Digital Communications

McGraw-Hill Series in Electrical and Computer Engineering

SENIOR CONSULTING EDITOR

Stephen W. Director, *University of Michigan, Ann Arbor*

Circuits and Systems
Communications and Signal Processing
Computer Engineering
Control Theory and Robotics
Electromagnetics
Electronics and VLSI Circuits
Introductory
Power
Antennas, Microwaves, and Radar

PREVIOUS CONSULTING EDITORS

Digital Communications

Fourth Edition

JOHN G. PROAKIS

Department of Electrical and Computer Engineering
Northeastern University

Boston Burr Ridge, IL Dubuque, IA Madison, WI New York San Francisco St. Louis
Bangkok Bogotá Caracas Lisbon London Madrid Mexico City Milan
New Delhi Seoul Singapore
Sydney Taipei Toronto

McGraw-Hill Higher Education

A Division of The ***McGraw-Hill*** *Companies*

DIGITAL COMMUNICATIONS

Published by McGraw-Hill, an imprint of The McGraw-Hill Companies, Inc., 1221 Avenue of the Americas, New York, NY, 10020.

This book is printed on acid-free paper.

2 3 4 5 6 7 8 9 0 DOC/DOC 0 9 8 7 6 5 4 3 2 1

ISBN 0-07-232111-3

Publisher: *Thomas Casson*
Sponsoring editor: *Catherine Fields Shultz*
Developmental editor: *Emily J. Gray*
Marketing manager: *John Wannemacher*
Project manager: *Craig S. Leonard*
Production supervisor: *Rose Hepburn*
Senior designer: *Kiera Cunningham*
New media: *Christopher Styles*
Compositor: *Interactive Composition Corporation*
Typeface: *10.5/12 Times Roman*
Printer: *Quebecor World Fairfield, Inc.*

Library of Congress Cataloging-in-Publication Data
Proakis, John G.
Digital communications / John G. Proakis.–4th ed.
p. cm.
ISBN 0-07-232111-3
1. Digital communications. I. Title.
TK5103.7.P76 2000
621.382–dc21 00-025305

www.mhhe.com

ABOUT THE AUTHOR

JOHN G. PROAKIS has been on the faculty of Northeastern University since September of 1969. During the period 1982–1997, he held the administrative positions of Chairman of the Department of Electrical and Computer Engineering, Associate Dean of Research and Graduate Studies, and as Interim Dean of the College of Engineering. He has also served on the staffs of GTE Laboratories and the MIT Lincoln Laboratory.

Dr. Proakis received the BSEE Degree from the University of Cincinnati, the MSEE Degree from MIT, and the Ph.D. in Engineering from Harvard University. His professional experience and research interests are in the general areas of digital communications and digital signal processing, about which he has written extensively.

To
Felia, George, and Elena

BRIEF CONTENTS

CONTENTS

PREFACE

The fourth edition of *Digital Communications* has undergone a minor revision. Several new topics have been added, including serial and parallel concatenated codes, punctured convolutional codes, turbo TCM and turbo equalization, and spatial multiplexing. Since this is an introductory-level text, the treatment of these topics is limited in scope.

The book is designed to serve as a text for a first-year graduate-level course for students in electrical engineering. It is also designed to serve as a text for self-study and as a reference book for the practicing engineer involved in the design of digital communications systems. As a background, I presume that the reader has a thorough understanding of basic calculus and elementary linear systems theory and some prior knowledge of probability and stochastic processes.

Chapter 1 is an introduction to the subject, including a historical perspective and a description of channel characteristics and channel models.

Chapter 2 contains a review of the basic elements of probability and stochastic processes. It deals with a number of probability distribution functions and moments that are used throughout the book. It also includes the derivation of the Chernoff bound, which is useful in obtaining bounds on the performance of digital communications systems.

Chapter 3 treats source coding for discrete and analog sources. Emphasis is placed on scalar and vector quantization techniques, and comparisons are made with basic results from rate-distortion theory.

In **Chapter 4**, the reader is introduced to the representation of digitally modulated signals and to the characterization of narrowband signals and systems. Also treated in this chapter are the spectral characteristics of digitally modulated signals. New material has been added on a linear representation of CPM signals.

Chapter 5 treats the design of modulation and optimum demodulation and detection methods for digital communications over an additive white Gaussian noise channel. Emphasis is placed on the evaluation of the error rate performance for the various digital signaling techniques and on the channel bandwidth requirements of the corresponding signals.

Chapter 6 is devoted to carrier phase estimation and time synchronization methods based on the maximum-likelihood criterion. Both decision-directed and non-decision-directed methods are described.

Chapter 7 treats the topics of channel capacity for several different channel models and random coding.

Chapter 8 treats linear block and convolutional codes. The new topics added to the chapter include serial and parallel interleaved concatenated block and convolutional codes, punctured and rate-compatible convolutional codes, the soft-output Viterbi algorithm (SOVA), and turbo TCM.

Chapter 9 is focused on signal design for bandlimited channels. This chapter includes the topics of partial response signals and run-length-limited codes for spectral shaping.

Chapter 10 treats the problem of demodulation and detection of signals corrupted by intersymbol interference. The emphasis is on optimum and suboptimum equalization methods and their performance. New topics added to the chapter include Tomlinson-Harashima precoding, reduced complexity maximum-likelihood detectors, and turbo equalization.

Chapter 11 treats adaptive channel equalization. The LMS and recursive least-squares algorithms are described, together with their performance characteristics. This chapter also includes a treatment of blind equalization algorithms. New topics added include the tap-leakage algorithm and methods for accelerating the initial convergence of the LMS algorithm.

Chapter 12 treats multichannel and multicarrier modulation. The latter subject is particularly appropriate in view of several important applications that have been developed over the past two decades.

Chapter 13 is devoted to spread spectrum signals and systems. The benefits of coding in the design of spread spectrum signals is emphasized throughout this chapter.

Chapter 14 treats communication through fading channels. Several channel fading statistical models are considered, with emphasis placed on Rayleigh fading and Nakagami fading. Trellis coding for fading channels is also included in this chapter. New material added includes a brief treatment of fading and multipath characteristics of mobile radio channels, receiver structures for fading multipath channels with intersymbol interference, and spatial multiplexing using multiple transmit and receive antennas.

Chapter 15 treats multiuser communications. The emphasis is on code-division multiple access (CDMA), signal detection and random access methods, such as ALOHA and carrier-sense multiple access (CSMA).

With 15 chapters and a variety of topics, the instructor has the flexibility to design either a one- or two-semester course. Chapters 3 through 6 provide a basic treatment of digital modulation/demodulation and detection methods. Channel coding, treated in Chapters 7 and 8, can be included along with modulation and demodulation in a one-semester course. The topics of channel equalization, fading channels, spread spectrum, and multiuser communications can be covered in a second-semester course.

Throughout my professional career, I have had the opportunity to work with and learn from a number of people whom I should like to publicly acknowledge. These include Dr. R. Price, P.R. Drouilhet, Jr., and Dr. P.E. Green, Jr., who introduced me to various aspects of digital communications through fading multipath channels and multichannel signal transmission during my employment at the MIT Lincoln Laboratory. I am also indebted to Professor D.W. Tufts, who supervised my Ph.D. dissertation at Harvard University and who introduced me to the problems of signal design and equalization for band-limited channels. Over the years, I have had the pleasure of working on a variety of research projects in collaboration with colleagues at GTE and Stein Associates, including Dr. S. Stein, Dr. B. Barrow, Dr. A.A. Giordano, Dr. A.H. Levesque, Dr. R. Greenspan, Dr. D. Freeman, P.H.Anderson, D. Gooding, and J. Lindholm. At Northeastern University, I have had the benefit of collaborating with Dr. M. Salehi, Dr. M. Stojanovic, and Dr. D. Brady. Dr. T. Schonhoff provided the graphs illustrating the spectral characteristics of CPFSK, and H. Gibbons provided the data for the graphs in Chapter 14 that show the performance of PSK and DPSK with diversity. The assistance of these colleagues is greatly appreciated.

McGraw-Hill and I would like to thank the following reviewers of this edition for their valuable suggestions: William E. Ryan, *University of Arizona*; Tan Wong, *University of Florida*; and Raymond Pickholtz, *George Washington University*.

Finally, I wish to express my appreciation to Gloria Doukakis, for typing the manuscript of this edition, and to Apostolos Rizos for preparing the Solutions Manual.

1

Introduction

In this book, we present the basic principles that underlie the analysis and design of digital communication systems. The subject of digital communications involves the transmission of information in digital form from a source that generates the information to one or more destinations. Of particular importance in the analysis and design of communication systems are the characteristics of the physical channels through which the information is transmitted. The characteristics of the channel generally affect the design of the basic building blocks of the communication system. Below, we describe the elements of a communication system and their functions.

1.1 ELEMENTS OF A DIGITAL COMMUNICATION SYSTEM

Figure 1.1–1 illustrates the functional diagram and the basic elements of a digital communication system. The source output may be either an analog signal, such as an audio or video signal, or a digital signal, such as the output of a teletype machine, that is discrete in time and has a finite number of output characters. In a digital communication system, the messages produced by the source are converted into a sequence of binary digits. Ideally, we should like to represent the source output (message) by as few binary digits as possible. In other words, we seek an efficient representation of the source output that results in little or no redundancy. The process of efficiently converting the output of either an analog or digital source into a sequence of binary digits is called *source encoding* or *data compression.*

The sequence of binary digits from the source encoder, which we call the *information sequence*, is passed to the *channel encoder*. The purpose of the channel encoder is to introduce, in a controlled manner, some redundancy in the binary information sequence that can be used at the receiver to overcome the effects of noise and interference encountered in the transmission of the signal

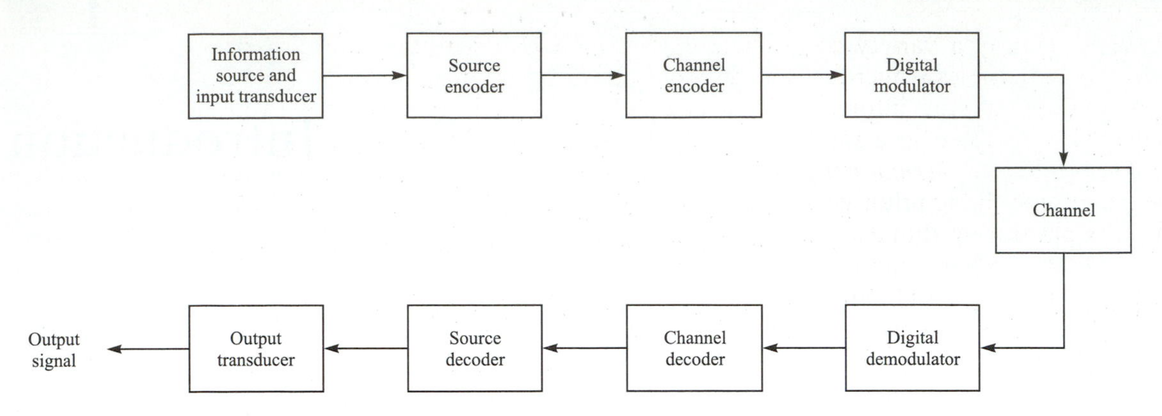

FIGURE 1.1–1
Basic elements of a digital communication system.

through the channel. Thus, the added redundancy serves to increase the reliability of the received data and improves the fidelity of the received signal. In effect, redundancy in the information sequence aids the receiver in decoding the desired information sequence. For example, a (trivial) form of encoding of the binary information sequence is simply to repeat each binary digit m times, where m is some positive integer. More sophisticated (nontrivial) encoding involves taking k information bits at a time and mapping each k-bit sequence into a unique n-bit sequence, called a *code word*. The amount of redundancy introduced by encoding the data in this manner is measured by the ratio n/k. The reciprocal of this ratio, namely k/n, is called the rate of the code or, simply, the *code rate*.

The binary sequence at the output of the channel encoder is passed to the *digital modulator*, which serves as the interface to the communication channel. Since nearly all the communication channels encountered in practice are capable of transmitting electrical signals (waveforms), the primary purpose of the digital modulator is to map the binary information sequence into signal waveforms. To elaborate on this point, let us suppose that the coded information sequence is to be transmitted one bit at a time at some uniform rate R bits per second (bits/s). The digital modulator may simply map the binary digit 0 into a waveform $s_0(t)$ and the binary digit 1 into a waveform $s_1(t)$. In this manner, each bit from the channel encoder is transmitted separately. We call this *binary modulation*. Alternatively, the modulator may transmit b coded information bits at a time by using $M = 2^b$ distinct waveforms $s_i(t)$, $i = 0, 1, \ldots, M - 1$, one waveform for each of the 2^b possible b-bit sequences. We call this *M-ary modulation* ($M > 2$). Note that a new b-bit sequence enters the modulator every b/R seconds. Hence, when the channel bit rate R is fixed, the amount of time available to transmit one of the M waveforms corresponding to a b-bit sequence is b times the time period in a system that uses binary modulation.

The *communication channel* is the physical medium that is used to send the signal from the transmitter to the receiver. In wireless transmission, the channel may be the atmosphere (free space). On the other hand, telephone channels usually employ a variety of physical media, including wire lines, optical fiber cables, and wireless (microwave radio). Whatever the physical medium used for transmission of the information, the essential feature is that the transmitted signal is corrupted in a random manner by a variety of possible mechanisms, such as additive *thermal noise* generated by electronic devices; man-made noise, e.g., automobile ignition noise; and atmospheric noise, e.g., electrical lightning discharges during thunderstorms.

At the receiving end of a digital communication system, the *digital demodulator* processes the channel-corrupted transmitted waveform and reduces the waveforms to a sequence of numbers that represent estimates of the transmitted data symbols (binary or M-ary). This sequence of numbers is passed to the channel decoder, which attempts to reconstruct the original information sequence from knowledge of the code used by the channel encoder and the redundancy contained in the received data.

A measure of how well the demodulator and decoder perform is the frequency with which errors occur in the decoded sequence. More precisely, the average probability of a bit-error at the output of the decoder is a measure of the performance of the demodulator–decoder combination. In general, the probability of error is a function of the code characteristics, the types of waveforms used to transmit the information over the channel, the transmitter power, the characteristics of the channel (i.e., the amount of noise, the nature of the interference), and the method of demodulation and decoding. These items and their effect on performance will be discussed in detail in subsequent chapters.

As a final step, when an analog output is desired, the source decoder accepts the output sequence from the channel decoder and, from knowledge of the source encoding method used, attempts to reconstruct the original signal from the source. Because of channel decoding errors and possible distortion introduced by the source encoder, and perhaps, the source decoder, the signal at the output of the source decoder is an approximation to the original source output. The difference or some function of the difference between the original signal and the reconstructed signal is a measure of the distortion introduced by the digital communication system.

1.2
COMMUNICATION CHANNELS AND THEIR CHARACTERISTICS

As indicated in the preceding discussion, the communication channel provides the connection between the transmitter and the receiver. The physical channel may be a pair of wires that carry the electrical signal, or an optical fiber that carries the information on a modulated light beam, or an underwater ocean channel in which the information is transmitted acoustically, or free space over which the information-bearing signal is radiated by use of an antenna. Other media that can be characterized as communication channels are data storage media, such as magnetic tape, magnetic disks, and optical disks.

One common problem in signal transmission through any channel is additive noise. In general, additive noise is generated internally by components such as resistors and solid-state devices used to implement the communication system. This is sometimes called *thermal noise*. Other sources of noise and interference may arise externally to the system, such as interference from other users of the channel. When such noise and interference occupy the same frequency band as the desired signal, their effect can be minimized by the proper design of the transmitted signal and its demodulator at the receiver. Other types of signal degradations that may be encountered in transmission over the channel are signal attenuation, amplitude and phase distortion, and multipath distortion.

The effects of noise may be minimized by increasing the power in the transmitted signal. However, equipment and other practical constraints limit the power level in the transmitted signal. Another basic limitation is the available channel bandwidth. A bandwidth constraint is usually due to the physical limitations of the medium and the electronic components used to implement the transmitter and the receiver. These two limitations constrain the amount of data that can be transmitted reliably over any communication channel as we shall observe in later chapters. Below, we describe some of the important characteristics of several communication channels.

Wireline channels. The telephone network makes extensive use of wire lines for voice signal transmission, as well as data and video transmission. Twisted-pair wire lines and coaxial cable are basically guided electromagnetic channels that provide relatively modest bandwidths. Telephone wire generally used to connect a customer to a central office has a bandwidth of several hundred kilohertz (kHz). On the other hand, coaxial cable has a usable bandwidth of several megahertz (MHz). Figure 1.2–1 illustrates the frequency range of guided electromagnetic channels, which include waveguides and optical fibers.

Signals transmitted through such channels are distorted in both amplitude and phase and further corrupted by additive noise. Twisted-pair wireline channels are also prone to crosstalk interference from physically adjacent channels. Because wireline channels carry a large percentage of our daily communications around the country and the world, much research has been performed on the characterization of their transmission properties and on methods for mitigating the amplitude and phase distortion encountered in signal transmission. In Chapter 9, we describe methods for designing optimum transmitted signals and their demodulation; in Chapters 10 and 11, we consider the design of channel equalizers that compensate for amplitude and phase distortion on these channels.

Fiber-optic channels. Optical fibers offer the communication system designer a channel bandwidth that is several orders of magnitude larger than coaxial cable channels. During the past two decades, optical fiber cables have been developed that have a relatively low signal attenuation, and highly reliable photonic devices have been developed for signal generation and signal detection. These technological advances have resulted in a rapid deployment of optical fiber channels, both

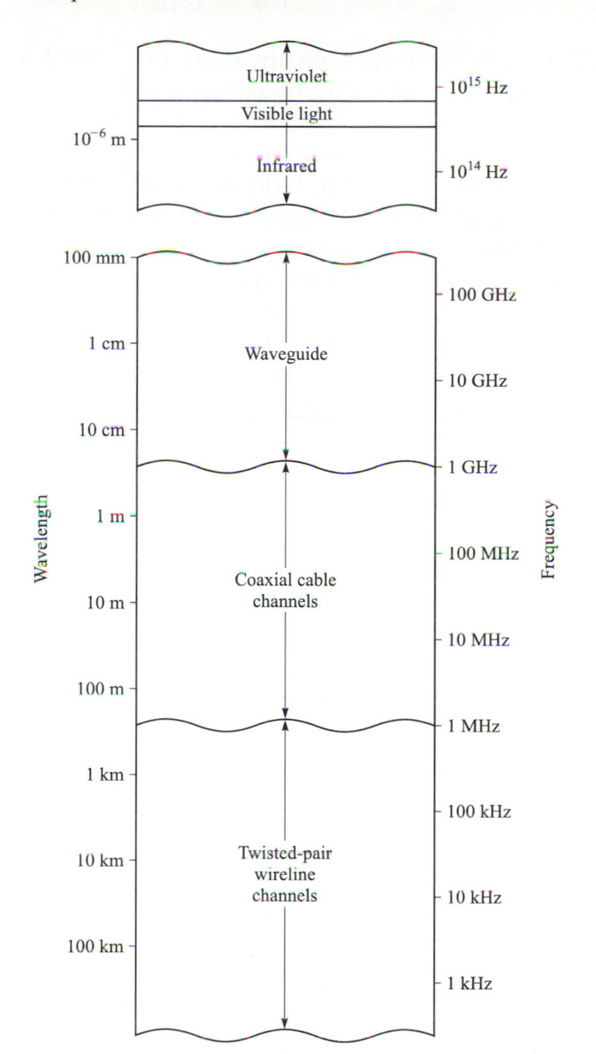

FIGURE 1.2–1
Frequency range for guided wire channel.

in domestic telecommunication systems as well as for trans-Atlantic and trans-Pacific communication. With the large bandwidth available on fiber-optic channels, it is possible for telephone companies to offer subscribers a wide array of telecommunication services, including voice, data, facsimile, and video.

The transmitter or modulator in a fiber-optic communication system is a light source, either a light-emitting diode (LED) or a laser. Information is transmitted by varying (modulating) the intensity of the light source with the message signal. The light propagates through the fiber as a light wave and is amplified periodically (in the case of digital transmission, it is detected and regenerated by repeaters) along the transmission path to compensate for signal attenuation. At the receiver, the light intensity is detected by a photodiode, whose output is an electrical signal that varies in direct proportion to the power of the light impinging on the photodiode. Sources of noise in fiber-optic channels are photodiodes and electronic amplifiers.

Wireless electromagnetic channels. In wireless communication systems, electromagnetic energy is coupled to the propagation medium by an antenna which serves as the radiator. The physical size and the configuration of the antenna depend primarily on the frequency of operation. To obtain efficient radiation of electromagnetic energy, the antenna must be longer than $\frac{1}{10}$ of the wavelength. Consequently, a radio station transmitting in the amplitude-modulated (AM) frequency band, say at $f_c = 1$ MHz [corresponding to a wavelength of $\lambda = c/f_c =$ 300 meters (m)], requires an antenna of at least 30 m. Other important characteristics and attributes of antennas for wireless transmission are described in Chapter 5.

Figure 1.2–2 illustrates the various frequency bands of the electromagnetic spectrum. The mode of propagation of electromagnetic waves in the atmosphere and in free space may be subdivided into three categories, namely, ground-wave propagation, sky-wave propagation, and line-of-sight (LOS) propagation. In the very low frequency (VLF) and audio frequency bands, where the wavelengths exceed 10 km, the earth and the ionosphere act as a waveguide for electromagnetic wave propagation. In these frequency ranges, communication signals practically propagate around the globe. For this reason, these frequency bands are primarily used to provide navigational aids from shore to ships around the world. The channel bandwidths available in these frequency bands are relatively small (usually 1–10 percent of the center frequency), and hence the information that is transmitted through these channels is of relatively slow speed and generally confined to digital transmission. A dominant type of noise at these frequencies is generated from thunderstorm activity around the globe, especially in tropical regions. Interference results from the many users of these frequency bands.

Ground-wave propagation, as illustrated in Fig. 1.2–3, is the dominant mode of propagation for frequencies in the medium frequency (MF) band (0.3–3 MHz). This is the frequency band used for AM broadcasting and maritime radio broadcasting. In AM broadcasting, the range with ground-wave propagation of even the more powerful radio stations is limited to about 150 km. Atmospheric noise, man-made noise, and thermal noise from electronic components at the receiver are dominant disturbances for signal transmission in the MF band.

Sky-wave propagation, as illustrated in Fig. 1.2–4, results from transmitted signals being reflected (bent or refracted) from the ionosphere, which consists of several layers of charged particles ranging in altitude from 50 to 400 km above the surface of the earth. During the daytime hours, the heating of the lower atmosphere by the sun causes the formation of the lower layers at altitudes below 120 km. These lower layers, especially the D-layer, serve to absorb frequencies below 2 MHz, thus severely limiting sky-wave propagation of AM radio broadcast. However, during the nighttime hours, the electron density in the lower layers of the ionosphere drops sharply and the frequency absorption

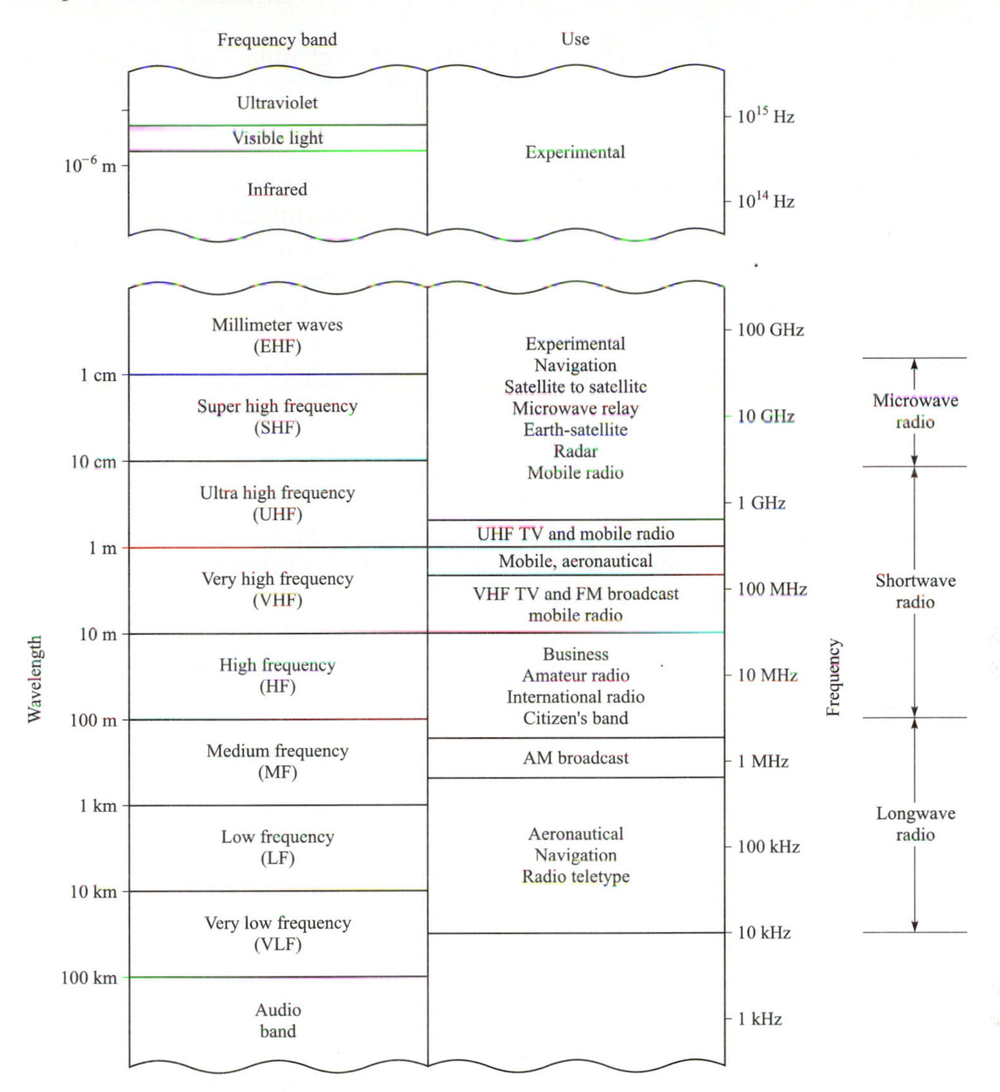

FIGURE 1.2–2
Frequency range for wireless electromagnetic channels. [*Adapted from Carlson* (1975), 2nd edition, © *McGraw-Hill Book Company Co. Reprinted with permission of the publisher*.]

that occurs during the daytime is significantly reduced. As a consequence, powerful AM radio broadcast stations can propagate over large distances via sky wave over the F-layer of the ionosphere, which ranges from 140 to 400 km above the surface of the earth.

A frequently occurring problem with electromagnetic wave propagation via sky wave in the high frequency (HF) range is *signal multipath*. Signal multipath occurs when the transmitted signal arrives at the receiver via multiple propaga-

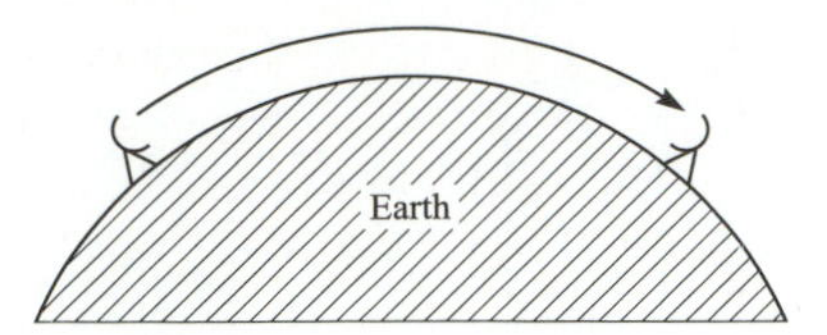

FIGURE 1.2–3
Illustration of ground-wave propagation.

tion paths at different delays. It generally results in intersymbol interference in a digital communication system. Moreover, the signal components arriving via different propagation paths may add destructively, resulting in a phenomenon called *signal fading*, which most people have experienced when listening to a distant radio station at night when sky wave is the dominant propagation mode. Additive noise in the HF range is a combination of atmospheric noise and thermal noise.

Sky-wave ionospheric propagation ceases to exist at frequencies above approximately 30 MHz, which is the end of the HF band. However, it is possible to have ionospheric scatter propagation at frequencies in the range 30–60 MHz, resulting from signal scattering from the lower ionosphere. It is also possible to communicate over distances of several hundred miles by use of tropospheric scattering at frequencies in the range 40–300 MHz. Troposcatter results from signal scattering due to particles in the atmosphere at altitudes of 10 miles or less. Generally, ionospheric scatter and tropospheric scatter involve large signal propagation losses and require a large amount of transmitter power and relatively large antennas.

Frequencies above 30 MHz propagate through the ionosphere with relatively little loss and make satellite and extraterrestrial communications possible. Hence, at frequencies in the very high frequency (VHF) band and higher, the dominant mode of electromagnetic propagation is LOS propagation. For terrestrial communication systems, this means that the transmitter and receiver antennas must be in direct LOS with relatively little or no obstruction. For this reason, television stations transmitting in the VHF and ultra high frequency (UHF) bands mount their antennas on high towers to achieve a broad coverage area.

In general, the coverage area for LOS propagation is limited by the curvature of the earth. If the transmitting antenna is mounted at a height h m above the surface of the earth, the distance to the radio horizon, assuming no physical obstructions such as mountains, is approximately $d = \sqrt{15h}$ km. For example,

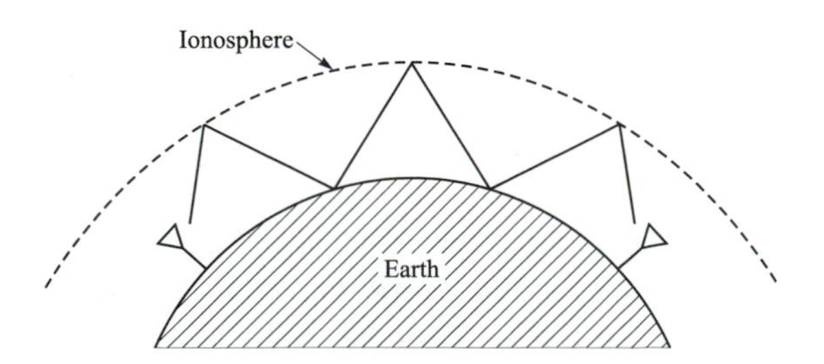

FIGURE 1.2–4
Illustration of sky-wave propagation.

a television antenna mounted on a tower of 300 m in height provides a coverage of approximately 67 km. As another example, microwave radio relay systems used extensively for telephone and video transmission at frequencies above 1 gigahertz (GHz) have antennas mounted on tall towers or on the top of tall buildings.

The dominant noise limiting the performance of a communication system in VHF and UHF ranges is thermal noise generated in the receiver front end and cosmic noise picked up by the antenna. At frequencies in the super high frequency (SHF) band above 10 GHz, atmospheric conditions play a major role in signal propagation. For example, at 10 GHz, the attenuation ranges from about 0.003 decibels per kilometer (dB/km) in light rain to about 0.3 dB/km in heavy rain. At 100 GHz, the attenuation ranges from about 0.1 dB/km in light rain to about 6 dB/km in heavy rain. Hence, in this frequency range, heavy rain introduces extremely high propagation losses that can result in service outages (total breakdown in the communication system).

At frequencies above the extremely high frequency (EHF) band, we have the infrared and visible light regions of the electromagnetic spectrum, which can be used to provide LOS optical communication in free space. To date, these frequency bands have been used in experimental communication systems, such as satellite-to-satellite links.

Underwater acoustic channels. Over the past few decades, ocean exploration activity has been steadily increasing. Coupled with this increase is the need to transmit data, collected by sensors placed under water, to the surface of the ocean. From there, it is possible to relay the data via a satellite to a data collection center.

Electromagnetic waves do not propagate over long distances under water except at extremely low frequencies. However, the transmission of signals at such low frequencies is prohibitively expensive because of the large and powerful transmitters required. The attenuation of electromagnetic waves in water can be expressed in terms of the *skin depth*, which is the distance a signal is attenuated by $1/e$. For seawater, the skin depth $\delta = 250/\sqrt{f}$, where f is expressed in Hz and δ is in m. For example, at 10 kHz, the skin depth is 2.5 m. In contrast, acoustic signals propagate over distances of tens and even hundreds of kilometers.

An underwater acoustic channel is characterized as a multipath channel due to signal reflections from the surface and the bottom of the sea. Because of wave motion, the signal multipath components undergo time-varying propagation delays that result in signal fading. In addition, there is frequency-dependent attenuation, which is approximately proportional to the square of the signal frequency. The sound velocity is nominally about 1500 m/s, but the actual value will vary either above or below the nominal value depending on the depth at which the signal propagates.

Ambient ocean acoustic noise is caused by shrimp, fish, and various mammals. Near harbors, there is also man-made acoustic noise in addition to the ambient noise. In spite of this hostile environment, it is possible to design and implement efficient and highly reliable underwater acoustic communication systems for transmitting digital signals over large distances.

Storage channels. Information storage and retrieval systems constitute a very significant part of data-handling activities on a daily basis. Magnetic tape, including digital audiotape and videotape, magnetic disks used for storing large amounts of computer data, optical disks used for computer data storage, and compact disks are examples of data storage systems that can be characterized as communication channels. The process of storing data on a magnetic tape or a magnetic or optical disk is equivalent to transmitting a signal over a telephone or a radio channel. The readback process and the signal processing involved in storage systems to recover the stored information are equivalent to the functions performed by a receiver in a telephone or radio communication system to recover the transmitted information.

Additive noise generated by the electronic components and interference from adjacent tracks is generally present in the readback signal of a storage system, just as is the case in a telephone or a radio communication system.

The amount of data that can be stored is generally limited by the size of the disk or tape and the density (number of bits stored per square inch) that can be achieved by the write/read electronic systems and heads. For example, a packing density of 10^9 bits per square inch has been demonstrated in magnetic disk storage systems. The speed at which data can be written on a disk or tape and the speed at which it can be read back are also limited by the associated mechanical and electrical subsystems that constitute an information storage system.

Channel coding and modulation are essential components of a well-designed digital magnetic or optical storage system. In the readback process, the signal is demodulated and the added redundancy introduced by the channel encoder is used to correct errors in the readback signal.

1.3 MATHEMATICAL MODELS FOR COMMUNICATION CHANNELS

In the design of communication systems for transmitting information through physical channels, we find it convenient to construct mathematical models that reflect the most important characteristics of the transmission medium. Then, the mathematical model for the channel is used in the design of the channel encoder and modulator at the transmitter and the demodulator and channel decoder at the receiver. Below, we provide a brief description of the channel models that are frequently used to characterize many of the physical channels that we encounter in practice.

The additive noise channel. The simplest mathematical model for a communication channel is the additive noise channel, illustrated in Fig. 1.3–1. In this model, the transmitted signal $s(t)$ is corrupted by an additive random noise

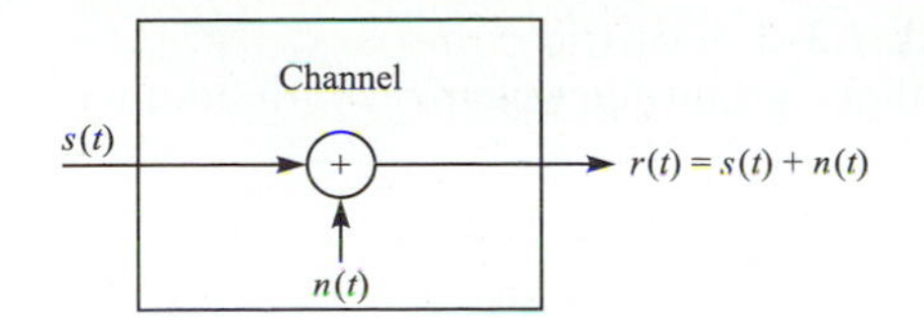

FIGURE 1.3–1
The additive noise channel.

process $n(t)$. Physically, the additive noise process may arise from electronic components and amplifiers at the receiver of the communication system or from interference encountered in transmission (as in the case of radio signal transmission).

If the noise is introduced primarily by electronic components and amplifiers at the receiver, it may be characterized as thermal noise. This type of noise is characterized statistically as a *Gaussian noise process*. Hence, the resulting mathematical model for the channel is usually called the *additive Gaussian noise channel*. Because this channel model applies to a broad class of physical communication channels and because of its mathematical tractability, this is the predominant channel model used in our communication system analysis and design. Channel attenuation is easily incorporated into the model. When the signal undergoes attenuation in transmission through the channel, the received signal is

$$r(t) = \alpha s(t) + n(t) \tag{1.3–1}$$

where α is the attenuation factor.

The linear filter channel. In some physical channels, such as wireline telephone channels, filters are used to ensure that the transmitted signals do not exceed specified bandwidth limitations and thus do not interfere with one another. Such channels are generally characterized mathematically as linear filter channels with additive noise, as illustrated in Fig. 1.3–2. Hence, if the channel input is the signal $s(t)$, the channel output is the signal

$$\begin{aligned} r(t) &= s(t) \star c(t) + n(t) \\ &= \int_{-\infty}^{\infty} c(\tau) s(t-\tau)\, d\tau + n(t) \end{aligned} \tag{1.3–2}$$

where $c(t)$ is the impulse response of the linear filter and $\star$ denotes convolution.

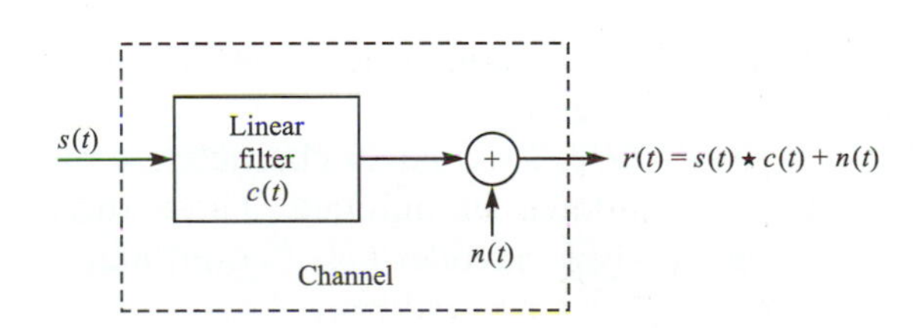

FIGURE 1.3–2
The linear filter channel with additive noise.

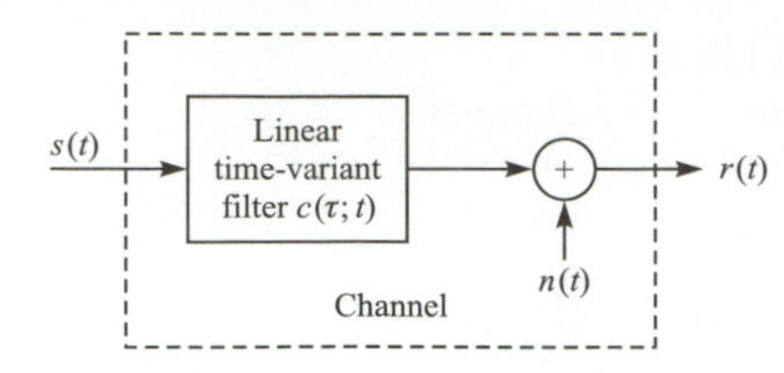

FIGURE 1.3–3
Linear time-variant filter channel with additive noise.

The linear time-variant filter channel. Physical channels such as underwater acoustic channels and ionospheric radio channels that result in time-variant multipath propagation of the transmitted signal may be characterized mathematically as time-variant linear filters. Such linear filters are characterized by a time-variant channel impulse response $c(\tau; t)$, where $c(\tau; t)$ is the response of the channel at time t due to an impulse applied at time $t - \tau$. Thus, τ represents the "age" (elapsed-time) variable. The linear time-variant filter channel with additive noise is illustrated in Fig. 1.3–3. For an input signal $s(t)$, the channel output signal is

$$\begin{aligned} r(t) &= s(t) \star c(\tau; t) + n(t) \\ &= \int_{-\infty}^{\infty} c(\tau; t) s(t - \tau)\, d\tau + n(t) \end{aligned} \tag{1.3–3}$$

A good model for multipath signal propagation through physical channels, such as the ionosphere (at frequencies below 30 MHz) and mobile cellular radio channels, is a special case of (1.3–3) in which the time-variant impulse response has the form

$$c(\tau; t) = \sum_{k=1}^{L} a_k(t) \delta(\tau - \tau_k) \tag{1.3–4}$$

where the $\{a_k(t)\}$ represents the possibly time-variant attenuation factor for the L multipath propagation paths and $\{\tau_k\}$ are the corresponding time delays. If (1.3–4) is substituted into (1.3–3), the received signal has the form

$$r(t) = \sum_{k=1}^{L} a_k(t) s(t - \tau_k) + n(t) \tag{1.3–5}$$

Hence, the received signal consists of L multipath components, where each component is attenuated by $\{a_k(t)\}$ and delayed by $\{\tau_k\}$.

The three mathematical models described above adequately characterize the great majority of the physical channels encountered in practice. These three channel models are used in this text for the analysis and design of communication systems.

1.4
A HISTORICAL PERSPECTIVE IN THE DEVELOPMENT OF DIGITAL COMMUNICATIONS

It is remarkable that the earliest form of electrical communication, namely *telegraphy*, was a digital communication system. The electric telegraph was developed by Samuel Morse and was demonstrated in 1837. Morse devised the variable-length binary code in which letters of the English alphabet are represented by a sequence of dots and dashes (code words). In this code, more frequently occurring letters are represented by short code words, while letters occurring less frequently are represented by longer code words. Thus, the *Morse code* was the precursor of the variable-length source coding methods described in Chapter 3.

Nearly 40 years later, in 1875, Emile Baudot devised a code for telegraphy in which every letter was encoded into fixed-length binary code words of length 5. In the *Baudot code*, binary code elements are of equal length and designated as mark and space.

Although Morse is responsible for the development of the first electrical digital communication system (telegraphy), the beginnings of what we now regard as modern digital communications stem from the work of Nyquist (1924), who investigated the problem of determining the maximum signaling rate that can be used over a telegraph channel of a given bandwidth without intersymbol interference. He formulated a model of a telegraph system in which a transmitted signal has the general form

$$s(t) = \sum_n a_n g(t - nT) \tag{1.4–1}$$

where $g(t)$ represents a basic pulse shape and $\{a_n\}$ is the binary data sequence of $\{\pm 1\}$ transmitted at a rate of $1/T$ bits/s. Nyquist set out to determine the optimum pulse shape that was band-limited to W Hz and maximized the bit rate under the constraint that the pulse caused no intersymbol interference at the sampling time $k/T, k = 0, \pm 1, \pm 2, \ldots$. His studies led him to conclude that the maximum pulse rate is $2W$ pulses/s. This rate is now called the *Nyquist rate*. Moreover, this pulse rate can be achieved by using the pulses $g(t) = (\sin 2\pi Wt)/2\pi Wt$. This pulse shape allows recovery of the data without intersymbol interference at the sampling instants. Nyquist's result is equivalent to a version of the sampling theorem for band-limited signals, which was later stated precisely by Shannon (1948b). The sampling theorem states that a signal of bandwidth W can be reconstructed from samples taken at the Nyquist rate of $2W$ samples/s using the interpolation formula

$$s(t) = \sum_n s\left(\frac{n}{2W}\right) \frac{\sin[2\pi W(t - n/2W)]}{2\pi W(t - n/2W)} \tag{1.4–2}$$

In light of Nyquist's work, Hartley (1928) considered the issue of the amount of data that can be transmitted reliably over a band-limited channel when multiple amplitude levels are used. Because of the presence of noise and other interference, Hartley postulated that the receiver can reliably estimate the received signal amplitude to some accuracy, say A_δ. This investigation led Hartley to conclude that there is a maximum data rate that can be communicated reliably over a band-limited channel when the maximum signal amplitude is limited to A_{max} (fixed power constraint) and the amplitude resolution is A_δ.

Another significant advance in the development of communications was the work of Kolmogorov (1939) and Wiener (1942), who considered the problem of estimating a desired signal waveform $s(t)$ in the presence of additive noise $n(t)$, based on observation of the received signal $r(t) = s(t) + n(t)$. This problem arises in signal demodulation. Kolmogorov and Wiener determined the linear filter whose output is the best mean-square approximation to the desired signal $s(t)$. The resulting filter is called the *optimum linear (Kolmogorov–Wiener) filter*.

Hartley's and Nyquist's results on the maximum transmission rate of digital information were precursors to the work of Shannon (1948a,b), who established the mathematical foundations for information transmission and derived the fundamental limits for digital communication systems. In his pioneering work, Shannon formulated the basic problem of reliable transmission of information in statistical terms, using probabilistic models for information sources and communication channels. Based on such a statistical formulation, he adopted a logarithmic measure for the information content of a source. He also demonstrated that the effect of a transmitter power constraint, a bandwidth constraint, and additive noise can be associated with the channel and incorporated into a single parameter, called the *channel capacity*. For example, in the case of an additive white (spectrally flat) Gaussian noise interference, an ideal band-limited channel of bandwidth W has a capacity C given by

$$C = W \log_2\left(1 + \frac{P}{WN_0}\right) \qquad \text{bits/s} \tag{1.4–3}$$

where P is the average transmitted power and N_0 is the power spectral density of the additive noise. The significance of the channel capacity is as follows: If the information rate R from the source is less than $C(R < C)$, then it is theoretically possible to achieve reliable (error-free) transmission through the channel by appropriate coding. On the other hand, if $R > C$, reliable transmission is not possible regardless of the amount of signal processing performed at the transmitter and receiver. Thus, Shannon established basic limits on communication of information and gave birth to a new field that is now called *information theory*.

Another important contribution to the field of digital communication is the work of Kotelnikov (1947), who provided a coherent analysis of the various digital communication systems based on a geometrical approach. Kotelnikov's approach was later expanded by Wozencraft and Jacobs (1965).

Following Shannon's publications, came the classic work of Hamming (1950) on error-detecting and error-correcting codes to combat the detrimental effects of channel noise. Hamming's work stimulated many researchers in the years that followed, and a variety of new and powerful codes were discovered, many of which are used today in the implementation of modern communication systems.

The increase in demand for data transmission during the last three to four decades, coupled with the development of more sophisticated integrated circuits, has led to the development of very efficient and more reliable digital communication systems. In the course of these developments, Shannon's original results and the generalization of his results on maximum transmission limits over a channel and on bounds on the performance achieved have served as benchmarks for any given communication system design. The theoretical limits derived by Shannon and other researchers that contributed to the development of information theory serve as an ultimate goal in the continuing efforts to design and develop more efficient digital communication systems.

There have been many new advances in the area of digital communications following the early work of Shannon, Kotelnikov, and Hamming. Some of the most notable advances are the following:

- The development of new block codes by Muller (1954), Reed (1954), Reed and Solomon (1960), Bose and Ray-Chaudhuri (1960a,b), and Goppa (1970, 1971).
- The development of concatenated codes by Forney (1966a).
- The development of computationally efficient decoding of Bose–Chaudhuri-Hocquenghem (BCH) codes, e.g., the Berlekamp–Massey algorithm (see Chien, 1964; Berlekamp, 1968).
- The development of convolutional codes and decoding algorithms by Wozencraft and Reiffen (1961), Fano (1963), Zigangirov (1966), Jelinek (1969), Forney (1970b, 1972, 1974), and Viterbi (1967, 1971).
- The development of trellis-coded modulation by Ungerboeck (1982), Forney et al. (1984), Wei (1987), and others.
- The development of efficient source encodings algorithms for data compression, such as those devised by Ziv and Lempel (1977, 1978), and Linde et al. (1980).
- The development of turbo codes and iterative decoding by Berrou et al. (1993).

1.5 OVERVIEW OF THE BOOK

Chapter 2 presents a brief review of the basic notions in the theory of probability and random processes. Our primary objectives in this chapter are to present results that are used throughout the book and to establish some necessary notation.

In Chapter 3, we provide an introduction to source coding for discrete and analog sources. Included in this chapter are the Huffman coding algorithm and the Lempel–Ziv algorithm for discrete sources, and scalar and vector quantization techniques for analog sources.

Chapter 4 treats the characterization of communication signals and systems from a mathematical viewpoint. Included in this chapter is a geometric representation of signal waveforms used for digital communications.

Chapter 5–8 are focused on modulation/demodulation and channel coding/decoding for the additive white Gaussian noise channel. The emphasis is on optimum demodulation and decoding techniques and their performance.

The design of efficient modulators and demodulators for linear filter channels with distortion is treated in Chapters 9–11. The focus is on signal design and on channel equalization methods to compensate for the channel distortion.

The final four chapters treat several more specialized topics. Chapter 12 treats multichannel and multicarrier communication systems. Chapter 13 is focused on spread spectrum signals for digital communications and their performance characteristics. Chapter 14 provides an in-depth treatment of communication through fading multipath channels. Included in this treatment is a description of channel characterization, signal design and demodulation techniques and their performance, and coding/decoding techniques and their performance. The last chapter of the book is focused on multiuser communication systems and multiple access methods.

1.6 BIBLIOGRAPHICAL NOTES AND REFERENCES

There are several historical treatments regarding the development of radio and telecommunications during the past century. These may be found in the books by McMahon (1984), Millman (1984), and Ryder and Fink (1984). We have already cited the classical works of Nyquist (1924), Hartley (1928), Kotelnikov (1947), Shannon (1948), and Hamming (1950), as well as some of the more important advances that have occurred in the field since 1950. The collected papers by Shannon have been published by IEEE Press in a book edited by Sloane and Wyner (1993) and previously in Russia in a book edited by Dobrushin and Lupanov (1963). Other collected works published by the IEEE Press that might be of interest to the reader are *Key Papers in the Development of Coding Theory*, edited by Berlekamp (1974), and *Key Papers in the Development of Information Theory*, edited by Slepian (1974).

2

Probability and Stochastic Processes

The theory of probability and stochastic processes is an essential mathematical tool in the design of digital communication systems. This subject is important in the statistical modeling of sources that generate the information, in the digitization of the source output, in the characterization of the channel through which the digital information is transmitted, in the design of the receiver that processes the information-bearing signal from the channel, and in the evaluation of the performance of the communication system. Our coverage of this rich and interesting subject is brief and limited in scope. We present a number of definitions and basic concepts in the theory of probability and stochastic processes, and we derive several results that are important in the design of efficient digital communication systems and in the evaluation of their performance.

We anticipate that most readers have had some prior exposure to the theory of probability and stochastic processes, so that our treatment serves primarily as a review. Some readers, however, who have had no previous exposure may find the presentation in this chapter extremely brief. These readers will benefit from additional reading of engineering-level treatments of the subject found in the texts by Davenport and Root (1958), Davenport (1970), Papoulis (1984), Helstrom (1991), Stark and Woods (1994), and Leon-Garcia (1994).

2.1 PROBABILITY

Let us consider an experiment, such as the rolling of a die, with a number of possible outcomes. The sample space S of the experiment consists of the set of all possible outcomes. In the case of the die,

$$S = \{1, 2, 3, 4, 5, 6\} \qquad (2.1\text{–}1)$$

where the integers $1, \ldots, 6$ represent the number of dots on the six faces of the die. These six possible outcomes are the sample points of the experiment. An event is a subset of S, and may consist of any number of sample points. For example, the event A defined as

$$A = \{2, 4\} \tag{2.1–2}$$

consists of the outcomes 2 and 4. The complement of the event A, denoted by $\bar{A}$, consists of all the sample points in S that are not in A and, hence,

$$\bar{A} = \{1, 3, 5, 6\} \tag{2.1–3}$$

Two events are said to be mutually exclusive if they have no sample points in common—that is, if the occurrence of one event excludes the occurrence of the other. For example, if A is defined as in Equation 2.1–2 and the event B is defined as

$$B = \{1, 3, 6\} \tag{2.1–4}$$

then A and B are mutually exclusive extents. Similarly, A and $\bar{A}$ are mutually exclusive events.

The union (sum) of two events in an event that consists of all the sample points in the two events. For example, if B is the event defined in Equation 2.1–4 and C is the event defined as

$$C = \{1, 2, 3\} \tag{2.1–5}$$

then the union of B and C, denoted by $B \cup C$, is the event

$$D = B \cup C = \{1, 2, 3, 6\} \tag{2.1–6}$$

Similarly, $A \cup \bar{A} = S$, where S is the entire sample space or the certain event. On the other hand, the intersection of two events is an event that consists of the points that are common to the two events. Thus, if $E = B \cap C$ represents the intersection of the events B and C, defined by Equations 2.1–4 and 2.1–5, respectively, then

$$E = \{1, 3\}$$

When the events are mutually exclusive, the intersection is the null event, denoted as $\varnothing$. For example, $A \cap B = \varnothing$, and $A \cap \bar{A} = \varnothing$. The definitions of union and intersection are extended to more than two events in a straightforward manner.

Associated with each event A contained in S is its probability $P(A)$. In the assignment of probabilities to events, we adopt an axiomatic viewpoint. That is, we postulate that the probability of the event A satisfies the condition $P(A) \geqslant 0$. We also postulate that the probability of the sample space (certain event) is $P(S) = 1$. The third postulate deals with the probability of mutually exclusive events. Suppose that A_i, $i = 1, 2, \ldots$, are a (possibly infinite) number of events in the sample space S such that

$$A_i \cap A_j = \varnothing; \qquad i \neq j = 1, 2, \ldots$$

Then the probability of the union of these mutually exclusive events satisfies the condition

$$P\left(\bigcup_i A_i\right) = \sum_i P(A_i) \tag{2.1–7}$$

For example, in a roll of a fair die, each possible outcome is assigned the probability $\frac{1}{6}$. The event A defined by Equation 2.1–2 consists of two mutually exclusive subevents or outcomes, and, hence, $P(A) = \frac{2}{6} = \frac{1}{3}$. Also, the probability of the event $A \cup B$, where A and B are the mutually exclusive events defined by Equations 2.1–2 and 2.1–4, respectively, is $P(A) + P(B) = \frac{1}{3} + \frac{1}{2} = \frac{5}{6}$.

Joint events and joint probabilities. Instead of dealing with a single experiment, let us perform two experiments and consider their outcomes. For example, the two experiments may be two separate tosses of a single die or a single toss of two dice. In either case, the sample space S consists of the 36 two-tuples (i, j) where $i, j = 1, 2, \ldots, 6$. If the dice are fair, each point in the sample space is assigned the probability $\frac{1}{36}$. We may now consider joint events, such as $\{i$ is even, $j = 3\}$, and determine the associated probabilities of such events from knowledge of the probabilities of the sample points.

In general, if one experiment has the possible outcomes A_i, $i = 1, 2, \ldots, n$, and the second experiment has the possible outcomes B_j, $j = 1, 2, \ldots, m$, then the combined experiment has the possible joint outcomes (A_i, B_j), $i = 1, 2, \ldots, n$, $j = 1, 2, \ldots, m$. Associated with each joint outcome (A_i, B_j) is the joint probability $P(A_i, B_j)$ which satisfies the condition

$$0 \leqslant P(A_i, B_j) \leqslant 1$$

Assuming that the outcomes B_j, $j = 1, 2, \ldots, m$, are mutually exclusive, it follows that

$$\sum_{j=1}^{m} P(A_i, B_j) = P(A_i) \tag{2.1–8}$$

Similarly, if the outcomes A_i, $i = 1, 2, \ldots, n$, are mutually exclusive then

$$\sum_{i=1}^{n} P(A_i, B_j) = P(B_j) \tag{2.1–9}$$

Furthermore, if all the outcomes of the two experiments are mutually exclusive, then

$$\sum_{i=1}^{n} \sum_{j=1}^{m} P(A_i, B_j) = 1 \tag{2.1–10}$$

The generalization of the above treatment to more than two experiments is straightforward.

Conditional probabilities. Consider a combined experiment in which a joint event occurs with probability $P(A, B)$. Suppose that the event B has occurred and we wish to determine the probability of occurrence of the event A. This is called the *conditional probability* of the event A given the occurrence of the event B and is defined as

$$P(A|B) = \frac{P(A, B)}{P(B)} \tag{2.1–11}$$

provided $P(B) > 0$. In a similar manner, the probability of the event B conditioned on the occurrence of the event A is defined as

$$P(B|A) = \frac{P(A, B)}{P(A)} \tag{2.1–12}$$

provided $P(A) > 0$. The relations in Equations 2.1–11 and 2.1–12 may also be expressed as

$$P(A, B) = P(A|B)P(B) = P(B|A)P(A) \tag{2.1–13}$$

The relations in Equations 2.1–11 to 2.1–13 also apply to a single experiment in which A and B are any two events defined on the sample space S and $P(A, B)$ is interpreted as the probability of $A \cap B$. That is, $P(A, B)$ denotes the simultaneous occurrence of A and B. For example, consider the events B and C given by Equations 2.1–4 and 2.1–5, respectively, for the single toss of a die. The joint even consists of the sample points $\{1, 3\}$. The conditional probability of the event C given that B occurred is

$$P(C|B) = \frac{\frac{2}{6}}{\frac{3}{6}} = \frac{2}{3}$$

In a single experiment, we observe that when two events A and B are mutually exclusive, $A \cap B = \varnothing$ and, hence, $P(A|B) = 0$. Also, if A is a subset of B, then $A \cap B = A$ and, hence,

$$P(A|B) = \frac{P(A)}{P(B)}$$

On the other hand, if B is a subset of A, we have $A \cap B = B$ and, hence,

$$P(A|B) = \frac{P(B)}{P(B)} = 1$$

An extremely useful relationship for conditional probabilities is Bayes' theorem, which states that if A_i, $i = 1, 2, \ldots, n$, are mutually exclusive events such that

$$\bigcup_{i=1}^{n} A_i = S$$

and B is an arbitrary event with nonzero probability then

$$P(A_i|B) = \frac{P(A_i, B)}{P(B)} = \frac{P(B|A_i)P(A_i)}{\sum_{j=1}^{n} P(B|A_j)P(A_j)} \tag{2.1–14}$$

We use this formula in Chapter 5 to derive the structure of the optimum receiver for a digital communication system in which the events A_j, $i = 1, 2, \ldots, n$, represent the possible transmitted messages in a given time interval; $P(A_i)$ represent their a priori probabilities; B represents the received signal, which consists of the transmitted message (one of the A_i) corrupted by noise; and $P(A_i|B)$ is the a posteriori probability of A_i conditioned on having observed the received signal B.

Statistical independence. The statistical independence of two or more events is another important concept in probability theory. It usually arises when we consider two or more experiments or repeated trials of a single experiment. To explain this concept, we consider two events A and B and their conditional probability $P(A|B)$, which is the probability of occurrence of A given that B has occurred. Suppose that the occurrence of A does not depend on the occurrence of B. That is,

$$P(A|B) = P(A) \tag{2.1–15}$$

Substitution of Equation 2.1–15 into Equation 2.1–13 yields the result

$$P(A, B) = P(A)P(B) \tag{2.1–16}$$

That is, the joint probability of the events A and B factors into the product of the elementary or marginal probabilities $P(A)$ and $P(B)$. When the events A and B satisfy the relation in Equation 2.1–16, they are said to be *statistically independent*.

For example, consider two successive experiments in tossing a die. Let A represent the even-numbered sample points $\{2, 4, 6\}$ in the first toss and B represent the even-numbered possible outcomes $\{2, 4, 6\}$ in the second toss. In a fair die, we assign the probabilities $P(A) = \frac{1}{2}$ and $P(B) = \frac{1}{2}$. Now, the joint probability of the joint event "even-numbered outcome on the first toss and even-numbered outcome on the second toss" is just the probability of the nine pairs of outcomes (i, j), $i = 2, 4, 6$, $j = 2, 4, 6$, which is $\frac{1}{4}$. Also,

$$P(A, B) = P(A)P(B) = \tfrac{1}{4}$$

Thus, the events A and B are statistically independent. Similarly, we may say that the outcomes of the two experiments are statistically independent.

The definition of statistical independence can be extended to three or more events. Three statistically independent events A_1, A_2, and A_3 must satisfy the following conditions:

$$\begin{aligned} P(A_1, A_2) &= P(A_1)P(A_2) \\ P(A_1, A_3) &= P(A_1)P(A_3) \\ P(A_2, A_3) &= P(A_2)P(A_3) \\ P(A_1, A_2, A_3) &= P(A_1)P(A_2)P(A_3) \end{aligned} \tag{2.1–17}$$

In the general case, the events A_j, $i = 1, 2, \ldots, n$, are statistically independent provided that the probabilities of the joint events taken $2, 3, 4, \ldots,$ and n at a time factor into the product of the probabilities of the individual events.

2.1.1 Random Variables, Probability Distributions, and Probability Densities

Given an experiment having a sample space S and elements $s \in S$, we define a function $X(s)$ whose domain is S and whose range is a set of numbers on the real line. The function $X(s)$ is called a *random variable*. For example, if we flip a coin, the possible outcomes are head (H) and tail (T), so S contains two points labeled H and T. Suppose we define a function $X(s)$ such that

$$X(s) = \begin{cases} 1 & (s = \mathrm{H}) \\ -1 & (s = \mathrm{T}) \end{cases} \tag{2.1–18}$$

Thus we have mapped the two possible outcomes of the coin-flipping experiment into the two points (± 1) on the real line. Another experiment is the toss of a die with possible outcomes $S = \{1, 2, 3, 4, 5, 6\}$. A random variable defined on this sample space may be $X(s) = s$, in which case the outcomes of the experiment are mapped into the integers $1, \ldots, 6$, or, perhaps, $X(s) = s^2$, in which case the possible outcomes are mapped into the integers $\{1, 4, 9, 16, 25, 36\}$. These are examples of discrete random variables.

Although we have used as examples experiments that have a finite set of possible outcomes, there are many physical systems (experiments) that generate continuous outputs (outcomes). For example, the noise voltage generated by an electronic amplifier has a continuous amplitude. Consequently, the sample space S of voltage amplitudes $v \in S$ is continuous and so is the mapping $X(v) = v$. In such a case, the random variable† X is said to be a *continuous random variable*.

Given a random variable X, let us consider the event $\{X \leqslant x\}$ where x is any real number in the interval $(-\infty, \infty)$. We write the probability of this event as $P(X \leqslant x)$ and denote it simply by $F(x)$, i.e.,

$$F(x) = P(X \leqslant x), \qquad -\infty < x < \infty \tag{2.1–19}$$

†The random variable $X(s)$ will be written simply as X.

The function $F(x)$ is called the *probability distribution function* of the random variable X. It is also called the *cumulative distribution function* (CDF). Since $F(x)$ is a probability, its range is limited to the interval $0 \leqslant F(x) \leqslant 1$. In fact, $F(-\infty) = 0$ and $F(\infty) = 1$. For example, the discrete random variable generated by flipping a fair coin and defined by Equation 2.1–18 has the CDF shown in Figure 2.1–1a. There are two discontinuities or jumps in $F(x)$, one at $x = -1$ and one at $x = 1$. Similarly, the random variable $X(s) = s$ generated by tossing a fair die has the CDF shown in Figure 2.1–1b. In this case $F(x)$ has six jumps, one at each of the points $x = 1, \ldots, 6$.

The CDF of a continuous random variable typically appears as shown in Figure 2.1–2. This is a smooth, nondecreasing function of x. In some practical problems, we may also encounter a random variable of a mixed type. The CDF of such a random variable is a smooth, nondecreasing function in certain parts of the real line and contains jumps at a number of discrete values of x. An example of such a CDF is illustrated in Figure 2.1–3.

The derivative of the CDF $F(x)$, denoted as $p(x)$, is called the *probability density function* (PDF) of the random variable X. Thus, we have

$$p(x) = \frac{dF(x)}{dx}, \qquad -\infty < x < \infty \tag{2.1–20}$$

or, equivalently

$$F(x) = \int_{-\infty}^{x} p(u)\, du, \qquad -\infty < x < \infty \tag{2.1–21}$$

Since $F(x)$ is a nondecreasing function, it follows that $p(x) \geqslant 0$. When the random variable is discrete or of a mixed type, the PDF contains impulses at the points of discontinuity of $F(x)$. In such cases, the discrete part of $p(x)$ may be expressed as

$$p(x) = \sum_{i=1}^{n} P(X = x_i)\delta(x - x_i) \tag{2.1–22}$$

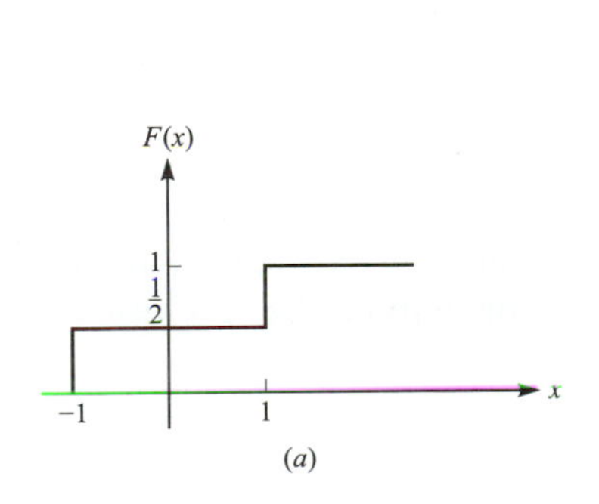

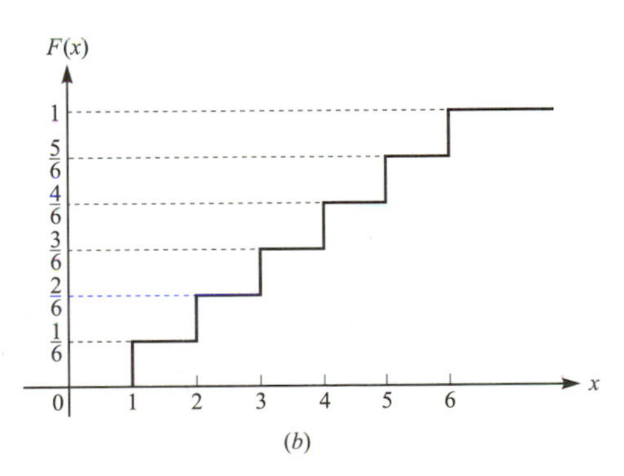

FIGURE 2.1–1
Examples of the cumulative distribution functions of two discrete random variables.

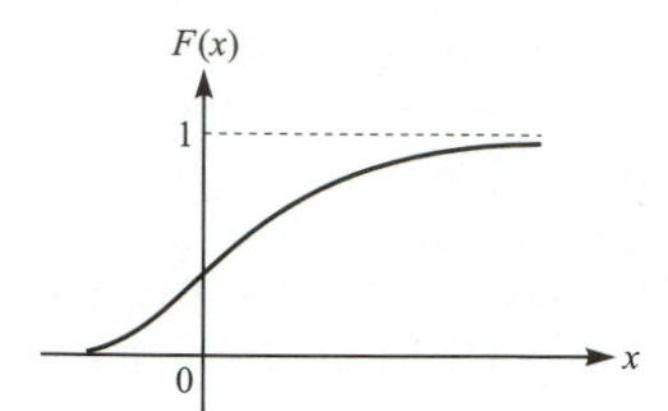

FIGURE 2.1–2
An example of the cumulative distribution function of a continuous random variable.

where x_i, $i = 1, 2, \ldots, n$, are the possible discrete values of the random variable; $P(X = x_i)$, $i = 2, \ldots, n$, are the probabilities; and $\delta(x)$ denotes an impulse at $x = 0$.

Often we are faced with the problem of determining the probability that a random variable X falls in an interval (x_1, x_2), where $x_2 > x_1$. To determine the probability of this event, let us begin with the event $\{X \leqslant x_2\}$. The event can always be expressed as the union of two mutually exclusive events $\{X \leqslant x_1\}$ and $\{x_1 < X \leqslant x_2\}$. Hence the probability of the event $\{X \leqslant x_2\}$ can be expressed as the sum of the probabilities of the mutually exclusive events. Thus we have

$$\begin{aligned} P(X \leqslant x_2) &= P(X \leqslant x_1) + P(x_1 < X \leqslant x_2) \\ F(x_2) &= F(x_1) + P(x_1 < X \leqslant x_2) \end{aligned}$$

or, equivalently,

$$\begin{aligned} P(x_1 < X \leqslant x_2) &= F(x_2) - F(x_1) \\ &= \int_{x_1}^{x_2} p(x)\, dx \end{aligned} \tag{2.1–23}$$

In other words, the probability of the event $\{x_1 < X \leqslant x_2\}$ is simply the area under the PDF in the range $x_1 < X \leqslant x_2$.

Multiple random variables, joint probability distributions, and joint probability densities. In dealing with combined experiments or repeated trials of a single experiment, we encounter multiple random variables and their CDFs and PDFs. Multiple random variables are basically multidimensional functions defined on a sample space of a combined experiment. Let us begin with two random variables

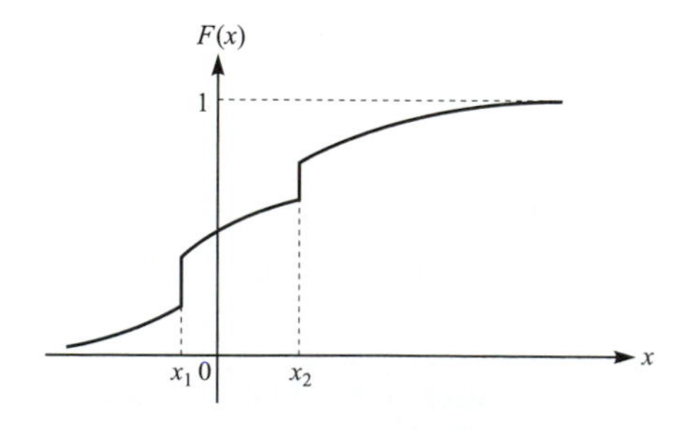

FIGURE 2.1–3
An example of the cumulative distribution function of a random variable of a mixed type.

X_1 and X_2, each of which may be continuous, discrete, or mixed. The joint CDF for the two random variables is defined as

$$\begin{aligned} F(x_1, x_2) &= P(X_1 \leqslant x_1, X_2 \leqslant x_2) \\ &= \int_{-\infty}^{x_1} \int_{-\infty}^{x_2} p(u_1, u_2)\, du_1\, du_2 \end{aligned} \tag{2.1–24}$$

where $p(x_1, x_2)$ is the joint PDF. The latter may also be expressed in the form

$$p(x_1, x_2) = \frac{\partial^2}{\partial x_1\, \partial x_2} F(x_1, x_2) \tag{2.1–25}$$

When the joint PDF $p(x_1, x_2)$ is integrated over one of the variables, we obtain the PDF of the other variable. That is,

$$\begin{aligned} \int_{-\infty}^{\infty} p(x_1, x_2)\, dx_1 &= p(x_2) \\ \int_{-\infty}^{\infty} p(x_1, x_2)\, dx_2 &= p(x_1) \end{aligned} \tag{2.1–26}$$

The PDFs $p(x_1)$ and $p(x_2)$ obtained from integrating over one of the variables are called *marginal PDFs*. Furthermore, if $p(x_1, x_2)$ is integrated over both variables, we obtain

$$\int_{-\infty}^{\infty} \int_{-\infty}^{\infty} p(x_1, x_2)\, dx_1\, dx_2 = F(\infty, \infty) = 1 \tag{2.1–27}$$

We also note that $F(-\infty, -\infty) = F(-\infty, x_2) = F(x_1, -\infty) = 0$.

The generalization of the above expressions to multidimensional random variables is straightforward. Suppose that X_i, $i = 1, 2, \ldots, n$, are random variables with a joint CDF defined as

$$\begin{aligned} F(x_1, x_2, \ldots, x_n) &= P(X_1 \leqslant x_1, X_2 \leqslant x_2, \ldots, X_n \leqslant x_n) \\ &= \int_{-\infty}^{x_1} \int_{-\infty}^{x_2} \cdots \int_{-\infty}^{x_n} p(u_1, u_2, \ldots, u_n)\, du_1\, du_2 \cdots du_n \end{aligned} \tag{2.1–28}$$

where $p(x_1, x_2, \ldots, x_n)$ is the joint PDF. By taking the partial derivatives of $F(x_1, x_2, \ldots, x_n)$ given by Equation 2.1–28, we obtain

$$p(x_1, x_2, \ldots, x_n) = \frac{\partial^n}{\partial x_1 \partial x_2 \cdots \partial x_n} F(x_1, x_2, \ldots, x_n) \tag{2.1–29}$$

Any number of variables in $p(x_1, x_2, \ldots, x_n)$ can be eliminated by integrating over these variables. For example, integration over x_2 and x_3 yields

$$\int_{-\infty}^{\infty} \int_{-\infty}^{\infty} p(x_1, x_2, x_3, \ldots, x_n)\, dx_2\, dx_3 = p(x_1, x_4, \ldots, x_n) \tag{2.1–30}$$

It also follows that $F(x_1, \infty, \infty, x_4, \ldots, x_n) = F(x_1, x_4, x_5, \ldots, x_n)$ and

$$F(x_1, -\infty - \infty, x_4, \ldots, x_n) = 0$$

Conditional probability distribution functions. Let us consider two random variables X_1 and X_2 with joint PDF $p(x_1, x_2)$. Suppose that we wish to determine the probability that the random variable $X_1 \leqslant x_1$ conditioned on

$$x_2 - \Delta x_2 < X_2 \leqslant x_2$$

where Δx_2 is some positive increment. That is, we wish to determine the probability of the event $(X_1 \leqslant x_1 | x_2 - \Delta x_2 < X_2 \leqslant x_2)$. Using the relations established earlier for the conditional probability of an event, the probability of the event $(X_1 \leqslant x_1 | x_2 - \Delta x_2 < X_2 \leqslant x_2)$ can be expressed as the probability of the joint event $(X_1 \leqslant x_1, x_2 - \Delta x_2 < X_2 \leqslant x_2)$ divided by the probability of the event $(x_2 - \Delta x_2 < X_2 \leqslant x_2)$. Thus

$$\begin{aligned} P(X_1 \leqslant x_1 | x_2 - \Delta x_2 < X_2 \leqslant x_2) &= \frac{\int_{-\infty}^{x_1} \int_{x_2 - \Delta x_2}^{x_2} p(u_1, u_2)\, du_1\, du_2}{\int_{x_2 - \Delta x_2}^{x_2} p(u_2)\, du_2} \\ &= \frac{F(x_1, x_2) - F(x_1, x_2 - \Delta x_2)}{F(x_2) - F(x_2 - \Delta x_2)} \end{aligned} \tag{2.1–31}$$

Assuming that the PDFs $p(x_1, x_2)$ and $p(x_2)$ are continuous functions over the interval $(x_2 - \Delta x_2, x_2)$, we may divide both numerator and denominator in Equation 2.1–31 by Δx_2 and take the limit as $\Delta x_2 \to 0$. Thus we obtain

$$\begin{aligned} P(X_1 \leqslant x_1 | X_2 = x_2) \equiv F(x_1 | x_2) &= \frac{\partial F(x_1, x_2)/\partial x_2}{\partial F(x_2)/\partial x_2} \\ &= \frac{\partial[\int_{-\infty}^{x_1} \int_{-\infty}^{x_2} p(u_1, u_2)\, du_1\, du_2]/\partial x_2}{\partial[\int_{-\infty}^{x_2} p(u_2)\, du_2]/\partial x_2} \\ &= \frac{\int_{-\infty}^{x_1} p(u_1, x_2) du_1}{p(x_2)} \end{aligned} \tag{2.1–32}$$

which is the conditional CDF of the random variable X_1 given the random variable X_2. We observe that $F(-\infty | x_2) = 0$ and $F(\infty | x_2) = 1$. By differentiating Equation 2.1–32 with respect to x_1, we obtain the corresponding PDF $p(x_1 | x_2)$ in the form

$$p(x_1 | x_2) = \frac{p(x_1, x_2)}{p(x_2)} \tag{2.1–33}$$

Alternatively, we may express the joint PDF $p(x_1, x_2)$ in terms of the conditional PDFs, $p(x_1 | x_2)$ or $p(x_2 | x_1)$, as

$$\begin{aligned} p(x_1, x_2) &= p(x_1 | x_2) p(x_2) \\ &= p(x_2 | x_1) p(x_1) \end{aligned} \tag{2.1–34}$$

The extension of the relations given above to multidimensional random variables is also easily accomplished. Beginning with the joint PDF of the random variables X_i, $i = 1, 2, \ldots, n$, we may write

$$p(x_1, x_2, \ldots, x_n) = p(x_1, x_2, \ldots, x_k | x_{k+1}, \ldots, x_n) p(x_{k+1}, \ldots, x_n) \quad (2.1\text{–}35)$$

where k is any integer in the range $1 < k < n$. The joint conditional CDF corresponding to the PDF $p(x_1, x_2, \ldots, x_k | x_{k+1}, \ldots, x_n)$ is

$$F(x_1, x_2, \ldots, x_k | x_{k+1}, \ldots, x_n) = \frac{\int_{-\infty}^{x_1} \cdots \int_{-\infty}^{x_k} p(u_1, u_2, \ldots, u_k, x_{k+1}, \ldots, x_n)\, du_1\, du_2 \cdots du_k}{p(x_{k+1}, \ldots, x_n)} \quad (2.1\text{–}36)$$

This conditional CDF satisfies the properties previously established for these functions, such as

$$F(\infty, x_2, \ldots, x_k | x_{k+1}, \ldots x_n) = F(x_2, x_3, \ldots, x_k | x_{k+1}, \ldots, x_n)$$
$$F(-\infty, x_2, \ldots, x_k | x_{k+1}, \ldots, x_n) = 0$$

Statistically independent random variables. We have already defined statistical independence of two or more events of a sample space S. The concept of statistical independence can be extended to random variables defined on a sample space generated by a combined experiment or by repeated trials of a single experiment. If the experiments result in mutually exclusive outcomes, the probability of an outcome in one experiment is independent of an outcome in any other experiment. That is, the joint probability of the outcomes factors into a product of the probabilities corresponding to each outcome. Consequently, the random variables corresponding to the outcomes in these experiments are independent in the sense that their joint PDF factors into a product of marginal PDFs. Hence the multidimensional random variables are statistically independent if and only if

$$F(x_1, x_2, \ldots, x_n) = F(x_1)F(x_2) \cdots F(x_n) \quad (2.1\text{–}37)$$

or, alternatively

$$p(x_1, x_2, \ldots, x_n) = p(x_1)p(x_2) \cdots p(x_n) \quad (2.1\text{–}38)$$

2.1.2 Functions of Random Variables

A problem that arises frequently in practical applications of probability is the following. Given a random variable X, which is characterized by its PDF $p(x)$, determine the PDF of the random variable $Y = g(X)$, where $g(X)$ is some given function of X. When the mapping g from X to Y is one-to-one, the determination of $p(y)$ is relatively straightforward. However, when the mapping is not one-to-one, as is the case, for example, when $Y = X^2$, we must be very careful in our derivation of $p(y)$.

EXAMPLE 2.1–1. Consider the random variable Y defined as

$$Y = aX + b \tag{2.1–39}$$

where a and b are constants. We assume that $a > 0$. If $a < 0$, the approach is similar (see Problem 2.3). We note that this mapping, illustrated in Figure 2.1–4a, is linear and monotonic. Let $F_X(x)$ and $F_Y(y)$ denote the CDFs for X and Y, respectively.† Then

$$\begin{aligned} F_Y(y) &= P(Y \leqslant y) = P(aX + b \leqslant y) = P\left(X \leqslant \frac{y-b}{a}\right) \\ &= \int_{-\infty}^{(y-b)/a} p_X(x)\, dx = F_X\left(\frac{y-b}{a}\right) \end{aligned} \tag{2.1–40}$$

By differentiating Equation 2.1–40 with respect to y, we obtain the relationship between the respective PDFs. It is

$$p_Y(y) = \frac{1}{a} p_X\left(\frac{y-b}{a}\right) \tag{2.1–41}$$

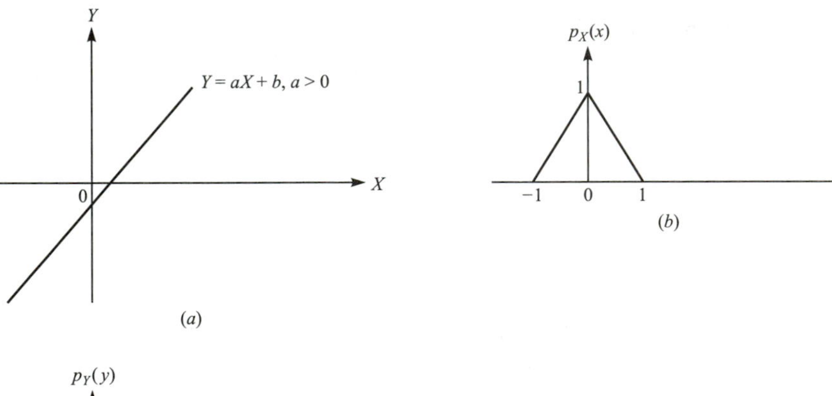

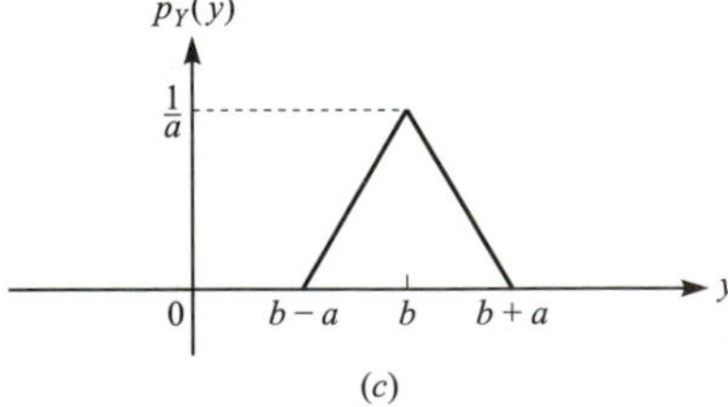

FIGURE 2.1–4
A linear transformation of a random variable X and an example of the corresponding PDFs of X and Y.

---

† To avoid confusion in changing variables, subscripts are used in the respective PDFs and CDFs.

Thus Equations 2.1–40 and 2.1–41 specify the CDF and PDF of the random variable Y in terms of the CDF and PDF of the random variable X for the linear transformation in Equation 2.1–39. To illustrate this mapping for a specific PDF $p_X(x)$, consider the one shown in Figure 2.1–4b. The PDF $p_Y(y)$ that results from the mapping in Equation 2.1–39 is shown in Figure 2.1–4c.

EXAMPLE 2.1–2. Consider the random variable Y defined as

$$Y = aX^3 + b, \qquad a > 0 \tag{2.1–42}$$

As in Example 2.1–1, the mapping between X and Y is one-to-one. Hence

$$\begin{aligned} F_Y(y) &= P(Y \leqslant y) = P(aX^3 + b \leqslant y) \\ &= P\left[X \leqslant \left(\frac{y-b}{a}\right)^{1/3}\right] = F_X\left[\left(\frac{y-b}{a}\right)^{1/3}\right] \end{aligned} \tag{2.1–43}$$

Differentiation of Equation 2.1–43 with respect to y yields the desired relationship between the two PDFs as

$$p_Y(y) = \frac{1}{3a[(y-b)/a]^{2/3}} p_X\left[\left(\frac{y-b}{a}\right)^{1/3}\right] \tag{2.1–44}$$

EXAMPLE 2.1–3. The random variable Y is defined as

$$Y = aX^2 + b, \qquad a > 0 \tag{2.1–45}$$

In contrast to Examples 2.1–1 and 2.1–2, the mapping between X and Y, illustrated in Figure 2.1–5, is not one-to-one. To determine the CDF of Y, we observe that

$$\begin{aligned} F_Y(y) &= P(Y \leqslant y) = P(aX^2 + b \leqslant y) \\ &= P\left(|X| \leqslant \sqrt{\frac{y-b}{a}}\right) \end{aligned}$$

Hence

$$F_Y(y) = F_x\left(\sqrt{\frac{y-b}{a}}\right) - F_X\left(-\sqrt{\frac{y-b}{a}}\right) \tag{2.1–46}$$

Differentiating Equation 2.1–46 with respect to y, we obtain the PDF of Y in terms of the PDF of X in the form

$$p_Y(y) = \frac{p_X[\sqrt{(y-b)/a}]}{2a\sqrt{[(y-b)/a]}} + \frac{p_X[-\sqrt{(y-b)/a}]}{2a\sqrt{[(y-b)/a]}} \tag{2.1–47}$$

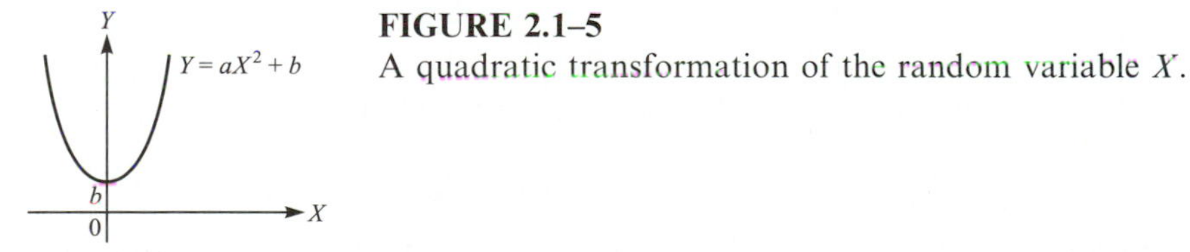

FIGURE 2.1–5
A quadratic transformation of the random variable X.

In Example 2.1–3, we observe that the equation $g(x) = ax^2 + b = y$ has two real solutions,

$$x_1 = \sqrt{\frac{y-b}{a}}, \qquad x_2 = -\sqrt{\frac{y-b}{a}}$$

and that $p_Y(y)$ consists of two terms corresponding to these two solutions. That is,

$$p_Y(y) = \frac{p_X[x_1 = \sqrt{(y-b)/a}]}{|g'[x_1 = \sqrt{(y-b)/a}]|} + \frac{p_X[x_2 = -\sqrt{(y-b)/a}]}{|g'[x_2 = -\sqrt{(y-b)/a}]|} \tag{2.1–48}$$

where $g'(x)$ denotes the first derivative of $g(x)$.

In the general case, suppose that $x_1, x_2, \ldots, x_n$ are the real roots of the equation $g(x) = y$. Then the PDF of the random variable $Y = g(x)$ may be expressed as

$$p_Y(y) = \sum_{i=1}^{n} \frac{p_X(x_i)}{|g'(x_i)|} \tag{2.1–49}$$

where the roots x_i, $i = 2, \ldots, n$, are functions of y.

Now let us consider functions of multidimensional random variables. Suppose that X_i, $i = 1, 2, \ldots, n$, are random variables with joint PDF $p_X(x_1, x_2, \ldots, x_n)$ and let Y_i, $i = 1, 2, \ldots n$, be another set of n random variables related to the X_i by the functions

$$Y_i = g_i(X_1, X_2, \ldots, X_n), \qquad i = 1, 2, \ldots, n \tag{2.1–50}$$

We assume that the $g_i(X_1, X_2, \ldots, X_n)$, $i = 1, 2, \ldots, n$, are single-valued functions with continuous partial derivatives and invertible. By "invertible" we mean that the X_i, $i = 1, 2, \ldots, n$, can be expressed as functions of Y_i, $i = 1, 2, \ldots, n$, in the form

$$X_i = g_i^{-1}(Y_1, Y_2, \ldots, Y_n), \qquad i = 1, 2, \ldots, n \tag{2.1–51}$$

where the inverse functions are also assumed to be single-valued with continuous partial derivatives. The problem is to determine the joint PDF of Y_i, $i = 1, 2, \ldots, n$, denoted by $p_Y(y_1, y_2, \ldots, y_n)$, given the joint PDF $p_X(x_1, x_2, \ldots, x_n)$.

To determine the desired relation, let R_X be the region in the n-dimensional space of the random variables X_i, $i = 1, 2, \ldots, n$, and let R_Y be the (one-to-one) mapping of R_X defined by the functions $Y_i = g_i(X_1, X_2, \ldots, X_n)$. Clearly

$$\begin{aligned}\iint_{R_Y} \cdots \int p_Y(y_1, y_2, \ldots, y_n)\, dy_1\, dy_2 \cdots dy_n \\ = \iint_{R_X} \cdots \int p_X(x_1, x_2, \ldots, x_n)\, dx_1\, dx_2 \cdots dx_n\end{aligned} \tag{2.1–52}$$

By making a change in variables in the multiple integral on the right-hand side of Equation 2.1–52 with the substitution

$$x_i = g_i^{-1}(y_1, y_2, \ldots, y_n) \equiv g_i^{-1}, \qquad i = 1, 2, \ldots, n$$

we obtain

$$\iint_{R_Y} \cdots \int p_Y(y_1, y_2, \ldots, y_n)\, dy_1\, dy_2 \ldots dy_n$$
$$= \iint_{R_Y} \cdots \int p_X(x_1 = g_1^{-1}, x_2 = g_2^{-1}, \ldots, x_n = g_n^{-1})|J| dy_1 dy_2 \ldots dy_n \tag{2.1–53}$$

where J denotes the Jacobian of the transformation, defined by the determinant

$$J = \begin{vmatrix} \dfrac{\partial g_1^{-1}}{\partial y_1} & \dfrac{\partial g_2^{-1}}{\partial y_2} & \cdots & \dfrac{\partial g_n^{-1}}{\partial y_1} \\ \vdots & \vdots & & \vdots \\ \dfrac{\partial g_1^{-1}}{\partial y_n} & \dfrac{\partial g_2^{-1}}{\partial y_n} & \cdots & \dfrac{\partial g_1^{-1}}{\partial y_n} \end{vmatrix} \tag{2.1–54}$$

Consequently, the desired relation for the joint PDF of the Y_i, $i = 1, 2, \ldots, n$, is

$$p_Y(y_1, y_2, \ldots, y_n) = p_X(x_1 = g_1^{-1}, x_2 = g_2^{-1}, \ldots, x_n = g_n^{-1})|J| \tag{2.1–55}$$

EXAMPLE 2.1–4. An important functional relation between two sets of n-dimensional random variables that frequently arises in practice is the linear transformation

$$Y_i = \sum_{j=1}^{n} a_{ij} X_j, \qquad i = 1, 2, \ldots, n \tag{2.1–56}$$

where the $\{a_{ij}\}$ are constants. It is convenient to employ the matrix form for the transformation, which is

$$\mathbf{Y} = \mathbf{AX} \tag{2.1–57}$$

where $\mathbf{X}$ and $\mathbf{Y}$ are n-dimensional vectors and $\mathbf{A}$ is an $n \times n$ matrix. We assume that $\mathbf{A}$ is nonsingular. Then $\mathbf{A}$ is invertible and, hence,

$$\mathbf{X} = \mathbf{A}^{-1}\mathbf{Y} \tag{2.1–58}$$

Equivalently, we have

$$X_i = \sum_{j=1}^{n} b_{ij} Y_j, \qquad i = 1, 2, \ldots, n \tag{2.1–59}$$

where $\{b_{ij}\}$ are the elements of the inverse matrix $\mathbf{A}^{-1}$. The Jacobian of this transformation is $J = 1/\det \mathbf{A}$. Hence

$$p_Y(y_1, y_2, \ldots, y_n)$$
$$= p_X\left(x_1 = \sum_{j=1}^{n} b_{1j}, y_j, x_2 = \sum_{j=1}^{n} b_{2j}y_j, \ldots, x_n = \sum_{j=1}^{n} b_{nj}y_j\right) \frac{1}{|\det \mathbf{A}|} \tag{2.1–60}$$

2.1.3 Statistical Averages of Random Variables

Averages play an important role in the characterization of the outcomes of experiments and the random variables defined on the sample space of the experiments. Of particular interest are the first and second moments of a single random variable and the joint moments, such as the correlation and covariance, between any pair of random variables in a multidimensional set of random variables. Also of great importance are the characteristic function for a single random variable and the joint characteristic function for a multidimensional set of random variables. This section is devoted to the definition of these important statistical averages.

First we consider a single random variable X characterized by its PDF $p(x)$. The *mean* or *expected value* of X is defined as

$$E(X) \equiv m_x = \int_{-\infty}^{\infty} xp(x)\, dx \tag{2.1–61}$$

where $E(\ )$ denotes expectation (statistical averaging). This is the first moment of the random variable X. In general, the nth moment is defined as

$$E(X^n) = \int_{-\infty}^{\infty} x^n p(x)\, dx \tag{2.1–62}$$

Now, suppose that we define a random variable $Y = g(X)$, where $g(X)$ is some arbitrary function of the random variable X. The expected value of Y is

$$E(Y) = E[g(X)] = \int_{-\infty}^{\infty} g(x)p(x)\, dx \tag{2.1–63}$$

In particular, if $Y = (X - m_x)^n$ where m_x is the mean value of X, then

$$E(Y) = E[(X - m_x)^n] = \int_{-\infty}^{\infty} (x - m_x)^n p(x)\, dx \tag{2.1–64}$$

This expected value is called the nth *central moment* of the random variable X, because it is a moment taken relative to the mean. When $n = 2$, the central moment is called the *variance* of the random variable and denoted as σ_x^2. That is,

$$\sigma_x^2 = \int_{-\infty}^{\infty} (x - m_x)^2 p(x)\, dx \tag{2.1–65}$$

This parameter provides a measure of the dispersion of the random variable X. By expanding the term $(x - m_x)^2$ in the integral of Equation 2.1–65 and noting that the expected value of a constant is equal to the constant, we obtain the expression that relates the variance to the first and second moments, namely,

$$\begin{aligned} \sigma_x^2 &= E(X^2) - [E(X)]^2 \\ &= E(X^2) - m_x^2 \end{aligned} \tag{2.1–66}$$

In the case of two random variables, X_1 and X_2, with joint PDF $p(x_1, x_2)$, we define the *joint moment* as

$$E(X_1^k X_2^n) = \int_{-\infty}^{\infty} \int_{-\infty}^{\infty} x_1^k x_2^n p(x_1, x_2)\, dx_1\, dx_2 \tag{2.1–67}$$

and the *joint central moment* as

$$\begin{aligned} &E[(X_1 - m_1)^k (X_2 - m_2)^n] \\ &\quad = \int_{-\infty}^{\infty} \int_{-\infty}^{\infty} (x_1 - m_1)^k (x_2 - m_2)^n p(x_1, x_2) dx_1 dx_2 \end{aligned} \tag{2.1–68}$$

where $m_i = E(X_i)$. Of particular importance to us are the joint moment and joint central moment correspondong to $k = n = 1$. These joint moments are called the *correlation* and the *covariance* of the random variables X_1 and X_2, respectively.

In considering multidimensional random variables, we can define joint moments of any order. However, the moments that are most useful in practical applications are the correlations and covariances between pairs of random variables. To elaborate, suppose that X_i, $i = 1, 2, \ldots, n$, are random variables with joint PDF $p(x_1, x_2, \ldots, x_n)$. Let $p(x_i, x_j)$ be the joint PDF of the random variables X_i and X_j. Then the correlation between X_i and X_j is given by the joint moment

$$E(X_i X_j) = \int_{-\infty}^{\infty} \int_{-\infty}^{\infty} x_i x_j (p(x_i, x_j)\, dx_i\, dx_j \tag{2.1–69}$$

and the covariance of X_i and X_j is

$$\begin{aligned} \mu_{ij} &\equiv E[(X_i - m_i)(X_j - m_j)] \\ &= \int_{-\infty}^{\infty} \int_{-\infty}^{\infty} (x_i - m_i)(x_j - m_j) p(x_i, x_j)\, dx_i\, dx_j \\ &= \int_{-\infty}^{\infty} \int_{-\infty}^{\infty} x_i x_j p(x_i, x_j)\, dx_i\, dx_j - m_i m_j \\ &= E(X_i X_j) - m_i m_j \end{aligned} \tag{2.1–70}$$

The $n \times n$ matrix with elements μ_{ij} is called the *covariance matrix* of the random variables X_i, $i = 1, 2, \ldots, n$. We shall encounter the covariance matrix in our discussion of jointly Gaussian random variables in Section 2.1.4.

Two random variables are said to be *uncorrelated* if $E(X_iX_j) = E(X_i)E(X_j) = m_im_j$. In that case, the covariance $\mu_{ij} = 0$. We note that when X_i and X_j are statistically independent, they are also uncorrelated. However, if X_i and X_j are uncorrelated, they are not necessarily statistically independent.

Two random variables are said to be *orthogonal* if $E(X_iX_j) = 0$. We note that this condition holds when X_i and X_j are uncorrelated and either one or both of the random variables have zero mean.

Characteristic functions. The *characteristic function* of a random variable X is defined as the statistical average

$$E(e^{jvX}) \equiv \psi(jv) = \int_{-\infty}^{\infty} e^{jvx} p(x)\, dx \tag{2.1–71}$$

where the variable v is real and $j = \sqrt{-1}$. We note that $\psi(jv)$ may be described as the Fourier transform† of the PDF $p(x)$. Hence the inverse Fourier transform is

$$p(x) = \frac{1}{2\pi}\int_{-\infty}^{\infty} \psi(jv)e^{-jvx}\, dv \tag{2.1–72}$$

One useful property of the characteristic function is its relation to the moments of the random variable. We note that the first derivative of Equation 2.1–71 with respect to v yields

$$\frac{d\psi(jv)}{dv} = j\int_{-\infty}^{\infty} xe^{jvx} p(x)\, dx$$

By evaluating the derivative at $v = 0$, we obtain the first moment (mean)

$$E(X) = m_x = -j\frac{d\psi(jv)}{dv}\bigg|_{v=0} \tag{2.1–73}$$

The differentiation process can be repeated, so that the nth derivative of $\psi(jv)$ evaluated at $v = 0$ yields the nth moment

$$E(X^n) = (-j)^n \frac{d^n\psi(jv)}{dv^n}\bigg|_{v=0} \tag{2.1–74}$$

Thus the moments of a random variable can be determined from the characteristic function. On the other hand, suppose that the characteristic function can be expanded in a Taylor series about the point $v = 0$. That is,

$$\psi(jv) = \sum_{n=0}^{\infty}\left[\frac{d^n\psi(jv)}{dv^n}\right]_{v=0}\frac{v^n}{n!} \tag{2.1–75}$$

† Usually the Fourier transform of a function $g(u)$ is defined as $G(v) = \int_{-\infty}^{\infty} g(u)e^{-juv}\, du$, which differs from Equation 2.1–71 by the negative sign in the exponential. This is a trivial difference, however, so we call the integral in Equation 2.1–71 a Fourier transform.

Using the relation in Equation 2.1–74 to eliminate the derivative in Equation 2.1–75, we obtain an expression for the characteristic function in terms of its moments in the form

$$\psi(jv) = \sum_{n=0}^{\infty} E(X^n)\frac{(jv)^n}{n!} \tag{2.1–76}$$

The characteristic function provides a simple method for determining the PDF of a sum of statistically independent random variables. To illustrate this point, let X_i, $i = 1, 2, \ldots, n$, be a set of n statistically independent random variables and let

$$Y = \sum_{i=1}^{n} X_i \tag{2.1–77}$$

The problem is to determine the PDF of Y. We shall determine the PDF of Y by first finding its characteristic function and then computing the inverse Fourier transform. Thus

$$\begin{aligned}
\psi_Y(jv) &= E(e^{jvY}) \\
&= E\left[\exp\left(jv\sum_{i=1}^{n} X_i\right)\right] \\
&= E\left[\prod_{i=1}^{n}(e^{jvX_i})\right] \\
&= \int_{-\infty}^{\infty}\cdots\int_{-\infty}^{\infty}\left(\prod_{i=1}^{n} e^{jvx_i}\right)p(x_1, x_2, \ldots, x_n)dx_1dx_2\cdots dx_n
\end{aligned} \tag{2.1–78}$$

Since the random variables are statistically independent, $p(x_1, x_2, \ldots, x_n) = p(x_1)p(x_2)\ldots p(x_n)$, and, hence, the nth order integral in Equation 2.1–78 reduces to a product of n single integrals, each corresponding to the characteristic function of one of the X_i. Hence,

$$\psi_Y(jv) = \prod_{i=1}^{n} \psi_{X_i}(jv) \tag{2.1–79}$$

If, in addition to their statistical independence, the X_i are identically distributed, then all the $\psi_{Xi}(jv)$ are identical. Consequently,

$$\psi_Y(jv) = [\psi_X(jv)]^n \tag{2.1–80}$$

Finally, the PDF of Y is determined from the inverse Fourier transform of $\psi_Y(jv)$, given by Equation 2.1–72.

Since the characteristic function of the sum of n statistically independent random variables is equal to the product of the characteristic functions of the individual random variables X_i, $i = 2, \ldots, n$, it follows that, in the transform domain, the PDF of Y is the n-fold convolution of the PDFs of the X_i.

Usually the n-fold convolution is more difficult to perform than the characteristic function method described above in determining the PDF of Y.

When working with n-dimensional random variables, it is appropriate to define an n-dimensional Fourier transform of the joint PDF. In particular, if X_i, $i = 1, 2, \ldots, n$, are random variables with PDF $p(x_1, x_2, \ldots, x_n)$, the *n-dimensional characteristic function* is defined as

$$\begin{aligned}\psi(jv_1, jv_2, \ldots, jv_n) &= E\left[\exp\left(j\sum_{i=1}^{n} v_i X_i\right)\right] \\ &= \int_{-\infty}^{\infty} \cdots \int_{-\infty}^{\infty} \exp\left(j\sum_{i=1}^{n} v_i x_i\right) p(x_1, x_2, \ldots, x_n)\, dx_1\, dx_2 \cdots dx_n \end{aligned} \tag{2.1–81}$$

Of special interest is the two-dimensional characteristic function

$$\psi(jv_1, jv_2) = \int_{-\infty}^{\infty}\int_{-\infty}^{\infty} e^{j(v_1 x_1 + v_2 x_2)} p(x_1, x_2) dx_1\, dx_2 \tag{2.1–82}$$

We observe that the partial derivatives of $\psi(jv_1, jv_2)$ with respect to v_1 and v_2 can be used to generate the joint moments. For example, it is easy to show that

$$E(X_1 X_2) = -\left.\frac{\partial^2 \psi(jv_1, jv_2)}{\partial v_1 \partial v_2}\right|_{v_1 = v_2 = 0} \tag{2.1–83}$$

Higher-order moments are generated in a straightforward manner.

2.1.4 Some Useful Probability Distributions

In subsequent chapters, we shall encounter several different types of random variables. In this section we list these frequently encountered random variables, their PDFs, their CDFs, and their moments. We begin with the binomial distribution, which is the distribution of a discrete random variable, and then we present the distributions of several continuous random variables.

Binomial distribution. Let X be a discrete random variable that has two possible values, say $X = 1$ or $X = 0$, with probabilities p and $1 - p$, respectively. The PDF of X is shown in Figure 2.1–6. Now, suppose that

$$Y = \sum_{i=1}^{n} X_i$$

where the X_i, $i = 1, 2, \ldots, n$, are statistically independent and identically distributed random variables with the PDF shown in Figure 2.1–6. What is the probability distribution function of Y?

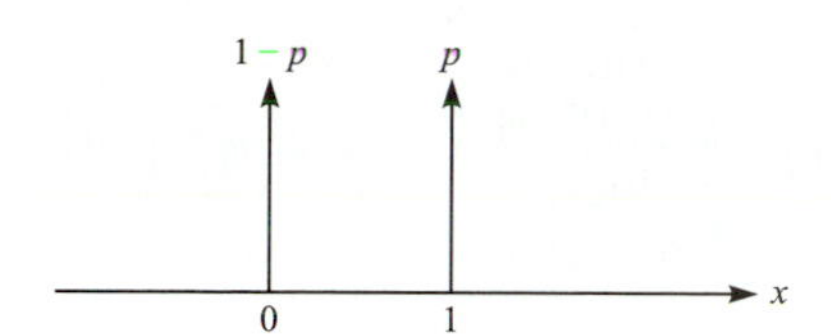

FIGURE 2.1–6
The probability distribution function of X.

To answer this question, we observe that the range of Y is the set of integers from 0 to n. The probability that $Y = 0$ is simply the probability that all the $X_i = 0$. Since the X_i are statistically independent,

$$P(Y = 0) = (1 - p)^n$$

The probability that $Y = 1$ is simply the probability that one $X_i = 1$ and the rest of the $X_i = 0$. Since this event can occur in n different ways,

$$P(Y = 1) = np(1 - p)^{n-1}$$

To generalize, the probability that $Y = k$ is the probability that k of the X_i are equal to one and $n - k$ are equal to zero. Since there are

$$\binom{n}{k} \equiv \frac{n!}{k!(n-k)!} \tag{2.1–84}$$

different combinations that result in the event $\{Y = k\}$, it follows that

$$P(Y = k) = \binom{n}{k} p^k (1 - p)^{n-k} \tag{2.1–85}$$

where $\binom{n}{k}$ is the binomial coefficient. Consequently, the PDF of Y may be expressed as

$$\begin{aligned} p(y) &= \sum_{k=0}^{n} P(Y = k)\delta(y - k) \\ &= \sum_{k=0}^{n} \binom{n}{k} p^k (1 - p)^{n-k} \delta(y - k) \end{aligned} \tag{2.1–86}$$

The CDF of Y is

$$\begin{aligned} F(y) &= P(Y \leqslant y) \\ &= \sum_{k=0}^{[y]} \binom{n}{k} p^k (1 - p)^{n-k} \end{aligned} \tag{2.1–87}$$

where [y] denotes the largest integer m such that $m \leqslant y$. The CDF in Equation 2.1–87 characterizes a binomially distributed random variable.

The first two moments of Y are

$$E(Y) = np$$
$$E(Y^2) = np(1-p) + n^2p^2 \tag{2.1–88}$$
$$\sigma^2 = np(1-p)$$

and the characteristic function is

$$\psi(jv) = (1 - p + pe^{jv})^n \tag{2.1–89}$$

Uniform distribution. The PDF and CDF of a uniformly distributed random variable X are shown in Figure 2.1–7. The first two moments of X are

$$E(X) = \tfrac{1}{2}(a+b)$$
$$E(X^2) = \tfrac{1}{3}(a^2 + b^2 + ab) \tag{2.1–90}$$
$$\sigma^2 = \tfrac{1}{12}(a-b)^2$$

and the characteristic function is

$$\psi(jv) = \frac{e^{jvb} - e^{jva}}{jv(b-a)} \tag{2.1–91}$$

Gaussian (normal) distribution. The PDF of a Gaussian or normally distributed random variable is

$$p(x) = \frac{1}{\sqrt{2\pi}\sigma} e^{-(x-m_x)^2/2\sigma^2} \tag{2.1–92}$$

where m_x is the mean and σ^2 is the variance of the random variable. The CDF is

$$\begin{aligned} F(x) &= \int_{-\infty}^{x} p(u)\, du \\ &= \frac{1}{\sqrt{2\pi}\sigma} \int_{-\infty}^{x} e^{-(u-m_x)^2/2\sigma^2}\, du \\ &= \frac{1}{2}\frac{2}{\sqrt{\pi}} \int_{-\infty}^{(x-m_x)/\sqrt{2}\sigma} e^{-t^2}\, dt \\ &= \tfrac{1}{2} + \tfrac{1}{2}\text{erf}\left(\frac{x-m_x}{\sqrt{2}\sigma}\right) \end{aligned} \tag{2.1–93}$$

where erf(x) denotes the error function, defined as

$$\text{erf}(x) = \frac{2}{\sqrt{\pi}} \int_0^x e^{-t^2}\, dt \tag{2.1–94}$$

The PDF and CDF are illustrated in Figure 2.1–8.

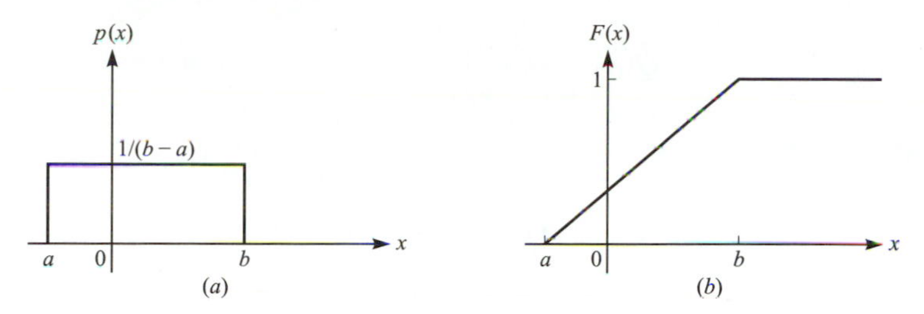

FIGURE 2.1–7
The PDF and CDF of a uniformly distributed random variable.

The CDF $F(x)$ may also be expressed in terms of the complementary error function. That is,

$$F(x) = 1 - \tfrac{1}{2}\mathrm{erfc}\left(\frac{x - m_x}{\sqrt{2}\sigma}\right)$$

where

$$\begin{aligned}\mathrm{erfc}(x) &= \frac{2}{\sqrt{\pi}}\int_x^\infty e^{-t^2}\,dt \\ &= 1 - \mathrm{erf}(x) \qquad (2.1\text{–}95)\end{aligned}$$

We note that $\mathrm{erf}(-x) = -\mathrm{erf}(x)$, $\mathrm{erfc}(-x) = 2 - \mathrm{erfc}(x)$, $\mathrm{erf}(0) = \mathrm{erfc}(\infty) = 0$, and $\mathrm{erf}(\infty) = \mathrm{erfc}(0) = 1$. For $x > m_x$, the complementary error function is proportional to the area under the tail of the Gaussian PDF. For large values of x, the complementary error function $\mathrm{erfc}(x)$ may be approximated by the asymptotic series

$$\mathrm{erfc}(x) = \frac{e^{-x^2}}{x\sqrt{\pi}}\left(1 - \frac{1}{2x^2} + \frac{1\cdot 3}{2^2 x^4} - \frac{1\cdot 3\cdot 5}{2^3 x^6} + \cdots\right) \qquad (2.1\text{–}96)$$

where the approximation error is less than the last term used.

The function that is frequently used for the area under the tail of the Gaussian PDF is denoted by $Q(x)$ and defined as

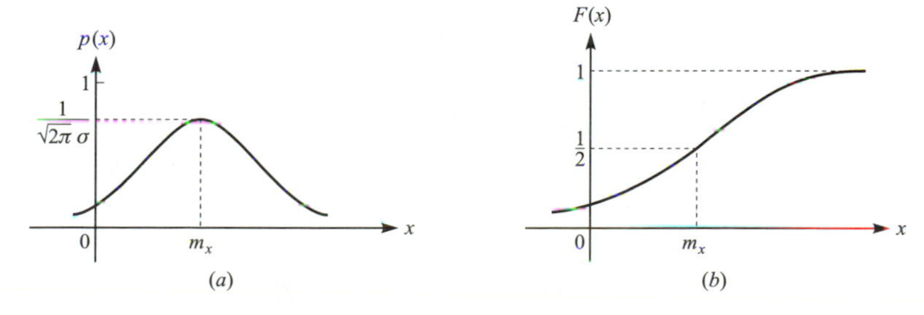

FIGURE 2.1–8
The PDF and CDF of a Gaussian-distributed random variable.

$$Q(x) = \frac{1}{\sqrt{2\pi}} \int_x^\infty e^{-t^2/2}\, dt, \qquad x \geqslant 0 \tag{2.1–97}$$

By comparing Equations 2.1–95 and 2.1–97, we find

$$Q(x) = \tfrac{1}{2}\,\mathrm{erfc}\left(\frac{x}{\sqrt{2}}\right) \tag{2.1–98}$$

The characteristic function of a Gaussian random variable with mean m_x and variance σ^2 is

$$\begin{aligned} \psi(jv) &= \int_{-\infty}^{\infty} e^{jvx} \left[\frac{1}{\sqrt{2\pi}\sigma} e^{-(x-m_x)^2/2\sigma^2} \right] dx \\ &= e^{jvm_x - (1/2)v^2\sigma^2} \end{aligned} \tag{2.1–99}$$

The central moments of a Gaussian random variable are

$$E[(X-m_x)^k] \equiv \mu_k = \begin{cases} 1\cdot 3 \cdots (k-1)\sigma^k & \text{(even } k\text{)} \\ 0 & \text{(odd } k\text{)} \end{cases} \tag{2.1–100}$$

and the ordinary moments may be expressed in terms of the central moments as

$$E(X^k) = \sum_{i=0}^{k} \binom{k}{i} m_x^i \mu_{k-i} \tag{2.1–101}$$

The sum of n statistically independent Gaussian random variables is also a Gaussian random variable. To demonstrate this point, let

$$Y = \sum_{i=1}^{n} X_i \tag{2.1–102}$$

where the X_i, $i = 1, 2, \ldots, n$, are statistically independent Gaussian random variables with means m_i and variances σ_i^2. Using the result in Equation 2.1–79, we find that the characteristic function of Y is

$$\begin{aligned} \psi_Y(jv) &= \prod_{i=1}^{n} \psi_{X_i}(jv) \\ &= \prod_{i=1}^{n} e^{jvm_i - v^2\sigma_i^2/2} \\ &= e^{jvm_y - v^2\sigma_y^2/2} \end{aligned} \tag{2.1–103}$$

where

$$m_y = \sum_{i=1}^{n} m_i, \qquad \sigma_y^2 = \sum_{i=1}^{n} \sigma_i^2 \tag{2.1–104}$$

Therefore, Y is Gaussian-distributed with mean m_y and variance σ_y^2.

Chi-square distribution. A chi-square-distributed random variable is related to a Gaussian-distributed random variable in the sense that the former can be viewed as a transformation of the latter. To be specific, let $Y = X^2$, where X is a Gaussian random variable. Then Y has a chi-square distribution. We distinguish between two types of chi-square distributions. The first is called a *central chi-square distribution* and is obtained when X has zero mean. The second is called a *noncentral chi-square distribution*, and is obtained when X has a nonzero mean.

First we consider the central chi-square distribution. Let X be Gaussian-distributed with zero mean and variance σ^2. Since $Y = X^2$, the result given in 2.1–47 applies directly with $a = 1$ and $b = 0$. Thus we obtain the PDF of Y in the form

$$p_Y(y) = \frac{1}{\sqrt{2\pi y}\sigma} e^{-y/2\sigma^2}, \qquad y \geqslant 0 \tag{2.1–105}$$

The CDF of Y is

$$\begin{aligned} F_Y(y) &= \int_0^y p_Y(u)\, du \\ &= \frac{1}{\sqrt{2\pi}\sigma} \int_0^y \frac{1}{\sqrt{u}} e^{-u/2\sigma^2}\, du \end{aligned} \tag{2.1–106}$$

which cannot be expressed in closed form. The characteristic function, however, can be determined in closed form. It is

$$\psi(jv) = \frac{1}{(1 - j2v\sigma^2)^{1/2}} \tag{2.1–107}$$

Now, suppose that the random variable Y is defined as

$$Y = \sum_{i=1}^{n} X_i^2 \tag{2.1–108}$$

where the X_i, $i = 1, 2, \ldots, n$, are statistically independent and identically distributed Gaussian random variables with zero mean and variance σ^2. As a consequence of the statistical independence of the X_i, the characteristic function of Y is

$$\psi_Y(jv) = \frac{1}{(1 - j2v\sigma^2)^{n/2}} \tag{2.1–109}$$

The inverse transform of this characteristic function yields the PDF

$$p_Y(y) = \frac{1}{\sigma^n 2^{n/2} \Gamma(\frac{1}{2}n)} y^{n/2-1} e^{-y/2\sigma^2}, \qquad y \geqslant 0 \tag{2.1–110}$$

where $\Gamma(p)$ is the gamma function, defined as

$$\begin{aligned} \Gamma(p) &= \int_0^{\infty} t^{p-1} e^{-t}\, dt, \qquad p > 0 \\ \Gamma(p) &= (p-1)!, \qquad p \text{ an integer} > 0 \\ \Gamma(\tfrac{1}{2}) &= \sqrt{\pi} \qquad \Gamma(\tfrac{3}{2}) = \tfrac{1}{2}\sqrt{\pi} \end{aligned} \tag{2.1–111}$$

This PDF, which is a generalization of Equation 2.1–105, is called a *chi-square* (or *gamma*) *PDF with n degrees of freedom*. It is illustrated in Figure 2.1–9. The case $n = 2$ yields the exponential distribution.

The first two moments of Y are

$$\begin{aligned} E(Y) &= n\sigma^2 \\ E(Y^2) &= 2n\sigma^4 + n^2\sigma^4 \\ \sigma_y^2 &= 2n\sigma^4 \end{aligned} \tag{2.1–112}$$

The CDF of Y is

$$F_Y(y) = \int_0^y \frac{1}{\sigma^n 2^{n/2} \Gamma(\frac{1}{2}n)} u^{n/2-1} e^{-u/2\sigma^2}\, du, \qquad y \geqslant 0 \tag{2.1–113}$$

This integral can be easily manipulated into the form of the incomplete gamma function, which is tabulated by Pearson (1965). When n is even, the integral in Equation 2.1–113 can be expressed in closed form. Specifically, let $m = \frac{1}{2}n$, where m is an integer. Then, by repeated integration by parts, we obtain

$$F_Y(y) = 1 - e^{-y/2\sigma^2} \sum_{k=0}^{m-1} \frac{1}{k!} \left(\frac{y}{2\sigma^2} \right)^k, \qquad y \geqslant 0 \tag{2.1–114}$$

Let us now consider a noncentral chi-square distribution, which results from squaring a Gaussian random variable having a nonzero mean. If X is Gaussian with mean m_x and variance σ^2, the random variable $Y = X^2$ has the PDF

$$p_Y(y) = \frac{1}{\sqrt{2\pi y}\,\sigma} e^{-(y+m_x^2)/2\sigma^2} \cosh\left(\frac{\sqrt{y}\, m_x}{\sigma^2} \right), \qquad y \geqslant 0 \tag{2.1–115}$$

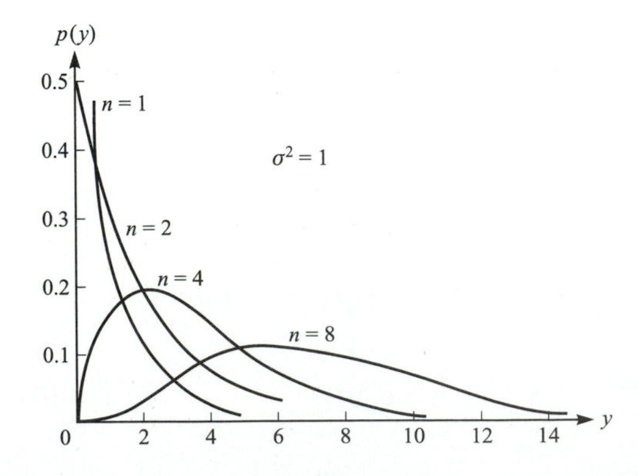

FIGURE 2.1–9
The PDF of a chi-square-distributed random variable for several degrees of freedom.

which is obtained by applying the result in Equation 2.1–47 to the Gaussian PDF given by Equation 2.1–92. The characteristic function corresponding to this PDF is

$$\psi_Y(jv) = \frac{1}{(1 - j2v\sigma^2)^{1/2}} e^{jm_x^2 v/(1-j2v\sigma^2)} \tag{2.1–116}$$

To generalize these results, let Y be the sum of squares of Gaussian random variables as defined by Equation 2.1–108. The X_i, $i = 1, 2, \ldots, n$, are assumed to be statistically independent with means m_i, $i = 1, 2, \ldots, n$, and identical variances equal to σ^2. Then the characteristic function of Y, obtained from Equation 2.1–116 by applying the relation in Equation 2.1–79, is

$$\psi_Y(jv) = \frac{1}{(1 - j2v\sigma^2)^{n/2}} \exp\left(\frac{jv \sum_{i=1}^{n} m_i^2}{1 - j2v\sigma^2} \right) \tag{2.1–117}$$

This characteristic function can be inverse-Fourier-transformed to yield the PDF

$$p_Y(y) = \frac{1}{2\sigma^2} \left(\frac{y}{s^2} \right)^{(n-2)/4} e^{-(s^2+y)/2\sigma^2} I_{n/2-1}\left(\sqrt{y} \frac{s}{\sigma^2} \right), \qquad y \geqslant 0 \tag{2.1–118}$$

where, by definition,

$$s^2 = \sum_{i=1}^{n} m_i^2 \tag{2.1–119}$$

and $I_\alpha(x)$ is the αth-order modified Bessel function of the first kind, which may be represented by the infinite series

$$I_\alpha(x) = \sum_{k=0}^{\infty} \frac{(x/2)^{\alpha+2k}}{k!\,\Gamma(\alpha + k + 1)}, \qquad x \geqslant 0 \tag{2.1–120}$$

The PDF given by Equation 2.1–118 is called the *noncentral chi-square PDF with n degrees of freedom*. The parameter s^2 is called the *noncentrality parameter of the distribution*.

The CDF of the noncentral chi-square with n degrees of freedom is

$$F_Y(y) = \int_0^y \frac{1}{2\sigma^2} \left(\frac{u}{s^2} \right)^{(n-2)/4} e^{-(s^2+u)/2\sigma^2} I_{n/2-1}\left(\sqrt{u} \frac{s}{\sigma^2} \right) du \tag{2.1–121}$$

There is no closed-form expression for this integral. However, when $m = \frac{1}{2}n$ is an integer, the CDF can be expressed in terms of the generalized Marcum's Q function, which is defined as

$$Q_m(a,b) = \int_b^{\infty} x\left(\frac{x}{a}\right)^{m-1} e^{-(x^2+a^2)/2} I_{m-1}(ax)\,dx$$
$$= Q_1(a,b) + e^{-(a^2+b^2)/2} \sum_{k=1}^{m-1} \left(\frac{b}{a}\right)^k I_k(ab) \tag{2.1–122}$$

where

$$Q_1(a,b) = e^{-(a^2+b^2)/2} \sum_{k=0}^{\infty} \left(\frac{a}{b}\right)^k I_k(ab), \qquad b > a > 0 \tag{2.1–123}$$

If we change the variable of integration in Equation 2.1–121 from u to x, where

$$x^2 = \frac{u}{\sigma^2}$$

and let $a^2 = s^2/\sigma^2$, then it is easily shown that

$$F_Y(y) = 1 - Q_m\left(\frac{s}{\sigma}, \frac{\sqrt{y}}{\sigma}\right) \tag{2.1–124}$$

Finally, we state that the first two moments of a noncentral chi-square-distributed random variable are

$$\begin{aligned} E(Y) &= n\sigma^2 + s^2 \\ E(Y^2) &= 2n\sigma^4 + 4\sigma^2 s^2 + (n\sigma^2 + s^2)^2 \\ \sigma_y^2 &= 2n\sigma^4 + 4\sigma^2 s^2 \end{aligned} \tag{2.1–125}$$

Rayleigh distribution. The Rayleigh distribution is frequently used to model the statistics of signals transmitted through radio channels such as cellular radio. This distribution is closely related to the central chi-square distribution. To illustrate this point, let $Y = X_1^2 + X_2^2$ where X_1 and X_2 are zero-mean statistically independent Gaussian random variables, each having a variance σ^2 From the discussion above, it follows that Y is chi-square-distributed with two degrees of freedom. Hence, the PDF of Y is

$$p_Y(y) = \frac{1}{2\sigma^2} e^{-y/2\sigma^2}, \qquad y \geqslant 0 \tag{2.1–126}$$

Now, suppose we define a new random variable

$$R = \sqrt{X_1^2 + X_2^2} = \sqrt{Y} \tag{2.1–127}$$

Making a simple change of variable in the PDF of Equation 2.1–126, we obtain the PDF of R in the form

$$p_R(r) = \frac{r}{\sigma^2} e^{-r^2/2\sigma^2}, \qquad r \geqslant 0 \tag{2.1–128}$$

This is the PDF of a Rayleigh-distributed random variable. The corresponding CDF if

$$\begin{aligned} F_R(r) &= \int_0^r \frac{u}{\sigma^2} e^{-u^2/2\sigma^2}\, du \\ &= 1 - e^{-r^2/2\sigma^2}, \qquad r \geqslant 0 \end{aligned} \tag{2.1–129}$$

The moments of R are

$$E(R^k) = (2\sigma^2)^{k/2}\Gamma(1 + \tfrac{1}{2}k) \tag{2.1–130}$$

and the variance is

$$\sigma_r^2 = (2 - \tfrac{1}{2}\pi)\sigma^2 \tag{2.1–131}$$

The characteristic function of the Rayleigh-distributed random variable is

$$\psi_R(jv) = \int_0^\infty \frac{r}{\sigma^2} e^{-r^2/2\sigma^2} e^{jvr}\, dr \tag{2.1–132}$$

This integral may be expressed as

$$\begin{aligned} \psi_R(jv) &= \int_0^\infty \frac{r}{\sigma^2} e^{-r^2/2\sigma^2} \cos vr\, dr + j\int_0^\infty \frac{r}{\sigma^2} e^{-r^2/2\sigma^2} \sin vr\, dr \\ &= {}_1F_1(1, \tfrac{1}{2}; -\tfrac{1}{2}v^2\sigma^2) + j\sqrt{\tfrac{1}{2}\pi}\, v\sigma^2 e^{-v^2\sigma^2/2} \end{aligned} \tag{2.1–133}$$

where ${}_1F_1(1, \frac{1}{2}; -a)$ is the confluent hypergeometric function, which is defined as

$${}_1F_1(\alpha, \beta; x) = \sum_{k=0}^{\infty} \frac{\Gamma(\alpha + k)\Gamma(\beta)x^k}{\Gamma(\alpha)\Gamma(\beta + k)k!}, \qquad \beta \neq 0, -1, -2, \ldots \tag{2.1–134}$$

Beaulieu (1990) has shown that ${}_1F_1(1, \frac{1}{2}; -a)$ may be expressed as

$${}_1F_1(1, \tfrac{1}{2}; -a) = -e^{-a} \sum_{k=0}^{\infty} \frac{a^k}{(2k-1)k!} \tag{2.1–135}$$

As a generalization of the above expression, consider the random variable

$$R = \sqrt{\sum_{i=1}^{n} X_i^2} \tag{2.1–136}$$

where the X_i, $i = 1, 2, \ldots n$, are statistically independent, identically distributed zero mean Gaussian random variables. The random variable R has a generalized Rayleigh distribution. Clearly, $Y = R^2$ is chi-square-distributed with n degrees of freedom. Its PDF is given by Equation 2.1–110. A simple change in variable in Equation 2.1–110 yields the PDF of R in the form

$$p_R(r) = \frac{r^{n-1}}{2^{(n-2)/2}\sigma^n\Gamma(\frac{1}{2}n)} e^{-r^2/2\sigma^2}, \qquad r \geqslant 0 \tag{2.1–137}$$

As a consequence of the functional relationship between the central chi-square and the Rayleigh distributions, the corresponding CDFs are similar. Thus, for any n, the CDF of R can be put in the form of the incomplete gamma function. In the special case when n is even, i.e., $n = 2m$, the CDF of R can be expressed in the closed form

$$F_R(r) = 1 - e^{-r^2/2\sigma^2} \sum_{k=0}^{m-1} \frac{1}{k!}\left(\frac{r^2}{2\sigma^2}\right)^k, \qquad r \geqslant 0 \tag{2.1–138}$$

Finally, we state that the kth moment of R is

$$E(R^k) = (2\sigma^2)^{k/2} \frac{\Gamma[\frac{1}{2}(n+k)]}{\Gamma(\frac{1}{2}n)}, \qquad k \geqslant 0 \tag{2.1–139}$$

which holds for any integer n.

Rice distribution. Just as the Rayleigh distribution is related to the central chi-square distribution, the Rice distribution is related to the noncentral chi-square distribution. To illustrate this relation, let $Y = X_1^2 + X_2^2$, where X_1 and X_2 are statistically independent Gaussian random variables with means m_i, $i = 1, 2$, and common variance σ^2. From the previous discussion, we know that Y has a noncentral chi-square distribution with noncentrality parameter $s^2 = m_1^2 + m_2^2$. The PDF of Y, obtained from Equation 2.1–118 for $n = 2$, is

$$p_Y(y) = \frac{1}{2\sigma^2} e^{-(s^2+y)/2\sigma^2} I_0\left(\sqrt{y}\frac{s}{\sigma^2}\right), \qquad y \geqslant 0 \tag{2.1–140}$$

Now, we define a new random variable $R = \sqrt{Y}$. The PDF of R, obtained from Equation 2.1–140 by a simple change of variable, is

$$p_R(r) = \frac{r}{\sigma^2} e^{-(r^2+s^2)/2\sigma^2} I_0\left(\frac{rs}{\sigma^2}\right), \qquad r \geqslant 0 \tag{2.1–141}$$

This is the PDF of a Ricean-distributed random variable. As will be shown in Chapter 5, this PDF characterizes the statistics of the envelope of a signal corrupted by additive narrowband Gaussian noise. It is also used to model the signal statistics of signals transmitted through some radio channels. The CDF of R is easily obtained by specializing the results in Equation 2.1–124 to the case $m = 1$. This yields

$$F_R(r) = 1 - Q_1\left(\frac{s}{\sigma}, \frac{r}{\sigma}\right), \qquad r \geqslant 0 \tag{2.1–142}$$

where $Q_1(a, b)$ is defined by Equation 2.1–123.

As a generalization of the expressions given above, let R be defined as in Equation 2.1–136 where the X_i, $i = 1, 2, \ldots, n$ are statistically independent Gaussian random variables with means m_i, $1, 2, \ldots, n$, and identical variances equal to σ^2. The random variable $R^2 = Y$ has a noncentral chi-square distribution with n degrees of freedom and noncentrality parameter s^2 given by Equation 2.1–119. Its PDF is given by Equation 2.1–118. Hence the PDF of R is

$$p_R(r) = \frac{r^{n/2}}{\sigma^2 s^{(n-2)/2}} e^{-(r^2+s^2)/2\sigma^2} I_{n/2-1}\left(\frac{rs}{\sigma^2}\right), \qquad r \geqslant 0 \qquad (2.1\text{–}143)$$

and the corresponding CDF is

$$F_R(r) = P(R \leqslant r) = P(\sqrt{Y} \leqslant r) = P(Y \leqslant r^2) = F_Y(r^2) \qquad (2.1\text{–}144)$$

where $F_Y(r^2)$ is given by Equation 2.1–121. In the special case where $m = \frac{1}{2}n$ is an integer, we have

$$F_R(r) = 1 - Q_m\left(\frac{s}{\sigma}, \frac{r}{\sigma}\right), \qquad r \geqslant 0 \qquad (2.1\text{–}145)$$

which follows from Equation 2.1–124. Finally, we state that the kth moment of R is

$$E(R^k) = (2\sigma^2)^{k/2} e^{-s^2/2\sigma^2} \frac{\Gamma[\frac{1}{2}(n+k)]}{\Gamma(\frac{1}{2}n)} {}_1F_1\left(\frac{n+k}{2}, \frac{n}{2}; \frac{s^2}{2\sigma^2}\right); \qquad k \geqslant 0 \qquad (2.1\text{–}146)$$

where ${}_1F_1\ (\alpha, \beta; x)$ is the confluent hypergeometric function.

Nakagami m-distribution. Both the Rayleigh distribution and the Rice distribution are frequently used to describe the statistical fluctuations of signals received from a multipath fading channel. These channel models are considered in Chapter 14. Another distribution that is frequently used to characterize the statistics of signals transmitted through multipath fading channels is the Nakagami m-distribution. The PDF for this distribution is given by Nakagami (1960) as

$$p_R(r) = \frac{2}{\Gamma(m)} \left(\frac{m}{\Omega}\right)^m r^{2m-1} e^{-mr^2/\Omega} \qquad (2.1\text{–}147)$$

where Ω is defined as

$$\Omega = E(R^2)$$

and the parameter m is defined as the ratio of moments, called the *fading figure*,

$$m = \frac{\Omega^2}{E[(R^2 - \Omega)^2]}, \qquad m \geqslant \tfrac{1}{2} \qquad (2.1\text{–}148)$$

A normalized version of Equation 2.1–147 may be obtained by defining another random variable $X = R/\sqrt{\Omega}$ (see Problem 2–15). The nth moment of R is

$$E(R^n) = \frac{\Gamma(m + \frac{1}{2}n)}{\Gamma(m)} \left(\frac{\Omega}{m}\right)^{n/2}$$

By setting $m = 1$, we observe that Equation 2.1–147 reduces to a Rayleigh PDF. For values of m in the range $\frac{1}{2} \leqslant m \leqslant 1$, we obtain PDFs that have larger tails than a Rayleigh-distributed random variable. For values of $m > 1$, the tail of the PDF decays faster than that of the Rayleigh. Figure 2.1–10 illustrates the PDFs for different values of m.

Lognormal distribution. Suppose that a random variable X is normally distributed with mean m and variance σ^2. Let us define a new random variable R that is related to X through the transformation $X = \ln R$. Then the PDF of R is

$$p(r) = \begin{cases} \dfrac{1}{\sqrt{2\pi}\sigma r} e^{-(\ln r - m)^2/2\sigma^2} & (r \geqslant 0) \\ 0 & (r < 0) \end{cases} \tag{2.1–149}$$

The lognormal distribution is suitable for modeling the effect of *shadowing* of the signal due to large obstructions, such as tall buildings, in mobile radio communications.

Multivariate Gaussian distribution. Of the many multivariate or multidimensional distributions that can be defined, the multivariate Gaussian distribution is the most important and the one most likely to be encountered in practice. We shall briefly introduce this distribution and state its basic properties.

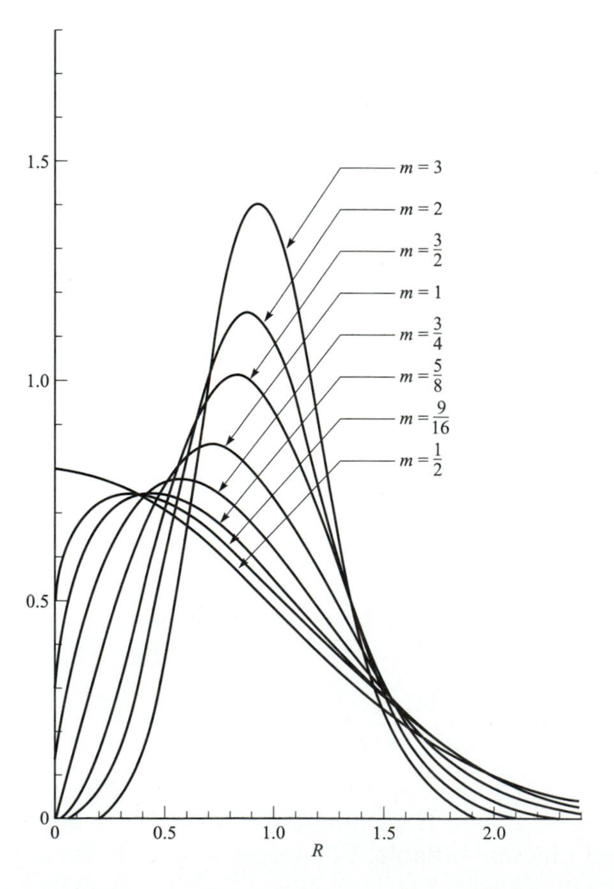

FIGURE 2.1–10
The PDF for the Nakagami-m distribution, shown with $\Omega = 1$. m is the fading figure. (*Miyagaki et al., 1978.*)

Let us assume that X_i, $i = 1, 2, \ldots, n$, are Gaussian random variables with means m_i, $i = 1, 2, \ldots, n$; variances σ_i^2, $i = 1, 2, \ldots, n$; and covariances μ_{ij}, $i, j = 1, 2, \ldots, n$. Clearly, $\mu_{ii} = \sigma_i^2$, $i = 1, 2, \ldots, n$. Let $\mathbf{M}$ denote the $n \times n$ covariance matrix with elements $\{\mu_{ij}\}$, let $\mathbf{X}$ denote the $n \times 1$ column vector of random variables, and let $\mathbf{m}_x$ denote the $n \times 1$ column vector of mean values m_i, $i, 2, \ldots, n$. The joint PDF of the Gaussian random variables X_i, $i = 1, 2, \ldots, n$, is defined as

$$p(x_1, x_2, \ldots, x_n) = \frac{1}{(2\pi)^{n/2}(\det \mathbf{M})^{1/2}} \exp[-\tfrac{1}{2}(\mathbf{x} - \mathbf{m}_x)'\mathbf{M}^{-1}(\mathbf{x} - \mathbf{m}_x)] \tag{2.1–150}$$

where $\mathbf{M}^{-1}$ denotes the inverse of $\mathbf{M}$ and $\mathbf{x}'$ denotes the transpose of $\mathbf{x}$.

The characteristic function corresponding to this n-dimensional joint PDF is

$$\psi(j\mathbf{v}) = E(e^{j\mathbf{v}'\mathbf{x}})$$

where $\mathbf{v}$ is an n-dimensional vector with elements v_i, $i = 1, 2, \ldots, n$. Evaluation of this n-dimensional Fourier transform yields the result

$$\psi(j\mathbf{v}) = \exp(j\mathbf{m}_x'\mathbf{v} - \tfrac{1}{2}\mathbf{v}'\mathbf{M}\mathbf{v}) \tag{2.1–151}$$

An important special case of Equation 2.1–150 is the bivariate or two-dimensional Gaussian PDF. The mean $\mathbf{m}_x$ and the covariance matrix $\mathbf{M}$ for this case are

$$\mathbf{m}_x = \begin{bmatrix} m_1 \\ m_2 \end{bmatrix}, \qquad \mathbf{M} = \begin{bmatrix} \sigma_1^2 & \mu_{12} \\ \mu_{12} & \sigma_2^2 \end{bmatrix} \tag{2.1–152}$$

where the joint central moment μ_{12} is defined as

$$\mu_{12} = E[(X_1 - m_1)(X_2 - m_2)]$$

It is convenient to define a normalized covariance

$$\rho_{ij} = \frac{\mu_{ij}}{\sigma_i \sigma_j}, \qquad i \neq j \tag{2.1–153}$$

where ρ_{ij} satisfies the condition $0 \leqslant |\rho_{ij}| \leqslant 1$. When dealing with the two-dimensional case, it is customary to drop the subscripts on μ_{12} and ρ_{12}. Hence the covariance matrix is expressed as

$$\mathbf{M} = \begin{bmatrix} \sigma_1^2 & \rho\sigma_1\sigma_2 \\ \rho\sigma_1\sigma_2 & \sigma_2^2 \end{bmatrix} \tag{2.1–154}$$

Its inverse is

$$\mathbf{M}^{-1} = \frac{1}{\sigma_1^2\sigma_2^2(1 - \rho^2)} \begin{bmatrix} \sigma_2^2 & -\rho\sigma_1\sigma_2 \\ -\rho\sigma_1\sigma_2 & \sigma_1^2 \end{bmatrix} \tag{2.1–155}$$

and $\det \mathbf{M} = \sigma_1^2\sigma_2^2(1 - \rho^2)$. Substitution for $\mathbf{M}^{-1}$ into Equation 2.1–150 yields the desired bivariate Gaussian PDF in the form

$$p(x_1, x_2) = \frac{1}{2\pi\sigma_1\sigma_2\sqrt{1-\rho^2}} \times \exp\left[-\frac{\sigma_2^2(x_1-m_1)^2 - 2\rho\sigma_1\sigma_2(x_1-m_1)(x_2-m_2)+\sigma_1^2(x_2-m_2)^2}{2\sigma_1^2\sigma_2^2(1-\rho^2)}\right] \tag{2.1–156}$$

We note that when $\rho = 0$, the joint PDF $p(x_1, x_2)$ in Equation 2.1–156 factors into the product $p(x_1)p(x_2)$, where $p(x_i)$, $i = 1, 2$, are the marginal PDFs. Since ρ is a measure of the correlation between X_1 and X_2, we have shown that when the Gaussian random variables X_1 and X_2 are uncorrelated, they are also statistically independent. This is an important property of Gaussian random variables, which does not hold in general for other distributions. It extends to n-dimensional Gaussian random variables in a straightforward manner. That is, if $\rho_{ij} = 0$ for $i \neq j$, then the random variables X_i, $i = 1, 2, \ldots, n$, are uncorrelated and, hence, statistically independent.

Now, let us consider a linear transformation of n Gaussian random variables X_i, $i = 1, 2, \ldots, n$, with mean vector $\mathbf{m}_x$ and covariance matrix $\mathbf{M}$. Let

$$\mathbf{Y} = \mathbf{AX} \tag{2.1–157}$$

where $\mathbf{A}$ is a nonsingular matrix. As shown previously, the Jacobian of this transformation is $J = 1/\det \mathbf{A}$. Since $\mathbf{X} = \mathbf{A}^{-1}\mathbf{Y}$, we may substitute for $\mathbf{X}$ in Equation 2.1–150 and, thus, we obtain the joint PDF of $\mathbf{Y}$ in the form

$$\begin{aligned} p(\mathbf{y}) &= \frac{1}{(2\pi)^{n/2}(\det \mathbf{M})^{1/2} \det \mathbf{A}} \exp[-\tfrac{1}{2}(\mathbf{A}^{-1}\mathbf{y} - \mathbf{m}_x)'\mathbf{M}^{-1}(\mathbf{A}^{-1}\mathbf{y} - \mathbf{m}_x)] \\ &= \frac{1}{(2\pi)^{n/2}(\det \mathbf{Q})^{1/2}} \exp[-\tfrac{1}{2}(\mathbf{y} - \mathbf{m}_y)'\mathbf{Q}^{-1}(\mathbf{y} - \mathbf{m}_y)] \end{aligned} \tag{2.1–158}$$

where the vector $\mathbf{m}_y$ and the matrix $\mathbf{Q}$ are defined as

$$\begin{aligned} \mathbf{m}_y &= \mathbf{Am}_x \\ \mathbf{Q} &= \mathbf{AMA}' \end{aligned} \tag{2.1–159}$$

Thus we have shown that a linear transformation of a set of jointly Gaussian random variables results in another set of jointly Gaussian random variables.

Suppose that we wish to perform a linear transformation that results in n statistically independent Gaussian random variables. How should the matrix $\mathbf{A}$ be selected? From our previous discussion, we know that the Gaussian random variables are statistically independent if they are pairwise-uncorrelated, i.e., if the covariance matrix $\mathbf{Q}$ is diagonal. Therefore, we must have

$$\mathbf{AMA}' = \mathbf{D} \tag{2.1–160}$$

where $\mathbf{D}$ is a diagonal matrix. The matrix $\mathbf{M}$ is a covariance matrix; hence, it is positive-definite. One solution is to select $\mathbf{A}$ to be an orthogonal matrix ($\mathbf{A}' = \mathbf{A}^{-1}$) consisting of columns that are the eigenvectors of the covariance matrix $\mathbf{M}$. Then $\mathbf{D}$ is a diagonal matrix with diagonal elements equal to the eigenvalues of $\mathbf{M}$.

EXAMPLE 2.1–5. Consider the bivariate Gaussian PDF with covariance matrix

$$\mathbf{M} = \begin{bmatrix} 1 & \frac{1}{2} \\ \frac{1}{2} & 1 \end{bmatrix}$$

Let us determine the transformation $\mathbf{A}$ that will result in uncorrelated random variables. First, we solve for the eigenvalues of $\mathbf{M}$. The characteristic equation is

$$\det(\mathbf{M} - \lambda\mathbf{I}) = 0$$
$$(1-\lambda)^2 - \tfrac{1}{4} = 0$$
$$\lambda = \tfrac{3}{2}, \tfrac{1}{2}$$

Next we determine the two eigenvectors. If $\mathbf{a}$ denotes an eigenvector, we have

$$(\mathbf{M} - \lambda\mathbf{I})\mathbf{a} = 0$$

With $\lambda_1 = \frac{3}{2}$ and $\lambda_2 = \frac{1}{2}$, we obtain the eigenvectors

$$\mathbf{a}_1 = \begin{bmatrix} \sqrt{\frac{1}{2}} \\ \sqrt{\frac{1}{2}} \end{bmatrix} \qquad \mathbf{a}_2 = \begin{bmatrix} \sqrt{\frac{1}{2}} \\ -\sqrt{\frac{1}{2}} \end{bmatrix}$$

Therefore,

$$\mathbf{A} = \sqrt{\tfrac{1}{2}}\begin{bmatrix} 1 & 1 \\ 1 & -1 \end{bmatrix}$$

It is easily verified that $\mathbf{A}^{-1} = \mathbf{A}'$ and that

$$\mathbf{AMA}' = \mathbf{D}$$

where the diagonal elements of $\mathbf{D}$ are $\frac{3}{2}$ and $\frac{1}{2}$.

2.1.5 Upper Bounds on the Tail Probability

In evaluating the performance of a digital communication system, it is often necessary to determine the area under the tail of the PDF. We refer to this area as the *tail probability*. In this section, we present two upper bounds on the tail probability. The first, obtained from the Chebyshev inequality, is rather loose. The second, called the *Chernoff bound*, is much tighter.

Chebyshev inequality. Suppose that X is an arbitrary random variable with finite mean m_x and finite variance σ_x^2. For any positive number δ,

$$P(|X - m_x| \geqslant \delta) \leqslant \frac{\sigma_x^2}{\delta^2} \tag{2.1–161}$$

This relation is called the *Chebyshev inequality*. The proof of this bound is relatively simple. We have

$$\sigma_x^2 = \int_{-\infty}^{\infty} (x - m_x)^2 p(x)\, dx \geqslant \int_{|x-m_x| \geqslant \delta} (x - m_x)^2 p(x)\, dx$$

$$\geqslant \delta^2 \int_{|x-m_x| \geqslant \delta} p(x)\, dx = \delta^2 P(|X - m_x| \geqslant \delta)$$

Thus the validity of the inequality is established.

It is apparent that the Chebyshev inequality is simply an upper bound on the area under the tails of the PDF $p(y)$, where $Y = X - m_x$, i.e., the area of $p(y)$ in the intervals $(\infty, -\delta)$ and (δ, ∞). Hence, the Chebyshev inequality may be expressed as

$$1 - [F_Y(\delta) - F_Y(-\delta)] \leqslant \frac{\sigma_x^2}{\delta^2} \tag{2.1–162}$$

or, equivalently, as

$$1 - [F_x(m_x + \delta) - F_x(m_x - \delta)] \leqslant \frac{\sigma_x^2}{\delta^2} \tag{2.1–163}$$

There is another way to view the Chebyshev bound. Working with the zero mean random variable $Y = X - m_x$, for convenience, suppose we define a function $g(Y)$ as

$$g(Y) = \begin{cases} 1 & (|Y| \geqslant \delta) \\ 0 & (|Y| < \delta) \end{cases} \tag{2.1–164}$$

Since $g(Y)$ is either 0 or 1 with probabilities $P(|Y| < \delta)$ and $P(|Y| \geqslant \delta)$, respectively, its mean value is

$$E[g(Y)] = P(|Y| \geqslant \delta) \tag{2.1–165}$$

Now suppose that we upper-bound $g(Y)$ by the quadratic $(Y/\delta)^2$, i.e.,

$$g(Y) \leqslant \left(\frac{Y}{\delta}\right)^2 \tag{2.1–166}$$

The graph of $g(Y)$ and the upper bound are shown in Figure 2.1–11. It follows that

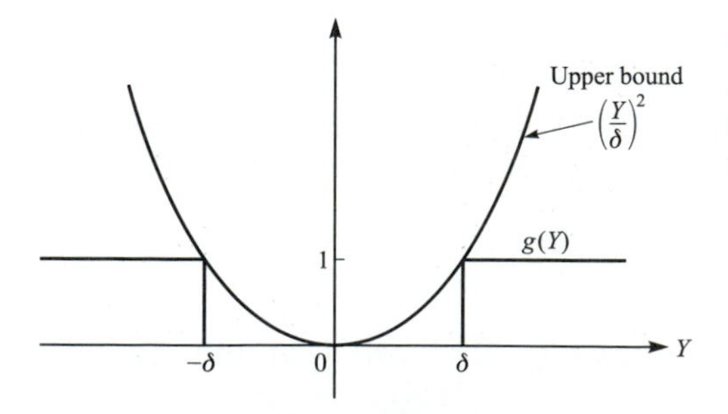

FIGURE 2.1–11
A quadratic upper bound on $g(Y)$ used in obtaining the tail probability (Chebyshev bound).

$$E[g(Y)] \leqslant E\left(\frac{Y^2}{\delta^2}\right) = \frac{E(Y^2)}{\delta^2} = \frac{\sigma_y^2}{\delta^2} = \frac{\sigma_x^2}{\delta^2}$$

Since $E[g(Y)]$ is the tail probability, as seen from Equation 2.1–165, we have obtained the Chebyshev bound.

For many practical applications, the Chebyshev bound is extremely loose. The reason for this may be attributed to the looseness of the quadratic $(Y/\delta)^2$ in overbounding $g(Y)$. There are certainly many other functions that can be used to overbound $g(Y)$. Below, we use an exponential bound to derive an upper bound on the tail probability that is extremely tight.

Chernoff bound. The Chebyshev bound given above involves the area under the two tails of the PDF. In some applications we are interested only in the area under one tail, either in the interval (δ, ∞) or in the interval $(-\infty, \delta)$. In such a case we can obtain an extremely tight upper bound by overbounding the function $g(Y)$ by an exponential having a parameter that can be optimized to yield as tight an upper bound as possible. Specifically, we consider the tail probability in the interval (δ, ∞). The function $g(Y)$ is overbounded as

$$g(Y) \leqslant e^{\nu(Y-\delta)} \tag{2.1–167}$$

where $g(Y)$ is now defined as

$$g(Y) = \begin{cases} 1 & (Y \geqslant \delta) \\ 0 & (Y < \delta) \end{cases} \tag{2.1–168}$$

and $\nu \geqslant 0$ is the parameter to be optimized. The graph of $g(Y)$ and the exponential upper bound are shown in Figure 2.1–12.

The expected value of $g(Y)$ is

$$E[g(Y)] = P(Y \geqslant \delta) \leqslant E(e^{\nu(Y-\delta)}) \tag{2.1–169}$$

This bound is valid for any $\nu \geqslant 0$. The tightest upper bound is obtained by selecting the value of ν that minimizes $E(e^{\nu(Y-\delta)})$. A necessary condition for a minimum is

$$\frac{d}{d\nu} E(e^{\nu(Y-\delta)}) = 0 \tag{2.1–170}$$

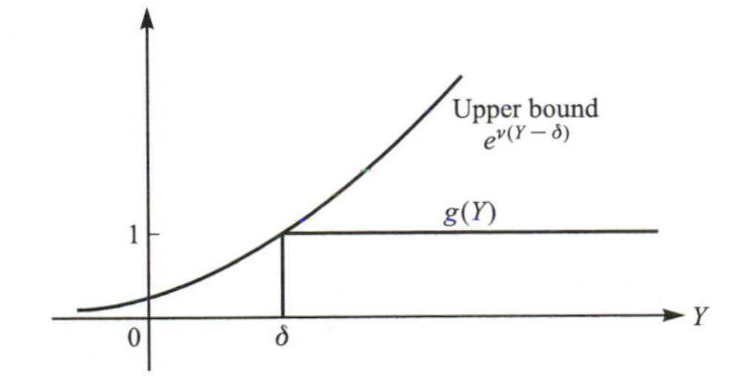

FIGURE 2.1–12
An exponential upper bound on $g(Y)$ used in obtaining the tail probability. (Chernoff bound).

But the order of differentiation and expectation can be interchanged, so that

$$\begin{aligned}\frac{d}{d\nu}E(e^{\nu(Y-\delta)}) &= E\left(\frac{d}{d\nu}e^{\nu(Y-\delta)}\right)\\ &= E[(Y-\delta)e^{\nu(Y-\delta)}]\\ &= e^{-\nu\delta}[E(Ye^{\nu Y})-\delta E(e^{\nu Y})]=0\end{aligned}$$

Therefore the value of ν that gives the tightest upper bound is the solution to the equation

$$E(Ye^{\nu Y})-\delta E(e^{\nu Y})=0 \tag{2.1–171}$$

Let $\hat{\nu}$ be the solution of Equation 2.1–171. Then, from Equation 2.1–169, the upper bound on the one-sided tail probability is

$$P(Y \geqslant \delta) \leqslant e^{-\hat{\nu}\delta}E(e^{\hat{\nu}Y}) \tag{2.1–172}$$

This is the Chernoff bound for the upper tail probability for a discrete or a continuous random variable having a zero mean.† This bound may be used to show that $Q(x) \leqslant e^{-x^2/2}$, where $Q(x)$ is the area in the tail of the Gaussian PDF (see Problem 2–18).

An upper bound on the lower tail probability can be obtained in a similar manner, with the result that

$$P(Y \leqslant \delta) \leqslant e^{-\hat{\nu}\delta}E(e^{\hat{\nu}Y}) \tag{2.1–173}$$

where $\hat{\nu}$ is the solution to Equation 2.1–171 and $\delta < 0$.

EXAMPLE 2.1–6. Consider the (Laplace) PDF

$$p(y)=\tfrac{1}{2}e^{-|y|} \tag{2.1–174}$$

which is illustrated in Figure 2.1–13. Let us evaluate the upper tail probability from the Chernoff bound and compare it with the true tail probability, which is

$$P(Y \geqslant \delta)=\int_{\delta}^{\infty}\tfrac{1}{2}e^{-y}\,dy=\tfrac{1}{2}e^{-\delta} \tag{2.1–175}$$

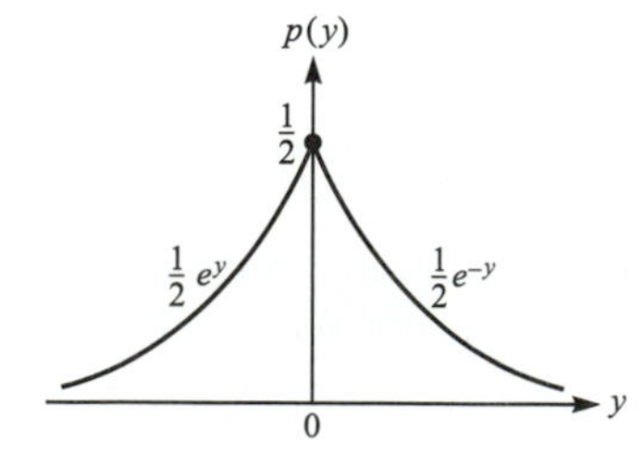

FIGURE 2.1–13
The PDF of a Laplace-distributed random variable.

† Note that $E(e^{\nu Y})$ for real ν is not the characteristic function of Y. It is called the *moment generating function of* Y.

To solve Equation 2.1–171 for $\hat{\nu}$, we must determine the moments $E(Ye^{\nu Y})$ and $E(e^{\nu Y})$. For the PDF in Equation 2.1–174, we find that

$$E(Ye^{\nu Y}) = \frac{2\nu}{(\nu+1)^2(\nu-1)^2}$$
$$E(e^{\nu Y}) = \frac{1}{(1+\nu)(1-\nu)} \tag{2.1–176}$$

Substituting these moments into Equation 2.1–171, we obtain the quadratic equation

$$\nu^2\delta + 2\nu - \delta = 0$$

which has the solutions

$$\hat{\nu} = \frac{-1 \pm \sqrt{1+\delta^2}}{\delta} \tag{2.1–177}$$

Since $\hat{\nu}$ must be positive, one of the two solutions is discarded. Thus

$$\hat{\nu} = \frac{-1 + \sqrt{1+\delta^2}}{\delta} \tag{2.1–178}$$

Finally, we evaluate the upper bound in Equation 2.1–172 by eliminating $E(e^{\hat{\nu}Y})$ using the second relation in Equation 2.1–176 and by substituting for $\hat{\nu}$ from Equation 2.1–178. The result is

$$P(Y \geqslant \delta) \leqslant \frac{\delta^2}{2(-1+\sqrt{1+\delta^2})} e^{1-\sqrt{1+\delta^2}} \tag{2.1–179}$$

For $\delta \gg 1$, Equation 2.1–179 reduces to

$$P(Y \geqslant \delta) \leqslant \frac{\delta}{2} e^{-\delta} \tag{2.1–180}$$

We note that the Chernoff bound decreases exponentially as δ increases. Consequently, it approximates closely the exact tail probability given by Equation 2.1–175. In contrast, the Chebyshev upper bound for the upper tail probability obtained by taking one-half of the probability in the two tails (due to symmetry in the PDF) is

$$P(Y \geqslant \delta) \leqslant \frac{1}{\delta^2}$$

Hence, this bound is extremely loose.

When the random variable has a nonzero mean, the Chernoff bound can be extended as we now demonstrate. If $Y = X - m_x$, we have

$$P(Y \geqslant \delta) = P(X - m_x \geqslant \delta) = P(X \geqslant m_x + \delta) = P(X \geqslant \delta_m)$$

where, by definition, $\delta_m = m_x + \delta$. Since $\delta > 0$, it follows that $\delta_m > m_x$. Let $g(X)$ be defined as

$$g(X) = \begin{cases} 1 & (X \geqslant \delta_m) \\ 0 & (X < \delta_m) \end{cases} \tag{2.1–181}$$

and upper-bounded as

$$g(X) \leqslant e^{\nu(X-\delta_m)} \tag{2.1–182}$$

From this point, the derivation parallels the steps contained in Equations 2.1–169 – 2.1–172. The final result is

$$P(X \geqslant \delta_m) \leqslant e^{-\hat{\nu}\delta_m} E(e^{\hat{\nu}X}) \tag{2.1–183}$$

where $\delta_m > m_x$ and $\hat{\nu}$ is the solution to the equation

$$E(Xe^{\nu X}) - \delta_m E(e^{\nu X}) = 0 \tag{2.1–184}$$

In a similar manner, we can obtain the Chernoff bound for the lower tail probability. For $\delta < 0$, we have

$$P(X - m_x \leqslant \delta) = P(X \leqslant m_x + \delta) = P(X \leqslant \delta_m) \leqslant E(e^{\nu(X-\delta_m)}) \tag{2.1–185}$$

From our previous development, it is apparent that Equation 2.1–185 results in the bound

$$P(X \leqslant \delta_m) \leqslant e^{-\hat{\nu}\delta_m} E(e^{\hat{\nu}X}) \tag{2.1–186}$$

where $\delta_m < m_x$ and $\hat{\nu}$ is the solution to Equation 2.1–184.

2.1.6 Sums of Random Variables and the Central Limit Theorem

We have previously considered the problem of determining the PDF of a sum of n statistically independent random variables. In this section, we again consider the sum of statistically independent random variables, but our approach is different and is independent of the particular PDF of the random variables in the sum. To be specific, suppose that X_i, $i = 1, 2, \ldots, n$, are statistically independent and identically distributed random variables, each having a finite mean m_x and a finite variance σ_x^2. Let Y be defined as the normalized sum, called the *sample mean*:

$$Y = \frac{1}{n}\sum_{i=1}^{n} X_i \tag{2.1–187}$$

First we shall determine upper bounds on the tail probabilities of Y and then we shall prove a very important theorem regarding the PDF of Y in the limit as $n \to \infty$.

The random variable Y defined in Equation 2.1–187 is frequently encountered in estimating the mean of a random variable X from a number of observations X_i, $i = 1, 2, \ldots, n$. In other words, the X_i, $i = 1, 2, \ldots,$ may be considered as independent samples drawn from a distribution $F_X(x)$, and Y is the estimate of the mean m_x.

The mean of Y is

$$E(Y) = m_y = \frac{1}{n}\sum_{i=1}^{n} E(X_i)$$
$$= m_x$$

The variance of Y is

$$\begin{aligned}
\sigma_y^2 &= E(Y^2) - m_y^2 = E(Y^2) - m_x^2 \\
&= \frac{1}{n^2}\sum_{i=1}^{n}\sum_{j=1}^{n} E(X_i X_j) - m_x^2 \\
&= \frac{1}{n^2}\sum_{i=1}^{n} E(X_i^2) + \frac{1}{n^2}\sum_{i=1}^{n}\sum_{\substack{j=1 \\ i\neq j}}^{n} E(X_i)E(X_j) - m_x^2 \\
&= \frac{1}{n}(\sigma_x^2 + m_x^2) + \frac{1}{n^2} n(n-1)m_x^2 - m_x^2 \\
&= \frac{\sigma_x^2}{n}
\end{aligned}$$

When Y is viewed as an estimate for the mean m_x, we note that its expected value is equal to m_x and its variance decreases inversely with the number of samples n. As n approaches infinity, the variance σ_y^2 approaches zero. An estimate of a parameter (in this case the mean m_x) that satisfies the conditions that its expected value converges to the true value of the parameter and the variance converges to zero as $n \to \infty$ is said to be a *consistent estimate*.

The tail probability of the random variable Y can be upper-bounded by use of the bounds presented in Section 2.1.5. The Chebyshev inequality applied to Y is

$$\begin{aligned}
P(|Y - m_y| \geqslant \delta) &\leqslant \frac{\sigma_y^2}{\delta^2} \\
P\left(\left|\frac{1}{n}\sum_{i=1}^{n} X_i - m_x\right| \geqslant \delta\right) &\leqslant \frac{\sigma_x^2}{n\delta^2}
\end{aligned} \tag{2.1–188}$$

In the limit as $n \to \infty$, Equation 2.1–188 becomes

$$\lim_{n\to\infty} P\left(\left|\frac{1}{n}\sum_{i=1}^{n} X_i - m_x\right| \geqslant \delta\right) = 0 \tag{2.1–189}$$

Therefore, the probability that the estimate of the mean differs from the true mean m_x by more than $\delta(\delta > 0)$ approaches zero as n approaches infinity. This statement is a form of the law of large numbers. Since the upper bound converges to zero relatively slowly, i.e., inversely with n, the expression in Equation 2.1–188 is called the *weak law of large numbers*.

The Chernoff bound applied to the random variable Y yields an exponential dependence of n, and thus provides a tighter upper bound on the one-sided tail probability. Following the procedure developed in Section 2.1.5, we can determine that the tail probability for y is

$$P(Y - m_y \geqslant \delta) = P\left(\frac{1}{n}\sum_{i=1}^{n} X_i - m_x \geqslant \delta\right)$$

$$= P\left(\sum_{i=1}^{n} X_i \geqslant n\delta_m\right) \leqslant E\left\{\exp\left[\nu\left(\sum_{i=1}^{n} X_i - n\delta_m\right)\right]\right\} \tag{2.1–190}$$

where $\delta_m = m_x + \delta$ and $\delta > 0$. But the X_i, $i = 1, 2, \ldots, n$, are statistically independent and identically distributed. Hence,

$$E\left\{\exp\left[\nu\left(\sum_{i=1}^{n} X_i - n\delta_m\right)\right]\right\} = e^{-\nu n\delta_m} E\left[\exp\left(\nu \sum_{i=1}^{n} X_i\right)\right]$$

$$= e^{-\nu n\delta_m} \prod_{i=1}^{n} E(e^{\nu X_i}) \tag{2.1–191}$$

$$= [e^{-\nu\delta_m} E(e^{\nu X})]^n$$

where X denotes any one of the X_i. The parameter ν that yields the tightest upper bound is obtained by differentiating Equation 2.1–191 and setting the derivative equal to zero. This yields the equation

$$E(Xe^{\nu X}) - \delta_m E(e^{\nu X}) = 0 \tag{2.1–192}$$

Let the solution of Equation 2.1–192 be denoted by $\hat{\nu}$. Then, the bound on the upper tail probability is

$$P\left(\frac{1}{n}\sum_{i=1}^{n} X_i \geqslant \delta_m\right) \leqslant [e^{-\hat{\nu}\delta_m} E(e^{\hat{\nu} X})]^n, \qquad \delta_m > m_x \tag{2.1–193}$$

In a similar manner, we find that the lower tail probability is upper-bounded as

$$P(Y \leqslant \delta_m) \leqslant [e^{-\hat{\nu}\delta_m} E(e^{\hat{\nu} X})]^n, \qquad \delta_m < m_x \tag{2.1–194}$$

where $\hat{\nu}$ is the solution to Equation 2.1–192.

EXAMPLE 2.1–7. Let X_i, $i = 1, 2, \ldots n$, be a set of statistically independent random variables defined as

$$X_i = \begin{cases} 1 & \text{(with probability } p < \frac{1}{2}) \\ -1 & \text{(with probability } 1 - p) \end{cases}$$

We wish to determine a tight upper bound on the probability that the sum of the X_i is greater than zero. Since $p < \frac{1}{2}$, we note that the sum will have a negative value for the mean; hence we seek the upper tail probability. With $\delta_m = 0$ in Equation 2.1–193, we have

$$P\left(\sum_{i=1}^{n} X_i \geqslant 0\right) \leqslant [E(e^{\hat{\nu}X})]^n \tag{2.1–195}$$

where $\hat{\nu}$ is the solution to the equation

$$E(Xe^{\nu X}) = 0 \tag{2.1–196}$$

Now

$$E(Xe^{\nu X}) = -(1-p)e^{-\nu} + pe^{\nu} = 0$$

Hence

$$\hat{\nu} = \ln\left(\sqrt{\frac{1-p}{p}}\right) \tag{2.1–197}$$

Furthermore,

$$E(e^{\hat{\nu}X}) = pe^{\hat{\nu}} + (1-p)e^{-\hat{\nu}}$$

Therefore the bound in Equation 2.1–195 becomes

$$\begin{aligned} P\left(\sum_{i=1}^{n} X_i \geqslant 0\right) &\leqslant [pe^{\hat{\nu}} + (1-p)e^{-\hat{\nu}}]^n \\ &\leqslant \left[p\sqrt{\frac{1-p}{p}} + (1-p)\sqrt{\frac{p}{1-p}}\right]^n \\ &\leqslant [4p(1-p)]^{n/2} \end{aligned} \tag{2.1–198}$$

We observe that the upper bound decays exponentially with n, as expected. In contrast, if the Chebyshev bound were evaluated, the tail probability would decrease inversely with n.

Central limit theorem. We conclude this section with an extremely useful theorem concering the CDF of a sum of random variables in the limit as the number of terms in the sum approaches infinity. There are several versions of this theorem. We shall prove the theorem for the case in which the random variables X_i, $i = 1, 2, \ldots, n$, being summed are statistically independent and identically distributed, each having a finite mean m_x and a finite variance σ_x^2. For convenience, we define the normalized random variable

$$U_i = \frac{X_i - m_x}{\sigma_x}, \qquad i = 1, 2, \ldots, n$$

Thus U_i has a zero mean and unit variance. Now, let

$$Y = \frac{1}{\sqrt{n}} \sum_{i=1}^{n} U_i \tag{2.1–199}$$

Since each term in the sum has a zero mean and a unit variance, it follows that the normalized (by $1\sqrt{n}$) random variable Y has zero mean and unit variance. We wish to determine the CDF of Y in the limit as $n \to \infty$.

The characteristic function of Y is

$$\psi_Y(jv) = E(e^{jvY}) = E\left[\exp\left(\frac{jv\sum_{i=1}^{n} U_i}{\sqrt{n}}\right)\right]$$
$$= \prod_{i=1}^{n} \psi_{U_i}\left(\frac{jv}{\sqrt{n}}\right) \tag{2.1–200}$$
$$= \psi_U\left(\frac{jv}{\sqrt{n}}\right)\Bigg]^n$$

where U denotes any of the U_i, which are identically distributed. Now, let us expand the characteristic function of U in a Taylor series. The expansion yields

$$\psi_U\left(j\frac{v}{\sqrt{n}}\right) = 1 + j\frac{v}{\sqrt{n}}E(U) - \frac{v^2}{n2!}E(U^2) + \frac{(jv)^3}{(\sqrt{n})^3 3!}E(U^3) - \cdots \tag{2.1–201}$$

Since $E(U) = 0$ and $E(U^2) = 1$, Equation 2.1–201 simplifies to

$$\psi_U\left(\frac{jv}{\sqrt{n}}\right) = 1 - \frac{v^2}{2n} + \frac{1}{n}R(v, n) \tag{2.1–202}$$

where $R(v, n)/n$ denotes the remainder. We note that $R(v, n)$ approaches zero as $n \to \infty$. Substitution of Equation 2.1–202 into Equation 2.1–200 yields the characteristic function of Y in the form

$$\psi_Y(jv) = \left[1 - \frac{v^2}{2n} + \frac{R(v, n)}{n}\right]^n \tag{2.1–203}$$

Taking the natural logarithm of Equation 2.1–203, we obtain

$$\ln \psi_Y(jv) = n \ln\left[1 - \frac{v^2}{2n} + \frac{R(v, n)}{n}\right] \tag{2.1–204}$$

For small values of x, $\ln(1 + x)$ can be expanded in the power series

$$\ln(1 + x) = x - \tfrac{1}{2}x^2 + \tfrac{1}{3}x^3 - \cdots$$

This expansion applied to Equation 2.1–204 yields

$$\ln \psi_Y(jv) = n\left[-\frac{v^2}{2n} + \frac{R(v, n)}{n} - \frac{1}{2}\left(-\frac{v^2}{2n} + \frac{R(v, n)}{n}\right)^2 + \cdots\right] \tag{2.1–205}$$

Finally, when we take the limit as $n \to \infty$, Equation 2.1–205 reduces to $\lim_{n\to\infty} \ln \psi_Y(jv) = -\frac{1}{2}v^2$, or, equivalently,

$$\lim_{n\to\infty} \psi_Y(jv) = e^{-v^2/2} \tag{2.1–206}$$

But, this is just the characteristic function of a Gaussian random variable with zero mean and unit variance. Thus we have the important result that the sum of statistically independent and identically distributed random variables with finite mean and variance approaches a Gaussian CDF as $n \to \infty$. This result is known as the *central limit theorem*.

Although we assumed that the random variables in the sum are identically distributed, the assumption can be relaxed provided that additional restrictions are imposed on the properties of the random variables. There is one variation of the theorem, for example, in which the assumption of identically distributed random variables is abandoned in favor of a condition on the third absolute moment of the random variables in the sum. For a discussion of this and other variations of the central limit theorem, the reader is referred to the book by Cramér (1946).

2.2 STOCHASTIC PROCESSES

Many of the random phenomena that occur in nature are functions of time. For example, the meteorological phenomena such as the random fluctuations in air temperature and air pressure are functions of time. The thermal noise voltages generated in the resistors of an electronic device such as a radio receiver are also a function of time. Similarly, the signal at the output of a source that generates information is characterized as a random signal that varies with time. An audio signal that is transmitted over a telephone channel is an example of such a signal. All these are examples of stochastic (random) processes. In our study of digital communications, we encounter stochastic processes in the characterization and modeling of signals generated by information sources, in the characterization of communication channels used to transmit the information, in the characterization of noise generated in a receiver, and in the design of the optimum receiver for processing the received random signal.

At any given time instant, the value of a stochastic process, whether it is the value of the noise voltage generated by a resistor or the amplitude of the signal generated by an audio source, is a random variable indexed by the parameter t. We shall denote such a process by $X(t)$. In general, the parameter t is continuous, whereas X may be either continuous or discrete, depending on the characteristics of the source that generates the stochastic process.

The noise voltage generated by a single resistor or a single information source represents a single realization of the stochastic process. Hence, it is called a *sample function* of the stochastic process. The set of all possible sample functions, e.g., the set of all noise voltage waveforms generated by resistors, constitutes an ensemble of sample functions or, equivalently, the stochastic process $X(t)$. In general, the number of sample functions in the ensemble is assumed to be extremely large; often it is infinite.

Having defined a stochastic process $X(t)$ as an ensemble of sample functions, we may consider the values of the process at any set of time instants $t_1 > t_2 > t_3 > \cdots > t_n$ where n is any positive integer. In general, the random variables $X_{t_i} \equiv X(t_i)$, $i = 1, 2, \ldots, n$, are characterized statistically by their joint PDF $p(x_{t_1}, x_{t_2}, \ldots, x_{t_n})$. Furthermore, all the probabilistic relations defined in Section 2.1 for multidimensional random variables carry over to the random variables X_{t_i}, $i = 1, 2, \ldots, n$.

Stationary stochastic processes. As indicated above, the random variables X_{t_i}, $i = 1, 2, \ldots, n$, obtained from the stochastic process $X(t)$ for any set of time instants $t_1 > t_2 > t_3 > \cdots > t_n$ and any n are characterized statistically by the joint PDF $p(x_{t_1}, x_{t_2}, \ldots, x_{t_n})$. Let us consider another set of n random variables $X_{t_i+t} \equiv X(t_i + t)$, $i = 1, 2, \ldots, n$, where t is an arbitrary time shift. These random variables are characterized by the joint PDF $p(x_{t_1+t}, x_{t_2+t}, \ldots, x_{t_n+t})$. The joint PDFs of the random variables X_{t_i} and X_{t_i+t}, $i = 1, 2, \ldots, n$, may or may not be identical. When they are identical, i.e., when

$$p(x_{t_1}, x_{t_2}, \ldots, x_{t_n}) = p(x_{t_1+t}, x_{t_2+t}, \ldots, x_{t_n+t}) \tag{2.2–1}$$

for all t and all n, the stochastic process is said to be *stationary in the strict sense*. That is, the statistics of a stationary stochastic process are invariant to any translation of the time axis. On the other hand, when the joint PDFs are different, the stochastic process is *nonstationary*.

2.2.1 Statistical Averages

Just as we have defined statistical averages for random variables, we may similarly define statistical averages for a stochastic process. Such averages are also called *ensemble averages*. Let $X(t)$ denote a random process and let $X_{t_i} \equiv X(t_i)$. The *nth moment* of the random variable X_{t_i} is defined as

$$E(X_{t_i}^n) = \int_{-\infty}^{\infty} x_{t_i}^n p(x_{t_i})\, dx_{t_i} \tag{2.2–2}$$

In general, the value of the nth moment will depend on the time instant t_i if the PDF of X_{t_i} depends on t_i. When the process is stationary, however, $p(x_{t_i+t}) = p(x_{t_i})$ for all t. Hence, the PDF is independent of time, and, as a consequence, the nth moment is independent of time.

Next we consider the two random variables $X_{t_i} \equiv X(t_i)$, $i = 1, 2$. The correlation between X_{t_1} and X_{t_2} is measured by the joint moment

$$E(X_{t_1} X_{t_2}) = \int_{-\infty}^{\infty} \int_{-\infty}^{\infty} x_{t_1} x_{t_2} p(x_{t_1}, x_{t_2})\, dx_{t_1}\, dx_{t_2} \tag{2.2–3}$$

Since this joint moment depends on the time instants t_1 and t_2, it is denoted by $\phi(t_1, t_2)$. The function $\phi(t_1, t_2)$ is called the *autocorrelation function* of the stochastic process. When the process $X(t)$ is stationary, the joint PDF of the pair (X_{t_1}, X_{t_2}) is identical to the joint PDF of the pair (X_{t_1+t}, X_{t_2+t}) for any arbitrary t.

This implies that the autocorrelation function of $X(t)$ does not depend on the specific time instants t_1 and t_2, but, instead, it depends on the time difference $t_1 - t_2$. Thus, for a stationary stochastic process, the joint moment in Equation 2.2–3 is

$$E(X_{t_1} X_{t_2}) = \phi(t_1, t_2) = \phi(t_1 - t_2) = \phi(\tau) \tag{2.2–4}$$

where $\tau = t_1 - t_2$ or, equivalently, $t_2 = t_1 - \tau$. If we let $t_2 = t_1 + \tau$, we have

$$\phi(-\tau) = E(X_{t_1} X_{t_1+\tau}) = E(X_{t_1'} X_{t_1'-\tau}) = \phi(\tau)$$

Therefore, $\phi(\tau)$ is an even function. We also note that $\phi(0) = E(X_t^2)$ denotes the average power in the process $X(t)$.

There exist nonstationary processes with the property that the mean value of the process is independent of time (a constant) and where the autocorrelation function satisfies the condition that $\phi(t_1, t_2) = \phi(t_1 - t_2)$. Such a process is called *wide-sense stationary*. Consequently, wide-sense stationarity is a less stringent condition than strict-sense stationarity. When reference is made to a stationary stochastic process in any subsequent discussion in which correlation functions are involved, the less stringent condition (wide-sense stationarity) is implied.

Related to the autocorrelation function is the autocovariance function of a stochastic process, which is defined as

$$\begin{aligned} \mu(t_1, t_2) &= E\{[X_{t_1} - m(t_1)][X_{t_2} - m(t_2)]\} \\ &= \phi(t_1, t_2) - m(t_1)m(t_2) \end{aligned} \tag{2.2–5}$$

where $m(t_1)$ and $m(t_2)$ are the means of X_{t_1} and X_{t_2}, respectively. When the process is stationary, the autocovariance function simplifies to

$$\mu(t_1, t_2) = \mu(t_1 - t_2) = \mu(\tau) = \phi(\tau) - m^2 \tag{2.2–6}$$

where $\tau = t_1 - t_2$.

Higher-order joint moments of two or more random variables derived from a stochastic process $X(t)$ are defined in an obvious manner. We note that for a Gaussian random process, higher-order moments can be expressed in terms of first and second moments. Consequently, a Gaussian random process is completely characterized by its first two moments.

Averages for a Gaussian process. Suppose that $X(t)$ is a Gaussian random process. Hence, at time instants $t = t_i$, $i = 1, 2, \ldots, n$, the random variables X_{t_i}, $i = 1, 2, \ldots, n$, are jointly Gaussian with mean values $m(t_i)$, $i = 1, 2, \ldots, n$, and autocovariances

$$\mu(t_i, t_j) = E[(X_{t_i} - m(t_i))(X_{t_j} - m(t_j))], \qquad i, j = 1, 2, \ldots, n \tag{2.2–7}$$

If we denote the $n \times n$ covariance matrix with elements $\mu(t_i, t_j)$ by $\mathbf{M}$ and the vector of mean values by $\mathbf{m}_x$, then the joint PDF of the random variables X_{t_i}, $i = 1, 2, \ldots, n$, is given by Equation 2.1–150.

If the Gaussian process is stationary, then $m(t_i) = m$ for all t_i and $\mu(t_i, t_j) = \mu(t_i - t_j)$. Again, we note that the Gaussian random process is completely specified by the mean and autocovariance functions. Since the joint Gaussian PDF depends only on these two moments, it follows that if the Gaussian process is wide-sense stationary, it is also strict-sense stationary. Of course, the converse is always true for any stochastic process.

Averages for joint stochastic processes. Let $X(t)$ and $Y(t)$ denote two stochastic processes and let $X_{t_i} \equiv X(t_i)$, $i = 1, 2, \ldots, n$, and $Y_{t'_j} \equiv Y(t'_j)$, $j = 1, 2, \ldots, m$, represent the random variables at times $t_1 > t_2 > t_3 > \cdots > t_n$ and $t'_1 > t'_2 > \cdots > t'_m$, respectively. The two processes are characterized statistically by their joint PDF

$$p(x_{t_1}, x_{t_2}, \ldots, x_{t_n}, y_{t'_1}, y_{t'_2}, \ldots, y_{t'_m})$$

for any set of time instants $t_1, t_2, \ldots, t_n, t'_1, t'_2, \ldots, t'_m$ and for any positive integer values of n and m.

The *cross-correlation function* of $X(t)$ and $Y(t)$, denoted by $\phi_{xy}(t_1, t_2)$, is defined as the joint moment

$$\phi_{xy}(t_1, t_2) = E(X_{t_1} Y_{t_2}) = \int_{-\infty}^{\infty} \int_{-\infty}^{\infty} x_{t_1} y_{t_2} p(x_{t_1}, y_{t_2})\, dx_{t_1}\, dy_{t_2} \tag{2.2–8}$$

and the *cross-covariance* is

$$\mu_{xy}(t_1, t_2) = \phi_{xy}(t_1, t_2) - m_x(t_1) m_y(t_2) \tag{2.2–9}$$

When the processes are jointly and individually stationary, we have $\phi_{xy}(t_1, t_2) = \phi_{xy}(t_1 - t_2)$ and $\mu_{xy}(t_1, t_2) = \mu_{xy}(t_1 - t_2)$. In this case, we note that

$$\phi_{xy}(-\tau) = E(X_{t_1} Y_{t_1+\tau}) = E(X_{t'_1-\tau} Y_{t'_1}) = \phi_{yx}(\tau) \tag{2.2–10}$$

The stochastic processes $X(t)$ and $Y(t)$ are said to be *statistically independent* if and only if

$$p(x_{t_1}, x_{t_2}, \ldots, x_{t_n}, y_{t'_1}, y_{t'_2}, \ldots, y_{t'_m}) = p(x_{t_1}, x_{t_2}, \ldots, x_{t_n}) p(y_{t'_1}, y_{t'_2}, \ldots, y_{t'_m})$$

for all choices of t_i and t'_i and for all positive integers n and m. The processes are said to be *uncorrelated* if

$$\phi_{xy}(t_1, t_2) = E(X_{t_1}) E(Y_{t_2})$$

Hence

$$\mu_{xy}(t_1, t_2) = 0$$

A *complex-valued stochastic process* $Z(t)$ is defined as

$$Z(t) = X(t) + jY(t) \tag{2.2–11}$$

where $X(t)$ and $Y(t)$ are stochastic processes. The joint PDF of the random variables $Z_{t_i} \equiv Z(t_i)$, $i = 1, 2, \ldots, n$, is given by the joint PDF of the components (X_{t_i}, Y_{t_i}), $i = 1, 2, \ldots, n$. Thus, the PDF that characterizes Z_{t_i}, $i = 1, 2, \ldots, n$, is

$$p(x_{t_1}, x_{t_2}, \ldots, x_{t_n}, y_{t_1}, y_{t_2}, \ldots y_{t_n})$$

The complex-valued stochastic process $Z(t)$ is encountered in the representation of narrowband band-pass noise in terms of its equivalent low-pass components. An important characteristic of such a process it its autocorrelation function. The function is defined as

$$\begin{aligned}\phi_{zz}(t_1, t_2) &= \tfrac{1}{2}E(Z_{t_1} Z_{t_2}^*) \\ &= \tfrac{1}{2}E[(X_{t_1} + jY_{t_1})(X_{t_2} - jY_{t_2})] \\ &= \tfrac{1}{2}\{\phi_{xx}(t_1, t_2) + \phi_{yy}(t_1, t_2) + j[\phi_{yx}(t_1, t_2) - \phi_{xy}(t_1, t_2)]\}\end{aligned} \tag{2.2–12}$$

where $\phi_{xx}(t_1, t_2)$ and $\phi_{yy}(t_1, t_2)$ are the autocorrelation functions of $X(t)$ and $Y(t)$, respectively, and $\phi_{yx}(t_1, t_2)$ and $\phi_{xy}(t_1, t_2)$ are the cross-correlation functions. The factor of $\frac{1}{2}$ in the definition of the autocorrelation function of a complex-valued stochastic process is an arbitrary but mathematically convenient normalization factor, as we will demonstrate in our treatment of such processes in Chapter 4.

When the processes $X(t)$ and $Y(t)$ are jointly and individually stationary, the autocorrelation function of $Z(t)$ becomes

$$\phi_{zz}(t_1, t_2) = \phi_{zz}(t_1 - t_2) = \phi_{zz}(\tau)$$

where $t_2 = t_1 - \tau$. Also, the complex conjugate of Equation 2.2–12 is

$$\phi_{zz}^*(\tau) = \tfrac{1}{2}E(Z_{t_1}^* Z_{t_1-\tau}) = \tfrac{1}{2}E(Z_{t_1'+\tau}^* Z_{t_1'}) = \phi_{zz}(-\tau) \tag{2.2–13}$$

Hence, $\phi_{zz}(\tau) = \phi_{zz}^*(-\tau)$.

Now, suppose that $Z(t) = X(t) + jY(t)$ and $W(t) = U(t) + jV(t)$ are two complex-valued stochastic processes. The cross-correlation function of $Z(t)$ and $W(t)$ is defined as

$$\begin{aligned}\phi_{zw}(t_1, t_2) &= \tfrac{1}{2}E(Z_{t_1} W_{t_2}^*) \\ &= \tfrac{1}{2}E[(X_{t_1} + jY_{t_1})(U_{t_2} - jV_{t_2})] \\ &= \tfrac{1}{2}\{\phi_{xu}(t_1, t_2) + \phi_{yv}(t_1, t_2) + j[\phi_{yu}(t_1, t_2) - \phi_{xv}(t_1, t_2)]\}\end{aligned} \tag{2.2–14}$$

When $X(t)$, $Y(t)$, $U(t)$, and $V(t)$ are pairwise-stationary, the cross-correlation functions in Equation 2.2–14 become functions of the time difference $\tau = t_1 - t_2$. Furthermore,

$$\phi_{zw}^*(\tau) = \tfrac{1}{2}E(Z_{t_1}^* W_{t_1-\tau}) = \tfrac{1}{2}E(Z_{t_1'+\tau}^* W_{t_i'}) = \phi_{wz}(-\tau) \tag{2.2–15}$$

2.2.2 Power Density Spectrum

The frequency content of a signal is a very basic characteristic that distinguishes one signal from another. In general, a signal can be classified as having either a finite (nonzero) average power (infinite energy) or finite energy. The frequency content of a finite energy signal is obtained as the Fourier transform of the corresponding time function. If the signal is periodic, its energy is infinite and, consequently, its Fourier transform does not exist. The mechanism for dealing with periodic signals is to represent them in a Fourier series. With such a representation, the Fourier coefficients determine the distribution of power at the various discrete frequency components.

A stationary stochastic process is an infinite energy signal, and, hence, its Fourier transform does not exist. The spectral characteristic of a stochastic signal is obtained by computing the Fourier transform of the autocorrelation function. That is, the distribution of power with frequency is given by the function

$$\Phi(f) = \int_{-\infty}^{\infty} \phi(\tau) e^{-j2\pi f\tau} \, d\tau \tag{2.2–16}$$

The inverse Fourier transform relationship is

$$\phi(\tau) = \int_{-\infty}^{\infty} \Phi(f) e^{j2\pi f\tau} \, df \tag{2.2–17}$$

We observe that

$$\begin{aligned} \phi(0) &= \int_{-\infty}^{\infty} \Phi(f) \, df \\ &= E(|X_t|^2) \geqslant 0 \end{aligned} \tag{2.2–18}$$

Since $\phi(0)$ represents the average power of the stochastic signal, which is the area under $\Phi(f)$, $\Phi(f)$ is the distribution of power as a function of frequency. Therefore, $\Phi(f)$ is called the *power density spectrum* of the stochastic process.

If the stochastic process is real, $\phi(\tau)$ is real and even, and, hence $\Phi(f)$ is real and even. On the other hand, if the process is complex, $\phi(\tau) = \phi^*(-\tau)$ and, hence

$$\begin{aligned} \Phi^*(f) &= \int_{-\infty}^{\infty} \phi^*(\tau) e^{j2\pi f\tau} \, d\tau = \int_{-\infty}^{\infty} \phi^*(-\tau) e^{-j2\pi f\tau} \, d\tau \\ &= \int_{-\infty}^{\infty} \phi(\tau) e^{-j2\pi f\tau} \, d\tau = \Phi(f) \end{aligned} \tag{2.2–19}$$

Therefore, $\Phi(f)$ is real.

The definition of a power density spectrum can be extended to two jointly stationary stochastic processes $X(t)$ and $Y(t)$, which have a cross-correlation function $\phi_{xy}(\tau)$. The Fourier transform of $\phi_{xy}(\tau)$, i.e.,

$$\Phi_{xy}(f) = \int_{-\infty}^{\infty} \phi_{xy}(\tau) e^{-j2\pi f\tau} \, d\tau \tag{2.2–20}$$

is called the *cross-power density spectrum*. If we conjugate both sides of Equation 2.2–20, we have

$$\begin{aligned} \Phi_{xy}^*(f) &= \int_{-\infty}^{\infty} \phi_{xy}^*(\tau) e^{j2\pi f\tau} d\tau = \int_{-\infty}^{\infty} \phi_{xy}^*(-\tau) e^{-j2\pi f\tau} \, d\tau \\ &= \int_{-\infty}^{\infty} \phi_{yx}(\tau) e^{-j2\pi f\tau} \, d\tau = \Phi_{yx}(f) \end{aligned} \tag{2.2–21}$$

This relation holds in general. However, if $X(t)$ and $Y(t)$ are real stochastic processes,

$$\Phi^*_{xy}(f) = \int_{-\infty}^{\infty} \phi_{xy}(\tau) e^{j2\pi f \tau}\, d\tau = \Phi_{xy}(-f) \tag{2.2–22}$$

By combining the result in Equation 2.2–21 with the result in Equation 2.2–22, we find that the cross-power density spectrum of two real processes satisfies the condition

$$\Phi_{yx}(f) = \Phi_{xy}(-f) \tag{2.2–23}$$

2.2.3 Response of a Linear Time-Invariant System to a Random Input Signal

Consider a linear time-invariant system (filter) that is characterized by its impulse response $h(t)$ or, equivalently, by its frequency response $H(f)$, where $h(t)$ and $H(f)$ are a Fourier transform pair. Let $x(t)$ be the input signal to the system and let $y(t)$ denote the output signal. The output of the system may be expressed in terms of the convolution integral as

$$y(t) = \int_{-\infty}^{\infty} h(\tau) x(t-\tau)\, d\tau \tag{2.2–24}$$

Now, suppose that $x(t)$ is a sample function of a stationary stochastic process $X(t)$. Then, the output $y(t)$ is a sample function of a stochastic process $Y(t)$. We wish to determine the mean and autocorrelation functions of the output.

Since convolution is a linear operation performed on the input signal $x(t)$, the expected value of the integral is equal to the integral of the expected value. Thus, the mean value of $Y(t)$ is

$$\begin{aligned} m_y = E[Y(t)] &= \int_{-\infty}^{\infty} h(\tau) E[X(t-\tau)]\, d\tau \\ &= m_x \int_{-\infty}^{\infty} h(\tau) d\tau = m_x H(0) \end{aligned} \tag{2.2–25}$$

where $H(0)$ is the frequency response of the linear system at $f = 0$. Hence, the mean value of the output process is a constant.

The autocorrelation function of the output is

$$\begin{aligned} \phi_{yy}(t_1, t_2) &= \tfrac{1}{2} E(Y_{t_1} Y^*_{t_2}) \\ &= \frac{1}{2} \int_{-\infty}^{\infty} \int_{-\infty}^{\infty} h(\beta) h^*(\alpha) E[X(t_1 - \beta) X^*(t_2 - \alpha)] d\alpha\, d\beta \\ &= \int_{-\infty}^{\infty} \int_{-\infty}^{\infty} h(\beta) h^*(\alpha) \phi_{xx}(t_1 - t_2 + \alpha - \beta)\, d\alpha\, d\beta \end{aligned}$$

The last step indicates that the double integral is a function of the time difference $t_1 - t_2$. In other words, if the input process is stationary, the output is also stationary. Hence

$$\phi_{yy}(\tau) = \int_{-\infty}^{\infty}\int_{-\infty}^{\infty} h^*(\alpha)h(\beta)\phi_{xx}(\tau + \alpha - \beta)d\alpha\, d\beta \tag{2.2–26}$$

By evaluating the Fourier transform of both sides of Equation 2.2–26, we obtain the power density spectum of the output process in the form

$$\begin{aligned}\Phi_{yy}(f) &= \int_{-\infty}^{\infty} \phi_{yy}(\tau)e^{-j2\pi f\tau}d\tau \\ &= \int_{-\infty}^{\infty}\int_{-\infty}^{\infty}\int_{-\infty}^{\infty} h^*(\alpha)h(\beta)\phi_{xx}(\tau + \alpha - \beta)e^{-j2\pi f\tau}d\tau\, d\alpha\, d\beta \\ &= \Phi_{xx}(f)|H(f)|^2\end{aligned} \tag{2.2–27}$$

Thus, we have the important result that the power density spectrum of the output signal is the product of the power density spectrum of the input multiplied by the magnitude squared of the frequency response of the system.

When the autocorrelation function $\phi_{yy}(\tau)$ is desired, it is usually easier to determine the power density spectrum $\Phi_{yy}(f)$ and then to compute the inverse transform. Thus, we have

$$\begin{aligned}\phi_{yy}(\tau) &= \int_{-\infty}^{\infty} \Phi_{yy}(f)e^{j2\pi f\tau}\, df \\ &= \int_{-\infty}^{\infty} \Phi_{xx}(f)|H(f)|^2 e^{j2\pi f\tau}\, df\end{aligned} \tag{2.2–28}$$

We observe that the average power in the output signal is

$$\phi_{yy}(0) = \int_{-\infty}^{\infty} \Phi_{xx}(f)|H(f)|^2\, df \tag{2.2–29}$$

Since $\phi_{yy}(0) = E(|Y_t)^2)$, it follows that

$$\int_{-\infty}^{\infty} \Phi_{xx}(f)|H(f)|^2\, df \geqslant 0$$

Suppose we let $|H(f)|^2 = 1$ for any arbitrarily small interval $f_1 \leqslant f \leqslant f_2$, and $H(f) = 0$ outside this interval. Then,

$$\int_{f_1}^{f_2} \Phi_{xx}(f)\, df \geqslant 0$$

But this is possible if and only if $\Phi_{xx}(f) \geqslant 0$ for all f.

EXAMPLE 2.2–1. Suppose that the low-pass filter illustrated in Figure 2.2–1 is excited by a stochastic process $x(t)$ having a power density spectrum

$$\Phi_{xx}(f) = \tfrac{1}{2}N_0, \qquad \text{for all } f$$

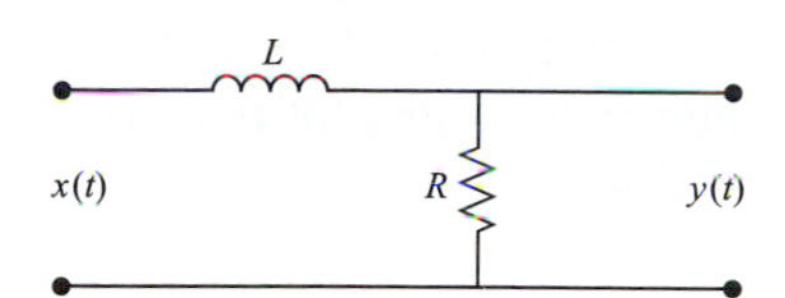

FIGURE 2.2–1
An example of a low-pass filter.

A stochastic process having a flat power density spectrum is called *white noise*. Let us determine the power density spectrum of the output process. The transfer function of the low-pass filter is

$$H(f) = \frac{R}{R + j2\pi fL} = \frac{1}{1 + j2\pi fL/R}$$

and, hence,

$$|H(f)|^2 = \frac{1}{1 + (2\pi L/R)^2 f^2} \tag{2.2–30}$$

The power density spectrum of the output process is

$$\Phi_{yy}(f) = \frac{N_0}{2} \frac{1}{1 + (2\pi L/R)^2 f^2} \tag{2.2–31}$$

This power density spectrum is illustrated in Figure 2.2–2. Its inverse Fourier transform yields the autocorrelation function

$$\begin{aligned} \phi_{yy}(\tau) &= \int_{-\infty}^{\infty} \frac{N_0}{2} \frac{1}{1 + (2\pi L/R)^2 f^2} e^{j2\pi f\tau}\, df \\ &= \frac{RN_0}{4L} e^{-(R/L)|\tau|} \end{aligned} \tag{2.2–32}$$

The autocorrelation function $\phi_{yy}(\tau)$ is shown in Figure 2.2–3. We observe that the second moment of the process $Y(t)$ is $\phi_{yy}(0) = RN_0/4L$.

As a final exercise, we determine the cross-correlation function between $y(t)$ and $x(t)$, where $x(t)$ denotes the input and $y(t)$ denotes the output of the linear system. We have

$$\begin{aligned} \phi_{yx}(t_1, t_2) &= \tfrac{1}{2}E(Y_{t_1} X_{t_2}^*) = \frac{1}{2}\int_{-\infty}^{\infty} h(\alpha) E[X(t_1 - \alpha) X^*(t_2)] \quad d\alpha \\ &= \int_{-\infty}^{\infty} h(\alpha)\phi_{xx}(t_1 - t_2 - \alpha) d\alpha = \phi_{yx}(t_1 - t_2) \end{aligned}$$

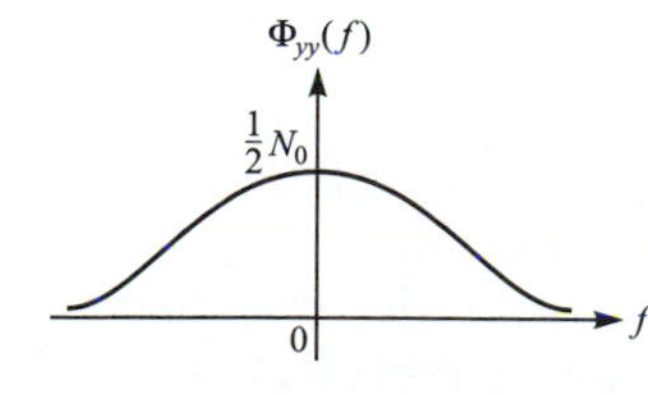

FIGURE 2.2–2
The power density spectrum of the low-pass filter output when the input is white noise.

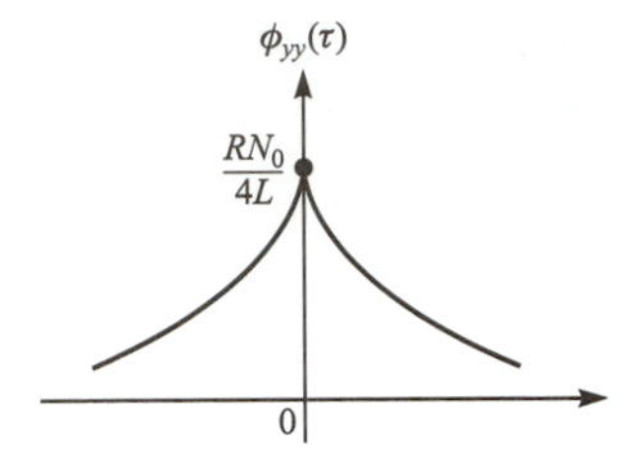

FIGURE 2.2–3
The autocorrelation function of the output of the low-pass filter for a white-noise input.

Hence, the stochastic processes $X(t)$ and $Y(t)$ are jointly stationary. With $t_1 - t_2 = \tau$, we have

$$\phi_{yx}(\tau) = \int_{-\infty}^{\infty} h(\alpha)\phi_{xx}(\tau - \alpha)d\alpha \tag{2.2–33}$$

Note that the integral in Equation 2.2–33 is a convolution integral. Hence in the frequency domain the relation 2.2–33 becomes

$$\Phi_{yx}(f) = \Phi_{xx}(f)H(f) \tag{2.2–34}$$

We observe that if the input process is white noise, the cross correlation of the input with the output of the system yields the impulse response $h(t)$ to within a scale factor.

2.2.4 Sampling Theorem for Band-Limited Stochastic Processes

Recall that a deterministic signal $s(t)$ that has a Fourier transform $S(f)$ is called band-limited if $S(f) = 0$ for $|f| > W$, where W is the highest frequency contained in $s(t)$. Such a signal is uniquely represented by samples of $s(t)$ taken at a rate of $f_s \geqslant 2W$ samples/s. The minimum rate $f_N = 2W$ samples/s is called the *Nyquist rate*. Sampling below the Nyquist rate results in frequency aliasing.

The band-limited signal sampled at the Nyquist rate can be reconstructed from its samples by use of the interpolation formula

$$s(t) = \sum_{n=-\infty}^{\infty} s\left(\frac{n}{2W}\right) \frac{\sin\left[2\pi W\left(t - \frac{n}{2W}\right)\right]}{2\pi W\left(t - \frac{n}{2W}\right)} \tag{2.2–35}$$

where $\{s(n/2W)\}$ are the samples of $s(t)$ taken at $t = n/2W$, $n = 0, \pm 1, \pm 2, \ldots$. Equivalently, $s(t)$ can be reconstructed by passing the sampled signal through an ideal low-pass filter with impulse response $h(t) = (\sin 2\pi Wt)/2\pi Wt$. Figure 2.2–4 illustrates the signal reconstruction process based on ideal interpolation.

A stationary stochastic process $X(t)$ is said to be *band-limited* if its power density spectrum $\Phi(f) = 0$ for $|f| > W$. Since $\Phi(f)$ is the Fourier transform of the autocorrelation function $\phi(\tau)$, it follows that $\phi(\tau)$ can be represented as

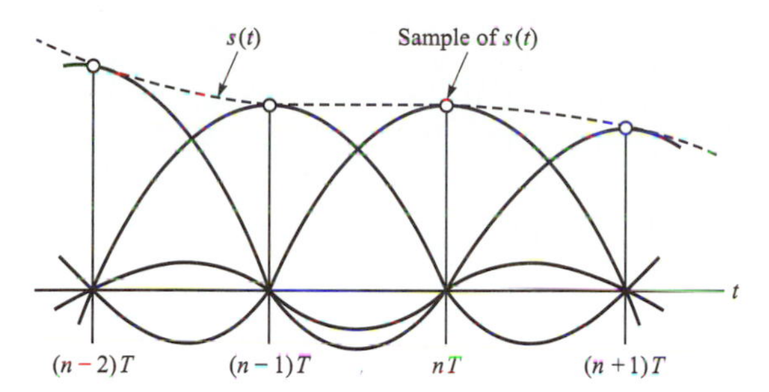

FIGURE 2.2–4
Signal reconstruction based on ideal interpolation.

$$\phi(\tau) = \sum_{n=-\infty}^{\infty} \phi\left(\frac{n}{2W}\right) \frac{\sin\left[2\pi W\left(\tau - \frac{n}{2W}\right)\right]}{2\pi W\left(\tau - \frac{n}{2W}\right)} \tag{2.2–36}$$

where $\{\phi(n/2W)\}$ are samples of $\phi(\tau)$ taken at $\tau = n/2W$, $n = 0, \pm 1, \pm 2, \ldots$.

Now, if $X(t)$ is a band-limited stationary stochastic process, then $X(t)$ can be represented as

$$X(t) = \sum_{n=-\infty}^{\infty} X\left(\frac{n}{2W}\right) \frac{\sin\left[2\pi W\left(t - \frac{n}{2W}\right)\right]}{2\pi W\left(t - \frac{n}{2W}\right)} \tag{2.2–37}$$

where $\{X(n/2W)\}$ are samples of $X(t)$ taken at $t = n/2W$, $n = 0, \pm 1, \pm 2, \ldots$. This is the sampling representation for a stationary stochastic process. The samples are random variables that are described statistically by appropriate joint probability density functions. The signal representation in Equation 2.2–37 is easily established by showing that (Problem 2.17)

$$E\left\{\left|X(t) - \sum_{n=-\infty}^{\infty} X\left(\frac{n}{2W}\right) \frac{\sin\left[2\pi W\left(t - \frac{n}{2W}\right)\right]}{2\pi W\left(t - \frac{n}{2W}\right)}\right|^2\right\} = 0 \tag{2.2–38}$$

Hence, equality between the sampling representation and the stochastic process $X(t)$ holds in the sense that the mean square error is zero.

2.2.5 Discrete-Time Stochastic Signals and Systems

The characterization of continuous-time stochastic signals given above can be easily carried over to discrete-time stochastic signals. Such signals are usually obtained by uniformly sampling a continuous-time stochastic process.

A discrete-time stochastic process $X(n)$ consists of an ensemble of sample sequences $\{x(n)\}$. The statistical properties of $X(n)$ are similar to the characterization of $X(t)$ with the restriction that n is now an integer (time) variable. Hence, the *mth moment* of $X(n)$ is defined as

$$E[X_n^m] = \int_{-\infty}^{\infty} X_n^m p(X_n)\, dX_n \tag{2.2–39}$$

and the *autocorrelation sequence* is

$$\phi(n, k) = \tfrac{1}{2}E(X_n X_k^*) = \tfrac{1}{2}\int_{-\infty}^{\infty}\int_{-\infty}^{\infty} X_n X_k^* p(X_n, X_k)\, dX_n\, dX_k \tag{2.2–40}$$

Similarly, the autocovariance sequence is

$$\mu(n, k) = \phi(n, k) - \tfrac{1}{2}E(X_n)E(X_k^*) \tag{2.2–41}$$

For a stationary process, we have $\phi(n, k) \equiv \phi(n - k)$, $\mu(n, k) \equiv \mu(n - k)$, and

$$\mu(n - k) = \phi(n - k) - \tfrac{1}{2}|m_x|^2 \tag{2.2–42}$$

where $m_x = E(X_n)$ is the mean value.

As in the case of continuous-time stochastic processes, a discrete-time stationary process has infinite energy but a finite average power, which is given as

$$E(|X_n|^2) = \phi(0) \tag{2.2–43}$$

The power density spectrum for the discrete-time process is obtained by computing the Fourier transform of $\phi(n)$. Since $\phi(n)$ is a discrete-time sequence, the Fourier transform is defined as

$$\Phi(f) = \sum_{n=-\infty}^{\infty} \phi(n) e^{-j2\pi f n} \tag{2.2–44}$$

and the inverse transform relationship is

$$\phi(n) = \int_{-1/2}^{1/2} \Phi(f) e^{j2\pi f n}\, df \tag{2.2–45}$$

We make the observation that the power density spectrum $\Phi(f)$ is periodic with a period $f_p = 1$. In other words, $\Phi(f + k) = \Phi(f)$ for $k = \pm 1, \pm 2, \ldots$. This is a characteristic of the Fourier transform of any discrete-time sequence such as $\phi(n)$.

Finally, let us consider the response of a discrete-time, linear time-invariant system to a stationary stochastic input signal. The system is characterized in the time domain by its unit sample response $h(n)$ and in the frequency domain by the frequency response $H(f)$, where

$$H(f) = \sum_{n=-\infty}^{\infty} h(n) e^{-j2\pi f n} \tag{2.2–46}$$

The response of the system to the stationary stochastic input signal $X(n)$ is given by the convolution sum

$$y(n) = \sum_{k=-\infty}^{\infty} h(k) x(n - k) \tag{2.2–47}$$

The mean value of the output of the system is

$$\begin{aligned} m_y = E[y(n)] &= \sum_{k=-\infty}^{\infty} h(k)E[x(n-k)] \\ &= m_x \sum_{k=-\infty}^{\infty} h(k) = m_x H(0) \end{aligned} \tag{2.2–48}$$

where $H(0)$ is the zero frequency [direct current (DC)] gain of the system.

The autocorrelation sequence for the output process is

$$\begin{aligned} \phi_{yy}(k) &= \tfrac{1}{2}E[y^*(n)y(n+k)] \\ &= \tfrac{1}{2}\sum_{i=-\infty}^{\infty}\sum_{j=-\infty}^{\infty} h^*(i)h(j)E[x^*(n-i)x(n+k-j)] \\ &= \sum_{i=-\infty}^{\infty}\sum_{j=-\infty}^{\infty} h^*(i)h(j)\phi_{xx}(k-j+i) \end{aligned} \tag{2.2–49}$$

This is the general form for the autocorrelation sequence of the system output in terms of the autocorrelation of the system input and the unit sample response of the system. By taking the Fourier transform of $\phi_{yy}(k)$ and substituting the relation in Equation 2.2–49, we obtain the corresponding frequency domain relationship

$$\Phi_{yy}(f) = \Phi_{xx}(f)|H(f)|^2 \tag{2.2–50}$$

which is identical to Equation 2.2–27 except that in Equation 2.2–50 the power density spectra $\Phi_{yy}(f)$ and $\Phi_{xx}(f)$ and the frequency response $H(f)$ are periodic functions of frequency with period $f_p = 1$.

2.2.6 Cyclostationary Processes

In dealing with signals that carry digital information we encounter stochastic processes that have statistical averages that are periodic. To be specific, let us consider a stochastic process of the form

$$X(t) = \sum_{n=-\infty}^{\infty} a_n g(t-nT) \tag{2.2–51}$$

where $\{a_n\}$ is a (discrete-time) sequence of random variables with mean $m_a = E(a_n)$ for all n and autocorrelation sequence $\phi_{aa}(k) = \frac{1}{2}E(a_n^* a_{n+k})$. The signal $g(t)$ is deterministic. The stochastic process $X(t)$ represents the signal for several different types of linear modulation techniques which are introduced in Chapter 4. The sequence $\{a_n\}$ represents the digital information sequence (of symbols) that is transmitted over the communication channel and $1/T$ represents the rate of transmission of the information symbols.

Let us determine the mean and autocorrelation function of $X(t)$. First, the mean value is

$$\begin{aligned} E[X(t)] &= \sum_{n=-\infty}^{\infty} E(a_n)g(t-nT) \\ &= m_a \sum_{n=-\infty}^{\infty} g(t-nT) \end{aligned} \tag{2.5–52}$$

We observe that the mean is time-varying. In fact, it is periodic with period T.

The autocorrelation function of $X(t)$ is

$$\begin{aligned} \phi_{xx}(t+\tau, t) &= \tfrac{1}{2}E[X(t+\tau)X^*(t)] \\ &= \tfrac{1}{2}\sum_{n=-\infty}^{\infty}\sum_{m=-\infty}^{\infty} E(a_n^* a_m)g^*(t-nT)g(t+\tau-mT) \\ &= \sum_{n=-\infty}^{\infty}\sum_{m=-\infty}^{\infty} \phi_{aa}(m-n)g^*(t-nT)g(t+\tau-mT) \end{aligned} \tag{2.2–53}$$

Again, we observe that

$$\phi_{xx}(t+\tau+kT, t+kT) = \phi_{xx}(t+\tau, t) \tag{2.2–54}$$

for $k = \pm 1, \pm 2, \ldots$. Hence, the autocorrelation function of $X(t)$ is also periodic with period T.

Such a stochastic process is called *cyclostationary* or *periodically stationary*. Since the autocorrelation function depends on both the variables t and τ, its frequency domain representation requires the use of a two-dimensional Fourier transform.

Since it is highly desirable to characterize such signals by their power density spectrum, an alternative approach is to compute the *time-average autocorrelation function* over a single period, defined as

$$\bar{\phi}_{xx}(\tau) = \frac{1}{T}\int_{-T/2}^{T/2} \phi_{xx}(t+\tau, t)\, dt \tag{2.2–55}$$

Thus, we eliminate the time dependence by dealing with the average autocorrelation function. Now, the Fourier transform of $\bar{\phi}_{xx}(\tau)$ yields the *average power density spectrum* of the cyclostationary stochastic process. This approach allows us to simply characterize cyclostationary processes in the frequency domain in terms of the power spectrum. That is, the power density spectrum is

$$\Phi_{xx}(f) = \int_{-\infty}^{\infty} \bar{\phi}_{xx}(\tau)e^{-j2\pi f\tau}\, d\tau \tag{2.2–56}$$

2.3 BIBLIOGRAPHICAL NOTES AND REFERENCES

In this chapter we have provided a review of basic concepts and definitions in the theory of probability and stochastic processes. As stated in the opening paragraph, this theory is an important mathematical tool in the statistical modeling of information sources, communication channels, and in the design of digital communication systems. Of particular importance in the evaluation of communication system performance is the Chernoff bound. This bound is frequently used in bounding the probability of error of digital communication systems that employ coding in the transmission of information. Our coverage also highlighted a number of probability distributions and their properties, which are frequently encountered in the design of digital communication systems.

The texts by Davenport and Root (1958), Davenport (1970), Papoulis (1984) Peebles (1987), Helstrom (1991), Stark and Woods (1994), and Leon-Garcia (1994) provide engineering-oriented treatments of probability and stochastic processes. A more mathematical treatment of probability theory may be found in the text by Loève (1955). Finally, we cite the book by Miller (1964), which treats multidimensional Gaussian distributions.

PROBLEMS

2.1 One experiment has four mutually exclusive outcomes A_i, $i = 1, 2, 3, 4$, and a second experiment has three mutually exclusive outcomes B_j, $j = 1, 2, 3$. The joint probabilities $P(A_i, B_j)$ are

$$\begin{array}{lll} P(A_1, B_1) = 0.10, & P(A_1, B_2) = 0.08, & P(A_1, B_3) = 0.13 \\ P(A_2, B_1) = 0.05, & P(A_2, B_2) = 0.03, & P(A_2, B_3) = 0.09 \\ P(A_3, B_1) = 0.05, & P(A_3, B_2) = 0.12, & P(A_3, B_3) = 0.14 \\ P(A_4, B_1) = 0.11, & P(A_4, B_2) = 0.04, & P(A_4, B_3) = 0.06 \end{array}$$

Determine the probabilities $P(A_i)$, $i = 1, 2, 3, 4$, and $P(B_j)$, $j = 1, 2, 3$.

2.2 The random variables X_i, $i = 1, 2, \ldots, n$, have joint PDF $p(x_1, x_2, \ldots, x_n)$. Prove that

$$p(x_1, x_2, x_3, \ldots, x_n) = p(x_n|x_{n-1}, \ldots, x_1)p(x_{n-1}|x_{n-2}, \ldots, x_1) \cdots p(x_3|x_2, x_1)p(x_2|x_1)p(x_1)$$

2.3 The PDF of a random variable X is $p(x)$. A random variable Y is defined as

$$Y = aX + b$$

where $a < 0$. Determine the PDF of Y in terms of the PDF of X.

2.4 Suppose that X is a Gaussian random variable with zero mean and unit variance. Let

$$Y = aX^3 + b, \qquad a > 0$$

Determine and plot the PDF of Y.

2.5 *a*) Let X_r and X_i be statistically independent zero-mean Gaussian random variables with identical variance. Show that a (rotational) transformation of the form

$$Y_r + jY_i = (X_r + jX_i)e^{j\phi}$$

results in another pair (Y_r, Y_i) of Gaussian random variables that have the same joint PDF as the pair (X_r, X_i).

b) Note that

$$\begin{bmatrix} Y_r \\ Y_i \end{bmatrix} = \mathbf{A}\begin{bmatrix} X_r \\ X_i \end{bmatrix}$$

where **A** is a 2×2 matrix. As a generalization of the two-dimensional transformation of the Gaussian random variables considered in (*a*), what property must the linear transformation **A** satisfy if the PDFs for **X** and **Y**, where $\mathbf{Y} = \mathbf{AX}$, $\mathbf{X} = (X_1 X_2 \cdots X_n)$ and $\mathbf{Y} = (Y_1 Y_2 \cdots Y_n)$ are identical?

2.6 The random variable Y is defined as

$$Y = \sum_{i=1}^{n} X_i$$

where the X_i, $i = 1, 2, \ldots, n$, are statistically independent random variables with

$$X_i = \begin{cases} 1 & \text{(with probability } p) \\ 0 & \text{(with probability } 1 - p) \end{cases}$$

a) Determine the characteristic function of Y.

b) From the characteristic function, determine the moments $E(Y)$ and $E(Y^2)$.

2.7 The four random variables X_1, X_2, X_3, X_4 are zero-mean jointly Gaussian random variables with covariance $\mu_{ij} = E(X_i X_j)$ and characteristic function $\psi(jv_1, jv_2, jv_3, jv_4)$. Show that

$$E(X_1 X_2 X_3 X_4) = \mu_{12}\mu_{34} + \mu_{13}\mu_{24} + \mu_{14}\mu_{23}$$

2.8 From the characteristic functions for the central chi-square and noncentral chi-square random variables given by Equations 2.1–109 and 2.1–117, respectively, determine the corresponding first and second moments given by Equations 2.1–112 and 2.1–125.

2.9 The PDF of a Cauchy distributed random variable X is

$$p(x) = \frac{a/\pi}{x^2 + a^2}, \qquad -\infty < x < \infty$$

a) Determine the mean and variance of X.

b) Determine the characteristic function of X.

2.10 The random variable Y is defined as

$$Y = \frac{1}{n}\sum_{i=1}^{n} X_i$$

where X_i, $i = 1, 2, \ldots, n$, are statistically independent and identically distributed random variables each of which has the Cauchy PDF given in Problem 2.9.

a) Determine the characteristic function of Y.
b) Determine the PDF of Y.
c) Consider the PDF of Y in the limit as $n \to \infty$. Does the central limit theorem hold? Explain your answer.

2.11 Assume that random processes $X(t)$ and $Y(t)$ are individually and jointly stationary.

a) Determine the autocorrelation function of $Z(t) = X(t) + Y(t)$.
b) Determine the autocorrelation function of $Z(t)$ when $X(t)$ and $Y(t)$ are uncorrelated.
c) Determine the autocorrelation function of $Z(t)$ when $X(t)$ and $Y(t)$ are uncorrelated and have zero means.

2.12 The autocorrelation function of a stochastic process $X(t)$ is

$$\phi_{xx}(\tau) = \tfrac{1}{2}N_0\delta(\tau)$$

Such a process is called *white noise*. Suppose $x(t)$ is the input to an ideal band-pass filter having the frequency response characteristic shown in Figure P2.12. Determine the total noise power at the output of the filter.

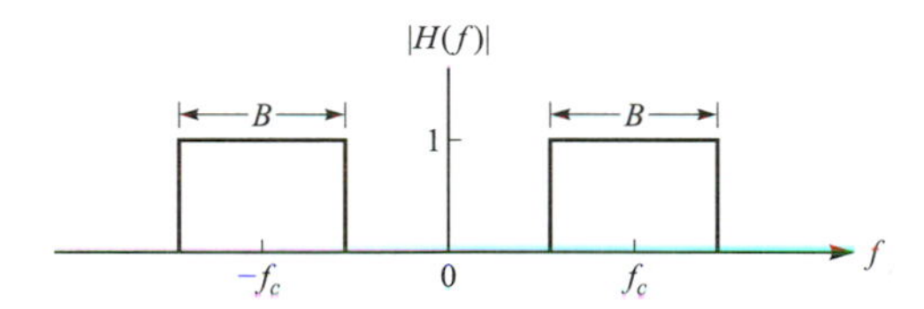

FIGURE P2.12

2.13 The covariance matrix of three random variables X_1, X_2, and X_3 is

$$\begin{bmatrix} \mu_{11} & 0 & \mu_{13} \\ 0 & \mu_{22} & 0 \\ \mu_{31} & 0 & \mu_{33} \end{bmatrix}$$

The linear transformation $\mathbf{Y} = \mathbf{AX}$ is made where

$$\mathbf{A} = \begin{bmatrix} 1 & 0 & 0 \\ 0 & 2 & 0 \\ 1 & 0 & 1 \end{bmatrix}$$

Determine the covariance matrix of $\mathbf{Y}$.

2.14 Let $X(t)$ be a stationary real normal process with zero mean. Let a new process $Y(t)$ be defined by

$$Y(t) = X^2(t)$$

Determine the autocorrelation function of $Y(t)$ in terms of the autocorrelation function of $X(t)$. *Hint*: Use the result on Gaussian variables derived in Problem 2.7.

2.15 For the Nakagami PDF, given by Equation 2.1–147, define the normalized random variable $X = R/\sqrt{\Omega}$. Determine the PDF of X.

2.16 The input $X(t)$ in the circuit shown in Figure P2.16 is a stochastic process with $E[X(t)] = 0$ and $\phi_{xx}(\tau) = \sigma^2\delta(\tau)$, i.e., $X(t)$ is a white noise process.
a) Determine the spectral density $\Phi_{yy}(f)$.
b) Determine $\phi_{yy}(\tau)$ and $E[Y^2(t)]$.

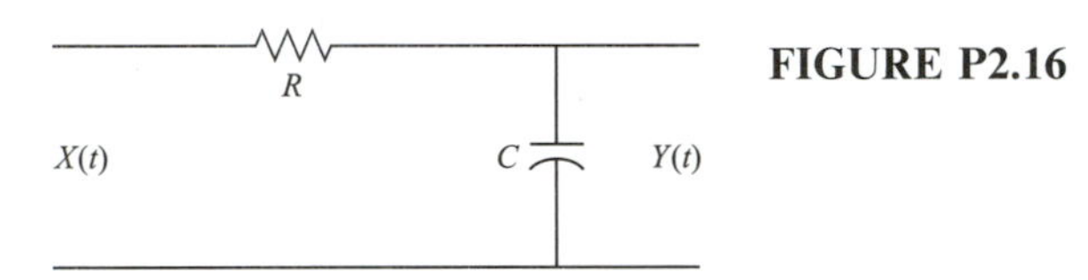

FIGURE P2.16

2.17 Demonstrate the validity of Equation 2.2–38.

2.18 Use the Chernoff bound to show that $Q(x) \leqslant e^{-x^2/2}$ where $Q(x)$ is defined by Equation 2.1–97.

2.19 Determine the mean, the autocorrelation sequence, and the power density spectrum of the output of a system with unit sample response

$$h(n) = \begin{cases} 1 & (n = 0) \\ -2 & (n = 1) \\ 1 & (n = 2) \\ 0 & (\text{otherwise}) \end{cases}$$

when the input $x(n)$ is a white-noise process with variance σ_x^2.

2.20 The autocorrelation sequence of a discrete-time stochastic process is $\phi(k) = (\frac{1}{2})^{|k|}$. Determine its power density spectrum.

2.21 A discrete-time stochastic process $X(n) \equiv X(nT)$ is obtained by periodic sampling of a continuous-time zero-mean stationary process $X(t)$ where T is the sampling interval, i.e., $f_s = 1/T$ is the sampling rate.
a) Determine the relationship between the autocorrelation function of $X(t)$ and the autocorrelation sequence of $X(n)$.
b) Express the power density spectrum of $X(n)$ in terms of the power density spectrum of the process $X(t)$.
c) Determine the conditions under which the power density spectrum of $X(n)$ is equal to the power density spectrum of $X(t)$

2.22 Consider a band-limited zero-mean stationary stochastic $X(t)$ with power density spectrum

$$\Phi(f) = \begin{cases} 1 & (|f| \leq W) \\ 0 & (|f| > W) \end{cases}$$

$X(t)$ is sampled at a rate $f_s = 1/T$ to yield a discrete-time process $X(n) \equiv X(nT)$.

a) Determine the expression for the autocorrelation sequence of $X(n)$.

b) Determine the minimum value of T that results in a white (spectrally flat) sequence.

c) Repeat (*b*) if the power density spectrum of $X(t)$ is

$$\Phi(f) = \begin{cases} 1 - |f|/W & (|f| \leq W) \\ 0 & (|f| > W) \end{cases}$$

2.23 Show that the functions

$$f_k(t) = \frac{\sin\left[2\pi W\left(t - \dfrac{k}{2W}\right)\right]}{2\pi W\left(t - \dfrac{k}{2W}\right)}, \qquad k = 0, \pm 1, \pm 2, \ldots$$

are orthogonal over the interval $[-\infty, \infty]$, i.e.,

$$\int_{-\infty}^{\infty} f_k(t) f_j(t)\, dt = \begin{cases} 1/2W & (k = j) \\ 0 & (k \neq j) \end{cases}$$

Therefore, the sampling theorem reconstruction formula may be viewed as a series expansion of the band-limited signal $s(t)$, where the weights are samples of $s(t)$ and the $\{f_k(t)\}$ are the set of orthogonal functions used in the series expansion.

2.24 The noise equivalent bandwidth of a system is defined as

$$B_{eq} = \frac{1}{G} \int_0^{\infty} |H(f)|^2 df$$

where $G = \max |H(f)|^2$. Using this definition, determine the noise equivalent bandwidth of the ideal band-pass filter shown in Figure P2.12 and the low-pass system shown in Figure P2.16.

3

Source Coding

Communication systems are designed to transmit the information generated by a source to some destination. Information sources may take a variety of different forms. For example, in radio broadcasting, the source is generally an audio source (voice or music). In TV broadcasting, the information source is a video source whose output is a moving image. The outputs of these sources are analog signals and, hence, the sources are called *analog sources*. In contrast, computers and storage devices, such as magnetic or optical disks, produce discrete outputs (usually binary or ASCII characters) and, hence, they are called *discrete sources*.

Whether a source is analog or discrete, a digital communication system is designed to transmit information in digital form. Consequently, the output of the source must be converted to a format that can be transmitted digitally. This conversion of the source output to a digital form is generally performed by the source encoder, whose output may be assumed to be a sequence of binary digits.

In this chapter, we treat source encoding based on mathematical models of information sources and provide a quantitative measure of the information emitted by a source. We consider the encoding of discrete sources first and then we discuss the encoding of analog sources. We begin by developing mathematical models for information sources.

3.1
MATHEMATICAL MODELS FOR INFORMATION SOURCES

Any information source produces an output that is random, i.e., the source output is characterized in statistical terms. Otherwise, if the source output were known exactly, there would be no need to transmit it. In this section, we consider both discrete and analog information sources, and we postulate mathematical models for each type of source.

The simplest type of discrete source is one that emits a sequence of letters selected from a finite alphabet. For example, a *binary source* emits a binary sequence of the form 100101110..., where the alphabet consists of the two letters {0, 1}. More generally, a discrete information source with an alphabet of L possible letters, say $\{x_1, x_2, \ldots, x_L\}$, emits a sequence of letters selected from the alphabet.

To construct a mathematical model for a discrete source, we assume that each letter in the alphabet $\{x_1, x_2, \ldots, x_L\}$ has a given probability p_k of occurrence. That is,

$$p_k = P(X = x_k), \qquad 1 \leqslant k \leqslant L$$

where

$$\sum_{k=1}^{L} p_k = 1$$

We consider two mathematical models of discrete sources. In the first, we assume that the output sequence from the source is statistically independent. That is, the current output letter is statistically independent from all past and future outputs. A source whose output satisfies the condition of statistical independence among output letters in the sequence is said to be *memoryless*. Such a source is called a *discrete memoryless source* (DMS).

If the discrete source output is statistically dependent, as, for example, English text, we may construct a mathematical model based on statistical stationarity. By definition, a discrete source is said to be *stationary* if the joint probabilities of two sequences of length n, say $a_1, a_2, \ldots, a_n$ and $a_{1+m}, a_{2+m}, \ldots, a_{n+m}$, are identical for all $n \geqslant 1$ and for all shifts m. In other words, the joint probabilities for any arbitrary length sequence of source outputs are invariant under a shift in the time origin.

An *analog* source has an output waveform $x(t)$ that is a sample function of a stochastic process $X(t)$. We assume that $X(t)$ is a stationary stochastic process with autocorrelation function $\phi_{xx}(\tau)$ and power spectral density $\Phi_{xx}(f)$. When $X(t)$ is a band-limited stochastic process, i.e., $\Phi_{xx}(f) = 0$ for $|f| \geqslant W$, the sampling theorem may be used to represent $X(t)$ as

$$X(t) = \sum_{n=-\infty}^{\infty} X\left(\frac{n}{2W}\right) \frac{\sin\left[2\pi W\left(t - \frac{n}{2W}\right)\right]}{2\pi W\left(t - \frac{n}{2W}\right)} \tag{3.1–1}$$

where $\{X(n/2W)\}$ denote the samples of the process $X(t)$ taken at the sampling (Nyquist) rate of $f_s = 2W$ samples/s. Thus, by applying the sampling theorem, we may convert the output of an analog source into an equivalent discrete-time source. Then, the source output is characterized statistically by the joint PDF $p(x_1, x_2, \ldots, x_m)$ for all $m \geqslant 1$, where $X_n = X(n/2W)$, $1 \leqslant n \leqslant m$, are the random variables corresponding to the samples of $X(t)$.

We note that the output samples $\{X(n/2W)\}$ from the stationary sources are generally continuous, and, hence, they cannot be represented in digital form without some loss in precision. For example, we may quantize each sample to a set of discrete values, but the quantization process results in loss of precision, and, consequently, the original signal cannot be reconstructed exactly from the quantized sample values. Later in this chapter, we shall consider the distortion resulting from quantization of the samples from an analog source.

3.2
A LOGARITHMIC MEASURE OF INFORMATION

To develop an appropriate measure of information, let us consider two discrete random variables with possible outcomes x_i, $i = 1, 2, \ldots n$, and y_i, $i = 1, 2, \ldots, m$, respectively. Suppose we observe some outcome $Y = y_j$ and we wish to determine, quantitatively, the amount of information that the occurrence of the event $Y = y_j$ provides about the event $X = x_i$, $i = 1, 2, \ldots, n$. We observe that when X and Y are statistically independent, the occurrence of $Y = y_j$ provides no information about the occurrence of the event $X = x_i$. On the other hand, when X and Y are fully dependent such that the occurrence of $Y = y_j$ determines the occurrence of $X = x_i$, the information content is simply that provided by the event $X = x_i$. A suitable measure that satisfies these conditions is the logarithm of the ratio of the conditional probability

$$P(X = x_i | Y = y_j) \equiv P(x_i | y_j)$$

divided by the probability

$$P(X = x_i) \equiv P(x_i)$$

That is, the information content provided by the occurrence of the event $Y = y_j$ about the event $X = x_i$ is defined as

$$I(x_i; y_j) = \log \frac{P(x_i | y_i)}{P(x_i)} \tag{3.2–1}$$

$I(x_i; y_j)$ is called the *mutual information* between x_i and y_j.

The units of $I(x_i; y_j)$ are determined by the base of the logarithm, which is usually selected as either 2 or e. When the base of the logarithm is 2, the units of $I(x_i; y_j)$ are bits, and when the base is e, the units of $I(x_i; y_j)$ are called *nats* (natural units). (The standard abbreviation for $\log_e$ is ln.) Since

$$\ln a = \ln 2 \log_2 a = 0.69315 \log_2 a$$

the information measured in nats is equal to ln 2 times the information measured in bits.

When the random variables X and Y are statistically independent, $P(x_i | y_j) = P(x_i)$ and, hence, $I(x_i; y_j) = 0$. On the other hand, when the occurrence of the event $Y = y_j$ uniquely determines the occurrence of the event $X = x_i$, the conditional probability in the numerator of Equation 3.2–1 is unity and, hence,

$$I(x_i; y_j) = \log \frac{1}{P(x_i)} = -\log P(x_i) \tag{3.2–2}$$

But Equation 3.2–2 is just the information of the event $X = x_j$. For this reason, it is called the *self-information* of the event $X = x_i$ and it is denoted as

$$I(x_i) = \log \frac{1}{P(x_i)} = -\log P(x_i) \tag{3.2–3}$$

We note that a high-probability event conveys less information than a low-probability event. In fact, if there is only a single event x with probability $P(x) = 1$, then $I(x) = 0$. To demonstrate further that the logarithmic measure of information content is the appropriate one for digital communications, let us consider the following example.

EXAMPLE 3.2–1. Suppose we have a discrete information source that emits a binary digit, either 0 or 1, with equal probability every τ_s seconds. The information content of each output from the source is

$$\begin{aligned} I(x_i) &= -\log_2 P(x_i), \qquad x_i = 0, 1 \\ &= -\log_2 \tfrac{1}{2} = 1 \text{ bit} \end{aligned}$$

Now suppose that successive outputs from the source are statistically independent, i.e., the source is memoryless. Let us consider a block of k binary digits from the source that occurs in a time interval $k\tau_s$. There are $M = 2^k$ possible k-bit blocks, each of which is equally probable with probability $1/M = 2^{-k}$. The self-information of a k-bit block is

$$I(x_i') = -\log_2 2^{-k} = k \text{ bits}$$

emitted in a time interval $k\tau_s$. Thus the logarithmic measure of information content possesses the desired additivity property when a number of source outputs is considered as a block.

Now let us return to the definition of mutual information given in Equation 3.2–1 and multiply the numerator and denominator of the ratio of probabilities by $P(y_i)$. Since

$$\frac{P(x_i|y_j)}{P(x_i)} = \frac{P(x_i|y_j)P(y_j)}{P(x_i)P(y_j)} = \frac{P(x_i, y_j)}{P(x_i)P(y_i)} = \frac{P(y_j|x_i)}{P(y_j)}$$

we conclude that

$$I(x_i; y_j) = I(y_j; x_i) \tag{3.2–4}$$

Therefore the information provided by the occurrence of the event $Y = y_j$ about the event $X = x_i$ is identical to the information provided by the occurrence of the event $X = x_i$ about the event $Y = y_j$.

EXAMPLE 3.2-2. Suppose that X and Y are binary-valued {0, 1} random variables that represent the input and output of a binary-input, binary-output channel. The input symbols are equally likely and the output symbols depend on the input according to the conditional probabilities

$$P(Y=0|X=0)=1-p_0$$
$$P(Y=1|X=0)=p_0$$
$$P(Y=1|X=1)=1-p_1$$
$$P(Y=0|X=1)=p_1$$

Let us determine the mutual information about the occurrence of the events $X=0$ and $X=1$, given that $Y=0$.

From the probabilities given above, we obtain

$$\begin{aligned} P(Y=0) &= P(Y=0|X=0)P(X=0)+P(Y=0|X=1)P(X=1) \\ &= \tfrac{1}{2}(1-p_0+p_1) \\ P(Y=1) &= P(Y=1|X=0)P(X=0)+P(Y=1|X=1)P(X=1) \\ &= \tfrac{1}{2}(1-p_1+p_0) \end{aligned}$$

Then, the mutual information about the occurrence of the event $X=0$, given that $Y=0$ is observed, is

$$I(x_1; y_1) = I(0; 0) = \log_2 \frac{P(Y=0|X=0)}{P(Y=0)} = \log_2 \frac{2(1-p_0)}{1-p_0+p_1}$$

Similarly, given that $Y=0$ is observed, the mutual information about the occurrence of the event $X=1$ is

$$I(x_2; y_1) \equiv I(1; 0) = \log_2 \frac{2p_1}{1-p_0+p_1}$$

Let us consider some special cases: First, if $p_0=p_1=0$, the channel is called *noiseless* and

$$I(0; 0) = \log_2 2 = 1 \text{ bit}$$

Hence, the output specifies the input with certainty. On the other hand, if $p_0=p_1=\frac{1}{2}$, the channel is *useless* because

$$I(0; 0) = \log_2 1 = 0$$

However, if $p_0=p_1=\frac{1}{4}$, then

$$I(0; 0) = \log_2 \tfrac{3}{2} = 0.587$$
$$I(0; 1) = \log_2 \tfrac{1}{2} = -1 \text{ bit}$$

In addition to the definition of mutual information and self-information, it is useful to define the *conditional self-information* as

$$I(x_i|y_j) = \log \frac{1}{P(x_i|y_j)} = -\log P(x_i|y_j) \qquad (3.2\text{–}5)$$

Then, by combining Equations 3.2–1, 3.2–3, and 3.2–5, we obtain the relationship

$$I(x_i; y_j) = I(x_i) - I(x_i|y_j) \qquad (3.2\text{–}6)$$

We interpret $I(x_i|y_j)$ as the self-information about the event $X=x_i$ after having observed the event $Y=y_j$. Since both $I(x_i) \geqslant 0$ and $I(x_i|y_j) \geqslant 0$, it follows that $I(x_i; y_j) < 0$ when $I(x_i|y_j) > I(x_i)$, and $I(x_i; y_j) > 0$ when $I(x_i|y_j) < I(x_i)$. Hence, the mutual information between a pair of events can be either positive, or negative, or zero.

3.2.1 Average Mutual Information and Entropy

Having defined the mutual information associated with the pair of events (x_i, y_j), which are possible outcomes of the two random variables X and Y, we can obtain the average value of the mutual information by simply weighting $I(x_i; y_j)$ by the probability of occurrence of the joint event and summing over all possible joint events. Thus, we obtain

$$\begin{aligned} I(X; Y) &= \sum_{i=1}^{n}\sum_{j=1}^{m} P(x_i, y_j) I(x_i; y_j) \\ &= \sum_{i=1}^{n}\sum_{j=1}^{m} P(x_i, y_j) \log \frac{P(x_i, y_j)}{P(x_i)P(y_j)} \end{aligned} \tag{3.2–7}$$

as the average mutual information betwen X and Y. We observe that $I(X; Y) = 0$ when X and Y are statistically independent. An important characteristic of the average mutual information is that $I(X; Y) \geqslant 0$ (see Problem 3–4).

Similarly, we define the average self-information, denoted by $H(X)$, as

$$\begin{aligned} H(X) &= \sum_{i=1}^{n} P(x_i) I(x_i) \\ &= -\sum_{i=1}^{n} P(x_i) \log P(x_i) \end{aligned} \tag{3.2–8}$$

When X represents the alphabet of possible output letters from a source, $H(X)$ represents the average self-information per source letter, and it is called the *entropy*† of the source. In the special case in which the letters from the source are equally probable, $P(x_i) = 1/n$ for all i, and, hence,

$$\begin{aligned} H(X) &= -\sum_{i=1}^{n} \frac{1}{n} \log \frac{1}{n} \\ &= \log n \end{aligned} \tag{3.2–9}$$

In general, $H(X) \leqslant \log n$ (see Problem 3–5) for any given set of source letter probabilities. In other words, *the entropy of a discrete source is a maximum when the output letters are equally probable.*

---

† The term *entropy* is taken from statistical mechanics (thermodynamics), where a function similar to equation 3.2–8 is called (thermodynamic) entropy.

EXAMPLE 3.2–3. Consider a source that emits a sequence of statistically independent letters, where each output letter is either 0 with probability q or 1 with probability $1 - q$. The entropy of this source is

$$H(X) \equiv H(q) = -q \log q - (1 - q) \log(1 - q) \tag{3.2–10}$$

The binary entropy function $H(q)$ is illustrated in Figure 3.2–1. We observe that the maximum value of the entropy function occurs at $q = \frac{1}{2}$ where $H(\frac{1}{2}) = 1$.

The average conditional self-information is called the *conditional entropy* and is defined as

$$H(X|Y) = \sum_{i=1}^{n} \sum_{j=1}^{m} P(x_i, y_j) \log \frac{1}{P(x_i|y_j)} \tag{3.2–11}$$

We interpret $H(X|Y)$ as the information or uncertainty in X after Y is observed. By combining Equations 3.2–7, 3.2–8, and 3.2–11 we obtain the relationship

$$I(X; Y) = H(X) - H(X|Y) \tag{3.2–12}$$

Since $I(X; Y) \geqslant 0$, it follows that $H(X) \geqslant H(X|Y)$, with equality if and only if X and Y are statistically independent. If we interpret $H(X|Y)$ as the average amount of (conditional self-information) uncertainty in X after we observe Y, and $H(X)$ as the average amount of uncertainty (self-information) prior to the observation, then $I(X; Y)$ is the average amount of (mutual information) uncertainty provided about the set X by the observation of the set Y. Since $H(X) \geqslant H(X|Y)$, it is clear that conditioning on the observation Y does not increase the entropy.

EXAMPLE 3.2–4. Let us evaluate the $H(X|Y)$ and $I(X; Y)$ for the binary-input, binary-output channel treated previously in Example 3.2–2 for the case where $p_0 = p_1 = p$. Let the probabilities of the input symbols be $P(X = 0) = q$ and $P(X = 1) = 1 - q$. Then the entropy is

$$H(X) \equiv H(q) = -q \log q - (1 - q) \log(1 - q)$$

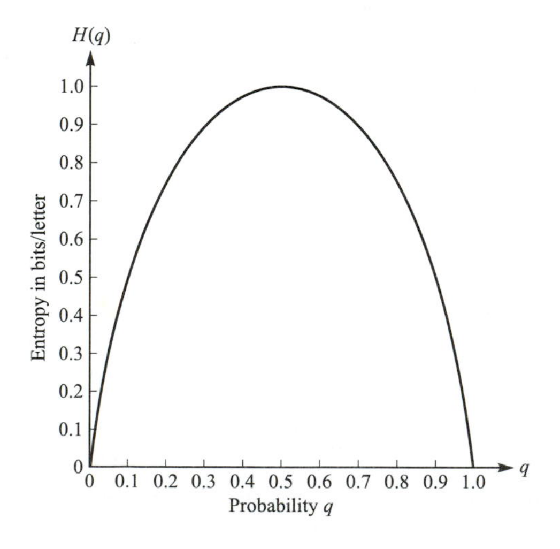

FIGURE 3.2–1
Binary entropy function.

where $H(q)$ is the binary entropy function and the conditional entropy $H(X|Y)$ is defined by Equation 3.2–11. A plot of $H(X|Y)$ as a function of q with p as a parameter is shown in Figure 3.2–2. The average mutual information $I(X; Y)$ is plotted in Figure 3.2–3.

As in the preceding example, when the conditional entropy $H(X|Y)$ is viewed in terms of a channel whose input is X and whose output is Y, $H(X|Y)$ is called the *equivocation* and is interpreted as the amount of average uncertainty remaining in X after observation of Y.

The results given above can be generalized to more than two random variables. In particular, suppose we have a block of k random variables $X_1X_2\cdots X_k$, with joint probability $P(x_1x_2\cdots x_k) \equiv P(X_1 = x_1, X_2 = x_2, \ldots, X_k = x_k)$. Then, the entropy for the block is defined as

$$H(X_1X_2\cdots X_k) = -\sum_{j_1=1}^{n_1}\sum_{j_2=1}^{n_2}\cdots\sum_{j_k=1}^{n_k} P(x_{j_1}x_{j_2}\cdots x_{j_k})\log P(x_{j_1}x_{j_2}\cdots x_{j_k}) \tag{3.2–13}$$

Since the joint probability $P(x_1x_2\cdots x_k)$ can be factored as

$$P(x_1x_2\cdots x_k) = P(x_1)P(x_2|x_1)P(x_3|x_1x_2)\cdots P(x_k|x_1x_2\cdots x_{k-1}) \tag{3.2–14}$$

it follows that

$$\begin{aligned} H(X_1X_2X_3\cdots X_k) &= H(X_1) + H(X_2|X_1) + H(X_3|X_1X_2) \\ &\quad + \cdots + H(X_k|X_1\cdots X_{k-1}) \\ &= \sum_{i=1}^{k} H(X_i|X_1X_2\cdots H_{i-1}) \end{aligned} \tag{3.2–15}$$

By applying the result $H(X) \geqslant H(X|Y)$, where $X = X_m$ and $Y = X_1X_2\cdots X_{m-1}$, in Equation 3.2–15 we obtain

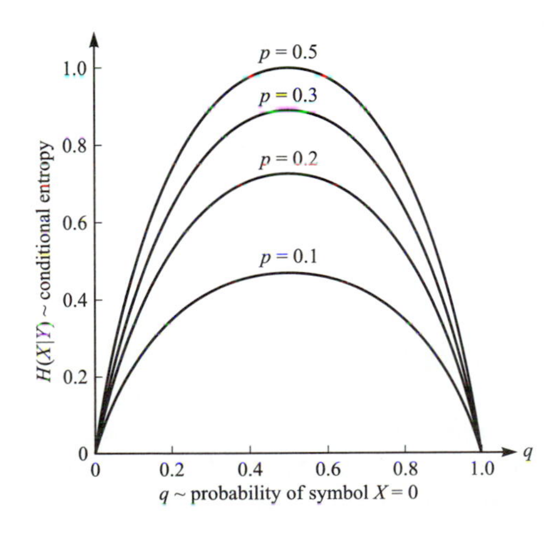

FIGURE 3.2–2
Conditional entropy for binary-input, binary-output symmetric channel.

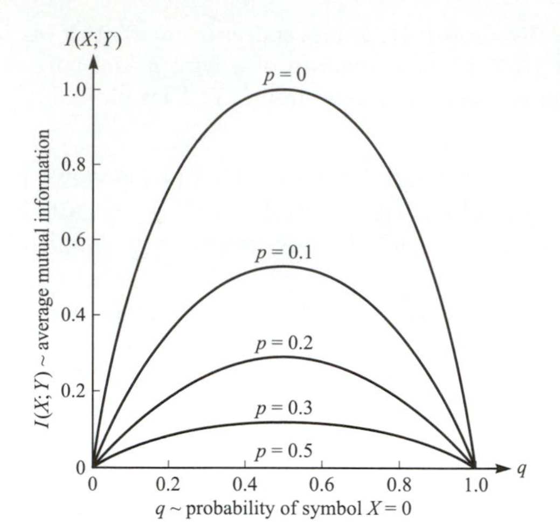

FIGURE 3.2–3
Average mutual information for binary-input, binary-output symmetric channel.

$$H(X_1X_2\cdots X_k) \leqslant \sum_{m=1}^{k} H(X_m) \tag{3.2–16}$$

with equality if and only if the random variables $X_1, X_2, \ldots, X_k$ are statistically independent.

3.2.2 Information Measures for Continuous Random Variables

The definition of mutual information given above for discrete random variables may be extended in a straightforward manner to continuous random variables. In particular, if X and Y are random variables with joint PDF $p(x, y)$ and marginal PDFs $p(x)$ and $p(y)$, the average mutual information between X and Y is defined as

$$I(X;\, Y) = \int_{-\infty}^{\infty}\int_{-\infty}^{\infty} p(x)p(y|x) \log \frac{p(y|x)p(x)}{p(x)p(y)}\, dx\, dy \tag{3.2–17}$$

Although the definition of the average mutual information carries over to continuous random variables, the concept of self-information does not. The problem is that a continuous random variable requires an infinite number of binary digits to represent it exactly. Hence, its self-information is infinite and, therefore, its entropy is also infinite. Nevertheless, we shall define a quantity that we call the *differential entropy* of the continuous random variable X as

$$H(X) = -\int_{-\infty}^{\infty} p(x) \log p(x)\, dx \tag{3.2–18}$$

We emphasize that this quantity does *not* have the physical meaning of self-information, although it may appear to be a natural extension of the definition of entropy for a discrete random variable (see Problem 3–6).

By defining the average conditional entropy of X given Y as

$$H(X|Y) = -\int_{-\infty}^{\infty}\int_{-\infty}^{\infty} p(x, y)\log p(x|y)\, dx\, dy \tag{3.2–19}$$

the average mutual information may be expressed as

$$I(X; Y) = H(X) - H(X|Y)$$

or, alternatively, as

$$I(X; Y) = H(Y) - H(Y|X)$$

In some cases of practical interest, the random variable X is discrete and Y is continuous. To be specific, suppose that X has possible outcomes x_i, $i = 1, 2, \ldots, n$, and Y is described by its marginal PDF $p(y)$. When X and Y are statistically dependent, we may express $p(y)$ as

$$p(y) = \sum_{i=1}^{n} p(y|x_i)P(x_i)$$

The mutual information provided about the event $X = x_i$ by the occurrence of the event $Y = y$ is

$$\begin{aligned} I(x_i; y) &= \log\frac{p(y|x_i)P(x_i)}{p(y)P(x_i)} \\ &= \log\frac{p(y|x_i)}{p(y)} \end{aligned} \tag{3.2–20}$$

Then, the average mutual information between X and Y is

$$I(X; Y) = \sum_{i=1}^{n}\int_{-\infty}^{\infty} p(y|x_i)P(x_i)\log\frac{p(y|x_i)}{p(y)}\, dy \tag{3.2–21}$$

EXAMPLE 3.2-5. Suppose that X is a discrete random variable with two equally probable outcomes $x_1 = A$ and $x_2 = -A$. Let the conditional PDFs $p(y|x_i)$, $i = 1, 2$, be Gaussian with mean x_i and variance σ^2. That is,

$$\begin{aligned} p(y|A) &= \frac{1}{\sqrt{2\pi}\sigma} e^{-(y-A)^2/2\sigma^2} \\ p(y|-A) &= \frac{1}{\sqrt{2\pi}\sigma} e^{-(y+A)^2/2\sigma^2} \end{aligned} \tag{3.2–22}$$

The average mutual information obtained from Equation 3.2–21 becomes

$$I(X; Y) = \frac{1}{2}\int_{-\infty}^{\infty}\left[p(y|A)\log\frac{p(y|A)}{p(y)} + p(y|-A)\log\frac{p(y|-A)}{p(y)}\right]dy \tag{3.2–23}$$

$$p(y) = \tfrac{1}{2}[p(y|A) + p(y|-A)] \tag{3.2–24}$$

In Chapter 7, it will be shown that the average mutual information $I(X; Y)$ given by Equation 3.2–23 represents the channel capacity of a binary-input additive white Gaussian noise channel.

3.3
CODING FOR DISCRETE SOURCES

In Section 3.2 we introduced a measure for the information content associated with a discrete random variable X. When X is the output of a discrete source, the entropy $H(X)$ of the source represents the average amount of information emitted by the source. In this section, we consider the process of encoding the output of a source, i.e., the process of representing the source output by a sequence of binary digits. A measure of the efficiency of a source-encoding method can be obtained by comparing the average number of binary digits per output letter from the source to the entropy $H(X)$.

The encoding of a discrete source having a finite alphabet size may appear, at first glance, to be a relatively simple problem. However, this is true only when the source is memoryless, i.e., when successive symbols from the source are statistically independent and each symbol is encoded separately. The discrete memoryless source (DMS) is by far the simplest model that can be devised for a physical source. Few physical sources, however, closely fit this idealized mathematical model. For example, successive output letters from a machine printing English text are expected to be statistically dependent. On the other hand, if the machine output is a computer program coded in Fortran, the sequence of output letters is expected to exhibit a much smaller dependence. In any case, we shall demonstrate that it is always more efficient to encode blocks of symbols instead of encoding each symbol separately. By making the block size sufficiently large, the average number of binary digits per output letter from the source can be made arbitrarily close to the entropy of the source.

3.3.1 Coding for Discrete Memoryless Sources

Suppose that a DMS produces an output letter or symbol every τ_s seconds. Each symbol is selected from a finite alphabet of symbols x_i, $i = 1, 2, \ldots, L$, occurring with probabilities $P(x_i)$, $i = 1, 2, \ldots, L$. The entropy of the DMS in bits per source symbol is

$$H(X) = -\sum_{i=1}^{L} P(x_i) \log_2 P(x_i) \leqslant \log_2 L \tag{3.3–1}$$

where equality holds when the symbols are equally probable. The average number of bits per source symbol is $H(X)$ and the source rate in bits/s is defined as $H(X)/\tau_s$.

Fixed-length code words. First we consider a block encoding scheme that assigns a unique set of R binary digits to each symbol. Since there are L possible symbols, the number of binary digits per symbol required for unique encoding when L is a power of 2 is

$$R = \log_2 L \tag{3.3–2}$$

and, when L is not a power of 2, it is

$$R = \lfloor \log_2 L \rfloor + 1 \tag{3.3–3}$$

where $\lfloor x \rfloor$ denotes the largest integer less than x. The code rate R in bits per symbol is now R and, since $H(X) \leqslant \log_2 L$, it follows that $R \geqslant H(X)$.

The efficiency of the encoding for the DMS is defined as the ratio $H(X)/R$. We observe that when L is a power of 2 and the source letters are equally probable, $R = H(X)$. Hence, a fixed-length code of R bits per symbol attains 100 percent efficiency. However, if L is not a power of 2 but the source symbols are still equally probable, R differs from $H(X)$ by at most 1 bit per symbol. When $\log_2 L \gg 1$, the efficiency of this encoding scheme is high. On the other hand, when L is small, the efficiency of the fixed-length code can be increased by encoding a sequence of J symbols at a time. To accomplish the desired encoding, we require L^J unique code words. By using sequences of N binary digits, we can accommodate 2^N possible code words. N must be selected such that

$$N \geqslant J \log_2 L$$

Hence, the minimum integer value of N required is

$$N = \lfloor J \log_2 L \rfloor + 1 \tag{3.3–4}$$

Now the average number of bits per source symbol is $N/J = R$, and, thus, the inefficiency has been reduced by approximately a factor of $1/J$ relative to the symbol-by-symbol encoding described above. By making J sufficiently large, the efficiency of the encoding procedure, measured by the ratio $JH(X)/N$, can be made as close to unity as desired.

The encoding methods described above introduce no distortion since the encoding of source symbols or blocks of symbols into code words is unique. This type of encoding is called *noiseless*.

Now, suppose we attempt to reduce the code rate R by relaxing the condition that the encoding process be unique. For example, suppose that only a fraction of the L^J blocks of symbols is encoded uniquely. To be specific, let us select the $2^N - 1$ most probable J-symbol blocks and encode each of them uniquely, while the remaining $L^J - (2^N - 1)$ J-symbol blocks are represented by the single remaining code word. This procedure results in a decoding failure or (distortion) probability of error every time a low probability block is mapped into this single code word. Let P_e denote this probability of error. Based on this block encoding procedure, Shannon (1948a) proved the following source coding theorem.

SOURCE CODING THEOREM I. Let X be the ensemble of letters from a DMS with finite entropy $H(X)$. Blocks of J symbols from the source are encoded into code words of length N from a binary alphabet. For any $\varepsilon > 0$, the probability P_e of a block decoding failure can be made arbitrarily small if

$$R \equiv \frac{N}{J} \geqslant H(X) + \varepsilon \tag{3.3–5}$$

and J is sufficiently large. Conversely, if

$$R \leqslant H(X) - \varepsilon \tag{3.3-6}$$

then P_e becomes arbitrarily close to 1 as J is made sufficiently large.

From this theorem, we observe that the average number of bits per symbol required to encode the output of a DMS with arbitrarily small probability of decoding failure is lower-bounded by the source entropy $H(X)$. On the other hand, if $R < H(X)$, the decoding failure rate approaches 100 percent as J is arbitrarily increased.

Variable-length code words. When the source symbols are not equally probable, a more efficient encoding method is to use variable-length code words. An example of such encoding is the Morse code, which dates back to the nineteenth century. In the Morse code, the letters that occur more frequently are assigned short code words and those that occur infrequently are assigned long code words. Following this general philosophy, we may use the probabilities of occurrence of the different source letters in the selection of the code words. The problem is to devise a method for selecting and assigning the code words to source letters. This type of encoding is called *entropy coding*.

For example, suppose that a DMS with output letters a_1, a_2, a_3, a_4 and corresponding probabilities $P(a_1) = \frac{1}{2}$, $P(a_2) = \frac{1}{4}$, and $P(a_3) = P(a_4) = \frac{1}{8}$ is encoded as shown in Table 3.3–1. Code I is a variable-length code that has a basic flaw. To see the flaw, suppose we are presented with the sequence 001001 . . . . Clearly, the first symbol corresponding to 00 is a_2. However, the next four bits are ambiguous (not uniquely decodable). They may be decoded either as a_4a_3 or as $a_1a_2a_1$. Perhaps, the ambiguity can be resolved by waiting for additional bits, but such a decoding delay is highly undesirable. We shall only consider codes that are decodable *instantaneously*, that is, without any decoding delay.

Code II in Table 3.3–1 is *uniquely decodable* and *instantaneously decodable*. It is convenient to represent the code words in this code graphically as terminal nodes of a tree, as shown in Figure 3.3–1. We observe that the digit 0 indicates the end of a code word for the first three code words. This characteristic plus the fact that no code word is longer than three binary digits makes this code instantaneously decodable. Note that no code word in this code is a prefix of any other

TABLE 3.3–1
Variable-length codes

Letter	$P(a_k)$	Code I	Code II	Code III
a_1	$\frac{1}{2}$	1	0	0
a_2	$\frac{1}{4}$	00	10	01
a_3	$\frac{1}{8}$	01	110	011
a_4	$\frac{1}{8}$	10	111	111

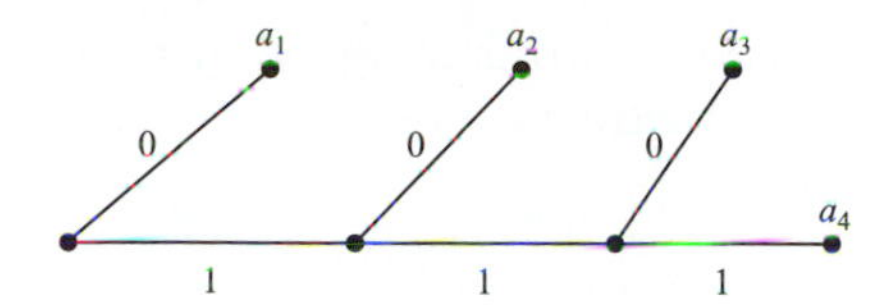

FIGURE 3.3–1
Code tree for code II in Table 3.3–1.

code word. In general, the *prefix condition* requires that for a given code word $\mathbf{C}_k$ of length k having elements $(b_1, b_2, \ldots, b_k)$, there is no other code word of length $l < k$ with elements $(b_1, b_2, \ldots, b_l)$ for $1 \leqslant l \leqslant k-1$. In other words, there is no code word of length $l < k$ that is identical to the first l binary digits of another code word of length $k > l$. This property makes the code words instantaneously decodable.

Code III given in Table 3.3–1 has the tree structures shown in Figure 3.3–2. We note that in this case the code is uniquely decodable but *not* instantaneously decodable. Clearly, this code does *not* satisfy the prefix condition.

Our main objective is to devise a systematic procedure for constructing uniquely decodable variable-length codes that are efficient in the sense that the average number of bits per source letter, defined as the quantity

$$\bar{R} = \sum_{k=1}^{L} n_k P(a_k) \tag{3.3–7}$$

is minimized. The conditions for the existence of a code that satisfies the prefix condition are given by the Kraft inequality.

Kraft inequality. A necessary and sufficient condition for the existence of a binary code with code words having lengths $n_1 \leqslant n_2 \leqslant \cdots \leqslant n_L$ that satisfy the prefix condition is

$$\sum_{k=1}^{L} 2^{-n_k} \leqslant 1 \tag{3.3–8}$$

First, we prove that Equation 3.3–8 is a sufficient condition for the existence of a code that satisfies the prefix condition. To construct such a code, we begin with a full binary tree of order $n = n_L$ that has 2^n terminal nodes and two nodes of order k stemming from each node of order $k-1$, for each k, $1 \leqslant k \leqslant n$. Let us select any node of order n_1 as the first code word $\mathbf{C}_1$. This choice eliminates 2^{n-n_1} terminal nodes (or the fraction 2^{-n_1} of the 2^n terminal nodes). From the remaining available nodes of order n_2, we select one node for the second code word $\mathbf{C}_2$. This choice eliminates 2^{n-n_2} terminal nodes (or the fraction 2^{-n_2} of the 2^n term-

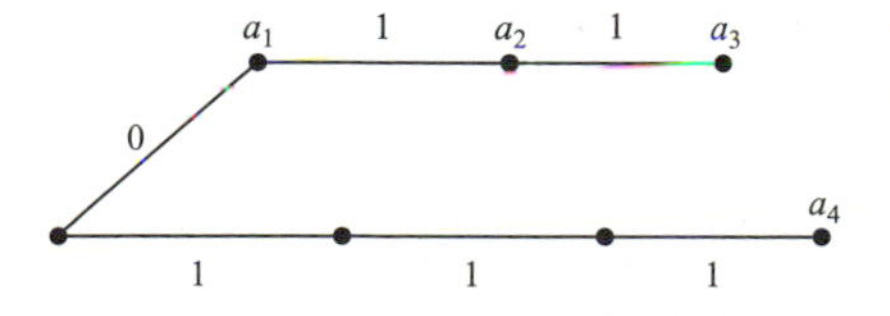

FIGURE 3.3–2
Code tree for code III in Table 3.3–1.

inal nodes). This process continues until the last code word is assigned at terminal node $n = n_L$. Since, at the node of order $j < L$, the fraction of the number of terminal nodes eliminated is

$$\sum_{k=1}^{j} 2^{-n_k} < \sum_{k=1}^{L} 2^{-n_k} \leqslant 1$$

there is always a node of order $k > j$ available to be assigned to the next code word. Thus, we have constructed a code tree that is embedded in the full tree of 2^n nodes as illustrated in Figure 3.3–3, for a tree having 16 terminal nodes and a source output consisting of five letters with $n_1 = 1, n_2 = 2, n_3 = 3$, and $n_4 = n_5 = 4$.

To prove that Equation 3.3–8 is a necessary condition, we observe that in the code tree of order $n = n_L$, the number of terminal nodes eliminated from the total number of 2^n terminal nodes is

$$\sum_{k=1}^{L} 2^{n-n_k} \leqslant 2^n$$

Hence,

$$\sum_{k=1}^{L} 2^{-n_k} \leqslant 1$$

and the proof of Equation 3.3–8 is complete.

The Kraft inequality may be used to prove the following (noiseless) source coding theorem, which applies to codes that satisfy the prefix condition.

SOURCE CODING THEOREM II. Let X be the ensemble of letters from a DMS with finite entropy $H(X)$ and output letters x_k, $1 \leqslant k \leqslant L$, with corresponding probabilities of occurrence p_k, $1 \leqslant k \leqslant L$. It is possible to construct a code that satisfies the prefix condition and has an average length $\bar{R}$ that satisfies the inequalities

$$H(X) \leqslant \bar{R} < H(X) + 1 \tag{3.3–9}$$

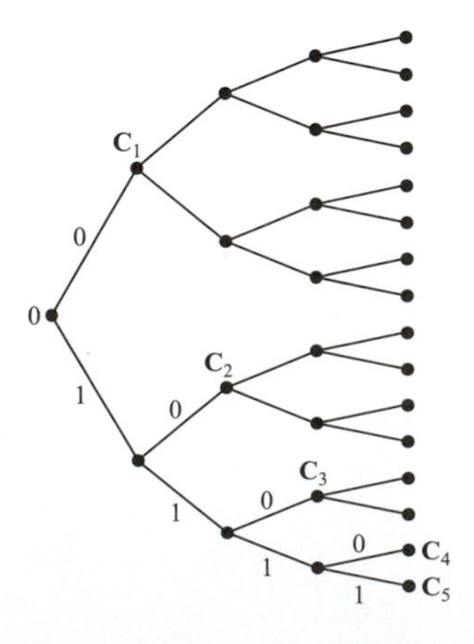

FIGURE 3.3–3
Construction of a binary tree code embedded in a full tree.

To establish the lower bound in Equation 3.3–9, we note that for code words that have length n_k, $1 \leqslant k \leqslant L$, the difference $H(X) - \bar{R}$ may be expressed as

$$\begin{aligned} H(X) - \bar{R} &= \sum_{k=1}^{L} p_k \log_2 \frac{1}{p_k} - \sum_{k=1}^{L} p_k n_k \\ &= \sum_{k=1}^{L} p_k \log_2 \frac{2^{-n_k}}{p_k} \end{aligned} \tag{3.3–10}$$

Use of the inequality $\ln x \leqslant x - 1$ in Equation 3.3–10 yields

$$\begin{aligned} H(X) - \bar{R} &\leqslant (\log_2 e) \sum_{k=1}^{L} p_k \left(\frac{2^{-n_k}}{p_k} - 1 \right) \\ &\leqslant (\log_2 e) \left(\sum_{k=1}^{L} 2^{-n_k} - 1 \right) \leqslant 0 \end{aligned}$$

where the last inequality follows from the Kraft inequality. Equality holds if and only if $p_k = 2^{-n_k}$ for $1 \leqslant k \leqslant L$.

The upper bound in Equation 3.3–9 may be established under the constraint that n_k, $1 \leqslant k \leqslant L$, are integers, by selecting the $\{n_k\}$ such that $2^{-n_k} \leqslant p_k < 2^{-n_k+1}$. But if the terms $p_k \geqslant 2^{-n_k}$ are summed over $1 \leqslant k \leqslant L$, we obtain the Kraft inequality, for which we have demonstrated that there exists a code that satisfies the prefix condition. On the other hand, if we take the logarithm of $p_k < 2^{-n_k+1}$, we obtain

$$\log p_k < -n_k + 1$$

or, equivalently,

$$n_k < 1 - \log p_k \tag{3.3–11}$$

If we multiply both sides of Equation 3.3–11 by p_k and sum over $1 \leqslant k \leqslant L$, we obtain the desired upper bound given in Equation 3.3–9. This completes the proof of Equation 3.3–9.

We have now established that variable-length codes that satisfy the prefix condition are efficient source codes for any DMS with source symbols that are not equally probable. Let us now describe an algorithm for constructing such codes.

Huffman coding algorithm. Huffman (1952) devised a variable-length encoding algorithm, based on the source letter probabilities $P(x_i)$, $i = 1, 2, \ldots, L$. This algorithm is optimum in the sense that the average number of binary digits required to represent the source symbols is a minimum, subject to the constraint that the code words satisfy the prefix condition, as defined above, which allows the received sequence to be uniquely and instantaneously decodable. We illustrate this encoding algorithm by means of two examples.

EXAMPLE 3.3-1. Consider a DMS with seven possible symbols $x_1, x_2, \ldots, x_7$ having the probabilities of occurrence illustrated in Figure 3.3–4. We have ordered the source symbols in decreasing order of the probabilities, i.e., $P(x_1) > P(x_2) > \cdots > P(x_7)$. We begin the encoding process with the two least probable symbols x_6 and x_7. These two symbols are tied together as shown in Figure 3.3–4, with the upper branch assigned a 0 and the lower branch assigned a 1. The probabilities of these two branches are added together at the node where the two branches meet to yield the probability 0.01. Now we have the source symbols $x_1, \ldots, x_5$ plus a new symbol, say x_6', obtained by combining x_6 and x_7. The next step is to join the two least probable symbols from the set $x_1, x_2, x_3, x_4, x_5, x_6'$. These are x_5 and x_6', which have a combined probability of 0.05. The branch from x_5 is assigned a 0 and the branch from x_6' is assigned a 1. This procedure continues until we exhaust the set of possible source letters. The result is a code tree with branches that contain the desired code words. The code words are obtained by beginning at the rightmost node in the tree and proceeding to the left. The resulting code words are listed in Figure 3.3–4. The average number of binary digits per symbol for this code is $\bar{R} = 2.21$ bits per symbol. The entropy of the source is 2.11 bits per symbol.

We make the observation that the code is not necessarily unique. For example, at the next to the last step in the encoding procedure, we have a tie between x_1 and x_3', since these symbols are equally probable. At this point, we chose to pair x_1 with x_2. An alternative is to pair x_2 with x_3'. If we choose this pairing, the resulting code is illustrated in Figure 3.3–5. The average number of bits per source symbol for this code is also 2.21. Hence, the resulting codes are equally efficient. Secondly, the assignment of a 0 to the upper branch and a 1 to the lower (less probable) branch is arbitrary. We may simply reverse the assignment of a 0 and 1 and still obtain an efficient code satisfying the prefix condition.

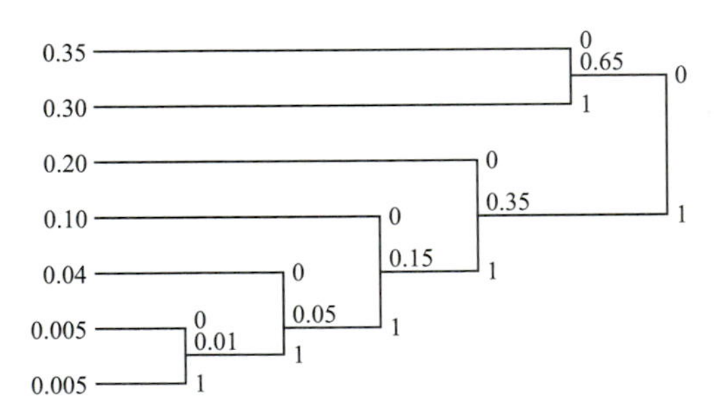

FIGURE 3.3–4
An example of variable-length source encoding for a DMS.

Letter	Probability	Self-information	Code
x_1	0.35	1.5146	00
x_2	0.30	1.7370	01
x_3	0.20	2.3219	10
x_4	0.10	3.3219	110
x_5	0.04	4.6439	1110
x_6	0.005	7.6439	11110
x_7	0.005	7.6439	11111

$H(X) = 2.11$ $\bar{R} = 2.21$

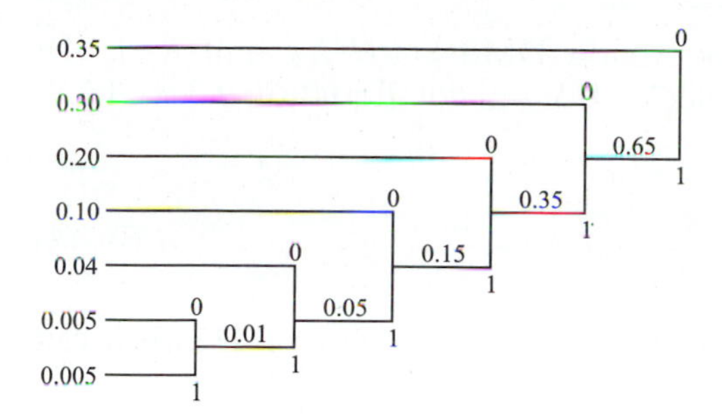

FIGURE 3.3–5
An alternative code for the DMS in Example 3.3–1.

Letter	Code
x_1	0
x_2	10
x_3	110
x_4	1110
x_5	11110
x_6	111110
x_7	111111

$\bar{R} = 2.21$

EXAMPLE 3.3–2. As a second example, let us determine the Huffman code for the output of a DMS illustrated in Figure 3.3–6. The entropy of this source is $H(X) = 2.63$ bits per symbol. The Huffman code as illustrated in Figure 3.3–6 has an average length of $\bar{R} = 2.70$ bits per symbol. Hence, its efficiency is 0.97.

The variable-length encoding (Huffman) algorithm described in the above examples generates a prefix code having an $\bar{R}$ that satisfies Equation 3.3–9. However, instead of encoding on a symbol-by-symbol basis, a more efficient procedure is to encode blocks of J symbols at a time. In such a case, the bounds in Equation 3.3–9 of source coding theorem II become

$$JH(X) \leqslant \bar{R}_J < JH(X) + 1, \tag{3.3–12}$$

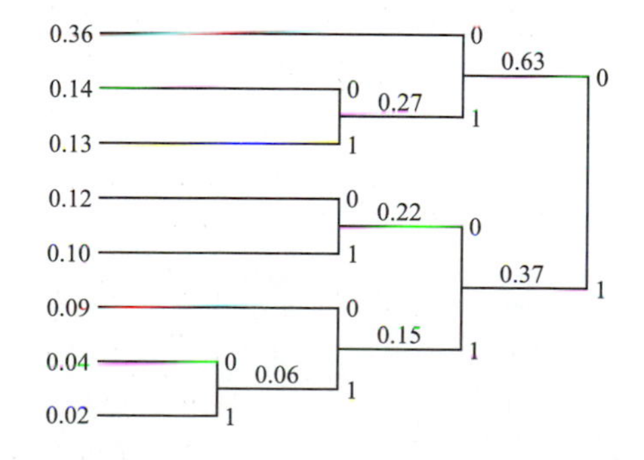

Letter	Code
x_1	00
x_2	010
x_3	011
x_4	100
x_5	101
x_6	110
x_7	1110
x_8	1111
$H(X) = 2.63$	$\bar{R} = 2.70$

FIGURE 3.3–6
Huffman code for Example 3.3–2.

since the entropy of a J-symbol block from a DMS is $JH(X)$, and $\bar{R}_J$ is the average number of bits per J-symbol blocks. If we divide Equation 3.3–12 by J, we obtain

$$H(X) \leqslant \frac{\bar{R}_J}{J} < H(X) + \frac{1}{J} \tag{3.3–13}$$

where $\bar{R}_J/J \equiv \bar{R}$ is the average number of bits per source symbol. Hence $\bar{R}$ can be made as close to $H(X)$ as desired by selecting J sufficiently large.

EXAMPLE 3.3–3. The output of a DMS consists of *letters* x_1, x_2, and x_3 with probabilities 0.45, 0.35, and 0.20, respectively. The entropy of this source is $H(X) = 1.518$ bits per symbol. The Huffman code for this source, given in Table 3.3–2, requires $\bar{R}_1 = 1.55$ bits per symbol and results in an efficiency of 97.9 percent. If pairs of symbols are encoded by means of the Huffman algorithm, the resulting code is as given in Table 3.3–3. The entropy of the source output for pairs of letters is $2H(X) =$ 3.036 bits per symbol pair. On the other hand, the Huffman code requires $\bar{R}_2 =$

TABLE 3.3–2
Huffman code for Example 3.3–3

Letter	Probability	Self-information	Code
x_1	0.45	1.156	1
x_2	0.35	1.520	00
x_3	0.20	2.330	01

$H(X) = 1.518$ bits/letter
$\bar{R}_1 = 1.55$ bits/letter
Efficiency = 97.9%

TABLE 3.3–3
Huffman code for encoding pairs of letters

Letter pair	Probability	Self-information	Code
x_1x_1	0.2025	2.312	10
x_1x_2	0.1575	2.676	001
x_2x_1	0.1575	2.676	010
x_2x_2	0.1225	3.039	011
x_1x_3	0.09	3.486	111
x_3x_1	0.09	3.486	0000
x_2x_3	0.07	3.850	0001
x_3x_2	0.07	3.850	1100
x_3x_3	0.04	4.660	1101

$2H(X) = 3.036$ bits/letter pair
$\bar{R}_2 = 3.0675$ bits/letter pair
$\frac{1}{2}\bar{R}_2 = 1.534$ bits/letter
Efficiency = 99.0%

3.0675 bits per symbol pair. Thus, the efficiency of the encoding increases to $2H(X)/\bar{R}_2 = 0.990$ or, equivalently, to 99.0 percent.

In summary, we have demonstrated that efficient encoding for a DMS may be done on a symbol-by-symbol basis using a variable-length code based on the Huffman algorithm. Furthermore, the efficiency of the encoding procedure is increased by encoding blocks of J symbols at a time. Thus, the output of a DMS with entropy $H(X)$ may be encoded by a variable-length code with an average number of bits per source letter that approaches $H(X)$ as closely as desired.

3.3.2 Discrete Stationary Sources

In the previous section, we described the efficient encoding of the output of a DMS. In this section, we consider discrete sources for which the sequence of output letters is statistically dependent. We limit our treatment to sources that are statistically stationary.

Let us evaluate the entropy of any sequence of letters from a stationary source. From the definition in Equation 3.2–13 and the result given in Equation 3.2–15, the entropy of a block of random variables $X_1 X_2 \cdots X_k$ is

$$H(X_1, X_2 \cdots X_k) = \sum_{i=1}^{k} H(X_i | X_1 X_2 \cdots X_{i-1}) \tag{3.3–14}$$

where $H(X_i | X_1 X_2 \cdots X_{i-1})$ is the conditional entropy of the ith symbol from the source given the previous $i-1$ symbols. The entropy per letter for the k-symbol block is defined as

$$H_k(X) = \frac{1}{k} H(X_1 X_2 \cdots X_k) \tag{3.3–15}$$

We define the information content of a stationary source as the entropy per letter in Equation 3.3–15 in the limit as $k \to \infty$. That is,

$$H_\infty(X) - \lim_{k \to \infty} H_k(X) = \lim_{k \to \infty} \frac{1}{k} H(X_1 X_2 \cdots X_k) \tag{3.3–16}$$

The existence of this limit is established below.

As an alternative, we may define the entropy per letter from the source in terms of the conditional entropy $H(X_k | X_1 X_2 \cdots X_{k-1})$ in the limit as k approaches infinity. Fortunately, this limit also exists and is identical to the limit in Equation 3.3–16. That is,

$$H_\infty(X) = \lim_{k \to \infty} H(X_k | X_1 X_2 \cdots X_{k-1}) \tag{3.3–17}$$

This result is also established below. Our development follows the approach in Gallager (1968).

First, we show that

$$H(X_k|X_1X_2\cdots X_{k-1}) \leqslant H(X_{k-1}|X_1X_2\cdots X_{k-2}) \qquad (3.3\text{–}18)$$

for $k \geqslant 2$. From our previous result that conditioning on a random variable cannot increase entropy, we have

$$H(X_k|X_1X_2\cdots X_{k-1}) \leqslant H(X_k|X_2X_3\cdots X_{k-1}) \qquad (3.3\text{–}19)$$

From the stationarity of the source, we have

$$H(X_k|X_2X_3\cdots X_{k-1}) = H(X_{k-1}|X_1X_2\cdots X_{k-2}) \qquad (3.3\text{–}20)$$

Hence, Equation 3.3–18 follows immediately. This result demonstrates that $H(X_k|X_1X_2\cdots X_{k-1})$ is a nonincreasing sequence in k.

Second, we have the result

$$H_k(X) \geqslant H(X_k|X_1X_2\cdots X_{k-1}) \qquad (3.3\text{–}21)$$

which follows immediately from Equations 3.3–14 and 3.3–15 and the fact that the last term in the sum of Equation 3.3–14 is a lower bound on each of the other $k-1$ terms.

Third, from the definition of $H_k(X)$, we may write

$$\begin{aligned} H_k(X) &= \frac{1}{k}[H(X_1X_2\cdots X_{k-1}) + H(X_k|X_1\cdots X_{k-1})] \\ &= \frac{1}{k}[(k-1)H_{k-1}(X) + H(X_k|X_1\cdots X_{k-1})] \\ &\leqslant \frac{k-1}{k}H_{k-1}(X) + \frac{1}{k}H_k(X) \end{aligned}$$

which reduces to

$$H_k(X) \leqslant H_{k-1}(X) \qquad (3.3\text{–}22)$$

Hence, $H_k(X)$ is a nonincreasing sequence in k.

Since $H_k(X)$ and the conditional entropy $H(X_k|X_1\cdots X_{k-1})$ are both nonnegative and nonincreasing with k, both limits must exist. Their limiting forms can be established by using Equations 3.3–14 and 3.3–15 to express $H_{k+j}(X)$ as

$$\begin{aligned} H_{k+j}(X) = &\frac{1}{k+j}H(X_1X_2\cdots X_{k-1}) \\ &+ \frac{1}{k+j}[H(X_k|X_1\cdots X_{k-1}) + H(X_{k+1}|X_1\cdots X_k) \\ &+ \cdots + H(X_{k+j}|X_1\cdots X_{k+j-1})] \end{aligned}$$

Since the conditional entropy is nonincreasing, the first term in the square brackets serves as an upper bound on the other terms. Hence,

$$H_{k+j}(X) \leqslant \frac{1}{k+j}H(X_1X_2\cdots X_{k-1}) + \frac{j+1}{k+j}H(X_k|X_1X_2\cdots X_{k-1}) \qquad (3.3\text{–}23)$$

For a fixed k, the limit of Equation 3.3–23 as $j \to \infty$ yields

$$H_\infty(X) \leqslant H(X_k | X_1 X_2 \cdots X_{k-1}) \tag{3.3–24}$$

But Equation 3.3–24 is valid for all k; hence, it is valid for $k \to \infty$. Therefore,

$$H_\infty(X) \leqslant \lim_{k\to\infty} H(X_k | X_1 X_2 \cdots X_{k-1}) \tag{3.3–25}$$

On the other hand, from Equation 3.3–21, we obtain in the limit as $k \to \infty$,

$$H_\infty(X) \geqslant \lim_{k\to\infty} H(X_k | X_1 X_2 \cdots X_{k-1}) \tag{3.3–26}$$

which establishes Equation 3.3–17.

Now suppose we have a discrete stationary source that emits J letters with $H_J(X)$ as the entropy per letter. We can encode the sequence of J letters with a variable-length Huffman code that satisfies the prefix condition by following the procedure described in the previous section. The resulting code has an average number of bits for the J-letter block that satisfies the condition

$$H(X_1 \cdots X_J) \leqslant \bar{R}_J < H(X_1 \cdots X_J) + 1 \tag{3.3–27}$$

By dividing each term of Equation 3.3–27 by J, we obtain the bounds on the average number $\bar{R} = \bar{R}_J / J$ of bits per source letter as

$$H_J(X) \leqslant \bar{R} < H_J(X) + \frac{1}{J} \tag{3.3–28}$$

By increasing the block size J, we can approach $H_J(X)$ arbitrarily closely, and in the limit as $J \to \infty$, $\bar{R}$ satisfies

$$H_\infty(X) \leqslant \bar{R} < H_\infty(X) + \varepsilon \tag{3.3–29}$$

where ε approaches zero as $1/J$. Thus, efficient encoding of stationary sources is accomplished by encoding large blocks of symbols into code words. We should emphasize, however, that the design of the Huffman code requires knowledge of the joint PDF for the J-symbol blocks.

3.3.3 The Lempel–Ziv Algorithm

From our preceding discussion, we have observed that the Huffman coding algorithm yields optimal source codes in the sense that the code words satisfy the prefix condition and the average block length is a minimum. To design a Huffman code for a DMS, we need to know the probabilities of occurrence of all the source letters. In the case of a discrete source with memory, we must know the joint probabilities of blocks of length $n \geqslant 2$. However, in practice, the statistics of a source output are often unknown. In principle, it is possible to estimate the probabilities of the discrete source output by simply observing a long information sequence emitted by the source and obtaining the probabilities empirically. Except for the estimation of the marginal probabilities $\{p_k\}$, corresponding to the frequency of occurrence of the individual source output letters, the computational complexity involved in estimating joint probabilities is extremely high. Consequently, the application of the Huffman coding method to source coding for many real sources with memory is generally impractical.

In contrast to the Huffman coding algorithm, the Lempel–Ziv source coding algorithm is designed to be independent of the source statistics. Hence, the Lempel–Ziv algorithm belongs to the class of *universal source coding algorithms*. It is a variable-to-fixed-length algorithm, where the encoding is performed as described below.

In the Lempel–Ziv algorithm, the sequence at the output of the discrete source is parsed into variable-length blocks, which are called *phrases*. A new phrase is introduced every time a block of letters from the source differs from some previous phrase in the last letter. The phrases are listed in a dictionary, which stores the location of the existing phrases. In encoding a new phrase, we simply specify the location of the existing phrase in the dictionary and append the new letter.

As an example, consider the binary sequence

10101101001001110101000011001110101100011011

Parsing the sequence as described above produces the following phrases:

1, 0, 10, 11, 01, 00, 100, 111, 010, 1000, 011, 001, 110, 101, 10001, 1011

We observe that each phrase in the sequence is a concatenation of a previous phrase with a new output letter from the source. To encode the phrases, we construct a dictionary as shown in Table 3.3–4. The dictionary locations are numbered consecutively, beginning with 1 and counting up, in this case to 16,

TABLE 3.3–4
Dictionary for Lempel-Ziv algorithm

	Dictionary location	Dictionary contents	Code word
1	0001	1	00001
2	0010	0	00000
3	0011	10	00010
4	0100	11	00011
5	0101	01	00101
6	0110	00	00100
7	0111	100	00110
8	1000	111	01001
9	1001	010	01010
10	1010	1000	01110
11	1011	011	01011
12	1100	001	01101
13	1101	110	01000
14	1110	101	00111
15	1111	10001	10101
16		1011	11101

which is the number of phrases in the sequence. The different phrases corresponding to each location are also listed, as shown. The code words are determined by listing the dictionary location (in binary form) of the previous phrase that matches the new phrase in all but the last location. Then, the new output letter is appended to the dictionary location of the previous phrase. Initially, the location 0000 is used to encode a phrase that has not appeared previously.

The source decoder for the code constructs an identical copy of the dictionary at the receiving end of the communication system and decodes the received sequence in step with the transmitted data sequence.

It should be observed that the table encoded 44 source bits into 16 code words of 5 bits each, resulting in 80 coded bits. Hence, the algorithm provided no data compression at all. However, the inefficiency is due to the fact that the sequence we have considered is very short. As the sequence is increased in length, the encoding procedure becomes more efficient and results in a compressed sequence at the output of the source.

How do we select the overall length of the table? In general, no matter how large the table is, it will eventually overflow. To solve the overflow problem, the source encoder and source decoder must use an identical procedure to remove phrases from the respective dictionaries that are not useful and substitute new phrases in their place.

The Lempel–Ziv algorithm is widely used in the compression of computer files. The "compress" and "uncompress" utilities under the UNIX© operating system and numerous algorithms under the MS-DOS operating system are implementations of various versions of this algorithm.

3.4 CODING FOR ANALOG SOURCES—OPTIMUM QUANTIZATION

As indicated in Section 3.1, an analog source emits a message waveform $x(t)$ that is a sample function of a stochastic process $X(t)$. When $X(t)$ is a band-limited, stationary stochastic process, the sampling theorem allows us to represent $X(t)$ by a sequence of uniform samples taken at the Nyquist rate.

By applying the sampling theorem, the output of an analog source is converted to an equivalent discrete-time sequence of samples. The samples are then quantized in amplitude and encoded. One type of simple encoding is to represent each discrete amplitude level by a sequence of binary digits. Hence, if we have L levels, we need $R = \log_2 L$ bits per sample if L is a power of 2, or $R = \lfloor \log_2 L \rfloor + 1$ if L is not a power of 2. On the other hand, if the levels are not equally probable, and the probabilities of the output levels are known, we may use Huffman coding (also called *entropy coding*) to improve the efficiency of the encoding process.

Quantization of the amplitudes of the sampled signal results in data compression, but it also introduces some distortion of the waveform or a loss of signal fidelity. The minimization of this distortion is considered in this section.

Many of the results given in this section apply directly to a discrete-time, continuous amplitude, memoryless Gaussian source. Such a source serves as a good model for the residual error in a number of source coding methods described in Section 3.5.

3.4.1 Rate-Distortion Function

Let us begin the discussion of signal quantization by considering the distortion introduced when the samples from the information source are quantized to a fixed number of bits. By the term *distortion*, we mean some measure of the difference between the actual source samples $\{x_k\}$ and the corresponding quantized values $\{\tilde{x}_k\}$ which we denote by $d(x_k, \tilde{x}_k)$. For example, a commonly used distortion measure is the *squared-error distortion*, defined as

$$d(x_k, \tilde{x}_k) = (x_k - \tilde{x}_k)^2 \tag{3.4–1}$$

which is used to characterize the quantization error in pulse-code modulation (PCM) in Section 3.5.1. Other distortion measures may take the general form

$$d(x_k, \tilde{x}_k) = |x - \tilde{x}_k|^p \tag{3.4–2}$$

where p takes values from the set of positive integers. The case $p = 2$ has the advantage of being mathematically tractable.

If $d(x_k, \tilde{x}_k)$ is the distortion measure per letter, the distortion between a sequence of n samples $\mathbf{X}_n$ and the corresponding n quantized values $\tilde{\mathbf{X}}_n$ is the average over the n source output samples, i.e.,

$$d(\mathbf{X}_n, \tilde{\mathbf{X}}_n) = \frac{1}{n}\sum_{k=1}^{n} d(x_k, \tilde{x}_k) \tag{3.4–3}$$

The source output is a random process, and, hence, the n samples in $\mathbf{X}_n$ are random variables. Therefore, $d(\mathbf{X}_n, \tilde{\mathbf{X}}_n)$ is a random variable. Its expected value is defined as the distortion D, i.e.,

$$D = E[d(\mathbf{X}_n, \tilde{\mathbf{X}}_n)] = \frac{1}{n}\sum_{k=1}^{n} E[d(x_k, \tilde{x}_k)] = E[d(x, \tilde{x})] \tag{3.4–4}$$

where the last step follows from the assumption that the source output process is stationary.

Now suppose we have a memoryless source with a continuous-amplitude output $\mathbf{X}$ that has a PDF $p(x)$, a quantized amplitude output alphabet $\tilde{\mathbf{X}}$, and a per letter distortion measure $d(x, \tilde{x})$, where $x \in \mathbf{X}$ and $\tilde{x} \in \tilde{\mathbf{X}}$. Then, the minimum rate in bits per source output that is required to represent the output $\mathbf{X}$ of the memoryless source with a distortion less than or equal to D is called the *rate-distortion function $R(D)$* and is defined as

$$R(D) = \min_{p(\tilde{x}|x):E[d(\mathbf{X},\tilde{\mathbf{X}})] \leqslant D} I(\mathbf{X}; \tilde{\mathbf{X}}) \tag{3.4–5}$$

where $I(\mathbf{X}; \tilde{\mathbf{X}})$ is the average mutual information between $\mathbf{X}$ and $\tilde{\mathbf{X}}$. In general, the rate $R(D)$ decreases as D increases or, conversely, $R(D)$ increases as D decreases.

One interesting model of a continuous-amplitude, memoryless information source is the Gaussian source model. In this case, Shannon proved the following fundamental theorem on the rate-distortion function.

THEOREM: RATE-DISTORTION FUNCTION FOR A MEMORYLESS GAUSSIAN SOURCE (SHANNON, 1959a). The minimum information rate necessary to represent the output of a discrete-time, continuous-amplitude memoryless Gaussian source based on a mean-square-error distortion measure per symbol (single letter distortion measure) is

$$R_g(D) = \begin{cases} \frac{1}{2}\log_2(\sigma_x^2/D) & (0 \leqslant D \leqslant \sigma_x^2) \\ 0 & (D > \sigma_x^2) \end{cases} \tag{3.4–6}$$

where σ_x^2 is the variance of the Gaussian source output.

We should note that Equation 3.4–6 implies that no information need be transmitted when the distortion $D \geqslant \sigma_x^2$. Specifically, $D = \sigma_x^2$ can be obtained by using zeros in the reconstruction of the signal. For $D > \sigma_x^2$, we can use statistically independent, zero-mean Gaussian noise samples with a variance of $D - \sigma_x^2$ for the reconstruction. $R_g(D)$ is plotted in Figure 3.4–1.

The rate-distortion function $R(D)$ of a source is associated with the following basic source coding theorem in information theory.

THEOREM: SOURCE CODING WITH A DISTORTION MEASURE (SHANNON, 1959a). There exists an encoding scheme that maps the source output into code words such that for any given distortion D, the minimum rate $R(D)$ bits per symbol (sample) is sufficient to reconstruct the source output with an average distortion that is arbitrarily close to D.

It is clear, therefore, that the rate-distortion function $R(D)$ for any source represents a lower bound on the source rate that is possible for a given level of distortion.

Let us return to the result in Equation 3.4–6 for the rate-distortion function of a memoryless Gaussian source. If we reverse the functional dependence between D and R, we may express D in terms of R as

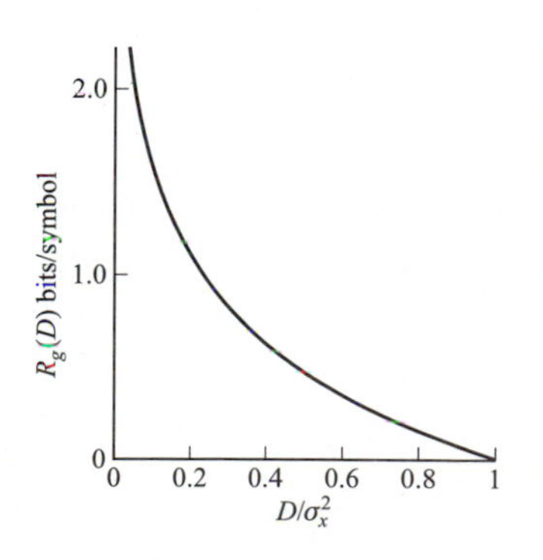

FIGURE 3.4–1
Rate-distortion function for a continuous-amplitude memoryless Gaussian source.

$$D_g(R) = 2^{-2R}\sigma_x^2 \tag{3.4–7}$$

This function is called the *distortion-rate function* for the discrete-time, memoryless Gaussian source.

When we express the distortion in Equation 3.4–7 in dB, we obtain

$$10\log_{10} D_g(R) = -6R + 10\log_{10}\sigma_x^2 \tag{3.4–8}$$

Note that the mean-square-error distortion decreases at the rate of 6 dB/bit.

Explicit results on the rate-distortion functions for memoryless non-Gaussian sources are not available. However, there are useful upper and lower bounds on the rate-distortion function for any discrete-time, continuous-amplitude, memoryless source. An upper bound is given by the following theorem.

THEOREM: UPPER BOUND ON $R(D)$. The rate-distortion function of a memoryless, continuous-amplitude source with zero mean and finite variance σ_x^2 with respect to the mean-square-error distortion measure is upper-bounded as

$$R(D) \leqslant \tfrac{1}{2}\log_2 \frac{\sigma_x^2}{D}, \qquad 0 \leqslant D \leqslant \sigma_x^2 \tag{3.4–9}$$

A proof of this theorem is given by Berger (1971). It implies that the Gaussian source requires the maximum rate among all other sources for a specified level of mean-square-error distortion. Thus, the rate distortion $R(D)$ of any continuous-amplitude, memoryless source with zero mean and finite variance σ_x^2 satisfies the condition $R(D) \leqslant R_g(D)$. Similarly, the distortion-rate function of the same source satisfies the condition

$$D(R) \leqslant D_g(R) = 2^{-2R}\sigma_x^2 \tag{3.4–10}$$

A lower bound on the rate-distortion function also exists. This is called the *Shannon lower bound* for a mean-square-error distortion measure, and is given as

$$R^*(D) = H(X) - \tfrac{1}{2}\log_2 2\pi e D \tag{3.4–11}$$

where $H(X)$ is the differential entropy of the continuous-amplitude, memoryless source. The distortion-rate function corresponding to Equation 3.4–11 is

$$D^*(R) = \frac{1}{2\pi e} 2^{-2[R-H(X)]} \tag{3.4–12}$$

Therefore, the rate-distortion function for any continuous-amplitude, memoryless source is bounded from above and below as

$$R^*(D) \leqslant R(D) \leqslant R_g(D) \tag{3.4–13}$$

and the corresponding distortion-rate function is bounded as

$$D^*(R) \leqslant D(R) \leqslant D_g(R) \tag{3.4–14}$$

The differential entropy of the memoryless Gaussian source is

$$H_g(X) = \tfrac{1}{2}\log_2 2\pi e\sigma_x^2 \tag{3.4–15}$$

so that the lower bound $R^*(D)$ in Equation 3.4–11 reduces to $R_g(D)$. Now, if we express $D^*(R)$ in terms of decibels and normalize it by setting $\sigma_x^2 = 1$ [or dividing $D^*(R)$ by σ_x^2], we obtain from Equation 3.4–12

$$10 \log_{10} D^*(R) = -6R - 6[H_g(X) - H(X)] \tag{3.4–16}$$

or, equivalently,

$$\begin{aligned} 10 \log_{10} \frac{D_g(R)}{D^*(R)} &= 6[H_g(X) - H(X)] \, \text{dB} \\ &= 6[R_g(D) - R^*(D)] \, \text{dB} \end{aligned} \tag{3.4–17}$$

The relations in Equations 3.4–16 and 3.4–17 allow us to compare the lower bound in the distortion with the upper bound which is the distortion for the Gaussian source. We note that $D^*(R)$ also decreases at -6 dB/bit. We should also mention that the differential entropy $H(X)$ is upper-bounded by $H_g(X)$, as shown by Shannon (1948b).

Table 3.4–1 lists four PDFs that are models commonly used for source signal distributions. The table shows the differential entropies, the differences in rates in bits per sample, and the difference in distortion between the upper and lower bounds. Note that the gamma PDF shows the greatest deviation from the Gaussian. The Laplacian PDF is the most similar to the Gaussian, and the uniform PDF ranks second of the PDFs shown in the table. These results provide some benchmarks on the difference between the upper and lower bounds on distortion and rate.

Before concluding this section, let us consider a band-limited Gaussian source with spectral density

$$\Phi(f) = \begin{cases} \sigma_x^2/2W & (|f| \leqslant W) \\ 0 & (|f| > W) \end{cases} \tag{3.4–18}$$

TABLE 3.4–1
Differential entropies and rate-distortion comparisons of four common PDFs for signal models

PDF	$p(x)$	$H(X)$	$R_g(D) - R^*(D)$ (bits/sample)	$D_g(R) - D^*(R)$ (dB)
Gaussian	$\frac{1}{\sqrt{2\pi}\sigma_x} e^{-x^2/2\sigma_x^2}$	$\frac{1}{2}\log_2(2\pi e \sigma_x^2)$	0	0
Uniform	$\frac{1}{2\sqrt{3}\sigma_x}, \ |x| \leqslant \sqrt{3}\sigma_x$	$\frac{1}{2}\log_2(12\sigma_x^2)$	0.255	1.53
Laplacian	$\frac{1}{\sqrt{2}\sigma_x} e^{-\sqrt{2}|x|/\sigma_x}$	$\frac{1}{2}\log_2(2e^2\sigma_x^2)$	0.104	0.62
Gamma	$\frac{\sqrt[4]{3}}{\sqrt{8\pi\sigma_x|x|}} e^{-\sqrt{3}|x|/2\sigma_x}$	$\frac{1}{2}\log_2(4\pi e^{0.423}\sigma_x^2/3)$	0.709	4.25

When the output of this source is sampled at the Nyquist rate, the samples are uncorrelated and, since the source is Gaussian, they are also statistically independent. Hence, the equivalent discrete-time Gaussian source is memoryless. The rate-distortion function for each sample is given by Equation 3.4–6. Therefore, the rate-distortion function for the band-limited white Gaussian source in bits/s is

$$R_g(D) = W \log_2 \frac{\sigma_x^2}{D} \qquad 0 \leqslant D \leqslant \sigma_x^2 \tag{3.4–19}$$

The corresponding distortion-rate function is

$$D_g(R) = 2^{-R/W} \sigma_x^2 \tag{3.4–20}$$

which, when expressed in decibels and normalized by σ_x^2, becomes

$$10 \log \frac{D_g(R)}{\sigma_x^2} = -\frac{3R}{W} \tag{3.4–21}$$

The more general case in which the Gaussian process is neither white nor band-limited has been treated by Gallager (1968) and Goblick and Holsinger (1967).

3.4.2 Scalar Quantization

In source encoding, the quantizer can be optimized if we know the PDF of the signal amplitude at the input to the quantizer. For example, suppose that the sequence $\{x_n\}$ at the input to the quantizer has a PDF $p(x)$ and let $L = 2^R$ be the desired number of levels. We wish to design the optimum scalar quantizer that minimizes some function of the quantization error $q = \tilde{x} - x$, where $\tilde{x}$ is the quantized value of x. To elaborate, suppose that $f(\tilde{x} - x)$ denotes the desired function of the error. Then, the distortion resulting from quantization of the signal amplitude is

$$D = \int_{-\infty}^{\infty} f(\tilde{x} - x) p(x)\, dx \tag{3.4–22}$$

In general, an optimum quantizer is one that minimizes D by optimally selecting the output levels and the corresponding input range of each output level. This optimization problem has been considered by Lloyd (1982) and Max (1960), and the resulting optimum quantizer is usually called the *Lloyd–Max quantizer*.

For a uniform quantizer, the output levels are specified as $\tilde{x}_k = \frac{1}{2}(2k - 1)\Delta$, corresponding to an input signal amplitude in the range $(k - 1)\Delta \leqslant x < k\Delta$, where Δ is the step size. When the uniform quantizer is symmetric with an even number of levels, the average distortion in Equation 3.4–22 may be expressed as

$$D = 2 \sum_{k=1}^{L/2-1} \int_{(k-1)\Delta}^{k\Delta} f\left(\tfrac{1}{2}(2k-1)\Delta - x\right)p(x)\,dx + 2\int_{(L/2-1)\Delta}^{\infty} f\left(\tfrac{1}{2}(L-1)\Delta - x\right)p(x)\,dx \tag{3.4–23}$$

In this case, the minimization of D is carried out with respect to the step-size parameter Δ. By differentiating D with respect to Δ, we obtain

$$\sum_{k=1}^{L/2-1} (2k-1) \int_{(k-1)\Delta}^{k\Delta} f\left(\tfrac{1}{2}(2k-1)\Delta - x\right)p(x)\,dx + (L-1)\int_{-(L/2-1)\Delta}^{\infty} f'\left(\tfrac{1}{2}(L-1)\Delta - x\right)p(x)\,dx = 0 \tag{3.4–24}$$

where $f'(x)$ denotes the derivative of $f(x)$.

By selecting the error criterion function $f(x)$, the solution of Equation 3.4–24 for the optimum step size can be obtained numerically on a digital computer for any given PDF $p(x)$. For the mean-square-error criterion, for which $f(x) = x^2$, Max (1960) evaluated the optimum step size Δ_{opt} and the minimum mean square error when the PDF $p(x)$ is zero-mean Gaussian with unit variance. Some of these results are given in Table 3.4–2. We observe that the minimum mean-square-error distortion $D_{\min}$ decreases by a little more than 5 dB for each doubling of the number of levels L. Hence, each additional bit that is employed in a uniform quantizer with optimum step size Δ_{opt} for a Gaussian-distributed signal amplitude reduces the distortion by more than 5 dB.

By relaxing the constraint that the quantizer be uniform, the distortion can be reduced further. In this case, we let the output level be $\tilde{x} = \tilde{x}_k$ when the input signal amplitude is in the range $x_{k-1} \leqslant x < x_k$. For an L-level quantizer, the end points are $x_0 = -\infty$ and $x_L = \infty$. The resulting distortion is

$$D = \sum_{k=1}^{L} \int_{x_{k-1}}^{x_k} f(\tilde{x}_k - x)p(x)\,dx \tag{3.4–25}$$

which is now minimized by optimally selecting the $\{\tilde{x}_k\}$ and $\{x_k\}$.

TABLE 3.4–2
Optimum step sizes for uniform quantization of a Gaussian random variable

Number of output levels	Optimum step size Δ_{opt}	Minimum MSE $D_{\min}$	$10 \log D_{\min}$ (dB)
2	1.596	0.3634	−4.4
4	0.9957	0.1188	−9.25
8	0.5860	0.03744	−14.27
16	0.3352	0.01154	−19.38
32	0.1881	0.00349	−24.57

The necessary conditions for a minimum distortion are obtained by differentiating D with respect to the $\{x_k\}$ and $\{\tilde{x}_k\}$. The result of this minimization is the pair of equations

$$f(\tilde{x}_k - x_k) = f(\tilde{x}_{k+1} - x_k), \qquad k = 1, 2, \ldots, L-1 \tag{3.4–26}$$

$$\int_{x_{k-1}}^{x_k} f'(\tilde{x}_k - x)p(x)\,dx = 0, \qquad k = 1, 2, \ldots, L \tag{3.4–27}$$

As a special case, we again consider minimizing the mean square value of the distortion. In this case, $f(x) = x^2$ and, hence, Equation 3.4–26 becomes

$$x_k = \tfrac{1}{2}(\tilde{x}_k + \tilde{x}_{k+1}), \qquad k = 1, 2, \ldots, L-1 \tag{3.4–28}$$

which is the midpoint between $\tilde{x}_k$ and $\tilde{x}_{k+1}$. The corresponding equations determining $\{\tilde{x}_k\}$ are

$$\int_{x_{k-1}}^{x_k} (\tilde{x}_k - x)p(x)\,dx = 0, \qquad k = 1, 2, \ldots, L \tag{3.4–29}$$

Thus, $\tilde{x}_k$ is the centroid of the area of $p(x)$ between x_{k-1} and x_k. These equations may be solved numerically for any given $p(x)$.

Tables 3.4–3 and 3.4–4 give the results of this optimization obtained by Max (1960) for the optimum four-level and eight-level quantizers of a Gaussian-distributed signal amplitude having zero mean and unit variance. In Table 3.4–5, we compare the minimum mean square distortion of a uniform quantizer to that of a nonuniform quantizer for the Gaussian-distributed signal amplitude. From the results of this table, we observe that the difference in the performance of the two types of quantizers is relatively small for small values of R (less than 0.5 dB for $R \leqslant 3$), but it increases as R increases. For example, at $R = 5$, the nonuniform quantizer is approximately 1.5 dB better than the uniform quantizer.

It is instructive to plot the minimum distortion as a function of the bit rate $R = \log_2 L$ bits per source sample (letter) for both the uniform and nonuniform quantizers. These curves are illustrated in Figure 3.4–2. The functional dependence of the distortion D on the bit rate R may be expressed as $D(R)$, the distortion-rate function. We observe that the distortion-rate function for the

TABLE 3.4–3
Optimum four-level quantizer for a Gaussian random variable

Level k	x_k	$\tilde{x}_k$
1	−0.9816	−1.510
2	0.0	−0.4528
3	0.9816	0.4528
4	∞	1.510

$D_{\min} = 0.1175$
$10 \log D_{\min} = -9.3$ dB

TABLE 3.4–4
Optimum eight-level quantizer for a Gaussian random variable (Max, 1960)

Level k	x_k	$\tilde{x}_k$
1	−1.748	−2.152
2	−1.050	−1.344
3	−0.5006	−0.7560
4	0	−0.2451
5	0.5006	0.2451
6	1.050	0.7560
7	1.748	1.344
8	∞	2.152

$D_{\min} = 0.03454$
$10 \log D_{\min} = -14.62$ dB

TABLE 3.4–5
Comparison of optimum uniform and nonuniform quantizers for a Gaussian random variable (Max, 1960; Paez and Glisson, 1972)

R (bits/sample)	$10 \log_{10} D_{\min}$	
	Uniform (dB)	Nonuniform (dB)
1	−4.4	−4.4
2	−9.25	−9.30
3	−14.27	−14.62
4	−19.38	−20.22
5	−24.57	−26.02
6	−29.83	−31.89
7	−35.13	−37.81

optimum nonuniform quantizer falls below that of the optimum uniform quantizer.

Since any quantizer reduces a continuous amplitude source into a discrete amplitude source, we may treat the discrete amplitude as letters, say $\tilde{X} = \{\tilde{x}_k, 1 \leqslant k \leqslant L\}$, with associated probabilities $\{p_k\}$. If the signal amplitudes are statistically independent, the discrete source is memoryless and, hence, its entropy is

$$H(\tilde{X}) = -\sum_{k=1}^{L} p_k \log_2 p_k \tag{3.4–30}$$

For example, the optimum four-level nonuniform quantizer for the Gaussian-distributed signal amplitude results in the probabilities $p_1 = p_4 = 0.1635$ for the two outer levels and $p_2 = p_3 = 0.3365$ for the two inner levels. The entropy for the discrete source is $H(\tilde{X}) = 1.911$ bits per letter. Hence, with entropy coding (Huffman coding) of blocks of output letters, we can achieve

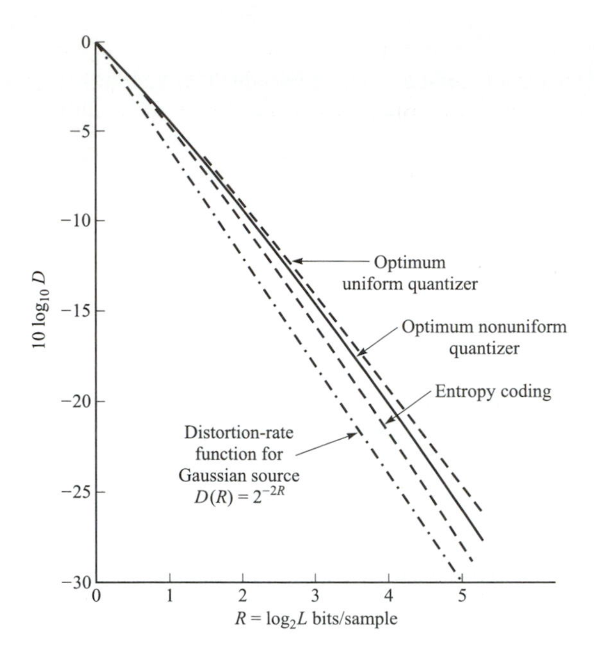

FIGURE 3.4–2
Distortion versus rate curves for discrete-time memoryless Gaussian source.

the minimum distortion of −9.30 dB with 1.911 bits per letter instead of 2 bits per letter. Max (1960) has given the entropy for the discrete source letters resulting from quantization. Table 3.4–6 lists the values of the entropy for the nonuniform quantizer. These values are also plotted in Figure 3.4–2 and labeled *entropy coding*.

From this discussion, we conclude that the quantizer can be optimized when the PDF of the continuous source output is known. The optimum quantizer of $L = 2^R$ levels results in a minimum distortion of $D(R)$, where $R = \log_2 L$ bits per sample. Thus, this distortion can be achieved by simply representing each quantized sample by R bits. However, more efficient encoding is possible. The discrete source output that results from quantization is characterized by a set of probabilities $\{p_k\}$ that can be used to design efficient variable-length codes for the

TABLE 3.4–6
Entropy of the output of an optimum nonuniform quantizer for a Gaussian random variable (Max, 1960)

$\bar{R}$ (bits/sample)	Entropy (bits/letter)	Distortion $10 \log_{10} D_{\min}$
1	1.0	−4.4
2	1.911	−9.30
3	2.825	−14.62
4	3.765	−20.22
5	4.730	−26.02

source output (entropy coding). The efficiency of any encoding method can be compared with the distortion-rate function or, equivalently, the rate-distortion function for the discrete-time, continuous-amplitude source that is characterized by the given PDF.

If we compare the performance of the optimum nonuniform quantizer with the distortion-rate function, we find, for example, that at a distortion of −26 dB, entropy coding is 0.41 bits per sample more than the minimum rate given by Equation 3.4–8, and simple block coding of each letter requires 0.68 bits per sample more than the minimum rate. We also observe that the distortion-rate functions for the optimal uniform and nonuniform quantizers for the Gaussian source approach the slope of −6 dB/bit asymptotically for large R.

3.4.3 Vector Quantization

In the previous section, we considered the quantization of the output signal from a continuous-amplitude source when the quantization is performed on a sample-by-sample basis, i.e., by scalar quantization. In this section, we consider the joint quantization of a block of signal samples or a block of signal parameters. This type of quantization is called *block* or *vector quantization*. It is widely used in speech coding for digital cellular systems.

A fundamental result of rate-distortion theory is that better performance can be achieved by quantizing vectors instead of scalars, even if the continuous-amplitude source is memoryless. If, in addition, the signal samples or signal parameters are statistically dependent, we can exploit the dependency by jointly quantizing blocks of samples or parameters and, thus, achieve an even greater efficiency (lower bit rate) compared with that which is achieved by scalar quantization.

The vector quantization problem may be formulated as follows. We have an n-dimensional vector $\mathbf{X} = [x_1 x_2 \cdots x_n]$ with real-valued, continuous-amplitude components $\{x_k, 1 \leqslant k \leqslant n\}$ that are described by a joint PDF $p(x_1, x_2, \ldots, x_n)$. The vector $\mathbf{X}$ is quantized into another n-dimensional vector $\tilde{\mathbf{X}}$ with components $\{\tilde{x}_k, 1 \leqslant k \leqslant n\}$. We express the quantization as $Q(\cdot)$, so that

$$\tilde{\mathbf{X}} = Q(\mathbf{X}) \qquad (3.4\text{–}31)$$

where $\tilde{\mathbf{X}}$ is the output of the vector quantizer when the input vector is $\mathbf{X}$.

Basically, vector quantization of blocks of data may be viewed as a pattern recognition problem involving the classification of blocks of data into a discrete number of categories or *cells* in a way that optimizes some fidelity criterion, such as mean-square error distortion. For example, let us consider the quantization of two-dimensional vectors $\mathbf{X} = [x_1 x_2]$. The two-dimensional space is partitioned into cells as illustrated in Figure 3.4–3, where we have arbitrarily selected hexagonal-shaped cells $\{C_k\}$. All input vectors that fall in cell C_k are quantized into the vector $\tilde{\mathbf{X}}_k$, which is shown in Figure 3.4–3 as the center of the hexagon. In this example, there are $L = 37$ vectors, one for each of the 37 cells into which the two-

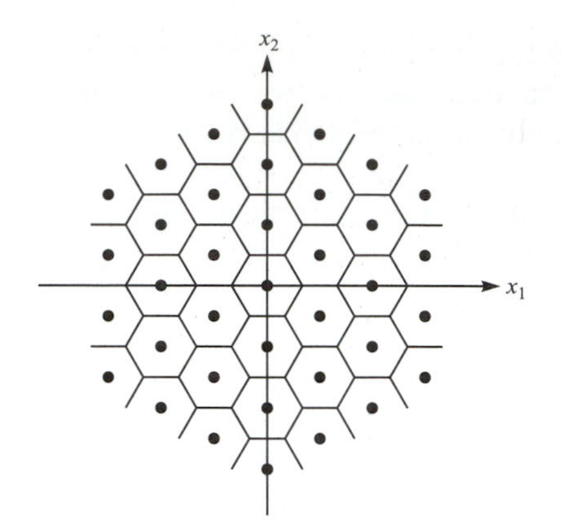

FIGURE 3.4–3
An example of quantization in two-dimensional space.

dimensional space has been partitioned. We denote the set of possible output vectors as$\{\tilde{\mathbf{X}}_k, 1 \leqslant k \leqslant L\}$.

In general, quantization of the n-dimensional vector $\mathbf{X}$ into an n-dimensional vector $\tilde{\mathbf{X}}$ introduces a quantization error or a distortion $d(\mathbf{X}, \tilde{\mathbf{X}})$. The average distortion over the set of input vectors $\mathbf{X}$ is

$$\begin{aligned} D &= \sum_{k=1}^{L} P(\mathbf{X} \in C_k) E[d(\mathbf{X}, \tilde{\mathbf{X}}_k) | \mathbf{X} \in C_k] \\ &= \sum_{k=1}^{L} P(\mathbf{X} \in C_k) \int_{\mathbf{X} \in C_k} d(\mathbf{X}, \tilde{\mathbf{X}}_k) p(\mathbf{X})\, d\mathbf{X} \end{aligned} \tag{3.4–32}$$

where $P(\mathbf{X} \in C_k)$ is the probability that the vector $\mathbf{X}$ falls in the cell C_k and $p(\mathbf{X})$ is the joint PDF of the n random variables. As in the case of scalar quantization, we can minimize D by selecting the cells $\{C_k, 1 \leqslant k \leqslant L\}$ for a given PDF $p(\mathbf{X})$.

A commonly used distortion measure for vector quantization is the mean square error (l_2 norm) defined as

$$d_2(\mathbf{X}, \tilde{\mathbf{X}}) = \frac{1}{n}(\mathbf{X} - \tilde{\mathbf{X}})'(\mathbf{X} - \tilde{\mathbf{X}}) = \frac{1}{n}\sum_{k=1}^{n}(x_k - \tilde{x}_k)^2 \tag{3.4–33}$$

or, more generally, the weighted mean square error

$$d_{2W}(\mathbf{X}, \tilde{\mathbf{X}}) = (\mathbf{X} - \tilde{\mathbf{X}})'\mathbf{W}(\mathbf{X} - \tilde{\mathbf{X}}) \tag{3.4–34}$$

where $\mathbf{W}$ is a positive-definite weighting matrix. Usually, $\mathbf{W}$ is selected to be the inverse of the covariance matrix of the input data vector $\mathbf{X}$.

Other distortion measures that are sometimes used are special cases of the l_p norm defined as

$$d_p(\mathbf{X}, \tilde{\mathbf{X}}) = \frac{1}{n}\sum_{k=1}^{n} |x_k - \tilde{x}_k|^p \tag{3.4–35}$$

The special case $p = 1$ is often used as an alternative to $p = 2$.

Vector quantization is not limited to quantizing a block of signal samples of a source waveform. It can also be applied to quantizing a set of parameters extracted from the data. For example, in linear predictive coding (LPC), described in Section 3.5.3, the parameters extracted from the signal are the prediction coefficients, which are the coefficients in the all-pole filter model for the source that generates the observed data. These parameters can be considered as a block and quantized as a block by application of some appropriate distortion measure. In the case of speech encoding, an appropriate distortion measure, proposed by Itakura and Saito (1968, 1975), is the weighted square error where the weighting matrix $\mathbf{W}$ is selected to be the normalized autocorrelation matrix $\mathbf{\Phi}$ of the observed data.

In speech processing, an alternative set of parameters that may be quantized as a block and transmitted to the receiver is the set of reflection coefficients $\{a_{ii}, 1 \leqslant i \leqslant m\}$. Yet another set of parameters that is sometimes used for vector quantization in linear predictive coding of speech comprises the log-area ratios $\{r_k\}$, which are defined in terms of the reflection coefficients as

$$r_k = \log \frac{1 + a_{kk}}{1 - a_{kk}}, \qquad 1 \leqslant k \leqslant m \tag{3.4–36}$$

Now, let us return to the mathematical formulation of vector quantization and let us consider the partitioning of the n-dimensional space into L cells $\{C_k, 1 \leqslant k \leqslant L\}$ so that the average distortion is minimized over all L-level quantizers. There are two conditions for optimality. The first is that the optimal quantizer employs a nearest-neighbor selection rule, which may be expressed mathematically as

$$Q(\mathbf{X}) = \tilde{\mathbf{X}}_k$$

if and only if

$$d(\mathbf{X}, \tilde{\mathbf{X}}_k) \leqslant d(\mathbf{X}, \tilde{\mathbf{X}}_j), \qquad k \neq j, \qquad 1 \leqslant j \leqslant L \tag{3.4–37}$$

The second condition necessary for optimality is that each output vector $\tilde{\mathbf{X}}_k$ be chosen to minimize the average distortion in cell C_k. In other words, $\tilde{\mathbf{X}}_k$ is the vector in C_k that minimizes

$$D_k = E[d(\mathbf{X}, \tilde{\mathbf{X}}) | \mathbf{X} \in C_k] = \int_{\mathbf{X} \in C_k} d(\mathbf{X}, \tilde{\mathbf{X}}) p(\mathbf{X}) \, d\mathbf{X} \tag{3.4–38}$$

The vector $\tilde{\mathbf{X}}_k$ that minimizes D_k is called the *centroid* of the cell. Thus, these conditions for optimality can be applied to partition the n-dimensional space into cells $\{C_k, 1 \leqslant k \leqslant L\}$ when the joint PDF $p(\mathbf{X})$ is known. It is clear that these two conditions represent the generalization of the optimum scalar quantization problem to the n-dimensional vector quantization problem. In general, we expect the code vectors to be closer together in regions where the joint PDF is large and farther apart in regions where $p(\mathbf{X})$ is small.

As an upper bound on the distortion of a vector quantizer, we may use the distortion of the optimal scalar quantizer, which can be applied to each component of the vector as described in the previous section. On the other hand, the best performance that can be achieved by optimum vector quantization is given by the rate-distortion function or, equivalently, the distortion-rate function.

The distortion-rate function, which was introduced in the previous section, may be defined in the context of vector quantization as follows. Suppose we form a vector $\mathbf{X}$ of dimension n from n consecutive samples $\{x_m\}$. The vector $\mathbf{X}$ is then quantized to form $\tilde{\mathbf{X}} = Q(\mathbf{X})$, where $\tilde{\mathbf{X}}$ is a vector from the set of $\{\tilde{\mathbf{X}}_k,\ 1 \leqslant k \leqslant L\}$. As described above, the average distortion D resulting from representing $\mathbf{X}$ by $\tilde{\mathbf{X}}$ is $E[d(\mathbf{X}, \tilde{\mathbf{X}})]$, where $d(\mathbf{X}, \tilde{\mathbf{X}})$ is the distortion per dimension, e.g.,

$$d(\mathbf{X}, \tilde{\mathbf{X}}) = \frac{1}{n}\sum_{k=1}^{n}(x_k - \tilde{x}_k)^2$$

The vectors $\{\tilde{\mathbf{X}}_k,\ 1 \leqslant k \leqslant L\}$ can be transmitted at an average bit rate of

$$R = \frac{H(\tilde{\mathbf{X}})}{n} \quad \text{bits per sample} \tag{3.4–39}$$

where $H(\tilde{\mathbf{X}})$ is the entropy of the quantized source output defined as

$$H(\tilde{\mathbf{X}}) = -\sum_{i=1}^{L} p(\tilde{\mathbf{X}}_i)\log_2 P(\tilde{\mathbf{X}}_i) \tag{3.4–40}$$

For a given average rate R, the minimum achievable distortion $D_n(R)$ is

$$D_n(R) = \min_{Q(\mathbf{X})} E[d(\mathbf{X}, \tilde{\mathbf{X}})] \tag{3.4–41}$$

where $R \geqslant H(\tilde{\mathbf{X}})/n$ and the minimum in Equation 3.4–41 is taken over all possible mappings $Q(\mathbf{X})$. In the limit as the number of dimensions n is allowed to approach infinity, we obtain

$$D(R) = \lim_{n\to\infty} D_n(R) \tag{3.4–42}$$

where $D(R)$ is the distortion-rate function that was introduced in the previous section. It is apparent from this development that the distortion-rate function can be approached arbitrarily closely by increasing the size n of the vectors.

The development above is predicated on the assumption that the joint PDF $p(\mathbf{X})$ of the data vector is known. However, in practice, the joint PDF $p(\mathbf{X})$ of the data may not be known. In such a case, it is possible to select the quantized output vectors adaptively from a set of training vectors $\mathbf{X}(m)$. Specifically, suppose that we are given a set of M training vectors where M is much greater than $L(M \gg L)$. An iterative clustering algorithm, called the *K-means algorithm*, where in our case $K = L$, can be applied to the training vectors. This algorithm iteratively subdivides the M training vectors into L clusters such that the two necessary conditions for optimality are satisfied. The K-means algorithm may be described as follows (Makhoul et al., 1985).

K-MEANS ALGORITHM

Step 1 Initialize by setting the iteration number $i = 0$. Choose a set of output vectors $\tilde{\mathbf{X}}_k(0)$, $1 \leqslant k \leqslant L$.

Step 2 Classify the training vectors $\{\mathbf{X}(m),\ 1 \leqslant m \leqslant M\}$ into the clusters $\{C_k\}$ by applying the nearest-neighbor rule

$$\mathbf{X} \in C_k(i) \quad \text{iff} \quad d(\mathbf{X}, \tilde{\mathbf{X}}_k(i)) \leqslant d(\mathbf{X}, \tilde{\mathbf{X}}_j(i)) \quad \text{for all } k \neq j$$

Step 3 Recompute (set i to $i + 1$) the output vectors of every cluster by computing the centroid

$$\tilde{\mathbf{X}}_k(i) = \frac{1}{M_k} \sum_{\mathbf{x} \in C_k} \mathbf{X}(m), \qquad 1 \leqslant k \leqslant L$$

of the training vectors that fall in each cluster. Also, compute the resulting average distortion $D(i)$ at the ith iteration.

Step 4 Terminate the test if the change $D(i-1) - D(i)$ in the average distortion is relatively small. Otherwise, go to Step 2.

The *K*-means algorithm converges to a local minimum (see Anderberg, 1973; Linde et al., 1980). By beginning the algorithm with different sets of initial output vectors $\{\tilde{\mathbf{X}}_k(0)\}$ and each time performing the optimization described in the *K*-means algorithm, it is possible to find a global optimum. However, the computational burden of this search procedure may limit the search to a few initializations.

Once we have selected the output vectors $\{\tilde{\mathbf{X}}_k,\ 1 \leqslant k \leqslant L\}$ we have established what is called a *code book*. Each input signal vector $\mathbf{X}(m)$ is quantized to the output vector that is nearest to it according to the distortion measure that is adopted. If the computation involves evaluating the distance between $\mathbf{X}(m)$ and each of the L possible output vectors $\{\tilde{\mathbf{X}}_k\}$, the procedure constitutes a *full search*. If we assume that each computation requires n multiplications and additions, the computational requirement for a full search is

$$\mathcal{C} = nL \tag{3.4–43}$$

multiplication and additions per input vector.

If we select L to be a power of 2, then $\log_2 L$ is the number of bits required to represent each vector. Now, if R denotes the bit rate per sample [per component or dimension of $\mathbf{X}(m)$], we have $nR = \log_2 L$, and, hence, the computational cost is

$$\mathcal{C} = n2^{nR} \tag{3.4–44}$$

Note that the number of computations grows exponentially with the dimensionality parameter n and the bit rate R per dimension.

The computational cost associated with a full search can be reduced by slightly suboptimum algorithms (see Chang et al., 1984; Gersho, 1982). A particularly simple approach is to construct the code book based on a binary tree search. The binary tree search is a hierarchical clustering method for partitioning the n-dimensional space in a way that reduces the computational cost of the search to be proportional to $\log_2 L$. This method begins by subdividing the n-

dimensional training vectors into two regions, using the K-means algorithm with $K = 2$. Thus, we obtain two regions and the corresponding centroids, say $\tilde{\mathbf{X}}_1$ and $\tilde{\mathbf{X}}_2$. In the next step, the points that fall into the first region are further subdivided into two regions by using the K-means algorithm with $K = 2$, and, thus, we obtain two centroids, say $\tilde{\mathbf{X}}_{11}$ and $\tilde{\mathbf{X}}_{12}$. This procedure is repeated for the second region to yield two other centroids $\tilde{\mathbf{X}}_{21}$ and $\tilde{\mathbf{X}}_{22}$. Thus, the n-dimensional region is subdivided into four regions, where each region has a corresponding centroid. This process is repeated until we have subdivided the n-dimensional space into $L = 2^{nR}$ regions, where nR is the number of bits per code vector. The corresponding code vectors may be viewed as terminal nodes in a binary tree, as shown in Figure 3.4–4.

Given a signal vector $\mathbf{X}(m)$, the search begins by comparing $\mathbf{X}(m)$ with the centroids $\tilde{\mathbf{X}}_1$ and $\tilde{\mathbf{X}}_2$. If $d(\mathbf{X}(m), \tilde{\mathbf{X}}_1) < d(\mathbf{X}(m), \tilde{\mathbf{X}}_2)$, we eliminate the half of the tree stemming from $\tilde{\mathbf{X}}_2$. Then, we compute the distortions $d(\mathbf{X}(m), \tilde{\mathbf{X}}_{11})$ and $d(\mathbf{X}(m), \tilde{\mathbf{X}}_{12})$. If $d(\mathbf{X}(m), \tilde{\mathbf{X}}_{11}) < d(\mathbf{X}(m), \tilde{\mathbf{X}}_{12})$, we eliminate the part of the tree stemming from $\tilde{\mathbf{X}}_{12}$ and continue the binary search along $\tilde{\mathbf{X}}_{11}$. The search terminates when we reach a terminal node.

The computational cost of the binary tree search is

$$\mathcal{C} = 2n \log_2 L = 2n^2 R$$

which is linear in R (the bit rate per dimension) compared to the exponential cost for the full search. Although the computational cost has been significantly reduced, the memory required to store the (centroid) vectors has increased from nL to approximately $2nL$, due to the fact that we now have to store the vectors at the intermediate nodes in addition to the vectors at the terminal nodes.

The binary tree search algorithm generates a *uniform tree*. In general, the resulting code book will be suboptimum in the sense that the code words result in more distortion compared to the code words generated by the method corre-

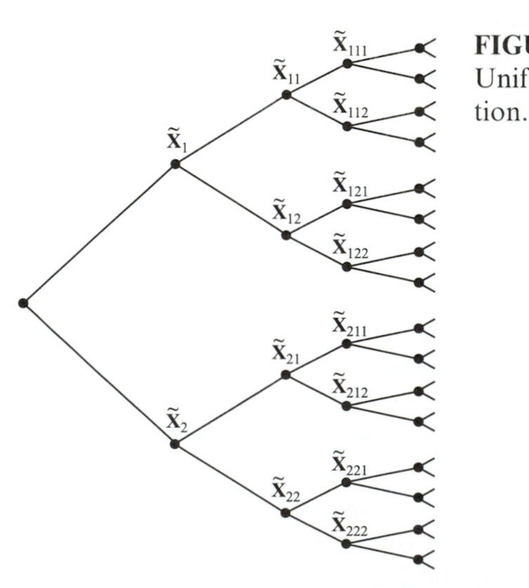

FIGURE 3.4–4
Uniform tree for binary search vector quantization.

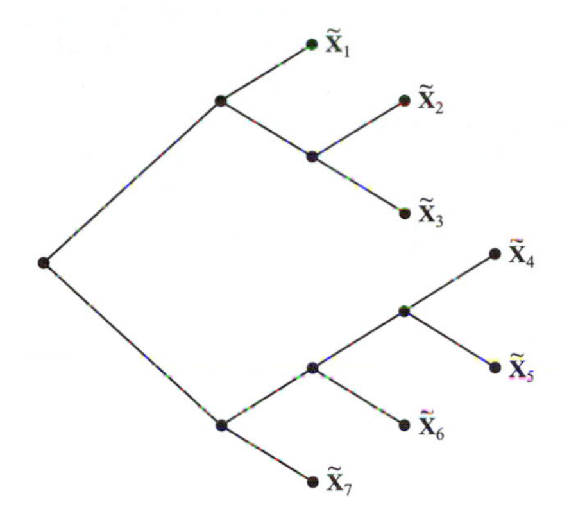

FIGURE 3.4–5
Nonuniform tree for binary search vector quantization.

sponding to a full search. An improvement in performance may be obtained by removing the restriction that the tree be uniform. In particular, a code book resulting in lower distortion is obtained by subdividing the cluster of test vectors having the largest total distortion at each step in the process. Thus, in the first step, the n-dimensional space is divided into two regions. In the second step, we select the cluster with the larger distortion and subdivide it. Now, we have three clusters. The next subdivision is performed on the cluster having the largest distortion. Thus we obtain four clusters and we repeat the process. The net result is that we generate a *nonuniform code* tree as illustrated in Figure 3.4–5 for the case $L = 7$. Note that L is no longer constrained to be a power of 2.

In order to demonstrate the benefits of vector quantization compared with scalar quantization, we present the following example taken from Makhoul et al. (1985).

EXAMPLE 3.4–1. Let x_1 and x_2 be two random variables with a uniform joint PDF

$$p(x_1, x_2) \equiv p(\mathbf{X}) = \begin{cases} \dfrac{1}{ab} & (\mathbf{X} \in \mathbf{C}) \\ 0 & (\text{otherwise}) \end{cases} \tag{3.4–45}$$

where **C** is the rectangular region illustrated in Figure 3.4–6. Note that the rectangle is rotated by 45° relative to the horizontal axis. Also shown in Figure 3.4–6 are the marginal densities $p(x_1)$ and $p(x_2)$.

If we quantize x_1 and x_2 separately by using uniform intervals of length Δ, the number of levels needed is

$$L_1 = L_2 = \frac{a+b}{\sqrt{2}\Delta} \tag{3.4–46}$$

Hence, the number of bits needed for coding the vector $\mathbf{X} = [x_1\ x_2]$ is

$$\begin{aligned} R_x &= R_1 + R_2 = \log_2 L_1 + \log_2 L_2 \\ R_x &= \log_2 \frac{(a+b)^2}{2\Delta^2} \end{aligned} \tag{3.4–47}$$

Thus, scalar quantization of each component is equivalent to vector quantization with the total number of levels

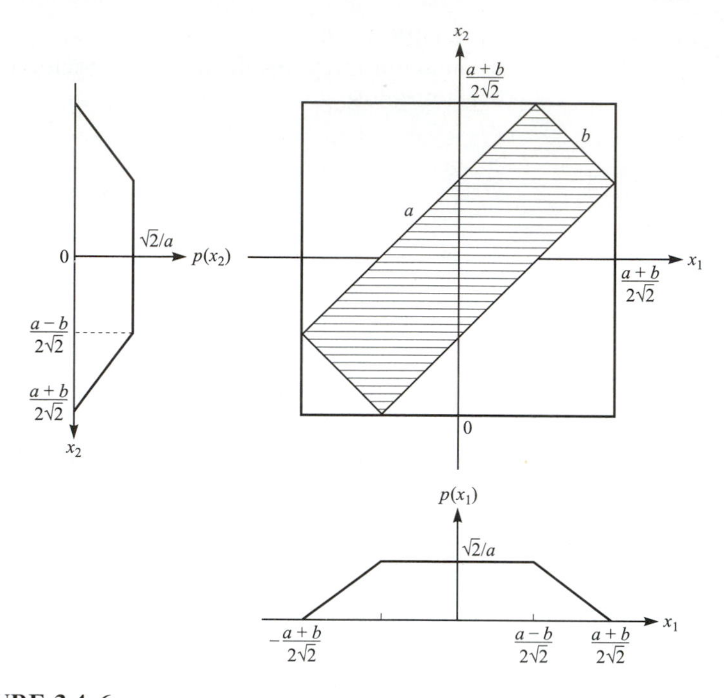

FIGURE 3.4–6
A uniform PDF in two dimensions. (*Makhoul et al., 1985.*)

$$L_x = L_1 L_2 = \frac{(a+b)^2}{2\Delta^2} \tag{3.4–48}$$

We observe that this approach is equivalent to covering the large square that encloses the rectangle by square cells, where each cell represents one of the L_x quantized regions. Since $p(\mathbf{X}) = 0$ except for $\mathbf{X} \in C$, this encoding is wasteful and results in an increase of the bit rate.

If we were to cover only the region for which $p(\mathbf{X}) \neq 0$ with squares having area Δ^2, the total number of levels that will result is the area of the rectangle divided by Δ^2, i.e.,

$$L'_x = \frac{ab}{\Delta^2} \tag{3.4–49}$$

Therefore, the difference in bit rate between the scalar and vector quantization methods is

$$R_x - R'_x = \log_2 \frac{(a+b)^2}{2ab} \tag{3.4–50}$$

For instance, if $a = 4b$, the difference in bit rate is

$$R_x - R'_x = 1.64 \text{ bits per vector}$$

Thus, vector quantization is 0.82 bits per sample better for the same distortion.

It is interesting to note that a linear transformation (rotation by 45°) will decorrelate x_1 and x_2 and render the two random variables statistically independent. Then scalar quantization and vector quantization achieve the same efficiency. Although a linear transformation can decorrelate a vector of random variables, it does not result in statistically independent random variables, in general. Consequently, vector quantization will always equal or exceed the performance of scalar quantization (see Problem 3–40).

Vector quantization has been applied to several types of speech encoding methods including both waveform and model-based methods which are treated in Section 3.5. In model-based methods such as LPC, vector quantization has made possible the coding of speech at rates below 1000 bits/s (see Buzo et al., 1980; Roucos et al., 1982; Paul, 1983). When applied to waveform encoding methods, it is possible to obtain good quality speech at 16,000 bits/s, or, equivalently, at $R = 2$ bits per sample. With additional computational complexity, it may be possible in the future to implement waveform encoders producing good-quality speech at a rate of $R = 1$ bit per sample.

3.5 CODING TECHNIQUES FOR ANALOG SOURCES

A number of coding techniques for analog sources have been developed over the past 40 years. Most of these have been applied to the encoding of speech and images. In this section, we briefly describe several of these methods and use speech encoding as an example in assessing their performance.

It is convenient to subdivide analog source encoding methods into three types. One type is called *temporal waveform coding*. In this type of encoding, the source encoder is designed to represent digitally the temporal characteristics of the source waveform. A second type of source encoding is *spectral waveform coding*. The signal waveform is usually subdivided into different frequency bands, and either the time waveform in each band or its spectral characteristics are encoded for transmission. The third type of source encoding is based on a mathematical model of the source and is called *model-based coding*.

3.5.1 Temporal Waveform Coding

There are several analog source coding techniques that are designed to represent the time-domain characteristics of the signal. The most commonly used methods are described in this section.

***Pulse-code modulation*† *(PCM)*.** Let $x(t)$ denote a sample function emitted by a source and let x_n denote the samples taken at a sampling rate $f_s \geqslant 2W$, where W is the highest frequency in the spectrum of $x(t)$. In PCM, each sample

† PCM, DPCM, and ADPCM are source coding techniques. They are not digital modulation methods.

of the signal is quantized to one of 2^R amplitude levels, where R is the number of binary digits used to represent each sample. Thus the rate from the source is Rf_s bits/s.

The quantization process may be modeled mathematically as

$$\tilde{x}_n = x_n + q_n \tag{3.5–1}$$

where $\tilde{x}_n$ represents the quantized value of x_n and q_n represents the quantization error, which we treat as an additive noise. Assuming that a uniform quantizer is used, having the input–output characteristic illustrated in Figure 3.5–1, the quantization noise is well characterized statistically by the uniform PDF

$$p(q) = \frac{1}{\Delta}, \qquad -\tfrac{1}{2}\Delta \leqslant q \leqslant \tfrac{1}{2}\Delta \tag{3.5–2}$$

where the step size of the quantizer is $\Delta = 2^{-R}$. The mean square value of the quantization error is

$$E(q^2) = \tfrac{1}{12}\Delta^2 = \tfrac{1}{12} \times 2^{-2R} \tag{3.5–3}$$

Measured in decibels, the mean square value of the noise is

$$10 \log \tfrac{1}{12}\Delta^2 = 10 \log\left(\tfrac{1}{12} \times 2^{-2R}\right) = -6R - 10.8\,\text{dB} \tag{3.5–4}$$

We observe that the quantization noise decreases by 6 dB/bit used in the quantizer. For example, a 7-bit quantizer results in a quantization noise power of −52.8 dB.

Many source signals such as speech waveforms have the characteristic that small signal amplitudes occur more frequently than large ones. However, a uniform quantizer provides the same spacing between successive levels throughout

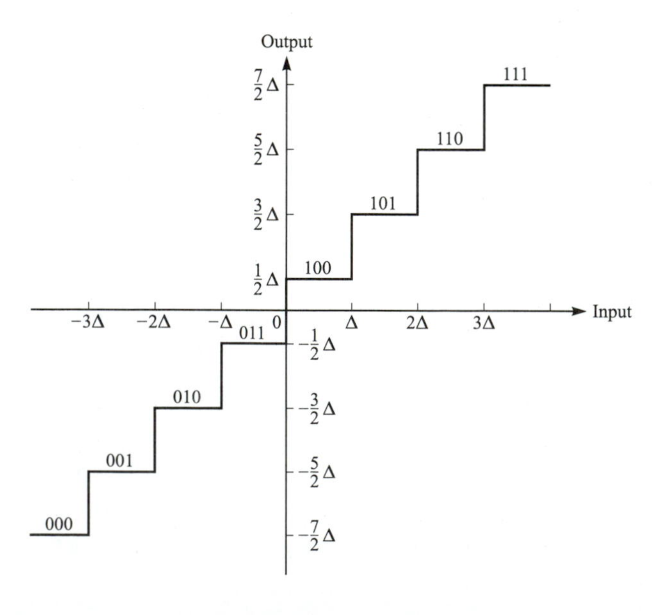

FIGURE 3.5–1
Input–output characteristic for a uniform quantizer.

the entire dynamic range of the signal. A better approach is to employ a nonuniform quantizer. A nonuniform quantizer characteristic is usually obtained by passing the signal through a non-linear device that compresses the signal amplitude, followed by a uniform quantizer. For example, a logarithmic compressor has an input–output magnitude characteristic of the form

$$|y| = \frac{\log(1 + \mu|x|)}{\log(1 + \mu)} \tag{3.5–5}$$

where $|x| \leqslant 1$ is the magnitude of the input, $|y|$ is the magnitude of the output, and μ is a parameter that is selected to give the desired compression characteristic. Figure 3.5–2 illustrates this compression relationship for several values of μ. The value $\mu = 0$ corresponds to no compression.

In the encoding of speech waveforms, for example, the value of $\mu = 255$ has been adopted as a standard in the United States and Canada. This value results in about a 24-dB reduction in the quantization noise power relative to uniform quantization, as shown by Jayant (1974). Consequently, a 7-bit quantizer used in conjunction with a $\mu = 255$ logarithmic compressor produces a quantization noise power of approximately −77 dB compared with the −53 dB for uniform quantization.

In the reconstruction of the signal from the quantized values, the inverse logarithmic relation is used to expand the signal amplitude. The combined compressor–expandor pair is termed a *compandor*.

Differential pulse-code modulation (DPCM). In PCM, each sample of the waveform is encoded independently of all the others. However, most source signals sampled at the Nyquist rate or faster exhibit significant correlation between successive samples. In other words, the average change in amplitude between successive samples is relatively small. Consequently, an encoding scheme that exploits the redundancy in the samples will result in a lower bit rate for the source output.

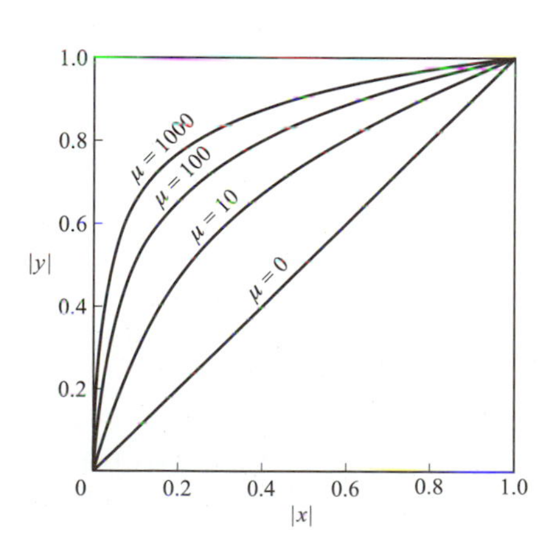

FIGURE 3.5–2
Input–output magnitude characteristic for a logarithmic compressor.

A relatively simple solution is to encode the differences between successive samples rather than the samples themselves. Since differences between samples are expected to be smaller than the actual sampled amplitudes, fewer bits are required to represent the differences. A refinement of this general approach is to predict the current sample based on the previous p samples. To be specific, let x_n denote the current sample from the source and let $\hat{x}_n$ denote the predicted value of x_n, defined as

$$\hat{x}_n = \sum_{i=1}^{p} a_i x_{n-i} \tag{3.5–6}$$

Thus $\hat{x}_n$ is a weighted linear combination of the past p samples and the $\{a_i\}$ are the predictor coefficients. The $\{a_i\}$ are selected to minimize some function of the error between x_n and $\hat{x}_n$.

A mathematically and practically convenient error function is the mean square error (MSE). With the MSE as the performance index for the predictor, we select the $\{a_i\}$ to minimize

$$\begin{aligned}\mathcal{E}_p = E(e_n^2) &= E\left[\left(x_n - \sum_{i=1}^{p} a_i x_{n-i}\right)^2\right] \\ &= E(x_n^2) - 2\sum_{i=1}^{p} a_i E(x_n x_{n-i}) + \sum_{i=1}^{p}\sum_{j=1}^{p} a_i a_j E(x_{n-i} x_{n-j})\end{aligned} \tag{3.5–7}$$

Assuming that the source output is (wide-sense) stationary, we may express Equation 3.5–7 as

$$\mathcal{E}_p = \phi(0) - 2\sum_{i=1}^{p} a_i \phi(i) + \sum_{i=1}^{p}\sum_{j=1}^{p} a_i a_j \phi(i-j) \tag{3.5–8}$$

where $\phi(m)$ is the autocorrelation function of the sampled signal sequence x_n. Minimization of $\mathcal{E}_p$ with respect to the predictor coefficients $\{a_i\}$ results in the set of linear equations

$$\sum_{i=1}^{p} a_i \phi(i-j) = \phi(j), \qquad j = 1, 2, \ldots, p \tag{3.5–9}$$

Thus, the values of the predictor coefficients are established. When the autocorrelation function $\phi(n)$ is not known a priori, it may be estimated from the samples $\{x_n\}$ using the relation†

$$\hat{\phi}(n) = \frac{1}{N}\sum_{i=1}^{N-n} x_i x_{i+n}, \qquad n = 0, 1, 2, \ldots, p \tag{3.5–10}$$

† The estimation of the autocorrelation function from a finite number of observations $\{x_i\}$ is a separate issue, which is beyond the scope of this discussion. The estimate in Equation 3.5–10 is one that is frequently used in practice.

and the estimate $\hat{\phi}(n)$ is used in Equation 3.5–9 to solve for the coefficients $\{a_i\}$. Note that the normalization factor of $1/N$ in Equation 3.5–10 drops out when $\hat{\phi}(n)$ is substituted in Equation 3.5–9.

The linear equations 3.5–9 for the predictor coefficients are called the *normal equations* or the *Yule–Walker equations*. There is an algorithm developed by Levinson (1947) and Durbin (1959) for solving these equations efficiently. It is described in Appendix A. We shall deal with the solution in greater detail in the subsequent discussion on linear predictive coding.

Having described the method for determining the predictor coefficients, let us now consider the block diagram of a practical DPCM system, shown in Figure 3.5–3a. In this configuration, the predictor is implemented with the feedback loop around the quantizer. The input to the predictor is denoted by $\tilde{x}_n$, which represents the signal sample x_n modified by the quantization process, and the output of the predictor is

$$\hat{\tilde{x}}_n = \sum_{i=1}^{p} a_i \tilde{x}_{n-i} \tag{3.5–11}$$

The difference

$$e_n = x_n - \hat{\tilde{x}}_n \tag{3.5–12}$$

is the input to the quantizer and $\tilde{e}_n$ denotes the output. Each value of the quantized prediction error $\tilde{e}_n$ is encoded into a sequence of binary digits and

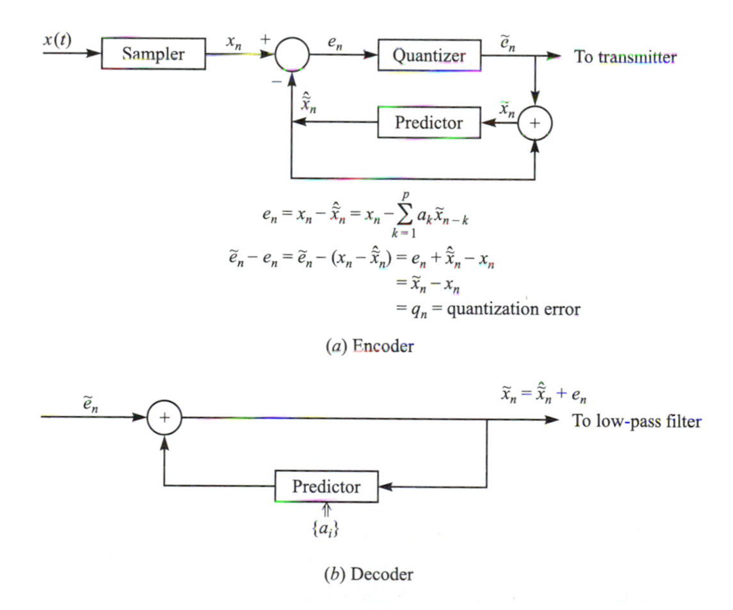

FIGURE 3.5–3
(a) Block diagram of a DPCM encoder. **(b)** DPCM decoder at the receiver.

transmitted over the channel to the destination. The quantized error $\tilde{e}_n$ is also added to the predicted value $\hat{\tilde{x}}_n$ to yield $\tilde{x}_n$.

At the destination, the same predictor that was used at the transmitting end is synthesized and its output $\hat{\tilde{x}}_n$ is added to $\tilde{e}_n$ to yield $\tilde{x}_n$. The signal $\tilde{x}_n$ is the desired excitation for the predictor and also the desired output sequence from which the reconstructed signal $\tilde{x}(t)$ is obtained by filtering, as shown in Figure 3.5–3b.

The use of feedback around the quantizer, as described above, ensures that the error in $\tilde{x}_n$ is simply the quantization error $q_n = \tilde{e}_n - e_n$ and that there is no accumulation of previous quantization errors in the implementation of the decoder. That is,

$$\begin{aligned} q_n &= \tilde{e}_n - e_n \\ &= \tilde{e}_n - (x_n - \hat{\tilde{x}}_n) \\ &= \tilde{x}_n - x_n \end{aligned} \tag{3.5–13}$$

Hence $\tilde{x}_n = x_n + q_n$. This means that the quantized sample $\tilde{x}_n$ differs from the input x_n by the quantization error q_n independent of the predictor used. Therefore, the quantization errors do not accumulate.

In the DPCM system illustrated in Figure 3.5–3, the estimate or predicted value $\hat{\tilde{x}}_n$ of the signal sample x_n is obtained by taking a linear combination of past values $\tilde{x}_{n-k}$, $k = 1, 2, \ldots, p$, as indicated by Equation 3.5–11. An improvement in the quality of the estimate is obtained by including linearly filtered past values of the quantized error. Specifically, the $\hat{\tilde{x}}_n$ estimate may be expressed as

$$\hat{\tilde{x}}_n = \sum_{i=1}^{p} a_i \tilde{x}_{n-i} + \sum_{i=1}^{m} b_i \tilde{e}_{n-i} \tag{3.5–14}$$

where $\{b_i\}$ are the coefficients of the filter for the quantized error sequence $\tilde{e}_n$. The block diagrams of the encoder at the transmitter and the decoder at the receiver are shown in Figure 3.5–4. The two sets of coefficients $\{a_i\}$ and $\{b_i\}$ are selected to minimize some function of the error $e_n = x_n - \hat{\tilde{x}}_n$, such as the mean square error.

Adaptive PCM and DPCM. Many real sources are quasi-stationary in nature. One aspect of the quasi-stationary characteristic is that the variance and the autocorrelation function of the source output vary slowly with time. PCM and DPCM encoders, however, are designed on the basis that the source output is stationary. The efficiency and performance of these encoders can be improved by having them adapt to the slowly time variant statistics of the source.

In both PCM and DPCM, the quantization error q_n resulting from a uniform quantizer operating on a quasi-stationary input signal will have a time-variant variance (quantization noise power). One improvement that reduces the dynamic range of the quantization noise is the use of an adaptive quantizer. Although the quantizer can be made adaptive in different ways, a relatively simple method is to use a uniform quantizer that varies its step size in accordance with the variance

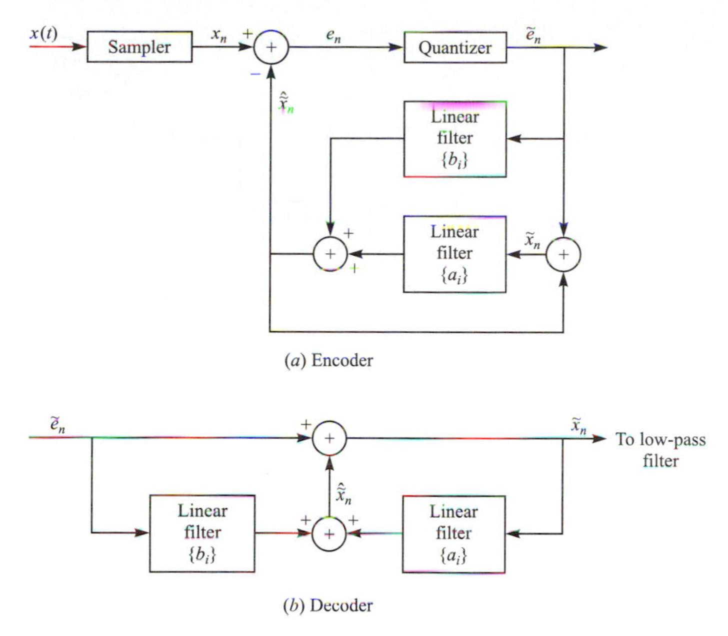

FIGURE 3.5–4
DPCM modified by the addition of linearly filtered error sequence.

of the past signal samples. For example, a short-term running estimate of the variance of x_n can be computed from the input sequence $\{x_n\}$ and the step size can be adjusted on the basis of such an estimate. In its simplest form, the algorithm for the step-size adjustment employs only the previous signal sample. Such an algorithm has been successfully used by Jayant (1974) in the encoding of speech signals. Figure 3.5–5 illustrates such a (3-bit) quantizer in which the step size is adjusted recursively according to the relation

$$\Delta_{n+1} = \Delta_n M(n) \tag{3.5–15}$$

where $M(n)$ is a factor, whose value depends on the quantizer level for the sample x_n, and Δ_n is the step size of the quantizer for processing x_n. Values of the multiplication factors optimized for speech encoding have been given by Jayant (1974). These values are displayed in Table 3.5–1 for 2-, 3-, and 4-bit adaptive quantization.

In DPCM, the predictor can also be made adaptive when the source output is quasi-stationary. The coefficients of the predictor can be changed periodically to reflect the changing signal statistics of the source. The linear equations given by Equation 3.5–9 still apply, with the short-term estimate of the autocorrelation function of x_n substituted in place of the ensemble correlation function. The predictor coefficients thus determined may be transmitted along with the quantized error $\tilde{e}(n)$ to the receiver, which implements the same predictor. Unfortunately, the transmission of the predictor coefficients results in a higher

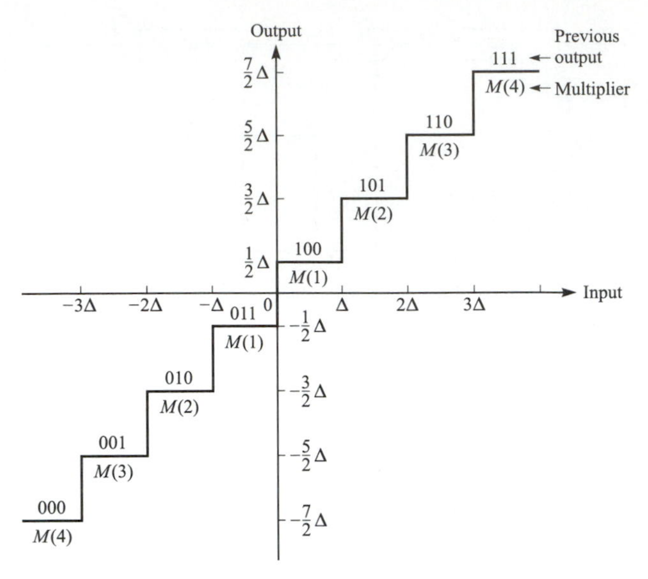

FIGURE 3.5–5
Example of a quantizer with an adaptive step size. (*Jayant, 1974.*)

bit rate over the channel, offsetting, in part, the lower data rate achieved by having a quantizer with fewer bits (fewer levels) to handle the reduced dynamic range in the error e_n resulting from adaptive prediction.

As an alternative, the predictor at the receiver may compute its own prediction coefficients from $\tilde{e}_n$ and $\tilde{x}_n$, where

$$\tilde{x}_n = \tilde{e}_n + \sum_{i=1}^{p} a_i \tilde{x}_{n-i} \tag{3.5–16}$$

If we neglect the quantization noise, $\tilde{x}_n$ is equivalent to x_n. Hence, $\tilde{x}_n$ may be used to estimate the autocorrelation function $\phi(n)$ at the receiver, and the resulting

TABLE 3.5–1
Multiplication factors for adaptive step size adjustment (Jayant, 1974)

	PCM			DPCM		
	2	**3**	**4**	**2**	**3**	**4**
$M(1)$	0.60	0.85	0.80	0.80	0.90	0.90
$M(2)$	2.20	1.00	0.80	1.60	0.90	0.90
$M(3)$		1.00	0.80		1.25	0.90
$M(4)$		1.50	0.80		1.70	0.90
$M(5)$			1.20			1.20
$M(6)$			1.60			1.60
$M(7)$			2.00			2.00
$M(8)$			2.40			2.40

estimates can be used in Equation 3.5–9 in place of $\phi(n)$ to solve for the predictor coefficients. For sufficiently fine quantization, the difference between x_n and $\tilde{x}_n$ is very small. Hence, the estimate of $\phi(n)$ obtained from $\tilde{x}_n$ is usually adequate for determining the predictor coefficients. Implemented in this manner, the adaptive predictor results in a lower source data rate.

Instead of using the block processing approach for determining the predictor coefficients $\{a_i\}$ as described above, we may adapt the predictor coefficients on a sample-by-sample basis by using a gradient-type algorithm, similar in form to the adaptive gradient equalization algorithm that is described in Chapter 11. Similar gradient-type algorithms have also been devised for adapting the filter coefficients $\{a_i\}$ and $\{b_i\}$ of the DPCM system shown in Figure 3.5–4. For details on such algorithms, the reader may refer to the book by Jayant and Noll (1984).

Delta modulation (DM). Delta modulation may be viewed as a simplified form of DPCM in which a two-level (1-bit) quantizer is used in conjunction with a fixed first-order predictor. The block diagram of a DM encoder–decoder is shown in Figure 3.5–6a. We note that

$$\hat{\tilde{x}}_n = \tilde{x}_{n-1} = \hat{\tilde{x}}_{n-1} + \tilde{e}_{n-1} \tag{3.5–17}$$

Since

$$\begin{aligned} q_n &= \tilde{e}_n - e_n \\ &= \tilde{e}_n - (x_n - \hat{\tilde{x}}_n) \end{aligned}$$

It follows that

$$\hat{\tilde{x}}_n = x_{n-1} + q_{n-1}$$

Thus the estimated (predicted) value of x_n is really the previous sample x_{n-1} modified by the quantization noise q_{n-1}. We also note that the difference equation 3.5–17 represents an integrator with an input $\tilde{e}_n$. Hence, an equivalent realization of the one-step predictor is an accumulator with an input equal to the quantized error signal $\tilde{e}_n$. In general, the quantized error signal is scaled by some value, say Δ_1, which is called the *step size*. This equivalent realization is illustrated in Figure 3.5–6b. In effect, the encoder shown in Figure 3.5–6 approximates a waveform $x(t)$ by a linear staircase function. In order for the approximation to be relatively good, the waveform $x(t)$ must change slowly relative to the sampling rate. This requirement implies that the sampling rate must be several (a factor of at least 5) times the Nyquist rate.

At any given sampling rate, the performance of the DM encoder is limited by two types of distortion, as illustrated in Figure 3.5–7. One is called *slope-overload distortion*. It is due to the use of a step size Δ_1 that is too small to follow portions of the waveform that have a steep slope. The second type of distortion, called *granular noise*, results from using a step size that is too large in parts of the waveform having a small slope. The need to minimize both of these two types of distortion results in conflicting requirements in the selection of the step size Δ_1.

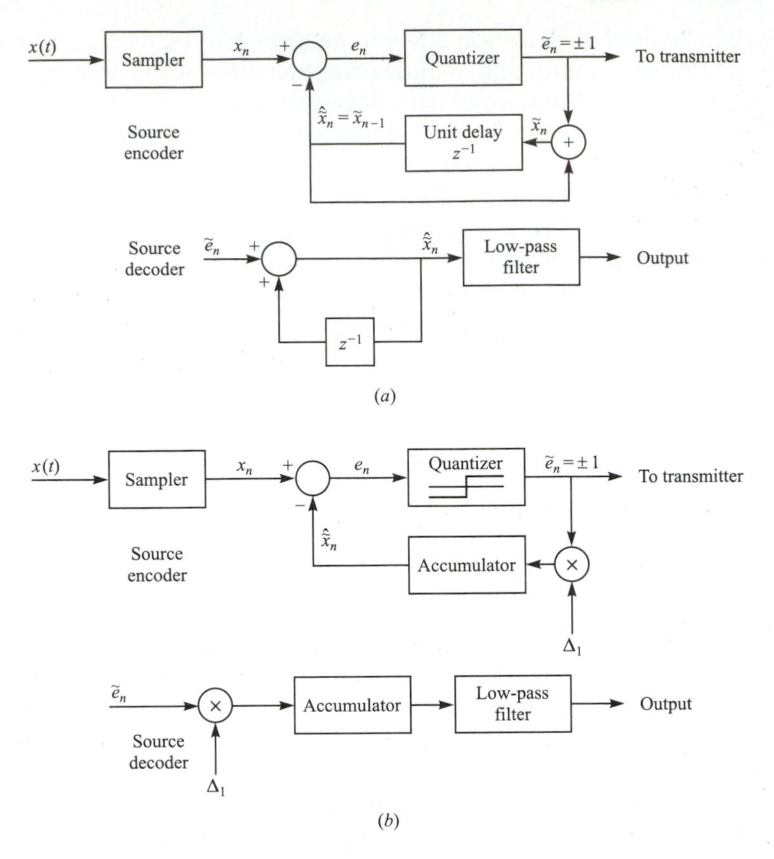

FIGURE 3.5–6
(a) Block diagram of a delta modulation system. **(b)** An equivalent realization of a delta modulation system.

One solution is to select Δ_1 to minimize the sum of the mean square values of these two distortions.

Even when Δ_1 is optimized to minimize the total mean square value of the slope-overload distortion and the granular noise, the performance of the DM encoder may still be less than satisfactory. An alternative solution is to employ a variable step size that adapts itself to the short-term characteristics of the source signal. That is, the step size is increased when the waveform has a steep slope and decreased when the waveform has a relatively small slope. This adaptive characteristic is illustrated in Figure 3.5–8.

A variety of methods can be used to adaptively set the step size in every iteration. The quantized error sequence $\tilde{e}_n$ provides a good indication of the slope characteristics of the waveform being encoded. When the quantized error $\tilde{e}_n$ is changing signs between successive iterations, this is an indication that the slope of the waveform in that locality is relatively small. On the other hand, when the waveform has a steep slope, successive values of the error $\tilde{e}_n$ are expected to have

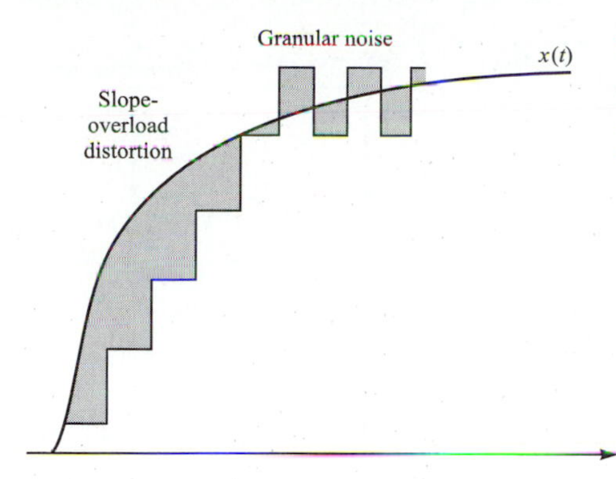

FIGURE 3.5–7
An example of slope-overload distortion and granular noise in a delta modulation encoder.

identical signs. From these observations, it is possible to devise algorithms that decrease or increase the step size depending on successive values of $\tilde{e}_n$. A relatively simple rule devised by Jayant (1970) is to adaptively vary the step size according to the relation

$$\Delta_n = \Delta_{n-1} K^{\tilde{e}_n \cdot \tilde{e}_{n-1}}, \qquad n = 1, 2, \ldots$$

where $K \geqslant 1$ is a constant that is selected to minimize the total distortion. A block diagram of a DM encoder–decoder that incorporates this adaptive algorithm is illustrated in Figure 3.5–9.

Several other variations of adaptive DM encoding have been investigated and described in the technical literature. A particularly effective and popular technique first proposed by Greefkes (1970) is called *continuously variable slope delta modulation* (CVSD). In CVSD the adaptive step-size parameter may be expressed as

$$\Delta_n = \alpha \Delta_{n-1} + k_1$$

if $\tilde{e}_n$, $\tilde{e}_{n-1}$, and $\tilde{e}_{n-2}$ have the same sign; otherwise,

$$\Delta_n = \alpha \Delta_{n-1} = k_2$$

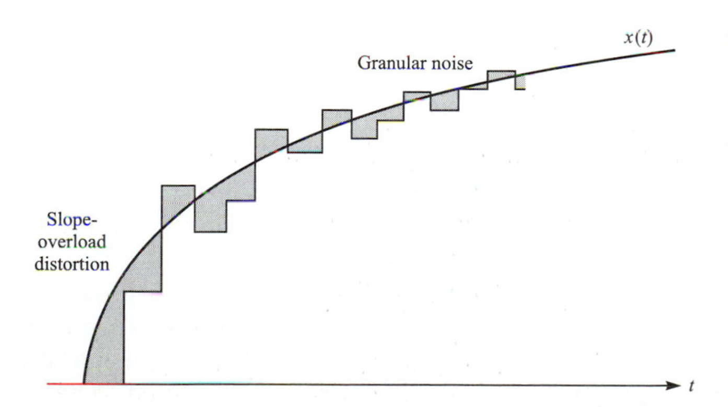

FIGURE 3.5–8
An example of variable-step-size delta modulation encoding.

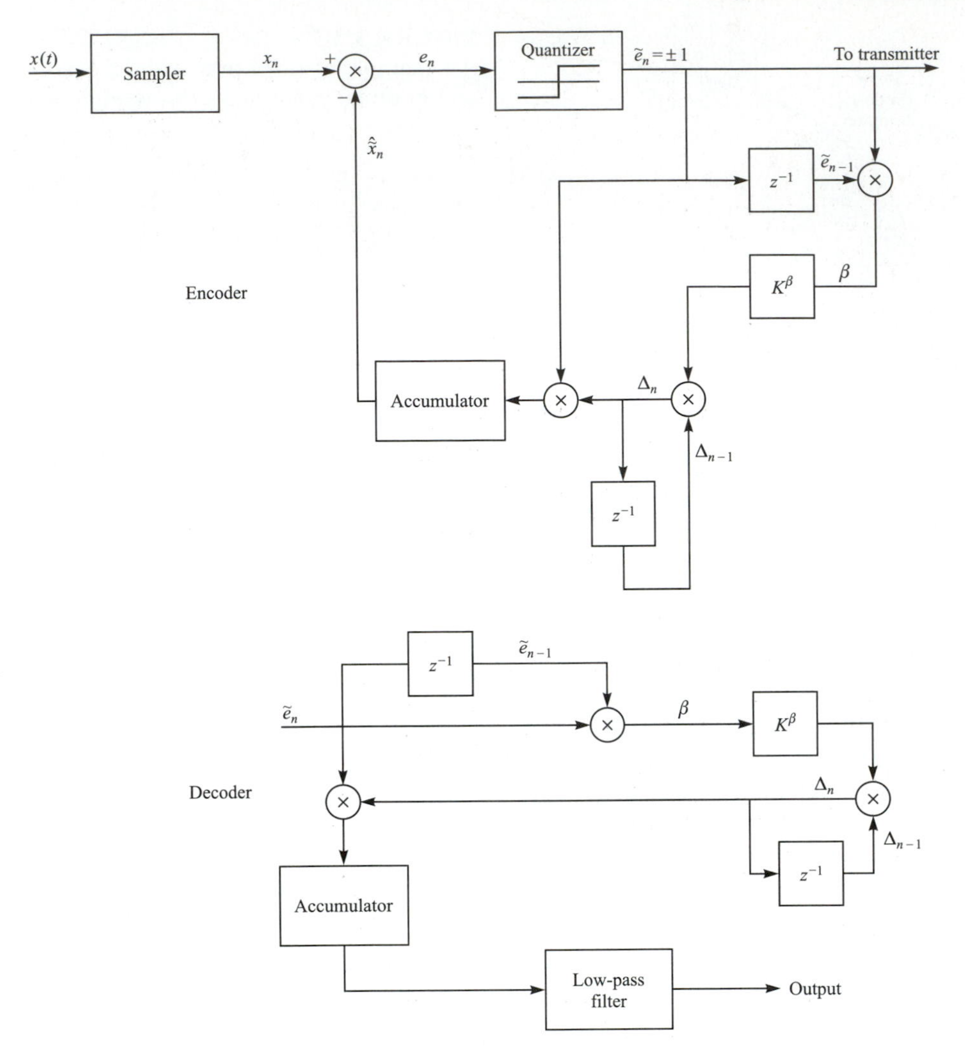

FIGURE 3.5–9
An example of a delta modulation system with adaptive step size.

The parameters α, k_1, and k_2 are selected such that $0 < \alpha < 1$ and $k_1 \gg k_2 > 0$. For more discussion on this and other variations of adaptive DM, the interested reader is referred to the papers by Jayant (1974) and Flanagan et al. (1979), which contain extensive references.

PCM, DPCM, adaptive PCM, and adaptive DPCM and DM are all source encoding techniques that attempt to faithfully represent the output waveform from the source. The following class of waveform encoding methods is based on a spectral decomposition of the source signal.

3.5.2 Spectral Waveform Coding

In this section, we briefly describe waveform coding methods that filter the source output signal into a number of frequency bands or subbands and separately encode the signal in each subband. The waveform encoding may be performed either on the time-domain waveforms in each subband or on the frequency-domain representation of the corresponding time-domain waveform in each subband.

Subband coding. In subband coding (SBC) of speech and image signals, the signal is divided into a small number of subbands and the time waveform in each subband is encoded separately. In speech coding, for example, the lower-frequency bands contain most of the spectral energy in voiced speech. In addition, quantization noise is more noticeable to the ear in the lower-frequency bands. Consequently, more bits are used for the lower-band signals and fewer are used for the higher-frequency bands.

Filter design is particularly important in achieving good performance in SBC. In practice, quadrature–mirror filters (QMFs) are generally used because they yield an alias-free response due to their perfect reconstruction property (see Vaidyanathan, 1993). By using QMFs in subband coding, the lower-frequency band is repeatedly subdivided by factors of 2, thus creating octave-band filters. The output of each QMF is decimated by a factor of 2, in order to reduce the sampling rate. For example, suppose that the bandwidth of a speech signal extends to 3200 Hz. The first pair of QMFs divides the spectrum into the low (0–1600 Hz) and high (1600–3200 Hz) bands. Then, the low band is split into low (0–800 Hz) and high (800–1600 Hz) bands by the use of another pair of QMFs. A third subdivision by another pair of QMFs can split the 0–800 Hz band into low (0–400 Hz) and high (400–800 Hz) bands. Thus, with three pairs of QMFs, we have obtained signals in the frequency bands 0–400, 400–800, 800–1600, and 1600–3200 Hz. The time-domain signal in each subband may now be encoded with different precision. In practice, adaptive PCM has been used for waveform encoding of the signal in each subband.

Adaptive Transform Coding. In adaptive transform coding (ATC), the source signal is sampled and subdivided into frames of N_f samples, and the data in each frame is transformed into the spectral domain for coding and transmission. At the source decoder, each frame of spectral samples is transformed back into the time domain and the signal is synthesized from the time-domain samples and passed through a digital-to-analog (D/A) converter. To achieve coding efficiency, we assign more bits to the more important spectral coefficients and fewer bits to the less important spectral coefficients. In addition, by designing an adaptive allocation in the assignment of the total number of bits to the spectral coefficients, we can adapt to possibly changing statistics of the source signal.

An objective in selecting the transformation from the time domain to the frequency domain is to achieve uncorrelated spectral samples. In this sense, the

Karhunen–Loéve transform (KLT) is optimal in that it yields spectral values that are uncorrelated, but the KLT is generally difficult to compute (see Wintz, 1972). The DFT and the *discrete cosine transform* (DCT) are viable alternatives, although they are suboptimum. Of these two, the DCT yields good performance compared with the KLT, and is generally used in practice (see Campanella and Robinson, 1971; Zelinsky and Noll, 1977).

In speech coding using ATC, it is possible to attain communication-quality speech at a rate of about 9600 bits/s.

3.5.3 Model-Based Source Coding

In contrast to the waveform encoding methods described above, model-based source coding represents a completely different approach. In this, the source is modeled as a linear system (filter) that, when excited by an appropriate input signal, results in the observed source output. Instead of transmitting the samples of the source waveform to the receiver, the parameters of the linear system are transmitted along with an appropriate excitation signal. If the number of parameters is sufficiently small, the model-based methods provide a large compression of the data.

The most widely used model-based coding method is called *linear predictive coding* (LPC). In this, the sampled sequence, denoted by x_n, $n = 0, 1, \ldots, N-1$, is assumed to have been generated by an all-pole (discrete-time) filter having the transfer function

$$H(z) = \frac{G}{1 - \sum_{k=1}^{p} a_k z^{-k}} \tag{3.5–18}$$

Appropriate excitation functions are an impulse, a sequence of impulses, or a sequence of white noise with unit variance. In any case, suppose that the input sequence is denoted by v_n, $n = 0, 1, 2, \ldots$. Then the output sequence of the all-pole model satisfies the difference equation

$$x_n = \sum_{k=1}^{p} a_k x_{n-k} + G v_n, \qquad n = 0, 1, 2, \ldots \tag{3.5–19}$$

In general, the observed source output x_n, $n = 0, 1, 2, \ldots, N-1$, does not satisfy the difference equation 3.5–19; only its model does. If the input is a white-noise sequence or an impulse, we may form an estimate (or prediction) of x_n by the weighted linear combination

$$\hat{x}_n = \sum_{k=1}^{p} a_k x_{n-k}, \qquad n > 0 \tag{3.5–20}$$

The difference between x_n and $\hat{x}_n$, namely,

$$e_n = x_n - \hat{x}_n = x_n - \sum_{k=1}^{p} a_k x_{n-k} \tag{3.5–21}$$

represents the error between the observed value x_n and the estimated (predicted) value $\hat{x}_n$. The filter coefficients $\{a_k\}$ can be selected to minimize the mean square value of this error.

Suppose for the moment that the input $\{v_n\}$ is a white-noise sequence. Then, the filter output x_n is a random sequence and so is the difference $e_n = x_n - \hat{x}_n$. The ensemble average of the squared error is

$$\begin{aligned}\mathcal{E}_p &= E(e_n^2) \\ &= E\left[\left(x_n - \sum_{k=1}^{p} a_k x_{n-k}\right)^2\right] \\ &= \phi(0) - 2\sum_{k-1}^{p} a_k\phi(k) + \sum_{k=1}^{p}\sum_{m=1}^{p} a_k a_m \phi(k-m)\end{aligned} \tag{3.5–22}$$

where $\phi(m)$ is the autocorrelation function of the sequence $x_n, n = 0, 1, \dots, N-1$. But $\mathcal{E}_p$ is identical to the MSE given by Equation 3.5–8 for a predictor used in DPCM. Consequently, minimization of $\mathcal{E}_p$ in Equation 3.5–22 yields the set of normal equations given previously by Equation 3.5–9. To completely specify the filter $H(z)$, we must also determine the filter gain G. From Equation 3.5–19, we have

$$E[(Gv_n)^2] = G^2 E(v_n^2) = G^2 = E\left[\left(x_n - \sum_{k=1}^{p} a_k x_{n-k}\right)^2\right] = \mathcal{E}_p \tag{3.5–23}$$

where $\mathcal{E}_p$ is the residual MSE obtained from Equation 3.5–22 by substituting the optimum prediction coefficients, which result from the solution of Equation 3.5–9. With this substitution, the expression for $\mathcal{E}_p$ and, hence, G^2 simplifies to

$$\mathcal{E}_p = G^2 = \phi(0) - \sum_{k=1}^{p} a_k \phi(k) \tag{3.5–24}$$

In practice, we do not usually know a priori the true autocorrelation function of the source output. Hence, in place of $\phi(n)$, we substitute an estimate $\hat{\phi}(n)$ as given by Equation 3.5–10, which is obtained from the set of samples $x_n, n = 0, 1, \dots, N-1$, emitted by the source.

As indicated previously, the Levinson–Durbin algorithm derived in Appendix A may be used to solve for the predictor coefficients $\{a_k\}$ recursively, beginning with a first-order predictor and iterating the order of the predictor up to order p. The recursive equations for the $\{a_k\}$ may be expressed as

$$\begin{aligned}
a_{ii} &= \frac{\hat{\phi}(i) - \sum_{k=1}^{i-1} a_{i-1k}\hat{\phi}(i-k)}{\hat{\mathcal{E}}_{i-1}}, \qquad i = 2, 3, \ldots, p \\
a_{ik} &= a_{i-1k} - a_{ii}a_{i-1i-k}, \qquad 1 \leqslant k \leqslant i-1 \\
\hat{\mathcal{E}}_i &= (1 - a_{ii})\hat{\mathcal{E}}_{i-1} \\
a_{11} &= \frac{\hat{\phi}(1)}{\hat{\phi}(0)} \\
\hat{\mathcal{E}}_0 &= \hat{\phi}(0)
\end{aligned} \tag{3.5–25}$$

where $a_{ik}, k = 1, 2, \ldots, i$, are the coefficients of the ith-order predictor. The desired coefficients for the predictor of order p are

$$a_k \equiv a_{pk}, \qquad k = 1, 2, \ldots, p \tag{3.5–26}$$

and the residual MSE is

$$\begin{aligned}
\hat{\mathcal{E}} = G^2 &= \hat{\phi}(0) - \sum_{k=1}^{p} a_k \hat{\phi}(k) \\
&= \hat{\phi}(0)\prod_{i=1}^{p}(1 - a_{ii}^2)
\end{aligned} \tag{3.5–27}$$

We observe that the recursive relations in Equation 3.5–25 give us not only the coefficients of the predictor for order p, but also the predictor coefficients of all orders less than p.

The residual MSE $\hat{\mathcal{E}}_i$, $i = 1, 2, \ldots, p$, forms a monotone decreasing sequence, i.e., $\hat{\mathcal{E}}_p \leqslant \hat{\mathcal{E}}_{p-1} \leqslant \cdots \leqslant \hat{\mathcal{E}}_1 \leqslant \hat{\mathcal{E}}_0$, and the prediction coefficients a_{ii} satisfy the condition

$$|a_{ii}| < 1, \qquad i = 1, 2, \ldots, p \tag{3.5–28}$$

This condition is necessary and sufficient for all the poles of $H(z)$ to be inside the unit circle. Thus Equation 3.5–28 ensures that the model is stable.

LPC has been successfully used in the modeling of a speech source. In this case, the coefficients a_{ii}, $i = 1, 2, \ldots, p$, are called *reflection coefficients* as a consequence of their correspondence to the reflection coefficients in the acoustic tube model of the vocal tract (see Rabiner and Schafer, 1978; Deller et al., 2000).

Once the predictor coefficients and the gain G have been estimated from the source output $\{x_n\}$, each parameter is coded into a sequence of binary digits and transmitted to the receiver. Source decoding or waveform synthesis may be accomplished at the receiver as illustrated in Figure 3.5–10. The signal generator is used to produce the excitation function $\{v_n\}$, which is scaled by G to produce the desired input to the all-pole filter model $H(z)$ synthesized from the received prediction coefficients. The analog signal may be reconstructed by passing the output sequence from $H(z)$ through an analog filter that basically performs the function of interpolating the signal between sample points. In this realization of

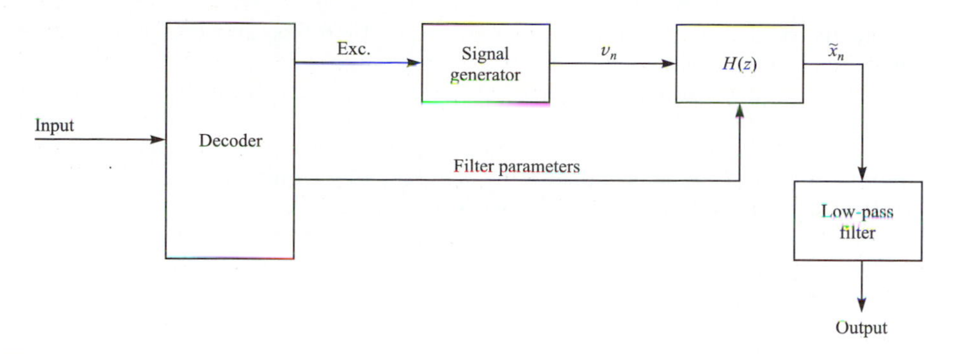

FIGURE 3.5–10
Block diagram of a waveform synthesizer (source decoder) for an LPC system.

the waveform synthesizer, the excitation function and the gain parameter must be transmitted along with the prediction coefficients to the receiver.

When the source output is stationary, the filter parameters need to be determined only once. However, the statistics of most sources encountered in practice are at best quasi-stationary. Under these circumstances, it is necessary to periodically obtain new estimates of the filter coefficients, the gain G, and the type of excitation function, and to transmit these estimates to the receiver.

EXAMPLE 3.5-1. The block diagram shown in Fig. 3.5–11 illustrates a model for a speech source. There are two mutually exclusive excitation functions to model voiced and unvoiced speech sounds. On a short-time basis, voiced speech is periodic with a fundamental frequency f_0 or a pitch period $1/f_0$ that depends on the speaker. Thus voiced speech is generated by exciting an all-pole filter model of the vocal tract by a periodic impulse train with a period equal to the desired pitch period. Unvoiced speech sounds are generated by exciting the all-pole filter model by the output of a random-noise generator. The speech encoder at the transmitter must determine the proper excitation function, the pitch period of voiced speech, the gain parameter G, and the prediction coefficients. These parameters are encoded into binary digits and transmitted to the receiver. Typically, the voiced and unvoiced information requires

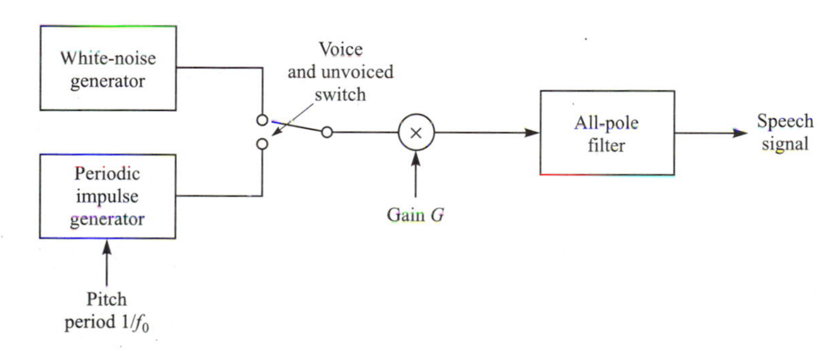

FIGURE 3.5–11
Block diagram model of the generation of a speech signal.

1 bit, the pitch period is adequately represented by 6 bits, and the gain parameter may be represented by 5 bits after its dynamic range is compressed logarithmically. The prediction coefficients require 8–10 bits per coefficient for adequate representation (see Rabiner and Schafer, 1978). The reason for such high accuracy is that relatively small changes in the prediction coefficients result in a large change in the pole positions of the filter model $H(z)$. The accuracy requirements may be lessened by transmitting the reflection coefficients a_{ii}, which have a smaller dynamic range. These are adequately represented by 6 bits. Thus, for a predictor or order $p = 10$ [five poles in $H(z)$], the total number of bits is 72. Because of the quasi-stationary nature of the speech signal, the linear system model must be changed periodically, typically once every 15–30 ms. Consequently, the bit rate from the source encoder is in the range 4800–2400 bits/s.

When the reflection coefficients are transmitted to the decoder, it is not necessary to recompute the prediction coefficients in order to realize the speech synthesizer. Instead, the synthesis is performed by realizing a lattice filter, shown in Figure 3.5–12, which utilizes the reflection coefficients directly and which is equivalent to the linear prediction filter.

The linear all-pole filter model, for which the filter coefficients are estimated via linear prediction, is by far the simplest linear model for a source. A more general source model is a linear filter that contains both poles and zeros. In a pole–zero model, the source output x_n satisfies the difference equation

$$x_n = \sum_{k=1}^{p} a_k x_{n-k} + \sum_{k=0}^{q} b_k v_{n-k}$$

where v_n is the input excitation sequence. The problem now is to estimate the filter parameters $\{a_k\}$ and $\{b_k\}$ from the data x_i, $i = 0, 1, \ldots, N-1$, emitted by the source. However, the MSE criterion applied to the minimization of the error $e_n = x_n - \hat{x}_n$, where $\hat{x}_n$ is an estimate of x_n, results in a set of non-linear equations for the parameters $\{a_k\}$ and $\{b_k\}$. Consequently, the evaluation of the $\{a_k\}$ and $\{b_k\}$ becomes tedious and difficult mathematically. To avoid having to solve the non-linear equations, a number of suboptimum methods have been devised for pole–zero modeling. A discussion of these techniques would lead to us too far afield, however.

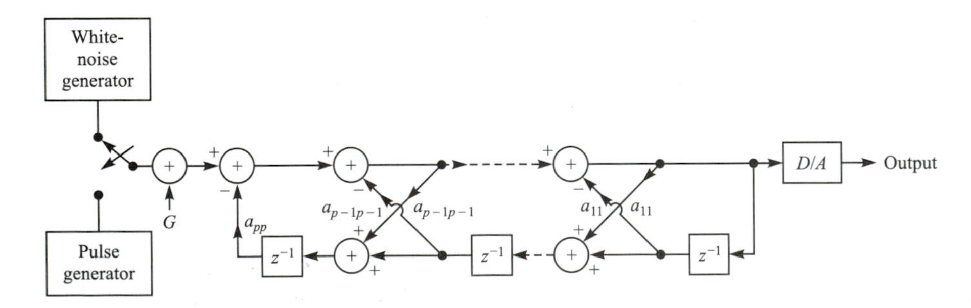

FIGURE 3.5–12
All-pole lattice filter for synthesizing the speech signal.

LPC as described above forms the basis for more complex model-based source encoding methods. When applied to speech coding, the model-based methods are generally called *vocoders* (for voice coders). In addition to the conventional LPC vocoder described above, other types of vocoders that have been implemented include the residual excited LPC (RELP) vocoder, the multipulse LPC vocoder, the code-excited LPC (CELP) vocoder, and the vector-sum-excited LPC (VSELP) vocoder. The CELP and VSELP vocoders employ vector-quantized excitation code books to achieve communication-quality speech at low bit rates. In these vocoders, the excitation sequence is selected from a code book of zero-mean Gaussian sequences. Typically, the excitation sequences consist of 40 samples each and the code book contains 1024 sequences. Consequently, each excitation vector is specified by 10 bits. CELP is widely used to achieve communication-quality speech transmission in digital cellular communication systems at rates below 10,000 bits/s.

Before concluding this section, we consider the application of waveform encoding and LPC to the encoding of speech signals and compare the bit rates of these coding techniques.

Encoding methods applied to speech signals. The transmission of speech signals over telephone lines, radio channels, and satellite channels constitutes by far the largest part of our daily communications. It is understandable, therefore, that over the past three decades more research has been performed on speech encoding than on any other type of information-bearing signal. In fact, all the encoding techniques described in this section have been applied to the encoding of speech signals. It is appropriate, therefore, to compare the efficiency of these methods in terms of the bit rate required to transmit the speech signal.

The speech signal is assumed to be band-limited to the frequency range 200–3200 Hz and sampled at a nominal rate of 8000 samples/s for all encoders except DM, where the sample rate f_s is identical to the bit rate. For an LPC encoder, the parameters given in Example 3.5–1 are assumed.

Table 3.5–2 summarizes the main characteristics of the encoding methods described in this section and the required bit rate. In terms of the quality of the speech signal synthesized at the receiver from the (error-free) binary sequence, all the waveform encoding methods (PCM, DPCM, ADPCM, DM, ADM) provide telephone (toll) quality speech. In other words, a listener would have difficulty discerning the difference between the digitized speech and the analog speech waveform. ADPCM and ADM are particularly efficient waveform encoding techniques. With CVSD, it is possible to operate down to 9600 bits/s with some noticeable waveform distortion. In fact, at rates below 16,000 bits/s, the distortion produced by waveform encoders increases significantly. Consequently, these techniques are not used below 9600 bits/s.

For rates below 9600 bits/s, encoding techniques, such as LPC, that are based on linear models of the source are usually employed. The synthesized speech obtained from this class of encoding techniques is intelligible. However, the speech signal has a synthetic quality and there is noticeable distortion.

TABLE 3.5–2
Encoding techniques applied to speech signals

Encoding method	Quantizer	Coder	Transmission rate (bits/s)
PCM	Linear	12 bits	96,000
Log PCM	Logarithmic	7–8 bits	56,000–64,000
DPCM	Logarithmic	4–6 bits	32,000–48,000
ADPCM	Adaptive	3–4 bits	24,000-32,000
DM	Binary	1 bit	32,000–64,000
ADM	Adaptive binary	1 bit	16,000–32,000
LPC/CELP			2400–9600

3.6 BIBLIOGRAPHICAL NOTES AND REFERENCES

Source coding has been an area of intense research activity since the publication of Shannon's classic papers in 1948 and the paper by Huffman (1952). Over the years, major advances have been made in the development of highly efficient source data compression algorithms. Of particular significance is the research on universal source coding and universal quantization published by Ziv (1985), Ziv and Lempel (1977, 1978), Davisson (1973), Gray (1975), and Davisson et al. (1981).

Treatments of rate-distortion theory are found in the books by Gallager (1968), Berger (1971), Viterbi and Omura (1979), Blahut (1987), and Gray (1990). Our treatment of rate-distortion functions was focused on speech processing applications. For practical applications of rate-distortion theory to image and video compression, the reader is referred to the *IEEE Signal Processing Magazine*, November 1998. The paper by Berger and Gibson (1998) on lossy source coding provides an overview of the major developments on this topic over the past 50 years.

Much work has been done over the past several decades on speech encoding methods. Our treatment provides an overview of this important topic. A more comprehensive treatment is given in the books by Rabiner and Schafer (1978), Jayant and Noll (1984), and Deller et al. (2000). In addition to these texts, there have been special issues of the *IEEE Transactions on Communications* (April 1979 and April 1982) and the *IEEE Journal on Selected Areas in Communications* (February 1988) devoted to speech encoding. We should also mention the publication by IEEE Press of a book containing reprints of published papers on waveform quantization and coding, edited by Jayant (1976), and the more recent survey paper by Jayant (1990).

Over the past decade, we have also seen a number of important developments in vector quantization. Our treatment of this topic was based on the tutorial paper by Makhoul et al. (1985). A comprehensive treatment of vector quantization and signal compression is provided in the book of Gersho and Gray (1992). The survey paper by Gray and Neuhoff (1998) describes the numerous advances that have been made on the topic of quantization over the past 50 years and includes a list of over 500 references.

PROBLEMS

3.1 Consider the joint experiment described in Problem 2.1 with the given joint probabilities $P(A_i, B_j)$. Suppose we observe the outcomes A_i, $i = 1, 2, 3, 4$, of experiment A.

a) Determine the mutual information $I(B_j; A_i)$ for $j = 1, 2, 3$ and $i = 1, 2, 3, 4$, in bits.

b) Determine the average mutual information $I(B; A)$.

3.2 Suppose the outcomes B_j, $j = 1, 2, 3$, in Problem 3.1 represent the three possible output letters from the DMS. Determine the entropy of the source.

3.3 Prove that $\ln u \leqslant u - 1$ and also demonstrate the validity of this inequality by plotting $\ln u$ and $u - 1$ on the same graph.

3.4 X and Y are two discrete random variables with probabilities

$$P(X = x, Y = y) \equiv P(x, y)$$

Show that $I(X; Y) \geqslant 0$, with equality if and only if X and Y are statistically independent.

[*Hint:* Use the inequality $\ln u \leqslant u - 1$, for $0 < u < 1$, to show that $-I(X; Y) \leqslant 0$.]

3.5 The output of a DMS consists of the possible letters $x_1, x_2, \ldots, x_n$, which occur with probabilities $p_1, p_2, \ldots, p_n$, respectively. Prove that the entropy $H(X)$ of the source is at most $\log n$.

3.6 Determine the differential entropy $H(X)$ of the uniformly distributed random variable X with PDF

$$p(x) = \begin{cases} a^{-1} & (0 \leqslant x \leqslant a) \\ 0 & (\text{otherwise}) \end{cases}$$

for the following three cases:

a) $a = 1$.

b) $a = 4$.

c) $a = \frac{1}{4}$.

Observe from these results that $H(X)$ is not an absolute measure, but only a relative measure of randomness.

3.7 A DMS has an alphabet of eight letters, x_i, $i = 1, 2, \ldots, 8$, with probabilities 0.25, 0.20, 0.15, 0.12, 0.10, 0.08, 0.05, and 0.05.

a) Use the Huffman encoding procedure to determine a binary code for the source output.

b) Determine the average number $\bar{R}$ of binary digits per source letter.
c) Determine the entropy of the source and compare it with $\bar{R}$.

3.8 A DMS has an alphabet of five letters, x_i, $i = 1, 2, \ldots, 5$, each occurring with probability $\frac{1}{5}$. Evaluate the efficiency of a fixed-length binary code in which
a) Each letter is encoded separately into a binary sequence.
b) Two letters at a time are encoded into a binary sequence.
c) Three letters at a time are encoded into a binary sequence.

3.9 Recall Equation 3.2–6:

$$I(x_i; y_j) = I(x_i) - I(x_i|y_j)$$

Prove that
a) $I(x_i; y_j) = I(y_j) - I(y_j|x_i)$.
b) $I(x_i; y_j) = I(x_i) + I(y_j) - I(x_i y_j)$, where $I(x_i y_j) = -\log P(x_i, y_j)$.

3.10 Let X be a geometrically distributed random variable; that is,

$$P(X = k) = p(1-p)^{k-1}, \qquad k = 1, 2, 3, \ldots$$

a) Find the entropy of X.
b) Knowing that $X > K$, where K is a positive integer, what is the entropy of X?

3.11 Let X and Y denote two jointly distributed discrete valued random variables.
a) Show that

$$H(X) = -\sum_{x,y} P(x, y) \log P(x)$$

$$H(Y) = -\sum_{x,y} P(x, y) \log P(y)$$

b) Use the above result to show that

$$H(X, Y) \leqslant H(X) + H(Y)$$

When does equality hold?
c) Show that

$$H(X|Y) \leqslant H(X)$$

with equality if and only if X and Y are independent.

3.12 Two binary random variables X and Y are distributed according to the joint distributions $P(X = Y = 0) = P(X = 0, Y = 1) = P(X = Y = 1) = \frac{1}{3}$. Compute $H(X)$, $H(Y)$, $H(X|Y)$, $H(Y|X)$, and $H(X, Y)$.

3.13 A Markov process is a process with one-step memory, i.e., a process such that

$$p(x_n|x_{n-1}, x_{n-2}, x_{n-3}, \ldots) = p(x_n|x_{n-1})$$

for all n. Show that, for a stationary Markov process, the entropy rate is given by $H(X_n|X_{n-1})$.

3.14 Let $Y = g(X)$, where g denotes a deterministic function. Show that, in general, $H(Y) \leqslant H(X)$. When does equality hold?

3.15 Show that $I(X;Y) = H(X) + H(Y) - H(XY)$.

3.16 Show that, for statistically independent events,

$$H(X_1 X_2 \cdots X_n) = \sum_{i=1}^{n} H(X_i)$$

3.17 For a noiseless channel, show that $H(X|Y) = 0$.

3.18 Show that

$$I(X_3; X_2|X_1) = H(X_3|X_1) - H(X_3|X_1 X_2)$$

and that

$$H(X_3|X_1) \geqslant H(X_3|X_1 X_2)$$

3.19 Let X be a random variable with PDF $p_x(x)$ and let $Y = aX + b$ be a linear transformation of X, where a and b are two constants. Determine the differential entropy $H(Y)$ in terms of $H(X)$.

3.20 The outputs x_1, x_2, and x_3 of a DMS with corresponding probabilities $p_1 = 0.45$, $p_2 = 0.35$, $p_3 = 0.20$ are transformed by the linear transformation $Y = aX + b$, where a and b are constants. Determine the entropy $H(Y)$ and comment on what effect the transformation has had on the entropy of X.

3.21 The optimum four-level nonuniform quantizer for a Gaussian-distributed signal amplitude results in the four levels a_1, a_2, a_3, and a_4, with corresponding probabilities of occurrence $p_1 = p_2 = 0.3365$ and $p_3 = p_4 = 0.1635$.

a) Design a Huffman code that encodes a single level at a time and determine the average bit rate.

b) Design a Huffman code that encodes two output levels at a time and determine the average bit rate.

c) What is the minimum rate obtained by encoding J output levels at a time as $J \to \infty$?

3.22 A first-order Markov source is characterized by the state probabilities $P(x_i)$, $i = 1, 2, \ldots, L$, and the transition probabilities $P(x_k|x_i)$, $k = 1, 2, \ldots, L$, and $k \neq i$ The entropy of the Markov source is

$$H(X) = \sum_{k=1}^{L} P(x_k) H(X|x_k)$$

where $H(X|x_k)$ is the entropy conditioned on the source being in state x_k.

Determine the entropy of the binary, first-order Markov source shown in Figure P3.22, which has the transition probabilities $P(x_2|x_1) = 0.2$ and $P(x_1|x_2) = 0.3$. {Note that the conditional entropies $H(X|x_1)$ and $H(X|x_2)$ are given by the binary entropy functions $H[P(x_2|x_1)]$ and $H[P(x_1|x_2)]$, respectively.} How does the entropy of the Markov source compare with the entropy of a binary DMS with the same output letter probabilities $P(x_1)$ and $P(x_2)$?

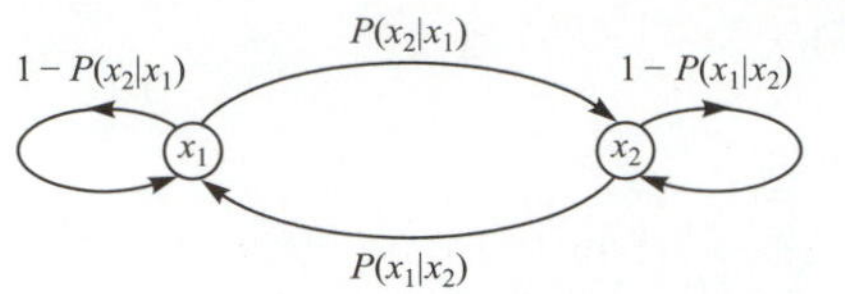

$1-P(x_2|x_1)$ $P(x_2|x_1)$ $1-P(x_1|x_2)$

x_1 x_2

$P(x_1|x_2)$

FIGURE P3.22

3.23 A memoryless source has the alphabet $\mathcal{A} = \{-5, -3, -1, 0, 1, 3, 5\}$, with corresponding probabilities $\{0.05, 0.1, 0.1, 0.15, 0.05, 0.25, 0.3\}$.

a) Find the entropy of the source.

b) Assuming that the source is quantized according to the quantization rule

$$\begin{aligned} q(-5) &= q(-3) = 4 \\ q(-1) &= q(0) = q(1) = 0 \\ q(3) &= q(5) = 4 \end{aligned}$$

find the entropy of the quantized source.

3.24 Design a *ternary* Huffman code, using 0, 1, and 2 as letters, for a source with output alphabet probabilities given by $\{0.05, 0.1, 0.15, 0.17, 0.18, 0.22, 0.13\}$. What is the resulting average code-word length? Compare the average code-word length with the entropy of the source. (In what base would you compute the logarithms in the expression for the entropy for a meaningful comparison?)

3.25 Find the Lempel–Ziv source code for the binary source sequence

000100100000011000010000000100000010100001000000110100000001100

Recover the original sequence back from the Lempel–Ziv source code.
[*Hint*: You require two passes of the binary sequence to decide on the size of the dictionary.]

3.26 Find the differential entropy of the continuous random variable X in the following cases:

a) X is an exponential random variable with parameter $\lambda > 0$, i.e.,

$$f_X(x) = \begin{cases} \lambda^{-1} e^{-x/\lambda} & (x>0) \\ 0 & (\text{otherwise}) \end{cases}$$

b) X is a Laplacian random variable with parameter $\lambda > 0$, i.e.,

$$f_X(x) = \frac{1}{2\lambda} e^{-|x|/\lambda}$$

c) X is a triangular random variable with parameter $\lambda > 0$, i.e.,

$$p_X(x) = \begin{cases} (x+\lambda)/\lambda^2 & (-\lambda \leqslant x \leqslant 0) \\ (-x+\lambda)/\lambda^2 & (0 < x \leqslant \lambda) \\ 0 & (\text{otherwise}) \end{cases}$$

3.27 It can be shown that the rate-distortion function for a Laplacian source, $p_X(x) = (2\lambda)^{-1} e^{-|x|/\lambda}$ with an absolute value of error-distortion measure $d(x, \hat{x}) = |x - \hat{x}|$ is given by

$$R(D) = \begin{cases} \log(\lambda/D) & (0 \leqslant D \leqslant \lambda) \\ 0 & (D>\lambda) \end{cases}$$

(see Berger, 1971).

a) How many bits per sample are required to represent the outputs of this source with an average distortion not exceeding $\frac{1}{2}\lambda$?

b) Plot $R(D)$ for three different values of λ and discuss the effect of changes in λ on these plots.

3.28 It can be shown that if X is a zero-mean continuous random variable with variance σ^2, its rate-distortion function, subject to squared error distortion measure, satisfies the lower and upper bounds given by the inequalities

$$H(X) - \tfrac{1}{2}\log 2\pi e D \leqslant R(D) \leqslant \tfrac{1}{2}\log\tfrac{1}{2}\sigma^2$$

where $H(X)$ denotes the differential entropy of the random variable X (see Cover and Thomas, 1991).

a) Show that, for a Gaussian random variable, the lower and upper bounds coincide.

b) Plot the lower and upper bounds for a Laplacian source with $\sigma = 1$.

c) Plot the lower and upper bounds for a triangular source with $\sigma = 1$.

3.29 A stationary random process has an autocorrelation function given by $R_X(\tau) = \frac{1}{2}A^2 e^{-|\tau|}\cos 2\pi f_0 \tau$ and it is known that the random process never exceeds 6 in magnitude. Assuming $A = 6$, how many quantization levels are required to guarantee a signal-to-quantization noise ratio of at least 60 dB?

3.30 An additive white Gaussian noise channel has the output $Y = X + G$, where X is the channel input and G is the noise with probability density function

$$p(n) = \frac{1}{\sqrt{2\pi}\sigma_n} = e^{-n^2/2\sigma_n^2}$$

If X is a white Gaussian input with $E(X) = 0$ and $E(X^2) = \sigma_x^2$, determine

a) The conditional differential entropy $H(X|G)$.

b) The average mutual information $I(X; Y)$.

3.31 A DMS has an alphabet of eight letters, x_i, $i = 1, 2, \ldots, 8$, with probabilities given in Problem 3.7. Use the Huffman encoding procedure to determine a ternary code (using symbols 0, 1, and 2) for encoding the source output.
[*Hint*: Add a symbol x_9 with probability $p_9 = 0$, and group three symbols at a time.]

3.32 Determine whether there exists a binary code with code-word lengths $(n_1, n_2, n_3, n_4) = (1, 2, 2, 3)$ that satisfy the prefix condition.

3.33 Consider a binary block code with 2^n code words of the same length n. Show that the Kraft inequality is satisfied for such a code.

3.34 Show that the entropy of an n-dimensional Gaussian vector $\mathbf{X} = [x_1\ x_2 \cdots x_n]$ with zero mean and covariance matrix $\mathbf{M}$ is

$$H(\mathbf{X}) = \tfrac{1}{2}\log_2(2\pi e)^n|\mathbf{M}|$$

3.35 Consider a DMS with output bits (0, 1) that are equiprobable. Define the distortion measure as $D = P_e$, where P_e is the probability of error in transmitting the binary symbols to the user over a BSC. Then the rate-distortion function is (Berger, 1971)

$$R(D) = 1 + D\log_2 D + (1 - D)\log_2(1 - D), \qquad 0 \leqslant D = P_e \leqslant \tfrac{1}{2}$$

Plot $R(D)$ for $0 \leqslant D \leqslant \frac{1}{2}$.

3.36 Evaluate the rate-distortion function for an M-ary symmetric channel given as

$$R(D) = \log_2 M + D\log_2 D + (1 - D)\log_2 \frac{1 - D}{M - 1}$$

for $M = 2, 4, 8$, and 16; where the distortion is the probability of error.

3.37 Consider the use of the weighted mean-square-error (MSE) distortion measure defined as

$$d_w(\mathbf{X}, \tilde{\mathbf{X}}) = (\mathbf{X} - \tilde{\mathbf{X}})'\mathbf{W}(\mathbf{X} - \tilde{\mathbf{X}})$$

where $\mathbf{W}$ is a symmetric, positive-definitive weighting matrix. By factorizing $\mathbf{W}$ as $\mathbf{W} = \mathbf{P}'\mathbf{P}$, show that $d_w(\mathbf{X}, \tilde{\mathbf{X}})$ is equivalent to an unweighted MSE distortion measure $d_2(\mathbf{X}', \tilde{\mathbf{X}}')$ involving transformed vectors $\mathbf{X}'$ and $\tilde{\mathbf{X}}'$.

3.38 Consider a stationary stochastic signal sequence $\{X(n)\}$ with zero mean and autocorrelation sequence

$$\phi(n) = \begin{cases} 1 & (n=0) \\ \frac{1}{2} & (n=\pm 1) \\ 0 & (\text{otherwise}) \end{cases}$$

a) Determine the prediction coefficient of the first-order minimum MSE predictor for $\{X(n)\}$ given by

$$\hat{x}(n) = a_1 x(n - 1)$$

and the corresponding minimum mean square error $\mathcal{E}_1$.

b) Repeat (*a*) for the second-order predictor

$$\hat{x}(n) = a_1 x(n - 1) + a_2 x(n - 2)$$

3.39 Consider the encoding of the random variables x_1 and x_2 that are characterized by the joint PDF $p(x_1, x_2)$ given by

$$p(x_1, x_2) = \begin{cases} 15/7ab & (x_1, x_2 \in C) \\ 0 & (\text{otherwise}) \end{cases}$$

as shown in Figure P3.39. Evaluate the bit rates required for uniform quantization of x_1 and x_2 separately (scalar quantization) and combined (vector) quantization of (x_1, x_2). Determine the difference in bit rate when $a = 4b$.

3.40 Consider the encoding of two random variables X and Y that are uniformly distributed on the region between two squares as shown in Figure P3.40.

a) Find $p_X(x)$ and $p_Y(y)$.

b) Assume that each of the random variables X and Y are quantized using four-level uniform quantizers. What is the resulting distortion? What is the resulting number of bits per (X, Y) pair?

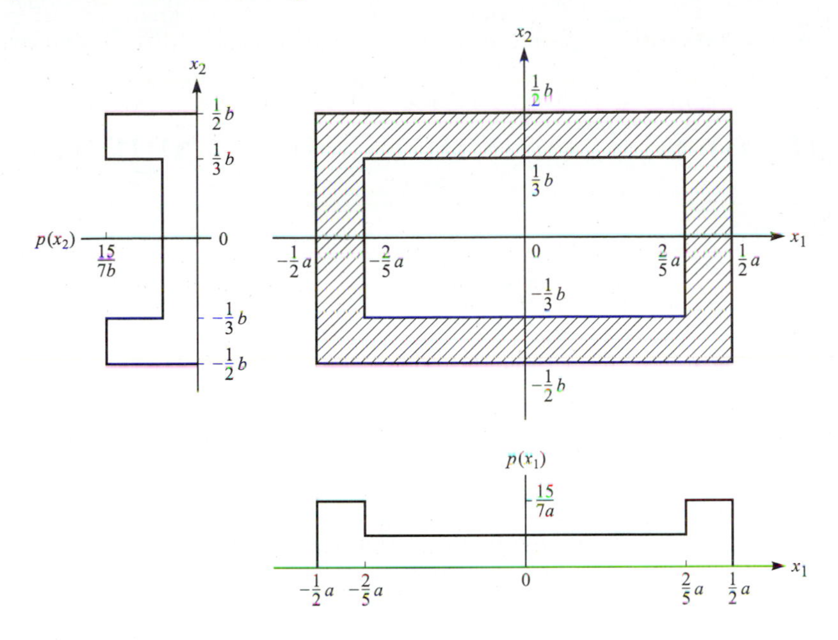

FIGURE P3.39

c) Now assume that instead of scalar quantizers for X and Y, we employ a vector quantizer to achieve the same level of distortion as in (*b*). What is the resulting number of bits per source output pair (X, Y)?

3.41 Two random variables X and Y are uniformly distributed on the square shown in Figure P3.41.

a) Find $p_X(x)$ and $p_Y(y)$.

b) Assume that each of the random variables X and Y are quantized using four-level uniform quantizers. What is the resulting distortion? What is the resulting number of bits per (X, Y) pair?

c) Now assume that, instead of scalar quantizers for X and Y, we employ a vector quantizer with the same number of bits per source output pair (X, Y) as in (*b*). What is the resulting distortion for this vector quantizer?

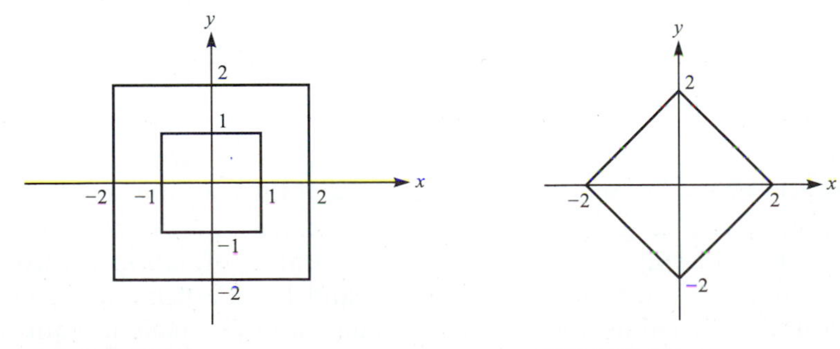

FIGURE P3.40 **FIGURE P3.41**

4

Characterization of Communication Signals and Systems

Signals can be categorized in a number of different ways, such as random versus deterministic, discrete time versus continuous time, discrete amplitude versus continuous amplitude, low-pass versus band-pass, finite energy versus infinite energy, finite average power versus infinite average power. In this chapter, we treat the characterization of signals and systems that are usually encountered in the transmission of digital information over a communication channel. In particular, we introduce the representation of various forms of digitally modulated signals and describe their spectral characteristics.

We begin with the characterization of band-pass signals and systems, including the mathematical representation of band-pass stationary stochastic processes. Then, we present a vector space representation of signals. We conclude with the representation of digitally modulated signals and their spectral characteristics.

4.1
REPRESENTATION OF BAND-PASS SIGNALS AND SYSTEMS

Many digital information-bearing signals are transmitted by some type of carrier modulation. The channel over which the signal is transmitted is limited in bandwidth to an interval of frequencies centered about the carrier, as in double-sideband modulation, or adjacent to the carrier, as in single-sideband modulation. Signals and channels (systems) that satisfy the condition that their bandwidth is much smaller than the carrier frequency are termed *narrowband band-pass signals and channels (systems)*. The modulation performed at the transmitting end of the communication system to generate the band-pass signal and the demodulation performed at the receiving end to recover the digital information involve frequency translations. With no loss of generality and for mathematical convenience, it is desirable to reduce all band-pass signals and channels to equivalent low-pass signals and channels. As a consequence, the results of the performance of the various modulation and demodulation techniques presented in the subsequent chapters are independent of carrier frequencies and channel frequency bands. The representation of band-pass signals and systems in terms of equivalent low-pass waveforms and the characterization of band-pass stationary stochastic processes are the main topics of this section.

4.1.1 Representation of Band-Pass Signals

Suppose that a real-valued signal $s(t)$ has a frequency content concentrated in a narrow band of frequencies in the vicinity of a frequency f_c, as shown in Figure 4.1–1. Our objective is to develop a mathematical representation of such signals. First, we construct a signal that contains only the positive frequencies in $s(t)$. Such a signal may be expressed as

$$S_+(f) = 2u(f)S(f) \tag{4.1–1}$$

where $S(f)$ is the Fourier transform of $s(t)$ and $u(f)$ is the unit step function. The equivalent time-domain expression for Equation 4.1–1 is

$$\begin{aligned} s_+(t) &= \int_{-\infty}^{\infty} S_+(f)e^{j2\pi ft}\,df \\ &= F^{-1}[2u(f)] \star F^{-1}[S(f)] \end{aligned} \tag{4.1–2}$$

The signal $s_+(t)$ is called the *analytic signal* or the *pre-envelope* of $s(t)$. We note that $F^{-1}[S(f)] = s(t)$ and

$$F^{-1}[2u(f)] = \delta(t) + \frac{j}{\pi t} \tag{4.1–3}$$

Hence,

$$\begin{aligned} s_+(t) &= \left[\delta(t) + \frac{j}{\pi t}\right] \star s(t) \\ &= s(t) + j\frac{1}{\pi t} \star s(t) \end{aligned} \tag{4.1–4}$$

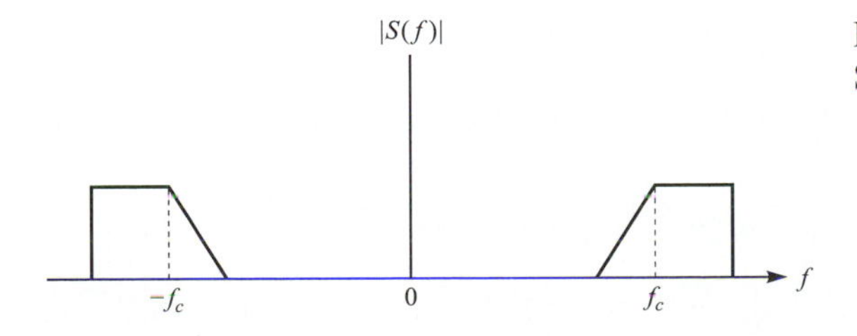

FIGURE 4.1–1
Spectrum of a band-pass signal.

We define $\hat{s}(t)$ as

$$\begin{aligned}\hat{s}(t) &= \frac{1}{\pi t} \star s(t) \\ &= \frac{1}{\pi}\int_{-\infty}^{\infty} \frac{s(\tau)}{t-\tau}\, d\tau\end{aligned} \tag{4.1–5}$$

The signal $\hat{s}(t)$ may be viewed as the output of the filter with impulse response

$$h(t) = \frac{1}{\pi t}, \qquad -\infty < t < \infty \tag{4.1–6}$$

when excited by the input signal $s(t)$. Such a filter is called a *Hilbert transformer*. The frequency response of this filter is simply

$$\begin{aligned}H(f) &= \int_{-\infty}^{\infty} h(t)e^{-j2\pi ft}\, dt \\ &= \frac{1}{\pi}\int_{-\infty}^{\infty} \frac{1}{t} e^{-j2\pi ft}\, dt \\ &= \begin{cases} -j & (f > 0) \\ 0 & (f = 0) \\ j & (f < 0) \end{cases}\end{aligned} \tag{4.1–7}$$

We observe that $|H(f)| = 1$ and that the phase response $\Theta(f) = -\frac{1}{2}\pi$ for $f > 0$ and $\Theta(f) = \frac{1}{2}\pi$ for $f < 0$. Therefore, this filter is basically a 90° phase shifter for all frequencies in the input signal.

The analytic signal $s_+(t)$ is a band-pass signal. We may obtain an equivalent low-pass representation by performing a frequency translation of $S_+(f)$. Thus, we define $S_l(f)$ as

$$S_l(f) = S_+(f + f_c) \tag{4.1–8}$$

The equivalent time-domain relation is

$$\begin{aligned}s_l(t) &= s_+(t)e^{-j2\pi f_c t} \\ &= [s(t) + j\hat{s}(t)]e^{-j2\pi f_c t}\end{aligned} \tag{4.1–9}$$

or, equivalently,

$$s(t) + j\hat{s}(t) = s_l(t)e^{j2\pi f_c t} \tag{4.1–10}$$

In general, the signal $s_l(t)$ is complex-valued (see Problem 4.5) and may be expressed as

$$s_l(t) = x(t) + jy(t) \tag{4.1–11}$$

If we substitute for $s_l(t)$ in Equation 4.1–11 and equate real and imaginary parts on each side, we obtain the relations

$$s(t) = x(t)\cos 2\pi f_c t - y(t)\sin 2\pi f_c t \tag{4.1–12}$$

$$\hat{s}(t) = x(t)\sin 2\pi f_c t + y(t)\cos 2\pi f_c t \tag{4.1–13}$$

The expression 4.1–12 is the desired form for the representation of a band-pass signal. The low-frequency signal components $x(t)$ and $y(t)$ may be viewed as amplitude modulations impressed on the carrier components $\cos 2\pi f_c t$ and $\sin 2\pi f_c t$, respectively. Since these carrier components are in phase quadrature, $x(t)$ and $y(t)$ are called the *quadrature components* of the band-pass signal $s(t)$.

Another representation of the signal in Equation 4.1–12 is

$$\begin{aligned} s(t) &= \text{Re}\{[x(t) + jy(t)]e^{j2\pi f_c t}\} \\ &= \text{Re}[s_l(t)e^{j2\pi f_c t}] \end{aligned} \tag{4.1–14}$$

where Re denotes the real part of the complex-valued quantity in the brackets following. The low-pass signal $s_l(t)$ is usually called the *complex envelope* of the real signal $s(t)$ and is basically the *equivalent low-pass signal*.

Finally, a third possible representation of a band-pass signal is obtained by expressing $s_l(t)$ as

$$s_l(t) = a(t)e^{j\theta(t)} \tag{4.1–15}$$

where

$$a(t) = \sqrt{x^2(t) + y^2(t)} \tag{4.1–16}$$

$$\theta(t) = \tan^{-1}\frac{y(t)}{x(t)} \tag{4.1–17}$$

Then

$$\begin{aligned} s(t) &= \text{Re}[s_l(t)e^{j2\pi f_c t}] \\ &= \text{Re}[a(t)e^{j[2\pi f_c t + \theta(t)]}] \\ &= a(t)\cos[2\pi f_c t + \theta(t)] \end{aligned} \tag{4.1–18}$$

The signal $a(t)$ is called the *envelope* of $s(t)$, and $\theta(t)$ is called the *phase* of $s(t)$. Therefore, Equations 4.1–12, 4.1–14, and 4.1–18 are equivalent representations of band-pass signals.

The Fourier transform of $s(t)$ is

$$S(f) = \int_{-\infty}^{\infty} s(t)e^{-j2\pi ft}\,dt = \int_{-\infty}^{\infty} \{\text{Re}[s_l(t)e^{j2\pi f_c t}]\}e^{-j2\pi ft}\,dt \tag{4.1–19}$$

Use of the identity

$$\text{Re}(\xi) = \tfrac{1}{2}(\xi + \xi^*) \tag{4.1–20}$$

in Equation 4.1–19 yields the result

$$\begin{aligned} S(f) &= \frac{1}{2}\int_{-\infty}^{\infty} [s_l(t)e^{j2\pi f_c t} + s_l^*(t)e^{-j2\pi f_c t}]e^{-j2\pi ft}\,dt \\ &= \tfrac{1}{2}[S_l(f - f_c) + S_l^*(-f - f_c)] \end{aligned} \tag{4.1–21}$$

where $S_l(f)$ is the Fourier transform of $s_l(t)$. This is the basic relationship between the spectrum of the real band-pass signal $S(f)$ and the spectrum of the equivalent low-pass signal $S_l(f)$.

The energy in the signal $s(t)$ is defined as

$$\begin{aligned}\mathcal{E} &= \int_{-\infty}^{\infty} s^2(t)\,dt \\ &= \int_{-\infty}^{\infty} \{\mathrm{Re}[s_l(t)e^{j2\pi f_c t}]\}^2\,dt\end{aligned} \tag{4.1–22}$$

When the identity in Equation 4.1–20 is used in Equation 4.1–22, we obtain the following result:

$$\begin{aligned}\mathcal{E} =& \frac{1}{2}\int_{-\infty}^{\infty} |s_l(t)|^2\,dt \\ &+ \frac{1}{2}\int_{-\infty}^{\infty} |s_l(t)|^2 \cos[4\pi f_c t + 2\theta(t)]\,dt\end{aligned} \tag{4.1–23}$$

Consider the second integral in Equation 4.1–23. Since the signal $s(t)$ is narrow-band, the real envelope $a(t) \equiv |s_l(t)|$ or, equivalently, $a^2(t)$ varies slowly relative to the rapid variations exhibited by the cosine function. A graphical illustration of the integrand in the second integral of Equation 4.1–23 is shown in Figure 4.1–2. The value of the integral is just the net area under the cosine function modulated by $a^2(t)$. Since the modulating waveform $a^2(t)$ varies slowly relative to the cosine function, the net area contributed by the second integral is very small relative to the value of the first integral in Equation 4.1–23 and, hence, it can be neglected. Thus, for all practical purposes, the energy in the band-pass signal $s(t)$, expressed in terms of the equivalent low-pass signal $s_l(t)$, is

$$\mathcal{E} = \frac{1}{2}\int_{-\infty}^{\infty} |s_l(t)|^2\,dt \tag{4.1–24}$$

where $|s_l(t)|$ is just the envelope $a(t)$ of $s(t)$.

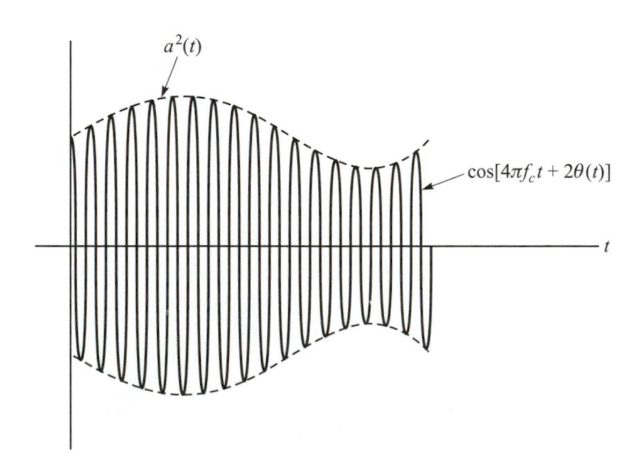

FIGURE 4.1–2
The signal $a^2(t)\cos[4\pi f_c t + 2\theta(t)]$.

4.1.2 Representation of Linear Band-Pass Systems

A linear filter or system may be described either by its impulse response $h(t)$ or by its frequency response $H(f)$, which is the Fourier transform of $h(t)$. Since $h(t)$ is real,

$$H^*(-f) = H(f) \tag{4.1–25}$$

Let us define $H_l(f - f_c)$ as

$$H_l(f - f_c) = \begin{cases} H(f) & (f > 0) \\ 0 & (f < 0) \end{cases} \tag{4.1–26}$$

Then

$$H_l^*(-f - f_c) = \begin{cases} 0 & (f > 0) \\ H^*(-f) & (f < 0) \end{cases} \tag{4.1–27}$$

Using Equation 4.1–25, we have

$$H(f) = H_l(f - f_c) + H_l^*(-f - f_c) \tag{4.1–28}$$

which resembles Equation 4.1–21 except for the factor $\frac{1}{2}$. The inverse transform of $H(f)$ in Equation 4.1–28 yields $h(t)$ in the form

$$\begin{aligned} h(t) &= h_l(t)e^{j2\pi f_c t} + h_l^*(t)e^{-j2\pi f_c t} \\ &= 2\,\mathrm{Re}[h_l(t)e^{j2\pi f_c t}] \end{aligned} \tag{4.1–29}$$

where $h_l(t)$ is the inverse Fourier transform of $H_l(f)$. In general, the impulse response $h_l(t)$ of the equivalent low-pass system is complex-valued.

4.1.3 Response of a Band-Pass System to a Band-Pass Signal

In Sections 4.1.1 and 4.1.2, we have shown that narrowband band-pass signals and systems can be represented by equivalent low-pass signals and systems. In this section, we demonstrate that the output of a band-pass system to a band-pass input signal is simply obtained from the equivalent low-pass input signal and the equivalent low-pass impulse response of the system.

Suppose that $s(t)$ is a narrowband band-pass signal and $s_l(t)$ is the equivalent low-pass signal. This signal excites a narrowband band-pass system characterized by its band-pass impulse response $h(t)$ or by its equivalent low-pass impulse response $h_l(t)$. The output of the band-pass system is also a band-pass signal, and, therefore, it can be expressed in the form

$$r(t) = \mathrm{Re}[r_l(t)e^{j2\pi f_c t}] \tag{4.1–30}$$

where $r(t)$ is related to the input signal $s(t)$ and the impulse response $h(t)$ by the convolution integral

$$r(t) = \int_{-\infty}^{\infty} s(\tau)h(t-\tau)\,d\tau \tag{4.1–31}$$

Equivalently, the output of the system, expressed in the frequency domain, is

$$R(f) = S(f)H(f) \tag{4.1–32}$$

Substituting from Equation 4.1–21 for $S(f)$ and from Equation 4.1–28 for $H(f)$, we obtain the result

$$R(f) = \tfrac{1}{2}[S_l(f-f_c) + S_l^*(-f-f_c)][H_l(f-f_c) + H_l^*(-f-f_c)] \tag{4.1–33}$$

When $s(t)$ is a narrowband signal and $h(t)$ is the impulse response of a narrowband system, $S_l(f-f_c) \approx 0$ and $H_l(f-f_c) = 0$ for $f < 0$. It follows from this narrowband condition that

$$S_l(f-f_c)H_l^*(-f-f_c) = 0, \qquad S_l^*(-f-f_c)H_l(f-f_c) = 0$$

Therefore, Equation 4.1–33 simplifies to

$$\begin{aligned} R(f) &= \tfrac{1}{2}[S_l(f-f_c)H_l(f-f_c) + S_l^*(-f-f_c)H_l^*(-f-f_c)] \\ &= \tfrac{1}{2}[R_l(f-f_c) + R_l^*(-f-f_c)] \end{aligned} \tag{4.1–34}$$

where

$$R_l(f) = S_l(f)H_l(f) \tag{4.1–35}$$

is the output spectrum of the equivalent low-pass system excited by the equivalent low-pass signal. It is clear that the time-domain relation for the output $r_l(t)$ is given by the convolution of $s_l(t)$ with $h_l(t)$. That is,

$$r_l(t) = \int_{-\infty}^{\infty} s_l(\tau)h_l(t-\tau)\,d\tau \tag{4.1–36}$$

The combination of Equation 4.1–36 with Equation 4.1–30 gives the relationship between the band-pass output signal $r(t)$ and the equivalent low-pass time functions $s_l(t)$ and $h_l(t)$. This simple relationship allows us to ignore any linear frequency translations encountered in the modulation of a signal for purposes of matching its spectral content to the frequency allocation of a particular channel. Thus, for mathematical convenience, we shall deal only with the transmission of equivalent low-pass signals through equivalent low-pass channels.

4.1.4 Representation of Band-Pass Stationary Stochastic Processes

The representation of band-pass signals presented in Section 4.1.1 applied to deterministic signals. In this section, we extend the representation to sample functions of a band-pass stationary stochastic process. In particular, we derive the important relations between the correlation functions and power spectra of the band-pass signal and the correlation functions and power spectra of the equivalent low-pass signal.

Suppose that $n(t)$ is a sample function of a wide-sense stationary stochastic process with zero mean and power spectral density $\Phi_{nn}(f)$. The power spectral density is assumed to be zero outside of an interval of frequencies centered around $\pm f_c$, where f_c is termed the *carrier frequency*. The stochastic process $n(t)$ is said to be a *narrowband band-pass process* if the width of the spectral density is much smaller than f_c. Under this condition, a sample function of the process $n(t)$ can be represented by any of the three equivalent forms given in Section 4.1.1, namely,

$$n(t) = a(t)\cos[2\pi f_c t + \theta(t)] \tag{4.1–37}$$

$$= x(t)\cos 2\pi f_c t - y(t)\sin 2\pi f_c t \tag{4.1–38}$$

$$= \mathrm{Re}[z(t)e^{j2\pi f_c t}] \tag{4.1–39}$$

where $a(t)$ is the envelope and $\theta(t)$ is the phase of the real-valued signal, $x(t)$ and $y(t)$ are the quadrature components of $n(t)$, and $z(t)$ is called the *complex envelope of* $n(t)$.

Let us consider the form given by Equation 4.1–38 in more detail. First, we observe that if $n(t)$ is zero mean, then $x(t)$ and $y(t)$ must also have zero mean values. In addition, the stationarity of $n(t)$ implies that the autocorrelation and cross-correlation functions of $x(t)$ and $y(t)$ satisfy the following properties:

$$\phi_{xx}(\tau) = \phi_{yy}(\tau) \tag{4.1–40}$$

$$\phi_{xy}(\tau) = -\phi_{yx}(\tau) \tag{4.1–41}$$

That these two properties follow from the stationarity of $n(t)$ is now demonstrated. The autocorrelation function $\phi_{nn}(\tau)$ of $n(t)$ is

$$\begin{aligned} E[n(t)n(t+\tau)] &= E\{[x(t)\cos 2\pi f_c t - y(t)\sin 2\pi f_c t] \\ &\quad \times [x(t+\tau)\cos 2\pi f_c(t+\tau) \\ &\quad - y(t+\tau)\sin 2\pi f_c(t+\tau)]\} \\ &= \phi_{xx}(\tau)\cos 2\pi f_c t \cos 2\pi f_c(t+\tau) \\ &\quad + \phi_{yy}(\tau)\sin 2\pi f_c t \sin 2\pi f_c(t+\tau) \\ &\quad - \phi_{xy}(\tau)\sin 2\pi f_c t \cos 2\pi f_c(t+\tau) \\ &\quad - \phi_{yx}(\tau)\cos 2\pi f_c t \sin 2\pi f_c(t+\tau) \end{aligned} \tag{4.1–42}$$

Use of the trigonometric identities

$$\begin{aligned} \cos A \cos B &= \tfrac{1}{2}[\cos(A-B) + \cos(A+B)] \\ \sin A \sin B &= \tfrac{1}{2}[\cos(A-B) - \cos(A+B)] \\ \sin A \cos B &= \tfrac{1}{2}[\sin(A-B) + \sin(A+B)] \end{aligned} \tag{4.1–43}$$

in Equation 4.1–42 yields the result

$$
\begin{aligned}
E[n(t)n(t+\tau)] = {} & \tfrac{1}{2}[\phi_{xx}(\tau)+\phi_{yy}(\tau)]\cos 2\pi f_c\tau \\
& + \tfrac{1}{2}[\phi_{xx}(\tau)-\phi_{yy}(\tau)]\cos 2\pi f_c(2t+\tau) \\
& - \tfrac{1}{2}[\phi_{yx}(\tau)-\phi_{xy}(\tau)]\sin 2\pi f_c\tau \\
& - \tfrac{1}{2}[\phi_{yx}(\tau)+\phi_{xy}(\tau)]\sin 2\pi f_c(2t+\tau)
\end{aligned}
\qquad (4.1\text{–}44)
$$

Since $n(t)$ is stationary, the right-hand side of Equation 4.1–44 must be independent of t. But this condition can only be satisfied if Equations 4.1–40 and 4.1–41 hold. As a consequence, Equation 4.1–44 reduces to

$$
\phi_{nn}(\tau) = \phi_{xx}(\tau)\cos 2\pi f_c\tau - \phi_{yx}(\tau)\sin 2\pi f_c\tau \qquad (4.1\text{–}45)
$$

We note that the relation between the autocorrelation function $\phi_{nn}(\tau)$ of the band-pass process and the autocorrelation and cross-correlation functions $\phi_{xx}(\tau)$ and $\phi_{yx}(\tau)$ of the quadrature components is identical in form to Equation 4.1–38, which expresses the band-pass process in terms of the quadrature components.

The autocorrelation function of the equivalent low-pass process

$$
z(t) = x(t) + jy(t) \qquad (4.1\text{–}46)
$$

is defined as

$$
\phi_{zz}(\tau) = \tfrac{1}{2}E[z^*(t)z(t+\tau)] \qquad (4.1\text{–}47)
$$

Substituting Equation 4.1–46 into Equation 4.1–47 and performing the expectation operation, we obtain

$$
\phi_{zz}(\tau) = \tfrac{1}{2}[\phi_{xx}(\tau)+\phi_{yy}(\tau)-j\phi_{xy}(\tau)+j\phi_{yx}(\tau)] \qquad (4.1\text{–}48)
$$

Now if the symmetry properties given in Equations 4.1–40 and 4.1–41 are used in Equation 4.1–48, we obtain

$$
\phi_{zz}(\tau) = \phi_{xx}(\tau) + j\phi_{yx}(\tau) \qquad (4.1\text{–}49)
$$

which relates the autocorrelation function of the complex envelope to the autocorrelation and cross-correlation functions of the quadrature components. Finally, we incorporate the result given by Equation 4.1–49 into Equation 4.1–45, and we have

$$
\phi_{nn}(\tau) = \mathrm{Re}[\phi_{zz}(\tau)e^{j2\pi f_c\tau}] \qquad (4.1\text{–}50)
$$

Thus, the autocorrelation function $\phi_{nn}(\tau)$ of the band-pass stochastic process is uniquely determined from the autocorrelation function $\phi_{zz}(\tau)$ of the equivalent low-pass process $z(t)$ and the carrier frequency f_c.

The power density spectrum $\Phi_{nn}(f)$ of the stochastic proces $n(t)$ is the Fourier transform of $\phi_{nn}(\tau)$. Hence,

$$\begin{aligned}\Phi_{nn}(f) &= \int_{-\infty}^{\infty} \{Re[\phi_{zz}(\tau)e^{j2\pi f_c \tau}]\} e^{-j2\pi f \tau} d\tau \\ &= \tfrac{1}{2}[\Phi_{zz}(f - f_c) + \Phi_{zz}(-f - f_c)]\end{aligned} \tag{4.1–51}$$

where $\Phi_{zz}(f)$ is the power density spectrum of the equivalent low-pass process $z(t)$. Since the autocorrelation function of $z(t)$ satisfies the property $\phi_{zz}(\tau) = \phi_{zz}^*(-\tau)$, it follows that $\Phi_{zz}(f)$ is a real-valued function of frequency.

Properties of the quadrature components. It was just demonstrated above that the cross-correlation function of the quadrature components $x(t)$ and $y(t)$ of the band-pass stationary stochastic process $n(t)$ satisfies the symmetry condition in Equation 4.1–41. Furthermore, any cross-correlation function satisfies the condition

$$\phi_{yx}(\tau) = \phi_{xy}(-\tau) \tag{4.1–52}$$

From these two conditions, we conclude that

$$\phi_{xy}(\tau) = -\phi_{xy}(-\tau) \tag{4.1–53}$$

That is, $\phi_{xy}(\tau)$ is an odd function of τ. Consequently, $\phi_{xy}(0) = 0$, and, hence, $x(t)$ and $y(t)$ are uncorrelated (for $\tau = 0$, only). Of course, this does not mean that the processes $x(t)$ and $y(t + \tau)$ are uncorrelated for all τ, since that would imply that $\phi_{xy}(\tau) = 0$ for all τ. If, indeed, $\phi_{xy}(\tau) = 0$ for all τ, then $\phi_{zz}(\tau)$ is real and the power spectral density $\Phi_{zz}(f)$ satisfies the condition

$$\Phi_{zz}(f) = \Phi_{zz}(-f) \tag{4.1–54}$$

and vice versa. That is, $\Phi_{zz}(f)$ is symmetric about $f = 0$.

In the special case in which the stationary stochastic process $n(t)$ is Gaussian, the quadrature components $x(t)$ and $y(t + \tau)$ are jointly Gaussian. Moreover, for $\tau = 0$, they are statistically independent, and, hence, their joint probability density function is

$$p(x, y) = \frac{1}{2\pi\sigma^2} e^{-(x^2+y^2)/2\sigma^2} \tag{4.1–55}$$

where the variance σ^2 is defined as $\sigma^2 = \phi_{xx}(0) = \phi_{yy}(0) = \phi_{nn}(0)$.

Representation of white noise. White noise is a stochastic process that is defined to have a flat (constant) power spectral density over the entire frequency range. This type of noise cannot be expressed in terms of quadrature components, as a result of its wideband character.

In problems concerned with the demodulation of narrowband signals in noise, it is mathematically convenient to model the additive noise process as white and to represent the noise in terms of quadrature components. This can be accomplished by postulating that the signals and noise at the receiving terminal have passed through an ideal band-pass filter, having a passband that includes the spectrum of the signals but is much wider. Such a filter will introduce

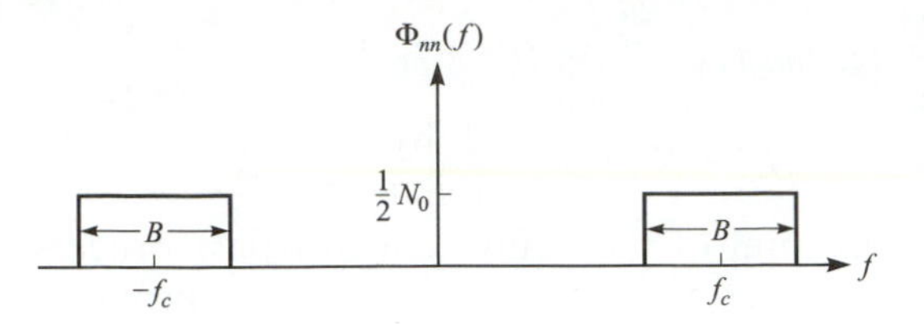

FIGURE 4.1–3
Band-pass noise with a flat spectrum.

negligible, if any, distortion on the signal, but it does eliminate the noise frequency components outside of the passband.

The noise resulting from passing the white noise process through a spectrally flat (ideal) band-pass filter is termed *band-pass white noise* and has the power spectral density depicted in Figure 4.1–3. Band-pass white noise can be represented by any of the forms given in Equations 4.1–37 to 4.1–39. The equivalent low-pass noise $z(t)$ has a power spectral density

$$\Phi_{zz}(f) = \begin{cases} N_0 & (|f| \leqslant \frac{1}{2}B) \\ 0 & (|f| > \frac{1}{2}B) \end{cases} \tag{4.1–56}$$

and its autocorrelation function is

$$\phi_{zz}(\tau) = N_0 \frac{\sin \pi B \tau}{\pi \tau} \tag{4.1–57}$$

The limiting form of $\phi_{zz}(\tau)$ as B approaches infinity is

$$\phi_{zz}(\tau) = N_0 \delta(\tau) \tag{4.1–58}$$

The power spectral density for white noise and band-pass white noise is symmetric about $f = 0$, so $\phi_{yx}(\tau) = 0$ for all τ. Therefore,

$$\phi_{zz}(\tau) = \phi_{xx}(\tau) = \phi_{yy}(\tau) \tag{4.1–59}$$

That is, the quadrature components $x(t)$ and $y(t)$ are uncorrelated for all time shifts τ and the autocorrelation functions of $z(t)$, $x(t)$, and $y(t)$ are all equal.

4.2
SIGNAL SPACE REPRESENTATIONS

In this section, we demonstrate that signals have characteristics that are similar to vectors and develop a vector representation for signal waveforms. We begin with some basic definitions and concepts involving vectors.

4.2.1 Vector Space Concepts

A vector $\mathbf{v}$ in an n-dimensional space is characterized by its n components $[v_1\ v_2 \cdots v_n]$. It may also be represented as a linear combination of *unit vectors* or *basis vectors* $\mathbf{e}_i$, $1 \leqslant i \leqslant n$, i.e.,

$$\mathbf{v} = \sum_{i=1}^{n} v_i \mathbf{e}_i \tag{4.2–1}$$

where, by definition, a unit vector has length unity and v_i is the projection of the vector $\mathbf{v}$ onto the unit vector $\mathbf{e}_i$.

The *inner product* of two n-dimensional vectors $\mathbf{v}_1 = [v_{11}\ v_{12} \cdots v_{1n}]$ and $\mathbf{v}_2 = [v_{21}\ v_{22} \cdots v_{2n}]$ is defined as

$$\mathbf{v}_1 \cdot \mathbf{v}_2 = \sum_{i=1}^{n} v_{1i} v_{2i} \tag{4.2–2}$$

Two vectors $\mathbf{v}_1$ and $\mathbf{v}_2$ are orthogonal if $\mathbf{v}_1 \cdot \mathbf{v}_2 = 0$. More generally, a set of m vectors $\mathbf{v}_k$, $1 \leqslant k \leqslant m$, are othogonal if

$$\mathbf{v}_i \cdot \mathbf{v}_j = 0 \tag{4.2–3}$$

for all $1 \leqslant i, j \leqslant m$, and $i \neq j$.

The *norm* of a vector $\mathbf{v}$ is denoted by $\|\mathbf{v}\|$ and is defined as

$$\|\mathbf{v}\| = (\mathbf{v} \cdot \mathbf{v})^{1/2} = \sqrt{\sum_{i=1}^{n} v_i^2} \tag{4.2–4}$$

which is simply its length. A set of m vectors is said to be *orthonormal* if the vectors are orthogonal and each vector has a unit norm. A set of m vectors is said to be *linearly independent* if no one vector can be represented as a linear combination of the remaining vectors.

Two n-dimensional vectors $\mathbf{v}_1$ and $\mathbf{v}_2$ satisfy the *triangle inequality*

$$\|\mathbf{v}_1 + \mathbf{v}_2\| \leqslant \|\mathbf{v}_1\| + \|\mathbf{v}_2\| \tag{4.2–5}$$

with equality if $\mathbf{v}_1$ and $\mathbf{v}_2$ are in the same direction, i.e., $\mathbf{v}_1 = a\mathbf{v}_2$ where a is a positive real scalar. From the triangle inequality there follows the *Cauchy–Schwarz inequality*

$$|\mathbf{v}_1 \cdot \mathbf{v}_2| \leqslant \|\mathbf{v}_1\| + \|\mathbf{v}_2\| \tag{4.2–6}$$

with equality if $\mathbf{v}_1 = a\mathbf{v}_2$. The norm square of the sum of two vectors may be expressed as

$$\|\mathbf{v}_1 + \mathbf{v}_2\|^2 = \|\mathbf{v}_1\|^2 + \|\mathbf{v}_2\|^2 + 2\mathbf{v}_1 \cdot \mathbf{v}_2 \tag{4.2–7}$$

If $\mathbf{v}_1$ and $\mathbf{v}_2$ are orthogonal, then $\mathbf{v}_1 \cdot \mathbf{v}_2 = 0$ and, hence,

$$\|\mathbf{v}_1 + \mathbf{v}_2\|^2 = \|\mathbf{v}_1\|^2 + \|\mathbf{v}_2\|^2 \tag{4.2–8}$$

This is the Pythagorean relation for two orthogonal n-dimensional vectors.

From matrix algebra, we recall that a linear transformation in an n-dimensional vector space is a matrix transformation of the form

$$\mathbf{v}' = \mathbf{A}\mathbf{v} \tag{4.2–9}$$

where the matrix $\mathbf{A}$ transforms the vector $\mathbf{v}$ into some vector $\mathbf{v}'$. In the special case where $\mathbf{v}' = \lambda\mathbf{v}$, i.e.,

$$\mathbf{A}\mathbf{v} = \lambda\mathbf{v} \tag{4.2–10}$$

where λ is some (positive or negative) scalar, the vector $\mathbf{v}$ is called an *eigenvector* of the transformation and λ is the corresponding *eigenvalue*.

Finally, let us review the Gram–Schmidt procedure for constructing a set of orthonormal vectors from a set of n-dimensional vectors $\mathbf{v}_i$, $1 \leqslant i \leqslant m$. We begin by arbitrarily selecting a vector from the set, say $\mathbf{v}_1$. By normalizing its length, we obtain the first vector, say

$$\mathbf{u}_1 = \frac{\mathbf{v}_1}{\|\mathbf{v}_1\|} \tag{4.2–11}$$

Next, we may select $\mathbf{v}_2$ and, first, subtract the projection of $\mathbf{v}_2$ onto $\mathbf{u}_1$. Thus, we obtain

$$\mathbf{u}_2' = \mathbf{v}_2 - (\mathbf{v}_2 \cdot \mathbf{u}_1)\mathbf{u}_1 \tag{4.2–12}$$

Then, we normalize the vector $\mathbf{u}_2'$ to unit length. This yields

$$\mathbf{u}_2 = \frac{\mathbf{u}_2'}{\|\mathbf{u}_2'\|} \tag{4.2–13}$$

The procedure continues by selecting $\mathbf{v}_3$ and subtracting the projections of $\mathbf{v}_3$ into $\mathbf{u}_1$ and $\mathbf{u}_2$. Thus, we have

$$\mathbf{u}_3' = \mathbf{v}_3 - (\mathbf{v}_3 \cdot \mathbf{u}_1)\mathbf{u}_1 - (\mathbf{v}_3 \cdot \mathbf{u}_2)\mathbf{u}_2 \tag{4.2–14}$$

Then, the orthonormal vector $\mathbf{u}_3$ is

$$\mathbf{u}_3 = \frac{\mathbf{u}_3'}{\|\mathbf{u}_3'\|} \tag{4.2–15}$$

By continuing this procedure, we construct a set of n_1 orthonormal vectors, where $n_1 \leqslant n$, in general. If $m < n$, then $n_1 \leqslant m$, and if $m \geqslant n$, then $n_1 \leqslant n$.

4.2.2 Signal Space Concepts

As in the case of vectors, we may develop a parallel treatment for a set of signals defined on some interval $[a, b]$. The *inner product* of two generally complex-valued signals $x_1(t)$ and $x_2(t)$ is denoted by $\langle x_1(t), x_2(t)\rangle$ and defined as

$$\langle x_1(t), x_2(t)\rangle = \int_a^b x_1(t)x_2^*(t)\,dt \tag{4.2–16}$$

The signals are orthogonal if their inner product is zero.

The *norm* of a signal is defined as

$$\|x(t)\| = \left(\int_a^b |x(t)|^2 \, dt \right)^{1/2} \tag{4.2–17}$$

A set of m signals are *orthonormal* if they are orthogonal and their norms are all unity. A set of m signals is *linearly independent*, if no signal can be represented as a linear combination of the remaining signals.

The *triangle inequality* for two signals is simply

$$\|x_1(t) + x_2(t)\| \leqslant \|x_1(t)\| + \|x_2(t)\| \tag{4.2–18}$$

and the *Cauchy–Schwarz inequality* is

$$\left| \int_a^b x_1(t) x_2^*(t) \, dt \right| \leqslant \left| \int_a^b |x_1(t)|^2 \, dt \right|^{1/2} \left| \int_a^b |x_2(t)|^2 \, dt \right|^{1/2} \tag{4.2–19}$$

with equality when $x_2(t) = a x_1(t)$, where a is any complex number.

4.2.3 Orthogonal Expansions of Signals

In this section, we develop a vector representation for signal waveforms, and, thus, we demonstrate an equivalence between a signal waveform and its vector representation.

Suppose that $s(t)$ is a deterministic, real-valued signal with finite energy

$$\mathcal{E}_s = \int_{-\infty}^{\infty} [s(t)]^2 \, dt \tag{4.2–20}$$

Furthermore, suppose that there exists a set of functions $\{f_n(t), \; n = 1, 2, \ldots, K\}$ that are orthonormal in the sense that

$$\int_{-\infty}^{\infty} f_n(t) f_m(t) \, dt = \begin{cases} 0 & (m \neq n) \\ 1 & (m = n) \end{cases} \tag{4.2–21}$$

We may approximate the signal $s(t)$ by a weighted linear combination of these functions, i.e.,

$$\hat{s}(t) = \sum_{k=1}^{K} s_k f_k(t) \tag{4.2–22}$$

where $\{s_k, \; 1 \leqslant k \leqslant K\}$ are the coefficients in the approximation of $s(t)$. The approximation error incurred is

$$e(t) = s(t) - \hat{s}(t) \tag{4.2–23}$$

Let us select the coefficients $\{s_k\}$ so as to minimize the energy $\mathcal{E}_e$ of the approximation error. Thus,

$$\mathcal{E}_e = \int_{-\infty}^{\infty} [s(t) - \hat{s}(t)]^2 \, dt$$
$$= \int_{-\infty}^{\infty} \left[s(t) - \sum_{k=1}^{K} s_k f_k(t) \right]^2 dt \tag{4.2–24}$$

The optimum coefficients in the series expansion of $s(t)$ may be found by differentiating Equation 4.2–24 with respect to each of the coefficients $\{s_k\}$ and setting the first derivatives to zero. Alternatively, we may use a well-known result from estimation theory based on the mean-square-error criterion, which, simply stated, is that the minimum of $\mathcal{E}_e$ with respect to the $\{s_k\}$ is obtained when the error is orthogonal to each of the functions in the series expansion. Thus,

$$\int_{-\infty}^{\infty} \left[s(t) - \sum_{k=1}^{K} s_k f_k(t) \right] f_n(t) \, dt = 0, \qquad n = 1, 2, \ldots, K \tag{4.2–25}$$

Since the functions $\{f_n(t)\}$ are orthonormal, Equation 4.2–25 reduces to

$$s_n = \int_{-\infty}^{\infty} s(t) f_n(t) \, dt, \qquad n = 1, 2, \ldots, K \tag{4.2–26}$$

Thus, the coefficients are obtained by projecting the signal $s(t)$ onto each of the functions $\{f_n(t)\}$. Consequently, $\hat{s}(t)$ is the projection of $s(t)$ onto the K-dimensional signal space spanned by the functions $\{f_n(t)\}$. The minimum mean square approximation error iss

$$\mathcal{E}_{\min} = \int_{-\infty}^{\infty} e(t) s(t) \, dt$$
$$= \int_{-\infty}^{\infty} [s(t)]^2 \, dt - \int_{-\infty}^{\infty} \sum_{k=1}^{K} s_k f_k(t) s(t) \, dt \tag{4.2–27}$$
$$= \mathcal{E}_s - \sum_{k=1}^{K} s_k^2$$

which is nonnegative, by definition.

When the minimum mean square approximation error $\mathcal{E}_{\min} = 0$,

$$\mathcal{E}_s = \sum_{k=1}^{K} s_k^2 = \int_{-\infty}^{\infty} [s(t)]^2 \, dt \tag{4.2–28}$$

Under the condition that $\mathcal{E}_{\min} = 0$, we may express $s(t)$ as

$$s(t) = \sum_{k=1}^{K} s_k f_k(t) \tag{4.2–29}$$

where it is understood that equality of $s(t)$ to its series expansion holds in the sense that the approximation error has zero energy.

When every finite energy signal can be represented by a series expansion of the form in Equation 4.2–29 for which $\mathcal{E}_{\min} = 0$, the set of orthonormal functions $\{f_n(t)\}$ is said to be *complete*.

EXAMPLE 4.2–1: TRIGONOMETRIC FOURIER SERIES. Consider a finite energy signal $s(t)$ that is zero everywhere except in the range $0 \leqslant t \leqslant T$ and has a finite number of discontinuities in this interval. Its periodic extension can be represented in a Fourier series as

$$s(t) = \sum_{k=0}^{\infty} \left(a_k \cos \frac{2\pi k t}{T} + b_k \sin \frac{2\pi k t}{T} \right) \tag{4.2–30}$$

where the coefficients $\{a_k, b_k\}$ that minimize the mean square error are given by

$$\begin{aligned} a_k &= \frac{1}{\sqrt{T}} \int_0^T s(t) \cos \frac{2\pi k t}{T} \, dt \\ b_k &= \frac{1}{\sqrt{T}} \int_0^T s(t) \sin \frac{2\pi k t}{T} \, dt \end{aligned} \tag{4.2–31}$$

The set of trigonometric functions $\{\sqrt{2/T} \cos 2\pi kt/T, \sqrt{2/T} \sin 2\pi kt/T\}$ is complete, and, hence, the series expansion results in zero mean square error. These properties are easily established from the development given above.

Gram–Schmidt procedure. Now suppose that we have a set of finite energy signal waveforms $\{s_i(t), i = 1, 2, \ldots, M\}$ and we wish to construct a set of orthonormal waveforms. The Gram–Schmidt orthogonalization procedure allows us to construct such a set. We begin with the first waveform $s_1(t)$, which is assumed to have energy $\mathcal{E}_1$. The first orthonormal waveform is simply constructed as

$$f_1(t) = \frac{s_1(t)}{\sqrt{\mathcal{E}_1}} \tag{4.2–32}$$

Thus, $f_1(t)$ is simply $s_1(t)$ normalized to unit energy.

The second waveform is constructed from $s_2(t)$ by first computing the projection of $f_1(t)$ onto $s_2(t)$, which is

$$c_{12} = \int_{-\infty}^{\infty} s_2(t) f_1(t) \, dt \tag{4.2–33}$$

Then, $c_{12} f_1(t)$ is subtracted from $s_2(t)$ to yield

$$f_2'(t) = s_2(t) - c_{12} f_1(t) \tag{4.2–34}$$

This waveform is orthogonal to $f_1(t)$, but it does not have unit energy. If $\mathcal{E}_2$ denotes the energy of $f_2'(t)$, the normalized waveform that is orthogonal to $f_1(t)$ is

$$f_2(t) = \frac{f_2'(t)}{\sqrt{\mathcal{E}_2}} \tag{4.2–35}$$

In general, the orthogonalization of the kth function leads to

$$f_k(t) = \frac{f'_k(t)}{\sqrt{\mathcal{E}_k}} \tag{4.2–36}$$

where

$$f'_k(t) = s_k(T) - \sum_{i=1}^{k-1} c_{ik} f_i(t) \tag{4.2–37}$$

and

$$c_{ik} = \int_{-\infty}^{\infty} s_k(t) f_i(t)\, dt, \qquad i = 1, 2, \ldots, k-1 \tag{4.2–38}$$

Thus, the orthogonalization process is continued until all the M signal waveforms $\{s_i(t)\}$ have been exhausted and $N \leqslant M$ orthonormal waveforms have been constructed. The dimensionality N of the signal space will be equal to M if all the signal waveforms are linearly independent, i.e., none of the signal waveforms is a linear combination of the other signal waveforms.

EXAMPLE 4.2–2. Let us apply the Gram–Schmidt procedure to the set of four waveforms illustrated in Figure 4.2–1a. The waveform $s_1(t)$ has energy $\mathcal{E}_1 = 2$, so that $f_1(t) = \sqrt{\frac{1}{2}} s_1(t)$. Next, we observe that $c_{12} = 0$; hence, $s_2(t)$ and $f_1(t)$ are orthogonal. Therefore, $f_2(t) = s_2(t)/\sqrt{\mathcal{E}_2} = \sqrt{\frac{1}{2}} s_2(t)$. To obtain $f_3(t)$, we compute c_{13} and c_{23}, which are $c_{13} = \sqrt{2}$ and $c_{23} = 0$. Thus,

$$f'_3(t) = s_3(t) - \sqrt{2} f_1(t) = \begin{cases} -1 & (2 \leqslant t \leqslant 3) \\ 0 & (\text{otherwise}) \end{cases}$$

Since $f'_3(t)$ has unit energy, it follows that $f_3(t) = f'_3(t)$. In determining $f_4(t)$, we find that $c_{14} = -\sqrt{2}$, $c_{24} = 0$, and $c_{34} = 1$. Hence,

$$f'_4(t) = s_4(t) + \sqrt{2} f_1(t) - f_3(t) = 0$$

Consequently, $s_4(t)$ is a linear combination of $f_1(t)$ and $f_3(t)$ and, hence, $f_4(t) = 0$. The three orthonormal functions are illustrated in Figure 4.2–1b.

Once we have constructed the set of orthonormal waveforms $\{f_n(t)\}$, we can express the M signals $\{s_n(t)\}$ as linear combinations of the $\{f_n(t)\}$. Thus, we may write

$$s_k(t) = \sum_{k=1}^{N} s_{kn} f_n(t), \qquad k = 1, 2, \ldots, M \tag{4.2–39}$$

and

$$\mathcal{E}_k = \int_{-\infty}^{\infty} [s_k(t)]^2\, dt = \sum_{n=1}^{N} s_{kn}^2 = \|\mathbf{s}_k\|^2 \tag{4.2–40}$$

Based on the expression in Equation 4.2–39, each signal may be represented by the vector

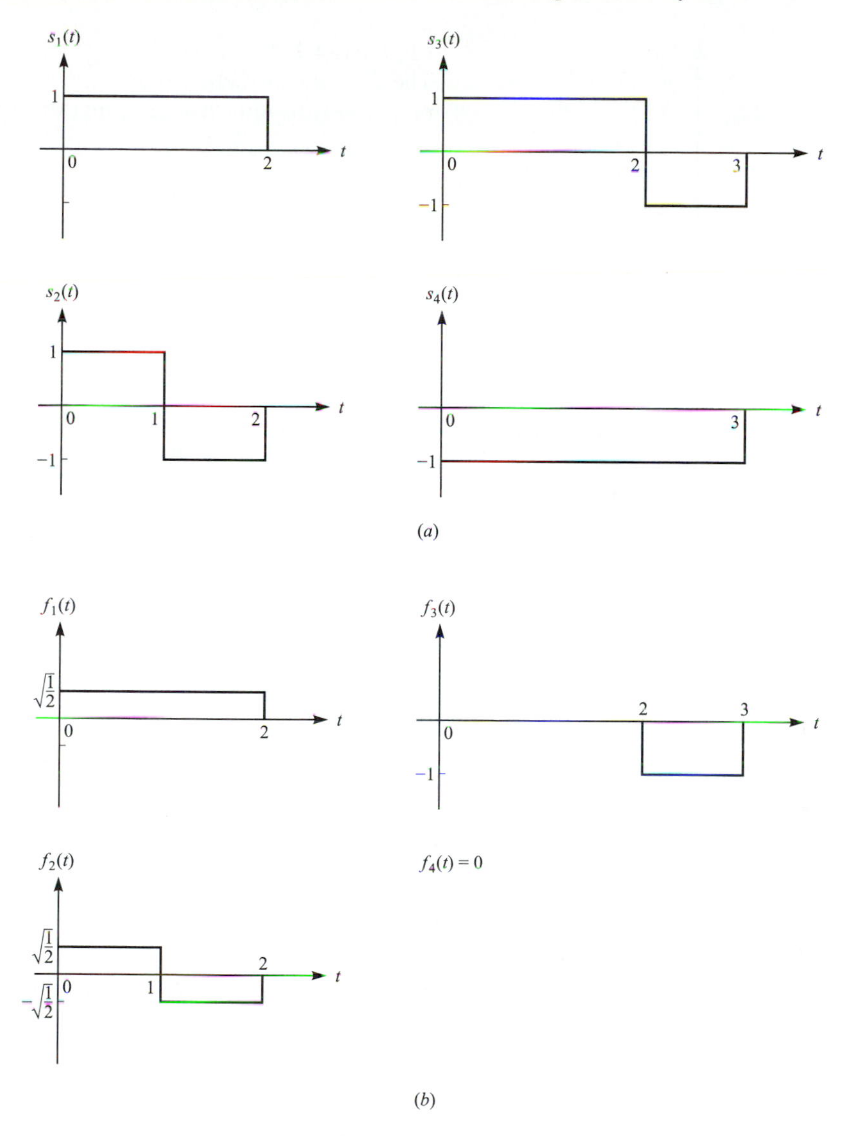

FIGURE 4.2–1
Gram–Schmidt orthogonalization of the signals $\{s_i(t), i = 1, 2, 3, 4\}$ and the corresponding orthogonal signals.

$$\mathbf{s}_k = [s_{k1}\, s_{k2} \cdots s_{kN}] \tag{4.2–41}$$

or, equivalently, as a point in the N-dimensional signal space with coordinates $\{s_{ki},\ i = 1, 2, \ldots, N\}$. The energy in the kth signal is simply the square of the length of the vector or, equivalently, the square of the Euclidean distance from the origin to the point in the N-dimensional space. Thus, any signal can be represented geometrically as a point in the signal space spanned by the orthonormal functions $\{f_n(t)\}$.

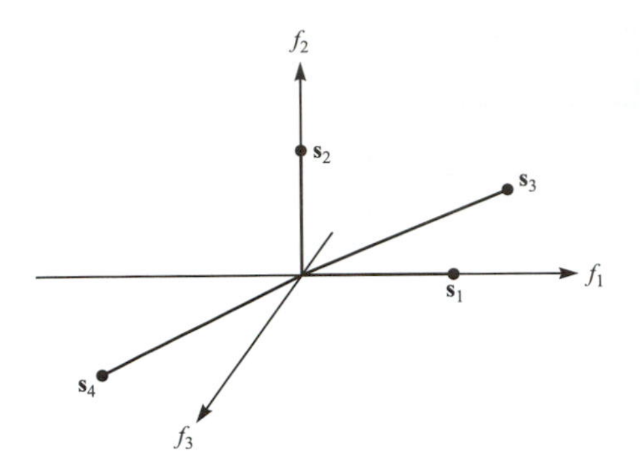

FIGURE 4.2–2
The four signal vectors represented as points in three-dimensional function space.

EXAMPLE 4.2–3. Let us obtain the vector representation of the four signals shown in Figure 4.2–1a by using the orthonormal set of functions in Fig. 4.2–1b. Since the dimensionality of the signal space is $N = 3$, each signal is described by three components. The signal $s_1(t)$ is characterized by the vector $\mathbf{s}_1 = (\sqrt{2}, 0, 0)$. Similarly, the signals $s_2(t)$, $s_3(t)$, and $s_4(t)$ are characterized by the vectors $\mathbf{s}_2 = (0, \sqrt{2}, 0)$, $\mathbf{s}_3 = (\sqrt{2}, 0, 1)$, and $\mathbf{s}_4 = (-\sqrt{2}, 0, 1)$, respectively. These vectors are shown in Figure 4.2–2. Their lengths are $\|\mathbf{s}_1\| = \sqrt{2}$, $\|\mathbf{s}_2\| = \sqrt{2}$, $\|\mathbf{s}_3\| = \sqrt{3}$, and $\|\mathbf{s}_4\| = \sqrt{3}$, and the corresponding signal energies are $\mathcal{E}_k = \|\mathbf{s}_k\|^2$, $k = 1, 2, 3, 4$.

We have demonstrated that a set of M finite energy waveforms $\{s_n(t)\}$ can be represented by a weighted linear combination of orthonormal functions $\{f_n(t)\}$ of dimensionality $N \leqslant M$. The functions $\{f_n(t)\}$ are obtained by applying the Gram–Schmidt orthogonalization procedure on $\{s_n(t)\}$. It should be emphasized, however, that the functions $\{f_n(t)\}$ obtained from the Gram–Schmidt procedure are not unique. If we alter the order in which the orthogonalization of the signals $\{s_n(t)\}$ is performed, the orthonormal waveforms will be different and the corresponding vector representation of the signals $\{s_n(t)\}$ will depend on the choice of the orthonormal functions $\{f_n(t)\}$. Nevertheless, the vectors $\{\mathbf{s}_n\}$ will retain their geometrical configuration and their lengths will be invariant to the choice of orthonormal functions $\{f_n(t)\}$.

EXAMPLE 4.2–4. An alternative set of orthonormal functions for the four signals in Figure 4.2–1a is illustrated in Figure 4.2–3a. By using these functions to expand $\{s_n(t)\}$, we obtain the corresponding vectors $\mathbf{s}_1 = (1, 1, 0)$, $\mathbf{s}_2 = (1, -1, 0)$, $\mathbf{s}_3 = (1, 1, -1)$, and $\mathbf{s}_4 = (-1, -1, -1)$, which are shown in Figure 4.2–3b. Note that the vector lengths are identical to those obtained from the orthonormal functions $\{f_n(t)\}$.

The orthogonal expansions described above were developed for real-valued signal waveforms. The extension to complex-valued signal waveforms is left as an exercise for the reader (see Problems 4.6 and 4.7).

Finally, let us consider the case in which the signal waveforms are band-pass and represented as

$$s_m(t) = \operatorname{Re}[s_{lm}(t)e^{j2\pi f_c t}], \qquad m = 1, 2, \ldots, M \qquad (4.2\text{–}42)$$

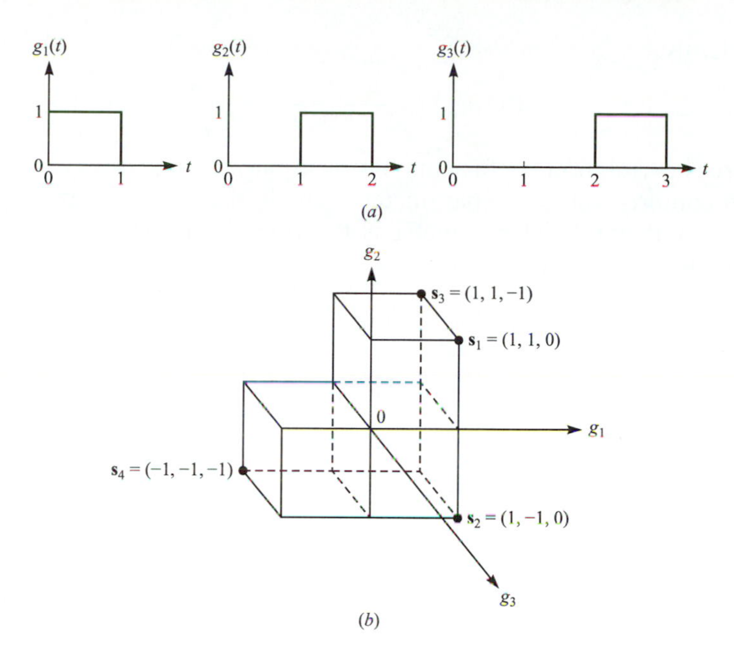

FIGURE 4.2–3
An alternative set of orthonormal functions for the four signals in Figure 4.2–1a and the corresponding signal points.

where $\{s_{lm}(t)\}$ denote the equivalent low-pass signals. Recall that the signal energies may be expressed either in terms of $s_m(t)$ or $s_{lm(t)}$, as

$$\begin{aligned}\mathcal{E}_m &= \int_{-\infty}^{\infty} s_m^2(t)\, dt \\ &= \frac{1}{2}\int_{-\infty}^{\infty} |s_{lm}(t)|^2\, dt\end{aligned} \tag{4.2–43}$$

The similarity between any pair of signal waveforms, say $s_m(t)$ and $s_k(t)$, is measured by the normalized cross correlation

$$\frac{1}{\sqrt{\mathcal{E}_m\mathcal{E}_k}}\int_{-\infty}^{\infty} s_m(t)s_k(t)\, dt = \mathrm{Re}\left\{\frac{1}{2\sqrt{\mathcal{E}_m\mathcal{E}_k}}\int_{-\infty}^{\infty} s_{lm}(t)s_{lk}^*(t)\, dt\right\} \tag{4.2–44}$$

We define the complex-valued cross-correlation coefficient ρ_{km} as

$$\rho_{km} = \frac{1}{2\sqrt{\mathcal{E}_m\mathcal{E}_k}}\int_{-\infty}^{\infty} s_{lm}^*(t)s_{lk}(t)\, dt \tag{4.2–45}$$

Then,

$$\mathrm{Re}(\rho_{km}) = \frac{1}{\sqrt{\mathcal{E}_m\mathcal{E}_k}}\int_{-\infty}^{\infty} s_m(t)s_k(t)\, dt \tag{4.2–46}$$

or, equivalently,

$$\mathrm{Re}(\rho_{km}) = \frac{\mathbf{s}_m \cdot \mathbf{s}_k}{\|\mathbf{s}_m\| \|\mathbf{s}_k\|} = \frac{\mathbf{s}_m \cdot \mathbf{s}_k}{\sqrt{\mathcal{E}_m \mathcal{E}_k}} \tag{4.2–47}$$

The cross-correlation coefficients between pairs of signal waveforms or signal vectors comprise one set of parameters that characterize the similarity of a set of signals. Another related parameter is the Euclidean distance $d_{km}^{(e)}$ between a pair of signals, defined as

$$\begin{aligned} d_{km}^{(e)} &= \|\mathbf{s}_m - \mathbf{s}_k\| \\ &= \left\{ \int_{-\infty}^{\infty} [s_m(t) - s_k(t)]^2 \, dt \right\}^{1/2} \\ &= \{\mathcal{E}_m + \mathcal{E}_k - 2\sqrt{\mathcal{E}_m \mathcal{E}_k} \mathrm{Re}(\rho_{km})\}^{1/2} \end{aligned} \tag{4.2–48}$$

When $\mathcal{E}_m = \mathcal{E}_k = \mathcal{E}$ for all m and k, this expression simplifies to

$$d_{km}^{(e)} = \{2\mathcal{E}[1 - \mathrm{Re}(\rho_{km})]\}^{1/2} \tag{4.2–49}$$

Thus, the Euclidean distance is an alternative measure of the similarity (or dissimilarity) of the set of signal waveforms or the corresponding signal vectors.

In the following section, we describe digitally modulated signals and make use of the signal space representation for such signals. We shall observe that digitally modulated signals, which are classified as linear, are conveniently expanded in terms of two orthonormal basis functions of the form

$$\begin{aligned} f_1(t) &= \sqrt{\frac{2}{T}} \cos 2\pi f_c t \\ f_2(t) &= -\sqrt{\frac{2}{T}} \sin 2\pi f_c t \end{aligned} \tag{4.2–50}$$

Hence, if $s_{lm}(t)$ is expressed as $s_{lm}(t) = x_l(t) + jy_l(t)$, it follows that $s_m(t)$ in Equation 4.2–42 may be expressed as

$$s_m(t) = x_l(t) f_1(t) + y_l(t) f_2(t) \tag{4.2–51}$$

where $x_l(t)$ and $y_l(t)$ represent the signal modulations.

4.3 REPRESENTATION OF DIGITALLY MODULATED SIGNALS

In the transmission of digital information over a communication channel, the modulator is the interface device that maps the digital information into analog waveforms that match the characteristics of the channel. The mapping is generally performed by taking blocks of $k = \log_2 M$ binary digits at a time from the information sequence $\{a_n\}$ and selecting one of $M = 2^k$ deterministic, finite energy waveforms $\{s_m(t), m = 1, 2, \ldots, M\}$ for transmission over the channel.

When the mapping from the digital sequence $\{a_n\}$ to waveforms is performed under the constraint that a waveform transmitted in any time interval depends on one or more previously transmitted waveforms, the modulator is said to have *memory*. On the other hand, when the mapping from the sequence $\{a_n\}$ to the waveforms $\{s_m(t)\}$ is performed without any constraint on previously transmitted waveforms, the modulator is called *memoryless*.

In addition to classifying the modulator as either memoryless or having memory, we may classify it as either *linear* or *non-linear*. Linearity of a modulation method requires that the principle of superposition applies in the mapping of the digital sequence into successive waveforms. In non-linear modulation, the superposition principle does not apply to signals transmitted in successive time intervals. We shall begin by describing memoryless modulation methods.

4.3.1 Memoryless Modulation Methods

As indicated above, the modulator in a digital communication system maps a sequence of binary digits into a set of corresponding signal waveforms. These waveforms may differ in either amplitude or in phase or in frequency, or some combination of two or more signal parameters. We consider each of these signal types separately, beginning with digital pulse amplitude modulation (PAM). In all cases, we assume that the sequence of binary digits at the input to the modulator occurs at a rate of R bits/s.

Pulse-amplitude-modulated (PAM) signals. In digital PAM, the signal waveforms may be represented as

$$\begin{aligned} s_m(t) &= \mathrm{Re}[A_m g(t) e^{j2\pi f_c t}] \\ &= A_m g(t) \cos 2\pi f_c t, \qquad m = 1, 2, \ldots, M, \qquad 0 \leqslant t \leqslant T \end{aligned} \tag{4.3–1}$$

where $\{A_m, 1 \leqslant m \leqslant M\}$ denote the set of M possible amplitudes corresponding to $M = 2^k$ possible k-bit blocks of *symbols*. The signal amplitudes A_m take the discrete values (levels)

$$A_m = (2m - 1 - M)d, \qquad m = 1, 2, \ldots, M \tag{4.3–2}$$

where $2d$ is the distance between adjacent signal amplitudes. The waveform $g(t)$ is a real-valued signal pulse whose shape influences the spectrum of the transmitted signal, as we shall observe later. The symbol rate for the PAM signal is R/k. This is the rate at which changes occur in the amplitude of the carrier to reflect the transmission of new information. The time interval $T_b = 1/R$ is called the *bit interval* and the time interval $T = k/R = kT_b$ is called the *symbol interval*.

The M PAM signals have energies

$$\begin{aligned}\mathcal{E}_m &= \int_0^T s_m^2(t)\,dt \\ &= \tfrac{1}{2}A_m^2 \int_0^T g^2(t)\,dt \\ &= \tfrac{1}{2}A_m^2 \mathcal{E}_g\end{aligned} \tag{4.3–3}$$

where $\mathcal{E}_g$ denotes the energy in the pulse $g(t)$. Clearly, these signals are one-dimensional ($N = 1$), and, hence, are represented by the general form

$$s_m(t) = s_m f(t) \tag{4.3–4}$$

where $f(t)$ is defined as the unit-energy signal waveform given as

$$f(t) = \sqrt{\frac{2}{\mathcal{E}_g}} g(t) \cos 2\pi f_c t \tag{4.3–5}$$

and

$$s_m = A_m \sqrt{\tfrac{1}{2}\mathcal{E}_g}, \qquad m = 1, 2, \ldots, M \tag{4.3–6}$$

The corresponding signal space diagrams for $M = 2$, $M = 4$, and $M = 8$ are shown in Figure 4.3–1. Digital PAM is also called *amplitude-shift keying* (ASK).

The mapping or assignment of k information bits to the $M = 2^k$ possible signal amplitudes may be done in a number of ways. The preferred assignment is one in which the adjacent signal amplitudes differ by one binary digit as illustrated in Figure 4.3–1. This mapping is called *Gray encoding*. It is important in the demodulation of the signal because the most likely errors caused by noise involve the erroneous selection of an adjacent amplitude to the transmitted signal amplitude. In such a case, only a single bit error occurs in the k-bit sequence.

We note that the Euclidean distance between any pair of signal points is

$$\begin{aligned}d_{mn}^{(e)} &= \sqrt{(s_m - s_n)^2} \\ &= \sqrt{\tfrac{1}{2}\mathcal{E}_g}\,|A_m - A_n| \\ &= d\sqrt{2\mathcal{E}_g}\,|m - n|\end{aligned} \tag{4.3–7}$$

Hence, the distance between a pair of adjacent signal points, i.e., the minimum Euclidean distance, is

$$d_{\min}^{(e)} = d\sqrt{2\mathcal{E}_g} \tag{4.3–8}$$

The carrier-modulated PAM signal represented by Equation 4.3–1 is a double-sideband (DSB) signal and requires twice the channel bandwidth of the equivalent low-pass signal for transmission. Alternatively, we may use single-sideband (SSB) PAM, which has the representation (lower or upper sideband):

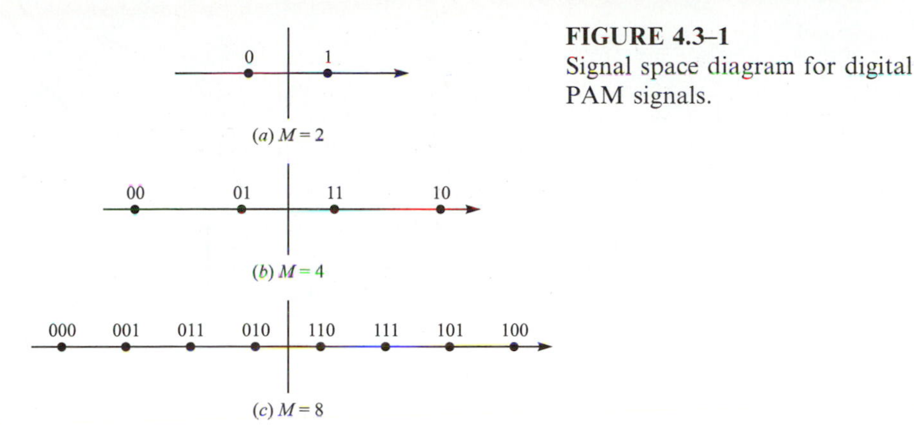

FIGURE 4.3–1
Signal space diagram for digital PAM signals.

$$s_m(t) = \text{Re}\{A_m[g(t) \pm j\hat{g}(t)]e^{j2\pi f_c t}\}, \qquad m = 1, 2, \ldots, M \tag{4.3–9}$$

where $\hat{g}(t)$ is the Hilbert transform of $g(t)$. Thus, the bandwidth of the SSB signal is half that of the DSB signal.

The digital PAM signal is also appropriate for transmission over a channel that does not require carrier modulation. In this case, the signal waveform may be simply represented as

$$s_m(t) = A_m g(t), \qquad m = 1, 2, \ldots, M \tag{4.3–10}$$

This is now called a *baseband* signal. For example a four-amplitude level baseband PAM signal is illustrated in Figure 4.3–2a. The carrier-modulated version of the signal is shown in Figure 4.3–2b.

In the special case of $M = 2$ signals, the binary PAM waveforms have the special property that

$$s_1(t) = -s_2(t)$$

Hence, these two signals have the same energy and a cross-correlation coefficient of -1. Such signals are called *antipodal.*

Phase-modulated signals. In digital phase modulation, the M signal waveforms are represented as

$$\begin{aligned} s_m(t) &= \text{Re}[g(t)e^{j2\pi(m-1)/M}e^{j2f_c t}], \qquad m = 1, 2, \ldots, M, \qquad 0 \leqslant t \leqslant T \\ &= g(t)\cos\left[2\pi f_c t + \frac{2\pi}{M}(m-1)\right] \\ &= g(t)\cos\frac{2\pi}{M}(m-1)\cos 2\pi f_c t - g(t)\sin\frac{2\pi}{M}(m-1)\sin 2\pi f_c t \end{aligned} \tag{4.3–11}$$

where $g(t)$ is the signal pulse shape and $\theta_m = 2\pi(m-1)/M$, $m = 1, 2, \ldots, M$, are the M possible phases of the carrier that convey the transmitted information. Digital phase modulation is usually called *phase-shift keying* (PSK).

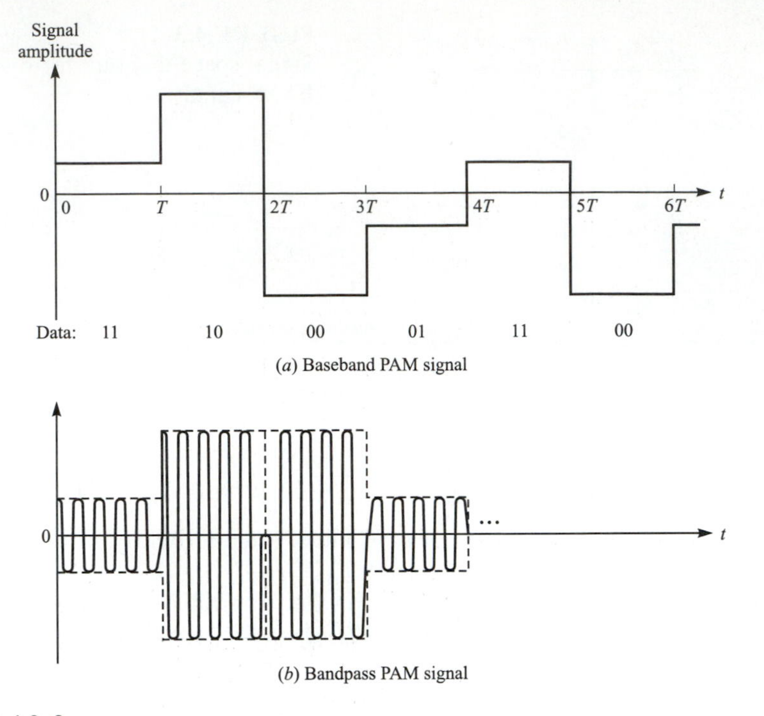

(*a*) Baseband PAM signal

(*b*) Bandpass PAM signal

FIGURE 4.3–2
Baseband and band-pass PAM signals.

We note that these signal waveforms have equal energy, i.e.,

$$\begin{aligned}\mathcal{E} &= \int_0^T s_m^2(t)\,dt \\ &= \frac{1}{2}\int_0^T g^2(t)\,dt = \tfrac{1}{2}\mathcal{E}_g\end{aligned} \tag{4.3–12}$$

Furthermore, the signal waveforms may be represented as a linear combination of two orthonormal signal waveforms, $f_1(t)$ and $f_2(t)$, i.e.,

$$s_m(t) = s_{m1}f_1(t) + s_{m2}f_2(t) \tag{4.3–13}$$

where

$$f_1(t) = \sqrt{\frac{2}{\mathcal{E}_g}}g(t)\cos 2\pi f_c t \tag{4.3–14}$$

$$f_2(t) = -\sqrt{\frac{2}{\mathcal{E}_g}}g(t)\sin 2\pi f_c t \tag{4.3–15}$$

and the two-dimensional vectors $\mathbf{s}_m = [s_{m1}\ s_{m2}]$ are given by

$$\mathbf{s}_m = \left[\sqrt{\frac{\mathcal{E}_g}{2}}\cos\frac{2\pi}{M}(m-1) \quad \sqrt{\frac{\mathcal{E}_g}{2}}\sin\frac{2\pi}{M}(m-1)\right], \qquad m = 1, 2, \ldots, M \tag{4.3–16}$$

Signal space diagrams for $M = 2$, 4, and 8 are shown in Figure 4.3–3. We note that $M = 2$ corresponds to one-dimensional signals, which are identical to binary PAM signals.

As is the case of PAM, the mapping or assignment of k information bits to the $M = 2^k$ possible phases may be done in a number of ways. The preferred assignment is Gray encoding, so that the most likely errors caused by noise will result in a single bit error in the k-bit symbol.

The Euclidean distance between signal points is

$$\begin{aligned} d_{mn}^{(e)} &= \|\mathbf{s}_m - \mathbf{s}_n\| \\ &= \left\{\mathcal{E}_g\left[1 - \cos\frac{2\pi}{M}(m-n)\right]\right\}^{1/2} \end{aligned} \tag{4.3–17}$$

The minimum Euclidean distance corresponds to the case in which $|m - n| = 1$, i.e., adjacent signal phases. In this case,

$$d_{\min}^{(e)} = \sqrt{\mathcal{E}_g\left(1 - \cos\frac{2\pi}{M}\right)} \tag{4.3–18}$$

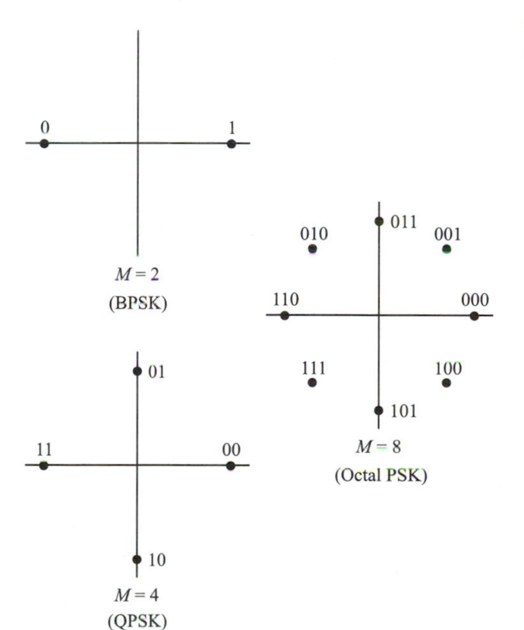

FIGURE 4.3–3
Signal space diagrams for PSK signals.

A variant of 4-phase PSK (QPSK), called $\pi/4$-QPSK is obtained by introducing an additional $\pi/4$ phase shift in the carrier phase in each symbol interval. This phase shift facilitates symbol synchronization.

Quadrature amplitude modulation. The bandwidth efficiency of PAM/SSB can also be obtained by simultaneously impressing two separate k-bit symbols from the information sequence $\{a_n\}$ on two quadrature carriers $\cos 2\pi f_c t$ and $\sin 2\pi f_c t$. The resulting modulation technique is called quadrature PAM or QAM, and the corresponding signal waveforms may be expressed as

$$\begin{aligned} s_m(t) &= \mathrm{Re}[(A_{mc} + jA_{ms})g(t)e^{j2\pi f_c t}], \qquad m = 1, 2, \ldots, M, \qquad 0 \leqslant t \leqslant T \\ &= A_{mc}g(t)\cos 2\pi f_c t - A_{ms}g(t)\sin 2\pi f_c t \end{aligned} \tag{4.3–19}$$

where A_{mc} and A_{ms} are the information-bearing signal amplitudes of the quadrature carriers and $g(t)$ is the signal pulse.

Alternatively, the QAM signal waveforms may be expressed as

$$\begin{aligned} s_m(t) &= \mathrm{Re}[V_m e^{j\theta m} g(t) e^{j2\pi f_c t}] \\ &= V_m g(t)\cos(2\pi f_c t + \theta_m) \end{aligned} \tag{4.3–20}$$

where $V_m = \sqrt{A_{mc}^2 + A_{ms}^2}$ and $\theta_m = \tan^{-1}(A_{ms}/A_{mc})$. From this expression, it is apparent that the QAM signal waveforms may be viewed as combined amplitude and phase modulation.

In fact, we may select any combination of M_1-level PAM and M_2-phase PSK to construct an $M = M_1 M_2$ combined PAM–PSK signal constellation. If $M_1 = 2^n$ and $M_2 = 2^m$, the combined PAM–PSK signal constellation results in the simultaneous transmission of $m + n = \log M_1 M_2$ binary digits occurring at a symbol rate $R/(m + n)$. Examples of signal space diagrams for combined PAM–PSK are shown in Figure 4.3–4, for $M = 8$ and $M = 16$.

As in the case of PSK signals, the QAM signal waveforms may be represented as a linear combination of two orthonormal signal waveforms, $f_1(t)$ and $f_2(t)$, i.e.,

$$s_m(t) = s_{m1} f_1(t) + s_{m2} f_2(t) \tag{4.3–21}$$

where

$$\begin{aligned} f_1(t) &= \sqrt{\frac{2}{\mathcal{E}_g}}\, g(t)\cos 2\pi f_c t \\ f_2(t) &= -\sqrt{\frac{2}{\mathcal{E}_g}}\, g(t)\sin 2\pi f_c t \end{aligned} \tag{4.3–22}$$

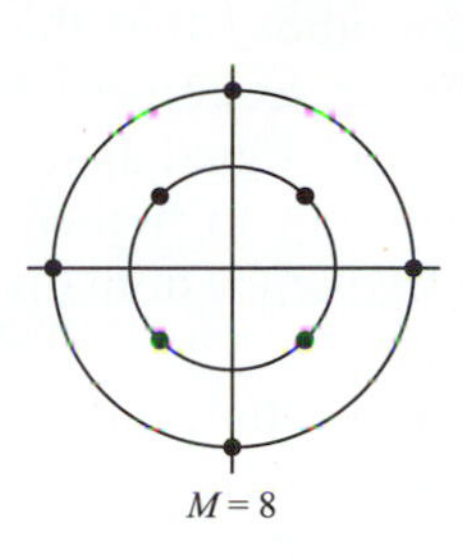

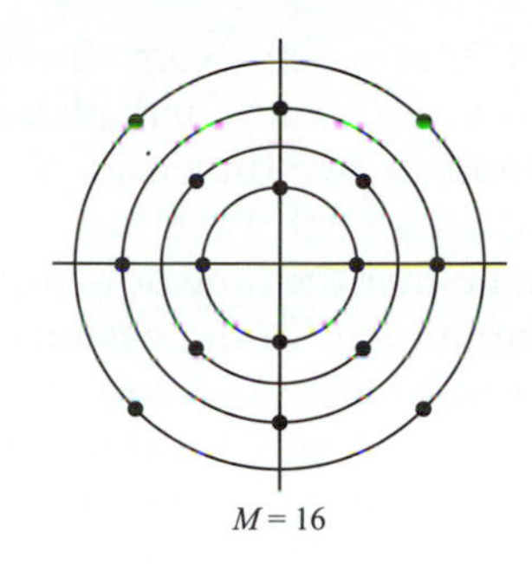

FIGURE 4.3–4
Examples of combined PAM–PSK signal space diagrams.

and

$$\begin{aligned} \mathbf{s}_m &= [s_{m1} \quad s_{m2}] \\ &= [A_{mc}\sqrt{\tfrac{1}{2}\mathcal{E}_g} \quad A_{ms}\sqrt{\tfrac{1}{2}\mathcal{E}_g}] \end{aligned} \tag{4.3–23}$$

where $\mathcal{E}_g$ is the energy of the signal pulse $g(t)$.

The Euclidean distance between any pair of signal vectors is

$$\begin{aligned} d_{mn}^{(e)} &= \|\mathbf{s}_m - \mathbf{s}_n\| \\ &= \sqrt{\tfrac{1}{2}\mathcal{E}_g[(A_{mc} - A_{nc})^2 + (A_{ms} - A_{ns})^2]} \end{aligned} \tag{4.3–24}$$

In the special case where the signal amplitudes take the set of discrete values $\{(2m - 1 - M)d,\ m = 1, 2, \ldots, M\}$, the signal space diagram is rectangular, as shown in Figure 4.3–5. In this case, the Euclidean distance between adjacent points, i.e., the minimum distance, is

$$d_{\min}^{(e)} = d\sqrt{2\mathcal{E}_g} \tag{4.3–25}$$

which is the same result as for PAM.

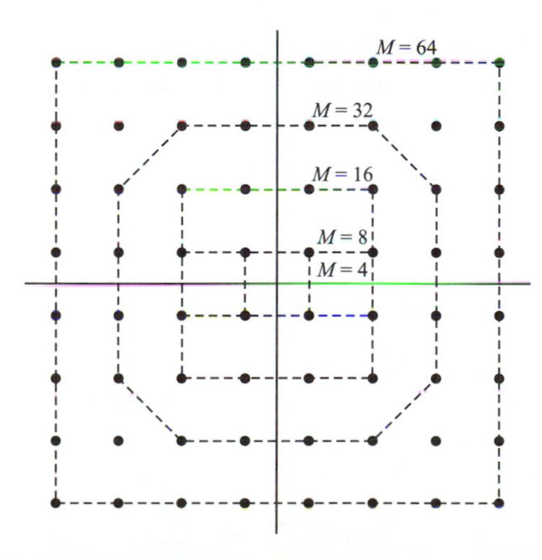

FIGURE 4.3–5
Several signal space diagrams for rectangular QAM.

Multidimensional signals. It is apparent from the discussion above that the digital modulation of the carrier amplitude and phase allows us to construct signal waveforms that correspond to two-dimensional vectors and signal space diagrams. If we wish to construct signal waveforms corresponding to higher-dimensional vectors, we may use either the time domain or the frequency domain or both in order to increase the number of dimensions.

Suppose we have N-dimensional signal vectors. For any N, we may subdivide a time interval of length $T_1 = NT$ into N subintervals of length $T = T_1/N$. In each subinterval of length T, we may use binary PAM (a one-dimensional signal) to transmit an element of the N-dimensional signal vector. Thus, the N time slots are used to transmit the N-dimensional signal vector. If N is even, a time slot of length T may be used to simultaneously transmit two components of the N-dimensional vector by modulating the amplitude of quadrature carriers independently by the corresponding components. In this manner, the N-dimensional signal vector is transmitted in $\frac{1}{2}NT$ seconds ($\frac{1}{2}N$ time slots).

Alternatively, a frequency band of width $N\Delta f$ may be subdivided into N frequency slots each of width Δf. An N-dimensional signal vector can be transmitted over the channel by simultaneously modulating the amplitude of N carriers, one in each of the N frequency slots. Care must be taken to provide sufficient frequency separation Δf between successive carriers so that there is no cross-talk interference among the signals on the N carriers. If quadrature carriers are used in each frequency slot, the N-dimensional vector (even N) may be transmitted in $\frac{1}{2}N$ frequency slots, thus reducing the channel bandwidth utilization by a factor of 2.

More generally, we may use both the time and frequency domains jointly to transmit an N-dimensional signal vector. For example, Figure 4.3–6 illustrates a subdivision of the time and frequency axes into 12 slots. Thus, an $N = 12$-dimensional signal vector may be transmitted by PAM or an $N = 24$-dimensional signal vector may be transmitted by use of two quadrature carriers (QAM) in each slot.

Orthogonal multidimensional signals. As a special case of the construction of multidimensional signals, let us consider the construction of M equal-energy orthogonal signal waveforms that differ in frequency and are represented as

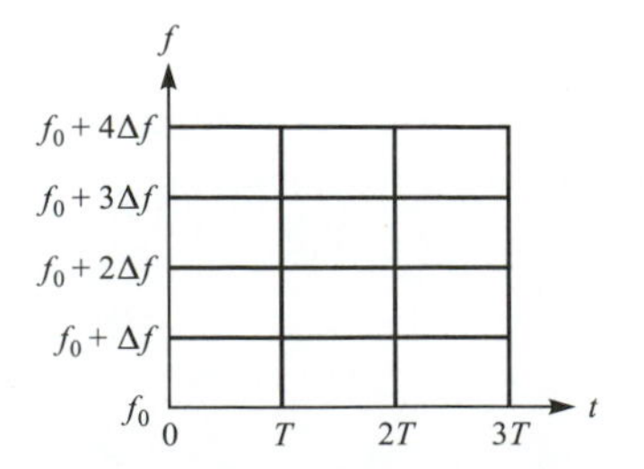

FIGURE 4.3–6
Subdivision of time and frequency axes into distinct slots.

$$
\begin{aligned}
s_m(t) &= \text{Re}[s_{lm}(t)e^{j2\pi f_c t}], \qquad m = 1, 2, \ldots, M, \qquad 0 \leqslant t \leqslant T \\
&= \sqrt{\frac{2\mathcal{E}}{T}} \cos[2\pi f_c t + 2\pi m \, \Delta f \; t]
\end{aligned}
\tag{4.3–26}
$$

where the equivalent low-pass signal waveforms are defined as

$$
s_{lm}(t) = \sqrt{\frac{2\mathcal{E}}{T}} e^{j2\pi m \, \Delta f \, t}, \qquad m = 1, 2, \ldots, M, \qquad 0 \leqslant t \leqslant T \tag{4.3–27}
$$

This type of frequency modulation is called *frequency-shift keying* (FSK).

These waveforms are characterized as having equal energy and cross-correlation coefficients

$$
\begin{aligned}
\rho_{km} &= \frac{2\mathcal{E}/T}{2\mathcal{E}} \int_0^T e^{j2\pi(m-k)\,\Delta f \, t} \, dt \\
&= \frac{\sin \pi T(m-k)\,\Delta f}{\pi T(m-k)\,\Delta f} e^{j\pi T(m-k)\,\Delta f}
\end{aligned}
\tag{4.3–28}
$$

The real part of ρ_{km} is

$$
\begin{aligned}
\rho_r \equiv \text{Re}(\rho_{km}) &= \frac{\sin[\pi T(m-k)\,\Delta f]}{\pi T(m-k)\Delta f} \cos[\pi T(m-k)\Delta f] \\
&= \frac{\sin[2\pi T(m-k)\,\Delta f]}{2\pi T(m-k)\,\Delta f}
\end{aligned}
\tag{4.3–29}
$$

First, we observe that $\text{Re}(\rho_{km}) = 0$ when $\Delta f = 1/2T$ and $m \neq k$. Since $|m-k| = 1$ corresponds to adjacent frequency slots, $\Delta f = 1/2T$ represents the minimum frequency separation between adjacent signals for orthogonality of the M signals. Plots of $\text{Re}(\rho_{km})$ versus Δf and $|\rho_{km}|$ versus Δf are shown in Figure 4.3–7. Note that $|\rho_{km}| = 0$ for multiples of $1/T$ whereas $\text{Re}(\rho_{km}) = 0$ for multiples of $1/2T$.

For the case in which $\Delta f = 1/2T$, the M FSK signals are equivalent to the N-dimensional vectors

$$
\begin{aligned}
\mathbf{s}_1 &= [\sqrt{\mathcal{E}} \quad 0 \quad 0 \quad \cdots \quad 0 \quad 0] \\
\mathbf{s}_2 &= [0 \quad \sqrt{\mathcal{E}} \quad 0 \quad \cdots \quad 0 \quad 0] \\
&\vdots \qquad \vdots \\
\mathbf{s}_N &= [0 \quad 0 \quad 0 \quad \cdots \quad 0, \sqrt{\mathcal{E}}]
\end{aligned}
\tag{4.3–30}
$$

where $N = M$. The distance between pairs of signals is

$$
d_{km}^{(e)} = \sqrt{2\mathcal{E}}, \qquad \text{for all } m, k \tag{4.3–31}
$$

which is also the minimum distance. Figure 4.3–8 illustrates the signal space diagram for $M = N = 2$ and $M = N = 3$.

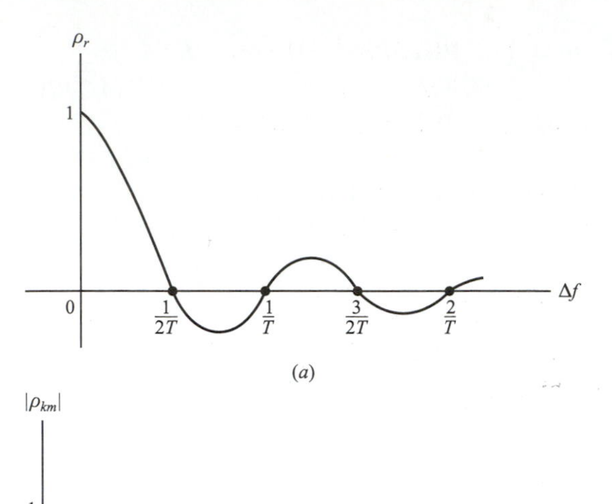

FIGURE 4.3–7
Cross-correlation coefficient as a function of frequency separation for FSK signals.

Biorthogonal Signals. A set of M biorthogonal signals can be constructed from $\frac{1}{2}M$ orthogonal signals by simply including the negatives of the orthogonal signals. Thus, we require $N = \frac{1}{2}M$ dimensions for the construction of a set of M biorthogonal signals. Figure 4.3–9 illustrates the biorthogonal signals for $M = 4$ and 6.

We note that the correlation between any pair of waveforms is either $\rho_r = -1$ or 0. The corresponding distances are $d = 2\sqrt{\mathcal{E}}$ or $\sqrt{2\mathcal{E}}$, with the latter being the minimum distance.

Simplex signals. Suppose we have a set of M orthogonal waveforms $\{s_m(t)\}$ or, equivalently, their vector representation $\{\mathbf{s}_m\}$. Their mean is

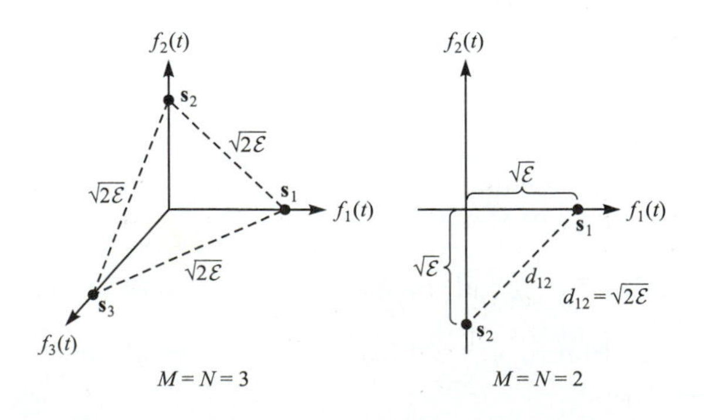

FIGURE 4.3–8
Orthogonal signals for $M = N = 3$ and $M = N = 2$.

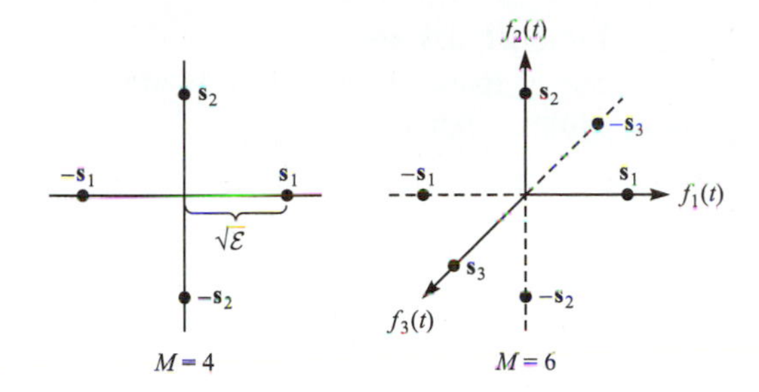

FIGURE 4.3–9
Signal space diagrams for $M = 4$ and $M = 6$ biorthogonal signals.

$$\bar{\mathbf{s}} = \frac{1}{M}\sum_{m=1}^{M} \mathbf{s}_m \tag{4.3–32}$$

Now, let us construct another set of M signals by subtracting the mean from each of the M orthogonal signals. Thus,

$$\mathbf{s}'_m = \mathbf{s}_m - \bar{\mathbf{s}}, \qquad m = 1, 2, \ldots, M \tag{4.3–33}$$

The effect of the subtraction is to translate the origin of the m orthogonal signals to the point $\bar{\mathbf{s}}$.

The resulting signal waveforms are called *simplex signals* and have the following properties. First, the energy per waveform is

$$\begin{aligned} \|\mathbf{s}'_m\|^2 &= \|\mathbf{s}_m - \bar{\mathbf{s}}\|^2 \\ &= \mathcal{E} - \frac{2}{M}\mathcal{E} + \frac{1}{M}\mathcal{E} \\ &= \mathcal{E}\left(1 - \frac{1}{M}\right) \end{aligned} \tag{4.3–34}$$

Second, the cross correlation of any pair of signals is

$$\begin{aligned} \text{Re}(\rho_{mn}) &= \frac{\mathbf{s}'_m \cdot \mathbf{s}'_n}{\|\mathbf{s}'_m\|\|\mathbf{s}'_n\|} \\ &= \frac{-1/M}{1 - 1/M} = -\frac{1}{M-1} \end{aligned} \tag{4.3–35}$$

for all m, n. Hence, the set of simplex waveforms is *equally correlated* and requires less energy, by the factor $1 - 1/M$, than the set of orthogonal waveforms. Since only the origin was translated, the distance between any pair of signal points is maintained at $d = \sqrt{2\mathcal{E}}$, which is the same as the distance between any pair of orthogonal signals.

Figure 4.3–10 illustrates the simplex signals for $M = 2, 3$, and 4. Note that the signal dimensionality is $N = M - 1$.

Signal waveforms from binary codes. A set of M signaling waveforms can be generated from a set of M binary code words of the form

$$\mathbf{C}_m = [c_{m1}\, c_{m2} \cdots c_{mN}], \qquad m = 1, 2, \ldots, M \tag{4.3–36}$$

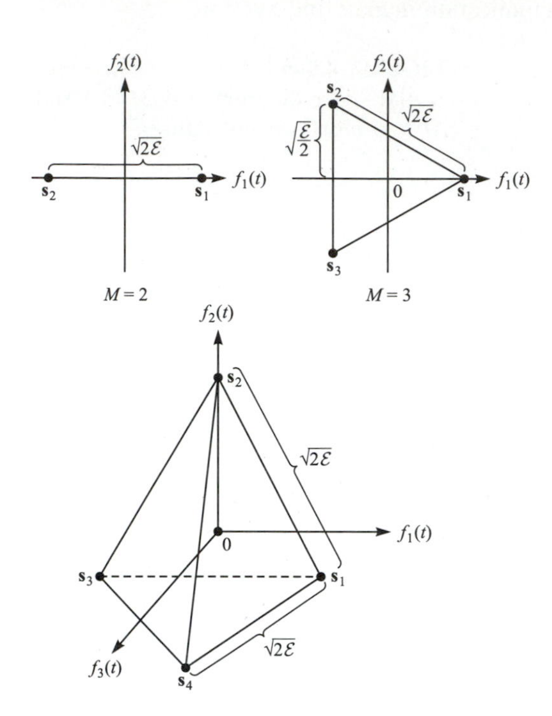

FIGURE 4.3–10
Signal space diagrams for M-ary simplex signals.

where $c_{mj} = 0$ or 1 for all m and j. Each component of a code word is mapped into an elementary binary PSK waveform as follows:

$$
\begin{aligned}
c_{mj} = 1 \Rightarrow s_{mj}(t) &= \sqrt{\frac{2\mathcal{E}_c}{T_c}} \cos 2\pi f_c t, \qquad 0 \leqslant t \leqslant T_c \\
c_{mj} = 0 \Rightarrow s_{mj}(t) &= -\sqrt{\frac{2\mathcal{E}_c}{T_c}} \cos 2\pi f_c t, \qquad 0 \leqslant t \leqslant T_c
\end{aligned}
\tag{4.3–37}
$$

where $T_c = T/N$ and $\mathcal{E}_c = \mathcal{E}/N$. Thus, the M code words $\{C_m\}$ are mapped into a set of M waveforms $\{s_m(t)\}$.

The waveforms can be represented in vector form as

$$
\mathbf{s}_m = [s_{m1}\, s_{m2} \cdots s_{mN}], \qquad m = 1, 2, \ldots, M \tag{4.3–38}
$$

where $s_{mj} = \pm\sqrt{\mathcal{E}/N}$ for all m and j. N is called the block length of the code, and it is also the dimension of the M waveforms.

We note that there are 2^N possible waveforms that can be constructed from the 2^N possible binary code words. We may select a subset of $M < 2^N$ signal waveforms for transmission of the information. We also observe that the 2^N possible signal points correspond to the vertices of an N-dimensional hypercube with its center at the origin. Figure 4.3–11 illustrates the signal points in $N = 2$ and 3 dimensions.

Each of the M waveforms has energy $\mathcal{E}$. The cross correlation between any pair of waveforms depends on how we select the M waveforms from the 2^N

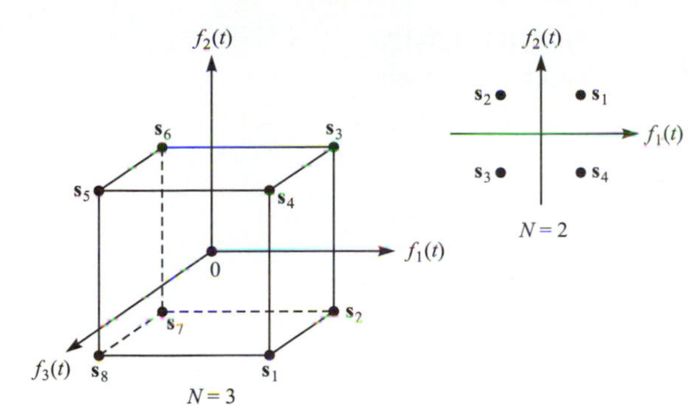

FIGURE 4.3–11
Signal space diagrams for signals generated from binary codes.

possible waveforms. This topic is treated in Chapter 7. Clearly, any adjacent signal points have a cross-correlation coefficient

$$\rho_r = \frac{\mathcal{E}(1 - 2/N)}{\mathcal{E}} = \frac{N - 2}{N} \tag{4.3–39}$$

and a corresponding distance of

$$\begin{aligned} d^{(e)} &= \sqrt{2\mathcal{E}(1 - \rho_r)} \\ &= \sqrt{4\mathcal{E}/N} \end{aligned} \tag{4.3–40}$$

This concludes our discussion of memoryless modulation signals.

4.3.2 Linear Modulation with Memory

The modulation signals introduced in the previous section were classified as memoryless, because there was no dependence between signals transmitted in non-overlapping symbol intervals. In this section, we present some modulation signals in which there is dependence between the signals transmitted in successive symbol intervals. This signal dependence is usually introduced for the purpose of shaping the spectrum of the transmitted signal so that it matches the spectral characteristics of the channel. Signal dependence between signals transmitted in different signal intervals is generally accomplished by encoding the data sequence at the input to the modulator by means of a *modulation code*, as described in Chapter 9.

In this section, we shall present examples of modulation signals with memory and characterize their memory in terms of Markov chains. We shall confine our treatment to baseband signals. The generalization to band-pass signals is relatively straightforward.

Figure 4.3–12 illustrates three different baseband signals and the corresponding data sequence. The first signal, called NRZ, is the simplest. The binary information digit 1 is represented by a rectangular pulse of polarity A and the binary digit 0 is represented by a rectangular pulse of polarity $-A$. Hence, the

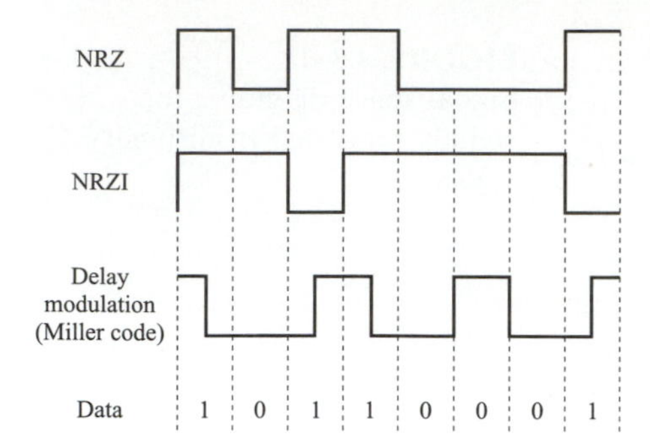

FIGURE 4.3–12
Baseband signals.

NRZ modulation is memoryless and is equivalent to a binary PAM or a binary PSK signal in a carrier-modulated system.

The NRZI signal is different from the NRZ signal in that transitions from one amplitude level to another occur only when a 1 is transmitted. The amplitude level remains unchanged when a zero is transmitted. This type of signal encoding is called *differential encoding*. The encoding operation is described mathematically by the relation

$$b_k = a_k \oplus b_{k-1} \tag{4.3–41}$$

where $\{a_k\}$ is the binary information sequence into the encoder, $\{b_k\}$ is the output sequence of the encoder, and $\oplus$ denotes addition modulo 2. When $b_k = 1$, the transmitted waveform is a rectangular pulse of amplitude A, and when $b_k = 0$, the transmitted waveform is a rectangular pulse of amplitude $-A$. Hence, the output of the encoder is mapped into one of two waveforms in exactly the same manner as for the NRZ signal.

The differential encoding operation introduces memory in the signal. The combination of the encoder and the modulator operations may be represented by a state diagram (a Markov chain) as shown in Figure 4.3–13. The state diagram may be described by two transition matrices corresponding to the two possible input bits $\{0, 1\}$. We note that when $a_k = 0$, the encoder stays in the same state. Hence, the state transition matrix for a zero is simply

$$\mathbf{T}_1 = \begin{bmatrix} 1 & 0 \\ 0 & 1 \end{bmatrix} \tag{4.3–42}$$

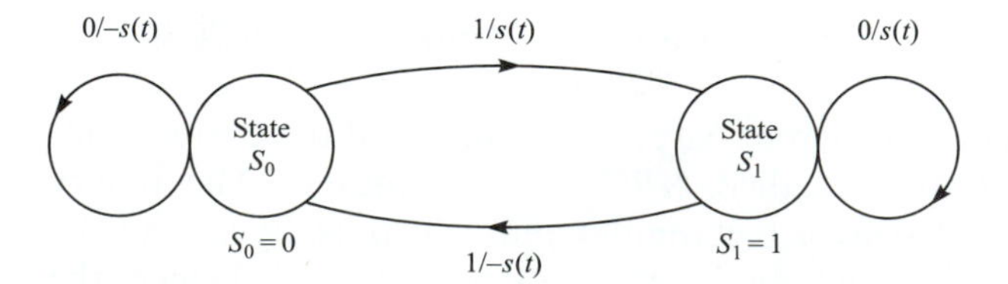

FIGURE 4.3–13
State diagram for the NRZI signal.

where $t_{ij} = 1$ if a_k results in a transition from state i to state j, $i = 1, 2$, and $j = 1, 2$; otherwise, $t_{ij} = 0$. Similarly, the state transition matrix for $a_k = 1$ is

$$\mathbf{T}_2 = \begin{bmatrix} 0 & 1 \\ 1 & 0 \end{bmatrix} \tag{4.3–43}$$

Thus, these two state transition matrices characterize the NRZI signal.

Another way to display the memory introduced by the precoding operation is by means of a trellis diagram. The trellis diagram for the NRZI signal is illustrated in Figure 4.3–14. The trellis provides exactly the same information concerning the signal dependence as the state diagram, but also depicts a time evolution of the state transitions.

The signal generated by delay modulation also has memory. As shown in Chapter 9, delay modulation is equivalent to encoding the data sequence by a run-length-limited code called a *Miller code* and using NRZI to transmit the encoded data. This type of digital modulation has been used extensively for digital magnetic recording and in carrier modulation systems employing binary PSK. The signal may be described by a state diagram that has four states as shown in Figure 4.3–15a. There are two elementary waveforms $s_1(t)$ and $s_2(t)$ and their negatives $-s_1(t)$ and $-s_2(t)$, which are used for transmitting the binary information. These waveforms are illustrated in Figure 4.3–15b. The mapping from bits to corresponding waveforms is illustrated in the state diagram. The state transition matrices that characterize the memory of this encoding and modulation method are easily obtained from the state diagram in Figure 4.3–15. When $a_k = 0$, we have

$$\mathbf{T}_1 = \begin{bmatrix} 0 & 0 & 0 & 1 \\ 0 & 0 & 0 & 1 \\ 1 & 0 & 0 & 0 \\ 1 & 0 & 0 & 0 \end{bmatrix} \tag{4.3–44}$$

and when $a_{k=1}$, the transition matrix is

$$\mathbf{T}_2 = \begin{bmatrix} 0 & 1 & 0 & 0 \\ 0 & 0 & 1 & 0 \\ 0 & 1 & 0 & 0 \\ 0 & 0 & 1 & 0 \end{bmatrix} \tag{4.3–45}$$

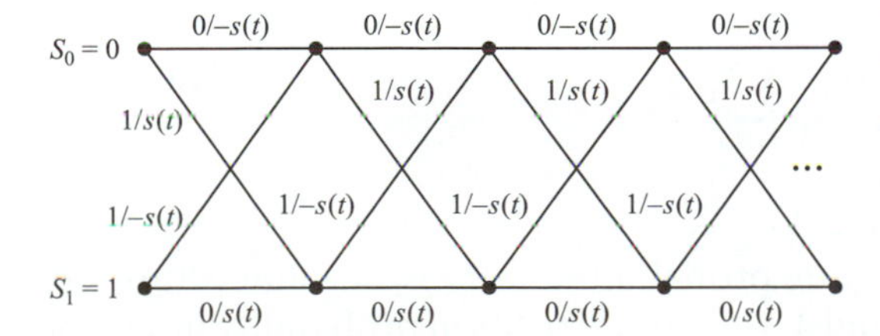

FIGURE 4.3–14
The trellis diagram for the NRZI signal.

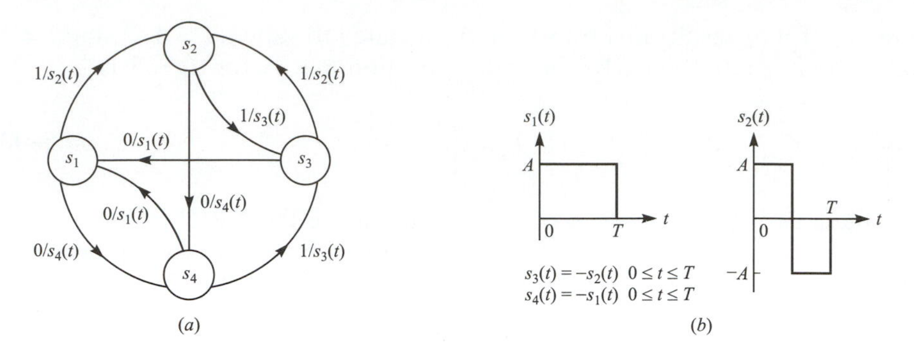

FIGURE 4.3–15
State diagram (**a**) and basic waveforms (**b**) for delay modulated (Miller-encoded) signal.

Thus, these two 4×4 state transition matrices characterize the state diagram for the Miller-encoded signal.

Modulation techniques with memory such as NRZI and Miller coding are generally characaterized by a K-state Markov chain with *stationary state probabilities* $\{p_i, i = 1, 2, \ldots, K\}$ and *transition probabilities* $\{p_{ij}, i, j = 1, 2, \ldots, K\}$. Associated with each transition is a signal waveform $s_j(t)$, $j = 1, 2, \ldots, K$. Thus, the transition probability p_{ij} denotes the probability that signal waveform $s_j(t)$ is transmitted in a given signaling interval after the transmission of the signal waveform $s_i(t)$ in the previous signaling interval. The transition probabilities may be arranged in matrix form as

$$\mathbf{P} = \begin{bmatrix} p_{11} & p_{12} & \cdots & p_{1K} \\ p_{21} & p_{22} & \cdots & p_{2K} \\ \vdots & \vdots & & \vdots \\ p_{K1} & p_{K2} & \cdots & p_{KK} \end{bmatrix} \tag{4.3–46}$$

where $\mathbf{P}$ is called the *transition probability matrix*.

The transition probability matrix is easily obtained from the transition matrices $\{\mathbf{T}_i\}$ and the corresponding probabilities of occurrence of the input bits (or, equivalently, the stationary state transition probabilities $\{p_i\}$). The general relationship may be expressed as

$$\mathbf{P} = \sum_{i=1}^{2} q_i \mathbf{T}_i \tag{4.3–47}$$

where $q_1 = P(a_k = 0)$ and $q_2 = P(a_k = 1)$.

For the NRZI signal with equal state probabilities $p_1 = p_2 = \frac{1}{2}$ and transition matrices given by Equations 4.3–42 and 4.3–43, the transition probability matrix is

$$\mathbf{P} = \begin{bmatrix} \frac{1}{2} & \frac{1}{2} \\ \frac{1}{2} & \frac{1}{2} \end{bmatrix} \tag{4.3–48}$$

Similarly, the transition probability matrix for the Miller-coded signal with equally likely symbols ($q_1 = q_2 = \frac{1}{2}$ or, equivalently, $p_1 = p_2 = p_3 = p_4 = \frac{1}{4}$) is

$$\mathbf{P} = \begin{bmatrix} 0 & \frac{1}{2} & 0 & \frac{1}{2} \\ 0 & 0 & \frac{1}{2} & \frac{1}{2} \\ \frac{1}{2} & \frac{1}{2} & 0 & 0 \\ \frac{1}{2} & 0 & \frac{1}{2} & 0 \end{bmatrix} \tag{4.3–49}$$

The transition probability matrix is useful in the determination of the spectral characteristics of digital modulation techniques with memory, as we shall observe in Section 4.4.

4.3.3 Non-linear Modulation Methods with Memory—CPFSK and CPM

In this section, we consider a class of digital modulation methods in which the phase of the signal is constrained to be continuous. This constraint results in a phase or frequency modulator that has memory. The modulation method is also non-linear.

Continuous-phase FSK (CPFSK). A conventional FSK signal is generated by shifting the carrier by an amount $f_n = \frac{1}{2}\Delta f\, I_n$, $I_n = \pm 1, \pm 3, \ldots, \pm(M-1)$, to reflect the digital information that is being transmitted. This type of FSK signal was described in Section 4.3.1, and it is memoryless. The switching from one frequency to another may be accomplished by having $M = 2^k$ separate oscillators tuned to the desired frequencies and selecting one of the M frequencies according to the particular k-bit symbol that is to be transmitted in a signal interval of duration $T = k/R$ seconds. However, such abrupt switching from one oscillator output to another in successive signaling intervals results in relatively large spectral side lobes outside of the main spectral band of the signal and, consequently, this method requires a large frequency band for transmission of the signal.

To avoid the use of signals having large spectral side lobes, the information-bearing signal frequency modulates a single carrier whose frequency is changed continuously. The resulting frequency-modulated signal is phase-continuous and, hence, it is called *continuous-phase* FSK (CPFSK). This type of FSK signal has memory because the phase of the carrier is constrained to be continuous.

In order to represent a CPFSK signal, we begin with a PAM signal

$$d(t) = \sum_n I_n g(t - nT) \tag{4.3–50}$$

where $\{I_n\}$ denotes the sequence of amplitudes obtained by mapping k-bit blocks of binary digits from the information sequence $\{a_n\}$ into the amplitude levels ± 1,

$\pm 3, \ldots, \pm(M-1)$ and $g(t)$ is a rectangular pulse of amplitude $1/2T$ and duration T seconds. The signal $d(t)$ is used to frequency-modulate the carrier. Consequently, the equivalent low-pass waveform $v(t)$ is expressed as

$$v(t) = \sqrt{\frac{2\mathcal{E}}{T}} \exp\left\{ j\left[4\pi T f_d \int_{-\infty}^{t} d(\tau)\, d\tau + \phi_0 \right]\right\} \tag{4.3–51}$$

where f_d is the *peak frequency deviation* and ϕ_0 is the initial phase of the carrier.

The carrier-modulated signal corresponding to Equation 4.3–51 may be expressed as

$$s(t) = \sqrt{\frac{2\mathcal{E}}{T}} \cos[2\pi f_c t + \phi(t; \mathbf{I}) + \phi_0] \tag{4.3–52}$$

where $\phi(t; \mathbf{I})$ represents the time-varying phase of the carrier, which is defined as

$$\begin{aligned} \phi(t; \mathbf{I}) &= 4\pi T f_d \int_{-\infty}^{t} d(\tau)\, d\tau \\ &= 4\pi T f_d \int_{-\infty}^{t} \left[\sum_n I_n g(\tau - nT) \right] d\tau \end{aligned} \tag{4.3–53}$$

Note that, although $d(t)$ contains discontinuities, the integral of $d(t)$ is continuous. Hence, we have a continuous-phase signal. The phase of the carrier in the interval $nT \leqslant t \leqslant (n+1)T$ is determined by integrating Equation 4.3–53. Thus,

$$\begin{aligned} \phi(t; \mathbf{I}) &= 2\pi f_d T \sum_{k=-\infty}^{n-1} I_k + 2\pi f_d (t - nT) I_n \\ &= \theta_n + 2\pi h I_n q(t - nT) \end{aligned} \tag{4.3–54}$$

where h, θ_n, and $q(t)$ are defined as

$$h = 2 f_d T \tag{4.3–55}$$

$$\theta_n = \pi h \sum_{k=-\infty}^{n-1} I_k \tag{4.3–56}$$

$$q(t) = \begin{cases} 0 & (t < 0) \\ t/2T & (0 \leqslant t \leqslant T) \\ \frac{1}{2} & (t > T) \end{cases} \tag{4.3–57}$$

We observe that θ_n represents the accumulation (memory) of all symbols up to time $(n-1)T$. The parameter h is called the *modulation index*.

Continuous-phase modulation (CPM). When expressed in the form of Equation 4.3–54, CPFSK becomes a special case of a general class of continuous-phase modulated (CPM) signals in which the carrier phase is

$$\phi(t; \mathbf{I}) = 2\pi \sum_{k=-\infty}^{n} I_k h_k q(t - kT), \qquad nT \leqslant t \leqslant (n+1)T \tag{4.3–58}$$

where $\{I_k\}$ is the sequence of M-ary information symbols selected from the alphabet $\pm1, \pm3, \ldots, \pm(M-1)$, $\{h_k\}$ is a sequence of modulation indices, and $q(t)$ is some normalized waveform shape.

When $h_k = h$ for all k, the modulation index is fixed for all symbols. When the modulation index varies from one symbol to another, the CPM signal is called *multi-h*. In such a case, the $\{h_k\}$ are made to vary in a cyclic manner through a set of indices.

The waveform $q(t)$ may be represented in general as the integral of some pulse $g(t)$, i.e.,

$$q(t) = \int_0^t g(\tau)\, d\tau \tag{4.3–59}$$

If $g(t) = 0$ for $t > T$, the CPM signal is called *full response CPM*. If $g(t) \neq 0$ for $t > T$, the modulated signal is called *partial response CPM*. Figure 4.3–16 illustrates several pulse shapes for $g(t)$, and the corresponding $q(t)$. It is apparent that an infinite variety of CPM signals can be generated by choosing different pulse shapes $g(t)$ and by varying the modulation index h and the alphabet size M.

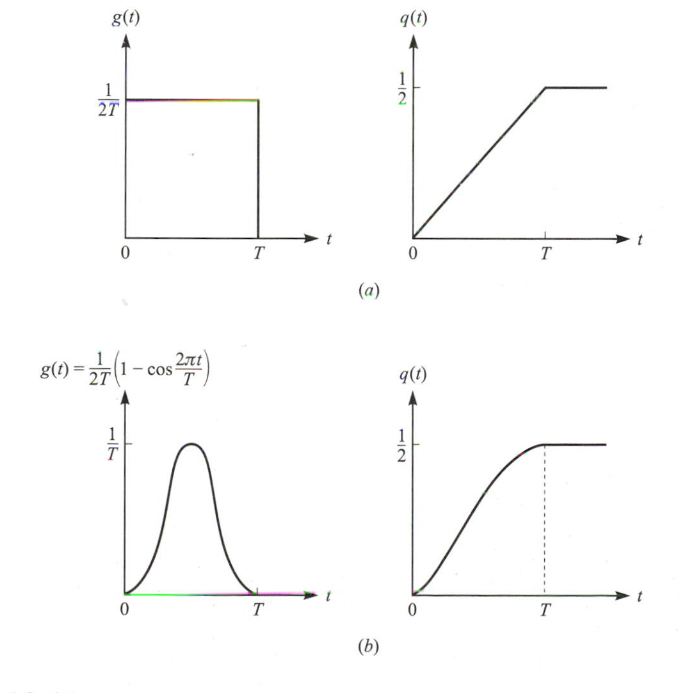

FIGURE 4.3–16
Pulse shapes for full response CPM (**a**, **b**) and partial response CPM (**c**, **d**, **e**).

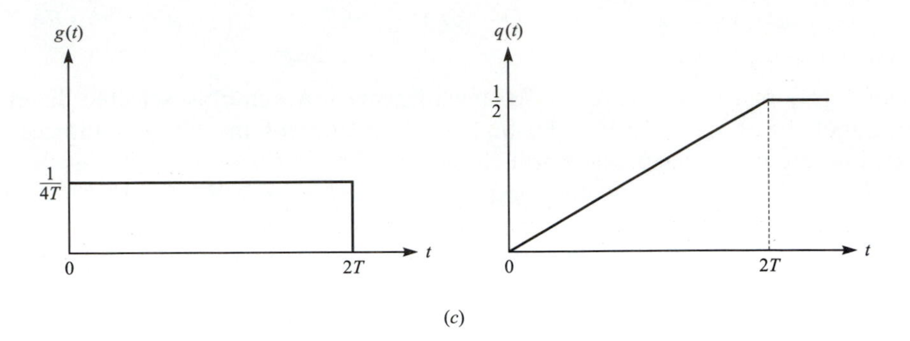

(c)

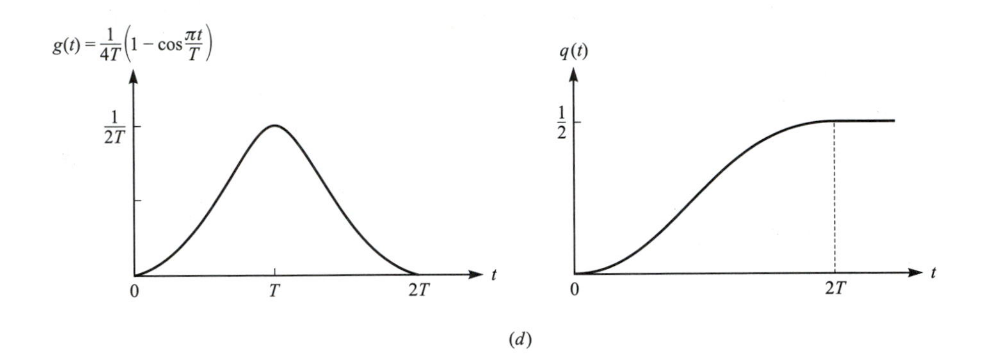

(d)

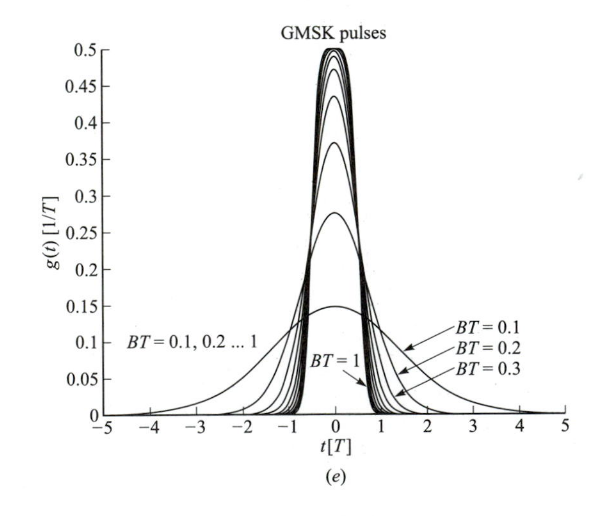

(e)

FIGURE 4.3–16
(*Continued*)

We note that the CPM signal has memory that is introduced through the phase continuity. For $L > 1$, additional memory is introduced in the CPM signal by the pulse $g(t)$.

Three popular pulse shapes are given in Table 4.3–1. LREC denotes a rectangular pulse of duration LT, where L is a positive integer. In this case, $L = 1$ results in a CPFSK signal, with the pulse as shown in Figure 4.3–16a. The LREC pulse for $L = 2$ is shown in Figure 4.3–16c. LRC denotes a raised cosine pulse of duration LT. The LRC pulses corresponding to $L = 1$ and $L = 2$ are shown in Figure 4.3–16b and d, respectively. The third pulse given in Table 4.3–1 is called a Gaussian minimum-shift keying (GMSK) pulse with bandwidth parameter B, which represents the -3 dB bandwidth of the Gaussian pulse. Figure 4.3–16e illustrates a set of GMSK pulses with time-bandwidth products BT ranging from 0.1 to 1. We observe that the pulse duration increases as the bandwidth of the pulse decreases, as expected. In practical applications, the pulse is usually truncated to some specified fixed duration. GMSK with $BT = 0.3$ is used in the European digital cellular communication system, called GSM. From Figure 4.3–16e we observe that when $BT = 0.3$, the GMSK pulse may be truncated at $|t| = 1.5T$ with a relatively small error incurred for $t > 1.5T$.

It is instructive to sketch the set of phase trajectories $\phi(t; \mathbf{I})$ generated by all possible values of the information sequence $\{I_n\}$. For example, in the case of CPFSK with binary symbols $I_n = \pm 1$, the set of phase trajectories beginning at time $t = 0$ is shown in Figure 4.3–17. For comparison, the phase trajectories for quaternary CPFSK are illustrated in Figure 4.3–18. These phase diagrams are called *phase trees*. We observe that the phase trees for CPFSK are piecewise linear as a consequence of the fact that the pulse $g(t)$ is rectangular. Smoother phase trajectories and phase trees are obtained by using pulses that do not contain discontinuities, such as the class of raised cosine pulses. For example, a phase trajectory generated by the sequence $(1, -1, -1, -1, 1, 1, -1, 1)$ for a partial response, raised cosine pulse of length $3T$ is illustrated in Figure 4.3–19. For comparison, the corresponding phase trajectory generated by CPFSK is also shown.

TABLE 4.3–1
Some commonly used CPM pulse shapes

LREC	$g(t) = \begin{cases} \dfrac{1}{2LT} & (0 \leqslant 1 \leqslant LT) \\ 0 & \text{(otherwise)} \end{cases}$
LRC	$g(t) = \begin{cases} \dfrac{1}{2LT}\left(1 - \cos\dfrac{2\pi t}{LT}\right) & (0 \leqslant 1 \leqslant LT) \\ 0 & \text{(otherwise)} \end{cases}$
GMSK	$g(t) = \{Q[2\pi B(t - \frac{T}{2})/(\ln 2)^{1/2}] - Q[2\pi B(t + \frac{T}{2})/(\ln 2)^{1/2}]\}$ $Q(t) = \int_t^\infty \dfrac{1}{\sqrt{2\pi}} e^{-x^2/2}\, dt$

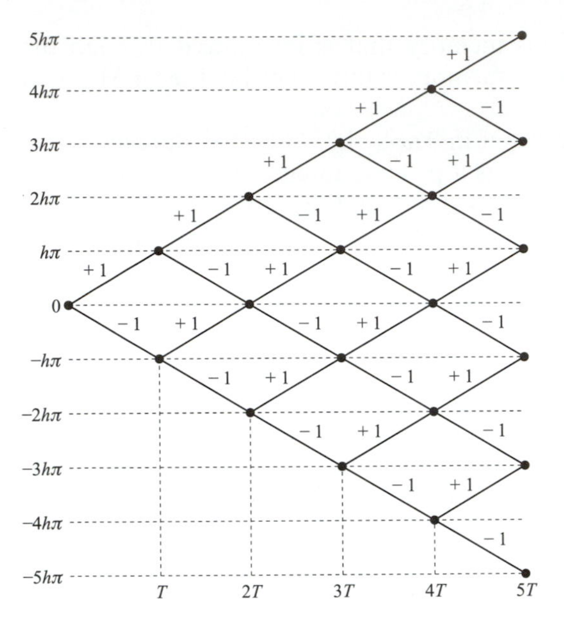

FIGURE 4.3–17
Phase trajectory for binary CPFSK.

The phase trees shown in these figures grow with time. However, the phase of the carrier is unique only in the range from $\phi = 0$ to $\phi = 2\pi$ or, equivalently, from $\phi = -\pi$ to $\phi = \pi$. When the phase trajectories are plotted modulo 2π, say in the range $(-\pi, \pi)$, the phase tree collapses into a structure called a *phase trellis*. To properly view the phase trellis diagram, we may plot the two quadrature components $x_c(t; \mathbf{I}) = \cos \phi(t; \mathbf{I})$ and $x_s(t; \mathbf{I}) = \sin \phi(t; \mathbf{I})$ as functions of time. Thus, we generate a three-dimensional plot in which the quadrature components x_c and x_s appear on the surface of a cylinder of unit radius. For example, Figure 4.3–20 illustrates the phase trellis or phase cylinder obtained with binary modulation, a modulation index $h = \frac{1}{2}$, and a raised cosine pulse of length $3T$.

Simpler representations for the phase trajectories can be obtained by displaying only the terminal values of the signal phase at the time instants $t = nT$. In this case, we restrict the modulation index of the CPM signal to be rational. In particular, let us assume that $h = m/p$, where m and p are relatively prime integers. Then, a full response CPM signal at the time instants $t = nT$ will have the terminal phase states

$$\Theta_s = \left\{0, \frac{\pi m}{p}, \frac{2\pi m}{p}, \dots, \frac{(p-1)\pi m}{p}\right\} \tag{4.3–60}$$

when m is even and

$$\Theta_s = \left\{0, \frac{\pi m}{p}, \frac{2\pi m}{p}, \dots, \frac{(2p-1)\pi m}{p}\right\} \tag{4.3–61}$$

when m is odd. Hence, there are p terminal phase states when m is even and $2p$ states when m is odd. On the other hand, when the pulse shape extends over L

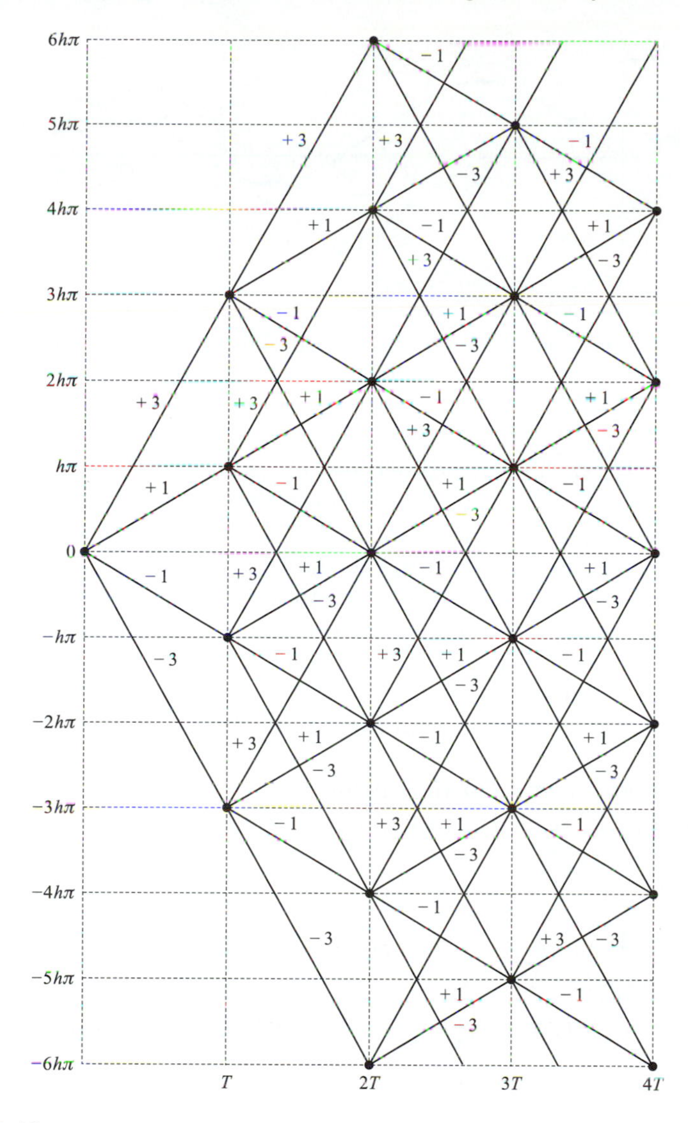

FIGURE 4.3–18
Phase trajectory for quaternary CPFSK.

symbol intervals (partial response CPM), the number of phase states may increase up to a maximum of S_t, where

$$S_t = \begin{cases} pM^{L-1} & \text{(even } m\text{)} \\ 2pM^{L-1} & \text{(odd } m\text{)} \end{cases} \tag{4.3–62}$$

where M is the alphabet size. For example, the binary CPFSK signal (full response, rectangular pulse) with $h = \frac{1}{2}$, has $S_t = 4$ (terminal) phase states. The

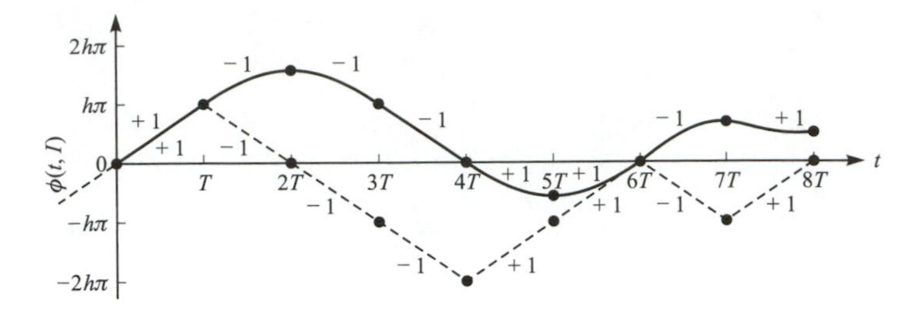

FIGURE 4.3–19
Phase trajectories for binary CPFSK (dashed) and binary, partial response CPM based on raised cosine pulse of length $3T$ (solid). [*From Sundberg (1986), © 1986 IEEE.*]

state trellis for this signal is illustrated in Figure 4.3–21. We emphasize that the phase transitions from one state to another are not true phase trajectories. They represent phase transitions for the (terminal) states at the time instants $t = nT$.

An alternative representation to the state trellis is the state diagram, which also illustrates the state transitions at the time instants $t = nT$. This is an even more compact representation of the CPM signal characteristics. Only the possible (terminal) phase states and their transitions are displayed in the state diagram. Time does not appear explicitly as a variable. For example, the state diagram for the CPFSK signal with $h = \frac{1}{2}$ is shown in Figure 4.3–22.

Minimum-shift keying (MSK). MSK is a special form of binary CPFSK (and, therefore, CPM) in which the modulation index $h = \frac{1}{2}$. The phase of the carrier in the interval $nT \leqslant t \leqslant (n+1)T$ is

$$\begin{aligned} \phi(t; \mathbf{I}) &= \tfrac{1}{2}\pi \sum_{k=-\infty}^{n-1} I_k + \pi I_n q(t - nT) \\ &= \theta_n + \tfrac{1}{2}\pi I_n\left(\frac{t - nT}{T}\right), \qquad nT \leqslant t \leqslant (n+1)T \end{aligned} \tag{4.3–63}$$

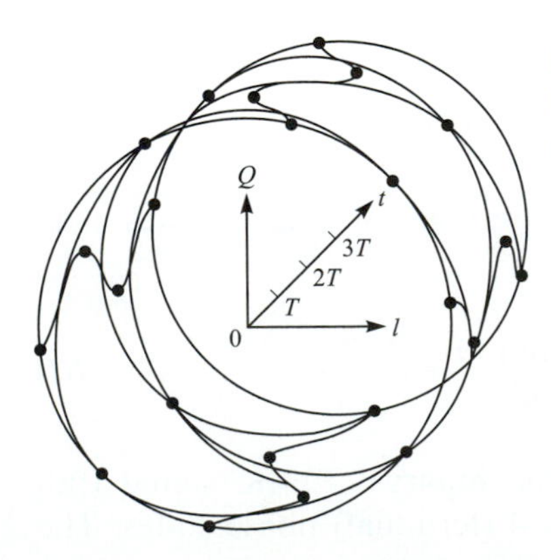

FIGURE 4.3–20
Phase cylinder for binary CPM with $h = \frac{1}{2}$ and a raised cosine pulse of length $3T$. [*From Sundberg (1986), © 1986 IEEE.*]

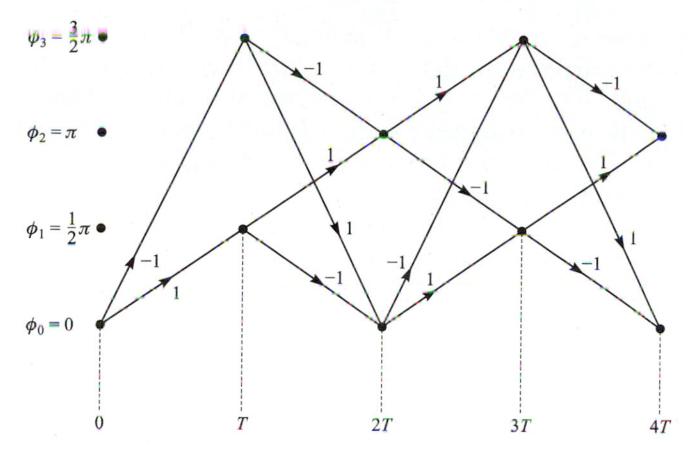

FIGURE 4.3–21
State trellis for binary CPFSK with $h = \frac{1}{2}$.

and the modulated carrier signal is

$$\begin{aligned} s(t) &= A\cos\left[2\pi f_c t + \theta_n + \tfrac{1}{2}\pi I_n\left(\frac{t-nT}{T}\right)\right] \\ &= A\cos\left[2\pi\left(f_c + \frac{1}{4T}I_n\right)t - \tfrac{1}{2}n\pi I_n + \theta_n\right], \qquad nT \leqslant t \leqslant (n+1)T \end{aligned} \tag{4.3–64}$$

The expression 4.3–64 indicates that the binary CPFSK signal can be expressed as a sinusoid having one of two possible frequencies in the interval $nT \leqslant t \leqslant (n+1)T$. If we define these frequencies as

$$\begin{aligned} f_1 &= f_c - \frac{1}{4T} \\ f_2 &= f_c + \frac{1}{4T} \end{aligned} \tag{4.3–65}$$

then the binary CPFSK signal given by Equation 4.3–64 may be written in the form

$$s_i(t) = A\cos[2\pi f_i t + \theta_n + \tfrac{1}{2}n\pi(-1)^{i-1}], \qquad i = 1, 2 \tag{4.3–66}$$

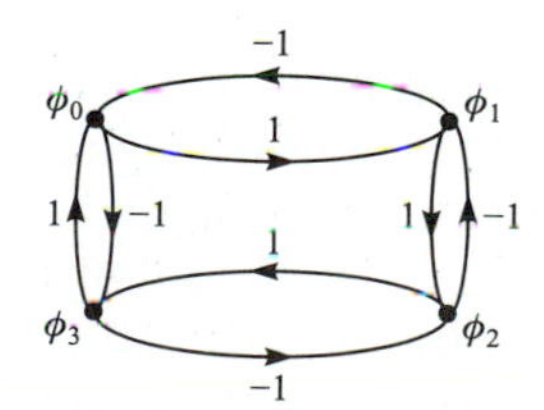

FIGURE 4.3–22
State diagram for binary CPFSK with $h = \frac{1}{2}$.

The frequency separation $\Delta f = f_2 - f_1 = 1/2T$. Recall that $\Delta f = 1/2T$ is the minimum frequency separation that is necessary to ensure the orthogonality of the signals $s_1(t)$ and $s_2(t)$ over a signaling interval of length T. This explains why binary CPFSK with $h = \frac{1}{2}$ is called minimum-shift keying (MSK). The phase in the nth signaling interval is the phase state of the signal that results in phase continuity between adjacent intervals.

MSK may also be represented as a form of four-phase PSK. Specifically, we may express the equivalent low-pass digitally modulated signal in the form (see Problem 4.14)

$$v(t) = \sum_{n=-\infty}^{\infty} [I_{2n} g(t - 2nT) - jI_{2n+1} g(t - 2nT - T)] \tag{4.3–67}$$

where $g(t)$ is a sinusoidal pulse defined as

$$g(t) = \begin{cases} \sin \dfrac{\pi t}{2T} & (0 \leqslant t \leqslant 2T) \\ 0 & (\text{otherwise}) \end{cases} \tag{4.3–68}$$

Thus, this type of signal is viewed as a four-phase PSK signal in which the pulse shape is one-half cycle of a sinusoid. The even-numbered binary-valued (± 1) symbols $\{I_{2n}\}$ of the information sequence $\{I_n\}$ are transmitted via the cosine of the carrier, while the odd-numbered symbols $\{I_{2n+1}\}$ are transmitted via the sine of the carrier. The transmission rate on the two orthogonal carrier components is $1/2T$ bits/s so that the combined transmission rate is $1/T$ bits/s. Note that the bit transitions on the sine and cosine carrier components are staggered or offset in time by T seconds. For this reason, the signal

$$\begin{aligned} s(t) = A\Bigg\{&\left[\sum_{n=-\infty}^{\infty} I_{2n} g(t - 2nT)\right] \cos 2\pi f_c t \\ &+ \left[\sum_{n=-\infty}^{\infty} I_{2n+1} g(t - 2nT - T)\right] \sin 2\pi f_c t\Bigg\} \end{aligned} \tag{4.3–69}$$

is called *offset quadrature PSK (OQPSK)* or *staggered quadrature PSK (SQPSK)*.

Figure 4.3–23 illustrates the representation of an MSK signal as two staggered quadrature-modulated binary PSK signals. The corresponding sum of the two quadrature signals is a constant amplitude, frequency-modulated signal.

It is also interesting to compare the waveforms for MSK with offset QPSK in which the pulse $g(t)$ is rectangular for $0 \leqslant t \leqslant 2T$, and with conventional quadrature (four-phase) PSK (QPSK) in which the pulse $g(t)$ is rectangular for $0 \leqslant t \leqslant 2T$. Clearly, all three of the modulation methods result in identical data rates. The MSK signal has continuous phase. The offset QPSK signal with a rectangular pulse is basically two binary PSK signals for which the phase transitions are staggered in time by T seconds. Thus, the signal contains phase jumps of $\pm 90^\circ$ that may occur as often as every T seconds. On the other hand, the conventional four-phase PSK signal with constant amplitude will

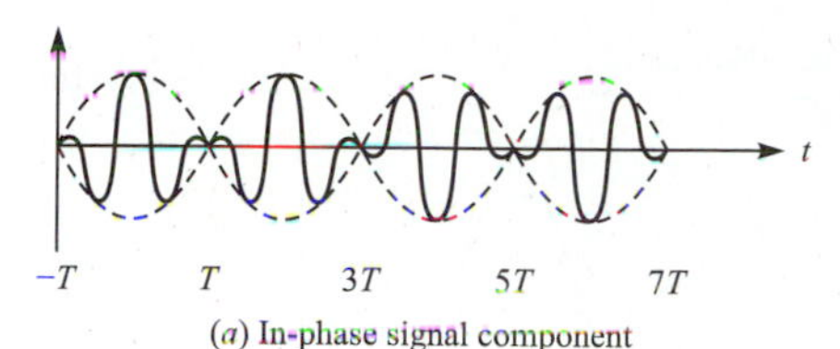

(a) In-phase signal component

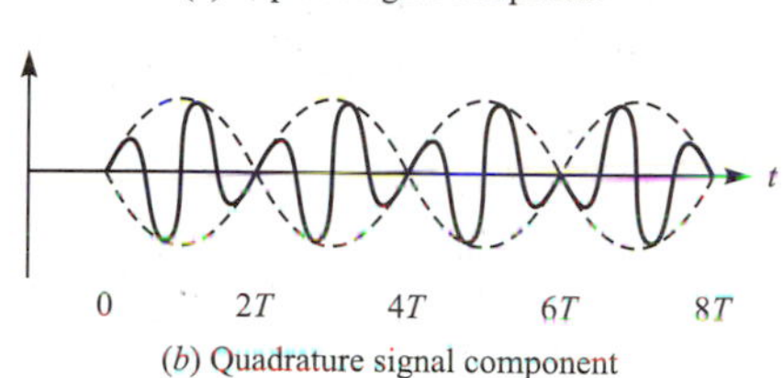

(b) Quadrature signal component

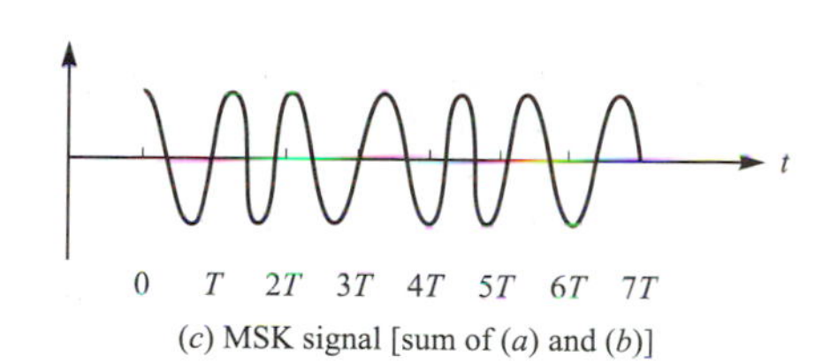

(c) MSK signal [sum of (a) and (b)]

FIGURE 4.3–23
Representation of an MSK signal as a form of two staggered binary PSK signals, each with a sinusoidal envelope.

contain phase jumps of $\pm 180^\circ$ or $\pm 90^\circ$ every $2T$ seconds. An illustration of these three signal types is given in Figure 4.3–24.

Signal space diagrams for CPM. In general, continuous-phase signals cannot be represented by discrete points in signal space as in the case of PAM, PSK, and QAM, because the phase of the carrier is time-variant. Instead, a continuous-phase signal is described by the various paths or trajectories from one phase state to another. For a constant-amplitude CPM signal, the various trajactories form a circle.

For example, Figure 4.3–25 illustrates the signal space (phase trajectory) diagram for CPFSK signals with $h = \frac{1}{4}$, $h = \frac{1}{3}$, $h = \frac{1}{2}$, and $h = \frac{2}{3}$. The beginning and ending points of these phase trajectories are marked in the figure by dots. Note that the length of the phase trajectory increases with an increase in h. An increase in h also results in an increase of the signal bandwidth, as demonstrated in the following section.

A linear representation of CPM. As described above, CPM is a non-linear modulation technique with memory. However, CPM may also be represented as a linear superposition of signal waveforms. Such a representation provides an alternative method for generating the modulated signal at the transmitter and/or demodulating the signal at the receiver.

Following the development originally given by Laurent (1986), we demonstrate that binary CPM may be represented by a linear superposition of a finite

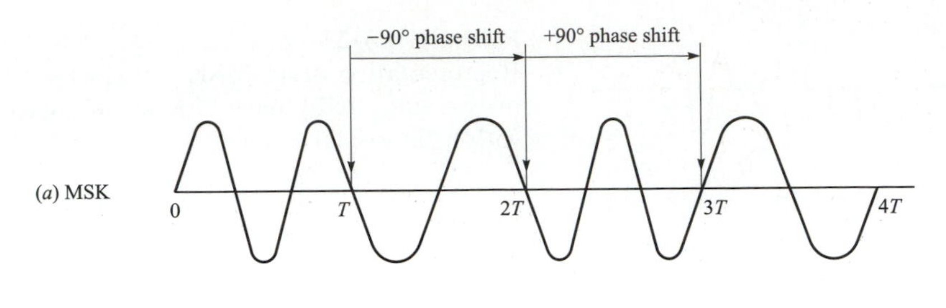

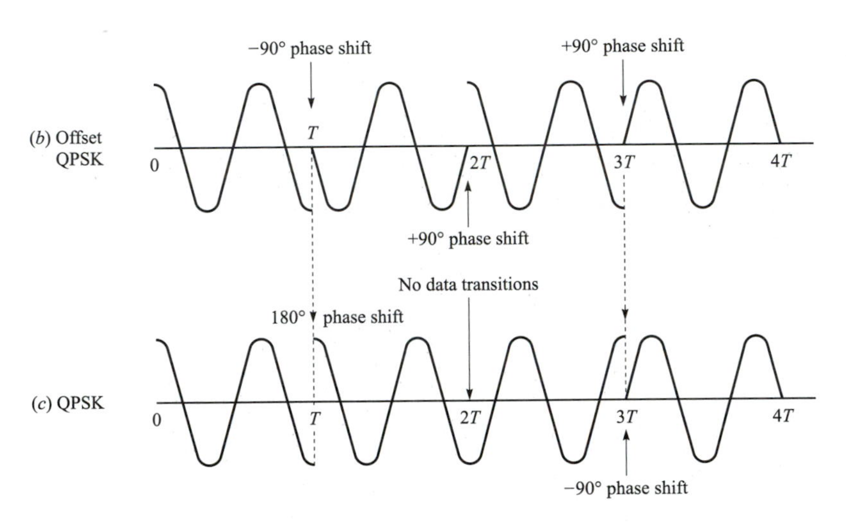

FIGURE 4.3–24
Signal waveforms for (**a**) MSK, (**b**) offset QPSK (rectangular pulse), and (**c**) conventional QPSK (rectangular pulse). [*From Gronemeyer and McBride (1976); © 1976 IEEE.*]

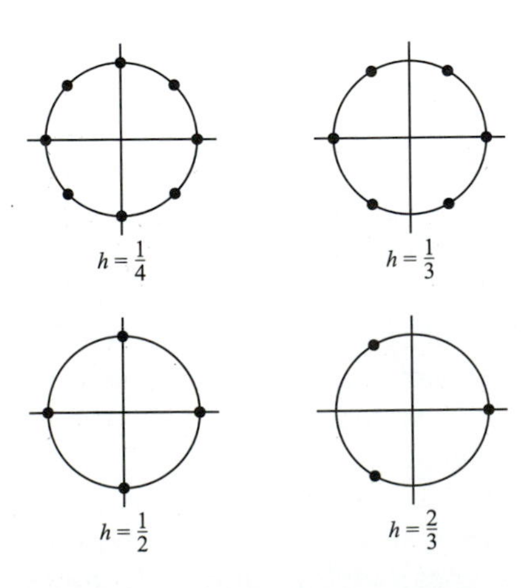

FIGURE 4.3–25
Signal space diagram for CPFSK.

number of amplitude-modulated pulses, provided that the pulse $g(t)$ is of finite duration LT, where T is the bit interval. We begin with the equivalent low-pass representation of CPM, which is

$$v(t) = \sqrt{\frac{2\mathcal{E}}{T}} e^{j\phi(t;\mathbf{I})}, \qquad nT \leqslant t \leqslant (n+1)T \tag{4.3–70}$$

where

$$\begin{aligned} \phi(t;\mathbf{I}) &= 2\pi h \sum_{k=-\infty}^{n} I_k q(t-kT), \qquad nT \leqslant t \leqslant (n+1)T \\ &= \pi h \sum_{k=-\infty}^{n-L} I_k + 2\pi h \sum_{k=n-L+1}^{n} I_k q(t-kT) \end{aligned} \tag{4.3–71}$$

and $q(t)$ is the integral of the pulse $g(t)$, as previously defined in Equation 4.3–59.

The exponential term $\exp[j\phi(t;\mathbf{I})]$ may be expressed as

$$\exp[j\phi(t;\mathbf{I})] = \exp\left(j\pi h \sum_{k=-\infty}^{n-L} I_k \right) \prod_{k=0}^{L-1} \exp\{ j2\pi h I_{n-k} q[t-(n-k)T] \} \tag{4.3–72}$$

Note that the first term on the right-hand side of Equation 4.3–72 represents the cumulative phase up to the information symbol I_{n-L}, and the second term consists of a product of L phase terms. Assuming that the modulation index h is not an integer and the data symbols are binary, i.e., $I_k = \pm 1$, the kth phase term may be expressed as

$$\begin{aligned} \exp\{j2\pi h I_{n-k} q[t-(n-k)T]\} &= \frac{\sin \pi h}{\sin \pi h} \exp\{j2\pi h I_{n-k} q[t-(n-k)T]\} \\ &= \frac{\sin\{\pi h - 2\pi h q[t-(n-k)]T\}}{\sin \pi h} \\ &\quad + \exp(j\pi h I_{n-k}) \frac{\sin\{2\pi h q[t-(n-k)T]\}}{\sin \pi h} \end{aligned} \tag{4.3–73}$$

It is convenient to define the signal pulse $s_0(t)$ as

$$s_0(t) = \begin{cases} \dfrac{\sin 2\pi h q(t)}{\sin \pi h} & (0 \leqslant t < LT) \\ \dfrac{\sin[\pi h - 2\pi h q(t-LT)]}{\sin \pi h} & (LT \leqslant t \leqslant 2LT) \\ 0 & (\text{otherwise}) \end{cases} \tag{4.3–74}$$

Then,

$$\exp[j\phi(t;\mathbf{I})] = \exp\left(j\pi h \sum_{k=-\infty}^{n-L} I_k \right) \prod_{k=0}^{L-1} \{s_0[t+(k+L-n)T] + \exp(j\pi h I_{n-k}) s_0[t-(k-n)T]\} \tag{4.3–75}$$

By performing the multiplication over the L terms in the product, we obtain a sum of 2^L terms, where 2^{L-1} terms are distinct and the other 2^{L-1} terms are time-shifted versions of the distinct terms. The final result may be expressed as

$$\exp[j\phi(t;\mathbf{I})] = \sum_n \sum_{k=0}^{2^{L-1}-1} e^{j\pi h A_{k,n}} c_k(t-nT) \tag{4.3–76}$$

where the pulses $c_k(t)$, for $0 \leqslant k \leqslant 2^{L-1}-1$, are defined as

$$c_k(t) = s_0(t)\prod_{n=1}^{L-1} s_0[t+(n+La_{k,n})T], \qquad 0 \leqslant t \leqslant T \cdot \min_{\{n\}} [L(2-a_{k,n})-n] \tag{4.3–77}$$

Note that each pulse is weighted by a complex coefficient $\exp(j\pi h A_{k,n})$, where

$$A_{k,n} = \sum_{m=-\infty}^{n} I_m - \sum_{m=1}^{L-1} I_{n-m} a_{k,m} \tag{4.3–78}$$

and the $\{a_{k,n} = 0 \text{ or } 1\}$ are the coefficients in the binary representation of the index k, i.e.,

$$k = \sum_{m=1}^{L-1} 2^{m-1} a_{k,m}, \qquad k = 0, 1, \ldots, 2^{L-1}-1 \tag{4.3–79}$$

Thus, the binary CPM signal is expressed as a weighted sum of 2^{L-1} real-valued pulses $\{c_k(t)\}$.

In this representation of CPM as a superposition of amplitude-modulated pulses, the pulse $c_0(t)$ is the most important component, because its duration is the longest and it contains the most significant part of the signal energy. Consequently, a simple approximation to a CPM signal is a partial response PAM signal having $c_0(t)$ as the basic pulse shape.

The focus for the above development was binary PCM. A representation of M-ary CPM as a superposition of PAM waveforms has been described by Mengali and Morelli (1995).

EXAMPLE 4.3–1. As a special case, let us consider the MSK signal, for which $h = \frac{1}{2}$ and $g(t)$ is a rectangular pulse of duration T. In this case,

$$\begin{aligned}\phi(t;\mathbf{I}) &= \frac{\pi}{2}\sum_{k=-\infty}^{n-1} I_k + \pi I_n q(t-nT) \\ &= \theta_n + \frac{\pi}{2} I_n \left(\frac{t-nT}{T}\right), \qquad nT \leqslant t \leqslant (n+1)T\end{aligned}$$

and

$$\exp[j\phi(t;\mathbf{I})] = \sum_n b_n c_0(t-nT)$$

where

$$c_0(t) = \begin{cases} \sin \frac{\pi t}{2T} & (0 \leqslant t \leqslant 2T) \\ 0 & (\text{otherwise}) \end{cases}$$

and

$$b_n = e^{j\pi A_{0,n}/2} = e^{j\pi(\theta_n + I_n)/2}$$

The complex-valued modified data sequence $\{b_n\}$ may be expressed recursively as

$$b_n = jb_{n-1}I_n$$

so that b_n alternates in taking real and imaginary values. By separating the real and the imaginary components, we obtain the equivalent low-pass signal representation given by Equations 4.3–67 and 4.3–68.

Multiamplitude CPM. Multiamplitude CPM is a generalization of ordinary CPM in which the signal amplitude is allowed to vary over a set of amplitude values while the phase of the signal is constrained to be continuous. For example, let us consider a two-amplitude CPFSK signal, which may be represented as

$$s(t) = 2A\cos[2\pi f_c t + \phi_2(t; \mathbf{I})] + A\cos[2\pi f_c t + \phi_1(t; \mathbf{J})] \tag{4.3–80}$$

where

$$\phi_2(t; \mathbf{I}) = \pi h \sum_{k=-\infty}^{n-1} I_k + \frac{\pi h I_n(t - nT)}{T}, \qquad nT \leqslant t \leqslant (n+1)T \tag{4.3–81}$$

$$\phi_1(t; \mathbf{J}) = \pi h \sum_{k=-\infty}^{n-1} J_k + \frac{\pi h J_n(t - nT)}{T}, \qquad nT \leqslant t \leqslant (n+1)T \tag{4.3–82}$$

The information is conveyed by the symbol sequences $\{I_n\}$ and $\{J_n\}$, which are related to two independent binary information sequences $\{a_n\}$ and $\{b_n\}$ that take values $\{0, 1\}$. We observe that the signal in Equation 4.3–80 is a superposition of two CPFSK signals of different amplitude. However, the sequences $\{I_n\}$ are $\{J_n\}$ are not statistically independent, but are constrained in order to achieve phase continuity in the superposition of the two components.

To elaborate, let us consider the case where $h = \frac{1}{2}$, so that we have the superposition of two MSK signals. At the symbol transition points, the two amplitude components are either in phase or 180° out of phase. The phase change in the signal is determined by the phase of the larger amplitude component, while the amplitude change is determined by the smaller component. Thus, the smaller component is constrained such that at the start and end of each symbol interval, it is either in phase or 180° out of phase with the larger component, independent of its phase. Under this constraint, the symbol sequences $\{I_n\}$ and $\{J_n\}$ may be expressed as

$$\begin{aligned} I_n &= 2a_n - 1 \\ J_n &= I_n(1 - 2b_n) = I_n\left(1 - \frac{b_n}{h}\right) \end{aligned} \tag{4.3–83}$$

TABLE 4.3–2

a_n	b_n	I_n	J_n	**Amplitude–phase relations**
0	0	−1	−1	Amplitude is constant; phase decreases
0	1	−1	1	Amplitude changes; phase decreases
1	0	1	1	Amplitude is constant; phase increases
1	1	1	−1	Amplitude changes; phase increases

These relationships are summarized in Table 4.3–2.

As a generalization, a multiamplitude CPFSK signal with n components may be expressed as

$$s(t) = 2^{N-1}\cos[2\pi f_c t + \phi_N(t; \mathbf{I})] + \sum_{m=1}^{N-1} 2^{m-1}\cos[2\pi f_c t + \phi_m(t; \mathbf{J}_m)] \quad (4.3\text{–}84)$$

where

$$\phi_N(t; \mathbf{I}) = \pi h I_n \frac{t - nT}{T} + \pi h \sum_{k=-\infty}^{n-1} I_k, \qquad nT \leqslant t \leqslant (n+1)T \quad (4.3\text{–}85)$$

and

$$\begin{aligned}\phi_m(t; \mathbf{J}_m) = {} & I_n \pi[h + \tfrac{1}{2}(J_{mn} + 1)]\frac{t - nT}{T} \\ & + \sum_{k=-\infty}^{n-1} \pi I_k[h + \tfrac{1}{2}(J_{mk} + 1)], \qquad nT \leqslant t \leqslant (n+1)T\end{aligned} \quad (4.3\text{–}86)$$

The sequences $\{I_n\}$ and $\{J_{mn}\}$ are statistically independent, binary-valued sequences that take values from the set $\{1, -1\}$.

From Equations 4.3–85 and 4.3–86, we observe that each component in the sum will be either in phase or 180° out of phase with the largest component at the end of the nth symbol interval, i.e., at $t - (n+1)T$. Thus, the signal states are specified by an amplitude level from the set of amplitudes $\{1, 3, 5, \ldots, 2^N - 1\}$ and a phase level from the set $\{0, \pi\theta, 2\pi\theta, \ldots, 2\pi - \pi h\}$. The phase constraint is required to maintain the phase continuity of the CPM signal.

Figure 4.3–26 illustrates the signal space diagrams for two-amplitude ($N = 2$) CPFSK with $h = \frac{1}{4}, \frac{1}{3}, \frac{1}{2}$, and $\frac{2}{3}$. The signal space diagrams for three-component ($N = 3$) CPFSK are shown in Figure 4.3–27. In this case, there are *four* amplitude levels. The number of states depends on the modulation index h as well as N. Note that the beginning and ending points of the phase trajectories are marked by dots.

Additional multiamplitude CPM signal formats may be obtained by using pulse shapes other than rectangular, as well as signal pulses that span more than one symbol (partial response).

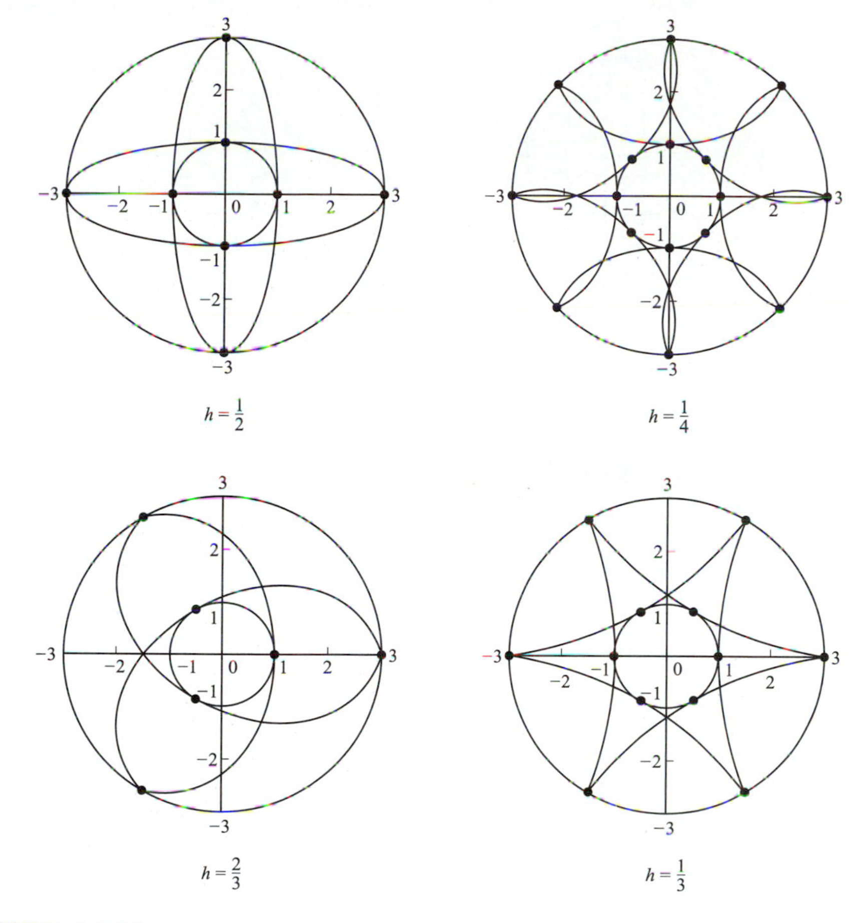

FIGURE 4.3–26
Signal space diagrams for two-component CPFSK.

4.4
SPECTRAL CHARACTERISTICS OF DIGITALLY MODULATED SIGNALS

In most digital communication systems, the available channel bandwidth is limited. Consequently, the system designer must consider the constraints imposed by the channel bandwidth limitation in the selection of the modulation technique used to transmit the information. For this reason, it is important for us to determine the spectral content of the digitally modulated signals described in Section 4.3.

Since the information sequence is random, a digitally modulated signal is a stochastic process. We are interested in determining the power density spectrum of such a process. From the power density spectrum, we can determine the

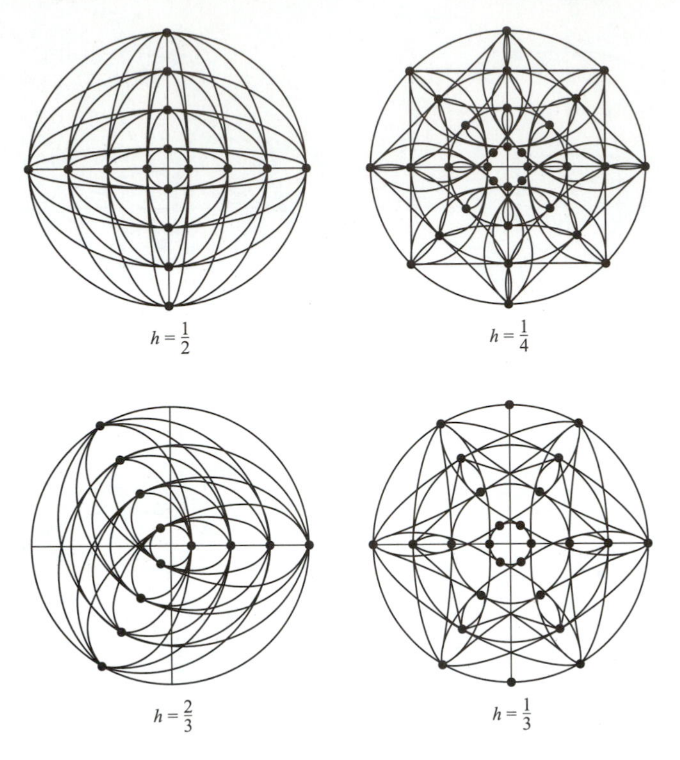

FIGURE 4.3–27
Signal space diagrams for three-component CPFSK.

channel bandwidth required to transmit the information-bearing signal. Below, we first derive the spectral characteristics of the class of linearly modulated signals. Then, we consider the non-linear CPFSK, CPM, and baseband modulated signals with memory.

4.4.1 Power Spectra of Linearly Modulated Signals

Beginning with the form

$$s(t) = \mathrm{Re}[v(t)e^{j2\pi f_c t}]$$

which relates the band-pass signal $s(t)$ to the equivalent low-pass signal $v(t)$, we may express the autocorrelation function of $s(t)$ as

$$\phi_{ss}(\tau) = \mathrm{Re}[\phi_{vv}(\tau)e^{j2\pi f_c \tau}] \tag{4.4–1}$$

where $\phi_{vv}(\tau)$ is the autocorrelation function of the equivalent low-pass signal $v(t)$. The Fourier transform of Equation 4.4–1 yields the desired expression for the power density spectrum $\Phi_{ss}(f)$ in the form

$$\Phi_{ss}(f) = \tfrac{1}{2}[\Phi_{vv}(f - f_c) + \Phi_{vv}(-f - f_c)] \tag{4.4–2}$$

where $\Phi_{vv}(f)$ is the power density spectrum of $v(t)$. It suffices to determine the autocorrelation function and the power density spectrum of the equivalent low-pass signal $v(t)$.

First we consider the linear digital modulation methods for which $v(t)$ is represented in the general form

$$v(t) = \sum_{n=-\infty}^{\infty} I_n g(t - nT) \tag{4.4–3}$$

where the transmission rate is $1/T = R/k$ symbols/s and $\{I_n\}$ represents the sequence of symbols that results from mapping k-bit blocks into corresponding signal points selected from the appropriate signal space diagram. Observe that in PAM, the sequence $\{I_n\}$ is real and corresponds to the amplitude values of the transmitted signal, but in PSK, QAM, and combined PAM–PSK, the sequence $\{I_n\}$ is complex-valued, since the signal points have a two-dimensional representation.

The autocorrelation function of $v(t)$ is

$$\begin{aligned} \phi_{vv}(t+\tau; t) &= \tfrac{1}{2}E[v^*(t)v(t+\tau)] \\ &= \tfrac{1}{2}\sum_{n=-\infty}^{\infty}\sum_{m=-\infty}^{\infty} E[I_n^* I_m] g^*(t-nT)g(t+\tau-mT) \end{aligned} \tag{4.4–4}$$

We assume that the sequence of information symbols $\{I_n\}$ is wide-sense stationary with mean μ_i and autocorrelation function

$$\phi_{ii}(m) = \tfrac{1}{2}E[I_n^* I_{n+m}] \tag{4.4–5}$$

Hence Equation 4.4–4 can be expressed as

$$\begin{aligned} \phi_{vv}(t+\tau; t) &= \sum_{n=-\infty}^{\infty}\sum_{m=-\infty}^{\infty} \phi_{ii}(m-n) g^*(t-nT)g(t+\tau-mT) \\ &= \sum_{m=-\infty}^{\infty} \phi_{ii}(m) \sum_{n=-\infty}^{\infty} g^*(t-nT)g(t+\tau-nT-mT) \end{aligned} \tag{4.4–6}$$

The second summation in Equation 4.4–6, namely

$$\sum_{n=-\infty}^{\infty} g^*(t-nT)g(t+\tau-nT-mT)$$

is periodic in the t variable with period T. Consequently, $\phi_{vv}(t+\tau; t)$ is also periodic in the t variable with period T. That is,

$$\phi_{vv}(t+T+\tau; t+T) = \phi_{vv}(t+\tau; t) \tag{4.4–7}$$

In addition, the mean value of $v(t)$, which is

$$E[v(t)] = \mu_i \sum_{n=-\infty}^{\infty} g(t-nT) \tag{4.4–8}$$

is periodic with period T. Therefore $v(t)$ is a stochastic process having a periodic mean and autocorrelation function. Such a process is called a *cyclostationary process* or a *periodically stationary process in the wide sense*, as described in Section 2.2.6.

In order to compute the power density spectrum of a cyclostationary process, the dependence of $\phi_{vv}(t+\tau; t)$ on the t variable must be eliminated. This can be accomplished simply by averaging $\phi_{vv}(t+\tau; t)$ over a single period. Thus,

$$\begin{aligned}\bar{\phi}_{vv}(\tau) &= \frac{1}{T}\int_{-T/2}^{T/2} \phi_{vv}(t+\tau; t)\,dt \\ &= \sum_{m=-\infty}^{\infty} \phi_{ii}(m) \sum_{n=-\infty}^{\infty} \frac{1}{T}\int_{-T/2}^{T/2} g^*(t-nT)g(t+\tau-nT-mT)\,dt \\ &= \sum_{m=-\infty}^{\infty} \phi_{ii}(m) \sum_{n=-\infty}^{\infty} \frac{1}{T}\int_{-T/2-nT}^{T/2-nT} g^*(t)g(t+\tau-mT)\,dt\end{aligned} \tag{4.4–9}$$

We interpret the integral in Equation 4.4–9 as the time-autocorrelation function of $g(t)$ and define it as

$$\phi_{gg}(\tau) = \int_{-\infty}^{\infty} g^*(t)g(t+\tau)\,dt \tag{4.4–10}$$

Consequently Equation 4.4–9 can be expressed as

$$\bar{\phi}_{vv}(\tau) = \frac{1}{T}\sum_{m=-\infty}^{\infty} \phi_{ii}(m)\phi_{gg}(\tau - mT) \tag{4.4–11}$$

The Fourier transform of the relation in Equation 4.4–11 yields the (average) power density spectrum of $v(t)$ in the form

$$\Phi_{vv}(f) = \frac{1}{T}|G(f)|^2\Phi_{ii}(f) \tag{4.4–12}$$

where $G(f)$ is the Fourier transform of $g(t)$, and $\Phi_{ii}(f)$ denotes the power density spectrum of the information sequence, defined as

$$\Phi_{ii}(f) = \sum_{m=-\infty}^{\infty} \phi_{ii}(m)e^{-j2\pi fmT} \tag{4.4–13}$$

The result 4.4–12 illustrates the dependence of the power density spectrum of $v(t)$ on the spectral characteristics of the pulse $g(t)$ and the information sequence $\{I_n\}$. That is, the spectral characterstics of $v(t)$ can be controlled by design of the pulse shape $g(t)$ and by design of the correlation characteristics of the information sequence.

Whereas the dependence of $\Phi_{vv}(f)$ on $G(f)$ is easily understood upon observation of Equation 4.4–12, the effect of the correlation properties of the information sequence is more subtle. First of all, we note that for an arbitrary autocorrelation $\phi_{ii}(m)$ the corresponding power density spectrum $\Phi_{ii}(f)$ is periodic in frequency with period $1/T$. In fact, the expression 4.4–13 relating the spectrum $\Phi_{ii}(f)$ to the autocorrelation $\phi_{ii}(m)$ is in the form of an exponential Fourier series with the $\{\phi_{ii}(m)\}$ as the Fourier coefficients. As a consequence, the autocorrelation sequence $\phi_{ii}(m)$ is given by

$$\phi_{ii}(m) = T \int_{-1/2T}^{1/2T} \Phi_{ii}(f) e^{j2\pi fmT} \, df \tag{4.4–14}$$

Second, let us consider the case in which the information symbols in the sequence are real and mutually uncorrelated. In this case, the autocorrelation function $\phi_{ii}(m)$ can be expressed as

$$\phi_{ii}(m) = \begin{cases} \sigma_i^2 + \mu_i^2 & (m = 0) \\ \mu_i^2 & (m \neq 0) \end{cases} \tag{4.4–15}$$

where σ_i^2 denotes the variance of an information symbol. When Equation 4.4–15 is used to substitute for $\phi_{ii}(m)$ in Equation 4.4–13, we obtain

$$\Phi_{ii}(f) = \sigma_i^2 + \mu_i^2 \sum_{m=-\infty}^{\infty} e^{-j2\pi fmT} \tag{4.4–16}$$

The summation in Equation 4.4–16 is periodic with period $1/T$. It may be viewed as the exponential Fourier series of a periodic train of impulses with each impulse having an area $1/T$. Therefore Equation 4.4–16 can also be expressed in the form

$$\Phi_{ii}(f) = \sigma_i^2 + \frac{\mu_i^2}{T} \sum_{m=-\infty}^{\infty} \delta\left(f - \frac{m}{T}\right) \tag{4.4–17}$$

Substitution of Equation 4.4–17 into Equation 4.4–12 yields the desired result for the power density spectrum of $v(t)$ when the sequence of information symbols is uncorrelated. That is,

$$\Phi_{vv}(f) = \frac{\sigma_i^2}{T} |G(f)|^2 + \frac{\mu_i^2}{T^2} \sum_{m=-\infty}^{\infty} \left| G\left(\frac{m}{T}\right) \right|^2 \delta\left(f - \frac{m}{T}\right) \tag{4.4–18}$$

The expression 4.4–18 for the power density spectrum is purposely separated into two terms to emphasize the two different types of spectral components. The first term is the continuous spectrum, and its shape depends only on the spectral characteristic of the signal pulse $g(t)$. The second term consists of discrete frequency components spaced $1/T$ apart in frequency. Each spectral line has a power that is proportional to $|G(f)|^2$ evaluated at $f = m/T$. Note that the discrete frequency components vanish when the information symbols have zero mean, i.e., $\mu_i = 0$. This condition is usually desirable for the digital modulation techniques under consideration, and it is satisfied when the information symbols are equally likely and symmetrically positioned in the complex plane. Thus, the system designer can control the spectral characteristics of the digitally modulated signal by proper selection of the characteristics of the information sequence to be transmitted.

EXAMPLE 4.4–1. To illustrate the spectral shaping resulting from $g(t)$, consider the rectangular pulse shown in Figure 4.4–1a. The Fourier transform of $g(t)$ is

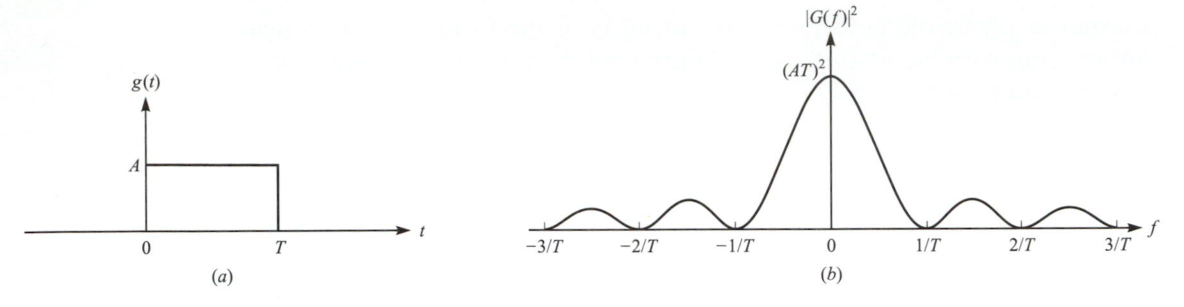

FIGURE 4.4–1
Rectangular pulse and its energy density spectrum $|G(f)|^2$.

$$G(f) = AT\frac{\sin \pi fT}{\pi fT}e^{-j\pi fT}$$

Hence

$$|G(f)|^2 = (AT)^2\left(\frac{\sin \pi fT}{\pi fT}\right)^2 \tag{4.4–19}$$

This spectrum is illustrated in Figure 4.4–1b. Note that it contains zeros at multiples of $1/T$ in frequency and that it decays inversely as the square of the frequency variable. As a consequence of the spectral zeros in $G(f)$, all but one of the discrete spectral components in Equation 4.4–18 vanish. Thus, upon substitution for $|G(f)|^2$ from Equation 4.4–19, Equation 4.4–18 reduces to

$$\Phi_{vv}(f) = \sigma_i^2 A^2 T\left(\frac{\sin \pi fT}{\pi fT}\right)^2 + A^2\mu_i^2\delta(f) \tag{4.4–20}$$

EXAMPLE 4.4–2. As a second illustration of the spectral shaping resulting from $g(t)$, we consider the raised cosine pulse

$$g(t) = \frac{A}{2}\left[1 + \cos\frac{2\pi}{T}\left(t - \frac{T}{2}\right)\right], \qquad 0 \leqslant t \leqslant T \tag{4.4–21}$$

This pulse is graphically illustrated in Figure 4.4–2a. Its Fourier transform is easily derived, and it may be expressed in the form

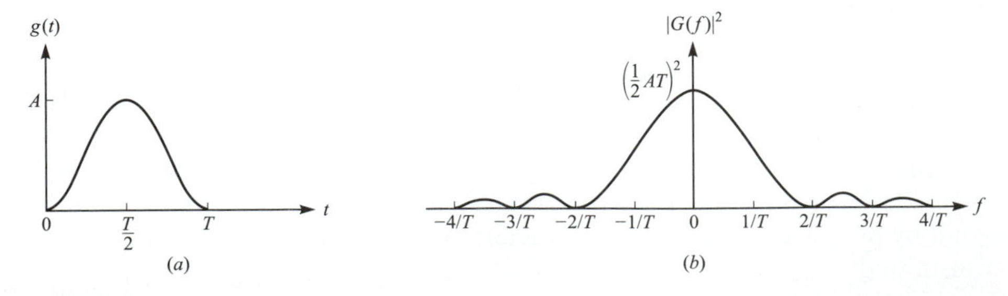

FIGURE 4.4–2
Raised cosine pulse and its energy density spectrum $|G(f)|^2$.

$$G(f) = \frac{AT}{2} \frac{\sin \pi f T}{\pi f T(1 - f^2 T^2)} e^{-j\pi f T} \tag{4.4–22}$$

The square of the magnitude of $G(f)$ is shown in Figure 4.4–2b. It is interesting to note that the spectrum has zeros at $f = n/T$, $n = \pm 2, \pm 3, \pm 4, \ldots$. Consequently, all the discrete spectral components in Equation 4.4–18, except the one at $f = 0$ and $f = \pm 1/T$, vanish. When compared with the spectrum of the rectangular pulse, the spectrum of the raised cosine pulse has a broader main lobe but the tails decay inversely as f^6.

EXAMPLE 4.4–3. To illustrate that spectral shaping can also be accomplished by operations performed on the input information sequence, we consider a binary sequence $\{b_n\}$ from which we form the symbols

$$I_n = b_n + b_{n-1} \tag{4.4–23}$$

The $\{b_n\}$ are assumed to be uncorrelated random variables, each having zero mean and unit variance. Then the autocorrelation function of the sequence $\{I_n\}$ is

$$\begin{aligned} \phi_{ii}(m) &= E(I_n I_{n+m}) \\ &= \begin{cases} 2 & (m = 0) \\ 1 & (m = \pm 1) \\ 0 & (\text{otherwise}) \end{cases} \end{aligned} \tag{4.4–24}$$

Hence, the power density spectrum of the input sequence is

$$\begin{aligned} \Phi_{ii}(f) &= 2(1 + \cos 2\pi f T) \\ &= 4 \cos^2 \pi f T \end{aligned} \tag{4.4–25}$$

and the corresponding power density spectrum for the (low-pass) modulated signal is

$$\Phi_{vv}(f) = \frac{4}{T} |G(f)|^2 \cos^2 \pi f T \tag{4.4–26}$$

4.4.2 Power Spectra of CPFSK and CPM Signals

In this section, we derive the power density spectrum for the class of constant amplitude CPM signals that were described in Section 4.3.3. We begin by computing the autocorrelation function and its Fourier transform, as was done in the case of linearly modulated signals.

The constant amplitude CPM signal is expressed as

$$s(t; \mathbf{I}) = A \cos[2\pi f_c t + \phi(t; \mathbf{I})] \tag{4.4–27}$$

where

$$\phi(t; \mathbf{I}) = 2\pi h \sum_{k=-\infty}^{\infty} I_k q(t - kT) \tag{4.4–28}$$

Each symbol in the sequence $\{I_n\}$ can take one of the M values $\{\pm 1, \pm 3, \ldots, \pm(M-1)\}$. These symbols are statistically independent and identically distributed with prior probabilities

$$P_n = P(I_k = n), \qquad n = \pm 1, \pm 3, \ldots, \pm(M-1) \tag{4.4–29}$$

where $\sum_n P_n = 1$. The pulse $g(t) = q'(t)$ is zero outside of the interval $[0, LT]$, $q(t) = 0$, $t < 0$, and $q(t) = \frac{1}{2}$ for $t > LT$.

The autocorrelation function of the equivalent low-pass signal

$$v(t) = e^{j\phi(t;\mathbf{I})}$$

is

$$\phi_{vv}(t+\tau; t) = \tfrac{1}{2}E\left[\exp\left(j2\pi h \sum_{k=-\infty}^{\infty} I_k[q(t+\tau-kT) - q(t-kT)]\right)\right] \tag{4.4–30}$$

First we express the sum in the exponent as a product of exponents. The result is

$$\phi_{vv}(t+\tau; t) = \tfrac{1}{2}E\left(\prod_{k=-\infty}^{\infty} \exp\{j2\pi h I_k[q(t+\tau-kT) - q(t-kT)]\}\right) \tag{4.4–31}$$

Next, we perform the expectation over the data symbols $\{I_k\}$. Since these symbols are statistically independent, we obtain

$$\phi_{vv}(t+\tau; t) = \tfrac{1}{2} \prod_{k=-\infty}^{\infty} \left(\sum_{\substack{n=-(M-1) \\ n \text{ odd}}}^{M-1} P_n \exp\{j2\pi h n[q(t+\tau-kT) = q(t-kT)]\} \right) \tag{4.4–32}$$

Finally, the average autocorrelation function is

$$\bar{\phi}_{vv}(\tau) = \frac{1}{T}\int_0^T \phi_{vv}(t+\tau; t)\, dt \tag{4.4–33}$$

Although Equation 4.4–32 implies that there is an infinite number of factors in the product, the pulse $g(t) = q'(t) = 0$ for $t < 0$ and $t > LT$, and $q(t) = 0$ for $t < 0$. Consequently only a finite number of terms in the product have nonzero exponents. Thus Equation 4.4–32 can be simplified considerably. In addition, if we let $\tau = \xi + mT$, where $0 \leqslant \xi < T$ and $m = 0, 1, \ldots,$ the average autocorrelation in Equation 4.4–33 reduces to

$$\bar{\phi}_{vv}(\xi + mT) = \frac{1}{2T}\int_0^T \prod_{k=1-L}^{m+1} \left(\sum_{\substack{n=-(M-1) \\ n \text{ odd}}}^{M-1} P_n \exp\{j2\pi h n[q(t+\xi-(k-m)T) - q(t-kT)]\} \right) \tag{4.4–34}$$

Let us focus on $\bar{\phi}_{vv}(\xi + mT)$ for $\xi + mT \geqslant LT$. In this case, Equation 4.4–34 may be expressed as

$$\bar{\phi}_{vv}(\xi + mT) = [\psi(jh)]^{m-L}\lambda(\xi), \qquad m \geqslant L, \qquad 0 \leqslant \xi < T \tag{4.4–35}$$

where $\psi(jh)$ is the characteristic function of the random sequence $\{I_n\}$, defined as

$$\begin{aligned}\psi(jh) &= E(e^{j\pi h I_n}) \\ &= \sum_{\substack{n=-(M-1) \\ n \text{ odd}}}^{M-1} P_n e^{j\pi h n}\end{aligned} \tag{4.4–36}$$

and $\lambda(\xi)$ is the remaining part of the average autocorrelation function, which may be expressed as

$$\begin{aligned}\lambda(\xi) = \frac{1}{2T}\int_0^T \prod_{k=1-L}^{0} &\left(\sum_{\substack{n=-(M-1) \\ n \text{ odd}}}^{M-1} P_n \exp\{j2\pi hn[\tfrac{1}{2} - q(t-kT)]\} \right) \\ \times \prod_{k=1-L}^{1} &\left(\sum_{\substack{n=-(M-1) \\ n \text{ odd}}}^{M-1} P_n \exp[j2\pi hnq(t+\xi-kT)] \right) dt, \qquad m \geqslant L\end{aligned} \tag{4.4–37}$$

Thus, $\bar{\phi}_{vv}(\tau)$ may be separated into a product of $\lambda(\xi)$ and $\psi(jh)$ as indicated in Equation 4.4–35 for $\tau = \xi + mT \geqslant LT$ and $0 \leqslant \xi < T$. This property is used below.

The Fourier transform of $\bar{\phi}_{vv}(\tau)$ yields the average power density spectrum as

$$\begin{aligned}\Phi_{vv}(f) &= \int_{\infty}^{\infty} \bar{\phi}_{vv}(\tau) e^{-j2\pi f\tau}\, dt \\ &= 2\,\mathrm{Re}\left[\int_0^{\infty} \bar{\phi}_{vv}(\tau) e^{-j2\pi f\tau}\, d\tau\right]\end{aligned} \tag{4.4–38}$$

But

$$\begin{aligned}\int_0^{\infty} \bar{\phi}_{vv}(\tau) e^{-j2\pi f\tau}\, d\tau = &\int_0^{LT} \bar{\phi}_{vv}(\tau) e^{-j2\pi f\tau}\, d\tau \\ &+ \int_{LT}^{\infty} \bar{\phi}_{vv}(\tau) e^{-j2\pi f\tau}\, d\tau\end{aligned} \tag{4.4–39}$$

With the aid of Equation 4.4–35, the integral in the range $LT \leqslant \tau < \infty$ may be expressed as

$$\int_{LT}^{\infty} \bar{\phi}_{vv}(\tau) e^{-2\pi f\tau}\, d\tau = \sum_{m=L}^{\infty} \int_{mT}^{(m+1)T} \bar{\phi}_{vv}(\tau) e^{-j2\pi f\tau}\, d\tau \tag{4.4–40}$$

Now, let $\tau = \xi + mT$. Then Equation 4.4–40 becomes

$$\begin{aligned}\int_{LT}^{\infty} \bar{\phi}_{vv}(\tau)e^{-j2\pi f\tau}\,dt &= \sum_{m=L}^{\infty} \int_0^T \bar{\phi}_{vv}(\xi + mT)e^{-j2\pi f(\xi+mT)}\,d\xi \\ &= \sum_{m=L}^{\infty} \int_0^T \lambda(\xi)[\psi(jh)]^{m-L}e^{-j2\pi f(\xi+mT)}\,d\xi \\ &= \sum_{n=0}^{\infty} \psi^n(jh)e^{-j2\pi fnT} \int_0^T \lambda(\xi)e^{-j2\pi f(\xi+LT)}\,d\xi\end{aligned} \tag{4.4–41}$$

A property of the characteristic function is $|\psi(jh)| \leqslant 1$. For values of h for which $|\psi(jh)| \leqslant 1$, the summation in Equation 4.4–41 converges and yields

$$\sum_{n=0}^{\infty} \psi^n(jh)e^{-j2\pi fnT} = \frac{1}{1 - \psi(jh)e^{-j2\pi fT}} \tag{4.4–42}$$

In this case, Equation 4.4–41 reduces to

$$\int_{LT}^{\infty} \bar{\phi}_{vv}(\tau)e^{-j2\pi f\tau}\,dt = \frac{1}{1 - \psi(jh)e^{-j2\pi fT}} \int_0^T \bar{\phi}_{vv}(\xi + LT)e^{-j2\pi f(\xi+LT)}\,d\xi \tag{4.4–43}$$

By combining Equations 4.4–38, 4.4–39, and 4.4–43, we obtain the power density spectrum of the CPM signal in the form

$$\Phi_{vv}(f) = 2\,\text{Re}\left[\int_0^{LT} \bar{\phi}_{vv}(\tau)e^{-j2\pi f\tau}\,d\tau + \frac{1}{1 - \psi(jh)e^{-j2\pi fT}} \int_{LT}^{(L+1)T} \bar{\phi}_{vv}(\tau)e^{-j2\pi f\tau}\,d\tau\right. \tag{4.4–44}$$

This is the desired result when $|\psi(jh)| < 1$. In general, the power density spectrum is evaluated numerically from Equation 4.4–44. The average autocorrelation function $\bar{\phi}_{vv}(\tau)$ for the range $0 \leqslant \tau \leqslant (L+1)T$ may be computed numerically from Equation 4.4–34.

For values of h for which $|\psi(jh)| = 1$, e.g., $h = K$, where K is an integer, we can set

$$\psi(jh) = e^{j2\pi\nu}, \qquad 0 \leqslant \nu < 1 \tag{4.4–45}$$

Then, the sum in Equation 4.4–41 becomes

$$\sum_{n=0}^{\infty} e^{-j2\pi T(f-\nu/T)n} = \tfrac{1}{2} + \frac{1}{2T}\sum_{n=-\infty}^{\infty} \delta\left(f - \frac{\nu}{T} - \frac{n}{T}\right) - j\tfrac{1}{2}\cot \pi T\left(f - \frac{\nu}{T}\right) \tag{4.4–46}$$

Thus, the power density spectrum now contains impulses located at frequencies

$$f_n = \frac{n+\nu}{T}, \qquad 0 \leqslant \nu < 1, \qquad n = 0, 1, 2, \ldots \tag{4.4–47}$$

The result 4.4–46 can be combined with Equations 4.4–41 and 4.4–39 to obtain the entire power density spectrum, which includes both a continuous spectrum component and a discrete spectrum component.

Let us return to the case for which $|\psi(jh)| < 1$. When symbols are equally probable, i.e.,

$$P_n = \frac{1}{M} \text{ for all } n$$

the characteristic function simplifies to the form

$$\begin{aligned}\psi(jh) &= \frac{1}{M} \sum_{\substack{n=-(M-1)\\ n \text{ odd}}}^{M-1} e^{j\pi hn} \\ &= \frac{1}{M} \frac{\sin M\pi h}{\sin \pi h}\end{aligned} \tag{4.4–48}$$

Note that in this case $\psi(jh)$ is real. The average autocorrelation function given by Equation 4.4–34 also simplifies in this case to

$$\bar{\phi}_{vv}(\tau) = \frac{1}{2T} \int_0^T \prod_{k=1-L}^{[\tau/T]} \frac{1}{M} \frac{\sin 2\pi h M[q(t+\tau-kT) - q(t-kT)]}{\sin 2\pi h[q(t+\tau-kT) - q(t-kT)]} dt \tag{4.4–49}$$

The corresponding expression for the power density spectrum reduces to

$$\begin{aligned}\Phi_{vv}(f) = 2\Bigg[&\int_0^{LT} \bar{\phi}_{vv}(\tau) \cos 2\pi f\tau \, d\tau \\ &+ \frac{1 - \psi(jh)\cos 2\pi fT}{1 + \psi^2(jh) - 2\psi(jh)\cos 2\pi fT} \int_{LT}^{(L+1)T} \bar{\phi}_{vv}(\tau) \cos 2\pi f\tau \, d\tau \\ &- \frac{\psi(jh)\sin 2\pi fT}{1 + \psi^2(jh) - 2\psi(jh)\cos 2\pi fT} \int_{LT}^{(L+1)T} \bar{\phi}_{vv}(\tau) \sin 2\pi f\tau \, d\tau\Bigg]\end{aligned} \tag{4.4–50}$$

Power density spectrum of CPFSK. A closed-form expression for the power density spectrum can be obtained from Equation 4.4–50 when the pulse shape $g(t)$ is rectangular and zero outside the interval $[0, T]$. In this case, $q(t)$ is linear for $0 \leqslant t \leqslant T$. The resulting power spectrum may be expressed as

$$\Phi_{vv}(f) = T\left[\frac{1}{M}\sum_{n=1}^{M} A_n^2(f) + \frac{2}{M^2}\sum_{n=1}^{M}\sum_{m=1}^{M} B_{nm}(f) A_n(f) A_m(f)\right] \tag{4.4–51}$$

where

$$A_n(f) = \frac{\sin \pi[fT - \frac{1}{2}(2n-1-M)h]}{\pi[fT - \frac{1}{2}(2n-1-M)h]}$$

$$B_{nm}(f) = \frac{\cos(2\pi fT - \alpha_{nm}) - \psi \cos \alpha_{nm}}{1 + \psi^2 - 2\psi \cos 2\pi fT} \quad (4.4\text{–}52)$$

$$\alpha_{nm} = \pi h(m + n - 1 - M)$$

$$\psi \equiv \psi(jh) = \frac{\sin M\pi h}{M \sin \pi h}$$

The power density spectrum of CPFSK for $M = 2, 4$, and 8 is plotted in Figures 4.4–3 to 4.4–5 as a function of the normalized frequency fT, with the modulation index $h = 2f_d T$ as a parameter. Note that only one-half of the bandwidth occupancy is shown in these graphs. The origin corresponds to the carrier f_c. The graphs illustrate that the spectrum of CPFSK is relatively smooth and well confined for $h < 1$. As h approaches unity, the spectra become very peaked and, for $h = 1$ when $|\psi| = 1$, we find that impulses occur at M frequencies. When $h > 1$, the spectrum becomes much broader. In communication systems where CPFSK is used, the modulation index is designed to conserve bandwidth, so that $h < 1$.

The special case of binary CPFSK with $h = \frac{1}{2}$ (or $f_d = 1/4T$) and $\psi = 0$ corresponds to MSK. In this case, the spectrum of the signal is

$$\Phi_{vv}(f) = \frac{16A^2 T}{\pi^2}\left(\frac{\cos 2\pi fT}{1 - 16f^2T^2}\right)^2 \quad (4.4\text{–}53)$$

where the signal amplitude $A = 1$ in Equation 4.4–52. In contrast the spectrum of four-phase offset (quadrature) PSK (OQPSK) with a rectangular pulse $g(t)$ of duration T is

$$\Phi_{vv}(f) = A^2 T\left(\frac{\sin \pi fT}{\pi fT}\right)^2 \quad (4.4\text{–}54)$$

If we compare these spectral characteristics, we should normalize the frequency variable by the bit rate or the bit interval T_b. Since MSK is binary FSK, it follows that $T = T_b$ in Equation 4.4–53. On the other hand, in OQPSK, $T = 2T_b$ so that Equation 4.4–54 becomes

$$\Phi_{vv}(f) = 2A^2 T_b\left(\frac{\sin 2\pi fT_b}{2\pi fT_b}\right)^2 \quad (4.4\text{–}55)$$

The spectra of the MSK and OQPSK signals are illustrated in Figure 4.4–6. Note that the main lobe of MSK is 50 percent wider than that for OQPSK. However, the side lobes in MSK fall off considerably faster. For example, if we compare the bandwidth W that contains 99 percent of the total power, we find that $W = 1.2T_b$ for MSK and $W \approx 8/T_b$ for OQPSK. Consequently, MSK has a narrower spectral occupancy when viewed in terms of fractional out-of-band

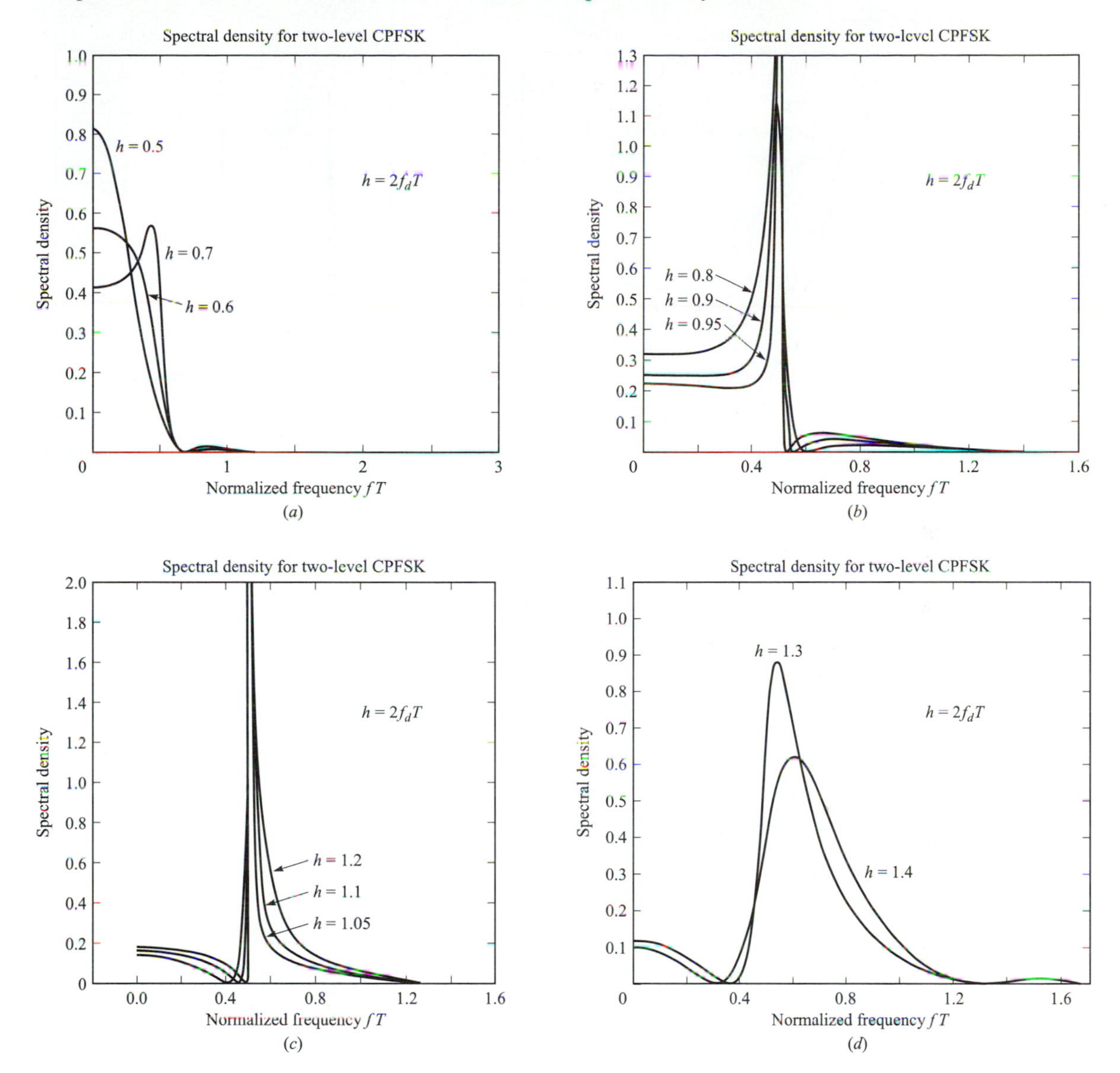

FIGURE 4.4–3
Power density spectrum of binary CPFSK.

power above $fT_b = 1$. Graphs for the fractional out-of-band power for OQPSK and MSK are shown in Figure 4.4–7. Note that MSK is significantly more bandwidth-efficient than QPSK. This efficiency accounts for the popularity of MSK in many digital communication systems.

Even greater bandwidth efficiency than MSK can be achieved by reducing the modulation index. However, the FSK signals will no longer be orthogonal and there will be an increase in the error probability.

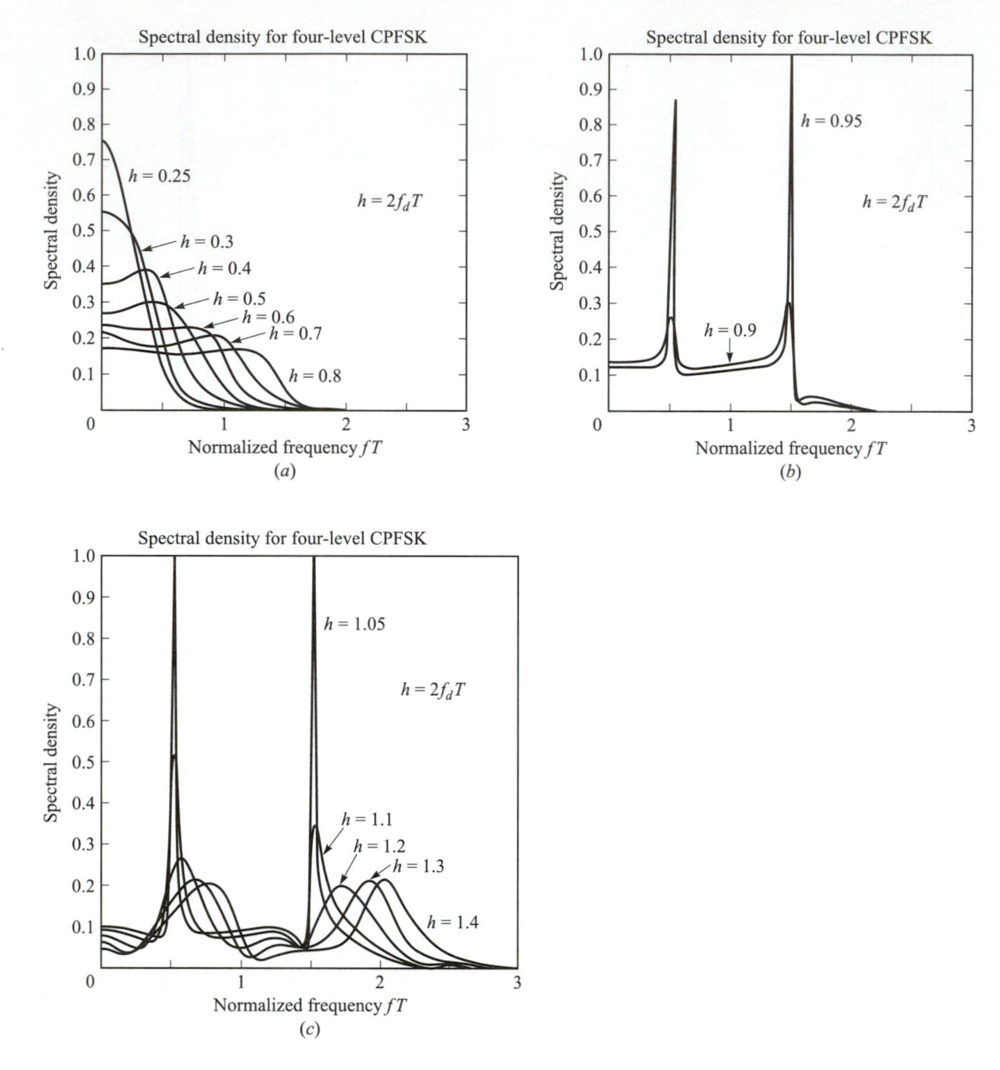

FIGURE 4.4–4
Power density spectrum of quaternary CPFSK.

Spectral characteristics of CPM. In general, the bandwidth occupancy of CPM depends on the choice of the modulation index h, the pulse shape $g(t)$, and the number of signals M. As we have observed for CPFSK, small values of h result in CPM signals with relatively small bandwidth occupancy, while large values of h result in signals with large bandwidth occupancy. This is also the case for the more general CPM signals.

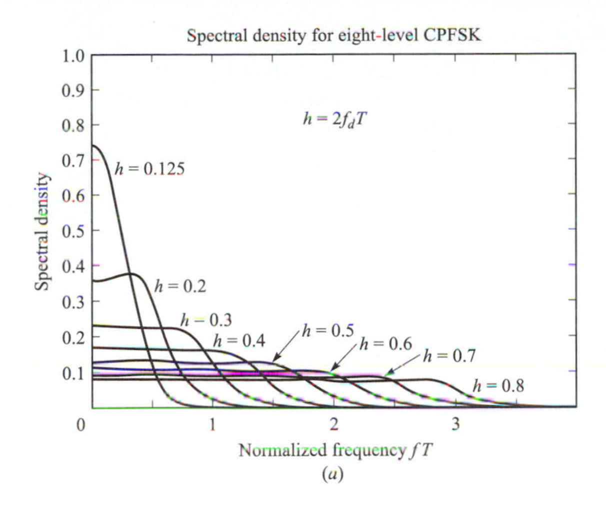

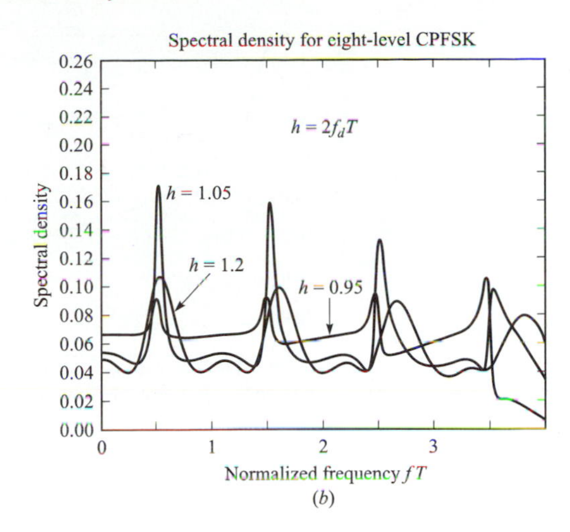

FIGURE 4.4–5
Power density spectrum of octal CPFSK.

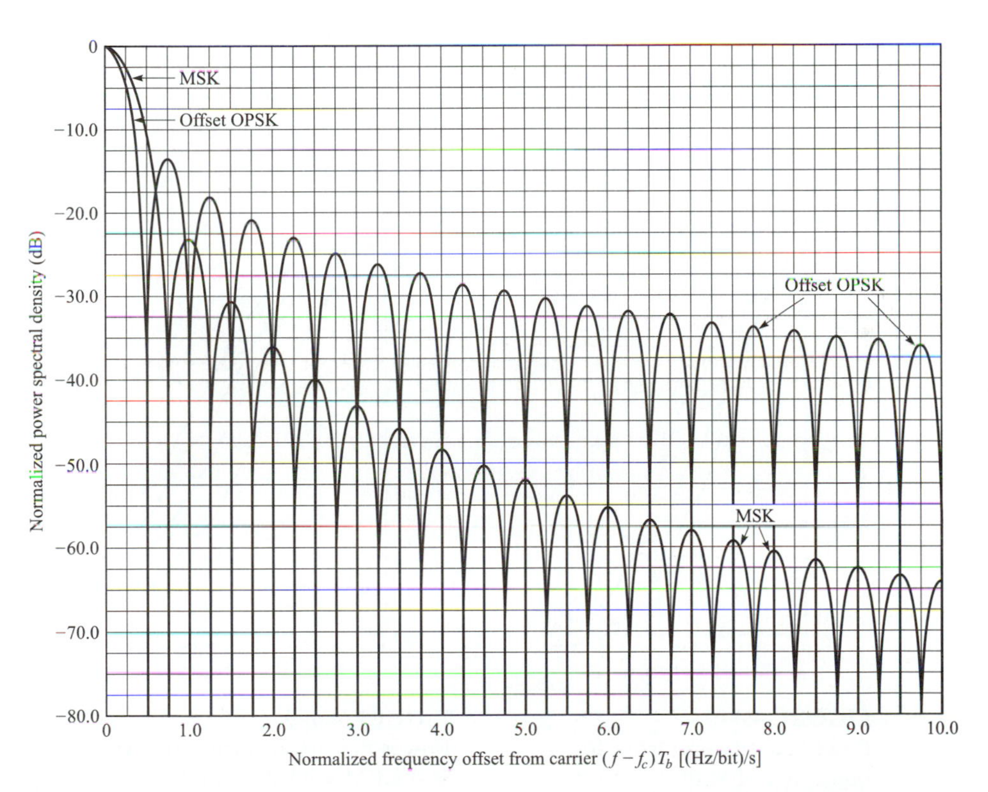

FIGURE 4.4–6
Power density spectra of MSK and offset QPSK. [*From Gronemeyer and McBride (1976); © IEEE.*]

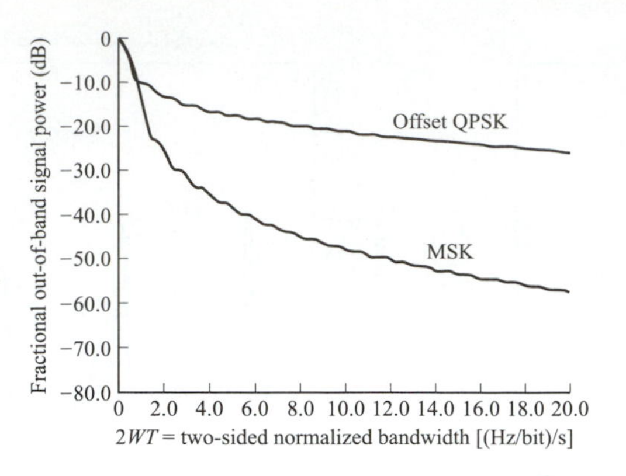

FIGURE 4.4–7
Fractional out-of-band power (normalized two-sided bandwidth $= 2WT$). [*From Gronemeyer and McBride (1976); © 1976 IEEE.*]

The use of smooth pulses such as raised cosine pulses of the form

$$g(t) = \begin{cases} \frac{1}{2LT}\left(1 - \cos\frac{2\pi t}{LT}\right) & (0 \leqslant t \leqslant LT) \\ 0 & (\text{otherwise}) \end{cases} \tag{4.4–56}$$

where $L = 1$ for full response and $L > 1$ for partial response, result in smaller bandwidth occupancy and, hence, greater bandwidth efficiency than the use of rectangular pulses. For example, Figure 4.4–8 illustrates the power density spectrum for binary CPM with different partial response raised cosine (LRC) pulses

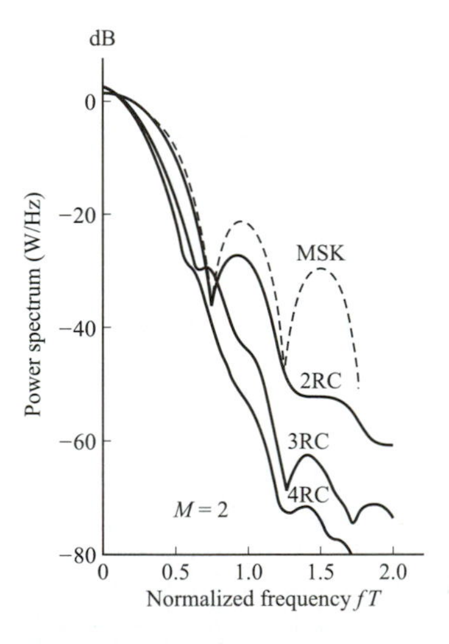

FIGURE 4.4–8
Power density spectrum for binary CPM with $h = \frac{1}{2}$ and different pulse shapes. [*From Aulin et al. (1981); © 1981 IEEE.*]

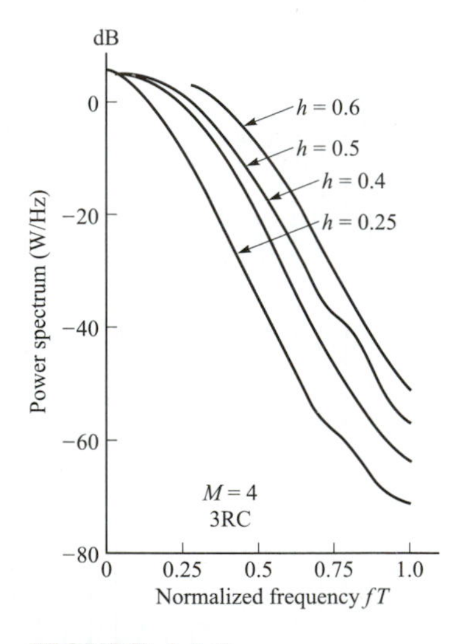

FIGURE 4.4–9
Power density spectrum for $M = 4$ CPM with $3RC$ and different modulation indices. [*From Aulin et al.* (1981); © *1981 IEEE.*]

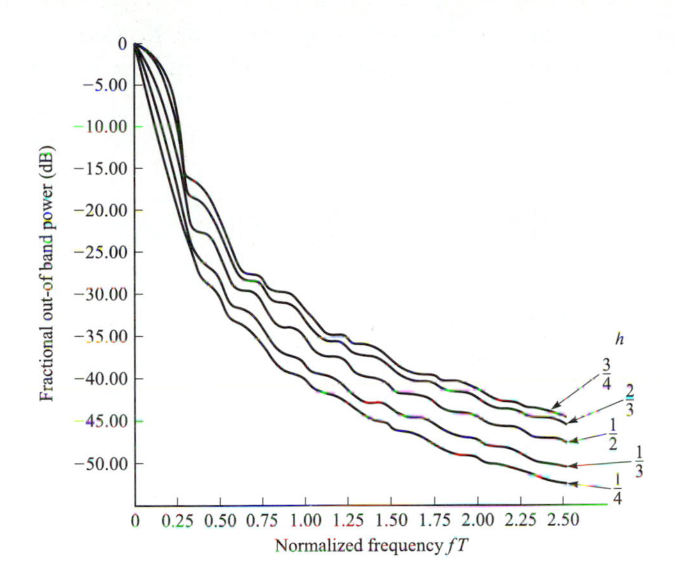

FIGURE 4.4–10
Fractional out-of-band power for two-component CPFSK. (*Mulligan, 1988.*)

when $h = \frac{1}{2}$. For comparison, the spectrum of binary CPFSK is also shown. Note that as L increases the pulse, $g(t)$ becomes smoother and the corresponding spectral occupancy of the signal is reduced.

The effect of varying the modulation index in a CPM signal is illustrated in Figure 4.4–9 for the case of $M = 4$ and a raised cosine pulse of the form given in Equation 4.4–56 with $L = 3$. Note that these spectral characteristics are similar to the ones illustrated previously for CPFSK, except that these spectra are narrower due to the use of a smoother pulse shape.

Finally, in Figure 4.4–10, we illustrate the fractional out-of-band power for two-amplitude CPFSK with several different values of h.

4.4.3 Power Spectra of Modulated Signals with Memory

In the last two sections, we have determined the spectral characteristics for the class of linearly modulated signals without memory and for the class of angle-modulated signals such as CPFSK and CPM, which are non-linear and possess memory. In this section, we consider the spectral characteristics of linearly modulated signals that have memory that can be modeled by a Markov chain. We have already encountered such signals in Section 4.3.2, where we described several types of baseband signals.

The power density spectrum of a digitally modulated signal that is generated by a Markov chain may be derived by following the basic procedure given in the previous section. Thus, we can determine the autocorrelation function and then evaluate its Fourier transform to obtain the power density spectrum. For signals that are generated by a Markov chain with transition probability matrix **P**, the power density spectrum of the modulated signal may be expressed in the general form (see Titsworth and Welch, 1961)

$$\begin{aligned}\Phi(f) = \frac{1}{T^2}\sum_{n=-\infty}^{\infty}\left|\sum_{i=1}^{K} p_i S_i\left(\frac{n}{T}\right)\right|^2 \delta\left(f-\frac{n}{T}\right) + \frac{1}{T}\sum_{i=1}^{K} p_i |S_i'(f)|^2 \\ + \frac{2}{T}\text{Re}\left[\sum_{i=1}^{K}\sum_{j=1}^{K} p_i S_i'^{*}(f) S_j'(f) P_{ij}(f)\right]\end{aligned} \tag{4.4–57}$$

where $S_i(f)$ is the Fourier transform of the signal waveform $s_i(t)$,

$$s_i'(t) = s_i(t) - \sum_{k=1}^{K} p_k s_k(t)$$

$P_{ij}(f)$ is the Fourier transform of the discrete-time sequence $p_{ij}(n)$, defined as

$$P_{ij}(f) = \sum_{n=1}^{\infty} p_{ij}(n) e^{-j2\pi nfT} \tag{4.4–58}$$

and K is the number of states of the modulator. The term $p_{ij}(n)$ denotes the probability that the signal $s_j(t)$ is transmitted n signaling intervals after the transmission of $s_i(t)$. Hence, $\{p_{ij}(n)\}$ are the transition probabilities in the transition probability matrix $\mathbf{P}^n$. Note that $p_{ij}(1) = p_{ij}$.

When there is no memory in the modulation method, the signal waveform transmitted on each signaling interval is independent of the waveforms transmitted in previous signaling intervals. The power density spectrum of the resultant signal may still be expressed in the form of Equation 4.4–57, if the transition probability matrix is replaced by

$$\mathbf{P} = \begin{bmatrix} p_1 & p_2 & \cdots & p_K \\ p_1 & p_2 & \cdots & p_K \\ \vdots & \vdots & & \vdots \\ p_1 & p_2 & \cdots & p_K \end{bmatrix} \tag{4.4–59}$$

and we impose the condition that $\mathbf{P}^n = \mathbf{P}$ for all $n \geqslant 1$. Under these conditions, the expression for the power density spectrum becomes a function of the stationary state probabilities $\{p_i\}$ only, and, hence, it reduces to the simpler form

$$\Phi(f) = \frac{1}{T^2} \sum_{n=-\infty}^{\infty} \left| \sum_{i=1}^{K} p_i S_i\left(\frac{n}{T}\right) \right|^2 \delta\left(f - \frac{n}{T}\right)$$
$$+ \frac{1}{T} \sum_{i=1}^{K} p_i(1 - p_i)|S_i(f)|^2 \tag{4.4–60}$$
$$- \frac{2}{T} \sum_{i=1}^{K} \sum_{\substack{j=1 \\ i<j}}^{K} p_i p_j \operatorname{Re}[S_i(f) S_j^*(f)]$$

We observe that our previous result for the power density spectrum of memoryless linear modulation given by Equation 4.4–18 may be viewed as a special case of Equation 4.4–60 in which all waveforms are identical except for a set of scale factors that convey the digital information (Problem 4.32).

We also make the observation that the first term in the expression for the power density spectrum given by either Equation 4.4–57 or 4.4–60 consists of discrete frequency components. This line spectrum vanishes when

$$\sum_{i=1}^{K} p_i S_i\left(\frac{n}{T}\right) = 0 \tag{4.4–61}$$

The condition 4.4–61 is usually imposed in the design of practical digital communication systems and is easily satisfied by an appropriate choice of signaling waveforms (Problem 4.33).

Now, let us determine the power density spectrum of the baseband-modulated signals described in Section 4.3.2. First, the NRZ signal is characterized by the two waveforms $s_1(t) = g(t)$ and $s_2(t) = -g(t)$, where $g(t)$ is a rectangular pulse of amplitude A. For $K = 2$, Equation 4.4–60 reduces to

$$\Phi(f) = \frac{(2p-1)^2}{T^2} \sum_{n=-\infty}^{\infty} \left| G\left(\frac{n}{T}\right) \right|^2 \delta\left(f - \frac{n}{T}\right) + \frac{4p(1-p)}{T} |G(f)|^2 \tag{4.4–62}$$

where

$$|G(f)|^2 = (AT)^2 \left(\frac{\sin \pi f T}{\pi f T} \right)^2 \tag{4.4–63}$$

Observe that when $p = \frac{1}{2}$, the line spectrum vanishes and $\Phi(f)$ reduces to

$$\Phi(f) = \frac{1}{T} |G(f)|^2 \tag{4.4–64}$$

The NRZI signal is characterized by the transition probability matrix

$$\mathbf{P} = \begin{bmatrix} \frac{1}{2} & \frac{1}{2} \\ \frac{1}{2} & \frac{1}{2} \end{bmatrix} \tag{4.4–65}$$

Notice that in this case $\mathbf{P}^n = \mathbf{P}$ for all $n \geqslant 1$. Hence, the special form for the power density spectrum given by Equation 4.4–62 applies to this modulation format as well. Consequently, the power density spectrum for the NRZI signal is identical to the spectrum of the NRZ signal.

Delay modulation has a transition probability matrix

$$\mathbf{P} = \begin{bmatrix} 0 & \frac{1}{2} & 0 & \frac{1}{2} \\ 0 & 0 & \frac{1}{2} & \frac{1}{2} \\ \frac{1}{2} & \frac{1}{2} & 0 & 0 \\ \frac{1}{2} & 0 & \frac{1}{2} & 0 \end{bmatrix} \tag{4.4–66}$$

and stationary state probabilities $p_i = \frac{1}{4}$ for $i = 1, 2, 3, 4$. Powers of $\mathbf{P}$ may be obtained by use of the relation

$$\mathbf{P}^4 \boldsymbol{\rho} = -\tfrac{1}{4}\boldsymbol{\rho} \tag{4.4–67}$$

where $\boldsymbol{\rho}$ is the signal correlation matrix with elements

$$\rho_{ij} = \frac{1}{T}\int_0^T s_i(t)s_j(t)\,dt \tag{4.4–68}$$

and where the four signals $\{s_i(t), i = 1, 2, 3, 4\}$ are shown in Figure 4.3–15. It is easily seen that

$$\boldsymbol{\rho} = \begin{bmatrix} 1 & 0 & 0 & -1 \\ 0 & 1 & -1 & 0 \\ 0 & -1 & 1 & 0 \\ -1 & 0 & 0 & 1 \end{bmatrix} \tag{4.4–69}$$

Consequently, powers of $\mathbf{P}$ can be generated from the relation

$$\mathbf{P}^{k+4}\boldsymbol{\rho} = -\tfrac{1}{4}\mathbf{P}^k\boldsymbol{\rho}, \qquad k > 1 \tag{4.4–70}$$

Use of Equations 4.4–66, 4.4–69, and 4.4–70 in Equation 4.4–57 yields the power density spectrum of delay modulation. It may be expressed in the form

$$\Phi(f) = \frac{1}{2\psi^2(17 + 8\cos 8\psi)}(23 - 2\cos\psi - 22\cos 2\psi - 12\cos 3\psi + 5\cos 4\psi + 12\cos 5\psi + 2\cos 6\psi - 8\cos 7\psi + 2\cos 8\psi) \tag{4.4–71}$$

where $\psi = \pi f T$.

The spectra of these baseband signals are illustrated in Figure 4.4–11. Observe that the spectra of the NRZ and NRZI signals peak at $f = 0$. Delay modulation has a narrower spectrum and a relatively small zero-frequency content. Its bandwidth occupancy is significantly smaller than that of the NRZ signal. These two characteristics make delay modulation an attractive choice for channels that do not pass DC, such as magnetic recording media.

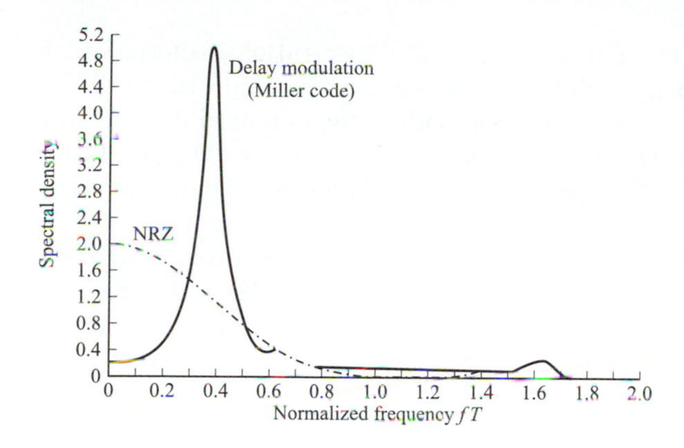

FIGURE 4.4–11
Power spectral density (one-sided) of Miller code (delay modulation) and NRZ/NRZI baseband signals. [*From Hecht and Guida (1969); © 1969 IEEE.*]

4.5
BIBLIOGRAPHICAL NOTES AND REFERENCES

The characteristics of signals and systems given in this chapter are very useful in the design of optimum modulation/demodulation and coding/decoding techniques for a variety of channel models. In particular, the digital modulation methods introduced in this chapter are widely used in digital communication systems. The next chapter is concerned with optimum demodulation techniques for these signals and their performance in an additive white Gaussian noise channel. A general reference for signal characterization is the book by Franks (1969).

Of particular importance in the design of digital communication systems are the spectral characteristics of the digitally modulated signals, which are presented in this chapter in some depth. Of these modulation techniques, CPM is one of the most important due to its efficient use of bandwidth. For this reason, it has been widely investigated by many researchers, and a large number of papers have been published in the technical literature. The most comprehensive treatment of CPM, including its performance and its spectral characteristics, can be found in the book by Anderson et al. (1986). In addition to this text, the tutorial paper by Sundberg (1986) presents the basic concepts and an overview of the performance characteristics of various CPM techniques.

The linear representation of CPM was developed by Laurent (1986) for binary modulation. It was extended to M-ary CPM signals by Mengali and Morelli (1995). Rimoldi (1988) showed that a CPM system can be decomposed into a continuous-phase and a memoryless modulator. This paper also contains over 100 references to published papers on this topic.

There are a large number of references dealing with the spectral characteristics of CPFSK and CPM. As a point of reference, we should mention that MSK was invented by Doelz and Heald in 1961. The early work on the power spectral density of CPFSK and CPM was done by Bennett and Rice (1963), Anderson and Salz (1965), and Bennett and Davey (1965). The book by

Lucky et al. (1968) also contains a treatment of the spectral characteristics of CPFSK. Most of the recent work is referenced in the paper by Sundberg (1986). We should also cite the special issue on bandwidth-efficient modulation and coding published by the *IEEE Transactions on Communications* (March 1981), which contains several papers on the spectral characteristics and performance of CPM.

The generalization of MSK to multiple amplitudes was investigated by Weber et al. (1978). The combination of multiple amplitudes with general CPM was proposed by Mulligan (1988) who investigated its spectral characteristics and its error probability performance in Gaussian noise with and without coding.

PROBLEMS

4.1 Prove the following properties of Hilbert transforms:

a) If $x(t) = x(-t)$, then $\hat{x}(t) = -\hat{x}(-t)$.

b) If $x(t) = -x(-t)$, then $\hat{x}(t) = \hat{x}(-t)$.

c) If $x(t) = \cos\omega_0 t$, then $\hat{x}(t) = \sin\omega_0 t$.

d) If $x(t) = \sin\omega_0 t$, then $\hat{x}(t) = -\cos\omega_0 t$.

e) $\hat{\hat{x}}(t) = -x(t)$.

f) $\int_{-\infty}^{\infty} x^2(t)\,dt = \int_{-\infty}^{\infty} \hat{x}^2(t)\,dt$.

g) $\int_{-\infty}^{\infty} x(t)\hat{x}(t)\,dt = 0$.

4.2 If $x(t)$ is a stationary random process with autocorrelation function $\phi_{ss}(\tau) = E[x(t)x(t+\tau)]$ and spectral density $\Phi_{xx}(f)$, then show that $\phi_{\hat{x}\hat{x}}(\tau) = \phi_{xx}(\tau)$, $\phi_{x\hat{x}}(\tau) = -\hat{\phi}_{xx}(\tau)$, and $\Phi_{\hat{x}\hat{x}}(f) = \Phi_{xx}(f)$.

4.3 Suppose that $n(t)$ is a zero-mean stationary narrowband process represented by either Equation 4.1–37, 4.1–38, or 4.1–39. The autocorrelation function of the equivalent low-pass process $z(t) = x(t) + jy(t)$ is defined as

$$\phi_{zz}(\tau) = \tfrac{1}{2}E[z^*(t)z(t+\tau)]$$

a) Show that

$$E[z(t)z(t+\tau)] = 0$$

b) Suppose $\phi_{zz}(\tau) = N_0\delta(\tau)$, and let

$$V = \int_0^T z(t)\,dt$$

Determine $E(V^2)$ and $E(VV^*) = E(|V|^2)$.

4.4 Determine the autocorrelation function of the stochastic process

$$x(t) = A\sin(2\pi f_c t + \theta)$$

where f_c is a constant and θ is a uniformly distributed phase, i.e.,

$$p(\theta) = \frac{1}{2\pi}, \qquad 0 \leqslant \theta \leqslant 2\pi$$

4.5 Prove that $s_l(t)$ is generally a complex-valued signal and give the condition under which it is real. Assume that $s(t)$ is a real-valued band-pass signal.

4.6 Suppose that $s(t)$ is either a real- or complex-valued signal that is represented as a linear combination of orthonormal functions $\{f_n(t)\}$, i.e.,

$$\hat{s}(t) = \sum_{k=1}^{K} s_k f_k(t)$$

where

$$\int_{-\infty}^{\infty} f_n(t) f_m^*(t)\, dt = \begin{cases} 0 & (m \neq n) \\ 1 & (m = n) \end{cases}$$

Determine the expressions for the coefficients $\{s_k\}$ in the expansion $\hat{s}_i(t)$ that minimize the energy

$$\mathcal{E}_e = \int_{-\infty}^{\infty} |s(t) - \hat{s}(t)|^2\, dt$$

and the corresponding residual error $\mathcal{E}_e$.

4.7 Suppose that a set of M signal waveforms $\{s_{lm}(t)\}$ are complex-valued. Derive the equations for the Gram–Schmidt procedure that will result in a set of $N \leqslant M$ orthonormal signal waveforms.

4.8 Determine the correlation coefficients ρ_{km} among the four signal waveforms $\{s_i(t)\}$ shown in Figure 4.2–1, and the corresponding Euclidean distances.

4.9 Consider a set of M orthogonal signal waveforms $s_m(t)$, $1 \leqslant m \leqslant M$, $0 \leqslant t \leqslant T$, all of which have the same energy $\mathcal{E}$. Determine a new set of M waveforms as

$$s'_m(t) = s_m(t) - \frac{1}{M}\sum_{k=1}^{M} s_k(t), \qquad 1 \leqslant m \leqslant M, \qquad 0 \leqslant t \leqslant T$$

Show that the M signal waveforms $\{s'_m(t)\}$ have equal energy, given by

$$\mathcal{E}' = (M-1)\mathcal{E}/M$$

and are equally correlated, with correlation coefficient

$$\rho_{mn} = \frac{1}{\mathcal{E}'}\int_0^T s'_m(t) s'_n(t)\, dt = -\frac{1}{M-1}$$

4.10 Consider the three waveforms $f_n(t)$ shown in Figure P4.10.

a) Show that these waveforms are orthonormal.

b) Express the waveform $x(t)$ as a linear combination of $f_n(t)$, $n = 1, 2, 3$, if

FIGURE P4.10

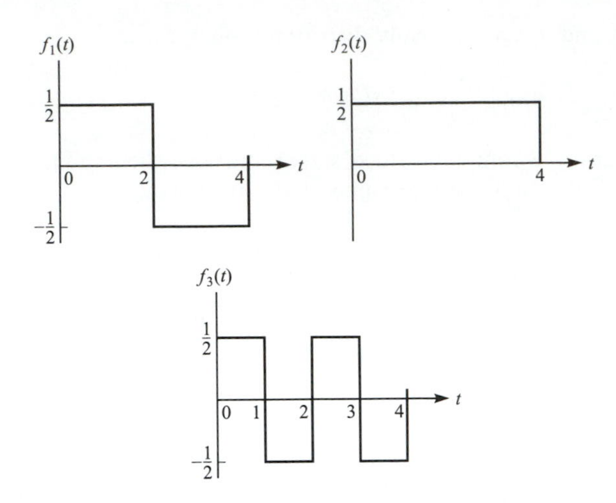

$$x(t) = \begin{cases} -1 & (0 \leqslant t < 1) \\ 1 & (1 \leqslant t < 3) \\ -1 & (3 \leqslant t < 4) \end{cases}$$

and determine the weighting coefficients.

4.11 Consider the four waveforms shown in Figure P4.11.

a) Determine the dimensionality of the waveforms and a set of basis functions.

b) Use the basis functions to represent the four waveforms by vectors $\mathbf{s}_1$, $\mathbf{s}_2$, $\mathbf{s}_3$, and $\mathbf{s}_4$.

c) Determine the minimum distance between any pair of vectors.

FIGURE P4.11

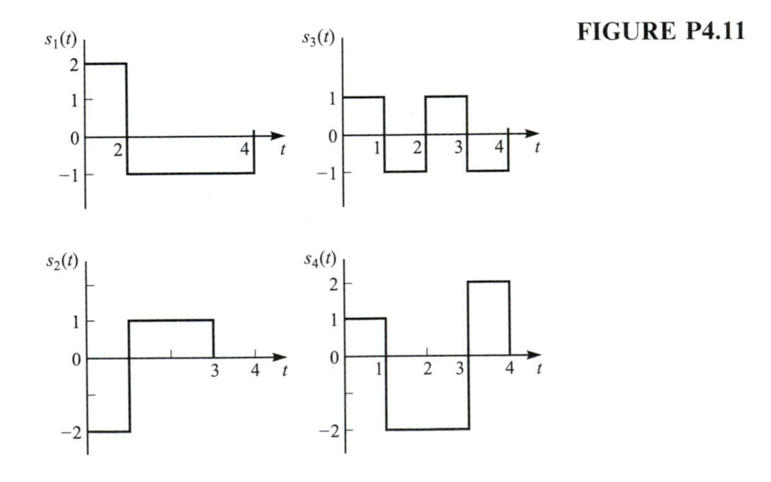

4.12 Determine a set of orthonormal functions for the four signals shown in Figure P4.12.

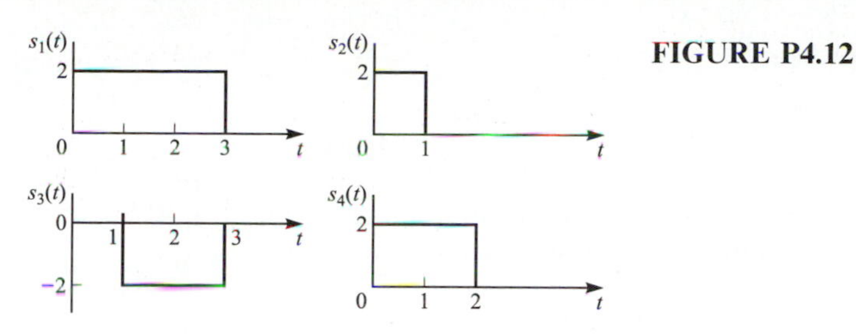

FIGURE P4.12

4.13 A low-pass Gaussian stochastic process $x(t)$ has a power spectral density

$$\Phi(f) = \begin{cases} N_0 & (|f| < B) \\ 0 & (|f| > B) \end{cases}$$

Determine the power spectral density and the autocorrelation function of $y(t) = x^2(t)$.

4.14 Consider an equivalent low-pass digitally modulated signal of the form

$$u(t) = \sum_n [a_n g(t - 2nT) - jb_n g(t - 2nT - T)]$$

where $\{a_n\}$ and $\{b_n\}$ are two sequences of statistically independent binary digits and $g(t)$ is a sinusoidal pulse defined as

$$g(t) = \begin{cases} \sin(\pi t/2T) & (0 < t < 2T) \\ 0 & (\text{otherwise}) \end{cases}$$

This type of signal is viewed as a four-phase PSK signal in which the pulse shape is one-half cycle of a sinusoid. Each of the information sequences $\{a_n\}$ and $\{b_n\}$ is transmitted at a rate of $1/2T$ bits/s and, hence, the combined transmission rate is $1/T$ bits/s. The two sequences are staggered in time by T seconds in transmission. Consequently, the signal $u(t)$ is called *staggered four-phase PSK*.

a) Show that the envelope $|u(t)|$ is a constant, independent of the information a_n on the in-phase component and information b_n on the quadrature component. In other words, the amplitude of the carrier used in transmitting the signal is constant.

b) Determine the power density spectrum of $u(t)$.

c) Compare the power density spectrum obtained from (b) with the power density spectrum of the MSK signal. What conclusion can you draw from this comparison?

4.15 Consider a four-phase PSK signal represented by the equivalent low-pass signal

$$u(t) = \sum_n I_n g(t - nT)$$

where I_n takes on one of the four possible values $\sqrt{\frac{1}{2}}(\pm 1 \pm j)$ with equal probability. The sequence of information symbols $\{I_n\}$ is statistically independent.

a) Determine and sketch the power density spectrum of $u(t)$ when

$$g(t) = \begin{cases} A & (0 \leqslant t \leqslant T) \\ 0 & (\text{otherwise}) \end{cases}$$

b) Repeat (a) when

$$g(t) = \begin{cases} A\sin(\pi t/T) & (0 \leqslant t \leqslant T) \\ 0 & (\text{otherwise}) \end{cases}$$

c) Compare the spectra obtained in (a) and (b) in terms of the 3-dB bandwidth and the bandwidth to the first spectral zero.

4.16 The random process $v(t)$ is defined as

$$v(t) = X\cos 2\pi f_c t - Y\sin 2\pi f_c t$$

where X and Y are random variables. Show that $v(t)$ is wide-sense stationary if and only if $E(X) = E(Y) = 0$, $E(X^2) = E(Y^2)$, and $E(XY) = 0$.

4.17 Carry out the Gram–Schmidt orthogonalization of the signals in Figure 4.2–1a in the order $s_4(t)$, $s_3(t)$, $s_1(t)$, and, thus, obtain a set of orthonormal functions $\{f_m(t)\}$. Then, determine the vector representation of the signals $\{s_n(t)\}$ by using the orthonormal functions $\{f_m(t)\}$. Also, determine the signal energies.

4.18 Determine the signal space representation of the four signals $s_k(t)$, $k = 1, 2, 3, 4$, shown in Figure P4.18, by using as basis functions the orthonormal functions $f_1(t)$ and $f_2(t)$. Plot the signal space diagram and show that this signal set is equivalent to that for a four-phase PSK signal.

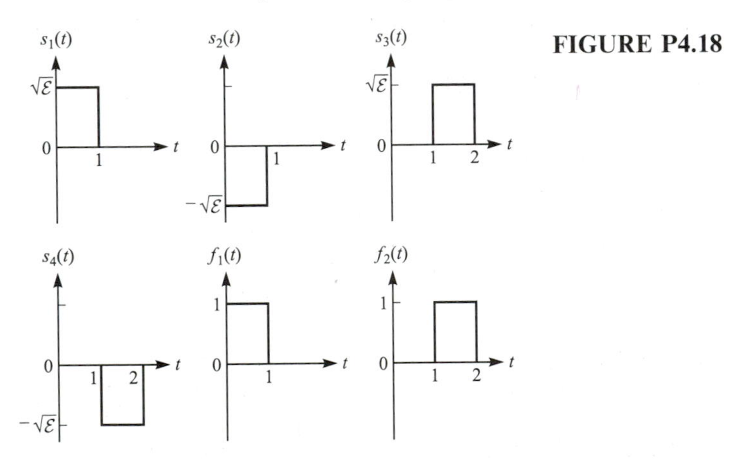

FIGURE P4.18

4.19 $\pi/4$-QPSK may be considered as two QPSK systems offset by $\pi/4$ radians.

a) Sketch the signal space diagram for a $\pi/4$-QPSK signal.

b) Using Gray encoding, label the signal points with the corresponding data bits.

4.20 The power density spectrum of the cyclostationary process

$$v(t) = \sum_{n=-\infty}^{\infty} I_n g(t - nT)$$

was derived in Section 4.4.1 by averaging the autocorrelation function $\phi_{vv}(t+\tau, t)$ over the period T of the process and then evaluating the Fourier transform of the average autocorrelation function. An alternative approach is to change the cyclostationary process into a stationary process $v_\Delta(t)$ by adding a random variable Δ, uniformly distributed over $0 \leqslant \Delta < T$, so that

$$v_\Delta(t) = \sum_{n=-\infty}^{\infty} I_n g(t - nT - \Delta)$$

and defining the spectral density of $v(t)$ as the Fourier transform of the autocorrelation function of the stationary process $v_\Delta(t)$. Derive the result in Equation 4.4–11 by evaluating the autocorrelation function of $v_\Delta(t)$ and its Fourier transform.

4.21 A PAM partial response signal (PRS) is generated as shown in Figure P4.21 by exciting an ideal low-pass filter of bandwidth W by the sequence

$$B_n = I_n + I_{n-1}$$

at a rate $1/T = 2W$ symbols/s. The sequence $\{I_n\}$ consists of binary digits selected independently from the alphabet $\{1, -1\}$ with equal probability. Hence, the filtered signal has the form

$$v(t) = \sum_{n=-\infty}^{\infty} B_n g(t - nT), \qquad T = \frac{1}{2W}$$

a) Sketch the signal space diagram for $v(t)$ and determine the probability of occurrence of each symbol.
b) Determine the autocorrelation and power density spectrum of the three-level sequence $\{B_n\}$.
c) The signal points of the sequence $\{B_n\}$ form a Markov chain. Sketch this Markov chain and indicate the transition probabilities among the states.

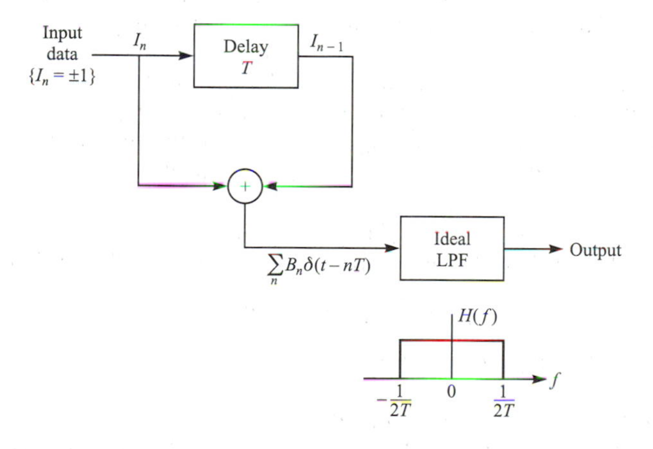

FIGURE P4.21

4.22 The low-pass equivalent representation of a PAM signal is

$$u(t) = \sum_n I_n g(t - nT)$$

Suppose $g(t)$ is a rectangular pulse and

$$I_n = a_n - a_{n-2}$$

where $\{a_n\}$ is a sequence of uncorrelated binary-valued $(1, -1)$ random variables that occur with equal probability.

a) Determine the autocorrelation function of the sequence $\{I_n\}$.
b) Determine the power density spectrum of $u(t)$.
c) Repeat (b) if the possible values of the a_n are (0, 1).

4.23 Show that $x(t) = s(t)\cos 2\pi f_c \pm \hat{s}(t)\sin 2\pi f_c t$ is a single-sideband signal, where $s(t)$ is band-limited to $B \leqslant f_c$ Hz and $\hat{s}(t)$ is its Hilbert transform.

4.24 Use the results in Section 4.4.3 to determine the power density spectrum of the binary FSK signals in which the waveforms are

$$s_i(t) = \sin \omega_i t, \qquad i = 1, 2, \qquad 0 \leqslant t \leqslant T$$

where $\omega_1 = n\pi/T$ and $\omega_2 = m\pi/T$, $n \neq m$, and m and n are arbitrary positive integers. Assume that $p_1 = p_2 = \frac{1}{2}$. Sketch the spectrum and compare this result with the spectrum of the MSK signal.

4.25 Use the results in Section 4.4.3 to determine the power density spectrum of multitone FSK (MFSK) signals for which the signal waveforms are

$$s_n(t) = \sin\frac{2\pi nt}{T}, \qquad n = 1, 2, \ldots, M, \quad 0 \leqslant t \leqslant T$$

Assume that the probabilities $p_n = 1/M$ for all n. Sketch the power spectral density.

4.26 A quadrature partial response signal (QPRS) is generated by two separate partial response signals of the type described in Problem 4.21 placed in phase quadrature. Hence, the QPRS is represented as

$$s(t) = \mathrm{Re}[v(t)e^{j2\pi f_c t}]$$

where

$$\begin{aligned} v(t) &= v_c(t) + jv_s(t) \\ &= \sum_n B_n g(t - nT) + j\sum_n C_n g(t - nT) \end{aligned}$$

and $B_n = I_n + I_{n-1}$ and $C_n = J_n + J_{n-1}$. The sequences $\{B_n\}$ and $\{C_n\}$ are independent and $I_n = \pm 1$, $J_n = \pm 1$ with equal probability.

a) Sketch the signal space diagram for the QPRS signal and determine the probability of occurrence of each symbol.
b) Determine the autocorrelations and power spectral density of $v_c(t)$, $v_s(t)$, and $v(t)$.
c) Sketch the Markov chain model and indicate the transition probabilities for the QPRS.

4.27 Determine the autocorrelation functions for the MSK and offset QPSK modulated signals based on the assumption that the information sequences for each of the two signals are uncorrelated and zero-mean.

4.28 Sketch the phase tree, the state trellis, and the state diagram for partial response CPM with $h = \frac{1}{2}$ and

$$g(t) = \begin{cases} 1/4T & (0 \leqslant t \leqslant 2T) \\ 0 & (\text{otherwise}) \end{cases}$$

4.29 Determine the number of terminal phase states in the state trellis diagram for
a) A full response binary CPFSK with either $h = \frac{2}{3}$ or $\frac{3}{4}$.
b) A partial response $L = 3$ binary CPFSK with either $h = \frac{2}{3}$ or $\frac{3}{4}$.

4.30 Show that 16 QAM can be represented as a superposition of two four-phase constant envelope signals where each component is amplified separately before summing, i.e.,

$$s(t) = G(A_n \cos 2\pi f_c t + B_n \sin 2\pi f_c t) + (C_n \cos 2\pi f_c t + D_n \sin 2\pi f_c t)$$

where $\{A_n\}$, $\{B_n\}$, $\{C_n\}$, and $\{D_n\}$ are statistically independent binary sequences with elements from the set $\{+1, -1\}$ and G is the amplifier gain. Thus, show that the resulting signal is equivalent to

$$s(t) = I_n \cos 2\pi f_c t + Q_n \sin 2\pi f_c t$$

and determine I_n and Q_n in terms of A_n, B_n, C_n, and D_n.

4.31 In the linear representation of CPM, show that the time durations of the 2^{L-1} pulses $\{c_k(t)\}$ are as follows:

$$\begin{aligned}
&c_0(t) = 0, && t < 0 \text{ and } t > (L+1)T \\
&c_1(t) = 0, && t < 0 \text{ and } t > (L-1)T \\
&c_2(t) = c_3(t) = 0, && t < 0 \text{ and } t > (L-2)T \\
&c_4(t) = c_5(t) = c_6(t) = c_7(t) = 0, && t < 0 \text{ and } t > (L-3)T \\
&\quad\vdots && \\
&c_{2^{L-2}}(t) = \cdots = c_{2^{L-1}}(t) = 0, && t < 0 \text{ and } t > T
\end{aligned}$$

4.32 Use the result in Equation 4.4–60 to derive the expression for the power density spectrum of memoryless linear modulation given by Equation 4.4–18 under the condition that

$$s_k(t) = I_k s(t), \qquad k = 1, 2, \ldots, K$$

where I_k is one of the K possible transmitted symbols that occur with equal probability.

4.33 Show that a sufficient condition for the absence of the line spectrum component in Equation 4.4–60 is

$$\sum_{i=1}^{K} p_i s_i(t) = 0$$

Is this condition necessary? Justify your answer.

4.34 The information sequence $\{a_n\}_{n=-\infty}^{\infty}$ is a sequence of iid random variables, each taking values +1 and −1 with equal probability. This sequence is to be transmitted at baseband by a biphase coding scheme, described by

$$s(t) = \sum_{n=-\infty}^{\infty} a_n g(t - nT)$$

where $g(t)$ is shown in Figure P4.34.

a) Find the power spectral density of $s(t)$.

b) Assume that it is desirable to have a zero in the power spectrum at $f = 1/T$. To this end, we use a precoding scheme by introducing $b_n = a_n + ka_{n-1}$, where k is some constant, and then transmit the $\{b_n\}$ sequence using the same $g(t)$. Is it possible to choose k to produce a frequency null at $f = 1/T$? If yes, what are the appropriate values and the resulting power spectrum?

c) Now assume we want to have zeros at all multiples of $f_0 = 1/4T$. Is it possible to have these zeros with an appropriate choice of k in the previous part? If not, then what kind of precoding do you suggest to result in the desired nulls?

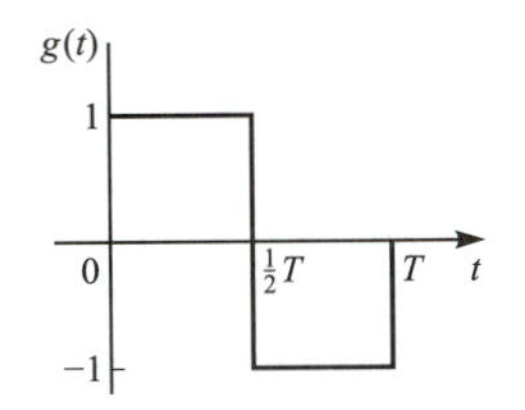

FIGURE P4.34

4.35 Starting with the definition of the transition probability matrix for delay modulation given in Equation 4.4–66, demonstrate that the relation

$$\mathbf{P}^4\boldsymbol{\rho} = -\tfrac{1}{4}\boldsymbol{\rho}$$

holds, and, hence,

$$\mathbf{P}^{k+4}\boldsymbol{\rho} = -\tfrac{1}{4}\mathbf{P}^k\boldsymbol{\rho}, \qquad k \geqslant 1$$

4.36 The two signal waveforms for binary FSK signal transmission with discontinuous phase are

$$s_0(t) = \sqrt{\frac{2\mathcal{E}_b}{T_b}} \cos\left[2\pi\left(f_c - \frac{\Delta f}{2}\right)t + \theta_0\right], \qquad 0 \leqslant t < T$$

$$s_1(t) = \sqrt{\frac{2\mathcal{E}_b}{T_b}} \cos\left[2\pi\left(f_c + \frac{\Delta f}{2}\right)t + \theta_1\right], \qquad 0 \leqslant t < T$$

where $\Delta f = 1/T \ll f_c$, and θ_0 and θ_1 are uniformly distributed random variables on the interval $(0, 2\pi)$. The signals $s_0(t)$ and $s_1(t)$ are equally probable.

a) Determine the power spectral density of the FSK signal.

b) Show that the power spectral density decays as $1/f^2$ for $f \gg f_c$.

5

Optimum Receivers for the Additive White Gaussian Noise Channel

In Chapter 4, we described various types of modulation methods that may be used to transmit digital information through a communication channel. As we have observed, the modulator at the transmitter performs the function of mapping the digital sequence into signal waveforms.

This chapter deals with the design and performance characteristics of optimum receivers for the various modulation methods, when the channel corrupts the transmitted signal by the addition of Gaussian noise. In Section 5.1, we first treat memoryless modulation signals, followed by modulation signals with memory. We evaluate the probability of error of the various modulation methods in Section 5.2. We treat the optimum receiver for CPM signals and its performance in Section 5.3. In Section 5.4, we derive the optimum receiver when the carrier phase of the signals is unknown at the receiver and is treated as a random variable. Finally, in Section 5.5, we consider the use of regenerative repeaters in signal transmission and carry out a link budget analysis for radio channels.

5.1 OPTIMUM RECEIVER FOR SIGNALS CORRUPTED BY ADDITIVE WHITE GAUSSIAN NOISE

Let us begin by developing a mathematical model for the signal at the input to the receiver. We assume that the transmitter sends digital information by use of M signal waveforms $\{s_m(t),\ m = 1, 2, \ldots, M\}$. Each waveform is transmitted within the symbol (signaling) interval of duration T. To be specific, we consider the transmission of information over the interval $0 \leqslant t \leqslant T$.

The channel is assumed to corrupt the signal by the addition of white Gaussian noise, as illustrated in Figure 5.1–1. Thus, the received signal in the interval $0 \leqslant t \leqslant T$ may be expressed as

$$r(t) = s_m(t) + n(t), \qquad 0 \leqslant t \leqslant T \tag{5.1–1}$$

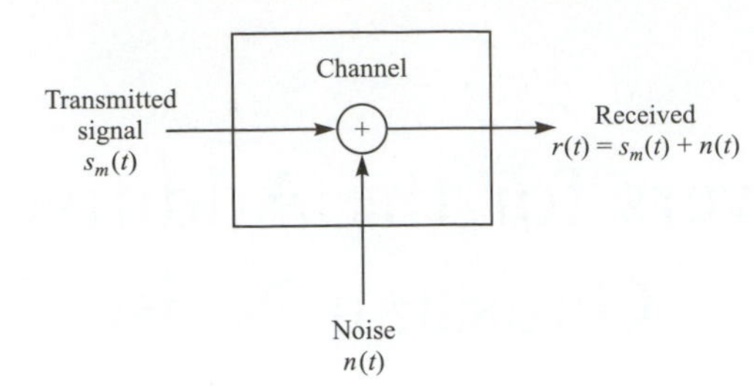

FIGURE 5.1–1
Model for received signal passed through an AWGN channel.

where $n(t)$ denotes a sample function of the additive white Gaussian noise (AWGN) process with power spectral density $\Phi_{nn}(f) = \frac{1}{2}N_0$ W/Hz. Based on the observation of $r(t)$ over the signal interval, we wish to design a receiver that is optimum in the sense that it minimizes the probability of making an error.

It is convenient to subdivide the receiver into two parts—the signal demodulator and the detector—as shown in Figure 5.1–2. The function of the signal demodulator is to convert the received waveform $r(t)$ into an N-dimensional vector $\mathbf{r} = [r_1 \; r_2 \cdots r_N]$, where N is the dimension of the transmitted signal waveforms. The function of the detector is to decide which of the M possible signal waveforms was transmitted based of the vector $\mathbf{r}$.

Two realizations of the signal demodulator are described in the next two sections. One is based on the use of signal correlators. The second is based on the use of matched filters. The optimum detector that follows the signal demodulator is designed to minimize the probability of error.

5.1.1 Correlation Demodulator

In this section, we describe a correlation demodulator that decomposes the received signal and the noise into N-dimensional vectors. In other words, the signal and the noise are expanded into a series of linearly weighted orthonormal basis functions $\{f_n(t)\}$. It is assumed that the N basis functions $\{f_n(t)\}$ span the signal space, so that every one of the possible transmitted signals of the set $\{s_m(t), 1 \leqslant m \leqslant M\}$ can be represented as a linear combination of $\{f_n(t)\}$. In the case of the noise, the functions $\{f_n(t)\}$ do not span the noise space. However, we show below that the noise terms that fall outside the signal space are irrelevant to the detection of the signal.

Suppose the received signal $r(t)$ is passed through a parallel bank of N cross correlators which basically compute the projection of $r(t)$ onto the N basis functions $\{f_n(t)\}$, as illustrated in Figure 5.1–3. Thus, we have

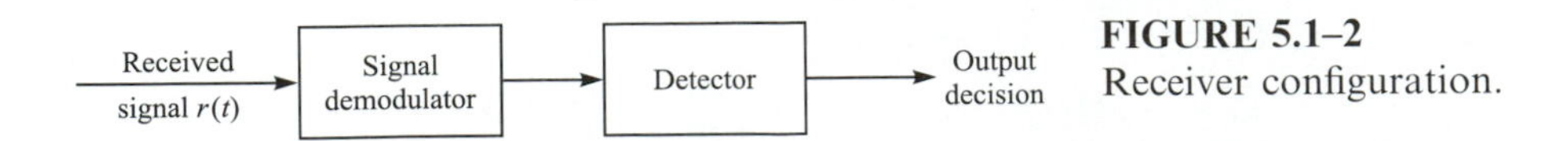

FIGURE 5.1–2
Receiver configuration.

$$\int_0^T r(t)f_k(t)\,dt = \int_0^T [s_m(t) + n(t)]f_k(t)\,dt$$
$$r_k = s_{mk} + n_k, \qquad k = 1, 2, \ldots, N \tag{5.1–2}$$

where

$$s_{mk} = \int_0^T s_m(t)f_k(t)\,dt, \qquad k = 1, 2, \ldots, N$$
$$n_k = \int_0^T n(t)f_k(t)\,dt, \qquad k = 1, 2, \ldots, N \tag{5.1–3}$$

The signal is now represented by the vector $\mathbf{s}_m$ with components s_{mk}, $k = 1, 2, \ldots, N$. Their values depend on which of the M signals was transmitted. The components $\{n_k\}$ are random variables that arise from the presence of the additive noise.

In fact, we can express the received signal $r(t)$ in the interval $0 \leqslant t \leqslant T$ as

$$r(t) = \sum_{k=1}^{N} s_{mk}f_k(t) + \sum_{k=1}^{N} n_k f_k(t) + n'(t)$$
$$= \sum_{k=1}^{N} r_k f_k(t) + n'(t) \tag{5.1–4}$$

The term $n'(t)$, defined as

$$n'(t) = n(t) - \sum_{k=1}^{N} n_k f_k(t) \tag{5.1–5}$$

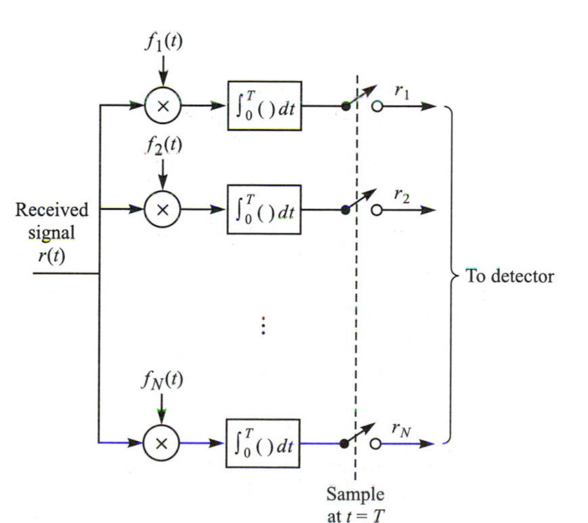

FIGURE 5.1–3
Correlation-type demodulator.

is a zero-mean Gaussian noise process that represents the difference between the original noise process $n(t)$ and the part corresponding to the projection of $n(t)$ onto the basis functions $\{f_k(t)\}$. We shall show below that $n'(t)$ is irrelevant to the decision as to which signal was transmitted. Consequently, the decision may be based entirely on the correlator output signal and noise components $r_k = s_{mk} + n_k$, $k = 1, 2, \ldots, N$.

Since the signals $\{s_m(t)\}$ are deterministic, the signal components are deterministic. The noise components $\{n_k\}$ are Gaussian, since each component can be viewed as the sampled output of a linear filter excited by Gaussian noise (see Section 5.1.2). Their mean values are

$$E(n_k) = \int_0^T E[n(t)]f_k(t)\,dt = 0 \tag{5.1–6}$$

for all n. Their covariances are

$$\begin{aligned} E(n_k n_m) &= \int_0^T \int_0^T E[n(t)n(\tau)]f_k(t)f_m(\tau)\,dt\,d\tau \\ &= \tfrac{1}{2}N_0 \int_0^T \int_0^T \delta(t-\tau)f_k(t)f_m(\tau)\,dt\,d\tau \\ &= \tfrac{1}{2}N_0 \int_0^T f_k(t)f_m(t)\,dt \\ &= \tfrac{1}{2}N_0\delta_{mk} \end{aligned} \tag{5.1–7}$$

where $\delta_{mk} = 1$ when $m = k$ and zero otherwise. Therefore, the N noise components $\{n_k\}$ are zero-mean uncorrelated Gaussian random variables with a common variance $\sigma_n^2 = \frac{1}{2}N_0$.

From the above development, it follows that the correlator outputs $\{r_k\}$ conditioned on the mth signal being transmitted are Gaussian random variables with mean

$$E(r_k) = E(s_{mk} + n_k) = s_{mk} \tag{5.1–8}$$

and equal variance

$$\sigma_r^2 = \sigma_n^2 = \tfrac{1}{2}N_0 \tag{5.1–9}$$

Since the noise components $\{n_k\}$ are uncorrelated Gaussian random variables, they are also statistically independent. As a consequence, the correlator outputs $\{r_k\}$ conditioned on the mth signal being transmitted are statistically independent Gaussian variables. Hence, the conditional probability density functions of the random variables $[r_1\ r_2 \cdots r_N] = \mathbf{r}$ are simply

$$p(\mathbf{r}|\mathbf{s}_m) = \prod_{k=1}^{N} p(r_k|s_{mk}), \qquad m = 1, 2, \ldots, M \tag{5.1–10}$$

where

$$p(r_k|s_{mk}) = \frac{1}{\sqrt{\pi N_0}} \exp\left[-\frac{(r_k - s_{mk})^2}{N_0}\right], \qquad k = 1, 2, \ldots, N \qquad (5.1\text{–}11)$$

By substituting Equation 5.1–11 into Equation 5.1–10, we obtain the joint conditional PDFs

$$p(\mathbf{r}|\mathbf{s}_m) = \frac{1}{(\pi N_0)^{N/2}} \exp\left[-\sum_{k=1}^{N} \frac{(r_k - s_{mk})^2}{N_0}\right], \qquad m = 1, 2, \ldots, M \quad (5.1\text{–}12)$$

As a final point we wish to show that the correlator outputs $(r_1, r_2, \ldots, r_N)$ are *sufficient statistics* for reaching a decision on which of the M signals was transmitted, i.e., that no additional relevant information can be extracted from the remaining noise process $n'(t)$. Indeed, $n'(t)$ is uncorrelated with the N correlator outputs $\{r_k\}$, i.e.,

$$\begin{aligned} E[n'(t)r_k] &= E[n'(t)]s_{mk} + E[n'(t)n_k] \\ &= E[n'(t)n_k] \\ &= E\left\{\left[n(t) - \sum_{j=1}^{N} n_j f_j(t)\right] n_k\right\} \\ &= \int_0^T E[n(t)n(\tau)] f_k(\tau)\, d\tau - \sum_{j=1}^{N} E(n_j n_k) f_j(t) \\ &= \tfrac{1}{2} N_0 f_k(t) - \tfrac{1}{2} N_0 f_k(t) = 0 \end{aligned} \qquad (5.1\text{–}13)$$

Since $n'(t)$ and $\{r_k\}$ are Gaussian and uncorrelated, they are also statistically independent. Consequently, $n'(t)$ does not contain any information that is relevant to the decision as to which signal waveform was transmitted. All the relevant information is contained in the correlator outputs $\{r_k\}$. Hence, $n'(t)$ may be ignored.

EXAMPLE 5.1–1. Consider an M-ary baseband PAM signal set in which the basic pulse shape $g(t)$ is rectangular as shown in Figure 5.1–4. The additive noise is a zero-mean white Gaussian noise process. Let us determine the basis function $f(t)$ and the output of the correlation-type demodulator. The energy in the rectangular pulse is

$$\mathcal{E}_g = \int_0^T g^2(t)\, dt = \int_0^T a^2\, dt = a^2 T$$

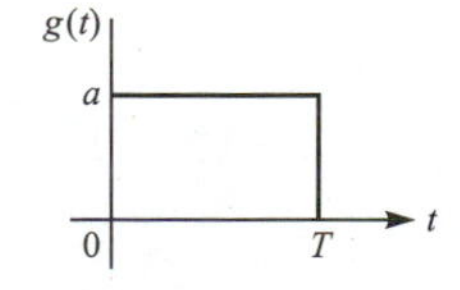

FIGURE 5.1–4
Signal pulse for Example 5.1–1.

Since the PAM signal set has dimension $N = 1$, there is only one basis function $f(t)$. This is given as

$$f(t) = \frac{1}{\sqrt{a^2 T}} g(t) = \begin{cases} 1/\sqrt{T} & (0 \leqslant t \leqslant T) \\ 0 & (\text{otherwise}) \end{cases}$$

The output of the correlation-type demodulator is

$$r = \int_0^T r(t) f(t)\, dt = \frac{1}{\sqrt{T}} \int_0^T r(t)\, dt$$

It is interesting to note that the correlator becomes a simple integrator when $f(t)$ is rectangular. If we substitute for $r(t)$, we obtain

$$\begin{aligned} r &= \frac{1}{\sqrt{T}} \left\{ \int_0^T [s_m(t) + n(t)] \right\} dt \\ &= \frac{1}{\sqrt{T}} \left[\int_0^T s_m(t)\, dt + \int_0^T n(t)\, dt \right] \\ r &= s_m + n \end{aligned}$$

where the noise term $E(n) = 0$ and

$$\begin{aligned} \sigma_n^2 &= E\left[\frac{1}{T} \int_0^T \int_0^T n(t) n(\tau)\, dt\, d\tau \right] \\ &= \frac{1}{T} \int_0^T \int_0^T E[n(t) n(\tau)]\, dt\, d\tau \\ &= \frac{N_0}{2T} \int_0^T \int_0^T \delta(t - \tau)\, dt\, d\tau = \tfrac{1}{2} N_0 \end{aligned}$$

The probability density function for the sampled output is

$$p(r|s_m) = \frac{1}{\sqrt{\pi N_0}} \exp\left[-\frac{(r - s_m)^2}{N_0} \right]$$

5.1.2 Matched-Filter Demodulator

Instead of using a bank of N correlators to generate the variables $\{r_k\}$, we may use a bank of N linear filters. To be specific, let us suppose that the impulse responses of the N filters are

$$h_k(t) = f_k(T - t), \qquad 0 \leqslant t \leqslant T \tag{5.1–14}$$

where $\{f_k(t)\}$ are the N basis functions and $h_k(t) = 0$ outside of the interval $0 \leqslant t \leqslant T$. The outputs of these filters are

$$\begin{aligned} y_k(t) &= \int_0^t r(\tau) h_k(t - \tau)\, d\tau \\ &= \int_0^t r(\tau) f_k(T - t + \tau)\, d\tau, \qquad k = 1, 2, \ldots, N \end{aligned} \tag{5.1–15}$$

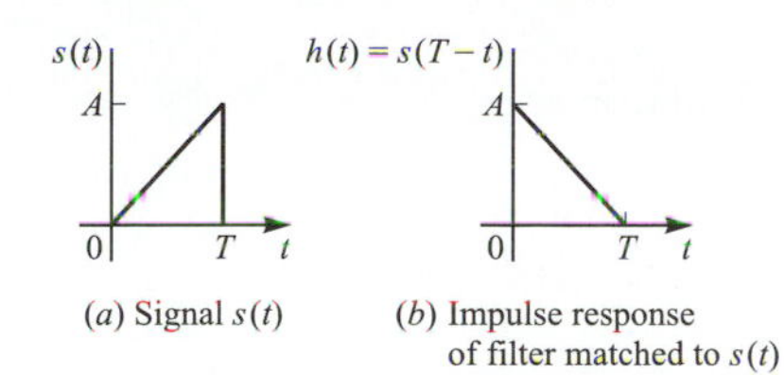

FIGURE 5.1–5
Signal $s(t)$ and filter matched to $s(t)$.

Now, if we sample the outputs of the filters at $t = T$, we obtain

$$y_k(T) = \int_0^T r(\tau) f_k(\tau)\, d\tau = r_k, \qquad k = 1, 2, \ldots, N \tag{5.1–16}$$

Hence, the sampled outputs of the filters at time $t = T$ are exactly the set of values $\{r_k\}$ obtained from the N linear correlators.

A filter whose impulse response $h(t) = s(T - t)$, where $s(t)$ is assumed to be confined to the time interval $0 \leqslant t \leqslant T$, is called the *matched filter* to the signal $s(t)$. An example of a signal and its matched filter are shown in Figure 5.1–5. The response of $h(t) = s(T - t)$ to the signal $s(t)$ is

$$y(t) = \int_0^t s(\tau) s(T - t + \tau)\, d\tau \tag{5.1–17}$$

which is basically the time-autocorrelation function of the signal $s(t)$. Figure 5.1–6 illustrates $y(t)$ for the triangular signal pulse shown in Figure 5.1–5. Note that the autocorrelation function $y(t)$ is an even function of t, which attains a peak at $t = T$.

In the case of the demodulator described above, the N matched filters are matched to the basis functions $\{f_k(t)\}$. Figure 5.1–7 illustrates the matched filter demodulator that generates the observed variables $\{r_k\}$.

Properties of the matched filter. A matched filter has some interesting properties. Let us prove the most important property, which may be stated as follows: If a signal $s(t)$ is corrupted by AWGN, the filter with an impulse response matched to $s(t)$ maximizes the output signal-to-noise ratio (SNR).

To prove this property, let us assume that the received signal $r(t)$ consists of the signal $s(t)$ and AWGN $n(t)$ which has zero-mean and power spectral density $\Phi_{nn}(f) = \frac{1}{2} N_0$ W/Hz. Suppose the signal $r(t)$ is passed through a filter with impulse response $h(t)$, $0 \leqslant t \leqslant T$, and its output is sampled at time $t = T$. The filter response to the signal and noise components is

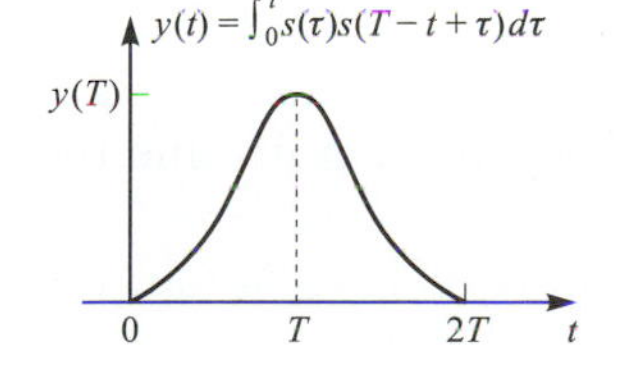

FIGURE 5.1–6
The matched filter output is the autocorrelation function of $s(t)$.

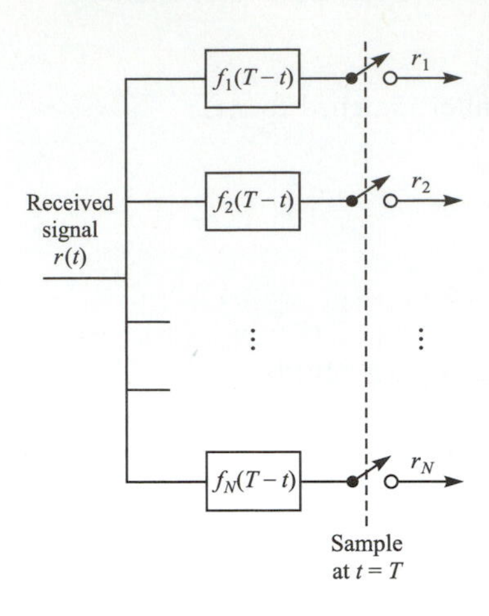

FIGURE 5.1–7
Matched filter demodulator.

$$\begin{aligned} y(t) &= \int_0^t r(\tau)h(t-\tau)\,d\tau \\ &= \int_0^t s(\tau)h(t-\tau)\,d\tau + \int_0^t n(\tau)h(t-\tau)\,d\tau \end{aligned} \tag{5.1–18}$$

At the sampling instant $t = T$, the signal and noise components are

$$\begin{aligned} y(T) &= \int_0^T s(\tau)h(T-\tau)\,d\tau + \int_0^T n(T)h(t-\tau)\,d\tau \\ &= y_s(T) + y_n(T) \end{aligned} \tag{5.1–19}$$

where $y_s(T)$ represents the signal component and $y_n(T)$ the noise component. The problem is to select the filter impulse response that maximizes the output signal-to-noise ratio (SNR_0) defined as

$$\text{SNR}_0 = \frac{y_s^2(T)}{E[y_n^2(T)]} \tag{5.1–20}$$

The denominator in Equation 5.1–20 is simply the variance of the noise term at the output of the filter. Let us evaluate $E[y_n^2(T)]$. We have

$$\begin{aligned} E[y_n^2(T)] &= \int_0^T \int_0^T E[n(\tau)n(t)]h(T-\tau)h(T-t)\,dt\,d\tau \\ &= \tfrac{1}{2}N_0 \int_0^T \int_0^T \delta(t-\tau)h(T-\tau)h(T-t)\,dt\,d\tau \\ &= \tfrac{1}{2}N_0 \int_0^T h^2(T-t)\,dt \end{aligned} \tag{5.1–21}$$

Note that the variance depends on the power spectral density of the noise and the energy in the impulse response $h(t)$.

By substituting for $y_s(T)$ and $E[y_n^2(T)]$ into Equation 5.1–20, we obtain the expression for the output SNR as

$$\text{SNR}_0 = \frac{[\int_0^T s(\tau)h(T-\tau)\,d\tau]^2}{\frac{1}{2}N_0\int_0^T h^2(T-t)\,dt} = \frac{[\int_0^T h(\tau)s(T-\tau)\,d\tau]^2}{\frac{1}{2}N_0\int_0^T h^2(T-t)\,dt} \tag{5.1–22}$$

Since the denominator of the SNR depends on the energy in $h(t)$, the maximum output SNR over $h(t)$ is obtained by maximizing the numerator subject to the constraint that the denominator is held constant. The maximization of the numerator is most easily performed by use of the Cauchy–Schwarz inequality, which states, in general, that if $g_1(t)$ and $g_2(t)$ are finite-energy signals, then

$$\left[\int_{-\infty}^{\infty} g_1(t)g_2(t)\,dt\right]^2 \leqslant \int_{-\infty}^{\infty} g_1^2(t)\,dt \int_{-\infty}^{\infty} g_2^2(t)\,dt \tag{5.1–23}$$

with equality when $g_1(t) = Cg_2(t)$ for any arbitrary constant C. If we set $g_1(t) = h(t)$ and $g_2(t) = s(T-t)$, it is clear that the SNR is maximized when $h(t) = Cs(T-t)$, i.e., $h(t)$ is matched to the signal $s(t)$. The scale factor C^2 drops out of the expression for the SNR, since it appears in both the numerator and the denominator.

The output (maximum) SNR obtained with the matched filter is

$$\begin{aligned}\text{SNR}_0 &= \frac{2}{N_0}\int_0^T s^2(t)\,dt \\ &= \frac{2\mathcal{E}}{N_0}\end{aligned} \tag{5.1–24}$$

Note that the output SNR from the matched filter depends on the energy of the waveform $s(t)$ but not on the detailed characteristics of $s(t)$. This is another interesting property of the matched filter.

Frequency-domain interpretation of the matched filter. The matched filter has an interesting frequency-domain interpretation. Since $h(t) = s(T-t)$, the Fourier transform of this relationship is

$$\begin{aligned}H(f) &= \int_0^T s(T-t)e^{-j2\pi ft}\,dt \\ &= \left[\int_0^T s(\tau)e^{j2\pi f\tau}\,d\tau\right]e^{-j2\pi fT} \\ &= S^*(f)e^{-j2\pi fT}\end{aligned} \tag{5.1–25}$$

We observe that the matched filter has a frequency response that is the complex conjugate of the transmitted signal spectrum multiplied by the phase factor $e^{-j2\pi fT}$, which represents the sampling delay of T. In other words, $|H(f)| = |S(f)|$, so that the magnitude response of the matched filter is identical to the transmitted signal spectrum. On the other hand, the phase of $H(f)$ is the negative of the phase of $S(f)$.

Now, if the signal $s(t)$ with spectrum $S(f)$ is passed through the matched filter, the filter output has a spectrum $Y(f) = |S(f)|^2 e^{-j2\pi fT}$. Hence, the output waveform is

$$
\begin{aligned}
y_s(t) &= \int_{-\infty}^{\infty} Y(f) e^{j2\pi ft}\, df \\
&= \int_{-\infty}^{\infty} |S(f)|^2 e^{-j2\pi fT} e^{j2\pi ft}\, df
\end{aligned}
\tag{5.1–26}
$$

By sampling the output of the matched filter at $t = T$, we obtain

$$
y_s(T) = \int_{-\infty}^{\infty} |S(f)|^2 df = \int_0^T s^2(t)\, dt = \mathcal{E} \tag{5.1–27}
$$

where the last step follows from Parseval's relation.

The noise at the output of the matched filter has a power spectral density

$$
\Phi_0(f) = \tfrac{1}{2}|H(f)|^2 N_0 \tag{5.1–28}
$$

Hence, the total noise power at the output of the matched filter is

$$
\begin{aligned}
P_n &= \int_{-\infty}^{\infty} \Phi_0(f)\, df \\
&= \tfrac{1}{2}N_0 \int_{-\infty}^{\infty} |H(f)|^2 df = \tfrac{1}{2}N_0 \int_{-\infty}^{\infty} |S(f)|^2 df = \tfrac{1}{2}\mathcal{E} N_0
\end{aligned}
\tag{5.1–29}
$$

The output SNR is simply the ratio of the signal power P_s, given by

$$
P_s = y_s^2(T) \tag{5.1–30}
$$

to the noise power P_n. Hence,

$$
\text{SNR}_0 = \frac{P_s}{P_n} = \frac{\mathcal{E}^2}{\frac{1}{2}\mathcal{E} N_0} = \frac{2\mathcal{E}}{N_0} \tag{5.1–31}
$$

which agrees with the result given by Equation 5.1–24.

EXAMPLE 5.1–2. $M = 4$ biorthogonal signals are constructed from the two orthogonal signals shown in Figure 5.1–8a for transmitting information over an AWGN channel. The noise is assumed to have zero-mean and power spectral density $\frac{1}{2}N_0$. Let us determine the basis functions for this signal set, the impulse responses of the matched-filter demodulators, and the output waveforms of the matched-filter demodulators when the transmitted signal is $s_1(t)$.

The $M = 4$ biorthogonal signals have dimensions $N = 2$. Hence, two basis functions are needed to represent the signals. From Figure 5.1–8a, we choose $f_1(t)$ and $f_2(t)$ as

$$
\begin{aligned}
f_1(t) &= \begin{cases} \sqrt{2/T} & (0 \leqslant t \leqslant \frac{1}{2}T) \\ 0 & (\text{otherwise}) \end{cases} \\
f_2(t) &= \begin{cases} \sqrt{2/T} & (\frac{1}{2}T \leqslant t \leqslant T) \\ 0 & (\text{otherwise}) \end{cases}
\end{aligned}
\tag{5.1–32}
$$

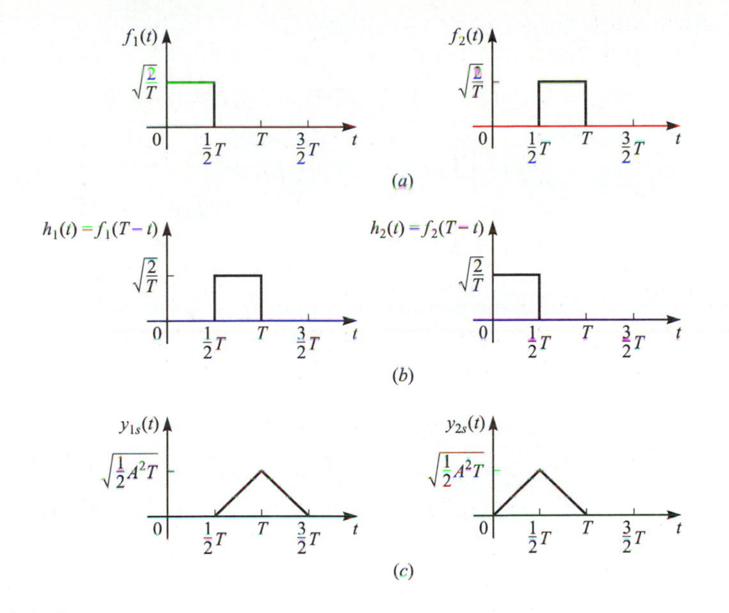

FIGURE 5.1–8
Basis functions and matched filter responses for Example 5.1–2.

The impulse responses of the two matched filters are

$$
\begin{aligned}
h_1(t) = f_1(T-t) &= \begin{cases} \sqrt{2/T} & \frac{1}{2}T \leqslant t \leqslant T) \\ 0 & \text{(otherwise)} \end{cases} \\
h_2(t) = f_2(T-t) &= \begin{cases} \sqrt{2/T} & (0 \leqslant t \leqslant \frac{1}{2}T) \\ 0 & \text{(otherwise)} \end{cases}
\end{aligned}
\tag{5.1–33}
$$

and are illustrated in Figure 5.1–8b.

If $s_1(t)$ is transmitted, the (noise-free) responses of the two matched filters are as shown in Figure 5.1–8c. Since $y_1(t)$ and $y_2(t)$ are sampled at $t = T$, we observe that $y_{1s}(T) = \sqrt{\frac{1}{2}A^2T}$ and $y_{2s}(T) = 0$. Note that $\frac{1}{2}A^2T = \mathcal{E}$, the signal energy. Hence, the received vector formed from the two matched filter outputs at the sampling instant $t = T$ is

$$
\mathbf{r} = [r_1 \quad r_2] = [\sqrt{\mathcal{E}} + n_1 \quad n_2] \tag{5.1–34}
$$

where $n_1 = y_{1n}(T)$ and $n_2 = y_{2n}(T)$ are the noise components at the outputs of the matched filters, given by

$$
y_{kn}(T) = \int_0^T n(t) f_k(t)\, dt, \qquad k = 1, 2 \tag{5.1–35}
$$

Clearly, $E(n_k) = E[y_{kn}(T)] = 0$. Their variance is

$$\begin{aligned}\sigma_n^2 = E[y_{kn}^2(T)] &= \int_0^T \int_0^T E[n(t)n(\tau)] f_k(t) f_k(\tau)\, dt\, d\tau \\ &= \tfrac{1}{2}N_0 \int_0^T \int_0^T \delta(t-\tau) f_k(\tau) f_k(t)\, dt\, d\tau \\ &= \tfrac{1}{2}N_0 \int_0^T f_k^2(t)\, dt = \tfrac{1}{2}N_0\end{aligned} \tag{5.1–36}$$

Observe that the SNR_0 for the first matched filter is

$$\mathrm{SNR}_0 = \frac{(\sqrt{\mathcal{E}})^2}{\frac{1}{2}N_0} = \frac{2\mathcal{E}}{N_0} \tag{5.1–37}$$

which agrees with our previous result. Also note that the four possible outputs of the two matched filters, corresponding to the four possible transmitted signals in Figure 5.1–8 are $(r_1, r_2) = (\sqrt{\mathcal{E}} + n_1, n_2)$, $(n_1, \sqrt{\mathcal{E}} + n_2)$, $(-\sqrt{\mathcal{E}} + n_1, n_2)$ and $(n_1, -\sqrt{\mathcal{E}} + n_2)$.

5.1.3 The Optimum Detector

We have demonstrated that, for a signal transmitted over an AWGN channel, either a correlation demodulator or a matched filter demodulator produces the vector $\mathbf{r} = [r_1\, r_2 \cdots r_N]$, which contains all the relevant information in the received signal waveform. In this section, we describe the optimum decision rule based on the observation vector $\mathbf{r}$. For this development, we assume that there is no memory in signals transmitted in successive signal intervals.

We wish to design a signal detector that makes a decision on the transmitted signal in each signal interval based on the observation of the vector $\mathbf{r}$ in each interval such that the probability of a correct decision is maximized. With this goal in mind, we consider a decision rule based on the computation of the *posterior probabilities* defined as

$$P(\text{signal } \mathbf{s}_m \text{ was transmitted} | \mathbf{r}), \qquad m = 1, 2, \ldots, M$$

which we abbreviate as $P(\mathbf{s}_m|\mathbf{r})$. The decision criterion is based on selecting the signal corresponding to the maximum of the set of posterior probabilities $\{P(\mathbf{s}_m|\mathbf{r})\}$. Later, we show that this criterion maximizes the probability of a correct decision and, hence, minimizes the probability of error. This decision criterion is called the *maximum a posteriori probability* (MAP) criterion.

Using Bayes' rule, the posterior probabilities may be expressed as

$$P(\mathbf{s}_m|\mathbf{r}) = \frac{p(\mathbf{r}|\mathbf{s}_m)P(\mathbf{s}_m)}{p(\mathbf{r})} \tag{5.1–38}$$

where $p(\mathbf{r}|\mathbf{s}_m)$ is the conditional PDF of the observed vector given $\mathbf{s}_m$, and $P(\mathbf{s}_m)$ is the *a priori probability* of the mth signal being transmitted. The denominator of Equation 5.1–38 may be expressed as

$$p(\mathbf{r}) = \sum_{m=1}^{M} p(\mathbf{r}|\mathbf{s}_m)P(\mathbf{s}_m) \tag{5.1–39}$$

From Equations 5.1–38 and 5.1–39, we observe that the computation of the posterior probabilities $P(\mathbf{s}_m|\mathbf{r})$ requires knowledge of the a priori probabilities $P(\mathbf{s}_m)$ and the conditional PDFs $p(\mathbf{r}|\mathbf{s}_m)$ for $m = 1, 2, \ldots, M$.

Some simplification occurs in the MAP criterion when the M signals are equally probable a priori, i.e., $P(\mathbf{s}_m) = 1/M$ for all M. Furthermore, we note that the denominator in Equation 5.1–38 is independent of which signal is transmitted. Consequently, the decision rule based on finding the signal that maximizes $P(\mathbf{s}_m|\mathbf{r})$ is equivalent to finding the signal that maximizes $p(\mathbf{r}|\mathbf{s}_m)$.

The conditional PDF $p(\mathbf{r}|\mathbf{s}_m)$ or any monotonic function of it is usually called the *likelihood function*. The decision criterion based on the maximum of $p(\mathbf{r}|\mathbf{s}_m)$ over the M signals is called the *maximum-likelihood* (ML) *criterion*. We observe that a detector based on the MAP criterion and one that is based on the ML criterion make the same decisions as long as the a priori probabilities $P(\mathbf{s}_m)$ are all equal, i.e., the signals $\{\mathbf{s}_m\}$ are equiprobable.

In the case of an AWGN channel, the likelihood function $p(\mathbf{r}|\mathbf{s}_m)$ is given by Equation 5.1–12. To simplify the computations, we may work with the natural logarithm of $p(\mathbf{r}|\mathbf{s}_m)$, which is a monotonic function. Thus,

$$\ln p(\mathbf{r}|\mathbf{s}_m) = -\tfrac{1}{2}N\ln(\pi N_0) - \frac{1}{N_0}\sum_{k=1}^{N}(r_k - s_{mk})^2 \tag{5.1–40}$$

The maximum of $\ln p(\mathbf{r}|\mathbf{s}_m)$ over $\mathbf{s}_m$ is equivalent to finding the signal $\mathbf{s}_m$ that minimizes the Euclidean distance

$$D(\mathbf{r}, \mathbf{s}_m) = \sum_{k=1}^{N}(r_k - s_{mk})^2 \tag{5.1–41}$$

We call $D(\mathbf{r}, \mathbf{s}_m)$, $m = 1, 2, \ldots, M$, the *distance metrics*. Hence, for the AWGN channel, the decision rule based on the ML criterion reduces to finding the signal $\mathbf{s}_m$ that is closest in distance to the received signal vector $\mathbf{r}$. We shall refer to this decision rule as *minimum distance detection*.

Another interpretation of the optimum decision rule based on the ML criterion is obtained by expanding the distance metrics in Equation 5.1–41 as

$$\begin{aligned} D(\mathbf{r}, \mathbf{s}_m) &= \sum_{n=1}^{N} r_n^2 - 2\sum_{n=1}^{N} r_n s_{mn} + \sum_{n=1}^{N} s_{mn}^2 \\ &= \|\mathbf{r}\|^2 - 2\mathbf{r}\cdot\mathbf{s}_m + \|\mathbf{s}_m\|^2, \qquad m = 1, 2, \ldots, M \end{aligned} \tag{5.1–42}$$

The term $\|\mathbf{r}\|^2$ is common to all distance metrics, and, hence, it may be ignored in the computations of the metrics. The result is a set of modified distance metrics

$$D'(\mathbf{r}, \mathbf{s}_m) = -2\mathbf{r}\cdot\mathbf{s}_m + \|\mathbf{s}_m\|^2 \tag{5.1–43}$$

Note that selecting the signal $\mathbf{s}_m$ that minimizes $D'(\mathbf{r}, \mathbf{s}_m)$ is equivalent to selecting the signal that maximizes the metric $C(\mathbf{r}, \mathbf{s}_m) = -D'(\mathbf{r}, \mathbf{s}_m)$, i.e.,

$$C(\mathbf{r}, \mathbf{s}_m) = 2\mathbf{r} \cdot \mathbf{s}_m - \|\mathbf{s}_m\|^2 \tag{5.1–44}$$

The term $\mathbf{r} \cdot \mathbf{s}_m$ represents the projection of the received signal vector onto each of the M possible transmitted signal vectors. The value of each of these projections is a measure of the correlation between the received vector and the mth signal. For this reason, we call $C(\mathbf{r}, \mathbf{s}_m)$, $m = 1, 2, \ldots, M$, the *correlation metrics* for deciding which of the M signals was transmitted. Finally, the terms $\|\mathbf{s}_m\|^2 = \mathcal{E}_m$, $m = 1, 2, \ldots, M$, may be viewed as bias terms that serve as compensation for signal sets that have unequal energies, such as PAM. If all signals have the same energy, $\|\mathbf{s}_m\|^2$ may also be ignored in the computation of the correlation metrics $C(\mathbf{r}, \mathbf{s}_m)$ and the distance metrics $D(\mathbf{r}, \mathbf{s}_m)$ or $D'(\mathbf{r}, \mathbf{s}_m)$.

It is easy to show (see Problem 5.5) that the correlation metrics $C(\mathbf{r}, \mathbf{s}_m)$ can also be expressed as

$$C(\mathbf{r}, \mathbf{s}_m) = 2\int_0^T r(t)s_m(t)\,dt - \mathcal{E}_m, \qquad m = 0, 1, \ldots, M \tag{5.1–45}$$

Therefore, these metrics can be generated by a demodulator that cross-correlates the received signal $r(t)$ with each of the M possible transmitted signals and adjusts each correlator output for the bias in the case of unequal signal energies. Equivalently, the received signal may be passed through a bank of M filters matched to the possible transmitted signals $\{s_m(t)\}$ and sampled at $t = T$, the end of the symbol interval. Consequently, the optimum receiver (demodulator and detector) can be implemented in the alternative configuration illustrated in Figure 5.1–9.

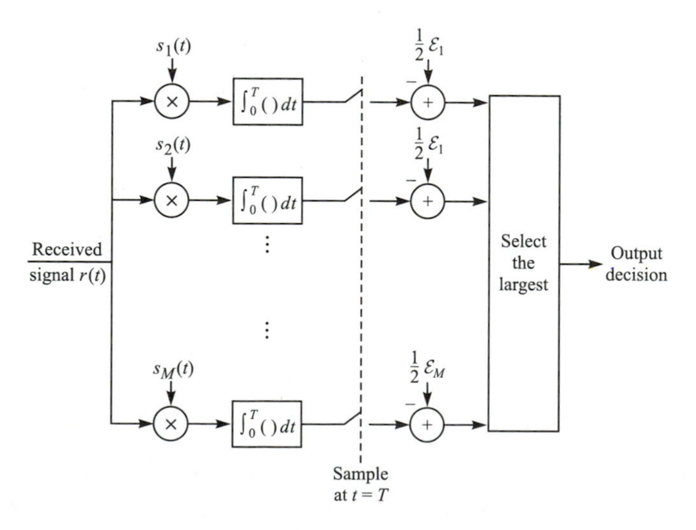

FIGURE 5.1–9
An alternative realization of the optimum AWGN receiver.

In summary, we have demonstrated that the optimum ML detector computes a set of M distances $D(\mathbf{r}, \mathbf{s}_m)$ or $D'(\mathbf{r}, \mathbf{s}_m)$ and selects the signal corresponding to the smallest (distance) metric. Equivalently, the optimum ML detector computes a set of M correlation metrics $C(\mathbf{r}, \mathbf{s}_m)$ and selects the signal corresponding to the largest correlation metric.

The above development for the optimum detector treated the important case in which all signals are equally probable. In this case, the MAP criterion is equivalent to the ML criterion. However, when the signals are not equally probable, the optimum MAP detector bases its decision on the probabilities $P(\mathbf{s}_m|\mathbf{r}), m = 1, 2, \ldots, M$, given by Equation 5.1–38 or, equivalently, on the *metrics*,

$$PM(\mathbf{r}, \mathbf{s}_m) = p(\mathbf{r}|\mathbf{s}_m)P(\mathbf{s}_m)$$

The following example illustrates this computation for binary PAM signals.

EXAMPLE 5.1–3. Consider the case of binary PAM signals in which the two possible signal points are $s_1 = -s_2 = \sqrt{\mathcal{E}_b}$, where $\mathcal{E}_b$ is the energy per bit. The prior probabilities are $P(s_1) = p$ and $P(s_2) = 1 - p$. Let us determine the metrics for the optimum MAP detector when the transmitted signal is corrupted with AWGN.

The received signal vector (one-dimensional) for binary PAM is

$$r = \pm\sqrt{\mathcal{E}_b} + y_n(T) \tag{5.1–46}$$

where $y_n(T)$ is a zero-mean Gaussian random variable with variance $\sigma_n^2 = \frac{1}{2}N_0$. Consequently, the conditional PDFs $p(r|s_m)$ for the two signals are

$$p(r|s_1) = \frac{1}{\sqrt{2\pi}\sigma_n} \exp\left[-\frac{(r - \sqrt{\mathcal{E}_b})^2}{2\sigma_n^2}\right] \tag{5.1–47}$$

$$p(r|s_2) = \frac{1}{\sqrt{2\pi}\sigma_n} \exp\left[-\frac{(r + \sqrt{\mathcal{E}_n})^2}{2\sigma_n^2}\right] \tag{5.1–48}$$

Then the metrics $PM(\mathbf{r}, \mathbf{s}_1)$ and $PM(\mathbf{r}, \mathbf{s}_2)$ are

$$\begin{aligned} PM(\mathbf{r}, \mathbf{s}_1) &= pp(r|s_1) \\ &= \frac{p}{\sqrt{2\pi}\sigma_n} \exp\left[-\frac{(r - \sqrt{\mathcal{E}_b})^2}{2\sigma_n^2}\right] \end{aligned} \tag{5.1–49}$$

$$PM(\mathbf{r}, \mathbf{s}_2) = \frac{1-p}{\sqrt{2\pi}\sigma_n} \exp\left[-\frac{(r + \sqrt{\mathcal{E}_b})^2}{2\sigma_n^2}\right] \tag{5.1–50}$$

If $PM(\mathbf{r}, \mathbf{s}_1) > PM(\mathbf{r}, \mathbf{s}_2)$, we select s_1 as the transmitted signal; otherwise, we select s_2. This decision rule may be expressed as

$$\frac{PM(\mathbf{r}, \mathbf{s}_1)}{PM(\mathbf{r}, \mathbf{s}_2)} \underset{s_2}{\overset{s_1}{\gtrless}} 1 \tag{5.1–51}$$

But

$$\frac{PM(\mathbf{r},\mathbf{s}_1)}{PM(\mathbf{r},\mathbf{s}_2)}=\frac{p}{1-p}\exp\left[\frac{(r+\sqrt{\mathcal{E}_b})^2-(r-\sqrt{\mathcal{E}_b})^2}{2\sigma_n^2}\right] \tag{5.1–52}$$

so that Equation 5.1–51 may be expressed as

$$\frac{(r+\sqrt{\mathcal{E}_b})^2-(r-\sqrt{\mathcal{E}_b})^2}{2\sigma_n^2}\underset{s_2}{\overset{s_1}{\gtrless}}\ln\frac{1-p}{p} \tag{5.1–53}$$

or equivalently,

$$\sqrt{\mathcal{E}_b}r\underset{s_2}{\overset{s_1}{\gtrless}}\tfrac{1}{2}\sigma_n^2\ln\frac{1-p}{p}=\tfrac{1}{4}N_0\ln\frac{1-p}{p} \tag{5.1–54}$$

This is the final form for the optimum detector. It computes the correlation metric $C(\mathbf{r},\mathbf{s}_1)=r\sqrt{\mathcal{E}_b}$ and compares it with threshold $\frac{1}{4}N_0\ln[(1-p)/p]$. Figure 5.1–10 illustrates the two signal points s_1 and s_2. The threshold, denoted by τ_h, divides the real line into two regions, say R_1 and R_2, where R_1 consists of the set of points that are greater than τ_h and R_2 consists of the set of points that are less than τ_h. If $r\sqrt{\mathcal{E}_b}>\tau_h$, the decision is made that s_1 was transmitted, and if $r\sqrt{\mathcal{E}_b}<\tau_h$, the decision is made that s_2 was transmitted. The threshold τ_h depends on N_0 and p. If $p=\frac{1}{2}$, $\tau_h=0$. If $p>\frac{1}{2}$, the signal point s_1 is more probable and, hence, $\tau_h<0$. In this case, the region R_1 is larger than R_2, so that s_1 is more likely to be selected than s_2. If $p<\frac{1}{2}$, the opposite is the case. Thus, the average probability of error is minimized.

It is interesting to note that in the case of unequal prior probabilities, it is necessary to know not only the values of the prior probabilities but also the value of the power spectral density N_0, or, equivalently, the noise-to-signal ratio $N_0/\mathcal{E}_b$, in order to compute the threshold. When $p=\frac{1}{2}$, the threshold is zero, and knowledge of N_0 is not required by the detector.

We conclude this section with the proof that the decision rule based on the maximum-likelihood criterion minimizes the probability of error when the M signals are equally probable a priori. Let us denote by R_m the region in the N-dimensional space for which we decide that signal $s_m(t)$ was transmitted when the vector $\mathbf{r}=[r_1\ r_2\cdots r_N]$ is received. The probability of a correct decision given that $s_m(t)$ was transmitted is

$$P(c|\mathbf{s}_m)=\int_{R_m}p(\mathbf{r}|\mathbf{s}_m)\,d\mathbf{r} \tag{5.1–55}$$

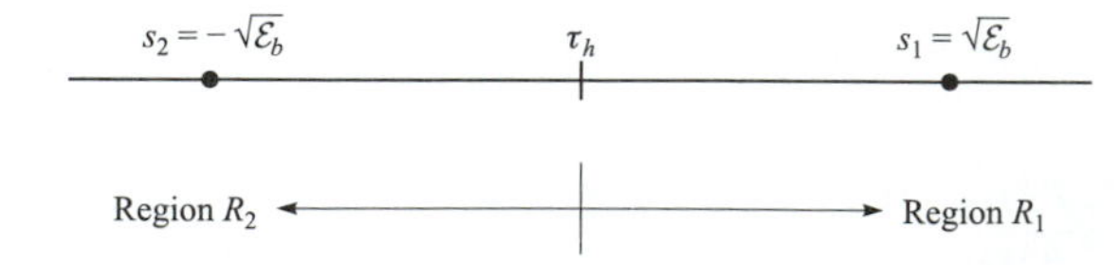

FIGURE 5.1–10
Signal space representation illustrating the operation of the optimum detector for binary (PAM) modulation.

The average probability of a correct decision is

$$\begin{aligned} P(c) &= \sum_{m=1}^{M} \frac{1}{M} P(c|\mathbf{s}_m) \\ &= \sum_{m=1}^{M} \frac{1}{M} \int_{R_m} p(\mathbf{r}|\mathbf{s}_m)\, d\mathbf{r} \end{aligned} \tag{5.1–56}$$

Note that $P(c)$ is maximized by selecting the signal $\mathbf{s}_m$ if $p(\mathbf{r}|\mathbf{s}_m)$ is larger than $p(\mathbf{r}|\mathbf{s}_k)$ for all $m \neq k$.

Similarly for the MAP criterion, when the M signals are not equally probable, the average probability of a correct decision is

$$P(c) = \sum_{m=1}^{M} \int_{R_m} P(\mathbf{s}_m|\mathbf{r}) p(\mathbf{r}) d\mathbf{r}$$

In order for $P(c)$ to be as large as possible, the points that are to be included in each particular region R_m are those for which $P(\mathbf{s}_m|\mathbf{r})$ exceeds all the other posterior probabilities.

5.1.4 The Maximum-Likelihood Sequence Detector

When the signal has no memory, the symbol-by-symbol detector described in the preceding section is optimum in the sense of minimizing the probability of a symbol error. On the other hand, when the transmitted signal has memory, i.e., the signals transmitted in successive symbol intervals are interdependent, the optimum detector is a detector that bases its decisions on observation of a sequence of received signals over successive signal intervals. Below, we describe two different types of detection algorithms. In this section, we describe a maximum-likelihood sequence detection algorithm that searches for the minimum Euclidean distance path through the trellis that characterizes the memory in the transmitted signal. In the following section, we describe a maximum a posteriori probability algorithm that makes decisions on a symbol-by-symbol basis, but each symbol decision is based on an observation of a sequence of received signal vectors.

To develop the maximum-likelihood sequence detection algorithm, let us consider, as an example, the NRZI signal described in Section 4.3.2. Its memory is characterized by the trellis shown in Figure 4.3–14. The signal transmitted in each signal interval is binary PAM. Hence, there are two possible transmitted signals corresponding to the signal points $s_1 = -s_2 = \sqrt{\mathcal{E}_b}$, where $\mathcal{E}_b$ is the energy per bit. The output of the matched-filter or correlation demodulator for binary PAM in the kth signal interval may be expressed as

$$r_k = \pm\sqrt{\mathcal{E}_b} + n_k \tag{5.1–57}$$

where n_k is a zero-mean Gaussian random variable with variance $\sigma_n^2 = N_0/2$. Consequently, the conditional PDFs for the two possible transmitted signals are

$$\begin{aligned} p(r_k|s_1) &= \frac{1}{\sqrt{2\pi}\sigma_n} \exp\left[-\frac{(r_k - \sqrt{\mathcal{E}_b})^2}{2\sigma_n^2}\right] \\ p(r_k|s_2) &= \frac{1}{\sqrt{2\pi}\sigma_n} \exp\left[-\frac{(r_k + \sqrt{\mathcal{E}_b})^2}{2\sigma_n^2}\right] \end{aligned} \tag{5.1–58}$$

Now, suppose we observe the sequence of matched-filter outputs $r_1, r_2, \ldots, r_K$. Since the channel noise is assumed to be white and Gaussian, and $f(t = iT)$, $f(t - jT)$ for $i \neq j$ are orthogonal, it follows that $E(n_k n_j) = 0, k \neq j$. Hence, the noise sequence $n_1, n_2, \ldots, n_K$ is also white. Consequently, for any given transmitted sequence $\mathbf{s}^{(m)}$, the joint PDF of $r_1, r_2, \ldots, r_K$ may be expressed as a product of K marginal PDFs, i.e.,

$$\begin{aligned} p(r_1, r_2, \ldots, r_K|\mathbf{s}^{(m)}) &= \prod_{k=1}^{K} p(r_k|s_k^{(m)}) \\ &= \prod_{k=1}^{K} \frac{1}{\sqrt{2\pi}\sigma_n} \exp\left[-\frac{(r_k - s_k^{(m)})^2}{2\sigma_n^2}\right] \\ &= \left(\frac{1}{\sqrt{2\pi}\sigma_n}\right)^K \exp\left[-\sum_{k=1}^{K}\frac{(r_k - s_k^{(m)})^2}{2\sigma_n^2}\right] \end{aligned} \tag{5.1–59}$$

where either $s_k = \sqrt{\mathcal{E}_b}$ or $s_k = -\sqrt{\mathcal{E}_b}$. Then, given the received sequence $r_1, r_2, \ldots, r_K$ at the output of the matched filter or correlation demodulator, the detector determines the sequence $\mathbf{s}^{(m)} = \{s_1^{(m)}, s_2^{(m)}, \ldots, s_K^{(m)}\}$ that maximizes the conditional PDF $p(r_1, r_2, \ldots, r_K|\mathbf{s}^{(m)})$. Such a detector is called the *maximum-likelihood* (ML) *sequence-detector*.

By taking the logarithm of Equation 5.1–59 and neglecting the terms that are independent of $(r_1, r_2, \ldots, r_K)$, we find that an equivalent ML sequence detector selects the sequence $\mathbf{s}^{(m)}$ that minimizes the *Euclidean distance metric*

$$D(\mathbf{r}, \mathbf{s}^{(m)}) = \sum_{k=1}^{K} (r_k - s_k^{(m)})^2 \tag{5.1–60}$$

In searching through the trellis for the sequence that minimizes the Euclidean distance $D(\mathbf{r}, \mathbf{s}^{(m)})$, it may appear that we must compute the distance $D(\mathbf{r}, \mathbf{s}^{(m)})$ for every possible sequence. For the NRZI example, which employs binary modulation, the total number of sequences is 2^K, where K is the number of outputs obtained from the demodulator. However, this is not the case. We may reduce the number of sequences in the trellis search by using the *Viterbi algorithm* to eliminate sequences as new data is received from the demodulator.

The Viterbi algorithm is a sequential trellis search algorithm for performing ML sequence detection. It is described in Chapter 8 as a decoding algorithm for convolutional codes. We describe it below in the context of the NRZI signal. We assume that the search process begins initially at state S_0. The corresponding trellis is shown in Figure 5.1–11.

At time $t = T$, we receive $r_1 = s_1^{(m)} + n$ from the demodulator, and at $t = 2T$, we receive $r_2 = s_2^{(m)} + n_2$. Since the signal memory is one bit, which we denote by $L = 1$, we observe that the trellis reaches its regular (steady state) form after two transitions. Thus, upon receipt of r_2 at $t = 2T$ (and thereafter), we observe that there are two signal paths entering each of the nodes and two signal paths leaving each node. The two paths entering node S_0 at $t = 2T$ correspond to the information bits (0, 0) and (1, 1) or, equivalently, to the signal points $(-\sqrt{\mathcal{E}_b}, -\sqrt{\mathcal{E}_b})$ and $(\sqrt{\mathcal{E}_b}, -\sqrt{\mathcal{E}_b})$, respectively. The two paths entering node S_1 at $t = 2T$ correspond to the information bits (0, 1) and (1, 0) or, equivalently, to the signal points $(-\sqrt{\mathcal{E}_b}, \sqrt{\mathcal{E}_b})$ and $(\sqrt{\mathcal{E}_b}, \sqrt{\mathcal{E}_b})$, respectively.

For the two paths entering node S_0, we compute the two Euclidean distance metrics

$$\begin{aligned} D_0(0, 0) &= (r_1 + \sqrt{\mathcal{E}_b})^2 + (r_2 + \sqrt{\mathcal{E}_b})^2 \\ D_0(1, 1) &= (r_1 - \sqrt{\mathcal{E}_b})^2 + (r_2 + \sqrt{\mathcal{E}_b})^2 \end{aligned} \tag{5.1–61}$$

by using the outputs r_1 and r_2 from the demodulator. The Viterbi algorithm compares these two metrics and discards the path having the larger (greater-distance) metric.† The other path with the lower metric is saved and is called the *survivor* at $t = 2T$. The elimination of one of the two paths may be done without compromising the optimality of the trellis search, because any extension of the path with the larger distance beyond $t = 2T$ will always have a larger metric than the survivor that is extended along the same path beyond $t = 2T$.

Similarly, for the two paths entering node S_1 at $t = 2T$, we compute the two Euclidean distance metrics

$$\begin{aligned} D_1(0, 1) &= (r_1 + \sqrt{\mathcal{E}_b})^2 + (r_2 - \sqrt{\mathcal{E}_b})^2 \\ D_1(1, 0) &= (r_1 - \sqrt{\mathcal{E}_b})^2 + (r_2 - \sqrt{\mathcal{E}_b})^2 \end{aligned} \tag{5.1–62}$$

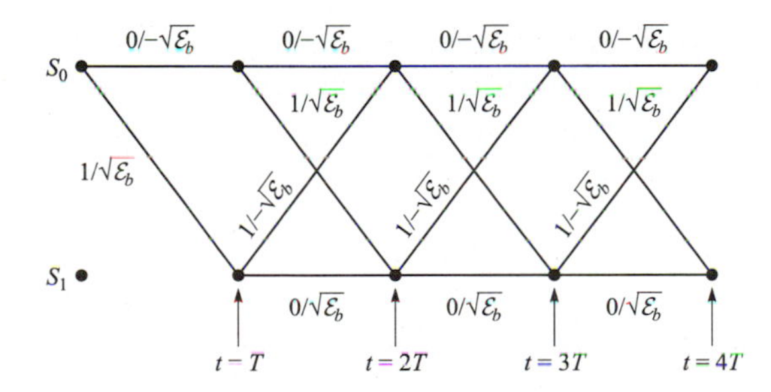

FIGURE 5.1–11
Trellis for NRZI signal.

†Note that, for NRZI, the reception of r_2 from the demodulator neither increases nor decreases the relative difference between the two metrics, $D_0(0, 0)$ and $D_0(1, 1)$. At this point, one may ponder on the implication of this observation. In any case, we continue with the description of the ML sequence detector based on the Viterbi algorithm.

by using the outputs r_1 and r_2 from the demodulator. The two metrics are compared and the signal path with the larger metric is eliminated. Thus, at $t = 2T$, we are left with two survivor paths, one at node S_0 and the other at node S_1, and their corresponding metrics. The signal paths at nodes S_0 and S_1 are then extended along the two survivor paths.

Upon receipt of r_3 at $t = 3T$, we compute the metrics of the two paths entering state S_0. Suppose the survivors at $t = 2T$ are the paths (0, 0) at S_0 and (0, 1) at S_1. Then, the two metrics for the paths entering S_0 at $t = 3T$ are

$$\begin{aligned} D_0(0, 0, 0) &= D_0(0, 0) + (r_3 + \sqrt{\mathcal{E}_b})^2 \\ D_0(0, 1, 1) &= D_1(0, 1) + (r_3 + \sqrt{\mathcal{E}_b})^2 \end{aligned} \tag{5.1–63}$$

These two metrics are compared and the path with the larger (greater-distance) metric is eliminated. Similarly, the metrics for the two paths entering S_1 at $t = 3T$ are

$$\begin{aligned} D_1(0, 0, 1) &= D_0(0, 0) + (r_3 - \sqrt{\mathcal{E}_b})^2 \\ D_1(0, 1, 0) &= D_1(0, 1) + (r_3 - \sqrt{\mathcal{E}_b})^2 \end{aligned} \tag{5.1–64}$$

These two metrics are compared and the path with the larger (greater-distance) metric is eliminated.

This process is continued as each new signal sample is received from the demodulator. Thus, the Viterbi algorithm computes two metrics for the two signal paths entering a node at each stage of the trellis search and eliminates one of the two paths at each node. The two survivor paths are then extended forward to the next state. Therefore, the number of paths searched in the trellis is reduced by a factor of 2 at each stage.

It is relatively easy to generalize the trellis search performed by the Viterbi algorithm for M-ary modulation. For example, delay modulation employs $M = 4$ signals and is characterized by the four-state trellis shown in Figure 5.1–12. We observe that each state has two signal paths entering and two signal paths leaving each node. The memory of the signal is $L = 1$. Hence, the Viterbi algorithm will have four survivors at each stage and their corresponding metrics. Two metrics corresponding to the two entering paths are computed at each node, and one of the two signal paths entering the node is eliminated at each state of the trellis. Thus, the Viterbi algorithm minimizes the number of trellis paths searched in performing ML sequence detection.

From the description of the Viterbi algorithm given above, it is unclear as to how decisions are made on the individual detected information symbols given the surviving sequences. If we have advanced to some stage, say K, where $K \gg L$ in the trellis, and we compare the surviving sequences, we shall find that with probability approaching one all surviving sequences will be identical in bit (or symbol) positions $K - 5L$ and less. In a practical implementation of the Viterbi algorithm, decisions on each information bit (or symbol) are forced after a delay of $5L$ bits (or symbols), and, hence, the surviving sequences are truncated to the $5L$ most recent bits (or symbols). Thus, a variable delay in bit or symbol detec-

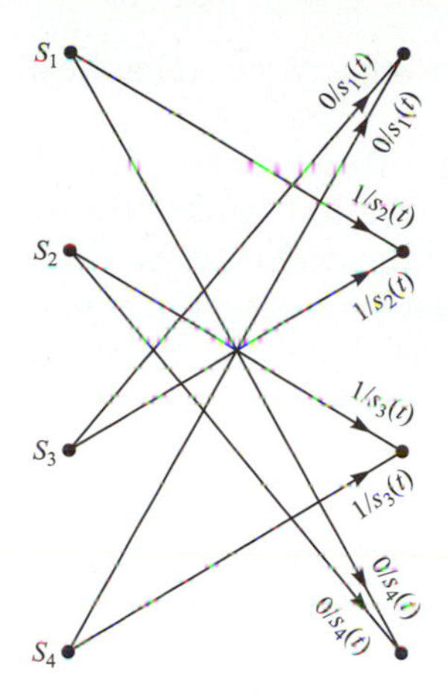

FIGURE 5.1–12
One stage of trellis diagram for delay modulation.

tion is avoided. The loss in performance resulting from the suboptimum detection procedure is negligible if the delay is at least $5L$.

EXAMPLE 5.1–4. Consider the decision rule for detecting the data sequence in an NRZI signal with a Viterbi algorithm having a delay of $5L$ bits. The trellis for the NRZI signal is shown in Figure 5.1–11. In this case, $L = 1$; hence the delay in bit detection is set to 5 bits. Hence, at $t = 6T$, we shall have two surviving sequences, one for each of the two states and the corresponding metrics $\mu_6(b_1, b_2, b_3, b_4, b_5, b_6)$ and $\mu_6(b_1', b_2', b_3', b_4', b_5', b_6')$. At this stage, with probability nearly equal to one, the bit b_1 will be the same as b_1'; that is, both surviving sequences will have a common first branch. If $b_1 \neq b_1'$, we may select the bit (b_1 or b_1') corresponding to the smaller of the two metrics. Then the first bit is dropped from the two surviving sequences. At $t = 7T$, the two metrics $\mu_7(b_2, b_3, b_4, b_5, b_6, b_7)$ and $\mu_7(b_2', b_3', b_4', b_5', b_6', b_7')$ will be used to determine the decision on bit b_2. This process continues at each stage of the search through the trellis for the minimum distance sequence. Thus the detection delay is fixed at 5 bits.†

5.1.5 A Symbol-by-Symbol MAP Detector for Signals with Memory

In contrast to the maximum-likelihood sequence detector for detecting the transmitted information, we now describe a detector that makes symbol-by-symbol decisions based on the computation of the maximum a posteriori probability (MAP) for each detected symbol. Hence, this detector is optimum in the sense that it minimizes the probability of a symbol error. The detection algorithm that is presented below is due to Abend and Fritchman (1970), who developed it as a

†One may have observed by now that the ML sequence detector and the symbol-by-symbol detector that ignores the memory in the NRZI signal reach the same decisions. Hence, there is no need for a decision delay. Nevertheless, the procedure described above applies in general.

detection algorithm for channels with intersymbol interference, i.e., channels with memory.

We illustrate the algorithm in the context of detecting a PAM signal with M possible levels. Suppose that it is desired to detect the information symbol transmitted in the kth signal interval, and let $r_1, r_2, \ldots, r_{k+D}$ be the observed received sequence, where D is the delay parameter which is chosen to exceed the signal memory, i.e., $D \geqslant L$, where L is the inherent memory in the signal. On the basis of the received sequence, we compute the posterior probabilities

$$P(s^{(k)} = A_m | r_{k+D}, r_{k+D-1}, \ldots, r_1) \tag{5.1–65}$$

for the M possible symbol values and choose the symbol with the largest probability. Since

$$P(s^{(k)} = A_m | r_{k+D}, \ldots, r_1) = \frac{p(r_{k+D}, \ldots, r_1 | s^{(k)} = A_m) P(s^{(k)} = A_m)}{p(r_{k+D}, r_{k+D-1}, \ldots, r_1)} \tag{5.1–66}$$

and since the denominator is common for all M probabilities, the MAP criterion is equivalent to choosing the value of $s^{(k)}$ that maximizes the numerator of Equation 5.1–66. Thus, the criterion for deciding on the transmitted symbol $s^{(k)}$ is

$$\tilde{s}^{(k)} = \arg\left\{ \max_{s^{(k)}} p(r_{k+D}, \ldots, r_1 | s^{(k)} = A_m) P(s^{(k)} = A_m) \right\} \tag{5.1–67}$$

When the symbols are equally probable, the probability $P(s^{(k)} = A_m)$ may be dropped from the computation.

The algorithm for computing the probabilities in Equation 5.1–67 recursively begins with the first symbol $s^{(1)}$. We have

$$\begin{aligned}
\tilde{s}^{(1)} &= \arg\left\{ \max_{s^{(1)}} p(r_{1+D}, \ldots, r_1 | s^{(1)} = A_m) P(s^{(1)} = A_m) \right\} \\
&= \arg\left\{ \max_{s^{(1)}} \sum_{s^{1+D}} \cdots \sum_{s^{(2)}} p(r_{1+D}, \ldots, r_1 | s^{(1+D)}, \ldots, s^{(1)}) P(s^{(1+D)}, \ldots, s^{(1)}) \right\} \\
&= \arg\left\{ \max_{s^{(1)}} \sum_{s^{(1+D)}} \cdots \sum_{s^{(2)}} p_1(s^{(1+D)}, \ldots, s^{(2)}, s^{(1)}) \right\}
\end{aligned} \tag{5.1–68}$$

where $\tilde{s}^{(1)}$ denotes the decision on $s^{(1)}$ and, for mathematical convenience, we have defined

$$p_1(s^{(1+D)}, \ldots, s^{(2)}, s^{(1)}) \equiv p(r_{1+D}, \ldots, r_1 | s^{(1+D)}, \ldots, s^{(1)}) P(s^{(1+D)}, \ldots, s^{(1)}) \tag{5.1–69}$$

The joint probability $P(s^{(1+D)}, \ldots, s^{(2)}, s^{(1)})$ may be omitted if the symbols are equally probable and statistically independent. As a consequence of the statistical independence of the additive noise sequence, we have

$$
\begin{aligned}
p(r_{1+D}, \ldots, r_1 | s^{(1+D)}, \ldots, s^{(1)}) \\
= p(r_{1+D} | s^{(1+D)}, \ldots, s^{(1+D-L)}) p(r_D | s^{(D)}, \ldots, s^{(D-L)}) \ldots \\
p(r_2 | s^{(2)}, s^{(1)}) p(r_1 | s^{(1)})
\end{aligned}
\tag{5.1–70}
$$

where we assume that $s^{(k)} = 0$ for $k \leqslant 0$.

For detection of the symbol $s^{(2)}$, we have

$$
\begin{aligned}
\tilde{s}^{(2)} &= \arg\left\{ \max_{s^{(2)}} p(r_{2+D}, \ldots, r_1 | s^{(2)} = A_m) P(s^{(2)} = A_m) \right\} \\
&= \arg\left\{ \max_{s^{(2)}} \sum_{s^{(2+D)}} \cdots \sum_{s^{(3)}} p(r_{2+D}, \ldots, r_1 | s^{(2+D)}, \ldots, s^{(2)}) P(s^{(2+D)}, \ldots, s^{(2)} \right\}
\end{aligned}
\tag{5.1–71}
$$

The joint conditional probability in the multiple summation can be expressed as

$$
\begin{aligned}
p(r_{2+D}, \ldots, r_1 | s^{(2+D)}, \ldots, s^{(2)}) \\
= p(r_{2+D} | s^{(2+D)}, \ldots, s^{(2+D-L)}) p(r_{1+D}, \ldots, r_1 | s^{(1+D)}, \ldots, s^{(2)})
\end{aligned}
\tag{5.1–72}
$$

Furthermore, the joint probability

$$
p(r_{1+D}, \ldots, r_1 | s^{(1+D)}, \ldots, s^{(2)}) P(s^{(1+D)}, \ldots, s^{(2)})
$$

can be obtained from the probabilities computed previously in the detection of $s^{(1)}$. That is,

$$
\begin{aligned}
p(r_{1+D}, \ldots, r_1 | s^{(1+D)}, \ldots, s^{(2)}) \\
= \sum_{s^{(1)}} p(r_{1+D}, \ldots, r_1 | s^{(1+D)}, \ldots, s^{(1)}) P(s^{(1+D)}, \ldots, s^{(1)}) \\
= \sum_{s^{(1)}} p_1(s^{(1+D)}, \ldots, s^{(2)}, s^{(1)})
\end{aligned}
\tag{5.1–73}
$$

Thus, by combining Equations 5.1–73 and 5.1–72 and then substituting into Equation 5.1–71, we obtain

$$
\tilde{s}^{(2)} = \arg\left\{ \max_{s^{(2)}} \sum_{s^{(2+D)}} \cdots \sum_{s^{(3)}} p_2(s^{(2+D)}, \ldots, s^{(3)}, s^{(2)}) \right\}
\tag{5.1–74}
$$

where, by definition,

$$
\begin{aligned}
p_2(s^{(2+D)}, \ldots, s^{(3)}, s^{(2)}) \\
= p(r_{2+D} | s^{(2+D)}, \ldots, s^{(2+D-L)}) P(s^{(2+D)}) \sum_{s^{(1)}} p_1(s^{(1+D)}, \ldots, s^{(2)}, s^{(1)})
\end{aligned}
\tag{5.1–75}
$$

In general, the recursive algorithm for detecting the symbol $s^{(k)}$ is as follows: upon reception of $r_{k+D}, \ldots, r_2, r_1$, we compute

$$\begin{aligned}\tilde{s}^{(k)} &= \arg\left\{\max_{s^{(k)}} p(r_{k+D}, \ldots, r_1 | s^{(k)}) P(s^{(k)})\right\} \\ &= \arg\left\{\max_{s^{(k)}} \sum_{s^{(k+D)}} \cdots \sum_{s^{(k+1)}} p_k(s^{(k+D)}, \ldots, s^{(k+1)}, s^{(k)})\right\}\end{aligned} \tag{5.1–76}$$

where, by definition,

$$\begin{aligned}&p_k(s^{(k+D)}, \ldots, s^{(k+1)}, s^{(k)}) \\ &\quad = p(r_{k+D} | s^{(k+D)}, \ldots, s^{(k+D-L)}) P(s^{(k+D)}) \sum_{s^{(k-1)}} p_{k-1}(s^{(k-1+D)}, \ldots, s^{(k-1)})\end{aligned} \tag{5.1–77}$$

Thus, the recursive nature of the algorithm is established by the relations 5.1–76 and 5.1–77.

The major problem with the algorithm is its computational complexity. In particular, the averaging performed over the symbols $s^{(k+D)}, \ldots, s^{(k+1)}, s^{(k)}$ in Equation 5.1–76 involves a large amount of computation per received signal, especially if the number M of amplitude levels $\{A_m\}$ is large. On the other hand, if M is small and the memory L is relatively short, this algorithm is easily implemented.

Symbol-by-symbol detection of signals with memory, based on the MAP criterion, is the basis for iterative decoding algorithms for convolutional codes and interleaved concatenated codes described in Chapter 8. These iterative decoding algorithms are based on the MAP algorithm described in the paper by Bahl et al. (1974), called the BCJR algorithm, which is different than the Abend-Fritchman MAP algorithm given above.

5.2 PERFORMANCE OF THE OPTIMUM RECEIVER FOR MEMORYLESS MODULATION

In this section, we evaluate the probability of error for the memoryless modulation signals described in Section 4.3.1. First, we consider binary PAM signals and then M-ary signals of various types.

5.2.1 Probability of Error for Binary Modulation

Let us consider binary PAM signals where the two signal waveforms are $s_1(t) = g(t)$ and $s_2(t) = -g(t)$, and $g(t)$ is an arbitrary pulse that is nonzero in the interval $0 \leqslant t \leqslant T_b$ and zero elsewhere.

Since $s_1(t) = -s_2(t)$, these signals are said to be *antipodal*. The energy in the pulse $g(t)$ is $\mathcal{E}_g$. As indicated in Section 4.3.1, PAM signals are one-dimensional,

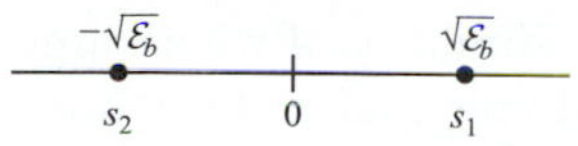

FIGURE 5.2–1
Signal points for binary antipodal signals.

and, hence, their geometric representation is simply the one-dimensional vector $s_1 = \sqrt{\mathcal{E}_b}$, $s_2 = -\sqrt{\mathcal{E}_b}$. Figure 5.2–1 illustrates the two signal points.

Let us assume that the two signals are equally likely and that signal $s_1(t)$ was transmitted. Then, the received signal from the (matched filter or correlation) demodulator is

$$r = s_1 + n = \sqrt{\mathcal{E}_b} + n \tag{5.2–1}$$

where n represents the additive Gaussian noise component, which has zero mean and variance $\sigma_n^2 = \frac{1}{2}N_0$. In this case, the decision rule based on the correlation metric given by Equation 5.1–44 compares r with the threshold zero. If $r > 0$, the decision is made in favor of $s_1(t)$, and if $r < 0$, the decision is made that $s_2(t)$ was transmitted. Clearly, the two conditional PDFs of r are

$$p(r|s_1) = \frac{1}{\sqrt{\pi N_0}} e^{-(r-\sqrt{\mathcal{E}_b})^2/N_0} \tag{5.2–2}$$

$$p(r|s_2) = \frac{1}{\sqrt{\pi N_0}} e^{-(r+\sqrt{\mathcal{E}_b})^2/N_0} \tag{5.2–3}$$

These two conditional PDFs are shown in Figure 5.2–2.

Given that $s_1(t)$ was transmitted, the probability of error is simply the probability that $r < 0$, i.e.,

$$\begin{aligned}
P(e|s_1) &= \int_{-\infty}^{0} p(r|s_1)\,dr \\
&= \frac{1}{\sqrt{\pi N_0}} \int_{-\infty}^{0} \exp\left[-\frac{(r-\sqrt{\mathcal{E}_b})^2}{N_0}\right] dr \\
&= \frac{1}{\sqrt{2\pi}} \int_{-\infty}^{-\sqrt{2\mathcal{E}_b/N_0}} e^{-x^2/2}\,dx \\
&= \frac{1}{\sqrt{2\pi}} \int_{\sqrt{2\mathcal{E}_b/N_0}}^{\infty} e^{-x^2/2}\,dx \\
&= Q\left(\sqrt{\frac{2\mathcal{E}_b}{N_0}}\right)
\end{aligned} \tag{5.2–4}$$

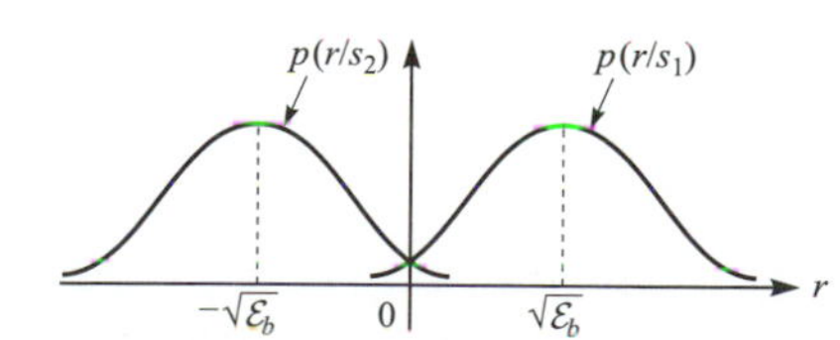

FIGURE 5.2–2
Conditional PDFs of two signals.

where $Q(x)$ is the Q-function defined in Equation 2.1–97. Similarly, if we assume that $s_2(t)$ was transmitted, $r = -\sqrt{\mathcal{E}_b} + n$ and the probability that $r > 0$ is also $P(e|s_2) = Q(\sqrt{2\mathcal{E}_b/N_0})$. Since the signals $s_1(t)$ and $s_2(t)$ are equally likely to be transmitted, the average probability of error is

$$\begin{aligned} P_b &= \tfrac{1}{2}P(e|s_1) + \tfrac{1}{2}P(e|s_2) \\ &= Q\left(\sqrt{\frac{2\mathcal{E}_b}{N_0}}\right) \end{aligned} \tag{5.2–5}$$

We should observe two important characteristics of this performance measure. First, we note that the probability of error depends only on the ratio $\mathcal{E}_b/N_0$ and not on any other detailed characteristics of the signals and the noise. Secondly, we note that $2\mathcal{E}_b/N_0$ is also the output SNR_0 from the matched-filter (and correlation) demodulator. The ratio $\mathcal{E}_b/N_0$ is usually called the *signal-to-noise ratio per bit*.

We also observe that the probability of error may be expressed in terms of the distance between the two signals s_1 and s_2. From Figure 5.2–1, we observe that the two signals are separated by the distance $d_{12} = 2\sqrt{\mathcal{E}_b}$. By substituting $\mathcal{E}_b = \frac{1}{4}d_{12}^2$ into Equation 5.2–5, we obtain

$$P_b = Q\left(\sqrt{\frac{d_{12}^2}{2N_0}}\right) \tag{5.2–6}$$

This expression illustrates the dependence of the error probability on the distance between the two signal points.

Next, let us evaluate the error probability for binary orthogonal signals. Recall that the signal vectors $\mathbf{s}_1$ and $\mathbf{s}_2$ are two-dimensional, as shown in Figure 5.2–3, and may be expressed, according to Equation 4.3–30, as

$$\begin{aligned} \mathbf{s}_1 &= [\sqrt{\mathcal{E}_b} \quad 0] \\ \mathbf{s}_2 &= [0 \quad \sqrt{\mathcal{E}_b}] \end{aligned} \tag{5.2–7}$$

where $\mathcal{E}_b$ denotes the energy for each of the waveforms. Note that the distance between these signal points is $d_{12} = \sqrt{2\mathcal{E}_b}$.

To evaluate the probability of error, let us assume that $\mathbf{s}_1$ was transmitted. Then, the received vector at the output of the demodulator is

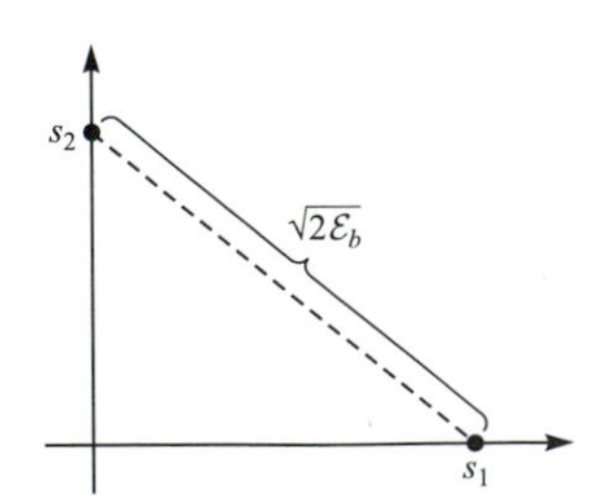

FIGURE 5.2–3
Signal points for binary orthogonal signals.

$$\mathbf{r} = [\sqrt{\mathcal{E}_b} + n_1 \; n_2] \tag{5.2–8}$$

We can now substitute for $\mathbf{r}$ into the correlation metrics given by Equation 5.1–44 to obtain $C(\mathbf{r}, \mathbf{s}_1)$ and $C(\mathbf{r}, \mathbf{s}_2)$. Then, the probability of error is the probability that $C(\mathbf{r}, \mathbf{s}_2) > C(\mathbf{r}, \mathbf{s}_1)$. Thus,

$$P(e|\mathbf{s}_1) = P[C(\mathbf{r}, \mathbf{s}_2) > C(\mathbf{r}_1, \mathbf{s}_1)] = P[n_2 - n_1 > \sqrt{\mathcal{E}_b}] \tag{5.2–9}$$

Since n_1 and n_2 are zero-mean statistically independent Gaussian random variables each with variance $\frac{1}{2}N_0$, the random variable $x = n_2 - n_1$ is zero-mean Gaussian with variance N_0. Hence,

$$\begin{aligned} P(n_2 - n_1 > \sqrt{\mathcal{E}_b}) &= \frac{1}{\sqrt{2\pi N_0}} \int_{\sqrt{\mathcal{E}_b}}^{\infty} e^{-x^2/2N_0}\, dx \\ &= \frac{1}{\sqrt{2\pi}} \int_{\sqrt{\mathcal{E}_b/N_0}}^{\infty} e^{-x^2/2}\, dx \\ &= Q\left(\sqrt{\frac{\mathcal{E}_b}{N_0}}\right) \end{aligned} \tag{5.2–10}$$

Because of symmetry, the same error probability is obtained when we assume that $\mathbf{s}_2$ is transmitted. Consequently, the average error probability for binary orthogonal signals is

$$P_b = Q\left(\sqrt{\frac{\mathcal{E}_b}{N_0}}\right) = Q(\sqrt{\gamma_b}) \tag{5.2–11}$$

where, by definition, γ_b is the SNR per bit.

If we compare the probability of error for binary antipodal signals with that for binary orthogonal signals, we find that orthogonal signals require a factor of 2 increase in energy to achieve the same error probability as antipodal signals. Since $10 \log_{10} 2 = 3$ dB, we say that orthogonal signals are 3 dB poorer than antipodal signals. The difference of 3 dB is simply due to the distance between the two signal points, which is $d_{12}^2 = 2\mathcal{E}_b$ for orthogonal signals, whereas $d_{12}^2 = 4\mathcal{E}_b$ for antipodal signals.

The error probability versus $10 \log_{10} \mathcal{E}_b/N_0$ for these two types of signals is shown in Figure 5.2–4. As observed from this figure, at any given error probability, the $\mathcal{E}_b/N_0$ required for orthogonal signals is 3 dB more than that for antipodal signals.

5.2.2 Probability of Error for M-ary Orthogonal Signals

For equal-energy orthogonal signals, the optimum detector selects the signal resulting in the largest cross correlation between the received vector $\mathbf{r}$ and each of the M possible transmitted signal vectors $\{\mathbf{s}_m\}$, i.e.,

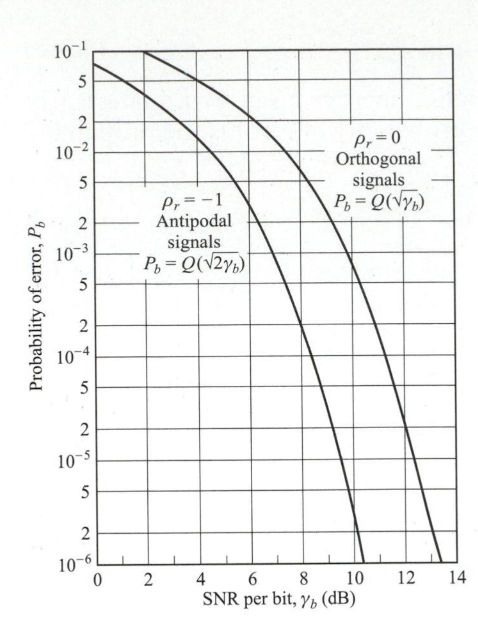

FIGURE 5.2–4
Probability of error for binary signals.

$$C(\mathbf{r}, \mathbf{s}_m) = \mathbf{r} \cdot \mathbf{s}_m = \sum_{k=1}^{M} r_k s_{mk}, \qquad m = 1, 2, \ldots, M \tag{5.2–12}$$

To evaluate the probability of error, let us suppose that the signal $\mathbf{s}_1$ is transmitted. Then the received signal vector is

$$\mathbf{r} = [\sqrt{\mathcal{E}_s} + n_1 \quad n_2 \quad n_3 \cdots n_M] \tag{5.2–13}$$

where $\mathcal{E}_s$ denotes the symbol energy and $n_1, n_2, \ldots, n_M$ are zero-mean, mutually statistically independent Gaussian random variables with equal variance $\sigma_n^2 = \frac{1}{2}N_0$. In this case, the outputs from the bank of M correlators are

$$\begin{aligned}
C(\mathbf{r}, \mathbf{s}_1) &= \sqrt{\mathcal{E}_s}(\sqrt{\mathcal{E}_s} + n_1) \\
C(\mathbf{r}, \mathbf{s}_2) &= \sqrt{\mathcal{E}_s} n_2 \\
\vdots \quad & \quad \vdots \\
C(\mathbf{r}, \mathbf{s}_M) &= \sqrt{\mathcal{E}_s} n_M
\end{aligned} \tag{5.2–14}$$

Note that the scale factor $\mathcal{E}_s$ may be eliminated from the correlator outputs by dividing each output by $\sqrt{\mathcal{E}_s}$. Then, with this normalization, the PDF of the first correlator output ($r_1 = \sqrt{\mathcal{E}_s} + n_1$) is

$$p_{r_1}(x_1) = \frac{1}{\sqrt{\pi N_0}} \exp\left[-\frac{(x_1 - \sqrt{\mathcal{E}_s})^2}{N_0}\right] \tag{5.2–15}$$

and the PDFs of the other $M-1$ correlator outputs are

$$p_{r_m}(x_m) = \frac{1}{\sqrt{\pi N_0}} e^{-x_m^2/N_0}, \qquad m = 2, 3, \ldots, M \tag{5.2–16}$$

It is mathematically convenient to first derive the probability that the detector makes a correct decision. This is the probability that r_1 is larger than each of the other $M-1$ correlator outputs $n_2, n_3, \ldots, n_M$. This probability may be expressed as

$$P_c = \int_{-\infty}^{\infty} P(n_2 < r_1, n_3 < r_1, \ldots, n_M < r_1 | r_1) p(r_1)\, dr_1 \tag{5.2–17}$$

where $P(n_2 < r_1, n_3 < r_1, \ldots, n_M < r_1 | r_1)$ denotes the joint probability that n_2, $n_3, \ldots, n_M$ are all less than r_1, conditioned on any given r_1. Then this joint probability is averaged over all r_1. Since the $\{r_m\}$ are statistically independent, the joint probabilty factors into a product of $M-1$ marginal probabilities of the form

$$\begin{aligned} P(n_m < r_1 | r_1) &= \int_{-\infty}^{r_1} p_{r_m}(x_m)\, dx_m, \qquad m = 2, 3, \ldots, M \\ &= \frac{1}{\sqrt{2\pi}} \int_{-\infty}^{r_1\sqrt{2/N_0}} e^{-x^2/2}\, dx \end{aligned} \tag{5.2–18}$$

These probabilities are identical for $m = 2, 3, \ldots, M$, and, hence, the joint probability under consideration is simply the result in Equation 5.2–18 raised to the $(M-1)$th power. Thus, the probability of a correct decision is

$$P_c = \int_{-\infty}^{\infty} \left(\frac{1}{2\pi} \int_{-\infty}^{r_1\sqrt{2/N_0}} e^{-x^2/2} dx \right)^{M-1} p(r_1)\, dr_1 \tag{5.2–19}$$

and the probability of a (k-bit) symbol error is

$$P_M = 1 - P_c \tag{5.2–20}$$

Therefore,

$$P_M = \frac{1}{\sqrt{2\pi}} \int_{-\infty}^{\infty} \left[1 - \left(\frac{1}{\sqrt{2\pi}} \int_{-\infty}^{y} e^{-x^2/2} dx \right)^{M-1} \right] \exp\left[-\frac{1}{2} \left(y - \sqrt{\frac{2\mathcal{E}_s}{N_0}} \right)^2 \right] dy \tag{5.2–21}$$

The same expression for the probability of error is obtained when any one of the other $M-1$ signals is transmitted. Since all the M signals are equally likely, the expression for P_M given in Equation 5.2–21 is the average probability of a symbol error. This expression can be evaluated numerically.

In comparing the performance of various digital modulation methods, it is desirable to have the probability of error expressed in terms of the SNR per bit, $\mathcal{E}_b/N_0$, instead of the SNR per symbol, $\mathcal{E}_s/N_0$. With $M = 2^k$, each symbol conveys k bits of information, and hence $\mathcal{E}_s = k\mathcal{E}_b$. Thus, Equation 5.2–21 may be expressed in terms of $\mathcal{E}_b/N_0$ by substituting for $\mathcal{E}_s$.

Sometimes, it is also desirable to convert the probability of a symbol error into an equivalent probability of a binary digit error. For equiprobable orthogonal signals, all symbol errors are equiprobable and occur with probability

$$\frac{P_M}{M-1} = \frac{P_M}{2^k - 1} \tag{5.2–22}$$

Furthermore, there are $\binom{k}{n}$ ways in which n bits out of k may be in error. Hence, the average number of bit errors per k-bit symbol is

$$\sum_{n=1}^{k} n \binom{k}{n} \frac{P_M}{2^k - 1} = k \frac{2^{k-1}}{2^k - 1} P_M \tag{5.2–23}$$

and the average bit error probability is just the result in Equation 5.2–23 divided by k, the number of bits per symbol. Thus,

$$P_b = \frac{2^{k-1}}{2^k - 1} P_M \approx \frac{P_M}{2}, \qquad k \gg 1 \tag{5.2–24}$$

The graphs of the probability of a binary digit error as a function of the SNR per bit, $\mathcal{E}_b/N_0$, are shown in Figure 5.2–5 for $M = 2, 4, 8, 16, 32$, and 64. This figure illustrates that, by increasing the number M of waveforms, one can reduce the SNR per bit required to achieve a given probability of a bit error. For example, to achieve a $P_b = 10^{-5}$, the required SNR per bit is a little more than 12 dB for $M = 2$, but if M is increased to 64 signal waveforms ($k = 6$ bits per symbol), the

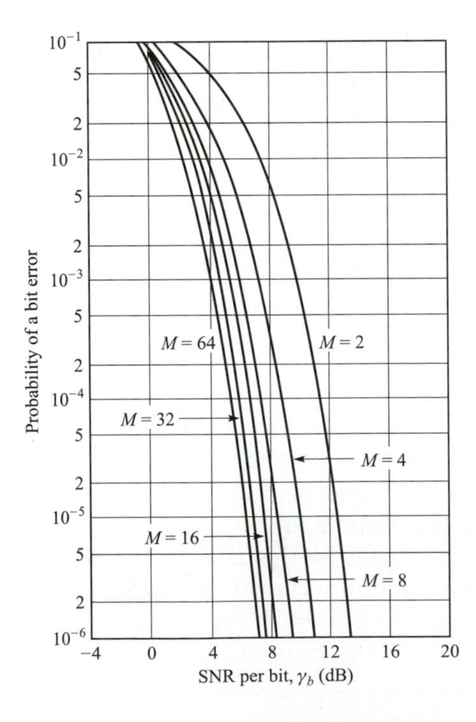

FIGURE 5.2–5
Probability of bit error for coherent detection of orthogonal signals.

required SNR per bit is approximately 6 dB. Thus, a savings of over 6 dB (a factor-of-4 reduction) is realized in transmitter power (or energy) required to achieve a $P_b = 10^{-5}$ by increasing M from $M = 2$ to $M = 64$.

What is the minimum required $\mathcal{E}_b/N_0$ to achieve an arbitrarily small probability of error as $M \to \infty$? This question is answered below.

A union bound on the probability of error. Let us investigate the effect of increasing M on the probability of error for orthogonal signals. To simplify the mathematical development, we first derive an upper bound on the probability of a symbol error that is much simpler than the exact form given in Equation 5.2–21.

Recall that the probability of error for binary orthogonal signals is given by Equation 5.2–11. Now, if we view the detector for M orthogonal signals as one that makes $M - 1$ binary decisions between the correlator output $C(\mathbf{r}, \mathbf{s}_1)$ that contains the signal and the other $M - 1$ correlator outputs $C(\mathbf{r}, \mathbf{s}_m)$, $m = 2, 3, \ldots, M$, the probability of error is upper-bounded by the *union bound* of the $M - 1$ events. That is, if E_i represents the event that $C(\mathbf{r}, \mathbf{s}_i) > C(\mathbf{r}, \mathbf{s}_1)$ for $i \neq 1$, then we have $P_M = P(\bigcup_{i=1}^{n} E_i) \leqslant \sum_{i=1}^{n} P(E_i)$. Hence,

$$P_M \leqslant (M-1)P_2 = (M-1)Q(\sqrt{\mathcal{E}_s/N_0}) < MQ(\sqrt{\mathcal{E}_s/N_0}) \qquad (5.2\text{–}25)$$

This bound can be simplified further by upper-bounding $Q\sqrt{\mathcal{E}_s/N_0})$. We have

$$Q(\sqrt{\mathcal{E}_s/N_0}) < e^{-\mathcal{E}_s/2N_0} \qquad (5.2\text{–}26)$$

Thus,

$$\begin{aligned} P_M &< Me^{-\mathcal{E}_s/2N_0} = 2^k e^{-k\mathcal{E}_b/2N_0} \\ &< e^{-k(\mathcal{E}_b/N_0 - 2\ln 2)/2} \end{aligned} \qquad (5.2\text{–}27)$$

As $k \to \infty$, or equivalently, as $M \to \infty$, the probability of error approaches zero exponentially, provided that $\mathcal{E}_b/N_0$ is greater than 2 ln 2, i.e.,

$$\frac{\mathcal{E}_b}{N_0} > 2\ln 2 = 1.39 \quad (1.42\,\text{dB}) \qquad (5.2\text{–}28)$$

The simple upper bound on the probability of error given by Equation 5.2–27 implies that, as long as SNR > 1.42 dB, we can achieve an arbitrarily low P_M. However, this union bound is not a very tight upper bound at a sufficiently low SNR due to the fact that the upper bound for the Q function in Equation 5.2–26 is loose. In fact, by more elaborate bounding techniques, it is shown in Chapter 7 that the upper bound in Equation 5.2–27 is sufficiently tight for $\mathcal{E}_b/N_0 > 4 \ln 2$. For $\mathcal{E}_b/N_0 < 4 \ln 2$, a tighter upper bound on P_M is

$$P_M < 2e^{-k(\sqrt{\mathcal{E}_b/N_0} - \sqrt{\ln 2})^2} \qquad (5.2\text{–}29)$$

Consequently, $P_m \to 0$ as $k \to \infty$, provided that

$$\frac{\mathcal{E}_b}{N_0} > \ln 2 = 0.693 \quad (-1.6\,\text{dB}) \qquad (5.2\text{–}30)$$

Hence, -1.6 dB is the minimum required SNR per bit to achieve an arbitrarily small probability of error in the limit as $k \to \infty$ $(M \to \infty)$. This minimum SNR per bit (-1.6 dB) is called the *Shannon limit* for an additive white Gaussian noise channel.

5.2.3 Probability of Error for M-ary Biorthogonal Signals

As indicated in Section 4.3, a set of $M = 2^k$ biorthogonal signals are constructed from $\frac{1}{2}M$ orthogonal signals by including the negatives of the orthogonal signals. Thus, we achieve a reduction in the complexity of the demodulator for the biorthogonal signals relative to that for orthogonal signals, since the former is implemented with $\frac{1}{2}M$ cross correlators or matched filters, whereas the latter requires M matched filters or cross correlators.

To evaluate the probability of error for the optimum detector, let us assume that the signal $s_1(t)$ corresponding to the vector $\mathbf{s}_1 = [\sqrt{\mathcal{E}_s}\ 0\ 0 \cdots 0]$ was transmitted. Then, the received signal vector is

$$\mathbf{r} = [\sqrt{\mathcal{E}_s} + n_1\ n_2 \cdots n_{M/2}] \tag{5.2–31}$$

where the $\{n_m\}$ are zero-mean, mutually statistically independent and identically distributed Gaussian random variables with variance $\sigma_n^2 = \frac{1}{2}N_0$. The optimum detector decides in favor of the signal corresponding to the largest in magnitude of the cross correlators

$$C(\mathbf{r}, \mathbf{s}_m) = \mathbf{r} \cdot \mathbf{s}_m = \sum_{k=1}^{M/2} r_k s_{mk}, \qquad m = 1, 2, \ldots, \tfrac{1}{2}M \tag{5.2–32}$$

while the sign of this largest term is used to decide whether $s_m(t)$ or $-s_m(t)$ was transmitted. According to this decision rule, the probability of a correct decision is equal to the probability that $r_1 = \sqrt{\mathcal{E}_s} + n_1 > 0$ and r_1 exceeds $|r_m| = |n_m|$ for $m = 2, 3, \ldots, \frac{1}{2}M$. But

$$P(|n_m| < r_1 | r_1 > 0) = \frac{1}{\sqrt{\pi N_0}} \int_{-r_1}^{r_1} e^{-x^2/N_0}\, dx = \frac{1}{\sqrt{2\pi}} \int_{-r_1/\sqrt{N_0/2}}^{r_1/\sqrt{N_0/2}} e^{-x^2/2}\, dx \tag{5.2–33}$$

Then, the probability of a correct decision is

$$P_c = \int_0^{\infty} \left(\frac{1}{\sqrt{2\pi}} \int_{-r_1/\sqrt{N_0/2}}^{r_1/\sqrt{N_0/2}} e^{-x^2/2} dx \right)^{M/2-1} p(r_1)\, dr_1$$

from which, upon substitution for $p(r_1)$, we obtain

$$P_c = \int_{-\sqrt{2\mathcal{E}_s/N_0}}^{\infty} \left(\frac{1}{\sqrt{2\pi}} \int_{-(v+\sqrt{2\mathcal{E}_s/N_0})}^{v+\sqrt{2\mathcal{E}_s/N_0}} e^{-x^2/2} dx \right)^{M/2-1} e^{-v^2/2} dv \tag{5.2–34}$$

where we have used the PDF of r_1 given in Equation 5.2–15. Finally, the probability of a symbol error $P_M = 1 - P_c$.

P_c, and hence, P_M may be evaluated numerically for different values of M from Equation 5.2–34. The graph shown in Figure 5.2–6 illustrates P_M as a function of $\mathcal{E}_b/N_0$, where $\mathcal{E}_s = k\mathcal{E}_b$, for $M = 2, 4, 8, 16$, and 32. We observe that this graph is similar to that for orthogonal signals (see Figure 5.2–5). However, in this case, the probability of error for $M = 4$ is greater than that for $M = 2$. This is due to the fact that we have plotted the symbol error probability P_M in Figure 5.2–6. If we plotted the equivalent bit error probability, we should find that the graphs for $M = 2$ and $M = 4$ coincide. As in the case of orthogonal signals, as $M \to \infty$ (or $k \to \infty$), the minimum required $\mathcal{E}_b/N_0$ to achieve an arbitrarily small probability of error is -1.6 dB, the Shannon limit.

5.2.4 Probability of Error for Simplex Signals

Next we consider the probability of error for M simplex signals. Recall from Section 4.3 that simplex signals are a set of M equally correlated signals with mutual cross-correlation coefficient $\rho_{mn} = -1/(M-1)$. These signals have the same minimum separation of $\sqrt{2\mathcal{E}_s}$ between adjacent signal points in M-dimensional space as orthogonal signals. They achieve this mutual separation with a transmitted energy of $\mathcal{E}_s(M-1)/M$, which is less than that required for ortho-

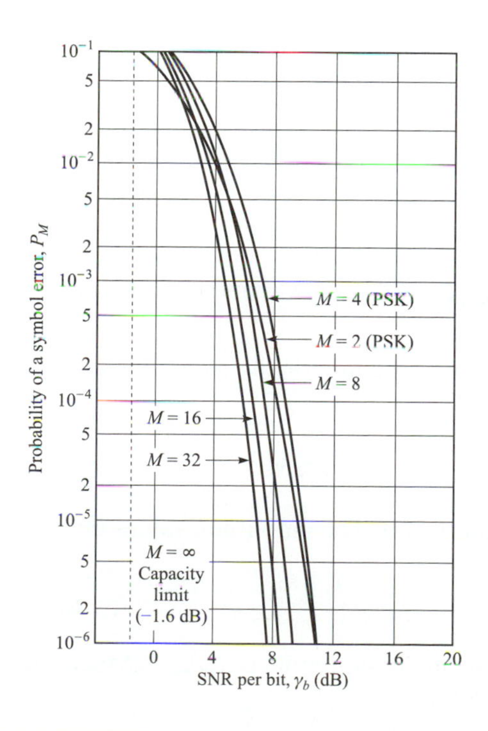

FIGURE 5.2–6
Probability of symbol error for biorthogonal signals.

gonal signals by a factor of $(M-1)/M$. Consequently, the probability of error for simplex signals is identical to the probability of error for orthogonal signals, but this performance is achieved with a saving of

$$10\log(1-\rho) = 10\log\frac{M}{M-1} \quad \text{dB} \tag{5.2–35}$$

in SNR. For $M=2$, the saving is 3 dB. However, as M is increased, the saving in SNR approaches 0 dB.

5.2.5 Probability of Error for M-ary Binary-Coded Signals

We have shown in Section 4.3 that binary-coded signal waveforms are represented by the signal vectors

$$\mathbf{s}_m = [s_{m1}\, s_{m2} \cdots s_{mN}], \qquad m = 1, 2, \ldots, M$$

where $s_{mj} = \pm\sqrt{\mathcal{E}/N}$ for all m and j. N is the block length of the code and is also the dimension of the M signal waveforms.

If $d_{\min}^{(e)}$ is the minimum Euclidean distance of the M signal waveforms, then the probability of a symbol error is upper-bounded as

$$\begin{aligned} P_M &< (M-1)P_b = (M-1)Q\left(\sqrt{\frac{(d_{\min}^{(e)})^2}{2N_0}}\right) \\ &< 2^k \exp\left[-\frac{(d_{\min}^{(e)})^2}{4N_0}\right] \end{aligned} \tag{5.2–36}$$

The value of the minimum Euclidean distance will depend on the selection of the code words, i.e., the design of the code.

5.2.6 Probability of Error for M-ary PAM

Recall that M-ary PAM signals are represented geometrically as M one-dimensional signal points with value

$$s_m = \sqrt{\tfrac{1}{2}\mathcal{E}_g}A_m, \qquad m = 1, 2, \ldots, M \tag{5.2–37}$$

where $\mathcal{E}_g$ is the energy of the basic signal pulse $g(t)$. The amplitude values may be expressed as

$$A_m = (2m-1-M)d, \qquad m = 1, 2, \ldots, M \tag{5.3–38}$$

where the Euclidean distance between adjacent signal points is $d\sqrt{2\mathcal{E}_g}$. Assuming equally probable signals, the average energy is

$$\begin{aligned}\mathcal{E}_{\text{av}} &= \frac{1}{M}\sum_{m=1}^{M}\mathcal{E}_m = \frac{d^2\mathcal{E}_g}{2M}\sum_{m=1}^{M}(2m-1-M)^2 \\ &= \frac{d^2\mathcal{E}_g}{2M}\left[\tfrac{1}{3}M(M^2-1)\right] = \tfrac{1}{6}(M^2-1)\,d^2\mathcal{E}_g\end{aligned} \tag{5.2–39}$$

Equivalently, we may characterize these signals in terms of their average power, which is

$$P_{\text{av}} = \frac{\mathcal{E}_{\text{av}}}{T} = \tfrac{1}{6}(M^2-1)\frac{d^2\mathcal{E}_g}{T} \tag{5.2–40}$$

The average probability of error for M-ary PAM can be determined from the decision rule that maximizes the correlation metrics given by Equation 5.2–44. Equivalently, the detector compares the demodulator output r with a set of $M-1$ thresholds, which are placed at the midpoints of successive amplitude levels, as shown in Figure 5.2–7. Thus, a decision is made in favor of the amplitude level that is closest to r.

The placing of the thresholds as shown in Figure 5.2–7 helps in evaluating the probability of error. We note that if the mth amplitude level is transmitted, the demodulator output is

$$r = s_m + n = \sqrt{\tfrac{1}{2}\mathcal{E}_g}A_m + n \tag{5.2–41}$$

where the noise variable n has zero-mean and variance $\sigma_n^2 = \frac{1}{2}N_0$. On the basis that all amplitude levels are equally likely a priori, the average probability of a symbol error is simply the probability that the noise variable n exceeds in magnitude one-half of the distance between levels. However, when either one of the two outside levels $\pm(M-1)$ is transmitted, an error can occur in one direction only. Thus, we have

$$\begin{aligned}P_M &= \frac{M-1}{M}P\left(|r-s_m| > d\sqrt{\tfrac{1}{2}\mathcal{E}_g}\right) \\ &= \frac{M-1}{M}\frac{2}{\sqrt{\pi N_0}}\int_{d\sqrt{\mathcal{E}_g/2}}^{\infty} e^{-x^2/N_0}dx \\ &= \frac{M-1}{M}\frac{2}{\sqrt{2\pi}}\int_{\sqrt{d^2\mathcal{E}_g/N_0}}^{\infty} e^{-x^2/2}dx \\ &= \frac{2(M-1)}{M}Q\left(\sqrt{\frac{d^2\mathcal{E}_g}{N_0}}\right)\end{aligned} \tag{5.2–42}$$

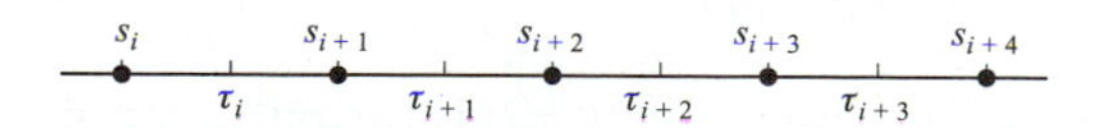

FIGURE 5.2–7
Placement of thresholds at midpoints of successive amplitude levels.

The error probability in Equation 5.2–42 can also be expressed in terms of the average transmitted power. From Equation 5.2–40, we note that

$$d^2 \mathcal{E}_g = \frac{6}{M^2 - 1} P_{\text{av}} T \tag{5.2–43}$$

By substituting for $d^2\mathcal{E}_g$ in Equation 5.2–42, we obtain the average probability of a symbol error for PAM in terms of the average power as

$$P_M = \frac{2(M-1)}{M} Q\left(\sqrt{\frac{6P_{\text{av}}T}{(M^2-1)N_0}}\right) \tag{5.2–44}$$

or, equivalently,

$$P_M = \frac{2(M-1)}{M} Q\left(\sqrt{\frac{6\mathcal{E}_{\text{av}}}{(M^2-1)N_0}}\right) \tag{5.2–45}$$

where $\mathcal{E}_{\text{av}} = P_{\text{av}} T$ is the average energy.

In plotting the probability of a symbol error for M-ary signals such as M-ary PAM, it is customary to use the SNR per bit as the basic parameter. Since $T = kT_b$ and $k = \log_2 M$, Equation 5.2–45 may be expressed as

$$P_M = \frac{2(M-1)}{M} Q\left(\sqrt{\frac{(6\log_2 M)\mathcal{E}_{b\,\text{av}}}{(M^2-1)N_0}}\right) \tag{5.2–46}$$

where $\mathcal{E}_{b\,\text{av}} = P_{\text{av}} T_b$ is the average bit energy and $\mathcal{E}_{b\,\text{av}}/N_0$ is the average SNR per bit. Figure 5.2–8 illustrates the probability of a symbol error as a function of $10 \log_{10} \mathcal{E}_{b\,\text{av}}/N_0$, with M as a parameter. Note that the case $M = 2$ corresponds to the error probability for binary antipodal signals. Also observe that the SNR per bit increases by over 4 dB for every factor-of-2 increase in M. For large M, the additional SNR per bit required to increase M by a factor of 2 approaches 6 dB.

5.2.7 Probability of Error for M-ary PSK

Recall from Section 4.3 that digital phase-modulated signal waveforms may be expressed as

$$\mathbf{s}_m(t) = g(t) \cos\left[2\pi f_c t + \frac{2\pi}{M}(m-1)\right], \qquad 1 \leqslant m \leqslant M, \qquad 0 \leqslant t \leqslant T \tag{5.2–47}$$

and have the vector representation

$$\mathbf{s}_m = \left[\sqrt{\mathcal{E}_s} \cos\frac{2\pi}{M}(m-1) \quad \sqrt{\mathcal{E}_s} \sin\frac{2\pi}{M}(m-1)\right] \tag{5.2–48}$$

where $\mathcal{E}_s = \frac{1}{2}\mathcal{E}_g$ is the energy in each of the waveforms and $g(t)$ is the pulse shape of the transmitted signal. Since the signal waveforms have equal energy, the

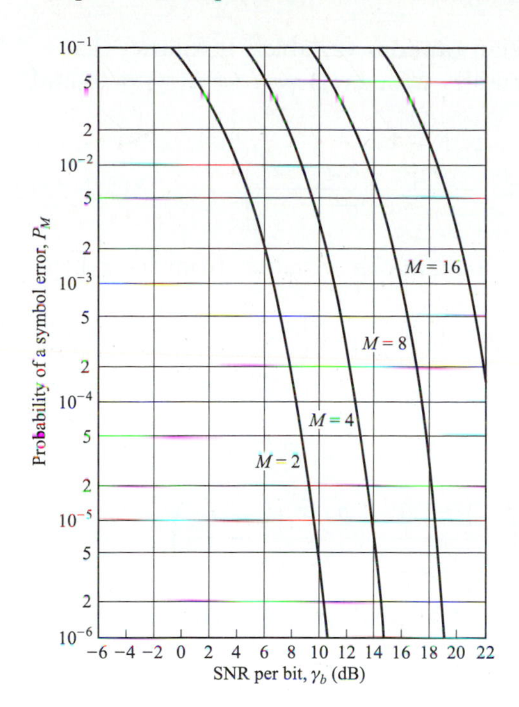

FIGURE 5.2–8
Probability of a symbol error for PAM.

optimum detector for the AWGN channel given by Equation 5.1–44 computes the correlation metrics

$$C(\mathbf{r}, \mathbf{s}_m) = \mathbf{r} \cdot \mathbf{s}_m, \qquad m = 1, 2, \ldots, M \tag{5.2–49}$$

In other words, the received signal vector $\mathbf{r} = [r_1 \;\; r_2]$ is projected onto each of the M possible signal vectors and a decision is made in favor of the signal with the largest projection.

The correlation detector described above is equivalent to a phase detector that computes the phase of the received signal from $\mathbf{r}$ and selects the signal vector $\mathbf{s}_m$ whose phase is closest to $\mathbf{r}$. Since the phase of $\mathbf{r}$ is

$$\Theta_r = \tan^{-1} \frac{r_2}{r_1} \tag{5.2–50}$$

we will determine the PDF of Θ_r, from which we shall compute the probability of error.

Let us consider the case in which the transmitted signal phase is $\Theta_r = 0$, corresponding to the signal $s_1(t)$. Hence, the transmitted signal vector is

$$\mathbf{s}_1 = [\sqrt{\mathcal{E}_s} \;\; 0] \tag{5.2–51}$$

and the received signal vector has components

$$\begin{aligned} r_1 &= \sqrt{\mathcal{E}_s} + n_1 \\ r_2 &= n_2 \end{aligned} \tag{5.2–52}$$

Because n_1 and n_2 are jointly Gaussian random variables, it follows that r_1 and r_2 are jointly Gaussian random variables with $E(r_1) = \sqrt{\mathcal{E}_s}$, $E(r_2) = 0$, and $\sigma_{r_1}^2 = \sigma_{r_2}^2 = \frac{1}{2}N_0 = \sigma_r^2$. Consequently,

$$p_r(r_1, r_2) = \frac{1}{2\pi\sigma_r^2} \exp\left[-\frac{(r_1 - \sqrt{\mathcal{E}_s})^2 + r_2^2}{2\sigma_r^2}\right] \tag{5.2–53}$$

The PDF of the phase Θ_r is obtained by a change in variables from (r_1, r_2) to

$$\begin{aligned} V &= \sqrt{r_1^2 + r_2^2} \\ \Theta_r &= \tan^{-1}\frac{r_2}{r_1} \end{aligned} \tag{5.2–54}$$

This yields the joint PDF

$$p_{V,\Theta_r}(V, \Theta_r) = \frac{V}{2\pi\sigma_r^2} \exp\left(-\frac{V^2 + \mathcal{E}_s - 2\sqrt{\mathcal{E}_s}V\cos\Theta_r}{2\sigma_r^2}\right)$$

Integration of $p_{V,\Theta_r}(V, \Theta_r)$ over the range of V yields $p_{\Theta_r}(\Theta_r)$. That is,

$$\begin{aligned} p_{\Theta_r}(\Theta_r) &= \int_0^\infty p_{V,\Theta_r}(V, \Theta_r)\, dV \\ &= \frac{1}{2\pi} e^{-\gamma_s \sin^2\Theta_r} \int_0^\infty V e^{-(V - \sqrt{2\gamma_s}\cos\Theta_r)^2/2}\, dV \end{aligned} \tag{5.2–55}$$

where for convenience, we have defined the symbol SNR as $\gamma_s = \mathcal{E}_s/N_0$. Figure 5.2–9 illustrates $p_{\Theta_r}(\Theta_r)$ for several values of the SNR parameter γ_s when the transmitted phase is zero. Note that $f_{\Theta_r}(\Theta_r)$ becomes narrower and more peaked about $\Theta_r = 0$ as the SNR γ_s increases.

When $s_1(t)$ is transmitted, a decision error is made if the noise causes the phase to fall outside the range $-\pi/M \leqslant \Theta_r \leqslant \pi/M$. Hence, the probability of a symbol error is

$$P_M = 1 - \int_{-\pi/M}^{\pi/M} p_{\Theta_r}(\Theta_r)\, d\Theta_r \tag{5.2–56}$$

In general, the integral of $p_{\Theta_r}(\Theta)$ does not reduce to a simple form and must be evaluated numerically, except for $M = 2$ and $M = 4$.

For binary phase modulation, the two signals $s_1(t)$ and $s_2(t)$ are antipodal, and, hence, the error probability is

$$P_2 = Q\left(\sqrt{\frac{2\mathcal{E}_b}{N_0}}\right) \tag{5.2–57}$$

When $M = 4$, we have in effect two binary phase-modulation signals in phase quadrature. Since there is no crosstalk or interference between the signals on the two quadrature carriers, the bit error probability is identical to that in Equation

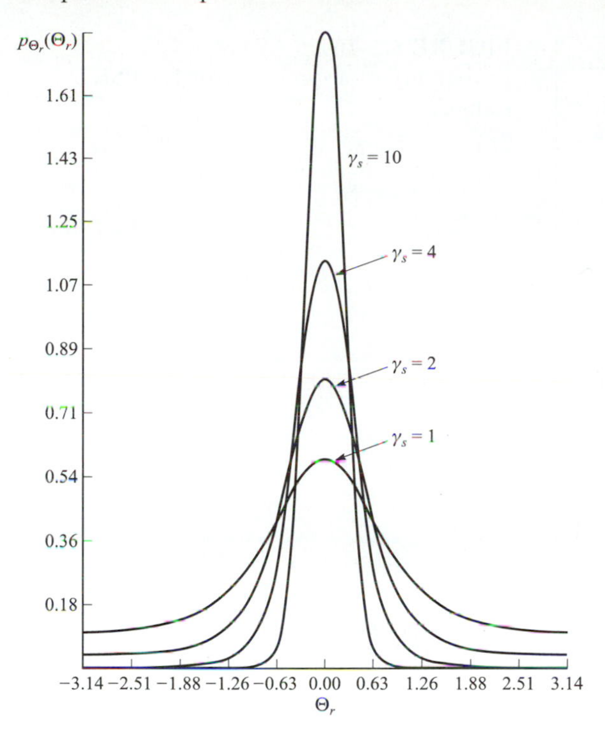

FIGURE 5.2–9
Probability density function $p_{\Theta_r}(\Theta_r)$ for $\gamma_s = 1, 2, 4$, and 10.

5.2–57. On the other hand, the symbol error probability for $M = 4$ is determined by noting that

$$P_c = (1 - P_2)^2 = \left[1 - Q\left(\sqrt{\frac{2\mathcal{E}_b}{N_0}}\right)\right]^2 \qquad (5.2\text{–}58)$$

where P_c is the probability of a correct decision for the 2-bit symbol. Equation 5.2–58 follows from the statistical independence of the noise on the quadrature carriers. Therefore, the symbol error probability for $M = 4$ is

$$\begin{aligned} P_4 &= 1 - P_c \\ &= 2Q\left(\sqrt{\frac{2\mathcal{E}_b}{N_0}}\right)\left[1 - \tfrac{1}{2}Q\left(\sqrt{\frac{2\mathcal{E}_b}{N_0}}\right)\right] \end{aligned} \qquad (5.2\text{–}59)$$

For $M > 4$, the symbol error probability P_M is obtained by numerically integrating Equation 5.2–56. Figure 5.2–10 illustrates this error probability as a function of the SNR per bit for $M = 2, 4, 8, 16$, and 32. The graphs clearly illustrate the penalty in SNR per bit as M increases beyond $M = 4$. For example, at $P_M = 10^{-5}$, the difference between $M = 4$ and $M = 8$ is approximately 4 dB, and the difference between $M = 8$ and $M = 16$ is approximately 5 dB. For large values of M, doubling the number of phases requires an additional 6 dB/bit to achieve the same performance.

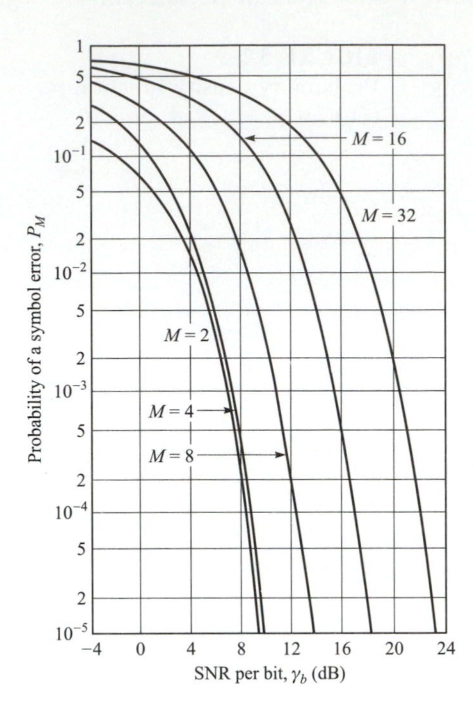

FIGURE 5.2–10
Probability of a symbol error for PSK signals.

An approximation to the error probability for large values of M and for large SNR may be obtained by first approximating $p_{\Theta_r}(\Theta)$. For $\mathcal{E}_s/N_0 \gg 1$ and $|\Theta_r| \leqslant \frac{1}{2}\pi$, $p_{\Theta_r}(\Theta_r)$ is well approximated as

$$p_{\Theta_r}(\Theta_r) \approx \sqrt{\frac{\gamma_s}{\pi}} \cos \Theta_r e^{-\gamma_s \sin^2 \Theta_r} \tag{5.2–60}$$

By substituting for $p_{\Theta_r}(\Theta_r)$ in Equation 5.2–56 and performing the change in variable from Θ_r to $u = \sqrt{\gamma_s} \sin \Theta_r$, we find that

$$\begin{aligned} P_M &\approx 1 - \int_{-\pi/M}^{\pi/M} \sqrt{\frac{\gamma_s}{\pi}} \cos \Theta_r e^{-\gamma_s \sin^2 \Theta_r} d\Theta_r \\ &\approx \frac{2}{\sqrt{\pi}} \int_{\sqrt{2\gamma_s} \sin \pi/M)}^{\infty} e^{-u^2} du \\ &= 2Q\left(\sqrt{2\gamma_s} \sin \frac{\pi}{M}\right) = 2Q\left(\sqrt{2k\gamma_b} \sin \frac{\pi}{M}\right) \end{aligned} \tag{5.2–61}$$

where $k = \log_2 M$ and $\gamma_s = k\gamma_b$. Note that this approximation† to the error probability is good for all values of M. For example, when $M = 2$ and $M = 4$,

†A better approximation of the error probability at low SNR is given in the paper by Lu et al. (1999).

we have $P_2 = P_4 = 2Q(\sqrt{2\gamma_s})$, which compares favorably (a factor-of-2 difference) with the exact probability given by Equation 5.2–57.

The equivalent bit error probability for M-ary PSK is rather tedious to derive due to its dependence on the mapping of k-bit symbols into the corresponding signal phases. When a Gray code is used in the mapping, two k-bit symbols corresponding to adjacent signal phases differ in only a single bit. Since the most probable errors due to noise result in the erroneous selection of an adjacent phase to the true phase, most k-bit symbol errors contain only a single-bit error. Hence, the equivalent bit error probability for M-ary PSK is well approximated as

$$P_b \approx \frac{1}{k} P_M \tag{5.2–62}$$

Our treatment of the demodulation of PSK signals assumed that the demodulator had a perfect estimate of the carrier phase available. In practice, however, the carrier phase is extracted from the received signal by performing some nonlinear operation that introduces a phase ambiguity. For example, in binary PSK, the signal is often squared in order to remove the modulation, and the double-frequency component that is generated is filtered and divided by 2 in frequency in order to extract an estimate of the carrier frequency and phase ϕ. These operations result in a phase ambiguity of 180° in the carrier phase. Similarly, in four-phase PSK, the received signal is raised to the fourth power in order to remove the digital modulation, and the resulting fourth harmonic of the carrier frequency is filtered and divided by 4 in order to extract the carrier component. These operations yield a carrier frequency component containing the estimate of the carrier phase ϕ, but there are phase ambiguities of ±90° and 180° in the phase estimate. Consequently, we do not have an absolute estimate of the carrier phase for demodulation.

The phase ambiguity problem resulting from the estimation of the carrier phase ϕ can be overcome by encoding the information in phase differences between successive signal transmissions as opposed to absolute phase encoding. For example, in binary PSK, the information bit 1 may be transmitted by shifting the phase of the carrier by 180° relative to the previous carrier phase, while the information bit 0 is transmitted by a zero phase shift relative to the phase in the previous signaling interval. In four-phase PSK, the relative phase shifts between successive intervals are 0, 90°, 180°, and −90°, corresponding to the information bits 00, 01, 11, and 10, respectively. The generalization to $M > 4$ phases is straightforward. The PSK signals resulting from the encoding process are said to be *differentially encoded*. The encoding is performed by a relatively simple logic circuit preceding the modulator.

Demodulation of the differentially encoded PSK signal is performed as described above, by ignoring the phase ambiguities. Thus, the received signal is demodulated and detected to one of the M possible transmitted phases in each signaling interval. Following the detector is a relatively simple phase comparator that compares the phases of the demodulated signal over two consecutive intervals in order to extract the information.

Coherent demodulation of differentially encoded PSK results in a higher probability of error than the error probability derived for absolute phase encoding. With differentially encoded PSK, an error in the demodulated phase of the signal in any given interval will usually result in decoding errors over two consecutive signaling intervals. This is especially the case for error probabilities below 0.1. Therefore, the probability of error in differentially encoded M-ary PSK is approximately twice the probability of error for M-ary PSK with absolute phase encoding. However, this factor-of-2 increase in the error probability translates into a relatively small loss in SNR.

5.2.8 Differential PSK (DPSK) and Its Performance

A differentially encoded phase-modulated signal also allows another type of demodulation that does not require the estimation of the carrier phase.† Instead, the received signal in any given signaling interval is compared to the phase of the received signal from the preceding signaling interval. To elaborate, suppose that we demodulate the differentially encoded signal by multiplying $r(t)$ by $\cos 2\pi f_c t$ and $\sin 2\pi f_c t$ integrating the two products over the interval T. At the kth signaling interval, the demodulator output is

$$\mathbf{r}_k = [\sqrt{\mathcal{E}_s}\cos(\theta_k - \phi) + n_{k_1} \quad \sqrt{\mathcal{E}_s}\sin(\theta_k - \phi) + n_{k_2}]$$

or, equivalently,

$$r_k = \sqrt{\mathcal{E}_s} e^{j(\theta_k - \phi)} + n_k \tag{5.2–63}$$

where θ_k is the phase angle of the transmitted signal at the kth signaling interval, ϕ is the carrier phase, and $n_k = n_{k_1} + jn_{k_2}$ is the noise vector. Similarly, the received signal vector at the output of the demodulator in the preceding signaling interval is

$$r_{k-1} = \sqrt{\mathcal{E}_s} e^{j(\theta_{k-1} - \phi)} + n_{k-1} \tag{5.2–64}$$

The decision variable for the phase detector is the phase difference between these two complex numbers. Equivalently, we can project r_k onto r_{k-1} and use the phase of the resulting complex number; that is,

$$r_k r_{k-1}^* = \mathcal{E}_s e^{j(\theta_k - \theta_{k-1})} + \sqrt{\mathcal{E}_s} e^{j(\theta_k - \phi)} n_{k-1}^* + \sqrt{\mathcal{E}_s} e^{-j(\theta_{k-1} - \phi)} n_k + n_k n_{k-1}^* \tag{5.2–65}$$

which, in the absence of noise, yields the phase difference $\theta_k - \theta_{k-1}$. Thus, the mean value of $r_k r_{k-1}^*$ is independent of the carrier phase. Differentially encoded PSK signaling that is demodulated and detected as described above is called *differential PSK* (DPSK).

†Because no phase estimation is required, DPSK is often considered to be a noncoherent communication technique. We take the view that DPSK represents a form of digital phase modulation in the extreme case where the phase estimate is derived only from the previous symbol interval.

The demodulation and detection of DPSK using matched filters is illustrated in Figure 5.2–11. If the pulse $g(t)$ is rectangular, the matched filters may be replaced by integrate-and-dump filters.

Let us now consider the evaluation of the error probability performance of a DPSK demodulator and detector. The derivation of the exact value of the probability of error for M-ary DPSK is extremely difficult, except for $M = 2$. The major difficulty is encountered in the determination of the PDF for the phase of the random variable $r_k r_{k-1}^*$, given by Equation 5.2–65. However, an approximation to the performance of DPSK is easily obtained, as we now demonstrate.

Without loss of generality, suppose the phase difference $\theta_k - \theta_{k-1} = 0$. Furthermore, the exponential factors $e^{-j(\theta_{k-1}-\phi)}$ and $e^{j(\theta_k-\phi)}$ in Equation 5.2–65 can be absorbed into the Gaussian noise components n_{k-1} and n_k, without changing their statistical properties. Therefore, $r_k r_{k-1}^*$ in Equation 5.2–65 can be expressed as

$$r_k r_{k-1}^* = \mathcal{E}_s + \sqrt{\mathcal{E}_s}(n_k + n_{k-1}^*) + n_k n_{k-1}^* \tag{5.2–66}$$

The complication in determining the PDF of the phase is the term $n_k n_{k-1}^*$. However, at SNRs of practical interest, the term $n_k n_{k-1}^*$ is small relative to the dominant noise term $\sqrt{\mathcal{E}_s}\,(n_k + n_{k-1}^*)$. If we neglect the term $n_k n_{k-1}^*$ and we also normalize $r_k r_{k-1}^*$ by dividing through by $\sqrt{\mathcal{E}_s}$, the new set of decision metrics becomes

$$\begin{aligned} x &= \sqrt{\mathcal{E}_s} + \mathrm{Re}(n_k + n_{k-1}^*) \\ y &= \mathrm{Im}(n_k + n_{k-1}^*) \end{aligned} \tag{5.2–67}$$

The variables x and y are uncorrelated Gaussian random variables with identical variances $\sigma_n^2 = N_0$. The phase is

$$\Theta_r = \tan^{-1}\frac{y}{x} \tag{5.2–68}$$

At this stage, we have a problem that is identical to the one we solved previously for phase-coherent demodulation. The only difference is that the noise variance is now twice as large as in the case of PSK. Thus we conclude that the performance of DPSK is 3 dB poorer than that for PSK. This result is relatively good

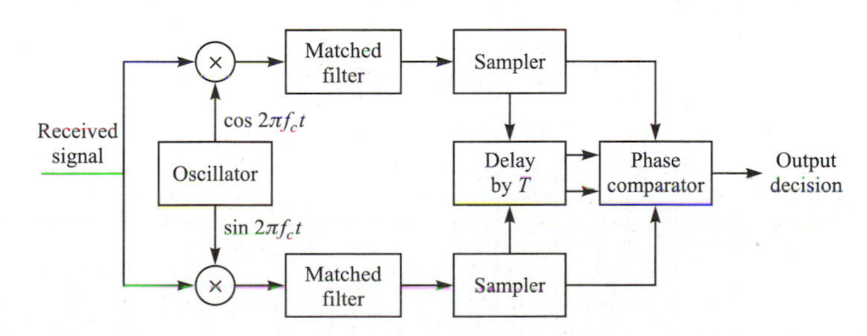

FIGURE 5.2-11
Block diagram of DPSK demodulator.

for $M \geqslant 4$, but it is pessimistic for $M = 2$ in the sense that the loss in binary DPSK relative to binary PSK is less than 3 dB at large SNR. This is demonstrated below.

In binary DPSK, the two possible transmitted phase differences are 0 and π rad. As a consequence, only the real part of $r_k r_{k-1}^*$ is needed for recovering the information. Using Equation 5.2–67, we express the real part as

$$\mathrm{Re}(r_k r_{k-1}^*) = \tfrac{1}{2}(r_k r_{k-1}^* + r_k^* r_{k-1})$$

Because the phase difference between the two successive signaling intervals is zero, an error is made if $\mathrm{Re}(r_k r_{k-1}^*) < 0$. The probability that $r_k r_{k-1}^* + r_k^* r_{k-1} < 0$ is a special case of a derivation, given in Appendix B concerned with the probability that a general quadratic form in complex-valued Gaussian random variables is less than zero. The general form for this probability is given by Equation B–21 of Appendix B, and it depends entirely on the first and second moments of the complex-valued Gaussian random variables r_k and r_{k-1}. Upon evaluating the moments and the parameters that are functions of the moments, we obtain the probability of error for binary DPSK in the form

$$P_b = \tfrac{1}{2} e^{-\mathcal{E}_b/N_0} \tag{5.2–69}$$

where $\mathcal{E}_b/N_0$ is the SNR per bit.

The graph is shown in Figure 5.2–12. Also shown in that illustration is the probability of error for binary, coherent PSK. We observe that at error prob-

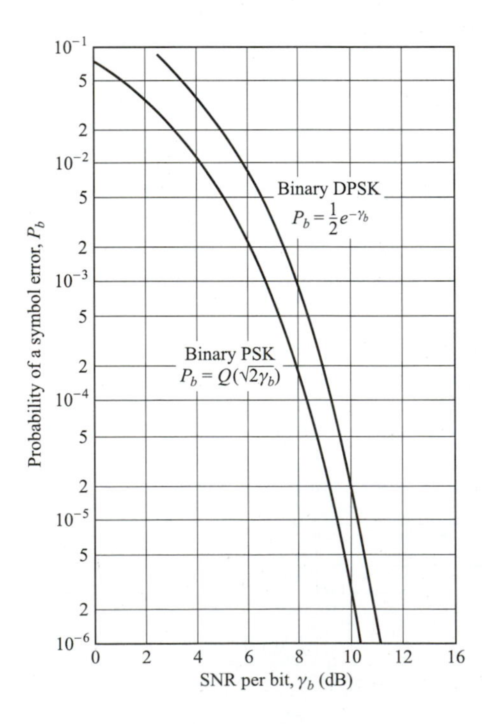

FIGURE 5.2–12
Probability of error for binary PSK and DPSK.

abilities of $P_b \leqslant 10^{-3}$ the difference in SNR between binary PSK and binary DPSK is less than 3 dB. In fact, at $P_b \leqslant 10^{-5}$, the difference in SNR is less than 1 dB.

The probability of a binary digit error for four-phase DPSK with Gray coding can be expressed in terms of well-known functions, but its derivation is quite involved. We simply state the result at this point and refer the interested reader to Appendix C for the details of derivation. It is expressed in the form

$$P_b = Q_1(a, b) - \tfrac{1}{2} I_0(ab) \exp[-\tfrac{1}{2}(a^2 + b^2)] \tag{5.2–70}$$

where $Q_1(a, b)$ is the Marcum Q function defined by Equations 2.1–122 and 2.1–123, $I_0(x)$ is the modified Bessel function of order zero, defined by Equation 2.1–120, and the parameters a and b are defined as

$$\begin{aligned} a &= \sqrt{2\gamma_b(1 - \sqrt{\tfrac{1}{2}})} \\ b &= \sqrt{2\gamma_b(1 + \sqrt{\tfrac{1}{2}})} \end{aligned} \tag{5.2–71}$$

Figure 5.2–13 illustrates the probability of a binary digit error for two- and four-phase DPSK and coherent PSK signaling obtained from evaluating the exact formulas derived in this section. Since binary DPSK is only slightly inferior to binary PSK at large SNR, and DPSK does not require an elaborate method for estimating the carrier phase, it is often used in digital communication systems.

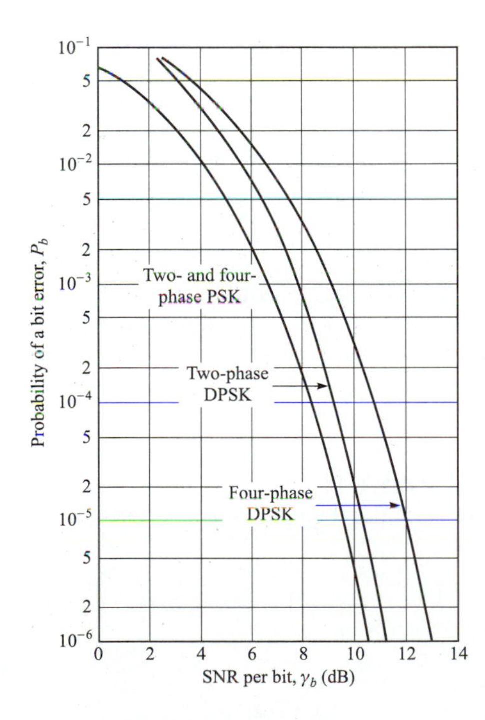

FIGURE 5.2–13
Probability of bit error for binary and four-phase PSK and DPSK.

On the other hand, four-phase DPSK is approximately 2.3 dB poorer in performance than four-phase PSK at large SNR. Consequently the choice between these two four-phase systems is not as clear-cut. One must weigh the 2.3 dB loss against the reduction in implementation complexity.

5.2.9 Probability of Error for QAM

Recall from Section 4.3 that QAM signal waveforms may be expressed as

$$s_m(t) = A_{mc}g(t)\cos 2\pi f_c t - A_{ms}g(t)\sin 2\pi f_c t, \qquad 0 \leqslant t \leqslant T \tag{5.2–72}$$

where A_{mc} and A_{ms} are the information-bearing signal amplitudes of the quadrature carriers and $g(t)$ is the signal pulse. The vector representation of these waveforms is

$$\mathbf{s}_m = [A_{mc}\sqrt{\tfrac{1}{2}\mathcal{E}_g} \quad A_{ms}\sqrt{\tfrac{1}{2}\mathcal{E}_g}] \tag{5.2–73}$$

To determine the probability of error for QAM, we must specify the signal point constellation. We begin with QAM signal sets that have $M = 4$ points. Figure 5.2–14 illustrates two four-point signal sets. The first is a four-phase modulated signal and the second is a QAM signal with two amplitude levels, labeled A_1 and A_2, and four phases. Because the probability of error is dominated by the minimum distance between pairs of signal points, let us impose the condition that $d_{\min}^{(e)} = 2A$ for both signal constellations and let us evaluate the average transmitter power, based on the premise that all signal points are equally probable. For the four-phase signal, we have

$$P_{\text{av}} = \tfrac{1}{4}(4)2A^2 = 2A^2 \tag{5.2–74}$$

For the two-amplitude, four-phase QAM, we place the points on circles of radii A and $\sqrt{3}A$. Thus, $d_{\min}^{(e)} = 2A$, and

$$P_{\text{av}} = \tfrac{1}{2}[2(3A^2) + 2A^2] = 2A^2 \tag{5.2–75}$$

which is the same average power as the $M = 4$-phase signal constellation. Hence, for all practical purposes, the error rate performance of the two signal sets is the same. In other words, there is no advantage of the two-amplitude QAM signal set over $M = 4$-phase modulation.

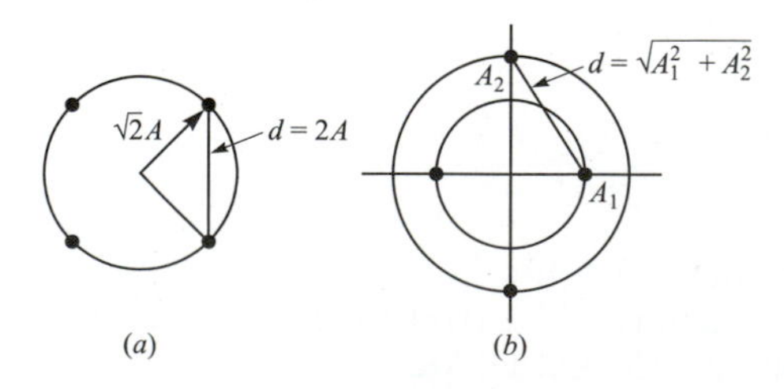

FIGURE 5.2–14
Two four-point signal constellations.

Next, let us consider $M = 8$ QAM. In this case, there are many possible signal constellations. We shall consider the four signal constellations shown in Figure 5.2–15, all of which consist of two amplitudes and have a minimum distance between signal points of $2A$. The coordinates (A_{mc}, A_{ms}) for each signal point, normalized by A, are given in the figure. Assuming that the signal points are equally probable, the average transmitted signal power is

$$\begin{aligned} P_{\text{av}} &= \frac{1}{M}\sum_{m=1}^{M}(A_{mc}^2 + A_{ms}^2) \\ &= \frac{A^2}{M}\sum_{m=1}^{M}(a_{mc}^2 + a_{ms}^2) \end{aligned} \tag{5.2–76}$$

where (a_{mc}, a_{ms}) are the coordinates of the signal points, normalized by A.

The two signal sets (a) and (c) in Figure 5.2–15 contain signal points that fall on a rectangular grid and have $P_{\text{av}} = 6A^2$. The signal set (b) requires an average transmitted power $P_{\text{av}} = 6.83A^2$, and (d) requires $P_{\text{av}} = 4.73A^2$. Therefore, the fourth signal set requires approximately 1 dB less power than the first two and 1.6 dB less power than the third to achieve the same probability of error. This signal constellation is known to be the best eight-point QAM constellation because it requires the least power for a given minimum distance between signal points.

For $M \geqslant 16$, there are many more possibilities for selecting the QAM signal points in the two-dimensional space. For example, we may choose a circular multiamplitude constellation for $M = 16$, as shown in Figure 4.3–4. In this case, the signal points at a given amplitude level are phase-rotated by $\frac{1}{4}\pi$ relative

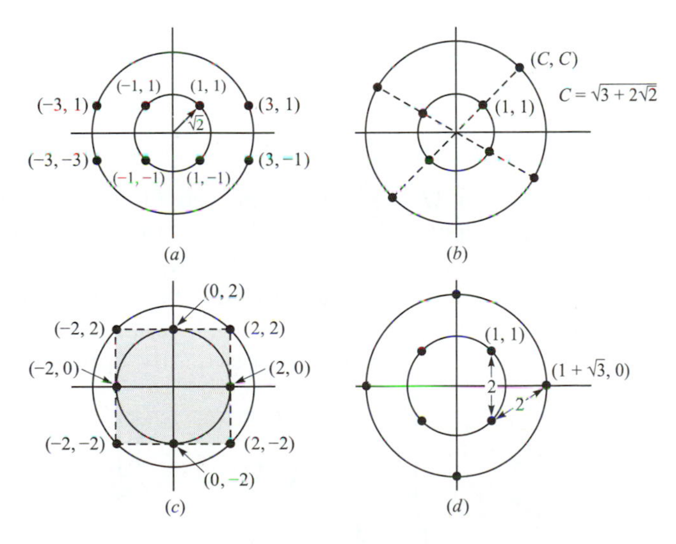

FIGURE 5.2–15
Four eight-point QAM signal constellations.

to the signal points at adjacent amplitude levels. This 16-QAM constellation is a generalization of the optimum 8-QAM constellation. However, the circular 16-QAM constellation is not the best 16-point QAM signal constellation for the AWGN channel.

Rectangular QAM signal constellations have the distinct advantage of being easily generated as two PAM signals impressed on phase-quadrature carriers. In addition, they are easily demodulated. Although they are not the best M-ary QAM signal constellations for $M \geqslant 16$, the average transmitted power required to achieve a given minimum distance is only slightly greater than the average power required for the best M-ary QAM signal constellation. For these reasons, rectangular M-ary QAM signals are most frequently used in practice.

For rectangular signal constellations in which $M = 2^k$ where k is even, the QAM signal constellation is equivalent to two PAM signals on quadrature carriers, each having $\sqrt{M} = 2^{k/2}$ signal points. Since the signals in the phase-quadrature components can be perfectly separated at the demodulator, the probability of error for QAM is easily determined from the probability of error for PAM. Specifically, the probability of a correct decision for the M-ary QAM system is

$$P_c = (1 - P_{\sqrt{M}})^2 \tag{5.2–77}$$

where $P_{\sqrt{M}}$ is the probability of error of an $\sqrt{M}$-ary PAM with one-half the average power in each quadrature signal of the equivalent QAM system. By appropriately modifying the probability of error for M-ary PAM, we obtain

$$P_{\sqrt{M}} = 2\left(1 - \frac{1}{\sqrt{M}}\right) Q\left(\sqrt{\frac{3}{M-1} \frac{\mathcal{E}_{\text{av}}}{N_0}}\right) \tag{5.2–78}$$

where $\mathcal{E}_{\text{av}}/N_0$ is the average SNR per symbol. Therefore, the probability of a symbol error for the M-ary QAM is

$$P_M = 1 - (1 - P_{\sqrt{M}})^2 \tag{5.2–79}$$

Note that this result is exact for $M = 2^k$ when k is even. On the other hand, when k is odd, there is no equivalent $\sqrt{M}$-ary PAM system. This is no problem, however, because it is rather easy to determine the error rate for a rectangular signal set. If we employ the optimum detector that bases its decisions on the optimum distance metrics given by Equation 5.1–43, it is relatively straightforward to show that the symbol error probability is tightly upper-bounded as

$$\begin{aligned} P_M &\leqslant 1 - \left[1 - 2Q\left(\sqrt{\frac{3\mathcal{E}_{\text{av}}}{(M-1)N_0}}\right)\right]^2 \\ &\leqslant 4Q\left(\sqrt{\frac{3k\mathcal{E}_{b\,\text{av}}}{(M-1)N_0}}\right) \end{aligned} \tag{5.2–80}$$

for any $k \geqslant 1$, where $\mathcal{E}_{b\text{av}}/N_0$ is the average SNR per bit. The probability of a symbol error is plotted in Figure 5.2–16 as a function of the average SNR per bit.

For nonrectangular QAM signal constellations, we may upper-bound the error probability by use of a union bound. An obvious upper bound is

$$P_M < (M-1)Q\left(\sqrt{[d_{\min}^{(e)}]^2/2N_0}\right)$$

where $d_{\min}^{(e)}$ is the minimum Euclidean distance between signal points. This bound may be loose when M is large. In such a case, we may approximate P_M by replacing $M-1$ by M_n, where M_n is the largest number of neighboring points that are at distance $d_{\min}^{(e)}$ from any constellation point.

It is interesting to compare the performance of QAM with that of PSK for any given signal size M, since both types of signals are two-dimensional. Recall that for M-ary PSK, the probability of a symbol error is approximated as

$$P_M \approx 2Q\left(\sqrt{2\gamma_s}\sin\frac{\pi}{M}\right) \qquad (5.2\text{–}81)$$

where γ_s is the SNR per symbol. For M-ary QAM, we may use the expression 5.2–78. Since the error probability is dominated by the argument of the Q function, we may simply compare the arguments of Q for the two signal formats. Thus, the ratio of these two arguments is

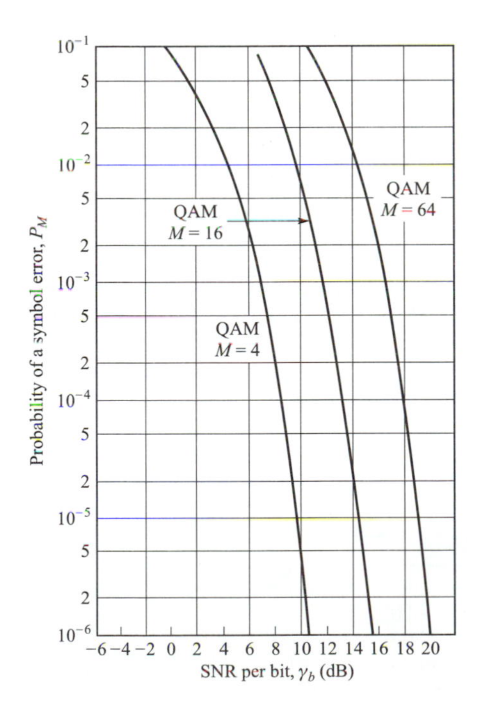

FIGURE 5.2–16
Probability of a symbol error for QAM.

$$\mathcal{R}_M = \frac{3/(M-1)}{2\sin^2(\pi/M)} \tag{5.2–82}$$

For example, when $M = 4$, we have $\mathcal{R}_M = 1$. Hence, 4-PSK and 4-QAM yield comparable performance for the same SNR per symbol. On the other hand, when $M > 4$, we find that $\mathcal{R}_M > 1$, so that M-ary QAM yields better performance than M-ary PSK. Table 5.2–1 illustrates the SNR advantage of QAM over PSK for several values of M. For example, we observe that 32-QAM has a 7-dB SNR advantage over 32-PSK.

5.2.10 Comparison of Digital Modulation Methods

The digital modulation methods described in this chapter can be compared in a number of ways. For example, one can compare them on the basis of the SNR required to achieve a specified probability of error. However, such a comparison would not be very meaningful, unless it were made on the basis of some constraint, such as a fixed data rate of transmission or, equivalently, on the basis of a fixed bandwidth. With this goal in mind, let us consider the bandwidth requirements for several modulation methods.

For multiphase signals, the channel bandwidth required is simply the bandwidth of the equivalent low-pass signal pulse $g(t)$, which depends on its detailed characteristics. For our purposes, we assume that $g(t)$ is a pulse of duration T and that its bandwidth W is approximately equal to the reciprocal of T. Thus, $W = 1/T$ and, since $T = k/R = (\log_2 M)/R$, it follows that

$$W = \frac{R}{\log_2 M} \tag{5.2–83}$$

Therefore, as M is increased, the channel bandwidth required, when the bit rate R is fixed, decreases. The bandwidth efficiency is measured by the bit rate to bandwidth ratio, which is

$$\frac{R}{W} = \log_2 M \tag{5.2–84}$$

TABLE 5.2–1
SNR advantage of M-ary QAM over M-ary PSK

M	$10 \log_{10} \mathcal{R}_M$
8	1.65
16	4.20
32	7.02
64	9.95

The bandwidth-efficient method for transmitting PAM is single-sideband. Then, the channel bandwidth required to transmit the signal is approximately equal to $1/2T$ and, since $T = k/R = (\log_2 M)/R$, it follows that

$$\frac{R}{W} = 2\log_2 M \tag{5.2–85}$$

This is a factor of 2 better than PSK.

In the case of QAM, we have two orthogonal carriers, with each carrier having a PAM signal. Thus, we double the rate relative to PAM. However, the QAM signal must be transmitted via double-sideband. Consequently, QAM and PAM have the same bandwidth efficiency when the bandwidth is referenced to the band-pass signal.

Orthogonal signals have totally different bandwidth requirements. If the $M = 2^k$ orthogonal signals are constructed by means of orthogonal carriers with minimum frequency separation of $1/2T$ for orthogonality, the bandwidth required for transmission of $k = \log_2 M$ information bits is

$$W = \frac{M}{2T} = \frac{M}{2(k/R)} = \frac{M}{2\log_2 M}R \tag{5.2–86}$$

In this case, the bandwidth increases as M increases. Similar relationships obtain for simplex and biorthogonal signals. In the case of biorthogonal signals, the required bandwidth is one-half of that for orthogonal signals.

A compact and meaningful comparison of these modulation methods is one based on the normalized data rate R/W (bits per second per hertz of bandwidth) versus the SNR per bit ($\mathcal{E}_b/N_0$) required to achieve a given error probability. Figure 5.2–17 illustrates the graph of R/W versus SNR per bit for PAM, QAM, PSK, and orthogonal signals, for the case in which the error probability is $P_M = 10^{-5}$. We observe that in the case of PAM, QAM, and PSK, increasing M results in a higher bit rate–to–bandwidth ratio R/W. However, the cost of achieving the higher data rate is an increase in the SNR per bit. Consequently, these modulation methods are appropriate for communication channels that are bandwidth limited, where we desire a bit rate–to–bandwidth ratio $R/W > 1$ and where there is sufficiently high SNR to support increases in M. Telephone channels and digital microwave radio channels are examples of such band-limited channels.

In contrast, M-ary orthogonal signals yield a bit rate–to–bandwidth ratio of $R/W \leqslant 1$. As M increases, R/W decreases due to an increase in the required channel bandwidth. However, the SNR per bit required to achieve a given error probability (in this case, $P_M = 10^{-5}$) decreases as M increases. Consequently, M-ary orthogonal signals are appropriate for power-limited channels that have sufficiently large bandwidth to accommodate a large number of signals. In this case, as $M \to \infty$, the error probability can be made as small as desired, provided that $\mathcal{E}_b/N_0 > 0.693$ (-1.6 dB). This is the minimum SNR per bit required to achieve reliable transmission in the limit as the channel bandwidth $W \to \infty$ and the corresponding bit rate–to–bandwidth ratio $R/W \to 0$.

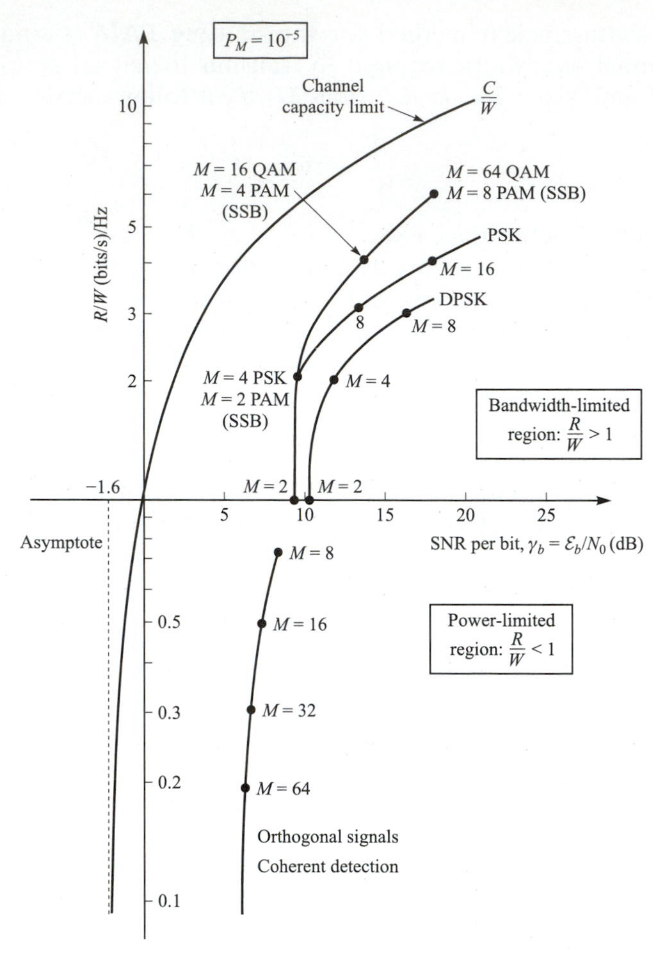

FIGURE 5.2–17
Comparison of several modulation methods at 10^{-5} symbol error probability.

Also shown in Figure 5.2–17 is the graph for the normalized capacity of the band-limited AWGN channel, which is due to Shannon (1948). The ratio C/W, where $C\,(= R)$ is the capacity in bits/s, represents the highest achievable bit rate–to–bandwidth ratio on this channel. Hence, it serves as the upper bound on the bandwidth efficiency of any type of modulation. This bound is derived in Chapter 7 and discussed in greater detail there.

5.3
OPTIMUM RECEIVER FOR CPM SIGNALS

We recall from Section 4.3 that CPM is a modulation method with memory. The memory results from the continuity of the transmitted carrier phase from one signal interval to the next. The transmitted CPM signal may be expressed as

$$s(t) = \sqrt{\frac{2\mathcal{E}}{T}} \cos[2\pi f_c t + \phi(t; \mathbf{I})] \tag{5.3–1}$$

where $\phi(t; \mathbf{I})$ is the carrier phase. The filtered received signal for an additive Gaussian noise channel is

$$r(t) = s(t) + n(t) \tag{5.3–2}$$

where

$$n(t) = n_c(t) \cos 2\pi f_c t - n_s(t) \sin 2\pi f_c t \tag{5.3–3}$$

5.3.1 Optimum Demodulation and Detection of CPM

The optimum receiver for this signal consists of a correlator followed by a maximum-likelihood sequence detector that searches the paths through the state trellis for the minimum Euclidean distance path. The Viterbi algorithm is an efficient method for performing this search. Let us establish the general state trellis structure for CPM and then describe the metric computations.

Recall that the carrier phase for a CPM signal with a fixed modulation index h may be expressed as

$$\begin{aligned} \phi(t; \mathbf{I}) &= 2\pi h \sum_{k=-\infty}^{n} I_k q(t - kT) \\ &= \pi h \sum_{k=-\infty}^{n-L} I_k + 2\pi h \sum_{k=n-L+1}^{n} I_k q(t - kT) \\ &= \theta_n + \theta(t; \mathbf{I}), \qquad nT \leqslant t \leqslant (n+1)T \end{aligned} \tag{5.3–4}$$

where we have assumed that $q(t) = 0$ for $t < 0$, $q(t) = \frac{1}{2}$ for $t \geqslant LT$, and

$$q(t) = \int_0^t g(\tau)\, d\tau \tag{5.3–5}$$

The signal pulse $g(t) = 0$ for $t < 0$ and $t \geqslant LT$. For $L = 1$, we have a full response CPM, and for $L > 1$, where L is a positive integer, we have a partial response CPM signal.

Now, when h is rational, i.e., $h = m/p$ where m and p are relatively prime positive integers, the CPM scheme can be represented by a trellis. In this case, there are p *phase states*

$$\Theta_s = \left\{0, \frac{\pi m}{p}, \frac{2\pi m}{p}, \ldots, \frac{(p-1)\pi m}{p}\right\} \tag{5.3–6}$$

when m is even, and $2p$ phase states

$$\Theta_s = \left\{0, \frac{\pi m}{p}, \ldots, \frac{(2p-1)\pi m}{p}\right\} \tag{5.3–7}$$

when m is odd. If $L = 1$, these are the only states in the trellis. On the other hand, if $L > 1$, we have an additional number of states due to the partial response character of the signal pulse $g(t)$. These additional states can be identified by expressing $\theta(t; \mathbf{I})$ given by Equation 5.3–4 as

$$\theta(t; \mathbf{I}) = 2\pi h \sum_{k=n-L+1}^{n-1} I_k q(t - kT) + 2\pi h I_n q(t - nT) \tag{5.3–8}$$

The first term on the right-hand side of Equation 5.3–8 depends on the information symbols $(I_{n-1}, I_{n-2}, \ldots, I_{n-L+1})$, which is called the *correlative state vector*, and represents the phase term corresponding to signal pulses that have not reached their final value. The second term in Equation 5.3–8 represents the phase contribution due to the most recent symbol I_n. Hence, the state of the CPM signal (or the modulator) at time $t = nT$ may be expressed as the combined *phase state* and *correlative state*, denoted as

$$S_n = \{\theta_n, I_{n-1}, I_{n-2}, \ldots, I_{n-L+1}\} \tag{5.3–9}$$

for a partial response signal pulse of length LT, where $L > 1$. In this case, the number of states is

$$N_s = \begin{cases} pM^{L-1} & (\text{even } m) \\ 2pM^{L-1} & (\text{odd } m) \end{cases} \tag{5.3–10}$$

when $h = m/p$.

Now, suppose the state of the modulator at $t = nT$ is S_n. The effect of the new symbol in the time interval $nT \leqslant t \leqslant (n+1)T$ is to change the state from S_n to S_{n+1}. Hence, at $t = (n+1)T$, the state becomes

$$S_{n+1} = (\theta_{n+1}, I_n, I_{n-1}, \ldots, I_{n-L+2})$$

where

$$\theta_{n+1} = \theta_n + \pi h I_{n-L+1}$$

EXAMPLE 5.3–1. Consider a binary CPM scheme with a modulation index $h = 3/4$ and a partial response pulse with $L = 2$. Let us determine the states S_n of the CPM scheme and sketch the phase tree and state trellis.

First, we note that there are $2p = 8$ phase states, namely,

$$\Theta_s = \{0, \pm\tfrac{1}{4}\pi, \pm\tfrac{1}{2}\pi, \pm\tfrac{3}{4}\pi, \pi\}$$

For each of these phase states, there are two states that result from the memory of the CPM scheme. Hence, the total number of states is $N_s = 16$, namely,

$(0,1), (0,-1), (\pi, 1), (\pi, -1), (\frac{1}{4}\pi, 1), (\frac{1}{4}\pi, -1), (\frac{1}{2}\pi, 1), (\frac{1}{2}\pi, -1),$
$(\frac{3}{4}\pi, 1), (\frac{3}{4}\pi, -1), (-\frac{1}{4}\pi, 1), (-\frac{1}{4}\pi, -1), (-\frac{1}{2}\pi, 1), (-\frac{1}{2}\pi, -1),$
$(-\frac{3}{4}\pi, 1), (-\frac{3}{4}\pi, -1)$

If the system is in phase state $\theta_n = -\frac{1}{4}\pi$ and $I_{n-1} = -1$, then

$$\begin{aligned}\theta_{n+1} &= \theta_n + \pi h I_{n-1}\\ &= -\tfrac{1}{4}\pi - \tfrac{3}{4}\pi = -\pi\end{aligned}$$

The state trellis is illustrated in Figure 5.3–1. A path through the state trellis corresponding to the sequence $(1, -1, -1, -1, 1, 1)$ is illustrated in Figure 5.3–2.

In order to sketch the phase tree, we must know the signal pulse shape $g(t)$. Figure 5.3–3 illustrates the phase tree when $g(t)$ is a rectangular pulse of duration $2T$, with initial state $(0, 1)$.

Having established the state trellis representation of CPM, let us now consider the metric computations performed in the Viterbi algorithm.

Metric computations. By referring back to the mathematical development for the derivation of the maximum likelihood demodulator given in Section

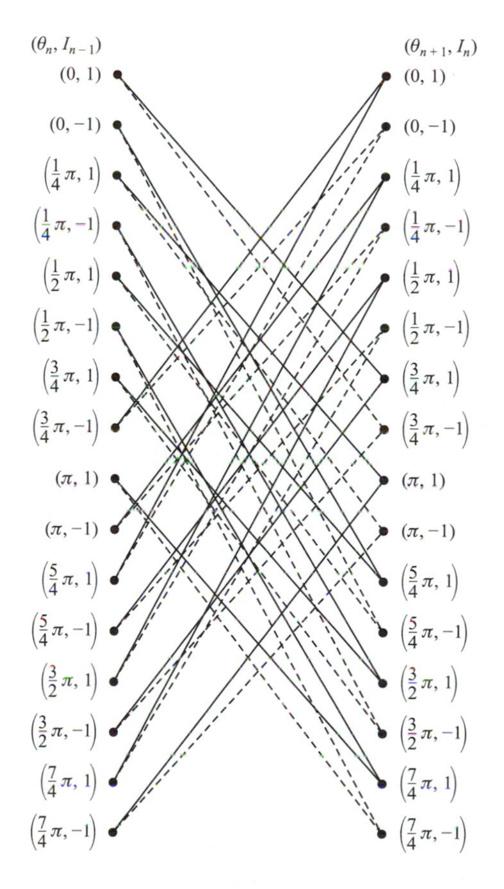

FIGURE 5.3–1
State trellis for partial response ($L = 2$) CPM with $h = \frac{3}{4}$.

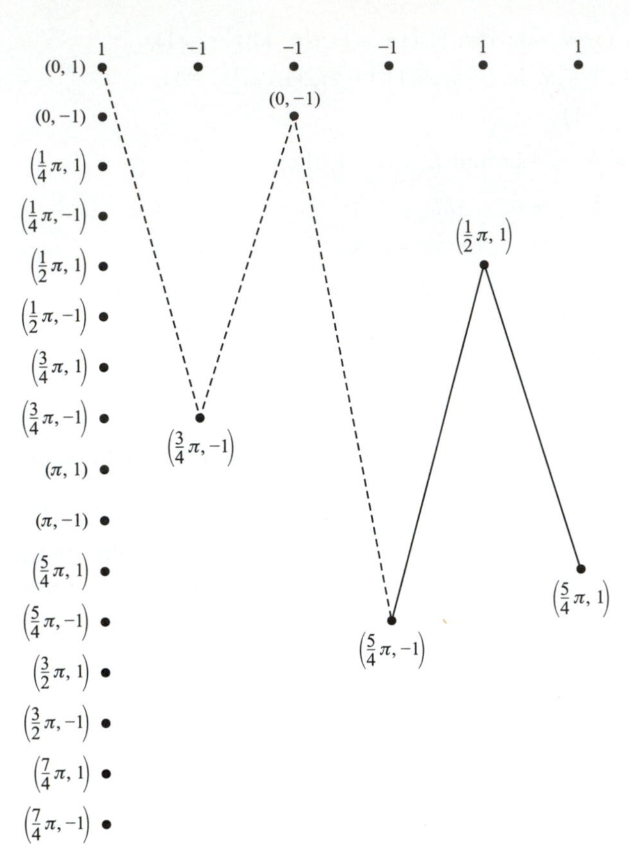

FIGURE 5.3–2
A single signal path through the trellis.

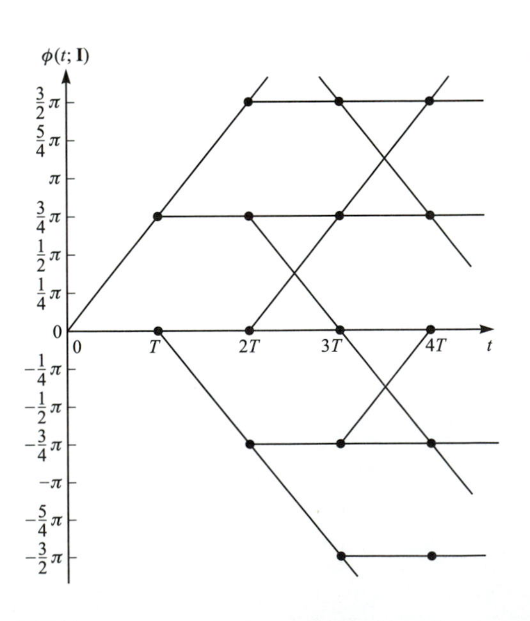

FIGURE 5.3–3
Phase tree for $L = 2$ partial response CPM with $h = \frac{3}{4}$

5.1.4, it is easy to show that the logarithm of the probability of the observed signal $r(t)$ conditioned on a particular sequence of transmitted symbols $\mathbf{I}$ is proportional to the cross-correlation metric

$$\begin{aligned} CM_n(\mathbf{I}) &= \int_{-\infty}^{(n+1)T} r(t)\cos[\omega_c t + \phi(t;\mathbf{I})]\,dt \\ &= CM_{n-1}(\mathbf{I}) + \int_{nT}^{(n+1)T} r(t)\cos[\omega_c t + \theta(t;\mathbf{I}) + \theta_n]\,dt \end{aligned} \tag{5.3–11}$$

The term $CM_{n-1}(\mathbf{I})$ represents the metrics for the surviving sequences up to time nT, and the term

$$v_n(\mathbf{I};\theta_n) = \int_{nT}^{(n+1)T} r(t)\cos[\omega_c t + \theta(t;\mathbf{I}) + \theta_n]\,dt \tag{5.3–12}$$

represents the additional increments to the metrics contributed by the signal in the time interval $nT \leqslant t \leqslant (n+1)T$. Note that there are M^L possible sequences $\mathbf{I} = (I_n, I_{n-1}, \ldots, I_{n-L+1})$ of symbols and p (or $2p$) possible phase states $\{\theta_n\}$. Therefore, there are pM^L (or $2pM^L$) different values of $v_n(\mathbf{I}, \theta_n)$ computed in each signal interval, and each value is used to increment the metrics corresponding to the pM^{L-1} surviving sequences from the previous signaling interval. A general block diagram that illustrates the computations of $v_n(\mathbf{I};\theta_n)$ for the Viterbi decoder is shown in Figure 5.3–4.

Note that the number of surviving sequences at each state of the Viterbi decoding process is pM^{L-1} (or $2pM^{L-1}$). For each surviving sequence, we have M new increments of $v_n(\mathbf{I};\theta_n)$ that are added to the existing metrics to yield pM^L (or $2pM^L$) sequences with pM^L (or $2pM^L$) metrics. However, this number is then reduced back to pM^{L-1} (or $2pM^{L-1}$) survivors with corresponding metrics by selecting the most probable sequence of the M sequences merging at each node of the trellis and discarding the other $M-1$ sequences.

5.3.2 Performance of CPM Signals

In evaluating the performance of CPM signals achieved with maximum-likelihood sequence detection, we must determine the minimum Euclidean

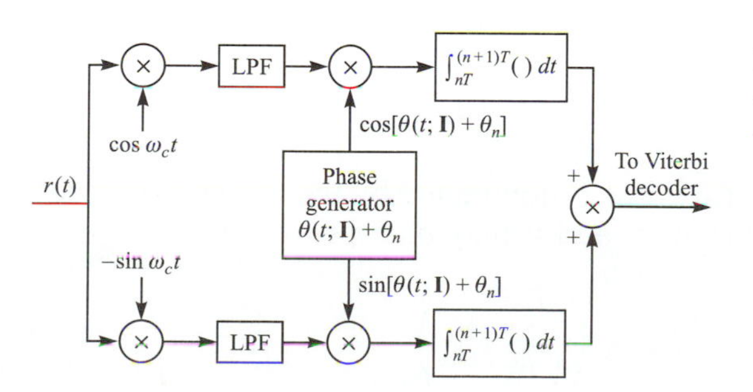

FIGURE 5.3–4
Computation of metric increments $v_n(\mathbf{I};\theta_n)$.

distance of paths through the trellis that separate at the node at $t = 0$ and remerge at a later time at the same node. The distance between two paths through the trellis is related to the corresponding signals as we now demonstrate.

Suppose that we have two signals $s_i(t)$ and $s_j(t)$ corresponding to two phase trajectories $\phi(t; \mathbf{I}_i)$ and $\phi(t; \mathbf{I}_j)$. The sequences $\mathbf{I}_i$ and $\mathbf{I}_j$ must be different in their first symbol. Then, the Euclidean distance between the two signals over an interval of length NT, where $1/T$ is the symbol rate, is defined as

$$\begin{aligned} d_{ij}^2 &= \int_0^{NT} [s_i(t) - s_j(t)]^2 \, dt \\ &= \int_0^{NT} s_i^2(t)\, dt + \int_0^{NT} s_j^2(t)\, dt - 2\int_0^{NT} s_i(t)s_j(t)\, dt \\ &= 2N\mathcal{E} - 2\frac{2\mathcal{E}}{T}\int_0^{NT} \cos[\omega_c t + \phi(t; \mathbf{I}_i)]\cos[\omega_c t + \phi(t; \mathbf{I}_j)]\, dt \\ &= 2N\mathcal{E} - \frac{2\mathcal{E}}{T}\int_0^{NT} \cos[\phi(t; \mathbf{I}_i) - \phi(t; \mathbf{I}_j)]\, dt \\ &= \frac{2\mathcal{E}}{T}\int_0^{NT} \{1 - \cos[\phi(t; \mathbf{I}_i) - \phi(t; \mathbf{I}_j)]\}\, dt \end{aligned} \tag{5.3–13}$$

Hence the Euclidean distance is related to the phase difference between the paths in the state trellis according to Equation 5.3–13.

It is desirable to express the distance d_{ij}^2 in terms of the bit energy. Since $\mathcal{E} = \mathcal{E}_b \log_2 M$, Equation 5.3–13 may be expressed as

$$d_{ij}^2 = 2\mathcal{E}_b \delta_{ij}^2 \tag{5.3–14}$$

where δ_{ij}^2 is defined as

$$\delta_{ij}^2 = \frac{\log_2 M}{T}\int_0^{NT} \{1 - \cos[\phi(t; \mathbf{I}_i) - \phi(t; \mathbf{I}_j)]\}\, dt \tag{5.3–15}$$

Furthermore, we observe that $\phi(t; \mathbf{I}_i) - \phi(t; \mathbf{I}_j) = \phi(t; \mathbf{I}_i - \mathbf{I}_j)$, so that, with $\boldsymbol{\xi} = \mathbf{I}_i - \mathbf{I}_j$, Equation 5.3–15 may be written as

$$\delta_{ij}^2 = \frac{\log_2 M}{T}\int_0^{NT} [1 - \cos\phi(t; \boldsymbol{\xi})]\, dt \tag{5.3–16}$$

where any element of $\boldsymbol{\xi}$ can take the values $0, \pm 2, \pm 4, \ldots, \pm 2(M-1)$, except that $\xi_0 \neq 0$.

The error rate performances for CPM is dominated by the term corresponding to the minimum Euclidean distance, and it may be expressed as

$$P_M = K_{\delta_{\min}} Q\left(\sqrt{\frac{\mathcal{E}_b}{N_0}\delta_{\min}^2}\right) \tag{5.3–17}$$

where $K_{\delta_{\min}}$ is the number of paths having the minimum distance

$$\begin{aligned}\delta_{\min}^2 &= \lim_{N\to\infty} \min_{i,j} \delta_{ij}^2 \\ &= \lim_{N\to\infty} \min_{i,j}\left\{\frac{\log_2 M}{T}\int_0^{NT}[1-\cos\phi(t;\mathbf{I}_i-\mathbf{I}_j)]\,dt\right\}\end{aligned} \tag{5.3–18}$$

We note that for conventional binary PSK with no memory, $N = 1$ and $\delta_{\min}^2 = \delta_{12}^2 = 2$. Hence, Equation 5.3–17 agrees with our previous result.

Since $\delta_{\min}^2$ characterizes the performance of CPM, we can investigate the effect on $\delta_{\min}^2$ resulting from varying the alphabet size M, the modulation index h, and the length of the transmitted pulse in partial response CPM.

First, we consider full response ($L = 1$) CPM. If we take $M = 2$ as a beginning, we note that the sequences

$$\begin{aligned}\mathbf{I}_i &= +1, -1, I_2, I_3 \\ \mathbf{I}_j &= -1, +1, I_2, I_3\end{aligned} \tag{5.3–19}$$

which differ for $k = 0, 1$ and agree for $k \geqslant 2$, result in two phase trajectories that merge after the second symbol. This corresponds to the difference sequence

$$\boldsymbol{\xi} = \{2, -2, 0, 0, \ldots\} \tag{5.3–20}$$

The Euclidean distance for this sequence is easily calculated from Equation 5.3–16, and provides an upper bound on $\delta_{\min}^2$. This upper bound for CPFSK with $M = 2$ is

$$d_B^2(h) = 2\left(1 - \frac{\sin 2\pi h}{2\pi h}\right), \qquad M = 2 \tag{5.3–21}$$

For example, where $h = \frac{1}{2}$, which corresponds to MSK, we have $d_B^2(\frac{1}{2}) = 2$, so that $\delta_{\min}^2(\frac{1}{2}) \leqslant 2$.

For $M > 2$ and full response CPM, it is also easily seen that phase trajectories merge at $t = 2T$. Hence, an upper bound on $\delta_{\min}^2$ can be obtained by considering the phase difference sequence $\boldsymbol{\xi} = \{\alpha, -\alpha, 0, 0, \ldots\}$ where $\alpha = \pm 2, \pm 4, \ldots, \pm 2(M-1)$. This sequence yields the upper bound for M-ary CPFSK as

$$d_B^2(h) = \min_{1\leqslant k\leqslant M-1}\left\{(2\log_2 M)\left(1 - \frac{\sin 2k\pi h}{2k\pi h}\right)\right\} \tag{5.3–22}$$

The graphs of $d_B^2(h)$ versus h for $M = 2, 4, 8, 16$ are shown in Figure 5.3–5. It is apparent from these graphs that large gains in performance can be achieved by increasing the alphabet size M. It must be remembered, however, that $\delta_{\min}^2(h) \leqslant d_B^2(h)$. That is, the upper bound may not be achievable for all values of h.

The minimum Euclidean distance $\delta_{\min}^2(h)$ has been determined, by evaluating Equation 5.3–16, for a variety of CPM signals by Aulin and Sundberg (1981). For example, Figure 5.3–6 illustrates the dependence of the Euclidean distance for binary CPFSK as a function of the modulation index h, with the number N

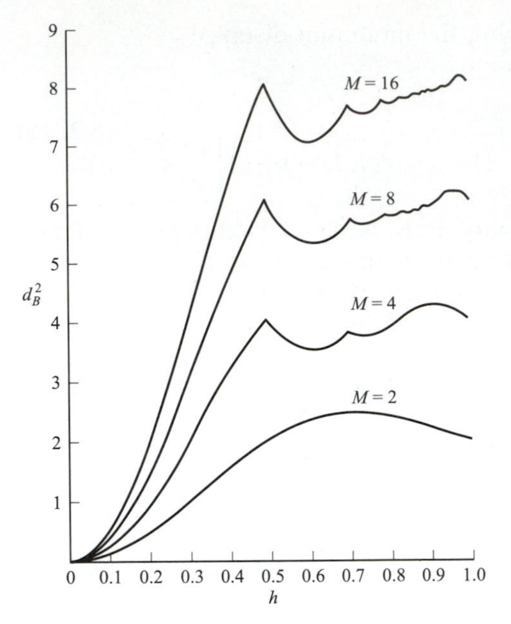

FIGURE 5.3–5
The upper bound d_B^2 as a function of the modulation index h for full response CPM with rectangular pulses. [*From Aulin and Sundberg (1984), © 1984 John Wiley Ltd. Reprinted with permission of the publisher.*]

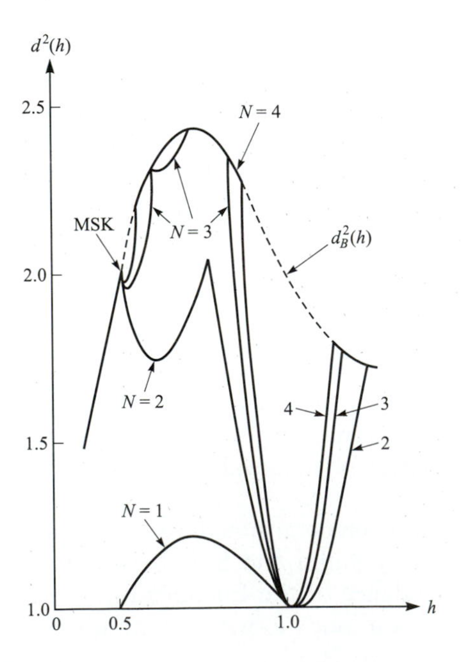

FIGURE 5.3–6
Squared minimum Euclidean distance as a function of the modulation index for binary CPFSK. The upper bound is d_B^2. [*From Aulin and Sundberg (1981), © 1981 IEEE.*]

of bit observation (decision) intervals ($N = 1, 2, 3, 4$) as a parameter. Also shown is the upper bound $d_B^2(h)$ given by Equation 5.3–21. In particular, we note that when $h = \frac{1}{2}$, $\delta_{\min}^2(\frac{1}{2}) = 2$, which is the same squared distance as PSK (binary or quaternary) with $N = 1$. On the other hand, the required observation interval for MSK is $N = 2$ intervals, for which we have $\delta_{\min}^2(\frac{1}{2}) = 2$. Hence, the performance of MSK with a Viterbi detector is comparable to (binary or quaternary) PSK as we have previously observed.

We also note from Figure 5.3–6 that the optimum modulation index for binary CPFSK is $h = 0.715$ when the observation interval is $N = 3$. This yields $\delta_{\min}^2(0.715) = 2.43$, or a gain of 0.85 dB relative to MSK.

Figure 5.3–7 illustrates the Euclidean distance as a function of h for $M = 4$ CPFSK, with the length of the observation interval N as a parameter. Also shown (as a dashed line where it is not reached) is the upper bound d_B^2 evaluated from Equation 5.3–22. Note that $\delta_{\min}^2$ achieves the upper bound for several values of h for some N. In particular, note that the maximum value of d_B^2, which occurs at $h \approx 0.9$, is approximately reached for $N = 8$ observed symbol intervals. The true maximum is achieved at $h = 0.914$ with $N = 9$. For this case,

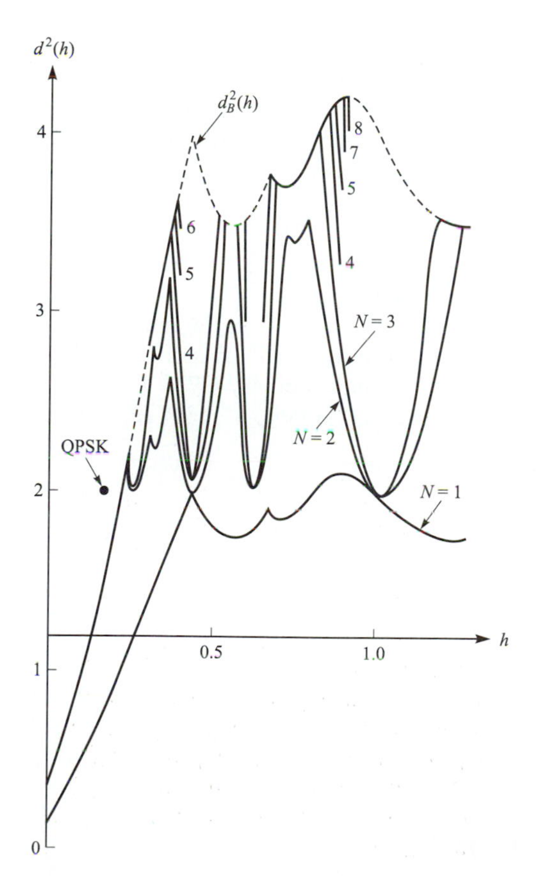

FIGURE 5.3–7
Squared minimum Euclidean distance as a function of the modulation index for quaternary CPFSK. The upper bound is d_B^2. [*From Aulin and Sundberg (1981), © 1981 IEEE.*]

$\delta^2_{\min}(0.914) = 4.2$, which represents a 3.2-dB gain over MSK. Also note that the Euclidean distance contains minima at $h = \frac{1}{3}, \frac{1}{2}, \frac{2}{3}, 1$, etc. These values of h are called *weak modulation indices* and should be avoided. Similar results are available for larger values of M and may be found in the paper by Aulin and Sundberg (1981) and the text by Anderson et al. (1986).

Large performance gains can also be achieved with maximum-likelihood sequence detection of CPM by using partial response signals. For example, the distance bound $d_B^2(h)$ for partial response, raised cosine pulses given by

$$g(t) = \begin{cases} \dfrac{1}{2LT}\left(1 - \cos\dfrac{2\pi t}{2LT}\right) & (0 \leqslant t \leqslant LT) \\ 0 & \text{(otherwise)} \end{cases} \tag{5.3–23}$$

is shown in Figure 5.3–8 for $M = 2$. Here, note that, as L increases, d_B^2 also achieves higher values. Clearly, the performance of CPM improves as the correlative memory L increases, but h must also be increased in order to achieve the larger values of d_B^2. Since a larger modulation index implies a larger bandwidth (for fixed L), while a larger memory length L (for fixed h) implies a smaller bandwidth, it is better to compare the Euclidean distance as a function of the normalized bandwidth $2WT_b$, where W is the 99 percent power bandwidth and T_b is the bit interval. Figure 5.3–9 illustrates this type of comparison with MSK used as a point of reference (0 dB). Note from this figure that there are several decibels to be gained by using partial response signals and higher signaling alphabets. The major price to be paid for this performance gain is the added exponentially increasing complexity in the implementation of the Viterbi detector.

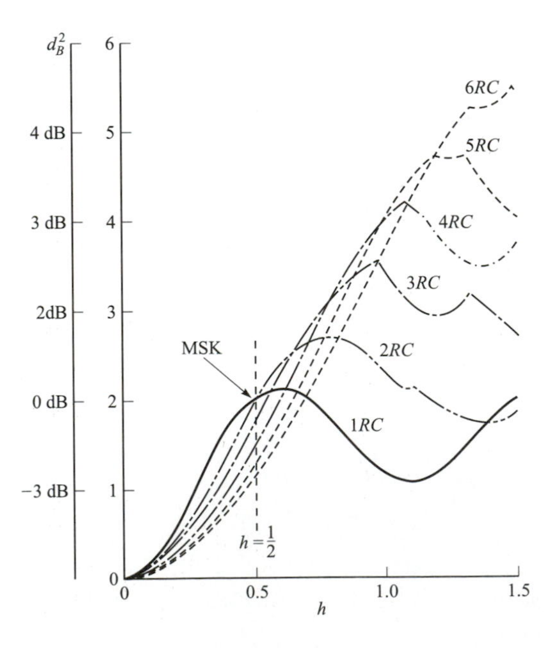

FIGURE 5.3–8
Upper bound d_B^2 on the minimum distance for partial response (raised cosine pulse) binary CPM. [*From Sundberg (1986), © 1986 IEEE.*]

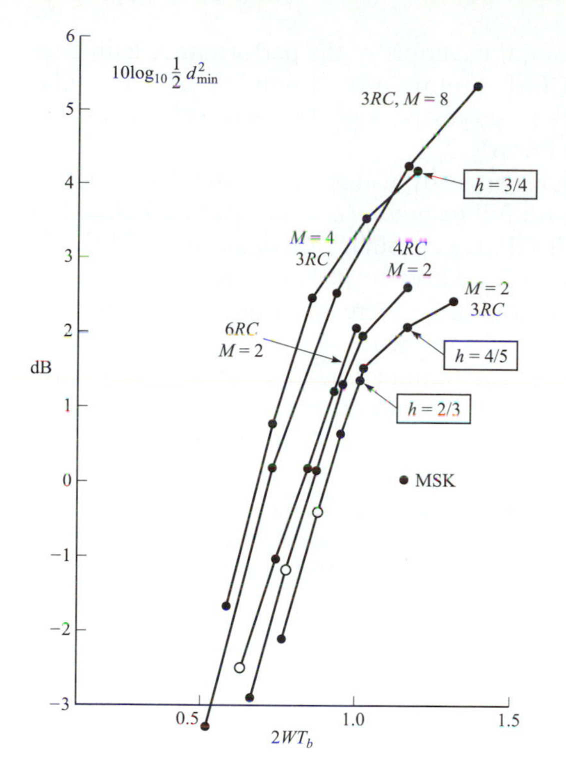

FIGURE 5.3–9
Power bandwidth tradeoff for partial response CPM signals with raised cosine pulses. W is the 99 percent in-band power bandwidth. [*From Sundberg (1986),* © *1986 IEEE.*]

The performance results shown in Figure 5.3–9 illustrate that a 3–4 dB gain relative to MSK can be easily obtained with relatively no increase in bandwidth by the use of raised cosine partial response CPM and $M = 4$. Although these results are for raised cosine signal pulses, similar gains can be achieved with other partial response pulse shapes. We emphasize that this gain in SNR is achieved by introducing memory into the signal modulation and exploiting the memory in the demodulation of the signal. No redundancy through coding has been introduced. In effect, the code has been built into the modulation and the trellis-type (Viterbi) decoding exploits the phase constraints in the CPM signal.

Additional gains in performance can be achieved by introducing additional redundancy through coding and increasing the alphabet size as a means of maintaining a fixed bandwidth. In particular, trellis-coded CPM using relatively simple convolution codes has been thoroughly investigated and many results are available in the technical literature. The Viterbi decoder for the convolutionally encoded CPM signal now exploits the memory inherent in the code and in the CPM signal. Performance gains of the order of 4–6 dB, relative to uncoded MSK with the same bandwidth, have been demonstrated by combining convolutional coding with CPM. Extensive numerical results for coded CPM are given by Lindell (1985).

Multi-h CPM. By varying the modulation index from one signaling interval to another, it is possible to increase the minimum Euclidean distance $\delta^2_{\min}$

between pairs of phase trajectories and, thus, improve the performance gain over constant-h CPM. Usually, multi-h CPM employs a fixed number H of modulation indices that are varied cyclically in successive signaling intervals. Thus, the phase of the signal varies piecewise linearly.

Significant gains in SNR are achievable by using only a small number of different values of h. For example, with full response ($L = 1$) CPM and $H = 2$, it is possible to obtain a gain of 3 dB relative to binary or quaternary PSK. By increasing H to $H = 4$, a gain of 4.5 dB relative to PSK can be obtained. The performance gain can also be increased with an increase in the signal alphabet. Table 5.3–1 lists the performance gains achieved with $M = 2, 4$, and 8 for several values of H. The upper bounds on the minimum Euclidean distance are also shown in Figure 5.3–10 for several values of M and H. Note that the major gain in performance is obtained when H is increased from $H = 1$ to $H = 2$. For $H > 2$, the additional gain is relatively small for small values of $\{h_i\}$. On the other hand, significant performance gains are achieved by increasing the alphabet size M.

The results shown above hold for full response CPM. One can also extend the use of multi-h CPM to partial response in an attempt to further improve performance. It is anticipated that such schemes will yield some additional performance gains, but numerical results on partial response, multi-h CPM are limited. The interested reader is referred to the paper by Aulin and Sundberg (1982b).

Multiamplitude CPM. Multiamplitude CPM (MACPM) is basically a combined amplitude and phase digital modulation scheme that allows us to increase the signaling alphabet relative to CPM in another dimension and, thus, to achieve higher data rates on a band-limited channel. Simultaneously, the com-

TABLE 5.3–1
Maximum values of the upper bound d_B^2 for multi-h linear phase CPM[a]

M	H	Max d_B^2	dB gain compared with MSK	h_1	h_2	h_3	h_4	$\bar{h}$
2	1	2.43	0.85	0.715				0.715
2	2	4.0	3.0	0.5	0.5			0.5
2	3	4.88	3.87	0.620	0.686	0.714		0.673
2	4	5.69	4.54	0.73	0.55	0.73	0.55	0.64
4	1	4.23	3.25	0.914				0.914
4	2	6.54	5.15	0.772	0.772			0.772
4	3	7.65	5.83	0.795	0.795	0.795		0.795
8	1	6.14	4.87	0.964				0.964
8	2	7.50	5.74	0.883	0.883			0.883
8	3	8.40	6.23	0.879	0.879	0.879		0.879

[a]From Aulin and Sundberg (1982b).

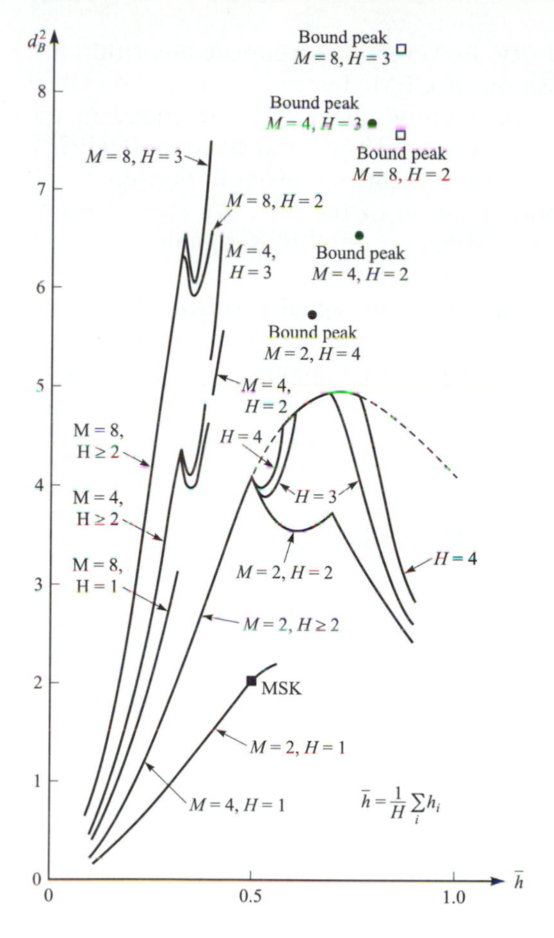

FIGURE 5.3–10
Upper bounds on minimum squared Euclidean distance for various M and H values. [*From Aulin and Sundberg (1982b),* © *1982 IEEE.*]

bination of multiple amplitude in conjunction with CPM results in a bandwidth-efficient modulation technique.

We have already observed the spectral characteristics of MACPM in Section 4.3. The performance characteristics of MACPM have been investigated by Mulligan (1988) for both uncoded and trellis-coded CPM. Of particular interest is the result that trellis-coded CPM with two amplitude levels achieves a gain of 3–4 dB relative to MSK without a significant increase in the signal bandwidth.

5.3.3 Symbol-by-Symbol Detection of CPM Signals

Besides the ML sequence detector, there are other types of detectors that can be used to recover the information sequence in a CPM signal. In this section, we consider symbol-by-symbol detectors. One type of symbol-by-symbol detector is the one described in Section 5.1–5, which exploits the memory of CPM by performing matched filtering or cross correlation over several signaling intervals.

Because of its computational complexity, however, this recursive algorithm has not been directly applied to the detection of CPM. Instead, two similar, albeit suboptimal, symbol-by-symbol detection methods have been described in the papers by deBuda (1972), Osborne and Luntz (1974), and Schonhoff (1976). One of these is functionally equivalent to the algorithm given in Section 5.1–5, and the second is a suboptimum approximation of the first. We shall describe these two methods in the context of demodulation of CPFSK signals, for which these detection algorithms have been applied directly.

To describe these methods, we assume that the signal is observed over the present signaling interval and D signaling intervals into the future in deciding on the information symbol transmitted in the present signaling interval. A block diagram of the demodulator, implemented as a bank of cross correlators, is shown in Figure 5.3–11. Recall that the transmitted CPFSK signal during the nth signaling interval is

$$s(t) = \mathrm{Re}[v(t)e^{j2\pi f_c t}]$$

where

$$v(t) = \exp\left\{ j\left[\frac{\pi h[t-(n-1)T]I_n}{T} + \pi h \sum_{k=0}^{n-1} I_k + \phi_0 \right] \right\}$$

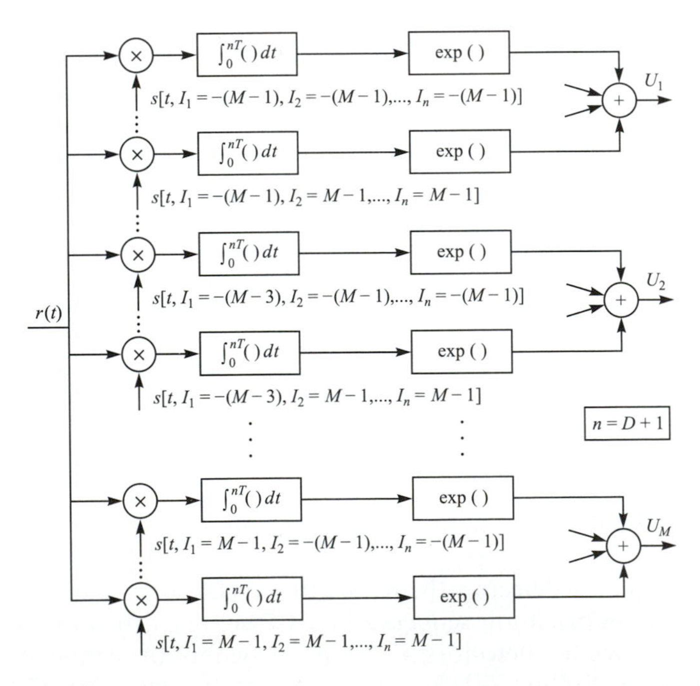

FIGURE 5.3–11
Block diagram of demodulator for detection of CPFSK.

$h = 2f_d T$ is the modulation index, f_d is the peak frequency deviation, and ϕ_0 is the initial phase angle of the carrier.

In detecting the symbol I_1 the cross correlations shown in Figure 5.3–11 are performed with the reference signals $s(t, I_1, I_2, \ldots, I_{1+D})$ for all M^{D+1} possible values of the symbols $I_1, I_2, \ldots, I_{1+D}$ transmitted over the $D+1$ signaling intervals. But these correlations in effect generate the variables $r_1, r_2, \ldots, r_{1+D}$, which in turn are the arguments of the exponentials that occur in the PDF

$$p(r_1, r_2, \ldots, r_{1+D} | I_1, I_2, \ldots, I_{1+D})$$

Finally, the summations over the M^D possible values of the symbols $I_2, I_3, \ldots, I_{1+D}$ represent the averaging of

$$p(r_1, r_2, \ldots, r_{1+D} | I_1, I_2, \ldots, I_{1+D}) P(I_1, I_2, \ldots, I_{1+D})$$

over the M^D possible values of these symbols. The M outputs of the demodulator constitute the decision variables from which the largest is selected to form the demodulated symbol. Consequently the metrics generated by the demodulator shown in Figure 5.3–11 are equivalent to the decision variables given by Equation 5.1–68 on which the decision on I_1 is based.

Signals received in subsequent signaling intervals are demodulated in the same manner. That is, the demodulator cross-correlates the signal received over $D+1$ signaling intervals with the M^{D+1} possible transmitted signals and forms the decision variables as illustrated in Figure 5.3–11. Thus the decision made on the mth signaling interval is based on the cross correlations performed over the signaling intervals $m, m+1, \ldots, m+D$. The initial phase in the correlation interval of duration $(D+1)T$ is assumed to be known. On the other hand, the algorithm described by Equations 5.1–76 and 5.1–77 involves an additional averaging operation over the previously detected symbols. In this respect, the demodulator shown in Figure 5.3–11 differs from the recursive algorithm described above. However, the difference is insignificant.

One suboptimum demodulation method that performs almost as well as the optimum method embodied in Figure 5.3–11 bases its decision on the largest output from the bank of M^{D+1} cross correlators. Thus the exponential functions and the summations are eliminated. But this method is equivalent to selecting the symbol I_m for which the probability density function $p(r_m, r_{m+1}, \ldots, r_{m+D} | I_m, I_{m+1}, \ldots, I_{m+D})$ is a maximum.

The performance of the detector shown in Figure 5.3–11 has been upper-bounded and evaluated numerically. Figure 5.3–12 illustrates the performance of binary CPFSK with $n = D+1$ as a parameter. The modulation index $h = 0.715$ used in generating these results minimizes the probability of error as shown by Schonhoff (1976). We note that an improvement of about 2.5 dB is obtained relative to orthogonal FSK ($n = 1$) by a demodulator that cross-correlates over two symbols. An additional gain of approximately 1.5 dB is obtained by extending the correlation time to three symbols. Further extension of the correlation time results in a relatively small additional gain.

Similar results are obtained with larger alphabet sizes. For example, Figures 5.3–13 and 5.3–14 illustrate the performance improvements for quaternary and

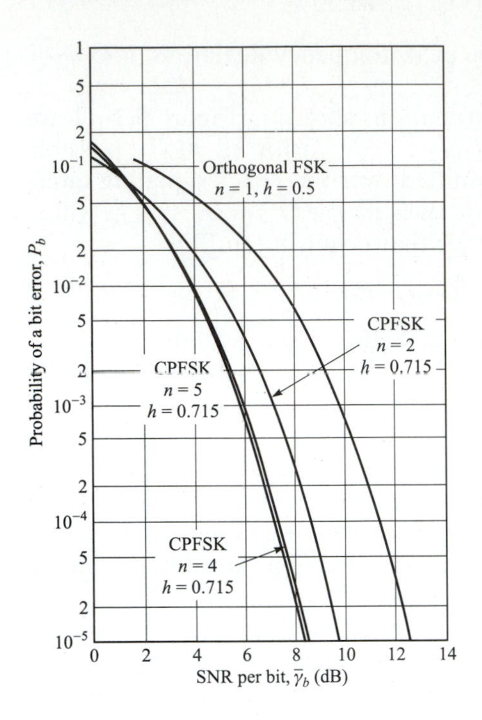

FIGURE 5.3–12
Performance of binary CPFSK with coherent detection.

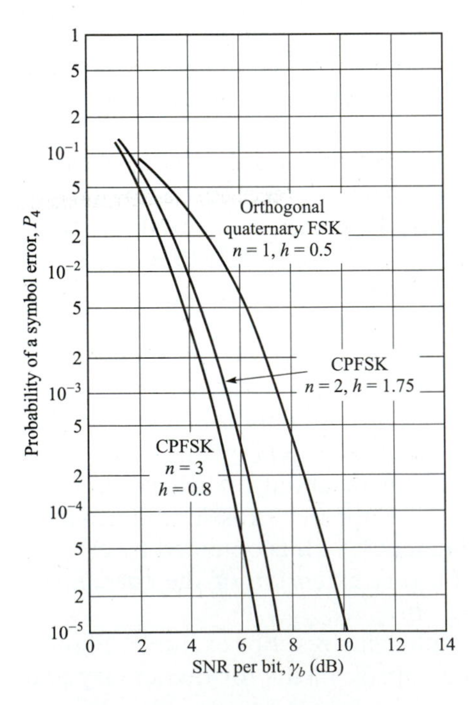

FIGURE 5.3–13
Performance of quaternary CPFSK with coherent detection.

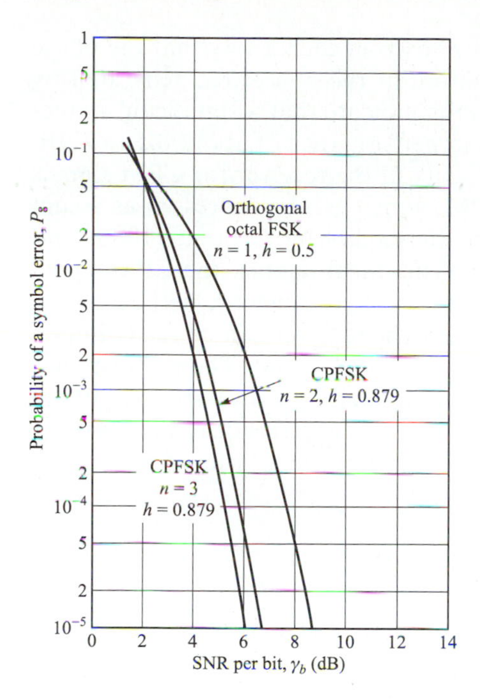

FIGURE 5.3–14
Performance of octal CPFSK with coherent detection.

octal CPFSK, respectively. The modulation indices given in these graphs are the ones that minimize the probability of a symbol error.

Instead of performing coherent detection, which requires knowledge of the carrier phase ϕ_0, we may assume that ϕ_0 is uniformly distributed over the interval 0 to 2π, and average over it in arriving at the decision variables. Thus coherent integration (cross correlation) is performed over the $n = D + 1$ signaling intervals, but the outputs of the correlators are envelope-detected. This is called *noncoherent detection* of CPFSK. In this detection scheme, performance is optimized by selecting n to be odd and making the decision on the middle symbol in the sequence of n symbols. The numerical results on the probability of error for noncoherent detection of CPFSK are similar to the results illustrated above for coherent detection. That is, a gain of 2–3 dB in performance is achieved by increasing the correlation interval from $n = 1$ to $n = 3$ and to $n = 5$.

5.3.4 Suboptimum Demodulation and Detection of CPM Signals

The high complexity inherent in the implementation of the maximum-likelihood sequence detector for CPM signals has been a motivating factor in the investigation of reduced-complexity detectors. Reduced-complexity Viterbi detectors were investigated by Svensson (1984), Svensson et al. (1984), Svensson and Sundberg (1983), Aulin et al. (1981), Simmons and Wittke (1983), Palenius and Svensson (1993), and Palenius (1991). The basic idea in achieving a reduced-complexity Viterbi detector is to design a receiver filter that has a shorter pulse than the transmitter. The receiver pulse $g_R(t)$ must be chosen in such a way that the phase tree generated by $g_R(t)$ is a good approximation of the phase tree generated by the transmitter pulse $g_T(t)$. Performance results indicate that a significant reduction in complexity can be achieved at a loss in performance of about 0.5 to 1 dB.

Another method for reducing the complexity of the receiver for CPM signals is to exploit the linear representation of CPM, which can be expressed as a sum of amplitude-modulated pulses as given in the papers by Laurent (1986) and Mengali and Morelli (1995). In many cases of practical interest the CPM signal can be approximated by a single amplitude-modulated pulse or, perhaps, by a sum of two amplitude-modulated pulses. Hence, the receiver can be easily implemented based on this linear representation of the CPM signal. The performance of such relatively simple receivers has been investigated by Kawas-Kaleh (1989). The results of this study indicate that such simplified receivers sacrifice little in performance but achieve a significant reduction in implementation complexity.

5.4
OPTIMUM RECEIVER FOR SIGNALS WITH RANDOM PHASE IN AWGN CHANNEL

In this section, we consider the design of the optimum receiver for carrier modulated signals when the carrier phase is unknown at the receiver and no attempt is made to estimate its value. Uncertainty in the carrier phase of the received signal may be due to one or more of the following reasons: First, the oscillators that are used at the transmitter and the receiver to generate the carrier signals are generally not phase synchronous. Second, the time delay in the propagation of the signal from the transmitter to the receiver is not generally known precisely. To elaborate on this point, a transmitted signal of the form

$$s(t) = \mathrm{Re}[g(t)e^{j2\pi f_c t}]$$

that propagates through a channel with delay t_0 will be received as

$$\begin{aligned} s(t - t_0) &= \mathrm{Re}[g(t - t_0)e^{j2\pi f_c (t-t_0)}] \\ &= \mathrm{Re}[g(t - t_0)e^{-j2\pi f_c t_0} e^{j2\pi f_c t}] \end{aligned}$$

The carrier phase shift due to the propagation delay t_0 is

$$\phi = -2\pi f_c t_0$$

Note that large changes in the carrier phase ϕ can occur due to relatively small changes in the propagation delay. For example, if the carrier frequency $f_c = 1$ MHz, an uncertainty or a change in the propagation delay of 0.5 μs will cause a phase uncertainty of π rad. In some channels (e.g., radio channels) the time delay in the propagation of the signal from the transmitter to the receiver may change rapidly and in an apparently random manner, so that the carrier phase of the received signal varies in an apparently random fashion.

In the absence of knowledge of the carrier phase, we may treat this signal parameter as a random variable and determine the form of the optimum receiver for recovering the transmitted information from the received signal. First, we treat the case of binary signals and, then, we consider M-ary signals.

5.4.1 Optimum Receiver for Binary Signals

We consider a binary communication system that uses the two carrier modulated signals $s_1(t)$ and $s_2(t)$ to transmit the information, where

$$s_m(t) = \text{Re}[s_{lm}(t)e^{j2\pi f_c t}], \qquad m = 1, 2, \qquad 0 \leqslant t \leqslant T \tag{5.4–1}$$

and $s_{lm}(t)$, $m = 1, 2$ are the equivalent low-pass signals. The two signals are assumed to have equal energy

$$\mathcal{E} = \int_0^T s_m^2(t)\, dt = \frac{1}{2}\int_0^T |s_{lm}(t)|^2\, dt \tag{5.4–2}$$

and are characterized by the complex-valued correlation coefficient

$$\rho_{12} \equiv \rho = \frac{1}{2\mathcal{E}}\int_0^T s_{l1}^*(t)s_{l2}(t)\, dt \tag{5.4–3}$$

The received signal is assumed to be a phase-shifted version of the transmitted signal and corrupted by the additive noise

$$\begin{aligned} n(t) &= \text{Re}\{[n_c(t) + jn_s(t)]e^{j2\pi f_c t}\} \\ &= \text{Re}[z(t)e^{j2\pi f_c t}] \end{aligned} \tag{5.4–4}$$

Hence, the received signal may be expressed as

$$r(t) = \text{Re}\{[s_{lm}(t)e^{j\phi} + z(t)]e^{j2\pi f_c t}\} \tag{5.4–5}$$

where

$$r_l(t) = s_{lm}(t)e^{j\phi} + z(t), \qquad 0 \leqslant t \leqslant T \tag{5.4–6}$$

is the equivalent low-pass received signal. This received signal is now passed through a demodulator whose sampled output at $t = T$ is passed to the detector.

The optimum demodulator. In Section 5.1.1, we demonstrated that if the received signal was correlated with a set of orthonormal functions $\{f_n(t)\}$ that spanned the signal space, the outputs from the bank of correlators provide a set of sufficient statistics for the detector to make a decision that minimizes the probability of error. We also demonstrated that a bank of matched filters could be substituted for the bank of correlators.

A similar orthonormal decomposition can also be employed for a received signal with an unknown carrier phase. However, it is mathematically convenient to deal with the equivalent low-pass signal and to specify the signal correlators or matched filters in terms of the equivalent low-pass signal waveforms.

To be specific, the impulse response $h_l(t)$ of a filter that is matched to the complex-valued equivalent low-pass signal $s_l(t)$, $0 \leqslant t \leqslant T$, is given as (see Problem 5.6)

$$h_l(t) = s_l^*(T - t) \tag{5.4–7}$$

and the output of such a filter at $t = T$ is simply

$$\int_0^T |s_l(t)|^2 \, dt = 2\mathcal{E} \tag{5.4–8}$$

where $\mathcal{E}$ is the signal energy. A similar result is obtained if the signal $s_l(t)$ is correlated with $s_l^*(t)$ and the correlator is sampled at $t = T$. Therefore, the optimum demodulator for the equivalent low-pass received signal $s_l(t)$ given in Equation 5.4–6 may be realized by two matched filters in parallel, one matched to $s_{l1}(t)$ and the other to $s_{l2}(t)$, and shown in Figure 5.4–1. The output of the matched filters or correlators at the sampling instant are the two complex numbers

$$r_m = r_{mc} + jr_{ms}, \qquad m = 1, 2 \tag{5.4–9}$$

Suppose that the transmitted signal is $s_1(t)$. Then, it is easily shown (see Problem 5.41) that

$$\begin{aligned} r_1 &= 2\mathcal{E} \cos\phi + n_{1c} + j(2\mathcal{E} \sin\phi + n_{1s}) \\ r_2 &= 2\mathcal{E}|\rho| \cos(\phi - \alpha_0) + n_{2c} + j[2\mathcal{E}|\rho| \sin(\phi - \alpha_0) + n_{2s}] \end{aligned} \tag{5.4–10}$$

where ρ is the complex-valued correlation coefficient of the two signals $s_{l1}(t)$ and $s_{l2}(t)$, which may be expressed as $\rho = |\rho| \exp(j\alpha_0)$. The random noise variables n_{1c}, n_{1s}, n_{2c}, and n_{2s} are jointly Gaussian, with zero-mean and equal variance.

The optimum detector. The optimum detector observes the random variables $[r_{1c} \; r_{1s} \; r_{2c} \; r_{2s}] = \mathbf{r}$, where $r_1 = r_{1c} + jr_{1s}$, and $r_2 = r_{2c} + jr_{2s}$, and bases its decision on the posterior probabilities $P(\mathbf{s}_m|\mathbf{r})$, $m = 1, 2$. These probabilities may be expressed as

$$P(\mathbf{s}_m|\mathbf{r}) = \frac{p(\mathbf{r}|\mathbf{s}_m)P(\mathbf{s}_m)}{p(\mathbf{r})}, \qquad m = 1, 2 \tag{5.4–11}$$

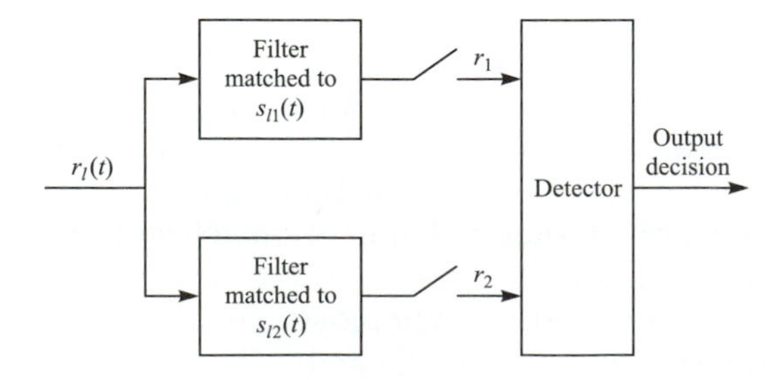

FIGURE 5.4–1
Optimum receiver for binary signals.

and, hence, the optimum decision rule may be expressed as

$$P(\mathbf{s}_1|\mathbf{r}) \underset{s_2}{\overset{s_1}{\gtrless}} P(\mathbf{s}_2|\mathbf{r})$$

or, equivalently,

$$\frac{p(\mathbf{r}|\mathbf{s}_1)}{p(\mathbf{r}|\mathbf{s}_2)} \underset{s_2}{\overset{s_1}{\gtrless}} \frac{P(\mathbf{s}_2)}{P(\mathbf{s}_1)} \tag{5.4–12}$$

The ratio of PDFs on the left-hand side of Equation 5.4–12 is the *likelihood ratio*, which we denote as

$$\Lambda(\mathbf{r}) = \frac{p(\mathbf{r}|\mathbf{s}_1)}{p(\mathbf{r}|\mathbf{s}_2)} \tag{5.4–13}$$

The right-hand side of Equation 5.4–12 is the ratio of the two prior probabilities, which takes the value of unity when the two signals are equally probable.

The probability density functions $p(\mathbf{r}|\mathbf{s}_1)$ and $p(\mathbf{r}|\mathbf{s}_2)$ can be obtained by averaging the PDFs $p(\mathbf{r}|\mathbf{s}_m, \phi)$ over the PDF of the random carrier phase, i.e.,

$$p(\mathbf{r}|\mathbf{s}_m) = \int_0^{2\pi} p(\mathbf{r}|\mathbf{s}_m, \phi)p(\phi)\, d\phi \tag{5.4–14}$$

We shall perform the integration indicated in Equation 5.4–14 for the special case in which the two signals are orthogonal, i.e., $\rho = 0$. In this case, the outputs of the demodulator are

$$\begin{aligned} r_1 &= r_{1c} + jr_{1s} \\ &= 2\mathcal{E}\cos\phi + n_{1c} + j(2\mathcal{E}\sin\phi + n_{1s}) \\ r_2 &= r_{2c} + jr_{2s} \\ &= n_{2c} + jn_{2s} \end{aligned} \tag{5.4–15}$$

where $(n_{1c}, n_{1s}, n_{2c}, n_{2s})$ are mutually uncorrelated and, hence, statistically independent, zero-mean Gaussian random variables (see Problem 5.40). Hence, the joint PDF of $\mathbf{r} = [r_{1c}\ r_{1s}\ r_{2c}\ r_{2s}]$ may be expressed as a product of the marginal PDFs. Consequently,

$$\begin{aligned} p(r_{1c}, r_{1s}|\mathbf{s}_1, \phi) &= \frac{1}{2\pi\sigma^2} \exp\left[-\frac{(r_{1c} - 2\mathcal{E}\cos\phi)^2 + (r_{1s} - 2\mathcal{E}\sin\phi)^2}{2\sigma^2}\right] \\ p(r_{2c}, r_{2s}) &= \frac{1}{2\pi\sigma^2} \exp\left(-\frac{r_{2c}^2 + r_{2s}^2}{2\sigma^2}\right) \end{aligned} \tag{5.4–16}$$

where $\sigma^2 = 2\mathcal{E}N_0$.

The uniform PDF for the carrier phase ϕ represents the most ignorance that can be exhibited by the detector. This is called the *least favorable PDF* for ϕ. With $p(\phi) = 1/2\pi$, $0 \leqslant \phi \leqslant 2\pi$, substituted into the integral in Equation 5.4–14, we obtain

$$\frac{1}{2\pi}\int_0^{2\pi} p(r_{1c}, r_{1s}|\mathbf{s}_1, \phi)\, d\phi$$
$$= \frac{1}{2\pi\sigma^2} \exp\left(-\frac{r_{1c}^2 + r_{1c}^2 + 4\mathcal{E}^2}{2\sigma^2}\right) \frac{1}{2\pi}\int_0^{2\pi} \exp\left[\frac{2\mathcal{E}(r_{1c}\cos\phi + r_{1s}\sin\phi)}{\sigma^2}\right] d\phi \tag{5.4–17}$$

But

$$\frac{1}{2\pi}\int_0^{2\pi} \exp\left[\frac{2\mathcal{E}(r_{1c}\cos\phi + r_{1s}\sin\phi)}{\sigma^2}\right] d\phi = I_0\left(\frac{2\mathcal{E}\sqrt{r_{1c}^2 + r_{1s}^2}}{\sigma^2}\right) \tag{5.4–18}$$

where $I_0(x)$ is the modified Bessel function of zeroth order, defined in Equation 2.1–120.

By performing a similar integration as in Equation 5.4–17 under the assumption that the signal $s_2(t)$ was transmitted, we obtain the result

$$p(r_{2c}, r_{2s}|\mathbf{s}_2) = \frac{1}{2\pi\sigma^2} \exp\left(-\frac{r_{2c}^2 + r_{2s}^2 + 4\mathcal{E}^2}{2\sigma^2}\right) I_0\left(\frac{2\mathcal{E}\sqrt{r_{2c}^2 + r_{2s}^2}}{\sigma^2}\right) \tag{5.4–19}$$

When we substitute these results into the likelihood ratio given by Equation 5.4–13, we obtain the result

$$\Lambda(\mathbf{r}) = \frac{I_0(2\mathcal{E}\sqrt{r_{1c}^2 + r_{1s}^2}/\sigma^2)}{I_0(2\mathcal{E}\sqrt{r_{2c}^2 + r_{2s}^2}/\sigma^2)} \underset{s_2}{\overset{s_1}{\gtrless}} \frac{P(\mathbf{s}_2)}{P(\mathbf{s}_1)} \tag{5.4–20}$$

Thus, the optimum detector computes the two envelopes $\sqrt{r_{1c}^2+r_{1s}^2}$ and $\sqrt{r_{2c}^2+r_{2s}^2}$ and the corresponding values of the Bessel function $I_0(2\mathcal{E}\sqrt{r_{1c}^2+r_{1s}^2}/\sigma^2)$ and $I_0(2\mathcal{E}\sqrt{r_{2c}^2+r_{2s}^2}/\sigma^2)$ to form the likelihood ratio. We observe that this computation requires knowledge of the noise variance σ^2. The likelihood ratio is then compared with the threshold $P(\mathbf{s}_2)/P(\mathbf{s}_1)$ to determine which signal was transmitted.

A significant simplification in the implementation of the optimum detector occurs when the two signals are equally probable. In such a case the threshold becomes unity, and, due to the monotonicity of the Bessel function shown in Figure 5.4–2, the optimum detection rule simplifies to

$$\sqrt{r_{1c}^2 + r_{1s}^2} \underset{s_2}{\overset{s_1}{\gtrless}} \sqrt{r_{2c}^2 + r_{2s}^2} \tag{5.4–21}$$

Thus, the optimum detector bases its decision on the two envelopes $\sqrt{r_{1c}^2+r_{1s}^2}$ and $\sqrt{r_{2c}^2+r_{2s}^2}$, and, hence, it is called an *envelope detector*.

We observe that the computation of the envelopes of the received signal samples at the output of the demodulator renders the carrier phase irrelevant in the decision as to which signal was transmitted. Equivalently, the decision may

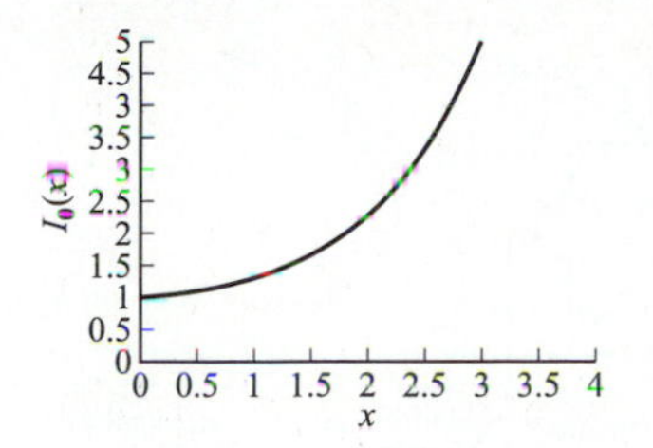

FIGURE 5.4–2
Graph of $I_0(x)$.

be based on the computation of the squared envelopes $r_{1c}^2 + r_{1s}^2$ and $r_{2c}^2 + r_{2s}^2$, in which case the detector is called a *square-law detector*.

Binary FSK signals are an example of binary orthogonal signals. Recall that in binary FSK we employ two different frequencies, say f_1 and $f_2 = f_1 + \Delta f$, to transmit a binary information sequence. The choice of minimum frequency separation $\Delta f = f_2 - f_1$ is considered below. Thus, the signal waveforms may be expressed as

$$
\begin{aligned}
s_1(t) &= \sqrt{2\mathcal{E}_b/T_b}\cos 2\pi f_1 t, \qquad 0 \leqslant t \leqslant T_b \\
s_2(t) &= \sqrt{2\mathcal{E}_b/T_b}\cos 2\pi f_2 t, \qquad 0 \leqslant t \leqslant T_b
\end{aligned}
\tag{5.4–22}
$$

and their equivalent low-pass counterparts are

$$
\begin{aligned}
s_{l1}(t) &= \sqrt{2\mathcal{E}_b/T_b}, \qquad 0 \leqslant t \leqslant T_b \\
s_{l2}(t) &= \sqrt{2\mathcal{E}_b/T_b}e^{j2\pi\Delta f t}, \qquad 0 \leqslant t \leqslant T_b
\end{aligned}
\tag{5.4–23}
$$

The received signal may be expressed as

$$
r(t) = \sqrt{\frac{2\mathcal{E}_b}{T_b}}\cos(2\pi f_m t + \phi_m) + n(t) \tag{5.4–24}
$$

where ϕ_m is the phase of the carrier frequency f_m. The demodulation of the real signal $r(t)$ may be accomplished, as shown in Figure 5.4–3, by using four correlators with the basis functions

$$
\begin{aligned}
f_{1m}(t) &= \sqrt{\frac{2}{T_b}}\cos[(2\pi f_1 + 2\pi m\,\Delta f)t], \qquad m = 0, 1 \\
f_{2m}(t) &= \sqrt{\frac{2}{T_b}}\sin[(2\pi f_1 + 2\pi m\,\Delta f)t], \qquad m = 0, 1
\end{aligned}
\tag{5.4–25}
$$

The four outputs of the correlators are sampled at the end of each signal interval and passed to the detector. If the mth signal is transmitted, the four samples at the detector may be expressed as

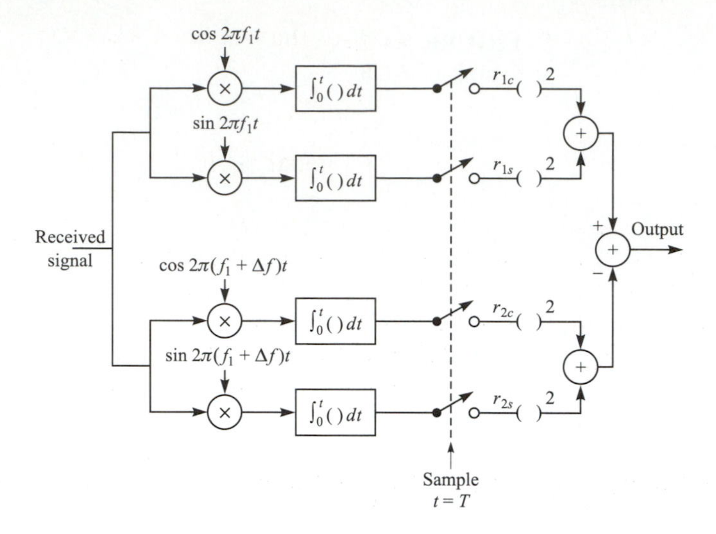

FIGURE 5.4–3
Demodulation and square-law detection of binary FSK signals.

$$\begin{aligned} r_{kc} &= \sqrt{\mathcal{E}_b}\left[\frac{\sin[2\pi(k-m)\,\Delta f\,T]}{2\pi(k-m)\,\Delta f\,T}\cos\phi_m \right.\\ &\quad \left. - \frac{\cos[2\pi(k-m)\,\Delta f\,T]-1}{2\pi(k-m)\,\Delta f\,T}\sin\phi_m\right] + n_{kc}, \qquad k, m = 1, 2 \\ r_{ks} &= \sqrt{\mathcal{E}_b}\left[\frac{\cos[2\pi(k-m)\,\Delta f\,T]-1}{2\pi(k-m)\,\Delta f\,T}\cos\phi_m \right.\\ &\quad \left. + \frac{\sin[2\pi(k-m)\,\Delta f\,T]}{2\pi(k-m)\,\Delta f\,T}\sin\phi_m\right] + n_{kc}, \qquad k, m = 1, 2 \end{aligned} \tag{5.4–26}$$

where n_{kc} and n_{ks} denote the Gaussian noise components in the sampled outputs.

We observe that when $k = m$, the sampled values to the detector are

$$\begin{aligned} r_{mc} &= \sqrt{\mathcal{E}_b}\cos\phi_m + n_{mc} \\ r_{ms} &= \sqrt{\mathcal{E}_b}\sin\phi_m + n_{ms} \end{aligned} \tag{5.4–27}$$

Furthermore, we observe that when $k \neq m$, the signal components in the samples r_{kc} and r_{ks} will vanish, independently of the values of the phase shifts ϕ_k, provided that the frequency separation between successive frequencies is $\Delta f = 1/T$. In such a case, the other two correlator outputs consist of noise only, i.e.,

$$r_{kc} = n_{kc}, \qquad r_{ks} = n_{ks}, \qquad k \neq m \tag{5.4–28}$$

With a frequency separation of $\Delta f = 1/T$, the relations 5.4–27 and 5.4–28 are consistent with the previous result 5.4–15 for the demodulator outputs. Therefore, we conclude that for envelope or square-law detection of FSK signals, the minimum frequency separation required for orthogonality of the signals is

$\Delta f = 1/T$. This separation is twice as large as that required when the detection is phase-coherent.

5.4.2 Optimum Receiver for M-ary Orthogonal Signals

The generalization of the optimum demodulator and detector to the case of M-ary orthogonal signals is straightforward. If the equal energy and equally probable signal waveforms are represented as

$$s_m(t) = \text{Re}[s_{lm}(t)e^{j2\pi f_c t}], \qquad m = 1, 2, \ldots, M, \qquad 0 \leqslant t \leqslant T \tag{5.4–29}$$

where $s_{lm}(t)$ are the equivalent low-pass signals, the optimum correlation-type or matched-filter-type demodulator produces the M complex-valued random variables

$$r_m = r_{mc} + jr_{ms} = \int_0^T r_l(t)s_{lm}^*(t)\,dt, \qquad m = 1, 2, \ldots, M \tag{5.4–30}$$

where $r_l(t)$ is the equivalent low-pass received signal. Then, the optimum detector, based on a random, uniformly distributed carrier phase, computes the M envelopes

$$|r_m| = \sqrt{r_{mc}^2 + r_{ms}^2}, \qquad m = 1, 2, \ldots, M \tag{5.4–31}$$

or, equivalently, the squared envelopes $|r_m|^2$, and selects the signal with the largest envelope (or squared envelope).

In the special case of M-ary orthogonal FSK signals, the optimum receiver has the structure illustrated in Figure 5.4–4. There are $2M$ correlators: two for each possible transmitted frequency. The minimum frequency separation between adjacent frequencies to maintain orthogonality is $\Delta f = 1/T$.

5.4.3 Probability of Error for Envelope Detection of M-ary Orthogonal Signals

Let us consider the transmission of M-ary orthogonal equal energy signals over an AWGN channel, which are envelope-detected at the receiver. We also assume that the M signals are equally probable a priori and that the signal $s_1(t)$ is transmitted in the signal interval $0 \leqslant t \leqslant T$.

The M decision metrics at the detector are the M envelopes

$$|r_m| = \sqrt{r_{mc}^2 + r_{ms}^2}, \qquad m = 1, 2, \ldots, M \tag{5.4–32}$$

where

$$\begin{aligned} r_{1c} &= \sqrt{\mathcal{E}_s}\cos\phi_1 + n_{1c} \\ r_{1s} &= \sqrt{\mathcal{E}_s}\sin\phi_1 + n_{1s} \end{aligned} \tag{5.4–33}$$

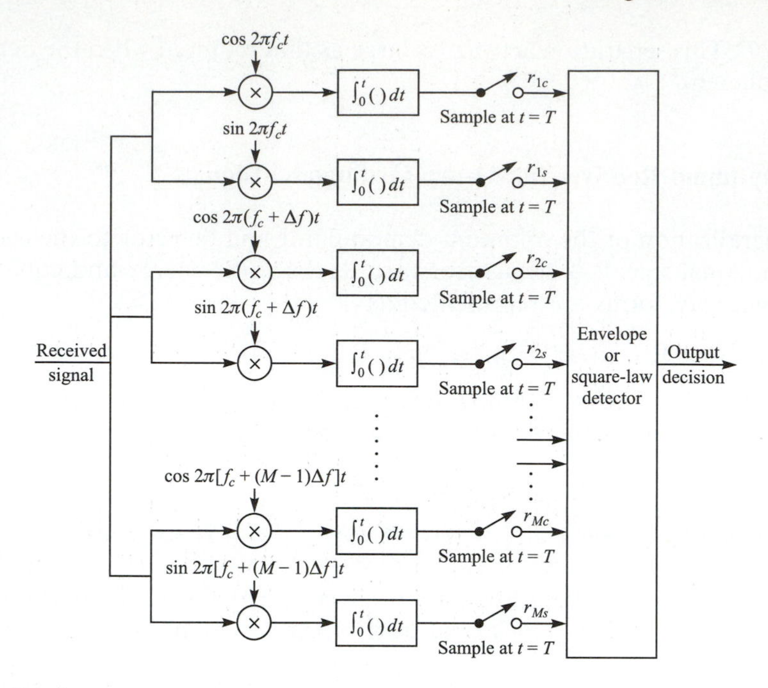

FIGURE 5.4–4
Demodulation of M-ary FSK signals for noncoherent detection.

and

$$r_{mc} = n_{mc}, \qquad r_{ms} = n_{ms}, \qquad m = 2, 3, \ldots, M \tag{5.4–34}$$

The additive noise components $\{n_{mc}\}$ and $\{n_{ms}\}$ are mutually statistically independent zero-mean Gaussian variables with equal variance $\sigma^2 = \frac{1}{2}N_0$. Thus the PDFs of the random variables at the input to the detector are

$$p_{\mathbf{r}_1}(r_{1c}, r_{1s}) = \frac{1}{2\pi\sigma^2} \exp\left(-\frac{r_{1c}^2 + r_{1s}^2 + \mathcal{E}_s}{2\sigma^2}\right) I_0\left(\frac{\sqrt{\mathcal{E}_s(r_{1c}^2 + r_{1s}^2)}}{\sigma^2}\right) \tag{5.4–35}$$

$$p_{\mathbf{r}_m}(r_{mc}, r_{ms}) = \frac{1}{2\pi\sigma^2} \exp\left(-\frac{r_{mc}^2 + r_{ms}^2}{2\sigma^2}\right), \qquad m = 2, 3, \ldots, M \tag{5.4–36}$$

Let us make a change in variables in the joint PDFs given by Equations 5.4–35 and 5.4–36. We define the normalized variables

$$\begin{aligned} R_m &= \frac{\sqrt{r_{mc}^2 + r_{ms}^2}}{\sigma} \\ \Theta_m &= \tan^{-1} \frac{r_{ms}}{r_{mc}} \end{aligned} \tag{5.4–37}$$

Clearly, $r_{mc} = \sigma R_m \cos \Theta_m$ and $r_{ms} = \sigma R_m \sin \Theta_m$. The Jacobian of this transformation is

$$|\mathbf{J}| = \begin{vmatrix} \sigma \cos \Theta_m & \sigma \sin \Theta_m \\ -\sigma R_m \sin \Theta_m & \sigma R_m \cos \Theta_m \end{vmatrix} = \sigma^2 R_m \tag{5.4–38}$$

Consequently,

$$p(R_1, \Theta_1) = \frac{R_1}{2\pi} \exp\left[-\frac{1}{2}\left(R_1^2 + 2\frac{\mathcal{E}_s}{N_0}\right)\right] I_0\left(\sqrt{\frac{2\mathcal{E}_s}{N_0}} R_1\right) \tag{5.4–39}$$

$$p(R_m, \Theta_m) = \frac{R_m}{2\pi} \exp(-\tfrac{1}{2} R_m^2), \qquad m = 2, 3, \ldots, M \tag{5.4–40}$$

Finally, by averaging $p(R_m, \Theta_m)$ over Θ_m, the factor of 2π is eliminated from Equations 5.4–39 and 5.4–40. Thus, we find that R_1 has a Rice probability distribution and R_m, $m = 2, 3, \ldots, M$, are each Rayleigh-distributed.

The probability of a correct decision is simply the probability that $R_1 > R_2$, and $R_1 > R_3, \ldots,$ and $R_1 > R_m$. Hence,

$$\begin{aligned} P_c &= P(R_2 < R_1, R_3 < R_1, \ldots, R_M < R_1) \\ &= \int_0^\infty P(R_2 < R_1, R_3 < R_1, \ldots, R_M < R_1 | R_1 = x) p_{R_1}(x)\, dx \end{aligned} \tag{5.4–41}$$

Because the random variables R_m, $m = 2, 3, \ldots, M$, are statistically independent and identically distributed, the joint probability in Equation 5.4–41 conditioned on R_1 factors into a product of $M - 1$ identical terms. Thus,

$$P_c = \int_0^\infty [P(R_2 < R_1 | R_1 = x)]^{M-1} p_{R_1}(x)\, dx \tag{5.4–42}$$

where

$$\begin{aligned} P(R_2 < R_1 | R_1 = x) &= \int_0^x p_{R_2}(r_2)\, dr_2 \\ &= 1 - e^{-x^2/2} \end{aligned} \tag{5.4–43}$$

The $(M - 1)$th power of Equation 5.4–43 may be expressed as

$$(1 - e^{-x^2/2})^{M-1} = \sum_{n=0}^{M-1} (-1)^n \binom{M-1}{n} e^{-nx^2/2} \tag{5.4–44}$$

Substitution of this result into Equation 5.4–42 and integration over x yields the probability of a correct decision as

$$P_c = \sum_{n=0}^{M-1} (-1)^n \binom{M-1}{n} \frac{1}{n+1} \exp\left[\frac{n\mathcal{E}_s}{(n+1)N_0}\right] \tag{5.4–45}$$

where $\mathcal{E}_s/N_0$ is the SNR per symbol. Then, the probability of a symbol error, which is $P_M = 1 - P_c$, becomes

$$P_M = \sum_{n=1}^{M-1} (-1)^{n+1} \binom{M-1}{n} \frac{1}{n+1} \exp\left[-\frac{nk\mathcal{E}_b}{(n+1)N_0}\right] \tag{5.4–46}$$

where $\mathcal{E}_b/N_0$ is the SNR per bit.

For binary orthogonal signals ($M = 2$), Equation 5.4–46 reduces to the simple form

$$P_2 = \tfrac{1}{2} e^{-\mathcal{E}_b/2N_0} \tag{5.4–47}$$

For $M > 2$, we may compute the probability of a bit error by making use of the relationship

$$P_b = \frac{2^{k-1}}{2^k - 1} P_M \tag{5.4–48}$$

which was established in Section 5.2. Figure 5.4–5 shows the bit-error probability as a function of the SNR per bit γ_b for $M = 2, 4, 8, 16$, and 32. Just as in the case of coherent detection of M-ary orthogonal signals (see Section 5.2.2), we observe that for any given bit-error probability, the SNR per bit decreases as M increases. It will be shown in Chapter 7 that, in the limit as $M \to \infty$ (or $k = \log_2 M \to \infty$), the probability of a bit error P_b can be made arbitrarily small provided that the SNR per bit is greater than the Shannon limit of -1.6 dB. The cost for increasing M is the bandwidth required to transmit the signals. For M-ary FSK, the frequency separation between adjacent frequencies is $\Delta f =$

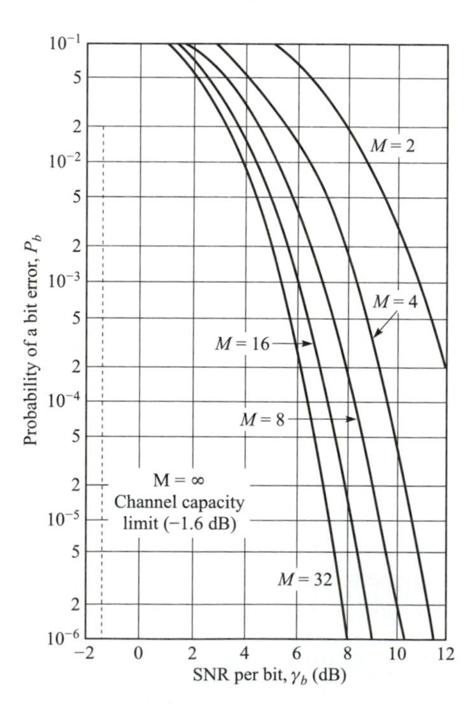

FIGURE 5.4–5
Probability of a bit error for noncoherent detection of orthogonal signals.

$1/T$ for signal orthogonality. The bandwidth required for the M signals is $W = M\,\Delta f = M/T$. Also, the bit rate is $R = k/T$, where $k = \log_2 M$. Therefore, the bit rate–to–bandwidth ratio is

$$\frac{R}{W} = \frac{\log_2 M}{M} \tag{5.4–49}$$

5.4.4 Probability of Error for Envelope Detection of Correlated Binary Signals

In this section, we consider the performance of the envelope detector for binary, equal-energy correlated signals. When the two signals are correlated, the input to the detector is the complex-valued random variables given by Equation 5.4–10. We assume that the detector bases its decision on the envelopes $|r_1|$ and $|r_2|$, which are correlated (statistically dependent). The marginal PDFs of $R_1 = |r_1|$ and $R_2 = |r_2|$ are Ricean distributed and may be expressed as

$$p(R_m) = \begin{cases} \dfrac{R_m}{2\mathcal{E}_s N_0} \exp\left(-\dfrac{R_m^2 + \beta_m^2}{4\mathcal{E} N_0}\right) I_0\left(\dfrac{\beta_m R_m}{2\mathcal{E} N_0}\right) & (R_m > 0) \\ 0 & (R_m < 0) \end{cases} \tag{5.4–50}$$

$m = 1, 2$, where $\beta_1 = 2\mathcal{E}$ and $\beta_2 = 2\mathcal{E}|\rho|$, based on the assumption that signal $s_1(t)$ was transmitted.

Since R_1 and R_2 are statistically dependent as a consequence of the non-orthogonality of the signals, the probability of error may be obtained by evaluating the double integral

$$P_b = P(R_2 > R_1) = \int_0^\infty \int_{x_1}^\infty p(x_1, x_2)\,dx_1\,dx_2 \tag{5.4–51}$$

where $p(x_1, x_2)$ is the joint PDF of the envelopes R_1 and R_2. This approach was first used by Helstrom (1955), who determined the joint PDF of R_1 and R_2 and evaluated the double integral in Equation 5.4–51.

An alternative approach is based on the observation that the probability of error may also be expressed as

$$P_b = P(R_2 > R_1) = P(R_2^2 > R_1^2) = P(R_2^2 - R_1^2 > 0) \tag{5.4–52}$$

But $R_2^2 - R_1^2$ is a special case of a general quadratic form in complex-valued Gaussian random variables, treated later in Appendix B. For the special case under consideration, the derivation yields the error probability in the form

$$P_b = Q_1(a, b) - \tfrac{1}{2} e^{-(a^2+b^2)/2} I_0(ab) \tag{5.4–53}$$

where

$$a = \sqrt{\frac{\mathcal{E}_b}{2N_0}(1 - \sqrt{1 - |\rho|^2})}$$
$$b = \sqrt{\frac{\mathcal{E}_b}{2N_0}(1 + \sqrt{1 - |\rho|^2})} \tag{5.4–54}$$

$Q_1(a, b)$ is the Marcum Q function defined in Equation 2.1–123 and $I_0(x)$ is the modified Bessel function of order zero.

The error probability P_b is illustrated in Figure 5.4–6 for several values of $|\rho|$. P_b is minimized when $\rho = 0$; that is, when the signals are orthogonal. For this case, $a = 0$, $b = \sqrt{\mathcal{E}_b/N_0}$, and Equation 5.4–53 reduces to

$$P_b = Q_1\left(0, \sqrt{\frac{\mathcal{E}_b}{N_0}}\right) - \tfrac{1}{2}e^{-\mathcal{E}_b/2N_0} \tag{5.4–55}$$

From the definition of $Q_1(a, b)$ in Equation 2.1–123, it follows that

$$Q_1\left(0, \sqrt{\frac{\mathcal{E}_b}{N_0}}\right) = e^{-\mathcal{E}_b/2N_0}$$

Substitution of these relations into Equation 5.4–55 yields the desired result given previously in Equation 5.4–47. On the other hand, when $|\rho| = 1$, the error probability in Equation 5.4–53 becomes $P_b = \frac{1}{2}$, as expected.

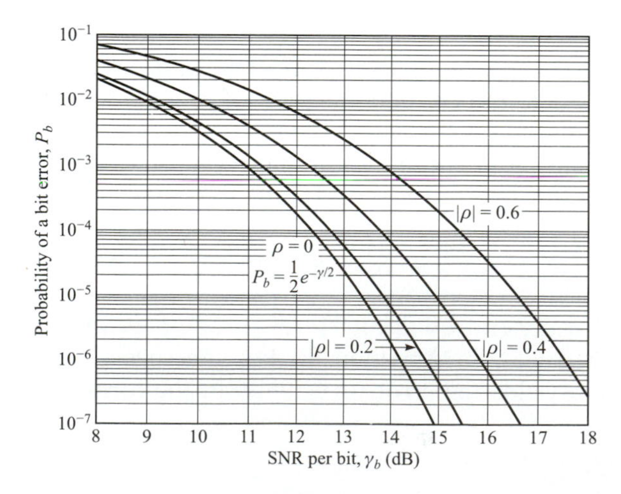

FIGURE 5.4–6
Probability of error for noncoherent detection of binary FSK.

5.5
PERFORMANCE ANALYSIS FOR WIRELINE AND RADIO COMMUNICATION SYSTEMS

In the transmission of digital signals through an AWGN channel, we have observed that the performance of the communication system, measured in terms of the probability of error, depends solely on the received SNR, $\mathcal{E}_b/N_0$, where $\mathcal{E}_b$ is the transmitted energy per bit and $\frac{1}{2}N_0$ is the power spectral density of the additive noise. Hence, the additive noise ultimately limits the performance of the communication system.

In addition to the additive noise, another factor that affects the performance of a communication system is channel attenuation. All physical channels, including wire lines and radio channels, are lossy. Hence, the signal is attenuated as it travels through the channel. The simple mathematical model for the attenuation shown in Figure 5.5–1 may be used for the channel. Consequently, if the transmitted signal is $s(t)$, the received signal, with $0 < \alpha \leqslant 1$ is

$$r(t) = \alpha s(t) + n(t) \tag{5.5–1}$$

Then, if the energy in the transmitted signal is $\mathcal{E}_b$, the energy in the received signal is $\alpha^2\mathcal{E}_b$. Consequently, the received signal has an SNR $\alpha^2\mathcal{E}_b/N_0$. Hence, the effect of signal attenuation is to reduce the energy in the received signal and thus to render the communication system more vulnerable to additive noise.

In analog communication systems, amplifiers called repeaters are used to periodically boost the signal strength in transmission through the channel. However, each amplifier also boosts the noise in the system. In contrast, digital communication systems allow us to detect and regenerate a clean (noise-free) signal in a transmission channel. Such devices, called *regenerative repeaters*, are frequently used in wireline and fiber-optic communication channels.

5.5.1 Regenerative Repeaters

The front end of each regenerative repeater consists of a demodulator/detector that demodulates and detects the transmitted digital information sequence sent by the preceding repeater. Once detected, the sequence is passed to the transmitter side of the repeater, which maps the sequence into signal waveforms that are transmitted to the next repeater. This type of repeater is called a regenerative repeater.

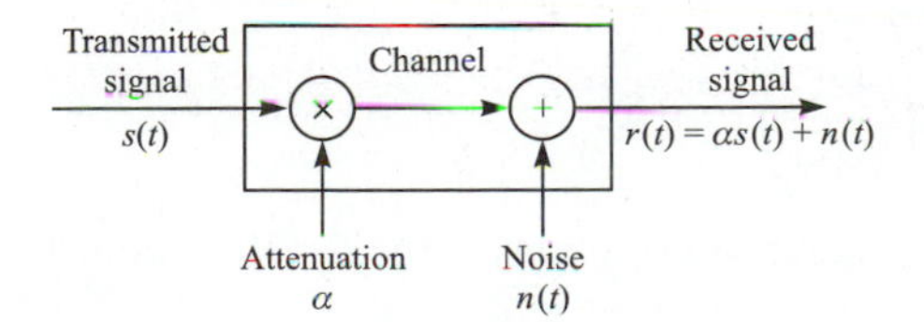

FIGURE 5.5–1
Mathematical model of channel with attenuation and additive noise.

Since a noise-free signal is regenerated at each repeater, the additive noise does not accumulate. However, when errors occur in the detector of a repeater, the errors are propagated forward to the following repeaters in the channel. To evaluate the effect of errors on the performance of the overall system, suppose that the modulation is binary PAM, so that the probability of a bit error for one hop (signal transmission from one repeater to the next repeater in the chain) is

$$P_b = Q\left(\sqrt{\frac{2\mathcal{E}_b}{N_0}}\right)$$

Since errors occur with low probability, we may ignore the probability that any one bit will be detected incorrectly more than once in transmission through a channel with K repeaters. Consequently, the number of errors will increase linearly with the number of regenerative repeaters in the channel, and therefore, the overall probability of error may be approximated as

$$P_b \approx KQ\left(\sqrt{\frac{2\mathcal{E}_b}{N_0}}\right) \tag{5.5–2}$$

In contrast, the use of K analog repeaters in the channel reduces the received SNR by K, and hence, the bit-error probability is

$$P_b \approx Q\left(\sqrt{\frac{2\mathcal{E}_b}{KN_0}}\right) \tag{5.5–3}$$

Clearly, for the same probability of error performance, the use of regenerative repeaters results in a significant saving in transmitter power compared with analog repeaters. Hence, in digital communication systems, regenerative repeaters are preferable. However, in wireline telephone channels that are used to transmit both analog and digital signals, analog repeaters are generally employed.

EXAMPLE 5.5–1. A binary digital communication system transmits data over a wireline channel of length 1000 km. Repeaters are used every 10 km to offset the effect of channel attenuation. Let us determine the $\mathcal{E}_b/N_0$ that is required to achieve a probability of a bit error of 10^{-5} if (a) analog repeaters are employed, and (b) regenerative repeaters are employed.

The number of repeaters used in the system is $K = 100$. If regenerative repeaters are used, the $\mathcal{E}_b/N_0$ obtained from Equation 5.5–2 is

$$10^{-5} = 100Q\left(\sqrt{\frac{2\mathcal{E}_b}{N_0}}\right)$$

$$10^{-7} = Q\left(\sqrt{\frac{2\mathcal{E}_b}{N_0}}\right)$$

which yields approximately 11.3 dB. If analog repeaters are used, the $\mathcal{E}_b/N_0$ obtained from Equation 5.5–3 is

$$10^{-5} = Q\left(\sqrt{\frac{2\mathcal{E}_b}{100N_0}}\right)$$

which yields $\mathcal{E}_b/N_0 \approx 29.6$ dB. Hence, the difference in the required SNR is about 18.3 dB, or approximately 70 times the transmitter power of the digital communication system.

5.5.2 Link Budget Analysis in Radio Communication Systems

In the design of radio communication systems that transmit over line-of-sight microwave channels and satellite channels, the system designer must specify the size of the transmit and receive antennas, the transmitted power, and the SNR required to achieve a given level of performance at some desired data rate. The system design procedure is relatively straightforward and is outlined below.

Let us begin with a transmit antenna that radiates isotropically in free space at a power level of P_T watts as shown in Figure 5.5–2. The power density at a distance d from the antenna is $P_T/4\pi d^2$ W/m^2. If the transmitting antenna has some directivity in a particular direction, the power density in that direction is increased by a factor called the antenna gain and denoted by G_T. In such a case, the power density at distance d is $P_T G_T/4\pi d^2$ W/m^2. The product $P_T G_T$ is usually called the *effective radiated power* (ERP or EIRP), which is basically the radiated power relative to an isotropic antenna, for which $G_T = 1$.

A receiving antenna pointed in the direction of the radiated power gathers a portion of the power that is proportional to its cross-sectional area. Hence, the received power extracted by the antenna may be expressed as

$$P_R = \frac{P_T G_T A_R}{4\pi d^2} \tag{5.5–4}$$

where A_R is the *effective area of the antenna*. From electromagnetic field theory, we obtain the basic relationship between the gain G_R of an antenna and its effective area as

$$A_R = \frac{G_R \lambda^2}{4\pi} \quad \text{m}^2 \tag{5.5–5}$$

where $\lambda = c/f$ is the wavelength of the transmitted signal, c is the speed of light (3×10^8 m/s), and f is the frequency of the transmitted signal.

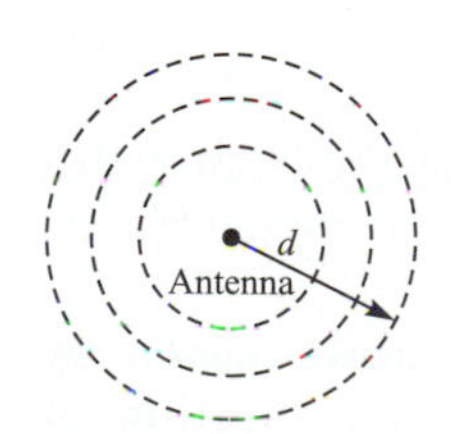

FIGURE 5.5–2
Isotropically radiating antenna.

If we substitute Equation 5.5–5 for A_R into Equation 5.5–4, we obtain an expression for the received power in the form

$$P_R = \frac{P_T G_T G_R}{(4\pi d/\lambda)^2} \tag{5.5–6}$$

The factor

$$L_s = \left(\frac{\lambda}{4\pi d}\right)^2 \tag{5.5–7}$$

is called the *free-space path loss*. If other losses, such as atmospheric losses, are encountered in the transmission of the signal, they may be accounted for by introducing an additional loss factor, say L_a. Therefore, the received power may be written in general as

$$P_R = P_T G_T G_R L_s L_a \tag{5.5–8}$$

As indicated above, the important characteristics of an antenna are its gain and its effective area. These generally depend on the wavelength of the radiated power and the physical dimensions of the antenna. For example, a parabolic (dish) antenna of diameter D has an effective area

$$A_R = \tfrac{1}{4}\pi D^2 \eta \tag{5.5–9}$$

where $\frac{1}{4}\pi D^2$ is the physical area and η is the *illumination efficiency factor*, which falls in the range $0.5 \leqslant \eta \leqslant 0.6$. Hence, the antenna gain for a parabolic antenna of diameter D is

$$G_R = \eta\left(\frac{\pi D}{\lambda}\right)^2 \tag{5.5–10}$$

As a second example, a horn antenna of physical area A has an efficiency factor of 0.8, an effective area of $A_R = 0.8A$, and an antenna gain of

$$G_R = \frac{10A}{\lambda^2} \tag{5.5–11}$$

Another parameter that is related to the gain (directivity) of an antenna is its beamwidth, which we denote as Θ_B and which is illustrated graphically in Figure 5.5–3. Usually, the beamwidth is measured as the -3 dB width of the antenna pattern. For example, the -3 dB beamwidth of a parabolic antenna is approximately

$$\Theta_B = 70(\lambda/D)^\circ \tag{5.5–12}$$

so that G_T is inversely proportional to Θ_B^2. That is, a decrease of the beamwidth by a factor of 2, which is obtained by doubling the diameter D, increases the antenna gain by a factor of 4 (6 dB).

Based on the general relationship for the received signal power given by Equation 5.5–8, the system designer can compute P_R from a specification of

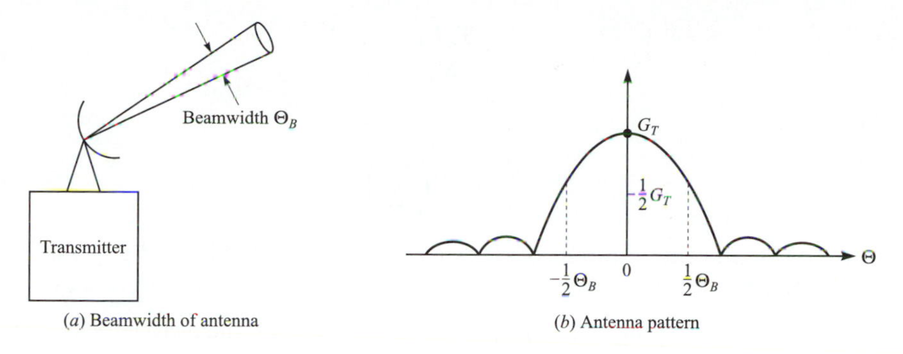

(a) Beamwidth of antenna
(b) Antenna pattern

FIGURE 5.5–3
Antenna beamwidth and pattern.

the antenna gains and the distance between the transmitter and the receiver. Such computations are usually done on a power basis, so that

$$(P_R)_{\text{dB}} = (P_T)_{\text{dB}} + (G_T)_{\text{dB}} + (G_R)_{\text{dB}} + (L_s)_{\text{dB}} + (L_a)_{\text{dB}} \tag{5.5–13}$$

EXAMPLE 5.5–2. Suppose that we have a satellite in geosynchronous orbit (36,000 km above the earth's surface) that radiates 100 W of power, i.e., 20 dB above 1 W (20 dBW). The transmit antenna has a gain of 17 dB, so that the ERP = 37 dBW. Also, suppose that the earth station employs a 3-m parabolic antenna and that the downlink is operating at a frequency of 4 GHz. The efficiency factor is $\eta = 0.5$. By substituting these numbers into Equation 5.5–10, we obtain the value of the antenna gain as 39 dB. The free-space path loss is

$$L_s = 195.6\,\text{dB}$$

No other losses are assumed. Therefore, the received signal power is

$$\begin{aligned}(P_R)_{\text{dB}} &= 20 + 17 + 39 - 195.6 \\ &= -119.6 \text{ dBW}\end{aligned}$$

or, equivalently,

$$P_R = 1.1 \times 10^{-12}\,\text{W}$$

To complete the link budget computation, we must consider the effect of the additive noise at the receiver front end. Thermal noise that arises at the receiver front end has a relatively flat power density spectrum up to about 10^{12} Hz, and is given as

$$N_0 = k_B T_0 \quad \text{W/Hz} \tag{5.5–14}$$

where k_B is Boltzmann's constant (1.38×10^{-23} W-s/K) and T_0 is the noise temperature in kelvin. Therefore, the total noise power in the signal bandwidth W is $N_0 W$.

The performance of the digital communication system is specified by the $\mathcal{E}_b/N_0$ required to keep the error rate performance below some given value. Since

$$\frac{\mathcal{E}_b}{N_0} = \frac{T_b P_R}{N_0} = \frac{1}{R}\frac{P_R}{N_0} \tag{5.5–15}$$

it follows that

$$\frac{P_R}{N_0} = R\left(\frac{\mathcal{E}_b}{N_0}\right)_{\text{req}} \tag{5.5–16}$$

where $(\mathcal{E}_b/N_0)_{\text{req}}$ is the required SNR per bit. Hence, if we have P_R/N_0 and the required SNR per bit, we can determine the maximum data rate that is possible.

EXAMPLE 5.5–3. For the link considered in Example 5.5–2, the received signal power is

$$P_R = 1.1 \times 10^{-12}\ \text{W} \quad (-119.6\ \text{dBW})$$

Now, suppose the receiver front end has a noise temperature of 300 K, which is typical for a receiver in the 4-GHz range. Then

$$N_0 = 4.1 \times 10^{-21}\ \text{W/Hz}$$

or, equivalently, -203.9 dBW/Hz. Therefore,

$$\frac{P_R}{N_0} = -119.6 + 203.9 = 84.3\ \text{dB Hz}$$

If the required SNR per bit is 10 dB, then, from Equation 5.5–16, we have the available rate as

$$\begin{aligned} R_{\text{dB}} &= 84.3 - 10 \\ &= 74.3\ \text{dB} \qquad (\text{with respect to 1 bit/s}) \end{aligned}$$

This corresponds to a rate of 26.9 megabits/s, which is equivalent to about 420 PCM channels, each operating at 64,000 bits/s.

It is a good idea to introduce some safety margin, which we shall call the *link margin* M_{dB}, in the above computations for the capacity of the communication link. Typically, this may be selected as $M_{\text{dB}} = 6$ dB. Then, the link budget computation for the link capacity may be expressed in the simple form

$$\begin{aligned} R_{\text{dB}} &= \left(\frac{P_R}{N_0}\right)_{\text{dB Hz}} - \left(\frac{\mathcal{E}_b}{N_0}\right)_{\text{req}} - M_{\text{dB}} \\ &= (P_T)_{\text{dBW}} + (G_T)_{\text{dB}} + (G_R)_{\text{dB}} \\ &\quad + (L_a)_{\text{dB}} + (L_s)_{\text{dB}} - \left(\frac{\mathcal{E}_b}{N_0}\right)_{\text{req}} - M_{\text{dB}} \end{aligned} \tag{5.5–17}$$

5.6 BIBLIOGRAPHICAL NOTES AND REFERENCES

In the derivation of the optimum demodulator for a signal corrupted by AWGN, we applied mathematical techniques that were originally used in deriving optimum receiver structures for radar signals. For example, the matched filter was first proposed by North (1943) for use in radar detection, and is sometimes called the North filter. An alternative method for deriving the optimum demodulator and detector is the Karhunen–Loeve expansion, which is described in the classical texts by Davenport and Root (1958), Helstrom (1968), and Van Trees (1968). Its use in radar detection theory is described in the paper by Kelly et al. (1960). These detection methods are based on the hypothesis testing methods developed by statisticians, e.g., Neyman and Pearson (1933) and Wald (1947).

The geometric approach to signal design and detection, which was presented in the context of digital modulation and which has its roots in Kotelnikov (1947) and Shannon's original work, is conceptually appealing and is now widely used since its use in the text by Wozencraft and Jacobs (1965).

Design and analysis of signal constellations for the AWGN channel have received considerable attention in the technical literature. Of particular significance is the performance analysis of two-dimensional (QAM) signal constellations that has been treated in the papers of Cahn (1960), Hancock and Lucky (1960). Campopiano and Glazer (1962), Lucky and Hancock (1962), Salz et al. (1971), Simon and Smith (1973), Thomas et al. (1974), and Foschini et al. (1974). Signal design based on multidimensional signal constellations has been described and analyzed in the paper by Gersho and Lawrence (1984).

The Viterbi algorithm was devised by Viterbi (1967) for the purpose of decoding convolutional codes. Its use as the optimal maximum-likelihood sequence detection algorithm for signals with memory was described by Forney (1972) and Omura (1971). Its use for carrier modulated signals was considered by Ungerboeck (1974) and MacKenchnie (1973). It was subsequently applied to the demodulation of CPM by Aulin and Sundberg (1981), Aulin et al. (1981), and Aulin (1980).

Our discussion of the demodulation and detection of signals with memory referenced journal papers published primarily in the United States. The author has recently learned that maximum-likelihood sequential detection algorithms for signals with memory (introduced by the channel through intersymbol interference) were also developed and published in Russia during the 1960s by D. Klovsky. An English translation of Klovsky's work is contained in his book coauthored with B. Nikolaev (1978). This book contains the first English language publication of the Klovsky–Nikolaev algorithm, which may be used in the detection of signals with memory.

PROBLEMS

5.1 A matched filter has the frequency response

$$H(f) = \frac{1 - e^{-j2\pi fT}}{j2\pi f}$$

a) Determine the impulse response $h(t)$ corresponding to $H(f)$

b) Determine the signal waveform to which the filter characteristic is matched.

5.2 Consider the signal

$$s(t) = \begin{cases} (A/T)t\cos 2\pi f_c t & (0 \leqslant t \leqslant T) \\ 0 & (\text{otherwise}) \end{cases}$$

a) Determine the impulse response of the matched filter for the signal.
b) Determine the output of the matched filter at $t = T$.
c) Suppose the signal $s(t)$ is passed through a correlator that correlates the input $s(t)$ with $s(t)$. Determine the value of the correlator output at $t = T$. Compare your result with that in (b).

5.3 This problem deals with the characteristics of a DPSK signal.
a) Suppose we wish to transmit the data sequence

1 1 0 1 0 0 0 1 0 1 1 0

by binary DPSK. Let $s(t) = A\cos(2\pi f_c t + \theta)$ represent the transmitted signal in any signaling interval of duration T. Give the phase of the transmitted signal for the data sequence. Begin with $\theta = 0$ for the phase of the first bit to be transmitted.
b) If the data sequence is uncorrelated, determine and sketch the power density spectrum of the signal transmitted by DPSK.

5.4 A binary digital communication system employs the signals

$$s_0(t) = 0, \qquad 0 \leqslant t \leqslant T$$
$$s_1(t) = A, \qquad 0 \leqslant t \leqslant T$$

for transmitting the information. This is called *on–off signaling*. The demodulator cross-correlates the received signal $r(t)$ with $s(t)$ and samples the output of the correlator at $t + T$.
a) Determine the optimum detector for an AWGN channel and the optimum threshold, assuming that the signals are equally probable.
b) Determine the probability of error as a function of the SNR. How does on–off signaling compare with antipodal signaling?

5.5 The correlation metrics given by Equation 5.1–44 are

$$C(\mathbf{r}, \mathbf{s}_m) = 2\sum_{n=1}^{N} r_n s_{mn} - \sum_{n=1}^{N} s_{mn}^2, \qquad m = 1, 2, \ldots, M$$

where

$$r_n = \int_0^T r(t) f_n(t)\, dt$$
$$s_{mn} = \int_0^T s_m(t) f_n(t)\, dt$$

Show that the correlation metrics are equivalent to the metrics

$$C(\mathbf{r}, \mathbf{s}_m) = 2\int_0^T r(t)s_m(t)\,dt - \int_0^T s_m^2(t)\,dt$$

5.6 Consider the equivalent low-pass (complex-valued) signal $s_l(t)$, $0 \leqslant t \leqslant T$, with energy

$$\mathcal{E} = \frac{1}{2}\int_0^T |s_l(t)|^2\,dt$$

Suppose that this signal is corrupted by AWGN, which is represented by its equivalent low-pass form $z(t)$. Hence, the observed signal is

$$r_l(t) = s_l(t) + z(t), \qquad 0 \leqslant t \leqslant T$$

The received signal is passed through a filter that has an (equivalent low-pass) impulse response $h_l(t)$. Determine $h_l(t)$ so that the filter maximizes the SNR at its output (at $t = T$).

5.7 Let $z(t) = x(t) + jy(t)$ be a complex-valued, zero-mean white Gaussian noise process with autocorrelation function $\phi_{zz}(\tau) = N_0\delta(\tau)$. Let $f_m(t)$, $m = 1, 2, \ldots, M$, be a set of M orthogonal equivalent low-pass waveforms defined on the interval $0 \leqslant t \leqslant T$. Define

$$N_{mr} = \text{Re}\left[\int_0^T z(t)f_m^*(t)\,dt\right], \qquad m = 1, 2, \ldots, M$$

a) Determine the variance of N_{mr}.
b) Show that $E(N_{mr}N_{kr}) = 0$ for $k \neq m$.

5.8 The two equivalent low-pass signals shown in Figure P5.8 are used to transmit a binary sequence over an additive white Gaussian noise channel. The received signal can be expressed as

$$r_l(t) = s_i(t) + z(t), \qquad 0 \leqslant t \leqslant T, \qquad i - 1, 2$$

where $z(t)$ is a zero-mean Gaussian noise process with autocorrelation function

$$\phi_{zz}(\tau) = \frac{1}{2}E[z^*(t)z(t+\tau)] = N_0\delta(\tau)$$

a) Determine the transmitted energy in $s_1(t)$ and $s_2(t)$ and the cross-correlation coefficient ρ_{12}.
b) Suppose the receiver is implemented by means of coherent detection using two matched filters, one matched to $s_1(t)$ and the other to $s_2(t)$. Sketch the equivalent low-pass impulse responses of the matched filters.

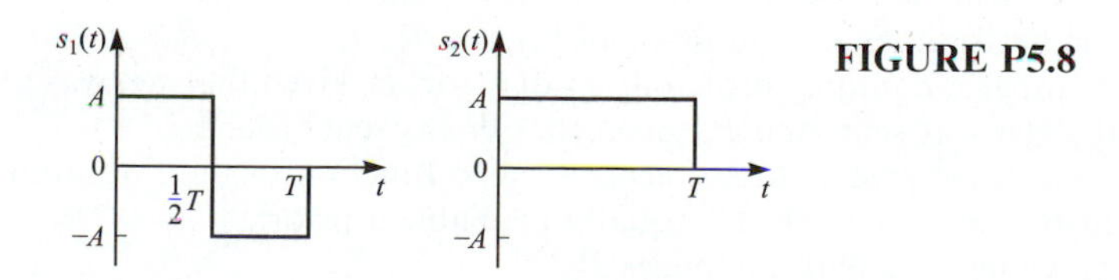

FIGURE P5.8

c) Sketch the noise-free response of the two matched filters when the transmitted signal is $s_2(t)$.
d) Suppose the receiver is implemented by means of two cross correlators (multipliers followed by integrators) in parallel. Sketch the output of each integrator as a function of time for the interval $0 \leqslant t \leqslant T$ when the transmitted signal is $s_2(t)$.
e) Compare the sketches in (c) and (d). Are they the same? Explain briefly.
f) From your knowledge of the signal characteristics, give the probability of error for this binary communication system.

5.9 Suppose that we have a complex-valued Gaussian random variable $z = x + jy$, where (x, y) are statistically independent variables with zero-mean and variance $E(x^2) = E(y^2) = \sigma^2$. Let

$$r = z + m, \qquad \text{where } m = m_r + jm_i$$

and define r as

$$r = a + jb$$

Clearly, $a = x + m_r$ and $b = y + m_i$. Determine the following probability density functions:
a) $p(a, b)$.
b) $p(u, \phi)$, where $u = \sqrt{a^2 + b^2}$ and $\phi = \tan^{-1} b/a$.
c) $p(u)$.
Note: In (b) it is convenient to define $\theta = \tan^{-1}(m_i/m_r)$ so that

$$m_r = \sqrt{m_r^2 + m_i^2}\cos\theta, \qquad m_i = \sqrt{m_r^2 + m_i^2}\sin\theta.$$

Furthermore, you must use the relation

$$\frac{1}{2\pi}\int_0^{2\pi} e^{\alpha\cos(\phi-\theta)}\,d\phi = I_0(\alpha) = \sum_{n=0}^{\infty}\frac{\alpha^{2n}}{2^{2n}(n!)^2}$$

where $I_0(\alpha)$ is the modified Bessel function of order zero.

5.10 A ternary communication system transmits one of three signals, $s(t)$, 0, or $-s(t)$, every T seconds. The received signal is either $r_l(t) = s(t) + z(t)$, $r_l(t) = z(t)$, or $r_l(t) = -s(t) + z(t)$, where $z(t)$ is white Gaussian noise with $E[z(t)] = 0$ and $\phi_{zz}(\tau) = \frac{1}{2}E[z(t)z^*(\tau)] = N_0\delta(t-\tau)$. The optimum receiver computes the correlation metric

$$U = \text{Re}\left[\int_0^T r_l(t)s^*(t)\,dt\right]$$

and compares U with a threshold A and a threshold $-A$. If $U > A$, the decision is made that $s(t)$ was sent. If $U < -A$, the decision is made in favor of $-s(t)$. If $-A < U < A$, the decision is made in favor of 0.
a) Determine the three conditional probabilities of error; P_e given that $s(t)$ was sent, P_e given that $-s(t)$ was sent, and P_e given that 0 was sent.
b) Determine the average probability of error P_e as a function of the threshold A, assuming that the three symbols are equally probable a priori.
c) Determine the value of A that minimizes P_e.

5.11 The two equivalent low-pass signals shown in Figure P5.11 are used to transmit a binary information sequence. The transmitted signals, which are equally probable, are corrupted by additive zero-mean white Gaussian noise having an equivalent low-pass representation $z(t)$ with an autocorrelation function

$$\begin{aligned}\phi_{zz}(\tau) &= \tfrac{1}{2}E[z^*(t)z(t+\tau)] \\ &= N_0\delta(\tau)\end{aligned}$$

a) What is the transmitted signal energy?
b) What is the probability of a binary digit error if coherent detection is employed at the receiver?
c) What is the probability of a binary digit error if noncoherent detection is employed at the receiver?

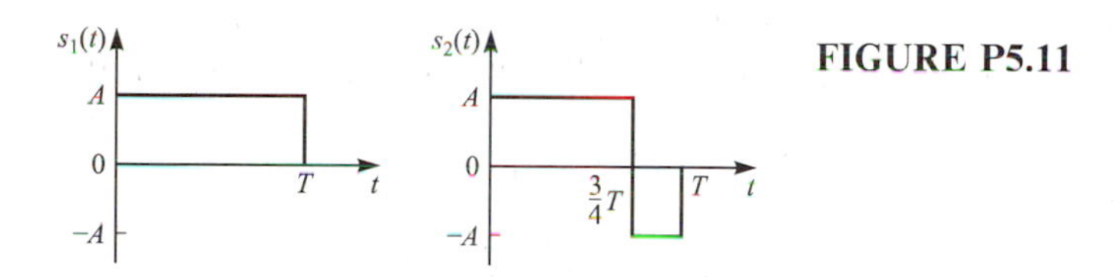

FIGURE P5.11

5.12 In Section 4.3.1 it wa shown that the minimum frequency separation for orthogonality of binary FSK signals with coherent detection is $\Delta f = 1/2T$. However, a lower error probability is possible with coherent detection of FSK if Δf is increased beyond $1/2T$. Show that the optimum value of Δf is $0.715/T$ and determine the probability of error for this value of Δf.

5.13 The equivalent low-pass waveforms for three signal sets are shown in Figure P5.13. Each set may be used to transmit one of four equally probable messages over an additive white Gaussian noise channel. The equivalent low-pass noise $z(t)$ has zero-mean and autocorrelation function $\phi_{zz}(\tau) = N_0\delta(\tau)$.

a) Classify the signal waveforms in sets I, II, and III. In other words, state the category or class to which each signal set belongs.
b) What is the *average* transmitted energy for each signal set?
c) For signal set I, specify the average probability of error if the signals are detected coherently.
d) For signal set II, give a union bound on the probability of a symbol error if the detection is performed (i) coherently and (ii) noncoherently.
e) Is it possible to use noncoherent detection on signal set III? Explain.
f) Which signal set or signal sets would you select if you wished to achieve a ratio of bit rate to bandwidth (R/W) of at least 2. *Briefly* explain your answer.

5.14 Consider a quaternary ($M = 4$) communication system that transmits, every T seconds, one of four equally probable signals: $s_1(t)$, $-s_1(t)$, $s_2(t)$, $-s_2(t)$. The signals $s_1(t)$ and $s_2(t)$ are orthogonal with equal energy. The additive noise is white Gaussian with zero-mean and autocorrelation function $\phi_{zz}(\tau) = N_0\delta(\tau)$. The demodulator consists of two filters matched to $s_1(t)$ and $s_2(t)$, and their outputs at the sampling instant are U_1 and U_2. The detector bases its decision on the following rule:

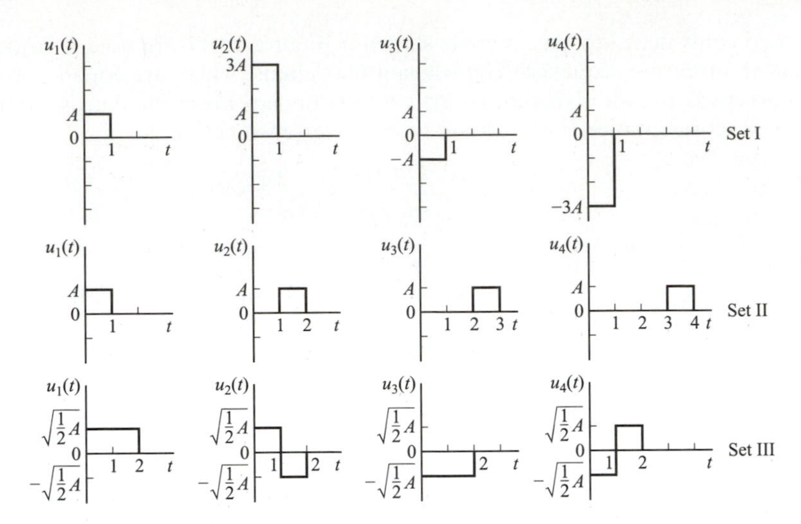

FIGURE P5.13

$$U_1 > |U_2| \Rightarrow s_1(t), \qquad U_1 < -|U_2| \Rightarrow -s_1(t)$$
$$U_2 > |U_1| \Rightarrow s_2(t), \qquad U_2 < -|U_1| \Rightarrow -s_2(t)$$

Since the signal set is biorthogonal, the error probability is given by $(1 - P_c)$ where P_c is given by Equation 5.2–34. Express this error probability in terms of a single integral and, thus, show that the symbol error probability for a biorthogonal signal set with $M = 4$ is identical to that for four-phase PSK. *Hint*: A change in variables from U_1 and U_2 to $W_1 = U_1 + U_2$ and $W_2 = U_1 - U_2$ simplifies the problem.

5.15 The input $s(t)$ to a band-pass filter is

$$s(t) = \text{Re}[s_0(t)e^{j2\pi f_c t}]$$

where $s_0(t)$ is a rectangular pulse as shown in Figure P5.15a.

a) Determine the output $y(t)$ of the band-pass filter for all $t \geqslant 0$ if the impulse response of the filter is

$$g(t) = \text{Re}[2h(t)e^{j2\pi f_c t}]$$

where $h(t)$ is an exponential as shown in Figure 5.15b.

b) Sketch the *equivalent low-pass output* of the filter.

c) When would you sample the output of the filter if you wished to have the maximum output at the sampling instant? What is the value of the maximum output?

d) Suppose that in addition to the input signal $s(t)$, there is additive white Gaussian noise

$$n(t) = \text{Re}[z(t)e^{j2\pi f_c t}]$$

where $\phi_{zz}(\tau) = N_0\delta(\tau)$. At the sampling instant determined in (c), the signal sample is corrupted by an additive Gaussian noise term. Determine its mean and variance.

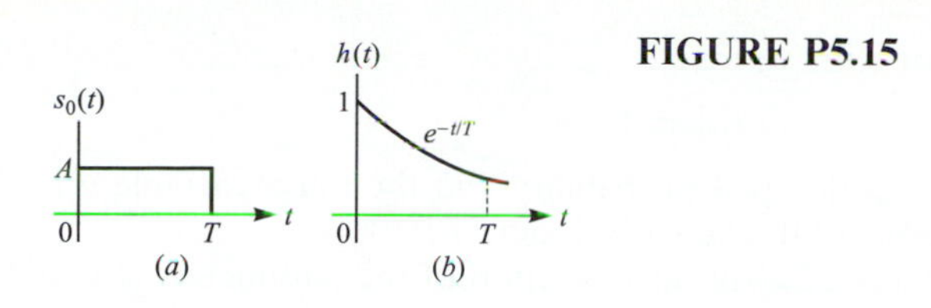

FIGURE P5.15

e) What is the signal-to-noise ratio γ of the sampled output?
f) Determine the signal-to-noise ratio when $h(t)$ is the matched filter to $s(t)$ and compare this result with the value of γ obtained in (e).

5.16 Consider the octal signal point constellations in Figure P5.16.
a) The nearest-neighbor signal points in the 8-QAM signal constellation are separated in distance by A units. Determine the radii a and b of the inner and outer circles.
b) The adjacent signal points in the 8-PSK are separated by a distance of A units. Determine the radius r of the circle.
c) Determine the average transmitter powers for the two signal constellations and compare the two powers. What is the relative power advantage of one constellation over the other? (Assume that all signal points are equally probable.)

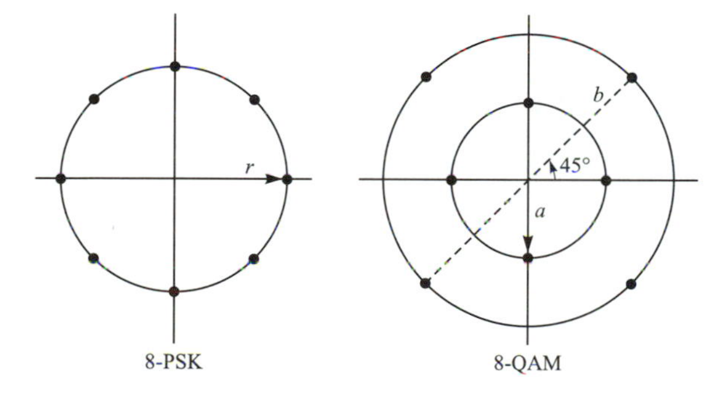

FIGURE P5.16

5.17 Consider the 8-point QAM signal constellation shown in Figure P5.16.
a) Is it possible to assign three data bits to each point of the signal constellation such that nearest (adjacent) points differ in only one bit position?
b) Determine the symbol rate if the desired bit rate is 90 Mbits/s.

5.18 Suppose that binary PSK is used for transmitting information over an AWGN with a power spectral density of $\frac{1}{2}N_0 = 10^{-10}$ W/Hz. The transmitted signal energy is $\mathcal{E}_b = \frac{1}{2}A^2T$, where T is the bit interval and A is the signal amplitude. Determine the signal amplitude required to achieve an error probability of 10^{-6} when the data rate is
a) 10 kbits/s.
b) 100 kbits/s.
c) 1 Mbit/s.

5.19 Consider a signal detector with an input

$$r = \pm A + n$$

where $+A$ and $-A$ occur with equal probability and the noise variable n is characterized by the (Laplacian) PDF shown in Figure P5.19.

a) Determine the probability of error as a function of the parameters A and σ.

b) Determine the SNR required to achieve an error probability of 10^{-5}. How does the SNR compare with the result for a Gaussian PDF?

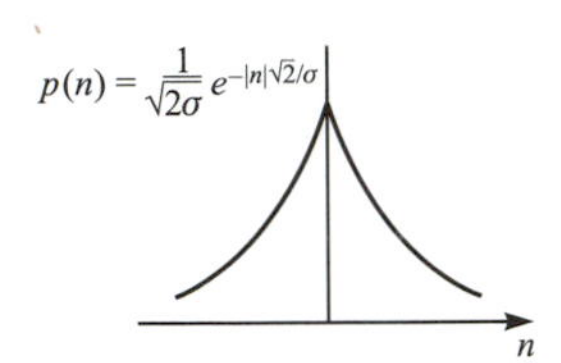

FIGURE P5.19

5.20 Consider the two 8-point QAM signal constellations shown in Figure P5.20. The minimum distance between adjacent points is $2A$. Determine the average transmitted power for each constellation, assuming that the signal points are equally probable. Which constellation is more power-efficient?

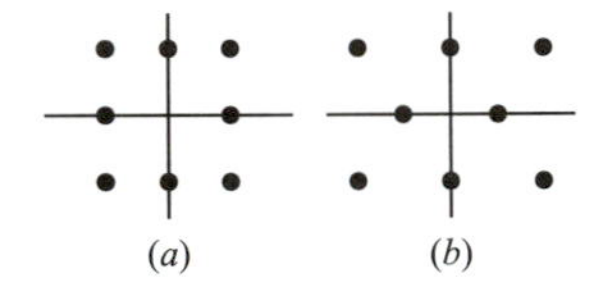

FIGURE P5.20

5.21 For the QAM signal constellation shown in Figure P5.21, determine the optimum decision boundaries for the detector, assuming that the SNR is sufficiently high so that errors only occur between adjacent points.

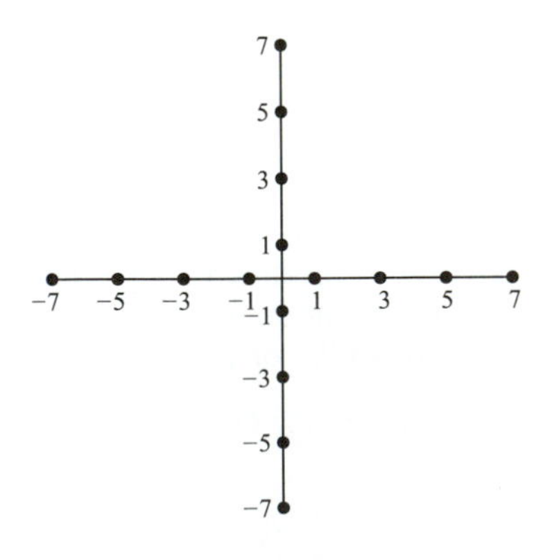

FIGURE P5.21

5.22 Specify a Gray code for the 16-QAM signal constellation shown in Figure P5.21.

5.23 Two quadrature carriers $\cos 2\pi f_c t$ and $\sin 2\pi f_c t$ are used to transmit digital information through an AWGN channel at two different data rates, 10 kbits/s and 100 kbits/s. Determine the relative amplitudes of the signals for the two carriers so that the $\mathcal{E}_b/N_b$ for the two channels is identical.

5.24 Three messages m_1, m_2, and m_3 are to be transmitted over an AWGN channel with noise power spectral density $\frac{1}{2}N_0$. The messages are

$$s_1(t) = \begin{cases} 1 & (0 \leqslant t \leqslant T) \\ 0 & (\text{otherwise}) \end{cases}$$

$$s_2(t) = -s_3(t) = \begin{cases} 1 & (0 \leqslant t \leqslant \frac{1}{2}T) \\ -1 & (\frac{1}{2}T \leqslant 0 \leqslant T) \\ 0 & (\text{otherwise}) \end{cases}$$

a) What is the dimensionality of the signal space?
b) Find an appropriate basis for the signal space. [*Hint*: You can find the basis without using the Gram–Schmidt procedure.]
c) Draw the signal constellation for this problem.
d) Derive and sketch the optimal decision regions R_1, R_2, and R_3.
e) Which of the three messages is more vulnerable to errors and why? In other words, which of $P(\text{error} \mid m_i \text{ transmitted})$, $i = 1, 2, 3$, is larger?

5.25 When the additive noise at the input to the demodulator is colored, the filter matched to the signal no longer maximizes the output SNR. In such a case we may consider the use of a prefilter that "whitens" the colored noise. The prefilter is followed by a filter matched to the prefiltered signal. Toward this end, consider the configuration shown in Figure P5.25.
a) Determine the frequency response characteristic of the prefilter that whitens the noise.
b) Determine the frequency response characteristic of the filter matched to $\tilde{s}(t)$.
c) Consider the prefilter and the matched filter as a single "generalized matched filter." What is the frequency response characteristic of this filter?
d) Determine the SNR at the input to the detector.

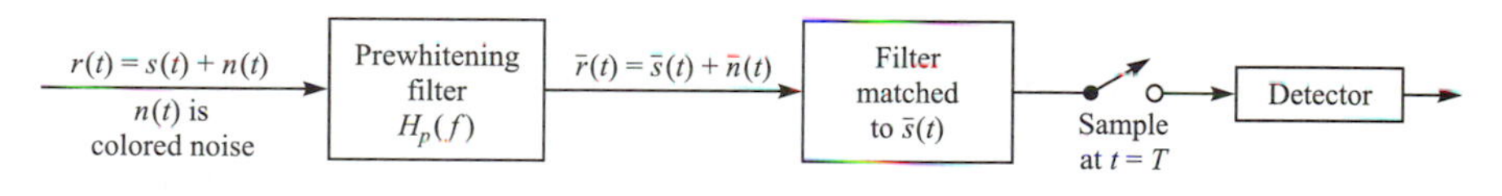

FIGURE P5.25

5.26 Consider a digital communication system that transmits information via QAM over a voice-band telephone channel at a rate of 2400 symbols/s. The additive noise is assumed to be white and Gaussian.
a) Determine the $\mathcal{E}_b/N_0$ required to achieve an error probability of 10^{-5} at 4800 bits/s.
b) Repeat (a) for a rate of 9600 bits/s.

c) Repeat (a) for a rate of 19,200 bits/s.
d) What conclusions do you reach from these results?

5.27 Consider the four-phase and eight-phase signal constellations shown in Figure P5.27. Determine the radii r_1 and r_2 of the circles such that the distance between two adjacent points in the two constellations is d. From this result, determine the additional transmitted energy required in the 8-PSK signal to achieve the same error probability as the four-phase signal at high SNR, where the probability of error is determined by errors in selecting adjacent points.

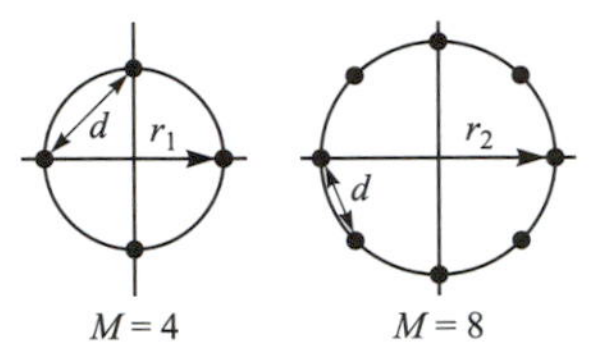

FIGURE P5.27

5.28 Digital information is to be transmitted by carrier modulation through an additive Gaussian noise channel with a bandwidth of 100 kHz and $N_0 = 10^{-10}$ W/Hz. Determine the maximum rate that can be transmitted through the channel for four-phase PSK, binary FSK, and four-frequency orthogonal FSK, which is detected noncoherently.

5.29 In a MSK signal, the initial state for the phase is either 0 or π rad. Determine the terminal phase state for the following four input pairs of input data:
a) 00.
b) 01.
c) 10.
d) 11.

5.30 A continuous-phase FSK signal with $h = \frac{1}{2}$ is represented as

$$s(t) = \pm\sqrt{\frac{2\mathcal{E}_b}{T_b}}\cos\left(\frac{\pi t}{2T_b}\right)\cos 2\pi f_c t \pm \sqrt{\frac{2\mathcal{E}_b}{T_b}}\sin\left(\frac{\pi t}{2T_b}\right)\sin 2\pi f_c t, \qquad 0 \leqslant t \leqslant 2T_b$$

where the $\pm$ signs depend on the information bits transmitted.
a) Show that this signal has constant amplitude.
b) Sketch a block diagram of the modulator for synthesizing the signal.
c) Sketch a block diagram of the demodulator and detector for recovering the information.

5.31 Sketch the state trellis and the state diagram for partial-response CPM with $h = \frac{1}{2}$ and

$$u(t) = \begin{cases} 1/4T & (0 \leqslant t \leqslant 2T) \\ 0 & (\text{otherwise}) \end{cases}$$

5.32 Determine the number of states in the state trellis diagram for
a) A full response binary CPFSK with either $h = \frac{2}{3}$ or $\frac{3}{4}$.
b) A partial-response $L = 3$ binary CPFSK with either $h = \frac{2}{3}$ or $\frac{3}{4}$.

5.33 Consider a biorthogonal signal set with $M = 8$ signal points. Determine a union bound for the probability of a symbol error as a function of $\mathcal{E}_b/N_0$. The signal points are equally likely a priori.

5.34 Consider an M-ary digital communication system where $M = 2^N$, and N is the dimension of the signal space. Suppose that the M signal vectors lie on the vertices of a hypercube that is centered at the origin. Determine the average probability of a symbol error as a function of $\mathcal{E}_s/N_0$ where $\mathcal{E}_s$ is the energy per symbol, $\frac{1}{2}N_0$ is the power spectral density of the AWGN, and all signal points are equally probable.

5.35 Consider the signal waveform

$$s(t) = \sum_{i=1}^{n} c_i p(t - iT_c)$$

where $p(t)$ is a rectangular pulse of unit amplitude and duration T_c. The $\{c_i\}$ may be viewed as a code vector $\mathbf{C} = [c_1 \; c_2 \cdots c_n]$, where the elements $c_i = \pm 1$. Show that the filter matched to the waveform $s(t)$ may be realized as a cascade of a filter matched to $p(t)$ followed by a discrete-time filter matched to the vector $\mathbf{C}$. Determine the value of the output of the matched filter at the sampling instant $t = nT_c$.

5.36 A speech signal is sampled at a rate of 8 kHz, logarithmically compressed and encoded into a PCM format using 8 bits per sample. The PCM data is transmitted through an AWGN baseband channel via M-level PAM. Determine the bandwidth required for transmission when
a) $M = 4$.
b) $M = 8$.
c) $M = 16$.

5.37 A Hadamard matrix is defined as a matrix whose elements are ± 1 and whose row vectors are pairwise orthogonal. In the case when n is a power of 2, an $n \times n$ Hadamard matrix is constructed by means of the recursion

$$\mathbf{H}_2 = \begin{bmatrix} 1 & 1 \\ 1 & -1 \end{bmatrix}, \qquad \mathbf{H}_{2n} = \begin{bmatrix} \mathbf{H}_n & \mathbf{H}_n \\ \mathbf{H}_n & -\mathbf{H}_n \end{bmatrix}$$

a) Let $\mathbf{C}_i$ denote the ith row of an $n \times n$ Hadamard matrix as defined above. Show that the waveforms constructed as

$$s_i(t) = \sum_{k=1}^{n} c_{ik} p(t - kT_c), \qquad i - 1, 2, \ldots, n$$

are orthogonal, where $p(t)$ is an arbitrary pulse confined to the time interval $0 \leqslant t \leqslant T_c$.
b) Show that the matched filters (or cross correlators) for the n waveforms $\{s_i(t)\}$ can be realized by a single filter (or correlator) matched to the pulse $p(t)$ followed by a set of n cross correlators using the code words $\{\mathbf{C}_i\}$.

5.38 The discrete sequence

$$r_k = \sqrt{\mathcal{E}_b} c_k + n_k, \qquad k = 1, 2, \ldots, n$$

represents the output sequence of samples from a demodulator, where $c_k = \pm 1$ are elements of one of two possible code words, $\mathbf{C}_1 = [1\ 1 \cdots 1]$ are $\mathbf{C}_2 = [1\ 1 \cdots 1\ \ -1 \cdots -1]$. The code word $\mathbf{C}_2$ has w elements that are $+1$ and $n - w$ elements that are -1, where w is some positive integer. The noise sequence $\{n_k\}$ is white Gaussian with variance σ^2.

a) What is the optimum maximum likelihood detector for the two possible transmitted signals?

b) Determine the probability of error as a function of the parameters $(\sigma^2, \mathcal{E}_b, w)$.

c) What is the value of w that minimizes the error probability?

5.39 Derive the outputs r_1 and r_2 shown in Figure 5.4–1 for a correlation-type demodulator. Assume that a signal $s_{l1}(t)$ is transmitted and that

$$r_l(t) = s_{l1}(t) e^{j\phi} + z(t)$$

where $z(t) = n_c(t) + jn_s(t)$ is the additive Gaussian noise.

5.40 Determine the covariances and variances of the Gaussian random noise variables n_{1c}, n_{2c}, n_{1s}, and n_{2s} in Equation 5.4–15 and the joint PDF.

5.41 Derive the matched filter outputs given by Equation 5.4–10.

5.42 In on–off keying of a carrier-modulated signal, the two possible signals are

$$s_0(t) = 0, \qquad 0 \leqslant t \leqslant T_b$$

$$s_1(t) \sqrt{\frac{2\mathcal{E}_b}{T_b}} \cos 2\pi f_c t, \qquad 0 \leqslant t \leqslant T_b$$

The corresponding received signals are

$$r(t) = n(t), \qquad 0 \leqslant t \leqslant T_b$$

$$r(t) = \sqrt{\frac{2\mathcal{E}_b}{T_b}} \cos(2\pi f_c t + \phi) + n(t), \qquad 0 \leqslant t \leqslant T_b$$

where ϕ is the carrier phase and $n(t)$ is AWGN.

a) Sketch a block diagram of the receiver (demodulator and detector) that employs noncoherent (envelope) detection.

b) Determine the PDFs for the two possible decision variables at the detector corresponding to the two possible received signals.

c) Derive the probability of error for the detector.

5.43 In two-phase DPSK, the received signal in one signaling interval is used as a phase reference for the received signal in the following signaling interval. The decision variable is

$$D = \text{Re}(V_m V_{m-1}^*) \underset{\text{“0”}}{\overset{\text{“1”}}{\gtrless}} 0$$

where

$$V_k = 2\mathcal{E}e^{j(\theta_k - \phi)} + N_k$$

represents the complex-valued output of the filter matched to the transmitted signal $u(t)$. N_k is a complex-valued Gaussian variable having zero-mean and statistically independent components.

a) Writing $V_k = X_k + jY_k$, show that D is equivalent to

$$d = \left[\tfrac{1}{2}(X_m + X_{m-1})\right]^2 + \left[\tfrac{1}{2}(Y_m + Y_{m-1})\right]^2 - \left[\tfrac{1}{2}(X_m - X_{m-1})\right]^2 - \left[\tfrac{1}{2}(Y_m - Y_{m-1})\right]^2$$

b) For mathematical convenience; suppose that $\theta_k = \theta_{k-1}$. Show that the random variables U_1, U_2, U_3, and U_4 are statistically independent Gaussian variables, where $U_1 = \frac{1}{2}(X_m + X_{m-1})$, $U_2 = \frac{1}{2}(Y_m + Y_{m-1})$, $U_3 = \frac{1}{2}(X_m - X_{m-1})$, and $U_4 = \frac{1}{2}(Y_m - Y_{m-1})$.

c) Define the random variables $W_1 = U_1^2 + U_2^2$ and $W_2 = U_3^2 + U_4^2$. Then

$$D = W_1 - W_2 \underset{\text{"0"}}{\overset{\text{"1"}}{\gtrless}} 0$$

Determine the probability density functions for W_1 and W_2.

d) Determine the probability of error P_b, where

$$P_b = P(D < 0) = P(W_1 - W_2 < 0) = \int_0^\infty P(W_2 > w_1 | w_1) p(w_1)\, dw_1$$

5.44 Recall that MSK can be represented as a four-phase offset PSK modulation having the low-pass equivalent form

$$v(t) = \sum_k [I_k u(t - 2kT_b) + jJ_k u(t - 2kT_b - T_b)]$$

where

$$u(t) = \begin{cases} \sin(\pi t/2T_b) & (0 \leqslant t \leqslant 2T_b) \\ 0 & (\text{otherwise}) \end{cases}$$

and $\{I_k\}$ and $\{J_k\}$ are sequences of information symbols (± 1).

a) Sketch the block diagram of an MSK demodulator for offset QPSK.

b) Evaluate the performance of the four-phase demodulator for AWGN if no account is taken of the memory in the modulation.

c) Compare the performance obtained in (b) with that for Viterbi decoding of the MSK signal.

d) The MSK signal is also equivalent to binary FSK. Determine the performance of noncoherent detection of the MSK signal. Compare your result with (b) and (c).

5.45 Consider a transmission line channel that employs $n - 1$ regenerative repeaters plus the terminal receiver in the transmission of binary information. Assume that the probability of error at the detector of each receiver is p and that errors among repeaters are statistically independent.

a) Show that the binary error probability at the terminal receiver is

$$P_n = \tfrac{1}{2}[1 - (1 - 2p)^n]$$

b) If $p = 10^{-6}$ and $n = 100$, determine an approximate value of P_n.

5.46 A digital communication system consists of a transmission line with 100 digital (regenerative) repeaters. Binary antipodal signals are used for transmitting the information. If the overall end-to-end error probability is 10^{-6}, determine the probability of error for each repeater and the required $\mathcal{E}_b/N_0$ to achieve this performance in AWGN.

5.47 A radio transmitter has a power output of $P_T = 1$ W at a frequency of 1 GHz. The transmitting and receiving antennas are parabolic dishes with diameter $D = 3$ m.

a) Determine the antenna gains.

b) Determine the EIRP for the transmitter.

c) The distance (free space) between the transmitting and receiving antennas is 20 km. Determine the signal power at the output of the receiving antenna in dBm.

5.48 A radio communication system transmits at a power level of 0.1 W at 1 GHz. The transmitting and receiving antennas are parabolic, each having a diameter of 1 m. The receiver is located 30 km from the transmitter.

a) Determine the gains of the transmitting and receiving antennas.

b) Determine the EIRP of the transmitted signal.

c) Determine the signal power from the receiving antenna.

5.49 A satellite in synchronous orbit is used to communicate with an earth station at a distance of 40,000 km. The satellite has an antenna with a gain of 15 dB and a transmitter power of 3 W. The earth station uses a 10-m parabolic antenna with an efficiency of 0.6. The frequency band is at $f = 10$ GHz. Determine the received power level at the output of the receiver antenna.

5.50 A spacecraft located 100,000 km from the earth is sending data at a rate of R bits/s. The frequency band is centered at 2 GHz and the transmitted power is 10 W. The earth station uses a parabolic antenna, 50 m in diameter, and the spacecraft has an antenna with a gain of 10 dB. The noise temperature of the receiver front end is $T_0 = 300$ K.

a) Determine the received power level.

b) If the desired $\mathcal{E}_b/N_0 = 10$ dB, determine the maximum bit rate that the spacecraft can transmit.

5.51 A satellite in geosynchronous orbit is used as a regenerative repeater in a digital communication system. Consider the satellite-to-earth link in which the satellite antenna has a gain of 6 dB and the earth station antenna has a gain of 50 dB. The downlink is operated at a center frequency of 4 GHz, and the signal bandwidth is 1 MHz. If the required $\mathcal{E}_b/N_0$ for reliable communication is 15 dB, determine the transmitted power for the satellite downlink. Assume that $N_0 = 4.1 \times 10^{-21}$ W/Hz.

6

Carrier and Symbol Synchronization

We have observed that in a digital communication system, the output of the demodulator must be sampled periodically, once per symbol interval, in order to recover the transmitted information. Since the propagation delay from the transmitter to the receiver is generally unknown at the receiver, symbol timing must be derived from the received signal in order to synchronously sample the output of the demodulator.

The propagation delay in the transmitted signal also results in a carrier offset, which must be estimated at the receiver if the detector is phase-coherent. In this chapter, we consider methods for deriving carrier and symbol synchronization at the receiver.

6.1 SIGNAL PARAMETER ESTIMATION

Let us begin by developing a mathematical model for the signal at the input to the receiver. We assume that the channel delays the signals transmitted through it and corrupts them by the addition of Gaussian noise. Hence, the received signal may be expressed as

$$r(t) = s(t-\tau) + n(t)$$

where

$$s(t) = \mathrm{Re}[s_l(t)e^{j2\pi f_c t}] \tag{6.1–1}$$

and where τ is the propagation delay and $s_l(t)$ is the equivalent low-pass signal.

The received signal may be expressed as

$$r(t) = \mathrm{Re}\{[s_l(t-\tau)e^{j\phi} + z(t)]e^{j2\pi f_c t}\} \tag{6.1–2}$$

where the carrier phase ϕ, due to the propagation delay τ, is $\phi = -2\pi f_c \tau$. Now, from this formulation, it may appear that there is only one signal parameter to be estimated, namely, the propagation delay, since one can determine ϕ from knowledge of f_c and τ. However, this is not the case. First of all, the oscillator that generates the carrier signal for demodulation at the receiver is generally not synchronous in phase with that at the transmitter. Furthermore, the two oscillators may be drifting slowly with time, perhaps in different directions. Consequently, the received carrier phase is not only dependent on the time delay τ. Furthermore, the precision to which one must synchronize in time for the purpose of demodulating the received signal depends on the symbol interval T. Usually, the estimation error in estimating τ must be a relatively small fraction of T. For example, ± 1 percent of T is adequate for practical applications. However, this level of precision is generally inadequate for estimating the carrier phase, even if ϕ depends only on τ. This is due to the fact that f_c is generally large, and, hence, a small estimation error in τ causes a large phase error.

In effect, we must estimate both parameters τ and ϕ in order to demodulate and coherently detect the received signal. Hence, we may express the received signal as

$$r(t) = s(t; \phi, \tau) + n(t) \tag{6.1–3}$$

where ϕ and τ represent the signal parameters to be estimated. To simplify the notation, we let $\boldsymbol{\psi}$ denote the parameter vector $\{\phi, \tau\}$, so that $s(t; \phi, \tau)$ is simply denoted by $s(t; \boldsymbol{\psi})$.

There are basically two criteria that are widely applied to signal parameter estimation: the *maximum-likelihood* (ML) criterion and the *maximum a posteriori probability* (MAP) criterion. In the MAP criterion, the signal parameter vector $\boldsymbol{\psi}$ is modeled as random and characterized by an a priori probability density function $p(\boldsymbol{\psi})$. In the maximum-likelihood criterion, the signal parameter vector $\boldsymbol{\psi}$ is treated as deterministic but unknown.

By performing an orthonormal expansion of $r(t)$ using N orthonormal functions $\{f_n(t)\}$, we may represent $r(t)$ by the vector of coefficients $[r_1\, r_2 \cdots r_N] \equiv \mathbf{r}$. The joint PDF of the random variables $[r_1\, r_2 \cdots r_N]$ in the expansion can be expressed as $p(\mathbf{r}|\boldsymbol{\psi})$. Then, the ML estimate of $\boldsymbol{\psi}$ is the value that maximizes $p(\mathbf{r}|\boldsymbol{\psi})$. On the other hand, the MAP estimate is the value of $\boldsymbol{\psi}$ that maximizes the a posteriori probability density function

$$p(\boldsymbol{\psi}|\mathbf{r}) = \frac{p(\mathbf{r}|\boldsymbol{\psi})p(\boldsymbol{\psi})}{p(\mathbf{r})} \tag{6.1–4}$$

We note that if there is no prior knowledge of the parameter vector $\boldsymbol{\psi}$, we may assume that $p(\boldsymbol{\psi})$ is uniform (constant) over the range of values of the parameters. In such a case, the value of $\boldsymbol{\psi}$ that maximizes $p(\mathbf{r}|\boldsymbol{\psi})$ also maximizes $p(\boldsymbol{\psi}|\mathbf{r})$. Therefore, the MAP and ML estimates are identical.

In our treatment of parameter estimation given below, we view the parameters ϕ and τ as unknown, but deterministic. Hence, we adopt the ML criterion for estimating them.

In the ML estimation of signal parameters, we require that the receiver extract the estimate by observing the received signal over a time interval $T_0 \geqslant T$, which is called the observation interval. Estimates obtained from a single observation interval are sometimes called one-shot estimates. In practice, however, the estimation is performed on a continuous basis by using tracking loops (either analog or digital) that continuously update the estimates. Nevertheless, one-shot estimates yield insight for tracking loop implementation. In addition, they prove useful in the analysis of the performance of ML estimation, and their performance can be related to that obtained with a tracking loop.

6.1.1 The Likelihood Function

Although it is possible to derive the parameter estimates based on the joint PDF of the random variables $[r_1\, r_2 \cdots r_N]$ obtained from the expansion of $r(t)$, it is convenient to deal directly with the signal waveforms when estimating their parameters. Hence, we shall develop a continuous-time equivalent of the maximization of $p(\mathbf{r}|\psi)$.

Since the additive noise $n(t)$ is white and zero-mean Gaussian, the joint PDF $p(\mathbf{r}|\psi)$ may be expressed as

$$p(\mathbf{r}|\psi) = \left(\frac{1}{\sqrt{2\pi}\sigma}\right)^N \exp\left\{-\sum_{n=1}^{N}\frac{[r_n - s_n(\psi)]^2}{2\sigma^2}\right\} \tag{6.1–5}$$

where

$$\begin{aligned} r_n &= \int_{T_0} r(t) f_n(t)\, dt \\ s_n(\psi) &= \int_{T_0} s(t;\psi) f_n(t)\, dt \end{aligned} \tag{6.1–6}$$

where T_0 represents the integration interval in the expansion of $r(t)$ and $s(t;\psi)$.

We note that the argument in the exponent may be expressed in terms of the signal waveforms $r(t)$ and $s(t;\psi)$, by substituting from Equation 6.1–6 into Equation 6.1–5. That is,

$$\lim_{N\to\infty}\frac{1}{2\sigma^2}\sum_{n=1}^{N}[r_n - s_n(\psi)]^2 = \frac{1}{N_0}\int_{T_0}[r(t) - s(t;\psi)]^2\, dt \tag{6.1–7}$$

where the proof is left as an exercise for the reader (see Problem 6.1). Now, the maximization of $p(\mathbf{r}|\psi)$ with respect to the signal parameters ψ is equivalent to the maximization of the *likelihood function*.

$$\Lambda(\psi) = \exp\left\{-\frac{1}{N_0}\int_{T_0}[r(t) - s(t;\psi)]^2\, dt\right\} \tag{6.1–8}$$

Below, we shall consider signal parameter estimation from the viewpoint of maximizing $\Lambda(\psi)$.

6.1.2 Carrier Recovery and Symbol Synchronization in Signal Demodulation

Symbol synchronization is required in every digital communication system which transmits information synchronously. Carrier recovery is required if the signal is detected coherently.

Figure 6.1–1 illustrates the block diagram of a binary PSK (or binary PAM) signal demodulator and detector. As shown, the carrier phase estimate $\hat{\phi}$ is used in generating the reference signal $g(t)\cos(2\pi f_c t + \hat{\phi})$ for the correlator. The symbol synchronizer controls the sampler and the output of the signal pulse generator. If the signal pulse is rectangular, then the signal generator can be eliminated.

The block diagram of an M-ary PSK demodulator is shown in Figure 6.1–2. In this case, two correlators (or matched filters) are required to correlate the received signal with the two quadrature carrier signals $g(t)\cos(2\pi f_c t + \hat{\phi})$ and $g(t)\sin(2\pi f_c t + \hat{\phi})$, where $\hat{\phi}$ is the carrier phase estimate. The detector is now a phase detector, which compares the received signal phases with the possible transmitted signal phases.

The block diagram of a PAM signal demodulator is shown in Figure 6.1–3. In this case, a single correlator is required, and the detector is an amplitude detector, which compares the received signal amplitude with the possible transmitted signal amplitudes. Note that we have included an automatic gain control (AGC) at the front end of the demodulator to eliminate channel gain variations, which would affect the amplitude detector. The AGC has a relatively long time constant, so that it does not respond to the signal amplitude variations that occur on a symbol-by-symbol basis. Instead, the AGC maintains a fixed average (signal plus noise) power at its output.

Finally, we illustrate the block diagram of a QAM demodulator in Figure 6.1–4. As in the case of PAM, an AGC is required to maintain a constant average power signal at the input to the demodulator. We observe that the demodulator

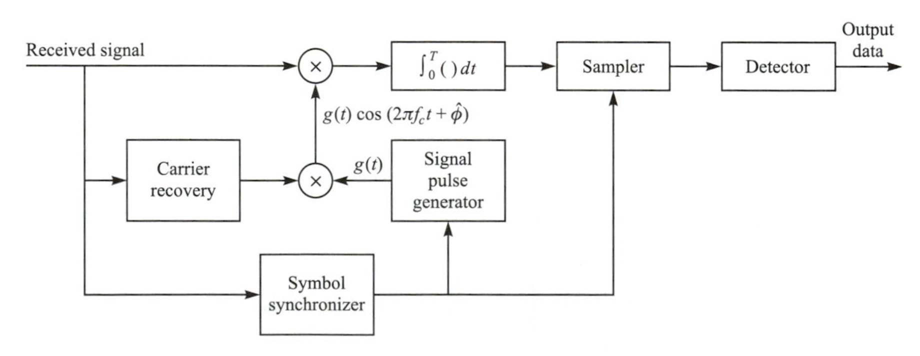

FIGURE 6.1–1
Block diagram of a binary PSK receiver.

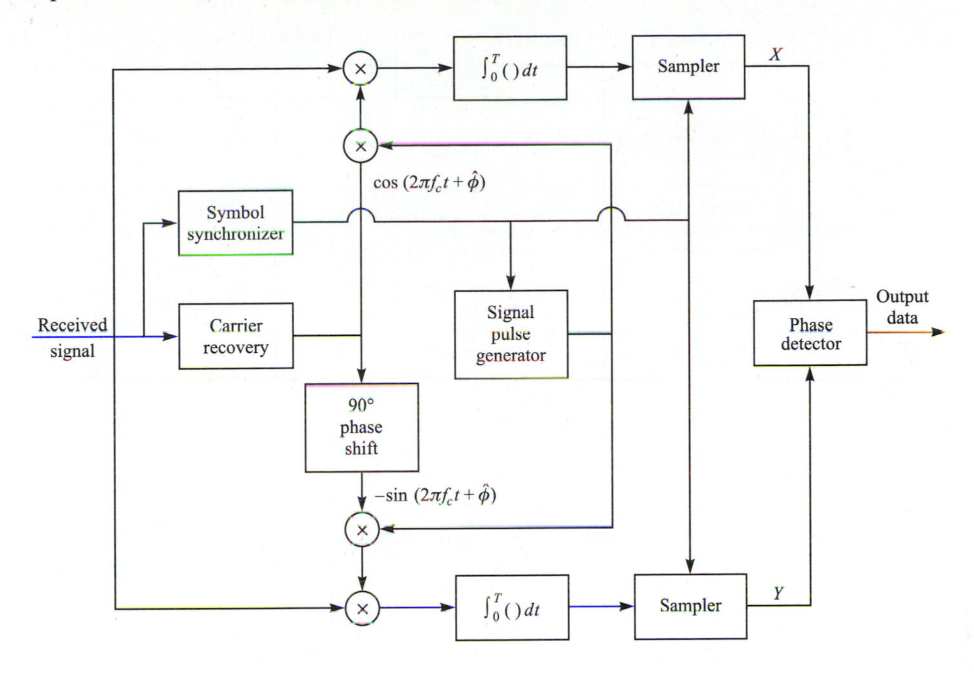

FIGURE 6.1–2
Block diagram of an M-ary PSK receiver.

is similar to a PSK demodulator, in that both generate in-phase and quadrature signal samples (X, Y) for the detector. In the case of QAM, the detector computes the Euclidean distance between the received noise-corrupted signal point and the M possible transmitted points, and selects the signal closest to the received point.

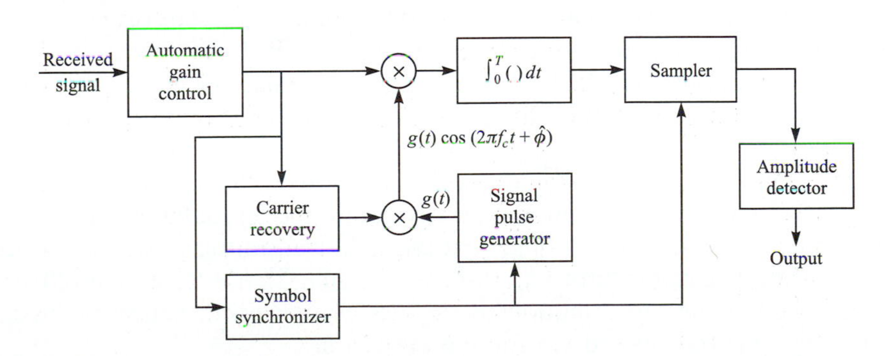

FIGURE 6.1–3
Block diagram of an M-ary PAM receiver.

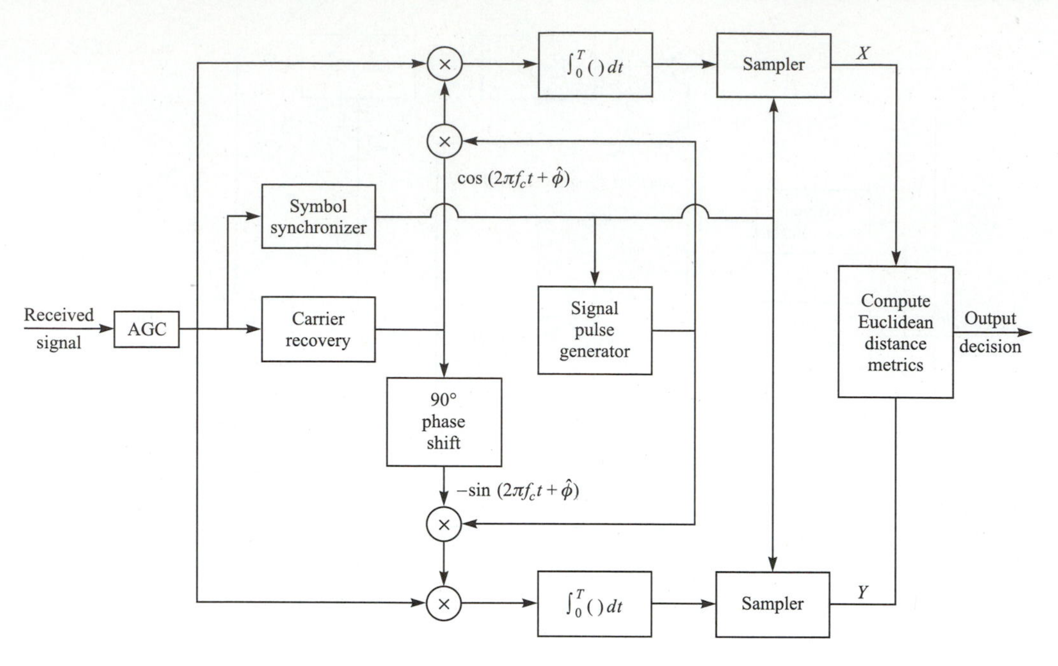

FIGURE 6.1–4
Block diagram of a QAM receiver.

6.2 CARRIER PHASE ESTIMATION

There are two basic approaches for dealing with carrier synchronization at the receiver. One is to multiplex, usually in frequency, a special signal, called a pilot signal, that allows the receiver to extract and, thus, to synchronize its local oscillator to the carrier frequency and phase of the received signal. When an unmodulated carrier component is transmitted along with the information-bearing signal, the receiver employs a phase-locked loop (PLL) to acquire and track the carrier component. The PLL is designed to have a narrow bandwidth so that it is not significantly affected by the presence of frequency components from the information-bearing signal.

The second approach, which appears to be more prevalent in practice, is to derive the carrier phase estimate directly from the modulated signal. This approach has the distinct advantage that the total transmitter power is allocated to the transmission of the information-bearing signal. In our treatment of carrier recovery, we confine our attention to the second approach; hence, we assume that the signal is transmitted via suppressed carrier.

In order to emphasize the importance of extracting an accurate phase estimate, let us consider the effect of a carrier phase error on the demodulation of a double-sideband, suppressed carrier (DSB/SC) signal. To be specific, suppose we have an amplitude-modulated signal of the form

$$s(t) = A(t)\cos(2\pi f_c t + \phi) \tag{6.2–1}$$

If we demodulate the signal by multiplying $s(t)$ with the carrier reference

$$c(t) = \cos(2\pi f_c t + \hat{\phi}) \tag{6.2–2}$$

we obtain

$$c(t)s(t) = \tfrac{1}{2}A(t)\cos(\phi - \hat{\phi}) + \tfrac{1}{2}A(t)\cos(4\pi f_c t + \phi + \hat{\phi})$$

The double-frequency component may be removed by passing the product signal $c(t)s(t)$ through a low-pass filter. This filtering yields the information-bearing signal

$$y(t) = \tfrac{1}{2}A(t)\cos(\phi - \hat{\phi}) \tag{6.2–3}$$

Note that the effect of the phase error $\phi - \hat{\phi}$ is to reduce the signal level in voltage by a factor $\cos(\phi - \hat{\phi})$ and in power by a factor $\cos^2(\phi - \hat{\phi})$. Hence, a phase error of 10° results in a signal power loss of 0.13 dB, and a phase error of 30° results in a signal power loss of 1.25 dB in an amplitude-modulated signal.

The effect of carrier phase errors in QAM and multiphase PSK is much more severe. The QAM and M-PSK signals may be represented as

$$s(t) = A(t)\cos(2\pi f_c t + \phi) - B(t)\sin(2\pi f_c t + \phi) \tag{6.2–4}$$

This signal is demodulated by the two quadrature carriers

$$\begin{aligned} c_c(t) &= \cos(2\pi f_c t + \hat{\phi}) \\ c_s(t) &= -\sin(2\pi f_c t + \hat{\phi}) \end{aligned} \tag{6.2–5}$$

Multiplication of $s(t)$ with $c_c(t)$ followed by low-pass filtering yields the in-phase component

$$y_I(t) = \tfrac{1}{2}A(t)\cos(\phi - \hat{\phi}) - \tfrac{1}{2}B(t)\sin(\phi - \hat{\phi}) \tag{6.2–6}$$

Similarly, multiplication of $s(t)$ by $c_s(t)$ followed by low-pass filtering yields the quadrature component

$$y_Q(t) = \tfrac{1}{2}B(t)\cos(\phi - \hat{\phi}) + \tfrac{1}{2}A(t)\sin(\phi - \hat{\phi}) \tag{6.2–7}$$

The expressions 6.2–6 and 6.2–7 clearly indicate that the phase error in the demodulation of QAM and M-PSK signals has a much more severe effect than in the demodulation of a PAM signal. Not only is there a reduction in the power of the desired signal component by a factor $\cos^2(\phi - \hat{\phi})$, but there is also crosstalk interference from the in-phase and quadrature components. Since the average power levels of $A(t)$ and $B(t)$ are similar, a small phase error causes a large degradation in performance. Hence, the phase accuracy requirements for QAM and multiphase coherent PSK are much higher than for DSB/SC PAM.

6.2.1 Maximum-Likelihood Carrier Phase Estimation

First, we derive the maximum-likelihood carrier phase estimate. For simplicity, we assume that the delay τ is known and, in particular, we set $\tau = 0$. The function to be maximized is the likelihood function given in Equation 6.1–8. With ϕ substituted for ψ, this function becomes

$$\begin{aligned}\Lambda(\phi) &= \exp\left\{-\frac{1}{N_0}\int_{T_0}[r(t)-s(t;\phi)]^2\,dt\right\} \\ &= \exp\left\{-\frac{1}{N_0}\int_{T_0} r^2(t)\,dt + \frac{2}{N_0}\int_{T_0} r(t)s(t;\phi)\,dt - \frac{1}{N_0}\int_{T_0} s^2(t;\phi)\,dt\right\}\end{aligned} \tag{6.2–8}$$

Note that the first term of the exponential factor does not involve the signal parameter ϕ. The third term, which contains the integral of $s^2(t;\phi)$, is a constant equal to the signal energy over the observation interval T_0 for any value of ϕ. Only the second term, which involves the cross correlation of the received signal $r(t)$ with the signal $s(t;\phi)$, depends on the choice of ϕ. Therefore, the likelihood function $\Lambda(\phi)$ may be expressed as

$$\Lambda(\phi) = C\exp\left[\frac{2}{N_0}\int_{T_0} r(t)s(t;\phi)\,dt\right] \tag{6.2–9}$$

where C is a constant independent of ϕ.

The ML estimate $\hat{\phi}_{ML}$ is the value of ϕ that maximizes $\Lambda(\phi)$ in Equation 6.2–9. Equivalently, the value $\hat{\phi}_{ML}$ also maximizes the logarithm of $\Lambda(\phi)$, i.e., the log-likelihood function

$$\Lambda_L(\phi) = \frac{2}{N_0}\int_{T_0} r(t)s(t;\phi)\,dt \tag{6.2–10}$$

Note that in defining $\Lambda_L(\phi)$ we have ignored the constant term $\ln C$.

EXAMPLE 6.2–1. As an example of the optimization to determine the carrier phase, let us consider the transmission of the unmodulated carrier $A\cos 2\pi f_c t$. The received signal is

$$r(t) = A\cos(2\pi f_c t + \phi) + n(t)$$

where ϕ is the unknown phase. We seek the value ϕ, say $\hat{\phi}_{ML}$, that maximizes

$$\Lambda_L(\phi) = \frac{2A}{N_0}\int_{T_0} r(t)\cos(2\pi f_c t + \phi)\,dt$$

A necessary condition for a maximum is that

$$\frac{d\Lambda_L(\phi)}{d\phi} = 0$$

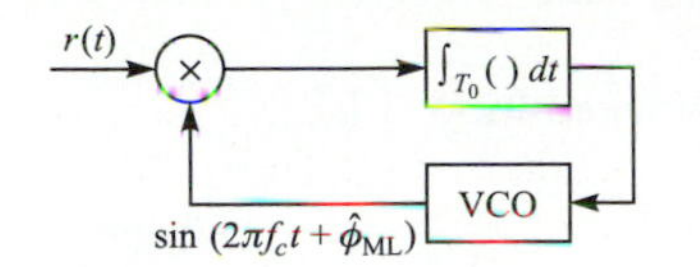

FIGURE 6.2–1
A PLL for obtaining the ML estimate of the phase of an unmodulated carrier.

This condition yields

$$\int_{T_0} r(t) \sin(2\pi f_c t + \hat{\phi}_{\mathrm{ML}})\, dt = 0 \tag{6.2–11}$$

or, equivalently,

$$\hat{\phi}_{\mathrm{ML}} = -\tan^{-1}\left[\int_{T_0} r(t) \sin 2\pi f_c t\, dt \Big/ \int_{T_0} r(t) \cos 2\pi f_c t\, dt\right] \tag{6.2–12}$$

We observe that the optimality condition given by Equation 6.2–11 implies the use of a loop to extract the estimate as illustrated in Figure 6.2–1. The loop filter is an integrator whose bandwidth is proportional to the reciprocal of the integration interval T_0. On the other hand, Equation 6.2–12 implies an implementation that uses quadrature carriers to cross-correlate with $r(t)$. Then $\hat{\phi}_{\mathrm{ML}}$ is the inverse tangent of the ratio of these two correlator outputs, as shown in Figure 6.2–2. Note that this estimation scheme yields $\hat{\phi}_{\mathrm{ML}}$ explicitly.

This example clearly demonstrates that the PLL provides the ML estimate of the phase of an unmodulated carrier.

6.2.2 The Phase-Locked Loop

The PLL basically consists of a multiplier, a loop filter, and a voltage-controlled oscillator (VCO), as shown in Figure 6.2–3. If we assume that the input to the PLL is the sinusoid $\cos(2\pi f_c t + \phi)$ and the output of the VCO is $\sin(2\pi f_c t + \hat{\phi})$, where $\hat{\phi}$ represents the estimate of ϕ, the product of these two signals is

$$\begin{aligned} e(t) &= \cos(2\pi f_c t + \phi) \sin(2\pi f_c t + \hat{\phi}) \\ &= \tfrac{1}{2}\sin(\hat{\phi} - \phi) + \tfrac{1}{2}\sin(4\pi f_c t + \phi + \hat{\phi}) \end{aligned} \tag{6.2–13}$$

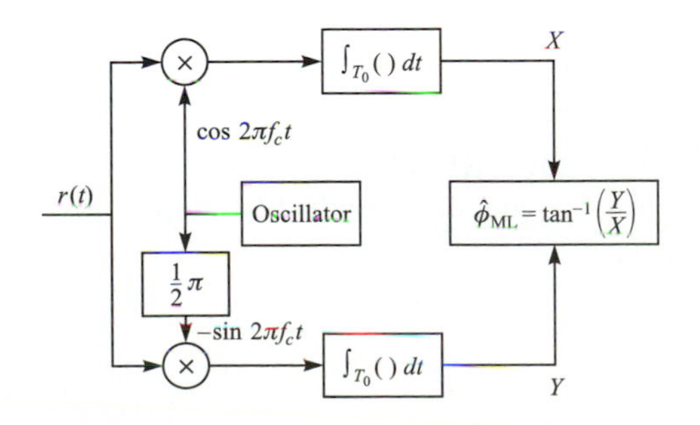

FIGURE 6.2–2
A (one-shot) ML estimate of the phase of an unmodulated carrier.

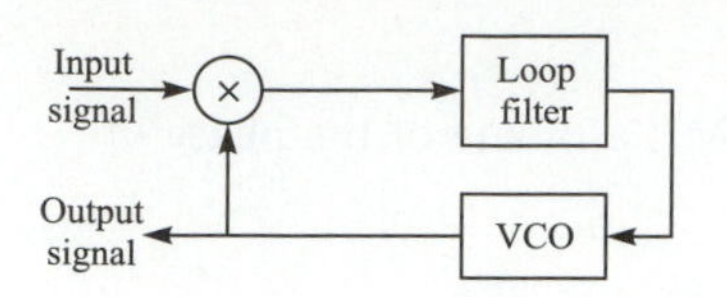

FIGURE 6.2–3
Basic elements of a phase-locked loop (PLL).

The loop filter is a low-pass filter that responds only to the low-frequency component $\frac{1}{2}\sin(\hat{\phi} - \phi)$ and removes the component at $2f_c$. This filter is usually selected to have the relatively simple transfer function

$$G(s) = \frac{1 + \tau_2 s}{1 + \tau_1 s} \tag{6.2–14}$$

where τ_1 and τ_2 are design parameters ($\tau_1 \gg \tau_2$) that control the bandwidth of the loop. A higher-order filter that contains additional poles may be used if necessary to obtain a better loop response.

The output of the loop filter provides the control voltage $v(t)$ for the VCO. The VCO is basically a sinusoidal signal generator with an instantaneous phase given by

$$2\pi f_c t + \hat{\phi}(t) = 2\pi f_c t + K \int_{-\infty}^{t} v(\tau)\, d\tau \tag{6.2–15}$$

where K is a gain constant in rad/V. Hence,

$$\hat{\phi}(t) = K \int_{-\infty}^{t} v(\tau)\, d\tau \tag{6.2–16}$$

By neglecting the double-frequency term resulting from the multiplication of the input signal with the output of the VCO, we may reduce the PLL into the equivalent closed-loop system model shown in Figure 6.2–4. The sine function of the phase difference $\hat{\phi} - \phi$ makes this system non-linear, and, as a consequence, the analysis of its performance in the presence of noise is somewhat involved, but, nevertheless, it is mathematically tractable for some simple loop filters.

In normal operation when the loop is tracking the phase of the incoming carrier, the phase error $\hat{\phi} - \phi$ is small and, hence,

$$\sin(\hat{\phi} - \phi) \approx \hat{\phi} - \phi \tag{6.2–17}$$

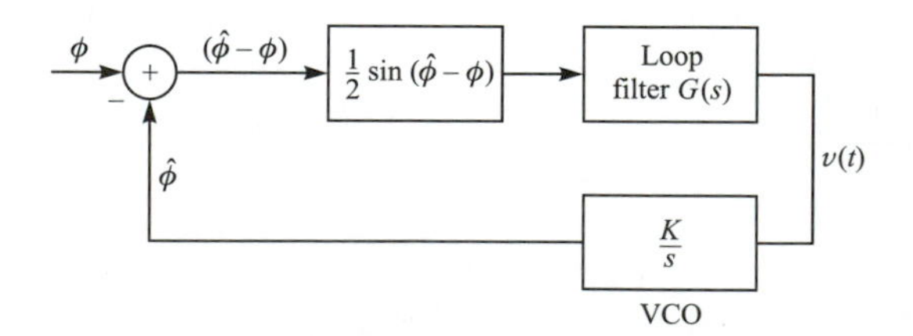

FIGURE 6.2–4
Model of phase-locked loop.

With this approximation, the PLL becomes linear and is characterized by the closed-loop transfer function

$$H(s) = \frac{KG(s)/s}{1 + KG(s)/s} \tag{6.2–18}$$

where the factor of $\frac{1}{2}$ has been absorbed into the gain parameter K. By substituting from Equation 6.2–14 for $G(s)$ into Equation 6.2–18, we obtain

$$H(s) = \frac{1 + \tau_2 s}{1 + (\tau_2 + 1/K)s + (\tau_1/K)s^2} \tag{6.2–19}$$

Hence, the closed-loop system for the linearized PLL is second-order when $G(s)$ is given by Equation 6.2–14. The parameter τ_2 controls the position of the zero, while K and τ_1 are used to control the position of the closed-loop system poles. It is customary to express the denominator of $H(s)$ in the standard form

$$D(s) = s^2 + 2\zeta\omega_n s + \omega_n^2 \tag{6.2–20}$$

where ζ is called the *loop damping factor* and ω_n is the natural frequency of the loop. In terms of the loop parameters, $\omega_n = \sqrt{K/\tau_1}$, and $\zeta = \omega_n(\tau_2 + 1/K)/2$, the closed-loop transfer function becomes

$$H(s) = \frac{(2\zeta\omega_n - \omega_n^2/K)s + \omega_n^2}{s^2 + 2\zeta\omega_n s + \omega_n^2} \tag{6.2–21}$$

The (one-sided) noise-equivalent bandwidth (see Problem 2.24) of the loop is

$$\begin{aligned} B_{\text{eq}} &= \frac{\tau_2^2(1/\tau_2^2 + K/\tau_1)}{4(\tau_2 + 1/K)} \\ &= \frac{1 + (\tau_2\omega_n)^2}{8\zeta/\omega_n} \end{aligned} \tag{6.2–22}$$

The magnitude response 20 log $|H(\omega)|$ as a function of the normalized frequency ω/ω_n is illustrated in Figure 6.2–5, with the damping factor ζ as a parameter and $\tau_1 \gg 1$. Note that $\zeta = 1$ results in a critically damped loop response, $\zeta < 1$ produces an underdamped response, and $\zeta > 1$ yields an overdamped response.

In practice, the selection of the bandwidth of the PLL involves a trade-off between speed of response and noise in the phase estimate, which is the topic considered below. On the one hand, it is desirable to select the bandwidth of the loop to be sufficiently wide to track any time variations in the phase of the received carrier. On the other hand, a wideband PLL allows more noise to pass into the loop, which corrupts the phase estimate. Below, we assess the effects of noise in the quality of the phase estimate.

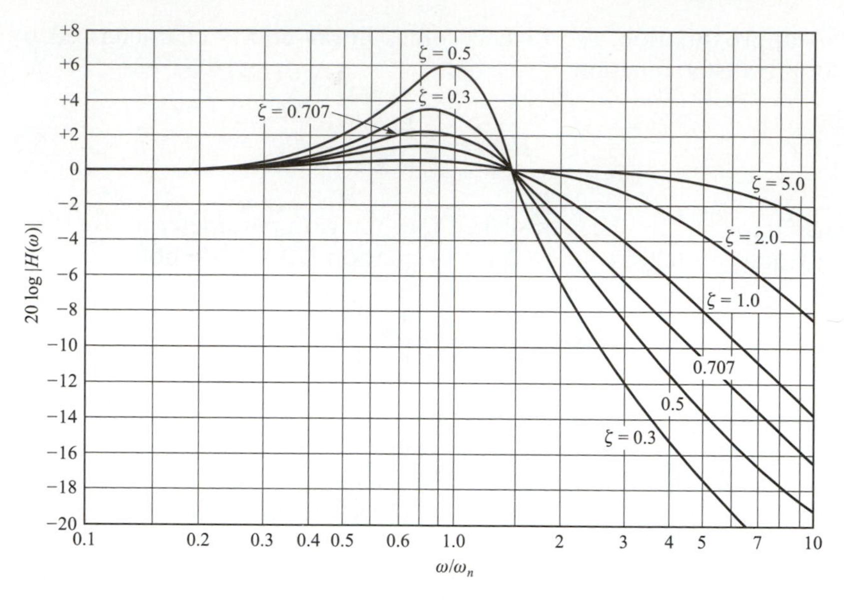

FIGURE 6.2–5
Frequency response of a second-order loop. [*From* Phaselock Techniques, *2nd edition, by F. M. Gardner, © 1979 by John Wiley and Sons, Inc. Reprinted with permission of the publisher.*]

6.2.3 Effect of Additive Noise on the Phase Estimate

In order to evaluate the effects of noise on the estimate of the carrier phase, let us assume that the noise at the input to the PLL is narrowband. For this analysis, we assume that the PLL is tracking a sinusoidal signal of the form

$$s(t) = A_c \cos[2\pi f_c t + \phi(t)] \tag{6.2–23}$$

that is corrupted by the additive narrowband noise

$$n(t) = x(t) \cos 2\pi f_c t - y(t) \sin 2\pi f_c t \tag{6.2–24}$$

The in-phase and quadrature components of the noise are assumed to be statistically independent, stationary Gaussian noise processes with (two-sided) power spectral density $\frac{1}{2}N_0$ W/Hz. By using simple trigonometric identities, the noise term in Equation 6.2–24 can be expressed as

$$n(t) = n_c(t) \cos[2\pi f_c t + \phi(t)] - n_s(t) \sin[2\pi f_c t + \phi(t)] \tag{6.2–25}$$

where

$$\begin{aligned} n_c(t) &= x(t) \cos \phi(t) + y(t) \sin \phi(t) \\ n_s(t) &= -x(t) \sin \phi(t) + y(t) \cos \phi(t) \end{aligned} \tag{6.2–26}$$

We note that

$$n_c(t) + jn_s(t) = [x(t) + jy(t)]e^{-j\phi(t)}$$

so that the quadrature components $n_c(t)$ and $n_s(t)$ have exactly the same statistical characteristics as $x(t)$ and $y(t)$.

If $s(t) + n(t)$ is multiplied by the output of the VCO and the double-frequency terms are neglected, the input to the loop filter is the noise-corrupted signal

$$\begin{aligned} e(t) &= A_c \sin \Delta\phi + n_c(t) \sin \Delta\phi - n_s(t) \cos \Delta\phi \\ &= A_c \sin \Delta\phi + n_1(t) \end{aligned} \tag{6.2–27}$$

where, by definition $\Delta\phi = \hat{\phi} - \phi$ is the phase error. Thus, we have the equivalent model for the PLL with additive noise as shown in Figure 6.2–6.

When the power $P_c = \frac{1}{2}A_c^2$ of the incoming signal is much larger than the noise power, we may linearize the PLL and, thus, easily determine the effect of the additive noise on the quality of the estimate $\hat{\phi}$. Under these conditions, the model for the linearized PLL with additive noise is illustrated in Figure 6.2–7. Note that the gain parameter A_c may be normalized to unity, provided that the noise terms are scaled by $1/A_c$, i.e., the noise terms become

$$n_2(t) = \frac{n_c(t)}{A_c} \sin \Delta\phi - \frac{n_s(t)}{A_c} \cos \Delta\phi \tag{6.2–28}$$

The noise term $n_2(t)$ is zero-mean Gaussian with a power spectral density $N_0/2A_c^2$. Since the noise $n_2(t)$ is additive at the input to the loop, the variance of the phase error $\Delta\phi$, which is also the variance of the VCO output phase, is

$$\begin{aligned} \sigma_{\hat{\phi}}^2 &= \frac{N_0}{2A_c^2} \int_{-\infty}^{\infty} |H(f)|^2 \, df \\ &= \frac{N_0}{A_c^2} \int_0^{\infty} |H(f)|^2 \, df \\ &= \frac{N_0 B_{\text{eq}}}{A_c^2} = \frac{N_0 B_{\text{eq}}}{A_c^2} \end{aligned} \tag{6.2–29}$$

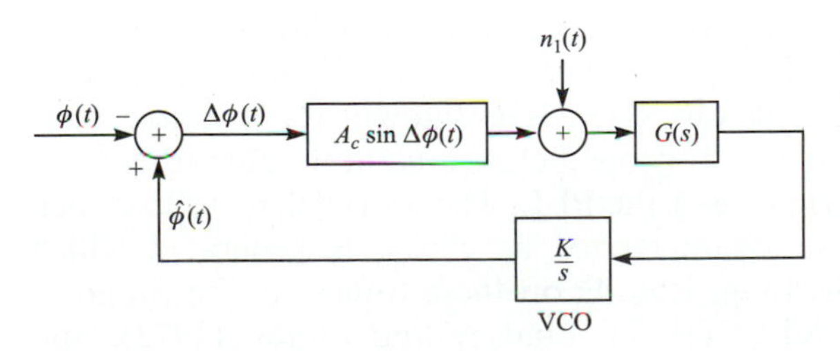

FIGURE 6.2–6
Equivalent PLL model with additive noise.

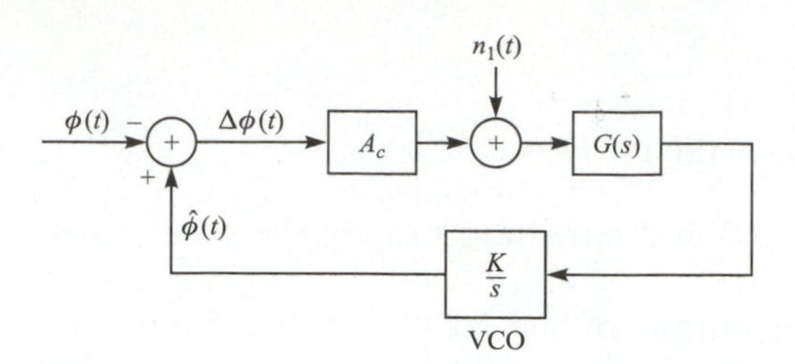

FIGURE 6.2–7
Linearized PLL model with additive noise.

where B_{eq} is the (one-sided) equivalent noise bandwidth of the loop, given in Equation 6.2–22. Note that $\sigma_{\hat{\phi}}^2$ is simply the ratio of total noise power within the bandwidth of the PLL divided by the signal power. Hence,

$$\sigma_{\hat{\phi}}^2 = \frac{1}{\gamma_L} \tag{6.2–30}$$

where γ_L is defined as the signal-to-noise ratio

$$\text{SNR} \equiv \gamma_L = \frac{A_c^2}{N_0 B_{eq}} \tag{6.2–31}$$

The expression for the variance $\sigma_{\hat{\phi}}^2$ of the VCO phase error applies to the case where the SNR is sufficiently high that the linear model for the PLL applies. An exact analysis based on the non-linear PLL is mathematically tractable when $G(s) = 1$, which results in a first-order loop. In this case, the probability density function for the phase error may be derived (see Viterbi, 1966) and has the form

$$p(\Delta\phi) = \frac{\exp(\gamma_L \cos \Delta\phi)}{2\pi I_0(\gamma_L)} \tag{6.2–32}$$

where γ_L is the SNR given by Equation 6.2–31 with B_{eq} being the appropriate noise bandwidth of the first-order loop, and $I_0(\cdot)$ is the modified Bessel function of order zero.

From the expression for $p(\Delta\phi)$, we may obtain the exact value of the variance for the phase error on a first-order PLL. This is plotted in Figure 6.2–8 as a function of $1/\gamma_L$. Also shown for comparison is the result obtained with the linearized PLL model. Note that the variance for the linear model is close to the exact variance for $\gamma_L > 3$. Hence, the linear model is adequate for practical purposes.

Approximate analyses of the statistical characteristics of the phase error for the non-linear PLL have also been performed. Of particular importance is the transient behavior of the PLL during initial acquisition. Another important problem is the behavior of PLL at low SNR. It is known, for example, that when the SNR at the input to the PLL drops below a certain value, there is a rapid deterioration in the performance of the PLL. The loop begins to lose lock and an impulsive type of noise, characterized as clicks, is generated which degrades the performance of the loop. Results on these topics can be found in the texts by Viterbi (1966), Lindsey (1972), Lindsey and Simon (1973), and

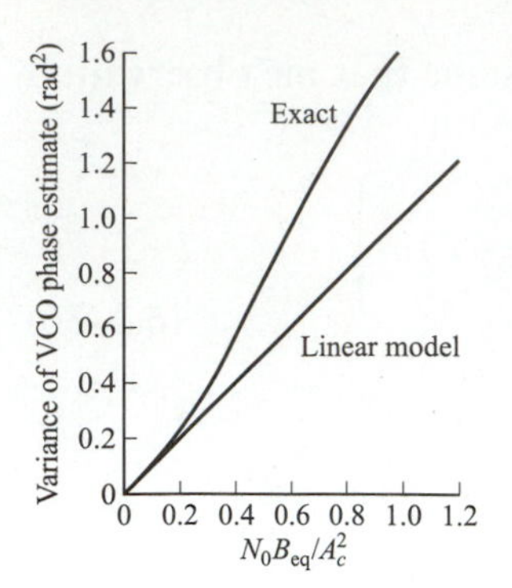

FIGURE 6.2–8
Comparison of VCO phase variance for exact and approximate (linear model) first-order PLL. [*From* Principles of Coherent Communication, *by A. J. Viterbi; © 1996 by McGraw-Hill Book Company. Reprinted with permission of the publisher.*]

Gardner (1979), and in the survey papers by Gupta (1975) and Lindsey and Chie (1981).

Up to this point, we have considered carrier phase estimation when the carrier signal is unmodulated. Below, we consider carrier phase recovery when the signal carries information.

6.2.4 Decision-Directed Loops

A problem arises in maximizing either Equation 6.2–9 or 6.2–10 when the signal $s(t;\phi)$ carries the information sequence $\{I_n\}$. In this case we can adopt one of two approaches: either we assume that $\{I_n\}$ is known or we treat $\{I_n\}$ as a random sequence and average over its statistics.

In decision-directed parameter estimation, we assume that the information sequence $\{I_n\}$ over the observation interval has been estimated and, in the absence of demodulation errors, $\tilde{I}_n = I_n$, where $\tilde{I}_n$ denotes the detected value of the information I_n. In this case $s(t;\phi)$ is completely known except for the carrier phase.

To be specific, let us consider the decision-directed phase estimate for the class of linear modulation techniques for which the received *equivalent low-pass signal* may be expressed as

$$r_l(t) = e^{-j\phi}\sum_n I_n g(t-nT) + z(t) = s_l(t)e^{-j\phi} + z(t) \tag{6.2–33}$$

where $s_l(t)$ is a known signal if the sequence $\{I_n\}$ is assumed known. The likelihood function and corresponding log-likelihood function for the equivalent low-pass signal are

$$\Lambda(\phi) = C\exp\left\{\mathrm{Re}\left[\frac{1}{N_0}\int_{T_0} r_l(t)s_l^*(t)e^{j\phi}dt\right]\right\} \tag{6.2–34}$$

$$\Lambda_L(\phi) = \mathrm{Re}\left\{\left[\frac{1}{N_0}\int_{T_0} r_l(t)s_l^*(t)\,dt\right]e^{j\phi}\right\} \tag{6.2–35}$$

If we substitute for $s_l(t)$ in Equation 6.2–35 and assume that the observation interval $T_0 = KT$, where K is a positive integer, we obtain

$$\begin{aligned}\Lambda_L(\phi) &= \text{Re}\left\{e^{j\phi}\frac{1}{N_0}\sum_{n=0}^{K-1} I_n^* \int_{nT}^{(n+1)T} r_l(t)g^*(t-nT)\,dt\right\} \\ &= \text{Re}\left\{e^{j\phi}\frac{1}{N_0}\sum_{n=0}^{K-1} I_n^* y_n\right\}\end{aligned} \tag{6.2–36}$$

where, by definition

$$y_n = \int_{nT}^{(n+1)T} r_l(t)g^*(t-nT)\,dt \tag{6.2–37}$$

Note that y_n is the output of the matched filter in the nth signal interval. The ML estimate of ϕ is easily found from Equation 6.2–36 by differentiating the log-likelihood

$$\Lambda_L(\phi) = \text{Re}\left(\frac{1}{N_0}\sum_{n=0}^{K-1} I_n^* y_n\right)\cos\phi - \text{Im}\left(\frac{1}{N_0}\sum_{n=0}^{K-1} I_n^* y_n\right)\sin\phi$$

with respect to ϕ and setting the derivative equal to zero. Thus, we obtain

$$\hat{\phi}_{ML} = -\tan^{-1}\left[\text{Im}\left(\sum_{n=0}^{K-1} I_n^* y_n\right)\Big/\text{Re}\left(\sum_{n=0}^{K-1} I_n^* y_n\right)\right] \tag{6.2–38}$$

We call $\hat{\phi}_{\text{ML}}$ in Equation 6.2–38 the *decision-directed* (or *decision-feedback) carrier phase estimate*. It is easily shown (Problem 6.10) that the mean value of $\hat{\phi}_{\text{ML}}$ is ϕ, so that the estimate is unbiased. Furthermore, the PDF of $\hat{\phi}_{\text{ML}}$ can be obtained (Problem 6.11) by using the procedure described in Section 5.2.7.

The block diagram of a double-sideband PAM signal receiver that incorporates the decision-directed carrier phase estimate given by Equation 6.2–38 is illustrated in Figure 6.2–9.

Another implementation of the PAM receiver that employs a decision-feedback PLL (DFPLL) for carrier phase estimation is shown in Figure 6.2–10. The received double-sideband PAM signal is given by $A(t)\cos(2\pi f_c t + \phi)$, where $A(t) = A_m g(t)$ and $g(t)$ is assumed to be a rectangular pulse of duration T. This received signal is multiplied by the quadrature carriers $c_c(t)$ and $c_s(t)$, as given by Equation 6.2–5, which are derived from the VCO. The product signal

$$\begin{aligned}r(t)\cos(2\pi f_c t + \hat{\phi}) &= \tfrac{1}{2}[A(t) + n_c(t)]\cos\Delta\phi \\ &\quad - \tfrac{1}{2}n_s(t)\sin\Delta\phi + \text{double-frequency terms}\end{aligned} \tag{6.2–39}$$

is used to recover the information carried by $A(t)$. The detector makes a decision on the symbol that is received every T seconds. Thus, in the absence of decision errors, it reconstructs $A(t)$ free of any noise. This reconstructed signal is used to multiply the product of the second quadrature multiplier, which has been

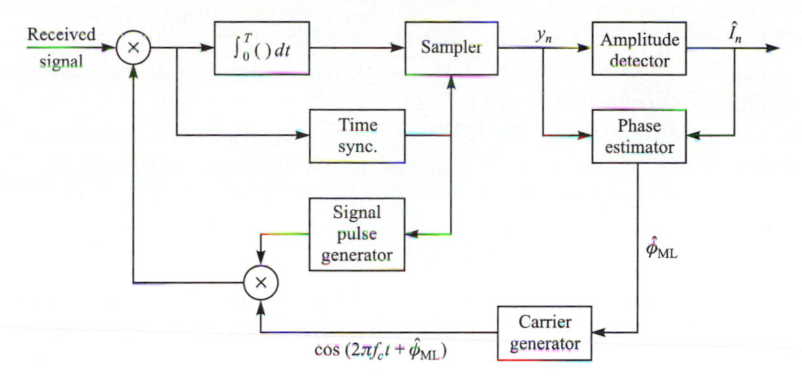

FIGURE 6.2–9
Block diagram of double-sideband PAM signal receiver with decision-directed carrier phase estimation.

delayed by T seconds to allow the demodulator to reach a decision. Thus, the input to the loop filter in the absence of decision errors is the error signal

$$\begin{aligned} e(t) &= \tfrac{1}{2}A(t)\{[A(t) + n_c(t)] \sin \Delta\phi - n_s(t) \cos \Delta\phi\} \\ &\quad + \text{double-frequency terms} \\ &= \tfrac{1}{2}A^2(t) \sin \Delta\phi + \tfrac{1}{2}A(t)[n_c(t) \sin \Delta\phi - n_s(t) \cos \Delta\phi] \\ &\quad + \text{double-frequency terms} \end{aligned} \tag{6.2–40}$$

The loop filter is low-pass and, hence, it rejects the double-frequency term in $e(t)$. The desired component is $A^2(t) \sin \Delta\phi$, which contains the phase error for driving the loop.

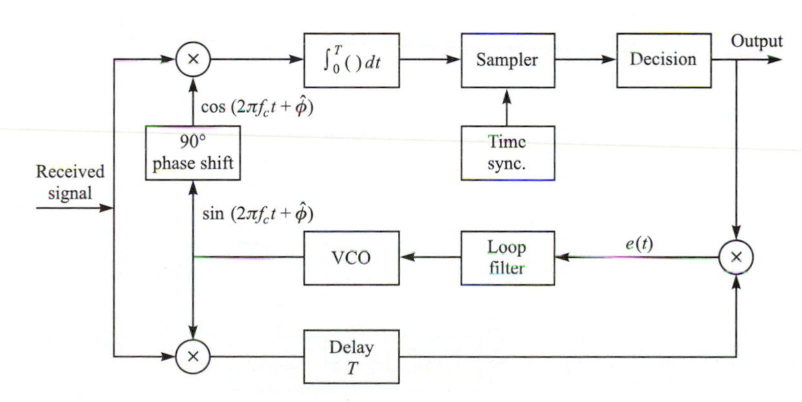

FIGURE 6.2–10
Carrier recovery with a decision-feedback PLL.

The ML estimate in Equation 6.2–38 is also appropriate for QAM. The block diagram of a QAM receiver that incorporates the decision-directed carrier phase estimate is shown in Figure 6.2–11.

In the case of M-ary PSK, the DFPLL has the configuration shown in Figure 6.2–12. The received signal is demodulated to yield the phase estimate

$$\hat{\theta}_m = \frac{2\pi}{M}(m-1)$$

which, in the absence of a decision error, is the transmitted signal phase θ_m. The two outputs of the quadrature multipliers are delayed by the symbol duration T and multiplied by $\cos\theta_m$ and $\sin\theta_m$ to yield

$$\begin{aligned} & r(t)\cos(2\pi f_c t + \hat{\phi})\sin\theta_m \\ & \quad = \tfrac{1}{2}[A\cos\theta_m + n_c(t)]\sin\theta_m\cos(\phi - \hat{\phi}) \\ & \qquad - \tfrac{1}{2}[A\sin\theta_m + n_s(t)]\sin\theta_m\sin(\phi - \hat{\phi}) \\ & \qquad + \text{double-frequency terms} \\ & r(t)\sin(2\pi f_c t + \hat{\phi})\cos\theta_m \\ & \quad = -\tfrac{1}{2}[A\cos\theta_m + n_c(t)]\cos\theta_m\sin(\phi - \hat{\phi}) \\ & \qquad - \tfrac{1}{2}[A\sin\theta_m + n_s(t)]\cos\theta_m\cos(\phi - \hat{\phi}) \\ & \qquad + \text{double-frequency terms} \end{aligned} \tag{6.2–41}$$

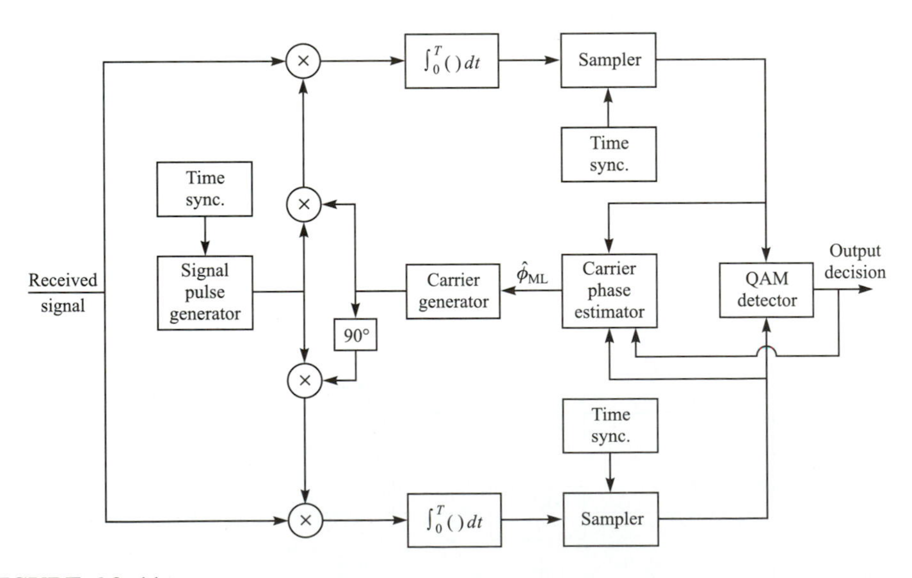

FIGURE 6.2–11
Block diagram of QAM signal receiver with decision-directed carrier phase estimation.

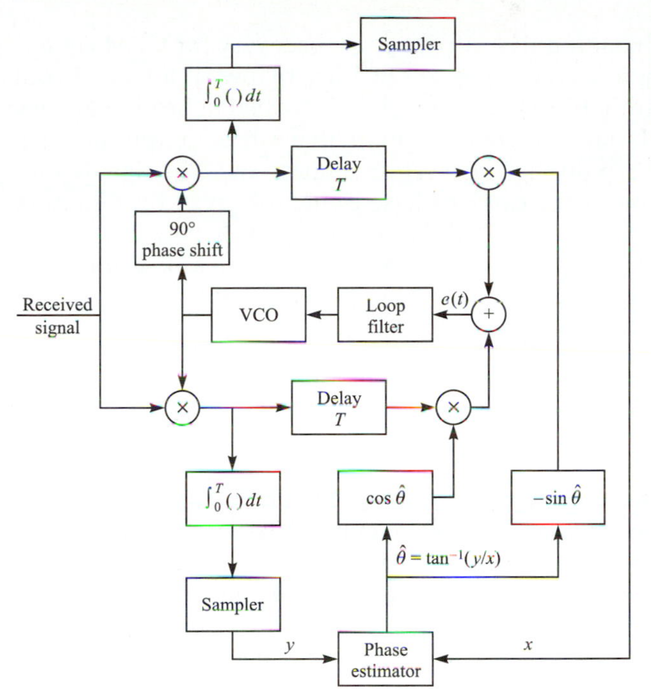

FIGURE 6.2–12 Carrier recovery for M-ary PSK using a decision-feedback PLL.

The two signals are added to generate the error signal

$$\begin{aligned} e(t) = &-\tfrac{1}{2}A\sin(\phi - \hat{\phi}) + \tfrac{1}{2}n_c(t)\sin(\phi - \hat{\phi} - \theta_m) \\ &+ \tfrac{1}{2}n_s(t)\cos(\phi - \hat{\phi} - \theta_m) + \text{double-frequency terms} \end{aligned} \tag{6.2–42}$$

This error signal is the input to the loop filter that provides the control signal for the VCO.

We observe that the two quadrature noise components in Equation 6.2–42 appear as additive terms. There is no term involving a product of two noise components as in an Mth-power law device, described in the next section. Consequently, there is no additional power loss associated with the decision-feedback PLL.

This M-phase tracking loop has a phase ambiguity of $360°/M$, necessitating the need to differentially encode the information sequence prior to transmission and differentially decode the received sequence after demodulation to recover the information.

The ML estimate in Equation 6.2–38 is also appropriate for QAM. The ML estimate for offset QPSK is also easily obtained (Problem 6.12) by maximizing the log-likelihood function in Equation 6.2–35, with $s_l(t)$ given as

$$s_l(t) = \sum_n I_n g(t - nT) + j\sum_n J_n g(t - nT - \tfrac{1}{2}T) \tag{6.2–43}$$

where $I_n = \pm 1$ and $J_n = \pm 1$.

Finally, we should also mention that carrier phase recovery for CPM signals can also be accomplished in a decision-directed manner by use of a PLL. From the optimum demodulator for CPM signals, which is described in Section 5.3, we can generate an error signal that is filtered in a loop filter whose output drives a PLL. Alternatively, we may exploit the linear representation of CPM signals and, thus, employ a generalization of the carrier phase estimator given by Equation 6.2–38, in which, the cross correlation of the received signal is performed with each of the pulses in the linear representation. A comprehensive description of carrier phase recover techniques for CPM is given in the book by Mengali and D'Andrea (1997).

6.2.5 Non-Decision-Directed Loops

Instead of using a decision-directed scheme to obtain the phase estimate, we may treat the data as random variables and simply average $\Lambda(\phi)$ over these random variables prior to maximization. In order to carry out this integration, we may use either the actual probability distribution function of the data, if it is known, or, perhaps, we may assume some probability distribution that might be a reasonable approximation to the true distribution. The following example illustrates the first approach.

EXAMPLE 6.2–2. Suppose the real signal $s(t)$ carries binary modulation. Then, in a signal interval, we have

$$s(t) = A\cos 2\pi f_c t, \qquad 0 \leqslant t \leqslant T$$

where $A = \pm 1$ with equal probability. Clearly, the PDF of A is given as

$$p(A) = \tfrac{1}{2}\delta(A-1) + \tfrac{1}{2}\delta(A+1)$$

Now, the likelihood function $\Lambda(\phi)$ given by Equation 6.2–9 is conditional on a given value of A and must be averaged over the two values. Thus,

$$\begin{aligned}\bar{\Lambda}(\phi) &= \int_{-\infty}^{\infty} \Lambda(\phi) p(A)\, dA \\ &= \tfrac{1}{2}\exp\left[\frac{2}{N_0}\int_0^T r(t)\cos(2\pi f_c t + \phi)\, dt\right] \\ &\quad + \tfrac{1}{2}\exp\left[-\frac{2}{N_0}\int_0^T r(t)\cos(2\pi f_c t + \phi)\, dt\right] \\ &= \cosh\left[\frac{2}{N_0}\int_0^T r(t)\cos(2\pi f_c t + \phi)\, dt\right]\end{aligned}$$

and the corresponding log-likelihood function is

$$\bar{\Lambda}_L(\phi) = \ln\cosh\left[\frac{2}{N_0}\int_0^T r(t)\cos(2\pi f_c t + \phi)\, dt\right] \qquad (6.2\text{–}44)$$

If we differentiate $\bar{\Lambda}_L(\phi)$ and set the derivative equal to zero, we obtain the ML estimate for the non-decision-directed estimate. Unfortunately, the functional relationship in Equation 6.2–44 is highly non-linear and, hence, an exact solution is difficult to obtain. On the other hand, approximations are possible. In particular,

$$\ln\cosh x = \begin{cases} \frac{1}{2}x^2 & (|x| \ll 1) \\ |x| & (|x| \gg 1) \end{cases} \tag{6.2–45}$$

With these approximations, the solution for ϕ becomes tractable.

In this example, we averaged over the two possible values of the information symbol. When the information symbols are M-valued, where M is large, the averaging operation yields highly non-linear functions of the parameter to be estimated. In such a case, we may simplify the problem by assuming that the information symbols are continuous random variables. For examples, we may assume that the symbols are zero-mean Gaussian. The following example illustrates this approximation and the resulting form for the average likelihood function.

EXAMPLE 6.2–3. Let us consider the same signal as in Example 6.2–2, but now we assume that the amplitude A is zero-mean Gaussian with unit variance. Thus,

$$p(A) = \frac{1}{\sqrt{2\pi}} e^{-A^2/2}$$

If we average $\Lambda(\phi)$ over the assumed PDF of A, we obtain the average likelihood $\bar{\Lambda}(\phi)$ in the form

$$\bar{\Lambda}(\phi) = C \exp\left\{\left[\frac{2}{N_0}\int_0^T r(t)\cos(2\pi f_c t + \phi)\,dt\right]^2\right\} \tag{6.2–46}$$

and the corresponding log-likelihood as

$$\bar{\Lambda}_L(\phi) = \left[\frac{2}{N_0}\int_0^T r(t)\cos(2\pi f_c t + \phi)\,dt\right]^2 \tag{6.2–47}$$

We can obtain the ML estimate of ϕ by differentiating $\bar{\Lambda}_L(\phi)$ and setting the derivative to zero.

It is interesting to note that the log-likelihood function is quadratic under the Gaussian assumption and that it is approximately quadratic, as indicated in Equation 6.2–45 for small values of the cross correlation of $r(t)$ with $s(t;\phi)$. In other words, if the cross correlation over a single interval is small, the Gaussian assumption for the distribution of the information symbols yields a good approximation to the log-likelihood function.

In view of these results, we may use the Gaussian approximation on all the symbols in the observation interval $T_0 = KT$. Specifically, we assume that the K information symbols are statistically independent and identically distributed. By averaging the likelihood function $\Lambda(\phi)$ over the Gaussian PDF for each of the K symbols in the interval $T_0 = KT$, we obtain the result

$$\bar{\Lambda}(\phi) = C \exp\left\{\sum_{n=0}^{K-1}\left[\frac{2}{N_0}\int_{nT}^{(n+1)T} r(t)\cos(2\pi f_c t + \phi)\,dt\right]^2\right\} \tag{6.2–48}$$

If we take the logarithm of Equation 6.2–48, differentiate the resulting log-likelihood function, and set the derivative equal to zero, we obtain the condition for the ML estimate as

$$\sum_{n=0}^{K-1} \int_{nT}^{(n+1)T} r(t)\cos(2\pi f_c t + \hat{\phi})\, dt \int_{nT}^{(n+1)T} r(t)\sin(2\pi f_c t + \hat{\phi})\, dt = 0 \qquad (6.2\text{–}49)$$

Although this equation can be manipulated further, its present form suggests the tracking loop configuration illustrated in Figure 6.2–13. This loop resembles a Costas loop, which is described below. We note that the multiplication of the two signals from the integrators destroys the sign carried by the information symbols. The summer plays the role of the loop filter. In a tracking loop configuration, the summer may be implemented either as a sliding-window digital filter (summer) or as a low-pass digital filter with exponential weighting of the past data.

In a similar manner, one can derive non-decision-directed ML phase estimates for QAM and M-PSK. The starting point is to average the likelihood function given by Equation 6.2–9 over the statistical characteristics of the data. Here again, we may use the Gaussian approximation (two-dimensional Gaussian for complex-valued information symbols) in averaging over the information sequence.

Squaring loop. The squaring loop is a non-decision-directed loop that is widely used in practice to establish the carrier phase of double-sideband suppressed carrier signals such as PAM. To describe its operation, consider the problem of estimating the carrier phase of the digitally modulated PAM signal of the form

$$s(t) = A(t)\cos(2\pi f_c t + \phi) \qquad (6.2\text{–}50)$$

where $A(t)$ carries the digital information. Note that $E[s(t)] = E[A(t)] = 0$ when the signal levels are symmetric about zero. Consequently, the average value of $s(t)$ does not produce any phase coherent frequency components at any frequency, including the carrier. One method for generating a carrier from the received signal is to square the signal and, thus, to generate a frequency component at $2f_c$, which can be used to drive a PLL tuned to $2f_c$. This method is illustrated in the block diagram shown in Figure 6.2–14.

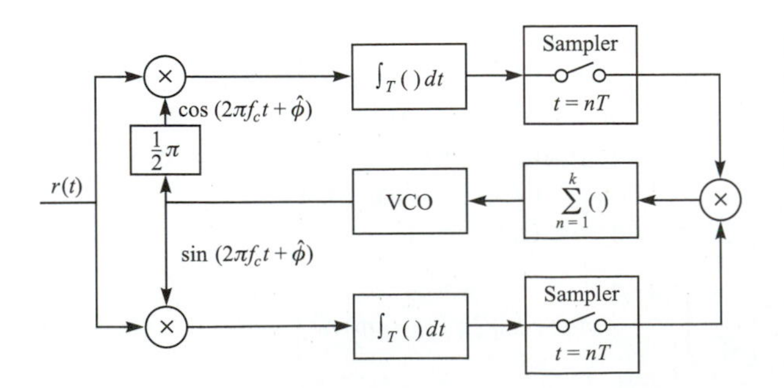

FIGURE 6.2–13
Non-decision-directed PLL for carrier phase estimation of PAM signals.

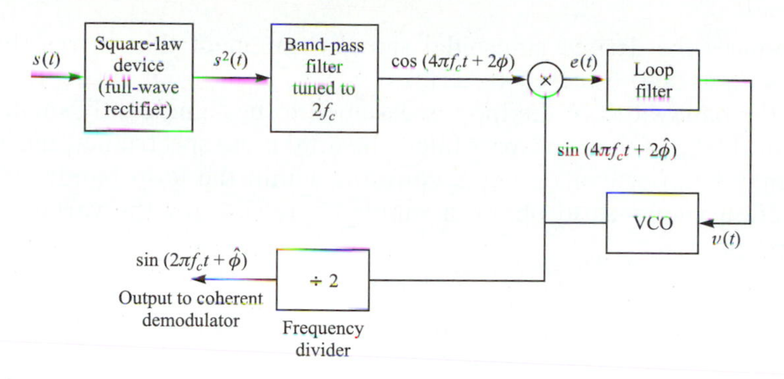

FIGURE 6.2–14
Carrier recover using a square-law device.

The output of the square-law device is

$$\begin{aligned} s^2(t) &= A^2(t)\cos^2(2\pi f_c t + \phi) \\ &= \tfrac{1}{2}A^2(t) + \tfrac{1}{2}A^2(t)\cos(4\pi f_c t + 2\phi) \end{aligned} \tag{6.2–51}$$

Since the modulation is a cyclostationary stochastic process, the expected value of $s^2(t)$ is

$$E[s^2(t)] = \tfrac{1}{2}E[A^2(t)] + \tfrac{1}{2}E[A^2(t)]\cos(4\pi f_c t + 2\phi) \tag{6.2–52}$$

Hence, there is power at the frequency $2f_c$.

If the output of the square-law device is passed through a band-pass filter tuned to the double-frequency term in Equation 6.2–51, the mean value of the filter is a sinusoid with frequency $2f_c$, phase 2ϕ, and amplitude $\frac{1}{2}E[A^2(t)]H(2f_c)$, where $H(2f_c)$ is the gain of the filter at $f = 2f_c$. Thus, the square-law device has produced a periodic component from the input signal $s(t)$. In effect, the squaring of $s(t)$ has removed the sign information contained in $A(t)$ and, thus, has resulted in phase-coherent frequency components at twice the carrier. The filtered frequency component at $2f_c$ is then used to drive the PLL.

The squaring operation leads to a noise enhancement that increases the noise power level at the input to the PLL and results in an increase in the variance of the phase error.

To elaborate on this point, let the input to the squarer be $s(t) + n(t)$, where $s(t)$ is given by Equation 6.2–50 and $n(t)$ represents the band-pass additive Gaussian noise process. By squaring $s(t) + n(t)$, we obtain

$$y(t) = s^2(t) + 2s(t)n(t) + n^2(t) \tag{6.2–53}$$

where $s^2(t)$ is the desired signal component and the other two components are the signal × noise and noise × noise terms. By computing the autocorrelation functions and power density spectra of these two noise components, one can easily show that both components have spectral power in the frequency band centered at $2f_c$. Consequently, the band-pass filter with bandwidth B_{bp} centered at $2f_c$, which produces the desired sinusoidal signal component that drives the PLL, also passes noise due to these two terms.

Since the bandwidth of the loop is designed to be significantly smaller than the bandwidth B_{bp} of the band-pass filter, the total noise spectrum at the input to the PLL may be approximated as a constant within the loop bandwidth. This approximation allows us to obtain a simple expression for the variance of the phase error as

$$\sigma_{\hat{\phi}}^2 = \frac{1}{\gamma_L S_L} \tag{6.2–54}$$

where S_L is called the squaring loss and is given by

$$S_L = \left(1 + \frac{B_{\text{bp}}/2B_{\text{eq}}}{\gamma_L}\right)^{-1} \tag{6.2–55}$$

Since $S_L < 1$, S_L^{-1} represents the increase in the variance of the phase error caused by the added noise (noise $\times$ noise terms) that results from the squarer. Note, for example, that when $\gamma_L = B_{\text{bp}}/2B_{\text{eq}}$, the loss is 3 dB.

Finally, we observe that the output of the VCO from the squaring loop must be frequency-divided by 2 to generate the phase-locked carrier for signal demodulation. It should be noted that the output of the frequency divider has a phase ambiguity of 180° relative to the phase of the received signal. For this reason, the binary data must be differentially encoded prior to transmission and differentially decoded at the receiver.

Costas loop. Another method for generating a properly phased carrier for a double-sideband suppressed carrier signal is illustrated by the block diagram shown in Figure 6.2–15. This scheme was developed by Costas (1956) and is called the *Costas loop*. The received signal is multiplied by $\cos(2\pi f_c t + \hat{\phi})$ and $\sin(2\pi f_c t + \hat{\phi})$, which are outputs from the VCO. The two products are

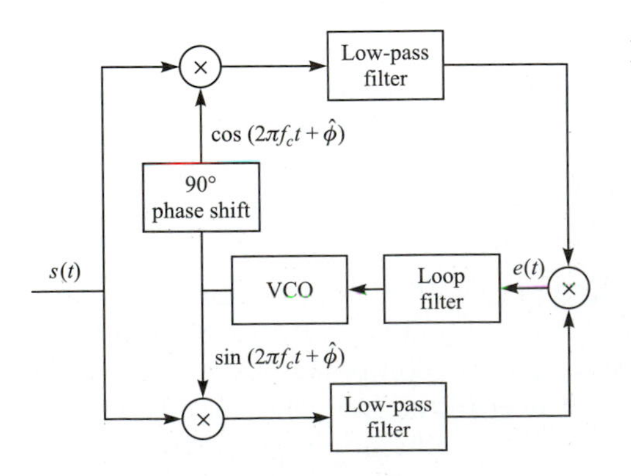

FIGURE 6.2–15
Block diagram of Costas loop.

$$\begin{aligned} y_c(t) &= [s(t) + n(t)] \cos(2\pi f_c t + \hat{\phi}) \\ &= \tfrac{1}{2}[A(t) + n_c(t)] \cos \Delta\phi + \tfrac{1}{2} n_s(t) \sin \Delta\phi \\ &\quad + \text{double-frequency terms} \\ y_s(t) &= [s(t) + n(t)] \sin(2\pi f_c t + \hat{\phi}) \\ &= \tfrac{1}{2}[A(t) + n_c(t)] \sin \Delta\phi - \tfrac{1}{2} n_s(t) \cos \Delta\phi \\ &\quad + \text{double-frequency terms} \end{aligned} \tag{6.2–56}$$

where the phase error $\Delta\phi = \hat{\phi} - \phi$. The double-frequency terms are eliminated by the low-pass filters following the multiplications.

An error signal is generated by multiplying the two outputs of the low-pass filters. Thus,

$$\begin{aligned} e(t) &= \tfrac{1}{8}\{[A(t) + n_c(t)]^2 - n_s^2(t)\} \sin(2\Delta\phi) \\ &\quad - \tfrac{1}{4} n_s(t)[A(t) + n_c(t)] \cos(2\Delta\phi) \end{aligned} \tag{6.2–57}$$

This error signal is filtered by the loop filter, whose output is the control voltage that drives the VCO. The reader should note the similarity of the Costas loop to the PLL shown in Figure 6.2–13.

We note that the error signal into the loop filter consists of the desired term $A^2(t) \sin 2(\hat{\phi} - \phi)$ plus terms that involve signal × noise and noise × noise. These terms are similar to the two noise terms at the input to the PLL for the squaring method. In fact, if the loop filter in the Costas loop is identical to that used in the squaring loop, the two loops are equivalent. Under this condition, the probability density function of the phase error and the performance of the two loops are identical.

It is interesting to note that the optimum low-pass filter for rejecting the double-frequency terms in the Costas loop is a filter matched to the signal pulse in the information-bearing signal. If matched filters are employed for the low-pass filters, their outputs could be sampled at the bit rate and at the end of each signal interval, and the discrete-time signal samples could be used to drive the loop. The use of the matched filter results in a smaller noise into the loop.

Finally, we note that, as in the squaring PLL, the output of the VCO contains a phase ambiguity of 180°, necessitating the need for differential encoding of the data prior to transmission and differential decoding at the demodulator.

Carrier estimation for multiple phase signals. When the digital information is transmitted via M-phase modulation of a carrier, the methods described above can be generalized to provide the properly phased carrier for demodulation. The received M-phase signal, excluding the additive noise, may be expressed as

$$s(t) = A \cos\left[2\pi f_c t + \phi + \frac{2\pi}{M}(m-1)\right], \qquad m = 1, 2, \ldots, M \tag{6.2–58}$$

where $2\pi(m-1)/M$ represents the information-bearing component of the signal phase. The problem in carrier recovery is to remove the information-bearing component and, thus, to obtain the unmodulated carrier $\cos(2\pi f_c t + \phi)$. One method by which this can be accomplished is illustrated in Figure 6.2–16, which represents a generalization of the squaring loop. The signal is passed through an Mth-power-law device, which generates a number of harmonics of f_c. The band-pass filter selects the harmonic $\cos(2\pi M f_c t + M\phi)$ for driving the PLL. The term

$$\frac{2\pi}{M}(m-1)M = 2\pi(m-1) \equiv 0 \pmod{2\pi}, \qquad m = 1, 2, \ldots, M$$

Thus, the information is removed. The VCO output is $\sin(2\pi M f_c t + M\hat{\phi})$, so this output is divided in frequency by M to yield $\sin(2\pi f_c t + \hat{\phi})$, and phase-shifted by $\frac{1}{2}\pi$ rad to yield $\cos(2\pi f_c t + \hat{\phi})$. These components are then fed to the demodulator. Although not explicitly shown, there is a phase ambiguity in these reference sinusoids of $360°/M$, which can be overcome by differential encoding of the data at the transmitter and differential decoding after demodulation at the receiver.

Just as in the case of the squaring PLL, the Mth-power PLL operates in the presence of noise that has been enhanced by the Mth-power-law device, which results in the output

$$y(t) = [s(t) + n(t)]^M$$

The variance of the phase error in the PLL resulting from the additive noise may be expressed in the simple form

$$\sigma^2_{\hat{\phi}} = \frac{S^{-1}_{\text{ML}}}{\gamma_L} \tag{6.2–59}$$

where γ_L is the loop SNR and S^{-1}_{ML} is the *M-phase power loss*. S_{ML} has been evaluated by Lindsey and Simon (1973) for $M = 4$ and 8.

Another method for carrier recovery in M-ary PSK is based on a generalization of the Costas loop. That method requires multiplying the received signal by M phase-shifted carriers of the form

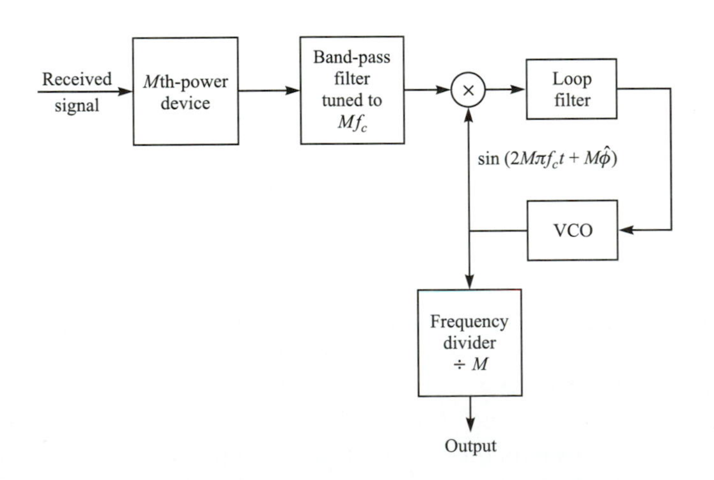

FIGURE 6.2–16
Carrier recovery with Mth-power-law device for M-ary PSK.

$$\sin\left[2\pi f_c t + \hat{\phi} + \frac{\pi}{M}(k-1)\right], \qquad k = 1, 2, \ldots, M$$

low-pass–filtering each product, and then multiplying the outputs of the low-pass filters to generate the error signal. The error signal excites the loop filter, which, in turn, provides the control signal for the VCO. This method is relatively complex to implement and, consequently, has not been generally used in practice.

Comparison of decision-directed with non-decision-directed loops. We note that the decision-feedback phase-locked loop (DFPLL) differs from the Costas loop only in the method by which $A(t)$ is rectified for the purpose of removing the modulation. In the Costas loop, each of the two quadrature signals used to rectify $A(t)$ is corrupted by noise. In the DFPLL, only one of the signals used to rectify $A(t)$ is corrupted by noise. On the other hand, the squaring loop is similar to the Costas loop in terms of the noise effect on the estimate $\hat{\phi}$. Consequently, the DFPLL is superior in performance to both the Costas loop and the squaring loop, provided that the demodulator is operating at error rates below 10^{-2} where an occasional decision error has a negligible effect on $\hat{\phi}$. Quantitative comparisons of the variance of the phase errors in a Costas loop to those in DFPLL have been made by Lindsey and Simon (1973), and show that the variance of the DFPLL is 4–10 times smaller for signal-to-noise ratios per bit above 0 dB.

6.3 SYMBOL TIMING ESTIMATION

In a digital communication system, the output of the demodulator must be sampled periodically at the symbol rate, at the precise sampling time instants $t_m = mT + \tau$, where T is the symbol interval and τ is a nominal time delay that accounts for the propagation time of the signal from the transmitter to the receiver. To perform this periodic sampling, we require a clock signal at the receiver. The process of extracting such a clock signal at the receiver is usually called *symbol synchronization* or *timing recovery*.

Timing recovery is one of the most critical functions that is performed at the receiver of a synchronous digital communication system. We should note that the receiver must know not only the frequency $(1/T)$ at which the outputs of the matched filters or correlators are sampled, but also where to take the samples within each symbol interval. The choice of sampling instant within the symbol interval of duration T is called the *timing phase*.

Symbol synchronization can be accomplished in one of several ways. In some communication systems, the transmitter and receiver clocks are synchronized to a master clock, which provides a very precise timing signal. In this case, the receiver must estimate and compensate for the relative time delay between the transmitted and received signals. Such may be the case for radio communication systems that operate in the very low frequency (VLF) band (below 30 kHz), where precise clock signals are transmitted from a master radio station.

Another method for achieving symbol synchronization is for the transmitter to simultaneously transmit the clock frequency $1/T$ or a multiple of $1/T$ along with the information signal. The receiver may simply employ a narrowband filter tuned to the transmitted clock frequency and, thus, extract the clock signal for sampling. This approach has the advantage of being simple to implement. There are several disadvantages, however. One is that the transmitter must allocate some of its available power to the transmission of the clock signal. Another is that some small fraction of the available channel bandwidth must be allocated for the transmission of the clock signal. In spite of these disadvantages, this method is frequently used in telephone transmission systems that employ large bandwidths to transmit the signals of many users. In such a case, the transmission of a clock signal is shared in the demodulation of the signals among the many users. Through this shared use of the clock signal, the penalty in the transmitter power and in bandwidth allocation is reduced proportionally by the number of users.

A clock signal can also be extracted from the received data signal. There are a number of different methods that can be used at the receiver to achieve self-synchronization. In this section, we treat both decision-directed and non-decision-directed methods.

6.3.1 Maximum-Likelihood Timing Estimation

Let us begin by obtaining the ML estimate of the time delay τ. If the signal is a baseband PAM waveform, it is represented as

$$r(t) = s(t; \tau) + n(t) \tag{6.3–1}$$

where

$$s(t; \tau) = \sum_n I_n g(t - nT - \tau) \tag{6.3–2}$$

As in the case of ML phase estimation, we distinguish between two types of timing estimators, decision-directed timing estimators and non-decision-directed estimators. In the former, the information symbols from the output of the demodulator are treated as the known transmitted sequence. In this case, the log-likelihood function has the form

$$\Lambda_L(\tau) = C_L \int_{T_0} r(t)s(t; \tau)\, dt \tag{6.3–3}$$

If we substitute Equation 6.3–2 into Equation 6.3–3, we obtain

$$\begin{aligned}\Lambda_L(\tau) &= C_L \sum_n I_n \int_{T_0} r(t)g(t - nT - \tau)\,dt \\ &= C_L \sum_n I_n y_n(\tau)\end{aligned} \tag{6.3–4}$$

where $y_n(t)$ is defined as

$$y_n(\tau) = \int_{T_0} r(t)g(t - nT - \tau)\,dt \tag{6.3–5}$$

A necessary condition for $\hat{\tau}$ to be the ML estimate of τ is that

$$\begin{aligned}\frac{d\Lambda_L(\tau)}{d\tau} &= \sum_n I_n \frac{d}{d\tau}\int_{T_0} r(t)g(t - nT - \tau)\,dt \\ &= \sum_n I_n \frac{d}{d\tau}[y_n(\tau)] = 0\end{aligned} \tag{6.3–6}$$

The result in Equation 6.3–6 suggest the implementation of the tracking loop shown in Figure 6.3–1. We should observe that the summation in the loop serves as the loop filter whose bandwidth is controlled by the length of the sliding window in the summation. The output of the loop filter drives the voltage-controlled clock (VCC), or voltage-controlled oscillator, which controls the sampling times for the input to the loop. Since the detected information sequence $\{I_n\}$ is used in the estimation of τ, the estimate is decision-directed.

The techniques described above for ML timing estimation of baseband PAM signals can be extended to carrier modulated signal formats such as QAM and PSK in a straightforward manner, by dealing with the equivalent low-pass form of the signals. Thus, the problem of ML estimation of symbol timing for carrier signals is very similar to the problem formulation for the baseband PAM signal.

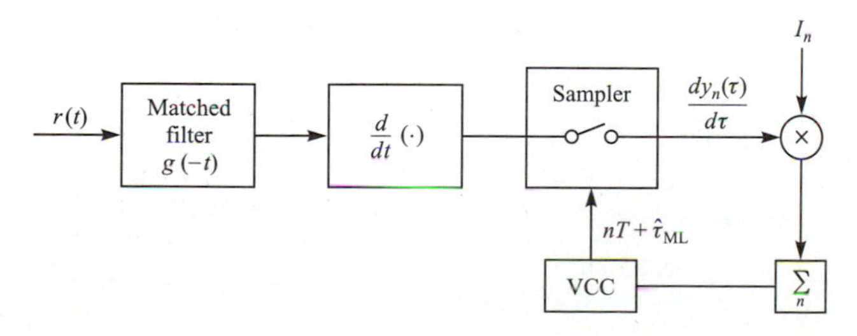

FIGURE 6.3–1
Decision-directed ML estimation of timing for baseband PAM.

6.3.2 Non-Decision-Directed Timing Estimation

A non-decision-directed timing estimate can be obtained by averaging the likelihood ratio $\Lambda(\tau)$ over the PDF of the information symbols, to obtain $\bar{\Lambda}(\tau)$, and then differentiating either $\bar{\Lambda}(\tau)$ or $\ln \bar{\Lambda}(\tau) = \bar{\Lambda}_L(\tau)$ to obtain the condition for the maximum-likelihood estimate $\hat{\tau}_{ML}$.

In the case of binary (baseband) PAM, where $I_n = \pm 1$ with equal probability, the average over the data yields

$$\bar{\Lambda}_L(\tau) = \sum_n \ln \cosh C y_n(\tau) \tag{6.3–7}$$

just as in the case of the phase estimator, Since $\ln \cosh x \approx \frac{1}{2}x^2$ for small x, the square-law approximation

$$\bar{\Lambda}_L(\tau) \approx \tfrac{1}{2}C^2 \sum_n y_n^2(\tau) \tag{6.3–8}$$

is appropriate for low signal-to-noise ratios. For multilevel PAM, we may approximate the statistical characteristics of the information symbols $\{I_n\}$ by the Gaussian PDF, with zero-mean and unit variance. When we average $\Lambda(\tau)$ over the Gaussian PDF, the logarithm of $\bar{\Lambda}(\tau)$ is identical to $\bar{\Lambda}_L(\tau)$ given by Equation 6.3–8. Consequently, the non-decision-directed estimate of τ may be obtained by differentiating Equation 6.3–8. The result is an approximation to the ML estimate of the delay time. The derivative of Equation 6.3–8 is

$$\frac{d}{d\tau} \sum_n y_n^2(\tau) = 2 \sum_n y_n(\tau) \frac{dy_n(\tau)}{d\tau} = 0 \tag{6.3–9}$$

where $y_n(\tau)$ is given by Equation 6.3–5.

An implementation of a tracking loop based on the derivative of $\bar{\Lambda}_L(\tau)$ given by Equation 6.3–7 is shown in Figure 6.3–2. Alternatively, an implementation of a tracking loop based on Equation 6.3–9 is illustrated in Figure 6.3–3. In both structures, we observe that the summation serves as the loop filter that drives the VCC. It is interesting to note the resemblance of the timing loop in Figure 6.3–3 to the Costas loop for phase estimation.

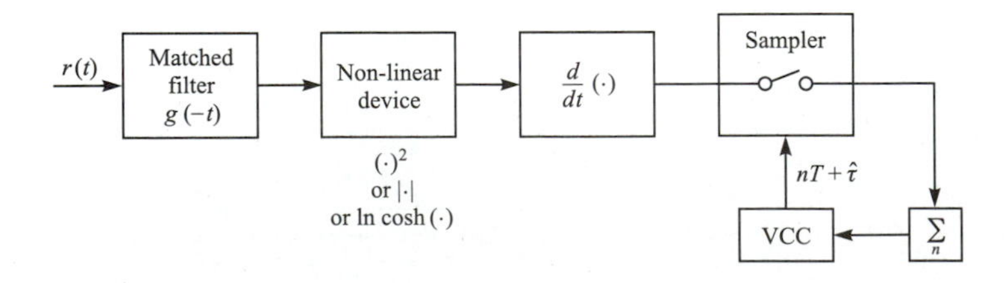

FIGURE 6.3–2
Non-decision-directed estimation of timing for binary baseband PAM.

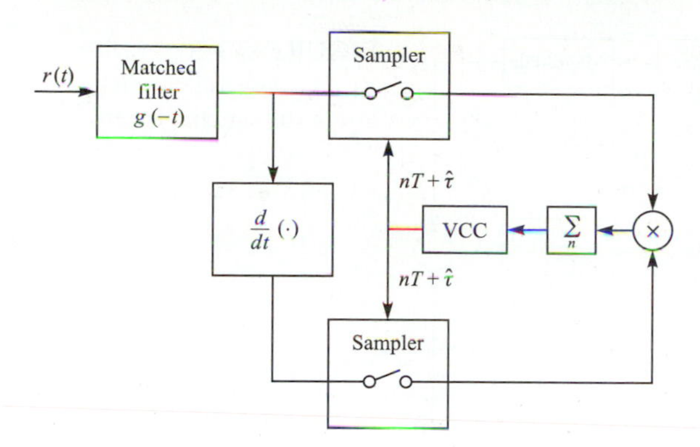

FIGURE 6.3–3
Non-decision-directed estimation of timing for baseband PAM.

Early–late gate synchronizers. Another non-decision-directed timing estimator exploits the symmetry properties of the signal at the output of the matched filter or correlator. To describe this method, let us consider the rectangular pulse $s(t)$, $0 \leqslant t \leqslant T$, shown in Figure 6.3–4a. The output of the filter matched to $s(t)$ attains its maximum value at time $t = T$, as shown in Figure 6.3–4b. Thus, the output of the matched filter is the time autocorrelation function of the pulse $s(t)$. Of course, this statement holds for any arbitrary pulse shape, so the approach that we describe applies in general to any signal pulse. Clearly, the proper time to sample the output of the matched filter for a maximum output is at $t = T$, i.e., at the peak of the correlation function.

In the presence of noise, the identification of the peak value of the signal is generally difficult. Instead of sampling the signal at the peak, suppose we sample early, at $t = T - \delta$ and late at $t = T + \delta$. The absolute values of the early samples $|y[m(T - \delta)]|$ and the late samples $|y[m(T + \delta)]|$ will be smaller (on the average in the presence of noise) than the samples of the peak value $|y(mT)|$. Since the autocorrelation function is even with respect to the optimum sampling time $t = T$, the absolute values of the correlation function at $t = T - \delta$ and $t = T + \delta$ are equal. Under this condition, the proper sampling time is the midpoint between $t = T - \delta$ and $t = T + \delta$. This condition forms the basis for the *early–late gate symbol synchronizer*.

Figure 6.3–5 illustrates the block diagram of an early–late gate synchronizer. In this figure, correlators are used in place of the equivalent matched filters. The two correlators integrate over the symbol interval T, but one correlator starts integrating δ seconds early relative to the estimated optimum sampling time and the other integrator starts integrating δ seconds late relative to the estimated

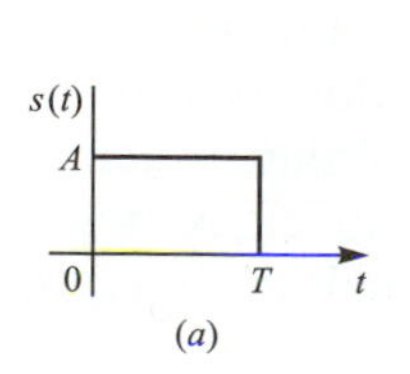

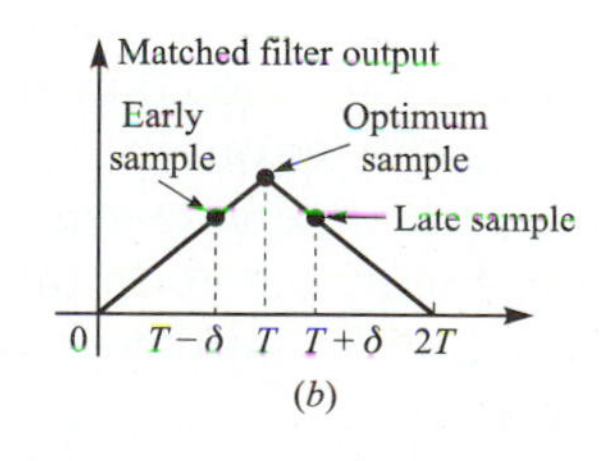

FIGURE 6.3–4
Rectangular signal pulse (a) and its matched filter output (b).

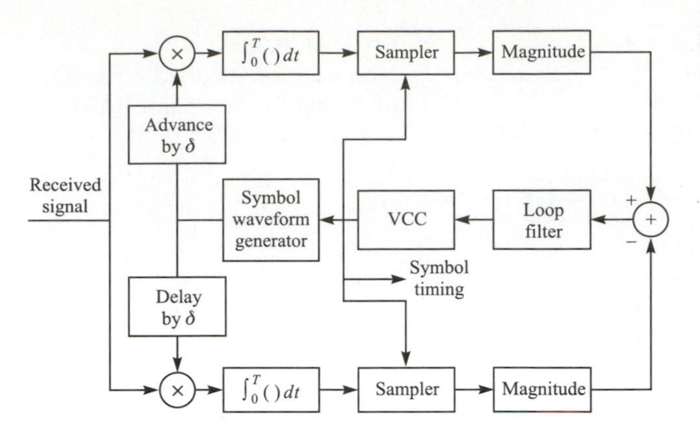

FIGURE 6.3–5
Block diagram of early–late gate synchronizer.

optimum sampling time. An error signal is formed by taking the difference beween the absolute values of the two correlator outputs. To smooth the noise corrupting the signal samples, the error signal is passed through a low-pass filter. If the timing is off relative to the optimum sampling time, the average error signal at the output of the low-pass filter is nonzero, and the clock signal is either retarded or advanced, depending on the sign of the error. Thus, the smoothed error signal is used to drive a VCC, whose output is the desired clock signal that is used for sampling. The output of the VCC is also used as a clock signal for a symbol waveform generator that puts out the same basic pulse waveform as that of the transmitting filter. This pulse waveform is advanced and delayed and then fed to the two correlators, as shown in Figure 6.3–5. Note that if the signal pulses are rectangular, there is no need for a signal pulse generator within the tracking loop.

We observe that the early–late gate synchronizer is basically a closed-loop control system whose bandwidth is relatively narrow compared to the symbol rate $1/T$. The bandwidth of the loop determines the quality of the timing estimate. A narrowband loop provides more averaging over the additive noise and, thus, improves the quality of the estimated sampling instants, provided that the channel propagation delay is constant and the clock oscillator at the transmitter is not drifting with time (or drifting very slowly with time). On the other hand, if the channel propagation delay is changing with time and/or the transmitter clock is also drifting with time, then the bandwidth of the loop must be increased to provide for faster tracking of time variations in symbol timing.

In the tracking mode, the two correlators are affected by adjacent symbols. However, if the sequence of information symbols has zero-mean, as is the case for PAM and some other signal modulations, the contribution to the output of the correlators from adjacent symbols averages out to zero in the low-pass filter.

An equivalent realization of the early–late gate synchronizer that is somewhat easier to implement is shown in Figure 6.3–6. In this case the clock signal from the VCC is advanced and delayed by δ, and these clock signals are used to sample the outputs of the two correlators.

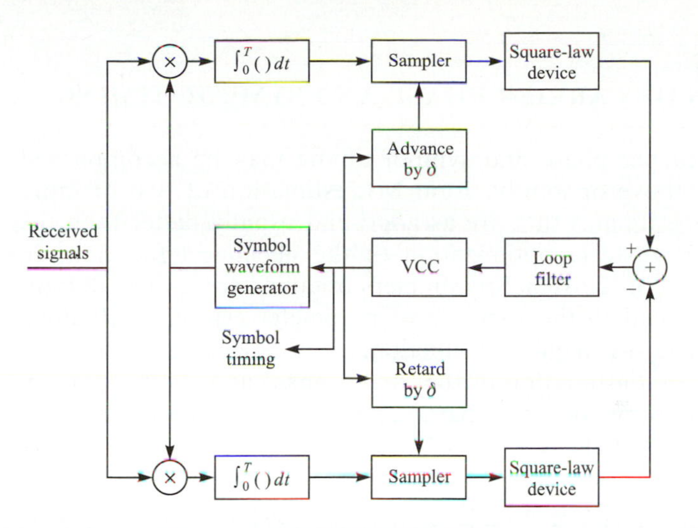

FIGURE 6.3–6
Block diagram of early–late gate synchronizer—an alternative form.

The early–late gate synchronizer described above is a non-decision-directed estimator of symbol timing that approximates the maximum-likelihood estimator. This assertion can be demonstrated by approximating the derivative of the log-likelihood function by the finite difference, i.e.,

$$\frac{d\bar{\Lambda}_L(\tau)}{d\tau} \approx \frac{\bar{\Lambda}_L(\tau+\delta) - \bar{\Lambda}_L(\tau-\delta)}{2\delta} \tag{6.3–10}$$

If we substitute for $\bar{\Lambda}_L(\tau)$ from Equation 6.3–8 into Equation 6.3–10, we obtain the approximation for the derivative as

$$\begin{aligned} \frac{d\bar{\Lambda}_L(\tau)}{d\tau} &= \frac{C^2}{4\delta}\sum_n \left[y_n^2(\tau+\delta) - y_n^2(\tau-\delta)\right] \\ &\approx \frac{C^2}{4\delta}\sum_n \left\{\left[\int_{T_0} r(t)g(t-nT-\tau-\delta)\,dt\right]^2 \right. \\ &\quad \left. - \left[\int_{T_0} r(t)g(t-nT-\tau+\delta)\,dt\right]^2\right\} \end{aligned} \tag{6.3–11}$$

But the mathematical expression in Equation 6.3–11 basically describes the functions performed by the early–late gate symbol synchronizers illustrated in Figures 6.3–5 and 6.3–6.

6.4 JOINT ESTIMATION OF CARRIER PHASE AND SYMBOL TIMING

The estimation of the carrier phase and symbol timing may be accomplished separately as described above or jointly. Joint ML estimation of two or more signal parameters yields estimates that are as good and usually better than the estimates obtained from separate optimization of the likelihood function. In other words, the variances of the signal parameters obtained from joint optimization are less than or equal to the variance of parameter estimates obtained from separately optimizing the likelihood function.

Let us consider the joint estimation of the carrier phase and symbol timing. The log-likelihood function for these two parameters may be expressed in terms of the equivalent low-pass signals as

$$\Lambda_L(\phi, \tau) = \text{Re}\left[\frac{1}{N_0}\int_{T_0} r(t)s_l^*(t; \phi, \tau)\, dt\right] \tag{6.4–1}$$

where $s_l(t; \phi, \tau)$ is the equivalent low-pass signal, which has the general form

$$s_l(t; \phi, \tau) = e^{-j\phi}\left[\sum_n I_n g(t - nT - \tau) + j\sum_n J_n w(t - nT - \tau)\right] \tag{6.4–2}$$

where $\{I_n\}$ and $\{J_n\}$ are the two information sequences.

We note that, for PAM, we may set $J_n = 0$ for all n, and the sequence $\{I_n\}$ is real. For QAM and PSK, we set $J_n = 0$ for all n and the sequence $\{I_n\}$ is complex-valued. For offset QPSK, both sequences $\{I_n\}$ and $\{J_n\}$ are nonzero and $w(t) = g(t - \frac{1}{2}T)$.

For decision-directed ML estimation of ϕ and τ, the log-likelihood function becomes

$$\Lambda_L(\phi, \tau) = \text{Re}\left\{\frac{e^{j\phi}}{N_0}\sum_n [I_n^* y_n(\tau) - jJ_n^* x_n(\tau)]\right\} \tag{6.4–3}$$

where

$$\begin{aligned} y_n(\tau) &= \int_{T_0} r(t)g^*(t - nT - \tau)\, dt \\ x_n(\tau) &= \int_{T_0} r(t)w^*(t - nT - \tau)\, dt \end{aligned} \tag{6.4–4}$$

Necessary conditions for the estimates of ϕ and τ to be the ML estimates are

$$\frac{\partial \Lambda_L(\phi, \tau)}{\partial \phi} = 0, \qquad \frac{\partial \Lambda_L(\phi, \tau)}{\partial \tau} = 0 \tag{6.4–5}$$

It is convenient to define

$$A(\tau) + jB(\tau) = \frac{1}{N_0} \sum [I_n^* y_n(\tau) - jJ_n^* x_n(\tau)] \tag{6.4–6}$$

With this definition, Equation 6.4–3 may be expressed in the simple form

$$\Lambda_L(\phi, \tau) = A(\tau)\cos\phi - B(\tau)\sin\phi \tag{6.4–7}$$

Now the conditions in Equation 6.4–5 for the joint ML estimates become

$$\frac{\partial \Lambda_L(\phi, \tau)}{\partial \phi} = -A(\tau)\sin\phi - B(\tau)\cos\phi = 0 \tag{6.4–8}$$

$$\frac{\partial \Lambda_L(\phi, \tau)}{\partial \tau} = \frac{\partial A(\tau)}{\partial \tau}\cos\phi - \frac{\partial B(\tau)}{\partial \tau}\sin\phi = 0 \tag{6.4–9}$$

From Equation 6.4–8, we obtain

$$\hat{\phi}_{\text{ML}} = -\tan^{-1}\left[\frac{B(\hat{\tau}_{\text{ML}})}{A(\hat{\tau}_{\text{ML}})}\right] \tag{6.4–10}$$

The solution to Equation 6.4–9 that incorporates Equation 6.4–10 is

$$\left[A(\tau)\frac{\partial A(\tau)}{\partial \tau} + B(\tau)\frac{\partial B(\tau)}{\partial \tau}\right]_{\tau=\hat{\tau}_{\text{ML}}} = 0 \tag{6.4–11}$$

The decision-directed tracking loop for QAM (or PSK) obtained from these equations is illustrated in Figure 6.4–1.

Offset QPSK requires a slightly more complex structure for joint estimation of ϕ and τ. The structure is easily derived from Equation 6.4–6 to 6.4–11.

In addition to the joint estimates given above, it is also possible to derive non-decision-directed estimates of the carrier phase and symbol timing, although we shall not pursue this approach.

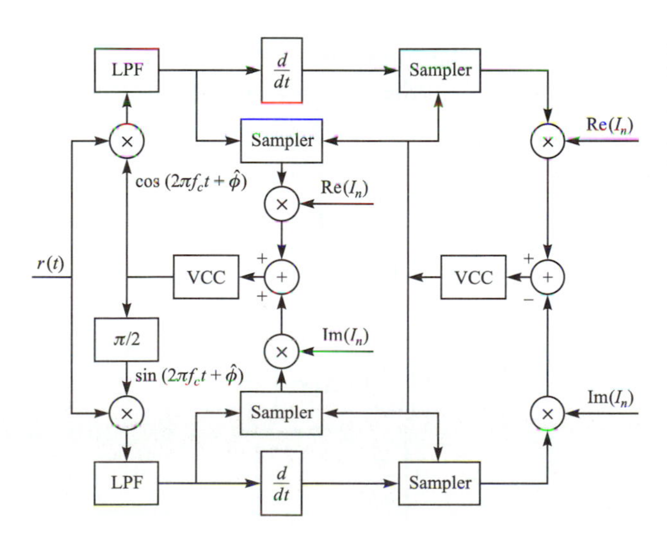

FIGURE 6.4–1
Decision-directed joint tracking loop for carrier phase and symbol timing in QAM and PSK.

We should also mention that one can combine the parameter estimation problem with the demodulation of the information sequence $\{I_n\}$. Thus, one can consider the joint maximum-likelihood estimation of $\{I_n\}$, the carrier phase ϕ, and the symbol timing parameter τ. Results on these joint estimation problems have appeared in the technical literature, e.g., Kobayashi (1971), Falconer (1976), and Falconer and Salz (1977).

6.5 PERFORMANCE CHARACTERISTICS OF ML ESTIMATORS

The quality of a signal parameter estimate is usually measured in terms of its bias and its variance. In order to define these terms, let us assume that we have a sequence of observations $[x_1 \ x_2 \ x_3 \cdots x_n] = \mathbf{x}$, with PDF $p(\mathbf{x}|\phi)$, from which we extract an estimate of a parameter ϕ. The bias of an estimate, say $\hat{\phi}(\mathbf{x})$, is defined as

$$\text{bias} = E[\hat{\phi}(\mathbf{x})] - \phi \tag{6.5–1}$$

where ϕ is the true value of the parameter. When $E[\hat{\phi}(\mathbf{x})] = \phi$, we say that the estimate is *unbiased*. The variance of the estimate $\hat{\phi}(x)$ is defined as

$$\sigma_{\hat{\phi}}^2 = E\{[\hat{\phi}(\mathbf{x})]^2\} - \{E[\hat{\phi}(\mathbf{x})]\}^2 \tag{6.5–2}$$

In general $\sigma_{\hat{\phi}}^2$ may be difficult to compute. However, a well-known result in parameter estimation (see Helstrom, 1968) is the Cramér–Rao lower bound on the mean square error defined as

$$E\{[\hat{\phi}(\mathbf{x}) - \phi]^2\} \geqslant \left\{\frac{\partial}{\partial\phi} E[\hat{\phi}(\mathbf{x})]\right\}^2 \bigg/ E\left\{\left[\frac{\partial}{\partial\phi}\ln p(\mathbf{x}|\phi)\right]^2\right\} \tag{6.5–3}$$

Note that when the estimate is unbiased, the numerator of Equation 6.5–3 is unity and the bound becomes a lower bound on the variance of $\sigma_{\hat{\phi}}^2$ of the estimate $\hat{\phi}(\mathbf{x})$, i.e.,

$$\sigma_{\hat{\phi}}^2 \geqslant 1 \bigg/ E\left\{\left[\frac{\partial}{\partial\phi}\ln p(\mathbf{x}|\phi)\right]^2\right\} \tag{6.5–4}$$

Since $\ln p(\mathbf{x}|\phi)$ differs from the log-likelihood function by a constant factor independent of ϕ, it follows that

$$\begin{aligned} E\left\{\left[\frac{\partial}{\partial\phi}\ln p(\mathbf{x}|\phi)\right]^2\right\} &= E\left\{\left[\frac{\partial}{\partial\phi}\ln \Lambda(\phi)\right]^2\right\} \\ &= -E\left\{\frac{\partial^2}{\partial\phi^2}\ln \Lambda(\phi)\right\} \end{aligned} \tag{6.5–5}$$

Therefore, the lower bound on the variance is

$$\sigma_{\hat{\phi}}^2 \geqslant 1 \bigg/ E\left\{\left[\frac{\partial}{\partial \phi} \ln \Lambda(\phi)\right]^2\right\} = -1 \bigg/ E\left[\frac{\partial^2}{\partial \phi^2} \ln \Lambda(\phi)\right] \tag{6.5–6}$$

This lower bound is a very useful result. It provides a benchmark for comparing the variance of any practical estimate to the lower bound. Any estimate that is unbiased and whose variance attains the lower bound is called an *efficient estimate*.

In general, efficient estimates are rare. When they exist, they are maximum-likelihood estimates. A well-known result from parameter estimation theory is that any ML parameter estimate is asymptotically (arbitrarily large number of observations) unbiased and efficient. To a large extent, these desirable properties constitute the importance of ML parameter estimates. It is also known that an ML estimate is asymptotically Gaussian distributed (with mean ϕ and variance equal to the lower bound given by Equation 6.5–6.)

In the case of the ML estimates described in this chapter for the two signal parameters, their variance is generally inversely proportional to the signal-to-noise ratio, or, equivalently, inversely proportional to the signal power multiplied by the observation interval T_0. Furthermore, the variance of the decision-directed estimates, at low error probabilities, are generally lower than the variance of non-decision-directed estimates. In fact, the performance of the ML decision-directed estimates for ϕ and τ attain the lower bound.

The following example is concerned with the evaluation of the Cramér–Rao lower bound for the ML estimate of the carrier phase.

EXAMPLE 6.5–1. The ML estimate of the phase of an unmodulated carrier was shown in Equation 6.2–11 to satisfy the condition

$$\int_{T_0} r(t) \sin(2\pi f_c t + \hat{\phi}_{\text{ML}})\, dt = 0 \tag{6.5–7}$$

where

$$\begin{aligned} r(t) &= s(t; \phi) + n(t) \\ &= A\cos(2\pi f_c t + \phi) + n(t) \end{aligned} \tag{6.5–8}$$

The condition in Equation 6.5–7 was derived by maximizing the log-likelihood function

$$\Lambda_L(\phi) = \frac{2}{N_0}\int_{T_0} r(t) s(t; \phi)\, dt \tag{6.5–9}$$

The variance of $\hat{\phi}_{\mathrm{ML}}$ is lower-bounded as

$$\begin{aligned}\sigma^2_{\hat{\phi}_{\mathrm{ML}}} &\geqslant \left\{\frac{2A}{N_0}\int_{T_0} E[r(t)]\cos(2\pi f_c t + \phi)\,dt\right\}^{-1} \\ &\geqslant \left\{\frac{A^2}{N_0}\int_{T_0} dt\right\}^{-1} = \frac{N_0}{A^2 T_0} \\ &\geqslant \frac{N_0/2T_0}{\frac{1}{2}A^2} = \frac{N_0 B_{\mathrm{eq}}}{\frac{1}{2}A^2}\end{aligned} \tag{6.5–10}$$

The factor $1/2T_0$ is simply the (one-sided) equivalent noise bandwidth of the ideal integrator, $A^2/2$ is the power in the sinusoidal signal, and $N_0 B_{\mathrm{eq}}$ is the total noise power.

From this example, we observe that the variance of the ML phase estimate is lower-bounded as

$$\sigma^2_{\hat{\phi}_{\mathrm{ML}}} \geqslant \frac{1}{\gamma_L} \tag{6.5–11}$$

where γ_L is the loop SNR. This is also the variance obtained for the phase estimate from a PLL with decision-directed estimation. As we have already observed, non-decision-directed estimates do not perform as well due to losses in the non-linearities required to remove the modulation, e.g., the squaring loss and the Mth-power loss.

Similar results can be obtained on the quality of the symbol timing estimates derived above. In addition to their dependence on the SNR, the quality of symbol timing estimates is a function of the signal pulse shape. For example, a pulse shape that is commonly used in practice is one that has a raised cosine spectrum (see Section 9.2). For such a pulse, the rms timing error ($\sigma_{\hat{\tau}}$) as a function of SNR is illustrated in Figure 6.5–1, for both decision-directed and non-decision-directed estimates. Note the significant improvement in performance of the decision-directed estimate compared with the non-decision-directed

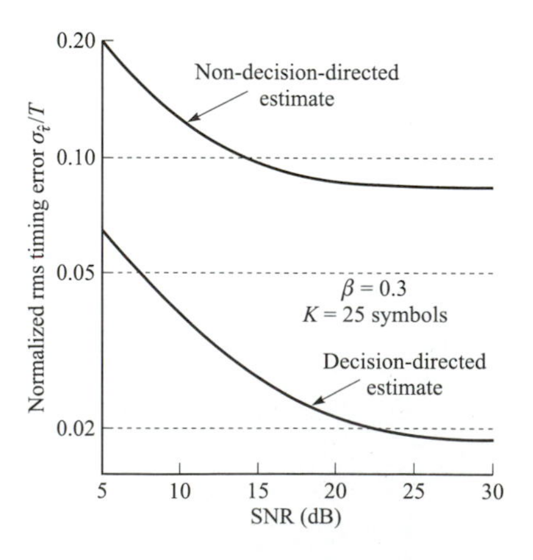

FIGURE 6.5–1
Performance of baseband symbol timing estimate for fixed signal and loop bandwidths. [*From* Synchronization Subsystems: Analysis and Design, *by L. Franks, 1983. Reprinted with permission of the author.*]

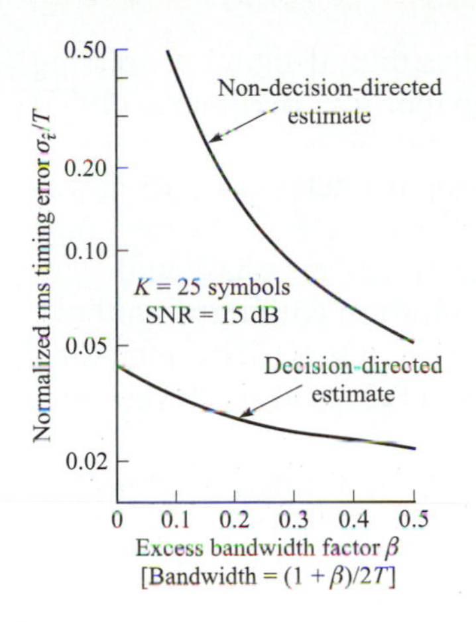

FIGURE 6.5–2
Performance of baseband symbol timing estimate for fixed SNR and fixed loop bandwidth. [*From* Synchronization Subsystems: Analysis and Design, *by L. Franks, 1983. Reprinted with permission of the author.*]

estimate. Now, if the bandwidth of the pulse is varied, the pulse shape is changed and, hence, the rms value of the timing error also changes. For example, when the bandwidth of the pulse that has a raised cosine spectrum is varied, the rms timing error varies as shown in Figure 6.5–2. Note that the error decreases as the bandwidth of the pulse increaes.

In conclusion, we have presented the ML method for signal parameter estimation and have applied it to the estimation of the carrier phase and symbol timing. We have also described their performance characteristics.

6.6 BIBLIOGRAPHICAL NOTES AND REFERENCES

Carrier recovery and timing synchronization are two topics that have been thoroughly investigated over the past three decades. The Costas loop was invented in 1956 and the decision-directed phase estimation methods were described in Proakis et al. (1964) and Natali and Walbesser (1969). The work on decision-directed estimation was motivated by earlier work of Price (1962a,b). Comprehensive treatments of phase-locked loops first appeared in the books by Viterbi (1966) and Gardner (1979). Books that cover carrier phase recovery and time synchronization techniques have been written by Stiffler (1971), Lindsey (1972), Lindsey and Simon (1973), Meyr and Ascheid (1990), Simon et al. (1995), Meyr et al. (1998), and Mengali and D'Andrea (1997).

A number of tutorial papers have appeared in IEEE journals on the PLL and on time synchronization. We cite, for example, the paper by Gupta (1975), which treats both analog and digital implementation of PLLs, and the paper by Lindsey and Chie (1981), which is devoted to the analysis of digital PLLs. In addition, the tutorial paper by Franks (1980) describes both carrier phase and symbol synchronization methods, including methods based on the maximum-likelihood estimation criterion. The paper by Franks is contained in a special issue of the *IEEE Transactions on Communications* (August 1980) devoted to synchronization. The paper by Mueller and Muller (1976) describes digital signal processing algorithms for extracting symbol timing and the paper by Bergmans (1995) evaluates the efficiency of data-aided timing recovery methods.

Aplication of the maximum-likelihood criterion to parameter estimation was first described in the context of radar parameter estimation (range and range rate). Subsequently, this optimal criterion was applied to carrier phase and symbol timing estimation as well as to joint parameter estimation with data symbols. Papers on these topics have been published by several researchers, including Falconer (1976), Mengali (1977), Falconer and Salz (1977), and Meyers and Franks (1980).

The Cramér–Rao lower bound on the variance of a parameter estimate is derived and evaluated in a number of standard texts on detection and estimation theory, such as Helstrom (1968) and Van Trees (1968). It is also described in several books on mathematical statistics, such as the book by Cramér (1946).

PROBLEMS

6.1 Prove the relation Equation 6.1–7.

6.2 Sketch the equivalent realization of the binary PSK receiver in Figure 6.1–1 that employs a matched filter instead of a correlator.

6.3 Suppose that the loop filter (see Equation 6.2–14) for a PLL has the transfer function

$$G(s) = \frac{1}{s + \sqrt{2}}$$

a) Determine the closed-loop transfer function $H(s)$ and indicate if the loop is stable.

b) Determine the damping factor and the natural frequency of the loop.

6.4 Consider the PLL for estimating the carrier phase of a signal in which the loop filter is specified as

$$G(s) = \frac{K}{1 + \tau_1 s}$$

a) Determine the closed-loop transfer function $H(s)$ and its gain at $f = 0$.

b) For what range of values of τ_1 and K is the loop stable?

6.5 The loop filter $G(s)$ in a PLL is implemented by the circuit shown in Figure P6.5. Determine the system function $G(s)$ and express the time constants τ_1 and τ_2 in terms of the circuit parameters.

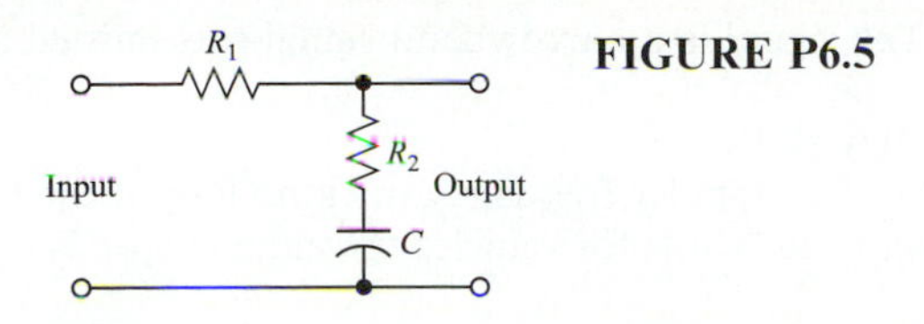

FIGURE P6.5

6.6 The loop filter $G(s)$ in a PLL is implemented with the active filter shown in Figure P6.6. Determine the system function $G(s)$ and express the time constants τ_1 and τ_2 in terms of the circuit parameters.

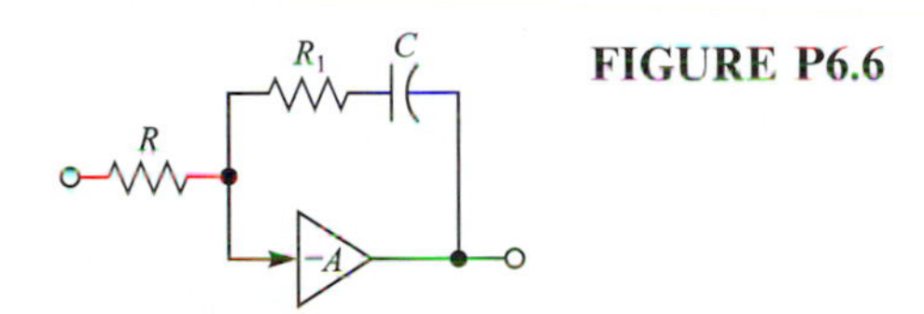

FIGURE P6.6

6.7 Show that the early–late gate synchronizer illustrated in Figure 6.3–5 is a close approximation to the timing recovery system illustrated in Figure P6.7.

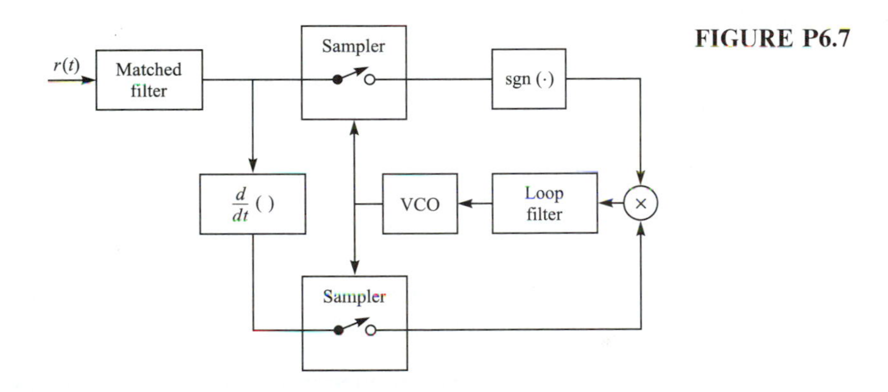

FIGURE P6.7

6.8 Based on an ML criterion, determine a carrier phase estimation method for binary on–off keying modulation.

6.9 In the transmission and reception of signals to and from moving vehicles, the transmitted signal frequency is shifted in direct proportion to the speed of the vehicle. The so-called *Doppler frequency shift* imparted to a signal that is received in a vehicle traveling at a velocity v relative to a (fixed) transmitter is given by the formula

$$f_D = \pm\frac{v}{\lambda}$$

where λ is the wavelength, and the sign depends on the direction (moving toward or moving away) that the vehicle is traveling relative to the transmitter. Suppose that a vehicle is traveling at a speed of 100 km/h relative to a base station in a mobile cellular communication system. The signal is a narrowband signal transmitted at a carrier frequency of 1 GHz.

a) Determine the Doppler frequency shift.

b) What should be the bandwidth of a Doppler frequency tracking loop if the loop is designed to track Doppler frequency shifts for vehicles traveling at speeds up to 100 km/h?

c) Suppose the transmitted signal bandwidth is 2 MHz centered at 1 GHz. Determine the Doppler frequency spread between the upper and lower frequencies in the signal.

6.10 Show that the mean value of the ML estimate in Equation 6.2–38 is ϕ, i.e., that the estimate is unbiased.

6.11 Determine the PDF of the ML phase estimate in Equation 6.2–38.

6.12 Determine the ML phase estimate for offset QPSK.

6.13 A single-sideband PAM signal may be represented as

$$u_m(t) = A_m[g_T(t)\cos 2\pi f_c t - \hat{g}_T(t)\sin 2\pi f_c t]$$

where $\hat{g}_T(t)$ is the Hilbert transform of $g_T(t)$ and A_m is the amplitude level that conveys the information. Demonstrate mathematically that a Costas loop can be used to demodulate the SSB PAM signal.

6.14 A carrier component is transmitted on the quadrature carrier in a communication system that transmits information via binary PSK. Hence, the received signal has the form

$$r(t) = \pm\sqrt{2P_s}\cos(2\pi f_c + \phi) + \sqrt{2P_c}\sin(2\pi f_c + \phi) + n(t)$$

where ϕ is the carrier phase and $n(t)$ is AWGN. The unmodulated carrier component is used as a pilot signal at the receiver to estimate the carrier phase.

a) Sketch a block diagram of the receiver, including the carrier phase estimator.

b) Illustrate mathematically the operations involved in the estimation of the carrier phase ϕ.

c) Express the probability of error for the detection of the binary PSK signal as a function of the total transmitted power $P_T = P_s + P_c$. What is the loss in performance due to the allocation of a portion of the transmitted power to the pilot signal? Evaluate the loss for $P_c/P_T = 0.1$.

6.15 Determine the signal and noise components at the input to a fourth-power ($M = 4$) PLL that is used to generate the carrier phase for demodulation of QPSK. By ignoring all noise components except those that are linear in the noise $n(t)$, determine the variance of the phase estimate at the output of the PLL.

6.16 The probability of error for binary PSK demodulation and detection when there is a carrier phase error ϕ_e is

$$P_2(\phi_e) = Q\left(\sqrt{\frac{2\mathcal{E}_b}{N_0}\cos^2\phi_e}\right)$$

Suppose that the phase error from the PLL is modeled as a zero-mean Gaussian random variable with variance $\sigma_\phi^2 \ll \pi$. Determine the expression for the average probability of error (in integral form).

6.17 Determine the ML estimate of the time delay τ for the QAM signal of the form

$$s(t) = \text{Re}[s_l(t; \tau)e^{j2\pi f_c t}]$$

where

$$s_l(t; \tau) = \sum_n I_n g(t - nT - \tau)$$

and $\{I_n\}$ is a sequence of complex-valued data.

6.18 Determine the joint ML estimate of τ and ϕ for a PAM signal.

6.19 Determine the joint ML estimate of τ and ϕ for offset QPSK.

7

Channel Capacity and Coding

In Chapter 5, we considered the problem of digital modulation by means of $M = 2^k$ signal waveforms, where each waveform conveys k bits of information. We observed that some modulation methods provide better performance than others. In particular, we demonstrated that orthogonal signaling waveforms allow us to make the probability of error arbitrarily small by letting the number of waveforms $M \to \infty$, provided that the SNR per bit $\gamma_b \geqslant -1.6$ dB. Thus, we can operate at the capacity of the additive white Gaussian noise channel in the limit as the bandwidth expansion factor $B_e = W/R \to \infty$. This is a heavy price to pay, because B_e grows exponentially with the block length k. Such inefficient use of channel bandwidth is highly undesirable.

In this and the following chapter, we consider signal waveforms generated from either binary or nonbinary sequences. The resulting waveforms are generally characterized by a bandwidth expansion factor that grows only linearly with k. Consequently, coded waveforms offer the potential for greater bandwidth efficiency than orthogonal M-ary waveforms. We shall observe that, in general, coded waveforms offer performance advantages not only in power-limited applications where $R/W < 1$, but also in bandwidth-limited systems where $R/W > 1$.

We begin by establishing several channel models that will be used to evaluate the benefits of channel coding, and we shall introduce the concept of channel capacity for the various channel models. Then, we treat the subject of code design for efficient communications.

7.1 CHANNEL MODELS AND CHANNEL CAPACITY

In the model of a digital communication system described in Section 1.1, we recall that the transmitter building blocks consist of the discrete-input, discrete-output channel encoder followed by the modulator. The function of the discrete channel encoder is to introduce, in a controlled manner, some redundancy in the binary information sequence, which can be used at the receiver to overcome the effects of noise and interference encountered in the transmission of the signal through the channel. The encoding process generally involves taking k information bits at a time and mapping each k-bit sequence into a unique n-bit sequence, called a *code word*. The amount of redundancy introduced by the encoding of the data in this manner is measured by the ratio n/k. The reciprocal of the ratio, namely k/n, is called the *code rate*.

The binary sequence at the output of the channel encoder is fed to the modulator, which serves as the interface to the communication channel. As we have discussed, the modulator may simply map each binary digit into one of two possible waveforms, i.e., a 0 is mapped into $s_1(t)$ and a 1 is mapped into $s_2(t)$. Alternatively, the modulator may transmit q-bit blocks at a time by using $M = 2^q$ possible waveforms.

At the receiving end of the digital communication system, the demodulator processes the channel-corrupted waveform and reduces each waveform to a scalar or a vector that represents an estimate of the transmitted data symbol (binary or M-ary). The detector, which follows the demodulator, may decide on whether the transmitted bit is a 0 or a 1. In such a case, the detector has made a *hard decision*. If we view the decision process at the detector as a form of quantization, we observe that a hard decision corresponds to binary quantization of the demodulator output. More generally, we may consider a detector that quantizes to $Q > 2$ levels, i.e., a Q-ary detector. If M-ary signals are used, then $Q \geqslant M$. In the extreme case when no quantization is performed, $Q = \infty$. In the case where $Q > M$, we say that the detector has made a *soft decision*.

The quantized output from the detector is then fed to the channel decoder, which exploits the available redundancy to correct for channel disturbances.

In the following sections, we describe three channel models that will be used to establish the maximum achievable bit rate for the channel.

7.1.1 Channel Models

In this section we describe channel models that will be useful in the design of codes. The simplest is the binary symmetric channel, which corresponds to the case with $M = 2$ and hard decisions at the detector.

Binary symmetric channel. Let us consider an additive noise channel and let the modulator and the demodulator/detector be included as parts of the channel. If the modulator employs binary waveforms and the detector makes hard decisions, then the composite channel, shown in Figure 7.1–1, has a discrete-time binary input sequence and a discrete-time binary output sequence. Such a composite channel is characterized by the set $X = \{0, 1\}$ of possible inputs, the set of $Y = \{0, 1\}$ of possible outputs, and a set of conditional probabilities that relate the possible outputs to the possible inputs. If the channel noise and other dis-

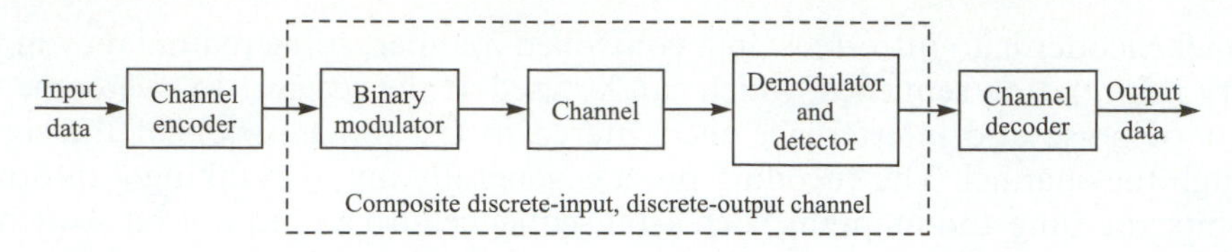

FIGURE 7.1–1
A composite discrete-input, discrete-output channel formed by including the modulator and the demodulator/detector as part of the channel.

turbances cause statistically independent errors in the transmitted binary sequence with average probability p, then

$$\begin{aligned} P(Y=0|X=1) &= P(Y=1|X=0) = p \\ P(Y=1|X=1) &= P(Y=0|X=0) = 1-p \end{aligned} \tag{7.1–1}$$

Thus, we have reduced the cascade of the binary modulator, the waveform channel, and the binary demodulator and detector into an equivalent discrete-time channel which is represented by the diagram shown in Figure 7.1–2. This binary-input, binary-output, symmetric channel is simply called a *binary symmetric channel* (BSC). Since each output bit from the channel depends only on the corresponding input bit, we say that the channel is memoryless.

Discrete memoryless channels. The BSC is a special case of a more general discrete-input, discrete-output channel. Suppose that the output symbols from the channel encoder are q-ary symbols, i.e., $X = \{x_0, x_1, \ldots, x_{q-1}\}$ and the output of the detector consists of Q-ary symbols, where $Q \geqslant M = 2^q$. If the channel and the modulation are memoryless, then the input–output characteristics of the composite channel, shown in Figure 7.1–1, are described by a set of qQ conditional probabilities

$$P(Y = y_i|X = x_j) \equiv P(y_i|x_j) \tag{7.1–2}$$

where $i = 0, 1, \ldots, Q-1$ and $j = 0, 1, \ldots, q-1$. Such a channel is called a *discrete memoryless channel* (DMC), and its graphical representation is shown in Figure 7.1–3. Hence, if the input to a DMC is a sequence of n symbols $u_1, u_2, \ldots, u_n$ selected from the alphabet X and the corresponding output is the

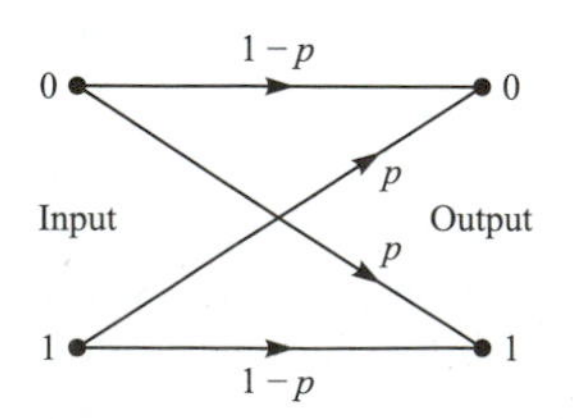

FIGURE 7.1–2
Binary symmetric channel.

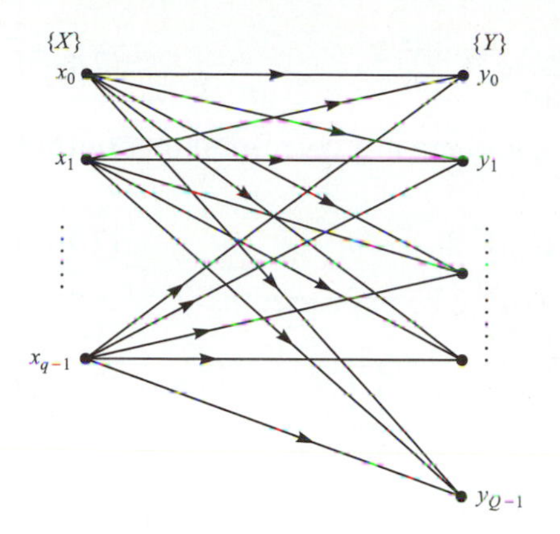

FIGURE 7.1–3
Discrete q-ary input, Q-ary output channel.

sequence $v_1, v_2, \ldots, v_n$ of symbols from the alphabet Y, the joint conditional probability is

$$P(Y_1 = v_1, Y_2 = v_2, \ldots, Y_n = v_n | X = u_1, \ldots, X = u_n) = \prod_{k=1}^{n} P(Y = v_k | X = u_k) \qquad (7.1\text{–}3)$$

This expression is simply a mathematical statement of the memoryless condition.

In general, the condition probabilities $\{P(y_i|x_j)\}$ that characterize a DMC can be arranged in the matrix form $\mathbf{P} = [p_{ji}]$, where, by definition, $p_{ji} \equiv P(y_i|x_j)$. $\mathbf{P}$ is called the *probability transition matrix* for the channel.

Discrete-input, continuous-output channel. Now, suppose that the input to the modulator comprises symbols selected from a finite and discrete input alphabet $X = \{x_0, x_1, \ldots, x_{q-1}\}$ and the output of the detector is unquantized ($Q = \infty$). Then, the input to the channel decoder can assume any value on the real line, i.e., $Y = \{-\infty, \infty\}$. This leads us to define a composite discrete-time memoryless channel that is characterized by the discrete input X, the continuous output Y, and the set of conditional probability density functions

$$p(y|X = x_k), \qquad k = 0, 1, \ldots, q-1$$

The most important channel of this type is the additive white Gaussian noise (AWGN) channel, for which

$$Y = X + G \qquad (7.1\text{–}4)$$

where G is a zero-mean Gaussian random variable with variance σ^2 and $X = x_k$, $k = 0, 1, \ldots, q-1$. For a given X, it follows that Y is Gaussian with mean x_k and variance σ^2. That is,

$$p(y|X = x_k) = \frac{1}{\sqrt{2\pi}\sigma} e^{-y-x_k)^2/2\sigma^2} \tag{7.1–5}$$

For any given input sequence, X_i, $i = 1, 2, \ldots, n$, there is a corresponding output sequence

$$Y_i = X_i + G_i, \qquad i = 1, 2, \ldots, n \tag{7.1–6}$$

The condition that the channel is memoryless may be expressed as

$$p(y_1, y_2, \ldots, y_n | X_1 = u_1, X_2 = u_2, \ldots, X_n = u_n) = \prod_{i=1}^{n} p(y_i | X_i = u_i) \tag{7.1–7}$$

Waveform channels. We may separate the modulator and demodulator from the physical channel, and consider a channel model in which the inputs are waveforms and the outputs are waveforms. Let us assume that such a channel has a given bandwidth W, with ideal frequency response $C(f) = 1$ within the bandwidth W, and the signal at its output is corrupted by additive white Gaussian noise. Suppose, that $x(t)$ is a band-limited input to such a channel and $y(t)$ is the corresponding output. Then,

$$y(t) = x(t) + n(t) \tag{7.1–8}$$

where $n(t)$ represents a sample function of the additive noise process. A suitable method for defining a set of probabilities that characterize the channel is to expand $x(t)$, $y(t)$, and $n(t)$ into a complete set of orthonormal functions. That is, we express $x(t)$, $y(t)$, and $n(t)$ in the form

$$\begin{aligned} y(t) &= \sum_i y_i f_i(t) \\ x(t) &= \sum_i x_i f_i(t) \\ n(t) &= \sum_i n_i f_i(t) \end{aligned} \tag{7.1–9}$$

where $\{y_i\}$, $\{x_i\}$, and $\{n_i\}$ are the sets of coefficients in the corresponding expansions, e.g.,

$$\begin{aligned} y_i &= \int_0^T y(t) f_i^*(t)\, dt \\ &= \int_0^T [x(t) + n(t)] f_i^*(t)\, dt \\ &= x_i + n_i \end{aligned} \tag{7.1–10}$$

The functions $\{f_i(t)\}$ form a complete orthonormal set over the interval $(0, T)$, i.e.,

$$\int_0^T f_i(t) f_j^*(t)\, dt = \delta_{ij} = \begin{cases} 1 & (i = j) \\ 0 & (i \neq j) \end{cases} \tag{7.1–11}$$

where δ_{ij} is the Kronecker delta function. Since the Gaussian noise is white, any complete set of orthonormal functions may be used in the expansions in Equation 7.1–9.

We may now use the coefficients in the expansion for characterizing the channel. Since

$$y_i = x_i + n_i$$

where n_i is Gaussian, it follows that

$$p(y_i|x_i) = \frac{1}{\sqrt{2\pi}\sigma_i} e^{-(y_i - x_i)^2/2\sigma_i^2}, \qquad i = 1, 2, \ldots \tag{7.1–12}$$

Since the functions $\{f_i(t)\}$ in the expansion are orthonormal, it follows that the $\{n_i\}$ are uncorrelated. Since they are Gaussian, they are also statistically independent. Hence,

$$p(y_1, y_2, \ldots, y_N|x_1, x_2, \ldots, x_N) = \prod_{i=1}^{N} p(y_i|x_i) \tag{7.1–13}$$

for any N. In this manner, the waveform channel is reduced to an equivalent discrete-time channel characterized by the conditional PDF given in Equation 7.1–12.

When the additive noise is white and Gaussian with spectral density $\frac{1}{2}N_0$, the variances $\sigma_i^2 = \frac{1}{2}N_0$ for all i in Equation 7.1–12. In this case, samples of $x(t)$ and $y(t)$ may be taken at the Nyquist rate of $2W$ samples/s, so that $x_i = x(i/2W)$ and $y_i = y(i/2W)$. Since the noise is white, the noise samples are statistically independent. Thus, Equations 7.1–12 and 7.1–13 describe the statistics of the sampled signal. We note that in a time interval of length T, there are $N = 2WT$ samples. This parameter is used below in obtaining the capacity of the band-limited AWGN waveform channel.

The choice of which channel model to use at any one time depends on our objectives. If we are interested in the design and analysis of the performance of the discrete channel encoder and decoder, it is appropriate to consider channel models in which the modulator and demodulator are a part of the composite channel. On the other hand, if our intent is to design and analyze the performance of the digital modulator and digital demodulator, we use a channel model for the waveform channel.

7.1.2 Channel Capacity

Now let us consider a DMC having an input alphabet $X = \{x_0, x_1, \ldots, x_{q-1}\}$, an output alphabet $Y = \{y_0, y_1, \ldots, y_{Q-1}\}$, and the set of transition probabilities $P(y_i|x_j)$ as defined in Equation 7.1–2. Suppose that the symbol x_j is transmitted and the symbol y_i is received. The mutual information provided about the event $X = x_j$ by the occurrence of the event $Y = y_i$ is $\log[P(y_i|x_j)/P(y_i)]$, where

$$P(y_i) \equiv P(Y = y_i) = \sum_{k=0}^{q-1} P(x_k)P(y_i|x_k) \tag{7.1–14}$$

Hence, the average mutual information provided by the output Y about the input X is

$$I(X; Y) = \sum_{j=0}^{q-1}\sum_{i=0}^{Q-1} P(x_j)P(y_i|x_j) \log \frac{P(y_i|x_j)}{P(y_i)} \tag{7.1–15}$$

The channel characteristics determine the transition probabilities $P(y_i|x_j)$, but the probabilities of the input symbols are under the control of the discrete channel encoder. The value of $I(X; Y)$ maximized over the set of input symbol probabilities $P(x_j)$ is a quantity that depends only on the characteristics of the DMC through the conditional probabilities $P(y_i|x_j)$. This quantity is called the *capacity* of the channel and is denoted by C. That is, the capacity of a DMC is defined as

$$\begin{aligned} C &= \max_{P(x_j)} I(X; Y) \\ &= \max_{P(x_j)} \sum_{j=0}^{q-1}\sum_{i=0}^{Q-1} P(x_j)P(y_i|x_j) \log \frac{P(y_i|x_j)}{P(y_i)} \end{aligned} \tag{7.1–16}$$

The maximization of $I(X; Y)$ is performed under the constraints that

$$P(x_j) \geqslant 0$$

$$\sum_{j=0}^{q-1} P(x_j) = 1$$

The units of C are bits per input symbol into the channel (bits per channel use) when the logarithm is base 2, and nats per input symbol when the natural logarithm (base e) is used. If a symbol enters the channel every τ_s seconds, the channel capacity in bits/s or nats/s is C/τ_s.

EXAMPLE 7.1–1. For the BSC with transition probabilities

$$P(0|1) = P(1|0) = p$$

the average mutual information is maximized when the input probabilities $P(0) = P(1) = \frac{1}{2}$. Thus, the capacity of the BSC is

$$C = p \log 2p + (1 - p) \log 2(1 - p) = 1 - H(p) \tag{7.1–17}$$

where $H(p)$ is the binary entropy function. A plot of C versus p is illustrated in Figure 7.1–4. Note that for $p = 0$, the capacity is 1 bit/channel use. On the other hand, for $p = \frac{1}{2}$, the mutual information between input and output is zero. Hence, the channel capacity is zero. For $\frac{1}{2} < p \leqslant 1$, we may reverse the position of 0 and 1 at the output of the BSC, so that C becomes symmetric with respect to the point $p = \frac{1}{2}$. In our treatment of binary modulation and demodulation given in Chapter 5, we showed that p is a monotonic function of the signal-to-noise ratio (SNR) as illustrated in Figure 7.1–5a. Consequently when C is plotted as a function of the SNR, it

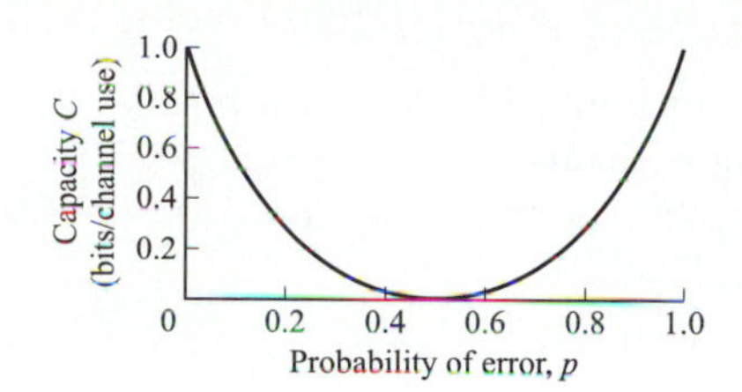

FIGURE 7.1–4
The capacity of a BSC as a function of the error probability p.

increases monotonically as the SNR increases. This characteristic behavior of C versus SNR is illustrated in Figure 7.1–5b.

Next let us consider the discrete-time AWGN memoryless channel described by the transition probability density functions defined by Equation 7.1–5. The capacity of this channel in bits per channel use is the maximum average mutual information between the discrete input $X = \{x_0, x_1, \ldots, x_{q-1}\}$ and the output $Y = \{-\infty, \infty\}$. That is,

$$C = \max_{P(x_i)} \sum_{i=0}^{q-1} \int_{-\infty}^{\infty} p(y|x_i)P(x_i) \log_2 \frac{p(y|x_i)}{p(y)} \, dy \tag{7.1–18}$$

where

$$p(y) = \sum_{k=0}^{q-1} p(y|x_k)P(x_k) \tag{7.1–19}$$

EXAMPLE 7.1–2. Let us consider a binary-input AWGN memoryless channel with possible inputs $X = A$ and $X = -A$. The average mutual information $I(X; Y)$ is maximized when the input probabilities are $P(X = A) = P(X = -A) = \frac{1}{2}$. Hence, the capacity of this channel in bits per channel use is

$$\begin{aligned} C &= \tfrac{1}{2} \int_{-\infty}^{\infty} p(y|A) \log_2 \frac{p(y|A)}{p(y)} \, dy \\ &\quad + \tfrac{1}{2} \int_{-\infty}^{\infty} p(y|-A) \log_2 \frac{p(y|-A)}{p(y)} \, dy \end{aligned} \tag{7.1–20}$$

Figure 7.1–6 illustrates C as a function of the ratio $A^2/2\sigma^2$. Note that C increases monotonically from 0 to 1 bit per symbol as this ratio increases.

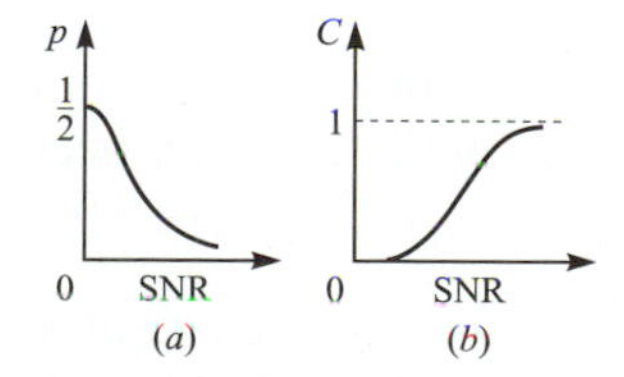

FIGURE 7.1–5
General behavior of error probability and channel capacity as a function of SNR.

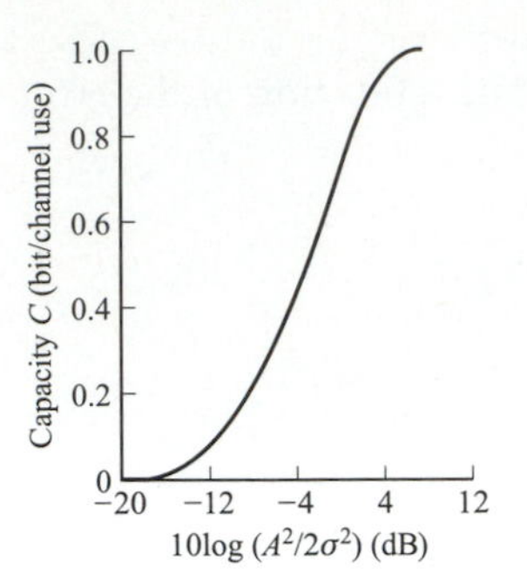

FIGURE 7.1–6
Channel capacity as a function of $A^2/2\sigma^2$ for a binary-input AWGN memoryless channel.

It is interesting to note that in the two channel models described above, the choice of equally probable input symbols maximizes the average mutual information. Thus, the capacity of the channel is obtained when the input symbols are equally probable. This is not always the solution for the capacity formulas given in Equations 7.1–16 and 7.1–18, however. Nothing can be said in general about the input probability assignment that maximizes the average mutual information. However, in the two channel models considered above, the channel transition probabilities exhibit a form of symmetry that results in the maximum of $I(X; Y)$ being obtained when the input symbols are equally probable. The symmetry condition can be expressed in terms of the elements of the probability transition matrix **P** of the channel. When each row of this matrix is a permutation of any other row and each column is a permutation of any other column, the probability transition matrix is symmetric and input symbols with equal probability maximize $I(X; Y)$.

In general, necessary and sufficient conditions for the set of input probabilities $\{P(x_j)\}$ to maximize $I(X; Y)$ and, thus, to achieve capacity on a DMC are that (Problem 7.1)

$$\begin{aligned} I(x_j; Y) &= C \qquad \text{for all } j \text{ with } P(x_j) > 0 \\ I(x_j; Y) &\leqslant C \qquad \text{for all } j \text{ with } P(x_j) = 0 \end{aligned} \tag{7.1–21}$$

where C is the capacity of the channel and

$$I(x_j; Y) = \sum_{i=0}^{Q-1} P(y_i|x_j) \log \frac{P(y_i|x_j)}{P(y_i)} \tag{7.1–22}$$

Usually, it is relatively easy to check if the equally probable set of input symbols satisfy the conditions 7.1–21. If they do not, then one must determine the set of unequal probabilities $\{P(x_j)\}$ that satisfy Equation 7.1–21.

Now let us consider a band-limited waveform channel with additive white Gaussian noise. Formally, the capacity of the channel per unit time has been defined by Shannon (1948b) as

$$C = \lim_{T\to\infty} \max_{p(x)} \frac{1}{T} I(X; Y) \tag{7.1–23}$$

where the average mutual information $I(X; Y)$ is given in Equation 3.2–17. Alternatively, we may use the samples or the coefficients $\{y_i\}$, $\{x_i\}$, and $\{n_i\}$ in the series expansions of $y(t)$, $x(t)$, and $n(t)$, respectively, to determine the average mutual information between $\mathbf{x}_N = [x_1\ x_2 \cdots x_N]$ and $\mathbf{y}_N = [y_1\ y_2 \cdots y_N]$, where $N = 2WT$, $y_i = x_i + n_i$, and $p(y_i|x_i)$ is given by Equation 7.1–12. The average mutual information between $\mathbf{x}_N$ and $\mathbf{y}_N$ for the AWGN channel is

$$\begin{aligned} I(\mathbf{X}_N; \mathbf{Y}_N) &= \int_{\mathbf{x}_N} \cdots \int \int_{\mathbf{y}_N} \cdots \int p(\mathbf{y}_N|\mathbf{x}_N) p(\mathbf{x}_N) \log \frac{p(\mathbf{y}_N|\mathbf{x}_N)}{p(\mathbf{y}_N)} \, d\mathbf{x}_N \, d\mathbf{y}_N \\ &= \sum_{i=1}^{N} \int_{-\infty}^{\infty} \int_{-\infty}^{\infty} p(y_i|x_i) p(x_i) \log \frac{p(y_i|x_i)}{p(y_i)} \, dy_i \, dx_i \end{aligned} \tag{7.1–24}$$

where

$$p(y_i|x_i) = \frac{1}{\sqrt{\pi N_0}} e^{-(y_i - x_i)^2/N_0} \tag{7.1–25}$$

The maximum of $I(X; Y)$ over the input PDFs $p(x_i)$ is obtained when the $\{x_i\}$ are statistically independent zero-mean Gaussian random variables, i.e.,

$$p(x_i) = \frac{1}{\sqrt{2\pi}\sigma_x} e^{-x_i^2/2\sigma_x^2} \tag{7.1–26}$$

where σ_x^2 is the variance of each x_i. Then, it follows from Equation 7.1–24 that

$$\begin{aligned} \max_{p(x)} I(\mathbf{X}_N; \mathbf{Y}_N) &= \sum_{i=1}^{N} \tfrac{1}{2} \log\left(1 + \frac{2\sigma_x^2}{N_0}\right) \\ &= \tfrac{1}{2} N \log\left(1 + \frac{2\sigma_x^2}{N_0}\right) \\ &= WT \log\left(1 + \frac{2\sigma_x^2}{N_0}\right) \end{aligned} \tag{7.1–27}$$

Suppose that we put a constraint on the average power in $x(t)$. That is,

$$\begin{aligned} P_{\text{av}} &= \frac{1}{T} \int_0^T E[x^2(t)] \, dt \\ &= \frac{1}{T} \sum_{i=1}^{N} E(x_i^2) \\ &= \frac{N\sigma_x^2}{T} \end{aligned} \tag{7.1–28}$$

Hence,

$$\sigma_x^2 = \frac{TP_{\text{av}}}{N} = \frac{P_{\text{av}}}{2W} \tag{7.1–29}$$

Substitution of this result into Equation 7.1–27 for σ_x^2 yields

$$\max_{p(x)} I(\mathbf{X}_N; \mathbf{Y}_N) = WT \log\left(1 + \frac{P_{\text{av}}}{WN_0}\right) \tag{7.1–30}$$

Finally, the channel capacity per unit time is obtained by dividing the result in Equation 7.1–30 by T. Thus

$$C = W \log\left(1 + \frac{P_{\text{av}}}{WN_0}\right) \tag{7.1–31}$$

This is the basic formula for the capacity of the band-limited AWGN waveform channel with a band-limited and average power-limited input. It was originally derived by Shannon (1948b).

A plot of the capacity in bits/s normalized by the bandwidth W is plotted in Figure 7.1–7 as a function of the ratio of signal power P_{av} to noise power WN_0. Note that the capacity increases monotonically with increasing SNR. Thus, for a fixed bandwidth, the capacity of the waveform channel increases with an increase in the transmitted signal power. On the other hand, if P_{av} is fixed, the capacity can be increased by increasing the bandwidth W. Figure 7.1–8 illustrates a graph of C versus W. Note that as W approaches infinity, the capacity of the channel approaches the asymptotic value

$$C_\infty = \frac{P_{\text{av}}}{N_0} \log_2 e = \frac{P_{\text{av}}}{N_0 \ln 2} \quad \text{bits/s} \tag{7.1–32}$$

It is instructive to express the normalized channel capacity C/W as a function of the SNR per bit. Since P_{av} represents the average transmitted power and C is the rate in bits/s, it follows that

$$P_{\text{av}} = C\mathcal{E}_b \tag{7.1–33}$$

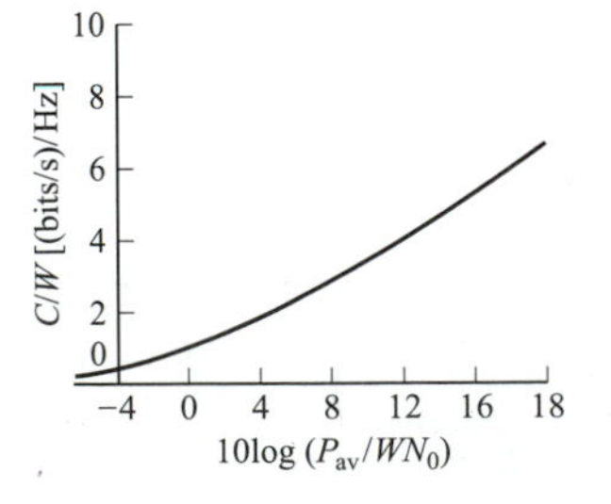

FIGURE 7.1–7
Normalized channel capacity as a function of SNR for band-limited AWGN channel.

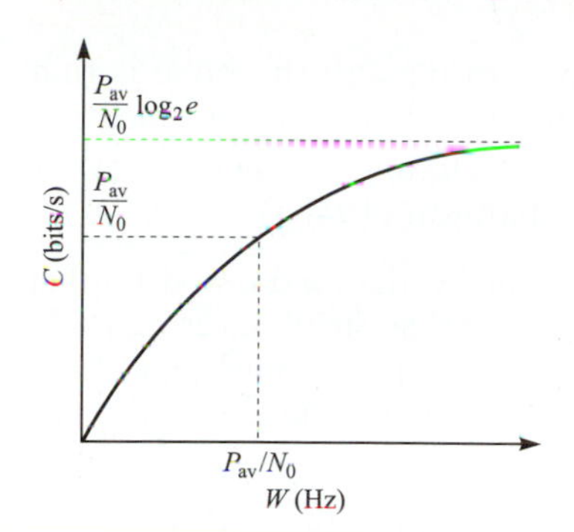

FIGURE 7.1–8
Channel capacity as a function of bandwidth with a fixed transmitted average power.

where $\mathcal{E}_b$ is the energy per bit. Hence, Equation 7.1–31 may be expressed as

$$\frac{C}{W} = \log_2\left(1 + \frac{C}{W}\frac{\mathcal{E}_b}{N_0}\right) \tag{7.1–34}$$

Consequently,

$$\frac{\mathcal{E}_b}{N_0} = \frac{2^{C/W} - 1}{C/W} \tag{7.1–35}$$

When $C/W = 1$, $\mathcal{E}_b/N_0 = 1$ (0 dB). As $C/W \to \infty$,

$$\begin{aligned}\frac{\mathcal{E}_b}{N_0} &\approx \frac{2^{C/W}}{C/W} \\ &\approx \exp\left(\frac{C}{W}\ln 2 - \ln\frac{C}{W}\right)\end{aligned} \tag{7.1–36}$$

Thus, $\mathcal{E}_b/N_0$ increases exponentially as $C/W \to \infty$. On the other hand, as $C/W \to 0$,

$$\frac{\mathcal{E}_b}{N_0} = \lim_{C/W \to 0} \frac{2^{C/W} - 1}{C/W} = \ln 2 \tag{7.1–37}$$

which is -1.6 dB. A plot C/W versus $\mathcal{E}_b/N_0$ is shown in Figure 5.2–17.

Thus, we have derived the channel capacities of three important channel models that are considered in this book. The first is the discrete-input, discrete-output channel, of which the BSC is a special case. The second is a discrete-input, continuous-output memoryless additive white Gaussian noise channel. From these two channel models, we can obtain benchmarks for the coded performance with hard- and soft-decision decoding in digital communication systems.

The third channel model focuses on the capacity in bits/s of a waveform channel. In this case, we assumed that we have a bandwidth limitation on the channel, an additive Gaussian noise that corrupts the signal, and an average power constraint at the transmitter. Under these conditions, we derived the result given in Equation 7.1–31.

The major significance of the channel capacity formulas given above is that they serve as upper limits on the transmission rate for reliable communication over a noisy channel. The fundamental rate that the channel capacity plays is given by the *noisy channel coding theorem* due to Shannon (1948a).

NOISY CHANNEL CODING THEOREM. There exist channel codes (and decoders) that make it possible to achieve reliable communication, with as small an error probability as desired, if the transmission rate $R < C$, where C is the channel capacity. If $R > C$, it is not possible to make the probability of error tend toward zero with any code.

In the following section, we explore the benefits of coding for the additive noise channel models described above, and use the channel capacity as the benchmark for assessing code performance.

7.1.3 Achieving Channel Capacity with Orthogonal Signals

In Section 5.2, we used a simple union bound to show that, for orthogonal signals, the probability of error can be made as small as desired by increasing the number M of waveforms, provided that $\mathcal{E}_b/N_0 > 2 \ln 2$. We indicated that the simple union bound does not produce the smallest lower bound on the SNR per bit. The problem is that the upper bound used in $Q(x)$ is very loose for small x.

An alternative approach is to use two different upper bounds for $Q(x)$, depending on the value of x. Beginning with Equation 5.2–21, we observe that

$$1 - [1 - Q(y)]^{M-1} \leqslant (M-1)Q(y) < Me^{-y^2/2} \tag{7.1–38}$$

This is just the union bound, which is tight when y is large, i.e., for $y > y_0$, where y_0 depends on M. When y is small, the union bound exceeds unity for large M. Since

$$1 - [1 - Q(y)]^{M-1} \leqslant 1 \tag{7.1–39}$$

for all y, we may use this bound for $y < y_0$ because it is tighter than the union bound. Thus Equation 5.2–21 may be upper-bounded as

$$P_M < \frac{1}{\sqrt{2\pi}} \int_{-\infty}^{y_0} e^{-(y-\sqrt{2\gamma})^2/2}\, dy + \frac{M}{\sqrt{2\pi}} \int_{y_0}^{\infty} e^{-y^2/2} e^{-(y-\sqrt{2\gamma})^2/2}\, dy \tag{7.1–40}$$

The value of y_0 that minimizes this upper bound is found by differentiating the right-hand side of Equation 7.1–40 and setting the derivative equal to zero. It is easily verified that the solution is

$$e^{y_0^2/2} = M \tag{7.1–41}$$

or, equivalently,

$$\begin{aligned} y_0 &= \sqrt{2 \ln M} = \sqrt{2 \ln 2 \log_2 M} \\ &= \sqrt{2k \ln 2} \end{aligned} \tag{7.1–42}$$

Having determined y_0, let us now compute simple exponential upper bounds for the integrals in Equation 7.1–40. For the first integral, we have

$$\begin{aligned} \frac{1}{\sqrt{2\pi}} \int_{-\infty}^{y_0} e^{-(y-\sqrt{2\gamma})^2/2} \, dy &= \frac{1}{\sqrt{\pi}} \int_{-\infty}^{-(\sqrt{2\gamma}-y_0)/\sqrt{2}} e^{-x^2} \, dx \\ &= Q(\sqrt{2\gamma} - y_0), \qquad y_0 \leqslant \sqrt{2\gamma} \\ &< e^{-(\sqrt{2\gamma}-y_0)^2/2}, \qquad y_0 \leqslant \sqrt{2\gamma} \end{aligned} \tag{7.1–43}$$

The second integral is upper-bounded as follows:

$$\begin{aligned} \frac{M}{\sqrt{2\pi}} \int_{y_0}^{\infty} e^{-y^2/2} e^{-(y-\sqrt{2\gamma})^2/2} \, dy &= \frac{M}{\sqrt{2\pi}} e^{-\gamma/2} \int_{y_0-\sqrt{\gamma/2}}^{\infty} e^{-x^2} \, dx \\ &< \begin{cases} M e^{-\gamma/2} & (y_0 \leqslant \sqrt{\frac{1}{2}\gamma}) \\ M e^{-\gamma/2} e^{-(y_0-\sqrt{\gamma/2})^2} & (y_0 \geqslant \sqrt{\frac{1}{2}\gamma}) \end{cases} \end{aligned} \tag{7.1–44}$$

Combining the bounds for the two integrals and substituting $e^{y_0^2/2}$ for M, we obtain

$$P_M < \begin{cases} e^{-(\sqrt{2\gamma}-y_0)^2/2} + e^{(y_0^2-\gamma)/2} & (0 \leqslant y_0 \leqslant (\sqrt{\frac{1}{2}\gamma}) \\ e^{-(\sqrt{2\gamma}-y_0)^2/2} + e^{(y_0^2-\gamma)/2} e^{-(y_0-\sqrt{\gamma/2})^2} & (\sqrt{\frac{1}{2}\gamma} \leqslant y_0 \leqslant \sqrt{2\gamma}) \end{cases} \tag{7.1–45}$$

In the range $0 \leqslant y_0 \leqslant \sqrt{\frac{1}{2}\gamma}$, the bound may be expressed as

$$P_M < e^{(y_0^2-\gamma)/2} (1 + e^{-(y_0-\sqrt{\gamma/2})^2}) < 2 e^{(y_0^2-\gamma)/2}, \qquad 0 \leqslant y_0 \leqslant \sqrt{\tfrac{1}{2}\gamma} \tag{7.1–46}$$

In the range $\sqrt{\gamma/2} \leqslant y_0 \leqslant \sqrt{2\gamma}$, the two terms in Equation 7.1–45 are identical. Hence,

$$P_M < 2e^{-(\sqrt{2\gamma}-y_0)^2/2}, \qquad \sqrt{\tfrac{1}{2}\gamma} \leqslant y_0 \leqslant \sqrt{2\gamma} \tag{7.1–47}$$

Now we substitute for y_0 and γ. Since $y_0 = 2 \ln M = \sqrt{2k \ln 2}$ and $\gamma = k\gamma_b$, the bounds in Equations 7.1–46 and 7.1–47 may be expressed as

$$P_M < \begin{cases} 2e^{-k(\gamma_b - 2\ln 2)/2} & (\ln M \leqslant \frac{1}{4}\gamma) \\ 2e^{-k(\sqrt{\gamma b} - \sqrt{\ln 2})^2} & (\frac{1}{4}\gamma \leqslant \ln M \leqslant \gamma) \end{cases} \tag{7.1–48}$$

The first upper bound coincides with the union bound presented earlier, but it is loose for large values of M. The second upper bound is better for large values of M. We note that $P_M \to 0$ as $k \to \infty$ ($M \to \infty$) provided that $\gamma_b > \ln 2$. But, $\ln 2$

is the limiting value of the SNR per bit required for reliable transmission when signaling at a rate equal to the capacity of the infinite-bandwidth AWGN channel as shown in Section 7.1.2. In fact, when the substitutions

$$\begin{aligned} y_0 &= \sqrt{2k \ln 2} = \sqrt{2RT \ln 2} \\ \gamma &= \frac{TP_{\text{av}}}{N_0} = TC_\infty \ln 2 \end{aligned} \tag{7.1–49}$$

are made into the two upper bounds given in Equations 7.1–46 and 7.1–47, where $C_\infty = P_{\text{av}}/(N_0 \ln 2)$ is the capacity of the infinite-bandwidth AWGN channel, the result is

$$P_M < \begin{cases} 2 \cdot 2^{-T(\frac{1}{2}C_\infty - R)} & (0 \leqslant \mathrm{R} \leqslant \frac{1}{4}C_\infty) \\ 2 \cdot 2^{-T(\sqrt{C_\infty} - \sqrt{R})^2} & (\frac{1}{4}C_\infty \leqslant R \leqslant C_\infty) \end{cases} \tag{7.1–50}$$

Thus we have expressed the bounds in terms of C_∞ and the bit rate in the channel. The first upper bound is appropriate for rates below $\frac{1}{4}C_\infty$, while the second is tighter than the first for rates between $\frac{1}{4}C_\infty$ and C_∞. Clearly, the probability of error can be made arbitrarily small by making $T \to \infty$ ($M \to \infty$ for fixed R), provided that $R < C_\infty = P_{\text{av}}/(N_0 \ln 2)$. Furthermore, we observe that the set of orthogonal waveforms achieves the channel capacity bound as $M \to \infty$, when the rate $R < C_\infty$.

7.1.4 Channel Reliability Functions

The exponential bounds on the error probability for M-ary orthogonal signals on an infinite-bandwidth AWGN channel given by Equation 7.1–50 may be expressed as

$$P_M < 2 \cdot 2^{-TE(R)} \tag{7.1–51}$$

The exponential factor

$$E(R) = \begin{cases} \frac{1}{2}C_\infty - R & (0 \leqslant R \leqslant \frac{1}{4}C_\infty) \\ (\sqrt{C_\infty} - \sqrt{R})^2 & (\frac{1}{4}C_\infty \leqslant R \leqslant C_\infty) \end{cases} \tag{7.1–52}$$

in Equation 7.1–51 is called the *channel reliability function* for the infinite-bandwidth AWGN channel. A plot of $E(R)/C_\infty$ is shown in Figure 7.1–9. Also shown is the exponential factor for the union bound on P_M, given by Equation 5.2–27, which may be expressed as

$$P_M \leqslant \tfrac{1}{2} \cdot 2^{-T(\frac{1}{2}C_\infty - R)}, \qquad 0 \leqslant R \leqslant \tfrac{1}{2}C_\infty \tag{7.1–53}$$

Clearly, the exponential factor in Equation 7.1–53 is not as tight as $E(R)$, due to the looseness of the union bound.

The bound given by Equations 7.1–51 and 7.1–52 has been shown by Gallager (1965) to be *exponentially tight*. This means that there does not exist

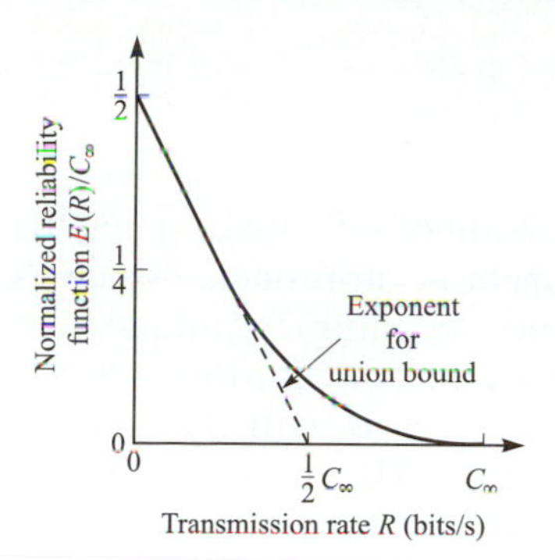

FIGURE 7.1–9
Channel reliability function for the infinite-bandwidth AWGN channel.

another reliability function, say $E_1(R)$, satisfying the condition $E_1(R) > E(R)$ for any R. Consequently, the error probability is bounded from above and below as

$$K_l 2^{-TE(R)} \leqslant P_M \leqslant K_u 2^{-TE(R)} \tag{7.1–54}$$

where the constants have only a weak dependence on T, i.e., they vary slowly with T.

Since orthogonal signals provide essentially the same performance as the optimum simplex signals for large M, the lower bound in Equation 7.1–54 applies for any signal set. Hence, the reliability function $E(R)$ given by Equation 7.1–52 determines the exponential characteristics of the error probability for digital signaling over the infinite-bandwidth AWGN channel.

Although the error probability can be made arbitrarily small by increasing the number of either orthogonal, biorthogonal, or simplex signals, with $R < C_\infty$, for a relatively modest number of signals, there is a large gap between the actual performance and the best achievable performance given by the channel capacity formula. For example, from Figure 5.2–17, we observe that a set of $M = 16$ orthogonal signals detected coherently requires an SNR per bit of approximately 7.5 dB to achieve a bit error rate of $P_e = 10^{-5}$. In contrast, the channel capacity formula indicates that for a $C/W = 0.5$, reliable transmission is possible with an SNR of −0.8 dB. This represents a rather large difference of 8.3 dB/ bit and serves as a motivation for searching for more efficient signaling waveforms. In this chapter and in Chapter 8, we demonstrate that coded waveforms can reduce this gap considerably.

Similar gaps in performance also exist in the bandwidth-limited region of Figure 5.2–17, where $R/W > 1$. In this region, however, we must be more clever in how we use coding to improve performance, because we cannot expand the bandwidth as in the power-limited region. The use of coding techniques for bandwidth-efficient communication is also treated in Chapter 8.

7.2 RANDOM SELECTION OF CODES

The design of coded modulation for efficient transmission of information may be divided into two basic approaches. One is the algebraic approach, which is primarily concerned with the design of coding and decoding techniques for specific classes of codes, such as cyclic block codes and convolutional codes. The second is the probabilistic approach, which is concerned with the analysis of the performance of a general class of coded signals. This approach yields bounds on the probability of error that can be attained for communication over a channel having some specified characteristic.

In this section, we adopt the probabilistic approach to coded modulation. The algebraic approach, based on block codes and on convolutional codes, is treated in Chapter 8.

7.2.1 Random Coding Based on M-ary Binary-Coded Signals

Let us consider a set of M coded signal waveforms constructed from a set of n-dimensional binary code words of the form

$$\mathbf{C}_i = [c_{i1} c_{i2} \cdots c_{in}], \qquad i = 1, 2, \ldots, M \tag{7.2–1}$$

where $c_{ij} = 0$ or 1. Each bit in the code word is mapped into a binary PSK waveform, so that the signal waveform corresponding to the code word $\mathbf{C}_i$ may be expressed as

$$s_i(t) = \sum_{j=1}^{n} s_{ij} f_j(t), \qquad i = 1, 2, \ldots, M \tag{7.2–2}$$

where

$$s_{ij} = \begin{cases} \sqrt{\mathcal{E}_c} & \text{when } c_{ij} = 1 \\ -\sqrt{\mathcal{E}_c} & \text{when } c_{ij} = 0 \end{cases} \tag{7.2–3}$$

and $\mathcal{E}_c$ is the energy per code bit. Thus, the waveforms $s_i(t)$ are equivalent to the n-dimensional vectors

$$\mathbf{s}_i = [s_{i1} \quad s_{i2} \quad \cdots \quad s_{in}], \qquad i = 1, 2, \ldots, M \tag{7.2–4}$$

which correspond to the vertices of a hypercube in n-dimensional space.

Now, suppose that the information rate into the encoder is R bits/s and we encode blocks of k bits at a time into one of the M waveforms. Hence, $k = RT$ and $M = 2^k = 2^{RT}$ signals are required. It is convenient to define a parameter D as

$$D = \frac{n}{T} \quad \text{dimensions/s} \tag{7.2–5}$$

Thus, $n = DT$ is the dimensionality of the signal space.

The hypercube has $2^n = 2^{DT}$ vertices, of which $M = 2^{RT}$ may be used to transmit the information. If we impose the condition that $D > R$, the fraction of the vertices that we use as signal points is

$$F = \frac{2^k}{2^n} = \frac{2^{RT}}{2^{DT}} = 2^{-(D-R)T} \tag{7.2–6}$$

Clearly, if $D > R$, we have $F \to 0$ as $T \to \infty$.

The question that we wish to pose is the following. Can we choose a subset $M = 2^{RT}$ vertices out of the $2^n = 2^{DT}$ available vertices such that the probability of error $P_e \to 0$ as $T \to \infty$ or, equivalently, as $n \to \infty$? Since the fraction F of vertices used approaches zero as $T \to \infty$, it should be possible to select M signal waveforms having a minimum distance that increases as $T \to \infty$ and, thus, $P_e \to 0$.

Instead of attempting to find a single set of M coded waveforms for which we compute the error probability, let us consider the ensemble of $(2^n)^M$ distinct ways in which we can select M vertices from the 2^n available vertices of the hypercube. Associated with each of the 2^{nM} selections, there is a communication system, consisting of a modulator, a channel, and a demodulator, that is optimum for the selected set of M waveforms. Thus, there are 2^{nM} communication systems, one for each choice of the M coded waveforms, as illustrated in Figure 7.2–1. Each communication system is characterized by its probability of error.

Suppose that our choice of M coded waveforms is based on random selection from the set of 2^{nM} possible sets of codes. Thus, the random selection of the mth code, denoted by $\{\mathbf{s}_i\}_m$, occurs with probability

$$P(\{\mathbf{s}_i\}_m) = 2^{-nM} \tag{7.2–7}$$

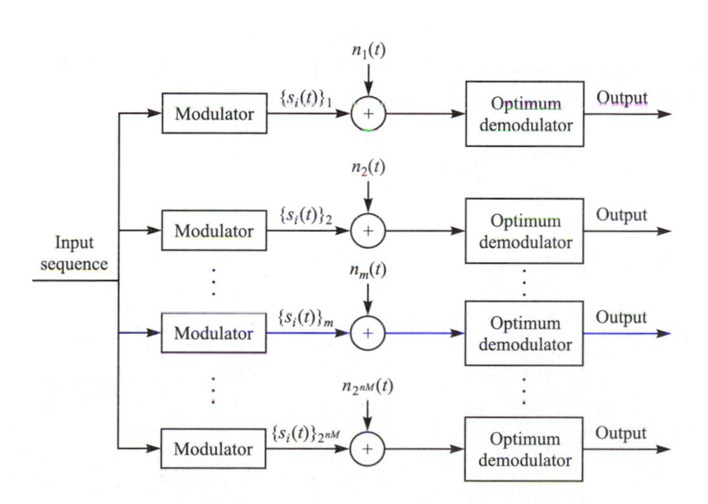

FIGURE 7.2–1
An ensemble of 2^{nM} communication systems. Each system employs a different set of M signals from the set of 2^{nM} possible choices.

and the corresponding conditional probability of error for this choice of coded signals is $P_e(\{\mathbf{s}_i\}_m)$. Then, the average probability of error over the ensemble of codes is

$$\begin{aligned}\bar{P}_e &= \sum_{m=1}^{2^{nM}} P_e(\{\mathbf{s}_i\}_m)P(\{\mathbf{s}_i\}_m) \\ &= 2^{-nM}\sum_{m=1}^{2^{nM}} P_e(\{\mathbf{s}_i\}_m)\end{aligned} \tag{7.2–8}$$

where the overbar on P_e denotes an average over the ensemble of codes.

It is clear that some choices of codes will result in a large probability of error. For example, the code that assigns all M k-bit sequences to the same vertex of the hypercube will result in a large probability of error. In such a case, $P_e(\{\mathbf{s}_i\}_m) > \bar{P}_e$. However, there will also be choices of codes for which $P_e(\{\mathbf{s}_i\}_m) < \bar{P}_e$. Consequently, if we obtain an upper bound on $\bar{P}_e$, this bound will also hold for those codes for which $P_e(\{\mathbf{s}_i\}_m) < \bar{P}_e$. Furthermore, if $\bar{P}_e \to 0$ as $T \to \infty$, then we conclude that, for these codes, $P_e(\{\mathbf{s}_i\}_m) \to 0$ as $T \to \infty$.

In order to determine an upper bound on $\bar{P}_e$, we consider the transmission of a k-bit message $\mathbf{X}_k \equiv [x_1x_2x_3 \cdots x_k]$, where $x_j = 0$ or 1 for $j = 1, 2, \ldots, k$. The conditional probability of error averaged over the ensemble of codes is

$$\overline{P_e(\mathbf{X}_k)} = \sum_{\text{all codes}} P_e(\mathbf{X}_k, \{\mathbf{s}_i\}_m)P(\{\mathbf{s}_i\}_m) \tag{7.2–9}$$

where $P_e(\mathbf{X}_k, \{\mathbf{s}_i\}_m)$ is the conditional probability of error for a given k-bit message $\mathbf{X}_k$, which is transmitted by use of the code $\{\mathbf{s}_i\}_m$. For the mth code, the probability of error $P_e(\mathbf{X}_k, \{\mathbf{s}_i\}_m)$ is upper-bounded as

$$P_e(\mathbf{X}_k, \{\mathbf{s}_i\}_m) \leqslant \sum_{\substack{l=1\\l\neq k}}^{M} P_{2m}(\mathbf{s}_l, \mathbf{s}_k) \tag{7.2–10}$$

where $P_{2m}(\mathbf{s}_i, \mathbf{s}_k)$ is the probability of error for a binary communication system that employs the signal vectors $\mathbf{s}_i$ and $\mathbf{s}_k$ to communicate one of two equally likely k-bit messages. Hence,

$$\overline{P_e(\mathbf{X}_k)} \leqslant \sum_{\text{all codes}} P(\{\mathbf{s}_i\}_m) \sum_{\substack{l=1\\l\neq k}}^{M} P_{2m}(\mathbf{s}_l, \mathbf{s}_k) \tag{7.2–11}$$

If we interchange the order of the summations in Equation 7.2–11, we obtain

$$\begin{aligned}\overline{P_e(\mathbf{X}_k)} &\leqslant \sum_{\substack{l=1\\l\neq k}}^{M}\left[\sum_{\text{all codes}} P(\{\mathbf{s}_i\}_m)P_{2m}(\mathbf{s}_l, \mathbf{s}_k)\right] \\ &\leqslant \sum_{\substack{l=1\\l\neq k}}^{M} \overline{P_2(\mathbf{s}_l, \mathbf{s}_k)}\end{aligned} \tag{7.2–12}$$

where $\overline{P_2(\mathbf{s}_l, \mathbf{s}_k)}$ represents the ensemble average of $P_{2m}(\mathbf{s}_l, \mathbf{s}_k)$ over the 2^{nM} codes or the 2^{nM} communication systems.

For the additive white Gaussian noise channel, the binary error probability $P_{2m}(\mathbf{s}_l, \mathbf{s}_k)$ is

$$P_{2m}(\mathbf{s}_l, \mathbf{s}_k) = Q\left(\sqrt{\frac{d_{lk}^2}{2N_0}}\right) \tag{7.2–13}$$

where $d_{lk}^2 = \|\mathbf{s}_l - \mathbf{s}_k\|^2$. If $\mathbf{s}_l$ and $\mathbf{s}_k$ differ in d coordinates,

$$d_{lk}^2 = \|\mathbf{s}_l - \mathbf{s}_k\|^2 = \sum_{j=1}^{n}(s_{lj} - s_{kj})^2 = d(2\sqrt{\mathcal{E}_c})^2 = 4d\mathcal{E}_c \tag{7.2–14}$$

Hence,

$$P_{2m}(\mathbf{s}_l, \mathbf{s}_k) = Q\left(\sqrt{\frac{2d\mathcal{E}_c}{N_0}}\right) \tag{7.2–15}$$

Now, we can average $P_{2m}(\mathbf{s}_l, \mathbf{s}_k)$ over the ensemble of codes. Since all the codes are equally probable, the signal vector $\mathbf{s}_l$ is equally likely to be any of the 2^n possible vertices of the hypercube and it is statistically independent of the signal vector $\mathbf{s}_k$. Therefore, $P(s_{li} = s_{ki}) = \frac{1}{2}$ and $P(s_{li} \neq s_{ki}) = \frac{1}{2}$, independently for all $i = 1, 2, \ldots, n$. Consequently, the probability that $\mathbf{s}_l$ and $\mathbf{s}_k$ differ in d positions is simply

$$P(d) = (\tfrac{1}{2})^n \binom{n}{d} \tag{7.2–16}$$

Hence, the expected value of $P_{2m}(\mathbf{s}_l, \mathbf{s}_k)$ over the ensemble of codes may be expressed as

$$\begin{aligned}\overline{P_2(\mathbf{s}_l, \mathbf{s}_k)} &= \sum_{d=0}^{n} P(d) Q\left(\sqrt{\frac{2d\mathcal{E}_c}{N_0}}\right) \\ &= \frac{1}{2^n}\sum_{d=0}^{n}\binom{n}{d} Q\left(\sqrt{\frac{2d\mathcal{E}_c}{N_0}}\right)\end{aligned} \tag{7.2–17}$$

The result 7.2–17 can be simplified if we upper-bound the Q function as

$$Q\left(\sqrt{\frac{2d\mathcal{E}_c}{N_0}}\right) < e^{-d\mathcal{E}_c/N_0}$$

Thus,

$$\begin{aligned}\overline{P_2(\mathbf{s}_l, \mathbf{s}_k)} &\leqslant 2^{-n} \sum_{d=0}^{n} \binom{n}{d} e^{-d\mathcal{E}_c/N_0} \\ &\leqslant 2^{-n}(1 + e^{-\mathcal{E}_c/N_0})^n \\ &\leqslant [\tfrac{1}{2}(1 + e^{-\mathcal{E}_c/N_0})]^n \end{aligned} \tag{7.2–18}$$

We observe that the right-hand side of Equation 7.2–18 is independent of the indices l and k. Hence, when we substitute the bound 7.2–18 into Equation 7.2–12, we obtain

$$\begin{aligned}\overline{P_e(\mathbf{X}_k)} &\leqslant \sum_{\substack{l=1\\ l\neq k}}^{M} \overline{P_2(\mathbf{s}_l, \mathbf{s}_k)} = (M-1)[\tfrac{1}{2}(1 + e^{-\mathcal{E}_c/N_0})]^n \\ &< M[\tfrac{1}{2}(1 + e^{-\mathcal{E}_c/N_0})]^n\end{aligned}$$

Finally, the unconditional average error probability $\bar{P}_e$ is obtained by averaging $\overline{P_e(X_k)}$ over all possible k-bit information sequences. Thus,

$$\begin{aligned}\bar{P}_e &= \sum_k \overline{P_e(\mathbf{X}_k)} P(\mathbf{X}_k) < M[\tfrac{1}{2}(1 + e^{-\mathcal{E}_c/N_0})]^n \sum_k P(\mathbf{X}_k) \\ &< M[\tfrac{1}{2}(1 + e^{-\mathcal{E}_c/N_0})]^n\end{aligned} \tag{7.2–19}$$

This result can be expressed in a more convenient form by first defining a parameter R_0, which is called the *cutoff rate* and has units of bits per dimension, as

$$\begin{aligned}R_0 &= \log_2 \frac{2}{1 + e^{-\mathcal{E}_c/N_0}} \\ &= 1 - \log_2(1 + e^{-\mathcal{E}_c/N_0}), \qquad \text{antipodal signaling}\end{aligned} \tag{7.2–20}$$

Then, Equation 7.2–19 becomes

$$\bar{P}_e < M 2^{-nR_0} = 2^{RT} 2^{-nR_0} \tag{7.2–21}$$

Since $n = DT$, Equation 7.2–21 may be expressed as

$$\bar{P}_e < 2^{-T(DR_0 - R)} \tag{7.2–22}$$

The parameter R_0 is plotted as a function of $\mathcal{E}_c/N_0$ in Figure 7.2–2. We observe that $0 \leqslant R_0 \leqslant 1$. Consequently, $\bar{P}_e \to 0$ as $T \to \infty$, provided that the information rate $R < DR_0$.

Alternatively, Equation 7.2–21 may be expressed as

$$\bar{P}_e < 2^{-n(R_0 - R/D)} \tag{7.2–23}$$

The ratio R/D also has units of bits per dimension and may be defined as

$$R_c = \frac{R}{D} = \frac{R}{n/T} = \frac{RT}{n} = \frac{k}{n} \tag{7.2–24}$$

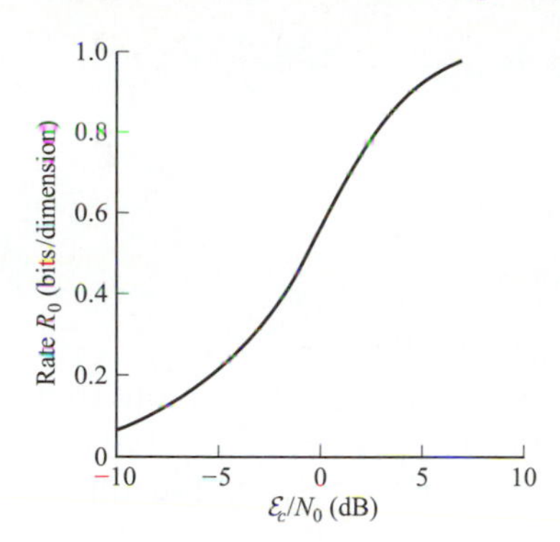

FIGURE 7.2–2
The cutoff rate R_0 as a function of the SNR per dimension in decibels.

Hence, R_c is the code rate and

$$\bar{P}_e < 2^{-n(R_0 - R_c)} \tag{7.2–25}$$

We conclude that when $R_c < R_0$, the average probability of error $\bar{P}_e \to 0$ as the code block length $n \to \infty$. Since the average value of the probability error can be made arbitrarily small as $n \to \infty$, it follows that there exist codes in the ensemble of 2^{nM} codes that have a probability of error no larger than $\bar{P}_e$.

From the derivation of the average error probability given above, we conclude that good codes exist. Although we do not normally select codes at random, it is interesting to consider the question of whether or not a randomly selected code is likely to be a good code. In fact, we can easily show that there are many good codes in the ensemble. First, we note that $\bar{P}_e$ is an ensemble average of error probabilities over all codes and that all these probabilities are obviously positive quantities. If a code is selected at random, the probability that its error probability $P_e > \alpha \bar{P}_e$ is less than $1/\alpha$. Consequently, no more than 10 percent of the codes have an error probability that exceeds $10\bar{P}_e$ and no more than 1 percent of the codes have an error probability that exceeds $100\bar{P}_e$.

We should emphasize that codes with error probabilities exceeding $\bar{P}_e$ are not necessarily poor codes. For example, suppose that an average error rate of $\bar{P}_e < 10^{-10}$ can be attained by using codes with dimensionality n_0 when $R_0 > R_c$. Then, if we select a code with error probability $1000\bar{P}_e = 10^{-7}$, we may compensate for this reduction in error probability by increasing n from n_0 to $n = 10n_0/7$. Thus, by a modest increase in dimensionality, we have a code with $\bar{P}_e < 10^{-10}$. In summary, good codes are abundant and, hence, they are easily found even by random selection.

It is also interesting to express the average error probability in Equation 7.2–25 in terms of the SNR per bit, γ_b. To accomplish this, we express the energy per signal waveform as

$$\mathcal{E} = n\mathcal{E}_c = k\mathcal{E}_b \tag{7.2–26}$$

Hence, $n = k\mathcal{E}_b/\mathcal{E}_c$. We also note that $R_c\mathcal{E}_b/\mathcal{E}_c = 1$. Therefore, Equation 7.2–25 may be expressed as

$$\bar{P}_e < 2^{-k(\gamma_b/\gamma_0 - 1)} \tag{7.2–27}$$

where γ_0 is a normalized SNR parameter, defined as

$$\begin{aligned}\gamma_0 &= \frac{R_c}{R_0}\gamma_b \\ &= \frac{R_c\gamma_b}{1-\log_2(1+e^{-R_c\gamma_b})}\end{aligned} \tag{7.2–28}$$

Now, we note that $\bar{P}_e \to 0$ as $k \to \infty$, provided that the SNR per bit, $\gamma_b > \gamma_0$.

The parameter γ_0 is plotted in Figure 7.2–3 as a function of $R_c\gamma_b$. Note that as $R_c\gamma_b \to 0$, $\gamma_0 \to 2\ln 2$. Consequently, the error probability for M-ary binary-coded signals is equivalent to the error probability obtained from the union bound for M-ary orthogonal signals, provided that the signal dimensionality is sufficiently large so that $\gamma_0 \approx 2\ln 2$.

The dimensionality parameter D that we introduced in Equation 7.2–5 is proportional to the channel bandwidth required to transmit the signals. Recall from the sampling theorem that a signal of bandwidth W may be represented by samples taken at a rate of $2W$ samples/s. Thus, in the time interval of length T there are $n = 2WT$ samples or, equivalently, n degrees of freedom (dimensions). Consequently, D may be equated with $2W$.

Finally, we note that the binary-coded signals considered in this section are appropriate when the SNR per dimension is small, e.g., $\mathcal{E}_c/N_0 < 10$. However, when $\mathcal{E}_c/N_0 > 10$, R_0 saturates at 1 bit per dimension. Since the code rate is restricted to be less than R_0, binary-coded signals become inefficient at $\mathcal{E}_c/N_0 > 10$. In such a case, we may use nonbinary-coded signals to achieve an increase in the number of bits per dimension. For example, multiple-amplitude coded signal sets can be constructed from nonbinary codes by mapping each code element into one of q possible amplitude levels (as in PAM). Such codes are considered below.

7.2.2 Random Coding Based on M-ary Multiamplitude Signals

Instead of constructing binary-coded signals, suppose we employ nonbinary codes with code words of the form given by Equation 7.2–1, where the code elements c_{ij} are selected from the set $\{0, 1, \ldots, q-1\}$. Each code element is mapped into one of q possible amplitude levels. Thus, we construct signals

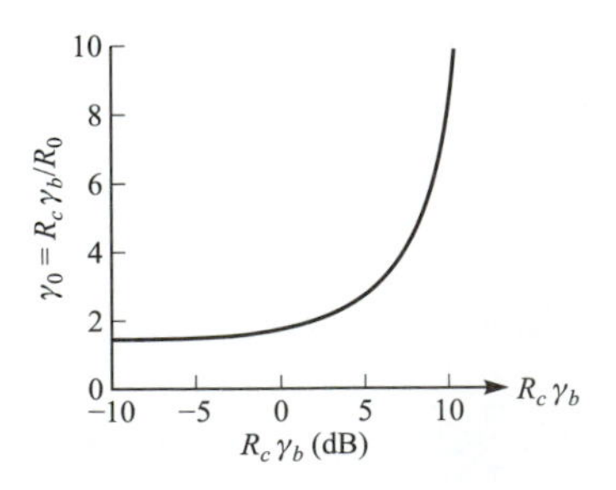

FIGURE 7.2–3
Lower bound on SNR per bit, γ_b, for binary antipodal signals.

corresponding to n-dimensional vectors $\{\mathbf{s}_i\}$ as in Equation 7.2–4, where the components $\{s_{ij}\}$ are selected from a multiamplitude set of q possible values. Now, we have q^n possible signals, from which we select $M = 2^{RT}$ signals to transmit k-bit blocks of information. The q amplitudes corresponding to the code elements $\{0, 1, \ldots, q-1\}$ may be denoted by $\{a_1, a_2, \ldots, a_q\}$, and they are assumed to be selected according to some specified probabilities $\{p_i\}$. The amplitude levels are assumed to be equally spaced over the interval $[-\sqrt{\mathcal{E}_c}, \sqrt{\mathcal{E}_c}]$. For example, Figure 7.2–4 illustrates the amplitude values for $q = 4$. In general, adjacent amplitude levels are separated by $2\sqrt{\mathcal{E}_c}/(q-1)$. This assignment guarantees not only that each component s_{ij} is peak-energy-limited to $\sqrt{\mathcal{E}_c}$, but, also, each code word is constrained in average energy to satisfy the condition

$$\|\mathbf{s}_i\|^2 < n\mathcal{E}_c \tag{7.2–29}$$

By repeating the derivation given above for random selection of codes in an AWGN channel, we find that the average probability of error is upper-bounded as

$$\bar{P}_e < M2^{-nR_0} = 2^{RT}2^{-nT_0} = 2^{-n(R_0 - R/D)} \tag{7.2–30}$$

where R_0 is defined as

$$R_0 = -\log_2\left(\sum_{l=1}^{q}\sum_{m=1}^{q} p_l p_m e^{-d_{lm}^2/4N_0}\right) \tag{7.2–31}$$

and

$$d_{lm} = |a_l - a_m|, \qquad l, m = 1, 2, \ldots, q \tag{7.2–32}$$

In the special case where all the amplitude levels are equally likely, $p_l = p_m = 1/q$ and Equation 7.2–31 reduces to

$$R_0 = -\log_2\left(\frac{1}{q^2}\sum_{l=1}^{q}\sum_{m=1}^{q} e^{-d_{lm}^2/4N_0}\right) \tag{7.2–33}$$

For example, where $q = 2$ and $a_1 = -\sqrt{\mathcal{E}_c}$, $a_2 = \sqrt{\mathcal{E}_c}$, we have $d_{11} = d_{22} = 0$, $d_{12} = d_{21} = 2\sqrt{\mathcal{E}_c}$, and, hence,

$$R_0 = \log_2 \frac{2}{1 + e^{-\mathcal{E}_c/N_0}}, \qquad q = 2$$

which agrees with our previous result. When $q = 4$, $a_1 = -\sqrt{\mathcal{E}_c}$, $a_2 = -\sqrt{\mathcal{E}_c}/3$, $a_3 = \sqrt{\mathcal{E}_c}/3$, and $a_4 = \sqrt{\mathcal{E}_c}$, we have $d_{mm} = 0$ for $m = 1, 2, 3, 4$, $d_{12} = d_{23} = d_{34} = d_{21} = d_{32} = d_{43} = 2\sqrt{\mathcal{E}_c}/3$, $d_{13} = d_{31} = d_{24} = d_{42} = 4\sqrt{\mathcal{E}_c}/3$, and $d_{14} = d_{41} = 2\sqrt{\mathcal{E}_c}$. Hence,

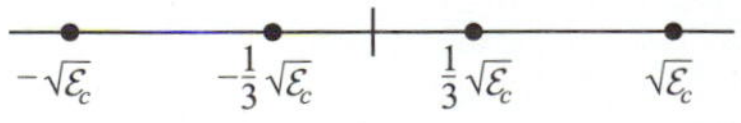

$-\sqrt{\mathcal{E}_c}$ $\quad -\frac{1}{3}\sqrt{\mathcal{E}_c}$ $\quad \frac{1}{3}\sqrt{\mathcal{E}_c}$ $\quad \sqrt{\mathcal{E}_c}$

FIGURE 7.2–4
Signal alphabet consisting of four amplitude levels.

$$R_0 = \log_2 \frac{8}{2 + 3e^{-\mathcal{E}_c/9N_0} + 2e^{-4\mathcal{E}_c/9N_0} + e^{-\mathcal{E}_c/N_0}}, \qquad q = 4 \tag{7.2–34}$$

Clearly, R_0 now saturates at 2 bits per dimension as $\mathcal{E}_c/N_0$ increases.

The graphs of R_0 as a function of $\mathcal{E}_c/N_0$ for equally spaced and equally likely amplitude levels are shown in Figure 7.2–5 for $q = 2, 3, 4, 8, 16, 32$, and 64. Note that the saturation level now occurs at $\log_2 q$ bits per dimension. Consequently, for high SNR, $\bar{P}_e \to 0$ as $n \to \infty$, provided that $R < DR_0 = 2WR_0$ bits/s.

If we remove the peak energy constraint on each of the elements, but retain the average energy constraint per code word as given by Equation 7.2–29, it is possible to obtain a larger upper bound on the number of bits per dimension. For this case, the result obtained by Shannon (1959b) is

$$\begin{aligned} R_0^* = {} & \frac{1}{2}\left[1 + \frac{\mathcal{E}_c}{N_0} - \sqrt{1 + \left(\frac{\mathcal{E}_c}{N_0}\right)^2}\,\right] \log_2 e \\ & + \tfrac{1}{2}\log_2\left[\frac{1}{2}\left(1 + \sqrt{1 + \left(\frac{\mathcal{E}_c}{N_0}\right)^2}\,\right)\right] \end{aligned} \tag{7.2–35}$$

The graph of R_0^* as a function of the SNR per dimension, $\mathcal{E}_c/N_0$, is also shown in Figure 7.2–5. It is clear that our selection of the equally spaced, equally likely amplitude levels that result in R_0 is suboptimum. However, these coded signals are easily generated and implemented in practice. This is an important advantage that justifies their use.

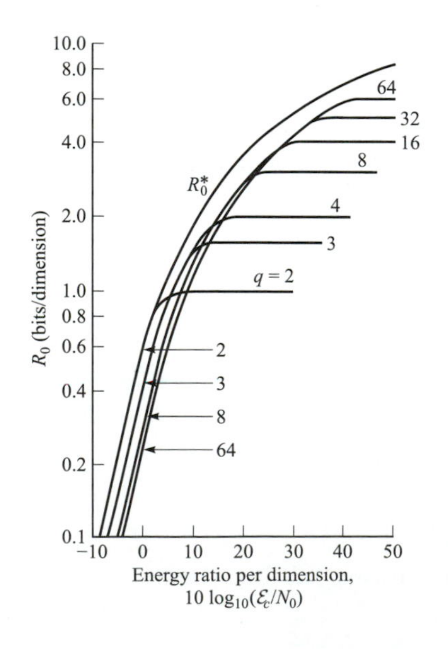

FIGURE 7.2–5
Cutoff rate R_0 for equally spaced q-level amplitude modulation with equal probabilities $p_c = 1/q$ [*From* Principles of Communication Engineering, *by J. M. Wozencraft and I. M. Jacobs, © 1965 by John Wiley and Sons, Inc. Reprinted with permission of the publisher.]*

7.2.3 Comparison of R_0^* with the Capacity of the AWGN Channel

The channel capacity of the band-limited additive white Gaussian noise channel with an average power constraint on the input signal was derived in Section 7.1.2, and is given by

$$C = W \log_2\left(1 + \frac{P_{\text{av}}}{WN_0}\right) \quad \text{bits/s} \tag{7.2–36}$$

where P_{av} is the average power of the input signal and W is the channel bandwidth. It is interesting to express the capacity of this channel in terms of bits per dimension and the average power in terms of energy per dimension. With $D = 2W$ and

$$\mathcal{E}_c = \frac{\mathcal{E}}{n} = \frac{P_{\text{av}} T}{n}$$

we have

$$P_{\text{av}} = \frac{n}{T}\mathcal{E}_c = D\mathcal{E}_c \tag{7.2–37}$$

By defining $C_n = C/2W = C/D$ and substituting for W and P_{av}, Equation 7.2–36 may be expressed as

$$\begin{aligned} C_n &= \tfrac{1}{2}\log_2\left(1 + 2\frac{\mathcal{E}_c}{N_0}\right) \\ &= \tfrac{1}{2}\log_2(1 + 2R_c\gamma_b) \quad \text{bits per dimension} \end{aligned} \tag{7.2–38}$$

This expression for the normalized capacity may be compared with R_0^*, as shown in Figure 7.2–6. Since C_n is the ultimate upper limit on the transmission rate R/D, $R_0^* < C_n$ as expected. We also observe that for small values of $\mathcal{E}_c/N_0$, the difference between R_0^* and C_n is approximately 3 dB. Therefore, the use of randomly selected, optimum average power-limited, multiamplitude signals yields a rate function R_0^* that is within 3 dB of the channel capacity. More elaborate bounding techniques are required to show that the probability of error can be made arbitrarily small when $R < DC_n = 2WC_n = C$.

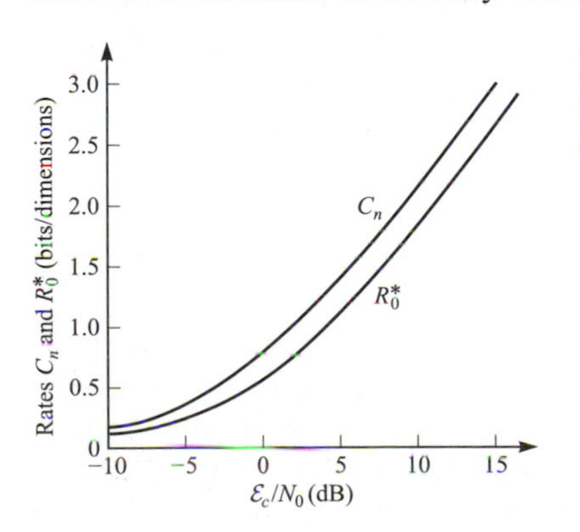

FIGURE 7.2–6
Comparison of cutoff rate R_0^* with the channel capacity for an AWGN channel.

7.3
COMMUNICATION SYSTEM DESIGN BASED ON THE CUTOFF RATE

In the foregoing discussion, we characterized coding and modulation performance in terms of the error probability, which is certainly a meaningful criterion for system design. However, in many cases, the computation of the error probability is extremely difficult, especially if non-linear operations such as signal quantization are performed in processing the signal at the receiver, or if the additive noise is non-Gaussian.

Instead of attempting to compute the exact probability of error for specific codes, we may use the ensemble average probability of error for randomly selected code words. The channel is assumed to have q input symbols $\{0, 1, \ldots, q-1\}$ and Q output symbols $\{0, 1, \ldots, Q-1\}$, and to be characterized by the transition probabilities $P(i|j)$, where $j = 0, 1, \ldots, q-1$ and $i = 0, 1, \ldots, Q-1$, with $Q \geqslant q$. The input symbols occur with probabilities $\{p_j\}$ and are assumed to be statistically independent. In addition, the noise on the channel is assumed to be statistically independent in time, so that there is no dependence among successive received symbols. Under these conditions, the ensemble average probability of error for randomly selected code words may be derived by applying the Chernoff bound (see Viterbi and Omura, 1979).

The general result that is obtained for the discrete memoryless channel is

$$\bar{P}_e < 2^{-n(R_Q - R/D)} \tag{7.3–1}$$

where n is the block length of the code, R is the information rate in bits/s, D is the number of dimensions per second, and R_Q is the cutoff rate for a quantizer with Q levels, defined as

$$R_Q = \max_{\{p_j\}} \left\{ -\log_2 \sum_{i=0}^{Q-1} \left[\sum_{j=0}^{q-1} p_j \sqrt{P(i|j)} \right]^2 \right\} \tag{7.3–2}$$

From the viewpoint of code design, the combination of modulator, waveform channel, and demodulator constitutes a discrete-time channel with q inputs and Q outputs. The transition probabilities $\{P(i|j)\}$ depend on the channel noise characteristics, the number of quantization levels, and the type of quantizer, e.g., uniform or nonuniform. For example, in the binary-input AWGN channel, the output of the correlator at the sampling instant may be expressed as

$$p(y|j) = \frac{1}{\sqrt{2\pi}\sigma} e^{-(y-m_j)^2/2\sigma^2}, \qquad j = 0, 1 \tag{7.3–3}$$

where $m_0 = -\sqrt{\mathcal{E}_c}$, $m_1 = \sqrt{\mathcal{E}_c}$, and $\sigma^2 = \frac{1}{2}N_0$. These two PDFs are shown in Figure 7.3–1. Also illustrated in the figure is a quantization scheme that subdivides the real line into five regions. From such a subdivision, we may compute the transition probabilities and optimally select the thresholds that subdivide the regions in a way that maximizes R_Q for any given Q. Thus,

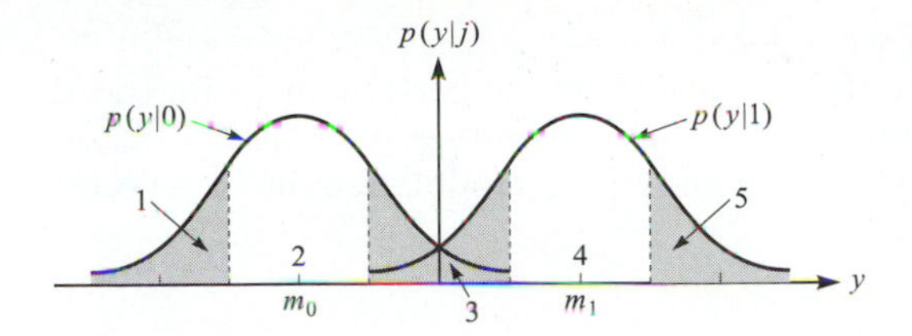

FIGURE 7.3–1
Example of quantization of the demodulator output into five levels.

$$P(i|j) = \int_{r_i} p(y|j)\, dy \qquad (7.3\text{–}4)$$

where the integral of $p(y|j)$ is evaluated over the region r_i that corresponds to the transition probability $P(i|j)$.

The value of the rate R_Q in the limit as $Q \to \infty$ yields the cutoff rate for the unquantized decoder. It is relatively straightforward to show that as $Q \to \infty$, the first summation (sum from $i = 0$ to $Q - 1$) in Equation 7.3–2 becomes an integral and the transition probabilities are replaced by the corresponding PDFs. Thus, when the channel consists of q discrete inputs and one continuous output y, which represents the unquantized output from a matched filter or a cross correlator in a system that employs either PSK or a multiamplitude (PAM) modulation, the cutoff rate is given by

$$R_0 = \max_{\{p_j\}} \left\{ -\log_2 \int_{-\infty}^{\infty} dy \left[\sum_{j=0}^{q-1} p_j \sqrt{p(y|j)} \right]^2 \right\} \qquad (7.3\text{–}5)$$

where p_j, $0 \leqslant j \leqslant q - 1$, is the probability of transmitting the jth symbol and $p(y|j)$ is the conditional probability density function of the output y from the matched filter or cross correlator when the jth signal is transmitted. This is the desired expression for unquantized (soft-decision) decoding.

We observe that when the input signal is binary PSK with $p_0 = p_1 = \frac{1}{2}$ and the noise is additive, white, and Gaussian, Equation 7.3–5 reduces to the familiar result given previously in Equation 7.2–20.

The general expressions in Equations 7.3–5 and 7.3–2 allow us to compare the performance of various receiver implementations based on a different number of quantization levels.

EXAMPLE 7.3–1. Let us compare the performance of a binary PSK input signal in an AWGN channel when the receiver quantizes the output to $Q = 2, 4$, and 8 levels. To simplify the optimization problem for the quantization of the signal at the output of the demodulator, the quantization levels are placed at $0, \pm\tau_h$, $\pm 2\tau_h, \ldots$, $\pm(2^{b-1} - 1)\tau_h$, where τ_h is the *quantizer step-size parameter*, which is to be selected, and b is the number of bits of the quantizer. A good strategy for the selection of τ_h is to choose it to minimize the SNR per bit γ_b that is required for operation at a code rate R_0. This implies that the step-size parameter must be optimized for every SNR, which in a practical implementation of the receiver means that the SNR must be measured. Fortunately, τ_h does not exhibit high sensitivity to small changes in SNR, so that it is possible to optimize τ_h for one SNR and obtain good performance for a wide range of SNRs about this nominal value by using a fixed τ_h.

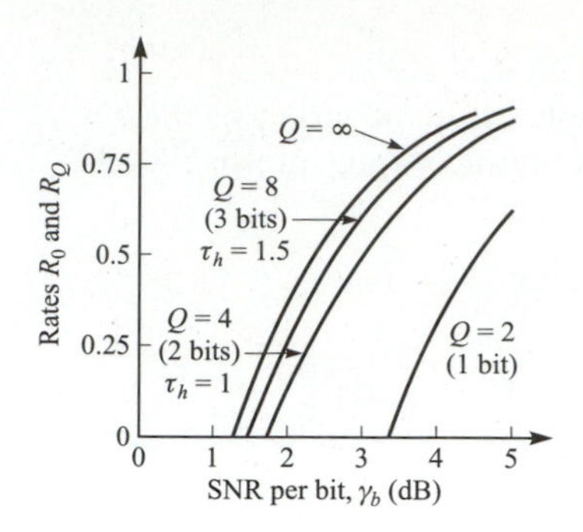

FIGURE 7.3–2
Effect of quantization on the performance of a coded communication system operating at a rate $R = R_0$ or $R = R_Q$, with binary PSK modulation on an AWGN channel.

Based on this approach, the expression for R_Q given by Equation 7.3–2 was evaluated for $b = 1$ (hard-decision decoding), 2, and 3 bits, corresponding to $Q = 2, 4$, and 8 levels of quantization. The results are plotted in Figure 7.3–2. The value of R_0 for unquantized soft-decision decoding, obtained by evaluating Equation 7.3–5 is also shown in Figure 7.3–2. We observe that 2-bit quantization with $\tau_h = 1.0$ gains about 1.4 dB over hard-decision decoding, and 3-bit quantization with $\tau_h = 0.5$ yields an additional 0.4 dB improvement. Thus, with a 3-bit quantizer, we are within 0.2 dB of the unquantized soft-decision decoding limit. Clearly, there is little to be gained by increasing the precision any further.

For binary signals ($q = 2$) and hard-decision decoding ($Q = 2$) the cutoff rate R_2 given in Equation 7.3–2 reduces to

$$R_2 = 1 - \log_2\left[1 + \sqrt{4p(1-p)}\right] \tag{7.3–6}$$

For binary PSK and AWGN, the probability of error is

$$p = Q\left(\sqrt{\frac{2\mathcal{E}_c}{N_0}}\right) = Q\left(\sqrt{2R_c\gamma_b}\right) \tag{7.3–7}$$

where $\gamma_b = \mathcal{E}_b/N_0$. We may set $R_2 = R_c$ in Equation 7.3–6 and solve Equation 7.3–6 for the value of γ_b that satisfies the equation with p given by Equation 7.3–7. For any code rate R_c, we interpret the value of γ_b that satisfies Equation 7.3–6 as the minimum SNR required to operate at R_c equal to the cutoff rate R_2.

Similarly, for binary PSK (antipodal signals) and soft-decision decoding ($Q = \infty$), the cutoff rate given by Equation 7.2–20 may be set to $R_0 = R_c$ and this equation may be solved for γ_b. The result is

$$\gamma_b = -\frac{1}{R_c}\ln(2^{1-R_c} - 1) \tag{7.3–8}$$

Again, for any code rate R_c, we interpret the value of γ_b obtained from Equation 7.3–8 as the minimum SNR per bit required to operate at a code rate equal to the cutoff rate R_0.

Figure 7.3–3 illustrates the plots of code rate versus the SNR per dimension for both hard-decision decoding and soft-decision decoding. The corresponding SNR per bit is $\gamma_b = (\mathcal{E}_c/N_0)/R_c$. We observe that for any given value of R_c, the

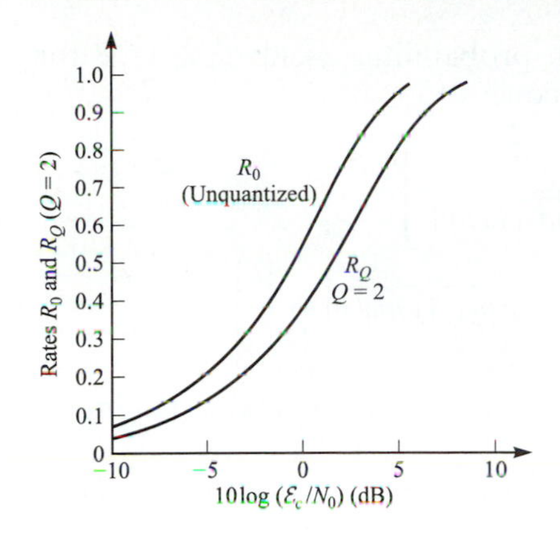

FIGURE 7.3–3
Comparison of R_0 (soft-decision decoding) and R_2 (hard-decision decoding) as a function of SNR per dimension.

difference between the SNR per bit for hard-decision decoding and soft-decision decoding is approximately 2 dB for the AWGN channel.

When a nonbinary code is used in conjunction with M-ary ($M = q$) signaling, the received signal at the output of the M matched filters may be represented by the vector $\mathbf{y} = [y_1 \; y_2 \cdots y_M]$. The cutoff rate for this M-input, M-output (unquantized) channel is

$$R_0 = \max_{\{p_j\}} \left\{ -\log_2 \sum_{j=0}^{M-1} \sum_{i=0}^{M-1} p_j p_i \int_{-\infty}^{\infty} \sqrt{p(\mathbf{y}|j) p(\mathbf{y}|i)}\, d\mathbf{y} \right\} \tag{7.3–9}$$

where $p(\mathbf{y}|j)$ is the conditional probability density function of the output vector $\mathbf{y}$ from the demodulator given that the jth signal was transmitted. Note that Equation 7.3–9 is similar in form to Equation 7.3–5 except that we now have an M-fold integral to perform because there are M outputs from the demodulator.

Let us assume that the M signals are orthogonal so that the M outputs conditioned on a particular input signal are statistically independent. As a consequence,

$$p(\mathbf{y}|j) = p_{s+n}(y_j) \prod_{\substack{i=0 \\ i \neq j}}^{M-1} p_n(y_i) \tag{7.3–10}$$

where $p_{s+n}(y_j)$ is the PDF of the matched filter output corresponding to the transmitted signal and $\{p_n(y_i)\}$ corresponds to the noise-only outputs from the other $M-1$ matched filters. When Equation 7.3–10 is incorporated into Equation 7.3–9, we obtain

$$R_0 = \max_{\{p_j\}} \left\{ -\log_2 \left[\sum_{j=0}^{M-1} p_j^2 + \sum_{j=0}^{M-1} \sum_{\substack{i=0 \\ i \neq j}}^{M-1} p_i p_j \left(\int_{-\infty}^{\infty} dy \sqrt{p_{s+n}(y) p_n(y)} \right)^2 \right] \right\} \tag{7.3–11}$$

The maximization of R_0 over the set of input probabilities yields $p_j = 1/M$ for $1 \leqslant j \leqslant M$. Consequently, Equation 7.3–11 reduces to

$$\begin{aligned} R_0 &= \log_2 \left\{ \frac{M}{1 + (M-1)[\int_{-\infty}^{\infty} \sqrt{p_{s+n}(y)p_n(y)}\, dy]^2} \right\} \\ &= \log_2 M - \log_2 \left\{ 1 + (M-1) \left[\int_{-\infty}^{\infty} \sqrt{p_{s+n}(y)p_n(y)}\, dy \right]^2 \right\} \end{aligned} \tag{7.3–12}$$

This is the desired result for the cutoff rate of an M-ary orthogonal signal input, M-ary vector output unquantized channel.

For phase coherent detection of the M-ary orthogonal signals the appropriate PDFs are

$$\begin{aligned} p_{s+n}(y) &= \frac{1}{\sqrt{2\pi}\sigma} e^{-(y-m)^2/2\sigma^2} \\ p_n(y) &= \frac{1}{\sqrt{2\pi}\sigma} e^{-y^2/2\sigma^2} \end{aligned} \tag{7.3–13}$$

where $m = \sqrt{\mathcal{E}}$ and $\sigma^2 = \frac{1}{2}N_0$. Substituting these relations into Equation 7.3–12 and evaluating the integral yields

$$\begin{aligned} R_0 &= \log_2 \left[\frac{M}{1 + (M-1)e^{-\mathcal{E}/2N_0}} \right] \\ &= \log_2 \left[\frac{M}{1 + (M-1)e^{-R_w \gamma_b/2}} \right] \end{aligned} \tag{7.3–14}$$

where $\mathcal{E}$ is the received energy per waveform, R_w is the information rate in bits per waveform, and $\gamma_b = \mathcal{E}_b/N_0$ is the SNR per bit.

We should emphasize that the rate parameter R_w has embedded in it the code rate R_c. For example, if $M = 2$ and the code is binary, then $R_w = R_c$. More generally, if the code is binary and $M = 2^\nu$, then each M-ary waveform conveys $R_w = \nu R_c$ bits of information. It is also interesting to note that if the code is binary and $M = 2$, then Equation 7.3–14 reduces to

$$R_0 = \log_2 \left(\frac{2}{1 + e^{-R_c \gamma_b/2}} \right), \qquad M = 2 \text{ orthogonal signals} \tag{7.3–15}$$

which is 3 dB worse than the cutoff rate for antipodal signals. If we set $R_w = R_0$ in Equation 7.3–14 and solve for γ_b, we obtain

$$\gamma_b = \frac{2}{R_0} \ln \left(\frac{M-1}{2^{-R_0} M - 1} \right) \tag{7.3–16}$$

Graphs of R_0 versus γ_b for several values of M are illustrated in Figure 7.3–4. Note that the curve for any value of M saturates at $R_0 = \log_2 M$.

It is also interesting to consider the limiting form of Equation 7.3–14 in the limit as $M \to \infty$. This yields

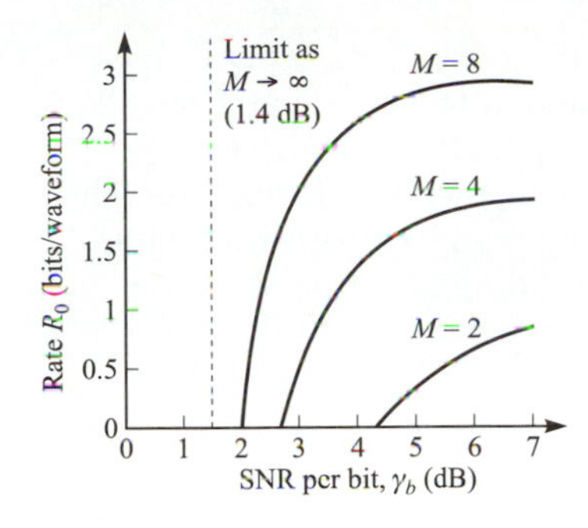

FIGURE 7.3–4
SNR per bit required to operate at a rate R_0 with M-ary orthogonal signals detected coherently in an AWGN channel.

$$\lim_{M\to\infty} R_0 = \frac{\mathcal{E}}{2N_0 \ln 2} \quad \text{bits per waveform} \tag{7.3–17}$$

Since $\mathcal{E} = P_{\text{av}}T$, where T is the time interval per waveform, it follows that

$$\lim_{M\to\infty} \frac{R_0}{T} = \frac{P_{\text{av}}}{2N_0 \ln 2} = \tfrac{1}{2}C_\infty \tag{7.3–18}$$

Hence, in the limit as $M \to \infty$, the cutoff rate is one-half of the capacity for the infinite bandwidth AWGN channel. Alternatively, the substitution of $\mathcal{E} = R_0\mathcal{E}_b$ into Equation 7.3–17 yields $\gamma_b = 2 \ln 2$ (1.4 dB), which is the minimum SNR required to operate at R_0 (as $M \to \infty$). Hence, signaling at a rate R_0 requires 3 dB more power than the Shannon limit.

The value of R_0 given in Equation 7.3–14 is based on the use of M-ary orthogonal signals, which are clearly suboptimal when M is small. If we attempt to maximize R_0 by selecting the best set of M waveforms, we should not be surprised to find that the simplex set of waveforms is optimum. In fact, R_0 for these optimum waveforms is simply given as

$$R_0 = \log_2\left[\frac{M}{1 + (M-1)e^{-M\mathcal{E}/2(M-1)N_0}}\right] \tag{7.3–19}$$

If we compare this expression with Equation 7.3–14, we observe that R_0 in Equation 7.3–19 simply reflects the fact that the simplex set is more energy-efficient by a factor $M/(M-1)$.

In the case of noncoherent detection, the probability density functions corresponding to signal-plus-noise and noise alone may be expressed as

$$\begin{aligned} p_{s+n}(y) &= ye^{-(y^2+a^2)/2} I_0(ay), \qquad y \geqslant 0 \\ p_n(y) &= ye^{-y^2/2}, \qquad y \geqslant 0 \end{aligned} \tag{7.3–20}$$

where, by definition, $a = \sqrt{2\mathcal{E}/N_0}$. The computation of R_0 given by Equation 7.3–12 does not yield a closed-form solution. Instead, the integral in Equation 7.3–12 must be evaluated numerically. Results for this case have been given by Jordan (1966) and Bucher (1980). For example, the (normalized) cutoff rate R_0 for M-ary orthogonal signals with noncoherent detection is shown in Figure 7.3–5

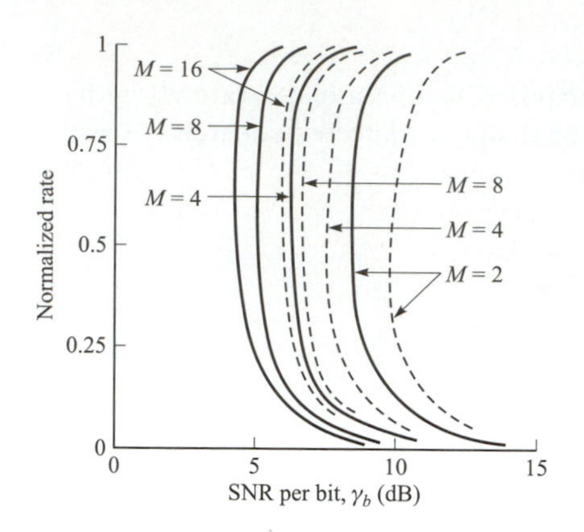

FIGURE 7.3–5
SNR per bit required to operate at a rate R_0 with M-ary orthogonal signals detected noncoherently in an AWGN channel. Graphs in broken lines represent performance with hard-decision decoding.

for $M = 2, 4, 8$, and 16. For purposes of comparison we also plot the cutoff rate for hard-decision decoding ($Q = M$) of the M-ary symbols. In this case, we have

$$R_Q = \log_2 \left\{ \frac{M}{[\sqrt{(1 - P_M)} + \sqrt{(M-1)P_M}]^2} \right\}, \qquad Q = M \qquad (7.3\text{–}21)$$

where P_M is the probability of a symbol error. For a relatively broad range of rates, the difference between soft- and hard-decision decoding is approximately 2 dB.

The most striking characteristic of the performance curves in Figure 7.3–5 is that there is an optimum code rate for any given M. Unlike the case of coherent detection, where the SNR per bit decreases monotonically with a decrease in code rate, the SNR per bit of noncoherent detection reaches a minimum in the vicinity of a normalized rate of 0.5 and increases for both high and low rates. The minimum is rather broad, so there is really a range of rates from 0.2 to 0.9 where the SNR per bit is within 1 dB of the minimum. This characteristic behavior in the performance with noncoherent detection is attributed to the non-linear characteristic of the detector.

7.4 BIBLIOGRAPHICAL NOTES AND REFERENCES

The pioneering work on channel characterization in terms of channel capacity and random coding was done by Shannon (1948a,b, 1949). Additional contributions were subsequently made by Gilbert (1952), Elias (1955), Gallager (1965), Wyner (1965), Shannon et al. (1967), Forney (1968), and Viterbi (1969). All these early publications are contained in the IEEE Press book entitled *Key Papers in the Development of Information Theory*, edited by Slepian (1974). The paper by Verdú (1998) in the 50th Anniversary Commemorative Issue of the *IEEE Transactions on Information Theory* gives a historical perspective of the numerous advances in information theory over the past 50 years.

The use of the cutoff rate parameter as a design criterion was proposed and developed by Wozencraft and Kennedy (1966) and by Wozencraft and Jacobs (1965). It was used by Jordan (1966) in the design of coded waveforms for M-ary orthogonal signals with coherent and noncoherent detection. Following these pioneering works, the cutoff rate has been widely used as a design criterion for coded signals in a variety of different channel conditions.

PROBLEMS

7.1 Show that the following two relations are necessary and sufficient conditions for the set of input probabilities $\{P(x_j)\}$ to maximize $I(X; Y)$ and, thus, to achieve capacity for a DMC:

$$I(x_j; Y) = C \qquad \text{for all } j \text{ with } P(x_j) > 0$$
$$I(x_j; Y) \leqslant C \qquad \text{for all } j \text{ with } P(x_j) = 0$$

where C is the capacity of the channel and

$$I(x_j; Y) = \sum_{i=0}^{Q-1} P(y_i|x_j) \log \frac{P(y_i|x_j)}{P(y_i)}$$

7.2 Figure P7.2 illustrates an M-ary symmetric DMC with transition probabilities $P(y|x) = 1 - p$ when $x = y = k$ for $k = 0, 1, \ldots, M-1$, and $P(y|x) = p/(M-1)$ when $x \neq y$.

a) Show that this channel satisfies the condition given in Problem 7.1 when $P(x_k) = 1/M$.

b) Determine and plot the channel capacity as a function of p.

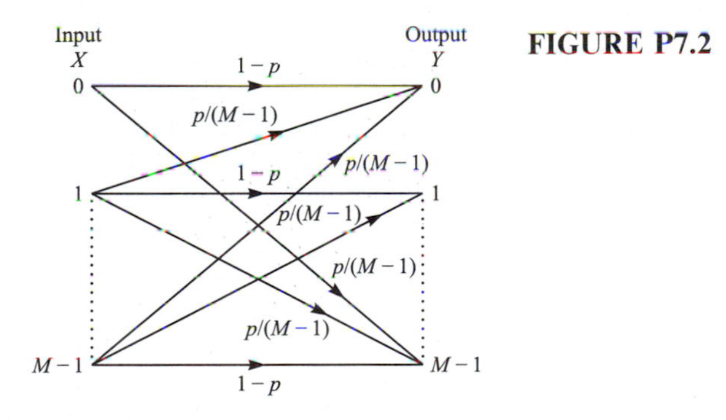

FIGURE P7.2

7.3 Determine the capacities of the channels shown in Figure P7.3.

7.4 Consider the two channels with the transition probabilities as shown in Figure P7.4. Determine if equally probable input symbols maximize the information rate through the channel.

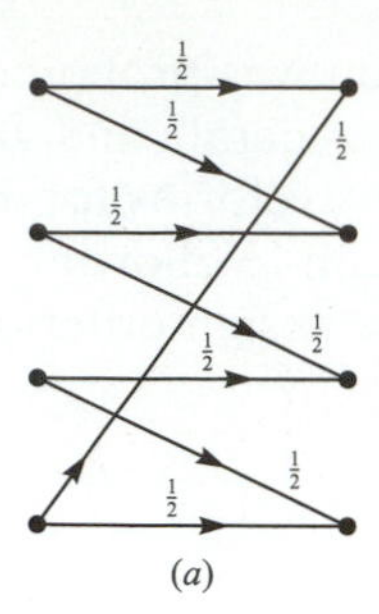

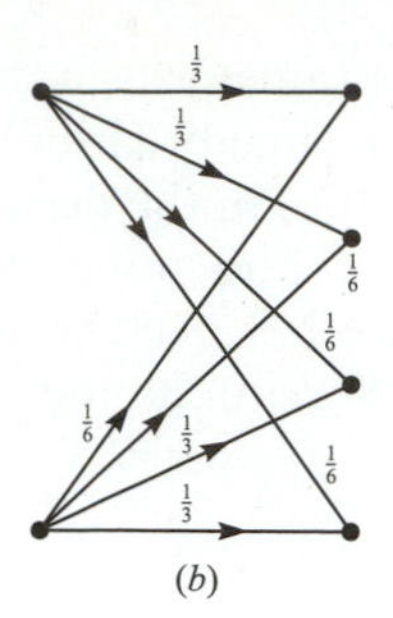

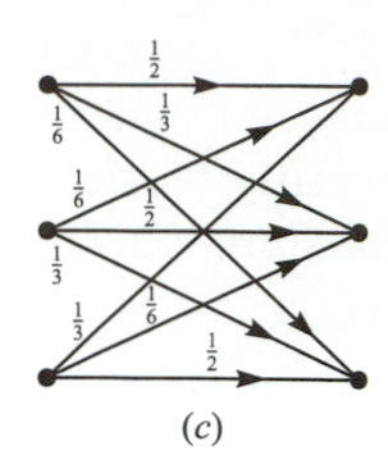

FIGURE P7.3

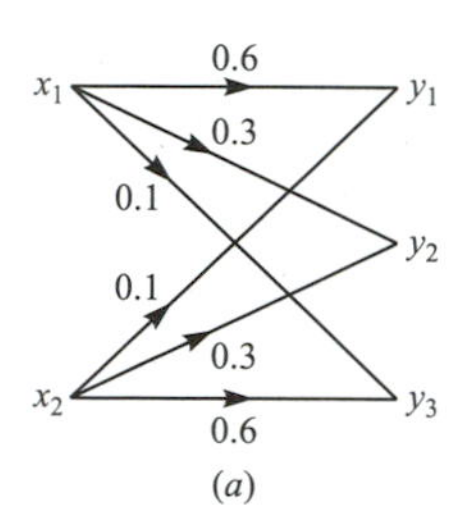

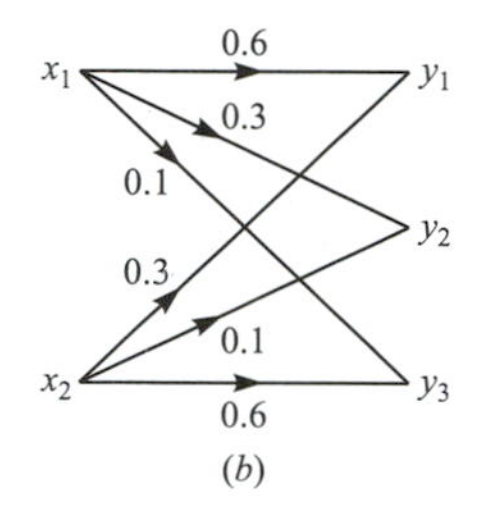

FIGURE P7.4

7.5 A telephone channel has a bandwidth $W = 3000$ Hz and a signal-to-noise power ratio of 400 (26 dB). Suppose we characterize the channel as a band-limited AWGN waveform channel with $P_{\mathrm{av}}/WN_0 = 400$.

a) Determine the capacity of the channel in bits/s.

b) Is the capacity of the channel sufficient to support the transmission of a speech signal that has been sampled and encoded by means of logarithmic PCM?

c) Usually, channel impairments other than additive noise limit the transmission rate over the telephone channel to less than the channel capacity of the equivalent band-limited AWGN channel considered in (a). Suppose that a transmission rate of $0.7C$ is achievable in practice without channel encoding. Which of the speech source encoding methods described in Section 3.5 provide sufficient compression to fit the bandwidth restrictions of the telephone channel?

7.6 Consider the binary-input, quaternary-output DMC shown in Figure P7.6.

a) Determine the capacity of the channel.

b) Show that this channel is equivalent to a BSC.

7.7 Determine the capacity for the channel shown in Figure P7.7.

7.8 Consider a BSC with crossover probability of error p. Suppose that R is the number of bits in a source code word that represents one of 2^R possible levels at the output of a quantizer.

a) Determine the probability that a code word transmitted over the BSC is received correctly.

b) Determine the probability of having at least one bit error in a code word transmitted over the BSC.

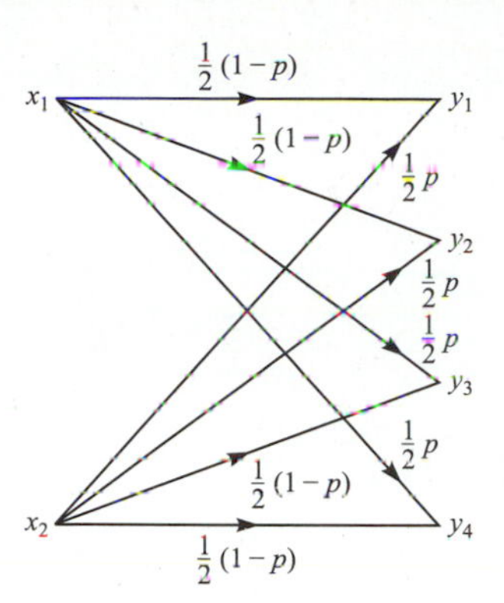

FIGURE P7.6

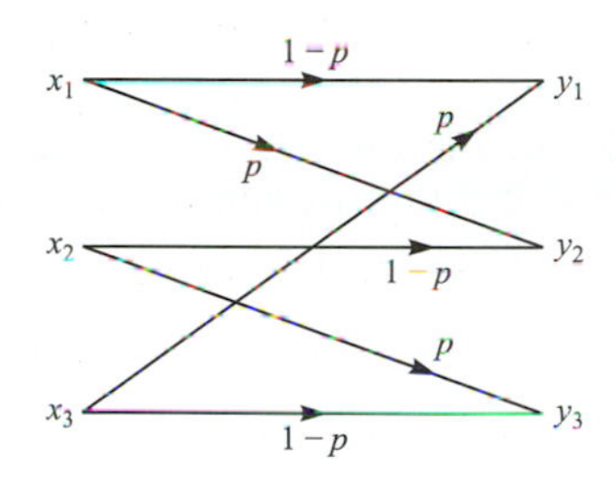

FIGURE P7.7

c) Determine the probability of having n_e or less bit errors in a code word.
d) Evaluate the probabilities in (a), (b), and (c) for $R = 5$, $p = 0.01$, and $n_e = 5$.

7.9 Show that, for a DMC, the average mutual information between a sequence $X_1 X_2 \cdots X_n$ of channel inputs and the corresponding channel outputs satisfy the condition

$$I(X_1 X_2 \cdots X_n; Y_1, Y_2, \ldots, Y_n) \leqslant \sum_{i=1}^{n} I(X_i; Y_i)$$

with equality if and only if the set of input symbols is statistically independent.

7.10 Figure P7.10 illustrates a binary erasure channel with transition probabilities $P(0|0) = P(1|1) = 1 - p$ and $P(e|0) = P(e|1) = p$. The probabilities for the input symbols are $P(X = 0) = \alpha$ and $P(X = 1) = 1 - \alpha$.
a) Determine the average mutual information $I(X; Y)$ in bits.
b) Determine the value of α that maximizes $I(X; Y)$, i.e., the channel capacity C in bits per channel use, and plot C as a function of p for the optimum value of α.
c) For the value of α found in (b), determine the mutual information $I(x; y) = I(0; 0)$, $I(1; 1)$, $I(0; e)$, and $I(1; e)$.

7.11 Consider the binary-input, ternary-output channel with transition probabilities shown in Figure P7.11, where e denotes an erasure. For the AWGN channel, α and p are defined as

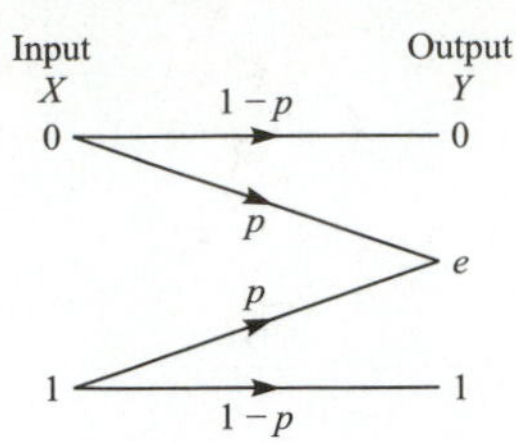

FIGURE P7.10

$$\alpha = \frac{1}{\sqrt{\pi N_0}} \int_{-\beta}^{\beta} e^{-(x+\sqrt{\mathcal{E}_c})^2/N_0}\, dx$$

$$p = \frac{1}{\sqrt{\pi N_0}} \int_{\beta}^{\infty} e^{-(x+\sqrt{\mathcal{E}_c})^2/N_0}\, dx$$

a) Determine R_Q for $Q = 3$ as a function of the probabilities α and p.

b) The rate parameter R_Q depends on the choice of the threshold β through the probabilities α and p. For any $\mathcal{E}_c/N_0$, the value of β that maximizes R_Q can be determined by trial and error. For example, it can be shown that for $\mathcal{E}_c/N_0$ below 0 dB, $\beta_{\text{opt}} = 0.65\sqrt{\frac{1}{2}N_0}$; for $1 \leqslant \mathcal{E}_c/N_0 \leqslant 10$, β_{opt} varies approximately linearly between $0.65\sqrt{\frac{1}{2}N_0}$ and $1.0\sqrt{\frac{1}{2}N_0}$. By using $\beta = 0.65\sqrt{\frac{1}{2}N_0}$ for the entire range of $\mathcal{E}_c/N_0$, plot R_Q versus $\mathcal{E}_c/N_0$ and compare this result with $R_Q(Q = \infty)$.

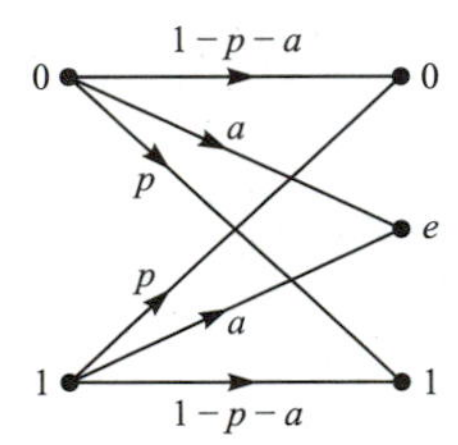

FIGURE P7.11

7.12 Find the capacity of the cascade connection of n binary-symmetric channels with the same crossover probability ε. What is the capacity when the number of channels goes to infinity?

7.13 Channels 1, 2, and 3 are shown in Figure P7.13.

a) Find the capacity of channel 1. What input distribution achieves capacity?

b) Find the capacity of channel 2. What input distribution achieves capacity?

c) Let C denote the capacity of the third channel and C_1 and C_2 represent the capacities of the first and second channel. Which of the following relations holds true and why?

$$C < \tfrac{1}{2}(C_1 + C_2) \tag{i}$$

$$C = \tfrac{1}{2}(C_1 + C_2) \tag{ii}$$

$$C > \tfrac{1}{2}(C_1 + C_2) \tag{iii}$$

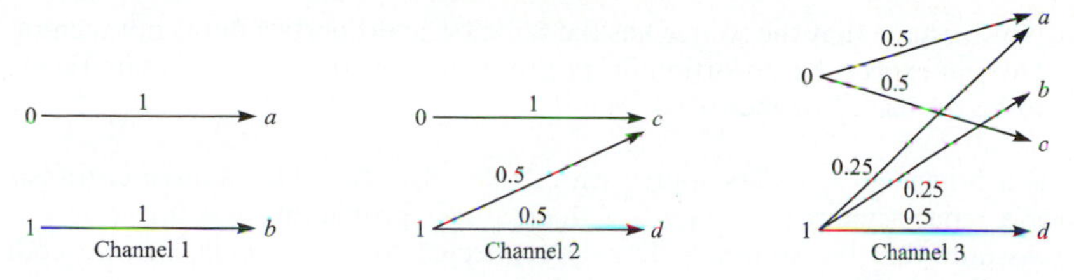

FIGURE P7.13

7.14 Let C denote the capacity of a discrete memoryless channel with input alphabet $\mathcal{X} = \{x_1, x_2, \ldots, x_N\}$ and output alphabet $\mathcal{Y} = \{y_1, y_2, \ldots, y_M\}$. Show that $C \leqslant \min\{\log M, \log N\}$.

7.15 The channel C (known as the Z channel) is shown in Figure P7.15.

a) Find the input probability distribution that achieves capacity.

b) What is the input distribution and capacity for the special cases $\varepsilon = 0$, $\varepsilon = 1$, and $\varepsilon = 0.5$?

c) Show that if n such channels are cascaded, the resulting channel will be equivalent to a Z channel with $\varepsilon_1 = \varepsilon^n$.

d) What is the capacity of the equivalent Z channel when $n \to \infty$?

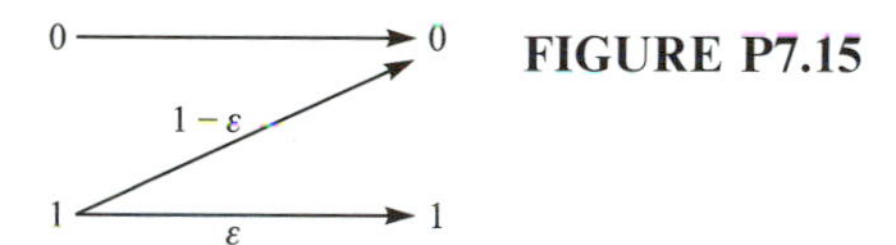

FIGURE P7.15

7.16 Find the capacity of an additive white Gaussian noise channel with a bandwidth 1 MHz, power 10 W, and noise power spectral density $\frac{1}{2}N_0 = 10^{-9}$ W/Hz.

7.17 Channel C_1 is an additive white Gaussian noise channel with a bandwidth W, average transmitter power P, and noise power spectral density $\frac{1}{2}N_0$. Channel C_2 is an additive Gaussian noise channel with the same bandwidth and power as channel C_1 but with noise power spectral density $\Phi_n(f)$. It is further assumed that the total noise power for both channels is the same; that is,

$$\int_{-W}^{W} \Phi_n(f)\, df = \int_{-W}^{W} \tfrac{1}{2}N_0\, df = N_0 W$$

Which channel do you think has a larger capacity? Give an intuitive reasoning.

7.18 A discrete-time memoryless Gaussian source with mean 0 and variance σ^2 is to be transmitted over a binary-symmetric channel with crossover probability p.

a) What is the minimum value of the distortion attainable at the destination (distortion is measured in mean-squared error)?

b) If the channel is a discrete-time memoryless additive Gaussian noise channel with input power P and noise power P_n, what is the minimum attainable distortion?

c) Now assume that the source has the same basic properties but is not memoryless. Do you expect the distortion in transmission over the binary-symmetric channel to be decreased or increased? Why?

7.19 X is a binary memoryless source with $P(X = 0) = 0.3$. This source is transmitted over a binary-symmetric channel with crossover probability $p = 0.1$.

a) Assume that the source is directly connected to the channel, i.e., no coding is employed. What is the error probability at the destination?

b) If coding is allowed, what is the minimum possible error probability in the reconstruction of the source?

c) For what values of p is reliable transmission possible (with coding, of course)?

7.20 Plot the capacity of an AWGN channel that employs binary antipodal signaling, with optimal bit-by-bit detection at the receiver, as a function of $\mathcal{E}_b/N_0$. On the same axis, plot the capacity of the same channel when binary orthogonal signaling is employed.

7.21 In a coded communication system, M messages $1, 2, \ldots, M = 2^k$ are transmitted by M *baseband* signals $x_1(t), x_2(t), \ldots, x_M(t)$, each of duration nT. The general form of $x_i(t)$ is given by

$$x_i(t) = \sum_{j=0}^{n-1} f_{ij}(t - jT)$$

where $f_{ij}(t)$ can be either of the two signals $f_1(t)$ or $f_2(t)$, where $f_1(t) = f_2(t) \equiv 0$ for all $t \notin [0, T]$. We further assume that $f_1(t)$ and $f_2(t)$ have equal energy $\mathcal{E}$ and the channel is ideal (no attenuation) with additive white Gaussian noise of power spectral density $\frac{1}{2}N_0$. This means that the received signal is $r(t) = x(t) + n(t)$, where $x(t)$ is one of the $x_i(t)$ and $n(t)$ represents the noise.

a) With $f_1(t) = -f_2(t)$, show that N, the dimensionality of the signal space, satisfies $N \leqslant n$.

b) Show that, in general, $N \leqslant 2n$.

c) With $M = 2$, show that, for general $f_1(t)$ and $f_2(t)$,

$$P(\text{error}|x_1(t) \text{ sent}) \leqslant \int \cdots \int_{R^N} \sqrt{p(\mathbf{r}|\mathbf{x}_1)p(\mathbf{r}|\mathbf{x}_2)}\, d\mathbf{r}$$

where $\mathbf{r}$, $\mathbf{x}_1$, and $\mathbf{x}_2$ are the vector representations of $r(t)$, $x_1(t)$, and $x_2(t)$ in the N-dimensional space.

d) Using the result of (c), show that, for general M,

$$P(\text{error}|x_m(t) \text{ sent}) \leqslant \sum_{\substack{1 \leqslant m' \leqslant M \\ m' \neq m}} \int \cdots \int_{R^N} \sqrt{p(\mathbf{r}|\mathbf{x}_m)p(\mathbf{r}|\mathbf{x}_{m'})}\, d\mathbf{r}$$

e) Show that

$$\int \cdots \int_{R^N} \sqrt{p(\mathbf{r}|\mathbf{x}_m)p(\mathbf{r}|\mathbf{x}_{m'})}\, d\mathbf{r} = \exp\left(-\frac{|\mathbf{x}_m - \mathbf{x}_{m'}|^2}{4N_0}\right)$$

and, therefore,

$$P(\text{error}|x_m(t) \text{ sent}) \leqslant \sum_{\substack{1 \leqslant m' \leqslant M \\ m' \neq m}} \exp\left(-\frac{|\mathbf{x}_m - \mathbf{x}_{m'}|^2}{4N_0}\right)$$

7.22 Use Equation 7.3–2 to show that for the binary symmetric channel (BSC), the cutoff rate is given as

$$R_Q|_{Q=2} \equiv R_2 = 1 - \log_2\left[1 + \sqrt{4p(1-p)}\right]$$

where p is the probability of error. Compute R_2 as a function of $\mathcal{E}_c/N_0$, assuming that the coded bits are transmitted by binary PSK (antipodal signals) where

$$p = Q\left(\sqrt{\frac{2\mathcal{E}_c}{N_0}}\right)$$

Also, plot R_0 for antipodal signals given by Equation 7.2–20 and, thus, compare R_0 with R_2. What is the loss in performance resulting from hard-decision decoding ($Q = 2$) compared with soft-decision decoding ($Q = \infty$)?

7.23 Using the expression for the cutoff R_2 for the BSC, given in Problem 7.22, plot R_2 as a function of $\mathcal{E}_c/N_0$ for the following binary modulation methods.

antipodal signals: $p = Q\left(\sqrt{\frac{2\mathcal{E}_c}{N_0}}\right)$

orthogonal signals : $p = Q\left(\sqrt{\frac{\mathcal{E}_c}{N_0}}\right)$

DPSK : $p = \frac{1}{2}e^{-\mathcal{E}_c/N_0}$

Comment on the difference in performance for the three modulation methods, as given by the cutoff rate.

7.24 Using the expression for the cutoff rate R_2 for the BSC, given in Problem 7.22, plot R_2 as a function of $\mathcal{E}_c/N_0$ for antipodal signals and also plot the channel capacity C for the BSC with antipodal signals. What is the difference in $\mathcal{E}_c/N_0$ between R_2 and C?

7.25 For M-ary PSK, equiprobable input symbols maximize the cutoff rate R_0. Show that

$$R_0 = \log_2 M - \log_2\left[\sum_{k=0}^{M-1} e^{-\|s_0 - s_k\|^2/4N_0}\right]$$

$$= \log_2 M - \log_2\left[\sum_{k=0}^{M-1} e^{-(\mathcal{E}_c/N_0)\sin^2(\pi k/M)}\right]$$

Plot R_0 as a function of $\mathcal{E}_c/N_0$ for $M = 2, 4, 8$, and 16.

8

Block and Convolutional Channel Codes

In Chapter 7, we treated channel coding and decoding from a general viewpoint, and showed that even randomly selected codes on the average yield performances close to the capacity of a channel. In the case of orthogonal signals, we demonstrated that the channel capacity limit can be achieved as the number of signals approaches infinity.

In this chapter, we describe specific codes and evaluate their performance for the additive white Gaussian noise channel. In particular, we treat two classes of codes, namely, linear block codes and convolutional codes. The code performance is evaluated for both hard-decision decoding and soft-decision decoding.

8.1 LINEAR BLOCK CODES

A block code consists of a set of fixed-length vectors called *code words*. The length of a code word is the number of elements in the vector and is denoted by n. The elements of a code word are selected from an alphabet of q symbols, or elements. When the alphabet consists of two symbols, 0 and 1, the code is a binary code and the elements of any code word are called bits. When the elements of a code word are selected from an alphabet having q symbols ($q > 2$), the code is nonbinary. It is interesting to note that when q is a power of 2, i.e., $q = 2^b$ where b is a positive integer, each q-ary symbol has an equivalent binary representation consisting of b bits, and, thus, a nonbinary code of block length N can be mapped into a binary code of block length $n = bN$.

There are 2^n possible code words in a binary block code of length n. From these 2^n code words, we may select $M = 2^k$ code words ($k < n$) to form a code. Thus, a block of k information bits is mapped into a code word of length n selected from the set of $M = 2^k$ code words. We refer to the resulting block code as an (n, k) code, and the ratio $k/n \equiv R_c$ is defined to be the *rate* of the code.

More generally, in a code having q symbols, there are q^n possible code words. A subset of $M = q^k$ code words may be selected to transmit k-symbol blocks of information.

Besides the code rate parameter R_c, an important parameter of a code word is its *weight*, which is simply the number of nonzero elements that it contains. In general, each code word has its own weight. The set of all weights in a code constitutes the *weight distribution* of the code. When all the M code words have equal weight, the code is called a *fixed-weight code* or a *constant-weight code*.

The encoding and decoding functions involve the arithmetic operations of addition and multiplication performed on code words. These arithmetic operations are performed according to the conventions of the algebraic field that has as its elements the symbols contained in the alphabet. For example, the symbols in a binary alphabet are 0 and 1; hence, the field has two elements. In general, a field F consists of a set of elements that has two arithmetic operations defined on its elements, namely, addition and multiplication, that satisfy the following properties (axioms).

Addition

1. The set F is closed under addition, i.e., if $a, b \in F$, then $a + b \in F$.
2. Addition is associative, i.e, if a, b, and c are elements of F, then $a + (b + c) = (a + b) + c$.
3. Addition is commutative, i.e., $a + b = b + a$.
4. The set contains an element called *zero* that satisfies the condition $a + 0 = a$.
5. Every element in the set has its own negative element. Hence, if b is an element, its negative is denoted by $-b$. The subtraction of two elements, such as $a - b$, is defined as $a + (-b)$.

Multiplication

1. The set F is closed under multiplication, i.e., if $a, b \in F$, then $ab \in F$.
2. Multiplication is associative, i.e., $a(bc) = (ab)c$.
3. Multiplication is commutative, i.e., $ab = ba$.
4. Multiplication is distributive over addition, i.e., $(a + b)c = ac + bc$.
5. The set F contains an element, called the *identity*, that satisfies the condition $a(1) = a$, for any element $a \in F$.
6. Every element of F, except zero, has an inverse. Hence, if $b \in F$ $(b \neq 0)$, then its inverse is defined as b^{-1}, and $bb^{-1} = 1$. The division of two elements, such as $a \div b$, is defined as ab^{-1}.

We are very familiar with the field of real numbers and the field of complex numbers. These fields have an infinite number of elements. However, as indicated above, codes are constructed from fields with a finite number of elements. A finite field with q elements is generally called a *Galois field* and denoted by GF(q),

Every field must have a *zero* element and a *one* element. Hence, the simplest field is GF(2). In general, when q is a prime, we can construct the finite field GF(q) consisting of the elements $\{0, 1, \ldots, q-1\}$. The addition and multiplication operations on the elements of GF(q) are defined modulo q and denoted as (mod q). For example, the addition and multiplication tables for GF(2) are

+	0	1
0	0	1
1	1	0

·	0	1
0	0	0
1	0	1

which are operations (mod 2). Similarly, the field GF(5) is a set consisting of the elements {0, 1, 2, 3, 4}. The addition and multiplication tables for GF(5) are

+	0	1	2	3	4
0	0	1	2	3	4
1	1	2	3	4	0
2	2	3	4	0	1
3	3	4	0	1	2
4	4	0	1	2	3

·	0	1	2	3	4
0	0	0	0	0	0
1	0	1	2	3	4
2	0	2	4	1	3
3	0	3	1	4	2
4	0	4	3	2	1

In general, the finite field GF(q) can be constructed only if q is a prime or a power of a prime. When q is a prime, multiplication and addition are based on modulo-q arithmetic as illustrated above. If $q = p^m$ where p is a prime and m is any positive integer, it is possible to extend the field GF(p) to the field GF(p^m). This is called the *extension field* of GF(p). Multiplication and addition of the elements in the extension field are based on modulo-p arithmetic.

With this brief introduction to the arithmetic operations that may be performed on the elements of code words, let us now consider some basic characteristics of block codes.

Suppose $\mathbf{C}_i$ and $\mathbf{C}_j$ are any two code words in an (n, k) block code. A measure of the difference between the code words is the number of corresponding elements or positions in which they differ. This measure is called the *Hamming distance* between the two code words and is denoted as d_{ij}. Clearly, d_{ij} for $i \neq j$ satisfies the condition $0 \leqslant d_{ij} \leqslant n$. The smallest value of the set $\{d_{ij}\}$ for the M code words is called the *minimum distance* of the code and is denoted as $d_{\min}$. Since the Hamming distance is a measure of the separation between pairs of code words, it is intimately related to the cross-correlation coefficient between corresponding pairs of waveforms generated from the code words. The relationship is discussed in Section 8.1.4

Besides characterizing a code as being binary or nonbinary, one can also describe it as either linear or non-linear. Suppose $\mathbf{C}_i$ and $\mathbf{C}_j$ are two code words in an (n, k) block code and let α_1 and α_2 be any two elements selected from the alphabet. Then the code is said to be linear if and only if $\alpha_1 \mathbf{C}_i + \alpha_2 \mathbf{C}_j$ is also a code word. This definition implies that a linear code must contain the all-zero code word. Consequently a constant-weight code is non-linear.

Suppose we have a binary linear block code, and let $\mathbf{C}_i$, $i = 1, 2, \ldots, M$, denote the M code words. For convenience, let $\mathbf{C}_1$ denote the all-zero code word, i.e., $\mathbf{C}_1 = [0\,0 \cdots 0]$, and let w_r denote the weight of the rth code word. It follows that w_r is the Hamming distance between the code words $\mathbf{C}_r$ and $\mathbf{C}_1$.

Thus, the distance $d_{1r} = w_r$. In general, the distance d_{ij} between any pair of code words $\mathbf{C}_i$ and $\mathbf{C}_j$ is simply equal to the weight of the code word formed by taking the difference between $\mathbf{C}_i$ and $\mathbf{C}_j$. Since the code is linear, the difference (equivalent to taking the modulo-2 sum for a binary code) between $\mathbf{C}_i$ and $\mathbf{C}_j$ is also a code word having a weight included in the set $\{w_r\}$. Hence, the weight distribution of a linear code completely characterizes the distance properties of the code. The minimum distance of the code is, therefore,

$$d_{\min} = \min_{r, r \neq 1}\{w_r\} \tag{8.1–1}$$

A number of elementary concepts from linear algebra are particularly useful in dealing with linear block codes. Specifically, the set of all n-tuples (vectors with n elements) form a vector space S. If we select a set of $k < n$ linearly independent vectors from S and from these construct the set of all linear combinations of these vectors, the resulting set forms a subspace of S, say S_c, of dimension k. Any set of k linearly independent vectors in the subspace S_c constitutes a basis. Now consider the set of vectors in S that are orthogonal to every vector in a basis for S_c (and, hence, orthogonal to all vectors in S_c). This set of vectors is also a subspace of S and is called the *null space* of S_c. If the dimension of S_c is k, the dimension of the null space is $n - k$.

Expressed in terms appropriate for binary block codes, the vector space S consists of the 2^n binary-valued n-tuples. The linear (n, k) code is a set of 2^k n-tuples called *code words*, which forms a subspace S_c over the field of two elements. Since there are 2^k code words in S_c, a basis for S_c has k code words. That is, k linearly independent code words are required to construct 2^k linear combinations, thus generating the entire code. The null space of S_c is another linear code, which consists of 2^{n-k} code words of block length n and $n - k$ information bits. Its dimension is $n - k$. In Section 8.1.1, we consider these relationships in greater detail.

8.1.1 The Generator Matrix and the Parity Check Matrix

Let $x_{m1}, x_{m2}, \ldots, x_{mk}$ denote the k information bits encoded into the code word $\mathbf{C}_m$. Throughout this chapter, we follow the established convention in coding of representing code words as row vectors. Thus, the vector of k information bits into the encoder is denoted by

$$\mathbf{X}_m = [x_{m1} \quad x_{m2} \quad \cdots \quad x_{mk}]$$

and the output of the encoder is the vector

$$\mathbf{C}_m = [c_{m1} \quad c_{m2} \quad \cdots \quad c_{mn}]$$

The encoding operation performed in a linear binary block encoder can be represented by a set of n equations of the form

$$c_{mj} = x_{m1}g_{1j} + x_{m2}g_{2j} + \cdots + x_{mk}g_{kj}, \qquad j = 1, 2, \ldots, n \tag{8.1–2}$$

where $g_{ij} = 0$ or 1 and $x_{mi}g_{ij}$ represents the product of x_{mi} and g_{ij}. The linear equations 8.1–2 may also be represented in a matrix form as

$$\mathbf{C}_m = \mathbf{X}_m \mathbf{G} \tag{8.1–3}$$

where $\mathbf{G}$, called the *generator matrix* of the code, is

$$\mathbf{G} = \begin{bmatrix} \leftarrow \mathbf{g}_1 \rightarrow \\ \leftarrow \mathbf{g}_2 \rightarrow \\ \vdots \\ \leftarrow \mathbf{g}_k \rightarrow \end{bmatrix} = \begin{bmatrix} g_{11} & g_{12} & \cdots & g_{1n} \\ g_{21} & g_{22} & \cdots & g_{2n} \\ \vdots & \vdots & & \vdots \\ g_{k1} & g_{k2} & \cdots & g_{kn} \end{bmatrix} \tag{8.1–4}$$

Note that any code word is simply a linear combination of the vectors $\{\mathbf{g}_i\}$ of $\mathbf{G}$, i.e.,

$$\mathbf{C}_m = x_{m1}\mathbf{g}_1 + x_{m2}\mathbf{g}_2 + \cdots + x_{mk}\mathbf{g}_k \tag{8.1–5}$$

Since the linear (n, k) code with 2^k code words is a subspace of dimension k, the row vectors $\{\mathbf{g}_i\}$ of the generator matrix $\mathbf{G}$ must be linearly independent, i.e., they must span a subspace of k dimensions. In other words, the $\{\mathbf{g}_i\}$ must be a basis for the (n, k) code. We note that the set of basis vectors is not unique, and, hence, $\mathbf{G}$ is not unique. We also note that, since the subspace has dimension k, the rank of $\mathbf{G}$ is k.

Any generator matrix of an (n, k) code can be reduced by row operations (and column permutations) to the "systematic form,"

$$\mathbf{G} = [\mathbf{I}_k | \mathbf{P}] = \left[\begin{array}{ccccc|cccc} 1 & 0 & 0 & \cdots & 0 & p_{11} & p_{12} & \cdots & p_{1n-k} \\ 0 & 1 & 0 & \cdots & 0 & p_{21} & p_{22} & \cdots & p_{2n-k} \\ \vdots & \vdots & \vdots & & \vdots & \vdots & \vdots & & \vdots \\ 0 & 0 & 0 & \cdots & 1 & p_{k1} & p_{k2} & \cdots & p_{kn-k} \end{array}\right] \tag{8.1–6}$$

where $\mathbf{I}_k$ is the $k \times k$ identity matrix and $\mathbf{P}$ is a $k \times (n-k)$ matrix that determines the $n - k$ redundant bits or parity check bits. Note that a generator matrix of the systematic form generates a linear block code in which the first k bits of each code word are identical to the information bits to be transmitted, and the remaining $n - k$ bits of each code word are linear combinations of the k information bits. These $n - k$ redundant bits are called *parity check bits*. The resulting (n, k) code is called a *systematic code*.

An (n, k) code generated by a generator matrix that is not in the systematic form 8.1–6 is called *nonsystematic*. However, such a generator matrix is equivalent to a generator matrix of the systematic form in the sense that one can be obtained from the other by elementary row operations and column permutations. The two (n, k) linear codes generated by the two equivalent generator matrices are said to be *equivalent*, and one can be obtained from the other by a permutation of the places of every element. Thus, every linear (n, k) code is equivalent to a linear systematic (n, k) code.

EXAMPLE 8.1–1. Consider a (7, 4) code with generator matrix

$$\mathbf{G} = \begin{bmatrix} 1 & 0 & 0 & 0 & 1 & 0 & 1 \\ 0 & 1 & 0 & 0 & 1 & 1 & 1 \\ 0 & 0 & 1 & 0 & 1 & 1 & 0 \\ 0 & 0 & 0 & 1 & 0 & 1 & 1 \end{bmatrix} = [\mathbf{I}_4|\mathbf{P}] \tag{8.1–7}$$

A typical code word may be expressed as

$$\mathbf{C}_m = [\, x_{m1} \quad x_{m2} \quad x_{m3} \quad x_{m4} \quad c_{m5} \quad c_{m6} \quad c_{m7}]$$

where the $\{x_{mj}\}$ represents the four information bits and the $\{c_{mj}\}$ represent the three parity check bits given by

$$\begin{aligned} c_{m5} &= x_{m1} + x_{m2} + x_{m3} \\ c_{m6} &= x_{m2} + x_{m3} + x_{m4} \\ c_{m7} &= x_{m1} + x_{m2} + x_{m4} \end{aligned} \tag{8.1–8}$$

A linear systematic (n, k) binary block encoder may be implemented by using a k-bit shift register and $n-k$ modulo-2 adders tied to the appropriate stages of the shift register. The $n-k$ adders generate the parity check bits, which are subsequently stored temporarily in a second shift register of length $n-k$. The k-bit block of information bits is shifted into the k-bit shift register and the $n-k$ parity check bits are computed. Then the k information bits followed by the $n-k$ parity check bits are shifted out of the two shift registers and fed to the modulator. This encoding is illustrated in Figure 8.1–1 for the (7, 4) code of Example 8.1–1.

Associated with any linear (n, k) code is the dual code of dimension $n-k$. The dual code is a linear $(n, n-k)$ code with 2^{n-k} code vectors, which is the null space of the (n, k) code. The generator matrix for the dual code, denoted by $\mathbf{H}$, consists of $n-k$ linearly independent code vectors selected from the null space. Any code word $\mathbf{C}_m$ of the (n, k) code is orthogonal to any code word in the dual code. Hence, any code word of the (n, k) code is orthogonal to every row of the matrix $\mathbf{H}$, i.e.,

$$\mathbf{C}_m\mathbf{H}' = \mathbf{0} \tag{8.1–9}$$

where $\mathbf{0}$ denotes an all-zero row vector with $n-k$ elements, and $\mathbf{C}_m$ is a code word of the (n, k). Since Equation 8.1–9 holds for every code word of the (n, k) code, it follows that

$$\mathbf{G}\mathbf{H}' = \mathbf{0} \tag{8.1–10}$$

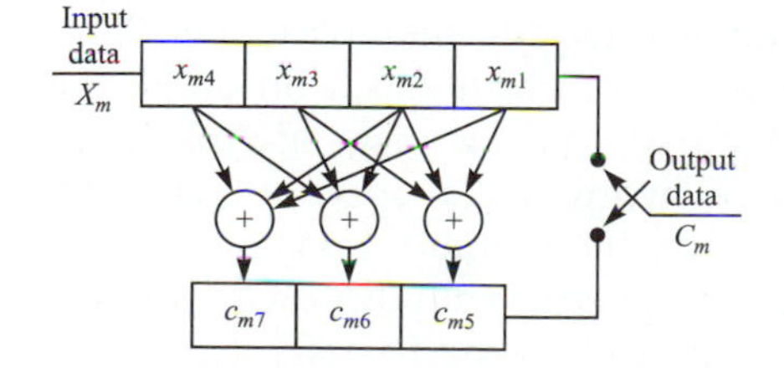

FIGURE 8.1–1
A linear shift register for generating a (7, 4) binary code.

where $\mathbf{0}$ is now a $k \times (n-k)$ matrix with all-zero elements.

Now suppose that the linear (n, k) code is systematic and its generator matrix $\mathbf{G}$ is given by the systematic form 8.1–6. Then, since $\mathbf{GH}' = \mathbf{0}$, it follows that

$$\mathbf{H} = [-\mathbf{P}' | \mathbf{I}_{n-k}] \tag{8.1–11}$$

The negative sign in Equation 8.1–11 may be dropped when dealing with binary codes, since modulo-2 subtraction is identical to modulo-2 addition.

EXAMPLE 8.1–2. For the systematic $(7, 4)$ code generated by matrix $\mathbf{G}$ given by Equation 8.1–7, we have, according to Equation 8.1–11, the matrix $\mathbf{H}$ in the form

$$\mathbf{H} = \begin{bmatrix} 1 & 1 & 1 & 0 & 1 & 0 & 0 \\ 0 & 1 & 1 & 1 & 0 & 1 & 0 \\ 1 & 1 & 0 & 1 & 0 & 0 & 1 \end{bmatrix} \tag{8.1–12}$$

Now, the product $\mathbf{C}_m\mathbf{H}'$ yields the three equations

$$\begin{aligned} x_{m1} + x_{m2} + x_{m3} + c_{m5} &= 0 \\ x_{m2} + x_{m3} + x_{m4} + c_{m6} &= 0 \\ x_{m1} + x_{m2} + x_{m4} + c_{m7} &= 0 \end{aligned} \tag{8.1–13}$$

Thus, we observe that the product $\mathbf{C}_m\mathbf{H}'$ is equivalent to adding the parity check bits to the corresponding linear combinations of the information bits used to compute $c_{mj}, j = 5, 6, 7$. That is, Equation 8.1–13 is equivalent to Equation 8.1–8. The matrix $\mathbf{H}$ may be used by the decoder to check that a received code word $\mathbf{Y}$ satisfies the condition 8.1–13, i.e., $\mathbf{YH}' = \mathbf{0}$. In so doing, the decoder checks the received parity check bits with the corresponding linear combination of the bits y_1, y_2, y_3, and y_4 that formed the parity check bits at the transmitter. It is, therefore, appropriate to call $\mathbf{H}$ the *parity check matrix* associated with the (n, k) code.

We make the following observation regarding the relation of the minimum distance of a code to its parity check matrix $\mathbf{H}$. The product $\mathbf{C}_m\mathbf{H}'$ with $\mathbf{C}_m \neq \mathbf{0}$ represents a linear combination of the n columns of $\mathbf{H}$. Since $\mathbf{C}_m\mathbf{H}' = \mathbf{0}$, the column vectors of $\mathbf{H}$ are linearly dependent. Suppose $\mathbf{C}_j$ denotes the minimum weight code word of a linear (n, k) code. It must satisfy the condition $\mathbf{C}_j\mathbf{H}' = \mathbf{0}$. Since the minimum weight is equal to the minimum distance, it follows that $d_{\min}$ of the columns of $\mathbf{H}$ are linearly dependent. Alternatively, we may say that no more than $d_{\min} - 1$ columns of $\mathbf{H}$ are linearly independent. Since the rank of $\mathbf{H}$ is at most $n - k$, we have $n - k \geqslant d_{\min} - 1$. Therefore, $d_{\min}$ is upper-bounded as

$$d_{\min} \leqslant n - k + 1 \tag{8.1–14}$$

Given a linear binary (n, k) code with minimum distance $d_{\min}$, we can construct a linear binary $(n + 1, k)$ code by appending one additional parity check bit to each code word. The check bit is usually selected to be a check bit on all the bits in the code word. Thus the added check bit is a 0 if the original code word has an even number of 1s and it is a 1 if the code word has an odd number of 1s. Consequently, if the minimum weight and, hence, the minimum distance of the code is odd, the added parity check bit increases the minimum distance by 1. We call the $(n + 1, k)$ code an *extended code*. Its parity check matrix is

$$\mathbf{H}_e = \left[\begin{array}{ccccc|c} & & & & & 0 \\ & & & & & 0 \\ & & \mathbf{H} & & & \vdots \\ & & & & & 0 \\ \hline 1 & 1 & 1 & \dots & 1 & 1 \end{array}\right] \tag{8.1–15}$$

where $\mathbf{H}$ is the parity check matrix of the original code.

A systematic (n, k) code can also be shortened by setting a number of the information bits to zero. That is, a linear (n, k) code consisting of k information bits and $n - k$ check bits can be shortened into an $(n - l, k - l)$ linear code by setting the first l bits to zero. These l bits are not transmitted. The $n - k$ check bits are computed in the usual manner, as in the original code. Since

$$\mathbf{C}_m = \mathbf{X}_m \mathbf{G}$$

the effect of setting the first l bits of $\mathbf{X}_m$ to zero is equivalent to reducing the number of rows of $\mathbf{G}$ by removing the first l rows. Equivalently, since

$$\mathbf{C}_m \mathbf{H}' = \mathbf{0}$$

we may remove the first l column of $\mathbf{H}$. The shortened $(n - l, k - l)$ code consists of 2^{k-l} code words. The minimum distance of these 2^{k-l} code words is at least as large as the minimum distance of the original (n, k) code.

8.1.2 Some Specific Linear Block Codes

In this subsection, we shall briefly describe three types of linear block codes that are frequently encountered in practice and list their important parameters.

Hamming codes. There are both binary and nonbinary Hamming codes. We limit our discussion to the properties of binary Hamming codes. These comprise a class of codes with the property that

$$(n, k) = (2^m - 1, 2^m - 1 - m) \tag{8.1–16}$$

where m is any positive integer. For example, if $m = 3$, we have a $(7, 4)$ code.

The parity check matrix $\mathbf{H}$ of a Hamming code has a special property that allows us to describe the code rather easily. Recall that the parity check matrix of an (n, k) code has $n - k$ rows and n columns. For the binary (n, k) Hamming code, the $n = 2^m - 1$ columns consist of all possible binary vectors with $n - k = m$ elements, except the all-zero vector. For example, the $(7, 4)$ code considered in Examples 8.1–1 and 8.1–2 is a Hamming code. Its parity check matrix consists of the seven column vectors (001), (010), (011), (100), (101), (110), (111).

If we desire to generate a systematic Hamming code, the parity check matrix $\mathbf{H}$ can be easily arranged in the systematic form of Equation 8.1–11. Then the corresponding generator matrix $\mathbf{G}$ can be constructed from Equation 8.1–11.

We make the observation that no two columns of **H** are linearly dependent, for otherwise the two columns would be identical. However, for $m > 1$, it is possible to find three columns of **H** that add to zero. Consequently, $d_{\min} = 3$ for an (n, k) Hamming code.

By adding an overall parity bit, a Hamming (n, k) code can be modified to yield an $(n + 1, k)$ code with $d_{\min} = 4$. On the other hand, an (n, k) Hamming code may be shortened to $(n - l, k - 1)$ by removing l rows of its generator matrix **G** or, equivalently, by removing l columns of its parity check matrix **H**.

The weight distribution for the class of Hamming (n, k) codes is known and is expressed in compact form by the weight enumerating polynomial

$$\begin{aligned} A(z) &= \sum_{i=0}^{n} A_i z^i \\ &= \frac{1}{n+1}\left[(1+z)^n + n(1+z)^{(n-1)/2}(1-z)^{(n+1)/2}\right] \end{aligned} \tag{8.1–17}$$

where A_i is the number of code words of weight i.

Hadamard codes. A Hadamard code is obtained by selecting as code words the rows of a Hadamard matrix. A Hadamard matrix $\mathbf{M}_n$ is an $n \times n$ matrix (n is an even integer) of 1s and 0s with the property that any row differs from any other row in exactly $\frac{1}{2}n$ positions.† One row of the matrix contains all zeros. The other rows contain $\frac{1}{2}n$ zeros and $\frac{1}{2}n$ ones.

For $n = 2$, the Hadamard matrix is

$$\mathbf{M}_2 = \begin{bmatrix} 0 & 0 \\ 0 & 1 \end{bmatrix} \tag{8.1–18}$$

Furthermore, from $\mathbf{M}_n$, we can generate the Hadamard matrix $\mathbf{M}_{2n}$ according to the relation

$$\mathbf{M}_{2n} = \begin{bmatrix} \mathbf{M}_n & \mathbf{M}_n \\ \mathbf{M}_n & \bar{\mathbf{M}}_n \end{bmatrix} \tag{8.1–19}$$

where $\bar{\mathbf{M}}_n$ denotes the complement (0s replaced by 1s and vice versa) of $\mathbf{M}_n$. Thus, by substituting Equation 8.1–18 into Equation 8.1–19, we obtain

$$\mathbf{M}_4 = \begin{bmatrix} 0 & 0 & 0 & 0 \\ 0 & 1 & 0 & 1 \\ 0 & 0 & 1 & 1 \\ 0 & 1 & 1 & 0 \end{bmatrix} \tag{8.1–20}$$

†Sometimes the elements of the Hadamard matrix are denoted by $+1$ and -1. Then the rows of the Hadamard matrix are mutually orthogonal. We also note that the $M = 2^k$ signal waveforms, constructed from Hadamard code words by mapping each bit in a code word into a binary PSK signal, are orthogonal.

The complement of $\mathbf{M}_4$ is

$$\bar{\mathbf{M}}_4 = \begin{bmatrix} 1 & 1 & 1 & 1 \\ 1 & 0 & 1 & 0 \\ 1 & 1 & 0 & 0 \\ 1 & 0 & 0 & 1 \end{bmatrix} \tag{8.1-21}$$

Now the rows of $\mathbf{M}_4$ and $\bar{\mathbf{M}}_4$ form a linear binary code of block length $n = 4$ having $2n = 8$ code words. The minimum distance of the code is $d_{\min} = \frac{1}{2}n = 2$.

By repeated application of Equation 8.1–19, we can generate Hadamard codes with block length $n = 2^m$, $k = \log_2 2n = \log_2 2^{m+1} = m + 1$, and $d_{\min} = \frac{1}{2}n = 2^{m-1}$, where m is a positive integer. In addition to the important special cases where $n = 2^m$, Hadamard codes of other block lengths are possible, but the codes are not linear.

Golay code. The Golay code is a binary linear (23, 12) code with $d_{\min} = 7$. The extended Golay code obtained by adding an overall parity to the (23, 12) is a binary linear (24, 12) code with $d_{\min} = 8$. Table 8.1–1 lists the weight distribution of the code words in the Golay (23, 12) and the extended Golay (24, 12) codes. We discuss the generation of the Golay code in Section 8.1.3.

8.1.3 Cyclic Codes

Cyclic codes are a subset of the class of linear codes that satisfy the following cyclic shift property: if $\mathbf{C} = [c_{n-1}c_{n-2} \cdots c_1 c_0]$ is a code word of a cyclic code, then $[c_{n-2}c_{n-3} \cdots c_0, c_{n-1}]$, obtained by a cyclic shift of the elements of $\mathbf{C}$, is also a code word. That is, all cyclic shifts of $\mathbf{C}$ are code words. As a consequence of the cyclic property, the codes possess a considerable amount of structure which can be

TABLE 8.1–1
Weight distribution of Golay (23, 12) and extended Golay (24, 12) codes

	Number of code words	
Weight	**(23, 12) code**	**(24, 12) code**
0	1	1
7	253	0
8	506	759
11	1288	0
12	1288	2576
15	506	0
16	253	759
23	1	0
24	0	1

exploited in the encoding and decoding operations. A number of efficient encoding and hard-decision decoding algorithms have been devised for cyclic codes that make it possible to implement long block codes with a large number of code words in practical communication systems. A description of specific algorithms is beyond the scope of this book. Our primary objective is to briefly describe a number of characteristics of cyclic codes.

In dealing with cyclic codes, it is convenient to associate with a code word $\mathbf{C} = [c_{n-1}c_{n-2} \cdots c_1 c_0]$ a polynomial $C(p)$ of degree $\leqslant n-1$, defined as

$$C(p) = c_{n-1}p^{n-1} + c_{n-2}p^{n-2} + \cdots + c_1 p + c_0 \tag{8.1–22}$$

For a binary code, each of the coefficients of the polynomial is either zero or one.

Now suppose we form the polynomial

$$pC(p) = c_{n-1}p^n + c_{n-2}p^{n-1} + \cdots + c_1 p^2 + c_0 p$$

This polynomial cannot represent a code word, since its degree may be equal to n (when $c_{n-1} = 1$). However, if we divide $pC(p)$ by $p^n + 1$, we obtain

$$\frac{pC(p)}{p^n + 1} = c_{n-1} + \frac{C_1(p)}{p^n + 1} \tag{8.1–23}$$

where

$$C_1(p) = c_{n-2}p^{n-1} + c_{n-3}p^{n-2} + \cdots + c_0 p + c_{n-1}$$

Note that the polynomial $C_1(p)$ represents the code word $\mathbf{C}_1 = [c_{n-2} \cdots c_0 c_{n-1}]$, which is just the code word $\mathbf{C}$ shifted cyclicly by one position. Since $C_1(p)$ is the remainder obtained by dividing $pC(p)$ by $p^n + 1$, we say that

$$C_1(p) = pC(p) \quad \mathrm{mod}(p^n + 1) \tag{8.1–24}$$

In a similar manner, if $C(p)$ represents a code word in a cyclic code, then $p^i C(p)$ mod $(p^n + 1)$ is also a code word of the cyclic code. Thus we may write

$$p^i C(p) = Q(p)(p^n + 1) + C_i(p) \tag{8.1–25}$$

where the remainder polynomial $C_i(p)$ represents a code word of the cyclic code and $Q(p)$ is the quotient.

We can generate a cyclic code by using a *generator polynomial* $g(p)$ of degree $n - k$. The generator polynomial of an (n, k) cyclic code is a factor of $p^n + 1$ and has the general form

$$g(p) = p^{n-k} + g_{n-k-1}p^{n-k-1} + \cdots + g_1 p + 1 \tag{8.1–26}$$

We also define a *message polynomial* $X(p)$ *as*

$$X(p) = x_{k-1}p^{k-1} + x_{k-2}p^{k-2} + \cdots + x_1 p + x_0 \tag{8.1–27}$$

where $[x_{k-1}x_{k-2} \cdots x_1 x_0]$ represent the k information bits. Clearly, the product $X(p)g(p)$ is a polynomial of degree less than or equal to $n - 1$, which may represent a code word. We note that there are 2^k polynomials $\{X_i(p)\}$, and, hence, there are 2^k possible code words that can be formed from a given $g(p)$.

Suppose we denote these code words as

$$C_m(p) = X_m(p)g(p), \qquad m = 1, 2, \ldots, 2^k \tag{8.1–28}$$

To show that the code words in Equation 8.1–28 satisfy the cyclic property, consider any code word $C(p)$ in Equation 8.1–28. A cyclic shift of $C(p)$ produces

$$C_1(p) = pC(p) + c_{n-1}(p^n + 1) \tag{8.1–29}$$

and, since $g(p)$ divides both $p^n + 1$ and $C(p)$, it also divides $C_1(p)$; i.e., $C_1(p)$ can be represented as

$$C_1(p) = X_1(p)g(p)$$

Therefore, a cyclic shift of any code word $C(p)$ generated by Equation 8.1–28 yields another code word.

From the above, we see that code words possessing the cyclic property can be generated by multiplying the 2^k message polynomials with a unique polynomial $g(p)$, called the generator polynomial of the (n, k) cyclic code, which divides $p^n + 1$ and has degree $n - k$. The cyclic code generated in this manner is a subspace S_c of the vector space S. The dimension of S_c is k.

EXAMPLE 8.1–3. Consider a code with block length $n = 7$. The polynomial $p^7 + 1$ has the following factors:

$$p^7 + 1 = (p + 1)(p^3 + p^2 + 1)(p^3 + p + 1) \tag{8.1–30}$$

To generate a (7, 4) cyclic code, we may take as a generator polynomial one of the following two polynomials:

$$\begin{aligned} g_1(p) &= p^3 + p^2 + 1 \\ g_2(p) &= p^3 + p + 1 \end{aligned} \tag{8.1–31}$$

The codes generated by $g_1(p)$ and $g_2(p)$ are equivalent. The code words in the (7, 4) code generated by $g_1(p) = p^3 + p^2 + 1$ are given in Table 8.1–2.

In general, the polynomial $p^n + 1$ may be factored as

$$p^n + 1 = g(p)h(p)$$

where $g(p)$ denotes the generator polynomial for the (n, k) cyclic code and $h(p)$ denotes the *parity polynomial* that has degree k. The latter may be used to generate the dual code.

For this purpose, we define the *reciprocal polynomial* of $h(p)$ as

$$\begin{aligned} p^k h(p^{-1}) &= p^k(p^{-k} + h_{k-1}p^{-k+1} + h_{k-2}p^{-k+2} + \cdots + h_1 p^{-1} + 1) \\ &= 1 + h_{k-1}p + h_{k-2}p^2 + \cdots + h_1 p^{k-1} + p^k \end{aligned} \tag{8.1–32}$$

Clearly, the reciprocal polynomial is also a factor of $p^n + 1$. Hence, $p^k h(p^{-1})$ is the generator polynomial of an $(n, n-k)$ cyclic code. This cyclic code is the dual code to the (n, k) code generated from $g(p)$. Thus, the $(n, n-k)$ dual code constitutes the null space of the (n, k) cyclic code.

TABLE 8.1–2
(7, 4) Cyclic code
Generator polynomial: $g_1(p)=p^3+p^2+1$

Information bits				Code words						
p^3	p^2	p^1	p^0	p^6	p^5	p^4	p^3	p^2	p^1	p^0
0	0	0	0	0	0	0	0	0	0	0
0	0	0	1	0	0	0	1	1	0	1
0	0	1	0	0	0	1	1	0	1	0
0	0	1	1	0	0	1	0	1	1	1
0	1	0	0	0	1	1	0	1	0	0
0	1	0	1	0	1	1	1	0	0	1
0	1	1	0	0	1	0	1	1	1	0
0	1	1	1	0	1	0	0	0	1	1
1	0	0	0	1	1	0	1	0	0	0
1	0	0	1	1	1	0	0	1	0	1
1	0	1	0	1	1	1	0	0	1	0
1	0	1	1	1	1	1	1	1	1	1
1	1	0	0	1	0	1	1	1	0	0
1	1	0	1	1	0	1	0	0	0	1
1	1	1	0	1	0	0	0	1	1	0
1	1	1	1	1	0	0	1	0	1	1

EXAMPLE 8.1–4. Let us consider the dual code to the (7, 4) cyclic code generated in Example 8.1–3. This dual code is a (7, 3) cyclic code associated with the parity polynomial

$$\begin{aligned} h_1(p) &= (p+1)(p^3+p+1) \\ &= p^4+p^3+p^2+1 \end{aligned} \tag{8.1–33}$$

The reciprocal polynomial is

$$p^4 h_1(p^{-1}) = 1+p+p^2+p^4$$

This polynomial generates the (7, 3) dual code given in Table 8.1–3. The reader can verify that the code words in the (7, 3) dual code are orthogonal to the code words in the (7, 4) cyclic code of Example 8.1–3. Note that neither the (7, 4) nor the (7, 3) codes are systematic.

It is desirable to show how a generator matrix can be obtained from the generator polynomial of a cyclic (n, k) code. A previously indicated, the generator matrix for an (n, k) code can be constructed from any set of k linearly independent code words. Hence, given the generator polynomial $g(p)$, an easily generated set of k linearly independent code words is the code words corresponding to the set of k linearly independent polynomials

$$p^{k-1}g(p), \quad p^{k-2}g(p), \quad \ldots, \quad pg(p), \quad g(p)$$

Since any polynomial of degree less than or equal to $n-1$ and divisible by $g(p)$ can be expressed as a linear combination of this set of polynomials, the set forms a basis of dimension k. Consequently, the code words associated with these polynomials form a basis of dimension k for the (n, k) cyclic code.

TABLE 8.1–3
(7, 3) Dual code
Generator polynomial: $p^4 h_1(p^{-1}) = p^4 + p^2 + p + 1$

Information bits			Code words						
p^2	p^1	p^0	p^6	p^5	p^4	p^3	p^2	p^1	p^0
0	0	0	0	0	0	0	0	0	0
0	0	1	0	0	1	0	1	1	1
0	1	0	0	1	0	1	1	1	0
0	1	1	0	1	1	1	0	0	1
1	0	0	1	0	1	1	1	0	0
1	0	1	1	0	0	1	0	1	1
1	1	0	1	1	1	0	0	1	0
1	1	1	1	1	0	0	1	0	1

EXAMPLE 8.1–5. The four rows of the generator matrix for the (7, 4) cyclic code with generator polynomial $g_1(p) = p^3 + p^2 + 1$ are obtained from the polynomials

$$p^i g_1(p) = p^{3+i} + p^{2+i} + p^i, \qquad i = 3, 2, 1, 0$$

It is easy to see that the generator matrix is

$$\mathbf{G}_1 = \begin{bmatrix} 1 & 1 & 0 & 1 & 0 & 0 & 0 \\ 0 & 1 & 1 & 0 & 1 & 0 & 0 \\ 0 & 0 & 1 & 1 & 0 & 1 & 0 \\ 0 & 0 & 0 & 1 & 1 & 0 & 1 \end{bmatrix} \qquad (8.1\text{–}34)$$

Similarly, the generator matrix for the (7, 4) cyclic code generated by the polynomial $g_2(p) = p^3 + p + 1$ is

$$\mathbf{G}_2 = \begin{bmatrix} 1 & 0 & 1 & 1 & 0 & 0 & 0 \\ 0 & 1 & 0 & 1 & 1 & 0 & 0 \\ 0 & 0 & 1 & 0 & 1 & 1 & 0 \\ 0 & 0 & 0 & 1 & 0 & 1 & 1 \end{bmatrix} \qquad (8.1\text{–}35)$$

The parity check matrices corresponding to $\mathbf{G}_1$ and $\mathbf{G}_2$ can be constructed in the same manner by using the respective reciprocal polynomials (Problem 8.8).

Note that the generator matrix obtained by this construction is not in systematic form. We can construct the generator matrix of a cyclic code in the systematic form $\mathbf{G} = [\mathbf{I}_k|\mathbf{P}]$ from the generator polynomial as follows. First, we observe that the lth row of $\mathbf{G}$ corresponds to a polynomial of the form $p^{n-l} + R_l(p)$, $l = 1, 2, \ldots, k$, where $R_l(p)$ is a polynomial of degree less than $n - k$. This form can be obtained by dividing p^{n-l} by $g(p)$. Thus, we have

$$\frac{p^{n-l}}{g(p)} = Q_l(p) + \frac{R_l(p)}{g(p)}, \qquad l = 1, 2, \ldots, k$$

or, equivalently,

$$p^{n-l} = Q_l(p)g(p) + R_l(p), \qquad l = 1, 2, \dots, k \tag{8.1–36}$$

where $Q_l(p)$ is the quotient. But $p^{n-l} + R_l(p)$ is a code word of the cyclic code since $p^{n-l} + R_l(p) = Q_l(p)g(p)$. Therefore the desired polynomial corresponding to the lth row of $\mathbf{G}$ is $p^{n-l} + R_l(p)$.

EXAMPLE 8.1–6. For the $(7, 4)$ cyclic code with generator polynomial $g_2(p) = p^3 + p + 1$, previously discussed in Example 8.1–5, we have

$$\begin{aligned} p^6 &= (p^3 + p + 1)g_2(p) + p^2 + 1 \\ p^5 &= (p^2 + 1)g_2(p) + p^2 + p + 1 \\ p^4 &= pg_2(p) + p^2 + p \\ p^3 &= g_2(p) + p + 1 \end{aligned}$$

Hence, the generator matrix of the code in systematic form is

$$\mathbf{G}_2 = \begin{bmatrix} 1 & 0 & 0 & 0 & 1 & 0 & 1 \\ 0 & 1 & 0 & 0 & 1 & 1 & 1 \\ 0 & 0 & 1 & 0 & 1 & 1 & 0 \\ 0 & 0 & 0 & 1 & 0 & 1 & 1 \end{bmatrix} \tag{8.1–37}$$

and the corresponding parity check matrix is

$$\mathbf{H}_2 = \begin{bmatrix} 1 & 1 & 1 & 0 & 1 & 0 & 0 \\ 0 & 1 & 1 & 1 & 0 & 1 & 0 \\ 1 & 1 & 0 & 1 & 0 & 0 & 1 \end{bmatrix} \tag{8.1–38}$$

It is left as an exercise for the reader to demonstrate that the generator matrix $\mathbf{G}_2$ given by Equation 8.1–35 and the systematic form given by Equation 8.1–37 generate the same set of code words (Problem 8.2).

The method for constructing the generator matrix $\mathbf{G}$ in systematic form according to Equation 8.1–36 also implies that a systematic code can be generated directly from the generator polynomial $g(p)$. Suppose that we multiply the message polynomial $X(p)$ by p^{n-k}. Thus, we obtain

$$p^{n-k}X(p) = x_{k-1}p^{n-1} + x_{k-2}p^{n-2} + \cdots + x_1 p^{n-k+1} + x_0 p^{n-k}$$

In a systematic code, this polynomial represents the first k bits in the code word $C(p)$. To this polynomial we must add a polynomial of degree less than $n - k$ representing the parity check bits. Now, if $p^{n-k}X(p)$ is divided by $g(p)$, the result is

$$\frac{p^{n-k}X(p)}{g(p)} = Q(p) + \frac{r(p)}{g(p)}$$

or, equivalently,

$$p^{n-k}X(p) = Q(p)g(p) + r(p) \tag{8.1–39}$$

where $r(p)$ has degree less than $n-k$. Clearly, $Q(p)g(p)$ is a code word of the cyclic code. Hence, by adding (modulo-2) $r(p)$ to both sides of Equation 8.1–39, we obtain the desired systematic code.

To summarize, the systematic code may be generated by

1. Multiplying the message polynomial $X(p)$ by p^{n-k}.
2. Dividing $p^{n-k}X(p)$ by $g(p)$ to obtain the remainder $r(p)$.
3. Adding $r(p)$ to $p^{n-k}X(p)$.

Below we demonstrate how these computations can be performed by using shift registers with feedback.

Since $p^n + 1 = g(p)h(p)$ or, equivalently, $g(p)h(p) = 0 \bmod (p^n + 1)$, we say that the polynomials $g(p)$ and $h(p)$ are *orthogonal*. Furthermore, the polynomials $p^i g(p)$ and $p^j h(p)$ are also orthogonal for all i and j. However, the vectors corresponding to the polynomials $g(p)$ and $h(p)$ are orthogonal only if the ordered elements of one of these vectors are reversed. The same statement applies to the vectors corresponding to $p^i g(p)$ and $p^j h(p)$. In fact, if the parity polynomial $h(p)$ is used as a generator for the $(n, n-k)$ dual code, the set of code words obtained just comprises the same code words generated by the reciprocal polynomial except that the code vectors are reversed. This implies that the generator matrix for the dual code obtained from the reciprocal polynomial $p^k h(p^{-1})$ can also be obtained indirectly from $h(p)$. Since the parity check matrix $\mathbf{H}$ for the (n, k) cyclic code is the generator matrix for the dual code, it follows that $\mathbf{H}$ can also be obtained from $h(p)$. The following example illustrates these relationships.

EXAMPLE 8.1–7. The dual code to the (7, 4) cyclic code generated by $g_1(p) = p^3 + p^2 + 1$ is the (7, 3) dual code that is generated by the reciprocal polynomial $p^4 h_1(p^{-1}) = p^4 + p^2 + p + 1$. However, we may also use $h_1(p)$ to obtain the generator matrix for the dual code. Then, the matrix corresponding to the polynomials $p^i h_1(p)$, $i = 2, 1, 0$, is

$$\mathbf{G}_{h1} = \begin{bmatrix} 1 & 1 & 1 & 0 & 1 & 0 & 0 \\ 0 & 1 & 1 & 1 & 0 & 1 & 0 \\ 0 & 0 & 1 & 1 & 1 & 0 & 1 \end{bmatrix}$$

The generator matrix for the (7, 3) dual code, which is the parity check matrix for the (7, 4) cyclic code, consists of the rows of $\mathbf{G}_{h1}$ taken in reverse order. Thus,

$$\mathbf{H}_1 = \begin{bmatrix} 0 & 0 & 1 & 0 & 1 & 1 & 1 \\ 0 & 1 & 0 & 1 & 1 & 1 & 0 \\ 1 & 0 & 1 & 1 & 1 & 0 & 0 \end{bmatrix}$$

The reader may verify that $\mathbf{G}_1\mathbf{H}_1' = 0$.

Note that the column vectors of $\mathbf{H}_1$ consist of all seven binary vectors of length 3, except the all-zero vector. But this is just the description of the parity check matrix for a (7, 4) Hamming code. Therefore, the (7, 4) cyclic code is equivalent to the (7, 4) Hamming code discussed previously in Examples 8.1–1 and 8.1–2.

Encoders for cyclic codes. The encoding operations for generating a cyclic code may be performed by a linear feedback shift register based on the use of either the generator polynomial or the parity polynomial. First, let us consider the use of $g(p)$.

As indicated above, the generation of a systematic cyclic code involves three steps, namely multiplying the message polynomial $X(p)$ by p^{n-k}, dividing the product by $g(p)$, and, finally, adding the remainder to $p^{n-k}X(p)$. Of these three steps, only the division if nontrivial.

The division of the polynomial $A(p) = p^{n-k}X(p)$ of degree $n-1$ by the polynomial

$$g(p) = g_{n-k}p^{n-k} + g_{n-k-1}p^{n-k-1} + \ldots + g_1 p + g_0$$

may be accomplished by the $(n-k)$ stage feedback shift register illustrated in Figure 8.1–2. Initially, the shift register contains all zeros. The coefficients of $A(p)$ are clocked into the shift register one (bit) coefficient at a time, beginning with the higher-order coefficients, i.e., with a_{n-1}, followed by a_{n-2}, and so on. After the kth shift, the first nonzero output of the quotient is $q_{k-1} = g_{n-k}a_{n-1}$. Subsequent outputs are generated as illustrated in Figure 8.1–2. For each output coefficient in the quotient, we must subtract the polynomial $g(p)$ multiplied by that coefficient, as in ordinary long division. The subtraction is performed by means of the feedback part of the shift register. Thus, the feedback shift register in Figure 8.1–2 performs division of two polynomials.

In our case, $g_{n-k} = g_0 = 1$, and, for binary codes, the arithmetic operations are performed in modulo-2 arithmetic. Consequently, the subtraction operations reduce to modulo-2 addition. Furthermore, we are only interested in generating the parity check bits for each code word, since the code is systematic. Consequently, the encoder for the cyclic code takes the form illustrated in Figure 8.1–3. The first k bits at the output of the encoder are simply the k information bits. These k bits are also clocked simultaneously into the shift register, since the switch 1 is in the closed position. Note that the polynomial multiplication of p^{n-k} with $X(p)$ is not performed explicitly. After the k information bits are all clocked into the encoder, the positions of the two switches are reversed. At this time, the contents of the shift register are simply the $n-k$ parity check bits, which correspond to the coefficients of the remainder polynomial. These $n-k$ bits are clocked out one at a time and sent to the modulator.

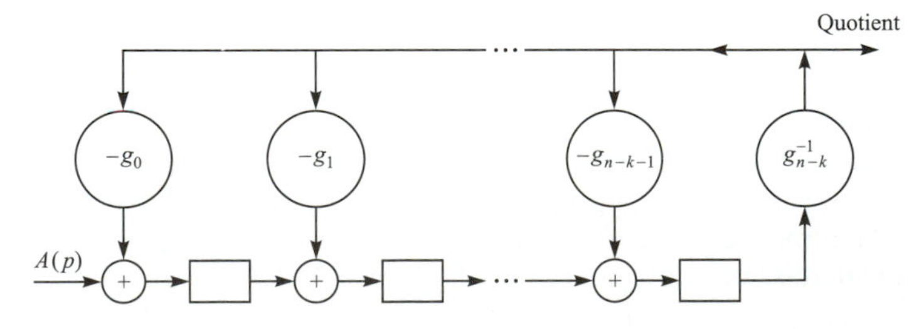

FIGURE 8.1–2
A feedback shift register for dividing the polynomial $A(p)$ by $g(p)$.

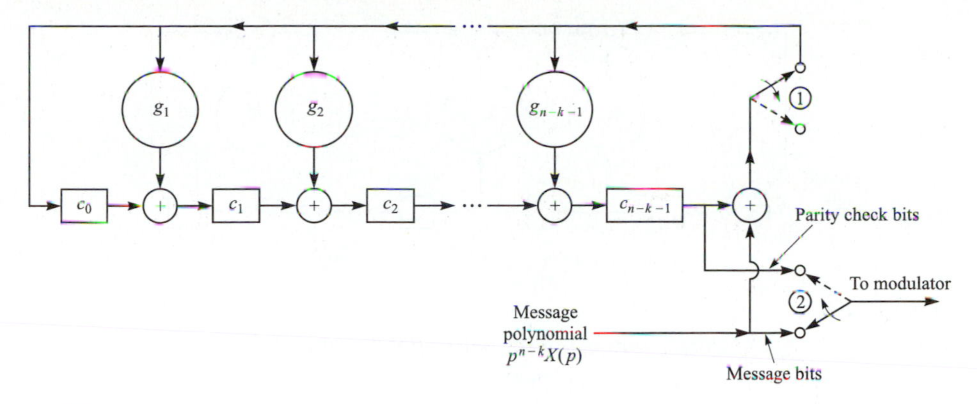

FIGURE 8.1–3
Encoding a cyclic code by use of the generator polynomial $g(p)$.

EXAMPLE 8.1–8. The shift register for encoding the (7, 4) cyclic code with generator polynomial $g(p) = p^3 + p + 1$ is illustrated in Figure 8.1–4. Suppose the input message bits are 0110. The contents of the shift register are as follows:

Input	Shift	Shift register contents		
	0	0	0	0
0	1	0	0	0
1	2	1	1	0
1	3	1	0	1
0	4	1	0	0

Hence, the three parity check bits are 100, which correspond to the code bits $c_5 = 0$, $c_6 = 0$, and $c_7 = 1$.

Instead of using the generator polynomial, we may implement the encoder for the cyclic code by making use of the parity polynomial

$$h(p) = p^k + h_{k-1}p^{k-1} + \cdots + h_1 p + 1$$

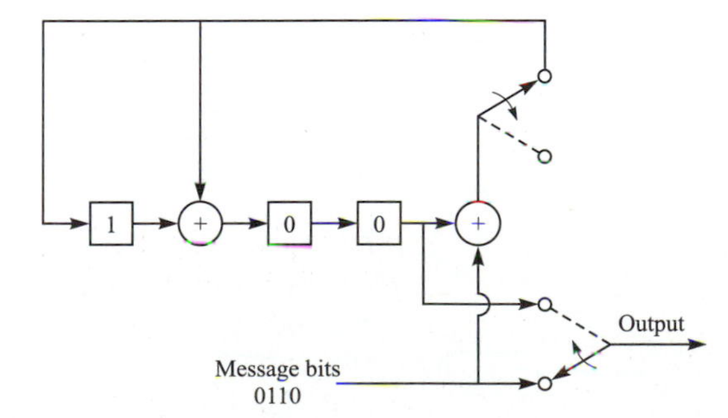

FIGURE 8.1–4
The encoder for the (7, 4) cyclic code with generator polynomial $g(p) = p^3 + p + 1$.

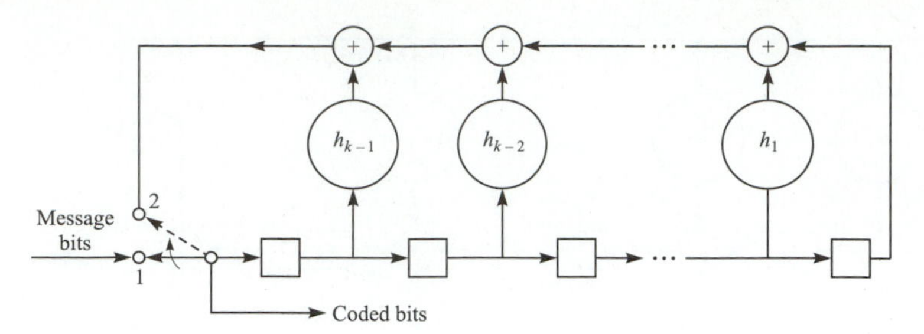

FIGURE 8.1–5
Encoder for an (n, k) cyclic code based on parity polynomial $h(p)$.

The encoder is shown in Figure 8.1–5. Initially, the k information bits are shifted into the shift register and simultaneously fed to the modulator. After all k information bits are in the shift register, the switch is thrown into position 2 and the shift register is clocked $n - k$ times to generate the $n - k$ parity check bits as illustrated in Figure 8.1–5.

EXAMPLE 8.1–9. The parity polynomial for the (7, 4) cyclic code generated by $g(p) = p^3 + p + 1$ is $h(p) = p^4 + p^2 + p + 1$. The encoder for this code based on the parity polynomial is illustrated in Figure 8.1–6. If the input to the encoder is the message bits 0110, the parity check bits are $c_5 = 0$, $c_6 = 0$, and $c_7 = 1$, as is easily verified.

It should be noted that the encoder based on the generator polynomial is simpler when $n - k < k\,(k > \frac{1}{2}n)$, i.e., for high-rate codes ($R_c > \frac{1}{2}$), while the encoder based on the parity polynomial is simpler when $k < n - k\,(k < \frac{1}{2}n)$, which corresponds to low-rate codes ($R_c < \frac{1}{2}$).

Cyclic Hamming codes. The class of cyclic codes include the Hamming codes, which have a block length $n = 2^m - 1$ and $n - k = m$ parity check bits, where m is any positive integer. The cyclic Hamming codes are equivalent to the Hamming codes described in Section 8.1.2.

Cyclic (23, 12) Golay code. The linear (23, 12) Golay code described in Section 8.1.2 can be generated as a cyclic code by means of the generator polynomial

$$g(p) = p^{11} + p^9 + p^7 + p^6 + p^5 + p + 1 \tag{8.1–40}$$

The code words have a minimum distance $d_{\min} = 7$.

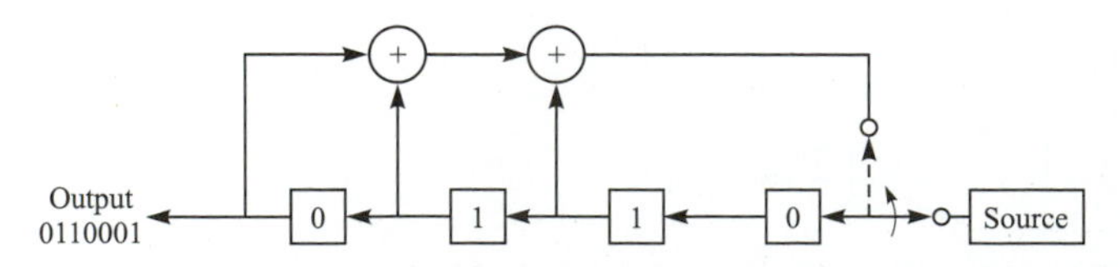

FIGURE 8.1–6
The encoder for the (7, 4) cyclic code based on the parity polynomial $h(p) = p^4 + p^2 + p + 1$.

Maximum-length shift-register codes. Maximum-length shift-register codes are a class of cyclic codes with

$$(n, k) = (2^m - 1, m) \tag{8.1-41}$$

where m is a positive integer. The code words are usually generated by means of an m-stage digital shift register with feedback, based on the parity polynomial. For each code word to be transmitted, the m information bits are loaded into the shift register, and the switch is thrown from position 1 to position 2. The contents of the shift register are shifted to the left one bit at a time for a total of $2^m - 1$ shifts. This operation generates a systematic code with the desired output length $n = 2^m - 1$. For example, the code words generated by the $m = 3$ stage shift register in Figure 8.1–7 are listed in Table 8.1–4.

Note that, with the exception of the all-zero code word, all the code words generated by the shift register are different cyclic shifts of a single code word. The reason for this structure is easily seen from the state diagram of the shift register, which is illustrated in Figure 8.1–8 for $m = 3$. When the shift register is loaded initially and shifted $2^m - 1$ times, it will cycle through all possible $2^m - 1$ states. Hence, the shift register is back to its original state in $2^m - 1$ shifts. Consequently, the output sequence is periodic with length $n = 2^m - 1$. Since there are $2^m - 1$ possible states, this length corresponds to the largest possible period. This explains why the $2^m - 1$ code words are different cyclic shifts of a single code word.

Maximum-length shift-register codes exist for any positive value of m. Table 8.1–5 lists the stages connected to the modulo-2 adder that result in a maximum-length shift register for $2 \leqslant m \leqslant 34$.

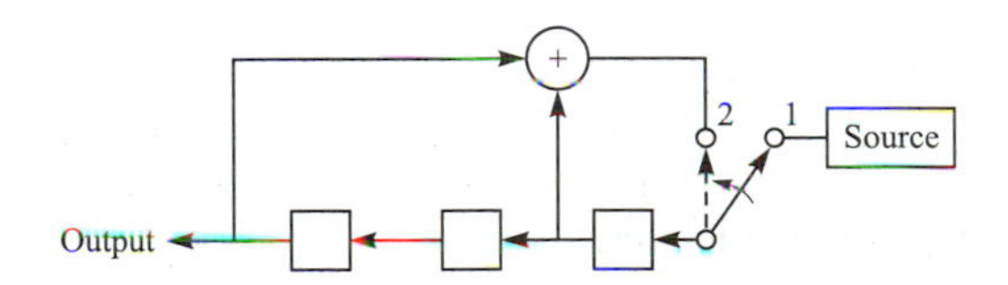

FIGURE 8.1–7
Three-stage ($m = 3$) shift register with feedback.

TABLE 8.1–4
Maximum-length shift-register code for $m = 3$

Information bits			Code words						
0	0	0	0	0	0	0	0	0	0
0	0	1	0	0	1	1	1	0	1
0	1	0	0	1	0	0	1	1	1
0	1	1	0	1	1	1	0	1	0
1	0	0	1	0	0	1	1	1	0
1	0	1	1	0	1	0	0	1	1
1	1	0	1	1	0	1	0	0	1
1	1	1	1	1	1	0	1	0	0

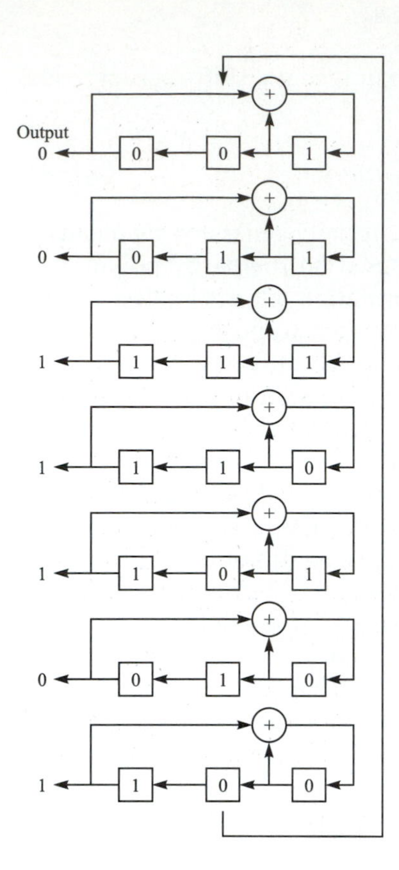

FIGURE 8.1–8
The seven states for the $m = 3$ maximum-length shift register.

TABLE 8.1–5
Shift-register connections for generating maximum-length sequences

m	Stages connected to modulo-2 adder	m	Stages connected to modulo-2 adder	m	Stages connected to modulo-2 adder
2	1, 2	13	1, 10, 11, 13	24	1, 18, 23, 24
3	1, 3	14	1, 5, 9, 14	25	1, 23
4	1, 4	15	1, 15	26	1, 21, 25, 26
5	1, 4	16	1, 5, 14, 16	27	1, 23, 26, 27
6	1, 6	17	1, 15	28	1, 26
7	1, 7	18	1, 12	29	1, 28
8	1, 5, 6, 7	19	1, 15, 18, 19	30	1, 8, 29, 30
9	1, 6	20	1, 18	31	1, 29
10	1, 8	21	1, 20	32	1, 11, 31, 32
11	1, 10	22	1, 22	33	1, 21
12	1, 7, 9, 12	23	1, 19	34	1, 8, 33, 34

Source: Forney (1970a).

Another characteristic of the code words in a maximum-length shift-register code is that each code word, with the exception of the all-zero code word, contains 2^{m-1} ones and $2^{m-1}-1$ zeros. Hence all these code words have identical weights, namely, $w = 2^{m-1}$. Since the code is linear, this weight is also the minimum distance of the code, i.e.,

$$d_{\min} = 2^{m-1}$$

Finally, note that the (7, 3) maximum-length shift-register code shown in Table 8.1–4 is identical to the (7, 3) code given in Table 8.1–3, which is the dual of the (7, 4) Hamming code given in Table 8.1–2. This is not a coincidence. The maximum-length shift-register codes are the dual codes of the cyclic Hamming $(2^m-1, 2^m-1-m)$ codes.

The shift register for generating the maximum-length code may also be used to generate a periodic binary sequence with period $n = 2^m - 1$. The binary periodic sequence exhibits a periodic autocorrelation $\phi(m)$ with values $\phi(m) = n$ for $m = 0, \pm n, \pm 2n, \ldots$, and $\phi(m) = -1$ for all other shifts as described in Section 13.2.4. This impulse-like autocorrelation implies that the power spectrum is nearly white and, hence, the sequence resembles white noise. As a consequence, maximum-length sequences are called pseudonoise (PN) sequences and find use in the scrambling of data and in the generation of spread spectrum signals.

Bose–Chaudhuri–Hocquenghem (BCH) codes. BCH codes comprise a large class of cyclic codes that include both binary and nonbinary alphabets. Binary BCH codes may be constructed with parameters

$$\begin{aligned} n &= 2^m - 1 \\ n - k &\leqslant mt \\ d_{\min} &= 2t + 1 \end{aligned} \tag{8.1–42}$$

where $m\,(m \geqslant 3)$ and t are arbitrary positive integers. Hence, this class of binary codes provides the communication system designer with a large selection of block lengths and code rates. Nonbinary BCH codes include the powerful Reed–Solomon codes that are described later.

The generator polynomials for BCH codes can be constructed from factors of $p^{2m-1} + 1$. Table 8.1–6 lists the coefficients of generator polynomials for BCH codes of block lengths $7 \leqslant n \leqslant 255$, corresponding to $3 \leqslant m \leqslant 8$. The coefficients are given in octal form, with the left-most digit corresponding to the highest-degree term of the generator polynomial. Thus, the coefficients of the generator polynomial for the (15.5) code are 2467, which in binary form is 10 100 110 111. Consequently, the generator polynomial is $g(p) = p^{10} + p^8 + p^5 + p^4 + p^2 + p + 1$.

A more extensive list of generator polynomials for BCH codes is given by Peterson and Weldon (1972), who tabulate the polynomial factors of $p^{2m-1} + 1$ for $m \leqslant 34$.

TABLE 8.1–6

Coefficients of generator polynomials (in octal form) for BCH codes of lengths $7 \leqslant n \leqslant 255$

n	k	t	$g(p)$
7	4	1	13
15	11	1	23
	7	2	721
	5	3	2467
31	26	1	45
	21	2	3551
	16	3	107657
	11	5	5423325
	6	7	313365047
63	57	1	103
	51	2	12471
	45	3	1701317
	39	4	166623567
	36	5	1033500423
	30	6	157464165547
	24	7	17323260404441
	18	10	1363026512351725
	16	11	6331141367235453
	10	13	472622305527250155
	7	15	5231045543503271737
127	120	1	211
	113	2	41567
	106	3	11554743
	99	4	3447023271
	92	5	624730022327
	85	6	130704476322273
	78	7	26230002166130115
	71	9	6255010713253127753
	64	10	1206534025570773100045
	57	11	335265252505705053517721
	50	13	54446512523314012421501421
	43	14	17721772213651227521220574343
	36	15	3146074666522075044764574721735
	29	21	403114461367670603667530141176155
	22	23	123376070404722522435445626637647043
	15	27	22057042445604554770523013762217604353
	8	31	7047264052751030651476224271567733130217
255	247	1	435
	239	2	267543
	231	3	156720665
	223	4	75626641375
	215	5	23157564726421
	207	6	16176560567636227

n	*k*	*t*	*g*(*p*)
	199	7	7633031270420722341
	191	8	2663470176115333714567
	187	9	52755313540001322236351
	179	10	22624710717340432416300455
	171	11	1541621421234235607706163067
	163	12	7500415510075602551574724514601
	155	13	3757513005407665015722506464677633
	147	14	1642130173537165525304165305441011711
	139	15	461401732060175561570722730247453567445
	131	18	215713331471510151261250277442142024165471
	123	19	120614052242066003717210326516141226272506267
	115	21	60526665572100247263636404600276352556313 472737
	107	22	22205772322066256312417300235347420176574 750154441
	99	23	10656667253473174222741416201574332252411 076432303431
	91	25	67502650303274441727236317247325110755507 62720724344561
	87	26	11013676341474323643523163430717204620672 2545273311721317
	79	27	66700035637657500020270344207366174621015 32671176654134 2355
	71	29	24024710520644321515554172112331163205444 250362557643221706035
	63	30	10754475055163544325315217357707003666111 7264552676136567025433 01
	55	31	73154252035011001330152753060320543254143 267550105570444260354736 17
	47	42	25335420170626465630330413774062331751233 341454460450050660245525431 73
	45	43	15202056055234161131101346376423701563670 02447076237303320215702505154 1
	37	45	51363302550670074141774474472454375304 20 7357061743234323476443547374030440 03
	29	47	30257155366730714655270640123613771153422 42324201174114060254757410403565037
	21	55	12562152570603326560017731536076121032273 41405653074542521153121614466513473725
	13	59	46417320050525645444265737142500660043306 77445476561403174677213570261344605005 47
	9	63	15726025217472463201031043255355134614162 36721204407454511276611554770556167751 6057

Source: Stenbit (1964), © 1964 IEEE.

8.1.4 Optimum Soft-Decision Decoding of Linear Block Codes

In this subsection, we derive the performance of linear binary block codes on an AWGN channel when optimum (unquantized) soft-decision decoding is employed at the receiver. The bits of a code word may be transmitted by any one of the binary signaling methods described in Chapter 5. For our purposes, we consider binary (or quaternary) coherent PSK, which is the most efficient method, and binary orthogonal FSK either with coherent detection or noncoherent detection.

Let $\mathcal{E}$ denote the transmitted signal energy per code word and let $\mathcal{E}_c$ denote the signal energy required to transmit a single element (bit) in the code word. Since there are n bits per code word, $\mathcal{E} = n\mathcal{E}_c$, and since each code word conveys k bits of information, the energy per information bit is

$$\mathcal{E}_b = \frac{\mathcal{E}}{k} = \frac{n}{k}\mathcal{E}_c = \frac{\mathcal{E}_c}{R_c} \tag{8.1–43}$$

The code words are assumed to be equally likely a priori with prior probability $1/M$.

Suppose the bits of a code word are transmitted by binary PSK. Thus each code word results in one of M signaling waveforms. From Chapter 5, we know that the optimum receiver, in the sense of minimizing the average probability of a code word error, for the AWGN channel, can be realized as a parallel bank of M filters matched to the M possible transmitted waveforms. The outputs of the M matched filters at the end of each signaling interval, which encompasses the transmission of n bits in the code word, are compared and the code word corresponding to the largest matched filter output is selected. Alternatively, M cross correlators can be employed. In either case, the receiver implementation can be simplified. That is, an equivalent optimum receiver can be realized by use of a single filter (or cross correlator) matched to the binary PSK waveform used to transmit each bit in the code word, followed by a decoder that forms the M decision variables corresponding to the M code words.

To be specific, let $r_j, j = 1, 2, \ldots, n$, represent the n sampled outputs of the matched filter for any particular code word. Since the signaling is binary coherent PSK, the output r_j may be expressed either as

$$r_j = \sqrt{\mathcal{E}_c} + n_j \tag{8.1–44}$$

when the jth bit of a code word is a 1, or as

$$r_j = -\sqrt{\mathcal{E}_c} + n_j \tag{8.1–45}$$

when the jth bit is a 0. The variables $\{n_j\}$ represent additive white Gaussian noise at the sampling instants. Each n_j has zero-mean and variance $\frac{1}{2}N_0$. From knowledge of the M possible transmitted code words and upon reception of $\{r_j\}$, the optimum decoder forms the M correlation metrics

$$CM_i = C(\mathbf{r}, \mathbf{C}_i) = \sum_{j=1}^{n} (2c_{ij} - 1) r_j, \qquad i = 1, 2, \ldots, M \tag{8.1–46}$$

where c_{ij} denotes the bit in the jth position of the ith code word. Thus, if $c_{ij} = 1$, the weighting factor $2c_{ij} - 1 = 1$, and if $c_{ij} = 0$, the weighting factor $2c_{ij} - 1 = -1$. In this manner, the weighting $2c_{ij} - 1$ aligns the signal components in $\{r_j\}$ such that the correlation metric corresponding to the actual transmitted code word will have a mean value $\sqrt{\mathcal{E}_c} n$, while the other $M - 1$ metrics will have smaller mean values.

Although the computations involved in forming the correlation metrics for soft-decision decoding according to Equation 8.1–46 are relatively simple, it may still be impractical to compute Equation 8.1–46 for all the possible code words when the number of code words is large, e.g., $M > 2^{10}$. In such a case it is still possible to implement soft-decision decoding using algorithms which employ techniques for discarding improbable code words without computing their entire correlation metrics as given by Equation 8.1–46. Several different types of soft-decision decoding algorithms have been described in the technical literature. The interested reader is referred to the papers by Forney (1966b), Weldon (1971), Chase (1972), Wainberg and Wolf (1973), Wolf (1978), and Matis and Modestino (1982).

In determining the probability of error for a linear block code, note that when such a code is employed on a binary-input, symmetric channel such as the AWGN channel with optimum soft-decision decoding, the error probability for the transmission of the mth code is the same for all m. Hence, we assume for simplicity that the all-zero code word $\mathbf{C}_1$ is transmitted. For correct decoding of $\mathbf{C}_1$, the correlation metric CM_1 must exceed all the other $M - 1$ correlation metrics CM_m, $m = 2, \ldots, M$. All the CM are Gaussian distributed. The mean value of CM_1 is $\sqrt{\mathcal{E}_c} n$, while the mean values of $CM_m, m = 2, \ldots, M$, is $\sqrt{\mathcal{E}_c} n(1 - 2w_m/n)$. The variance of each decision variable is $\frac{1}{2}N_0$. The derivation of the exact expression for the probability of correct decoding or, equivalently, the probability of a code word error is complicated by the correlations among the M correlation metrics. The cross-correlation coefficients between $\mathbf{C}_1$ and the other $M - 1$ code words are

$$\rho_m = 1 - \frac{2w_m}{n}, \qquad m = 2, \ldots, M \tag{8.1–47}$$

where w_m denotes the weight of the mth code word.

Instead of attempting to derive the exact error probability, we resort to a union bound. The probability that $CM_m > CM_1$ is

$$P_2(m) = Q\left(\sqrt{\frac{\mathcal{E}}{N_0}(1 - \rho_m)}\right) \tag{8.1–48}$$

where $\mathcal{E} = k\mathcal{E}_b$ is the transmitted energy per waveform. Substitution for ρ_m from Equation 8.1–47 and for $\mathcal{E}$ yields

$$P_2(m) = Q\left(\sqrt{\frac{2\mathcal{E}_b}{N_0} R_c w_m}\right) = Q(\sqrt{2\gamma_b R_c w_m}) \tag{8.1–49}$$

where γ_b is the SNR per bit and R_c is the code rate. Then the average probability of a code word error is bounded from above by the sum of the binary error events given by Equation 8.1–49. Thus,

$$P_M \leqslant \sum_{m=2}^{M} P_2(m) \leqslant \sum_{m=2}^{M} Q(\sqrt{2\gamma_b R_c w_m}) \tag{8.1–50}$$

The computation of the probability of error for soft-decision decoding according to Equation 8.1–50 requires knowledge of the weight distribution of the code. Weight distributions of many codes are given in a number of texts on coding theory, e.g., Berlekamp (1968) and MacWilliams and Sloane (1977).

A somewhat looser bound is obtained by noting that

$$Q(\sqrt{2\gamma_b R_c w_m}) \leqslant Q\sqrt{2\gamma_b R_c d_{\min}}) < \exp(-\gamma_b R_c d_{\min}) \tag{8.1–51}$$

Consequently,

$$P_M \leqslant (M-1)Q(\sqrt{2\gamma_b R_c d_{\min}}) < \exp(-\gamma_b R_c d_{\min} + k \ln 2) \tag{8.1–52}$$

This bound is particularly useful since it does not require knowledge of the weight distribution of the code. When the upper bound in Equation 8.1–52 is compared with the performance of an uncoded binary PSK system, which is upper-bounded as $\frac{1}{2}\exp(-\gamma_b)$, we find that coding yields a gain of approximately $10\log(R_c d_{\min} - k \ln 2/\gamma_b)$ dB. We may call this the *coding gain*. We note that its value depends on the code parameters and also on the SNR per bit γ_b.

The expression for the probability of error for equicorrelated waveforms that can be obtained for the simplex signals described in Section 5.2 gives us yet a third approximation to the error probabilities for coded waveforms. We know that the maximum cross-correlation coefficient between a pair of coded waveforms is

$$\rho_{\max} = 1 - \frac{2}{n} d_{\min} \tag{8.1–53}$$

If we assume as a worst case that all the M code words have a cross-correlation coefficient equal to $\rho_{\max}$, then the code word error probability can easily be approximated. Since some code words are separated by more than the minimum distance, the error probability evaluated for $\rho_r = \rho_{\max}$ is actually an upper bound. Thus,

$$P_M \leqslant 1 - \frac{1}{\sqrt{2\pi}} \int_{-\infty}^{\infty} e^{-v^2/2} \left(\frac{1}{\sqrt{2\pi}} \int_{-\infty}^{v+\sqrt{4\gamma_b R_c d_{\min}}} e^{-x^2/2} \, dx \right)^{M-1} dv \qquad (8.1\text{–}54)$$

The bounds on the performance of linear block codes given above are in terms of the block error or code word error probability. The evaluation of the equivalent bit error probability P_b is much more complicated. In general, when a block error is made, some of the k information bits in the block will be correct and some will be in error. For orthogonal waveforms, the conversion factor that multiplies P_M to yield P_b is $2^{k-1}/(2^k - 1)$. This factor is unity for $k = 1$ and approaches $\frac{1}{2}$ as k increases, which is equivalent to assuming that, on the average, half of the k bits will be in error when a block error occurs. The conversion factor for coded waveforms depends in a complicated way on the distance properties of the code, but is certainly no worse than assuming that, on the average, half of the k bits will be in error when a block error occurs. Consequently, $P_b \leqslant \frac{1}{2} P_M$.

The bounds on performance given by Equations 8.1–50, 8.1–52, and 8.1–54 also apply to the case in which a pair of bits of a code word are transmitted by quaternary PSK, since quaternary PSK may be viewed as being equivalent to two independent binary PSK waveforms transmitted in phase quadrature. Furthermore, the bounds in Equations 8.1–52 and 8.1–54, which depend only on the minimum distance of the code, apply also to non-linear binary block codes.

If binary orthogonal FSK is used to transmit each bit of a code word on the AWGN channel, the optimum receiver can be realized by means of two matched filters, one matched to the frequency corresponding to a transmission of a 0, and the other to the frequency corresponding to a transmission of a 1, followed by a decoder that forms the M correlation metrics corresponding to the M possible code words. The detection at the receiver may be coherent or noncoherent. In either case, let r_{0j} and r_{1j} denote the input samples to the combiner. The correlation metrics formed by the decoder may be expressed as

$$CM_i = \sum_{j=1}^{n} [c_{ij} r_{1j} + (1 - c_{ij}) r_{0j}], \qquad i = 1, 2, \ldots, M \qquad (8.1\text{–}55)$$

where c_{ij} represents the jth bit in the ith code word. The code word corresponding to the largest of the $\{CM_i\}$ is selected as the transmitted code word.

If the detection of the binary FSK waveforms is coherent, the random variables $\{r_{0j}\}$ and $\{r_{1j}\}$ are Gaussian and, hence, the correlation metrics $\{CM_i\}$ are also Gaussian. In this case, bounds on the performance of the code are easily obtained. To be specific, suppose that the all-zero code word $\mathbf{C}_1$ is transmitted. Then,

$$\left.\begin{aligned} r_{0j} &= \sqrt{\mathscr{E}_c} + n_{0j} \\ r_{1j} &= n_{1j} \end{aligned}\right\} \qquad j = 1, 2, \ldots, n \qquad (8.1\text{–}56)$$

where the $\{n_{ij}\}$, $i = 0, 1, j = 1, 2, \ldots, n$, are mutually statistically independent Gaussian random variables with zero-mean and variance $\frac{1}{2} N_0$. Consequently

CM_1 is Gaussian with mean $\sqrt{\mathcal{E}_c}n$ and variance $\frac{1}{2}nN_0$. On the other hand, the correlation metric CM_m, corresponding to the code word having weight w_m, is Gaussian with mean $\sqrt{\mathcal{E}_c}n(1 - w_m/n)$ and variance $\frac{1}{2}nN_0$. Since the $\{CM_m\}$ are correlated, we again resort to a union bound. The correlation coefficients are given by

$$\rho_m = 1 - \frac{w_m}{n} \tag{8.1–57}$$

Hence, the probability that $CM_m > CM_1$ is

$$P_2(m) = Q(\sqrt{\gamma_b R_c w_m}) \tag{8.1–58}$$

Comparison of this result with that given in Equation 8.1–49 for coherent PSK reveals that coherent PSK requires 3 dB less SNR to achieve the same performance. This is not surprising in a view of the fact that uncoded PSK is 3 dB better than binary orthogonal FSK with coherent detection. Hence, the advantage of PSK over FSK is maintained in the coded waveforms. We conclude, then, that the bounds given in Equations 8.1–50, 8.1–52, and 8.1–54 apply to coded waveforms transmitted by binary orthogonal coherent FSK with γ_b replaced by $\frac{1}{2}\gamma_b$.

If square-law detection of the binary orthogonal FSK signal is employed at the receiver, the performance is further degraded by the noncoherent combining loss, as shown in Chapter 12. Suppose again that the all-zero code word is transmitted. Then the correlation metrics are given by Equation 8.1–55, where the input variables to the decoder are now

$$\left.\begin{aligned} r_{0j} &= |\sqrt{\mathcal{E}_c} + N_{0j}|^2 \\ r_{1j} &= |N_{1j}|^2 \end{aligned}\right\} \quad j = 1, 2, \ldots, n \tag{8.1–59}$$

where $\{N_{0j}\}$ and $\{N_{1j}\}$ represent complex-valued mutually statistically independent Gaussian random variables with zero-mean and variance N_0. The correlation metric CM_1 is given as

$$CM_1 = \sum_{j=1}^{n} r_{0j} \tag{8.1–60}$$

while the correlation metric corresponding to the code word having weight w_m is statistically equivalent to the correlation metric of a code word in which $c_{mj} = 1$ for $1 \leqslant j \leqslant w_m$ and $c_{mj} = 0$ for $w_m + 1 \leqslant j \leqslant n$. Hence, CM_m may be expressed as

$$CM_m = \sum_{j=1}^{w_m} r_{1j} + \sum_{j=w_m+1}^{n} r_{0j} \tag{8.1–61}$$

The difference between CM_1 and CM_m is

$$CM_1 - CM_m = \sum_{j=1}^{w_m} (r_{0j} - r_{1j}) \tag{8.1–62}$$

and the probability of error is simply the probability that $CM_1 - CM_m < 0$. But this difference is a special case of the general quadratic form in complex-valued Gaussian random variables considered in Chapter 12 and Appendix B. The expression for the probability of error in deciding between CM_1 and CM_m is (see Section 12.1.1)

$$P_2(m) = \frac{1}{2^{2w_m-1}} \exp(-\tfrac{1}{2}\gamma_b R_c w_m) \sum_{i=0}^{w_m-1} K_i (\tfrac{1}{2}\gamma_b R_c w_m)^i \tag{8.1–63}$$

where, by definition,

$$K_i = \frac{1}{i!} \sum_{r=0}^{w_m-1-i} \binom{2w_m - 1}{r} \tag{8.1–64}$$

The union bound obtained by summing $P_2(m)$ over $2 \leqslant m \leqslant M$ provides us with an upper bound on the probability of a code word error.

As an alternative, we may use the minimum distance instead of the weight distribution to obtain the looser upper bound

$$P_M \leqslant \frac{M-1}{2^{2d_{\min}-1}} \exp(-\tfrac{1}{2}\gamma_b R_c d_{\min}) \sum_{i=0}^{d_{\min}-1} K_i (\tfrac{1}{2}\gamma_b R_c d_{\min})^i \tag{8.1–65}$$

A measure of the noncoherent combining loss inherent in the square-law detection and combining of the n elementary binary FSK waveforms in a code word can be obtained from Figure 12.1–1, where $d_{\min}$ is used in place of L. The loss obtained is relative to the case in which the n elementary binary FSK waveforms are first detected coherently and combined as in Equation 8.1–55 and then the sums are square-law-detected or envelope-detected to yield the M decision variables. The binary error probability for the latter case is

$$P_2(m) = \tfrac{1}{2} \exp(-\tfrac{1}{2}\gamma_b R_c w_m) \tag{8.1–66}$$

and, hence,

$$P_M \leqslant \sum_{m=2}^{M} P_2(m)$$

If $d_{\min}$ is used instead of the weight distribution, the union bound for the code word error probability in the latter case is

$$P_M \leqslant \tfrac{1}{2}(M-1) \exp(-\tfrac{1}{2}\gamma_b R_c d_{\min}) \tag{8.1–67}$$

The channel bandwidth required to transmit the coded waveforms can be determined as follows. If binary PSK is used to transmit each bit in a code word, the required bandwidth is approximately equal to the reciprocal of the time interval devoted to the transmission of each bit. For an information rate of R bits/s, the time available to transmit k information bits and $n-k$ redundant (parity) bits (n total bits) is $T = k/R$. Hence,

$$W = \frac{1}{T/n} = \frac{n}{k/R} = \frac{R}{R_c} \tag{8.1–68}$$

Therefore, the bandwidth expansion factor B_e for the coded waveform is

$$B_e = \frac{W}{R} = \frac{n}{k} = \frac{1}{R_c} \tag{8.1–69}$$

On the other hand, if binary FSK with noncoherent detection is employed for transmitting the bits in a code word, $W \approx 2n/T$, and, hence, the bandwidth expansion factor increases by approximately a factor of 2 relative to binary PSK. In any case, B_e increases inversely with the code rate, or, equivalently, it increases linearly with the block size n.

We are now in a position to compare the performance characteristics and bandwidth requirements of coded signaling waveforms with orthogonal signaling waveforms. A comparison of the expression for P_M given in Equation 5.2–21 for orthogonal waveforms and in Equation 8.1–54 for coded waveforms with coherent PSK indicates that the coded waveforms result in a loss of at most $10 \log(n/2d_{\min})$ dB relative to orthogonal waveforms having the same number of waveforms. On the other hand, if we compensate for the loss in SNR due to coding by increasing the number of code words so that coded transmission requires $M_c = 2^{k_c}$ waveforms and orthogonal signaling requires $M_o = 2^{k_o}$ waveforms, then (from the union bounds in Equations 5.2–27 and 8.1–52), the performance obtained with the two sets of signaling waveforms at high SNR is about equal if

$$k_o = 2R_c d_{\min} \tag{8.1–70}$$

Under this condition, the bandwidth expansion factor for orthogonal signaling can be expressed as

$$B_{eo} = \frac{M_o}{2 \log_2 M_o} = \frac{2^{k_o}}{2k_o} = \frac{2^{2R_c d_{\min}}}{4R_c d_{\min}} \tag{8.1–71}$$

while, for coded signaling waveforms, we have $B_{ec} = 1/R_c$. The ratio of B_{eo} given in Equation 8.1–71 to B_{ec}, which is

$$\frac{B_{eo}}{B_{ec}} = \frac{2^{2R_c d_{\min}}}{4d_{\min}} \tag{8.1–72}$$

provides a measure of the relative bandwidth between orthogonal signaling and signaling with coded coherent PSK waveforms.

For example, suppose we use a (63, 33) binary cyclic code that has a minimum distance $d_{\min} = 12$. The bandwidth ratio for orthogonal signaling relative to this code, given by Equation 8.1–72, is 127. This is indicative of the bandwidth efficiency obtained through coding relative to orthogonal signaling.

8.1.5 Hard-Decision Decoding of Linear Block Codes

The bounds given in Section 8.1.4 on the performance of coded signaling waveforms on the AWGN channel are based on the premise that the samples from the matched filter or cross correlator are not quantized. Although this processing yields the best performance, the basic limitation is the computational burden of forming M correlation metrics and comparing these to obtain the largest. The amount of computation becomes excessive when the number M of code words is large.

To reduce the computational burden, the analog samples can be quantized and the decoding operations are then performed digitally. In this subsection, we consider the extreme situation in which each sample corresponding to a single bit of a code word is quantized to two levels: zero and one. That is, a (hard) decision is made as to whether each transmitted bit in a code word is a 0 or a 1. The resulting discrete-time channel (consisting of the modulator, the AWGN channel, and the demodulator/detector) constitutes a BSC with crossover probability p. If coherent PSK is employed in transmitting and receiving the bits in each code word, then

$$p = Q\left(\sqrt{\frac{2\mathcal{E}_c}{N_0}}\right) = Q(\sqrt{2\gamma_b R_c}) \tag{8.1–73}$$

On the other hand, if FSK is used to transmit the bits in each code word, then

$$p = Q(\sqrt{\gamma_b R_c}) \tag{8.1–74}$$

for coherent detection and

$$p = \tfrac{1}{2}\exp(-\tfrac{1}{2}\gamma_b R_c) \tag{8.1–75}$$

for noncoherent detection.

Minimum-distance (maximum-likelihood) decoding. The n bits from the detector corresponding to a received code word are passed to the decoder, which compares the received code word with the M possible transmitted code words and decides in favor of the code word that is closest in Hamming distance (number of bit positions in which two code words differ) to the received code word. This minimum distance decoding rule is optimum in the sense that it results in a minimum probability of a code word error for the binary symmetric channel.

A conceptually simple, albeit computationally inefficient, method for decoding is to first add (modulo 2) the received code word vector to all the M possible transmitted code words $\mathbf{C}_i$ to obtain the error vectors $\mathbf{e}_i$. Hence, $\mathbf{e}_i$ represents the error event that must have occurred on the channel in order to transform the code word $\mathbf{C}_i$ into the particular received code word. The number of errors in transforming $\mathbf{C}_i$ into the received code word is just equal to the number of 1s in $\mathbf{e}_i$. Thus, if we simply compute the weight of each of the M error vectors $\{\mathbf{e}_i\}$ and decide in favor of the code word that results in the smallest weight error vector, we have, in effect, a realization of the minimum distance decoding rule.

A more efficient method for hard-decision decoding makes use of the parity check matrix $\mathbf{H}$. To elaborate, suppose that $\mathbf{C}_m$ is the transmitted code word and $\mathbf{Y}$ is the received code word at the output of the detector. In general, $\mathbf{Y}$ may be expressed as

$$\mathbf{Y} = \mathbf{C}_m + \mathbf{e}$$

where $\mathbf{e}$ denotes an arbitrary binary error vector. The product $\mathbf{YH}'$ yields

$$\begin{aligned}\mathbf{YH}' &= (\mathbf{C}_m + \mathbf{e})\mathbf{H}' \\ &= \mathbf{C}_m\mathbf{H}' + \mathbf{eH}' \\ &= \mathbf{eH}' = \mathbf{S}\end{aligned} \tag{8.1–76}$$

where the $(n-k)$-dimensional vector $\mathbf{S}$ is called the *syndrome of the error pattern*. In other words, the vector $\mathbf{S}$ has components that are zero for all parity check equations that are satisfied and nonzero for all parity check equations that are not satisfied. Thus, $\mathbf{S}$ contains the pattern of failures in the parity checks.

We emphasize that the syndrome $\mathbf{S}$ is a characteristic of the error pattern and not of the transmitted code word. Furthermore, we observe that there are 2^n possible error patterns and only 2^{n-k} syndromes. Consequently, different error patterns result in the same syndrome.

Suppose we construct a decoding table in which we list all the 2^k possible code words in the first row, beginning with the all-zero code word in the first (leftmost) column. This all-zero code word also represents the all-zero error pattern. We fill in the first column by listing first all $n-1$ error patterns $\{\mathbf{e}_i\}$ of weight 1. If $n < 2^{n-k}$, we may then list all double error patterns, then all triple error patterns, etc., until we have a total of 2^{n-k} entries in the first column. Thus, the number of rows that we can have is 2^{n-k}, which is equal to the number of syndromes. Next, we add each error pattern in the first column to the corresponding code words. Thus, we fill in the remainder of the $n \times (n-k)$ table as follows:

$$\begin{array}{lllll}
\mathbf{C}_1 & \mathbf{C}_2 & \mathbf{C}_3 & \cdots & \mathbf{C}_{2^k} \\
\mathbf{e}_2 & \mathbf{C}_2 + \mathbf{e}_2 & \mathbf{C}_3 + \mathbf{e}_2 & \cdots & \mathbf{C}_{2^k} + \mathbf{e}_2 \\
\mathbf{e}_3 & \mathbf{C}_2 + \mathbf{e}_3 & \mathbf{C}_3 + \mathbf{e}_3 & \cdots & \mathbf{C}_{2^k} + \mathbf{e}_3 \\
\vdots & \vdots & \vdots & & \vdots \\
\mathbf{e}_{2^{n-k}} & \mathbf{C}_2 + \mathbf{e}_{2^{n-k}} & \mathbf{C}_3 + \mathbf{e}_{2^{n-k}} & \cdots & \mathbf{C}_{2^k} + \mathbf{e}_{2^{n-k}}
\end{array}$$

This table is called a *standard array*. Each row, including the first, consists of k possible received code words that would result from the corresponding error pattern in the first column. Each row is called a *coset* and the first (leftmost) code word (or error pattern) is called a *coset leader*. Therefore, a coset consists of all the possible received code words resulting from a particular error pattern (coset leader).

EXAMPLE 8.1–10. Let us construct the standard array for the (5, 2) systematic code with generator matrix given by

$$\mathbf{G} = \begin{bmatrix} 1 & 0 & 1 & 0 & 1 \\ 0 & 1 & 0 & 1 & 1 \end{bmatrix}$$

This code has a minimum distance $d_{\min} = 3$. The standard array is given in Table 8.1–7. Note that in this code, the coset leaders consist of the all-zero error pattern, five error patterns of weight 1, and two error patterns of weight 2. Although many more double error patterns exist, there is only room for two to complete the table. These were selected such that their corresponding syndromes are distinct from those of the single error patterns.

Now, suppose that $\mathbf{e}_i$ is a coset leader and that $\mathbf{C}_m$ was the transmitted code word. Then, the error pattern $\mathbf{e}_i$ would result in the received code word

$$\mathbf{Y} = \mathbf{C}_m + \mathbf{e}_i$$

The syndrome is

$$\mathbf{S} = (\mathbf{C}_m + \mathbf{e}_i)\mathbf{H}' = \mathbf{C}_m\mathbf{H}' + \mathbf{e}_i\mathbf{H}' = \mathbf{e}_i\mathbf{H}'$$

Clearly, all received code words in the same coset have the same syndrome, since the latter depends only on the error pattern. Furthermore, each coset has a different syndrome. Having established this characteristic of the standard array, we may simply construct a syndrome decoding table in which we list the 2^{n-k} syndromes and the corresponding 2^{n-k} coset leaders that represent the minimum weight error patterns. Then, given a received code vector $\mathbf{Y}$, we compute the syndrome

$$\mathbf{S} = \mathbf{Y}\mathbf{H}'$$

For the computed $\mathbf{S}$, we find the corresponding (most likely) error vector, say $\hat{\mathbf{e}}_m$. This error vector is added to $\mathbf{Y}$ to yield the decoded word

$$\hat{\mathbf{C}}_m = \mathbf{Y} \oplus \hat{\mathbf{e}}_m$$

TABLE 8.1–7
Standard array for the (5, 2) code

Code words			
0 0 0 0 0	**0 1 0 1 1**	**1 0 1 0 1**	**1 1 1 1 0**
0 0 0 0 1	0 1 0 1 0	1 0 1 0 0	1 1 1 1 1
0 0 0 1 0	0 1 0 0 1	1 0 1 1 1	1 1 1 0 0
0 0 1 0 0	0 1 1 1 1	1 0 0 0 1	1 1 0 1 0
0 1 0 0 0	0 0 0 1 1	1 1 1 0 1	1 0 1 1 0
1 0 0 0 0	1 1 0 1 1	0 0 1 0 1	0 1 1 1 0
1 1 0 0 0	1 0 0 1 1	0 1 1 0 1	0 0 1 1 0
1 0 0 1 0	1 1 0 0 1	0 0 1 1 1	0 1 1 0 0

EXAMPLE 8.1–11. Consider the (5, 2) code with the standard array given in Table 8.1–7. The syndromes versus the most likely error patterns are given in Table 8.1–8. Now suppose the actual error vector on the channel is

$$\mathbf{e} = [1 \quad 0 \quad 1 \quad 0 \quad 0]$$

The syndrome computed for the error is $\mathbf{S} = [0 \ 0 \ 1]$. Hence, the error determined from the table is $\hat{\mathbf{e}} = [0 \ 0 \ 0 \ 0 \ 1]$. When $\hat{\mathbf{e}}$ is added to $\mathbf{Y}$, the result is a decoding error. In other words the (5, 2) code corrects all single errors and only two double errors, namely $[1 \ 1 \ 0 \ 0 \ 0]$ and $[1 \ 0 \ 0 \ 1 \ 0]$.

Syndrome decoding of cyclic codes. As described above, hard-decision decoding of a linear block code may be accomplished by first computing the syndrome $\mathbf{S} = \mathbf{YH}'$, then using a table lookup to find the most probable error pattern $\hat{\mathbf{e}}$ corresponding to the computed syndrome $\mathbf{S}$, and, finally, adding the error pattern $\hat{\mathbf{e}}$ to the received vector $\mathbf{Y}$ to obtain the most probable code word $\hat{\mathbf{C}}_m$. When the code is cyclic, the syndrome computation may be performed by a shift register similar in form to that used for encoding.

To elaborate, let us consider a systematic cyclic code and let us represent the received code vector $\mathbf{Y}$ by the polynomial $Y(p)$. In general, $\mathbf{Y} = \mathbf{C} + \mathbf{e}$, where $\mathbf{C}$ is the transmitted code word and $\mathbf{e}$ is the error vector. Hence, we have

$$\begin{aligned} Y(p) &= C(p) + e(p) \\ &= X(p)g(p) + e(p) \end{aligned} \tag{8.1–77}$$

Now, suppose we divide $Y(p)$ by the generator polynomial $g(p)$. This division will yield

$$\frac{Y(p)}{g(p)} = Q(p) + \frac{R(p)}{g(p)}$$

or, equivalently,

$$Y(p) = Q(p)g(p) + R(p) \tag{8.1–78}$$

TABLE 8.1–8
Syndrome table for the (5, 2) code

Syndrome	Error pattern
0 0 0	0 0 0 0 0
0 0 1	0 0 0 0 1
0 1 0	0 0 0 1 0
1 0 0	0 0 1 0 0
0 1 1	0 1 0 0 0
1 0 1	1 0 0 0 0
1 1 0	1 1 0 0 0
1 1 1	1 0 0 1 0

The remainder $R(p)$ is a polynomial of degree less than or equal to $n-k-1$. If we combine Equation 8.1–77 with Equation 8.1–78, we obtain

$$e(p) = [X(p) + Q(p)]g(p) + R(p) \tag{8.1–79}$$

This relationship illustrates that the remainder $R(p)$ obtained from dividing $Y(p)$ by $g(p)$ depends only on the error polynomial $e(p)$, and, hence, $R(p)$ is simply the syndrome associated with the error pattern $\mathbf{e}$. Therefore,

$$Y(p) = Q(p)g(p) + S(p) \tag{8.1–80}$$

where $S(p)$ is the syndrome polynomial of degree less than or equal to $n-k-1$. If $g(p)$ divides $Y(p)$ exactly, then $S(p) = 0$ and the received decoded word is $\hat{\mathbf{C}}_m = \mathbf{Y}$.

The division of $Y(p)$ by the generator polynomial $g(p)$ may be carried out by means of a shift register which performs division as described previously. First the received vector $\mathbf{Y}$ is shifted into an $(n-k)$-stage shift register as illustrated in Figure 8.1–9. Initially, all the shift-register contents are zero and the switch is closed in position 1. After the entire n-bit received vector has been shifted into the register, the contents of the $n-k$ stages constitute the syndrome with the order of the bits numbered as shown in Figure 8.1–9. These bits may be clocked out by throwing the switch into position 2. Given the syndrome from the $(n-k)$-stage shift register, a table lookup may be performed to identify the most probable error vector.

EXAMPLE 8.1–12. Let us consider the syndrome computation for the (7, 4) cyclic Hamming code generated by the polynomial $g(p) = p^3 + p + 1$. Suppose that the received vector is $\mathbf{Y} = [1\ 0\ 0\ 1\ 1\ 0\ 1]$. This is fed into the three-stage register shown in Figure 8.1–10. After seven shifts, the contents of the shift register are 110, which corresponds to the syndrom $\mathbf{S} = [0\ 1\ 1]$. The most probable error vector corresponding to this syndrome is $\mathbf{e} = [0\ 0\ 0\ 1\ 0\ 0\ 0]$ and, hence,

$$\hat{\mathbf{C}}_m = \mathbf{Y} + \mathbf{e} = [1\ 0\ 0\ 0\ 1\ 0\ 1]$$

The information bits are 1 0 0 0.

The table lookup decoding method using the syndrome is practical only when $n-k$ is small, e.g., $n-k < 10$. This method is impractical for many interesting and powerful codes. For example, if $n-k = 20$, the table has 2^{20} (approxi-

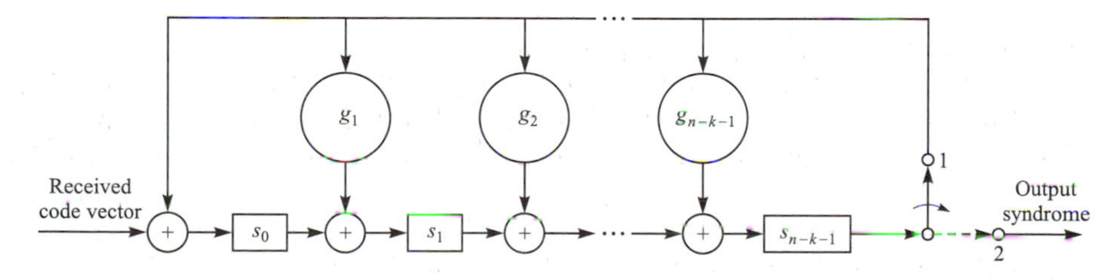

FIGURE 8.1–9
An $(n-k)$-stage shift register for computing the syndrome.

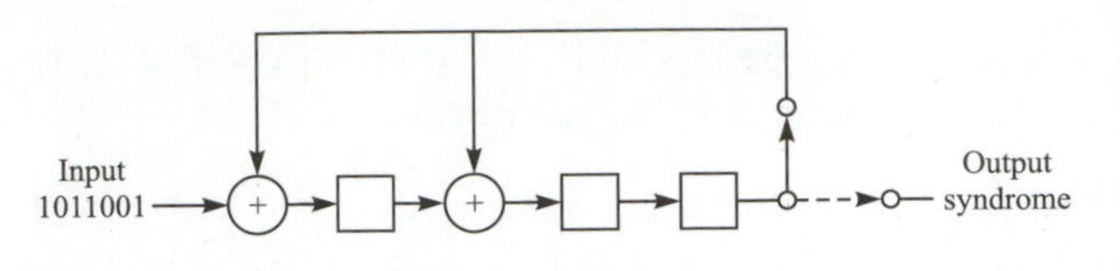

Shift	Register contents
0	000
1	100
2	010
3	001
4	010
5	101
6	100
7	110

FIGURE 8.1–10
Syndrome computation for the (7, 4) cyclic code with generator polynomial $g(p) = p^3 + p + 1$ and received vector $\mathbf{Y} = [1\,0\,0\,1\,1\,0\,1]$.

mately 1 million) entries. Such a large amount of storage and the time required to locate an entry in such a large table renders the table lookup decoding method impractical for long codes having large numbers of check bits.

More efficient and practical hard-decision decoding algorithms have been devised for the class of cyclic codes and, more specifically, the BCH codes. A description of these algorithms requires further development of computational methods with finite fields, which is beyond the scope of our treatment of coding theory. It suffices to indicate that efficient decoding algorithms exist which make it possible to implement long BCH codes with high redundancy in practical digital communication systems. The interested reader is referred to the texts of Peterson and Weldon (1972), Lin and Costello (1983), Blahut (1983), and Berlekamp (1968), and to the paper by Forney (1965).

Error detection and error correction capability. It is clear from the discussion above that when the syndrome consists of all zeros, the received code word is one of the 2^k possible transmitted code words. Since the minimum separation between a pair of code words is $d_{\min}$, it is possible for an error pattern of weight $d_{\min}$ to transform one of these 2^k code words in the code into another code word. When this happens, we have an *undetected error*. On the other hand, if the actual number of errors is less than $d_{\min}$, the syndrome will have a nonzero weight. When this occurs, we have detected the presence of one or more errors on the channel. Clearly, the (n, k) block code is capable of *detecting* $d_{\min} - 1$ errors. Error detection may be used in conjunction with an automatic repeat-request (ARQ) scheme for retransmission of the code word.

The *error correction capability* of a code also depends on the minimum distance. However, the number of correctable error patterns is limited by the number of possible syndromes or coset leaders in the standard array. To determine the error correction capability of an (n, k) code, it is convenient to view the 2^k code words as points in an n-dimensional space. If each code word is viewed as the center of a sphere of radius (Hamming distance) t, the largest value that t may have without intersection (or tangency) of any pair of the 2^k spheres is

$t = \lfloor \frac{1}{2}(d_{\min} - 1) \rfloor$, where $\lfloor x \rfloor$ denotes the largest integer contained in x. Within each sphere lie all the possible received code words of distance less than or equal to t from the valid code word. Consequently, any received code vector that falls within a sphere is decoded into the valid code word at the center of the sphere. This implies that an (n, k) code with minimum distance $d_{\min}$ is capable of correcting $t = \lfloor \frac{1}{2}(d_{\min} - 1) \rfloor$ errors. Figure 8.1–11 is a two-dimensional representation of the code words and the spheres.

As described above, a code may be used to detect $d_{\min} - 1$ errors or to correct $t = \lfloor \frac{1}{2}(d_{\min} - 1) \rfloor$ errors. Clearly, to correct t errors implies that we have detected t errors. However, it is also possible to detect more than t errors if we compromise in the error correction capability of the code. For example, a code with $d_{\min} = 7$ can correct $t = 3$ errors. If we wish to detect four errors, we can do so by reducing the radius of the sphere around each code word from 3 to 2. Thus, patterns with four errors are detectable, but only patterns of two errors are correctable. In other words, when only two errors occur, these are corrected, and when three or four errors occur, the receiver may ask for a retransmission. If more than four errors occur, they will go undetected if the code word falls within a sphere of radius 2. Similarly, for $d_{\min} = 7$, five errors can be detected and one error corrected. In general, a code with minimum distance $d_{\min}$ can detect e_d errors and correct e_c errors, where

$$e_d + e_c \leqslant d_{\min} - 1$$

and

$$e_c \leqslant e_d$$

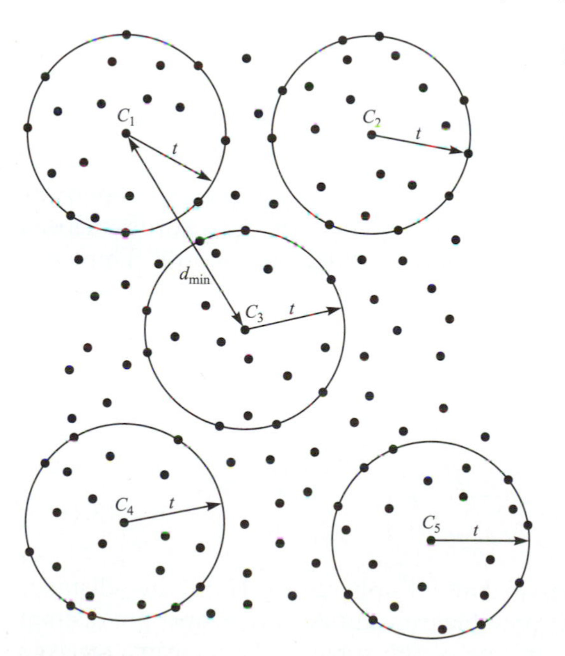

FIGURE 8.1–11
A representation of code words as centers of spheres of radius $t = \lfloor \frac{1}{2}(d_{\min} - 1) \rfloor$.

Probability of error based on error correction. We conclude this section with the derivation of the probability of error for hard-decision decoding of linear binary block codes based on error correction only.

From the above discussion, it is clear that the optimum decoder for a binary symmetric channel will decode correctly if (but not necessarily only if) the number of errors in a code word is less than half the minimum distance $d_{\min}$ of the code. That is, any number of errors up to

$$t = \lfloor \tfrac{1}{2}(d_{\min} - 1) \rfloor$$

are always correctable. Since the binary symmetric channel is memoryless, the bit errors occur independently. Hence, the probability of m errors in a block of n bits is

$$P(m, n) = \binom{n}{m} p^m (1-p)^{n-m} \tag{8.1–81}$$

and, therefore, the probability of a code word error is upper-bounded by the expression

$$P_M \leqslant \sum_{m=t+1}^{n} P(m, n) \tag{8.1–82}$$

Equality holds in Equation 8.1–82 if the linear block code is a perfect code. In order to describe the basic characteristics of a perfect code, suppose we place a sphere of radius t around each of the possible transmitted code words. Each sphere around a code word contains the set of all code words of Hamming distance less than or equal to t from the code word. Now, the number of code words in a sphere of radius $t = \lfloor \frac{1}{2}(d_{\min} - 1) \rfloor$ is

$$1 + \binom{n}{1} + \binom{n}{2} + \cdots + \binom{n}{t} = \sum_{i=0}^{t} \binom{n}{i}$$

Since there are $M = 2^k$ possible transmitted code words, there are 2^k nonoverlapping spheres each having a radius t. The total number of code words enclosed in the 2^k spheres cannot exceed the 2^n possible received code words. Thus, a t-error correcting code must satisfy the inequality

$$2^k \sum_{i=0}^{t} \binom{n}{i} \leqslant 2^n$$

or, equivalently,

$$2^{n-k} \geqslant \sum_{i=0}^{t} \binom{n}{i} \tag{8.1–83}$$

A perfect code has the property that all spheres of Hamming distance $t = \lfloor \frac{1}{2}(d_{\min} - 1) \rfloor$ around the $M = 2^k$ possible transmitted code words are disjoint and every received code word falls in one of the spheres. Thus, every received code word is at most at distance t from one of the possible transmitted code words and Equation 8.1–83 holds with equality. For such a code, all error patterns of weight less than or equal to t are corrected by the optimum (minimum distance) decoder. On the other hand, any error pattern of weight $t+1$ or greater cannot be corrected. Consequently, the expression for the error probability given in Equation 8.1–82 holds with equality. The Golay (23, 12) code, having $d_{\min} = 7$ and $t = 3$, is a perfect code. The Hamming codes, which have the parameters $n = 2^{n-k} - 1$, $d_{\min} = 3$, and $t = 1$, are also perfect codes. These two nontrivial codes and the trivial code consisting of two code words of odd length n and $d_{\min} = n$ are the only perfect binary block codes. These codes are optimum on the BSC in the sense that they result in a minimum error probability among all codes having the same block length and the same number of information bits.

The optimality property defined above also holds for quasiperfect codes. A quasiperfect code is characterized by the property that all spheres of Hamming radius t around the M possible transmitted code words are disjoint and every received code word is at most at distance $t+1$ from one of the possible transmitted code words. For such a code, all error patterns of weight less than or equal to t and some error patterns of weight $t+1$ are correctable, but any error pattern of weight $t+2$ or greater leads to incorrect decoding of the code word. Clearly, Equation 8.1–82 is an upper bound on the error probability and

$$P_M \geqslant \sum_{m=t+2}^{n} P(m, n) \tag{8.1–84}$$

is a lower bound.

A more precise measure of the performance for quasiperfect codes can be obtained by making use of the inequality in Equation 8.1–83. That is, the total number of code words outside the 2^k spheres of radius t is

$$N_{t+1} = 2^n - 2^k \sum_{i=0}^{t} \binom{n}{i}$$

If these code words are equally subdivided into 2^k sets and each set is associated with one of the 2^k spheres, then each sphere is enlarged by the addition of

$$\beta_{t+1} = 2^{n-k} - \sum_{i=0}^{t} \binom{n}{i} \tag{8.1–85}$$

code words having distance $t+1$ from the transmitted code word. Consequently, of the $\binom{n}{t+1}$ error patterns of distance $t+1$ from each code word, we can correct β_{t+1} error patterns. Thus, the error probability for decoding the quasiperfect code may be expressed as

$$P_M = \sum_{m=t+2}^{n} P(m, n) + \left[\binom{n}{t+1} - \beta_{t+1}\right] p^{t+1}(1-p)^{n-t-1} \tag{8.1–86}$$

There are many known quasiperfect codes, although they do not exist for all choices of n and k. Since such codes are optimum for the binary symmetric channel, any (n, k) linear block code must have an error probability that is at least as large as Equation 8.1–86. Consequently, Equation 8.1–86 is a lower bound on the probability of error for any (n, k) linear block code, where t is the largest integer such that $\beta_{t+1} \geqslant 0$.

Another pair of upper and lower bounds is obtained by considering two code words that differ by the minimum distance. First, we note that P_M cannot be less than the probability of erroneously decoding the transmitted code word as its nearest neighbor, which is at distance $d_{\min}$ from the transmitted code word. That is,

$$P_M \geqslant \sum_{m=[d_{\min}/2]+1}^{d_{\min}} \binom{d_{\min}}{m} p^m (1-p)^{d_{\min}-m} \tag{8.1–87}$$

On the other hand, P_M cannot be greater than $M-1$ times the probability of erroneously decoding the transmitted code word as its nearest neighbor, which is at distance $d_{\min}$ from the transmitted code word. That is a union bound, which is expressed as

$$P_M \leqslant (M-1) \sum_{m=[d_{\min}/2]+1}^{d_{\min}} \binom{d_{\min}}{m} p^m (1-p)^{d_{\min}-m} \tag{8.1–88}$$

When M is large, the lower bound in Equation 8.1–87 and the upper bound in Equation 8.1–88 are very loose.

A tight upper bound on P_M can be obtained by applying the Chernoff bound presented earlier in Section 2.1.6. We assume again that the all-zero code was transmitted. In comparing the received code word to the all-zero code word and to a code word of weight w_m, the probability of a decoding error, obtained from the Chernoff bound (Problem 8.22), is upper-bounded by the expression

$$P_2(w_m) \leqslant [4p(1-p)]^{w_m/2} \tag{8.1–89}$$

The union of these binary decisions yields the upper bound

$$P_M \leqslant \sum_{m=2}^{M} [4p(1-p)]^{w_m/2} \tag{8.1–90}$$

A simpler version of Equation 8.1–90 is obtained if we employ $d_{\min}$ in place of the weight distribution. That is,

$$P_M \leqslant (M-1)[4p(1-p)]^{d_{\min}/2} \tag{8.1–91}$$

Of course, Equation 8.1–90 is a tighter upper bound than Equation 8.1–91.

In Section 8.1.6, we compare the various bounds given above for a specific code, namely, the Golay (23, 12) code. In addition, we compare the error rate performance of hard-decision and soft-decision decoding.

8.1.6 Comparison of Performance between Hard-Decision and Soft-Decision Decoding

It is both interesting and instructive to compare the bounds on the error rate performance of linear block codes for soft-decision decoding and hard-decision decoding on an AWGN channel. For illustrative purposes, we shall use the Golay (23, 12) code, which has the relatively simple weight distribution given in Table 8.1–1. As stated previously, this code has a minimum distance $d_{\min} = 7$.

First we compute and compare the bounds on the error probability for hard-decision decoding. Since the Golay (23, 12) code is a perfect code, the exact error probability for hard-decision decoding is

$$\begin{aligned} P_M &= \sum_{m=4}^{23} \binom{23}{m} p^m (1-p)^{23-m} \\ &= 1 - \sum_{m=0}^{3} \binom{23}{m} p^m (1-p)^{23-m} \end{aligned} \tag{8.1–92}$$

where p is the probability of a binary digit error for the binary symmetric channel. Binary (or four-phase) coherent PSK is assumed to be the modulation/demodulation technique for the transmission and reception of the binary digits contained in each code word. Thus, the appropriate expression for p is given by Equation 8.1–73. In addition to the exact error probability given by Equation 8.1–92, we have the lower bound given by Equation 8.1–87 and the three upper bounds given by Equations 8.1–88, 8.1–90, and 8.1–91.

Numerical results obtained from these bounds are compared with the exact error probability in Figure 8.1–12. We observe that the lower bound is very loose. At $P_M = 10^{-5}$, the lower bound is off by approximately 2 dB from the exact error probability. At $P_M = 10^{-2}$, the difference increases to approximately 4 dB. Of the three upper bounds, the one given by Equation 8.1–88 is the tightest; it differs by less than 1 dB from the exact error probability at $P_M = 10^{-5}$. The Chernoff bound in Equation 8.1–90, which employs the weight distribution, is also relatively tight. Finally, the Chernoff bound that employs only the minimum distance of the code is the poorest of the three. At $P_M = 10^{-5}$, it differs from the exact error probability by approximately 2 dB. All three upper bounds are very loose for error rates above $P_M = 10^{-2}$.

It is also interesting to compare the performance between soft- and hard-decision decoding. For this comparison, we use the upper bounds on the error probability for soft-decision decoding given by Equation 8.1–52 and the exact error probability for hard-decision decoding given by Equation 8.1–92. Figure 8.1–13 illustrates these performance characteristics. We observe that the two bounds for soft-decision decoding differ by approximately 0.5 dB at $P_M = 10^{-6}$ and by approximately 1 dB at $P_M = 10^{-2}$. We also observe that the difference in performance between hard- and soft-decision decoding is approximately 2 dB in the range $10^{-2} < P_M < 10^{-6}$. In the range $P_M > 10^{-2}$, the curve of the error probability for hard-decision decoding crosses the curves

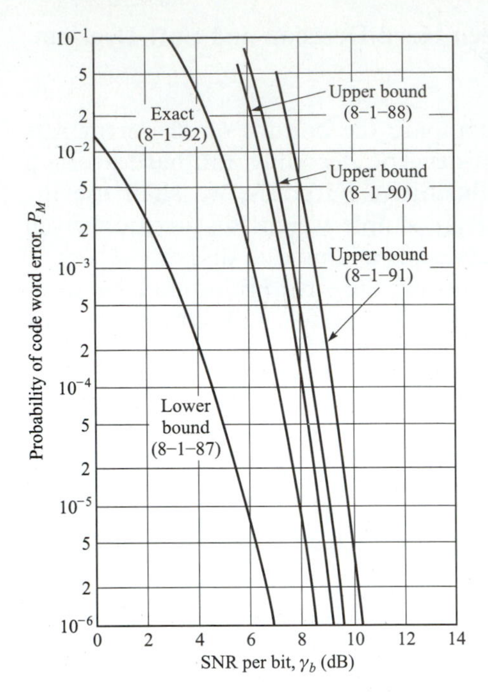

FIGURE 8.1–12
Comparison of bounds with exact error probability for hard-decision decoding of Golay (23, 12) code.

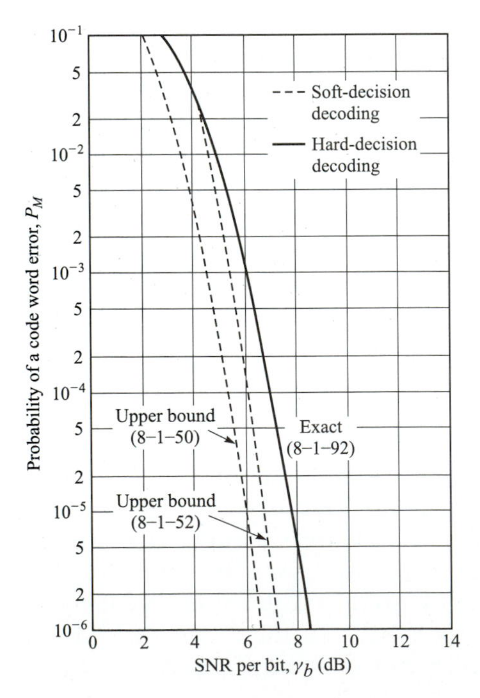

FIGURE 8.1–13
Comparison of soft-decision decoding with hard-decision decoding for the Golay (23, 12) code.

for the bounds. This behavior indicates that the bounds for soft-decision decoding are loose when $P_M > 10^{-2}$.

As we observed in Section 7.3, by comparing the cutoff rate for hard-decision decoding and for soft-decision decoding, the 2-dB difference in performance is a fundamental result that applies in general to coded digital communications over the AWGN channel. This 2-dB difference in performance is also obtained, as shown below, by computing the capacity of the AWGN channel with hard- and soft-decision decoding.

The channel capacity of the BSC in bits per code symbol, derived in Section 7.1.2, is

$$C = 1 + p\log_2 p + (1-p)\log_2(1-p) \tag{8.1–93}$$

where the probability of a bit error for binary, coherent PSK on an AWGN channel is given by Equation 8.1–73. Suppose we use Equation 8.1–73 for p, let $C = R_c$ in Equation 8.1–93, and then determine the value of γ_b that satisfies this equation. The result is shown in Figure 8.1–14 as a graph of R_c versus γ_b. For example, suppose that we are interested in using a code with rate $R_c = \frac{1}{2}$. For this code rate, note that the minimum SNR per bit required to achieve capacity with hard-decision decoding is approximately 1.6 dB.

What is the limit on the minimum SNR as the code rate approaches zero? For small values of R_c, the probability p can be approximated as

$$p \approx \tfrac{1}{2} - \sqrt{\gamma_b R_c/\pi} \tag{8.1–94}$$

When the expression for p is substituted into Equation 8.1–93 and the logarithms in Equation 8.1–93 are approximated by

$$\log_2(1+x) \simeq \frac{x - \frac{1}{2}x^2}{\ln 2}$$

the channel capacity formula reduces to

$$C = \frac{2}{\pi \ln 2}\gamma_b R_c \tag{8.1–95}$$

Now we set $C = R_c$. Thus, in the limit as R_c approaches zero, we obtain the result

$$\gamma_b = \tfrac{1}{2}\pi \ln 2 \qquad (0.37 \text{ dB}) \tag{8.1–96}$$

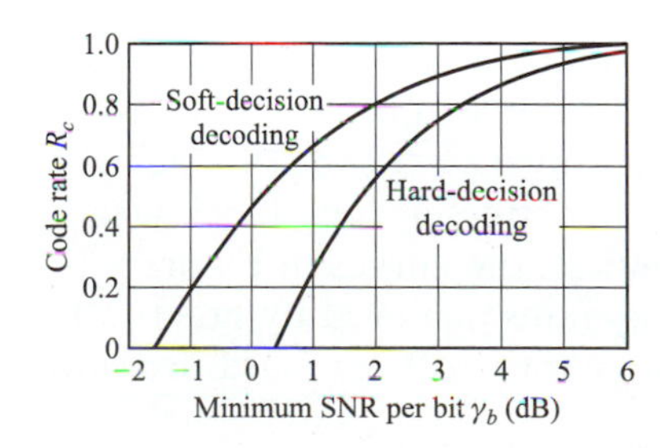

FIGURE 8.1–14
Code rate as a function of the minimum SNR per bit for soft- and hard-decision decoding.

The capacity of the binary-input AWGN channel with soft-decision decoding can be computed in a similar manner. The expression for the capacity in bits per code symbol, derived in Section 7.1.2, is

$$C = \frac{1}{2}\sum_{k=0}^{1}\int_{-\infty}^{\infty} p(y|k)\log_2 \frac{p(y|k)}{p(y)}\, dy \tag{8.1–97}$$

where $p(y|k)$, $k = 0, 1$, denote the probability density functions of the demodulator output conditioned on the transmitted bit being a 0 and a 1, respectively. For the AWGN channel, we have

$$p(y|k) = \frac{1}{\sqrt{2\pi}\sigma} e^{-(y-m_k)^2/2\sigma^2}, \qquad k = 0, 1 \tag{8.1–98}$$

where $m_0 = -\sqrt{\mathcal{E}_c}$, $m_1 = \sqrt{\mathcal{E}_c}, \sigma^2 = \frac{1}{2}N_0$, and $\mathcal{E}_c = R_c\mathcal{E}_b$. The unconditional probability density $p(y)$ is simply one-half of the sum of $p(y|1)$ and $p(y|0)$. As R_c approaches zero, the expression 8.9–97 for the channel capacity can be approximated as

$$C \approx \frac{\gamma_b R_c}{\ln 2} \tag{8.1–99}$$

Again, we set $C = R_c$. Thus, as $R_c \to 0$, the minimum SNR per bit to achieve capacity is

$$\gamma_b = \ln 2 \qquad (-1.6\,\text{dB}) \tag{8.1–100}$$

By using Equations 8.1–98 in Equation 8.1–97 and setting $C = R_c$, a numerical solution can be obtained for code rates in the range $0 \leqslant R_c \leqslant 1$. The result of this solution is also shown in Figure 8.1–14.

From the above, we observe that in the limit as R_c approaches zero, the difference in SNR γ_b between hard- and soft-decision decoding is $\frac{1}{2}\pi$, which is approximately 2 dB. On the other hand, as R_c increases toward unity, the difference in γ_b between these two decoding techniques decreases. For example, at $R_c = 0.8$, the difference is about 1.5 dB.

The curves in Figure 8.1–14 provide more information than just the difference in performance between soft- and hard-decision decoding. These curves also specify the minimum SNR per bit that is required for a given code rate. For example, a code rate of $R_c = 0.8$ can provide arbitrarily small error probability at an SNR per bit of 2 dB, when soft-decision decoding is used. By comparison, an uncoded binary PSK requires 9.6 dB to achieve an error probability of 10^{-5}. Hence, a 7.6-dB gain is possibly by employing a rate $R_c = \frac{4}{5}$ code. To achieve such a large coding gain usually implies the use of an extremely long block length code, and, generally, a complex decoder. Nevertheless, the curves in Figure 8.1–14 provide a benchmark for comparing the coding gains achieved by practically implementable codes with the ultimate limits for either soft- or hard-decision decoding.

8.1.7 Bounds on Minimum Distance of Linear Block Codes

The expressions for the probability of error derived in this chapter for soft-decision and hard-decision decoding of linear binary block codes clearly indicate the importance that the minimum distance parameter plays in the performance of the code. If we consider soft-decision decoding, for example, the upper bound on the error probability given by Equation 8.1–52 indicates that, for a given code rate $R_c = k/n$, the probability of error in an AWGN channel decreases exponentially with $d_{\min}$. When this bound is used in conjunction with the lower bound on $d_{\min}$ given below, we obtain an upper bound on P_M, the probability of a code word error. Similarly, we may use the upper bound given by Equation 8.1–82 for the probability of error for hard-decision decoding in conjunction with the lower bound on $d_{\min}$ to obtain an upper bound on the error probability for linear binary block codes on the binary symmetric channel.

On the other hand, an upper bound on $d_{\min}$ can be used to determine a lower bound on the probability of error achieved by the best code. For example, suppose that hard-decision decoding is employed. In this case, we have the two lower bounds on P_M given by Equations 8.1–86 and 8.1–87, with the former being the tighter. When either one of these two bounds is used in conjunction with an upper bound on $d_{\min}$, the result is a lower bound on P_M for the best (n, k) code. Thus, upper and lower bounds on $d_{\min}$ are important in assessing the capabilities of codes.

A simple upper bound on the minimum distance of an (n, k) binary or non-binary linear block code, called the Singleton bound, was given in Equation 8.1–14 as $d_{\min} \leqslant n - k + 1$. It is convenient to normalize this expression by the block size n. That is,

$$\frac{d_{\min}}{n} \leqslant (1 - R_c) + \frac{1}{n} \tag{8.1–101}$$

where R_c is the code rate. For large n, the factor $1/n$ can be neglected.

If a code has the largest possible distance, i.e., $d_{\min} = n - k + 1$, it is called a *maximum-distance-separable code*. Except for the trivial repetition-type codes, there are no binary maximum-separable codes. In fact, the upper bound in Equation 8.1–101 is extremely loose for binary codes. On the other hand, non-binary codes with $d_{\min} = n - k + 1$ do exist. For example, the Reed–Solomon codes, which comprise a subclass of BCH codes, are maximum-distance-separable.

In addition to the upper bound given above, there are several relatively tight bounds on the minimum distance of linear block codes. We shall briefly describe four important bounds, three of which are upper bounds and the other a lower bound. The derivations of these bounds are lengthy and are not of particular interest in our subsequent discussion. The interested reader may refer to chapter 4 of the book by Peterson and Weldon (1972) for those derivations.

One upper bound on the minimum distance can be obtained from the inequality in Equation 8.1–83. By taking the logarithm of both sides of Equation 8.1–83 and dividing by n, we obtain

$$1 - R_c \geqslant \frac{1}{n}\log_2 \sum_{i=0}^{t} \binom{n}{i} \tag{8.1–102}$$

Since the error-correcting capability of the code, measured by t, is related to the minimum distance, the above relation is an upper bound on the minimum distance. It is called the *Hamming upper bound.*

The asymptotic form of Equation 8.1–102 is obtained by letting $n \to \infty$. Now, for any n, let t_0 be the largest integer t for which (8.1–102) holds. Then, it can be shown (Peterson and Weldon, 1972) that as $n \to \infty$, the ratio t/n for any (n, k) block code cannot exceed t_0/n, where t_0/n satisfies the equation

$$1 - R_c = H(t_0/n) \tag{8.1–103}$$

and $H(x)$ is the binary entropy function defined by Equation 3.2–10.

The generalization of the Hamming bound to nonbinary codes is simply

$$1 - R_c \geqslant \frac{1}{n}\log_q \left[\sum_{i=0}^{t} \binom{n}{i}(q-1)^i\right] \tag{8.1–104}$$

Another upper bound, developed by Plotkin (1960), may be stated as follows. The number of check digits required to achieve a minimum distance $d_{\min}$ in an (n, k) linear block code satisfies the inequality

$$n - k \geqslant \frac{q d_{\min} - 1}{q - 1} - 1 - \log_q d_{\min} \tag{8.1–105}$$

where q is the alphabet size. When the code is binary, Equation 8.1–105 may be expressed as

$$\frac{d_{\min}}{n}\left(1 - \frac{1}{2d_{\min}}\log_2 d_{\min}\right) \leqslant \frac{1}{2}\left(1 - R_c + \frac{2}{n}\right)$$

In the limit as $n \to \infty$ with $d_{\min}/n \leqslant \frac{1}{2}$, Equation 8.1–105 reduces to

$$\frac{d_{\min}}{n} \leqslant \tfrac{1}{2}(1 - R_c) \tag{8.1–106}$$

Finally, there is another tight upper bound on the minimum distance obtained by Elias (Berlekamp, 1968). It may be expressed in its asymptotic form as

$$\frac{d_{\min}}{n} \leqslant 2A(1 - A) \tag{8.1–107}$$

where the parameter A is related to the code rate through the equation

$$R_c = 1 + A\log_2 A + (1 - A)\log_2(1 - A), \qquad 0 \leqslant A \leqslant \tfrac{1}{2} \tag{8.1–108}$$

Lower bounds on the minimum distance of (n, k) block codes also exist. In particular, binary block codes exist that have a normalized minimum distance that asymptotically satisfies the inequality

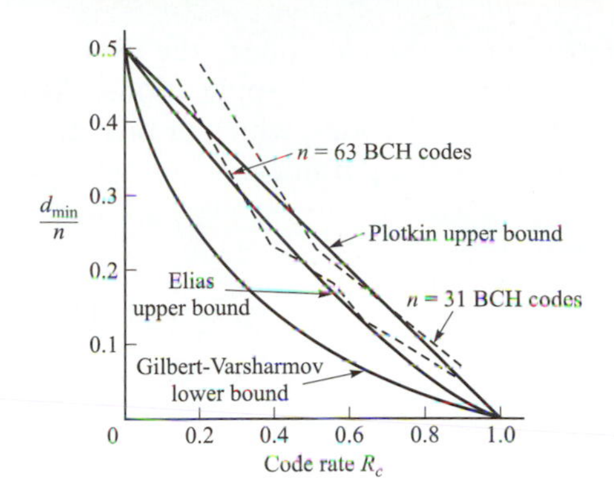

FIGURE 8.1–15
Upper and lower bounds on normalized minimum distance as a function of code rate.

$$\frac{d_{\min}}{n} \geqslant \alpha \tag{8.1–109}$$

where α is related to the code rate through the equation

$$\begin{aligned} R_c &= 1 - H(\alpha) \\ &= 1 + \alpha \log_2 \alpha + (1 - \alpha) \log_2 (1 - \alpha), \qquad 0 \leqslant \alpha \leqslant \tfrac{1}{2} \end{aligned} \tag{8.1–110}$$

This lower bound is a special case of a lower bound developed by Gilbert (1952) and Varsharmov (1957), which applies to nonbinary and binary block codes.

The asymptotic bounds given above are plotted in Figure 8.1–15 for binary codes. Also plotted in the figure for purposes of comparison are curves of the minimum distance as a function of code rate for BCH codes of block lengths $n = 31$ and 63. We observe that for $n = 31$ and 63, the normalized minimum distance falls well above the Varsharmov–Gilbert lower bound. As the block length n increases, the efficiency of the BCH codes diminishes. For example, when $n = 1023$, the curve for the normalized minimum distance falls close to the Varsharmov–Gilbert bound. As n increases beyond $n = 1023$, the normalized minimum distance of the BCH codes continues to decrease and falls below the Varsharmov–Gilbert bound. That is, $d_{\min}/n$ approaches zero as n tends to infinity. Consequently the BCH codes, which are the most important class of cyclic codes, are not very efficient at large block lengths.

8.1.8 Nonbinary Block Codes and Concatenated Block Codes

A nonbinary block code consists of a set of fixed-length code words in which the elements of the code words are selected from an alphabet of q symbols, denoted by $\{0, 1, 2, \ldots, q-1\}$. Usually, $q = 2^k$, so that k information bits are mapped into one of the q symbols. The length of the nonbinary code word is denoted by N and the number of information symbols encoded into a block of N symbols is denoted by K. The minimum distance of the nonbinary code is denoted by $D_{\min}$. A systematic (N, K) block code consists of K information symbols and $N - K$ parity check symbols.

Among the various types of nonbinary linear block codes, the Reed–Solomon codes are some of the most important for practical applications. As indicated previously, they comprise a subset of the BCH codes, which in turn are a class of cyclic codes. These codes are described by the parameters

$$\begin{aligned} N &= q - 1 = 2^k - 1 \\ K &= 1, 2, 3, \ldots, N - 1 \\ D_{\min} &= N - K + 1 \\ R_c &= K/N \end{aligned} \tag{8.1–111}$$

Such a code is guaranteed to correct up to

$$\begin{aligned} t &= \lfloor \tfrac{1}{2}(D_{\min} - 1) \rfloor \\ &= \lfloor \tfrac{1}{2}(N - K) \rfloor \end{aligned} \tag{8.1–112}$$

symbol errors. Of course, these codes may be extended or shortened in the manner described previously for binary block codes.

The weight distribution $\{A_i\}$ of the class of Reed–Solomon codes is known. The coefficients in the weight enumerating polynomial are given as

$$A_i = \binom{N}{i}(q-1)\sum_{j=0}^{i-D}(-1)^j \binom{i-1}{j} q^{i-j-D}, \qquad i \geqslant D_{\min} \tag{8.1–113}$$

where $D \equiv D_{\min}$ and $q = 2^k$.

One reason for the importance of the Reed–Solomon codes is their good distance properties. A second reason for their importance is the existence of efficient hard-decision decoding algorithms, which make it possible to implement relatively long codes in many practical applications where coding is desirable.

A nonbinary code is particularly matched to an M-ary modulation technique for transmitting the 2^k possible symbols. Specifically, M-ary orthogonal signaling, e.g., M-ary FSK, is frequently used. Each of the 2^k symbols in the q-ary alphabet is mapped to one of the $M = 2^k$ orthogonal signals. Thus, the transmission of a code word is accomplished by transmitting N orthogonal signals, where each signal is selected from the set of $M = 2^k$ possible signals.

The optimum demodulator for such a signal corrupted by AWGN consists of M matched filters (or cross correlators) whose outputs are passed to the decoder, either in the form of soft decisions or in the form of hard decisions. If hard decisions are made by the demodulator, the symbol error probability P_M and the code parameters are sufficient to characterize the performance of the decoder. In fact, the modulator, the AWGN channel, and the demodulator form an equivalent discrete (M-ary) input, discrete (M-ary) output, symmetric memoryless channel characterized by the transition probabilities $P_c = 1 - P_M$ and $P_M/(M-1)$. This channel model, which is illustrated in Figure 8.1–16, is a generalization of the BSC.

The performance of the hard-decision decoder may be characterized by the following upper bound on the code word error probability:

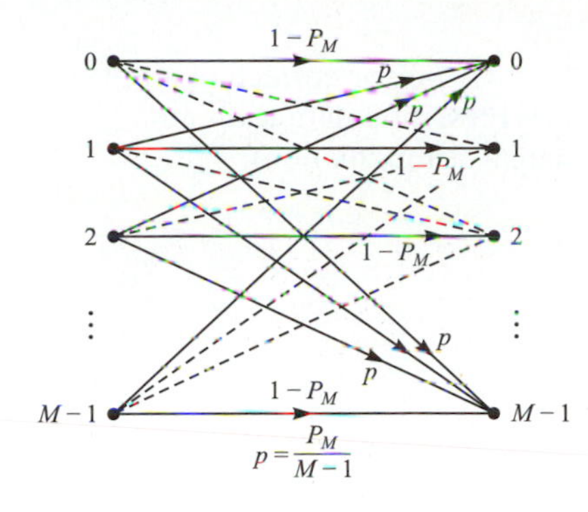

FIGURE 8.1–16
M-ary input, M-ary output, symmetric memoryless channel.

$$P_e \leqslant \sum_{i=t+1}^{N} \binom{N}{i} P_M^i (1 - P_M)^{N-i} \tag{8.1–114}$$

where t is the number of errors guaranteed to be corrected by the code.

When a code word error is made, the corresponding symbol error probability is

$$P_{es} = \frac{1}{N} \sum_{i=t+1}^{N} i \binom{N}{i} P_M^i (1 - P_M)^{N-i} \tag{8.1–115}$$

Furthermore, if the symbols are converted to binary digits, the bit error probability corresponding to Equation 8.1–115 is

$$P_{eb} = \frac{2^{k-1}}{2^k - 1} P_{es} \tag{8.1–116}$$

EXAMPLE 8.1–13. Let us evaluate the performance of an $N = 2^5 - 1 = 31$ Reed–Solomon code with $D_{\min} = 3, 5, 9$, and 17. The corresponding values of K are 29, 27, 23, and 15. The modulation is $M = 32$ orthogonal FSK with noncoherent detection at the receiver.

The probability of a symbol error is given by Equation 5.4–46 and may be expressed as

$$P_M = \frac{1}{M} e^{-\gamma} \sum_{n=2}^{M} (-1)^n \binom{M}{n} e^{\gamma/n} \tag{8.1–117}$$

where γ is the SNR per code symbol. By using Equation 8.1–117 in Equation 8.1–115 and combining the result with Equation 8.1–116, we obtain the bit error probability. The results of these computations are plotted in Figure 8.1–17. Note that the more powerful codes (large $D_{\min}$) give poorer performance at low SNR per bit than the weaker codes. On the other hand, at high SNR, the more powerful codes give better performance. Hence, there are crossovers among the various codes, as illustrated for example in Figure 8.1–17 for the $t = 1$ and $t = 8$ codes. Crossovers also occur among the $t = 1, 2$, and 4 codes at smaller values of SNR per bit. Similarly, the curves for $t = 4$ and 8 and for $t = 8$ and 2 cross in the region of high SNR. This is the characteristic behavior for noncoherent detection of the coded waveforms.

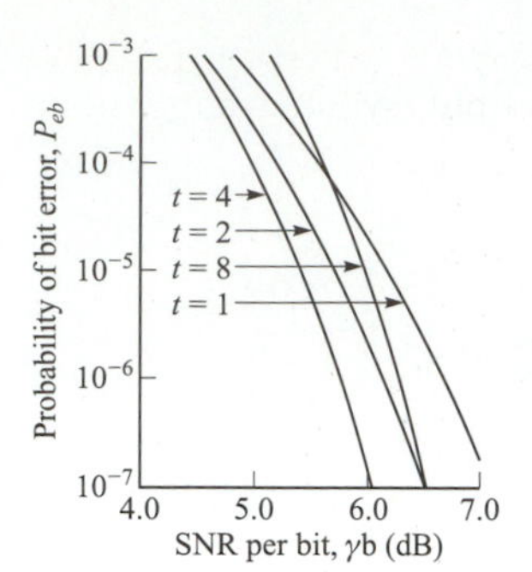

FIGURE 8.1–17
Performance of several $N = 31$, t-error correcting Reed–Solomon codes with 32-ary FSK modulation on an AWGN channel (noncoherent demodulation).

If the demodulator does not make a hard decision on each symbol, but, instead, passes the unquantized matched filter outputs to the decoder, soft-decision decoding can be performed. This decoding involves the formation of $q^K = 2^{kK}$ correlation metrics, where each metric corresponds to one of the q^K code words and consists of a sum of N matched filter outputs corresponding to the N code symbols. The matched filter outputs may be added coherently, or they may be envelope-detected and then added, or they may be square-law detected and then added. If coherent detection is used and the channel noise is AWGN, the computation of the probability of error is a straightforward extension of the binary case considerd in Section 8.1.4. On the other hand, when envelope detection or square-law detection and noncoherent combining are used to form the decision variables, the computation of the decoder performance is considerably more complicated.

Concatenated block codes. A concatenated code consists of two separate codes which are combined to form a larger code. Usually one code is selected to be nonbinary and the other is binary. The two codes are concatenated as illustrated in Figure 8.1–18. The nonbinary (N, K) code forms the outer code and the binary code forms the inner code. Code words are formed by subdividing a block of kK information bits into K groups, called *symbols*, where each symbol consists of k bits. The K k-bit symbols are encoded into N k-bit symbols by the outer encoder, as is usually done with a nonbinary code. The inner encoder takes each k-bit symbol and encodes it into a binary block code of length n. Thus we

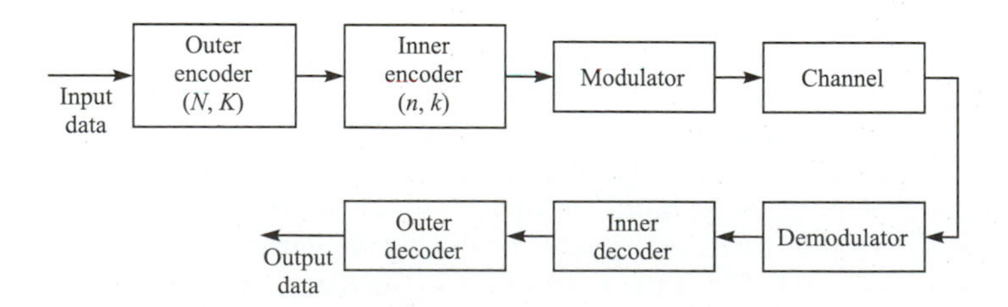

FIGURE 8.1–18
Block diagram of a communication system employing a concatenated code.

obtain a concatenated block code having a block length of Nn bits and containing kK infomration bits. That is, we have created an equivalent (Nn, Kk) long binary code. The bits in each code word are transmitted over the channel by means of PSK or, perhaps, by FSK.

We also indicate that the minimum distance of the concatenated code is $d_{\min}D_{\min}$, where $D_{\min}$ is the minimum distance of the outer code and $d_{\min}$ is the minimum distance of the inner code. Furthermore, the rate of the concatenated code is Kk/Nn, which is equal to the product of the two code rates.

A hard-decision decoder for a concatenated code is conveniently separated into an inner decoder and an outer decoder. The inner decoder takes the hard decisions on each group of n bits, corresponding to a code word of the inner code, and makes a decision on the k information bits based on maximum-likelihood (minimum-distance) decoding. These k bits represent one symbol of the outer code. When a block of N k-bit symbols is received from the inner decoder, the outer decoder makes a hard decision on the K k-bit symbols based on maximum-likelihood decoding.

Soft-decision decoding is also a possible alternative with a concatenated code. Usually, the soft-decision decoding is performed on the inner code, if it is selected to have relatively few code words, i.e., if 2^k is not too large. The outer code is usually decoded by means of hard-decision decoding, especially if the block length is long and there are many code words. On the other hand, there may be a significant gain in performance when soft-decision decoding is used on both the outer and inner codes, to justify the additional decoding complexity. This is the case in digital communications over fading channels, as we shall demonstrate in Chapter 14.

We conclude this subsection with the following example.

EXAMPLE 8.1–14. Suppose that the (7, 4) Hamming code described in Examples 8.1–1 and 8.1–2 is used as the inner code in a concatenated code in which the outer code is a Reed–Solomon code. Since $k = 4$, we select the length of the Reed–Solomon code to be $N = 2^4 - 1 = 5$. The number of information symbols K per outer code word may be selected over the range $1 \leqslant K \leqslant 14$ in order to achieve a desired code rate.

8.1.9 Interleaving of Coded Data for Channels with Burst Errors

Most of the well-known codes that have been devised for increasing the reliability in the transmission of information are effective when the errors caused by the channel are statistically independent. This is the case for the AWGN channel. However, there are channels that exhibit bursty error characteristics. One example is the class of channels characterized by multipath and fading, which is described in detail in Chapter 14. Signal fading due to time-variant multipath propagation often causes the signal to fall below the noise level, thus resulting in a large number of errors. A second example is the class of magnetic recording channels (tape or disk) in which defects in the recording media result in clusters of errors. Such error clusters are not usually corrected by codes that are optimally designed for statistically independent errors.

Considerable work has been done on the construction of codes that are capable of correcting burst errors. Probably the best known burst error correcting codes are the subclass of cyclic codes called Fire codes, named after P. Fire (1959), who discovered them. Another class of cyclic codes for burst error correction were subsequently discovered by Burton (1969).

A *burst* of errors of length b is defined as a sequence of b-bit errors, the first and last of which are 1s. The *burst error correction capability* of a code is defined as one less than the length of the shortest uncorrectable burst. It is relatively easy to show that a systematic (n, k) code, which has $n-k$ parity check bits, can correct bursts of length $b \leqslant \lfloor \frac{1}{2}(n-k) \rfloor$.

An effective method for dealing with burst error channels is to interleave the coded data in such a way that the bursty channel is transformed into a channel having independent errors. Thus, a code designed for independent channel errors (short bursts) is used.

A block diagram of a system that employs interleaving is shown in Figure 8.1–19. The encoded data is reordered by the interleaver and transmitted over the channel. At the receiver, after (either hard- or soft-decision) demodulation, the deinterleaver puts the data in proper sequence and passes it to the decoder. As a result of the interleaving/deinterleaving, error bursts are spread out in time so that errors within a code word appear to be independent.

The interleaver can take one of two forms: a block structure or a convolutional structure. A block *interleaver* formats the encoded data in a rectangular array of m rows and n columns. Usually, each row of the array constitutes a code word of length n. An *interleaver of degree m* consists of m rows (m code words) as illustrated in Figure 8.1–20. The bits are read out column-wise and transmitted over the channel. At the receiver, the deinterleaver stores the data in the same rectangular array format, but it is read out row-wise, one code word at a time. As a result of this reordering of the data during transmission, a burst of errors of length $l = mb$ is broken up into m bursts of length b. Thus, an (n, k) code that can handle burst errors of length $b \leqslant \lfloor \frac{1}{2}(n-k) \rfloor$ can be combined with an interleaver of degree m to create an interleaved (mn, mk) block code that can handle bursts of length mb.

A *convolutional interleaver* can be used in place of a block interleaver in much the same way. Convolutional interleavers are better matched for use with the

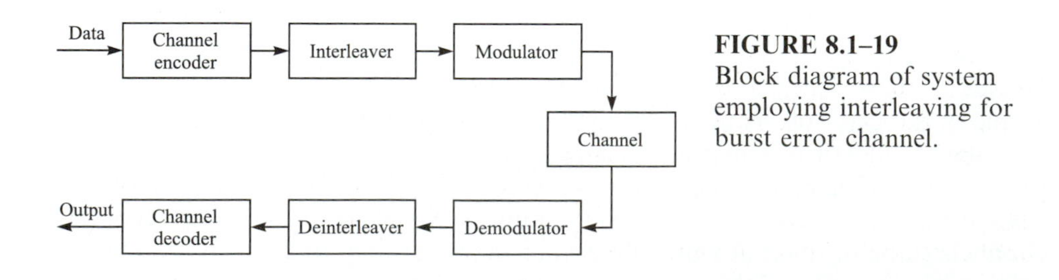

FIGURE 8.1–19
Block diagram of system employing interleaving for burst error channel.

Read out bits to modulator

Read in coded bits from encoder

1	8	15	22	29	36	⋯		$mn-6$
2	9	16	23	30	37	⋯		$mn-5$
3	10	17	24	31	38	⋯		$mn-4$
4	11	18	25	32	39	⋯		$mn-3$
5	12	19	26	33	40	⋯		$mn-2$
6	13	20	27	34	41	⋯		$mn-1$
7	14	21	28	35	42	⋯		mn

m rows

← $n-k$ parity bits → ← k data bits →

FIGURE 8.1–20
A block interleaver for coded data.

class of convolutional codes that is described in Section 8.2. Convolutional interleaver structures have been described by Ramsey (1970) and Forney (1971).

8.1.10 Serial and Parallel Concatenated Block Codes

An interleaver may be used in conjunction with a concatenated code to construct a code with extremely long code words. In a serially concatenated block code (SCBC) the interleaver is inserted between the two encoders as shown in Figure 8.1–21. Both codes are linear systematic binary codes. The outer code is a (p, k) code and the inner code is an (n, p) code. The block interleaver length is selected as $N = mp$, where m corresponds to the number of outer code words. The encoding and interleaving is performed as follows: mk information bits are encoded by the outer encoder to produce mp coded bits. These $N = mp$ coded bits are read out of the interleaver in different order according to the permutation algorithm of the interleaver. The mp bits at the output of the interleaver are fed to the inner encoder in blocks of length p. Therefore, a block of mk information bits is encoded by the SCBC into a block of mn bits. The resulting code rate is $R_c^s = k/n$, which is the product of the code rates of the inner and outer encoders. However, the block length of the SCBC is nm bits, which can be significantly larger than the block length of the conventional serial concatenation of the block codes without the use of the interleaver.

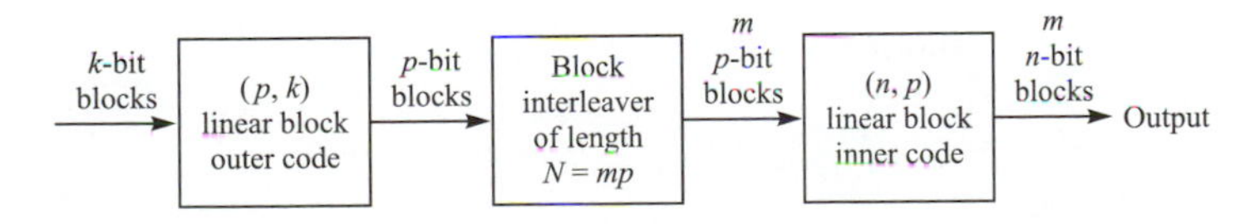

FIGURE 8.1–21
Serial concatenated block code (SCBC) with interleaver.

The block interleaver is usually implemented as a pseudorandom interleaver, i.e., an interleaver that pseudorandomly permutes the block of N bits. For purposes of analyzing the performance of SCBC, such an interleaver may be modeled as a *uniform interleaver*, which is defined as a device that maps a given input word of weight w into all distinct $\binom{N}{w}$ permutations with equal probability $1/\binom{N}{w}$.

By use of interleaving, parallel concatenated block codes (PCBC) can be constructed in a similar manner. Figure 8.1–22 illustrates the basic configuration of such an encoder based on two constituent binary codes. The constituent codes may be identical or different. The two encoders are systematic, binary linear encoders, denoted as (n_1, k) and (n_2, k). The pseudorandom block interleaver has length $N = k$, and, thus, the overall PCBC has block length $n_1 + n_2 - k$ and rate $k/(n_1 + n_2 - k)$, since the information bits are transmitted only once. More generally, we may encode mk bits $(m > 1)$ and, thus, use an interleaver of length $N = mk$. The design of interleavers for parallel concatenated codes is considered in a paper by Daneshgaran and Mondin (1999).

The use of an interleaver in the construction of SCBC and PCBC results in code words that are both large in block length and relatively sparse. Decoding of these types of codes is generally performed iteratively, using soft-in/soft-out maximum a posteriori probability (MAP) algorithms. An iterative MAP decoding algorithm for serially concatenated codes is described in the paper by Benedetto et al. (1998). Iterative MAP decoding algorithms for parallel concatenated codes have been described in a number of papers, including Berrou et al. (1993), Benedetto and Montorsi (1996), Hagenauer et al. (1996) and in the book by Heegard and Wicker (1999).

The combination of code concatenation with interleaving and iterative MAP decoding results in performance very close to the Shannon limit at moderate error rates, such as 10^{-4} to 10^{-5} (low SNR region). Concatenated codes are treated again in the next section, which deals with convolutional codes.

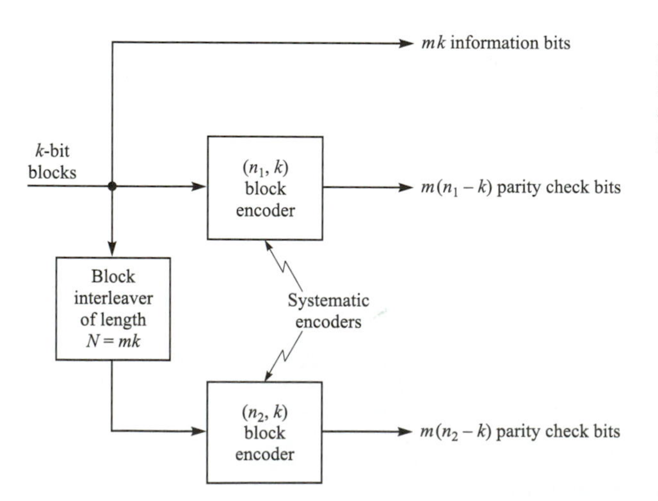

FIGURE 8.1–22
Parallel concatenated block code (PCBC) with interleaver.

8.2 CONVOLUTIONAL CODES

A convolutional code is generated by passing the information sequence to be transmitted through a linear finite-state shift register. In general, the shift register consists of K (k-bit) stages and n linear algebraic function generators, as shown in Figure 8.2–1. The input data to the encoder, which is assumed to be binary, is shifted into and along the shift register k bits at a time. The number of output bits for each k-bit input sequence is n bits. Consequently, the code rate is defined as $R_c = k/n$, consistent with the definition of the code rate for a block code. The parameter K is called the *constraint length* of the convolution code.†

One method for describing a convolutional code is to give its generator matrix, just as we did for block codes. In general, the generator matrix for a convolutional code is semi-infinite since the input sequence is semi-infinite in length. As an alternative to specifying the generator matrix, we shall use a functionally equivalent representation in which we specify a set of n vectors, one vector for each of the n modulo-2 adders. Each vector has Kk dimensions and contains the connections of the encoder to that modulo-2 adder. A 1 in the ith position of the vector indicates that the corresponding stage in the shift register is connected to the modulo-2 adder, and a 0 in a given position indicates that no connection exists between that stage and the modulo-2 adder.

To be specific, let us consider the binary convolutional encoder with constraint length $K = 3$, $k = 1$, and $n = 3$, which is shown in Figure 8.2–2. Initially, the shift register is assumed to be in the all-zero state. Suppose the first input bit

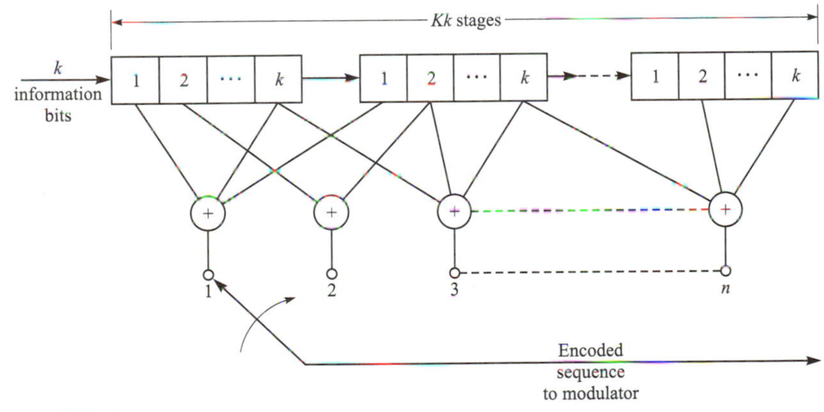

FIGURE 8.2–1
Convolutional encoder

†In many cases, the constraint length of the code is given in bits rather than k-bit bytes. Hence the shift register may be called an *L-stage shift register*, where $L = Kk$. Furthermore, L may not be a multiple of k, in general.

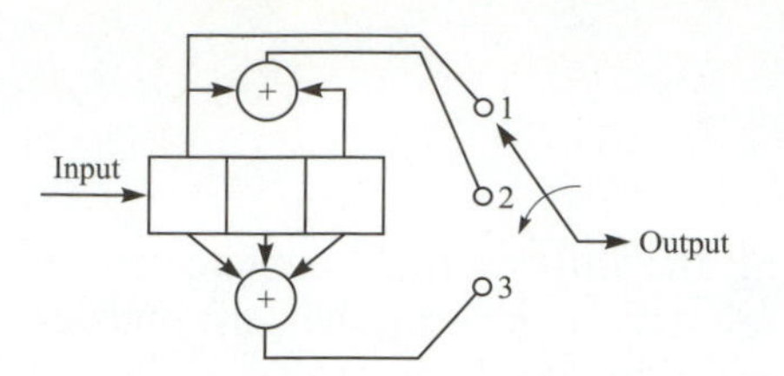

FIGURE 8.2–2
$K = 3, k = 1, n = 3$ convolutional encoder.

is a 1. Then the output sequence of 3 bits is 111. Suppose the second bit is a 0. The output sequence will then be 001. If the third bit is a 1, the output will be 100, and so on. Now, suppose we number the outputs of the function generators that generate each 3-bit output sequence as 1, 2, and 3, from top to bottom, and similarly number each corresponding function generator. Then, since only the first stage is connected to the first function generator (no modulo-2 adder is needed), the generator is

$$\mathbf{g}_1 = [100]$$

The second function generator is connected to stages 1 and 3. Hence

$$\mathbf{g}_2 = [101]$$

Finally,

$$\mathbf{g}_3 = [111]$$

The generators for this code are more conveniently given in octal form as (4, 5, 7). We conclude that, when $k = 1$, we require n generators, each of dimension K to specify the encoder.

For a rate k/n binary convolutional code with $k > 1$ and constraint length K, the n generators are Kk-dimensional vectors, as stated above. The following example illustrates the case in which $k = 2$ and $n = 3$.

EXAMPLE 8.2–1. Consider the rate 2/3 convolutional encoder illustrated in Figure 8.2–3. In this encoder, 2 bits at a time are shifted into it and 3 output bits are generated. The generators are

$$\mathbf{g}_1 = [1011], \qquad \mathbf{g}_2 = [1101], \qquad \mathbf{g}_3 = [1010]$$

In octal form, these generators are (13, 15, 12).

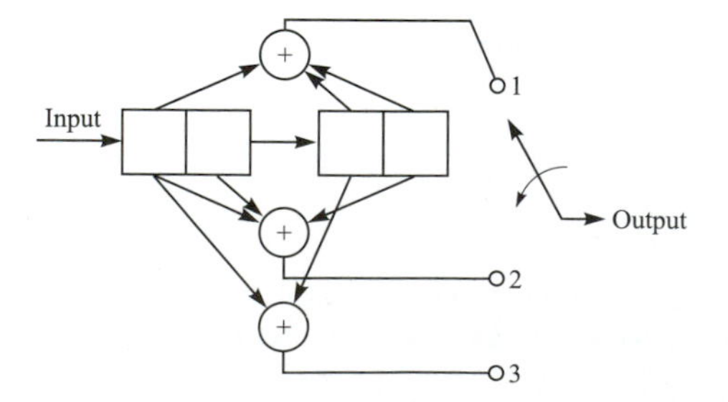

FIGURE 8.2–3
$K = 2, k = 2, n = 3$ convolutional encoder.

There are three alternative methods that are often used to describe a convolutional code. These are the tree diagram, the trellis diagram, and the state diagram. For example, the tree diagram for the convolutional encoder shown in Figure 8.2–2 is illustrated in Figure 8.2–4. Assuming that the encoder is in the all-zero state initially, the diagram shows that, if the first input bit is a 0, the output sequence is 000 and, if the first bit is a 1, the output sequence is 111. Now, if the first input bit is a 1 and the second bit is a 0, the second set of 3 output bits is 001. Continuing through the tree, we see that if the third bit is a 0, then the output is 011, while if the third bit is a 1, then the output is 100. Given that a particular sequence has taken us to a particular node in the tree, the branching rule is to follow the upper branch if the next input bit is a 0 and the lower branch if the bit is a 1. Thus, we trace a particular path through the tree that is determined by the input sequence.

Close observation of the tree that is generated by the convolutional encoder shown in Figure 8.2–2 reveals that the structure repeats itself after the third stage. This behavior is consistent with the fact that the constraint length $K = 3$. That is, the 3-bit output sequence at each stage is determined by the input bit and the two previous input bits, i.e., the 2 bits contained in the first two stages of the shift register. The bit in the last stage of the shift register is shifted out at the right and does not affect the output. Thus we may say that the 3-bit output sequence for each input bit is determined by the input bit and the four possible states of the shift register, denoted as $a = 00$, $b = 01$, $c = 10$, $d = 11$. If we label each node in the tree to correspond to the four possible states in the shift register, we find that at the third stage there are two nodes with the label a, two with the label b, two with label c, and two with the label d. Now we observe that all branches emanating from two nodes having the same label (same state) are identical in the sense that they generate identical output sequences. This means that the two nodes having the same label can be merged. If we do this

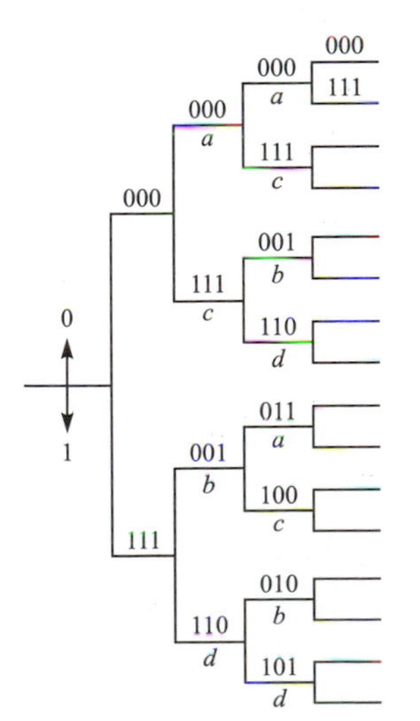

FIGURE 8.2–4
Tree diagram for rate 1/3, $K = 3$ convolutional code.

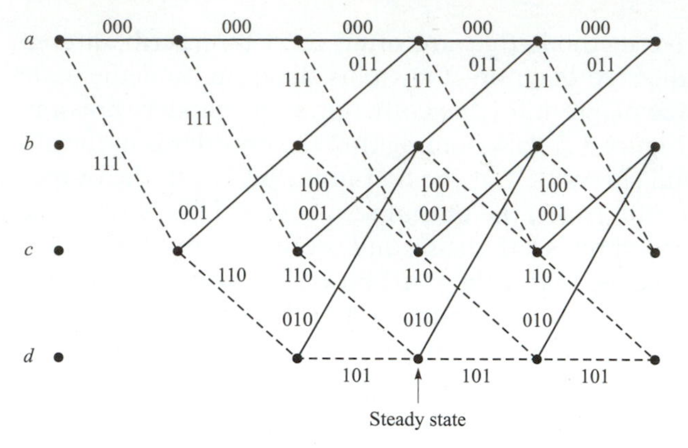

FIGURE 8.2–5
Trellis diagram for rate 1/3, $K = 3$ convolutional code.

to the tree shown in Figure 8.2–4, we obtain another diagram, which is more compact, namely, a *trellis*. For example, the trellis diagram for the convolutional encoder of Figure 8.2–2 is shown in Figure 8.2–5. In drawing this diagram, we use the convention that a solid line denotes the output generated by the input bit 0 and a dotted line the output generated by the input bit 1. In the example being considered, we observe that, after the initial transient, the trellis contains four nodes at each stage, corresponding to the four states of the shift register, a, b, c, and d. After the second stage, each node in the trellis has two incoming paths and two outgoing paths. Of the two outgoing paths, one corresponds to the input bit 0 and the other to the path followed if the input bit is a 1.

Since the output of the encoder is determined by the input and the state of the encoder, an even more compact diagram than the trellis is the state diagram. The state diagram is simply a graph of the possible states of the encoder and the possible transitions from one state to another. For example the state diagram for the encoder shown in Figure 8.2–2 is illustrated in Figure 8.2–6. This diagram shows that the possible transitions are

$$a \xrightarrow{0} a,\quad a \xrightarrow{1} c,\quad b \xrightarrow{0} a,\quad b \xrightarrow{1} c,\quad c \xrightarrow{0} b,\quad c \xrightarrow{1} d,\quad d \xrightarrow{0} b,\quad d \xrightarrow{1} d,$$

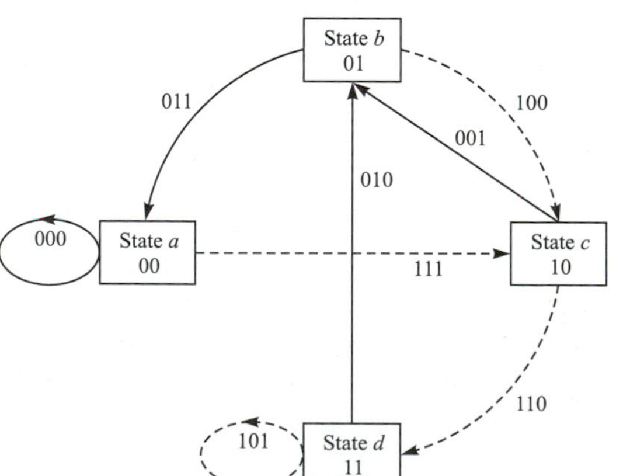

FIGURE 8.2–6
State diagram for rate 1/3, $K = 3$ convolutional code.

where $\alpha \xrightarrow{1} \beta$ denotes the transition from state α to β when the input bit is a 1. The 3 bits shown next to each branch in the state diagram represent the output bits. A dotted line in the graph indicates that the input bit is a 1, while the solid line indicates that the input bit is a 0.

EXAMPLE 8.2-2. Let us consider the $k = 2$, rate 2/3 convolutional code described in Example 8.2-1 and shown in Figure 8.2-3. The first two input bits may be 00, 01, 10, or 11. The corresponding output bits are 000, 010, 111, 101. When the next pair of input bits enters the encoder, the first pair is shifted to the second stage. The corresponding output bits depend on the pair of bits shifted into the second stage and the new pair of input bits. Hence, the tree diagram for this code, shown in Figure 8.2-7, has four branches per node, corresponding to the four possible pairs of input symbols. Since the constraint length of the code is $K = 2$, the tree begins to repeat after the second stage. As illustrated in Figure 8.2-7, all the branches emanating from nodes labeled a (state a) yield identical outputs. By merging the nodes having identical labels, we obtain the trellis, which is shown in Figure 8.2-8. Finally, the state diagram for this code is shown in Figure 8.2-9.

To generalize, we state that a rate k/n, constraint length K, convolutional code is characterized by 2^k branches emanating from each node of the tree

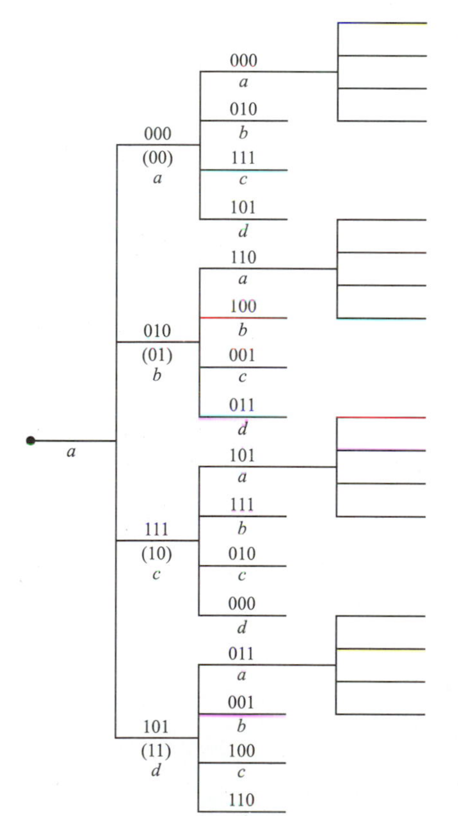

FIGURE 8.2-7
Tree diagram for $K = 2, k = 2, n = 3$ convolutional code.

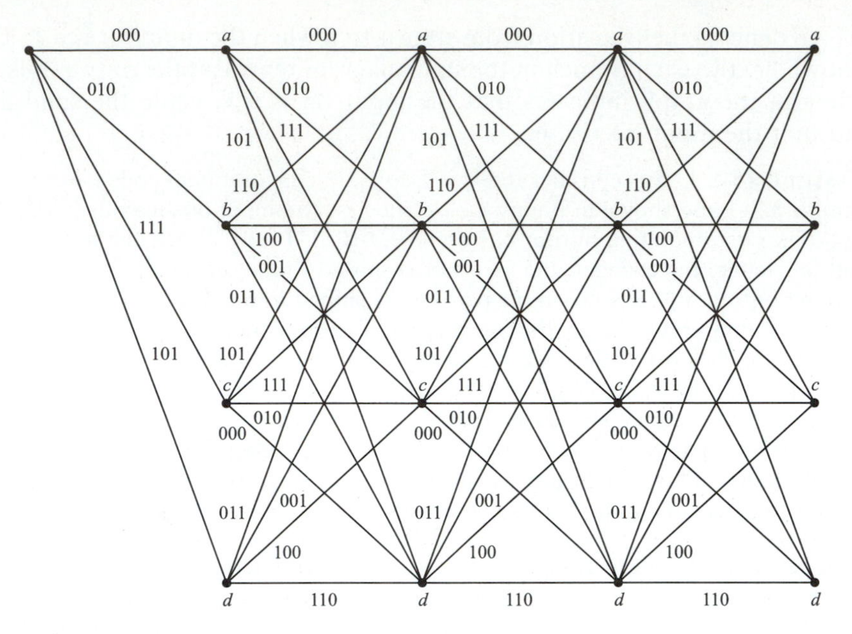

FIGURE 8.2–8
Trellis diagram for $K = 2, k = 2, n = 3$ convolutional code.

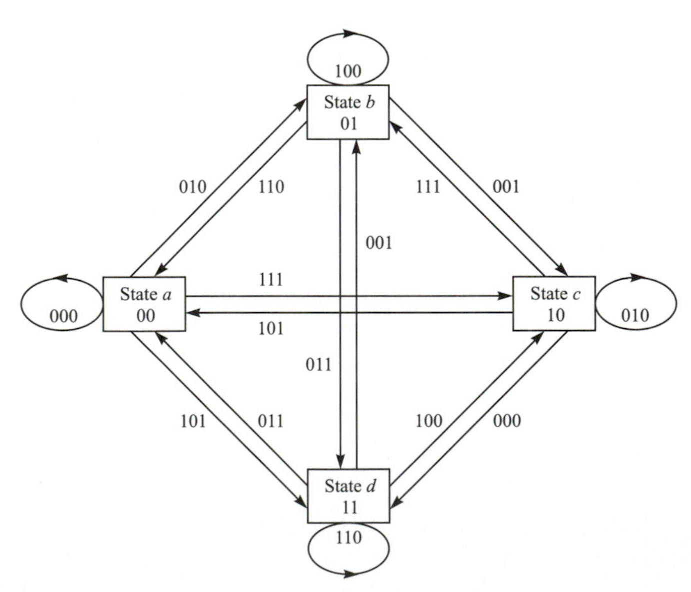

FIGURE 8.2–9
State diagram for $K = 2, k = 2, n = 3$ convolutional code.

diagram. The trellis and the state diagrams each have $2^{k(K-1)}$ possible states. There are 2^k branches entering each state and 2^k branches leaving each state (in the trellis and tree, this is true after the initial transient).

The three types of diagrams described above are also used to represent nonbinary convolutional codes. When the number of symbols in the code alphabet is $q = 2^k$, $k > 1$, the resulting nonbinary code may also be represented as an equivalent binary code. The following example considers a convolutional code of this type.

EXAMPLE 8.2–3. Let us consider the convolutional code generated by the encoder shown in Figure 8.2–10. This code may be described as a binary convolutional code with parameters $K = 2$, $k = 2$, $n = 4$, $R_c = 1/2$, and having the generators

$$\mathbf{g}_1 = [1010], \qquad \mathbf{g}_2 = [0101], \qquad \mathbf{g}_3 = [1110], \qquad \mathbf{g}_4 = [1001]$$

Except for the difference in rate, this code is similar in form to the rate 2/3, $k = 2$ convolutional code considered in Example 8.2–1.

Alternatively, the code generated by the encoder in Figure 8.2–10 may be described as a nonbinary ($q = 4$) code with one quaternary symbol as an input and two quaternary symbols as an output. In fact, if the output of the encoder is treated by the modulator and demodulator as q-ary ($q = 4$) symbols that are transmitted over the channel by means of some M-ary ($M = 4$) modulation technique, the code is appropriately viewed as nonbinary.

In any case, the tree, the trellis, and the stage diagrams are independent of how we view the code. That is, this particular code is characterized by a tree with four branches emanating from each node, or a trellis with four possible states and four branches entering and leaving each state, or, equivalently, by a state diagram having the same parameters as the trellis.

8.2.1 The Transfer Function of a Convolutional Code

The distance properties and the error rate performance of a convolutional code can be obtained from its state diagram. Since a convolutional code is linear, the set of Hamming distances of the code sequences generated up to some stage in the tree, from the all-zero code sequence, is the same as the set of distances of the

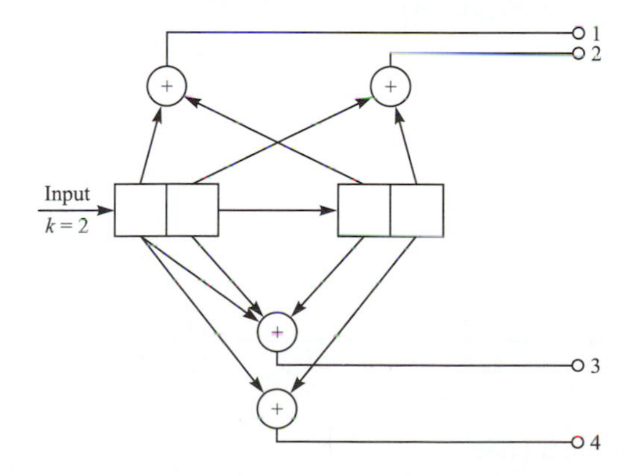

FIGURE 8.2–10
$K = 2, k = 2, n = 4$ convolutional encoder.

code sequences with respect to any other code sequence. Consequently, we assume without loss of generality that the all-zero code sequence is the input to the encoder.

The state diagram shown in Figure 8.2–6 will be used to demonstrate the method for obtaining the distance properties of a convolutional code. First, we label the branches of the state diagram as either $D^0 = 1$, D^1, D^2, or D^3, where the exponent of D denotes the Hamming distance between the sequence of output bits corresponding to each branch and the sequence of output bits corresponding to the all-zero branch. The self-loop at node a can be eliminated, since it contributes nothing to the distance properties of a code sequence relative to the all-zero code sequence. Furthermore, node a is split into two nodes, one of which represents the input and the other the output of the state diagram. Figure 8.2–11 illustrates the resulting diagram. We use this diagram, which now consists of five nodes because node a was split into two, to write the four state equations

$$\begin{aligned} X_c &= D^3 X_a + D X_b \\ X_b &= D X_c + D X_d \\ X_d &= D^2 X_c + D^2 X_d \\ X_e &= D^2 X_b \end{aligned} \tag{8.2–1}$$

The transfer function for the code is defined as $T(D) = X_e / X_a$. By solving the state equations given above, we obtain

$$\begin{aligned} T(D) &= \frac{D^6}{1 - 2D^2} \\ &= D^6 + 2D^8 + 4D^{10} + 8D^{12} + \cdots \\ &= \sum_{d=6}^{\infty} a_d D^d \end{aligned} \tag{8.2–2}$$

where, by definition,

$$a_d = \begin{cases} 2^{(d-6)/2} & \text{(even d)} \\ 0 & \text{(odd d)} \end{cases} \tag{8.2–3}$$

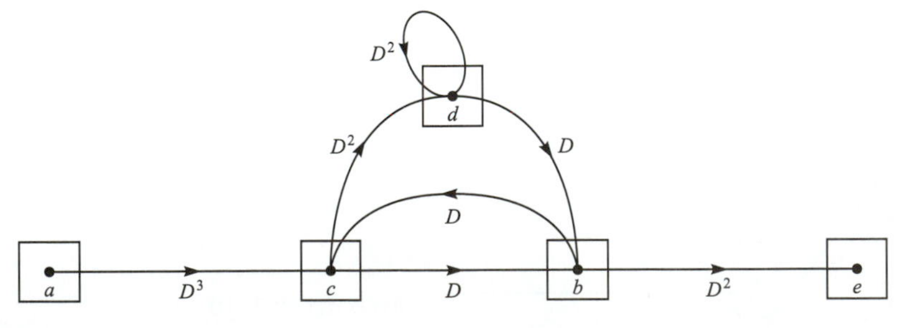

FIGURE 8.2–11
State diagram for rate 1/3, $K = 3$ convolutional code.

The transfer function for this code indicates that there is a single path of Hamming distance $d = 6$ from the all-zero path that merges with the all-zero path at a given node. From the state diagram shown in Figure 8.2–6 or the trellis diagram shown in Figure 8.2–5, it is observed that the $d = 6$ path is *acbe*. There is no other path from node a to node e having a distance $d = 6$. The second term in Equation 8.2–2 indicates that there are two paths from node a to node e having a distance $d = 8$. Again, from the state diagram or the trellis, we observe that these paths are *acdbe* and *acbcbe*. The third term in Equation 8.2–2 indicates that there are four paths of distance $d = 10$, and so forth. Thus the transfer function gives us the distance properties of the convolutional code. The minimum distance of the code is called the *minimum free distance* and denoted by d_{free}. In our example, $d_{\text{free}} = 6$.

The transfer function can be used to provide more detailed information than just the distance of the various paths. Suppose we introduce a factor N into all branch transitions caused by the input bit 1. Thus, as each branch is traversed, the cumulative exponent on N increases by one only if that branch transition is due to an input bit 1. Furthermore, we introduce a factor of J into each branch of the state diagram so that the exponent of J will serve as a counting variable to indicate the number of branches in any given path from node a to node e. For the rate 1/3 convolutional code in our example, the state diagram that incorporates the additional factors of J and N is shown in Figure 8.2–12.

The state equations for the state diagram shown in Figure 8.2–12 are

$$\begin{aligned} X_c &= JND^3 X_a + JNDX_b \\ X_b &= JDX_c + JDX_d \\ X_d &= JND^2 X_c + JND^2 X_d \\ X_e &= JD^2 X_b \end{aligned} \tag{8.2–4}$$

Upon solving these equations for the ratio X_e/X_a, we obtain the transfer function

$$\begin{aligned} T(D, N, J) &= \frac{J^3 ND^6}{1 - JND^2(1 + J)} \\ &= J^3 ND^6 + J^4 N^2 D^8 + J^5 N^2 D^8 + J^5 N^3 D^{10} \\ &\quad + 2J^6 N^3 D^{10} + J^7 N^3 D^{10} + \cdots \end{aligned} \tag{8.2–5}$$

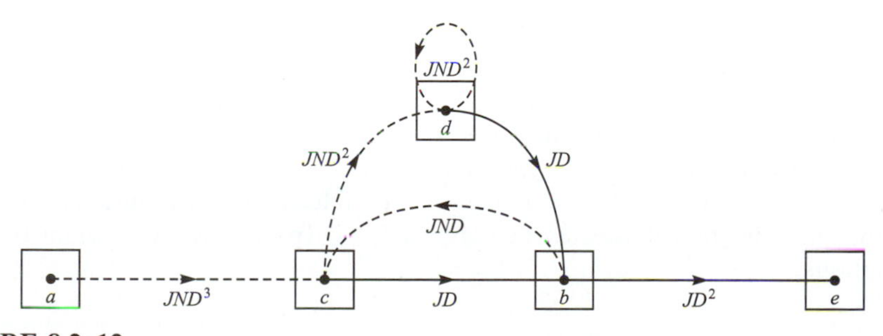

FIGURE 8.2–12
State diagram for rate 1/3, $K = 3$ convolutional code.

This form for the transfer functions gives the properties of all the paths in the convolutional code. That is, the first term in the expansion of $T(D, N, J)$ indicates that the distance $d = 6$ path is of length 3 and of the three information bits, one is a 1. The second and third terms in the expansion of $T(D, N, J)$ indicate that of the two $d = 8$ terms, one is of length 4 and the second has length 5. Two of the four information bits in the path having length 4 and two of the five information bits in the path having length 5 are 1s. Thus, the exponent of the factor J indicates the length of the path that merges with the all-zero path for the first time, the exponent of the factor N indicates the number of 1s in the information sequence for that path, and the exponent of D indicates the distance of the sequence of encoded bits for that path from the all-zero sequence.

The factor J is particularly important if we are transmitting a sequence of finite duration, say m bits. In such a case, the convolutional code is truncated after m nodes or m branches. This implies that the transfer function for the truncated code is obtained by truncating $T(D, N, J)$ at the term J^m. On the other hand, if we are transmitting an extremely long sequence, i.e., essentially an infinite-length sequence, we may wish to suppress the dependence of $T(D, N, J)$ on the parameter J. This is easily accomplished by setting $J = 1$. Hence, for the example given above, we have

$$\begin{aligned} T(D, N, 1) = T(D, N) &= \frac{ND^6}{1 - 2ND^2} \\ &= ND^6 + 2N^2D^8 + 4N^3D^{10} + \cdots \\ &= \sum_{d=6}^{\infty} a_d N^{(d-4)/2} D^d \end{aligned} \tag{8.2–6}$$

where the coefficients $\{a_d\}$ are defined by Equation 8.2–3.

The procedure outlined above for determining the transfer function of a binary convolutional code is easily extended to nonbinary codes. In the following example, we determine the transfer function of the nonbinary convolutional code previously introduced in Example 8.2–3.

EXAMPLE 8.2–4. The convolutional code shown in Figure 8.2–10 has the parameters $K = 2$, $k = 2$, $n = 4$. In this example, we have a choice of how we label distances and count errors, depending on whether we treat the code as binary or nonbinary. Suppose we treat the code as nonbinary. Thus, the input to the encoder and the output are treated as quaternary symbols. In particular, if we treat the input and output as quaternary symbols 00, 01, 10, and 11, the distance measured in symbols between the sequences 0111 and 0000 is 2. Furthermore, suppose that an input symbol 00 is decoded as the symbol 11; then we have made one symbol error. This convention applied to the convolutional code shown in Figure 8.2–10 results in the state diagram illustrated in Figure 8.2–13, from which we obtain the state equations

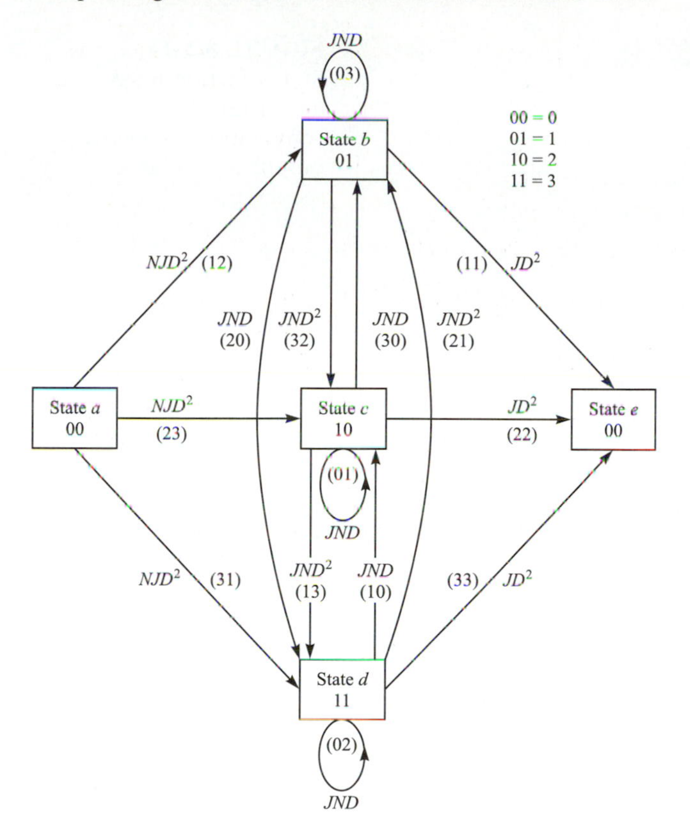

FIGURE 8.2–13
State diagram for $K = 2$, $k = 2$, rate 1/2 nonbinary code.

$$
\begin{aligned}
X_b &= NJD^2 X_a + NJDX_b + NJDX_c + NJD^2 X_d \\
X_c &= NJD^2 X_a + NJD^2 X_b + NJDX_c + NJDX_d \\
X_d &= NJD^2 X_a + NJDX_b + NJD^2 X_c + NJDX_d \\
X_e &= JD^2 (X_b + X_c + X_d)
\end{aligned}
\tag{8.2–7}
$$

Solution of these equations leads to the transfer function

$$
T(D, N, J) = \frac{3NJ^2 D^4}{1 - 2NJD - NJD^2} \tag{8.2–8}
$$

This expression for the transfer function is particularly appropriate when the quaternary symbols at the output of the encoder are mapped into a corresponding set of quaternary waveforms $s_m(t)$, $m = 1, 2, 3, 4$, e.g., four orthogonal waveforms. Thus, there is a one-to-one correspondence between code symbols and signal waveforms.

Alternatively, for example, the output of the encoder may be transmitted as a sequence of binary digits by means of binary PSK. In such a case, it is appropriate to measure distance in terms of bits. When this convention is employed, the state diagram is labeled as shown in Figure 8.2–14. Solution of the state equations obtained from this state diagram yields a transfer function that is different from the one given in Equation 8.2–8.

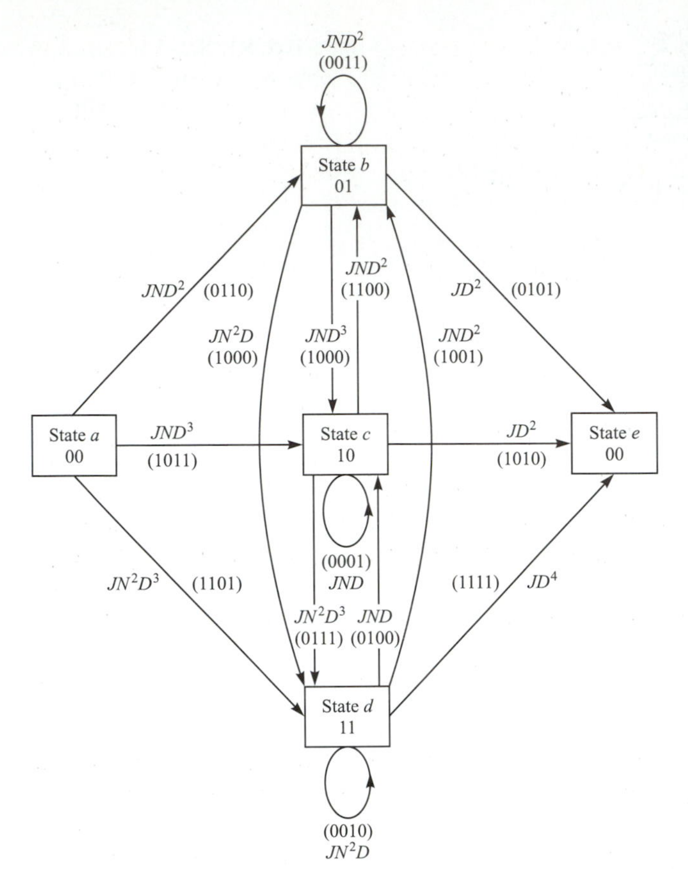

FIGURE 8.2–14
State diagram for $K = 2$, $k = 2$, rate 1/2 convolutional code with output treated as a binary sequence.

Some convolutional codes exhibit a characteristic behavior that is called *catastrophic error propagation*. When a code that has this characteristic is used on a binary symmetric channel, it is possible for a finite number of channel errors to cause an infinite number of decoding errors. Such a code can be identified from its state diagram. It will contain a zero-distance path (a path with multiplier $D^0 = 1$) from some nonzero state back to the same state. This means that one can loop around this zero-distance path an infinite number of times without increasing the distance relative to the all-zero path. But, if this self-loop corresponds to the transmission of a 1, the decoder will make an infinite number of errors. Since such codes are easily recognized, they are easily avoided in practice.

8.2.2 Optimum Decoding of Convolutional Codes—The Viterbi Algorithm

In the decoding of a block code for a memoryless channel, we computed the distances (Hamming distance for hard-decision decoding and Euclidean distance for soft-decision decoding) between the received code word and the 2^k possible transmitted code words. Then we selected the code word that was closest in distance to the received code word. This decision rule, which requires the computation of 2^k metrics, is optimum in the sense that it results in a minimum probability of error for the binary symmetric channel with $p < \frac{1}{2}$ and the additive white Gaussian noise channel.

Unlike a block code, which has a fixed length n, a convolutional encoder is basically a finite-state machine. Hence the optimum decoder is a maximum-likelihood sequence estimator (MLSE) of the type described in Section 5.1.4 for signals with memory, such as NRZI and CPM. Therefore, optimum decoding of a convolutional code involves a search through the trellis for the most probable sequence. Depending on whether the detector following the demodulator performs hard or soft decisions, the corresponding metric in the trellis search may be either a Hamming metric or a Euclidean metric, respectively. We elaborate below, using the trellis in Figure 8.2–5 for the convolutional code shown in Figure 8.2–2.

Consider the two paths in the trellis that begin at the initial state a and remerge at state a after three state transitions (three branches), corresponding to the two information sequences 000 and 100 and the transmitted sequences 000 000 000 and 111 001 011, respectively. We denote the transmitted bits by $\{c_{jm}, j = 1, 2, 3; m = 1, 2, 3\}$, where the index j indicates the jth branch and the index m the mth bit in that branch. Correspondingly, we define $\{r_{jm}, j = 1, 2, 3; m = 1, 2, 3\}$ as the output of the demodulator. If the decoder performs hard-decision decoding, the detector output for each transmitted bit is either 0 or 1. On the other hand, if soft-decision decoding is employed and the coded sequence is transmitted by binary coherent PSK, the input to the decoder is

$$r_{jm} = \sqrt{\mathcal{E}_c}(2c_{jm} - 1) + n_{jm} \tag{8.2–9}$$

where n_{jm} represents the additive noise and $\mathcal{E}_c$ is the transmitted signal energy for each code bit.

A metric is defined for the jth branch of the ith path through the trellis as the logarithm of the joint probability of the sequence $\{r_{jm}, m = 1, 2, 3\}$ conditioned on the transmitted sequence $\{c_{jm}^{(i)}, m = 1, 2, 3\}$ for the ith path. That is,

$$\mu_j^{(i)} = \log P(\mathbf{Y}_j | \mathbf{C}_j^{(i)}), \qquad j = 1, 2, 3, \ldots \tag{8.2–10}$$

Furthermore, a metric for the ith path consisting of B branches through the trellis is defined as

$$PM^{(i)} = \sum_{j=1}^{B} \mu_j^{(i)} \tag{8.2–11}$$

The criteron for deciding between two paths through the trellis is to select the one having the larger metric. This rule maximizes the probability of a correct decision or, equivalently, it minimizes the probability of error for the sequence of information bits. For example, suppose that hard-decision decoding is performed by the demodulator, yielding the received sequence {101 000 100}. Let $i = 0$ denote the three-branch all-zero path and $i = 1$ the second three-branch path that begins in the initial state a and remerges with the all-zero path at state a after three transitions. The metrics for these two paths are

$$\begin{aligned} PM^{(0)} &= 6\log(1-p) + 3\log p \\ PM^{(1)} &= 4\log(1-p) + 5\log p \end{aligned} \tag{8.2–12}$$

where p is the probability of a bit error. Assuming that $p < \frac{1}{2}$, we find that the metric $PM^{(0)}$ is larger than the metric $PM^{(1)}$. This result is consistent with the observation that the all-zero path is at Hamming distance $d = 3$ from the received sequence, while the $i = 1$ path is at Hamming distance $d = 5$ from the received path. Thus, the Hamming distance is an equivalent metric for hard-decision decoding.

Similarly, suppose that soft-decision decoding is employed and the channel adds white Gaussian noise to the signal. Then the demodulator output is described statistically by the probability density function

$$p(r_{jm}|c_{jm}^{(i)}) = \frac{1}{\sqrt{2\pi}\sigma} \exp\left\{-\frac{[r_{jm} - \sqrt{\mathcal{E}_c}(2c_{jm}^{(i)} - 1)]^2}{2\sigma^2}\right\} \tag{8.2–13}$$

where $\sigma^2 = \frac{1}{2}N_0$ is the variance of the additive Gaussian noise. If we neglect the terms that are common to all branch metrics, the branch metric for the jth branch of the ith path may be expressed as

$$\mu_j^{(i)} = \sum_{m=1}^{n} r_{jm}(2c_{jm}^{(i)} - 1) \tag{8.2–14}$$

where, in our example, $n = 3$. Thus the correlation metrics for the two paths under consideration are

$$\begin{aligned} CM^{(0)} &= \sum_{j=1}^{3}\sum_{m=1}^{3} r_{jm}(2c_{jm}^{(0)} - 1) \\ CM^{(1)} &= \sum_{j=1}^{3}\sum_{m=1}^{3} r_{jm}(2c_{jm}^{(1)} - 1) \end{aligned} \tag{8.2–15}$$

Having defined the branch metrics and path metrics computed by the decoder, we now consider the use of the Viterbi algorithm for optimum decoding of the convolutionally encoded information sequence. We consider the two paths described above, which merge at state a after three transitions. Note that any particular path through the trellis that stems from this node will add identical terms to the path metrics $CM^{(0)}$ and $CM^{(1)}$. Consequently, if $CM^{(0)} > CM^{(1)}$ at the merged node a after three transitions, $CM^{(0)}$ will continue to be larger than $CM^{(1)}$ for any path that stems from node a. This means that the path corresponding to $CM^{(1)}$ can be discarded from further consideration. The path corresponding to the metric $CM^{(0)}$ is the *survivor*. Similarly, one of the two paths that merge at state b can be eliminated on the basis of the two corresponding metrics. This procedure is repeated at state c and state d. As a result, after the first three transitions, there are four surviving paths, one terminating at each state, and a corresponding metric for each survivor. This procedure is repeated at each stage of the trellis as new signals are received in subsequent time intervals.

In general, when a binary convolutional code with $k = 1$ and constraint length K is decoded by means of the Viterbi algorithm, there are 2^{K-1} states. Hence, there are 2^{K-1} surviving paths at each stage and 2^{K-1} metrics, one for each surviving path. Furthermore, a binary convolutional code in which k bits at a time are shifted into an encoder that consists of K (k-bit) shift-register stages generates a trellis that has $2^{k(K-1)}$ states. Consequently, the decoding of such a code by means of the Viterbi algorithm requires keeping track of $2^{k(K-1)}$ surviving paths and $2^{k(K-1)}$ metrics. At each stage of the trellis, there are 2^k paths that merge at each node. Since each path that converges at a common node requires the computation of a metric, there are 2^k metrics computed for each node. Of the 2^k paths that merge at each node, only one survives, and this is the most-probable (minimum-distance) path. Thus, the number of computations in decoding performed at each stage increases exponentially with k and K. The exponential increase in computational burden limits the use of the Viterbi algorithm to relatively small values of K and k.

The decoding delay in decoding a long information sequence that has been convolutionally encoded is usually too long for most practical applications. Moreover, the memory required to store the entire length of surviving sequences is large and expensive. As indicated in Section 5.1.4, a solution to this problem is to modify the Viterbi algorithm in a way which results in a fixed decoding delay without significantly affecting the optimal performance of the algorithm. Recall that the modification is to retain at any given time t only the most recent δ decoded information bits (symbols) in each surviving sequence. As each new information bit (symbol) is received, a final decision is made on the bit (symbol) received δ branches back in the trellis, by comparing the metrics in the surviving sequences and deciding in favor of the bit in the sequence having the largest metric. If δ is chosen sufficiently large, all surviving sequences will contain the identical decoded bit (symbol) δ branches back in time. That is, with high probability, all surviving sequences at time t stem from the same node at $t - \delta$. It has been found experimentally (computer simulation) that a delay $\delta \geqslant 5K$ results in a negligible degradation in the performance relative to the optimum Viterbi algorithm.

8.2.3 Probability of Error for Soft-Decision Decoding

The topic of this subsection is the error rate performance of the Viterbi algorithm on an additive white Gaussian noise channel with soft-decision decoding.

In deriving the probability of error for convolutional codes, the linearity property for this class of codes is employed to simplify the derivation. That is, we assume that the all-zero sequence is transmitted and we determine the probability of error in deciding in favor of another sequence. The coded binary digits for the jth branch of the convolutional code, denoted as $\{c_{jm}, m = 1, 2, \ldots, n\}$

and defined in Section 8.2.2, are assumed to be transmitted by binary PSK (or four-phase PSK) and demodulated coherently. The output of the demodulator, which is the input to the Viterbi decoder, is the sequence $\{r_{jm}, m = 1, 2, \ldots, n; j = 1, 2, \ldots\}$ where r_{jm} is defined in Equation 8.2–9.

The Viterbi soft-decision decoder forms the branch metrics defined by Equation 8.2–14 and from these computes the path metrics

$$CM^{(i)} = \sum_{j=1}^{B} \mu_j^{(i)} = \sum_{j=1}^{B} \sum_{m=1}^{n} r_{jm}(2c_{jm}^{(i)} - 1) \tag{8.2–16}$$

where i denotes any one of the competing paths at each node and B is the number of branches (information symbols) in a path. For example, the all-zero path, denoted as $i = 0$, has a path metric

$$\begin{aligned} CM^{(0)} &= \sum_{j=1}^{B} \sum_{m=1}^{n} (-\sqrt{\mathcal{E}_c} + n_{jm})(-1) \\ &= \sqrt{\mathcal{E}_c} Bn + \sum_{j=1}^{B} \sum_{m=1}^{n} n_{jm} \end{aligned} \tag{8.2–17}$$

Since the convolutional code does not necessarily have a fixed length, we derive its performance from the probability of error for sequences that merge with the all-zero sequence for the first time at a given node in the trellis. In particular, we define the first-event error probability as the probability that another path that merges with the all-zero path at node B has a metric that exceeds the metric of the all-zero path for the first time. Suppose the incorrect path, call it $i = 1$, that merges with the all-zero path differs from the all-zero path in d bits, i.e., there are d 1s in the path $i = 1$ and the rest are 0s. The probability of error in the pairwise comparison of the metrics $CM^{(0)}$ and $CM^{(1)}$ is

$$\begin{aligned} P_2(d) &= P(CM^{(1)} \geqslant CM^{(0)}) = P(CM^{(1)} - CM^{(0)} \geqslant 0) \\ &= P\left[2 \sum_{j=1}^{B} \sum_{m=1}^{n} r_{jm}(c_{jm}^{(1)} - c_{jm}^{(0)}) \geqslant 0\right] \end{aligned} \tag{8.2–18}$$

Since the coded bits in the two paths are identical except in the d positions, Equation 8.2–18 can be written in the simpler form

$$P_2(d) = P\left(\sum_{l=1}^{d} r_l' \geqslant 0\right) \tag{8.2–19}$$

where the index l runs over the set of d bits in which the two paths differ and the set $\{r_l'\}$ represents the input to the decoder for these d bits.

The $\{r_l'\}$ are independent and identically distributed Gaussian random variables with mean $-\sqrt{\mathcal{E}_c}$ and variance $\frac{1}{2}N_0$. Consequently the probability of error in the pairwise comparison of these two paths that differ in d bits is

$$
\begin{aligned}
P_2(d) &= Q\left(\sqrt{\frac{2\mathcal{E}_c}{N_0}d}\right) \\
&= Q(\sqrt{2\gamma_b R_c d})
\end{aligned}
\tag{8.2–20}
$$

where $\gamma_b = \mathcal{E}_b/N_0$ is the received SNR per bit and R_c is the code rate.

Although we have derived the first-event error probability for a path of distance d from the all-zero path, there are many possible paths with different distances that merge with the all-zero path at a given node B. In fact, the transfer function $T(D)$ provides a complete description of all the possible paths that merge with the all-zero path at node B and their distances. Thus we can sum the error probability in Equation 8.2–20 over all possible path distances. Upon performing this summation, we obtain an upper bound on the first-event error probability in the form

$$
\begin{aligned}
P_e &\leqslant \sum_{d=d_{\text{free}}}^{\infty} a_d P_2(d) \\
&\leqslant \sum_{d=d_{\text{free}}}^{\infty} a_d Q(\sqrt{2\gamma_b R_c d})
\end{aligned}
\tag{8.2–21}
$$

where a_d denotes the number of paths of distance d from the all-zero path that merge with the all-zero path for the first time.

There are two reasons why Equation 8.2–21 is an upper bound on the first-event error probability. One is that the events that result in the error probabilities $\{P_2(d)\}$ are not disjoint. This can be seen from observation of the trellis. Second, by summing over all possible $d \geqslant d_{\text{free}}$, we have implicitly assumed that the convolutional code has infinite length. If the code is truncated periodically after B nodes, the upper bound in Equation 8.2–21 can be improved by summing the error events for $d_{\text{free}} \leqslant d \leqslant B$. This refinement has some merit in determining the performance of short convolutional codes, but the effect on performance is negligible when B is large.

The upper bound in Equation 8.2–21 can be expressed in a slightly different form if the Q function is upper-bounded by an exponential. That is,

$$
Q(\sqrt{2\gamma_b R_c d}) \leqslant e^{-\gamma_b R_c d} = D^d|_{D=e^{-\gamma_b R_c}} \tag{8.2–22}
$$

If we use Equation 8.2–22 in Equation 8.2–21, the upper bound on the first-event error probability can be expressed as

$$
P_e < T(D)|_{D=e^{-\gamma_b R_c}} \tag{8.2–23}
$$

Although the first-event error probability provides a measure of the performance of a convolutional code, a more useful measure of performance is the bit error probability. This probability can be upper-bounded by the procedure used in bounding the first-event error probability. Specifically, we know that when an incorrect path is selected, the information bits in which the selected path differs from the correct path will be decoded incorrectly. We also know that the exponents in the factor N contained in the transfer function $T(D, N)$ indicate the number of information bit errors (number of 1s) in selecting an incorrect path that merges with the all-zero path at some node B. If we multiply the pairwise error probability $P_2(d)$ by the number of incorrectly decoded information bits for the incorrect path at the node where they merge, we obtain the bit error rate for that path. The average bit error probability is upper-bounded by multiplying each pairwise error probability $P_2(d)$ by the corresponding number of incorrectly decoded information bits, for each possible incorrect path that merges with the correct path at the Bth node, and summing over all d.

The appropriate multiplication factors corresponding to the number of information bit errors for each incorrectly selected path may be obtained by differentiating $T(D, N)$ with respect to N. In general, $T(D, N)$ can be expressed as

$$T(D, N) = \sum_{d=d_{\text{free}}}^{\infty} a_d D^d N^{f(d)} \tag{8.2–24}$$

where $f(d)$ denotes the exponent of N as a function of d. Taking the derivative of $T(D, N)$ with respect to N and setting $N = 1$, we obtain

$$\begin{aligned} \left.\frac{dT(D, N)}{dN}\right|_{N=1} &= \sum_{d=d_{\text{free}}}^{\infty} a_d f(d) D^d \\ &= \sum_{d=d_{\text{free}}}^{\infty} \beta_d D^d \end{aligned} \tag{8.2–25}$$

where $\beta_d = a_d f(d)$. Thus the bit error probability for $k = 1$ is upper-bounded by

$$\begin{aligned} P_b &< \sum_{d=d_{\text{free}}}^{\infty} \beta_d P_2(d) \\ &< \sum_{d=d_{\text{free}}}^{\infty} \beta_d Q(\sqrt{2\gamma_b R_c d}) \end{aligned} \tag{8.2–26}$$

If the Q function is upper-bounded by an exponential as indicated in Equation 8.2–22, then Equation 8.2–26 can be expressed in the simple form

$$\begin{aligned} P_b &< \left.\sum_{d=d_{\text{free}}}^{\infty} \beta_d D^d\right|_{D=e^{-\gamma_b R_c}} \\ &< \left.\frac{dT(D, N)}{dN}\right|_{N=1,\, D=e^{-\gamma_b R_c}} \end{aligned} \tag{8.2–27}$$

If $k > 1$, the equivalent bit error probability is obtained by dividing Equations 8.2–26 and 8.2–27 by k.

The expressions for the probability of error given above are based on the assumption that the code bits are transmitted by binary coherent PSK. The results also hold for four-phase coherent PSK, since this modulation/demodulation technique is equivalent to two independent (phase-quadrature) binary PSK systems. Other modulation and demodulation techniques, such as coherent and noncoherent binary FSK, can be accommodated by recomputing the pairwise error probability $P_2(d)$. That is, a change in the modulation and demodulation technique used to transmit the coded information sequence affects only the computation of $P_2(d)$. Otherwise, the derivation for P_b remains the same.

Although the above derivation of the error probability for Viterbi decoding of a convolutional code applies to binary convolutional codes, it is relatively easy to generalize it to nonbinary convolutional codes in which each nonbinary symbol is mapped into a distinct waveform. In particular, the coefficients $\{\beta_d\}$ in the expansion of the derivative of $T(D, N)$, given in Equation 8.2–25, represent the number of symbol errors in two paths separated in distance (measured in terms of symbols) by d symbols. Again, we denote the probability of error in a pairwise comparison of two paths that are separated in distance by d as $P_2(d)$. Then the symbol error probability, for a k-bit symbol, is upper-bounded by

$$P_M \leqslant \sum_{d=d_{\text{free}}}^{\infty} \beta_d P_2(d)$$

The symbol error probability can be converted into an equivalent bit error probability. For example, if 2^k orthogonal waveforms are used to transmit the k-bit symbols, the equivalent bit error probability is P_M multiplied by a factor $2^{k-1}/(2^k - 1)$, as shown in Chapter 5.

8.2.4 Probability of Error for Hard-Decision Decoding

We now consider the performance achieved by the Viterbi decoding algorithm on a binary symmetric channel. For hard-decision decoding of the convolutional code, the metrics in the Viterbi algorithm are the Hamming distances between the received sequence and the $2^{k(K-1)}$ surviving sequences at each node of the trellis.

As in our treatment of soft-decision decoding, we begin by determining the first-event error probability. The all-zero path is assumed to be transmitted. Suppose that the path being compared with the all-zero path at some node B has distance d from the all-zero path. If d is odd, the all-zero path will be correctly selected if the number of errors in the received sequence is less than $\frac{1}{2}(d+1)$; otherwise, the incorrect path will be selected. Consequently, the probability of selecting the incorrect path is

$$P_2(d) = \sum_{k=(d+1)/2}^{d} \binom{d}{k} p^k (1-p)^{d-k} \tag{8.2–28}$$

where p is the probability of a bit error for the binary symmetric channel. If d is even, the incorrect path is selected when the number of errors exceeds $\frac{1}{2}d$. If the number of errors equals $\frac{1}{2}d$, there is a tie between the metrics in the two paths, which may be resolved by randomly selecting one of the paths; thus, an error occurs half the time. Consequently, the probability of selecting the incorrect path is

$$P_2(d) = \sum_{k=d/2+1}^{d} \binom{d}{k} p^k (1-p)^{d-k} + \frac{1}{2}\binom{d}{\frac{1}{2}d} p^{d/2}(1-p)^{d/2} \tag{8.2–29}$$

As indicated in Section 8.2.3, there are many possible paths with different distances that merge with the all-zero path at a given node. Therefore, there is no simple exact expression for the first-event error probability. However, we can overbound this error probability by the sum of the pairwise error probabilities $P_2(d)$ over all possible paths that merge with the all-zero path at the given node. Thus, we obtain the union bound

$$P_e < \sum_{d=d_{\text{free}}}^{\infty} a_d P_2(d) \tag{8.2–30}$$

where the coefficients $\{a_d\}$ represent the number of paths corresponding to the set of distances $\{d\}$. These coefficients are the coefficients in the expansion of the transfer function $T(D)$ or $T(D, N)$.

Instead of using the expressions for $P_2(d)$ given in Equations 8.2–28 and 8.2–29, we can use the Chernoff upper bound

$$P_2(d) < [4p(1-p)]^{d/2} \tag{8.2–31}$$

which was given in Section 8.1.5. Use of this bound in Equation 8.2–30 yields a looser upper bound on the first-event error probability, in the form

$$\begin{aligned} P_e &< \sum_{d=d_{\text{free}}}^{\infty} a_d [4p(1-p)]^{d/2} \\ &< T(D)|_{D=\sqrt{4p(1-p)}} \end{aligned} \tag{8.2–32}$$

Let us now determine the probability of a bit error. As in the case of soft-decision decoding, we make use of the fact that the exponents in the factors of N that appear in the transfer function $T(D, N)$ indicate the number of nonzero information bits that are in error when an incorrect path is selected over the all-zero path. By differentiating $T(D, N)$ with respect to N and setting $N = 1$, the exponents of N become multiplication factors of the corresponding error-event probabilities $P_2(d)$. Thus, we obtain the expression for the upper bound on the bit error probability, in the form

$$P_b < \sum_{d=d_{\text{free}}}^{\infty} \beta_d P_2(d) \tag{8.2–33}$$

where the $\{\beta_d\}$ are the coefficients in the expansion of the derivative of $T(D, N)$, evaluated at $N = 1$. For $P_2(d)$, we may use either the expressions given in

Equations 8.2–28 and 8.2–29 or the upper bound in Equation 8.2–31. If the latter is used, the upper bound on P_b can be expressed as

$$P_b < \left.\frac{dT(D, N)}{dN}\right|_{N=1, D=\sqrt{4p(1-p)}} \tag{8.2–34}$$

When $k > 1$, the results given in Equations 8.2–33 and 8.2–34 for P_b should be divided by k.

A comparison of the error probability for the rate 1/3, $K = 3$ convolutional code with soft-decision decoding and hard-decision decoding is made in Figure 8.2–15. Note that the Chernoff upper-bound given by Equation 8.2–34 is less than 1 dB above the tighter upper bound given by Equation 8.2–33 in conjunction with Equations 8.2–28 and 8.2–29. The advantage of the Chernoff bound is its computational simplicity. In comparing the performance between soft-decision and hard-decision decoding, note that the difference obtained from the upper bounds is approximately 2.5 dB for $10^{-6} \leqslant P_b \leqslant 10^{-2}$.

Finally, we should mention that the ensemble average error rate performance of a convolutional code on a discrete memoryless channel, just as in the case of a block code, can be expressed in terms of the cutoff rate parameter R_0 as (for the derivation, see Viterbi and Omura, 1979)

$$\bar{P}_b < \frac{(q-1)q^{-KR_0/R_c}}{(1 - q^{-(R_0-R_c)/R_c})^2}, \qquad R_c \leqslant R_0 \tag{8.2–35}$$

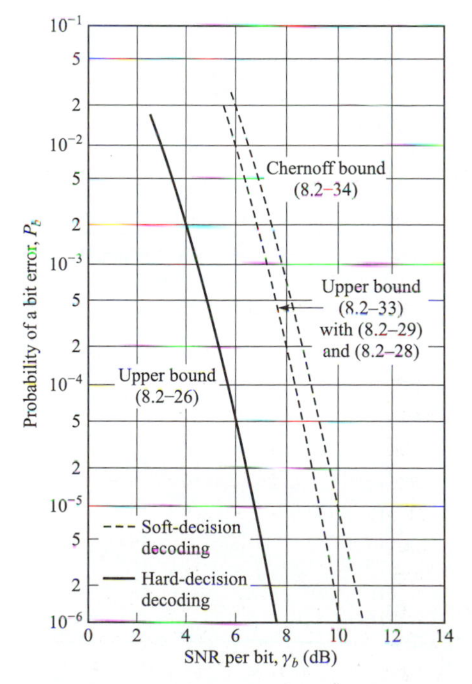

FIGURE 8.2–15
Comparison of soft-decision and hard-decision decoding for $K = 3, k = 1, n = 3$ convolutional code.

where q is the number of channel input symbols, K is the constraint length of the code, R_c is the code rate, and R_0 is the cutoff rate defined in Sections 7.2 and 8.1.6. Therefore, conclusions reached by computing R_0 for various channel conditions apply to both block codes and convolutional codes.

8.2.5 Distance Properties of Binary Convolutional Codes

In this subsection, we shall tabulate the minimum free distance and the generators for several binary, short-constraint-length convolutional codes for several code rates. These binary codes are optimal in the sense that, for a given rate and a given constraint length, they have the largest possible d_{free}. The generators and the corresponding values of d_{free} tabulated below have been obtained by Odenwalder (1970), Larsen (1973), Paaske (1974), and Daut et al. (1982) using computer search methods.

Heller (1968) has derived a relatively simple upper bound on the minimum free distance of a rate $1/n$ convolutional code. It is given by

$$d_{\text{free}} \leqslant \min_{l \geqslant 1} \left\lfloor \frac{2^{l-1}}{2^l - 1}(K + l - 1)n \right\rfloor \tag{8.2–36}$$

where $\lfloor x \rfloor$ denotes the largest integer contained in x. For purposes of comparison, this upper bound is also given in the tables for the rate $1/n$ codes. For rate k/n convolutional codes, Daut et al. (1982) has given a modification to Heller's bound. The values obtained from this upper bound for k/n are also tabulated.

Tables 8.2–1 to 8.2–7 list the parameters of rate $1/n$ convolutional codes for $n = 2, 3, \ldots, 8$. Tables 8.2–8 to 8.2–11 list the parameters of several rate k/n convolutional codes for $k \leqslant 4$ and $n \leqslant 8$.

TABLE 8.2–1
Rate 1/2 maximum free distance codes

Constraint length K	Generators in octal		d_{free}	Upper bound on d_{free}
3	5	7	5	5
4	15	17	6	6
5	23	35	7	8
6	53	75	8	8
7	133	171	10	10
8	247	371	10	11
9	561	753	12	12
10	1,167	1,545	12	13
11	2,335	3,661	14	14
12	4,335	5,723	15	15
13	10,533	17,661	16	16
14	21,675	27,123	16	17

Sources: Odenwalder (1970) and Larsen (1973).

TABLE 8.2–2
Rate 1/3 maximum free distance codes

Constraint length K	Generators in octal			d_{free}	Upper bound on d_{free}
3	5	7	7	8	8
4	13	15	17	10	10
5	25	33	37	12	12
6	47	53	75	13	13
7	133	145	175	15	15
8	225	331	367	16	16
9	557	663	711	18	18
10	1,117	1,365	1,633	20	20
11	2,353	2,671	3,175	22	22
12	4,767	5,723	6,265	24	24
13	10,533	10,675	17,661	24	24
14	21,645	35,661	37,133	26	26

Sources: Odenwalder (1970) and Larsen (1973).

TABLE 8.2–3
Rate 1/4 maximum free distance codes

Constraint length K	Generators in octal				d_{free}	Upper bound on d_{free}
3	5	7	7	7	10	10
4	13	15	15	17	13	15
5	25	27	33	37	16	16
6	53	67	71	75	18	18
7	135	135	147	163	20	20
8	235	275	313	357	22	22
9	463	535	733	745	24	24
10	1,117	1,365	1,633	1,653	27	27
11	2,327	2,353	2,671	3,175	29	29
12	4,767	5,723	6,265	7,455	32	32
13	11,145	12,477	15,537	16,727	33	33
14	21,113	23,175	35,527	35,537	36	36

Source: Larsen (1973).

TABLE 8.2–4
Rate 1/5 maximum free distance codes

Constraint length K	Generators in octal					d_{free}	Upper bound on d_{free}
3	7	7	7	5	5	13	13
4	17	17	13	15	15	16	16
5	37	27	33	25	35	20	20
6	75	71	73	65	57	22	22
7	175	131	135	135	147	25	25
8	257	233	323	271	357	28	28

Source: Daut et al. (1982).

TABLE 8.2–5
Rate 1/6 maximum free distance codes

Constraint length K	Generators in octal			d_{free}	Upper bound on d_{free}
3	7	7	7	16	16
	7	5	5		
4	17	17	13	20	20
	13	15	15		
5	37	35	27	24	24
	33	25	35		
6	73	75	55	27	27
	65	47	57		
7	173	151	135	30	30
	135	163	137		
8	253	375	331	34	34
	235	313	357		

Source: Daut et al. (1982).

TABLE 8.2–6
Rate 1/7 maximum free distance codes

Constraint length K	Generators in octal				d_{free}	Upper bound on d_{free}
3	7	7	7	7	18	18
	5	5	5			
4	17	17	13	13	23	23
	13	15	15			
5	35	27	25	27	28	28
	33	35	37			
6	53	75	65	75	32	32
	47	67	57			
7	165	145	173	135	36	36
	135	147	137			
8	275	253	375	331	40	40
	235	313	357			

Source: Daut et al. (1982).

8.2.6 Punctured Convolutional Codes

In some practical applications, there is a need to employ high-rate convolutional codes, e.g., rates of $(n-1)/n$. As we have observed, the trellis for such high-rate codes has 2^{n-1} branches that enter each state. Consequently, there are 2^{n-1} metric computations per state that must be performed in implementing the Viterbi

TABLE 8.2–7
Rate 1/8 maximum free distance codes

Constraint length K	Generators in octal				d_{free}	Upper bound on d_{free}
3	7	7	5	5	21	21
	5	7	7	7		
4	17	17	13	13	26	26
	13	15	15	17		
5	37	33	25	25	32	32
	35	33	27	37		
6	57	73	51	65	36	36
	75	47	67	57		
7	153	111	165	173	40	40
	135	135	147	137		
8	275	275	253	371	45	45
	331	235	313	357		

Source: Daut et al. (1982).

TABLE 8.2–8
Rate 2/3 maximum free distance codes

Constraint length K	Generators in octal			d_{free}	Upper bound on d_{free}
2	17	06	15	3	4
3	27	75	72	5	6
4	236	155	337	7	7

Source: Daut et al. (1982).

TABLE 8.2–9
Rate $k/5$ maximum free distance codes

Rate	Constraint length K	Generators in octal					d_{free}	Upper bound on d_{free}
2/5	2	17	07	11	12	04	6	6
	3	27	71	52	65	57	10	10
	4	247	366	171	266	373	12	12
3/5	2	35	23	75	61	47	5	5
4/5	2	237	274	156	255	337	3	4

Source: Daut et al. (1982).

TABLE 8.2–10
Rate $k/7$ maximum free distance codes

Rate	Constraint length K	Generators in octal				d_{free}	Upper bound on d_{free}
2/7	2	05	06	12	15	9	9
		15	13	17			
	3	33	55	72	47	14	14
		25	53	75			
	4	312	125	247	366	18	18
		171	266	373			
3/7	2	45	21	36	62	8	8
		57	43	71			
4/7	2	130	067	237	274	6	7
		156	255	337			

Source: Daut et al. (1982).

TABLE 8.2–11
Rates 3/4 and 3/8 maximum free distance codes

Rate	Constraint length K	Generators in octal				d_{free}	Upper bound on d_{free}
3/4	2	13	25	61	47	4	4
3/8	2	15	42	23	61	8	8
		51	36	75	47		

Source: Daut et al. (1982).

algorithm and as many comparisons of the updated metrics in order to select the best path at each state. Therefore, the implementation of the decoder of a high-rate code can be very complex.

The computational complexity inherent in the implementation of the decoder of a high-rate convolutional code can be avoided by designing the high-rate code from a low-rate code in which some of the coded bits are deleted from transmission. The deletion of selected coded bits at the output of a convolutional encoder is called *puncturing*. Thus, one can generate high-rate convolutional codes by puncturing rate $1/n$ codes with the result that the decoder maintains the low complexity of the rate $1/n$ code. We note, of course, that puncturing a code reduces the free distance of the rate $1/n$ code by some amount that depends on the degree of puncturing.

The puncturing process may be described as periodically deleting selected bits from the output of the encoder, thus, creating a periodically time varying trellis code. We begin with a rate $1/n$ parent code and define a *puncturing period* P, corresponding to P input information bits to the encoder. Hence, in one period, the encoder outputs nP coded bits. Associated with the nP encoded bits is a *puncturing matrix* $\mathbf{P}$ of the form

$$\mathbf{P} = \begin{bmatrix} p_{11} & p_{12} & \cdots & p_{1p} \\ p_{21} & p_{22} & \cdots & p_{2p} \\ \vdots & \vdots & & \vdots \\ p_{n1} & p_{n2} & \cdots & p_{np} \end{bmatrix} \tag{8.2–37}$$

where each column of $\mathbf{P}$ corresponds to the n possible output bits from the encoder for each input bit and each element of $\mathbf{P}$ is either 0 or 1. When $p_{ij} = 1$, the corresponding output from the encoder is transmitted. When $p_{ij} = 0$, the corresponding output bit from the encoder is deleted. Thus, the code rate is determined by the period P and the number of bits deleted.

If we delete N bits out of nP, the code rate is $P/(nP - N)$, where N may take any integer value in the range 0 to $(n-1)P - 1$. Hence, the achievable code rates are

$$R_c = \frac{P}{P+M}, \qquad M = 1, 2, \ldots (n-1)P \tag{8.2–38}$$

EXAMPLE 8.2–5. Let us construct a rate $\frac{3}{4}$ code by puncturing the output of the rate $\frac{1}{3}$, $K = 3$ encoder shown in Figure 8.2–2. There are many choices for P and M in Equation 8.2–38 to achieve the desired rate. We may take the smallest value of P, namely, $P = 3$. Then, out of every $nP = 9$ output bits, we delete $N = 5$ bits. Thus, we achieve a rate $\frac{3}{4}$ punctured convolutional code. As the puncturing matrix, we may select $\mathbf{P}$ as

$$\mathbf{P} = \begin{bmatrix} 1 & 1 & 1 \\ 1 & 0 & 0 \\ 0 & 0 & 0 \end{bmatrix} \tag{8.2–39}$$

Figure 8.2–16 illustrates the generation of the punctured code from the rate $\frac{1}{3}$ parent code. The corresponding trellis for the punctured code is also shown in Figure 8.2–16.

In the example given above, the puncturing matrix was selected arbitrarily. However, some puncturing matrices are better than others in that the trellis paths have better Hamming distance properties. A computer search is usually employed to find good puncturing matrices. Generally, the high-rate punctured convolutional codes generated in this manner have a free distance that is either equal to or 1 bit less than the best same high-rate convolutional code obtained directly without puncturing.

Yasuda et al. (1984), Hole (1988), Lee (1988), Haccoun and Bégin (1989), Bégin and Haccoun (1989), and Bégin et al. (1990) have investigated the construction and properties of small and large constraint length punctured convolutional codes generated from low-rate codes. In general, high-rate codes with good distance properties are obtained by puncturing rate $\frac{1}{2}$ maximum free distance codes. For example, in Table 8.2–12 we list the puncturing matrices for code rates of $\frac{2}{3} \leqslant R_c \leqslant \frac{7}{8}$ which are obtained by puncturing rate $\frac{1}{2}$ codes with constraint lengths $3 \leqslant K \leqslant 9$. The free distances of the punctured codes are also

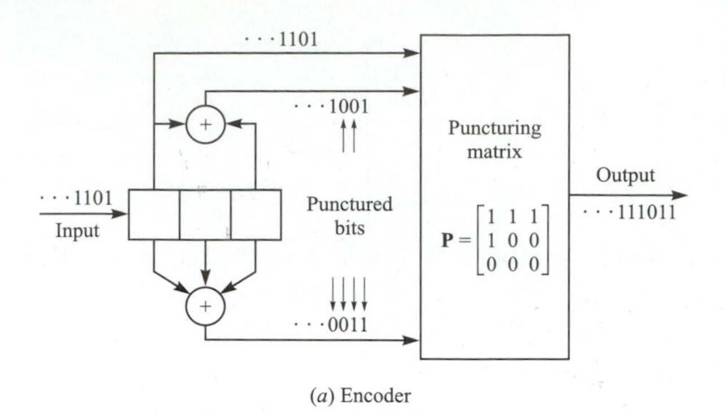

(*a*) Encoder

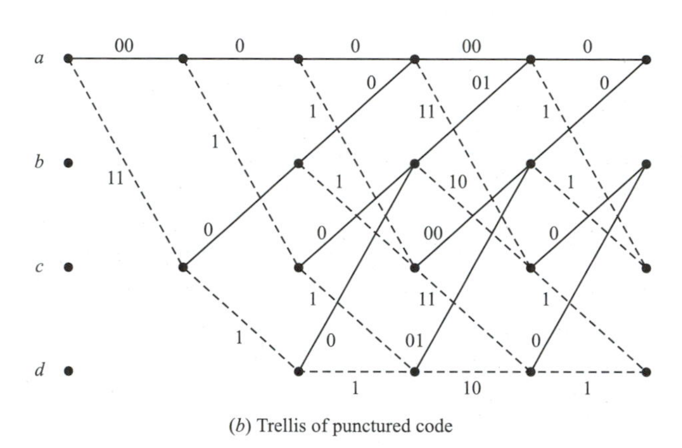

(*b*) Trellis of punctured code

FIGURE 8.2–16 Generation of a rate 3/4 punctured code from a rate 1/3 convolutional code.

given in the table. Punctured convolutional codes for additional rates and larger constraint lengths may be found in the papers referred to above.

The decoding of punctured convolutional codes is performed in the same manner as the decoding of the low-rate $1/n$ parent code, using the trellis of the $1/n$ code. The path metrics in the trellis for soft-decision decoding are computed in the conventional way as described previously. When one or more bits in a branch are punctured, the corresponding branch metric increment is computed based on the nonpunctured bits; thus, the punctured bits do not contribute to the branch metrics. Error events in a punctured code are generally longer than error events in the low rate $1/n$ parent code. Consequently, the decoder must wait longer than five constraint lengths before making final decisions on the received bits.

For soft-decision decoding, the performance of the punctured codes is given by the error probability (upper bound) expression in Equation 8.2–21 for the first event error probability and by the expression (upper bound)

$$P_b \leqslant \frac{1}{b} \sum_{d=d_{\text{free}}}^{\infty} \beta_d Q\left(\sqrt{2\gamma_b R_c d}\right) \tag{8.2–40}$$

for the bit error probability, where R_c is the code rate of the punctured code ($R_c = b/c$) and b is the number of information bits per branch.

TABLE 8.2–12
Puncturing matrices for code rates of $2/3 \leqslant R_c \leqslant 7/8$ from rate 1/2 code

K	Rate 2/3		Rate 3/4		Rate 4/5		Rate 5/6		Rate 6/7		Rate 7/8	
	P	d_{free}	P	d_{free}	P	d_{free}	P	d_{free}	P	d_{free}	P	d_{free}
3	10 11	3	101 110	3	1011 1100	2	10111 11000	2	101111 110000	2	1011111 1100000	2
4	11 10	4	110 101	4	1011 1100	3	10100 11011	3	100011 111100	2	1000010 1111101	2
5	11 10	4	101 110	3	1010 1101	3	10111 11000	3	101010 110101	3	1010011 1101100	3
6	10 11	6	100 111	4	1000 1111	4	10000 11111	4	110110 101001	3	1011101 1100010	3
7	11 10	6	110 101	5	1111 1000	4	11010 10101	4	111010 100101	3	1111010 1000101	3
8	10 11	7	110 101	6	1010 1101	5	11100 10011	4	101001 110110	4	1010100 1101011	4
9	11 10	7	111 100	6	1101 1010	5	10110 11001	5	110110 101001	4	1101011 1010100	4

An approach for the design of good punctured codes is to search and select puncturing matrices that yield the maximum free distance. A somewhat better approach is to determine the weight spectrum $\{\beta_d\}$ of the dominant terms of the punctured code and to calculate the corresponding bit error probability bound given by Equation 8.2–40. The code corresponding to the puncturing matrix that results in the best error rate performance may then be selected as the best punctured code, provided that it is not catastrophic. In general, in determining the weight spectrum for a punctured code, it is necessary to search through a larger number of paths over longer lengths than the underlying low rate $1/n$ parent code. Weight spectra for several punctured codes are given in the papers by Haccoun and Bégin (1989) and Bégin et al. (1990).

Rate-compatible punctured convolutional codes. In the transmission of compressed digital speech signals and in some other applications, there is a need to transmit some groups of information bits with more redundancy than others. In other words, the different groups of information bits require unequal error protection to be provided in the transmission of the information sequence, where the more important bits are transmitted with more redundancy. Instead of using separate codes to encode the different groups of bits, it is desirable to use a single code that has variable redundancy. This can be accomplished by puncturing the same low rate $1/n$ convolutional code by different amounts as described by Hagenauer (1988). The puncturing matrices are selected to satisfy a rate-compatibility criterion, where the basic requirement is that lower-rate codes (higher redundancy) should transmit the same coded bits as all higher-rate codes plus additional bits. The resulting codes obtained from a single rate $1/n$ convolutional code are called *rate-compatible punctured convolutional* (RCPC) *codes*.

EXAMPLE 8.2–6. From the rate $\frac{1}{3}$, $K = 4$ maximum free distance convolutional code, let us construct a RCPC code. The RCPC codes for this example are taken from the paper of Hagenauer (1988), who selected $P = 8$ and generated codes of rates ranging from $\frac{4}{11}$ to $\frac{8}{9}$. The puncturing matrices are listed in Table 8.2–13. Note that the rate $\frac{1}{2}$ code has a puncturing matrix with all zeros in the third row. Hence all bits from the third branch of the rate $\frac{1}{3}$ encoder are deleted. Higher code rates are obtained by deleting additional bits from the second branch of the rate $\frac{1}{3}$ encoder. However, note that when a 1 appears in a puncturing matrix of a high-rate code, a 1 also appears in the same position for all lower-rate codes.

In applying RCPC codes to systems that require unequal error protection of the information sequence, we may format the groups of bits into a frame structure, as suggested by Hagenauer et al. (1990) and illustrated in Figure 8.2–17, where, for example, three groups of bits of different lengths N_1, N_2, and N_3 are arranged in order of their corresponding specified error protection probabilities $p_1 > p_2 > p_3$. Each frame is terminated after the last group of information bits (N_3) by $K - 1$ zeros, which result in overhead bits that are used for the purpose of terminating the trellis in the all-zero state. We then select an appropriate set of RCPC codes that satisfy the error protection requirements, i.e., the specified error probabilities $\{p_k\}$. In our example, the group of bits will be encoded by the use of three puncturing matrices having period P corresponding to a set of RCPC codes generated from a rate $1/n$ code. Thus, the bits requiring the least protection are transmitted first, followed by the bits requiring the next higher level of protection, up to the group of bits requiring the highest level of protection, followed by the all-zero terminating sequence. All rate transitions occur within the frame without compromising the designed error rate performance requirements. As in the encoding, the bits within a frame are decoded by a single Viterbi algorithm using the trellis of the rate $1/n$ code and performing metric computations based on the appropriate puncturing matrix for each group of bits.

It can be shown (see Problem 8.39) that the average effective code rate of this scheme is

$$R_{\text{av}} = \frac{\sum_{j=1}^{J} N_j P}{\sum_{j=1}^{J} N_j(P + M_j) + (K - 1)(P + M_J)} \tag{8.2–41}$$

where J is the number of groups of bits in the frame, P is the period of the RCPC codes, and the second term in the denominator corresponds to the overhead code bits which are transmitted with the lowest code rate (highest redundancy).

TABLE 8.2–13
Rate-compatible punctured convolutional codes constructed from rate 1/3, $K = 4$ Code with $P = 8$
$R_c = P/(P + M)$, $M = 1, 2, 4, 6, 8, 10, 12, 14$

Rate	Puncturing matrix P
$\frac{1}{3}$	$\begin{bmatrix} 1&1&1&1&1&1&1&1 \\ 1&1&1&1&1&1&1&1 \\ 1&1&1&1&1&1&1&1 \end{bmatrix}$
$\frac{4}{11}$	$\begin{bmatrix} 1&1&1&1&1&1&1&1 \\ 1&1&1&1&1&1&1&1 \\ 1&1&1&0&1&1&1&0 \end{bmatrix}$
$\frac{2}{5}$	$\begin{bmatrix} 1&1&1&1&1&1&1&1 \\ 1&1&1&1&1&1&1&1 \\ 1&0&1&0&1&0&1&0 \end{bmatrix}$
$\frac{4}{9}$	$\begin{bmatrix} 1&1&1&1&1&1&1&1 \\ 1&1&1&1&1&1&1&1 \\ 1&0&0&0&1&0&0&0 \end{bmatrix}$
$\frac{1}{2}$	$\begin{bmatrix} 1&1&1&1&1&1&1&1 \\ 1&1&1&1&1&1&1&1 \\ 0&0&0&0&0&0&0&0 \end{bmatrix}$
$\frac{4}{7}$	$\begin{bmatrix} 1&1&1&1&1&1&1&1 \\ 1&1&1&0&1&1&1&0 \\ 0&0&0&0&0&0&0&0 \end{bmatrix}$
$\frac{4}{6}$	$\begin{bmatrix} 1&1&1&1&1&1&1&1 \\ 1&0&1&0&1&0&1&0 \\ 0&0&0&0&0&0&0&0 \end{bmatrix}$
$\frac{4}{5}$	$\begin{bmatrix} 1&1&1&1&1&1&1&1 \\ 1&0&0&0&1&0&0&0 \\ 0&0&0&0&0&0&0&0 \end{bmatrix}$
$\frac{8}{9}$	$\begin{bmatrix} 1&1&1&1&1&1&1&1 \\ 1&0&0&0&0&0&0&0 \\ 0&0&0&0&0&0&0&0 \end{bmatrix}$

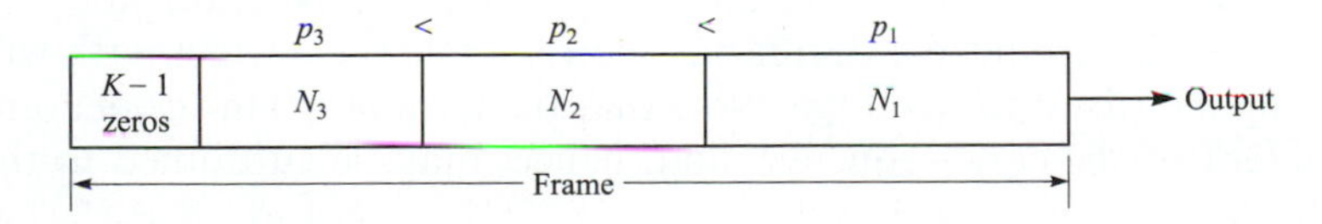

FIGURE 8.2–17
Frame structure for transmitting data with unequal error protection.

8.2.7 Other Decoding Algorithms for Convolutional Codes

The Viterbi algorithm described in Section 8.2.2 is the optimum decoding algorithm (in the sense of maximum-likelihood decoding of the entire sequence) for convolutional codes. However, it requires the computation of 2^{kK} metrics at each node of the trellis and the storage of $2^{k(K-1)}$ metrics and $2^{k(K-1)}$ surviving sequences, each of which may be about $5kK$ bits long. The computational burden and the storage required to implement the Viterbi algorithm make it impractical for convolutional codes with large constraint length.

Prior to the discovery of the optimum algorithm by Viterbi, a number of other algorithms had been proposed for decoding convolutional codes. The earliest was the sequential decoding algorithm originally proposed by Wozencraft (1957), further treated by Wozencraft and Reiffen (1961), and subsequently modified by Fano (1963).

Sequential decoding algorithm. The Fano sequential decoding algorithm searches for the most probable path through the tree or trellis by examining one path at a time. The increment added to the metric along each branch is proportional to the probability of the received signal for that branch, just as in Viterbi decoding, with the exception that an additional negative constant is added to each branch metric. The value of this constant is selected such that the metric for the correct path will increase on the average, while the metric for any incorrect path will decrease on the average. By comparing the metric of a candidate path with a moving (increasing) threshold, Fano's algorithm detects and discards incorrect paths.

To be more specific, let us consider a memoryless channel. The metric for the ith path through the tree or trellis from the first branch to branch B may be expressed as

$$CM^{(i)} = \sum_{j=1}^{B} \sum_{m=1}^{n} \mu_{jm}^{(i)} \tag{8.2–42}$$

where

$$\mu_{jm}^{(i)} = \log_2 \frac{p(r_{jm}|c_{jm}^{(i)})}{p(r_{jm})} - \mathcal{K} \tag{8.2–43}$$

In Equation 8.2–43, r_{jm} is the demodulator output sequence, $p(r_{jm}|c_{jm}^{(i)})$ denotes the PDF of r_{jm} conditional on the code bit $c_{jm}^{(i)}$ for the mth bit of the jth branch of the ith path, and $\mathcal{K}$ is a positive constant. $\mathcal{K}$ is selected as indicated above so that the incorrect paths will have a decreasing metric while the correct path will have an increasing metric on the average. Note that the term $p(r_{jm})$ in the denominator is independent of the code sequence, and, hence, may be subsumed in the constant factor.

The metric given by Equation 8.2–43 is generally applicable for either hard- or soft-decision decoding. However, it can be considerably simplified when hard-decision decoding is employed. Specifically, if we have a BSC with transition (error) probability p, the metric for each received bit, consistent with the form in Equation 8.2–43 is given by

$$\mu_{jm}^{(i)} = \begin{cases} \log_2[2(1-p)] - R_c & (\text{if } \tilde{r}_{jm} = c_{jm}^{(i)}) \\ \log_2 2p - R_c & (\text{if } \tilde{r} \neq c_{jm}^{(i)}) \end{cases} \tag{8.2–44}$$

where $\tilde{r}_{jm}$ is the hard-decision output from the demodulator, $c_{jm}^{(i)}$ is the mth code bit in the jth branch of the ith path in the tree, and R_c is the code rate. Note that this metric requires some (approximate) knowledge of the error probability.

EXAMPLE 8.2–7. Suppose we have a rate $R_c = 1/3$ binary convolutional code for transmitting information over a BSC with $p = 0.1$. By evaluating Equation 8.2–44 we find that

$$\mu_{jm}^{(i)} = \begin{cases} 0.52 & (\text{if } \tilde{r}_{jm} = c_{jm}^{(i)}) \\ -2.65 & (\text{if } \tilde{r}_{jm} \neq c_{jm}^{(i)}) \end{cases} \tag{8.2–45}$$

To simplify the computations, the metric in Equation 8.2–45 may be normalized. It is well approximated as

$$\mu_{jm}^{(i)} = \begin{cases} 1 & (\text{if } \tilde{r}_{jm} = c_{jm}^{(i)}) \\ -5 & (\text{if } \tilde{r}_{jm} \neq c_{jm}^{(i)}) \end{cases} \tag{8.2–46}$$

Since the code rate is 1/3, there are three output bits from the encoder for each input bit. Hence, the branch metric consistent with Equation 8.2–46 is

$$\mu_j^{(i)} = 3 - 6d$$

or, equivalently,

$$\mu_j^{(i)} = 1 - 2d \tag{8.2–47}$$

where d is the Hamming distance of the three received bits from the three branch bits. Thus, the metric $\mu_j^{(i)}$ is simply related to the Hamming distance between received bits and the code bits in the jth branch of the ith path.

Initially, the decoder may be forced to start on the correct path by the transmission of a few known bits of data. Then it proceeds forward from node to node, taking the most probable (largest metric) branch at each node and increasing the threshold such that the threshold is never more than some preselected value, say τ, below the metric. Now suppose that the additive noise (for soft-decision decoding) or demodulation errors resulting from noise on the channel (for hard-decision decoding) cause the decoder to take an incorrect path because it appears more probable than the correct path. This is illustrated in Figure 8.2–18. Since the metrics of an incorrect path decrease on the average, the metric will fall below the current threshold, say τ_0. When this occurs, the decoder backs up and takes alternative paths through the tree or trellis, in order of decreasing branch metrics, in an attempt to find another path that exceeds the threshold τ_0. If it is successful in finding an alternative path, it continues along that path, always selecting the most probable branch at each node. On the other

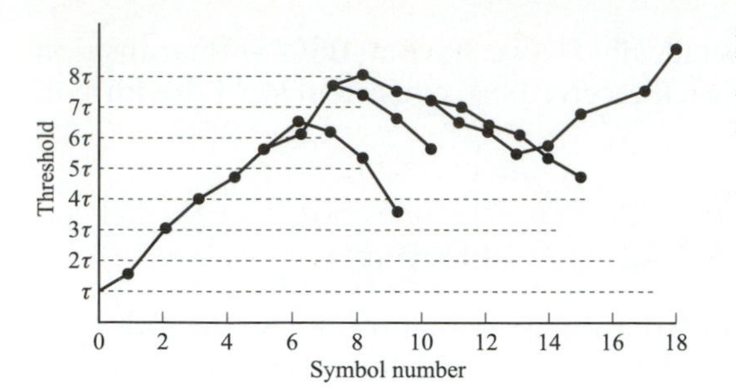

FIGURE 8.2–18
An example of the path search in sequential decoding. [*From Jordan (1966), © 1966 IEEE.]*

hand, if no path exists that exceeds the threshold τ_0, the threshold is reduced by an amount τ and the original path is retraced. If the original path does not stay above the new threshold, the decoder resumes its backward search for other paths. This procedure is repeated, with the threshold reduced by τ for each repetition, until the decoder finds a path that remains above the adjusted threshold. A simplified flow diagram of Fano's algorithm is shown in Figure 8.2–19.

The sequential decoding algorithm requires a buffer memory in the decoder to store incoming demodulated data during periods when the decoder is searching for alternate paths. When a search terminates, the decoder must be capable of processing demodulated bits sufficiently fast to empty the buffer prior to

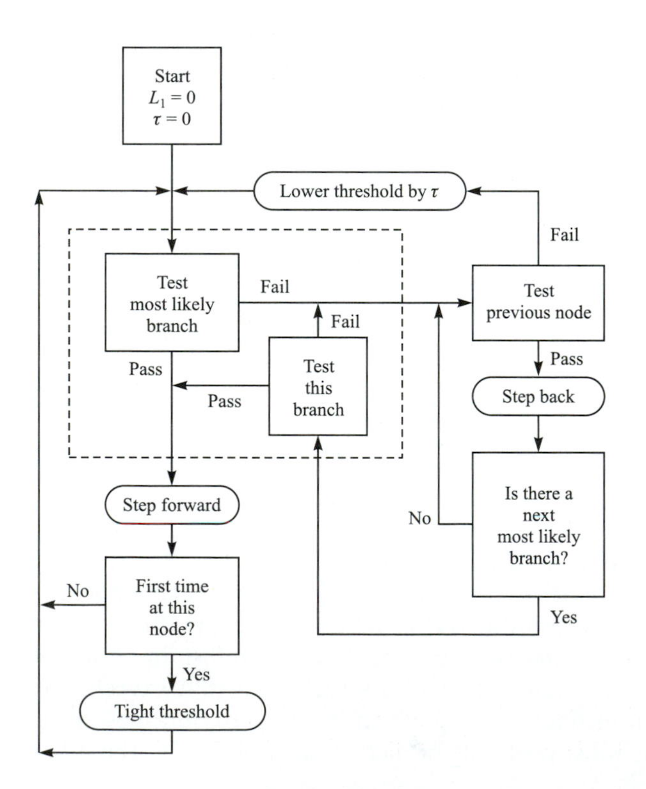

FIGURE 8.2–19
A simplified flow diagram of Fano's algorithm. [*From Jordan (1966), © 1966 IEEE.*]

commencing a new search. Occasionally, during extremely long searches, the buffer may overflow. This causes loss of data, a condition that can be remedied by retransmission of the lost information. In this regard, we should mention that the cutoff rate R_0 has special meaning in sequential decoding. It is the rate above which the average number of decoding operations per decoded digit becomes infinite, and it is termed the *computational cutoff rate* R_{comp}. In practice, sequential decoders usually operate at rates near R_0.

The Fano sequential decoding algorithm has been successfully implemented in several communication systems. Its error rate performance is comparable to that of Viterbi decoding. However, in comparison with Viterbi decoding, sequential decoding has a significantly larger decoding delay. On the positive side, sequential decoding requires less storage than Viterbi decoding and, hence, it appears attractive for convolutional codes with a large constraint length. The issues of computational complexity and storage requirements for sequential decoding are interesting and have been thoroughly investigated. For an analysis of these topics and other characteristics of the Fano algorithm, the interested reader may refer to Gallager (1968), Wozencraft and Jacobs (1965), Savage (1966), and Forney (1974).

Stack algorithm. Another type of sequential decoding algorithm, called a *stack algorithm*, has been proposed independently by Jelinek (1969) and Zigangirov (1966). In contrast to the Viterbi algorithm, which keeps track of $2^{(K-1)k}$ paths and corresponding metrics, the stack sequential decoding algorithm deals with fewer paths and their corresponding metrics. In a stack algorithm, the more probable paths are ordered according to their metrics, with the path at the top of the stack having the largest metric. At each step of the algorithm, only the path at the top of the stack is extended by one branch. This yields 2^k successors and their corresponding metrics. These 2^k successors along with the other paths are then reordered according to the values of the metrics, and all paths with metrics that fall below some preselected amount from the metric of the top path may be discarded. Then the process of extending the path with the largest metric is repeated. Figure 8.2–20 illustrates the first few steps in a stack algorithm.

It is apparent that when none of the 2^k extensions of the path with the largest metric remains at the top of the stack, the next step in the search involves the extension of another path that has climbed to the top of the stack. It follows that the algorithm does not necessarily advance by one branch through the trellis in every iteration. Consequently, some amount of storage must be provided for newly received signals and previously received signals in order to allow the algorithm to extend the search along one of the shorter paths, when such a path reaches the top of the stack.

In a comparison of the stack algorithm with the Viterbi algorithm, the stack algorithm requires fewer metric computations, but this computational saving is offset to a large extent by the computations involved in reordering the stack after every iteration. In comparison with the Fano algorithm, the stack algorithm is computationally simpler, since there is no retracing over the same path as is done

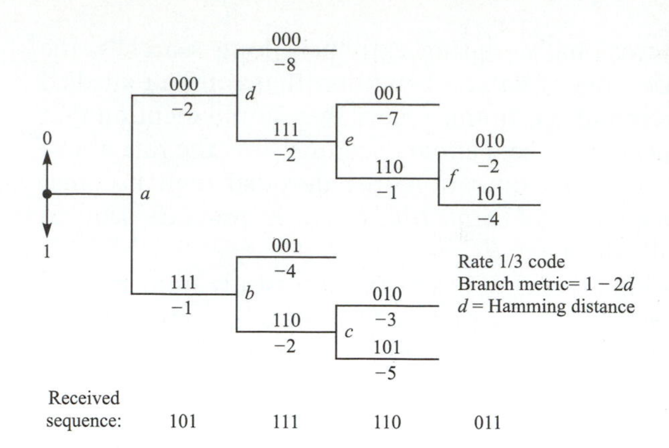

FIGURE 8.2–20
An example of the stack algorithm for decoding a rate 1/3 convolutional code.

Stack with accumulated path metrics

Step *a*	Step *b*	Step *c*	Step *d*	Step *e*	Step *f*
−1	−2	−3	−2	−1	−2
−3	−3	−3	−3	−3	−3
	−4	−4	−4	−4	−4
		−5	−5	−5	−4
			−8	−7	−5
				−8	−7
					−8

in the Fano algorithm. On the other hand, the stack algorithm requires more storage than the Fano algorithm.

Feedback decoding. A third alternative to the optimum Viterbi decoder is a method called *feedback decoding* (Heller, 1975), which has been applied to decoding for a BSC (hard-decision decoding). In feedback decoding, the decoder makes a hard decision on the information bit at stage j based on metrics computed from stage j to stage $j + m$, where m is a preselected positive integer. Thus, the decision on the information bit is either 0 or 1 depending on whether the minimum Hamming distance path that begins at stage j and ends at stage $j + m$ contains a 0 or 1 in the branch emanating from stage j. Once a decision is made on the information bit at stage j, only that part of the tree that stems from the bit selected at stage j is kept (half the paths emanating from node j) and the remaining part is discarded. This is the feedback feature of the decoder.

The next step is to extend the part of the tree that has survived to stage $j + 1 + m$ and consider the paths from stage $j + 1$ to $j + 1 + m$ in deciding on the bit at stage $j + 1$. Thus, this procedure is repeated at every stage. The parameter m is simply the number of stages in the tree that the decoder looks ahead before making a hard decision. Since a large value of m results in a large amount of storage, it is desirable to select m as small as possible. On the other hand, m must be sufficiently large to avoid a severe degradation in performance. To balance these two conflicting requirements, m is usually selected in the range $K \leqslant m \leqslant 2K$, where K is the constraint length. Note that this decoding delay is significantly smaller than the decoding delay in a Viterbi decoder, which is usually about $5K$.

EXAMPLE 8.2–8. Let us consider the use of a feedback decoder for the rate 1/3 convolutional code shown in Figure 8.2–2. Figure 8.2–21 illustrates the tree diagram and the operation of the feedback decoder for $m = 2$. That is, in decoding the bit at branch j, the decoder considers the paths at branches j, $j + 1$, and $j + 2$. Beginning with the first branch, the decoder computes eight metrics (Hamming distances) and decides that the bit for the first branch is 0 if the minimum distance path is contained in the upper part of the tree, and 1 if the minimum distance path is contained in the lower part of the tree. In this example, the received sequence for the first three branches is assumed to be 101111110, so that the minimum distance path is in the upper part of the tree. Hence, the first output bit is 0.

The next step is to extend the upper part of the tree (the part of the tree that has survived) by one branch, and to compute the eight metrics for branches 2, 3, and 4. For the assumed received sequence 111110011, the minimum-distance path is contained in the lower part of the section of the tree that survived from the first step. Hence, the second output bit is 1. The third step is to extend this lower part of the tree and to repeat the procedure described for the first two steps.

Instead of computing metrics as described above, a feedback decoder for the BSC may be efficiently implemented by computing the syndrome from the received sequence and using a table lookup method for correcting errors. This method is similar to the one described for decoding block codes. For some

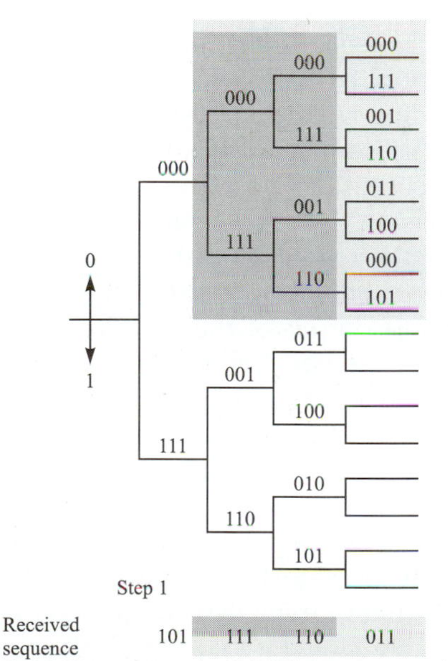

Step 1: Upper-tree metrics: 7,6,5,2*; lower-tree metrics: 5,4,3,4 → 0
Step 2: Upper-tree metrics: 7,6,5,6; lower-tree metrics: 3,6,1*,2 → 1

FIGURE 8.2–21
An example of feedback decoding for a rate 1/3 convolutional code.

convolutional codes, the feedback decoder simplifies to a form called a *majority logic decoder* or a *threshold decoder* (Massey, 1963; Heller, 1975).

Soft-output algorithms. The outputs of the Viterbi algorithm and the three algorithms described in this section are hard decisions. In some cases, it is desirable to have soft outputs from the decoder. This is the case if the decoding is being performed on an inner code in a concatenated code, where it is desirable to provide soft decisions to the input of the outer decoder. This is also the case in iterative decoding of concatenated codes, previously discussed in the context of block codes in Section 8.1.10, and further treated in the context of convolutional codes in Section 8.2.10.

The optimum metric that provides a measure of the reliability of symbol decisions is the a posteriori probability of the detected symbol conditioned on the received signal vector $\mathbf{r} = \{r_{jm}, m = 1, 2, \cdots, n; j = 1, 2, \cdots B\}$, where $\{r_{jm}\}$ is the sequence of soft outputs from the demodulator, n is the number of output symbols from the encoder for each k input symbols, and j is the branch index. For example, the output of the demodulator for a binary convolutional code and binary PSK modulation in an AWGN channel is

$$r_{jm} = (2c_{jm} - 1)\sqrt{\mathcal{E}_c} + n_{jm}$$

where $\{c_{jm} = 0, 1\}$ are the output bits from the encoder. Given the received vector $\mathbf{r}$, decisions on the transmitted information bits are based on the maximum a posteriori probability (MAP), which may be expressed as

$$P(x_i = 0|\mathbf{r}) = 1 - P(x_i = 1|\mathbf{r}) \tag{8.2–48}$$

where x_i denotes the ith information bit in the sequence. Thus, under the MAP criterion, a decision is made on a symbol-by-symbol basis by selecting the information symbol, or bit in this case, corresponding to the largest a posteriori probability. If the a posteriori probabilities for the possible transmitted symbols are nearly the same, the decision is unreliable. Hence, the a posteriori probability associated with the decided symbol (the hard decision) is the soft output from the decoder that provides a measure, or metric, for the reliability of the hard decision. Since the MAP criterion minimizes the probability of a symbol error, the a posteriori probability metric is the optimum soft output of the decoder.

An algorithm for recursively computing the a posteriori probabilities for each received symbol given the received signal sequence $\mathbf{r}$ from the demodulator has been described in the paper by Bahl, Cocke, Jelinek, and Raviv (1974). This symbol-by-symbol decoding algorithm, called the BCJR algorithm, is based on the MAP criterion and provides a hard decision on each received symbol and the a posteriori probability metric that serves as a measure for the reliability of the hard decision. The major disadvantage to the BCJR algorithm is its computational complexity, which is about three to four times larger than the computational complexity of the Viterbi algorithm.

In contrast to the MAP symbol-by-symbol detection criterion, the Viterbi algorithm selects the sequence that maximizes the probability $p(\mathbf{r}|\mathbf{x})$, where $\mathbf{x}$ is the vector of information bits. In this case, the soft output metric is the Euclidean distance associated with the sequence of received symbols, as opposed to the individual symbols. However, it is possible to derive symbol metrics from the sequence or path metrics. Hagenauer and Hoeher (1989) devised a soft-output Viterbi algorithm (SOVA) that provides a reliability metric for each decoded symbol. The SOVA is based on the observation that the probability that a hard decision on a given symbol at the output of the Viterbi algorithm is correct is proportional to the difference in path metrics between a surviving sequence and its associated nonsurviving sequences. This observation allows us to form an estimate of the error probability, or the probability of a correct decision, for each symbol by comparing the path metrics of the surviving path with the path metrics of nonsurviving paths.

For example, let us consider a binary convolutional code with binary PSK modulation. Since the Viterbi algorithm makes decisions with a decoding delay δ, at time $t = i + \delta$ the Viterbi decoder outputs the bit $\hat{x}_{is}$ from the most probable surviving sequence. When we trace back along the surviving path from t to $t - \delta$, we observe that we have discarded $\delta + 1$ paths. Let us consider the jth discarded path and its corresponding bit x_{ij} at time $t = i$. If $\hat{x}_{is} \neq x_{ij}$, let Δ_j ($\Delta_j \geqslant 0$) be equal to the difference in the path metrics between the surviving path and the jth discarded path. If $\hat{x}_{is} = x_{ij}$, let $\Delta_j = \infty$. This comparison is performed for all discarded paths. From the set $\{\Delta_j, j = 0, 1, 2, \cdots, \delta\}$ we select the smallest value, defined as $\Delta_{\min} = \min\{\Delta_0, \Delta_1, \cdots, \Delta_\delta\}$. Then, the probability of error for the bit $\hat{x}_{is}$ is approximated as

$$\hat{P}_e = \frac{1}{1 + e^{\Delta_{\min}}} \tag{8.2–49}$$

Note that if $\Delta_{\min}$ is very small, $\hat{P}_e \approx \frac{1}{2}$, so the decision on $\hat{x}_{is}$ is unreliable. Thus, $\hat{P}_e$ provides a reliability metric for the hard decisions at the output of the Viterbi algorithm. We note, however, that $\hat{P}_e$ is only an approximation to the true error probability. That is, $\hat{P}_e$ is not the optimum soft-output metric for the hard decisions at the output of the Viterbi algorithm. In fact, it has been observed in a paper by Wang and Wicker (1996) that $\hat{P}_e$ underestimates the true error probability at low SNR. Nevertheless, this soft-output metric from the Viterbi algorithm leads to a significant improvement in the performance of the decoder in a concatenated code.

From Equation 8.2–49 we can obtain an estimate of the probability of a correct decision as

$$\hat{P}_c = 1 - \hat{P}_e = \frac{e^{\Delta_{\min}}}{1 + e^{\Delta_{\min}}} \tag{8.2–50}$$

In the soft-in, soft-out iterative decoding algorithms discussed in Section 8.2.10, the soft-output metric used in the algorithms is usually the logarithm of $\hat{P}_c/\hat{P}_e$. That is,

$$L(x_i) \equiv \ln \frac{\hat{P}_c}{\hat{P}_e} = \Delta_{\min}(i) \tag{8.2–51}$$

where x_i is the ith information bit being decoded.

8.2.8 Practical Considerations in the Application of Convolutional Codes

Convolutional codes are widely used in many practical applications of communication system design. Viterbi decoding is predominantly used for short constraint lengths ($K \leqslant 10$), while sequential decoding is used for long-constraint-length codes, where the complexity of Viterbi decoding becomes prohibitive. The choice of constraint length is dictated by the desired coding gain.

From the error probability results for soft-decision decoding given by Equation 8.2–26 it is apparent that the coding gain achieved by a convolutional code over an uncoded binary PSK or QPSK system is

$$\text{Coding gain} \leqslant 10 \log_{10}(R_c d_{\text{free}})$$

We also know that the minimum free distance d_{free} can be increased either by decreasing the code rate or by increasing the constraint length, or both. Table 8.2–14 provides a list of upper bounds on the coding gain for several convolutional codes. For purposes of comparison, Table 8.2–15 lists the actual coding gains for several short-constraint-length convolutional codes with Viterbi decoding. It should be noted that the coding gain increases toward the asymptotic limit as the SNR per bit increases.

These results are based on soft-decision Viterbi decoding. If hard-decision decoding is used, the coding gains are reduced by approximately 2 dB for the AWGN channel.

Larger coding gains than those listed in Tables 8.2–14 and 8.2–15 are achieved by employing long-constraint-length convolutional codes, e.g., $K = 50$, and decoding such codes by sequential decoding. Invariably, sequential

TABLE 8.2–14
Upper bounds on coding gain for soft-decision decoding of some convolutional codes

Rate 1/2 codes			Rate 1/3 codes		
Constraint length K	d_{free}	**Upper bound, dB**	**Constraint length K**	d_{free}	**Upper bound, dB**
3	5	3.98	3	8	4.26
4	6	4.77	4	10	5.23
5	7	5.44	5	12	6.02
6	8	6.02	6	13	6.37
7	10	6.99	7	15	6.99
8	10	6.99	8	16	7.27
9	12	7.78	9	18	7.78
10	12	7.78	10	20	8.24

TABLE 8.2–15
Coding gain (dB) for soft-decision Viterbi decoding

P_b	$\mathcal{E}_b/N_0$ uncoded, dB	$R_c = 1/3$		$R_c = 1/2$			$R_c = 2/3$		$R_c = 3/4$	
		$K = 8$	$K = 8$	$K = 5$	$K = 6$	$K = 7$	$K = 6$	$K = 8$	$K = 6$	$K = 9$
10^{-3}	6.8	4.2	4.4	3.3	3.5	3.8	2.9	3.1	2.6	2.6
10^{-5}	9.6	5.7	5.9	4.3	4.6	5.1	4.2	4.6	3.6	4.2
10^{-7}	11.3	6.2	6.5	4.9	5.3	5.8	4.7	5.2	3.9	4.8

Source: Jacobs (1974); © IEEE.

decoders are implemented for hard-decision decoding to reduce complexity. Figure 8.2–22 illustrates the error rate performance of several constraint-length $K = 7$ convolutional codes for rates 1/2 and 1/3 and for sequential decoding (with hard decisions) of a rate 1/2 and a rate 1/3 constraint-length $K = 41$ convolutional codes. Note that the $K = 41$ codes achieve an error rate of 10^{-6} at 2.5 and 3 dB, which are within 4–4.5 dB of the channel capacity limit, i.e., in the vicinity of the cutoff rate limit. However, the rate 1/2 and rate 1/3, $K = 7$ codes with soft-decision Viterbi decoding operate at about 5 and 4.4 dB at 10^{-6}, respectively. These short-constraint-length codes achieve a coding gain of about 6 dB at 10^{-6}, while the long-constraint-length codes gain about 7.5–8 dB.

Two important issues in the implementation of Viterbi decoding are

1. The effect of path memory truncation, which is a desirable feature that ensures a fixed decoding delay.
2. The degree of quantization of the input signal to the Viterbi decoder.

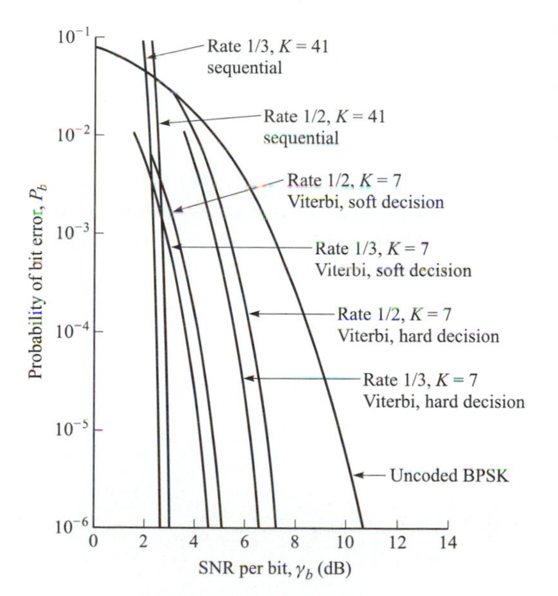

FIGURE 8.2–22
Performance of rate 1/2 and rate 1/3 Viterbi and sequential decoding. [*From Omura and Levitt (1982).* © *1982 IEEE.*]

As a rule of thumb, we stated that path memory truncation to about five constraint lengths has been found to result in negligible performance loss. Figure 8.2–23 illustrates the performance obtained by simulation for rate 1/2, constraint-lengths $K = 3$, 5, and 7 codes with memory path length of 32 bits. In addition to path memory truncation, the computations were performed with eight-level (three bits) quantized input signals from the demodulator. The broken curves are performance results obtained from the upper bound in the bit error rate given by Equation 8.2–26. Note that the simulation results are close to the theoretical upper bounds, which indicate that the degradation due to path memory truncation and quantization of the input signal has a minor effect on performance (0.20–0.30 dB).

Figure 8.2–24 illustrates the bit error rate performance obtained via simulation for hard-decision decoding of convolutional codes with $K = 3$–8. Note that with the $K = 8$ code, an error rate of 10^{-5} requires about 6 dB, which represents a coding gain of nearly 4 dB relative to uncoded QPSK.

The effect of input signal quantization is further illustrated in Figure 8.2–25 for a rate 1/2, $K = 5$ code. Note that 3-bit quantization (eight levels) is about 2 dB better than hard-decision decoding, which is the ultimate limit between soft-decision decoding and hard-decision decoding on the AWGN channel. The combined effect of signal quantization and path memory truncation for the rate 1/2, $K = 5$ code with 8-, 16-, and 32-bit path memories and either 1- or 3-bit quantization is shown in Figure 8.2–26. It is apparent from these results that a path memory as short as three constraint lengths does not seriously degrade performance.

When the signal from the demodulator is quantized to more than two levels, another problem that must be considered is the spacing between quantization levels. Figure 8.2–27 illustrates the simulation results for an eight-level uniform quantizer as a function of the quantizer threshold spacing. We observe that there is an optimum spacing between thresholds (approximately equal to 0.5).

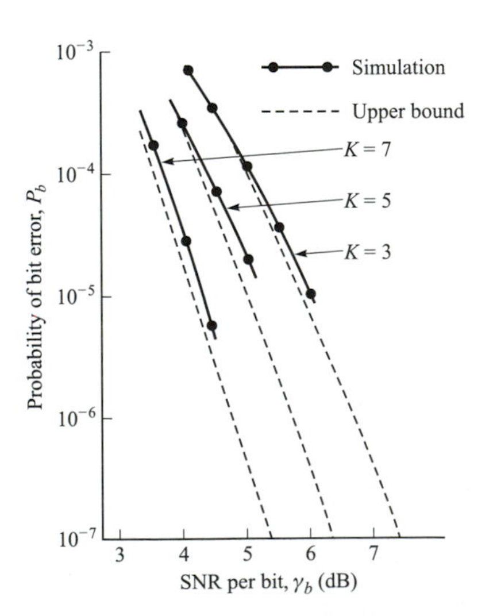

FIGURE 8.2–23
Bit error probability for rate 1/2 Viterbi decoding with eight-level quantized inputs to the decoder and 32-bit path memory. [*From Heller and Jacobs (1971).* © *1971 IEEE.*]

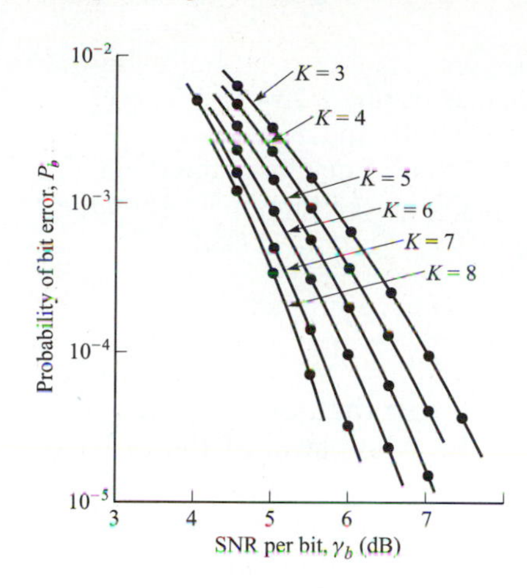

FIGURE 8.2–24
Performance of rate 1/2 codes with hard-decision Viterbi decoding and 32-bit path memory truncation. [*From Heller and Jacobs (1971).* © *1971 IEEE.*]

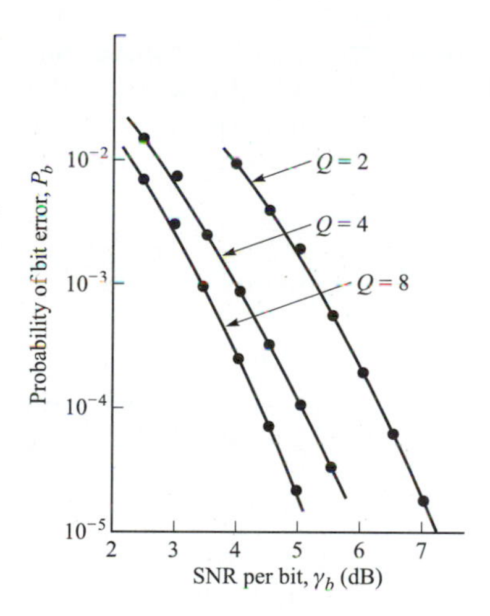

FIGURE 8.2–25
Performance of rate 1/2, $K = 5$ code with eight-, four-, and two-level quantization at the input to the Viterbi decoder. Path truncation length = 32 bits. [*From Heller and Jacobs (1971).* © *1971 IEEE.*]

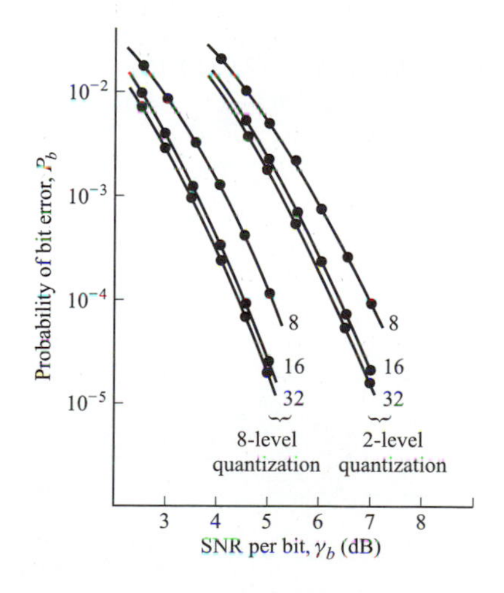

FIGURE 8.2–26
Performance of rate 1/2, $K = 5$ code with 32-, 16-, and 8-bit path memory truncation and eight- and two-level quantization. [*From Heller and Jacobs (1971).* © *1971 IEEE.*]

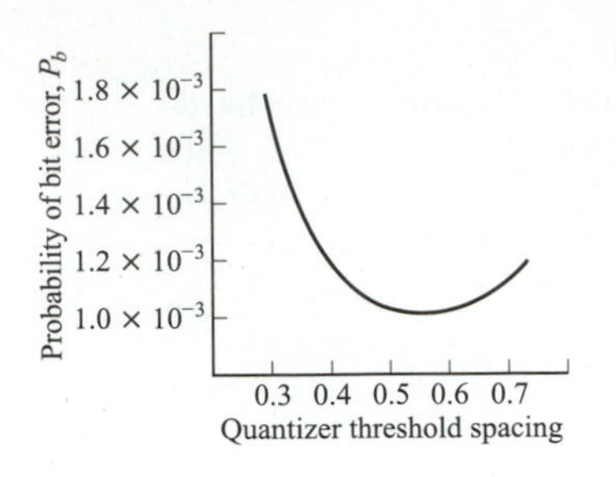

FIGURE 8.2–27
Error rate performance of rate 1/2, $K = 5$ Viterbi decoder for $\mathcal{E}_b/N_0 = 3.5$ dB and eight-level quantization as a function of quantizer threshold level spacing for equally spaced thresholds. [*From Heller and Jacobs (1971). © 1971 IEEE.*]

However, the optimum is sufficiently broad (0.4–0.7), so that, once it is set, there is little degradation resulting from variations in the AGC level of the order of ±20 percent.

Finally, we should point out some important results in the performance degradation due to carrier phase variations. Figure 8.2–28 illustrates the performance of a rate 1/2, $K = 7$ code with eight-level quantization and a carrier phase tracking loop SNR γ_L. Recall that in a PLL, the phase error has a variance that is inversely proportional to γ_L. The results in Figure 8.2–28 indicate that the degradation is large when the loop SNR is small ($\gamma_L < 12$ dB), and causes the error rate performance to bottom out at a relatively high error rate.

8.2.9 Nonbinary Dual-k Codes and Concatenated Codes

Our treatment of convolutional codes thus far has been focused primarily on binary codes. Binary codes are particularly suitable for channels in which binary

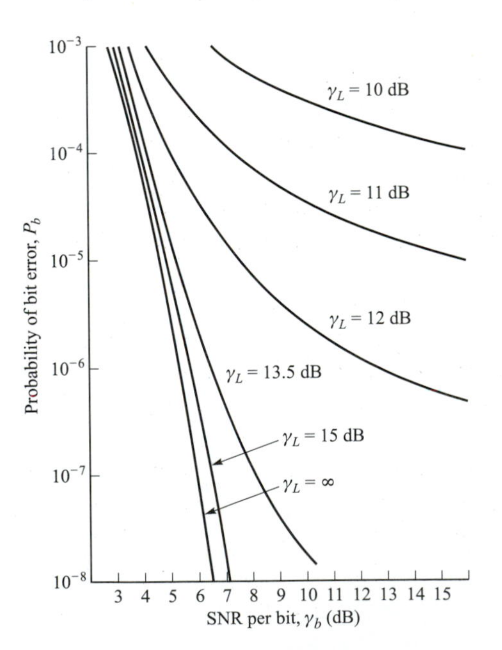

FIGURE 8.2–28
Performance of a rate 1/2, $K = 7$ code with Viterbi decoding and eight-level quantization as a function of the carrier phase tracking loop SNR γ_L [*From Heller and Jacobs (1971). © 1971 IEEE.*]

or quaternary PSK modulation and coherent demodulation is possible. However, there are many applications in which PSK modulation and coherent demodulation is not suitable or possible. In such cases, other modulation techniques, e.g., M-ary FSK, are employed in conjunction with noncoherent demodulation. Nonbinary codes are particularly matched to M-ary signals that are demodulated noncoherently.

In this subsection, we describe a class of nonbinary convolutional codes, called *dual-k codes*, that are easily decoded by means of the Viterbi algorithm using either soft-decision or hard-decision decoding. They are also suitable either as an outer code or as an inner code in a concatenated code, as will also be described below.

A dual-k rate 1/2 convolutional encoder may be represented as shown in Figure 8.2–29. It consists of two ($K = 2$) k-bit shift-register stages and $n = 2k$ function generators. Its output is two k-bit symbols. We note that the code considered in Example 8.2–3 is a dual-2 convolutional code.

The $2k$ function generators for the dual-k codes have been given by Viterbi and Jacobs (1975). These may be expressed in the form

$$
\begin{aligned}
\begin{bmatrix} \leftarrow \mathbf{g}_1 \rightarrow \\ \leftarrow \mathbf{g}_2 \rightarrow \\ \vdots \\ \leftarrow \mathbf{g}_k \rightarrow \end{bmatrix}
&= \left[\begin{array}{ccccc|ccccc}
1 & 0 & 0 & \cdots & 0 & 1 & 0 & 0 & \cdots & 0 \\
0 & 1 & 0 & \cdots & 0 & 0 & 1 & 0 & \cdots & 0 \\
\vdots & \vdots & \vdots & & \vdots & \vdots & \vdots & & & \vdots \\
0 & 0 & 0 & \cdots & 1 & 0 & 0 & \cdots & 0 & 1
\end{array}\right] = [\mathbf{I}_k \quad \mathbf{I}_k] \\
\begin{bmatrix} \leftarrow \mathbf{g}_{k+1} \rightarrow \\ \leftarrow \mathbf{g}_{k+2} \rightarrow \\ \vdots \\ \leftarrow \mathbf{g}_{2k} \rightarrow \end{bmatrix}
&= \left[\begin{array}{ccccccc|cccccc}
1 & 1 & 0 & 0 & \cdots & & 0 & 1 & 0 & 0 & \cdots & & 0 \\
0 & 0 & 1 & 0 & \cdots & & 0 & 0 & 1 & 0 & \cdots & & 0 \\
\vdots & \vdots & \vdots & \vdots & \cdots & & \vdots & \vdots & \vdots & \vdots & \cdots & & \vdots \\
0 & 0 & 0 & & \cdots & 0 & 1 & 0 & 0 & & \cdots & 1 & 0 \\
1 & 0 & 0 & & \cdots & 0 & 0 & 0 & 0 & & \cdots & 0 & 1
\end{array}\right] \\
&= \left[\begin{array}{ccccccc|c}
1 & 1 & 0 & 0 & \cdots & & 0 & \\
0 & 0 & 1 & 0 & \cdots & & 0 & \\
\vdots & \vdots & \vdots & \vdots & \cdots & & \vdots & \mathbf{I}_k \\
0 & 0 & 0 & & \cdots & 0 & 1 & \\
1 & 0 & 0 & & \cdots & 0 & 0 &
\end{array}\right]
\end{aligned} \tag{8.2–52}
$$

where $\mathbf{I}_k$ denotes the $k \times k$ identity matrix.

The general form for the transfer function of a rate 1/2 dual-k code has been derived by Odenwalder (1976). It is expressed as

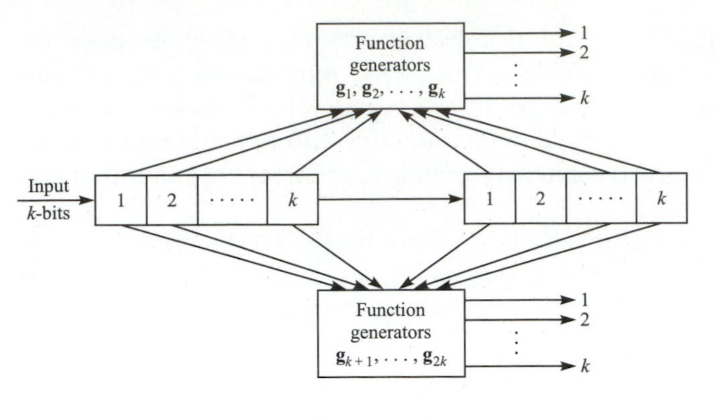

FIGURE 8.2–29
Encoder for rate 1/2 dual-k codes.

$$\begin{aligned} T(D, N, J) &= \frac{(2^k - 1)D^4 J^2 N}{1 - NJ[2D + (2^k - 3)D^2]} \\ &= \sum_{i=4}^{\infty} a_i D^i N^{f(i)} J^{h(i)} \end{aligned} \tag{8.2–53}$$

where D represents the Hamming distance for the q-ary ($q = 2^k$) symbols, the $f(i)$ exponent on N represents the number of information symbol errors that are produced in selecting a branch in the tree or trellis other than a corresponding branch on the all-zero path, and the $h(i)$ exponent on J is equal to the number of branches in a given path. Note that the minimum free distance is $d_{\text{free}} = 4$ symbols ($4k$ bits).

Lower-rate dual-k convolutional codes can be generated in a number of ways, the simplest of which is to repeat each symbol generated by the rate 1/2 code r times, where $r = 1, 2, \ldots, m$ ($r = 1$ corresponds to each symbol appearing once). If each symbol in any particular branch of the tree or trellis or state diagram is repeated r times, the effect is to increase the distance parameter from D to D^r. Consequently the transfer function for a rate $1/2r$ dual-k code is

$$T(D, N, J) = \frac{(2^k - 1)D^{4r} J^2 N}{1 - NJ[2D^r + (2^k - 3)D^{2r}]} \tag{8.2–54}$$

In the transmission of long information sequences, the path length parameter J in the transfer function may be suppressed by setting $J = 1$. The resulting transfer function $T(D, N)$ may be differentiated with respect to N, and N is set to unity. This yields

$$\begin{aligned} \left.\frac{dT(D, N)}{dN}\right|_{N=1} &= \frac{(2^k - 1)D^{4r}}{[1 - 2D^r - (2^k - 3)D^{2r}]^2} \\ &= \sum_{i=4r}^{\infty} \beta_i D^i \end{aligned} \tag{8.2–55}$$

where β_i represents the number of symbol errors associated with a path having distance D^i from the all-zero path, as described previously in Section 8.2.3. The expression in Equation 8.2–55 may be used to evaluate the error probability for dual-k codes under various channel conditions.

Performance of dual-k codes with M-ary modulation. Suppose that a dual-k code is used in conjunction with M-ary orthogonal signaling at the modulator, where $M = 2^k$. Each symbol from the encoder is mapped into one of the M possible orthogonal waveforms. The channel is assumed to add white Gaussian noise. The demodulator consists of M matched filters.

If the decoder performs hard-decision decoding, the performance of the code is determined by the symbol error probability P_M. This error probability has been computed in Chapter 5 for both coherent and noncoherent detection. From P_M, we can determine $P_2(d)$ according to Equation 8.2–28 or 8.2–29, which is the probability of error in a pairwise comparison of the all-zero path with a path that differs in d symbols. The probability of a bit error is upper-bounded as

$$P_b < \frac{2^{k-1}}{2^k - 1} \sum_{d=4r}^{\infty} \beta_d P_2(d) \tag{8.2–56}$$

The factor $2^{k-1}/(2^k - 1)$ is used to convert the symbol error probability to the bit error probability.

Instead of hard-decision decoding, suppose that the decoder performs soft-decision decoding using the output of a demodulator that employs a square-law detector. The expression for the bit error probability given by Equation 8.2–56 still applies, but now $P_2(d)$ is given by (see Section 12.1.1)

$$P_2(d) = \frac{1}{2^{2d-1}} \exp(-\tfrac{1}{2}\gamma_b R_c d) \sum_{i=0}^{d-1} K_i (\tfrac{1}{2}\gamma_b R_c d)^i \tag{8.2–57}$$

where

$$K_i = \frac{1}{i!} \sum_{l=0}^{d-1-i} \binom{2d-1}{l}$$

and $R_c = 1/2r$ is the code rate. This expression follows from the result 8.1–63.

Concatenated codes. In Section 8.1.8, we considered the concatenation of two block codes to form a long block code. Now that we have described convolutional codes, we broaden our viewpoint and consider the concatenation of a block code with a convolutional code or the concatenation of two convolutional codes.

In a conventional concatenated code, the outer code is usually chosen to be nonbinary, with each symbol selected from an alphabet of $q = 2^k$ symbols. This code may be a block code, such as a Reed–Solomon code, or a convolutional code, such as a dual-k code. The inner code may be either binary or nonbinary, and either a block or a convolutional code. For example, a Reed–Solomon code may be selected as the outer code and a dual-k code may be selected as the inner code. In such a concatenation scheme, the number of symbols in the outer

(Reed–Solomon) code q equals 2^k, so that each symbol of the outer code maps into a k-bit symbol of the inner dual-k code. M-ary orthogonal signals may be used to transmit the symbols.

The decoding of such concatenated codes may also take a variety of different forms. If the inner code is a convolutional code having a short constraint length, the Viterbi algorithm provides an efficient means for decoding, using either soft-decision or hard-decision decoding.

If the inner code is a block code, and the decoder for this code performs soft-decision decoding, the outer decoder may also perform soft-decision decoding using as inputs the metrics corresponding to each word of the inner code. On the other hand, the inner decoder may make a hard decision after receipt of the code word and feed the hard decisions to the outer decoder. Then the outer decoder must perform hard-decision decoding.

The following example describes a concatenated code in which the outer code is a convolutional code and the inner code is a block code.

EXAMPLE 8.2–9. Suppose we construct a concatenated code by selecting a dual-k code as the outer code and a Hadamard code as the inner code. To be specific, we select a rate 1/2 dual-5 code and a Hadamard (16, 5) inner code. The dual-5 rate 1/2 code has a minimum free distance $D_{\text{free}} = 4$ and the Hadamard code has a minimum distance $d_{\text{min}} = 8$. Hence, the concatenated code has an effective minimum distance of 32. Since there are 32 code words in the Hadamard code and 32 possible symbols in the outer code, in effect, each symbol from the outer code is mapped into one of the 32 Hadamard code words.

The probability of a symbol error in decoding the inner code may be determined from the results of the performance of block codes given in Sections 8.1.4 and 8.1.5 for soft-decision and hard-decision decoding, respectively. First, suppose that hard-decision decoding is performed in the inner decoder with the probability of a code word (symbol of outer code) error denoted as P_{32}, since $M = 32$. Then the performance of the outer code and, hence, the performance of the concatenated code is obtained by using this error probability in conjunction with the transfer function for the dual-5 code given by Equation 8.2–53.

On the other hand, if soft-decision decoding is used on both the outer and the inner codes, the soft-decision metric from each received Hadamard code word is passed to the Viterbi algorithm, which computes the accumulated metrics for the competing paths through the trellis. We shall give numerical results on the performance of concatenated codes of this type in our discussion of coding for Rayleigh fading channels.

8.2.10 Parallel and Serial Concatenated Convolutional Codes

In Section 8.1.10 we introduced serial and parallel concatenated block codes in which an interleaver is used to construct extremely long codes. In this section we consider the construction and decoding of concatenated codes with interleaving, using convolutional codes.

Parallel concatenated convolutional codes (PCCC) with interleaving, also called *turbo codes*, were introduced by Berrou et al. (1993). A basic turbo encoder, shown in Figure 8.2–30, is a recursive systematic encoder that employs two convolutional encoders in parallel, where the second encoder is preceded by an interleaver. The two recursive systematic convolutional encoders may be either identical or different. We observe that the nominal rate at the output of the turbo encoder is $R_c = 1/3$. However, by puncturing the parity check bits at the output of the binary convolutional encoders, we may achieve higher rates, such as rate 1/2 or 2/3. As in the case of concatenated block codes, the interleaver is usually selected to be a block pseudorandom interleaver that reorders the bits in the information sequence before feeding them to the second encoder. In effect, the use of two convolutional encoders in conjunction with the interleaver produces a code that contains very few code words of low weight. This characteristic does not necessarily imply that the free distance of the concatenated code is especially large. However, the use of the interleaver in conjunction with the two encoders results in code words that have relatively few nearest neighbors. That is, the code words are relatively sparse. Hence, the coding gain achieved by a turbo code is due in part to this feature, i.e., the reduction in the number of nearest neighboring code words that result from interleaving. Benedetto and Montorsi (1996) have shown that the number of nearest neighbors is reduced by a factor of N, where N is the block length of the interleaver.

The second important aspect in achieving good performance with turbo codes is the use of iterative decoding based on the MAP criterion. The basic configuration of the iterative decoder is illustrated in Figure 8.2–31. The iterative decoding algorithm usually employed in PCCC is the BCJR algorithm described in the paper by Bahl et al. (1974) or some variation of this algorithm. We observe from Figure 8.2–31 that the first decoder is provided with inputs from the demodulator corresponding to the information bits and the parity check bits $P1$. Similarly, the second decoder is provided with inputs from the demodulator corresponding to the information bits and the parity check bits $P2$. Each decoder

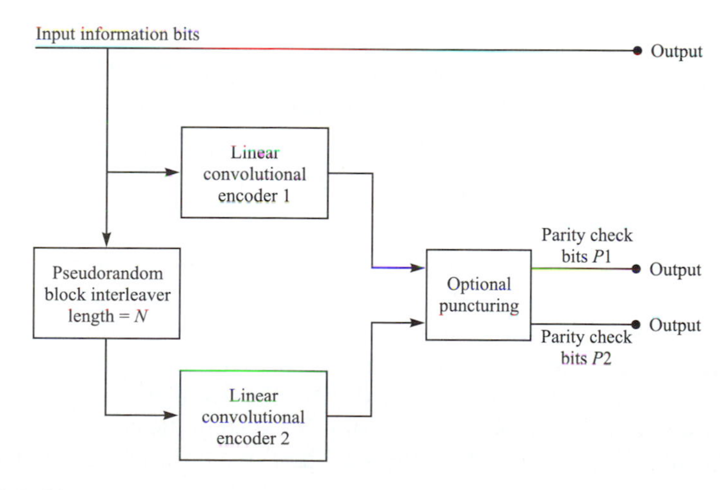

FIGURE 8.2–30
Encoder for parallel concatenated code (turbo code).

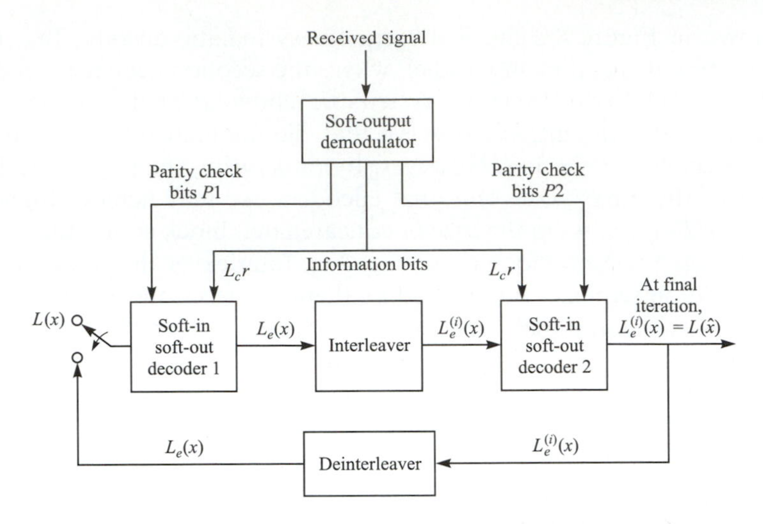

FIGURE 8.2–31
Block diagram of iterative decoder for a PCCC (turbo code).

attempts to decode the information sequence $\{x_i\}$ by computing the a posteriori probability for each information bit obtained from its respective parity bits and passing the soft-output information to the other decoder which uses this soft output as a priori information to improve its probability of correct detection. The soft-output information that is exchanged between the two decoders, as shown in Figure 8.2–31, is called the *extrinsic information.*

Let us define the output of the symbol-by-symbol MAP decoder as the a posteriori log-likelihood ratio for a transmitted information bit $x = 1$ and a transmitted information $x = -1$ as

$$L(\hat{x}) = \log \frac{P(x = 1|\mathbf{r})}{P(x = -1|\mathbf{r})}$$

where $\mathbf{r}$ is the vector of signal outputs from the demodulator. In computing $L(\hat{x})$, the decoder uses a priori information, defined as the log-likelihood ratio of a priori probabilities, i.e.,

$$L(x) = \log \frac{P(x = 1)}{P(x = -1)}$$

and channel values $L_c r$ at the output of the demodulator, where L_c is a constant† proportional to the received SNR. The decoder output consists of the soft outputs $L(\hat{x})$ for each information bit and corresponding soft outputs, called extrinsic information and denoted as $L_e(\hat{x})$, which represent the computed soft outputs for the information bits that are obtained from the decoder's received parity bits.

Hagenauer et al. (1996) have shown that for systematic codes, the soft output for the information bit x may be expressed as three additive terms, namely,

†It is easily shown that for the AWGN channel, $L_c = 4\mathcal{E}_c/N_0$, where $\mathcal{E}_c$ is the energy per code bit.

$$L(\hat{x}) = L_c r + L(x) + L_e(\hat{x}) \tag{8.2–58}$$

where the extrinsic information $L_e(\hat{x})$ is exchanged between the decoders. Hence, in the first iteration of the iterative decoding algorithm, decoder 1 computes the extrinsic information

$$L_{e1}(x) = L_1(\hat{x}) - [L_c r + L(x)] \tag{8.2–59}$$

where $L(x) = 0$ if, a priori, the information bits are equally probable. This extrinsic information from the first decoder, with appropriate interleaving, is passed to the decoder 2, which uses $L_{e1}(\hat{x})$, as the a priori value in place of $L(x)$ to compute $L_2(\hat{x})$. Hence, the extrinsic information value computed by decoder 2 for each information bit is

$$L_{e2}(\hat{x}) = L_2(\hat{x}) - [L_c r + L_{e1}(\hat{x})] \tag{8.2–60}$$

Then, decoder 1 will use the extrinsic information values $L_{e2}(\hat{x})$ as prior information in the second iteration. The computation is repeated in each iteration.

The iterative process is usually terminated after a predetermined number of iterations, when the soft-output values $L_2(\hat{x})$ stabilize and change little between successive iterations. In the final iteration, decoder 2 combines the two extrinsic information values in computing the soft output values

$$L_2(\hat{x}) = L_c r + L_{e1}(\hat{x}) + L_{e2}(\hat{x})$$

Typically, four iterations are adequate if the decoders are operating at a high enough SNR to achieve an error rate in the range 10^{-5} to 10^{-6}, whereas, about eight to ten iterations may be needed when the error rate is in the range of 10^{-5}, where the SNR is lower.

An important factor in the performance of the turbo code is the length of the interleaver, which is sometimes referred to as the *interleaver gain*. With a sufficiently large interleaver and iterative MAP decoding, the performance of a turbo code is very close to the Shannon limit. For example, a rate 1/2 turbo code of block length $N = 2^{16}$ with 18 iterations of decoding per bit, achieves an error probability of 10^{-5} at an SNR of 0.6 dB.

The major drawback with decoding turbo codes with large interleavers is decoding delay and the computational complexity inherent in the iterative decoding algorithm. In most data communication systems, however, the decoding delay is tolerable, and the additional computational complexity is usually justified by the significant coding gain that is achieved by the turbo code.

A second method for constructing concatenated convolutional codes with interleaving is serial concatenation. Benedetto et al. (1998) have investigated the construction and the performance of serial concatenated convolutional codes (SCCC) with interleaving and have developed an iterative decoding algorithm for such codes. In comparing the error rate performance of SCCC with PCCC (turbo codes), Benedetto et al. (1998) found that SCCC generally exhibit better performance than PCCC for error rates below 10^{-2}.

8.3 CODED MODULATION FOR BANDWIDTH-CONSTRAINED CHANNELS—TRELLIS-CODED MODULATION

In the treatment of block and convolutional codes in Sections 8.1 and 8.2, respectively, performance improvement was achieved by expanding the bandwidth of the transmitted signal by an amount equal to the reciprocal of the code rate. Recall for example that the improvement in performance achieved by an (n, k) binary block code with soft-decision decoding is approximately $10 \log_{10}(R_c d_{\min} - k \ln 2/\gamma_b)$ compared with uncoded binary or quaternary PSK. For example, when $\gamma_b = 10$, the (24, 12) extended Golay code gives a coding gain of 5 dB. This coding gain is achieved at a cost of doubling the bandwidth of the transmitted signal and, of course, at the additional cost in receiver implementation complexity. Thus, coding provides an effective method for trading bandwidth and implementation complexity against transmitter power. This situation applies to digital communication systems that are designed to operate in the power-limited region where $R/W < 1$.

In this section, we consider the use of coded signals for bandwidth-constrained channels. For such channels, the digital communication system is designed to use bandwidth-efficient multilevel amplitude and phase modulation, such as PAM, PSK, DPSK, or QAM, and operates in the region where $R/W > 1$. When coding is applied to the bandwidth-constrained channel, a performance gain is desired without expanding the signal bandwidth. This goal can be achieved by increasing the number of signals over the corresponding uncoded system to compensate for the redundancy introduced by the code.

For example, suppose that a system employing uncoded four-phase PSK modulation achieves an $R/W = 2$ (bits/s)/Hz at an error probability of 10^{-6}. For this error rate the SNR per bit is $\gamma_b = 10.5$ dB. We may try to reduce the SNR per bit by use of coded signals, but this must be done without expanding the bandwidth. If we choose a rate $R_c = 2/3$ code, it must be accompanied by an increase in the number of signal points from four (2 bits per symbol) to eight (3 bits per symbol). Thus, the rate 2/3 code used in conjunction with eight-phase PSK, for example, yields the same data throughput as uncoded four-phase PSK. However, we recall that an increase in the number of signal phases from four to eight requires an additional 4 dB approximately in signal power to maintain the same error rate. Hence, if coding is to provide a benefit, the performance gain of the rate 2/3 code must overcome this 4-dB penalty.

If the modulation is treated as a separate operation independent of the encoding, the use of very powerful codes (large-constraint-length convolutional codes or large-block-length block codes) is required to offset the loss and provide some significant coding gain. On the other hand, if the modulation is an integral part of the encoding process and is designed in conjunction with the code to increase the minimum Euclidean distance between pairs of coded signals, the loss from the expansion of the signal set is easily overcome and a significant coding gain is achieved with relatively simple codes. The key to this integrated modulation and coding approach is to devise an effective method for mapping the coded bits into signal points such that the minimum Euclidean distance is maximized. Such a method was developed by Ungerboeck (1982), based on the principle of *mapping by set partitioning*. We describe this principle by means of Examples 8.3–1 and 8.3–2.

Set partitioning. We begin with a given signal constellation, such as M-ary PAM, or QAM or PSK, and partition the constellation into subsets in a way that the minimum Euclidean distance between signal points in a subset is increased with each partition. The following two examples illustrate the set partitioning method proposed by Ungerboeck.

EXAMPLE 8.3–1: AN 8-PSK SIGNAL CONSTELLATION. Let us partition the eight-phase signal constellation shown in Figure 8.3–1 into subsets of increasing minimum Euclidean distance. In the eight-phase signal set, the signal points are located on a circle of radius $\sqrt{\mathcal{E}}$ and have a minimum distance separation of

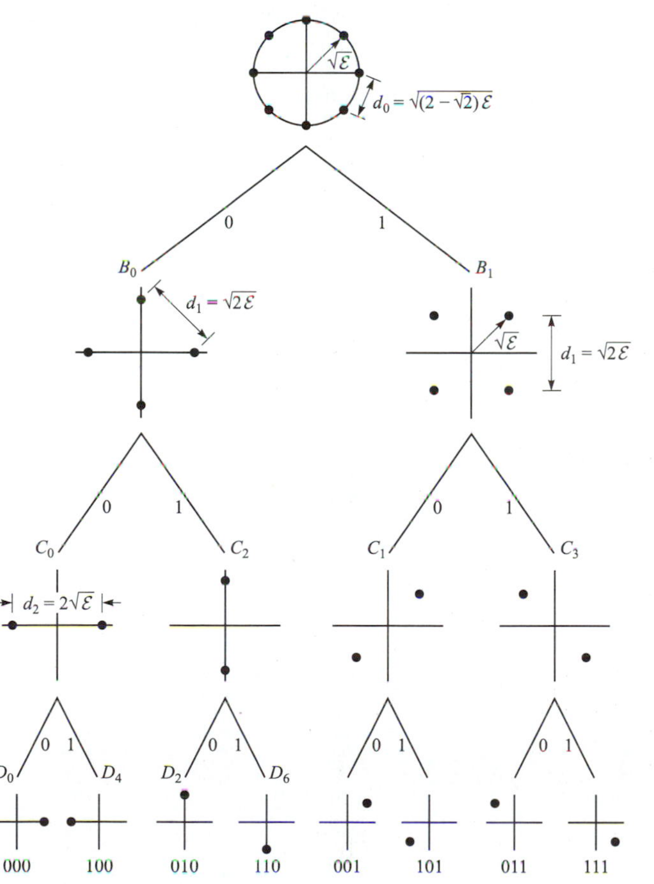

FIGURE 8.3–1
Set partitioning of an 8-PSK signal set.

$$d_0 = 2\sqrt{\mathcal{E}} \sin \tfrac{1}{8}\pi = \sqrt{(2 - \sqrt{2})\mathcal{E}} = 0.765\sqrt{\mathcal{E}}$$

In the first partitioning, the eight points are subdivided into two subsets of four points each, such that the minimum distance between points increases to $d_1 = \sqrt{2\mathcal{E}}$. In the second level of partitioning, each of the two subsets is subdivided into two subsets of two points, such that the minimum distance increases to $d_2 = 2\sqrt{\mathcal{E}}$. This results in four subsets of two points each.

Finally, the last stage of partitioning leads to eight subsets, where each subset contains a single point. Note that each level of partitioning increases the minimum Euclidean distance between signal points. The results of these three stages of partitioning are illustrated in Figure 8.3–1. The way in which the coded bits are mapped into the partitioned signal points is described below.

EXAMPLE 8.3–2: A 16-QAM SIGNAL CONSTELLATION. The 16-point rectangular signal constellation shown in Figure 8.3–2 is first divided into two subsets by assigning alternate points to each subset as illustrated in the figure. Thus, the distance between points is increased from $2\sqrt{\mathcal{E}}$ to $2\sqrt{2\mathcal{E}}$ by the first partitioning. Further partitioning of the two subsets leads to greater separation in Euclidean distance between signal points as illustrated in Figure 8.3–2. It is interesting to note that for the rectangular signal constellations, each level of partitioning increases the minimum Euclidean distance by $\sqrt{2}$, i.e., $d_{i+1}/d_i = \sqrt{2}$ for all i.

In these two examples, the partitioning was carried out to the limit where each subset contains only a single point. In general, this may not be necessary. For example, the 16-point QAM signal constellation may be partitioned only twice, to yield four subsets of four points each. Similarly, the eight-phase PSK signal constellation can be partitioned twice, to yield four subsets of two points each.

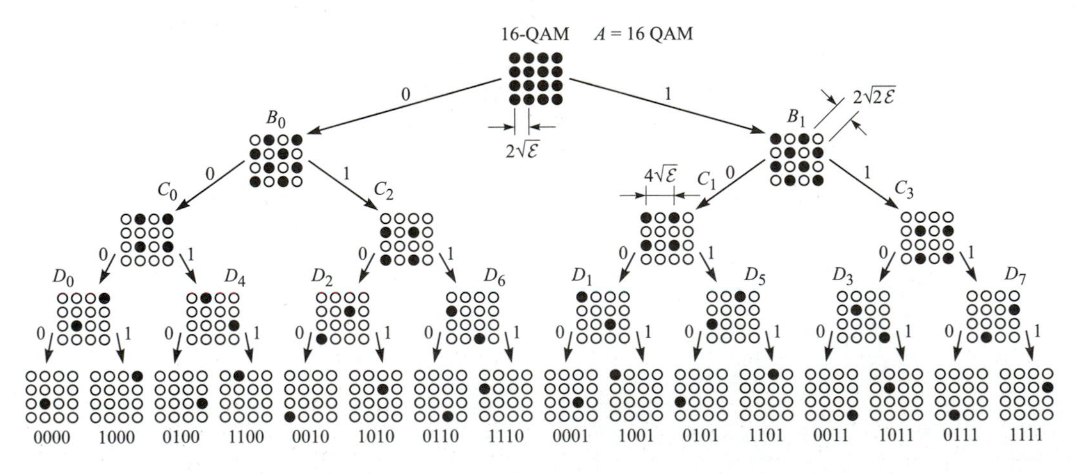

FIGURE 8.3–2
Set partitioning of 16-QAM signal.

Trellis-coded modulation (TCM). The degree to which the signal is partitioned depends on the characteristics of the code. In general, the encoding process is performed as illustrated in Figure 8.3–3. A block of m information bits is separated into two groups of length k_1 and k_2, respectively. The k_1 bits are encoded into n bits, while the k_2 bits are left uncoded. Then, the n bits from the encoder are used to select one of the possible subsets in the partitioned signal set, while the k_2 bits are used to select one of 2^{k_2} signal points in each subset. When $k_2 = 0$, all m information bits are encoded.

The assignment of signal subsets to state transitions in the trellis is based on three heuristic rules devised by Ungerboeck (1982). The rules are

1. Use all subsets with equal frequency in the trellis.
2. Transitions originating from the same state or merging into the same state in the trellis are assigned subsets that are separated by the largest Euclidean distance.
3. Parallel state transitions (when they occur) are assigned signal points separated by the largest Euclidean distance. Parallel transitions in the trellis are characteristic of TCM that contains one or more uncoded information bits.

EXAMPLE 8.3–3. Consider the use of the rate 1/2 convolutional encoder shown in Figure 8.3–4a to encode one information bit while the second information bit is left uncoded. This code results in the four-state trellis shown in Figure 8.3–4b. When used in conjunction with an eight-point signal constellation, such as eight-point PSK or QAM, the two encoded output bits are used to select one of the four subsets in the partitioned signal constellation, while the remaining information bit is used to select one of the two points within each subset. Let us use the eight-point PSK consellation to complete this example. The four subsets assigned to the trellis in Figure 8.3–4b correspond to the subsets labeled C_0, C_1, C_2, C_3 in Figure 8.3–1. Note that the Euclidean distance of points within any subset is $d_2 = 2\sqrt{\mathcal{E}}$ and the largest minimum distance between signal points in any pair of subsets is $d_1 = \sqrt{2\mathcal{E}}$. The mappings of the coded bits (c_2, c_1) and the uncoded bit c_3 to the state transitions, using the convention (c_3, c_2, c_1) are shown in Figure 8.3–4c. We note that each trellis state has two parallel transitions, corresponding to the two possible values of the uncoded bit. The phase assignments in the eight-point PSK constellation are shown in Figure 8.3–4d. It should be noted that the mapping of the bits (c_3, c_2, c_1) into the eight signal points in the constellation is not unique. Several other mappings are possible. For example, an equally good mapping is obtained if the four-point subsets B_0 and B_1 shown in Figure 8.3–1, are interchanged, so that the signal points in the subsets C_0, C_1, C_2, and C_3 will also change.

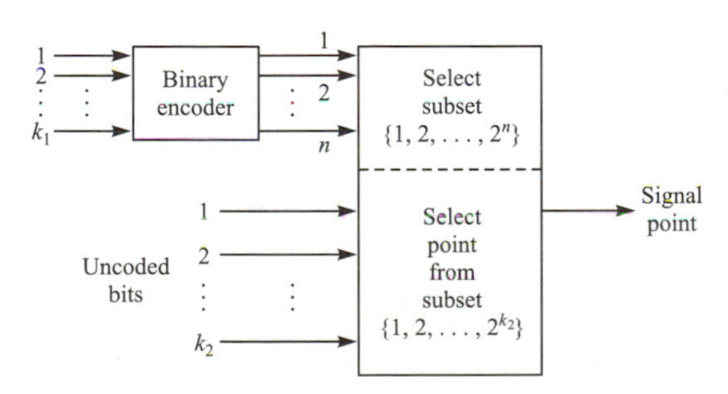

FIGURE 8.3–3
General structure of combined encoder/modulator.

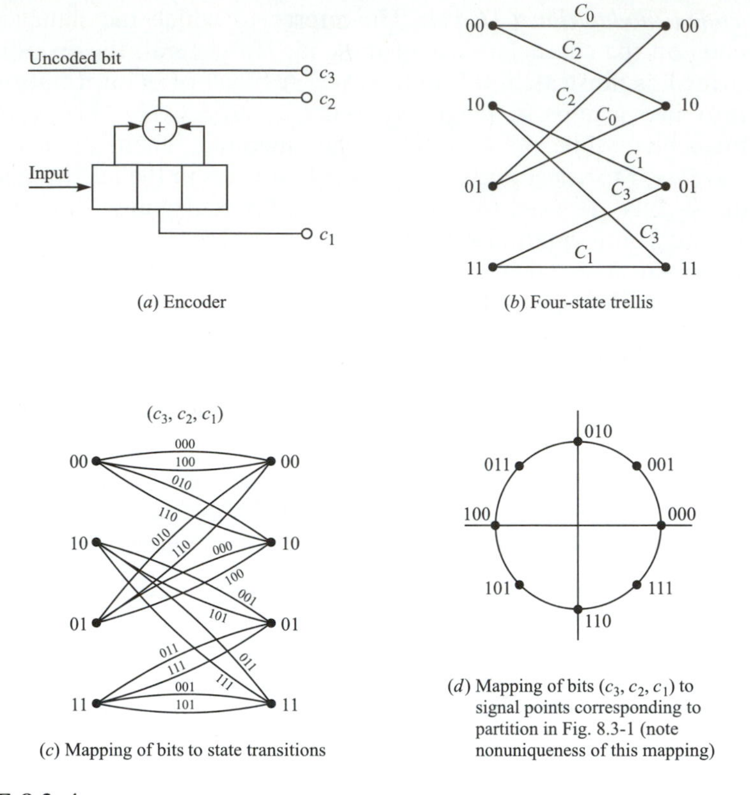

FIGURE 8.3–4
Four-state trellis-coded modulation with 8-PSK signal constellation.

In general, the number of states $S = 2^v$ in the code trellis is a function of the number of memory elements in the encoder. Hence, we may increase the number of trellis states while maintaining the same code rate. For example, Figure 8.3–5 illustrates a rate 2/3 code that has eight trellis states. In this case, both information bits are coded.

Let us now evaluate the performance of the trellis-coded 8-PSK and compare its performance with that of uncoded 4-*PSK*, which we use as a reference in

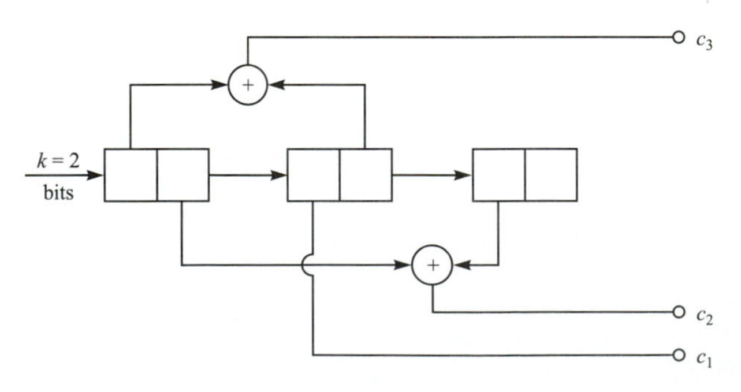

FIGURE 8.3–5
Rate $\frac{2}{3}$, eight-state trellis code.

measuring the coding gain of the trellis-coded modulation. Uncoded 4-*PSK* employs the signal points in either subset B_0 or B_1 of Figure 8.3–1, for which the minimum distance of the signal points is $\sqrt{2\mathcal{E}}$. Note that the 4-PSK signal corresponds to a trivial one-state trellis with four parallel state transitions, as shown in Figure 8.3–6. The subsets D_0, D_2, D_4, and D_6 in Figure 8.3–1 are used as the signal points for the purpose of illustration.

For the trellis-coded 8-PSK modulation, we use the four-state trellis shown in Figure 8.3–4b and c. We observe that each branch in the trellis corresponds to one of the four subsets C_0, C_1, C_2, or C_3. As indicated above, for the eight-point constellation, each of the subsets C_0, C_1, C_2, and C_3 contains two signal points. Hence, the state transition C_0 contains the two signal points corresponding to the bits $(c_3c_2c_1) = (000)$ and (100), or $(0, 4)$ in octal representation. Similarly, C_2 contains the two signal points corresponding to (010) and (110) or to (2, 6) in octal, C_1 contains the points corresponding to (001) and (101) or (1, 5) in octal, and C_3 contains the points corresponding to (011) and (111) or (3, 7) in octal. Thus, each transition in the four-state trellis contains two parallel paths, as previously indicated. As shown in Figure 8.3–6, any two signal paths that diverge from one state and remerge at the same state after more than one transition have a squared Euclidean distance of $d_0^2 + 2d_1^2 = d_0^2 + d_2^2$ between them. For example, the signal paths 0, 0, 0 and 2, 1, 2 are separated by $d_0^2 + d_2^2 = [(0.765)^2 + 4]\mathcal{E}$

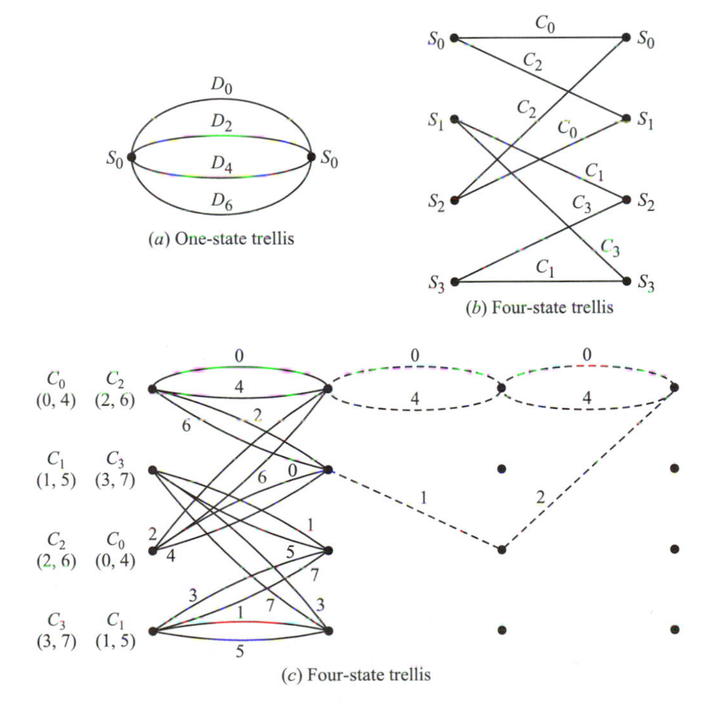

FIGURE 8.3–6
Uncoded 4-PSK and trellis-coded 8-PSK modulation.

$= 4.585\mathcal{E}$. On the other hand, the squared Euclidean distance between parallel transitions is $d_2^2 = 4\mathcal{E}$. Hence, the minimum Euclidean distance separation between paths that diverge from any state and remerge at the same state in the four-state trellis is $d_2 = 2\sqrt{\mathcal{E}}$. The minimum distance in the trellis code is called the *free Euclidean distance* and denoted by D_{fed}.

In the four-state trellis of Figure 8.3–6b, $D_{\text{fed}} = 2\sqrt{\mathcal{E}}$. When compared with the Euclidean distance $d_0 = \sqrt{2\mathcal{E}}$ for the uncoded 4-PSK modulation, we observe that the four-state trellis code gives a coding gain of 3 dB.

We should emphasize that the four-state trellis code illustrated in Figure 8.3–6b is optimum in the sense that it provides the largest free Euclidean distance. Clearly, many other four-state trellis codes can be constructed, including the one shown in Figure 8.3–7, which consists of four distinct transitions from each state to all other states. However, neither this code nor any of the other possible four-state trellis codes gives a larger D_{fed}.

In the four-state trellis code, the parallel transitions were separated by the Euclidean distance $2\sqrt{\mathcal{E}}$, which is also D_{fed}. Hence, the coding gain of 3 dB is limited by the distance of the parallel transitions. Larger gains in performance relative to uncoded 4-PSK can be achieved by using trellis codes with more states, which allow for the elimination of the parallel transitions. Thus, trellis codes with eight or more states would use distinct transitions to obtain a larger D_{fed}.

For example, in Figure 8.3–8, we illustrate an eight-state trellis code due to Ungerboeck (1982) for the 8-PSK signal constellation. The state transitions for maximizing the free Euclidean distance were determined from application of the three basic rules given above. In this case, note that the minimum squared Euclidean distance is

$$D_{\text{fed}}^2 = d_0^2 + 2d_1^2 = 4.585\mathcal{E}$$

which, when compared with $d_0^2 = 2\mathcal{E}$ for uncoded 4-PSK, represents a gain of 3.6 dB. Ungerboeck (1982, 1987) has also found rate 2/3 trellis codes with 16, 32, 64, 128, and 256 states that achieve coding gains ranging from 4 to 5.75 dB for 8-PSK modulation.

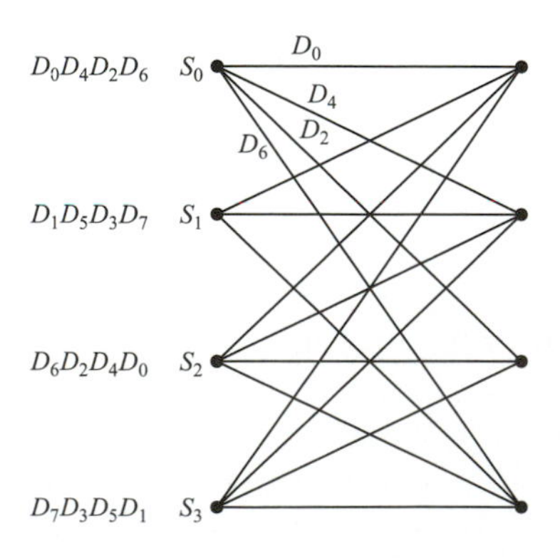

FIGURE 8.3–7
An alternative four-state trellis code.

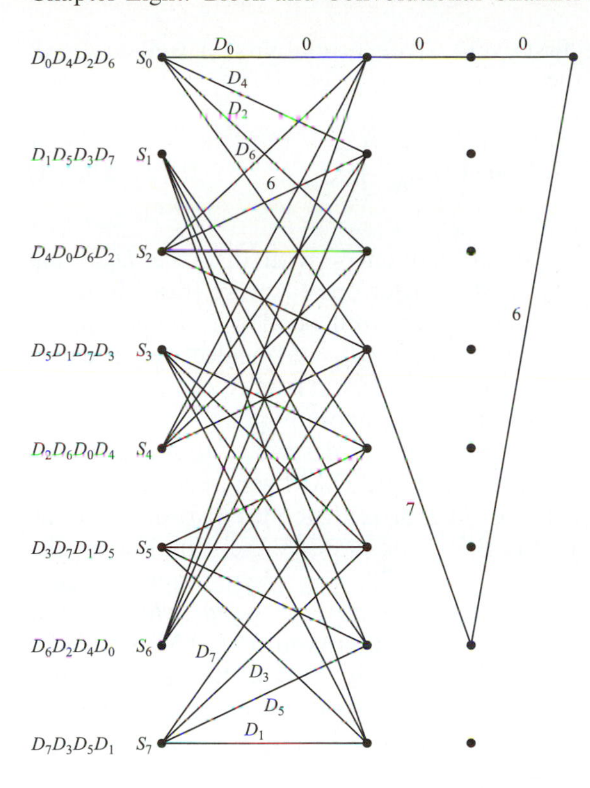

FIGURE 8.3–8
Eight-state trellis code for coded 8-PSK modulation.

The basic principle of set partitioning is easily extended to larger PSK signal constellations that yield greater bandwidth efficiency. For example, 3 (bits/s)/Hz can be achieved with either uncoded 8-PSK or with trellis-coded 16-PSK modulation. Ungerboeck (1987) has devised trellis codes and has evaluated the coding gains achieved by simple rate 1/2 and rate 2/3 convolutional codes for the 16-PSK signal constellations. The results are summarized below.

Soft-decision Viterbi decoding for trellis-coded modulation is accomplished in two steps. Since each branch in the trellis corresponds to a signal subset, the first step in decoding is to determine the best signal point within each subset, i.e., the point in each subset that is closest in distance to the received point. We may call this *subset decoding*. In the second step, the signal point selected from each subset and its squared distance metric are used for the corresponding branch in the Viterbi algorithm to determine the signal path through the code trellis that has the minimum sum of squared distances from the sequence of received (noisy channel output) signals.

The error rate performance of the trellis-coded signals in the presence of additive Gaussian noise can be evaluated by following the procedure described in Section 8.2 for convolutional codes. Recall that this procedure involves the computation of the probability of error for all different error events and summing these error event probabilities to obtain a union bound on the first-event error probability. Note, however, that, at high SNR, the first-event error probability is dominated by the leading term, which has the minimum distance D_{fed}.

Consequently, at high SNR, the first-event error probability is well approximated as

$$P_e \approx N_{\text{fed}} Q\left(\sqrt{\frac{D_{\text{fed}}^2}{2N_0}}\right) \tag{8.3–1}$$

where N_{fed} denotes the number of signal sequences with distance D_{fed} that diverge at any state and remerge at that state after one or more transitions.

In computing the coding gain achieved by trellis-coded modulation, we usually focus on the gain achieved by increasing D_{fed} and neglect the effect of N_{fed}. However, trellis codes with a large number of states may result in a large N_{fed} that cannot be ignored in assessing the overall coding gain.

In addition to the trellis-coded PSK modulations described above, powerful trellis codes have also been developed for PAM and QAM signal constellations. Of particular practical importance is the class of trellis-coded two-dimensional rectangular signal constellations. Figure 8.3–9 illustrates these signal constellations for M-QAM where $M = 16$, 32, 64, and 128. The $M = 32$ and 128 constellations have a cross pattern and are sometimes called *cross-constellations*. The underlying rectangular grid containing the signal points in M-QAM is called a *lattice of type* Z_2 (the subscript indicates the dimensionality of the space). When set partitioning is applied to this class of signal constellations, the minimum Euclidean distance between successive partitions is $d_{i+1}/d_i = \sqrt{2}$ for all i, as previously observed in Example 8.3–2.

Figure 8.3–10 illustrates an eight-state trellis code that can be used with any of the M-QAM rectangular signal constellations for which $M = 2^k$, where

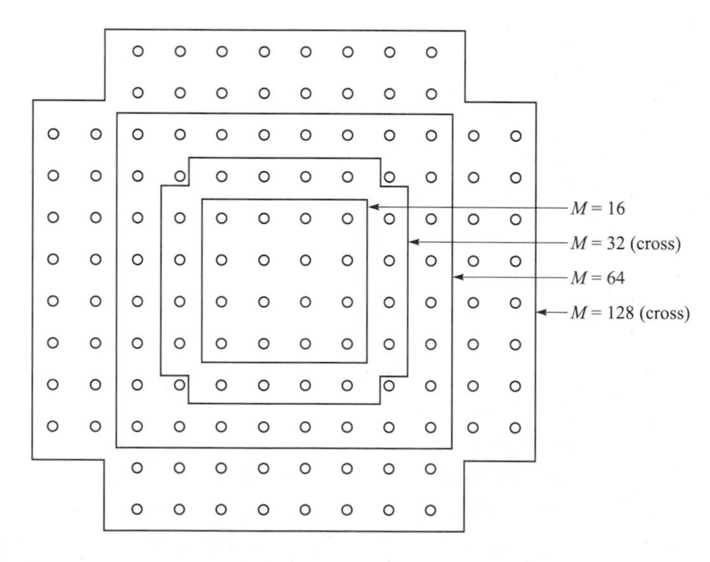

FIGURE 8.3–9
Rectangular two-dimensional (QAM) signal constellations.

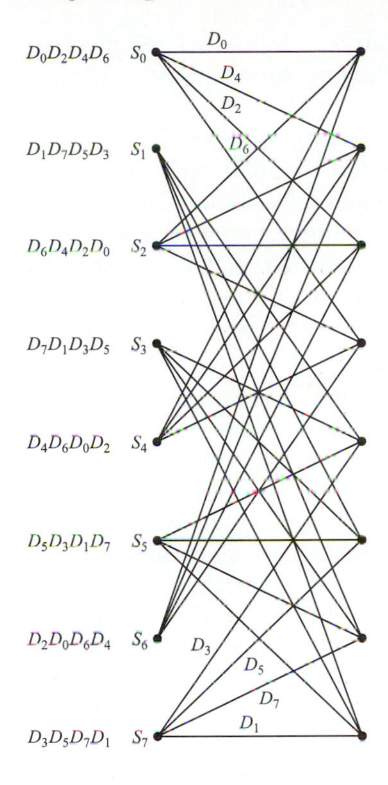

FIGURE 8.3–10
Eight-state trellis for rectangular QAM signal constellations.

$k = 4, 5, 6, \ldots$, etc. With the eight-state trellis, we associate eight signal subsets, so that any of the M-QAM signal sets for $M \geqslant 16$ are suitable. For $M = 2^{m+1}$, two input bits ($k_1 = 2$) are encoded into $n = 3\,(n = k_1 + 1)$ bits that are used to select one of the eight states corresponding to the eight subsets. The additional $k_2 = m - k_1$ input bits are used to select signal points within a subset, and result in parallel transitions in the eight-state trellis. Hence, 16-QAM involves two parallel transitions in each branch of the trellis. More generally, the choice of an $M = 2^{m+1}$-point QAM signal constellation implies that the eight-state trellis contains 2^{m-2} parallel transitions in each branch.

The assignment of signal subsets to transitions is based on the same set of basic (heuristic) rules described above for the 8-PSK signal constellation. Thus, the four (branches) transitions originating from or leading to the same state are assigned either the subsets D_0, D_2, D_4, D_6 or D_1, D_3, D_5, D_7. Parallel transitions are assigned signal points contained within the corresponding subsets. This eight-state trellis code provides a coding gain of 4 dB. The Euclidean distance of parallel transitions exceeds the free Euclidean distance, and, hence, the code performance is not limited by parallel transitions.

Larger size trellis codes for M-QAM provide even larger coding gains. For example, trellis codes with 2^{ν} states for an $M = 2^{m+1}$ QAM signal constellation can be constructed by convolutionally encoding k_1 input bits into $k_1 + 1$ output bits. Thus, a rate $R_c = k_1/(k_1 + 1)$ convolutional code is employed for this purpose. Usually, the choice of $k_1 = 2$ provides a significant fraction of the total coding gain that is achievable. The additional $k_2 = m - k_1$ input bits are uncoded and are transmitted in each signal interval by selecting signal points within a subset.

Tables 8.3–1 to 8.3–3, taken from the paper by Ungerboeck (1987), provide a summary of coding gains achievable with trellis-coded modulation. Table 8.3–1 summarizes the coding gains achieved for trellis-coded (one-dimensional) PAM modulation with rate 1/2 trellis codes. Note that the coding gain with a 128-state trellis code is 5.8 dB for octal PAM, which is close to the channel cutoff rate R_0 and less than 4 dB from the channel capacity limit for error rates in the range of 10^{-6}–10^{-8}. We should also observe that the number of paths N_{fed} with free Euclidean distance D_{fed} becomes large with an increase in the number of states.

Table 8.3–2 lists the coding gain for trellis-coded 16-PSK. Again, we observe that the coding gain for eight or more trellis stages exceeds 4 dB, relative to uncoded 8-PSK. A simple rate 1/2 code yields 5.33 dB gain with a 128-stage trellis.

Table 8.3–3 contains the coding gains obtained with trellis-coded QAM signals. Relatively simple rate 2/3 trellis codes yield a gain of 6 dB with 128 trellis stages for $m = 3$ and 4.

The results in these tables clearly illustrate the significant coding gains that are achievable with relatively simple trellis codes. A 6-dB coding gain is close to the cutoff rate R_0 for the signal sets under consideration. Additional gains that would lead to transmission in the vicinity of the channel capacity bound are difficult to attain without a significant increase in coding/decoding complexity. Continued partitioning of large signal sets quickly leads to signal point separation within any subset that exceeds the free Euclidean distance of the code. In such cases, parallel transitions are no longer the limiting factor on D_{fed}. Usually, a partition to eight subsets is sufficient to obtain a coding gain of 5–6 dB with simple rate 1/2 or rate 2/3 trellis codes with either 64 or 128 trellis stages, as indicated in Tables 8.3–1 to 8.3–3.

Convolutional encoders for the linear trellis codes listed in Tables 8.3–1 to 8.3–3 for the M-PAM, M-PSK, and M-QAM signal constellations are given in the papers by Ungerboeck (1982, 1987). The encoders may be realized either with feedback or without feedback. For example Figure 8.3–11 illustrates three feedback-free convolutional encoders corresponding to 4-, 8-, and 16-state trellis codes for 8-PSK and 16-QAM signal constellations. Equivalent realizations of these trellis codes based on systematic convolutional encoders with feedback are shown in Figure 8.3–12. Usually, the systematic convolutional encoders are preferred in practical applications.

A potential problem with linear trellis codes is that the modulated signal sets are not usually invariant to phase rotations. This poses a problem in practical applications where differential encoding is usually employed to avoid phase ambiguities when a receiver must recover the carrier phase after a temporary loss of signal. For two-dimensional signal constellations, it is possible to achieve 180° phase invariance by use of a linear trellis code. However, it is not possible to achieve 90° phase invariance with a linear code. In such a case, a non-linear code must be used. The problem of phase invariance and differential encoding/decod-

TABLE 8.3–1
Coding gains for trellis-coded PAM signals

Number of states	k_1	Code rate $\frac{k_1}{k_1+1}$	$m=1$ coding gain (dB) of 4-PAM versus uncoded 2-PAM	$m=2$ coding gain (dB) of 8-PAM versus uncoded 4-PAM	$m \to \infty$ asymptotic coding gain (dB)	$m \to \infty$ N_{fed}
4	1	1/2	2.55	3.31	3.52	4
8	1	1/2	3.01	3.77	3.97	4
16	1	1/2	3.42	4.18	4.39	8
32	1	1/2	4.15	4.91	5.11	12
64	1	1/2	4.47	5.23	5.44	36
128	1	1/2	5.05	5.81	6.02	66

Source: Ungerboeck (1987).

TABLE 8.3–2
Coding gains for trellis-coded 16-PSK modulation

Number of states	k_1	Code rate $\frac{k_1}{k_1+1}$	$m=3$ coding gain (dB) of 16-PSK versus uncoded 8-PSK	$m \to \infty$ N_{fed}
4	1	1/2	3.54	4
8	1	1/2	4.01	4
16	1	1/2	4.44	8
32	1	1/2	5.13	8
64	1	1/2	5.33	2
128	1	1/2	5.33	2
256	2	2/3	5.51	8

Source: Ungerboeck (1987).

TABLE 8.3–3
Coding gains for trellis-coded QAM modulation

Number of states	k_1	Code rate $\frac{k_1}{k_1+1}$	$m=3$ gain (dB) of 16-QAM versus uncoded 8-QAM	$m=4$ gain (dB) of 32-QAM versus uncoded 16-QAM	$m=5$ gain (dB) of 64-QAM versus uncoded 32-QAM	$m=\infty$ asymptotic coding gain (dB)	N_{fed}
4	1	1/2	3.01	3.01	2.80	3.01	4
8	2	2/3	3.98	3.98	3.77	3.98	16
16	2	2/3	4.77	4.77	4.56	4.77	56
32	2	2/3	4.77	4.77	4.56	4.77	16
64	2	2/3	5.44	5.44	5.23	5.44	56
128	2	2/3	6.02	6.02	5.81	6.02	344
256	2	2/3	6.02	6.02	5.81	6.02	44

Source: Ungerboeck (1987).

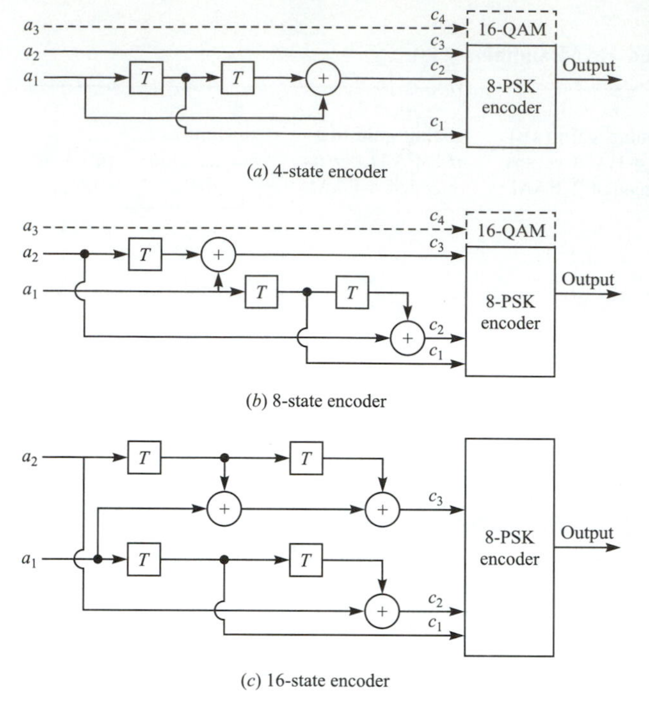

FIGURE 8.3–11 Minimal feedback-free convolutional encoders for 8-PSK and 16-QAM signals. [*From Ungerboeck (1982). © 1982 IEEE.*]

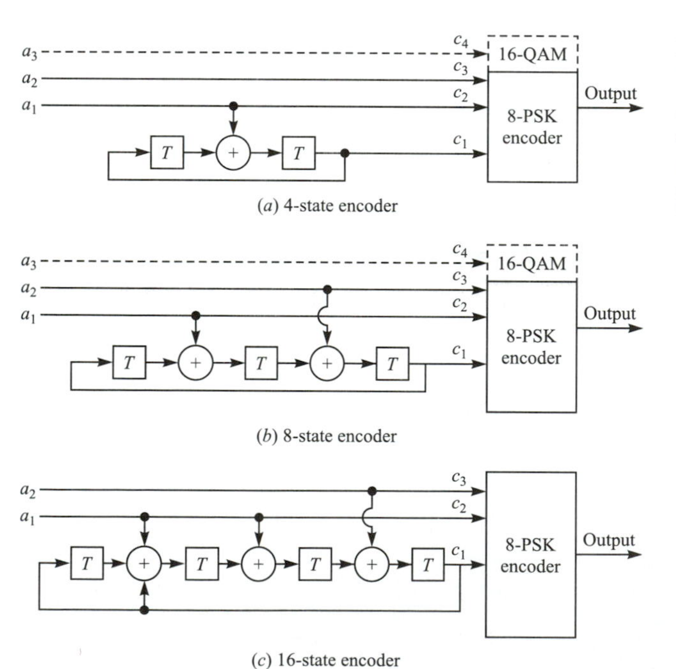

FIGURE 8.3–12 Equivalent realizations of systematic convolutional encoders with feedback for 8-PSK and 16-QAM. [*From Ungerboeck (1982). © 1982 IEEE.*

ing was solved by Wei (1984a,b), who devised linear and non-linear trellis codes that are rotationally invariant under either 180° or 90° phase rotations, respectively. For example, Figure 8.3–13 illustrates a non-linear eight-state convolutional encoder for a 32-QAM rectangular signal constellation that is invariant under 90° phase rotations. This trellis code has been adopted as an international standard (V.32 and V.33) for 9600 and 14,000 bits/s (high-speed) telephone line modems.

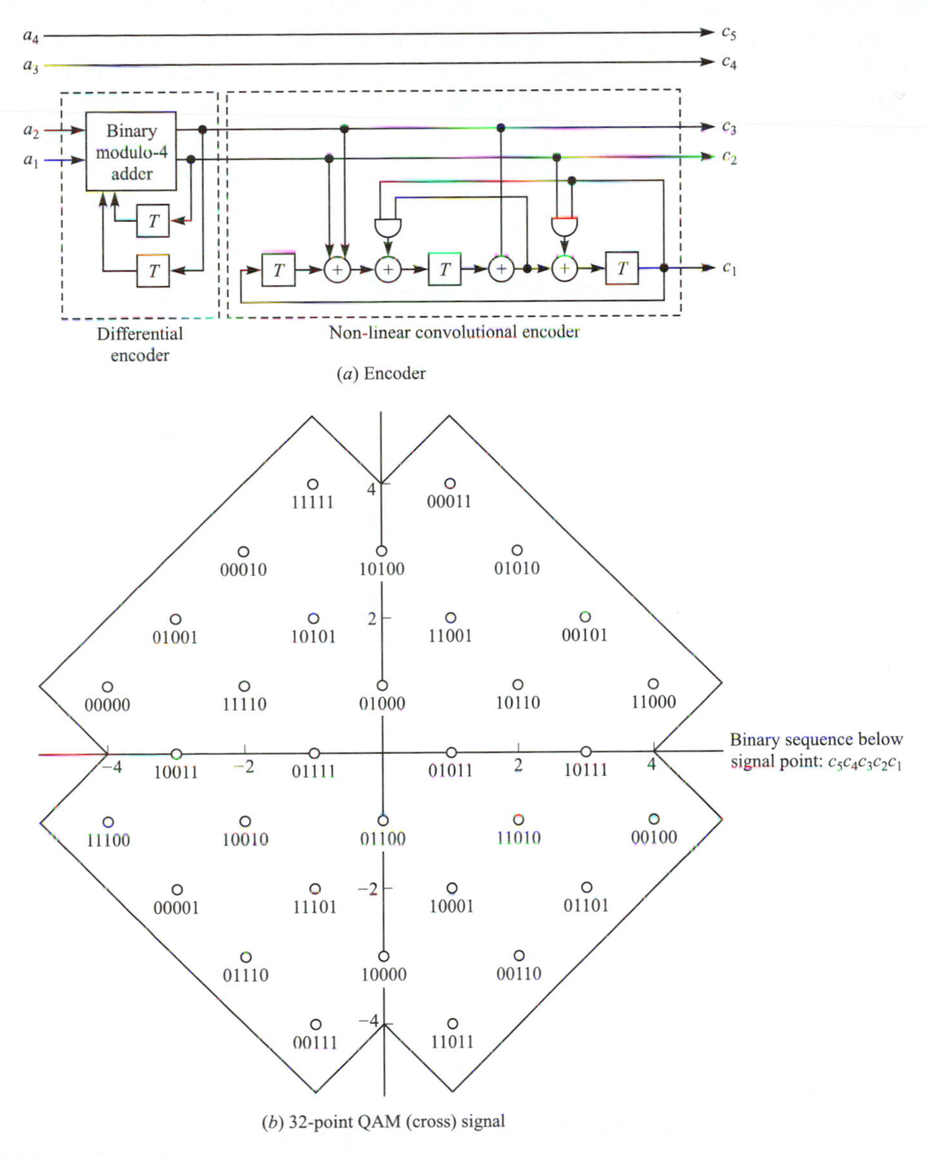

FIGURE 8.3–13
Eight-state non-linear convolutional encoder for 32-QAM signal set that exhibits invariance under 90° phase rotations.

Trellis-coded modulation schemes have also been developed for multidimensional signals. In practical systems, multidimensional signals are transmitted as a sequence of either one-dimensional (PAM) or two-dimensional (QAM) signals. Trellis codes based on 4-, 8-, and 16-dimensional signal constellations have been constructed, and some of these codes have been implemented in commercially available modems. A potential advantage of trellis-coded multidimensional signals is that we can use smaller constituent two-dimensional signal constellations that allow for a trade-off between coding gain and implementation complexity. For example, a 16-state linear four-dimensional code, also designed by Wei (1987), is currently used as one of the codes, for the V.34 telephone modem standard. The constituent two-dimensional signal constellation contains a maximum of 1664 signal points. The modem can transmit as many as 10 bits per symbol (eight uncoded bits) to achieve data rates as high as 33,600 bits/s. The papers by Wei (1987), Ungerboeck (1987), Gersho and Lawrence (1984), and Forney et al. (1984) treat multidimensional signal constellations for trellis-coded modulation.

Finally, we should mention that another design technique for trellis-coded modulation based on lattices and cosets of a sublattice has been described by Calderbank and Sloane (1987) and Forney (1988). This method for constructing trellis codes provides an alternative to the set partitioning method described above. However, the two methods are closely related. In this alternative method, a block of k_1 bits is fed to a convolutional encoder. Each block of k_1 input bits produces an output symbol that is a coset of the sublattice Λ', which is a subset of the chosen lattice. A second block of k_2 input bits is used to select one of the points in the coset at the output of the convolutional encoder. It is apparent that the cosets of the sublattice are akin to the subsets in set partitioning and the elements of the cosets are akin to the signal points within a subset. This method has led to the discovery of new powerful trellis codes involving larger signal constellations, many of which are listed in the paper by Calderbank and Sloane (1987).

Turbo-coded bandwidth efficient modulation. The performance of TCM can be further improved by code concatenation. There are several different methods described in the literature. We shall briefly describe two schemes for code concatenation using parallel concatenated codes, which we simply refer to as turbo coding.

In one scheme, described in the paper by Le Goff et al. (1994), the information sequence is fed to a binary turbo encoder that employs a parallel concatenation of a component convolutional code with interleaving to generate a systematic binary turbo code. As shown in Figure 8.3–14, the output of the turbo encoder is ultimately connected to the signal mapper after the binary sequence from the turbo code has been appropriately multiplexed, the parity bit sequence has been punctured to achieve the desired code rate, and the data and parity sequences have been interleaved. Gray mapping is typically used in mapping coded bits to modulation signal points, separately for the in-phase (I) and quadrature (Q) signal components.

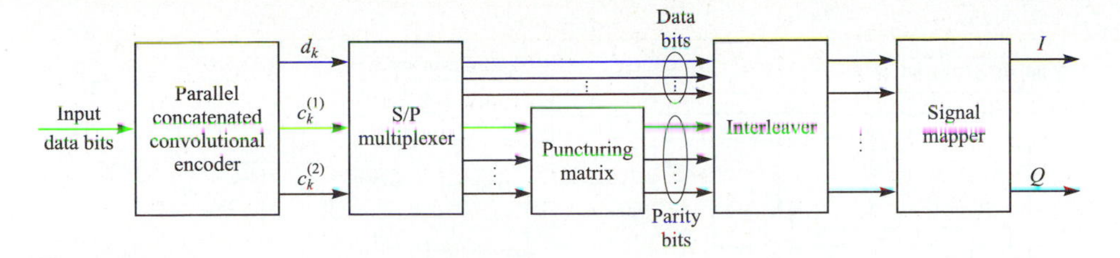

FIGURE 8.3–14
Encoder for concatenation of a PCCC (turbo code) with TCM.

Figure 8.3–15 illustrates the block diagram of the decoder for this turbo-coding scheme. Based on each received I and Q symbol, the receiver computes the logarithm of the likelihood ratio or the MAP of each systematic bit and each parity bit. After deinterleaving, depuncturing, and demultiplexing of these logarithmic metrics, the systematic and parity bit information are fed to the standard binary turbo decoder.

This scheme for constructing turbo-coded bandwidth efficient modulation imposes no constraints on the type or size of the signal constellation. In addition, this scheme can be matched to any conventional binary turbo code. In fact, this scheme is also suitable if the turbo code is replaced by a serially concatenated convolutional code.

A second scheme employs a conventional Ungerboeck trellis code with interleaving to yield a parallel concatenated TCM. The basic configuration of the turbo TCM encoder, as described in the paper by Robertson and Wörz (1998), is illustrated in Figure 8.3–16. To avoid a rate loss, the parity sequence is punctured, as described below, in such a way that all information bits are transmitted only once, and the parity bits from the two encoders are alternately punctured. The block interleaver operates on groups of $m-1$ information bits, where the signal constellation consists of 2^m signal points.

To illustrate the group interleaving and puncturing, let us consider a rate $R_c = \frac{2}{3}$ TCM code, a block interleaver of length $N = 6$, and 8-PSK modulation ($m = 3$). Hence, the number of information bits per block is $N(m-1) = 12$, and the interleaving is performed on pairs of information bits as shown in Figure 8.3–16 where, for example, a pair of bits in an even position (2, 4, 6) is mapped to

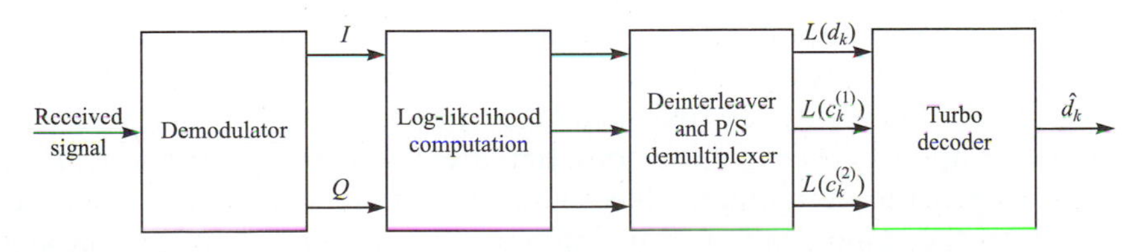

FIGURE 8.3–15
Decoder for concatenated PCCC/TCM code.

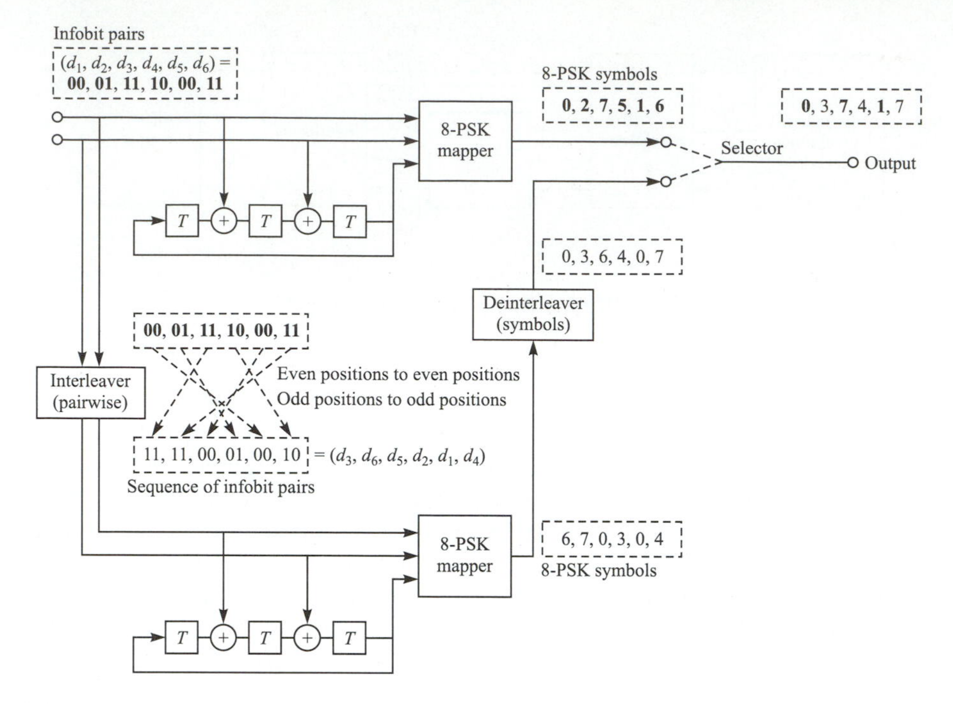

FIGURE 8.3–16
Turbo TCM encoder shown for 8-PSK with two-dimensional component codes memory 3. An example of interleaving with $N = 6$ is shown. Bold letters indicate that symbols or pairs of bits correspond to the upper encoder. [*From Robertson and Wörz (1998); © 1998 IEEE.*]

another even position and a pair of bits in an odd position is mapped to another odd position. The output of the second TCM encoder is deinterleaved symbolwise as illustrated in Figure 8.3–16, and the output symbol sequence is obtained by puncturing the two signal-point sequences, i.e., by selecting every other symbol from each of the two sequences. That is, we select the even-numbered symbols from the top symbol mapper and the odd-numbered symbols from the bottom symbol mapper. (In general, some of the information bits can remain uncoded, depending on the signal constellation and the signal mapping. In this example, both information bits are coded.)

A block diagram of the turbo decoder is shown in Figure 8.3–17. In the conventional binary iterative turbo decoder, each output of each component decoder is usually split into three parts, namely, the systematic part, the a priori part, and the extrinsic part, where only the latter is passed between the two decoders. In this TCM scheme, the systematic part cannot be separated from the extrinsic component, because the noise that affects the parity component also affects the systematic component due to the fact that both components are transmitted by the same symbol. This implies that the output of the decoders can only be split into two components, namely, the a priori information and the extrinsic-systematic information. Hence, each decoder passes the extrinsic-sys-

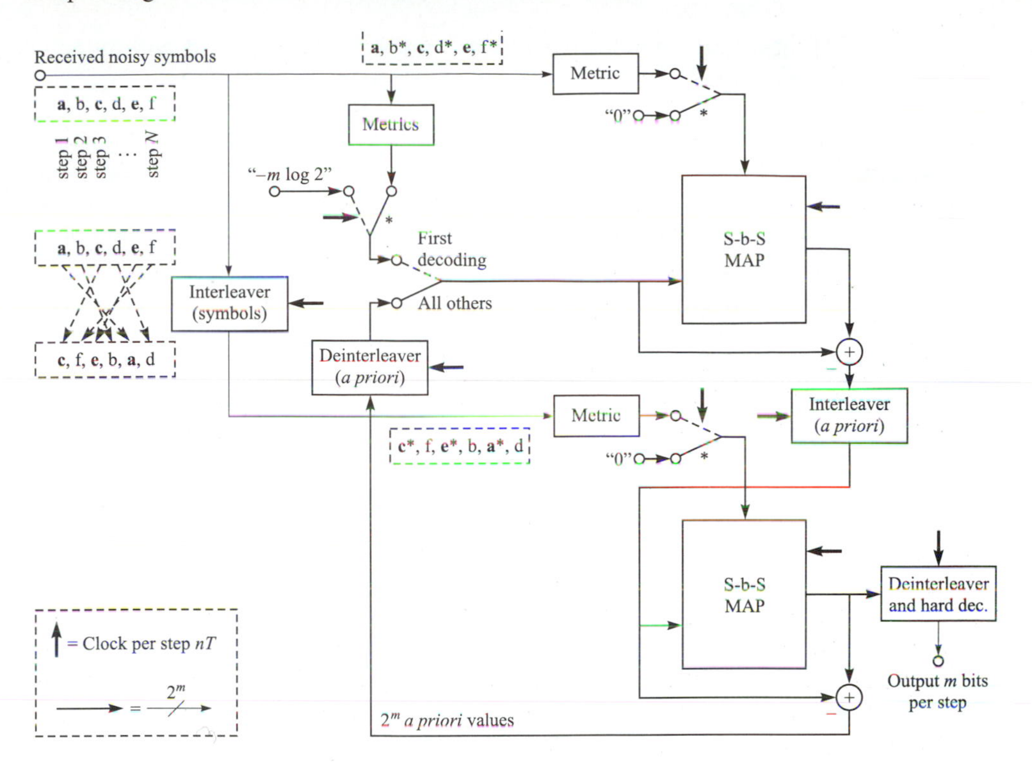

FIGURE 8.3–17
Turbo TCM decoder corresponding to the encoder in Figure 8.3–16. [*From Robertson and Wörz (1998); © 1998 IEEE.*]

tematic information to the other decoder. Each decoder ignores those symbols where the pertinent parity bit was not sent and obtains the systematic information through its a priori input. In the first iteration, the a priori input of the first decoder is initialized with the missing systematic information. Details of the iterative decoder computations are given in the paper by Robertson and Wörz (1998). An additional coding gain of about 1.7 dB has been achieved by use of a turbo TCM compared to conventional TCM, at error rates in the vicinity of 10^{-4}. This means that turbo TCM achieves a performance close to the Shannon capacity on an AWGN channel.

8.4 BIBLIOGRAPHICAL NOTES AND REFERENCES

The pioneering work on coding and coded waveforms for digital communications was done by Shannon (1948a,b), Hamming (1950), and Golay (1949). These works were rapidly followed with papers on code performance by

Gilbert (1952), new codes by Muller (1954) and Reed (1954), and coding techniques for noisy channels by Elias (1954, 1955) and Slepian (1956).

During the period 1960–1970, there were a number of significant contributions in the development of coding theory and decoding algorithms. In particular, we cite the papers by Reed and Solomon (1960) on Reed–Solomon codes, the papers by Hocquenghem (1959) and Bose and Ray-Chaudhuri (1960a,b) on BCH codes, and the Ph.D dissertation of Forney (1966a) on concatenated codes. These works were followed by the papers of Goppa (1970, 1971) on the construction of a new class of linear cyclic codes, now called Goppa codes (see also Berlekamp, 1973), and the paper of Justesen (1972) on a constructive technique for asymptotically good codes. During this period, work on decoding algorithms was primarily focused on BCH codes. The first decoding algorithm for binary BCH codes was developed by Peterson (1960). A number of refinements and generalizations by Chien (1964), Forney (1965), Massey (1965), and Berlekamp (1968) led to the development of a computationally efficient algorithm for BCH codes, which is described in detail by Lin and Costello (1983) and Wicker (1995). A treatment of Reed–Solomon codes is given in the book by Wicker and Bhargava (1994).

In parallel with these developments on block codes are the developments in convolutional codes, which were invented by Elias (1955). The major problem in convolutional coding was decoding. Wozencraft and Reiffen (1961) described a sequential decoding algorithm for convolutional codes. This algorithm was later modified and refined by Fano (1963), and it is now called the *Fano algorithm*. Subsequently, the stack algorithm was devised by Zigangirov (1966) and Jelinek (1969), and the Viterbi algorithm was devised by Viterbi (1967). The optimality and the relatively modest complexity for small constraint lengths have served to make the Viterbi algorithm the most popular in decoding of convolutional codes with $K \leqslant 10$.

One of the most important contributions in coding during the 1970s was the work of Ungerboeck and Csajka (1976) on coding for bandwidth-constrained channels. In this paper, it was demonstrated that a significant coding gain can be achieved through the introduction of redundancy in a bandwidth-constrained channel and trellis codes were described for achieving coding gains of 3–4 dB. This work has generated much interest among researchers and has led to a large number of publications over the past 15 years. A number of references can be found in the papers by Ungerboeck (1982, 1987) and Forney et al. (1984). The papers by Benedetto et al. (1988, 1994) focus on applications and performance evaluation. Additional papers on coded modulation for bandwidth-constrained channels may also be found in the Special Issue on Voiceband Telephone Data Transmission, *IEEE Journal on Selected Areas in Communication* (September 1984, August 1989, and December 1989). A comprehensive treatment of trellis-coded modulation is given in the book by Biglieri et al. (1991).

A major new advance in coding and decoding is the construction of parallel and serially concatenated codes with interleaving, and the decoding of such codes using iterative MAP algorithms. Both PCCC and SCCC have been shown to yield performance very close to the Shannon limit with iterative decoding.

PCCCs, called turbo codes, and the use of iterative decoding were first described in a paper by Berrou et al. (1993). Serially concatenated codes with interleaving and their performance have been treated in the paper by Benedetto et al. (1998). Turbo coding and decoding is also treated in the books by Heegard and Wicker (1999), Johannesson and Zigangirov (1999), and Schlegel (1997). Performance bounds for turbo codes are given in the paper by Duman and Salehi (1997).

In addition to the references given above on coding, decoding, and coded signal design, we should mention the collection of papers published by the IEEE Press entitled *Key Papers in the Development of Coding Theory*, edited by Berlekamp (1974). This book contains important papers that were published in the first 25 years of coding theory. We should also cite the Special Issue on Error-Correcting Codes, *IEEE Transactions on Communications* (October 1971). Finally, the survey papers by Calderbank (1998), Costello et al. (1998) and Forney and Ungerboeck (1998) highlight the major developments in coding and decoding over the past 50 years and include a large number of references.

PROBLEMS

8.1 The generator matrix for a linear binary code is

$$\mathbf{G} = \begin{bmatrix} 0 & 0 & 1 & 1 & 1 & 0 & 1 \\ 0 & 1 & 0 & 0 & 1 & 1 & 1 \\ 1 & 0 & 0 & 1 & 1 & 1 & 0 \end{bmatrix}$$

a) Express $\mathbf{G}$ in systematic $[\mathbf{I}|\mathbf{P}]$ form.
b) Determine the parity check matrix $\mathbf{H}$ for the code.
c) Construct the table of syndromes for the code.
d) Determine the minimum distance of the code.
e) Demonstrate that the code word corresponding to the information sequence 101 is orthogonal to $\mathbf{H}$.

8.2 List the code words generated by the matrices given in Equations 8.1–35 and 8.1–37, and, thus, demonstrate that these matrices generate the same set of code words.

8.3 The weight distribution of Hamming codes is known. Expressed as a polynomial in powers of x, the weight distribution for the binary Hamming codes of block length n is

$$\begin{aligned} A(x) &= \sum_{i=0}^{n} A_i x^i \\ &= \frac{1}{n+1}[(1+x)^n + n(1+x)^{(n-1)/2}(1-x)^{(n+1)/2}] \end{aligned}$$

where A_i is the number of code words of weight i. Use this formula to determine the weight distribution of the (7, 4) Hamming code and check your result with the list of code words given in Table 8.1–2.

8.4 The polynomial

$$g(p) = p^4 + p + 1$$

is the generator for the (15, 11) Hamming binary code.

a) Determine a generator matrix **G** for this code in systematic form.

b) Determine the generator polynomial for the dual code.

8.5 For the (7, 4) cyclic Hamming code with generator polynomial $g(p) = p^3 + p^2 + 1$, construct an (8, 4) extended Hamming code and list all the code words. What is $d_{\min}$ for the extended code?

8.6 An (8, 4) linear block code is constructed by shortening a (15, 11) Hamming code generated by the generator polynomial $g(p) = p^4 + p + 1$.

a) Construct the code words of the (8, 4) code and list them.

b) What is the minimum distance of the (8, 4) code?

8.7 The polynomial $p^{15} + 1$ when factored yields

$$p^{15} + 1 = (p^4 + p^3 + 1)(p^4 + p^3 + p^2 + p + 1) \times (p^4 + p + 1)(p^2 + p + 1)(p + 1)$$

a) Construct a systematic (15, 5) code using the generator polynomial

$$g(p) = (p^4 + p^3 + p^2 + p + 1)(p^4 + p + 1)(p^2 + p + 1)$$

b) What is the minimum distance of the code?

c) How many random errors per code word can be corrected?

d) How many errors can be detected by this code?

e) List the code words of a (15, 2) code constructed from the generator polynomial

$$g(p) = \frac{p^{15} + 1}{p^2 + p + 1}$$

and determine the minimum distance.

8.8 Construct the parity check matrices $\mathbf{H}_1$ and $\mathbf{H}_2$ corresponding to the generator matrices $\mathbf{G}_1$ and $\mathbf{G}_2$ given by Equations 8.1–34 and 8.1–35, respectively.

8.9 Construct an extended (8, 4) code from the (7, 4) Hamming code by specifying the generator matrix and the parity check matrix.

8.10 A systematic (6, 3) code has the generator matrix

$$\mathbf{G} = \begin{bmatrix} 1 & 0 & 0 & 1 & 1 & 0 \\ 0 & 1 & 0 & 0 & 1 & 1 \\ 0 & 0 & 1 & 1 & 0 & 1 \end{bmatrix}$$

Construct the standard array and determine the correctable error patterns and their corresponding syndromes.

8.11 Construct the standard array for the (7, 3) code with generator matrix

$$\mathbf{G} = \begin{bmatrix} 1 & 0 & 0 & 1 & 0 & 1 & 1 \\ 0 & 1 & 0 & 1 & 1 & 1 & 0 \\ 0 & 0 & 1 & 0 & 1 & 1 & 1 \end{bmatrix}$$

and determine the correctable patterns and their corresponding syndromes.

8.12 Determine the correctable error patterns (of least weight) and their syndromes for the systematic (7, 4) cyclic Hamming code.

8.13 Prove that if the sum of two error patterns $\mathbf{e}_1$ and $\mathbf{e}_2$ is a valid code word $\mathbf{C}_j$, then each pattern has the same syndrome.

8.14 Let $g(p) = p^8 + p^6 + p^4 + p^2 + 1$ be a polynomial over the binary field.
a) Find the lowest-rate cyclic code whose generator polynomial is $g(p)$. What is the rate of this code?
b) Find the minimum distance of the code found in (a).
c) What is the coding gain for the code found in (a)?

8.15 The polynomial $g(p) = p + 1$ over the binary field is considered.
a) Show that this polynomial can generate a cyclic code for any choice of n. Find the corresponding k.
b) Find the systematic form of **G** and **H** for the code generated by $g(p)$.
c) Can you say what type of code this generator polynomial generates?

8.16 Design a (6, 2) cyclic code by choosing the shortest possible generator polynomial.
a) Determine the generator matrix **G** (in the systematic form) for this code and find all possible code words.
b) How many errors can be corrected by this code?

8.17 Prove that any two n-tuples in the same row of a standard array add to produce a valid code word.

8.18 Beginning with a (15, 7) BCH code, construct a shortened (12, 4) code. Give the generator matrix for the shortened code.

8.19 In Section 8.1.2, it was indicated that when an (n, k) Hadamard code is mapped into waveforms by means of binary PSK, the corresponding $M = 2^k$ waveforms are orthogonal. Determine the bandwidth expansion factor for the M orthogonal waveforms and compare this with the bandwidth requirements of orthogonal FSK detected coherently.

8.20 Show that the signaling waveforms generated from a maximum-length shift-register code by mapping each bit in a code word into a binary PSK signal are equicorrelated with correlation coefficient $\rho_r = -1/(M-1)$, i.e., the M waveforms form a simplex set.

8.21 Compute the error probability obtained with a (7, 4) Hamming code on an AWGN channel, both for hard-decision and soft-decision decoding. Use Equations 8.1–50, 8.1–52, 8.1–82, 8.1–90, and 8.1–91.

8.22 Use the results in Section 2.1.6 to obtain the Chernoff bound for hard-decision decoding given by Equations 8.1–89 and 8.1–90. Assume that the all-zero code word is transmitted and determine an upper bound on the probability that code word C_m, having weight w_m, is selected. This occurs if $\frac{1}{2}w_m$ or more bits are in error. To apply the Chernoff bound, define a sequence of random variables $\{X_i\}$ as

$$X_i = \begin{cases} 1 & \text{(with probability } p) \\ -1 & \text{(with probability } 1-p) \end{cases}$$

where $i = 1, 2, \ldots, w_m$, and p is the probability of error. For the BSC, the $\{X_i\}$ are statistically independent.

8.23 A convolutional code is described by

$$\mathbf{g}_1 = [1 \quad 0 \quad 1], \qquad \mathbf{g}_2 = [1 \quad 1 \quad 1], \qquad \mathbf{g}_3 = [1 \quad 1 \quad 1]$$

a) Draw the encoder corresponding to this code.
b) Draw the state-transition diagram for this code.
c) Draw the trellis diagram for this code.
d) Find the transfer function and the free distance of this code.
e) Verify whether or not this code is catastrophic.

8.24 The convolutional code of Problem 8.23 is used for transmission over an AWGN channel with hard-decision decoding. The output of the demodulator detector is $(101001011110111\cdots)$. Using the Viterbi algorithm, find the transmitted sequence.

8.25 Repeat Problem 8.23 for a code with

$$\mathbf{g}_1 = [1 \quad 1 \quad 0], \qquad \mathbf{g}_2 = [1 \quad 0 \quad 1], \qquad \mathbf{g}_3 = [1 \quad 1 \quad 1]$$

8.26 The block diagram of a binary convolutional code is shown in Figure P8.26.
a) Draw the state diagram for the code.
b) Find the transfer function of the code, $T(D)$.
c) What is d_{free}, the minimum free distance of the code?
d) Assume that a message has been encoded by this code and transmitted over a binary-symmetric channel with an error probability of $p = 10^{-5}$. If the received sequence is $\mathbf{r} = (110, 110, 110, 111, 010, 101, 101)$, using the Viterbi algorithm, find the transmitted bit sequence.
e) Find an upper bound to the bit error probability of the code when the above binary-symmetric channel is employed. Make any reasonable approximation.

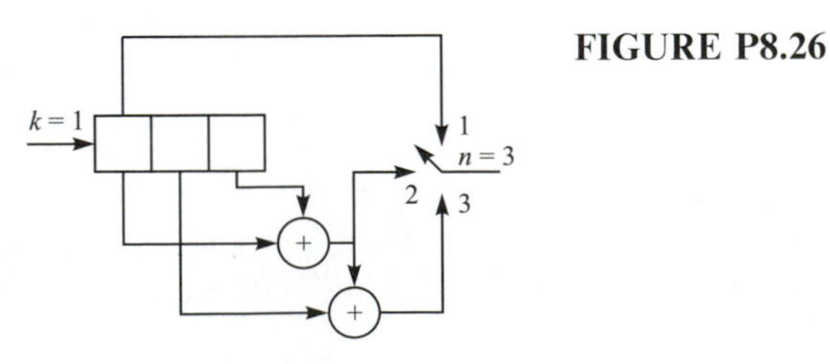

FIGURE P8.26

8.27 The block diagram of a (3, 1) convolutional code is shown in Figure P8.27.

a) Draw the state diagram of the code.

b) Find the transfer function $T(D)$ of the code.

c) Find the minimum free distance (d_{free}) of the code and show the corresponding path (at distance d_{free} from the all-zero code word) in the trellis.

d) Assume that four information bits (x_1, x_2, x_3, x_4), followed by two zero bits, have been encoded and sent via a binary-symmetric channel with crossover probability equal to 0.1. The received sequence is (111, 111, 111, 111, 111, 111). Use the Viterbi decoding algorithm to find the most likely data sequence.

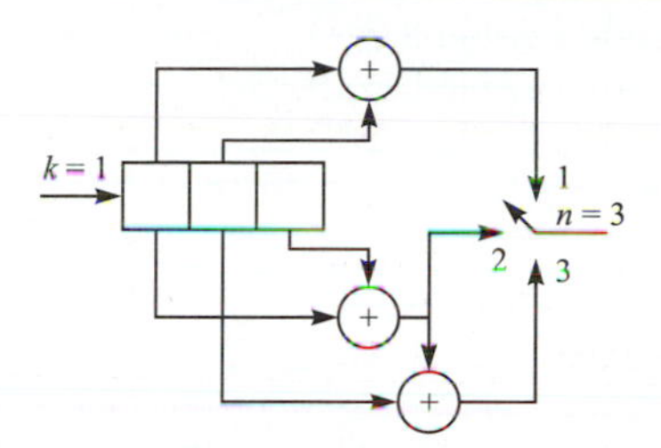

FIGURE P8.27

8.28 In the convolutional code generated by the encoder shown in Figure P8.28.

a) Find the transfer function of the code in the form $T(N, D)$.

b) Find d_{free} of the code.

c) If the code is used on a channel using hard-decision Viterbi decoding, assuming the crossover probability of the channel is $p = 10^{-6}$, use the hard-decision bound to find an upper bound on the average bit error probability of the code.

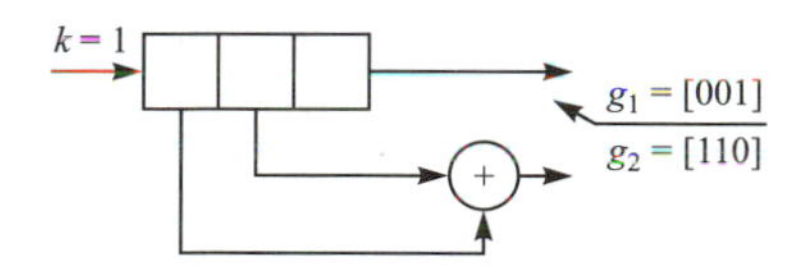

FIGURE P8.28

8.29 Figure P8.29 depicts a rate 1/2, constraint length $K = 2$, convolutional code.

a) Sketch the tree diagram, the trellis diagram, and the state diagram.

b) Solve for the transfer function $T(D, N, J)$ and, from this, specify the minimum free distance.

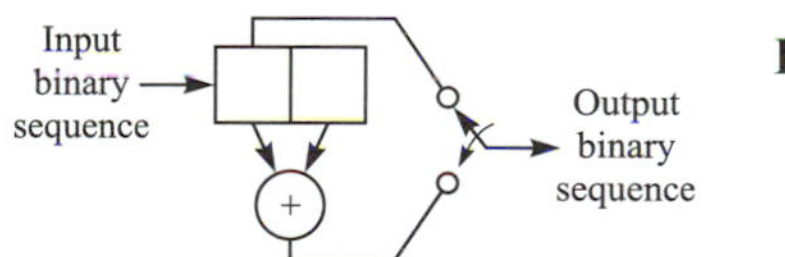

FIGURE P8.29

8.30 A rate 1/2, $K = 3$, binary convolutional encoder is shown in Figure P8.30.

a) Draw the tree diagram, the trellis diagram, and the state diagram.

b) Determine the transfer function $T(D, N, J)$ and, from this, specify the minimum free distance.

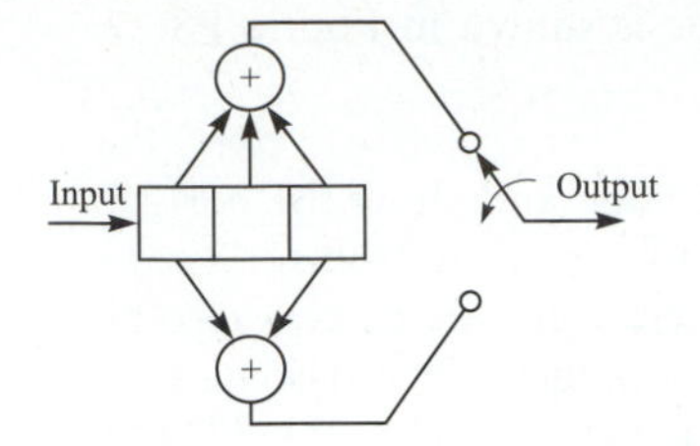

FIGURE P8.30

8.31 Sketch the convolutional encoders for the following codes:
a) Rate 1/2, $K = 5$, maximum free distance code (Table 8.2–1).
b) Rate 1/3, $K = 5$, maximum free distance code (Table 8.2–2).
c) Rate 2/3, $K = 2$, maximum free distance code (Table 8.2–8).

8.32 Draw the state diagram for the rate 2/3, $K = 2$, convolutional code indicated in Problem 8.31c and, for each transition, show the output sequence and the distance of the output sequence from the all-zero sequence.

8.33 Consider the $K = 3$, rate 1/2, convolutional code shown in Figure P8.30. Suppose that the code is used on a binary symmetric channel and the received sequence for the first eight branches is 0 0 0 1 1 0 0 0 0 0 0 0 1 0 0 1. Trace the decisions on a trellis diagram and label the survivors' Hamming distance metric at each node level. If a tie occurs in the metrics required for a decision, always choose the upper path (arbitrary choice).

8.34 Use the transfer function derived in Problem 8.30 for the $R_c = 1/2$, $K = 3$, convolutional code to compute the probability of a bit error for an AWGN channel with
a) Hard-decision decoding.
b) Soft-decision decoding.
Compare the performance by plotting the results of the computation on the same graph.

8.35 Use the generators given by Equation 8.2–36 to obtain the encoder for a dual-3, rate 1/2 convolutional code. Determine the state diagram and derive the transfer function $T(D, N, J)$.

8.36 Draw the state diagram for the convolutional code generated by the encoder shown in Figure P8.36 and, thus, determine if the code is catastrophic or noncatastrophic. Also, give an example of a rate 1/2, $K = 4$, convolutional encoder that exhibits catastrophic error propagation.

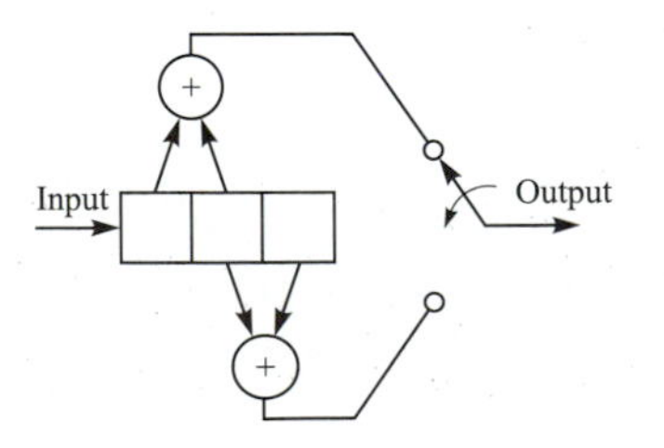

FIGURE P8.36

8.37 A trellis-coded signal is formed as shown in Figure P8.37 by encoding one bit by use of a rate 1/2 convolutional code, while three additional information bits are left uncoded. Perform the set partitioning of a 32-QAM (cross) constellation and indicate the subsets in the partition. By how much is the distance between adjacent signal points increased as a result of partitioning?

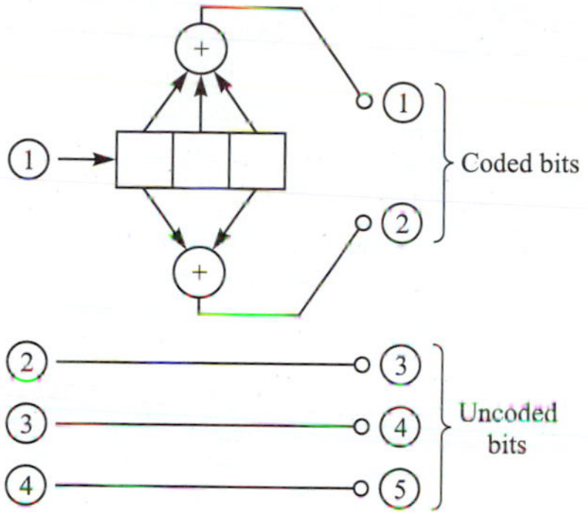

FIGURE P8.37

8.38 Let $\mathbf{x}_1$ and $\mathbf{x}_2$ be two code words of length n with distance d and assume that these two code words are transmitted via a binary-symmetric channel with crossover probability p. Let $P(d)$ denote the error probability in transmission of these two code words.

a) Show that

$$P(d) \leqslant \sum_{i=1}^{2^n} \sqrt{p(\mathbf{y}_i|\mathbf{x}_1)p(\mathbf{y}_i|\mathbf{x}_2)}$$

where the summation is over all binary sequences $\mathbf{y}_i$.

b) From the above, conclude that

$$P(d) \leqslant [4p(1-p)]^{d/2}$$

8.39 Prove Equation 8.2–41.

9

Signal Design for Band-Limited Channels

In previous chapters, we considered the transmission of digital information through an additive Gaussian noise channel. In effect, no bandwidth constraint was imposed on the signal design and the communication system design.

In this chapter, we consider the problem of signal design when the channel is band-limited to some specified bandwidth of W Hz. Under this condition, the channel may be modeled as a linear filter having an equivalent low-pass frequency response $C(f)$ that is zero for $|f| > W$.

The first topic that is treated is the design of the signal pulse $g(t)$ in a linearly modulated signal, represented as

$$v(t) = \sum_n I_n g(t - nT)$$

that efficiently utilizes the total available channel bandwidth W. We shall see that when the channel is ideal for $|f| \leqslant W$, a signal pulse can be designed that allows us to transmit at symbol rates comparable to or exceeding the channel bandwidth W. On the other hand, when the channel is not ideal, signal transmission at a symbol rate equal to or exceeding W results in intersymbol interference (ISI) among a number of adjacent symbols.

The second topic that is treated in this chapter is the use of coding to shape the spectrum of the transmitted signal and, thus, to avoid the problem of ISI.

We begin our discussion with a general characterization of band-limited linear filter channels.

9.1 CHARACTERIZATION OF BAND-LIMITED CHANNELS

Of the various channels available for digital communications, telephone channels are by far the most widely used. Such channels are characterized as *band-limited linear filters*. This is certainly the proper characterization when frequency-division multiplexing (FDM) is used as a means for establishing channels in the telephone network. Recent additions to the telephone network employ pulse-code modulation (PCM) for digitizing and encoding the analog signal and time-division multiplexing (TDM) for establishing multiple channels. Nevertheless, filtering is still used on the analog signal prior to sampling and encoding. Consequently, even though the present telephone network employs a mixture of FDM and TDM for transmission, the linear filter model for telephone channels is still appropriate.

For our purposes, a band-limited channel such as a telephone channel will be characterized as a linear filter having an equivalent low-pass frequency-response characteristic $C(f)$. Its equivalent low-pass impulse response is denoted by $c(t)$. Then, if a signal of the form

$$s(t) = \text{Re}[v(t)e^{j2\pi f_c t}] \tag{9.1–1}$$

is transmitted over a band-pass telephone channel, the equivalent low-pass received signal is

$$r_l(t) = \int_{-\infty}^{\infty} v(\tau)c(t-\tau)\,d\tau + z(t) \tag{9.1–2}$$

where the integral represents the convolution of $c(t)$ with $v(t)$, and $z(t)$ denotes the additive noise. Alternatively, the signal term can be represented in the frequency domain as $V(f)C(f)$, where $V(f)$ is the Fourier transform of $v(t)$.

If the channel is band-limited to W Hz, then $C(f) = 0$ for $|f| > W$. As a consequence, any frequency components in $V(f)$ above $|f| = W$ will not be passed by the channel. For this reason, we limit the bandwidth of the transmitted signal to W Hz also.

Within the bandwidth of the channel, we may express the frequency response $C(f)$ as

$$C(f) = |C(f)|e^{j\theta(f)} \tag{9.1–3}$$

where $|C(f)|$ is the amplitude-response characteristic and $\theta(f)$ is the phase-response characteristic. Furthermore, the envelope delay characteristic is defined as

$$\tau(f) = -\frac{1}{2\pi}\frac{d\theta(f)}{df} \tag{9.1–4}$$

A channel is said to be *nondistorting* or *ideal* if the amplitude response $|C(f)|$ is constant for all $|f| \leqslant W$ and $\theta(f)$ is a linear function of frequency, i.e., $\tau(f)$ is a constant for all $|f| \leqslant W$. On the other hand, if $|C(f)|$ is not constant for all $|f| \leqslant W$, we say that the channel *distorts the transmitted signal $V(f)$ in amplitude*, and, if $\tau(f)$ is not constant for all $|f| \leqslant W$, we say that the channel *distorts the signal $V(f)$ in delay*.

As a result of the amplitude and delay distortion caused by the nonideal channel frequency-response characteristic $C(f)$, a succession of pulses transmitted through the channel at rates comparable to the bandwidth W are smeared to the point that they are no longer distinguishable as well-defined pulses at the receiving terminal. Instead, they overlap, and, thus, we have intersymbol interference. As an example of the effect of delay distortion on a transmitted pulse, Figure 9.1–1a illustrates a band-limited pulse having zeros periodically spaced in time at points labeled $\pm T$, $\pm 2T$, etc. If information is conveyed by the pulse amplitude, as in PAM, for example, then one can transmit a sequence of pulses, each of which has a peak at the periodic zeros of the other pulses. However, transmission of the pulse through a channel modeled as having a linear envelope delay characteristic $\tau(f)$ [quadratic phase $\theta(f)$] results in the received pulse shown in Figure 9.1–1b having zero-crossings that are no longer periodically spaced. Consequently, a sequence of successive pulses would be smeared into one another and the peaks of the pulses would no longer be distinguishable. Thus, the channel delay distortion results in intesymbol interference. As will be discussed in Chapter 10, it is possible to compensate for the nonideal frequency-response characteristic of the channel by use of a filter or equalizer at the demodulator. Figure 9.1–1c illustrates the output of a linear equalizer that compensates for the linear distortion in the channel.

The extent of the intersymbol interference on a telephone channel can be appreciated by observing a frequency-response characteristic of the channel. Figure 9.1–2 illustrates the measured average amplitude and delay as functions of frequency for a medium-range (180–725 mi) telephone channel of the switched

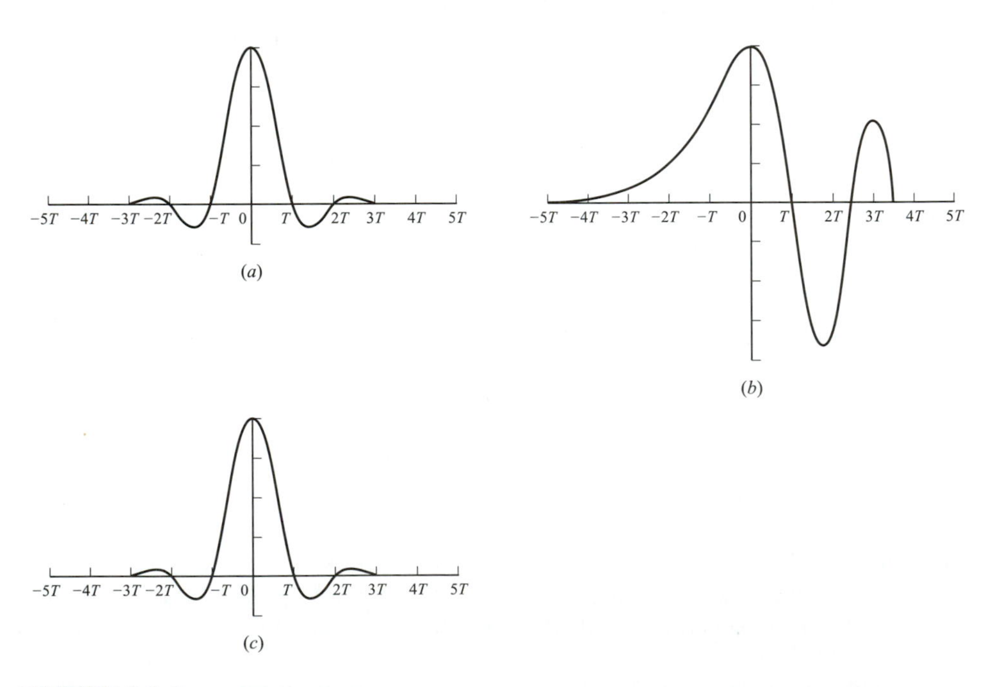

FIGURE 9.1–1
Effect of channel distortion: (**a**) channel input; (**b**) channel output; (**c**) equalizer output.

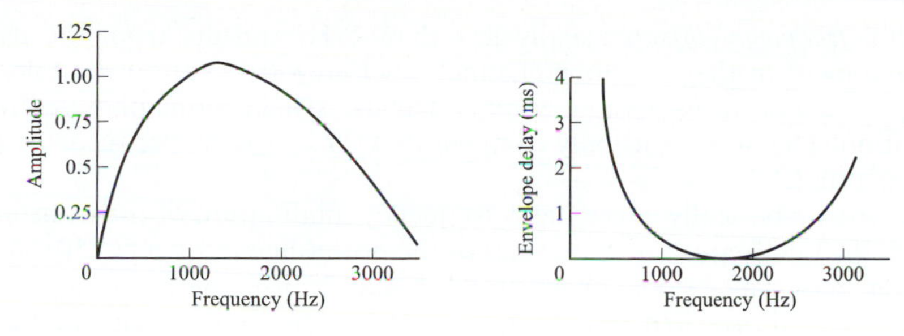

FIGURE 9.1–2
Average amplitude and delay characteristics of medium-range telephone channel.

telecommunications network as given by Duffy and Tratcher (1971). We observe that the usable band of the channel extends from about 300 Hz to about 3000 Hz. The corresponding impulse response of this average channel is shown in Figure 9.1–3. Its duration is about 10 ms. In comparison, the transmitted symbol rates on such a channel may be of the order of 2500 pulses or symbols per second. Hence, intersymbol interference might extend over 20–30 symbols.

In addition to linear distortion, signals transmitted through telephone channels are subject to other impairments, specifically non-linear distortion, frequency offset, phase jitter, impulse noise, and thermal noise.

Non-linear distortion in telephone channels arises from non-linearities in amplifiers and compandors used in the telephone system. This type of distortion is usually small and it is very difficult to correct.

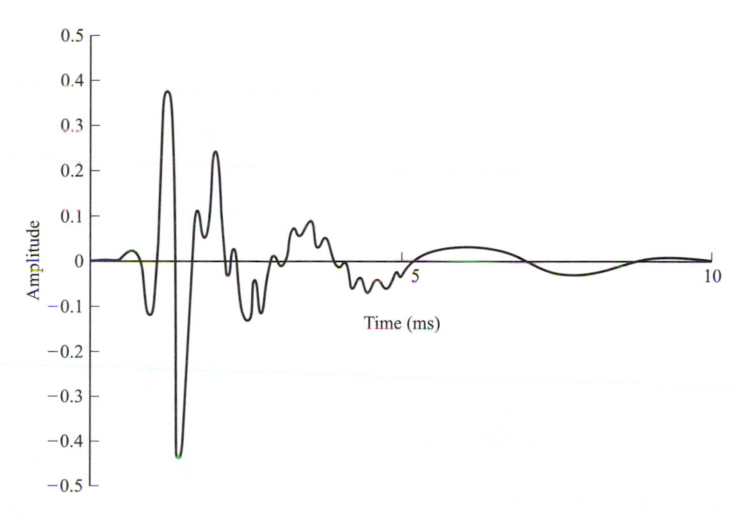

FIGURE 9.1–3
Impulse response of average channel with amplitude and delay shown in Figure 9.1–2.

A small *frequency offset*, usually less than 5 Hz, results from the use of carrier equipment in the telephone channel. Such an offset cannot be tolerated in high-speed digital transmission systems that use synchronous phase-coherent demodulation. The offset is usually compensated for by the carrier recovery loop in the demodulator.

Phase jitter is basically a low-index frequency modulation of the transmitted signal with the low-frequency harmonics of the power line frequency (50–60 Hz). Phase jitter poses a serious problem in digital transmission of high rates. However, it can be tracked and compensated for, to some extent, at the demodulator.

Impulse noise is an additive disturbance. It arises primarily from the switching equipment in the telephone system. *Thermal* (Gaussian) *noise* is also present at levels of 30 dB or more below the signal.

The degree to which one must be concerned with these channel impairments depends on the transmission rate over the channel and the modulation technique. For rates below 1800 bits/s ($R/W < 1$), one can choose a modulation technique, e.g., FSK, that is relatively insensitive to the amount of distortion encountered on typical telephone channels from all the sources listed above. For rates between 1800 and 2400 bits/s ($R/W \approx 1$), a more bandwidth-efficient modulation technique such as four-phase PSK is usually employed. At these rates, some form of compromise equalization is often employed to compensate for the average amplitude and delay distortion in the channel. In addition, the carrier recovery method is designed to compensate for the frequency offset. The other channel impairments are not that serious in their effects on the error rate performance at these rates. At transmission rates above 2400 bits/s ($R/W > 1$), bandwidth-efficient coded modulation techniques such as trellis-coded QAM, PAM, and PSK are employed. For such rates, special attention must be paid to linear distortion, frequency offset, and phase jitter. Linear distortion is usually compensated for by means of an adaptive equalizer. Phase jitter is handled by a combination of signal design and some type of phase compensation at the demodulator. At rates above 9600 bits/s, special attention must be paid not only to linear distortion, phase jitter, and frequency offset, but also to the other channel impairments mentioned above.

Unfortunately, a channel model that encompasses all the impairments listed above becomes difficult to analyze. For mathematical tractability the channel model that is adopted in this and the next two chapters is a linear filter that introduces amplitude and delay distortion and adds Gaussian noise.

Besides the telephone channels, there are other physical channels that exhibit some form of time dispersion and, thus, introduce intersymbol interference. Radio channels such as shortwave ionospheric channels (HF), tropospheric scatter channels, and mobile radio channels are examples of time-dispersive channels. In these channels, time dispersion and, hence, intersymbol interference are the result of multiple propagation paths with different path delays. The number of paths and the relative time delays among the paths vary with time, and, for this reason, these radio channels are usually called *time-variant multipath channels*. The time-variant multipath conditions give rise to a wide variety of frequency-response characteristics. Consequently the frequency-response characterization that is used for telephone channels is inappropriate for time-variant multipath channels. Instead, these radio channels are characterized statistically, as explained in more detail in Chapter 14, in terms of the scattering function, which, in brief, is a two-dimensional representation of the average received signal power as a function of relative time delay and Doppler frequency.

For illustrative purposes, a scattering function measured on a medium-range (150 mi) trophospheric scatter channel is shown in Figure 9.1–4. The total time duration (multipath spread) of the channel response is approximately 0.7 μs on the average, and the spread between "half-power points" in Doppler frequency is a little less than 1 Hz on the strongest path and somewhat larger on the other paths. Typically, if one is transmitting at a rate of 10^7 symbols/s over such a channel, the multipath spread of 0.7 μs will result in intersymbol interference that spans about seven symbols.

In this chapter, we deal exclusively with the linear time-invariant filter model for a band-limited channel. The adaptive equalization techniques presented in Chapters 10 and 11 for combating intersymbol interference are also applicable to time-variant multipath channels, under the condition that the time variations in the channel are relatively slow in comparison to the total channel bandwidth or, equivalently, to the symbol transmission rate over the channel.

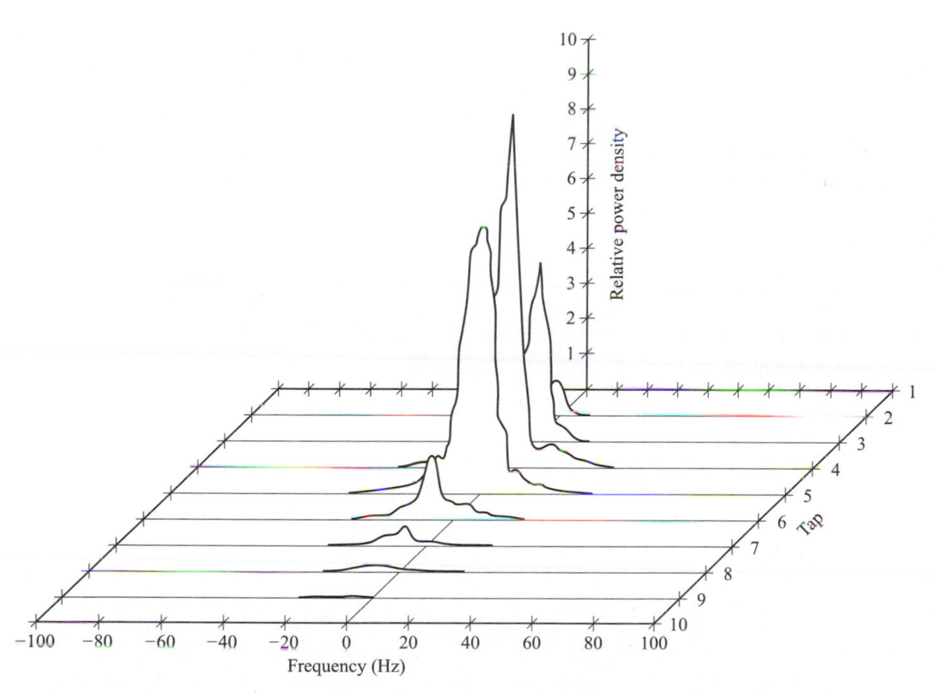

FIGURE 9.1–4
Scattering function of a medium-range tropospheric scatter channel.

9.2
SIGNAL DESIGN FOR BAND-LIMITED CHANNELS

It was shown in Chapter 4 that the equivalent low-pass transmitted signal for several different types of digital modulation techniques has the common form

$$v(t) = \sum_{n=0}^{\infty} I_n g(t - nT) \tag{9.2–1}$$

where $\{I_n\}$ represents the discrete information-bearing sequence of symbols and $g(t)$ is a pulse that, for the purposes of this discussion, is assumed to have a band-limited frequency-response characteristic $G(f)$, i.e., $G(f) = 0$ for $|f| > W$. This signal is transmitted over a channel having a frequency response $C(f)$, also limited to $|f| \leqslant W$. Consequently, the received signal can be represented as

$$r_l(t) = \sum_{n=0}^{\infty} I_n h(t - nT) + z(t) \tag{9.2–2}$$

where

$$h(t) = \int_{-\infty}^{\infty} g(\tau) c(t - \tau)\, d\tau \tag{9.2–3}$$

and $z(t)$ represents the additive white Gaussian noise.

Let us suppose that the received signal is passed first through a filter and then sampled at a rate $1/T$ samples/s. We shall show in a subsequent section that the optimum filter from the point of view of signal detection is one matched to the received pulse. That is, the frequency response of the receiving filter is $H^*(f)$. We denote the output of the receiving filter as

$$y(t) = \sum_{n=0}^{\infty} I_n x(t - nT) + \nu(t) \tag{9.2–4}$$

where $x(t)$ is the pulse representing the response of the receiving filter to the input pulse $h(t)$ and $\nu(t)$ is the response of the receiving filter to the noise $z(t)$.

Now, if $y(t)$ is sampled at times $t = kT + \tau_0$, $k = 0, 1, \ldots,$ we have

$$y(kT + \tau_0) \equiv y_k = \sum_{n=0}^{\infty} I_n x(kT - nT + \tau_0) + \nu(kT + \tau_0) \tag{9.2–5}$$

or, equivalently,

$$y_k = \sum_{n=0}^{\infty} I_n x_{k-n} + \nu_k, \qquad k = 0, 1, \ldots \tag{9.2–6}$$

where τ_0 is the transmission delay through the channel. The sample values can be expressed as

$$y_k = x_0 \left(I_k + \frac{1}{x_0} \sum_{\substack{n=0 \\ n \neq k}}^{\infty} I_n x_{k-n} \right) + \nu_k, \qquad k = 0, 1, \ldots \tag{9.2–7}$$

We regard x_0 as an arbitrary scale factor, which we arbitrarily set equal to unity for convenience. Then

$$y_k = I_k + \sum_{\substack{n=0 \\ n \neq k}}^{\infty} I_n x_{k-n} + \nu_k \tag{9.2–8}$$

The term I_k represents the desired information symbol at the kth sampling instant, the term

$$\sum_{\substack{n=0 \\ n \neq k}}^{\infty} I_n x_{k-n}$$

represents the ISI, and ν_k is the additive Gaussian noise variable at the kth sampling instant.

The amount of intersymbol interference and noise in a digital communication system can be viewed on an oscilloscope. For PAM signals, we can display the received signal $y(t)$ on the vertical input with the horizontal sweep rate set at $1/T$. The resulting oscilloscope display is called an *eye pattern* because of its resemblance to the human eye. For example, Figure 9.2–1 illustrates the eye patterns for binary and four-level PAM modulation. The effect of ISI is to cause the eye to close, thereby reducing the margin for additive noise to cause errors. Figure 9.2–2 graphically illustrates the effect of intersymbol interference in reducing the opening of a binary eye. Note that intersymbol interference distorts the position of the zero-crossings and causes a reduction in the eye opening. Thus, it causes the system to be more sensitive to a synchronization error.

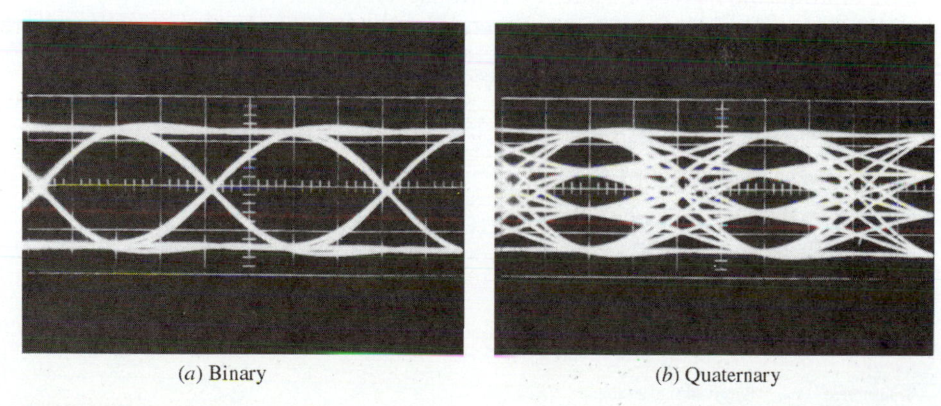

(*a*) Binary (*b*) Quaternary

FIGURE 9.2–1
Examples of eye patterns for binary and quaternary amplitude-shift keying (or PAM).

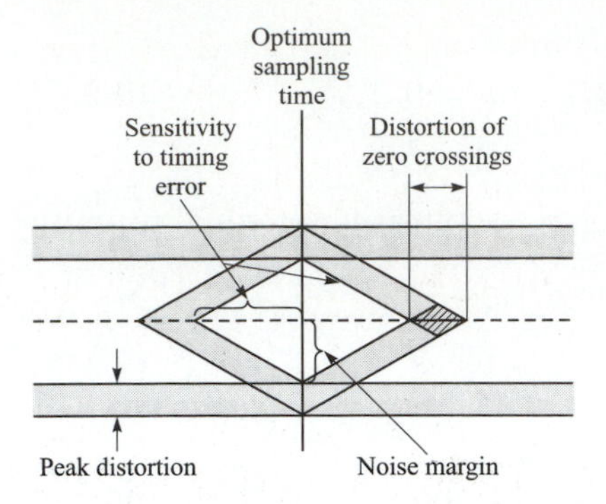

FIGURE 9.2–2
Effect of intersymbol interference on eye opening.

For PSK and QAM it is customary to display the "eye pattern" as a two-dimensional scatter diagram illustrating the sampled values $\{y_k\}$ that represent the decision variables at the sampling instants. Figure 9.2–3 illustrates such an eye pattern for an 8-PSK signal. In the absence of intersymbol interference and noise, the superimposed signals at the sampling instants would result in eight distinct points corresponding to the eight transmitted signal phases. Intersymbol interference and noise result in a deviation of the received samples $\{y_k\}$ from the desired 8-PSK signal. The larger the intersymbol interference and noise, the larger the scattering of the received signal samples relative to the transmitted signal points.

Below, we consider the problem of signal design under the condition that there is no intersymbol interference at the sampling instants.

9.2.1 Design of Band-Limited Signals for No Intersymbol Interference—The Nyquist Criterion

For the discussion in this section and in Section 9.2.2, we assume that the band-limited channel has ideal frequency-response characteristics, i.e., $C(f) = 1$ for $|f| \leqslant W$. Then the pulse $x(t)$ has a spectral characteristic $X(f) = |G(f)|^2$, where

$$x(t) = \int_{-W}^{W} X(f) e^{j2\pi ft}\, df \tag{9.2–9}$$

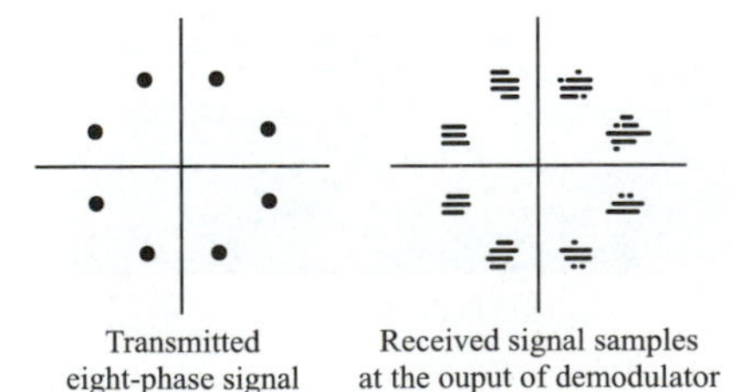

FIGURE 9.2–3
Two-dimensional digital "eye patterns."

We are interested in determining the spectral properties of the pulse $x(t)$ and, hence, the transmitted pulse $g(t)$, that results in no intersymbol interference. Since

$$y_k = I_k + \sum_{\substack{n=0 \\ n \neq k}}^{\infty} I_n x_{k-n} + \nu_k \tag{9.2–10}$$

the condition for no intersymbol interference is

$$x(t = kT) \equiv x_k = \begin{cases} 1 & (\mathrm{k}=0) \\ 0 & (\mathrm{k} \neq 0) \end{cases} \tag{9.2–11}$$

Below, we derive the necessary and sufficient condition on $X(f)$ in order for $x(t)$ to satisfy the above relation. This condition is known as the *Nyquist pulse-shaping criterion* or *Nyquist condition for zero ISI* and is stated in the following theorem.

THEOREM (NYQUIST). The necessary and sufficient condition for $x(t)$ to satisfy

$$x(nT) = \begin{cases} 1 & (\mathrm{n}=0) \\ 0 & (\mathrm{n} \neq 0) \end{cases} \tag{9.2–12}$$

is that its Fourier transform $X(f)$ satisfy

$$\sum_{m=-\infty}^{\infty} X(f + m/T) = T \tag{9.2–13}$$

Proof. In general, $x(t)$ is the inverse Fourier transform of $X(f)$. Hence,

$$x(t) = \int_{-\infty}^{\infty} X(f) e^{j2\pi f t} \, df \tag{9.2–14}$$

At the sampling instants $t = nT$, this relation becomes

$$x(nT) = \int_{-\infty}^{\infty} X(f) e^{j2\pi f nT} \, df \tag{9.2–15}$$

Let us break up the integral in Equation 9.2–15 into integrals covering the finite range of $1/T$. Thus, we obtain

$$\begin{aligned} x(nT) &= \sum_{m=-\infty}^{\infty} \int_{(2m-1)/2T}^{(2m+1)/2T} X(f) e^{j2\pi f nT} \, df \\ &= \sum_{m=-\infty}^{\infty} \int_{-1/2T}^{1/2T} X(f + m/T) e^{j2\pi f nT} \, dt \\ &= \int_{-1/2T}^{1/2T} \left[\sum_{m=-\infty}^{\infty} X(f + m/T) \right] e^{j2\pi f nT} \, df \\ &= \int_{-1/2T}^{1/2T} B(f) e^{j2\pi f nT} \, df \end{aligned} \tag{9.2–16}$$

where we have defined $B(f)$ as

$$B(f) = \sum_{m=-\infty}^{\infty} X(f + m/T) \tag{9.2–17}$$

Obviously $B(f)$ is a periodic function with period $1/T$, and, therefore, it can be expanded in terms of its Fourier series coefficients $\{b_n\}$ as

$$B(f) = \sum_{n=-\infty}^{\infty} b_n e^{j2\pi nfT} \tag{9.2–18}$$

where

$$b_n = T \int_{-1/2T}^{1/2T} B(f) e^{-j2\pi nfT}\, df \tag{9.2–19}$$

Comparing Equations 9.2–19 and 9.2–16, we obtain

$$b_n = Tx(-nT) \tag{9.2–20}$$

Therefore, the necessary and sufficient condition for Equation 9.2–10 to be satisfied is that

$$b_n = \begin{cases} T & (n = 0) \\ 0 & (n \neq 0) \end{cases} \tag{9.2–21}$$

which, when substituted into Equation 9.2–18, yields

$$B(f) = T \tag{9.2–22}$$

or, equivalently,

$$\sum_{m=-\infty}^{\infty} X(f + m/T) = T \tag{9.2–23}$$

This concludes the proof of the theorem.

Now suppose that the channel has a bandwidth of W. Then $C(f) \equiv 0$ for $|f| > W$ and, consequently, $X(f) = 0$ for $|f| > W$. We distinguish three cases.

1. When $T < 1/2W$, or, equivalently, $1/T > 2W$, since $B(f) = \sum_{n=-\infty}^{+\infty} X(f + n/T)$ consists of nonoverlapping replicas of $X(f)$, separated by $1/T$ as shown in Figure 9.2–4, there is no choice for $X(f)$ to ensure $B(f) \equiv T$ in this case and there is no way that we can design a system with no ISI.
2. When $T = 1/2W$, or, equivalently, $1/T = 2W$ (the Nyquist rate), the replications of $X(f)$, separated by $1/T$, are as shown in Figure 9.2–5. It is clear that in this case there exists only one $X(f)$ that results in $B(f) = T$, namely,

$$X(f) = \begin{cases} T & (|f| < W) \\ 0 & (\text{otherwise}) \end{cases} \tag{9.2–24}$$

which corresponds to the pulse

$$x(t) = \frac{\sin(\pi t/T)}{\pi t/T} \equiv \operatorname{sinc}\left(\frac{\pi t}{T}\right) \tag{9.2–25}$$

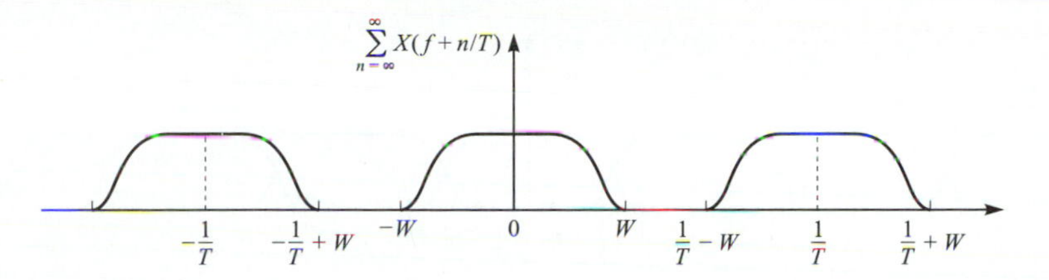

FIGURE 9.2–4
Plot of $B(f)$ for the case $T < 1/2W$.

This means that the smallest value of T for which transmission with zero ISI is possible is $T = 1/2W$, and for this value, $x(t)$ has to be a sinc function. The difficulty with this choice of $x(t)$ is that it is noncausal and, therefore, non-realizable. To make it realizable, usually a delayed version of it, i.e., $\text{sinc}[\pi(t - t_0)/T]$ is used and t_0 is chosen such that for $t < 0$, we have $\text{sinc}[\pi(t - t_0)/T] \approx 0$. Of course, with this choice of $x(t)$, the sampling time must also be shifted to $mT + t_0$. A second difficulty with this pulse shape is that its rate of convergence to zero is slow. The tails of $x(t)$ decay as $1/t$; consequently, a small mistiming error in sampling the output of the matched filter at the demodulator results in an infinite series of ISI components. Such a series is not absolutely summable because of the $1/t$ rate of decay of the pulse, and, hence, the sum of the resulting ISI does not converge.

3. When $T > 1/2W$, $B(f)$ consists of overlapping replications of $X(f)$ separated by $1/T$, as shown in Figure 9.2–6. In this case, there exist numerous choices for $X(f)$ such that $B(f) \equiv T$.

A particular pulse spectrum, for the $T > 1/2W$ case, that has desirable spectral properties and has been widely used in practice is the raised cosine spectrum. The raised cosine frequency characteristic is given as (see Problem 9.16)

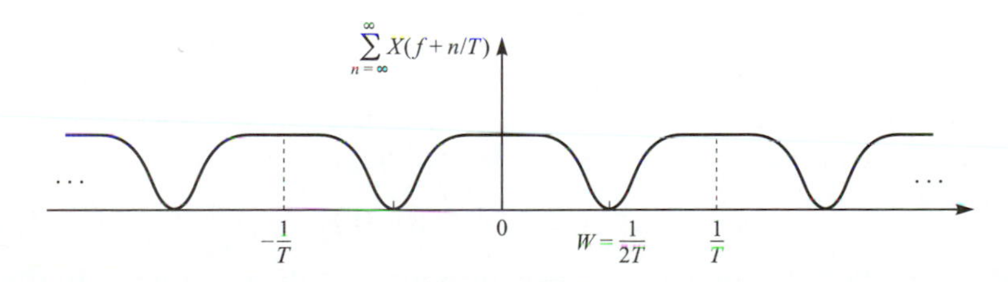

FIGURE 9.2–5
Plot of $B(f)$ for the case $T = 1/2W$.

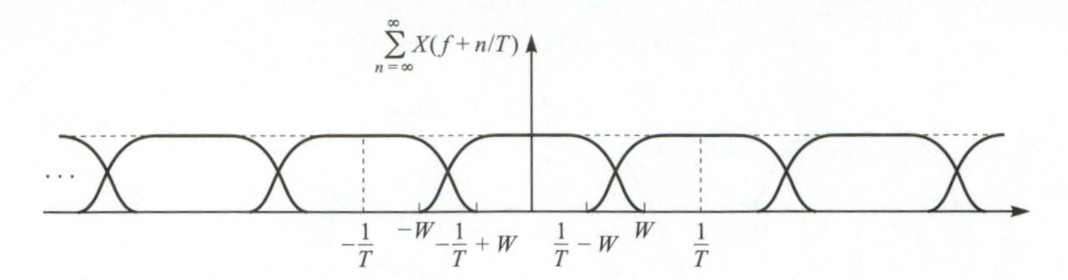

FIGURE 9.2–6
Plot of $B(f)$ for the case $T > 1/2W$.

$$X_{rc}(f) = \begin{cases} T & \left(0 \leqslant |f| \leqslant \dfrac{1-\beta}{2T}\right) \\ \dfrac{T}{2}\left\{1 + \cos\left[\dfrac{\pi T}{\beta}\left(|f| - \dfrac{1-\beta}{2T}\right)\right]\right\} & \left(\dfrac{1-\beta}{2T} \leqslant |f| \leqslant \dfrac{1+\beta}{2T}\right) \\ 0 & \left(|f| > \dfrac{1+\beta}{2T}\right) \end{cases} \tag{9.2–26}$$

where β is called the *roll-off factor* and takes values in the range $0 \leqslant \beta \leqslant 1$. The bandwidth occupied by the signal beyond the Nyquist frequency $1/2T$ is called the *excess bandwidth* and is usually expressed as a percentage of the Nyquist frequency. For example, when $\beta = \frac{1}{2}$, the excess bandwidth is 50 percent and when $\beta = 1$, the excess bandwidth is 100 percent. The pulse $x(t)$, having the raised cosine spectrum, is

$$\begin{aligned} x(t) &= \frac{\sin(\pi t/T)}{\pi t/T} \frac{\cos(\pi\beta t/T)}{1 - 4\beta^2 t^2/T^2} \\ &= \operatorname{sinc}(\pi t/T) \frac{\cos(\pi\beta t/T)}{1 - 4\beta^2 t^2/T^2} \end{aligned} \tag{9.2–27}$$

Note that $x(t)$ is normalized so that $x(0) = 1$. Figure 9.2–7 illustrates the raised cosine spectral characteristics and the corresponding pulses for $\beta = 0, \frac{1}{2}$, and 1. Note that for $\beta = 0$, the pulse reduces to $x(t) = \operatorname{sinc}(\pi t/T)$, and the symbol rate $1/T = 2W$. When $\beta = 1$, the symbol rate is $1/T = W$. In general, the tails of $x(t)$ decay as $1/t^3$ for $\beta > 0$. Consequently, a mistiming error in sampling leads to a series of ISI components that converges to a finite value.

Because of the smooth characteristics of the raised cosine spectrum, it is possible to design practical filters for the transmitter and the receiver that approximate the overall desired frequency response. In the special case where the channel is ideal, i.e., $C(f) = 1$, $|f| \leqslant W$, we have

$$X_{rc}(f) = G_T(f) G_R(f) \tag{9.2–28}$$

where $G_T(f)$ and $G_R(f)$ are the frequency responses of the two filters. In this case, if the receiver filter is matched to the transmitter filter, we have $X_{rc}(f) = G_T(f)G_R(f) = |G_T(f)|^2$. Ideally,

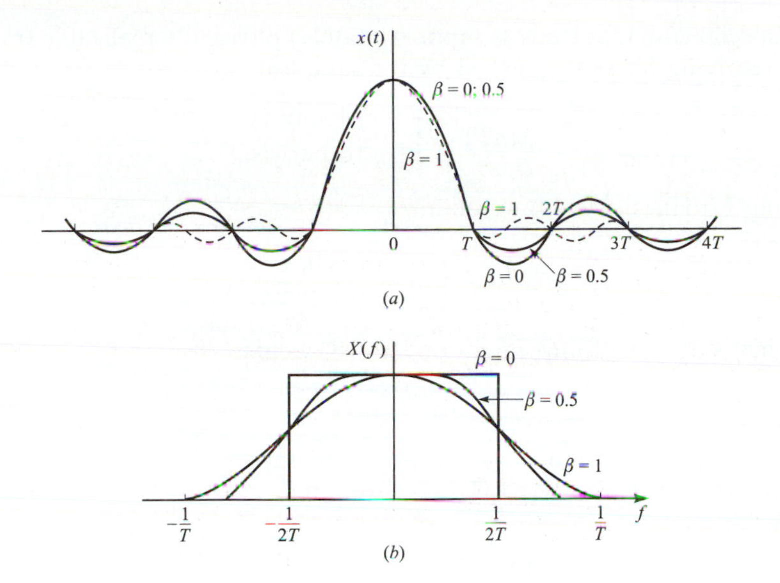

FIGURE 9.2–7
Pulses having a raised cosine spectrum.

$$G_T(f) = \sqrt{|X_{rc}(f)|}e^{-j2\pi f t_0} \tag{9.2–29}$$

and $G_R(f) = G_T^*(f)$, where t_0 is some nominal delay that is required to ensure physical realizability of the filter. Thus, the overall raised cosine spectral characteristic is split evenly between the transmitting filter and the receiving filter. Note also that an additional delay is necessary to ensure the physical realizability of the receiving filter.

9.2.2 Design of Band-Limited Signals with Controlled ISI—Partial-Response Signals

As we have observed from our discussion of signal design for zero ISI, it is necessary to reduce the symbol rate $1/T$ below the Nyquist rate of $2W$ symbols/s to realize practical transmitting and receiving filters. On the other hand, suppose we choose to relax the condition of zero ISI and, thus, achieve a symbol transmission rate of $2W$ symbols/s. By allowing for a controlled amount of ISI, we can achieve this symbol rate.

We have already seen that the condition for zero ISI is $x(nT) = 0$ for $n \neq 0$. However, suppose that we design the band-limited signal to have controlled ISI at one time instant. This means that we allow one additional nonzero value in the samples $\{x(nT)\}$. The ISI that we introduce is deterministic or "controlled" and, hence, it can be taken into account at the receiver, as discussed below.

One special case that leads to (approximately) physically realizable transmitting and receiving filters is specified by the samples†

$$x(nT) = \begin{cases} 1 & (n = 0, 1) \\ 0 & (\text{otherwise}) \end{cases} \tag{9.2–30}$$

Now, using Equation 9.2–20, we obtain

$$b_n = \begin{cases} T & (n = 0, -1) \\ 0 & (\text{otherwise}) \end{cases} \tag{9.2–31}$$

which, when substituted into Equation 9.2–18, yields

$$B(f) = T + Te^{-j2\pi fT} \tag{9.2–32}$$

As in the preceding section, it is impossible to satisfy the above equation for $T < 1/2W$. However, for $T = 1/2W$, we obtain

$$\begin{aligned} X(f) &= \begin{cases} \frac{1}{2W}(1 + e^{-j\pi f/W}) & (|f| < W) \\ 0 & (\text{otherwise}) \end{cases} \\ &= \begin{cases} \frac{1}{W}e^{-j\pi f/2W}\cos\frac{\pi f}{2W} & (|f| < W) \\ 0 & (\text{otherwise}) \end{cases} \end{aligned} \tag{9.2–33}$$

Therefore, $x(t)$ is given by

$$x(t) = \text{sinc}(2\pi Wt) + \text{sinc}[2\pi(Wt - \tfrac{1}{2})] \tag{9.2–34}$$

This pulse is called a *duobinary signal pulse*. It is illustrated along with its magnitude spectrum in Figure 9.2–8. Note that the spectrum decays to zero smoothly, which means that physically realizable filters can be designed that approximate this spectrum very closely. Thus, a symbol rate of $2W$ is achieved.

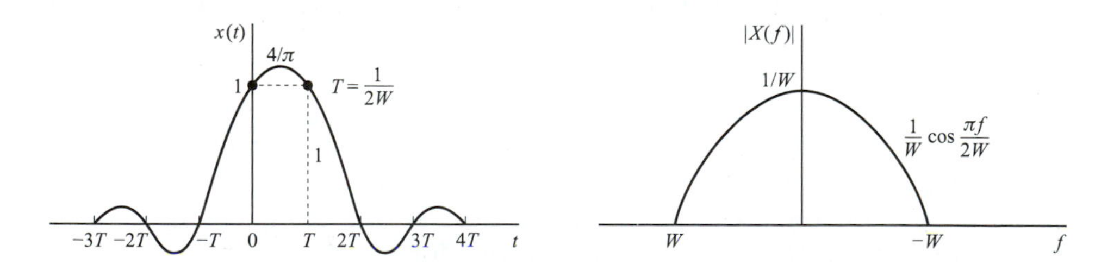

FIGURE 9.2–8
Time-domain and frequency-domain characteristics of a duobinary signal.

†It is convenient to deal with samples of $x(t)$ that are normalized to unity for $n = 0, 1$.

Another special case that leads to (approximately) physically realizable transmitting and receiving filters is specified by the samples

$$x\left(\frac{n}{2W}\right) = x(nT) = \begin{cases} 1 & (n = -1) \\ -1 & (n = 1) \\ 0 & (\text{otherwise}) \end{cases} \tag{9.2–35}$$

The corresponding pulse $x(t)$ is given as

$$x(t) = \operatorname{sinc}\frac{\pi(t+T)}{T} - \operatorname{sinc}\frac{\pi(t-T)}{T} \tag{9.2–36}$$

and its spectrum is

$$X(f) = \begin{cases} \dfrac{1}{2W}(e^{j\pi f/W} - e^{-j\pi f/W}) = \dfrac{j}{W}\sin\dfrac{\pi f}{W} & |f| \leqslant W \\ 0 & |f| > W \end{cases} \tag{9.2–37}$$

This pulse and its magnitude spectrum are illustrated in Figure 9.2–9. It is called a *modified duobinary signal pulse*. It is interesting to note that the spectrum of this signal has a zero at $f = 0$, making it suitable for transmission over a channel that does not pass DC.

One can obtain other interesting and physically realizable filter characteristics, as shown by Kretzmer (1966) and Lucky et al. (1968), by selecting different values for the samples $\{x(n/2W)\}$ and more than two nonzero samples. However, as we select more nonzero samples, the problem of unraveling the controlled ISI becomes more cumbersome and impractical.

In general, the class of band-limited signal pulses that have the form

$$x(t) = \sum_{n=-\infty}^{\infty} x\left(\frac{n}{2W}\right)\operatorname{sinc}\left[2\pi W\left(t - \frac{n}{2W}\right)\right] \tag{9.2–38}$$

and their corresponding spectra

$$X(f) = \begin{cases} \dfrac{1}{2W}\displaystyle\sum_{n=-\infty}^{\infty} x\left(\frac{n}{2W}\right)e^{-jn\pi f/W} & (|f| \leqslant W) \\ 0 & (|f| \leqslant W) \end{cases} \tag{9.2–39}$$

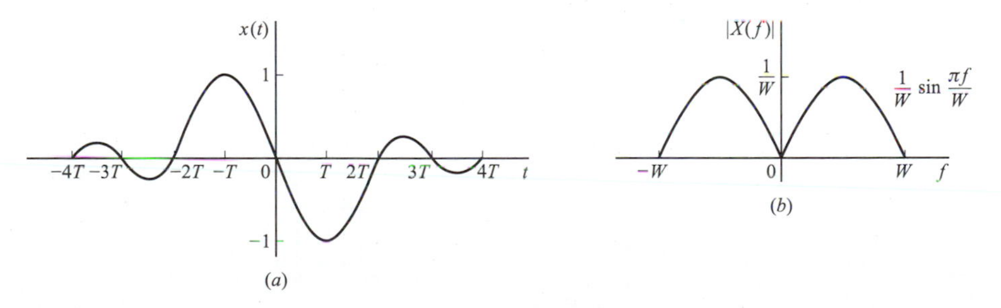

FIGURE 9.2–9
Time-domain and frequency-domain characteristics of a modified duobinary signal.

are called *partial-response signals* when controlled ISI is a purposely introduced by selecting two or more nonzero samples from the set $\{x(n/2W)\}$. The resulting signal pulses allow us to transmit information symbols at the Nyquist rate of $2W$ symbols/s. The detection of the received symbols in the presence of controlled ISI is described below.

Alternative characterization of partial-response signals. We conclude this subsection by presenting another interpretation of a partial-response signal. Suppose that the partial-response signal is generated, as shown in Figure 9.2–10, by passing the discrete-time sequence $\{I_n\}$ through a discrete-time filter with coefficients $x_n \equiv x(n/2W)$, $n = 0, 1, \ldots, N-1$, and using the output sequence $\{B_n\}$ from this filter to excite periodically with an input $B_n\delta(t - nT)$ an analog filter having an impulse response $\text{sinc}(2\pi Wt)$. The resulting output signal is identical to the partial-response signal given by Equation 9.2–38.

Since

$$B_n = \sum_{k=0}^{N-1} x_k I_{n-k} \tag{9.2–40}$$

the sequence of symbols $\{B_n\}$ is correlated as a consequence of the filtering performed on the sequence $\{I_n\}$. In fact, the autocorrelation function of the sequence $\{B_n\}$ is

$$\begin{aligned} \phi(m) &= E(B_n B_{n+m}) \\ &= \sum_{k=0}^{N-1} \sum_{l=0}^{N-1} x_k x_l E(I_{n-k} I_{n+m-l}) \end{aligned} \tag{9.2–41}$$

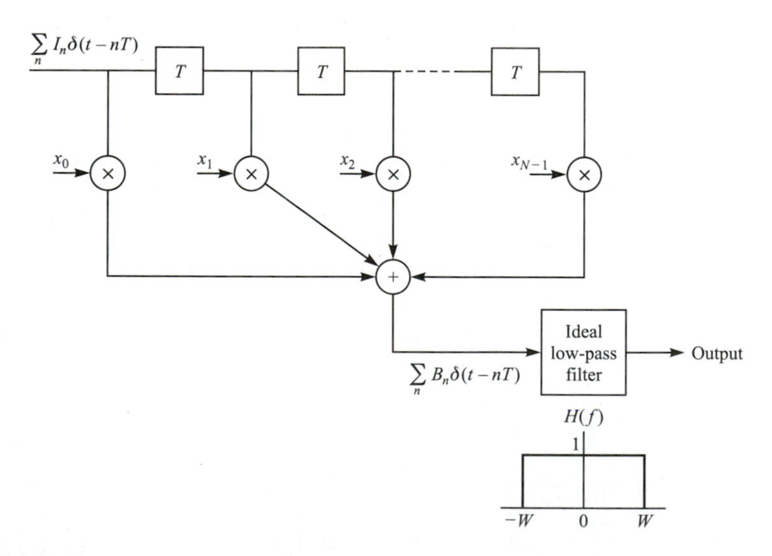

FIGURE 9.2–10
An alternative method for generating a partial-response signal.

When the input sequence is zero-mean and white,

$$E(I_{n-k}I_{n+m-l}) = \delta_{m+k-1} \tag{9.2–42}$$

where we have used the normalization $E(I_n^2) = 1$. Substitution of Equation 9.2–42, into Equation 9.2–41 yields the desired autocorrelation function for $\{B_n\}$ in the form

$$\phi(m) = \sum_{k=0}^{N-1-|m|} x_k x_{k+|m|}, \qquad m = 0, \pm 1, \ldots, \pm(N-1) \tag{9.2–43}$$

The corresponding power spectral density is

$$\begin{aligned}\Phi(f) &= \sum_{m=-(N-1)}^{N-1} \phi(m)e^{-j2\pi fmT} \\ &= \left|\sum_{m=0}^{N-1} x_m e^{-j2\pi fmT}\right|^2\end{aligned} \tag{9.2–44}$$

where $T = 1/2W$ and $|f| \leqslant 1/2T = W$. Thus, the partial-response signal designs provide spectral shaping of the signal transmitted through the channel.

9.2.3 Data Detection for Controlled ISI

In this section, we describe two methods for detecting the information symbols at the receiver when the received signal contains controlled ISI. One is a symbol-by-symbol detection method that is relatively easy to implement. The second method is based on the maximum-likelihood criterion for detecting a sequence of symbols. The latter method minimizes the probability of error but is a little more complex to implement. In particular, we consider the detection of the duobinary and the modified duobinary partial-response signals. In both cases, we assume that the desired spectral characteristic $X(f)$ for the partial-response signal is split evenly between the transmitting and receiving filters, i.e., $|G_T(f)| = |G_R(f)| = |X(f)|^{1/2}$. This treatment is based on PAM signals, but it is easily generalized to QAM and PSK.

Symbol-by-symbol suboptimum detection. For the duobinary signal pulse, $x(nT) = 1$, for $n = 0, 1$, and is zero otherwise. Hence, the samples at the output of the receiving filter (demodulator) have the form

$$y_m = B_m + \nu_m = I_m + I_{m-1} + \nu_m \tag{9.2–45}$$

where $\{I_m\}$ is the transmitted sequence of amplitudes and $\{\nu_m\}$ is a sequence of additive Gaussian noise samples. Let us ignore the noise for the moment and consider the binary case where $I_m = \pm 1$ with equal probability. Then B_m takes on one of three possible values, namely, $B_m = -2, 0, 2$ with corresponding probabilities 1/4, 1/2, 1/4. If I_{m-1} is the detected symbol from the $(m-1)$th signaling interval, its effect on B_m, the received signal in the mth signaling interval, can be eliminated by subtraction, thus allowing I_m to be detected. This process can be repeated sequentially for every received symbol.

The major problem with this procedure is that errors arising from the additive noise tend to propagate. For example, if I_{m-1} is in error, its effect on B_m is not eliminated but, in fact, is reinforced by the incorrect subtraction. Consequently, the detection of I_m is also likely to be in error.

Error propagation can be avoided by *precoding* the data at the transmitter instead of eliminating the controlled ISI by subtraction at the receiver. The precoding is performed on the binary data sequence prior to modulation. From the data sequence $\{D_n\}$ of 1s and 0s that is to be transmitted, a new sequence $\{P_n\}$, called the *precoded sequence*, is generated. For the duobinary signal, the precoded sequence is defined as

$$P_m = D_m \ominus P_{m-1}, \qquad m = 1, 2, \ldots \tag{9.2–46}$$

where $\ominus$ denotes modulo-2 subtraction.† Then we set $I_m = -1$ if $P_m = 0$ and $I_m = 1$ if $P_m = 1$, i.e., $I_m = 2P_m - 1$. Note that this precoding operation is identical to that described in Section 4.3.2 in the context of our discussion of an NRZI signal.

The noise-free samples at the output of the receiving filter are given by

$$\begin{aligned} B_m &= I_m + I_{m-1} \\ &= (2P_m - 1) + (2P_{m-1} - 1) \\ &= 2(P_m + P_{m-1} - 1) \end{aligned} \tag{9.2–47}$$

Consequently,

$$P_m + P_{m-1} = \tfrac{1}{2}B_m + 1 \tag{9.2–48}$$

Since $D_m = P_m \oplus P_{m-1}$, it follows that the data sequence D_m is obtained from B_m using the relation

$$D_m = \tfrac{1}{2}B_m + 1 \quad (\text{mod } 2) \tag{9.2–49}$$

Consequently, if $B_m = \pm 2$, then $D_m = 0$, and if $B_m = 0$, then $D_m = 1$. An example that illustrates the precoding and decoding operations is given in Table 9.2–1. In the presence of additive noise, the sampled outputs from the receiving filter are given by Equation 9.2–45. In this case $y_m = B_m + \nu_m$ is compared with the two thresholds set at $+1$ and -1. The data sequence $\{D_n\}$ is obtained according to the detection rule

$$D_m = \begin{cases} 1 & (|y_m| < 1) \\ 0 & (|y_m| \geqslant 1) \end{cases} \tag{9.2–50}$$

†Although this is identical to modulo-2 addition, it is convenient to view the precoding operation for duobinary in terms of modulo-2 subtraction.

TABLE 9.2–1
Binary signaling with duobinary pulses

Data sequence D_n		1	1	1	0	1	0	0	1	0	0	0	1	1	0	1
Precoded sequence P_n	0	1	0	1	1	0	0	0	1	1	1	1	0	1	1	0
Transmitted sequence I_n	−1	1	−1	1	1	−1	−1	−1	1	1	1	1	−1	1	1	−1
Received sequence B_n		0	0	0	2	0	−2	−2	0	2	2	2	0	0	2	0
Decoded sequence D_n		1	1	1	0	1	0	0	1	0	0	0	1	1	0	1

The extension from binary PAM to multilevel PAM signaling using the duobinary pulses is straightforward. In this case the M-level amplitude sequence $\{I_m\}$ results in a (noise-free) sequence

$$B_m = I_m + I_{m-1}, \qquad m = 1, 2, \ldots \tag{9.2–51}$$

which has $2M - 1$ possible equally spaced levels. The amplitude levels are determined from the relation

$$I_m = 2P_m - (M - 1) \tag{9.2–52}$$

where $\{P_m\}$ is the precoded sequence that is obtained from an M-level data sequence $\{D_m\}$ according to the relation

$$P_m = D_m \ominus P_{m-1} \quad (\text{mod } M) \tag{9.2–53}$$

where the possible values of the sequence $\{D_m\}$ are $0, 1, 2, \ldots, M - 1$.

In the absence of noise, the samples at the output of the receiving filter may be expressed as

$$B_m = I_m + I_{m-1} = 2[P_m + P_{m-1} - (M - 1)] \tag{9.2–54}$$

Hence,

$$P_m + P_{m-1} = \tfrac{1}{2}B_m + (M - 1) \tag{9.2–55}$$

Since $D_m = P_m + P_{m-1}$ (mod M), it follows that

$$D_m = \tfrac{1}{2}B_m + (M - 1) \quad (\text{mod } M) \tag{9.2–56}$$

An example illustrating multilevel precoding and decoding is given in Table 9.2–2.

In the presence of noise, the received signal-plus-noise is quantized to the nearest of the possible signal levels and the rule given above is used on the quantized values to recover the data sequence.

In the case of the modified duobinary pulse, the controlled ISI is specified by the values $x(n/2W) = -1$, for $n = 1$, $x(n/2W) = 1$ for $n = -1$, and zero otherwise. Consequently, the noise-free sampled output from the receiving filter is given as

TABLE 9.2–2
Four-level signal transmission with duobinary pulses

Data sequence D_m		0	0	1	3	1	2	0	3	3	2	0	1	0
Precoded sequence P_m	0	0	0	1	2	3	3	1	2	1	1	3	2	2
Transmitted sequence I_m	−3	−3	−3	−1	1	3	3	−1	1	−1	−1	3	1	1
Received sequence B_n		−6	−6	−4	0	4	6	2	0	0	−2	2	4	2
Decoded sequence D_m		0	0	1	3	1	2	0	3	3	2	0	1	0

$$B_m = I_m - I_{m-2} \tag{9.2–57}$$

where the M-level sequence $\{I_m\}$ is obtained by mapping a precoded sequence according to the Equation 9.2–52 and

$$P_m = D_m \oplus P_{m-2} \quad (\text{mod } M) \tag{9.2–58}$$

From these relations, it is easy to show that the detection rule for recovering the data sequence $\{D_m\}$ from $\{B_m\}$ in the absence of noise is

$$D_m = \tfrac{1}{2}B_m \quad (\text{mod } M) \tag{9.2–59}$$

As demonstrated above, the precoding of the data at the transmitter makes it possible to detect the received data on a symbol-by-symbol basis without having to look back at previously detected symbols. Thus, error propagation is avoided.

The symbol-by-symbol detection rule described above is not the optimum detection scheme for partial-response signals due to the memory inherent in the received signal. Nevertheless, symbol-by-symbol detection is relatively simple to implement and is used in many practical applications involving duobinary and modified duobinary pulse signals. Its performance is evaluated in the following section.

Maximum-likelihood sequence detection. It is clear from the above discussion that partial-response waveforms are signal waveforms with memory. This memory is conveniently represented by a trellis. For example, the trellis for the duobinary partial-response signal for binary data transmission is illustrated in Figure 9.2–11. For binary modulation, this trellis contains two states, corresponding to the two possible input values of I_m, i.e., $I_m = \pm 1$. Each branch in the trellis is labeled by two numbers. The first number on the left is the new data bit, i.e., $I_{m+1} = \pm 1$. This number determines the transition to the new state. The number on the right is the received signal level.

The duobinary signal has a memory of length $L = 1$. Hence, for binary modulation the trellis has $S_t = 2$ states. In general, for M-ary modulation, the number of trellis states is M^L.

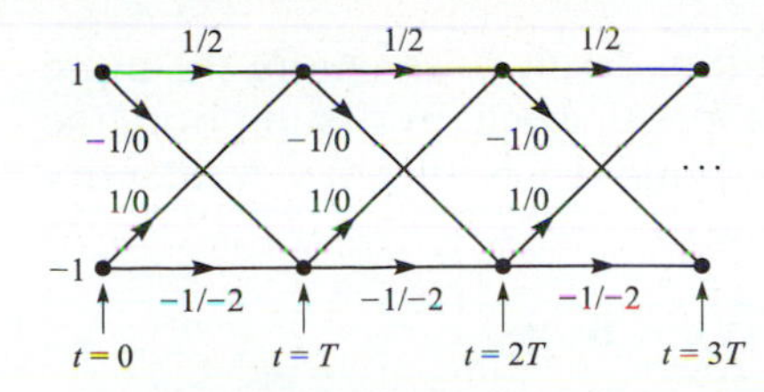

FIGURE 9.2–11
Trellis for duobinary partial-response signal.

The optimum maximum-likelihood (ML) sequence detector selects the most probable path through the trellis upon observing the received data sequence $\{y_m\}$ at the sampling instants $t = mT$, $m = 1, 2, \ldots$. In general, each node in the trellis will have M incoming paths and M corresponding metrics. One out of the M incoming paths is selected as the most probable, based on the values of the metrics and the other $M - 1$ paths and their metrics are discarded. The surviving path at each node is then extended to M new paths, one for each of the M possible input symbols, and the search process continues. This is basically the Viterbi algorithm for performing the trellis search.

For the class of partial-response signals, the received sequence $\{y_m, 1 \leqslant m \leqslant N\}$ is generally described statistically by the joint PDF $p(\mathbf{y}_N|\mathbf{I}_N)$, where $\mathbf{y}_N = [y_1 \; y_2 \; \cdots y_N]'$ and $\mathbf{I}_N = [I_1 \; I_2 \; \cdots I_N]'$ and $N > L$. When the additive noise is zero-mean Gaussian, $p(\mathbf{y}_N|\mathbf{I}_N)$ is a multivariate Gaussian PDF, i.e.,

$$f(\mathbf{y}_N|\mathbf{I}_N) = \frac{1}{(2\pi \det \mathbf{C})^{N/2}} \exp[-\tfrac{1}{2}(\mathbf{y}_N - \mathbf{B}_N)'\mathbf{C}^{-1}(\mathbf{y}_N - \mathbf{B}_N)] \tag{9.2–60}$$

where $\mathbf{B}_N = [B_1 \; B_2 \; \cdots \; B_N]'$ is the mean of the vector $\mathbf{y}_N$ and $\mathbf{C}$ is the $\mathrm{N} \times \mathrm{N}$ covariance matrix of $\mathbf{y}_N$. Then, the ML sequence detector selects the sequence through the trellis that maximizes the PDF $p(\mathbf{y}_N|\mathbf{I}_N)$.

The computation for finding the most probable sequence through the trellis is simplified by taking the natural logarithms of $p(\mathbf{y}_N|\mathbf{I}_N)$. Thus,

$$\ln p(\mathbf{y}_N|\mathbf{I}_N) = -\tfrac{1}{2}N \ln(2\pi \det \mathbf{C}) - \tfrac{1}{2}(\mathbf{y}_N - \mathbf{B}_N)'\mathbf{C}^{-1}(\mathbf{y}_N - \mathbf{B}_N) \tag{9.2–61}$$

Given the received sequence $\{y_m\}$, the data sequence $\{I_m\}$ that maximizes $\ln p(\mathbf{y}_N|\mathbf{I}_N)$ is identical to the sequence $\{I_N\}$ that minimizes $(\mathbf{y}_N - \mathbf{B}_N)'\mathbf{C}^{-1}(\mathbf{y}_N - \mathbf{B}_N)$, i.e.,

$$\hat{\mathbf{I}}_N = \arg\min_{\mathbf{I}_N}[(\mathbf{y}_N - \mathbf{B}_N)'\mathbf{C}^{-1}(\mathbf{y}_N - \mathbf{B}_N)] \tag{9.2–62}$$

The metric computations in the trellis search are complicated by the correlation of the noise samples at the output of the matched filter for the partial-response signal. For example, in the case of the duobinary signal waveform, the correlation of the noise sequence $\{\nu_m\}$ is over two successive signal samples. Hence, ν_m and ν_{m+k} are correlated for $k = 1$ and uncorrelated for $k > 1$. In general, a partial-response signal waveform with memory L will result in a correlated noise sequence at the output of the matched filter, which satisfies the condition $E[\nu_m \nu_{m+k}] = 0$ for $k > L$. In such a case, the Viterbi algorithm for performing the trellis search may be modified as described in Chapter 10.

Some simplification in the metric computations result if we ignore the noise correlation by assuming that $E(\nu_m \nu_{m+k}) = 0$ for $k > 0$. Then, by assumption, the covariance matrix $\mathbf{C} = \sigma_\nu^2 \mathbf{1}_N$, where $\sigma_\nu^2 = E[\nu_m^2]$ and $\mathbf{1}_N$ is the $N \times N$ identity matrix.† In this case, Equation 9.2–62 simplifies to

$$\begin{aligned}\hat{\mathbf{I}}_N &= \arg\min_{\mathbf{I}_N} [(\mathbf{y}_N - \mathbf{B}_N)'(\mathbf{y}_N - \mathbf{B}_N)] \\ &= \arg\min_{\mathbf{I}_N} \left[\sum_{m=1}^{N} \left(y_m - \sum_{k=0}^{L} x_k I_{m-k} \right)^2 \right] \end{aligned} \tag{9.2–63}$$

where

$$B_m = \sum_{k=0}^{L} x_k I_{m-k}$$

and $x_k = x(kT)$ are the sampled values of the partial-response signal waveform. In this case, the metric computations at each node of the trellis have the form

$$DM_m(\mathbf{I}_m) = DM_{m-1}(\mathbf{I}_{m-1}) + \left(y_m - \sum_{k=0}^{L} x_k I_{m-k} \right)^2 \tag{9.2–64}$$

where $DM_m(\mathbf{I}_m)$ are the distance metrics at time $t = mT$, $DM_{m-1}(\mathbf{I}_{m-1})$ are the distance metrics at time $t = (m-1)T$, and the second term on the right-hand side of Equation 9.2–64 represents the new increments to the metrics based on the new received sample y_m.

As indicated in Section 5.1.4, ML sequence detection introduces a variable delay in detecting each transmitted information symbol. In practice, the variable delay is avoided by truncating the surviving sequences to the N_t most recent symbols, where $N_t \gg 5L$, thus achieving a fixed delay. In the case that the M^L surviving sequences at time $t = mT$ disagree on the symbol I_{m-N_1}, the symbol in the most probable surviving sequence may be chosen. The loss in performance resulting from this truncation is negligible if $N_t > 5L$.

9.2.4 Signal Design for Channels with Distortion

In Sections 9.2.1 and 9.2.2, we described signal design criteria for the modulation filter at the transmitter and the demodulation filter at the receiver when the channel is ideal. In this section, we perform the signal design under the condition that the channel distorts the transmitted signal. We assume that the channel frequency-response $C(f)$ is known for $|f| \leqslant W$ and that $C(f) = 0$ for $|f| > W$. The filter responses $G_T(f)$ and $G_R(f)$ may be selected to minimize the error probability at the detector. The additive channel noise is assumed to be

†We are using $\mathbf{1}_N$ here to avoid confusion with $\mathbf{I}_N$.

Gaussian with power spectral density $\Phi_{nn}(f)$. Figure 9.2–12 illustrates the overall system under consideration.

For the signal component at the output of the demodulator, we must satisfy the condition

$$G_T(f)C(f)G_R(f) = X_d(f)e^{-j2\pi f t_0}, \qquad |f| \leqslant W \tag{9.2–65}$$

where $X_d(f)$ is the desired frequency response of the cascade of the modulator, channel, and demodulator, and t_0 is a time delay that is necessary to ensure the physical realizability of the modulation and demodulation filters. The desired frequency response $X_d(f)$ may be selected to yield either zero ISI or controlled ISI at the sampling instants. We shall consider the case of zero ISI by selecting $X_d(f) = X_{rc}(f)$, where $X_{rc}(f)$ is the raised cosine spectrum with an arbitrary roll-off factor.

The noise at the output of the demodulation filter may be expressed as

$$\nu(t) = \int_{-\infty}^{\infty} n(t-\tau)g_R(\tau)d\tau \tag{9.2–66}$$

where $n(t)$ is the input to the filter. Since $n(t)$ is zero-mean Gaussian, $\nu(t)$ is zero-mean Gaussian, with a power spectral density

$$\Phi_{\nu\nu}(f) = \Phi_{nn}(f)|G_R(f)|^2 \tag{9.2–67}$$

For simplicity, we consider binary PAM transmission. Then, the sampled output of the matched filter is

$$y_m = x_0 I_m + \nu_m = I_m + \nu_m \tag{9.2–68}$$

where x_0 is normalized† to unity, $I_m = \pm d$, and ν_m represents the noise term, which is zero-mean Gaussian with variance

$$\sigma_\nu^2 = \int_{-\infty}^{\infty} \Phi_{nn}(f)|G_R(f)|^2\, df \tag{9.2–69}$$

Consequently, the probability of error is

$$P_2 = \frac{1}{\sqrt{2\pi}} \int_{d/\sigma_\nu}^{\infty} e^{-y^2/2} dy = Q\left(\sqrt{\frac{d^2}{\sigma_\nu^2}}\right) \tag{9.2–70}$$

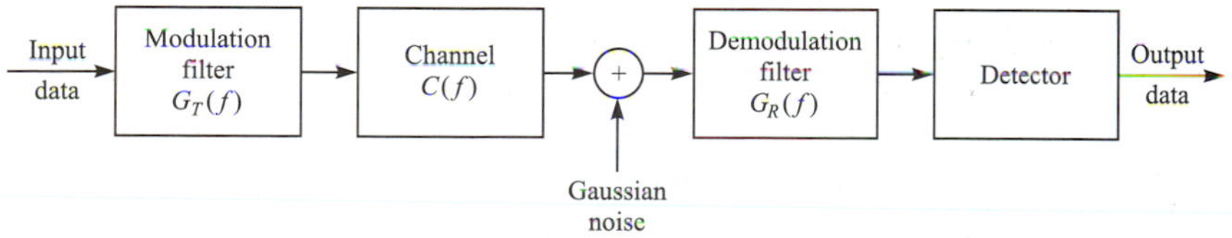

FIGURE 9.2–12
System model for the design of the modulation and demodulation filters.

---

†By setting $x_0 = 1$ and $I_m = \pm d$, the scaling by x_0 is incorporated into the parameter d.

The probability of error is minimized by maximizing the ratio d^2/σ_v^2 or, equivalently, by minimizing the noise-to-signal ratio σ_v^2/d^2.

Let us consider two possible solutions for the case in which the additive Gaussian noise is white, so that $\Phi_{nn}(f) = N_0/2$. First, suppose that we precompensate for the total channel distortion at the transmitter, so that the filter at the receiver is matched to the received signal. In this case, the transmitter and receiver filters have the magnitude characteristics

$$\begin{aligned} |G_T(f)| &= \frac{\sqrt{X_{rc}(f)}}{|C(f)|}, \qquad |f| \leqslant W \\ |G_R(f)| &= \sqrt{X_{rc}(f)}, \qquad |f| \leqslant W \end{aligned} \tag{9.2–71}$$

The phase characteristic of the channel frequency response $C(f)$ may also be compensated at the transmitter filter. For these filter characteristics, the average transmitted power is

$$\begin{aligned} P_{\text{av}} &= \frac{E(I_m^2)}{T} \int_{-W}^{W} g_T^2(t)\, dt = \frac{d^2}{T} \int_{-W}^{W} |G_T(f)|^2\, df \\ &= \frac{d^2}{T} \int_{-W}^{W} \frac{X_{rc}(f)}{|C(f)|^2}\, df \end{aligned} \tag{9.2–72}$$

and, hence,

$$d^2 = P_{\text{av}} T \left[\int_{-W}^{W} \frac{X_{rc}(f)}{|C(f)|^2}\, df \right]^{-1} \tag{9.2–73}$$

The noise at the output of the receiver filter is $\sigma_v^2 = N_0/2$ and, hence, the SNR at the detector is

$$\frac{d^2}{\sigma_v^2} = \frac{2P_{\text{av}}T}{N_0} \left[\int_{-W}^{W} \frac{X_{rc}(f)}{|C(f)|^2}\, df \right]^{-1} \tag{9.2–74}$$

As an alternative, suppose we split the channel compensation equally between the transmitter and receiver filters, i.e.,

$$\begin{aligned} |G_T(f)| &= \frac{\sqrt{X_{rc}(f)}}{|C(f)|^{1/2}}, \qquad |f| \leqslant W \\ |G_R(f)| &= \frac{\sqrt{X_{rc}(f)}}{|C(f)|^{1/2}} \qquad |f| \leqslant W \end{aligned} \tag{9.2–75}$$

The phase characteristic of $C(f)$ may also be split equally between the transmitter and receiver filters. In this case, the average transmitter power is

$$P_{av} = \frac{d^2}{T} \int_{-W}^{W} \frac{X_{rc}(f)}{|C(f)|}\, df \tag{9.2–76}$$

and the noise variance at the output of the receiver filter is

$$\sigma_v^2 = \frac{N_0}{2} \int_{-W}^{W} \frac{X_{rc}(f)}{|C(f)|} df \tag{9.2–77}$$

Hence, the SNR at the detector is

$$\frac{d^2}{\sigma_v^2} = \frac{2P_{\text{av}}T}{N_0} \left[\int_{-W}^{W} \frac{X_{rc}(f)}{|C(f)|} df \right]^{-2} \tag{9.2–78}$$

From Equations 9.2–73 and 9.2–78, we observe that when we express the SNR d^2/σ_v^2 in terms of the average transmitter power P_{av}, there is a loss incurred due to channel distortion. In the case of the filters given by Equation 9.2–71, the loss is

$$10 \log \int_{-W}^{W} \frac{X_{rc}(f)}{|C(f)|^2} df \tag{9.2–79}$$

and, in the case of the filters given by Equation 9.2–75, the loss is

$$10 \log \left[\int_{-W}^{W} \frac{X_{rc}(f)}{|C(f)|} df \right]^2 \tag{9.2–80}$$

We observe that when $C(f) = 1$ for $|f| \leqslant W$, the channel is ideal and

$$\int_{-W}^{W} X_{rc}(f)\, df = 1 \tag{9.2–81}$$

so that no loss is incurred. On the other hand, when there is amplitude distortion, $|C(f)| < 1$ for some range of frequencies in the band $|f| \leqslant W$ and, hence, there is a loss in SNR as given by Equations 9.2–79 and 9.2–80. The interested reader may show (see Problem 9.30) that the filters given by Equation 9.2–75 result in the smaller SNR loss.

EXAMPLE 9.2–1. Let us determine the transmitting and receiving filters given by Equation 9.2–75 for a binary communication system that transmits data at a rate of 4800 bits/s over a channel with frequency (magnitude) response

$$|C(f)| = \frac{1}{\sqrt{1 + (f/W)^2}}, \qquad |f| \leqslant W \tag{9.2–82}$$

where $W = 4800$ Hz. The additive noise is zero-mean white Gaussian with spectral density $\frac{1}{2}N_0 = 10^{-15}$ W/Hz.

Since $W = 1/T = 4800$, we use a signal pulse with a raised cosine spectrum and $\beta = 1$. Thus,

$$\begin{aligned} X_{rc}(f) &= \tfrac{1}{2}T[1 + \cos(\pi T|f|)] \\ &= T\cos^2\left(\frac{\pi|f|}{9600}\right) \end{aligned} \tag{9.2–83}$$

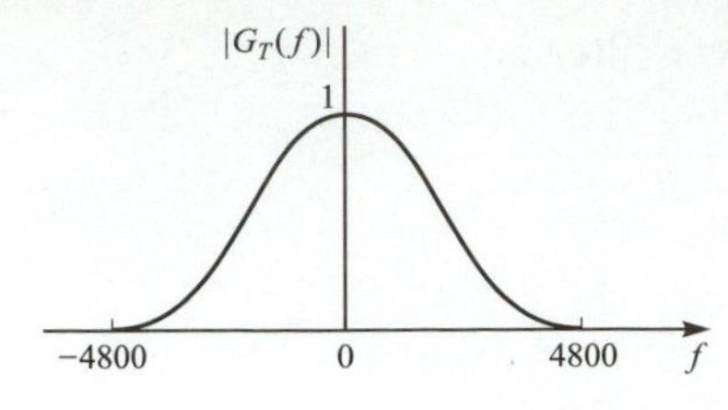

FIGURE 9.2–13
Frequency response of an optimum transmitter filter.

Then,

$$|G_T(f)| = |G_R(f)| = \left[1 + \left(\frac{f}{4800}\right)^2\right]^{1/4} \cos\left(\frac{\pi|f|}{9600}\right), \qquad |f| \leqslant 4800 \tag{9.2–84}$$

and $|G_T(f)| = |G_R(f)| = 0$, otherwise. Figure 9.2–13 illustrates the filter characteristic $G_T(f)$.

One can now use these filters to determine the amount of transmitted energy $\mathcal{E}$ required to achieve a specified error probability. This problem is left as an exercise for the reader.

9.3 PROBABILITY OF ERROR IN DETECTION OF PAM

In this section, we evaluate the performance of the receiver for demodulating and detecting an M-ary PAM signal in the presence of additive white Gaussian noise at its input. First, we consider the case in which the transmitter and receiver filters $G_T(f)$ and $G_R(f)$ are designed for zero ISI. Then, we consider the case in which $G_T(f)$ and $G_R(f)$ are designed such that $x(t) = g_T(t) \star g_R(t)$ is either a duobinary signal or a modified duobinary signal.

9.3.1 Probability of Error for Detection of PAM with Zero ISI

In the absence of ISI, the received signal sample at the output of the receiving matched filter has the form

$$y_m = x_0 I_m + \nu_m \tag{9.3–1}$$

where

$$x_0 = \int_{-W}^{W} |G_T(f)|^2 df = \mathcal{E}_g \tag{9.3–2}$$

and ν_m is the additive Gaussian noise that has zero-mean and variance

$$\sigma_\nu^2 = \tfrac{1}{2}\mathcal{E}_g N_0 \tag{9.3–3}$$

In general, I_m takes one of M possible equally spaced amplitude values with equal probability. Given a particular amplitude level, the problem is to determine the probability of error.

The problem of evaluating the probability of error for digital PAM in a band-limited, additive white Gaussian noise channel, in the absence of ISI, is identical to the evaluation of the error probability for M-ary PAM as given in Section 5.2. The final result that is obtained from the derivation is

$$P_M = \frac{2(M-1)}{M} Q\left(\sqrt{\frac{2\mathcal{E}_g}{N_0}}\right) \tag{9.3–4}$$

But $\mathcal{E}_g = 3\mathcal{E}_{\text{av}}/(M^2-1)$, $\mathcal{E}_{\text{av}} = k\mathcal{E}_{b\text{av}}$ is the average energy per symbol and $\mathcal{E}_{b\text{av}}$ is the average energy per bit. Hence,

$$P_M = \frac{2(M-1)}{M} Q\left(\sqrt{\frac{6(\log_2 M)\mathcal{E}_{b\text{av}}}{(M^2-1)N_0}}\right) \tag{9.3–5}$$

This is exactly the form for the probability of error of M-ary PAM derived in Section 5.2 (see Equation 5.2–46). In the treatment of PAM given in this chapter, we imposed the additional constraint that the transmitted signal is band-limited to the bandwidth allocated for the channel. Consequently, the transmitted signal pulses were designed to be band-limited and to have zero ISI.

In contrast, no bandwidth constraint was imposed on the PAM signals considered in Section 5.2. Nevertheless, the receivers (demodulators and detectors) in both cases are optimum (matched filters) for the corresponding transmitted signals. Consequently, no loss in error rate performance results from the bandwidth constraint when the signal pulse is designed for zero ISI and the channel does not distort the transmitted signal.

9.3.2 Probability of Error for Detection of Partial-Response Signals

In this section we determine the probability of error for detection of digital M-ary PAM signaling using duobinary and modified duobinary pulses. The channel is assumed to be an ideal band-limited channel with additive white Gaussian noise. The model for the communication system is shown in Figure 9.3–1.

We consider two types of detectors. The first is the symbol-by-symbol detector and the second is the optimum ML sequence detector described in the previous section.

Symbol-by-symbol detector. At the transmitter, the M-level data sequence $\{D_m\}$ is precoded as described previously. The precoder output is mapped into one of M possible amplitude levels. Then the transmitting filter with frequency response $G_T(f)$ has an output

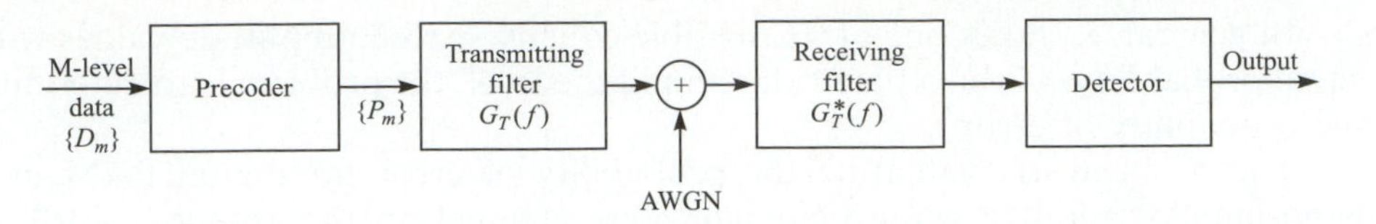

FIGURE 9.3–1
Block diagram of modulator and demodulator for partial-response signals.

$$v(t) = \sum_{n=-\infty}^{\infty} I_n g_T(t - nT) \qquad (9.3\text{–}6)$$

The partial-respone function $X(f)$ is divided equally between the transmitting and receiving filters. Hence, the receiving filter is matched to the transmitted pulse, and the cascade of the two filters results in the frequency characteristic

$$|G_T(f)G_R(f)| = |X(f)| \qquad (9.3\text{–}7)$$

The matched filter output is sampled at $t = nT = n/2W$ and the samples are fed to the decoder. For the duobinary signal, the output of the matched filter at the sampling instant may be expressed as

$$y_m = I_m + I_{m-1} + \nu_m = B_m + \nu_m \qquad (9.3\text{–}8)$$

where ν_m is the additive noise component. Similarly, the output of the matched filter for the modified duobinary signal is

$$y_m = I_m - I_{m-2} + \nu_m = B_m + \nu_m \qquad (9.3\text{–}9)$$

For binary transmission, let $I_m = \pm d$, where $2d$ is the distance between signal levels. Then, the corresponding values of B_m are $(2d, 0, -2d)$. For M-ary PAM signal transmission, where $I_m = \pm d, \pm 3d, \ldots, \pm(M-1)d$, the received signal levels are $B_m = 0, \pm 2d, \pm 4d, \ldots, \pm 2(M-1)d$. Hence, the number of received levels is $2M - 1$, and the scale factor d is equivalent to $x_0 = \mathcal{E}_g$.

The input transmitted symbols $\{I_m\}$ are assumed to be equally probable. Then, for duobinary and modified duobinary signals, it is easily demonstrated that, in the absence of noise, the received output levels have a (triangular) probability distribution of the form

$$P(B = 2md) = \frac{M - |m|}{M^2}, \qquad m = 0, \pm 1, \pm 2, \ldots, \pm(M-1) \qquad (9.3\text{–}10)$$

where B denotes the noise-free received level and $2d$ is the distance between any two adjacent received signal levels.

The channel corrupts the signal transmitted through it by the addition of white Gaussian noise with zero-mean and power spectral density $\frac{1}{2}N_0$.

We assume that a symbol error occurs whenever the magnitude of the additive noise exceeds the distance d. This assumption neglects the rare event that a large noise component with magnitude exceeding d may result in a received signal level that yields a correct symbol decision. The noise component ν_m is zero-mean Gaussian with variance

$$\begin{aligned}\sigma_\nu^2 &= \tfrac{1}{2}N_0 \int_{-W}^{W} |G_R(f)|^2 df \\ &= \tfrac{1}{2}N_0 \int_{-W}^{W} |X(f)| df = \frac{2N_0}{\pi}\end{aligned} \tag{9.3–11}$$

for both the duobinary and the modified duobinary signals. Hence, an upper bound on the symbol probability of error is

$$\begin{aligned}P_M &< \sum_{m=-(M-2)}^{M-2} P(|y - 2md| > d | B = 2md) P(B = 2md) \\ &\quad + 2P[y + 2(M-1)d > d | B = -2(M-2)d] P[B = -2(M-1)d] \\ &= P(|y| > d | B = 0) \left\{ 2 \sum_{m=0}^{M-1} P(B = 2md) - P(B = 0) - P[B = -2(M-1)d] \right\} \\ &= (1 - M^{-2}) P(|y| > d | B = 0)\end{aligned} \tag{9.3–12}$$

But

$$\begin{aligned}P(|y| > d | B = 0) &= \frac{2}{\sqrt{2\pi}\sigma_\nu} \int_d^\infty e^{-x^2/2\sigma_\nu^2} dx \\ &= 2Q\left(\sqrt{\frac{\pi d^2}{2N_0}}\right)\end{aligned} \tag{9.3–13}$$

Therefore, the average probability of a symbol error is upper-bounded as

$$P_M < 2(1 - M^{-2}) Q\left(\sqrt{\frac{\pi d^2}{2N_0}}\right) \tag{9.3–14}$$

The scale factor d in Equation 9.3–14 can be eliminated by expressing it in terms of the average power transmitted into the channel. For the M-ary PAM signal in which the transmitted levels are equally probable, the average power at the output of the transmitting filter is

$$P_{\text{av}} = \frac{E(I_m^2)}{T} \int_{-W}^{W} |G_T(f)|^2 \, df = \frac{E(I_m^2)}{T} \int_{-W}^{W} |X(f)| df = \frac{4}{\pi T} E(I_m^2) \tag{9.3–15}$$

where $E(I_m^2)$ is the mean square value of the M signal levels, which is

$$E(I_m^2) = \tfrac{1}{3} d^2 (M^2 - 1) \tag{9.3–16}$$

Therefore,

$$d^2 = \frac{3\pi P_{\text{av}} T}{4(M^2 - 1)} \tag{9.3–17}$$

By substituting the value of d^2 from Equation 9.3–17 into Equation 9.3–14, we obtain the upper bound on the symbol error probability as

$$P_M < 2\left(1 - \frac{1}{M^2}\right) Q\left(\sqrt{\left(\frac{\pi}{4}\right)^2 \frac{6}{M^2 - 1} \frac{\mathcal{E}_{\text{av}}}{N_0}}\right) \tag{9.3–18}$$

where $\mathcal{E}_{\text{av}}$ is the average energy per transmitted symbol, which can be also expressed in terms of the average bit energy as $\mathcal{E}_{\text{av}} = k\mathcal{E}_{b\text{av}} = (\log_2 M)\mathcal{E}_{b\text{av}}$.

The expression in Equation 9.3–18 for the probability of error of M-ary PAM holds for both duobinary and modified duobinary partial-response signals. If we compare this result with the error probability of M-ary PAM with zero ISI, which can be obtained by using a signal pulse with a raised cosine spectrum, we note that the performance of partial-response duobinary or modified duobinary has a loss of $(\frac{1}{4}\pi)^2$, or 2.1 dB. This loss in SNR is due to the fact that the detector for the partial-response signals makes decisions on a symbol-by-symbol basis, and ignoring the inherent memory contained in the received signal at the input to the detector.

Maximum-likelihood sequence detector. The ML sequence detector searches through the trellis for the most probable transmitted sequence $\{I_m\}$ as previously described in Section 9.2.3. At each stage of the search process the detector compares the metrics of paths that merge at each of the nodes and selects the path that is most probable at each node. The performance of the detector may be evaluated by determining the probability of error events, based on a Euclidean distance metric, as was done for soft-decision decoding of convolutional codes. The general derivation is given in Section 10.1.4. In the case of the duobinary and modified duobinary signals, it is demonstrated that the 2.1-dB loss inherent in the suboptimum symbol-by-symbol detector is completely recovered by the ML sequence detector.

9.4
MODULATION CODES FOR SPECTRUM SHAPING

We have observed that the power spectral density of a digital communication signal can be controlled and shaped by selecting the transmitted signal pulse $g(t)$ and by introducing correlation through coding, which is used to combat channel distortion and noise in transmission. Coding for spectrum shaping is introduced following the channel encoding so that the spectrum of the transmitted signal matches the spectral characteristics of a baseband or equivalent low-pass channel.

Codes that are used for spectrum shaping are generally called either *modulation codes*, or *line codes*, or *data translation codes*. Such codes generally place restrictions on the sequence of bits into the modulator and, thus, introduce correlation and, hence, memory into the transmitted signal. It is this type of coding that is treated in this section.

Modulation codes are usually employed in magnetic recording, in optical recording, and in digital communications over cable systems to achieve spectral shaping and to eliminate or minimize the DC content in the transmitted (or stored) baseband signal. In magnetic recording channels, the modulation code is designed to increase the distance between transitions in the recorded waveform and, thus, intersymbol interference effects are also reduced.

As an example of the use of a modulation code, let us consider a magnetic recording system, which consists of the elements shown in the block diagram of Figure 9.4–1. The binary data sequence to be stored is used to generate a write current. This current may be viewed as the output of the "modulator." The most commonly used method to map the information sequence into the write current waveform is NRZI, which was described in Section 4.3.2. Recall that in NRZI, a transition from one amplitude to another (A to $-A$ or $-A$ to A) occurs only when the information bit is a 1. No transition occurs when the information bit is a 0, i.e., the amplitude level remains the same as in the previous signal interval. The positive amplitude pulse results in magnetizing the medium in one (direction) polarity and the negative pulse magnetizes the medium in the opposite (direction) polarity.

Since the input data sequence is basically random with equally probable 1s and 0s, we shall encounter level transitions from A to $-A$ or $-A$ to A with probability 1/2 for every data bit. The readback signal for a positive transition ($-A$ to A) is a pulse that is well-modeled mathematically as

$$g(t) = \frac{1}{1 + (2t/T_{50})^2} \tag{9.4–1}$$

where T_{50} is defined as the width of the pulse at its 50 percent amplitude level, as shown in Figure 9.4–2. Similarly, the readback signal for a negative transition (A to $-A$) is the pulse $-g(t)$. The value of T_{50} is determined by the characteristics of the medium, the read/write heads, and the distance of the head to the medium.

Now, suppose we write a positive transition followed by a negative transition. Let's vary the time interval between the two transitions, which we denote as

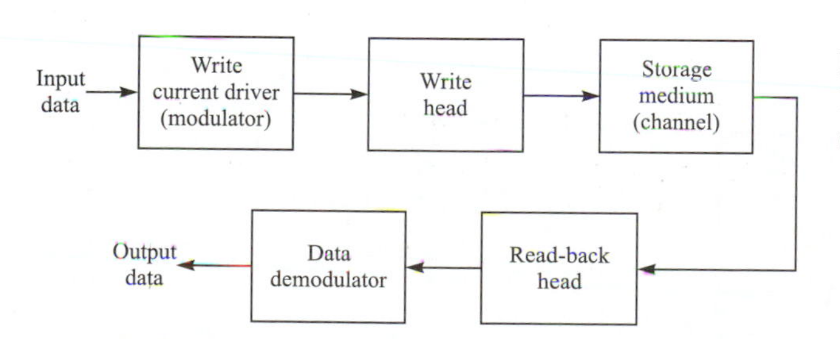

FIGURE 9.4–1
Block diagram of magnetic storage read/write system.

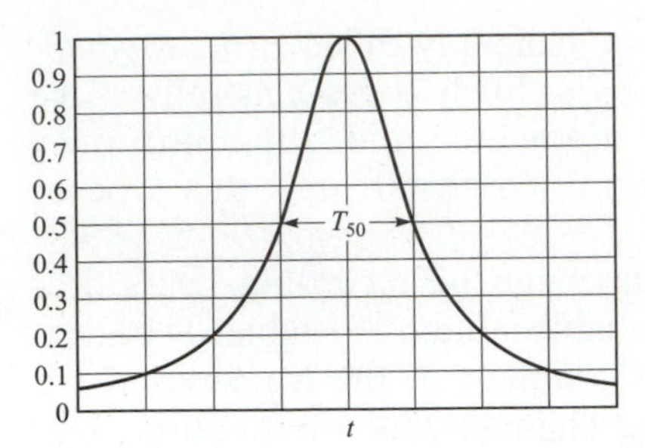

FIGURE 9.4–2
Readback pulse in magnetic recording system.

T_b (the bit time interval). Figure 9.4–3 illustrates the readback signal pulses, which are obtained by a superposition of $g(t)$ with $-g(t - T_b)$. The parameter $\Delta = T_{50}/T_b$ is defined as the *normalized density*. The closer the bit transitions (T_b small), the larger will be the value of the normalized density and, hence, the larger will be the packing density. We notice that as Δ is increased, the peak amplitudes of the readback signal are reduced and are also shifted in time from the desired time instants. In other words, the pulses interfere with one another, thus limiting the density with which we can write. This problem serves as a motivation to design modulation codes that take the original data sequence and transform (encode) it into another sequence that results in a write waveform in which amplitude transitions are spaced farther apart. For example, if we use NRZI, the encoded sequence into the modulator must contain one or more 0s between 1s.

The second problem encountered in magnetic recording is the need to avoid (or minimize) having a DC content in the modulated signal (the write current) due to the frequency-response characteristics of the readback system and associated electronics. This requirement also arises in digital communication over cable channels. This problem can be overcome by altering (encoding) the data sequence into the modulator. A class of codes that satisfy these objectives are the modulation codes described below.

Runlength-limited codes. Codes that have a restriction on the number of consecutive 1s or 0s in a sequence are generally called *runlength-limited codes.*

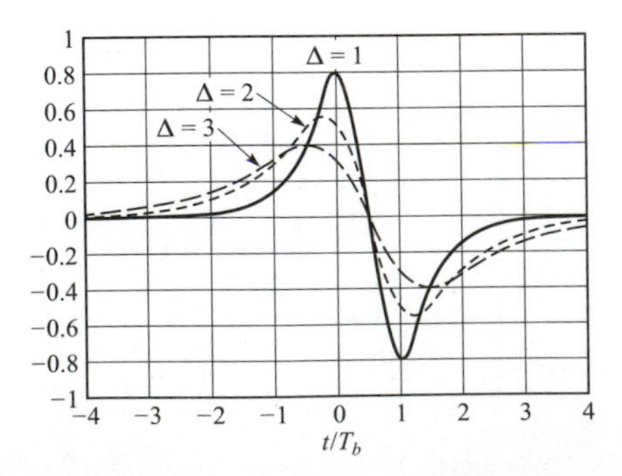

FIGURE 9.4–3
Readback signal response to a pulse.

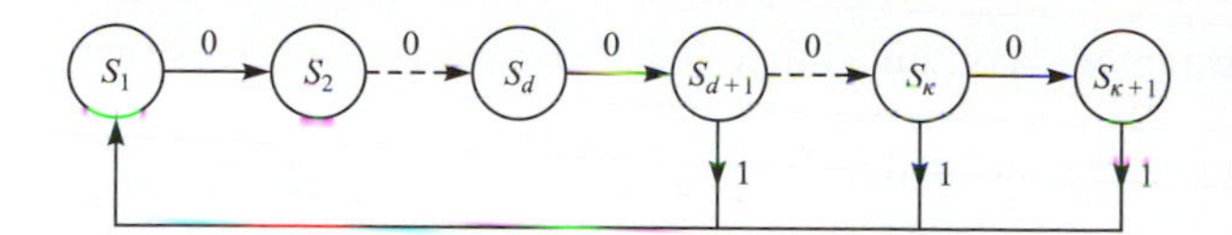

FIGURE 9.4–4
Finite-state sequential machine for a (d, κ)-coded sequence.

These codes are generally described by two parameters, say d and κ, where d denotes the minimum number of 0s between two 1s in a sequence, and κ denotes the maximum number of 0s between two 1s in a sequence. When used with NRZI modulation, the effect of placing d zeros between successive 1s is to spread the transitions farther apart, thus reducing the overlap in the channel response due to successive transitions and hence reducing the intersymbol interference. Setting an upper limit κ on the runlength of 0s ensures that transitions occur frequently enough so that symbol timing information can be recovered from the received modulated signal. Runlength-limited codes are usually called (d, κ) codes.†

The (d, κ) code sequence constraints may be represented by a finite-state sequential machine with $\kappa + 1$ states, denoted as S_i, $1 \leqslant i \leqslant \kappa + 1$, as shown in Figure 9.4–4. We observe that an output data bit 0 takes the sequence from state S_i to S_{i+1}, $i \leqslant \kappa$. The output data bit 1 takes the sequence to state S_1. The output bit from the encoder may be a 1 only when the sequence is in state S_i, $d + 1 \leqslant i \leqslant \kappa + 1$. When the sequence is in state $S_{\kappa+1}$, the output bit is always 1.

The finite-state sequential machine may also be represented by a *state transition matrix*, denoted as $\mathbf{D}$, which is a square $(\kappa + 1) \times (\kappa + 1)$ matrix with elements d_{ij}, where

$$d_{i1} = 1 \qquad (i \geqslant d + 1)$$
$$d_{ij} = \begin{cases} 1 & (j = i + 1) \\ 0 & (\text{otherwise}) \end{cases} \tag{9.4–2}$$

EXAMPLE 9.4–1. Let us determine the state transition matrix for a $(d, \kappa) = (1, 3)$ code. The $(1, 3)$ code has four states. From Figure 9.4–4, we obtain its state transition matrix, which is

$$\mathbf{D} = \begin{bmatrix} 0 & 1 & 0 & 0 \\ 1 & 0 & 1 & 0 \\ 1 & 0 & 0 & 1 \\ 1 & 0 & 0 & 0 \end{bmatrix} \tag{9.4–3}$$

An important parameter of any (d, κ) code is the number of sequences of a certain length, say n, that satisfy the (d, κ) constraints. As n is allowed to increase, the number of sequences $N(n)$ that satisfy the (d, κ) constraint also increases. The number of information bits that can be uniquely represented with $N(n)$ code sequences is

$$k = \lfloor \log_2 N(n) \rfloor$$

†In fact, they are usually called (d, k) codes, where k is the maximum runlength of zeros. We have substituted the Greek letter kappa κ for k, to avoid confusion with our previous use of k.

where $\lfloor x \rfloor$ denotes the largest integer contained in x. The maximum code rate is then $R_c = k/n$.

The capacity of a (d, κ) code is defined as

$$C(d, \kappa) = \lim_{n\to\infty} \frac{1}{n} \log_2 N(n) \tag{9.4–4}$$

Clearly, $C(d, \kappa)$ is the maximum possible rate that can be achieved with the (d, κ) constraints. Shannon (1948) showed that the capacity is given as

$$C(d, \kappa) = \log_2 \lambda_{\max} \tag{9.4–5}$$

where $\lambda_{\max}$ is the largest real eigenvalue of the state transition matrix $\mathbf{D}$.

EXAMPLE 9.4–2. Let us determine the capacity of a $(d, \kappa) = (1, 3)$ code. Using the state transition matrix given in Example 9.4–1 for the $(1, 3)$ code, we have

$$\det(\mathbf{D} - \lambda\mathbf{I}) = \det \begin{bmatrix} -\lambda & 1 & 0 & 0 \\ 1 & -\lambda & 1 & 0 \\ 1 & 0 & -\lambda & 1 \\ 1 & 0 & 0 & -\lambda \end{bmatrix} \tag{9.4–6}$$
$$= \lambda^4 - \lambda^2 - \lambda - 1 = 0$$

The maximum real root of this polynomial is found to be $\lambda_{\max} = 1.4656$. Therefore, the capacity $C(1, 3) = \log_2 \lambda_{\max} = 0.5515$.

The capacities of (d, κ) codes for $0 \leqslant d \leqslant 6$ and $2 \leqslant \kappa \leqslant 15$ are given in Table 9.4–1. We observe that $C(d, \kappa) < \frac{1}{2}$ for $d \geqslant 3$ and any value of κ. The most commonly used codes for magnetic recording employ $d \leqslant 2$; hence, their rate R_c is at least $\frac{1}{2}$.

TABLE 9.4–1
Capacity $C(d, \kappa)$ versus runlength parameters d and κ

κ	$d = 0$	$d = 1$	$d = 2$	$d = 3$	$d = 4$	$d = 5$	$d = 6$
2	.8791	.4057					
3	.9468	.5515	.2878				
4	.9752	.6174	.4057	.2232			
5	.9881	.6509	.4650	.3218	.1823		
6	.9942	.6690	.4979	.3746	.2269	.1542	
7	.9971	.6793	.5174	.4057	.3142	.2281	.1335
8	.9986	.6853	.5293	.4251	.3432	.2709	.1993
9	.9993	.6888	.5369	.4376	.3620	.2979	.2382
10	.9996	.6909	.5418	.4460	.3746	.3158	.2633
11	.9998	.6922	.5450	.4516	.3833	.3285	.2804
12	.9999	.6930	.5471	.4555	.3894	.3369	.2924
13	.9999	.6935	.5485	.4583	.3937	.3432	.3011
14	.9999	.6938	.5495	.4602	.3968	.3478	.3074
15	.9999	.6939	.5501	.4615	.3991	.3513	.3122
∞	1.000	.6942	.5515	.4650	.4057	.3620	.3282

Now let us turn our attention to the construction of some runlength-limited codes. In general, (d, κ) codes can be constructed either as fixed-length codes or as variable-length codes. In a fixed-length code, each bit or block of k bits is encoded into a block of $n > k$ bits.

In principle, the construction of a fixed-length code is straightforward. For a given block length n, we may select the subset of the 2^n code words that satisfy the specified runlength constraints. From this subset, we eliminate code words that do not satisfy the runlength constraints when concatenated. Thus, we obtain a set of code words that satisfy the constraints and can be used in the mapping of the input data bits to the encoder. The encoder and decoding operations can be performed by use of a lookup table.

EXAMPLE 9.4–3. Let us construct a $d = 0$, $\kappa = 2$ code of length $n = 3$, and determine its efficiency. By listing all the code words, we find that the following five code words satisfy the (0, 2) constraint: (010), (011), (101), (110), (111). We may select any four of these code words and use them to encode the pairs of data bits (00, 01, 10, 11). Thus, we have a rate $k/n = 2/3$ code that satisfies the (0, 2) constraint.

The fixed-length code in this example is not very efficient. The capacity is $C(0, 2) = 0.8791$, so that this code has an *efficiency* of

$$\text{Efficiency} = \frac{R_c}{C(d, \kappa)} = \frac{2/3}{0.8791} = 0.76$$

Surely, better (0, 2) codes can be constructed by increasing the block length n.

In the following example, we place no restriction on the maximum runlength of zeros.

EXAMPLE 9.4–4. Let us construct a $d = 1, \kappa = \infty$ code of length $n = 5$. In this case, we are placing no constraint on the number of consecutive zeros. To construct the code, we select from the set of 32 possible code words those that satisfy the $d = 1$ constraint. There are eight such code words, which implies that we can encode three information bits with each code word. The code is given in Table 9.4–2. Note that the first bit of each code word is a 0, whereas the last bit may be either 0 or 1. Consequently, the $d = 1$ constraint is satisfied when these code words are concatenated. This code has a rate $R_c = 3/5$. When compared with the capacity $C(1, \infty) = 0.6942$ obtained from Table 9.4–1, the code efficiency is 0.864, which is quite acceptable.

TABLE 9.4–2
Fixed length $d = 1$, $\kappa = \infty$ code

Input data bits	Output coded sequence
0 0 0	0 0 0 0 0
0 0 1	0 0 0 0 1
0 1 0	0 0 0 1 0
0 1 1	0 0 1 0 0
1 0 0	0 0 1 0 1
1 0 1	0 1 0 0 0
1 1 0	0 1 0 0 1
1 1 1	0 1 0 1 0

The code construction method described in the two examples above produces fixed-length (d, κ) codes that are *state-independent*. By state-independent, we mean that fixed-length code words can be concatenated without violating the (d, κ) constraints. In general, fixed-length state-independent (d, κ) codes require large block lengths, except in cases such as those in the examples above where d is small. Simpler (shorter-length) codes are generally possible by allowing for state-dependence and for variable-length code words. Below, we consider codes for which both the input blocks to the encoder and the output blocks may have variable length. For the code words to be uniquely decodable at the receiver, the variable-length code should satisfy the prefix condition, described in Chapter 3.

EXAMPLE 9.4–5. A very simple uniquely decodable variable-length $d = 0$, $\kappa = 2$ code is

$$0 \rightarrow 01$$
$$10 \rightarrow 10$$
$$11 \rightarrow 11$$

The code in the above example has a fixed output block size but a variable input block size. In general both the input and output blocks may be variable. The following example illustrates the latter case.

EXAMPLE 9.4–6. Let us construct a (2, 7) variable block size code. The solution to this code construction is certainly not unique, nor is it trivial. We picked this example because the (2, 7) code has been widely used by IBM in many of its disk storage systems. The code is listed in Table 9.4–3. We observe that the input data blocks of 2, 3, and 4 bits are mapped into output data blocks of 4, 6, and 8 bits, respectively. Hence, the code rate is $R_c = 1/2$. Since this is the code rate for all code words, the code is called a *fixed-rate* code. This code has an efficiency of 0.5/0.5174 = 0.966. Note that this code satisfies the prefix condition.

Another code that has been widely used in magnetic recording is the rate 1/2, $(d, \kappa) = (1, 3)$ code in Table 9.4–4. We observe that when the information bit is a 0, the first output bit is 1 if the previous input bit was 0, or a 0 if the previous input bit was a 1. When the information bit is a 1, the encoder output is 01.

TABLE 9.4–3
Code book for variable-length (2, 7) code

Input data bits	Output coded sequence
1 0	1 0 0 0
1 1	0 1 0 0
0 1 1	0 0 0 1 0 0
0 1 0	0 0 1 0 0 0
0 0 0	1 0 0 1 0 0
0 0 1 1	0 0 1 0 0 1 0 0
0 0 1 0	0 0 0 0 1 0 0 0

TABLE 9.4–4
Encoder for (1, 3) Miller code

Input data bits	Output coded sequence
0	$x\,0$
1	0 1

$x = 0$, if preceding input bit is 1
$x = 1$, if preceding input bit is 0

Decoding of this code is simple. The first bit of the 2-bit block is redundant and may be discarded. The second bit is the information bit. This code is usually called the *Miller code*. We observe that this is a state-dependent code, which is described by the state diagram shown in Figure 9.4–5. There are two states labeled S_1 and S_2 with transitions as shown in the figure. When the encoder is at state S_1, an input bit 1 results in the encoder staying in state S_1 and outputs 01. This is denoted as 1/01. If the input bit is a 0, the encoder enters state S_2 and outputs 00. This is denoted as 0/00. Similarly, if the encoder is in state S_2, an input bit 0 causes no transition and the encoder output is 10. On the other hand, if the input bit is a 1, the encoder enters state S_1 and outputs 01. Figure 9.4–6 shows the trellis for the Miller code.

The mapping of coded bits into signal waveforms. The output sequence from a (d, κ) encoder is mapped by the modulator into signal waveforms for transmission over the channel. If the binary digit 1 is mapped into a rectangular pulse of amplitude A and the binary digit 0 is mapped into a rectangular pulse of amplitude $-A$, the result is a (d, κ) coded NRZ modulated signal. Note that the duration of the rectangular pulses is $T_c = R_c/R_b = R_c T_b$, where R_b is the information (bit) rate into the encoder, T_b is the corresponding (uncoded) bit interval, and R_c is the code rate for the (d, κ) code.

When the (d, κ) code is a state-independent fixed-length code with code rate $R_c = k/n$, we may consider each n-bit block as generating one signal waveform of duration nT_c. Thus, we have $M = 2^k$ signal waveforms, one for each of the 2^k possible k-bit data blocks. These coded waveforms have the general form given by Equations 4.3–6 and 4.3–38. In this case, there is no dependence between the transmission of successive waveforms.

In contrast to the situation considered above, the modulation signal is no longer memoryless when NRZI is used and/or the (d, κ) code is state-dependent. Let us consider the effect of mapping the coded bits into an NRZI signal waveform.

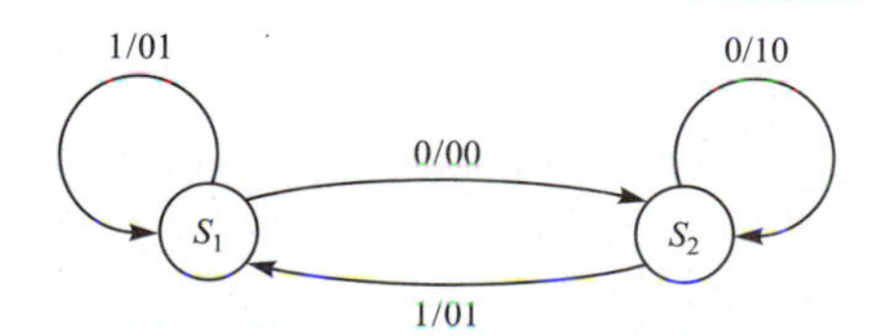

FIGURE 9.4–5
State diagram for $d = 1, \kappa = 3$ (Miller) code.

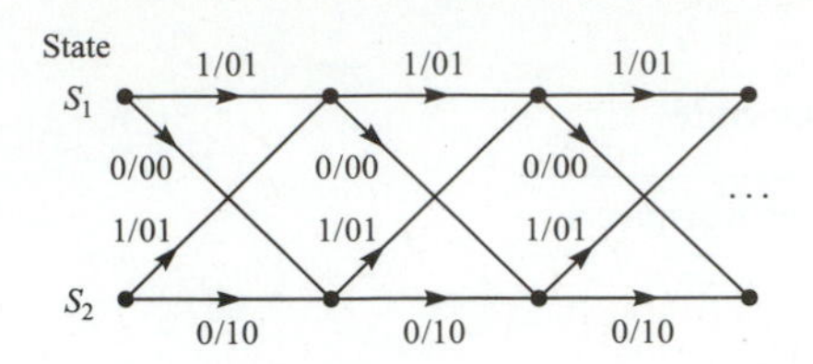

FIGURE 9.4–6
Trellis for $d = 1$, $\kappa = 3$ (Miller) code.

Recall that the state dependence in the NRZI signal is due to the differential encoding of the information sequence. The differential encoding is a form of precoding, which is described mathematically as

$$p_k = d_k \oplus p_{k-1}$$

where $\{d_k\}$ is the binary sequence into the precoder, $\{p_k\}$ is the output binary sequence from the precoder, and $\oplus$ denotes modulo-2 addition. This encoding is characterized by the state diagram shown in Figure 9.4–7a. Then, the sequence $\{p_k\}$ is transmitted by NRZ. Thus, when $p_k = 1$, the modulator output is a rectangular pulse of amplitude A, and when $p_k = 0$, the modulator output is a rectangular pulse of amplitude $-A$. When the signal waveforms are superimposed on the state diagram of Figure 9.4–7a, we obtain the corresponding state diagram shown in Figure 9.4–7b. The corresponding trellis is shown in Figure 9.4–7c.

When the output of a state-dependent (d, κ) encoder is followed by an NRZI modulator, we may simply combine the two-state diagrams into a single-state diagram for the (d, κ) code with precoding. A similar combination can be per-

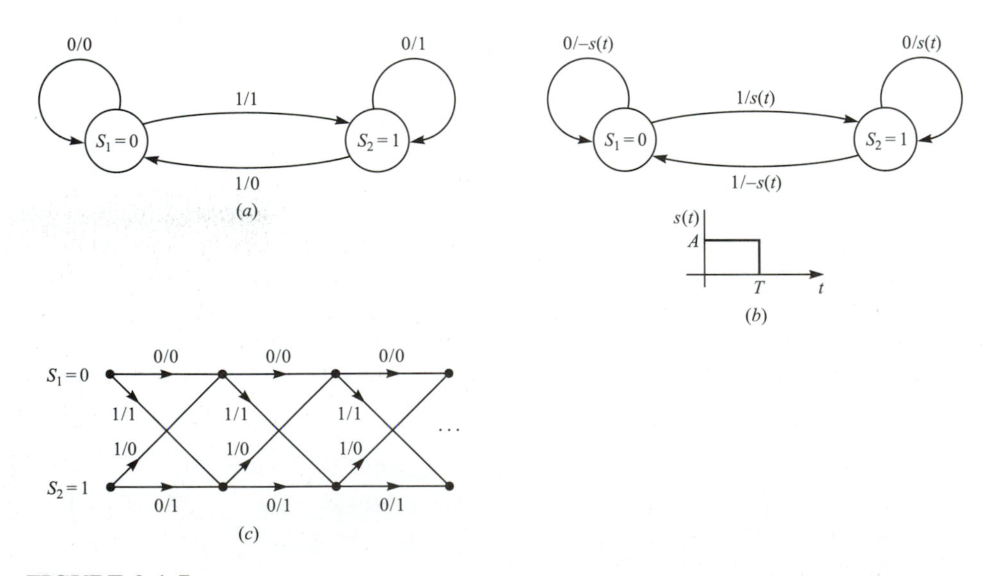

FIGURE 9.4–7
State and trellis diagrams for NRZI signal.

formed with the corresponding trellises. The following example illustrates the approach for the (1, 3) Miller code followed by NRZI modulation.

EXAMPLE 9.4–7. Let us determine the state diagram of the combined (1, 3) Miller code followed by the precoding inherent in NRZI modulation. Since the (1, 3) Miller code has two states and the precoder has two states, the state diagram for the combined encoder has four states, which we denote as $(S_M, S_N) = (\sigma_1, s_1)$, (σ_1, s_2), (σ_2, s_1), (σ_2, s_2), where $S_M = \{\sigma_1, \sigma_2\}$ represents the two states of the Miller code and $S_N = \{s_1, s_2\}$ represents the two states of the precoder for NRZI. For each data input bit into the Miller encoder, we obtain two output bits which are then precoded to yield two precoded output bits. The resulting state diagram is shown in Figure 9.4–8, where the first bit denotes the information bit into the Miller encoder and the next two bits represent the corresponding output of the precoder.

The trellis diagram for the Miller precoded sequence may be obtained directly from the combined state diagram or from a combination of the trellises of the two codes. The result of this combination is the four-state trellis, one stage of which is shown in Figure 9.4–9.

It is left as an exercise for the reader to show that the four signal waveforms obtained by mapping each pair of bits of the Miller-precoded sequence into an NRZ signal are biorthogonal and that the resulting modulated signal waveform is identical to the delay modulation that was described in Section 4.3.2.

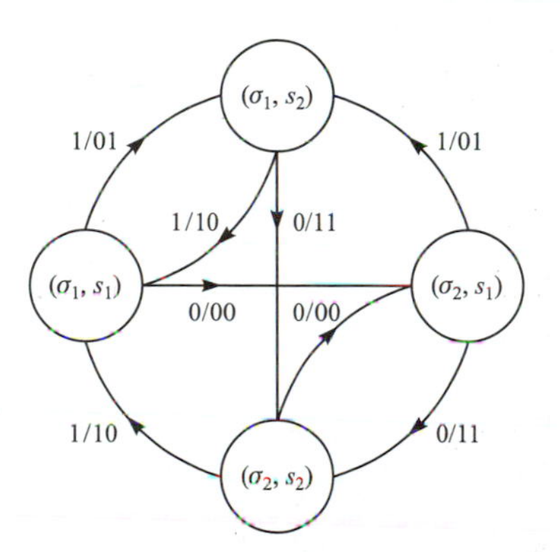

FIGURE 9.4–8
State diagram of the Miller code followed by the precoder.

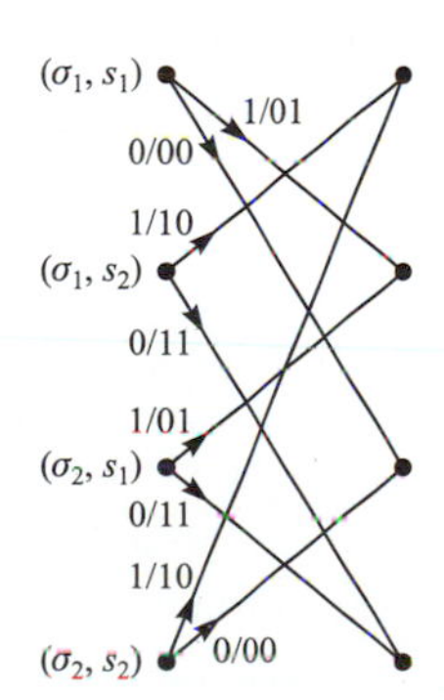

FIGURE 9.4–9
One stage of trellis diagram for the Miller code followed by the precoder.

From the state diagram of a state-dependent runlength-limited code, one can obtain the transition probability matrix, as described in Section 4.3.2. Then, the power spectral density of the code may be determined, as shown in Section 4.4.3.

9.5
BIBLIOGRAPHICAL NOTES AND REFERENCES

The pioneering work on signal design for bandwidth-constrained channels was done by Nyquist (1928). The use of binary partial-response signals was originally proposed by Lender (1963) and was later generalized by Kretzmer (1966). Other early work on problems dealing with intersymbol interference (ISI) and transmitter and receiver optimization with constraints on ISI was done by Gerst and Diamond (1961), Tufts (1965), Smith (1965), and Berger and Tufts (1967). "Faster than Nyquist" transmission has been studied by Mazo (1975) and Foschini (1984).

Modulation codes were also first introduced by Shannon (1948). Some of the early work on the construction of runlength-limited codes is found in the papers by Freiman and Wyner (1964), Gabor (1967), Franaszek (1968, 1969, 1970), Tang and Bahl (1970), and Jacoby (1977). Additional results may be found in papers by Adler et al. (1983) and Karabed and Siegel (1991). The motivation for most of the work on runlength-limited codes was provided by applications to magnetic and optical recording. A tutorial paper on runlength-limited codes has been published by Immink (1990). A tutorial treatment of modulation and coding for storage channels is given in the paper by Siegel and Wolf (1991). A recent survey paper on modulation and coding techniques for storage systems has been published by Immink et al. (1998).

PROBLEMS

9.1 A channel is said to be *distortionless* if the response $y(t)$ to an input $x(t)$ is $Kx(t-t_0)$, where K and t_0 are contants. Show that if the frequency response of the channel is $A(f)e^{j\theta(f)}$, where $A(f)$ and $\theta(f)$ are real, the necessary and sufficient conditions for distortionless transmission are $A(f)=K$ and $\theta(f)=2\pi f t_0 \pm n\pi$, $n=0,1,2,\ldots$.

9.2 The raised cosine spectral characteristic is given by Equation 9.2–26.

a) Show that the corresponding impulse response is

$$x(t)=\frac{\sin(\pi t/T)}{\pi t/T}\,\frac{\cos(\beta\pi t/T)}{1-4\beta^2t^2/T^2}$$

b) Determine the Hilbert transform of $x(t)$ when $\beta=1$.

c) Does $\hat{x}(t)$ possess the desirable properties of $x(t)$ that make it appropriate for data transmission? Explain.

d) Determine the envelope of the SSB suppressed-carrier signal generated from $x(t)$.

9.3 *a*) Show that (Poisson sum formula)

$$x(t) = \sum_{k=-\infty}^{\infty} g(t)h(t-kT) \Rightarrow X(f) = \frac{1}{T}\sum_{n=-\infty}^{\infty} H\left(\frac{n}{T}\right)G\left(f-\frac{n}{T}\right)$$

Hint: Make a Fourier-series expansion of the periodic factor

$$\sum_{k=-\infty}^{\infty} h(t-kT)$$

b) Using the result in (a), verify the following versions of the Poisson sum:

$$\sum_{k=-\infty}^{\infty} h(kT) = \frac{1}{T}\sum_{n=-\infty}^{\infty} H\left(\frac{n}{T}\right) \quad \text{(i)}$$

$$\sum_{k=-\infty}^{\infty} h(t-kT) = \frac{1}{T}\sum_{n=-\infty}^{\infty} H\left(\frac{n}{T}\right)\exp\left(\frac{j2\pi nt}{T}\right) \quad \text{(ii)}$$

$$\sum_{k=-\infty}^{\infty} h(kT)\exp(-j2\pi kTf) = \frac{1}{T}\sum_{n=-\infty}^{\infty} H\left(f-\frac{n}{T}\right) \quad \text{(iii)}$$

c) Derive the condition for no intersymbol interference (Nyquist criterion) by using the Poisson sum formula.

9.4 Suppose a digital communication system employs Gaussian-shaped pulses of the form

$$x(t) = \exp(-\pi a^2 t^2)$$

To reduce the level of intersymbol interference to a relatively small amount, we impose the condition that $x(T) = 0.01$, where T is the symbol interval. The bandwidth W of the pulse $x(t)$ is defined as that value of W for which $X(W)/X(0) = 0.01$, where $X(f)$ is the Fourier transform of $x(t)$. Determine the value of W and compare this value to that of raised cosine spectrum with 100 percent rolloff.

9.5 Show that the impulse response of a filter having a square-root raised cosine spectral characteristic is given as

$$x_{sr}(t) = \frac{(4\beta t/T)\cos[\pi(1+\beta)t/T] + \sin[\pi(1-\beta)t/T]}{(\pi t/T)[1-(4\beta t/T)^2]}$$

9.6 It is desired to implement a (discrete-time) finite impulse response (FIR) filter that provides square-root raised cosine spectral shaping. The coefficients of the FIR filter are the sampled values of the time response given in Problem 9.5, where the samples are taken at $t = kT/2$, for $k = 0, \pm 1, \pm 2, \cdots, \pm N$.

a) Determine the effect on the spectral characteristic resulting from the truncation of the filter response for $N = 10, 15,$ and 20 and roll-off factor $\beta = 1/2$, by computing their frequency response

$$X_{sr}(\omega) = \sum_{n=-N}^{N} x(nT_s)e^{-j\omega nT_s}$$

where $T_s = T/2$.

b) Plot the spectral characteristics of these three filters for $N = 10, 15$, and 20 and compare your results with the ideal square-root raised cosine spectrum.

9.7 Figure P9.7 illustrates a block diagram of a QAM or PSK modulator and demodulator (modem) in which the modulated signals are synthesized digitally and demodulated digitally. The FIR filters have square-root raised cosine spectral characteristics and employ a sampling rate of $2/T$, where the symbol rate $1/T =$ 2400 symbols/s. The FIR interpolators employ a sampling rate of $6/T$ and are designed as linear phase FIR filters that pass the desired signal spectrum.

a) Write a software program that implements the digital modulator in Figure P9.7 for the following parameters: roll-off factor $\beta = 0.25$, length of FIR shaping filter $= 21$, length of FIR interpolator $= 11$, carrier frequency $f_c = 1800$ Hz.

b) Generate 5000 samples of the digital signal sequence $x_d(n)$ and compute and plot the power spectral density of this modulated signal.

c) Repeat (b) for five more iterations and compute the average power spectrum over the total of six signal records. Comment on the results.

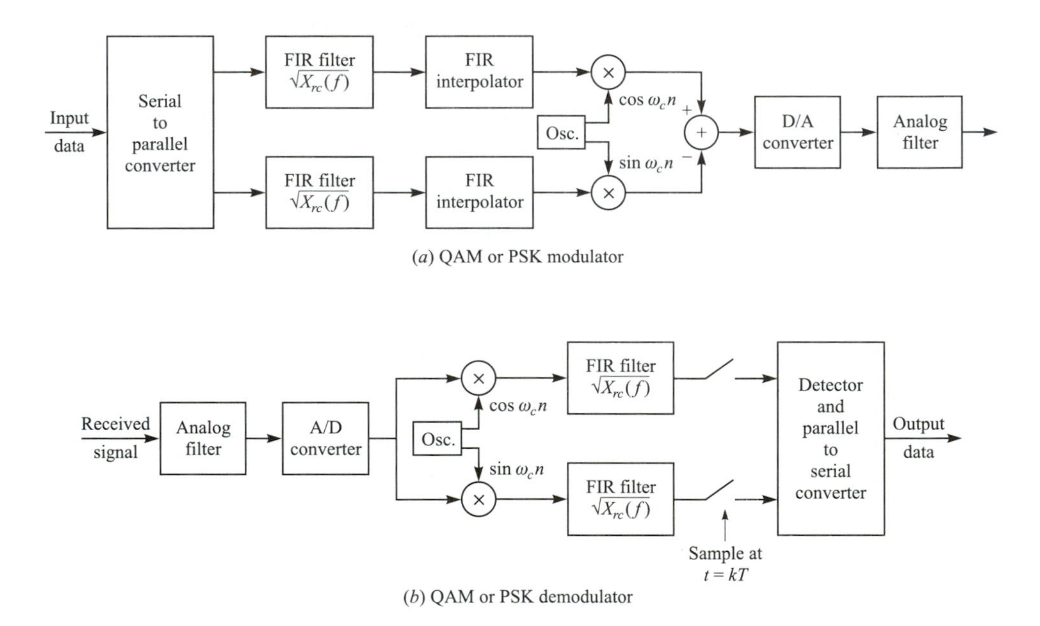

FIGURE P9.7

9.8 (Carrierless QAM or PSK modem) Consider the transmission of a QAM or M-ary PSK ($M \geqslant 4$) signal at a carrier frequency f_c, where the carrier is comparable to the bandwidth of the baseband signal. The band-pass signal may be represented as

$$s(t) = \text{Re}\left[\sum_n I_n g(t - nT) e^{j2\pi f_c t}\right]$$

a) Show that $s(t)$ can be expressed as

$$s(t) = \text{Re}\left[\sum_n I'_n Q(t - nT)\right]$$

where $Q(t)$ is defined as

$$Q(t) = q(t) + j\hat{q}(t)$$
$$q(t) = g(t)\cos 2\pi f_c t$$
$$\hat{q}(t) = g(t)\sin 2\pi f_c t$$

and I'_n is a phase rotated symbol, i.e., $I'_n = I_n e^{j2\pi f_c nT}$.

b) Using FIR filters with responses $q(t)$ and $\hat{q}(t)$, sketch the block diagram of the modulator and demodulator implementation that does not require the mixer to translate the signal to band-pass at the modulator and to baseband at the demodulator.

9.9 (Carrierless amplitude or phase [CAP] modulation) In some practical applications in wireline data transmission, the bandwidth of the signal to be transmitted is comparable to the carrier frequency. In such systems, it is possible to eliminate the step of mixing the baseband signal with the carrier component. Instead, the band-pass signal can be synthesized directly, by embedding the carrier component in the realization of the FIR shaping filters. Thus, the modem is realized as shown in the block diagram in Figure P9.9, where the FIR shaping filters have the impulse responses

$$q(t) = g(t)\cos 2\pi f_c t$$
$$\hat{q}(t) = g(t)\sin 2\pi f_c t$$

and $g(t)$ is a pulse that has a square-root raised cosine spectral characteristic.

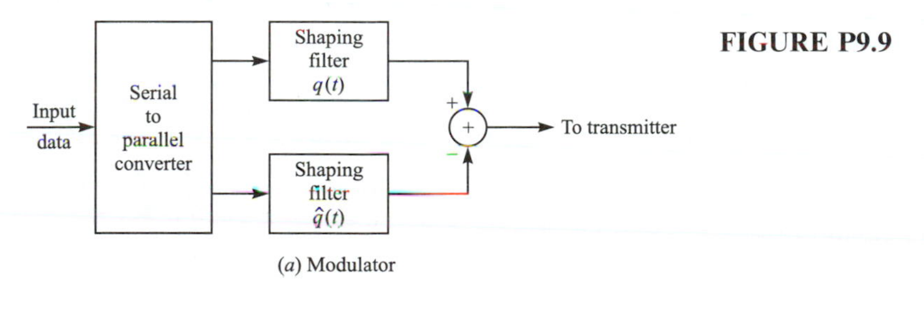

(*a*) Modulator

FIGURE P9.9

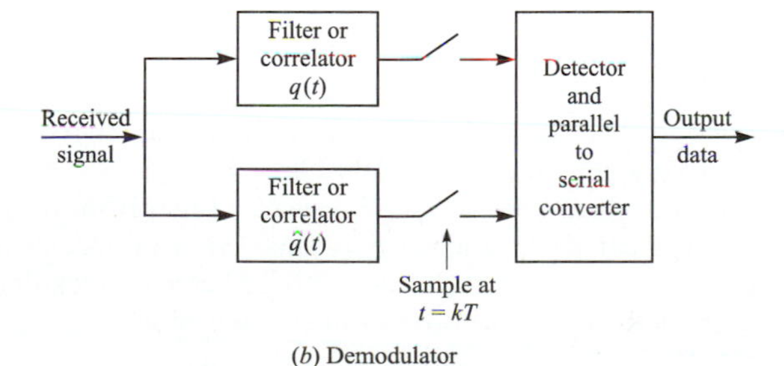

(*b*) Demodulator

a) Show that

$$\int_{-\infty}^{\infty} q(t)\hat{q}(t)\,dt = 0$$

and that this system can be used to transmit two-dimensional signal constellations.

b) Under what conditions is this CAP modem identical to the carrierless QAM/PSK modem treated in Problem 9.8.

9.10 A band-limited signal having bandwidth W can be represented as

$$x(t) = \sum_{n=-\infty}^{\infty} x_n \frac{\sin[2\pi W(t - n/2W)]}{2\pi W(t - n/2W)}$$

a) Determine the spectrum $X(f)$ and plot $|X(f)|$ for the following cases:

$$x_0 = 2, \quad x_1 = 1, \quad x_2 = -1, \quad x_n = 0, \quad n \neq 0, 1, 2 \qquad \text{(i)}$$
$$x_{-1} = -1, \quad x_0 = 2, \quad x_1 = -1, \quad x_n = 0, \quad n \neq -1, 0, 1 \qquad \text{(ii)}$$

b) Plot $x(t)$ for these two cases.

c) If these signals are used for binary signal transmission, determine the number of received levels possible at the sampling instants $t = nT = n/2W$ and the probabilities of occurrence of the received levels. Assume that the binary digits at the transmitter are equally probable.

9.11 A 4-kHz band-pass channel is to be used for transmission of data at a rate of 9600 bits/s. If $\frac{1}{2}N_0 = 10^{-10}$ W/Hz is the spectral density of the additive zero-mean Gaussian noise in the channel, design a QAM modulation and determine the average power that achieves a bit error probability of 10^{-6}. Use a signal pulse with a raised cosine spectrum having a roll-off factor of at least 50 percent.

9.12 Determine the bit rate that can be transmitted through a 4-kHz voice-band telephone (bandpass) channel if the following modulation methods are used:

a) Binary PAM.
b) Four-phase PSK.
c) 8-point QAM.
d) Binary orthogonal FSK, with noncoherent detection.
e) Orthogonal four-FSK with noncoherent detection.
f) Orthogonal 8-FSK with noncoherent detection.

For (*a*)–(*c*), assume that the transmitter pulse shape has a raised cosine spectrum with a 50 percent roll-off.

9.13 An ideal voice-band telephone line channel has a band-pass frequency-response characteristic spanning the frequency range 600–3000 Hz.

a) Design an $M = 4$ PSK (quadrature PSK or QPSK) system for transmitting data at a rate of 2400 bits/s and a carrier frequency $f_c = 1800$ Hz. For spectral shaping, use a raised cosine frequency-response characteristic. Sketch a block diagram of the system and describe the functional operation of each block.

b) Repeat (*a*) for a bit rate $R = 4800$ bits/s and a 8-QAM signal.

9.14 A voice-band telephone channel passes the frequencies in the band from 300 to 3300 Hz. It is desired to design a modem that transmits at a symbol rate of 2400 symbols/s, with the objective of achieving 9600 bits/s. Select an appropriate QAM signal constellation, carrier frequency, and the roll-off factor of a pulse with a raised cosine spectrum that utilizes the entire frequency band. Sketch the spectrum of the transmitted signal pulse and indicate the important frequencies.

9.15 A communication system for a voice-band (3 kHz) channel is designed for a received SNR at the detector of 30 dB when the transmitter power is $P_s = -3$ dBW. Determine the value of P_s if it is desired to expand the bandwidth of the system to 10 kHz, while maintaining the same SNR at the detector.

9.16 Show that a pulse having the raised-cosine spectrum given by Equation 9.2–26 satisfies the Nyquist criterion given by Equation 9.2–13 for any value of the roll-off factor β.

9.17 Show that, for any value of β, the raised cosine spectrum given by Equation 9.2–26 satisfies

$$\int_{-\infty}^{\infty} X_{rc}(f)df = 1$$

[*Hint*: Use the fact that $X_{rc}(f)$ satisfies the Nyquist criterion given by Equation 9.2–13.]

9.18 The Nyquist criterion gives the necessary and sufficient condition for the spectrum $X(f)$ of the pulse $x(t)$ that yields zero ISI. Prove that for any pulse that is band-limited to $|f| < 1/T$, the zero-ISI condition is satisfied if Re$[X(f)$, for $f > 0$, consists of a rectangular function plus an arbitrary odd function around $f = 1/2T$, and Im$[X(f)]$ is any arbitrary even function around $f = 1/2T$.

9.19 A voice-band telephone channel has a passband characteristic in the frequency range 300 Hz $< f <$ 3000 Hz.
a) Select a symbol rate and a power efficient constellation size to achieve 9600 bits/s signal transmission.
b) If a square-root raised cosine pulse is used for the transmitter pulse $g(t)$, select the roll-off factor. Assume that the channel has an ideal frequency-response characteristic.

9.20 Design an M-ary PAM system that transmits digital information over an ideal channel with bandwidth $W = 2400$ Hz. The bit rate is 14,400 bits/s. Specify the number of transmitted points, the number of received signal points using a duobinary signal pulse, and the required $\mathcal{E}_b$ to achieve an error probability of 10^{-6}. The additive noise is zero-mean Gaussian with a power spectral density of 10^{-4}W/Hz.

9.21 A binary PAM signal is generated by exciting a raised cosine roll-off filter with a 50 percent roll-off factor and is then DSB/SC amplitude-modulated on a sinusoidal carrier as illustrated in Figure P9.21. The bit rate is 2400 bits/s.

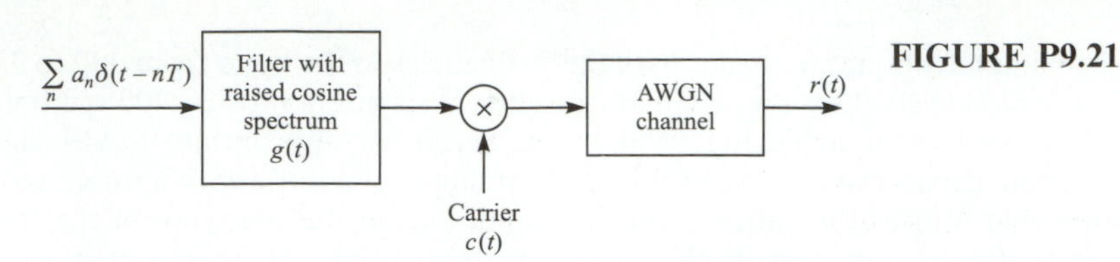

FIGURE P9.21

a) Determine the spectrum of the modulated binary PAM signal and sketch it.
b) Draw the block diagram illustrating the optimum demodulator/detector for the received signal, which is equal to the transmitted signal plus additive white Gaussian noise.

9.22 The elements of the sequence $\{a_n\}_{n=-\infty}^{\infty}$ are independent binary random variables taking values of ± 1 with equal probability. This data sequence is used to modulate the basic pulse $g(t)$ shown in Figure P9.22a. The modulated signal is

$$X(t) = \sum_{n=-\infty}^{+\infty} a_n g(t - nT)$$

a) Find the power spectral density of $X(t)$.
b) If $g_1(t)$ (shown in Figure 9.22b) is used instead of $g(t)$, how would the power spectrum in (a) change?
c) In (b) assume we want to have a null in the spectrum at $f = 1/3T$. This is done by a precoding of the form $b_n = a_n + \alpha a_{n-3}$. Find the α that provides the desired null.
d) Is it possible to employ a precoding of the form $b_n = a_n + \sum_{i=1}^{N} \alpha_i a_{n-i}$ for some finite N such that the final power spectrum will be identical to zero for $1/3T \leqslant |f| \leqslant 1/2T$? If yes, how? If no, why? [*Hint*: Use properties of analytic functions.]

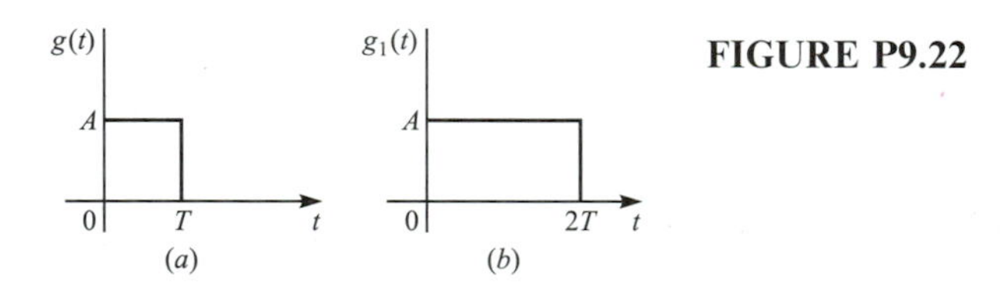

FIGURE P9.22

9.23 Consider the transmission of data via PAM over a voice-band telephone channel that has a bandwidth of 3000 Hz. Show how the symbol rate varies as a function of the excess bandwidth. In particular, determine the symbol rate for an excess bandwidth of 25, 33, 50, 67, 75 and 100 percent.

9.24 The binary sequence 10010110010 is the input to a precoder whose output is used to modulate a duobinary transmitting filter. Construct a table as in Table 9.2–1 showing the precoded sequence, the transmitted amplitude levels, the received signal levels, and the decoded sequence.

9.25 Repeat Problem 9.24 for a modified duobinary signal pulse.

9.26 A precoder for a partial response signal fails to work if the desired partial response at $n = 0$ is zero modulo M. For example, consider the desired response for $M = 2$:

$$x(nT) = \begin{cases} 2 & (n = 0) \\ 1 & (n = 1) \\ -1 & (n = 2) \\ 0 & (\text{otherwise}) \end{cases}$$

Show why this response cannot be precoded.

9.27 Consider the RC low-pass filter shown in Figure P9.27, where $\tau = RC = 10^{-6}$.
a) Determine and sketch the envelope (group) delay of the filter as a function of frequency.
b) Suppose that the input to the filter is a low-pass signal of bandwidth $\Delta f = 1$ kHz. Determine the effect of the RC filter on this signal.

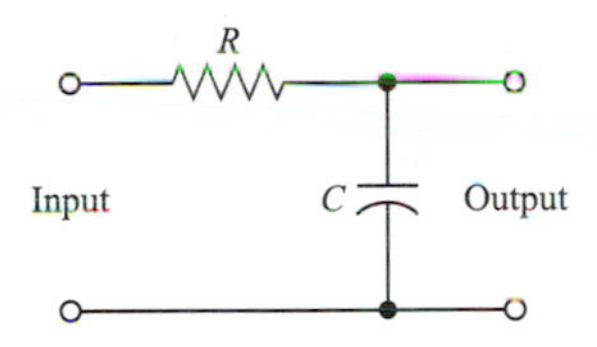

FIGURE P9.27

9.28 A microwave radio channel has a frequency response

$$C(f) = 1 + 0.3 \cos 2\pi f T$$

Determine the frequency-response characteristic of the transmitting and receiving filters that yield zero ISI at a rate of $1/T$ symbols/s and have a 50 percent excess bandwidth. Assume that the additive noise spectrum is flat.

9.29 $M = 4$ PAM modulation is used for transmitting at a bit rate of 9600 bits/s on a channel having a frequency response

$$C(f) = \frac{1}{1 + j(f/2400)}$$

for $|f| \leqslant 2400$, and $C(f) = 0$ otherwise. The additive noise is zero-mean white Gaussian with power spectral density $\frac{1}{2}N_0$ W/Hz. Determine the (magnitude) frequency-response characteristic of the optimum transmitting and receiving filters.

9.30 Use the Cauchy–Schwarz inequality to show that the transmitter and receiver filters given by Equation 9.2–75 minimize the noise-to signal ratio σ_v^2/d^2, where σ_v^2 is the noise power given by Equation 9.2–69, where $\Phi_{nn}(f) = N_0/2$.

9.31 Suppose that a channel frequency response is given as

$$C(f) = \begin{cases} 1 & (|f| \leqslant W/2) \\ \frac{1}{2} & \left(\frac{W}{2} < |f| < W\right) \end{cases}$$

Determine the loss in SNR incurred, as given by Equations 9.2–79 and 9.2–80, for the filters given by the corresponding equations 9.2–71 and 9.2–75, respectively. Which filters result in a smaller loss?

9.32 Determine the capacity of a (0, 1) runlength-limited code. Compare its capacity with that of a $(1, \infty)$ code and explain the relationship.

9.33 A ternary signal format is designed for a channel that does not pass DC. The binary input information sequence is transmitted by mapping a 1 into either a positive pulse or a negative pulse, and a zero is transmitted by the absence of a pulse. Hence, for the transmission of 1s, the polarity of the pulses alternate. This is called an AMI (alternate mark inversion) code. Determine the capacity of the code.

9.34 Give another description of the AMI code described in Problem 9.33 using the running digit sum (RDS) with the constraint that the RDS can take only the values 0 and $+1$.

9.35 (kBnT codes) From Problem 9.33 note that the AMI code is a "pseudo-ternary" code in that it transmits one bit per symbol using a ternary alphabet, which has the capacity of $\log_2 3 = 1.58$ bits. Such a code does not provide sufficient spectral shaping. Better spectral shaping is achieved by the class of block codes designated as kBnT, where k denotes the number of information bits and n denotes the number of ternary symbols per block. By selecting the largest k possible for each n, we obtain the following table:

k	n	**Code**
1	1	1B1T
3	2	3B2T
4	3	4B3T
6	4	6B4T

Determine the efficiency of these codes by computing the ratio of the code in bits per symbol divided by $\log_2 3$. Note that 1B1T is the AMI code.

9.36 This problem deals with the capacity of two (d, κ) codes.

a) Determine the capacity of a (d, κ) code that has the following state transition matrix:

$$\mathbf{D} = \begin{bmatrix} 1 & 1 \\ 1 & 0 \end{bmatrix}$$

b) Repeat (*a*) for

$$\mathbf{D} = \begin{bmatrix} 1 & 1 \\ 0 & 1 \end{bmatrix}$$

c) Comment on the differences between (*a*) and (*b*).

9.37 A simplified model of the telegraph code consists of two symbols (Blahut, 1990). A dot consists of one time unit of line closure followed by one time unit of line open. A dash consists of three units of the line closure followed by one time unit of line open.

a) Viewing this code as a constrained code with symbols of equal duration, give the constraints.
b) Determine the state transition matrix.
c) Determine the capacity.

9.38 Determine the state transition matrix for the runlength-constrained code described by the state diagram shown in Figure P9.38. Sketch the corresponding trellis.

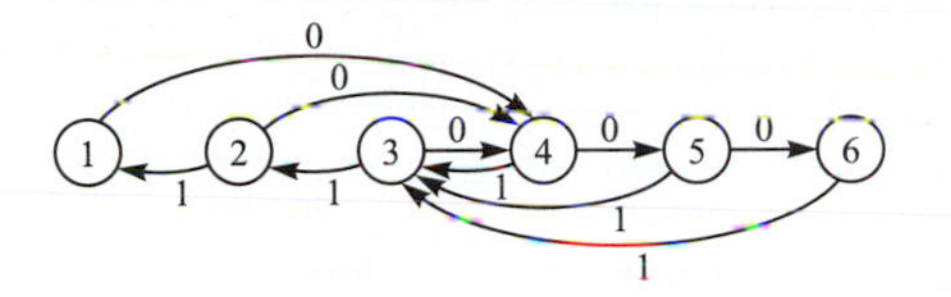

FIGURE P9.38

9.39 Determine the state transition matrix for the (2, 7) runlength-limited code specified by the state diagram shown in Figure P9.39.

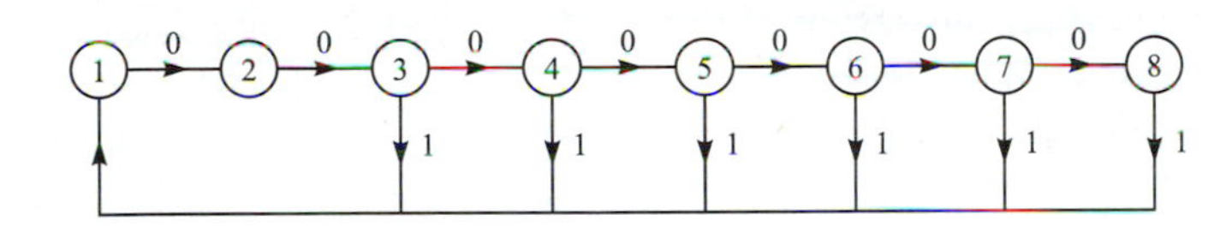

FIGURE P9.39

9.40 Show that the four signal waveforms obtained in the mapping of each pair of bits of the Miller precoded sequence into the NRZ signal are biorthogonal. Then, show that the resulting modulated signal waveform is identical to the delay modulation previously described in Section 4.3.2.

10

Communication Through Band-Limited Linear Filter Channels

In Chapter 9, we focused on the design of the modulator and demodulator filters for band-limited channels. The design procedure was based on the assumption that the (ideal or nonideal) channel response characteristic $C(f)$ was known a priori. However, in practical digital communication systems that are designed to transmit at high speed through band-limited channels, the frequency response $C(f)$ of the channel is not known with sufficient precision to design optimum filters for the modulator and demodulator. For example, in digital communication over the dial-up telephone network, the communication channel will be different every time we dial a number, because the channel route will be different. This is an example of a channel whose characteristics are unknown a priori. There are other types of channels, e.g., wireless channels such as radio channels and underwater acoustic channels, whose frequency-response characteristics are time-variant. For such channels, it is not possible to design optimum fixed demodulation filters.

In this chapter, we consider the problem of receiver design in the presence of channel distortion, which is not known a priori, and AWGN. The channel distortion results in intersymbol interference, which, if left uncompensated, causes high error rates. The solution to the ISI problem is to design a receiver that employs a means for compensating or reducing the ISI in the received signal. The compensator for the ISI is called an *equalizer*.

Several types of equalization methods are treated in this chapter. One is based on the maximum-likelihood (ML) sequence detection criterion, which is optimum from a probability of error viewpoint. A second equalization method is based on the use of a linear filter with adjustable coefficients. A third equalization method that is described exploits the use of previously detected symbols to suppress the ISI in the present symbol being detected, and it is called *decision-feedback equalization*. Finally, we consider reduced complexity maximum-likelihood detection methods. We begin with the derivation of the optimum detector for channels with ISI.

10.1
OPTIMUM RECEIVER FOR CHANNELS WITH ISI AND AWGN

In this section, we derive the structure of the optimum demodulator and detector for digital transmission through a nonideal band-limited channel with additive Gaussian noise. We begin with the transmitted (equivalent low-pass) signal given by Equation 9.2–1. The received (equivalent low-pass) signal is expressed as

$$r_l(t) = \sum_n I_n h(t - nT) + z(t) \tag{10.1–1}$$

where $h(t)$ represents the response of the channel to the input signal pulse $g(t)$ and $z(t)$ represents the additive white Gaussian noise.

First we demonstrate that the optimum demodulator can be realized as a filter matched to $h(t)$, followed by a sampler operating at the symbol rate $1/T$ and a subsequent processing algorithm for estimating the information sequence $\{I_n\}$ from the sample values. Consequently, the samples at the output of the matched filter are sufficient for the estimation of the sequence $\{I_n\}$.

10.1.1 Optimum Maximum-Likelihood Receiver

Let us expand the received signal $r_l(t)$ in the series

$$r_l(t) = \lim_{N\to\infty} \sum_{k=1}^{N} r_k f_k(t) \tag{10.1–2}$$

where $\{f_k(t)\}$ is a complete set of orthonormal functions and $\{r_k\}$ are the observable random variables obtained by projecting $r_l(t)$ onto the set $\{f_k(t)\}$. It is easily shown that

$$r_k = \sum_n I_n h_{kn} + z_k, \qquad k = 1, 2, \ldots \tag{10.1–3}$$

where h_{kn} is the value obtained from projecting $h(t - nT)$ onto $f_k(t)$, and z_k is the value obtained from projecting $z(t)$ onto $f_k(t)$. The sequence $\{z_k\}$ is Gaussian with zero-mean and covariance

$$\tfrac{1}{2}E(z_k^* z_m) = N_0 \delta_{km} \tag{10.1–4}$$

The joint probability density function of the random variables $\mathbf{r}_N \equiv [r_1 \ r_2 \cdots r_N]$ conditioned on the transmitted sequence $\mathbf{I}_p \equiv [I_1 \ I_2 \cdots I_p]$, where $p \leqslant N$, is

$$p(\mathbf{r}_N|\mathbf{I}_p) = \left(\frac{1}{2\pi N_0}\right)^N \exp\left(-\frac{1}{2N_0}\sum_{k=1}^{N}\left|r_k - \sum_n I_n h_{kn}\right|^2\right) \tag{10.1–5}$$

In the limit as the number N of observable random variables approaches infinity, the logarithm of $p(\mathbf{r}_N|\mathbf{I}_p)$ is proportional to the metrics $PM(\mathbf{I}_p)$, defined as

$$PM(\mathbf{I}_p) = -\int_{-\infty}^{\infty} \left| r_1(t) - \sum_n I_n h(t - nT) \right|^2 dt$$
$$= -\int_{-\infty}^{\infty} |r_l(t)|^2 dt + 2\text{Re} \sum_n \left[I_n^* \int_{-\infty}^{\infty} r_l(t) h^*(t - nT)\, dt \right] \qquad (10.1\text{–}6)$$
$$- \sum_n \sum_m I_n^* I_m \int_{-\infty}^{\infty} h^*(t - nT) h(t - mT)\, dt$$

The maximum-likelihood estimates of the symbols $I_1, I_2, \ldots, I_p$ are those that maximize this quantity. Note, however, that the integral of $|r_l(t)|^2$ is common to all metrics, and, hence, it may be discarded. The other integral involving $r(t)$ gives rise to the variables

$$y_n \equiv y(nT) = \int_{-\infty}^{\infty} r_l(t) h^*(t - nT)\, dt \qquad (10.1\text{–}7)$$

These variables can be generated by passing $r(t)$ through a filter matched to $h(t)$ and sampling the output at the symbol rate $1/T$. The samples $\{y_n\}$ form a set of sufficient statistics for the computation of $PM(\mathbf{I}_p)$ or, equivalently, of the correlation metrics

$$CM(\mathbf{I}_p) = 2\text{Re}\left(\sum_n I_n^* y_n \right) - \sum_n \sum_m I_n^* I_m x_{n-m} \qquad (10.1\text{–}8)$$

where, by definition, $x(t)$ is the response of the matched filter to $h(t)$ and

$$x_n \equiv x(nT) = \int_{-\infty}^{\infty} h^*(t) h(t + nT)\, dt \qquad (10.1\text{–}9)$$

Hence, $x(t)$ represents the output of a filter having an impulse response $h^*(-t)$ and an excitation $h(t)$. In other words, $x(t)$ represents the autocorrelation function of $h(t)$. Consequently, $\{x_n\}$ represents the samples of the autocorrelation function of $h(t)$, taken periodically at $1/T$. We are not particularly concerned with the noncausal characterstic of the filter matched to $h(t)$, since, in practice, we can introduce a sufficiently large delay to ensure causality of the matched filter.

If we substitute for $r_l(t)$ in Equation 10.1–7 using Equation 10.1–1, we obtain

$$y_k = \sum_n I_n x_{k-n} + \nu_k \qquad (10.1\text{–}10)$$

where ν_k denotes the additive noise sequence of the output of the matched filter, i.e.,

$$\nu_k = \int_{-\infty}^{\infty} z(t) h^*(t - kT)\, dt \qquad (10.1\text{–}11)$$

The output of the demodulator (matched filter) at the sampling instants is corrupted by ISI as indicated by Equation 10.1–10. In any practical system, it is reasonable to assume that the ISI affects a finite number of symbols. Hence, we may assume that $x_n = 0$ for $|n| > L$. Consequently, the ISI observed at the output of the demodulator may be viewed as the output of a finite state machine. This implies that the channel output with ISI may be represented by a trellis diagram, and the maximum-likelihood estimate of the information sequence $(I_1, I_2, \ldots, I_p)$ is simply the most probable path through the trellis given the received demodulator output sequence $\{y_n\}$. Clearly, the Viterbi algorithm provides an efficient means for performing the trellis search.

The metrics that are computed for the MLSE of the sequence $\{I_k\}$ are given by Equation 10.1–8. It can be seen that these metrics can be computed recursively in the Viterbi algorithm, according to the relation

$$CM_n(\mathbf{I}_n) = CM_{n-1}(\mathbf{I}_{n-1}) + \text{Re}\left[I_n^* \left(2y_n - x_0 I_n - 2 \sum_{m=1}^{L} x_m I_{n-m} \right) \right] \quad (10.1\text{–}12)$$

Figure 10.1–1 illustrates the block diagram of the optimum receiver for an AWGN channel with ISI.

10.1.2 A Discrete-Time Model for a Channel with ISI

In dealing with band-limited channels that result in ISI, it is convenient to develop an equivalent discrete-time model for the analog (continuous-time) system. Since the transmitter sends discrete-time symbols at a rate of $1/T$ symbols/s and the sampled output of the matched filter at the receiver is also a discrete-time signal with samples occurring at a rate of $1/T$ per second, it follows that the cascade of the analog filter at the transmitter with impulse response $g(t)$, the channel with impulse response $c(t)$, the matched filter at the receiver with impulse response $h^*(-t)$, and the sampler can be represented by an equivalent discrete-time tranversal filter having tap gain coefficients $\{x_k\}$. Consequently, we have an equivalent discrete-time transversal filter that spans a time interval of $2LT$ seconds. Its input is the sequence of information symbols $\{I_k\}$ and its output is the discrete-time sequence $\{y_k\}$ given by Equation 10.1–10. The equivalent discrete-time model is shown in Figure 10.1–2.

The major difficulty with this discrete-time model occurs in the evaluation of performance of the various equalization or estimation techniques that are dis-

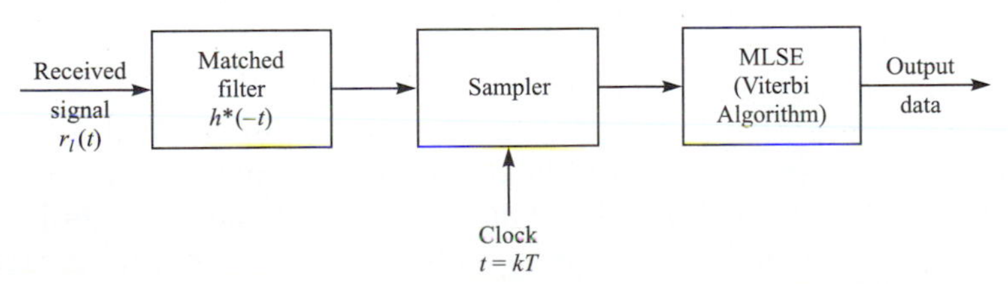

FIGURE 10.1–1
Optimum receiver for an AWGN channel with ISI.

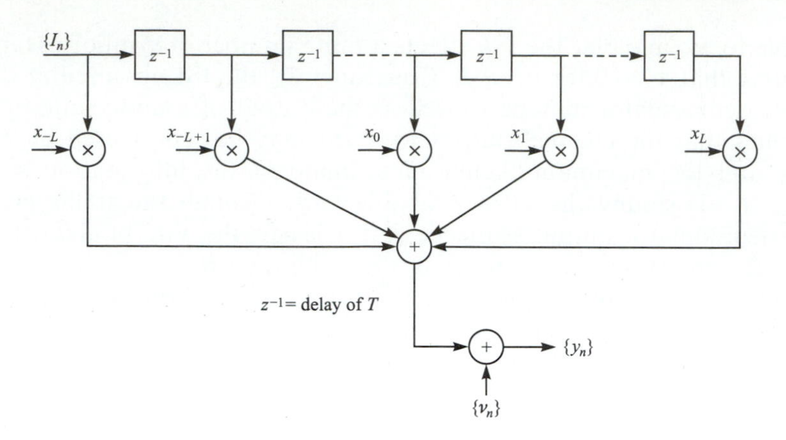

FIGURE 10.1–2
Equivalent discrete-time model of channel with intersymbol interference.

cussed in the following sections. The difficulty is caused by the correlations in the noise sequence $\{\nu_k\}$ at the output of the matched filter. That is, the set of noise variables $\{\nu_k\}$ is a Gaussian-distributed sequence with zero-mean and autocorrelation function (see Problem 10.5)

$$\tfrac{1}{2}E(\nu_k^*\nu_j) = \begin{cases} N_0 x_{j-k} & (|k-j| \leqslant L) \\ 0 & (\text{otherwise}) \end{cases} \tag{10.1–13}$$

Hence, the noise sequence is correlated unless $x_k = 0$, $k \neq 0$. Since it is more convenient to deal with the white noise sequence when calculating the error rate performance, it is desirable to whiten the noise sequence by further filtering the sequence $\{y_k\}$. A discrete-time noise-whitening filter is determined as follows.

Let $X(z)$ denote the (two-sided) z transform of the sampled autocorrelation function $\{x_k\}$, i.e.,

$$X(z) = \sum_{k=-L}^{L} x_k z^{-k} \tag{10.1–14}$$

Since $x_k = x_{-k}^*$, it follows that $X(z) = X^*(1/z^*)$ and the $2L$ roots of $X(z)$ have the symmetry that if ρ is a root, $1/\rho^*$ is also a root. Hence, $X(z)$ can be factored and expressed as

$$X(z) = F(z)F^*(z^{-1}) \tag{10.1–15}$$

where $F(z)$ is a polynomial of degree L having the roots $\rho_1, \rho_2, \ldots, \rho_L$ and $F^*(z^{-1})$ is a polynomial of degree L having the roots $1/\rho_1^*, 1/\rho_2^*, \ldots, 1/\rho_L^*$. Assuming that there are no roots on the unit circle, an appropriate noise-whitening filter has a z transform $1/F^*(z^{-1})$. Since there are 2^L possible choices for the roots of $F^*(z^{-1})$, each choice resulting in a filter characteristic that is identical in magnitude but different in phase from other choices of the roots, we propose to choose the unique $F^*(z^{-1})$ that results in an anticausal impulse response with poles corresponding to the zeros of $X(z)$ that are outside the unit circle. Such an anticausal filter is stable. Selecting the noise-whitening filter in this manner ensures that the resulting channel response, characterized by $F(z)$, is minimum phase. Consequently, passage of the sequence $\{y_k\}$ through the digital filter $1/F^*(z^{-1})$ results in an output sequence $\{v_k\}$ that can be expressed as

$$v_l = \sum_{n=0}^{L} f_n I_{k-n} + \eta_k \tag{10.1–16}$$

where $\{\eta_k\}$ is a white Gaussian noise sequence and $\{f_k\}$ is a set of tap coefficients of an equivalent discrete-time transversal filter having a transfer function $F(z)$. The cascade of the matched filter, the sampler, and the noise-whitening filter is called the *whitened matched filter* (WMF).

It is convenient to normalize the energy of $F(z)$ to unity, i.e.,

$$\sum_{n=0}^{L} |f_n|^2 = 1$$

The minimum-phase condition on $F(z)$ implies that the energy in the first M values of the impulse response $\{f_0, f_1, \ldots, f_M\}$ is a maximum for every M.

In summary, the cascade of the transmitting filter $g(t)$, the channel $c(t)$, the matched filter $h^*(-t)$, the sampler, and the discrete-time noise-whitening filter $1/F^*(z^{-1})$ can be represented as an equivalent discrete-time transversal filter having the set $\{f_k\}$ as its tap coefficients. The additive noise sequence $\{\eta_k\}$ corrupting the output of the discrete-time transversal filter is a white Gaussian noise sequence having zero-mean and variance N_0. Figure 10.1–3 illustrates the model of the equivalent discrete-time system with white noise. We refer to this model as the *equivalent discrete-time white noise filter model.*

EXAMPLE 10.1–1. Suppose that the transmitter signal pulse $g(t)$ has duration T and unit energy and the received signal pulse is $h(t) = g(t) + ag(t - T)$. Let us determine the equivalent discrete-time white noise filter model. The sampled autocorrelation function is given by

$$x_k = \begin{cases} a^* & (k = -1) \\ 1 + |a|^2 & (k = 0) \\ a & (k = 1) \end{cases} \tag{10.1–17}$$

The z transform of x_k is

$$\begin{aligned} X(z) &= \sum_{k=-1}^{1} x_k z^{-k} \\ &= a^* z + (1 + |a|^2) + az^{-1} \\ &= (az^{-1} + 1)(a^* z + 1) \end{aligned} \tag{10.1–18}$$

Under the assumption that $|a| < 1$, one chooses $F(z) = az^{-1} + 1$, so that the equivalent transversal filter consists of two taps having tap gain coefficients $f_0 = 1, f_1 = a$. Note that the correlation sequence $\{x_k\}$ may be expressed in terms of the $\{f_n\}$ as

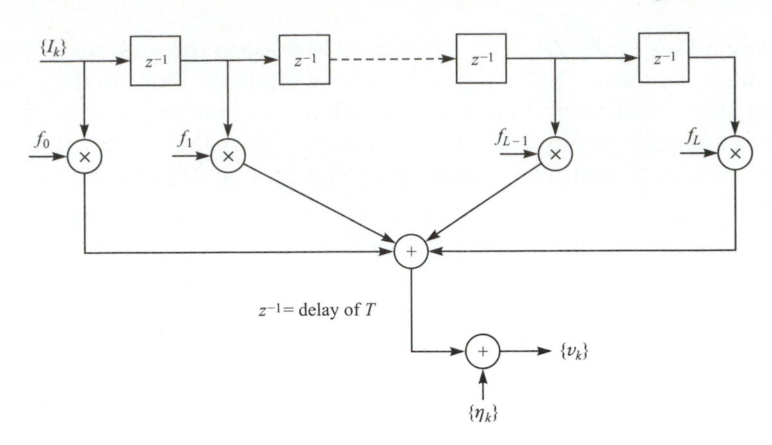

FIGURE 10.1–3
Equivalent discrete-time model of intersymbol interference channel with AWGN.

$$x_k = \sum_{n=0}^{L-k} f_n^* f_{n+k}, \qquad k = 0, 1, 2, \ldots, L \qquad (10.1\text{–}19)$$

When the channel impulse response is changing slowly with time, the matched filter at the receiver becomes a time-variable filter. In this case, the time variations of the channel/matched-filter pair result in a discrete-time filter with time-variable coefficients. As a consequence, we have time-variable intersymbol interference effects, which can be modeled by the filter illustrated in Figure 10.1–3, where the tap coefficients are slowly varying with time.

The discrete-time white noise linear filter model for the intersymbol interference effects that arise in high-speed digital transmission over nonideal band-limited channels will be used throughout the remainder of this chapter in our discussion of compensation techniques for the interference. In general, the compensation methods are called *equalization techniques* or *equalization algorithms*.

10.1.3 The Viterbi Algorithm for the Discrete-Time White Noise Filter Model

MLSE of the information sequence $\{I_k\}$ is most easily described in terms of the received sequence $\{v_k\}$ at the output of the whitening filter. In the presence of intersymbol interference that spans $L+1$ symbols (L interfering components), the MLSE criterion is equivalent to the problem of estimating the state of a discrete-time finite-state machine. The finite-state machine in this case is the equivalent discrete-time channel with coefficients $\{f_k\}$, and its state at any instant in time is given by the L most recent inputs, i.e., the state at time k is

$$S_k = (I_{k-1}, I_{k-2}, \ldots, I_{k-L}) \qquad (10.1\text{–}20)$$

where $I_k = 0$ for $k \leqslant 0$. Hence, if the information symbols are M-ary, the channel filter has M^L states. Consequently, the channel is described by an M^L-state trellis and the Viterbi algorithm may be used to determine the most probable path through the trellis.

The metrics used in the trellis search are akin to the metrics used in soft-decision decoding of convolutional codes. In brief, we begin with the samples $v_1, v_2, \ldots, v_{L+1}$, from which we compute the M^{L+1} metrics

$$\sum_{k=1}^{L+1} \ln p(v_k | I_k, I_{k-1}, \ldots, I_{k-L}) \tag{10.1–21}$$

The M^{L+1} possible sequences of $I_{L+1}, I_L, \ldots, I_2, I_1$ are subdivided into M^L groups corresponding to the M^L states $(I_{L+1}, I_L, \ldots, I_2)$. Note that the M sequences in each group (state) differ in I_1 and correspond to the paths through the trellis that merge at a single node. From the M sequences in each of the M^L states, we select the sequence with the largest probability (with respect to I_1) and assign to the surviving sequence the metric

$$\begin{aligned} PM_1(\mathbf{I}_{L+1}) &\equiv PM_1(I_{L+1}, I_L, \ldots, I_2) \\ &= \max_{I_1} \sum_{k=1}^{L+1} \ln p(v_k | I_k, I_{k-1}, \ldots, I_{k-L}) \end{aligned} \tag{10.1–22}$$

The $M-1$ remaining sequences from each of the M^L groups are discarded. Thus, we are left with M^L surviving sequences and their metrics.

Upon reception of v_{L+2}, the M^L surviving sequences are extended by one stage, and the corresponding M^{L+1} probabilities for the extended sequences are computed using the previous metrics and the new increment, which is $\ln p(v_{L+2} | I_{L+2}, I_{L+1}, \ldots, I_2)$. Again, the M^{L+1} sequences are subdivided into M^L groups corresponding to the M^L possible states $(I_{L+2}, \ldots, I_3)$ and the most probable sequence from each group is selected, while the other $M-1$ sequences are discarded.

The procedure described continues with the reception of subsequent signal samples. In general, upon reception of v_{L+k}, the metrics†

$$PM_k(\mathbf{I}_{L+K}) = \max_{I_k} [\ln p(v_{L+k} | I_{L+k}, \ldots, I_k) + PM_{k-1}(\mathbf{I}_{L+k-1}))] \tag{10.1–23}$$

that are computed give the probabilities of the M^L surviving sequences. Thus, as each signal sample is received, the Viterbi algorithm involves first the computation of the M^{L+1} probabilities

$$\ln p(v_{L+k} | I_{L+k}, \ldots, I_k) + PM_{k-1}(\mathbf{I}_{L+k-1}) \tag{10.1–24}$$

corresponding to the M^{L+1} sequences that form the continuations of the M^L surviving sequences from the previous stage of the process. Then the M^{L+1} sequences are subdivided into M^L groups, with each group containing M sequences that terminate in the same set of symbols $I_{L+k}, \ldots, I_{k+1}$ and differ in the symbol I_k. From each group of M sequences, we select the one having the

†We observe that the metrics $PM_k(\mathbf{I})$ are simply related to the Euclidean distance metrics $DM_k(\mathbf{I})$ when the additive noise is Gaussian.

largest probability as indicated by Equation 10.1–23, while the remaining $M-1$ sequences are discarded. Thus, we are left again with M^L sequences having the metrics $PM_k(\mathbf{I}_{L+k})$.

As indicated previously, the delay in detecting each information symbol is variable. In practice, the variable delay is avoided by truncating the surviving sequences to the q most recent symbols, where $q \gg L$, thus achieving a fixed delay. In the case that the M^L surviving sequences at time k disagree on the symbol I_{k-q}, the symbol in the most probable sequence may be chosen. The loss of performance resulting from this suboptimum decision procedure is negligible if $q \geqslant 5L$.

EXAMPLE 10.1–2. For illustrative purposes, suppose that a duobinary signal pulse is employed to transmit four-level ($M=4$) PAM. Thus, each symbol is a number selected from the set $\{-3, -1, 1, 1, 3\}$. The controlled intersymbol interference in this partial-response signal is represented by the equivalent discrete-time channel model shown in Figure 10.1–4. Suppose we have received v_1 and v_2, where

$$\begin{aligned} v_1 &= I_1 + \eta_1 \\ v_2 &= I_2 + I_1 + \eta_2 \end{aligned} \tag{10.1–25}$$

and $\{\eta_i\}$ is a sequence of statistically independent zero-mean Gaussian noise. We may now compute the 16 metrics

$$PM_1(I_2, I_1) = -\sum_{k=1}^{2}\left(v_k - \sum_{j=0}^{1} I_{k-j}\right)^2, \qquad I_1, I_2 = \pm 1, \pm 3 \tag{10.1–26}$$

where $I_k = 0$ for $k \leqslant 0$.

Note that any subsequently received signals $\{v_i\}$ do not involve I_1. Hence, at this stage, we may discard 12 of the 16 possible pairs $\{I_1, I_2\}$. This step is illustrated by the tree diagram shown in Figure 10.1–5. In other words, after computing the 16 metrics corresponding to the 16 paths in the tree diagram, we discard three out of the four paths that terminate with $I_2 = 3$ and save the most probable of these four. Thus, the metric for the surviving path is

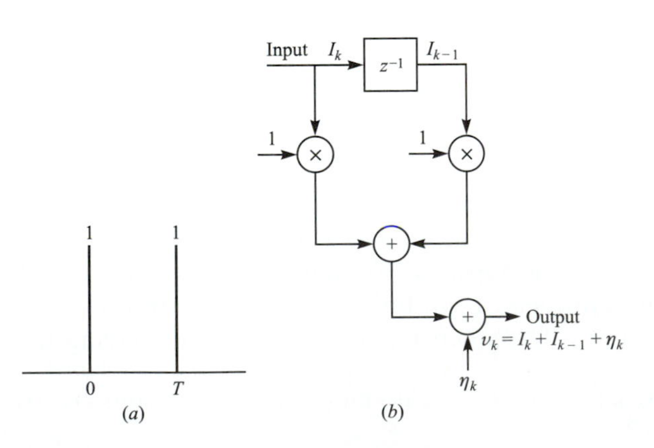

FIGURE 10.1–4
Equivalent discrete-time model for intersymbol interference resulting from a duobinary pulse.

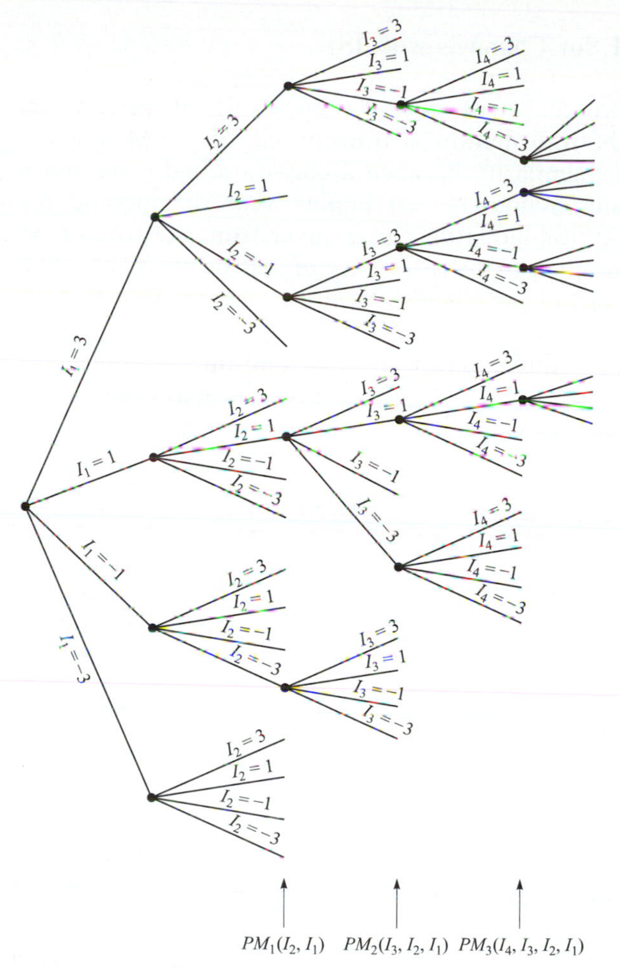

FIGURE 10.1–5
Tree diagram for Viterbi decoding of the duobinary pulse.

$$PM_1(I_2 = 3, I_1) = \max_{I_1}\left[-\sum_{k=1}^{2}\left(v_k - \sum_{j=0}^{1} I_{k-j}\right)^2\right]$$

The process is repeated for each set of four paths terminating with $I_2 = 1$, $I_2 = -1$, and $I_2 = -3$. Thus four paths and their corresponding metrics survive after v_1 and v_2 are received.

When v_3 is received, the four paths are extended as shown in Figure 10.1–5 to yield 16 paths and 16 corresponding metrics given by

$$PM_2(I_3, I_2, I_1) = PM_1(I_2, I_1) - \left(v_3 - \sum_{j=0}^{1} I_{3-j}\right)^2 \tag{10.1–27}$$

Of the four paths terminating with the $I_3 = 3$, we save the most probable. This procedure is again repeated for $I_3 = 1$, $I_3 = -1$, and $I_3 = -3$. Consequently, only four paths survive at this stage. The procedure is then repeated for each subsequently received signal v_k for $k > 3$.

10.1.4 Performance of MLSE for Channels with ISI

We shall now determine the probability of error for the MLSE of the received information sequence when the information is transmitted via PAM and the additive noise is Gaussian. The similarity between a convolutional code and a finite-duration intersymbol interference channel implies that the method for computing the error probability for the latter carries over from the former. In particular, the method for computing the performance of soft-decision decoding of a convolutional code by means of the Viterbi algorithm, described in Section 8.2.3, applies with some modification.

In PAM signaling with the additive Gaussian noise and intersymbol interference, the metrics used in the Viterbi algorithm may be expressed as in Equation 10.1–23, or, equivalently, as

$$PM_{k-L}(\mathbf{I}_k) = PM_{k-L-1}(\mathbf{I}_{k-1}) - \left(v_k - \sum_{j=0}^{L} f_j I_{k-j}\right)^2 \tag{10.1–28}$$

where the symbols $\{I_n\}$ may take the values $\pm d, \pm 3d, \ldots, \pm(M-1)d$, and $2d$ is the distance between successive levels. The trellis has M^L states, defined at time k as

$$S_k = (I_{k-1}, I_{k-2}, \ldots, I_{k-L}) \tag{10.1–29}$$

Let the estimated symbols from the Viterbi algorithm be denoted by $\{\tilde{I}_n\}$ and the corresponding estimated state at time k by

$$\tilde{S}_k = (\tilde{I}_{k-1}, \tilde{I}_{k-2}, \ldots, \tilde{I}_{k-L}) \tag{10.1–30}$$

Now suppose that the estimated path through the trellis diverges from the correct path at time k and remerges with the correct path at time $k+l$. Thus, $\tilde{S}_k = S_k$ and $\tilde{S}_{k+1} = S_{k+1}$, but $\tilde{S}_m \neq S_m$ for $k < m < k+l$. As in a convolutional code, we call this an *error event*. Since the channel spans $L+1$ symbols, it follows that $l \geqslant L+1$.

For such an error event, we have $\tilde{I}_k \neq I_k$ and $\tilde{I}_{k+l-L-1} \neq I_{k+l-L-1}$, but $\tilde{I}_m = I_m$ for $k-L \leqslant m \leqslant k-1$ and $k+l-L \leqslant m \leqslant k+l-1$. It is convenient to define an error vector $\boldsymbol{\varepsilon}$ corresponding to this error event as

$$\boldsymbol{\varepsilon} = [\varepsilon_k \quad \varepsilon_{k+1} \quad \cdots \quad \varepsilon_{k+l-L-1}] \tag{10.1–31}$$

where the components of $\boldsymbol{\varepsilon}$ are defined as

$$\varepsilon_j = \frac{1}{2d}(I_j - \tilde{I}_j), \qquad j = k, k+1, \ldots, k+l-L-1 \tag{10.1–32}$$

The normalization factor of $2d$ in Equation 10.1–32 results in elements ε_j that take on the values $0, \pm 1, \pm 2, \pm 3, \ldots, \pm(M-1)$. Moreover, the error vector is characterized by the properties that $\varepsilon_k \neq 0$, $\varepsilon_{k+l-L-1} \neq 0$, and there is no sequence of L consecutive elements that are zero. Associated with the error vector in Equation 10.1–31 is the polynomial of degree $l-L-1$,

$$\varepsilon(z) = \varepsilon_k + \varepsilon_{k+1}z^{-1} + \varepsilon_{k+2}z^{-2} + \cdots + \varepsilon_{k+l-L-1}z^{-(l-L-1)} \tag{10.1–33}$$

We wish to determine the probability of occurrence of the error event that begins at time k and is characterized by the error vector $\boldsymbol{\varepsilon}$ given in Equation 10.1–31 or, equivalently, by the polymonial given in Equation 10.1–33. To accomplish this, we follow the procedure developed by Forney (1972). Specifically, for the error event $\boldsymbol{\varepsilon}$ to occur, the following three subevents E_1, E_2, and E_3 must occur:

E_1: At time k, $\tilde{S}_k = S_k$.

E_2: The information symbols $I_k, I_{k+1}, \ldots, I_{k+l-L-1}$ when added to the scaled error sequence $2d(\varepsilon_k, \varepsilon_{k+1}, \ldots, \varepsilon_{k+1-L-1})$ must result in an allowable sequence, i.e., the sequence $\tilde{I}_k, \tilde{I}_{k+1}, \ldots, \tilde{I}_{k+l-L-1}$ must have values selected from $\pm d, \pm 3d, \pm \cdots \pm (M-1)d$.

E_3: For $k \leqslant m < k+l$, the sum of the branch metrics of the estimated path exceeds the sum of the branch metrics of the correct path.

The probability of occurrence of E_3 is

$$P(E_3) = P\left[\sum_{i=1}^{k+l-1}\left(v_i - \sum_{j=0}^{L} f_j \tilde{I}_{i-j}\right)^2 < \sum_{i=k}^{k+l-1}\left(v_i - \sum_{j=0}^{L} f_j I_{i-j}\right)^2\right] \tag{10.1–34}$$

But

$$v_i = \sum_{j=0}^{L} f_j I_{i-j} + \eta_i \tag{10.1–35}$$

where $\{\eta_i\}$ is a real-valued white Gaussian noise sequence. Substitution of Equation 10.1–35 into Equation 10.1–34 yields

$$\begin{aligned} P(E_3) &= P\left[\sum_{i=k}^{k+l-1}\left(\eta_i + 2d\sum_{j=0}^{L} f_j \varepsilon_{i-j}\right)^2 < \sum_{i=k}^{k+l-1}\eta_i^2\right] \\ &= P\left[4d\sum_{i=k}^{k+l-1}\eta_i\left(\sum_{j=0}^{L} f_j\varepsilon_{i-j}\right) < -4d^2\sum_{i=k}^{k+l-1}\left(\sum_{j=0}^{L} f_j\varepsilon_{i-j}\right)^2\right] \end{aligned} \tag{10.1–36}$$

where $\varepsilon_j = 0$ for $j < k$ and $j > k + l - L - 1$. If we define

$$\alpha_i = \sum_{j=0}^{L} f_j \varepsilon_{i-j} \tag{10.1–37}$$

then Equation 10.1–36 may be expressed as

$$P(E_3) = P\left(\sum_{i=k}^{k+l-1} \alpha_i \eta_i < -d \sum_{i=k}^{k+1-1} \alpha_i^2\right) \tag{10.1–38}$$

where the factor of $4d$ common to both terms has been dropped. Now Equation 10.1–38 is just the probability that a linear combination of statistically independent Gaussian random variables is less than some negative number. Thus

$$P(E_3) = Q\left(\sqrt{\frac{2d^2}{N_0}\sum_{i=k}^{k+l-1}\alpha_i^2}\right) \tag{10.1–39}$$

For convenience, we define

$$\delta^2(\boldsymbol{\varepsilon}) = \sum_{i=k}^{k+l-1}\alpha_i^2 = \sum_{i=k}^{k+l-1}\left(\sum_{j=0}^{L} f_j\varepsilon_{i-j}\right)^2 \tag{10.1–40}$$

where $\varepsilon_j = 0$ for $j < k$ and $j > k + l - L - 1$. Note that the $\{\alpha_i\}$ resulting from the convolution of $\{f_i\}$ with $\{\varepsilon_j\}$ are the coefficients of the polynomial

$$\begin{aligned}\alpha(z) &= F(z)\varepsilon(z)\\ &= \alpha_k + \alpha_{k+1}z^{-1} + \cdots + \alpha_{k+l-1}z^{-(l-1)}\end{aligned} \tag{10.1–41}$$

Furthermore, $\delta^2(\boldsymbol{\varepsilon})$ is simply equal to the coefficient of z^0 in the polynomial

$$\begin{aligned}\alpha(z)\alpha(z^{-1}) &= F(z)F(z^{-1})\varepsilon(z)\varepsilon(z^{-1})\\ &= X(z)\varepsilon(z)\varepsilon(z^{-1})\end{aligned} \tag{10.1–42}$$

We call $\delta^2(\boldsymbol{\varepsilon})$ the *Euclidean weight* of the error event $\boldsymbol{\varepsilon}$.

An alternative method for representing the result of convolving $\{f_j\}$ with $\{\varepsilon_j\}$ is the matrix form

$$\boldsymbol{\alpha} = \mathbf{ef}$$

where $\boldsymbol{\alpha}$ is an l-dimensional vector, $\mathbf{f}$ is an $(L+1)$-dimensional vector, and $\mathbf{e}$ is an $l \times (L+1)$ matrix defined as

$$\boldsymbol{\alpha} = \begin{bmatrix}\alpha_k\\ a_{k+1}\\ \vdots\\ \alpha_{k+l-1}\end{bmatrix}, \qquad \mathbf{f} = \begin{bmatrix}f_0\\ f_1\\ \vdots\\ f_L\end{bmatrix}$$

$$\mathbf{e} = \begin{bmatrix}\varepsilon_k & 0 & 0 & \cdots & 0 & \cdots & 0\\ \varepsilon_{k+1} & \varepsilon_k & 0 & \cdots & 0 & \cdots & 0\\ \varepsilon_{k+2} & \varepsilon_{k+1} & \varepsilon_k & \cdots & 0 & \cdots & 0\\ \vdots & \vdots & \vdots & & \vdots & & \vdots\\ \varepsilon_{k+l-1} & \cdots & \cdots & \cdots & \cdots & \cdots & \varepsilon_{k+l-L-1}\end{bmatrix} \tag{10.1–43}$$

Then

$$\begin{aligned}\delta^2(\boldsymbol{\varepsilon}) &= \boldsymbol{\alpha}'\boldsymbol{\alpha} \\ &= \mathbf{f}'\mathbf{e}'\mathbf{e}\mathbf{f} \\ &= \mathbf{f}'\mathbf{A}\mathbf{f}\end{aligned} \tag{10.1–44}$$

where $\mathbf{A}$ is an $(L+1) \times (L+1)$ matrix of the form

$$\mathbf{A} = \mathbf{e}'\mathbf{e} = \begin{bmatrix} \beta_0 & \beta_1 & \beta_2 & \cdots & \beta_L \\ \beta_1 & \beta_0 & \beta_1 & \cdots & \beta_{L-1} \\ \beta_2 & \beta_1 & \beta_0 & \beta_1 & \beta_{L-2} \\ \vdots & \vdots & \vdots & \vdots & \vdots \\ \beta_L & \cdots & \cdots & \cdots & \beta_0 \end{bmatrix} \tag{10.1–45}$$

and

$$\beta_m = \sum_{i=k}^{k+l-1-m} \varepsilon_i \varepsilon_{i+m} \tag{10.1–46}$$

We may use either Equations 10.1–40 and 10.1–41 or Equations 10.1–45 and 10.1–46 in evaluating the error rate performance. We consider these computations later. For now we conclude that the probability of the subevent E_3, given by Equations 10.1–39, may be expressed as

$$\begin{aligned}P(E_3) &= Q\left(\sqrt{\frac{2d^2}{N_0}\delta^2(\boldsymbol{\varepsilon})}\right) \\ &= Q\left(\sqrt{\frac{6}{M^2-1}\gamma_{\text{av}}\delta^2(\boldsymbol{\varepsilon})}\right)\end{aligned} \tag{10.1–47}$$

where we have used the relation

$$d^2 = \frac{3}{M^2-1} TP_{\text{av}} \tag{10.1–48}$$

to eliminate d^2 and $\gamma_{\text{av}} = TP_{\text{av}}/N_0$. Note that, in the absence of intersymbol interference, $\delta^2(\boldsymbol{\varepsilon}) = 1$ and $P(E_3)$ is proportional to the symbol error probability of M-ary PAM.

The probability of the subevent E_2 depends only on the statistical properties of the input sequence. We assume that the information symbols are equally probable and that the symbols in the transmitted sequence are statistically independent. Then, for an error of the form $|\varepsilon_i| = j, j = 1, 2, \ldots, M-1$, there are $M-j$ possible values of I_i such that

$$I_i = \tilde{I}_i + 2d\varepsilon_i$$

Hence

$$P(E_2) = \prod_{i=0}^{l-L-1} \frac{M - |\varepsilon_i|}{M} \tag{10.1–49}$$

The probability of the subevent E_1 is much more difficult to compute exactly because of its dependence on the subevent E_3. That is, we must compute $P(E_1|E_3)$. However, $P(E_1|E_3) = 1 - P_M$, where P_M is the symbol error probability. Hence $P(E_1|E_3)$ is well approximated (and upper-bounded) by unity for reasonably low symbol error probabilities. Therefore, the probability of the error event $\boldsymbol{\varepsilon}$ is well approximated and upper-bounded as

$$P(\boldsymbol{\varepsilon}) \leqslant Q\left(\sqrt{\frac{6}{M^2-1}\gamma_{\text{av}}\delta^2(\boldsymbol{\varepsilon})}\right) \prod_{i=0}^{l-L-1} \frac{M - |\varepsilon_i|}{M} \tag{10.1–50}$$

Let E be the set of all error events $\boldsymbol{\varepsilon}$ starting at time k and let $w(\boldsymbol{\varepsilon})$ be the corresponding number of nonzero components (Hamming weight or number of symbol errors) in each error event $\boldsymbol{\varepsilon}$. Then the probability of a symbol error is upper-bounded (union bound) as

$$\begin{aligned} P_M &\leqslant \sum_{\boldsymbol{\varepsilon} \in E} w(\boldsymbol{\varepsilon}) P(\boldsymbol{\varepsilon}) \\ &\leqslant \sum_{\boldsymbol{\varepsilon} \in E} w(\boldsymbol{\varepsilon}) Q\left(\sqrt{\frac{6}{M^2-1}\gamma_{\text{av}}\delta^2(\boldsymbol{\varepsilon})}\right) \prod_{i=0}^{l-L-1} \frac{M - |\varepsilon_i|}{M} \end{aligned} \tag{10.1–51}$$

Now let D be the set of all $\delta(\boldsymbol{\varepsilon})$. For each $\delta \in D$, let E_δ be the subset of error events for which $\delta(\boldsymbol{\varepsilon}) = \delta$. Then Equation 10.1–51 may be expressed as

$$\begin{aligned} P_M &\leqslant \sum_{\delta \in D} Q\left(\sqrt{\frac{6}{M^2-1}\gamma_{\text{av}}\delta^2}\right) \left[\sum_{\boldsymbol{\varepsilon} \in E_\delta} w(\boldsymbol{\varepsilon}) \prod_{i=0}^{l-L-1} \frac{M - |\varepsilon_i|}{M}\right] \\ &\leqslant \sum_{\delta \in D} K_\delta Q\left(\sqrt{\frac{6}{M^2-1}\gamma_{\text{av}}\delta^2}\right) \end{aligned} \tag{10.1–52}$$

where

$$K_\delta = \sum_{\boldsymbol{\varepsilon} \in E_\delta} w(\boldsymbol{\varepsilon}) \prod_{i=0}^{l-L-1} \frac{M - |\varepsilon_i|}{M} \tag{10.1–53}$$

The expression for the error probability in Equation 10.1–52 is similar to the form of the error probability for a convolutional code with soft-decision decoding given by Equation 8.2–26. The weighting factors $\{K_\delta\}$ may be determined by means of the error state diagram, which is akin to the state diagram of a convolutional encoder. This approach has been illustrated by Forney (1972) and Viterbi and Omura (1979).

In general, however, the use of the error state diagram for computing P_M is tedious. Instead, we may simplify the computation of P_M by focusing on the dominant term in the summation of Equation 10.1–52. Because of the exponential dependence of each term in the sum, the expression P_M is dominated by the term corresponding to the minimum value of δ, denoted as $\delta_{\min}$. Hence the symbol error probability may be approximated as

$$P_M \approx K_{\delta_{\min}} Q\left(\sqrt{\frac{6}{M^2-1}\gamma_{\mathrm{av}}\delta_{\min}^2}\right) \tag{10.1–54}$$

where

$$K_{\delta_{\min}} = \sum_{\boldsymbol{\varepsilon} \in E_{\delta_{\min}}} w(\boldsymbol{\varepsilon}) \prod_{i=0}^{l-L-1} \frac{M-|\varepsilon_i|}{M} \tag{10.1–55}$$

In general, $\delta_{\min}^2 \leqslant 1$. Hence, 10 log $\delta_{\min}^2$ represents the loss in SNR due to intersymbol interference.

The minimum value of δ may be determined either from Equation 10.1–40 or from evaluation of the quadratic form in Equation 10.1–44 for different error sequences. In the following two examples we use Equation 10.1–40.

EXAMPLE 10.1–3. Consider a two path channel ($L = 1$) with arbitrary coefficients f_0 and f_1 satisfying the constraint $f_0^2 + f_1^2 = 1$. The channel characteristic is

$$F(z) = f_0 + f_1 z^{-1} \tag{10.1–56}$$

For an error event of length n,

$$\varepsilon(z) = \varepsilon_0 + \varepsilon_1 z^{-1} + \cdots + \varepsilon_{n-1} z^{-(n-1)}, \qquad n \geqslant 1 \tag{10.1–57}$$

The product $\alpha(z) = F(z)\varepsilon(z)$ may be expressed as

$$\alpha(z) = \alpha_0 + \alpha_1 z^{-1} + \cdots + \alpha_n z^{-n} \tag{10.1–58}$$

where $\alpha_0 = \varepsilon_0 f_0$ and $\alpha_n = f_1 \varepsilon_{n-1}$. Since $\varepsilon_0 \neq 0$, $\varepsilon_{n-1} \neq 0$, and

$$\delta^2(\boldsymbol{\varepsilon}) = \sum_{k=0}^{n} \alpha_k^2 \tag{10.1–59}$$

it follows that

$$\delta_{\min}^2 \geqslant f_0^2 + f_1^2 = 1$$

Indeed, $\delta_{\min}^2 = 1$ when a single error occurs, i.e., $\varepsilon(z) = \varepsilon_0$. Thus, we conclude that there is no loss in SNR in maximum-likelihood sequence estimation of the information symbols when the channel dispersion has length 2.

EXAMPLE 10.1–4. The controlled intersymbol interference in a partial-response signal may be viewed as having been generated by a time-dispersive channel. Thus, the intersymbol interference from a duobinary pulse may be represented by the (normalized) channel characteristic

$$F(z) = \sqrt{\tfrac{1}{2}} + \sqrt{\tfrac{1}{2}} z^{-1} \tag{10.1–60}$$

Similarly, the representation for a modified duobinary pulse is

$$F(z)=\sqrt{\tfrac{1}{2}}-\sqrt{\tfrac{1}{2}}z^{-2} \qquad (10.1\text{–}61)$$

The minimum distance $\delta^2_{\min}=1$ for any error event of the form

$$\varepsilon(z)=\pm(1-z^{-1}-z^{-2}\cdots-z^{-(n-1)}), \qquad n\geqslant 1 \qquad (10.1\text{–}62)$$

for the channel given by Equation 10.1–60, since

$$\alpha(z)=\pm\sqrt{\tfrac{1}{2}}\mp\sqrt{\tfrac{1}{2}}z^{-n}$$

Similarly, when

$$\varepsilon(z)=\pm(1+z^{-2}+z^{-4}+\cdots+z^{-2(n-1)}), \qquad n\geqslant 1 \qquad (10.1\text{–}63)$$

$\delta^2_{\min}=1$ for the channel given by Equation 10.1–61 since

$$\alpha(z)=\pm\sqrt{\tfrac{1}{2}}\mp\sqrt{\tfrac{1}{2}}z^{-2n}$$

Hence the MLSE of these two partial-response signals result in no loss in SNR. In contrast, the suboptimum symbol-by-symbol detection described previously resulted in a 2.1-dB loss.

The constant $K_{\delta_{\min}}$ is easily evaluated for these two signals. With precoding, the number of output symbol errors (Hamming weight) associated with the error events in Equations 10.1–62 and 10.1–63 is two. Hence,

$$K_{\delta_{\min}}=2\sum_{n=1}^{\infty}\left(\frac{M-1}{M}\right)^n=2(M-1) \qquad (10.1\text{–}64)$$

On the other hand, without precoding, these error events result in n symbol errors, and, hence,

$$K_{\delta_{\min}}=2\sum_{n=1}^{\infty}n\left(\frac{M-1}{M}\right)^n=2M(M-1) \qquad (10.1\text{–}65)$$

As a final exercise, we consider the evaluation of $\delta^2_{\min}$ from the quadratic form in Equation 10.1–44. The matrix $\mathbf{A}$ of the quadratic form is positive-definite; hence, all its eigenvalues are positive. If $\{\mu_k(\boldsymbol{\varepsilon})\}$ are the eigenvalues and $\{\mathbf{v}_k(\boldsymbol{\varepsilon})\}$ are the corresponding orthonormal eigenvectors of $\mathbf{A}$ for an error event $\boldsymbol{\varepsilon}$, then the quadratic form in Equation 10.1–44 can be expressed as

$$\delta^2(\boldsymbol{\varepsilon})=\sum_{k=1}^{L+1}\mu_k(\boldsymbol{\varepsilon})[\mathbf{f}'\mathbf{v}_k(\boldsymbol{\varepsilon})]^2 \qquad (10.1\text{–}66)$$

In other words, $\delta^2(\boldsymbol{\varepsilon})$ is expressed as a linear combination of the squared projections of the channel vector $\mathbf{f}$ onto the eigenvectors of $\mathbf{A}$. Each squared projection of the sum is weighted by the corresponding eigenvalue $\mu_k(\boldsymbol{\varepsilon})$, $k=1,2,\ldots,L+1$. Then

$$\delta^2_{\min}=\min_{\boldsymbol{\varepsilon}}\delta^2(\boldsymbol{\varepsilon}) \qquad (10.1\text{–}67)$$

It is interesting to note that the worst channel characteristic of a given length $L+1$ can be obtained by finding the eigenvector corresponding to the minimum eigenvalue. Thus, if $\mu_{\min}(\boldsymbol{\varepsilon})$ is the minimum eigenvalue for a given error event $\boldsymbol{\varepsilon}$ and $\mathbf{v}_{\min}(\boldsymbol{\varepsilon})$ is the corresponding eigenvector, then

$$\mu_{\min} = \min_{\boldsymbol{\varepsilon}} \mu_{\min}(\boldsymbol{\varepsilon})$$

$$\mathbf{f} = \min_{\boldsymbol{\varepsilon}} \mathbf{v}_{\min}(\boldsymbol{\varepsilon})$$

and

$$\delta_{\min}^2 = \mu_{\min}$$

EXAMPLE 10.1–5. Let us determine the worst time-dispersive channel of length 3 ($L = 2$) by finding the minimum eigenvalue of $\mathbf{A}$ for different error events. Thus,

$$F(z) = f_0 + f_1 z^{-1} + f_2 z^{-2}$$

where f_0, f_1, and f_2 are the components of the eigenvector of $\mathbf{A}$ corresponding to the minimum eigenvalue. An error event of the form

$$\varepsilon(z) = 1 - z^{-1}$$

results in a matrix

$$\mathbf{A} = \begin{bmatrix} 2 & -1 & 0 \\ -1 & 2 & -1 \\ 0 & -1 & 2 \end{bmatrix}$$

which has the eigenvalues $\mu_1 = 2$, $\mu_2 = 2 + \sqrt{2}$, $\mu_3 = 2 - \sqrt{2}$. The eigenvector corresponding to μ_3 is

$$\mathbf{v}_3' = \begin{bmatrix} \frac{1}{2} & \sqrt{\frac{1}{2}} & \frac{1}{2} \end{bmatrix} \tag{10.1–68}$$

We may also consider the dual error event

$$\epsilon(z) = 1 + z^{-1}$$

which results in the matrix

$$\mathbf{A} = \begin{bmatrix} 2 & 1 & 0 \\ 1 & 2 & 1 \\ 0 & 1 & 2 \end{bmatrix}$$

This matrix has eigenvalues identical to those of the one for $\varepsilon(z) = 1 - z^{-1}$. The corresponding eigenvector for $\mu_3 = 2 - \sqrt{2}$ is

$$\mathbf{v}_3' = \begin{bmatrix} -\frac{1}{2} & \sqrt{\frac{1}{2}} & -\frac{1}{2} \end{bmatrix} \tag{10.1–69}$$

Any other error events lead to larger values for $\mu_{\min}$. Hence, $\mu_{\min} = 2 - \sqrt{2}$ and the worst-case channel is either

$$\begin{bmatrix} \frac{1}{2} & \sqrt{\frac{1}{2}} & \frac{1}{2} \end{bmatrix} \quad \text{or} \quad \begin{bmatrix} -\frac{1}{2} & \sqrt{\frac{1}{2}} & -\frac{1}{2} \end{bmatrix}$$

TABLE 10.1–1
Maximum performance loss and corresponding channel characteristics

Channel length $L+1$	Performance loss $-10\log\delta^2_{\min}$ dB	Minimum-distance channel
3	2.3	0.50, 0.71, 0.50
4	4.2	0.38,0.60,0.60,0.38
5	5.7	0.29, 0.50, 0.58, 0.50, 0.29
6	7.0	0.23, 0.42, 0.52, 0.52, 0.42, 0.23

The loss in SNR from the channel is

$$-10\log\delta^2_{\min} = -10\log\mu_{\min} = 2.3\,\text{dB}$$

Repetitions of the above computation for channels with $L = 3, 4$, and 5 yield the results given in Table 10.1–1.

10.2
LINEAR EQUALIZATION

The MLSE for a channel with ISI has a computational complexity that grows exponentially with the length of the channel time dispersion. If the size of the symbol alphabet is M and the number of interfering symbols contributing to ISI is L, the Viterbi algorithm computes M^{L+1} metrics for each new received symbol. In most channels of practical interest, such a large computational complexity is prohibitively expensive to implement.

In this and the following sections, we describe suboptimum channel equalization approaches to compensate for the ISI. One approach employs a linear transversal filter, which is described in this section. This filter structure has a computational complexity that is a linear function of the channel dispersion length L.

The linear filter most often used for equalization is the transversal filter shown in Figure 10.2–1. Its input is the sequence $\{v_k\}$ given in Equation 10.1–16 and its output in the estimate of the information sequence $\{I_k\}$. The estimate of the kth symbol may be expressed as

$$\hat{I}_k = \sum_{j=-K}^{K} c_j v_{k-j} \tag{10.2–1}$$

where $\{c_j\}$ are the $2K + 1$ complex-valued tap weight coefficients of the filter. The estimate $\hat{I}_k$ is quantized to the nearest (in distance) information symbol to form the decision $\tilde{I}_k$. If $\tilde{I}_k$ is not identical to the transmitted information symbol I_k, an error has been made.

Considerable research has been performed on the criterion for optimizing the filter coefficients $\{c_k\}$. Since the most meaningful measure of performance for a digital communication system is the average probability of error, it is desirable to

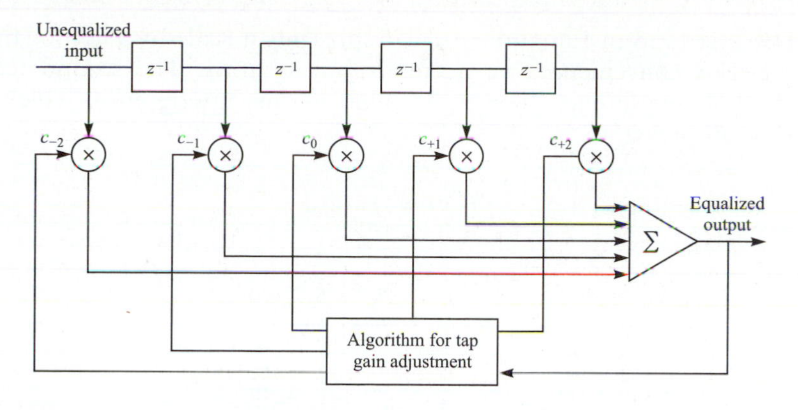

FIGURE 10.2–1
Linear transversal filter.

choose the coefficients to minimize this performance index. However, the probability of error is a highly non-linear function of $\{c_j\}$. Consequently, the probability of error as a performance index for optimizing the tap weight coefficients of the equalizer is computationally complex.

Two criteria have found widespread use in optimizing the equalizer coefficients $\{c_j\}$. One is the peak distortion criterion and the other is the mean-square-error criterion.

10.2.1 Peak Distortion Criterion

The peak distortion is simply defined as the worst-case intersymbol interference at the output of the equalizer. The minimization of this performance index is called the *peak distortion criterion*. First we consider the minimization of the peak distortion assuming that the equalizer has an infinite number of taps. Then we shall discuss the case in which the transversal equalizer spans a finite time duration.

We observe that the cascade of the discrete-time linear filter model having an impulse response $\{f_n\}$ and an equalizer having an impulse response $\{c_n\}$ can be represented by a single equivalent filter having the impulse response

$$q_n = \sum_{j=-\infty}^{\infty} c_j f_{n-j} \tag{10.2–2}$$

That is, $\{q_n\}$ is simply the convolution of $\{c_n\}$ and $\{f_n\}$. The equalizer is assumed to have an infinite number of taps. Its output at the kth sampling instant can be expressed in the form

$$\hat{I}_k = q_0 I_k + \sum_{n\neq k} I_n q_{k-n} + \sum_{j=-\infty}^{\infty} c_j \eta_{k-j} \tag{10.2–3}$$

The first term in Equation 10.2–3 represents a scaled version of the desired symbol. For convenience, we normalize q_0 to unity. The second term is the intersymbol interference. The peak value of this interference, which is called the *peak distortion*, is

$$\begin{aligned} \mathcal{D}(\mathbf{c}) &= \sum_{\substack{n=-\infty \\ n\neq 0}}^{\infty} |q_n| \\ &= \sum_{\substack{n=-\infty \\ n\neq 0}}^{\infty} \left| \sum_{j=-\infty}^{\infty} c_j f_{n-j} \right| \end{aligned} \tag{10.2–4}$$

Thus, $\mathcal{D}(\mathbf{c})$ is a function of the equalizer tap weights.

With an equalizer having an infinite number of taps, it is possible to select the tap weights so that $\mathcal{D}(\mathbf{c}) = 0$, i.e., $q_n = 0$ for all n except $n = 0$. That is, the intersymbol interference can be completely eliminated. The values of the tap weights for accomplishing this goal are determined from the condition

$$q_n = \sum_{j=-\infty}^{\infty} c_j f_{n-j} = \begin{cases} 1 & (n = 0) \\ 0 & (n \neq 0) \end{cases} \tag{10.2–5}$$

By taking the z transform of Equation 10.2–5, we obtain

$$Q(z) = C(z)F(z) = 1 \tag{10.2–6}$$

or, simply,

$$C(z) = \frac{1}{F(z)} \tag{10.2–7}$$

where $C(z)$ denotes the z transform of the $\{c_j\}$. Note that the equalizer, with transfer function $C(z)$, is simply the inverse filter to the linear filter model $F(z)$. In other words, complete elimination of the intersymbol interference requires the use of an inverse filter to $F(z)$. We call such a filter a *zero-forcing filter*. Figure 10.2–2 illustrates in block diagram the equivalent discrete-time channel and equalizer.

The cascade of the noise-whitening filter having the transfer function $1/F^*(z^{-1})$ and the zero-forcing equalizer having the transfer function $1/F(z)$ results in an equivalent zero-forcing equalizer having the transfer function

$$C'(z) = \frac{1}{F(z)F^*(z^{-1})} = \frac{1}{X(z)} \tag{10.2–8}$$

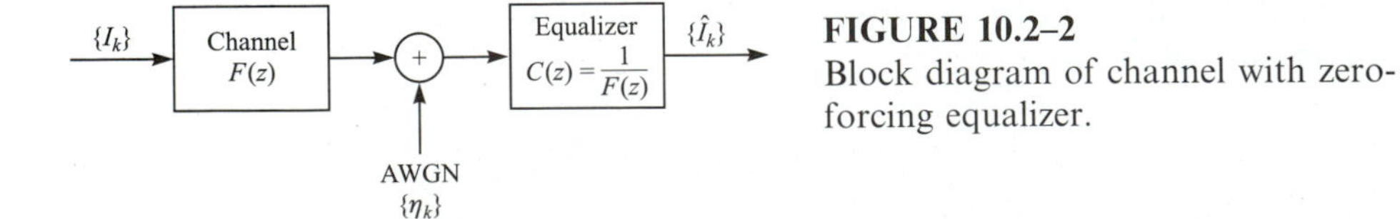

FIGURE 10.2–2
Block diagram of channel with zero-forcing equalizer.

as shown in Figure 10.2–3. This combined filter has as its input the sequence $\{y_k\}$ of samples from the matched filter, given by Equation 10.1–10. Its output consists of the desired symbols corrupted only by additive zero-mean Gaussian noise. The impulse response of the combined filter is

$$\begin{aligned} c'_k &= \frac{1}{2\pi j}\oint C'(z)z^{k-1}\,dz \\ &= \frac{1}{2\pi j}\oint \frac{z^{k-1}}{X(z)}\,dz \end{aligned} \tag{10.2–9}$$

where the integration is performed on a closed contour that lies within the region of convergence of $C'(z)$. Since $X(z)$ is a polynomial with $2L$ roots $(\rho_1, \rho_2, \ldots, \rho_L, 1/\rho_1^*, 1/\rho_2^*, \ldots, 1/\rho_L^*)$, it follows that $C'(z)$ must converge in an annular region in the z plane that includes the unit circle ($z = e^{j\theta}$). Consequently, the closed contour in the integral can be the unit circle.

The performance of the infinite-tap equalizer that completely eliminates the intersymbol interference can be expressed in terms of the SNR at its output. For mathematical convenience, we normalize the received signal energy to unity.† This implies that $q_0 = 1$ and that the expected value of $|I_k|^2$ is also unity. Then the SNR is simply the reciprocal of the noise variance σ_n^2 at the output of the equalizer.‡

The value of σ_n^2 can be simply determined by observing that the noise sequence $\{\nu_k\}$ at the input to the equivalent zero-forcing equalizer $C'(z)$ has zero-mean and a power spectral density

$$\Phi_{\nu\nu}(\omega) = N_0 X(e^{j\omega T}), \qquad |\omega| \leqslant \frac{\pi}{T} \tag{10.2–10}$$

where $X(e^{j\omega t})$ is obtained from $X(z)$ by the substitution $z = e^{j\omega T}$. Since $C'(z) = 1/X(z)$, it follows that the noise sequence at the output of the equalizer has a power spectral density

$$\Phi_{nn}(\omega) = \frac{N_0}{X(e^{j\omega T})}, \qquad |\omega| \leqslant \frac{\pi}{T} \tag{10.2–11}$$

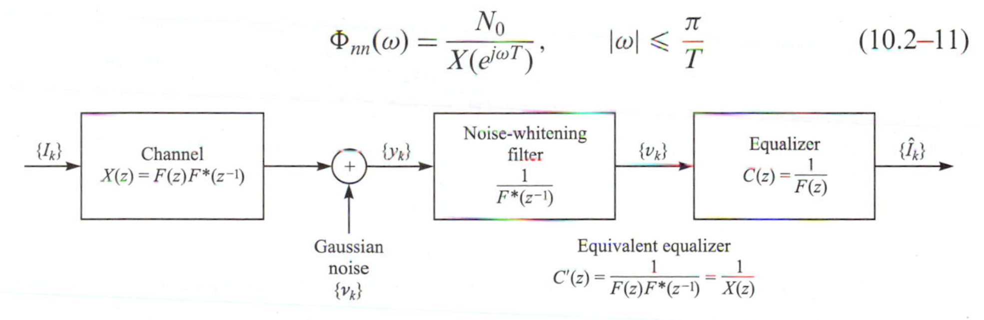

FIGURE 10.2–3
Block diagram of channel with equivalent zero-forcing equalizer.

---

†This normalization is used throughout this chapter for mathematical convenience.
‡If desired, one can multiply this normalized SNR at the output of the equalizer by the signal energy.

Consequently, the variance of the noise variable at the output of the equalizer is

$$\begin{aligned}\sigma_n^2 &= \frac{T}{2\pi}\int_{-\pi/T}^{\pi/T} \Phi_{nn}(\omega)\,d\omega \\ &= \frac{TN_0}{2\pi}\int_{-\pi/T}^{\pi/T} \frac{d\omega}{X(e^{j\omega T})}\end{aligned} \tag{10.2–12}$$

and the SNR for the zero-forcing equalizer is

$$\gamma_\infty = \frac{1}{\sigma_n^2} = \left[\frac{TN_0}{2\pi}\int_{-\pi/T}^{\pi/T} \frac{d\omega}{X(e^{j\omega T})}\right]^{-1} \tag{10.2–13}$$

where the subscript on γ indicates that the equalizer has an infinite number of taps.

The spectral characteristics $X(e^{j\omega T})$ corresponding to the Fourier transform of the sampled sequence $\{x_n\}$ has an interesting relationship to the analog filter $H(\omega)$ used at the receiver. Since

$$x_k = \int_{-\infty}^{\infty} h^*(t)h(t+kT)\,dt$$

use of Parseval's theorem yields

$$x_k = \frac{1}{2\pi}\int_{-\infty}^{\infty} |H(\omega)|^2 e^{j\omega kT}\,d\omega \tag{10.2–14}$$

where $H(\omega)$ is the Fourier transform of $h(t)$. But the integral in Equation 10.2–14 can be expressed in the form

$$x_k = \frac{1}{2\pi}\int_{-\pi/T}^{\pi/T}\left[\sum_{n=-\infty}^{\infty}\left|H\left(\omega+\frac{2\pi n}{T}\right)\right|^2\right]e^{j\omega kT}\,d\omega \tag{10.2–15}$$

Now, the Fourier transform of $\{x_k\}$ is

$$X(e^{j\omega T}) = \sum_{k=-\infty}^{\infty} x_k e^{-j\omega kT} \tag{10.2–16}$$

and the inverse transform yields

$$x_k = \frac{T}{2\pi}\int_{-\pi/T}^{\pi/T} X(e^{j\omega T})e^{j\omega kT}\,d\omega \tag{10.2–17}$$

From a comparison of Equations 10.2–15 and 10.2–17, we obtain the desired relationship beween $X(e^{j\omega T})$ and $H(\omega)$. That is,

$$X(e^{j\omega T}) = \frac{1}{T}\sum_{n=-\infty}^{\infty}\left|H\left(\omega+\frac{2\pi n}{T}\right)\right|^2, \qquad |\omega| \leqslant \frac{\pi}{T} \tag{10.2–18}$$

where the right-hand side of Equation 10.2–18 is called the *folded spectrum* of $|H(\omega)|^2$. We also observe that $|H(\omega)|^2 = X(\omega)$, where $X(\omega)$ is the Fourier transform of the waveform $x(t)$ and $x(t)$ is the response of the matched filter to the input pulse $h(t)$. Therefore the right-hand side of Equation 10.2–18 can also be expressed in terms of $X(\omega)$

Substitution for $X(e^{j\omega T})$ in Equation 10.2–13 using the result in Equation 10.2–18 yields the desired expression for the SNR in the form

$$\gamma_\infty = \left[\frac{T^2 N_0}{2\pi}\int_{-\pi/T}^{\pi/T}\frac{d\omega}{\sum_{n=-\infty}^{\infty}|H(\omega+2\pi n/T)|^2}\right]^{-1} \tag{10.2–19}$$

We observe that if the folded spectral characteristic of $H(\omega)$ possesses any zeros, the integrand becomes infinite and the SNR goes to zero. In other words, the performance of the equalizer is poor whenever the folded spectral characteristic possesses nulls or takes on small values. This behavior occurs primarily because the equalizer, in eliminating the intersymbol interference, enhances the additive noise. For example, if the channel contains a spectral null in its frequency response, the linear zero-forcing equalizer attempts to compensate for this by introducing an infinite gain at that frequency. But this compensates for the channel distortion at the expense of enhancing the additive noise. On the other hand, an ideal channel coupled with an appropriate signal design that results in no intersymbol interference will have a folded spectrum that satisfies the condition

$$\sum_{n=-\infty}^{\infty}\left|H\left(\omega+\frac{2\pi n}{T}\right)\right|^2 = T, \qquad |\omega| \leqslant \frac{\pi}{T} \tag{10.2–20}$$

In this case, the SNR achieves its maximum value, namely,

$$\gamma_\infty = \frac{1}{N_0} \tag{10.2–21}$$

Finite-length equalizer. Let us now turn our attention to an equalizer having $2K+1$ taps. Since $c_j = 0$ for $|j| > K$, the convolution of $\{f_n\}$ with $\{c_n\}$ is zero outside the range $-K \leqslant n \leqslant K+L-1$. That is, $q_n = 0$ for $n < -K$ and $n > K+L-1$. With q_0 normalized to unity, the peak distortion is

$$\mathcal{D}(\mathbf{c}) = \sum_{\substack{n=-K\\ n\neq 0}}^{K+L-1}|q_n| = \sum_{\substack{n=-K\\ n\neq 0}}^{K+L-1}\left|\sum_j c_j f_{n-j}\right| \tag{10.2–22}$$

Although the equalizer has $2K+1$ adjustable parameters, there are $2K+L$ nonzero values in the response $\{q_n\}$. Therefore, it is generally impossible to completely eliminate the intersymbol interference at the output of the equalizer. There is always some residual interference when the optimum coefficients are used. The problem is to minimize $\mathcal{D}(\mathbf{c})$ with respect to the coefficients $\{c_j\}$.

The peak distortion given by Equation 10.2–22 has been shown by Lucky (1965) to be a convex function of the coefficients $\{c_j\}$. That is, it possesses a global minimum and no local minima. Its minimization can be carried out numerically using, for example, the method of steepest descent. Little more can be said for the general solution to this minimization problem. However, for one special but important case, the solution for the minimization of $\mathcal{D}(\mathbf{c})$ is known. This is the case in which the distortion at the input to the equalizer, defined as

$$D_0 = \frac{1}{|f_0|} \sum_{n=1}^{L} |f_n| \tag{10.2–23}$$

is less than unity. This condition is equivalent to having the eye open prior to equalization. That is, the intersymbol interference is not severe enough to close the eye. Under this condition, the peak distortion $\mathcal{D}(\mathbf{c})$ is minimized by selecting the equalizer coefficients to force $q_n = 0$ for $1 \leqslant |n| \leqslant K$ and $q_0 = 1$. In other words, the general solution to the minimization of $\mathcal{D}(\mathbf{c})$, when $D_0 < 1$, is the zero-forcing solution for $\{q_n\}$ in the range $1 \leqslant |n| \leqslant K$. However, the values of $\{q_n\}$ for $K + 1 \leqslant n \leqslant K + L - 1$ are nonzero, in general. These nonzero values constitute the residual intersymbol interference at the output of the equalizer.

10.2.2 Mean-Square-Error (MSE) Criterion

In the MSE criterion, the tap weight coefficients $\{c_j\}$ of the equalizer are adjusted to minimize the mean square value of the error

$$\varepsilon_k = I_k - \hat{I}_k \tag{10.2–24}$$

where I_k is the information symbol transmitted in the kth signaling interval and $\hat{I}_k$ is the estimate of that symbol at the output of the equalizer, defined in Equation 10.2–1. When the information symbols $\{I_k\}$ are complex-valued, the performance index for the MSE criterion, denoted by J, is defined as

$$J = E|\varepsilon_k|^2 = E|I_k - \hat{I}_k|^2 \tag{10.2–25}$$

On the other hand, when the information symbols are real-valued, the performance index is simply the square of the real part of ε_k. In either case, J is a quadratic function of the equalizer coefficients $\{c_j\}$. In the following discussion, we consider the minimization of the complex-valued form given in Equation 10.2–25.

Infinite-length equalizer. First, we shall derive the tap weight coefficients that minimize J when the equalizer has an infinite number of taps. In this case, the estimate $\hat{I}_k$ is expressed as

$$\hat{I}_k = \sum_{j=-\infty}^{\infty} c_j v_{k-j} \tag{10.2–26}$$

Substitution of Equation 10.2–26 into the expression for J given in Equation 10.2–25 and expansion of the result yields a quadratic function of the coefficients $\{c_j\}$. This function can be easily minimized with respect to the $\{c_j\}$ to yield a set (infinite in number) of linear equations for the $\{c_j\}$. Alternatively, the set of linear equations can be obtained by invoking the orthogonality principle in mean square estimation. That is, we select the coefficients $\{c_j\}$ to render the error ε_k orthogonal to the signal sequence $\{v^*_{k-l}\}$ for $-\infty < l < \infty$. Thus,

$$E(\varepsilon_k v^*_{k-l}) = 0, \qquad -\infty < l < \infty \tag{10.2–27}$$

Substitution for ε_k in Equation 10.2–27 yields

$$E\left[\left(I_k - \sum_{j=-\infty}^{\infty} c_j v_{k-j}\right) v^*_{k-l}\right] = 0$$

or, equivalently,

$$\sum_{j=-\infty}^{\infty} c_j E(v_{k-j} v^*_{k-l}) = E(I_k v^*_{k-l}), \qquad -\infty < l < \infty \tag{10.2–28}$$

To evaluate the moments in Equation 10.2–28, we use the expression for v_k given in Equation 10.1–16. Thus, we obtain

$$\begin{aligned} E(v_{k-j} v^*_{k-l}) &= \sum_{n=0}^{L} f^*_n f_{n+l-j} + N_0 \delta_{lj} \\ &= \begin{cases} x_{l-j} + N_0 \delta_{ij} & (|l-j| \leqslant L) \\ 0 & (\text{otherwise}) \end{cases} \end{aligned} \tag{10.2–29}$$

and

$$E(I_k v^*_{k-l}) = \begin{cases} f^*_{-l} & (-L \leqslant l \leqslant 0) \\ 0 & (\text{otherwise}) \end{cases} \tag{10.2–30}$$

Now, if we substitute Equations 10.2–29 and 10.2–30 into Equation 10.2–28 and take the z transform of both sides of the resulting equation, we obtain

$$C(z)[F(z)F^*(z^{-1}) + N_0] = F^*(z^{-1}) \tag{10.2–31}$$

Therefore, the transfer function of the equalizer based on the MSE criterion is

$$C(z) = \frac{F^*(z^{-1})}{F(z)F^*(z^{-1}) + N_0} \tag{10.2–32}$$

When the noise-whitening filter is incorporated into $C(z)$, we obtain an equivalent equalizer having the transfer function

$$\begin{aligned} C'(z) &= \frac{1}{F(z)F^*(z^{-1}) + N_0} \\ &= \frac{1}{X(z) + N_0} \end{aligned} \tag{10.2–33}$$

We observe that the only difference between this expression for $C'(z)$ and the one based on the peak distortion criterion is the noise spectral density factor N_0 that appears in Equation 10.2–33. When N_0 is very small in comparison with the signal, the coefficients that minimize the peak distortion $\mathcal{D}(\mathbf{c})$ are approximately equal to the coefficients that minimize the MSE performance index J. That is, in the limit as $N_0 \to 0$, the two criteria yield the same solution for the tap weights. Consequently, when $N_0 = 0$, the minimization of the MSE results in complete elimination of the intersymbol interference. On the other hand, that is not the case when $N_0 \neq 0$. In general, when $N_0 \neq 0$, there is both residual intersymbol interference and additive noise at the output of the equalizer.

A measure of the residual intersymbol interference and additive noise is obtained by evaluating the minimum value of J, denoted by $J_{\min}$, when the transfer function $C(z)$ of the equalizer is given by Equation 10.2–32. Since $J = E|\varepsilon_k|^2 = E(\varepsilon_k I_k^*) - E(\varepsilon_k \hat{I}_k^*)$, and since $E(\varepsilon_k \hat{I}_k^*) = 0$ by virtue of the orthogonality conditions given in Equation 10.2–27, it follows that

$$\begin{aligned} J_{\min} &= E(\varepsilon_k I_k^*) \\ &= E|I_k|^2 - \sum_{j=-\infty}^{\infty} c_j E(v_{k-j} I_k^*) \\ &= 1 - \sum_{j=-\infty}^{\infty} c_j f_{-j} \end{aligned} \tag{10.2–34}$$

This particular form for $J_{\min}$ is not very informative. More insight on the performance of the equalizer as a function of the channel characteristics is obtained when the summation in Equation 10.2–34 is transformed into the frequency domain. This can be accomplished by first noting that the summation in Equation 10.2–34 is the convolution of $\{c_j\}$ with $\{f_j\}$, evaluated at a shift of zero. Thus, if $\{b_k\}$ denotes the convolution of these two sequences, the summation in Equation 10.2–34 is simply equal to b_0. Since the z transform of the sequence $\{b_k\}$ is

$$\begin{aligned} B(z) &= C(z)F(z) \\ &= \frac{F(z)F^*(z^{-1})}{F(z)F^*(z^{-1}) + N_0} \\ &= \frac{X(z)}{X(z) + N_0} \end{aligned} \tag{10.2–35}$$

the term b_0 is

$$\begin{aligned} b_0 &= \frac{1}{2\pi j} \oint \frac{B(z)}{z} dz \\ &= \frac{1}{2\pi j} \oint \frac{X(z)}{z[X(z) + N_0]} dz \end{aligned} \tag{10.2–36}$$

The contour integral in Equation 10.2–36 can be transformed into an equivalent line integral by the change of variable $z = e^{j\omega T}$. The result of this change of variable is

$$b_0 = \frac{T}{2\pi}\int_{-\pi/T}^{\pi/T} \frac{X(e^{j\omega T})}{X(e^{j\omega T}) + N_0} d\omega \tag{10.2–37}$$

Finally, substitution of the result in Equation 10.2–37 for the summation in Equation 10.2–34 yields the desired expression for the minimum MSE in the form

$$\begin{aligned} J_{\min} &= 1 - \frac{T}{2\pi}\int_{-\pi/T}^{\pi/T} \frac{X(e^{j\omega T})}{X(e^{j\omega T}) + N_0} d\omega \\ &= \frac{T}{2\pi}\int_{-\pi/T}^{\pi/T} \frac{N_0}{X(e^{j\omega T}) + N_0} d\omega \\ &= \frac{T}{2\pi}\int_{-\pi/T}^{\pi/T} \frac{N_0}{T^{-1}\sum_{n=-\infty}^{\infty}|H(\omega + 2\pi n/T)|^2 + N_0} d\omega \end{aligned} \tag{10.2–38}$$

In the absence of intersymbol interference, $X(e^{j\omega T}) = 1$ and, hence,

$$J_{\min} = \frac{N_0}{1 + N_0} \tag{10.2–39}$$

We observe that $0 \leqslant J_{\min} \leqslant 1$. Furthermore, the relationship between the output (normalized by the signal energy) SNR γ_∞ and $J_{\min}$ must be

$$\gamma_\infty = \frac{1 - J_{\min}}{J_{\min}} \tag{10.2–40}$$

More importantly, this relation between γ_∞ and $J_{\min}$ also holds when there is residual intersymbol interference in addition to the noise.

Finite-length equalizer. Let us now turn our attention to the case in which the transversal equalizer spans a finite time duration. The output of the equalizer in the kth signaling interval is

$$\hat{I}_k = \sum_{j=-K}^{K} c_j v_{k-j} \tag{10.2–41}$$

The MSE for the equalizer having $2K+1$ taps, denoted by $J(K)$, is

$$J(K) = E|I_k - \hat{I}_k|^2 = E\left|I_k - \sum_{j=-K}^{K} c_j v_{k-j}\right|^2 \tag{10.2–42}$$

Minimization of $J(K)$ with respect to the tap weights $\{c_j\}$ or, equivalently, forcing the error $\varepsilon_k = I_k - \hat{I}_k$ to be orthogonal to the signal samples v_{j-l}^*, $|l| \leqslant K$, yields the following set of simultaneous equations:

$$\sum_{j=-K}^{K} c_j \Gamma_{lj} = \xi_l, \qquad l = -K, \ldots, -1, 0, 1, \ldots, K \tag{10.2–43}$$

where

$$\Gamma_{lj} = \begin{cases} x_{l-j} + N_0\delta_{lj} & (|l-j| \leqslant L) \\ 0 & \text{(otherwise)} \end{cases} \tag{10.2–44}$$

and

$$\xi_l = \begin{cases} f_{-l}^* & (-L \leqslant l \leqslant 0) \\ 0 & \text{(otherwise)} \end{cases} \tag{10.2–45}$$

It is convenient to express the set of linear equations in matrix form. Thus,

$$\boldsymbol{\Gamma}\mathbf{C} = \boldsymbol{\xi} \tag{10.2–46}$$

where $\mathbf{C}$ denotes the column vector of $2K+1$ tap weight coefficients, $\boldsymbol{\Gamma}$ denotes the $(2K+1) \times (2K+1)$ Hermitian covariance matrix with elements Γ_{ij} and $\boldsymbol{\xi}$ is a $(2K+1)$-dimensional column vector with elements ξ_i. The solution of Equation 10.2–46 is

$$\mathbf{C}_{\text{opt}} = \boldsymbol{\Gamma}^{-1}\boldsymbol{\xi} \tag{10.2–47}$$

Thus, the solution for $\mathbf{C}_{\text{opt}}$ involves inverting the matrix $\boldsymbol{\Gamma}$. The optimum tap weight coefficients given by Equation 10.2–47 minimize the performance index $J(K)$, with the result that the minimum value of $J(K)$ is

$$\begin{aligned} J_{\min}(K) &= 1 - \sum_{j=-K}^{0} c_j f_{-j} \\ &= 1 - \boldsymbol{\xi}^{t*}\boldsymbol{\Gamma}^{-1}\boldsymbol{\xi} \end{aligned} \tag{10.2–48}$$

where $\boldsymbol{\xi}^t$ represents the transpose of the column vector $\boldsymbol{\xi}$. $J_{\min}(K)$ may be used in Equation 10.2–40 to compute the output SNR for the linear equalizer with $2K+1$ tap coefficients.

10.2.3 Performance Characteristics of the MSE Equalizer

In this section, we consider the performance characteristics of the linear equalizer that is optimized by using the MSE criterion. Both the minimum MSE and the probability of error are considered as performance measures for some specific channels. We begin by evaluating the minimum MSE $J_{\min}$ and the output SNR γ_∞ for two specific channels. Then, we consider the evaluation of the probability of error.

EXAMPLE 10.2–1. First, we consider an equivalent discrete-time channel model consisting of two components f_0 and f_1, which are normalized to $|f_0|^2 + |f_1|^2 = 1$. Then

$$F(z) = f_0 + f_1 z^{-1} \tag{10.2–49}$$

and

$$X(z) = f_0 f_1^* z + 1 + f_0^* f_1 z^{-1} \tag{10.2–50}$$

The corresponding frequency response is

$$\begin{aligned} X(e^{j\omega T}) &= f_0 f_1^* e^{j\omega T} + 1 + f_0^* f_1 e^{-j\omega T} \\ &= 1 + 2|f_0||f_1|\cos(\omega T + \theta) \end{aligned} \quad (10.2\text{–}51)$$

where θ is the angle of $f_0 f_1^*$. We note that this channel characteristic possesses a null at $\omega = \pi/T$ when $f_0 = f_1 = \sqrt{\frac{1}{2}}$.

A linear equalizer with an infinite number of taps, adjusted on the basis of the MSE criterion, will have the minimum MSE given by Equation 10.2–38. Evaluation of the integral in Equation 10.2–38 for the $X(e^{j\omega T})$ given in Equation 10.2–51 yields the result

$$\begin{aligned} J_{\min} &= \frac{N_0}{\sqrt{N_0^2 + 2N_0(|f_0|^2 + |f_1|^2) + (|f_0|^2 - |f_1|^2)^2}} \\ &= \frac{N_0}{\sqrt{N_0^2 + 2N_0 + (|f_0|^2 - |f_1|^2)^2}} \end{aligned} \quad (10.2\text{–}52)$$

Let us consider the special case in which $f_0 = f_1 = \sqrt{\frac{1}{2}}$. The minimum MSE is $J_{\min} = N_0/\sqrt{N_0^2 + 2N_0}$ and the corresponding output SNR is

$$\begin{aligned} \gamma_\infty &= \sqrt{1 + \frac{2}{N_0}} - 1 \\ &\approx \left(\frac{2}{N_0}\right)^{1/2}, \qquad N_0 \ll 1 \end{aligned} \quad (10.2\text{–}53)$$

This result should be compared with the output SNR of $1/N_0$ obtained in the case of no intersymbol interference. A significant loss in SNR occurs from this channel.

EXAMPLE 10.2–2. As a second example, we consider an exponentially decaying characteristic of the form

$$f_k = \sqrt{1 - a^2}\, a^k, \qquad k = 0, 1, \ldots$$

where $a < 1$. The Fourier transform of this sequence is

$$X(e^{j\omega T}) = \frac{1 - a^2}{1 + a^2 - 2a\cos\omega T} \quad (10.2\text{–}54)$$

which is a function that contains a minimum at $\omega = \pi/T$.

The output SNR for this channel is

$$\begin{aligned} \gamma_\infty &= \left(\sqrt{1 + 2N_0\frac{1 + a^2}{1 - a^2} + N_0^2} - 1\right)^{-1} \\ &\approx \frac{1 - a^2}{(1 + a^2)N_0}, \qquad N_0 \ll 1 \end{aligned} \quad (10.2\text{–}55)$$

Therefore, the loss in SNR due to the presence of the interference is

$$-10\log_{10}\left(\frac{1 - a^2}{1 + a^2}\right)$$

Probability of error performance of linear MSE equalizer. Above, we discussed the performance of the linear equalizer in terms of the minimum achievable MSE $J_{\min}$ and the output SNR γ that is related to $J_{\min}$ through the formula in Equation 10.2–40. Unfortunately, there is no simple relationship between these quantities and the probability of error. The reason is that the linear MSE equalizer contains some residual intersymbol interference at its output. This situation is unlike that of the infinitely long zero-forcing equalizer, for which there is no residual interference, but only Gaussian noise. The residual interference at the output of the MSE equalizer is not well characterized as an additional Gaussian noise term, and, hence, the output SNR does not translate easily into an equivalent error probability.

One approach to computing the error probability is a brute force method that yields an exact result. To illustrate this method, let us consider a PAM signal in which the information symbols are selected from the set of values $2n - M - 1, n = 1, 2, \ldots, M$, with equal probability. Now consider the decision on the symbol I_n. The estimate of I_n is

$$\hat{I}_n = q_0 I_n + \sum_{k \neq n} I_k q_{n-k} + \sum_{j=-K}^{K} c_j \eta_{n-j} \tag{10.2–56}$$

where $\{q_n\}$ represent the convolution of the impulse response of the equalizer and equivalent channel, i.e.,

$$q_n = \sum_{k=-K}^{K} c_k f_{n-k} \tag{10.2–57}$$

and the input signal to the equalizer is

$$v_k = \sum_{j=0}^{L} f_j I_{k-j} + \eta_k \tag{10.2–58}$$

The first term in the right-hand side of Equation 10.2–56 is the desired symbol, the middle term is the intersymbol interference, and the last term is the Gaussian noise. The variance of the noise is

$$\sigma_n^2 = N_0 \sum_{j=-K}^{K} c_j^2 \tag{10.2–59}$$

For an equalizer with $2K + 1$ taps and a channel response that spans $L + 1$ symbols, the number of symbols involved in the intersymbol interference is $2K + L$.

Define

$$\mathcal{D} = \sum_{k \neq n} I_k q_{n-k} \tag{10.2–60}$$

For a particular sequence of $2K + L$ information symbols, say the sequence $\mathbf{I}_J$, the intersymbol interference term $\mathcal{D} \equiv D_J$ is fixed. The probability of error for a fixed D_J is

$$\begin{aligned} P_M(D_J) &= 2\frac{M-1}{M} P(N + D_J > q_0) \\ &= \frac{2(M-1)}{M} Q\left(\sqrt{\frac{(q_0 - D_j)^2}{\sigma_n^2}}\right) \end{aligned} \tag{10.2–61}$$

where N denotes the additive noise term. The average probability of error is obtained by averaging $P_M(D_J)$ over all possible sequences $\mathbf{I}_J$. That is,

$$\begin{aligned} P_M &= \sum_{\mathbf{I}_J} P_M(D_J) P(\mathbf{I}_J) \\ &= \frac{2(M-1)}{M} \sum_{\mathbf{I}_J} Q\left(\sqrt{\frac{(q_0 - D_J)^2}{\sigma_n^2}}\right) P(\mathbf{I}_J) \end{aligned} \tag{10.2–62}$$

When all the sequences are equally likely,

$$P(\mathbf{I}_J) = \frac{1}{M^{2K+L}} \tag{10.2–63}$$

The conditional error probability terms $P_M(D_J)$ are dominated by the sequence that yields the largest value of D_J. This occurs when $I_n = \pm(M-1)$ and the signs of the information symbols match the signs of the corresponding $\{q_n\}$. Then,

$$D_J^* = (M-1) \sum_{k \neq 0} |q_k|$$

and

$$P_M(D_J^*) = \frac{2(M-1)}{M} Q\left(\sqrt{\frac{q_0^2}{\sigma_n^2}\left(1 - \frac{M-1}{q_0} \sum_{k \neq 0} |q_k|\right)^2}\right) \tag{10.2–64}$$

Thus, an upper bound on the average probability of error for equally likely symbol sequences is

$$P_M \leqslant P_M(D_J^*) \tag{10.2–65}$$

If the computation of the exact error probability in Equation 10.2–62 proves to be too cumbersome and too time consuming because of the large number of terms in the sum and if the upper bound is too loose, one can resort to one of a number of different approximate methods that have been devised, which are known to yield tight bounds on P_M. A discussion of these different approaches would take us too far afield. The interested reader is referred to the papers by Saltzberg (1968), Lugannani (1969), Ho and Yeh (1970), Shimbo and Celebiler (1971), Glave (1972), Yao (1972), and Yao and Tobin (1976).

As an illustration of the performance limitations of a linear equalizer in the presence of severe intersymbol interference, we show in Figure 10.2–4 the probability of error for binary (antipodal) signaling, as measured by Monte Carlo simulation, for the three discrete-time channel characteristics shown in Figure 10.2–5. For purposes of comparison, the performance obtained for a channel with no intersymbol interference is also illustrated in Figure 10.2–4. The equivalent discrete-time channel shown in Figure 10.2–5a is typical of the response of a good-quality telephone channel. In contrast, the equivalent discrete-time channel characteristics shown in Figure 10.2–5b and c result in severe intersymbol interference. The spectral characteristics $|X(e^{j\omega})|$ for the three channels, illustrated in Figure 10.2–6, clearly show that the channel in Figure 10.2–5c has the worst spectral characteristic. Hence the performance of the linear equalizer for this channel is the poorest of the three cases. Next in performance is the channel shown in Figure 10.2–5b, and finally, the best performance is obtained with the channel shown in Fig. 10.2–5a. In fact, the error rate of the latter is within 3 dB of the error rate achieved with no interference.

One conclusion reached from the results on output SNR γ_∞ and the limited probability of error results illustrated in Figure 10.2–4 is that a linear equalizer yields good performance on channels such as telephone lines, where the spectral

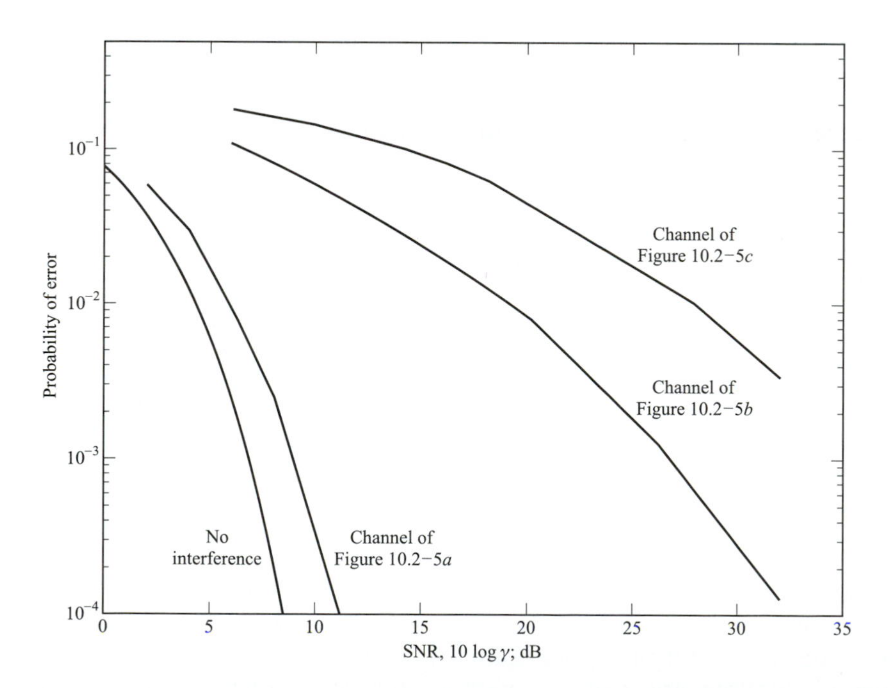

FIGURE 10.2–4
Error rate performance of linear MSE equalizer. Thirty-one taps in transversal equalizer. ($\gamma = \frac{1}{N_0}\sum_k |f_k|^2$).

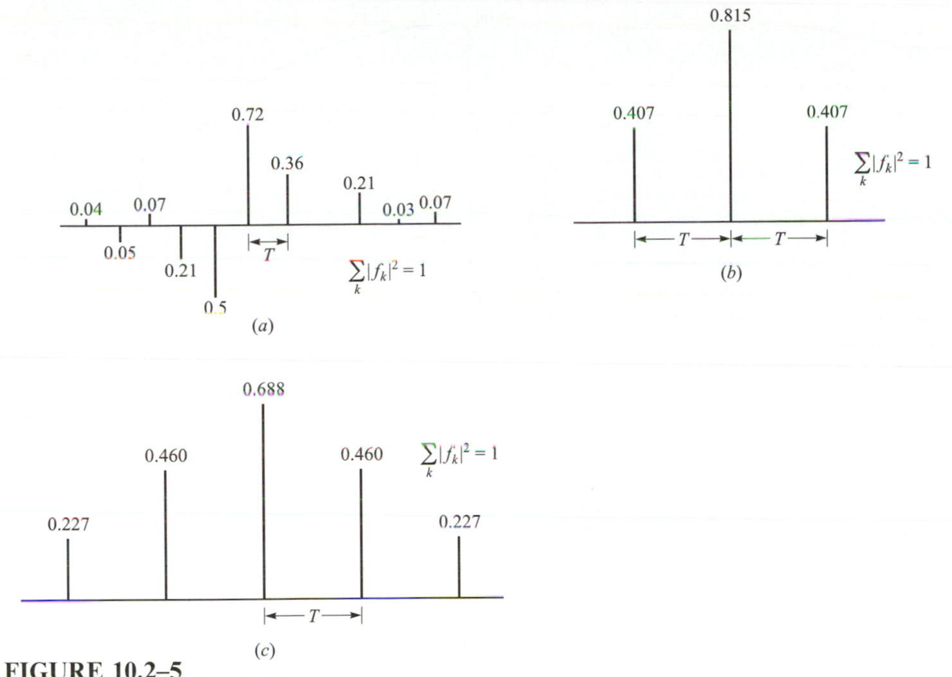

FIGURE 10.2–5
Three discrete-time channel characteristics

characteristics of the channels are well behaved and do not exhibit spectral nulls. On the other hand, a linear equalizer is inadequate as a compensator for the intersymbol interference on channels with spectral nulls, which may be encountered in radio transmission. In general, the channel spectral nulls result in a large noise enhancement at the output of the linear equalizer.

The basic limitation of the linear equalizer to cope with severe ISI has motivated a considerable amount of research into non-linear equalizers with low computational complexity. The decision-feedback equalizer described in Section 10.3 is shown to be an effective solution to this problem.

10.2.4 Fractionally Spaced Equalizers

In the linear equalizer structures that we have described in the previous section, the equalizer taps are spaced at the reciprocal of the symbol rate, i.e., at the reciprocal of the signaling rate $1/T$. This tap spacing is optimum if the equalizer is preceded by a filter matched to the channel distorted transmitted pulse. When the channel characteristics are unknown, the receiver filter is sometimes matched to the transmitted signal pulse and the sampling time is optimized for this suboptimum filter. In general, this approach leads to an equalizer performance that is very sensitive to the choice of sampling time.

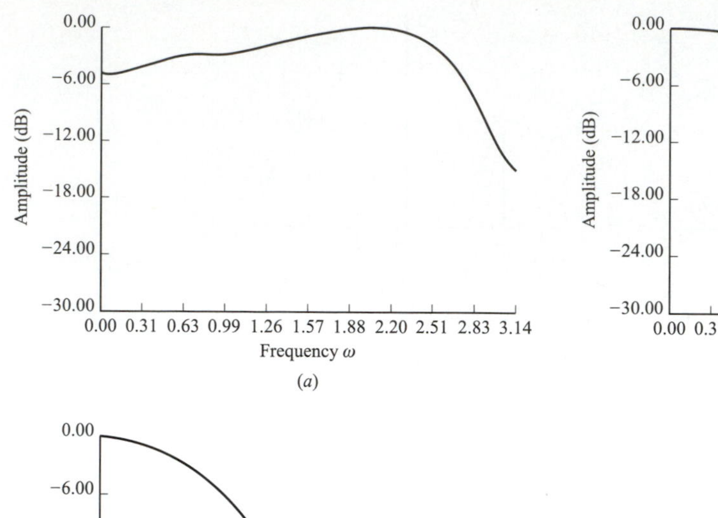

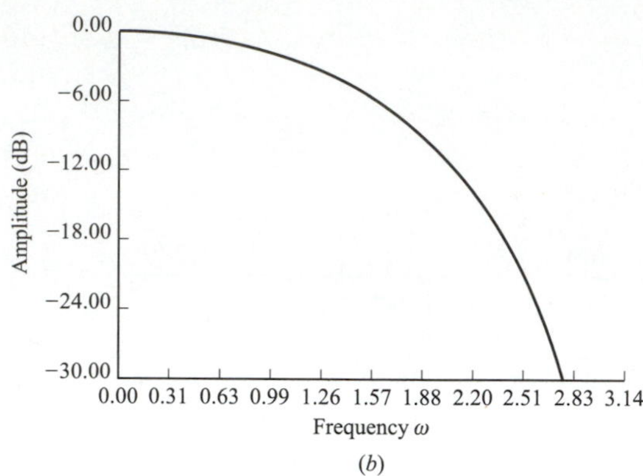

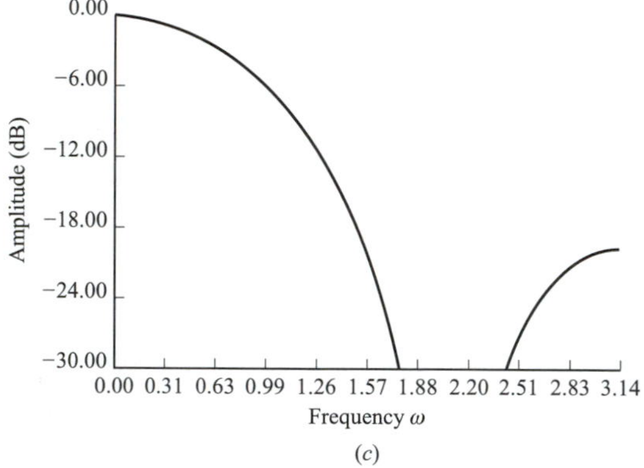

FIGURE 10.2–6
Amplitude spectra for the channels shown in Figure 10.2–5a, b, and c, respectively.

The limitations of the symbol rate equalizer are most easily evident in the frequency domain. From Equation 9.2–5, the spectrum of the signal at the input to the equalizer may be expressed as

$$Y_T(f) = \frac{1}{T}\sum_n X\left(f - \frac{n}{T}\right)e^{j2\pi(f-n/T)\tau_0} \tag{10.2–66}$$

where $Y_T(f)$ is the folded or aliased spectrum, where the folding frequency is $1/2T$. Note that the received signal spectrum is dependent on the choice of the sampling delay τ_0. The signal spectrum at the output of the equalizer is $C_T(f)Y_T(f)$, where

$$C_T(f) = \sum_{k=-K}^{K} c_k e^{-j2\pi fkT} \tag{10.2–67}$$

It is clear from these relationships that the symbol rate equalizer can only compensate for the frequency-response characteristics of the aliased received signal. It cannot compensate for the channel distortion inherent in $X(f)e^{j2\pi f\tau_0}$.

In contrast to the symbol rate equalizer, a *fractionally spaced equalizer* (FSE) is based on sampling the incoming signal at least as fast as the Nyquist rate. For example, if the transmitted signal consists of pulses having a raised cosine spectrum with a roll-off factor β, its spectrum extends to $F_{\max} = (1+\beta)/2T$. This signal can be sampled at the receiver at a rate

$$2F_{\max} = \frac{1+\beta}{T} \tag{10.2–68}$$

and then passed through an equalizer with tap spacing of $T/(1+\beta)$. For example, if $\beta = 1$, we would have a $\frac{1}{2}T$-spaced equalizer. If $\beta = 0.5$, we would have a $\frac{2}{3}T$-spaced equalizer, and so forth. In general, then, a digitally implemented fractionally spaced equalizer has tap spacing of MT/N where M and N are integers and $N > M$. Usually, a $\frac{1}{2}T$-spaced equalizer is used in many applications.

Since the frequency response of the FSE is

$$C_{T'}(f) = \sum_{k=-K}^{K} c_k e^{-j2\pi f k T'} \tag{10.2–69}$$

where $T' = MT/N$, it follows that $C_{T'}(f)$ can equalize the received signal spectrum beyond the Nyquist frequency $f = 1/2T$ to $f = (1+\beta)/2T = N/2MT$. The equalized spectrum is

$$\begin{aligned} C_{T'}(f)Y_{T'}(f) &= C_{T'}(f)\sum_n X\left(f - \frac{n}{T'}\right)e^{j2\pi(f-n/T')\tau_0} \\ &= C_{T'}(f)\sum_n X\left(f - \frac{nN}{MT}\right)e^{j2\pi(f-nN/MT)\tau_0} \end{aligned} \tag{10.2–70}$$

Since $X(f) = 0$ for $|f| > N/2MT$, Equation 10.2–70 may be expressed as

$$C_{T'}(f)Y_{T'}(f) = C_{T'}(f)X(f)e^{j2\pi f\tau_0}, \qquad |f| \leqslant \frac{1}{2T'} \tag{10.2–71}$$

Thus, we observe that the FSE compensates for the channel distortion in the received signal before the aliasing effects due to symbol rate sampling. In other words, $C_{T'}(f)$ can compensate for an arbitrary timing phase.

The FSE output is sampled at the symbol rate $1/T$ and has the spectrum

$$\sum_k C_{T'}\left(f - \frac{k}{T}\right)X\left(f - \frac{k}{T}\right)e^{j2\pi(f-k/T)\tau_0} \tag{10.2–72}$$

In effect, the optimum FSE is equivalent to the optimum linear receiver consisting of the matched filter followed by a symbol rate equalizer.

Let us now consider the adjustment of the tap coefficients in the FSE. The input to the FSE may be expressed as

$$y\left(\frac{kMT}{N}\right) = \sum_n I_n x\left(\frac{kMT}{N} - nT\right) + v\left(\frac{kMT}{N}\right) \tag{10.2–73}$$

In each symbol interval, the FSE produces an output of the form

$$\hat{I}_k = \sum_{n=-K}^{K} c_n y\left(kT - \frac{nMT}{N}\right) \tag{10.2–74}$$

where the coefficients of the equalizer are selected to minimize the MSE. This optimization leads to a set of linear equations for the equalizer coefficients that have the solution

$$\mathbf{C}_{\text{opt}} = \mathbf{A}^{-1}\boldsymbol{\alpha} \tag{10.2–75}$$

where $\mathbf{A}$ is the covariance matrix of the input data and $\boldsymbol{\alpha}$ is the vector of cross correlations. These equations are identical in form to those for the symbol rate equalizer, but there are some subtle differences. One is that $\mathbf{A}$ is Hermitian, but not Toeplitz. In addition, $\mathbf{A}$ exhibits periodicities that are inherent in a cyclostationary process, as shown by Qureshi (1985). As a result of the fractional spacing, some of the eigenvalues of $\mathbf{A}$ are nearly zero. Attempts have been made by Long et al. (1988a,b) to exploit this property in the coefficient adjustment.

An analysis of the performance of fractionally spaced equalizers, including their convergence properties, is given in a paper by Ungerboeck (1976). Simulation results demonstrating the effectiveness of the FSE over a symbol rate equalizer have also been given in the papers by Qureshi and Forney (1977) and Gitlin and Weinstein (1981). We cite two examples from these papers. First, Figure 10.2–7 illustrates the performance of the symbol rate equalizer and a $\frac{1}{2}T$-FSE for a channel with high-end amplitude distortion, whose characteristics are also shown in this figure. The symbol-spaced equalizer was preceded with a filter matched to the transmitted pulse that had a (square-root) raised cosine spectrum with a 20 percent roll-off ($\beta = 0.2$). The FSE did not have any filter preceding it. The symbol rate was 2400 symbols/s and the modulation was QAM. The received SNR was 30 dB. Both equalizers had 31 taps; hence, the $\frac{1}{2}T$-FSE spanned one-half of the time interval of the symbol rate equalizer. Nevertheless, the FSE outperformed the symbol rate equalizer when the latter was optimized at the best sampling time. Furthermore, the FSE did not exhibit any sensitivity to timing phase, as illustrated in Figure 10.2–7b.

Similar results were obtained by Gitlin and Weinstein. For a channel with poor envelope delay characteristics, the SNR performance of the symbol rate equalizer and a $\frac{1}{2}T$-FSE are illustrated in Figure 10.2–8. In this case, both equalizers had the same time span. The T-spaced equalizer had 24 taps while the FSE had 48 taps. The symbol rate was 2400 symbols/s and the data rate was 9600 bits/s with 16-QAM modulation. The signal pulse had a raised cosine spectrum with $\beta = 0.12$. Note again that the FSE outperformed the T-spaced equalizer by several decibels, even when the latter was adjusted for optimum sampling. The results in these two papers clearly demonstrate the superior performance achieved with a fractionally spaced equalizer.

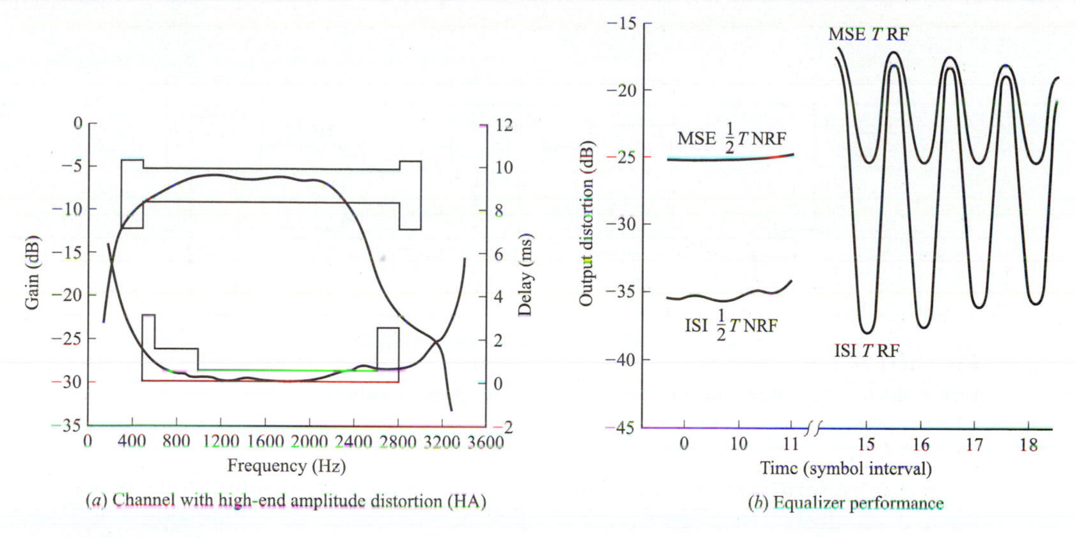

FIGURE 10.2–7
T and $\frac{1}{2}T$ equalizer performance as a function of timing phase for 2400 symbols per second. (NRF indicates no receiver filter.) [*From Qureshi and Forney (1977).* © *1977 IEEE.*]

10.2.5 Baseband and Passband Linear Equalizers

The linear equalizer treated above was described in terms of equivalent low-pass signals. However, in a practical implementation, the linear equalizer shown in Figure 10.2–1 can be realized either at baseband or at bandpass. For example, Figure 10.2–9 illustrates the demodulation of QAM or multiphase PSK by first translating the signal to baseband and equalizing the baseband signal with an equalizer having complex-valued coefficients. In effect, the equalizer with a complex-valued (in-phase and quadrature components) input is equivalent to four parallel equalizers with real-valued tap coefficients as shown in Figure 10.2–10. We generally refer to the equalizer in Figure 10.2–9 as a complex-valued *baseband equalizer*.

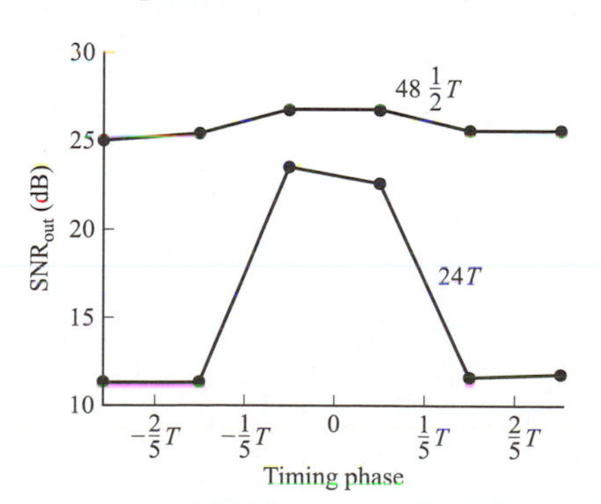

FIGURE 10.2–8
Performance of T and $\frac{1}{2}T$ equalizers as a function of timing phase for 2400 symbols/s 16-QAM on a channel with poor envelope delay. [*From Gitlin and Weinstein (1981). Reprinted with permission from* Bell System Technical Journal. © *1981 AT & T.*]

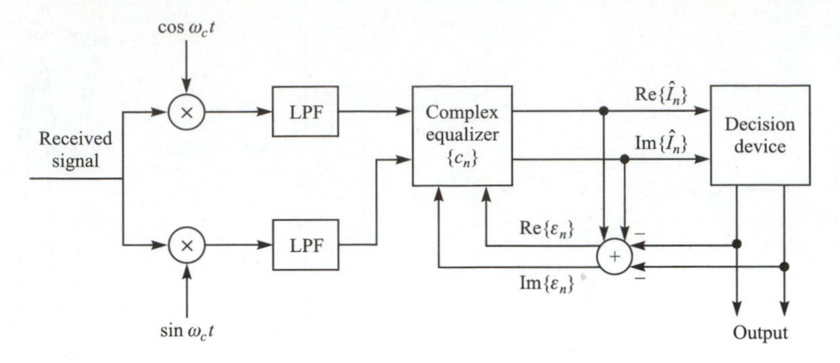

FIGURE 10.2–9
QAM and PSK signal demodulator with baseband equalizer.

As an alternative, we may equalize the signal at passband. This is accomplished as shown in Figure 10.2–11 for two-dimensional signal constellations such as QAM and PSK. The received signal is filtered and, in parallel, it is passed through a Hilbert transformer, called a *phase-splitting filter*. Thus, we have the equivalent of in-phase and quadrature components at passband, which are fed to a passband complex equalizer. We may call this equalizer structure a complex-valued *passband equalizer*. Following the equalization, the signal is down-converted to baseband and detected.

The complex-valued baseband equalizer may be implemented either as a symbol rate equalizer (SRE) or as a fractionally spaced equalizer (FSE), with the latter being preferable in view of its insensitivity to the sampling phase within a symbol interval.

The complex-valued passband equalizer must be an FSE, with samples of the received signal taken at some multiple of the symbol rate that exceeds the Nyquist rate.

An alternative passband FSE to the structure shown in Figure 10.2–11 is illustrated in Figure 10.2–12. In this FSE, real-valued samples of the received signal are taken at the Nyquist rate or faster and equalized at bandpass by a linear equalizer that has complex-valued coefficients. We note that this equalizer structure does not explicitly implement a Hilbert transformer to perform phase

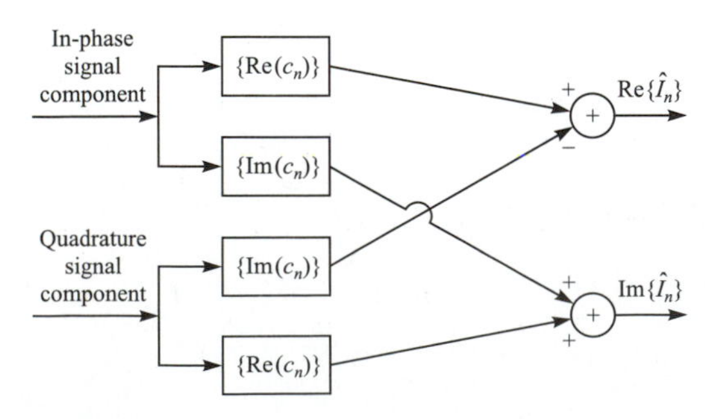

FIGURE 10.2–10
Complex-valued baseband equalizer for QAM and PSK signals.

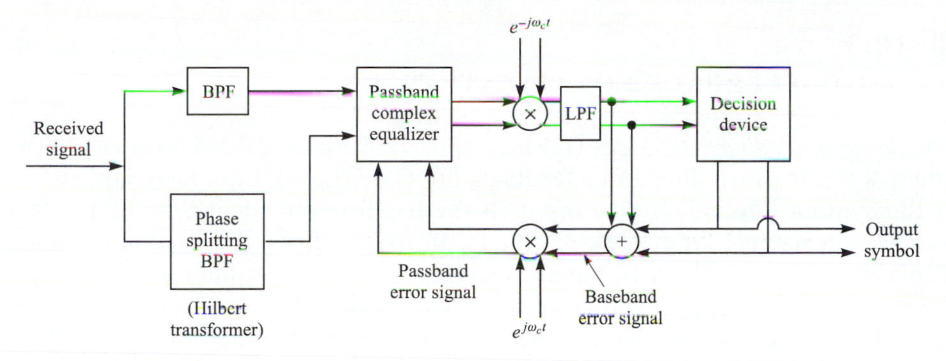

FIGURE 10.2–11
QAM or PSK signal equalization at passband.

splitting. Instead, the phase-splitting function is embedded in the equalizer coefficients and, thus, the Hilbert transform is avoided. This alternative passband FSE structure in Figure 10.2–12 has been called a *phase-splitting FSE* (PS-FSE). Its properties and its performance has been investigated by Mueller and Werner (1982), Im and Un (1987), and Ling and Qureshi (1990).

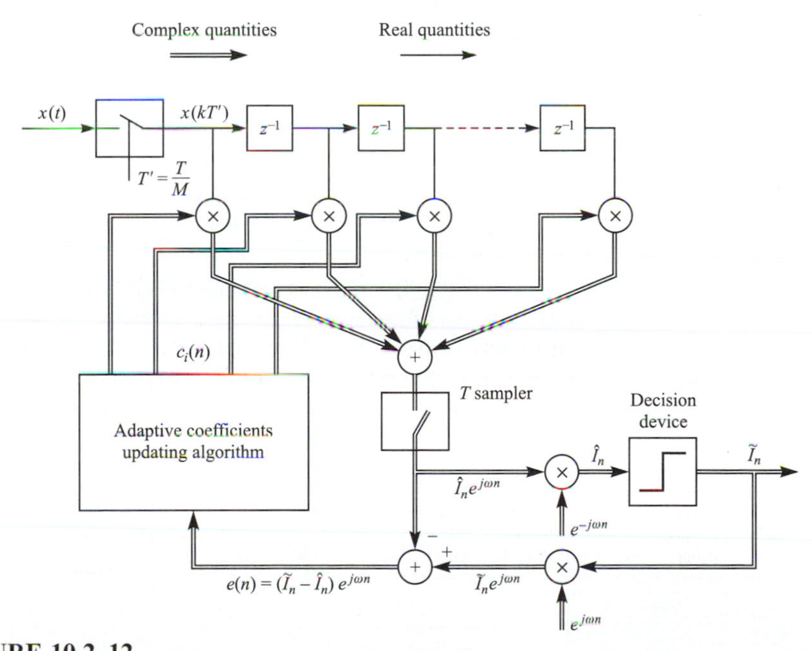

FIGURE 10.2–12
Structure of a phase-splitting fractionally spaced equalizer. [*From Ling and Qureshi (1990); © 1990 IEEE.*]

10.3 DECISION-FEEDBACK EQUALIZATION

The *decision-feedback equalizer* (DFE), depicted in Figure 10.3–1, consists of two filters, a feedforward filter and a feedback filter. As shown, both have taps spaced at the symbol interval T. The input to the feedforward section is the received signal sequence $\{v_k\}$. In this respect, the feedforward filter is identical to the linear transversal equalizer described in Section 10.2. The feedback filter has as its input the sequence of decisions on previously detected symbols. Functionally, the feedback filter is used to remove that part of the intersymbol interference from the present estimate caused by previously detected symbols.

In the case where the feedforward and feedback filters have infinite duration impulse respones, Price (1972) showed that the optimum feedforward filter in a zero-forcing DFE is the whitened matched filter described previously. In such a case, the coefficients of the feedback filter are simply related to the coefficients of $F(z)$.

Below, we apply the MSE criterion to optimize the coefficients of the two filters, with each filter having a finite number of coefficients.

10.3.1 Coefficient Optimization

From the description given above, it follows that the equalizer output can be expressed as

$$\hat{I}_k = \sum_{j=-K_1}^{0} c_j v_{k-j} + \sum_{j=1}^{K_2} c_j \tilde{I}_{k-j} \qquad (10.3\text{–}1)$$

where $\hat{I}_k$ is an estimate of the kth information symbol, $\{c_j\}$ are the tap coefficients of the filter, and $\{\tilde{I}_{k-1}, \ldots, \tilde{I}_{k-K_2}\}$ are previously detected symbols. The equalizer is assumed to have $(K_1 + 1)$ taps in its feedforward section and K_2 in its feedback section. It should be observed that this equalizer is non-linear because the feedback filter contains previously detected symbols $\{\tilde{I}_k\}$.

Both the peak distortion criterion and the MSE criterion result in a mathematically tractable optimization of the equalizer coefficients, as can be concluded from the papers by George et al. (1971), Price (1972), Salz (1973), and

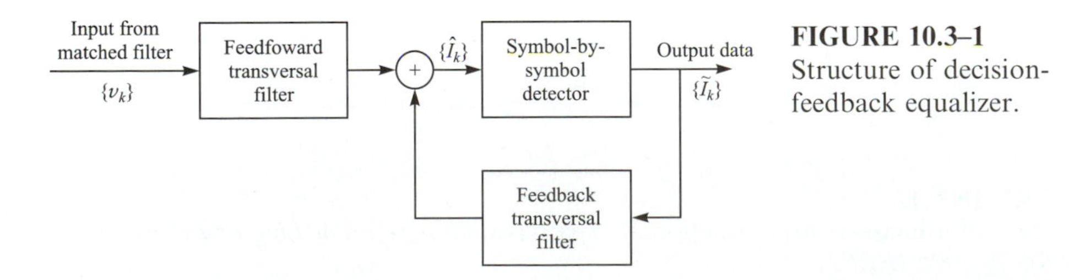

FIGURE 10.3–1
Structure of decision-feedback equalizer.

Proakis (1975). Since the MSE criterion is more prevalent in practice, we focus our attention on it. Based on the assumption that previously detected symbols in the feedback filter are correct, the minimization of MSE

$$J(K_1, K_2) = E|I_k - \hat{I}_k|^2 \tag{10.3–2}$$

leads to the following set of linear equations for the coefficients of the feedforward filter:

$$\sum_{j=-K_1}^{0} \psi_{lj} c_j = f_{-l}^*, \qquad l = -K_1, \ldots, -1, 0 \tag{10.3–3}$$

where

$$\psi_{lj} = \sum_{m=0}^{-l} f_m^* f_{m+l-j} + N_0 \delta_{lj}, \qquad l, j = -K_1, \ldots, -1, 0 \tag{10.3–4}$$

The coefficients of the feedback filter of the equalizer are given in terms of the coefficients of the feedforward section by the following expression:

$$c_k = -\sum_{j=-K_1}^{0} c_j f_{k-j}, \qquad k = 1, 2, \ldots, K_2 \tag{10.3–5}$$

The values of the feedback coefficients result in complete elimination of intersymbol interference from previously detected symbols, provided that previous decisions are correct and that $K_2 \geqslant L$ (see Problem 10.20).

10.3.2 Performance Characteristics of DFE

We now turn our attention to the performance achieved with decision-feedback equalization. The exact evaluation of the performance is complicated to some extent by occasional incorrect decisions made by the detector, which then propagate down the feedback section. In the absence of decision errors, the minimum MSE is given as

$$J_{\min}(K_1) = 1 - \sum_{j=-K_1}^{0} c_j f_{-j} \tag{10.3–6}$$

By going to the limit ($K_1 \to \infty$) of an infinite number of taps in the feedforward filter, we obtain the smallest achievable MSE, denoted as $J_{\min}$. With some effort $J_{\min}$ can be expressed in terms of the spectral characteristics of the channel and additive noise, as shown by Salz (1973). This more desirable form for $J_{\min}$ is

$$J_{\min} = \exp\left\{\frac{T}{2\pi}\int_{-\pi/T}^{\pi/T} \ln\left[\frac{N_0}{X(e^{j\omega T}) + N_0}\right] d\omega\right\} \tag{10.3–7}$$

The corresponding output SNR is

$$\gamma_\infty = \frac{1 - J_{\min}}{J_{\min}} = -1 + \exp\left\{\frac{T}{2\pi}\int_{-\pi/T}^{\pi/T} \ln\left[\frac{N_0 + X(e^{j\omega T})}{N_0}\right] d\omega\right\} \tag{10.3–8}$$

We observe again, that, in the absence of intersymbol interference, $X(e^{j\omega T}) = 1$, and hence, $J_{\min} = N_0/(1 + N_0)$. The corresponding output SNR is $\gamma_\infty = 1/N_0$.

EXAMPLE 10.3–1. It is interesting to compare the value of $J_{\min}$ for the decision-feedback equalizer with the value of $J_{\min}$ obtained with the linear MSE equalizer. For example, let us consider the discrete-time equivalent channel consisting of two taps f_0 and f_1. The minimum MSE for this channel is

$$\begin{aligned} J_{\min} &= \exp\left\{\frac{T}{2\pi}\int_{-\pi/T}^{\pi/T} \ln\left[\frac{N_0}{1 + N_0 + 2|f_0||f_1|\cos(\omega T + \theta)}\right] d\omega\right\} \\ &= N_0 \exp\left[-\frac{1}{2\pi}\int_{-\pi}^{\pi} \ln(1 + N_0 + 2|f_0||f_1|\cos\omega)\, d\omega\right] \\ &= \frac{2N_0}{1 + N_0 + \sqrt{(1 + N_0)^2 - 4|f_0 f_1|^2}} \end{aligned} \tag{10.3–9}$$

Note that $J_{\min}$ is maximized when $|f_0| = |f_1| = \sqrt{\frac{1}{2}}$. Then

$$J_{\min} = \frac{2N_0}{1 + N_0 + \sqrt{(1 + N_0)^2 - 1}} \approx 2N_0, \qquad N_0 \ll 1 \tag{10.3–10}$$

The corresponding output SNR is

$$\gamma_\infty \approx \frac{1}{2N_0}, \qquad N_0 \ll 1 \tag{10.3–11}$$

Therefore, there is a 3-dB degradation in output SNR due to the presence of intersymbol interference. In comparison, the performance loss for the linear equalizer is very severe. Its output SNR as given by Equalizer 10.2–53 is $\gamma_\infty \approx (2/N_0)^{1/2}$ for $N_0 \ll 1$.

EXAMPLE 10.3–2. Consider the exponentially decaying channel characteristic of the form

$$f_k = (1 - a^2)^{1/2} a^k, \qquad k = 0, 1, 2, \ldots \tag{10.3–12}$$

where $a < 1$. The output SNR of the decision-feedback equalizer is

$$\begin{aligned}
\gamma_\infty &= -1 + \exp\left\{\frac{1}{2\pi}\int_{-\pi}^{\pi} \ln\left[\frac{1+a^2+(1-a^2)/N_0 - 2a\cos\omega}{1+a^2-2a\cos\omega}\right]d\omega\right\} \\
&= -1 + \frac{1}{2N_0}\left\{1-a^2+N_0(1+a^2)+\sqrt{[1-a^2+N_0(1+a^2)]^2-4a^2N_0^2}\right\} \\
&\approx \frac{(1-a^2)[1+N_0(1+a^2)/(1-a^2)]-N_0}{N_0} \\
&\approx \frac{1-a^2}{N_0}, \qquad N_0 \ll 1 \qquad\qquad (10.3\text{–}13)
\end{aligned}$$

Thus, the loss in SNR is $10\log_{10}(1-a^2)$ dB. In comparison, the linear equalizer has a loss of $10\log_{10}[(1-a^2)/(1+a^2)]$ dB.

These results illustrate the superiority of the decision-feedback equalizer over the linear equalizer when the effect of decision errors on performance is neglected. It is apparent that a considerable gain in performance can be achieved relative to the linear equalizer by the inclusion of the decision-feedback section, which eliminates the intersymbol interference from previously detected symbols.

One method of assessing the effect of decision errors on the error rate performance of the decision-feedback equalizer is Monte Carlo simulation on a digital computer. For purposes of illustration, we offer the following results for binary PAM signaling through the equivalent discrete-time channel models shown in Figure 10.2–5b and c.

The results of the simulation are displayed in Figure 10.3–2. First of all, a comparison of these results with those presented in Figure 10.2–4 leads us to conclude that the decision-feedback equalizer yields a significant improvement in performance relative to the linear equalizer having the same number of taps. Second, these results indicate that there is still a significant degradation in performance of the decision-feedback equalizer due to the residual intersymbol interference, especially on channels with severe distortion such as the one shown in Figure 10.2–5c. Finally, the performance loss due to incorrect decisions being fed back is 2 dB, approximately, for the channel responses under consideration. Additional results on the probability of error for a decision-feedback equalizer with error propagation may be found in the papers by Duttweiler et al. (1974) and Beaulieu (1994).

The structure of the DFE that is analyzed above employs a T-spaced filter for the feedforward section. The optimality of such a structure is based on the assumption that the analog filter preceding the DFE is matched to the channel-corrupted pulse response and its output is sampled at the optimum time instant. In practice, the channel response is not known a priori, so it is not possible to design an ideal matched filter. In view of this difficulty, it is customary in practical applications to use a fractionally spaced feedforward filter. Of course, the feedback filter tap spacing remains at T. The use of the FSE for the feedforward filter eliminates the system sensitivity to a timing error.

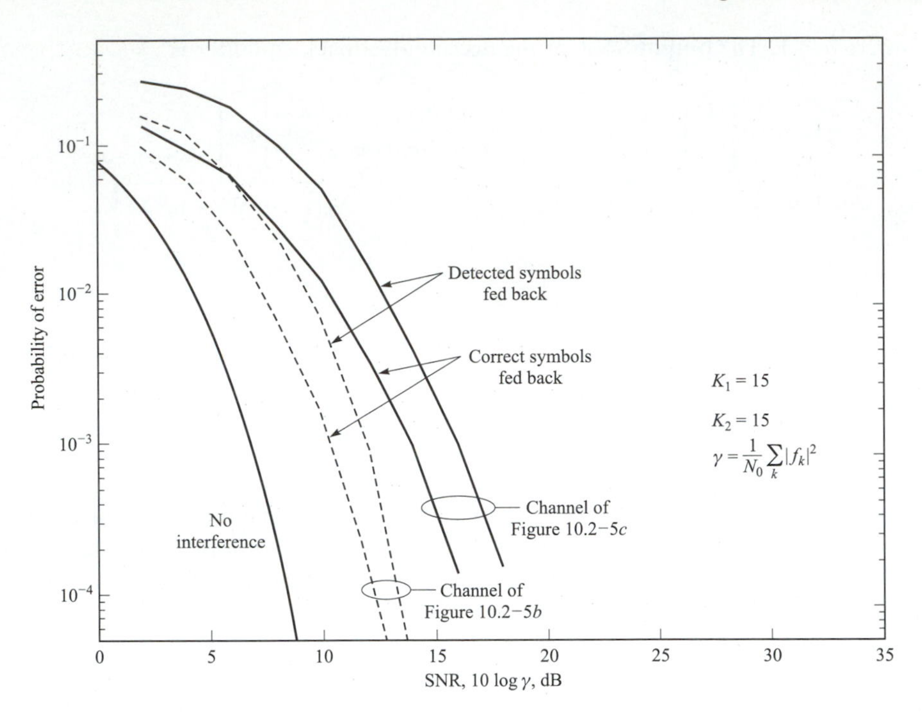

FIGURE 10.3–2
Performance of decision-feedback equalizer with and without error propagation.

Performance comparison with the MLSE. We conclude this subsection on the performance of the DFE by comparing its performance against that of the MLSE. For the two-path channel with $f_0 = f_1 = \sqrt{\frac{1}{2}}$, we have shown that the MLSE suffers no SNR loss while the decision-feedback equalizer suffers a 3-dB loss. On channels with more distortion, the SNR advantage of the MLSE over decision-feedback equalization is even greater. Figure 10.3–3 illustrates a comparison of the error rate performance of these two equalization techniques, obtained via Monte Carlo simulation, for binary PAM and the channel characteristics shown in Figure 10.2–5b and c. The error rate curves for the two methods have different slopes; hence the difference in SNR increases as the error probability decreases. As a benchmark, the error rate for the AWGN channel with no intersymbol interference is also shown in Figure 10.3–3.

10.3.3 Predictive Decision-Feedback Equalizer

Belfiore and Park (1979) proposed another DFE structure that is equivalent to the one shown in Figure 10.3–1 under the condition that the feedforward filter has an infinite number of taps. This structure consists of an FSE as a feedforward filter and a linear predictor as a feedback filter, as shown in the configura-

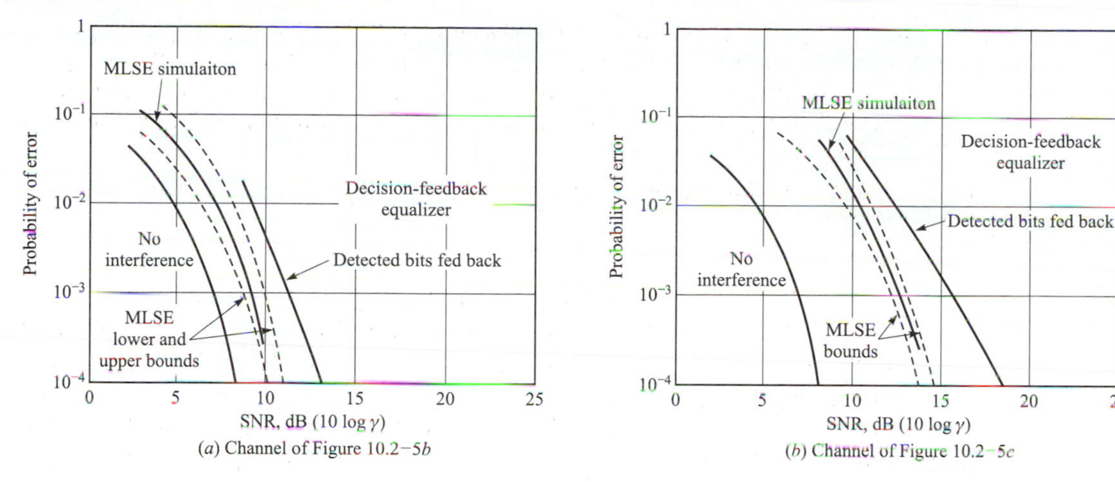

FIGURE 10.3–3
Comparison of performance between MLSE and decision-feedback equalization for channel characteristics shown (a) in Figure 10.2–5 b and (b) in Figure 10.2–5c.

tion given in Figure 10.3–4. Let us briefly consider the performance characteristics of this equalizer.

First of all, the noise at the output of the infinite length feedforward filter has the power spectral density

$$\frac{N_0 X(e^{j\omega T})}{|N_0 + X(e^{j\omega T})|^2}, \qquad |\omega| \leqslant \frac{\pi}{T} \tag{10.3–14}$$

The residual intersymbol interference has the power spectral density

$$\left|1 - \frac{X(e^{j\omega T})}{N_0 + X(e^{j\omega T})}\right|^2 = \frac{N_0^2}{|N_0 + X(e^{j\omega T})|^2}, \qquad |\omega| \leqslant \frac{\pi}{T} \tag{10.3–15}$$

The sum of these two spectra represents the power spectral density of the total noise and intersymbol interference at the output of the feedforward filter. Thus, on adding Equations 10.3–14 and 10.3–15, we obtain

$$|E_t(\omega)|^2 = \frac{N_0}{N_0 + X(e^{j\omega T})}, \qquad |\omega| \leqslant \frac{\pi}{T} \tag{10.3–16}$$

As we have observed previously, if $X(e^{j\omega T}) = 1$, the channel is ideal and, hence, it is not possible to reduce the MSE any further. On the other hand, if there is channel distortion, the power in the error sequence at the output of the feedforward filter can be reduced by means of linear prediction based on past values of the error sequence.

If $\mathcal{B}(\omega)$ represents the frequency response of the infinite length feedback predictor, i.e.,

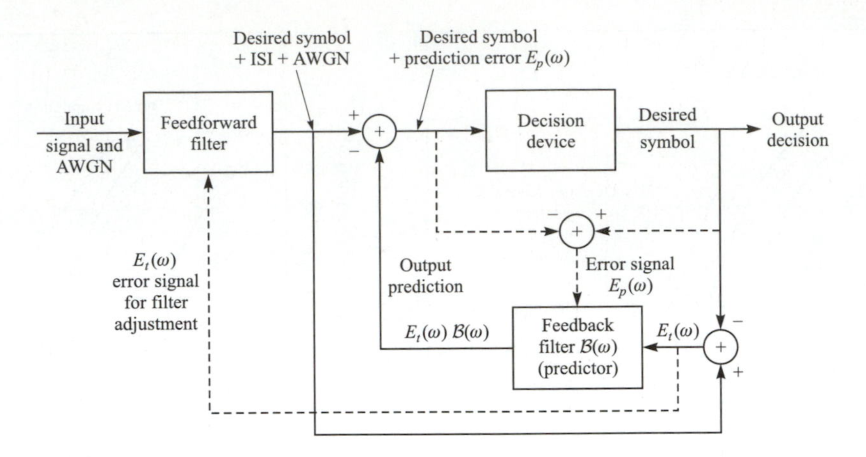

FIGURE 10.3–4
Block diagram of predictive DFE.

$$\mathcal{B}(\omega) = \sum_{n=1}^{\infty} b_n e^{-j\omega nT} \tag{10.3–17}$$

then the error at the output of the predictor is

$$E_p(\omega) = E_t(\omega) - E_t(\omega)\mathcal{B}(\omega) = E_t(\omega)[1 - \mathcal{B}(\omega)] \tag{10.3–18}$$

The minimization of the mean square value of this error, i.e.,

$$J = \frac{1}{2\pi}\int_{-\pi/T}^{\pi/T} |1 - \mathcal{B}(\omega)|^2 |E_t(\omega)|^2 d\omega \tag{10.3–19}$$

over the predictor coefficients $\{b_n\}$ yields the optimum predictor in the form

$$\mathcal{B}(\omega) = 1 - \frac{G(\omega)}{g_0} \tag{10.3–20}$$

where $G(\omega)$ is the solution to the spectral factorization

$$G(\omega)G^*(-\omega) = \frac{1}{|E_t(\omega)|^2} \tag{10.3–21}$$

and

$$G(\omega) = \sum_{n=0}^{\infty} g_n e^{-j\omega nT} \tag{10.3–22}$$

The output of the infinite length linear predictor is a white noise sequence with power spectral density $1/g_0^2$ and the corresponding minimum MSE is given by Equation 10.3–7. Therefore, the MSE performance of the infinite length predictive DFE is identical to the conventional DFE.

Although these two DFE structures result in equivalent performance if their lengths are infinite, the predictive DFE is suboptimum if the lengths of the two filters are finite. The reason for the optimality of the conventional DFE is relatively simple. The optimization of its tap coefficients in the feedforward and feedback filters is done jointly. Hence, it yields the minimum MSE. On the other hand, the optimizations of the feedforward filter and the feedback predictor in the predictive DFE are done separately. Hence, its MSE is at least as large as that of the conventional DFE. In spite of this suboptimality of the predictive DFE, it is suitable as an equalizer for trellis-coded signals, where the conventional DFE is not as suitable, as described in the next chapter. Finally, we should indicate that the feedforward filter in the predictive DFE is usually implemented as a fractionally spaced filter to avoid sensitivity to the sampling phase.

10.3.4 Equalization at the Transmitter—Tomlinson–Harashima Precoding

If the channel response is known to the transmitter, the equalizer can be placed at the transmitter end of the communication system. Thus, the noise enhancement that is generally inherent when the equalizer (linear or DFE) is placed at the receiver is avoided. In practice, however, channel characteristics generally vary with time, so it is cumbersome to place the entire equalizer at the transmitter.

In wireline channels, the channel characteristics do not vary significantly with time. Therefore, it is possible to place the feedback filter of the DFE at the transmitter and the feedforward filter at the receiver. This approach has the advantage that the problem of error propagation due to incorrect decisions in the feedback filter is completely eliminated. Thus, the tail (postcursors) in the channel response is cancelled without any penalty in the SNR. The linear fractionally spaced feedforward part of the DFE, which ideally is the WMF, compensates for ISI that results from any small time variation in the channel response. The synthesis of the feedback filter of the DFE at the transmitter side is usually performed after the response of the channel is measured at the receiver by the transmission of a channel probe signal and the receiver sends to the transmitter the coefficients of the feedback filter.

The one problem with this approach to implementing the DFE is that the signal points at the transmitter, after subtracting the postcursors of the ISI, generally have a larger dynamic range than the original signal constellation and, consequently, require a larger transmitter power. This problem can be avoided by precoding the information symbols prior to transmission as described by Tomlinson (1971) and Harashima and Miyakawa (1972).

We describe the precoding technique for a PAM signal constellation. Since a rectangular QAM signal constellation may be viewed as two PAM signal sets on quadrature carriers, the precoding is easily extended to QAM. For simplicity, we assume that the feedforward filter in the DFE is the WMF and that the channel response, characterized by the parameters $\{f_i, 0 \leqslant i \leqslant L\}$, is perfectly known to the transmitter and the receiver. The information symbols $\{I_k\}$ are assumed to take the values $\{\pm 1, \pm 3, \ldots, \pm(M-1)\}$.

In the precoding, the ISI due to the postcursors $\{f_i, 1 \leqslant i \leqslant L\}$ is subtracted from the symbol to be transmitted and, if the difference falls outside of the range $(-M, M]$, it is reduced to the range by subtracting an integer multiple of $2M$ from this difference. Hence, the precoder output may be expressed as

$$a_k = I_k - \sum_{j=1}^{L} f_j a_{k-j} + 2Mb_k \tag{10.3–23}$$

where $\{b_k\}$ represents the appropriate integer that brings $\{a_k\}$ to the desired range. In other words, $\{a_k\}$ is reduced to the desired range by performing a modulo-$2M$ operation. By using the z transform to describe the operation of the precoder, we have

$$A(z) = I(z) - [F(z) - 1]A(z) + 2MB(z) \tag{10.3–24}$$

where the channel coefficient f_0 is normalized to unity for convenience. Hence, the transmitted sequence is

$$A(z) = \frac{I(z) + 2MB(z)}{F(z)} \tag{10.3–25}$$

Since the channel response is $F(z)$, the received signal sequence may be expressed as

$$\begin{aligned} V(z) &= A(z) + W(z) \\ &= [I(z) + 2MB(z)] + W(z) \end{aligned} \tag{10.3–26}$$

where $W(z)$ represents the AWGN term. Therefore, the received data sequence term $I(z) + 2MB(z)$ at the input to the detector is free of ISI and $I(z)$ can be recovered from $V(z)$ by use of a symbol-by-symbol detector that decodes the symbols modulo-$2M$. Figure 10.3–5 illustrates the block diagram of the system that implements the precoder and the feedback filter of the DFE at the transmitter.

The placement of the feedback filter at the transmitter makes it possible to use the DFE in conjunction with trellis-coded modulation (TCM). Since the

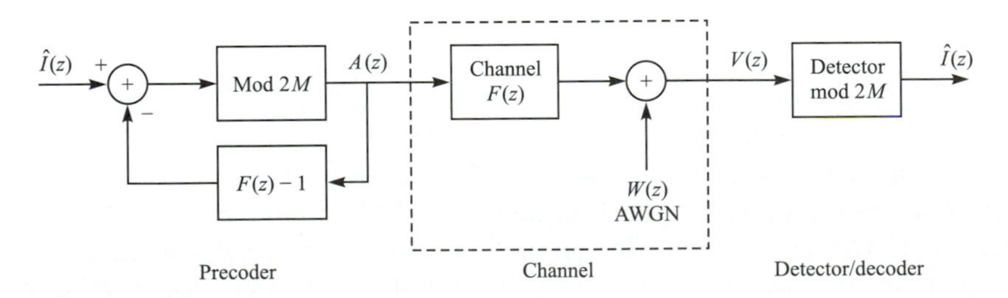

FIGURE 10.3–5
Tomlinson–Harashima precoding.

equalizer at the receiver is a linear filter, decisions from the output of the Viterbi (TCM) detector can be used to adjust the coefficients of the equalizer. In this case, the Viterbi detector performs the modulo-$2M$ operations in its metric computations.

10.4
REDUCED COMPLEXITY ML DETECTORS

The performance results of the three basic equalization methods described above, namely, MLSE, linear equalization (LE), and decision-feedback equalization (DFE), clearly show the superiority of MLSE in channels with severe ISI. Such channels are encountered in wireless communications and in high-density magnetic recording systems. The performance advantage of MLSE has motivated a significant amount of research on methods that retain the performance characteristics of MLSE, but do so at a reduced complexity.

The early work on the design of reduced complexity MLSE focused on methods that reduce the length of the ISI span by preprocessing the received signal prior to the maximum-likelihood detector. Falconer and Magee (1973) and Beare (1978) used a linear equalizer to reduce the span of the ISI to some small specified length prior to the Viterbi detector. Lee and Hill (1977) employed a DFE in place of the LE. Thus, the large ISI span in the channel is reduced to a sufficiently small length, called the *desired impulse response*, so that the complexity of the Viterbi detector following the LE or DFE is manageable. We may view this role of the LE or the DFE, prior to the Viterbi detector, as equalizing the channel response to a specified partial-response characteristic of short duration (the desired impulse response) which the Viterbi detector can handle with sufficiently small complexity. The choice of the desired impulse response is tailored to the ISI characteristics of the channel. This approach to reducing the complexity of the Viterbi detector has proved to be very effective in high-density magnetic recording systems, as illustrated in the papers by Siegel and Wolf (1991), Tyner and Proakis (1993), Moon and Carley (1988), and Proakis (1998).

Another general approach is to reduce the complexity of the Viterbi detector directly, by reducing the number of surviving sequences. The papers by Vermuelen and Hellman (1974), Fredricsson (1974), and Foschini (1977) describe algorithms that reduce the number of surviving sequences in the Viterbi detector. Other works on this class of methods include the papers by Clark et al. (1984, 1985) and Wesolowski (1987a).

The most effective approach in terms of performance for reducing the complexity of the Viterbi detector directly is the method described in the papers by Bergmans et al. (1987), Eyuboglu and Qureshi (1988), and Duel-Hallen and Heegard (1989). The filter preceding the Viterbi detector is the whitened matched filter (WMF) described previously. The WMF reduces the channel to one that has a minimum phase characteristic. The basic algorithm described in these papers for reducing the computational complexity of the Viterbi detector employs decision feedback within the Viterbi detector to reduce the effective length of the ISI from L terms to L_0 terms, where $L_0 < L$. This may be accomplished in one of two ways, as described by Bergmans et al. (1987), either by using "global feedback" or "local feedback" from preliminary decisions that are present in the Viterbi detector. The use of global feedback is illustrated in Figure 10.4–1, where preliminary decisions obtained by using the most probable surviving sequence from the Viterbi detector are used to synthesize the tail in the ISI due to the channel coefficients $(f_{L_0+1}, f_{L_0+2}, \ldots, f_{L-1}, f_L)$. Thus, for M-ary modulations, the computational complexity of the Viterbi detector is reduced from M^L to M^{L_0}, which amounts to a reduction by the factor M^{L-L_0}. The primary drawback of using global feedback is that if one or more of the symbols $\hat{I}_{k-L_0-1}$, ..., $\hat{I}_{k-L}$ in the most probable surviving sequence are incorrect, the subtraction of the tail in the ISI is also incorrect and, thus, the metric computations are corrupted by the residual ISI resulting from this imperfect cancellation.

To remedy this problem, one may use the preliminary decisions corresponding to each surviving sequence to cancel the ISI in the tail of the corresponding surviving sequence. Thus, the ISI will be perfectly cancelled when the correct sequence is among the surviving sequences, even if it is not the most probable sequence. Bergmans et al. (1987) described this approach as using "local feedback" to perform the tail cancellation.

It is interesting to note that if L_0 is selected as unity ($L_0 = 1$), the Viterbi detector reduces to the simple feedback filter of a conventional DFE. At the

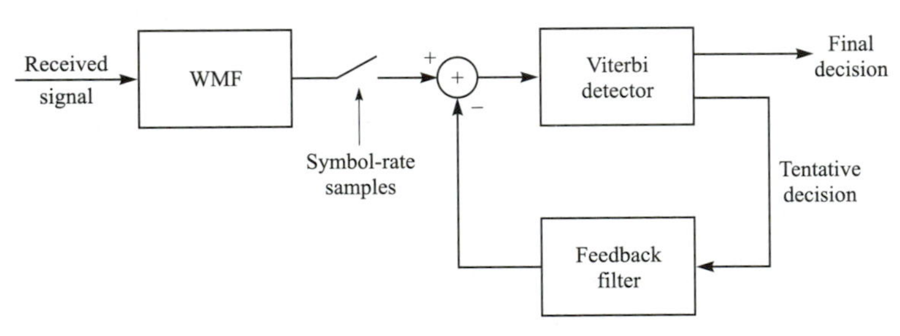

(a) Block diagram of symbol detector

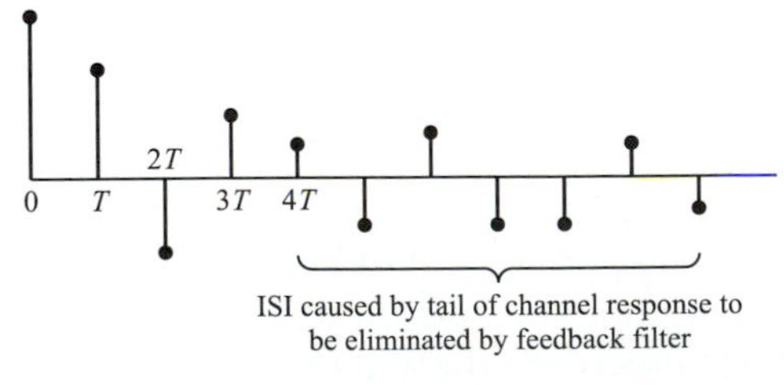

(b) Channel response

FIGURE 10.4–1
Reduced complexity ML sequence detector using feedback from the Viterbi detector.

other extreme, when $L_0 = L$, we have a full complexity Viterbi detector. The analytical and simulation results given in the paper by Bergmans et al. (1987) clearly illustrate that local feedback gives superior performance to global feedback.

10.5 ITERATIVE EQUALIZATION AND DECODING—TURBO EQUALIZATION

Iterative decoding and the turbo-coding principle that was described in Section 8.2.10 can be applied to channel equalization. Suppose the transmitter of a digital communication system employs a binary systematic convolutional encoder followed by a block interleaver and a modulator. The channel is a linear time-dispersive channel that introduces ISI. In such a case, we may view the channel as an inner encoder in a serially concatenated code. Hence, we can apply iterative decoding based on the MAP criterion.

The basic configuration of the iterative equalizer–decoder is shown in Figure 10.5–1. The input to the MAP equalizer is the sequence $\{v_k\}$ from the WMF. The equalizer computes the logarithm of the likelihood ratio of the coded bits, denoted as $L^E(\hat{x})$, which represents the a posteriori values of the coded bits. The outer decoder receives as an input the extrinsic part of $L^E(\hat{x})$, which is defined as

$$L_e^E(\hat{x}) = L^E(\hat{x}) - L_e^D(\hat{x}) \tag{10.5–1}$$

where $L_e^D(\hat{x})$ is the extrinsic part of the outer decoder output after interleaving. $L_e^E(\hat{x})$ is deinterleaved prior to being fed to the outer decoder.

The outer decoder computes the logarithm of the likelihood ratio of the coded bits, denoted by $L^D(\hat{x}')$ and the information bits, denoted as $L^D(\hat{I})$. The extrinsic part of $L^D(\hat{x}')$, denoted as $L_e^D(\hat{x}')$, is the incremental information about

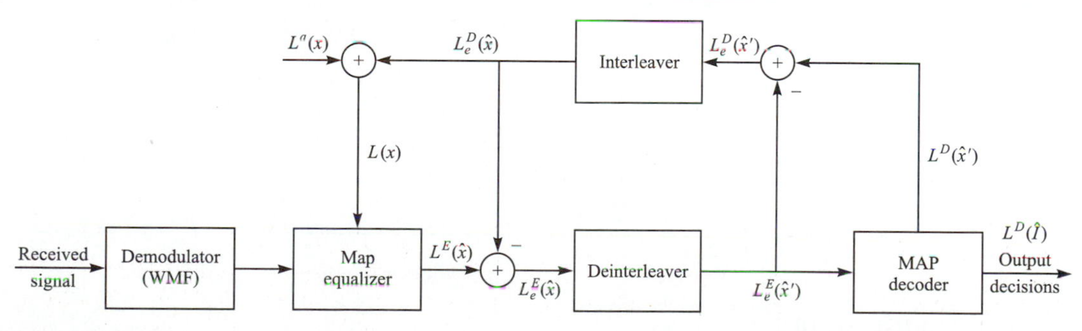

FIGURE 10.5–1
Iterative equalization and decoding.

the current bit obtained by the decoder after observing all the information for all the received bits. The extrinsic information is computed as

$$L_e^D(\hat{x}') = L^D(\hat{x}') - L_e^E(\hat{x}') \tag{10.5–2}$$

$L_e^E(\hat{x}')$ is interleaved to produce $L_e^E(\hat{x})$ and fed to the MAP equalizer. We emphasize the importance of feeding back only the extrinsic part $L_e^D(\hat{x})$, thus, minimizing the correlation between the a priori information used by the equalizer and previous equalizer outputs. Similarly, we reduce the a posteriori information $L^E(\hat{x})$ by the a priori information values $L_e^D(\hat{x})$ to obtain the extrinsic information value $L_e^E(\hat{x})$, which is fed to the outer decoder after deinterleaving.

The computation of the log-likelihood ratios is described in the paper by Bauch et al. (1997). The power of this iterative equalization–decoding scheme can be assessed from the performance results given in this paper. Figure 10.5–2 illustrates the bit error probability obtained through simulation of the five-tap time-invariant channel given in Figure 10.2–5c. The outer decoder used is a rate 1/2 recursive systematic convolutional code with constraint length $K = 5$. The interleaver used was a pseudorandom block interleaver of length $N = 4096$ bits. Binary PSK was used for modulation. The graph illustrates the performance gain as the number of iterations is increased. We observe that after eight iterations, the performance of the iterative equalizer–decoder is within 0.8 dB of the performance of the encoded data without ISI, at a bit error probability of 10^{-4}.

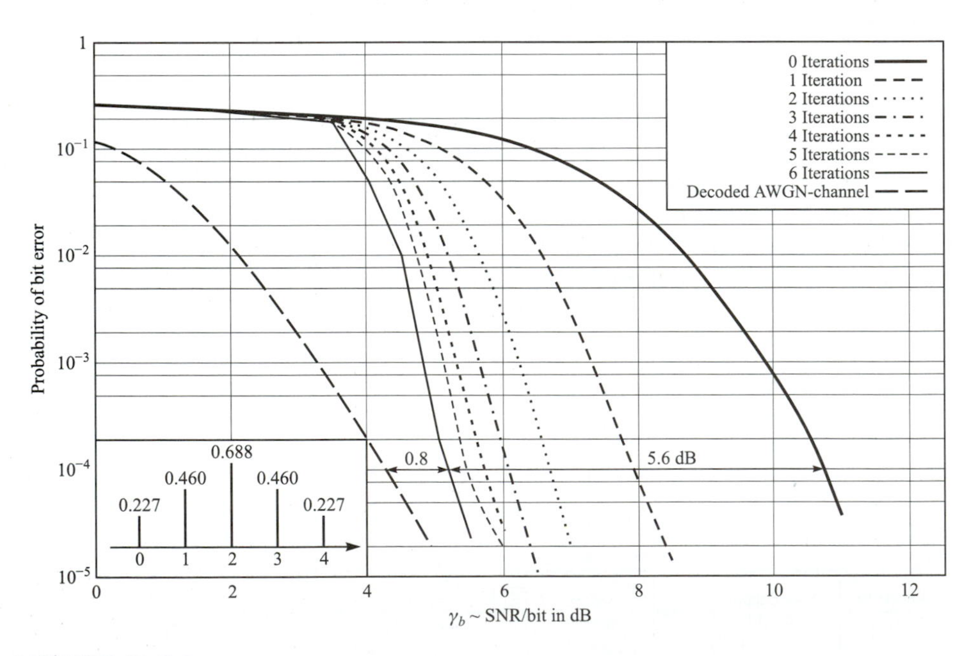

FIGURE 10.5–2
Channel taps and bit error rate for a time-invariant channel. [*From Bauch et al. (1997).*]

Hence, the iterative equalizer eliminates nearly the entire loss due to ISI. In contrast, the optimum (noniterative) Viterbi detector for this channel suffers a loss of approximately 7 dB, due to ISI, as can be observed from Figure 10.3–3b. Therefore, the iterative equalizer has achieved a performance gain of over 6 dB, aside from the coding gain due to the convolutional code. The performance of this method of iterative equalization has been evaluated for cellular radio channels by Bauch et al. (1998). An implementation of iterative equalization–decoding using non-linear circuits is described in a paper by Hagenauer et al. (1999).

An alternative approach to iterative equalization–decoding is to employ a parallel concatenated code (turbo code) followed by a block interleaver and a modulator at the transmitter side. The receiver employs a MAP equalizer followed by a turbo decoder. The extrinsic information generated by the turbo decoder is fed back to the MAP equalizer. Thus, we have an iterative equalizer–turbo decoder structure, which is called a turbo equalizer. Turob equalization is treated by Raphaeli and Zarai (1998) and Douillard et al. (1995).

10.6 BIBLIOGRAPHICAL NOTES AND REFERENCES

Channel equalization for digital communications was developed by Lucky (1965, 1966), who focused on linear equalizers that were optimized using the peak distortion criterion. The mean-square-error criterion for optimization of the equalizer coefficients was proposed by Widrow (1966).

Decision-feedback equalization was proposed and analyzed by Austin (1967). Analyses of the performance of the DFE can be found in the papers by Monsen (1971), George et al. (1971), Price (1972), Salz (1973), Duttweiler et al. (1974), and Altekar and Beaulieu (1993).

The use of the Viterbi algorithm as the optimal maximum-likelihood sequence estimator for symbols corrupted by ISI was proposed and analyzed by Forney (1972) and Omura (1971). Its use for carrier-modulated signals was considered by Ungerboeck (1974) and MacKenchnie (1973).

The use of iterative MAP algorithms in suppressing ISI in coded systems, called turbo equalization, represents a major new advance in suppression of intersymbol interference in signal transmission through band-limited channels. It is anticipated that iterative MAP equalization algorithms will be incorporated in future communication systems. The implementation of turbo equalization, described in the paper by Hagenauer et al. (1999), is the first attempt at implementing an iterative MAP equalization algorithm in a coded system.

PROBLEMS

10.1 In a binary PAM system, the input to the detector is

$$y_m = a_m + n_m + i_m$$

where $a_m = \pm 1$ is the desired signal, n_m is a zero-mean Gaussian random variable with variance σ_n^2, and i_m represents the ISI due to channel distortion. The ISI term is a random variable that takes the values $-\frac{1}{2}$, 0, and $\frac{1}{2}$ with probabilities $\frac{1}{4}$, $\frac{1}{2}$, and $\frac{1}{4}$, respectively. Determine the average probability of error as a function of σ_n^2.

10.2 In a binary PAM system, the clock that specifies the sampling of the correlator output is offset from the optimum sampling time by 10 percent.
a) If the signal pulse used is rectangular, determine the loss in SNR due to the mistiming.
b) Determine the amount of ISI introduced by the mistiming and determine its effect on performance.

10.3 The frequency-response characteristic of a low-pass channel can be approximated by

$$H(f) = \begin{cases} 1 + \alpha \cos 2\pi f t_0 & (|\alpha| < 1,\ |f| \leqslant W) \\ 0 & (\text{otherwise}) \end{cases}$$

where W is the channel bandwidth. An input signal $s(t)$ whose spectrum is band-limited to W Hz is passed through the channel.
a) Show that

$$y(t) = s(t) + \tfrac{1}{2}\alpha[s(t - t_0) + s(t + t_0)]$$

Thus, the channel produces a pair of echoes.
b) Suppose that the received signal $y(t)$ is passed through a filter matched to $s(t)$. Determine the output of the matched filter at $t = kT$, $k = 0, \pm 1, \pm 2, \ldots$, where T is the symbol duration.
c) What is the ISI pattern resulting from the channel if $t_0 = T$?

10.4 A wireline channel of length 1000 km is used to transmit data by means of binary PAM. Regenerative repeaters are spaced 50 km apart along the system. Each segment of the channel has an ideal (constant) frequency response over the frequency band $0 \leqslant f \leqslant 1200$ Hz and an attenuation of 1 dB/km. The channel noise is AWGN.
a) What is the highest bit rate that can be transmitted without ISI?
b) Determine the required $\mathcal{E}_b/N_0$ to achieve a bit error of $P_2 = 10^{-7}$ for each repeater.
c) Determine the transmitted power at each repeater to achieve the desired $\mathcal{E}_b/N_0$, where $N_0 = 4.1 \times 10^{-21}$ W/Hz.

10.5 Prove the relationship in Equation 10.1–13 for the autocorrelation of the noise at the output of the matched filter.

10.6 In the case of PAM with correlated noise, the correlation metrics in the Viterbi algorithm may be expressed in general as (Ungerboeck, 1974)

$$CM(\mathbf{I}) = 2\sum_{n} I_n r_n - \sum_{n}\sum_{m} I_n I_m x_{n-m}$$

where $x_n = x(nT)$ is the sampled signal output of the matched filter, $\{I_n\}$ is the data sequence, and $\{r_n\}$ is the received signal sequence at the output of the matched filter. Determine the metric for the duobinary signal.

10.7 Consider the use of a (square-root) raised cosine signal pulse with a roll-off factor of unity for transmission of binary PAM over an ideal band-limited channel that passes the pulse without distortion. Thus, the transmitted signal is

$$v(t) = \sum_{k=-\infty}^{\infty} I_k g_T(t - kT_b)$$

where the signal interval $T_b = \frac{1}{2}T$. Thus, the symbol rate is double of that for no ISI.

a) Determine the ISI values at the output of a matched filter demodulator.

b) Sketch the trellis for the maximum-likelihood sequence detector and label the states.

10.8 A binary antipodal signal is transmitted over a nonideal band-limited channel, which introduces ISI over two adjacent symbols. For an isolated transmitted signal pulse $s(t)$, the (noise-free) output of the demodulator is $\sqrt{\mathcal{E}_b}$ at $t = T$, $\sqrt{\mathcal{E}_b}/4$ at $t = 2T$, and zero for $t = kT$, $k > 2$, where $\mathcal{E}_b$ is the signal energy and T is the signaling interval.

a) Determine the average probability of error, assuming that the two signals are equally probable and the additive noise is white and Gaussian.

b) By plotting the error probability obtained in (a) and that for the case of no ISI, determine the relative difference in SNR of the error probability of 10^{-6}.

10.9 Derive the expression in Equation 10.3–5 for the coefficients in the feedback filter of the DFE.

10.10 Binary PAM is used to transmit information over an unequalized linear filter channel. When $a = 1$ is transmitted, the noise-free output of the demodulator is

$$x_m = \begin{cases} 0.3 & (m = 1) \\ 0.9 & (m = 0) \\ 0.3 & (m = -1) \\ 0 & (\text{otherwise}) \end{cases}$$

a) Design a three-tap zero-forcing linear equalizer so that the output is

$$q_m = \begin{cases} 1 & (m = 0) \\ 0 & (m = \pm 1) \end{cases}$$

b) Determine q_m for $m = \pm 2, \pm 3$, by convolving the impulse response of the equalizer with the channel response.

10.11 The transmission of a signal pulse with a raised cosine spectrum through a channel results in the following (noise-free) sampled output from the demodulator:

$$x_k = \begin{cases} -0.5 & (k = -2) \\ 0.1 & (k = -1) \\ 1 & (k - 0) \\ -0.2 & (k = 1) \\ 0.05 & (k = 2) \\ 0 & (\text{otherwise}) \end{cases}$$

a) Determine the tap coefficients of a three-tap linear equalizer based on the zero-forcing criterion.

b) For the coefficients determined in (a), determine the output of the equalizer for the case of the isolated pulse. Thus, determine the residual ISI and its span in time.

10.12 A nonideal band-limited channel introduces ISI over three successive symbols. The (noise-free) response of the matched filter demodulator sampled at the sampling time kT is

$$\int_{-\infty}^{\infty} s(t)s(t - kT)\,dt = \begin{cases} \mathcal{E}_b & (k = 0) \\ 0.9\mathcal{E}_b & (k = \pm 1) \\ 0.1\mathcal{E}_b & (k = \pm 2) \\ 0 & (\text{otherwise}) \end{cases}$$

a) Determine the tap coefficients of a three-tap linear equalizer that equalizes the channel (received signal) response to an equivalent partial-response (duobinary) signal

$$y_k = \begin{cases} \mathcal{E}_b & (k = 0, 1) \\ 0 & (\text{otherwise}) \end{cases}$$

b) Suppose that the linear equalizer in (a) is followed by a Viterbi sequence detector for the partial signal. Give an estimate of the error probability if the additive noise is white and Gaussian, with power spectral density $\frac{1}{2}N_0$ W/Hz.

10.13 Determine the tap weight coefficients of a three-tap zero-forcing equalizer if the ISI spans three symbols and is characterized by the values $x(0) = 1$, $x(-1) = 0.3$, $x(1) = 0.2$. Also determine the residual ISI at the output of the equalizer for the optimum tap coefficients.

10.14 In line-of-sight microwave radio transmission, the signal arrives at the receiver via two propagation paths: the direct path and a delayed path that occurs due to signal reflection from surrounding terrain. Suppose that the received signal has the form

$$r(t) = s(t) + \alpha s(t - T) + n(t)$$

where $s(t)$ is the transmitted signal, α is the attenuation ($\alpha < 1$) of the secondary path, and $n(t)$ is AWGN.

a) Determine the output of the demodulator at $t = T$ and $t = 2T$ that employs a filter matched to $s(t)$.

b) Determine the probability of error for a symbol-by-symbol detector if the transmitted signal is binary antipodal and the detector ignores the ISI.

c) What is the error rate performance of a simple (one-tap) DFE that estimates α and removes the ISI? Sketch the detector structure that employs a DFE.

10.15 Repeat Problem 10.10 using the MSE as the criterion for optimizing the tap coefficients. Assume that the noise power spectral density is 0.1 W/Hz.

10.16 In a magnetic recording channel, where the readback pulse resulting from a positive transition in the write current has the form

$$p(t) = \left[1 + \left(\frac{2t}{T_{50}}\right)^2\right]^{-1}$$

a linear equalizer is used to equalize the pulse to a partial response. The parameter T_{50} is defined as the width of the pulse at the 50 percent amplitude level. The bit rate is $1/T_b$ and the ratio of $T_{50}/T_b = \Delta$ is the normalized density of the recording. Suppose the pulse is equalized to the partial-response values

$$x(nT) = \begin{cases} 1 & (n = -1, 1) \\ 2 & (n = 0) \\ 0 & (\text{otherwise}) \end{cases}$$

where $x(t)$ represents the equalized pulse shape.

a) Determine the spectrum $X(f)$ of the band-limited equalized pulse.

b) Determine the possible output levels at the detector, assuming that successive transitions can occur at the rate $1/T_b$.

c) Determine the error rate performance of the symbol-by-symbol detector for this signal, assuming that the additive noise is zero-mean Gaussian with variance σ^2.

10.17 Sketch the trellis for the Viterbi detector of the equalized signal in Problem 10.16 and label all the states. Also, determine the minimum Euclidean distance between merging paths.

10.18 Consider the problem of equalizing the discrete-time equivalent channel shown in Figure P10.18. The information sequence $\{I_n\}$ is binary (± 1) and uncorrelated. The additive noise $\{v_n\}$ is white and real-valued, with variance N_0. The received sequence $\{y_n\}$ is processed by a linear three-tap equalizer that is optimized on the basis of the MSE criterion.

a) Determine the optimum coefficients of the equalizer as a function of N_0.

b) Determine the three eigenvalues λ_1, λ_2, and λ_3 of the covariance matrix $\mathbf{\Gamma}$ and the corresponding (normalized to unit length) eigenvectors $\mathbf{v}_1$, $\mathbf{v}_2$, $\mathbf{v}_3$.

c) Determine the minimum MSE for the three-tap equalizer as a function of N_0.

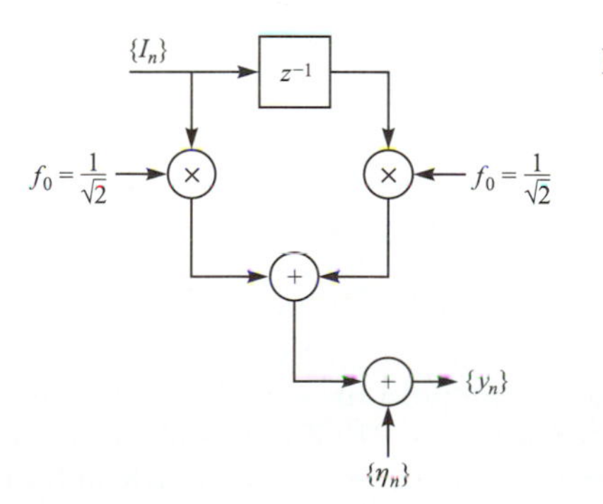

FIGURE P10.18

d) Determine the output SNR for the three-tap equalizer as a function of N_0. How does this compare with the output SNR for the infinite-tap equalizer? For example, evaluate the output SNR for these two equalizers when $N_0 = 0.1$.

10.19 Use the orthogonality principle to derive the equations for the coefficients in a decision-feedback equalizer based on the MSE criterion and given by Equation 10.3–3 and 10.3–5.

10.20 Suppose that the discrete-time model for the intersymbol interference is characterized by the tap coefficients $f_0, f_1, \ldots, f_L$. From the equations for the tap coefficients of a decision-feedback equalizer (DFE), show that only L taps are needed in the feedback filter of the DFE. That is, if $\{c_k\}$ are the coefficients of the feedback filter, then $c_k = 0$ for $k \geqslant L+1$.

10.21 Consider the channel model shown in Figure P10.21. $\{v_n\}$ is a real-valued white noise sequence with zero-mean and variance N_0. Suppose the channel is to be equalized by a DFE having a two-tap feedforward filter (c_0, c_{-1}) and a one-tap feedback filter (c_1). The $\{c_i\}$ are optimized using the MSE criterion.
a) Determine the optimum coefficients and their approximate values for $N_0 \ll 1$.
b) Determine the exact value of the minimum MSE and a first-order approximation appropriate to the case $N_0 \ll 1$.
c) Determine the exact value of the output SNR for the three-tap equalizer as a function of N_0 and a first-order approximation appropriate to the case $N_0 \ll 1$.
d) Compare the results in (b) and (c) with the performance of the infinite-tap DFE.
e) Evaluate and compare the exact values of the output SNR for the three-tap and infinite-tap DFE in the special cases where $N_0 = 0.1$ and 0.01. Comment on how well the three-tap equalizer performs relative to the infinite-tap equalizer.

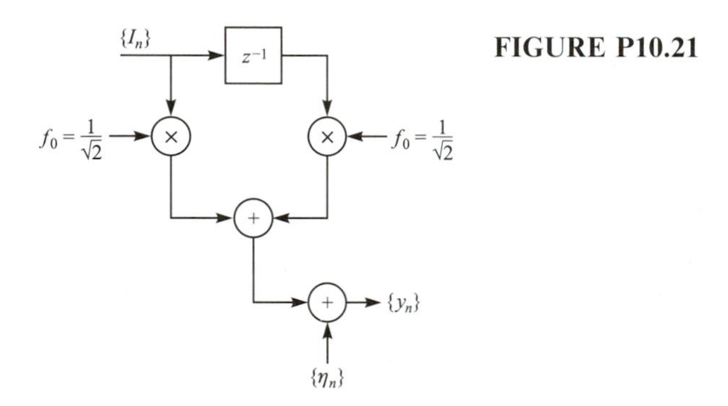

FIGURE P10.21

10.22 A pulse and its (raised cosine) spectral characteristic are shown in Figure P10.22. This pulse is used for transmitting digital information over a band-limited channel at a rate $1/T$ symbols/s.
a) What is the roll-off factor β?
b) What is the pulse rate?
c) The channel distorts the signal pulses. Suppose the sampled values of the filtered received pulse $x(t)$ are as shown in Figure P10.22c. It is obvious that there are five interfering signal components. Give the sequence of +1s and −1s that

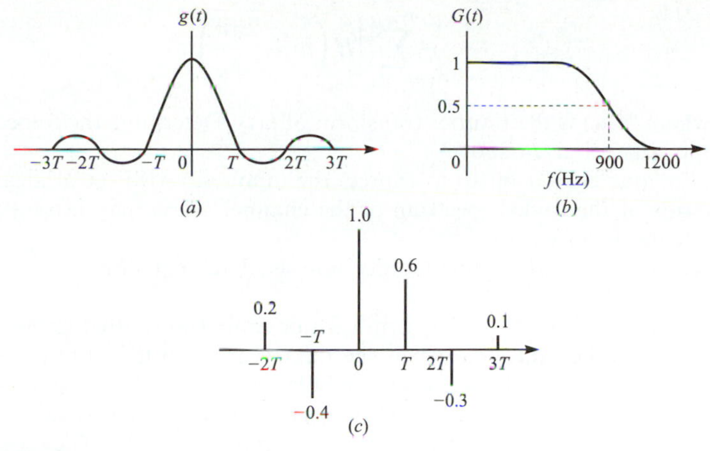

FIGURE P10.22

will cause the largest (destructive or constructive) interference and the corresponding value of the interference (the peak distortion).

d) What is the probability of occurrence of the worst sequence obtained in (c), assuming that all binary digits are equally probable and independent?

10.23 A time-dispersive channel having an impulse response $h(t)$ is used to transmit four-phase PSK at a rate $R = 1/T$ symbols/s. The equivalent discrete-time channel is shown in Figure P10.23. The sequence $\{\eta_k\}$ is a white noise sequence having zero-mean and variance $\sigma^2 = N_0$.

a) What is the sampled autocorrelation function sequence $\{x_k\}$ defined by

$$x_k = \int_{-\infty}^{\infty} h^*(t)h(t + kT)\, dt$$

for this channel?

b) The minimum MSE performance of a linear equalizer and a decision-feedback equalizer having an infinite number of taps depends on the *folded-spectrum of the channel*

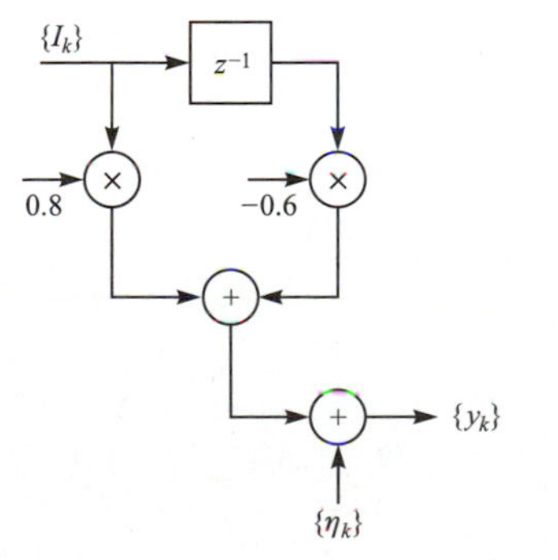

FIGURE P10.23

$$\frac{1}{T}\sum_{n=-\infty}^{\infty}\left|H\left(\omega+\frac{2\pi n}{T}\right)\right|^2$$

where $H(\omega)$ is the Fourier transform of $h(t)$. Determine the folded spectrum of the channel given above.

c) Use your answer in (b) to express the minimum MSE of a linear equalizer in terms of the folded spectrum of the channel. (You may leave your answer in integral form.)

d) Repeat (c) for an infinite-tap decision-feedback equalizer.

10.24 Consider a four-level PAM system with possible transmitted levels, 3, 1, −1, and −3. The channel through which the data is transmitted introduces intersymbol interference over two successive symbols. The equivalent discrete-time channel model is shown in Figure P10.24. $\{\eta_k\}$ is a sequence of real-valued independent zero-mean Gaussian noise variables with variance $\sigma^2 = N_0$. The received sequence is

$$y_1 = 0.8I_1 + n_1$$
$$y_2 = 0.8I_2 - 0.6I_1 + n_2$$
$$y_3 = 0.8I_3 - 0.6I_2 + n_3$$
$$\vdots$$
$$y_k = 0.8I_k - 0.6I_{k-1} + n_k$$

a) Sketch the tree structure, showing the possible signal sequences for the received signals y_1, y_2, and y_3.

b) Suppose the Viterbi algorithm is used to detect the information sequence. How many probabilities must be computed at each stage of the algorithm?

c) How many surviving sequences are there in the Viterbi algorithm for this channel?

d) Suppose that the received signals are

$$y_1 = 0.5, \qquad y_2 = 2.0, \qquad y_3 = -1.0$$

Determine the surviving sequences through stage y_3 and the corresponding metrics.

e) Give a tight upper bound for the probability of error for four-level PAM transmitted over this channel.

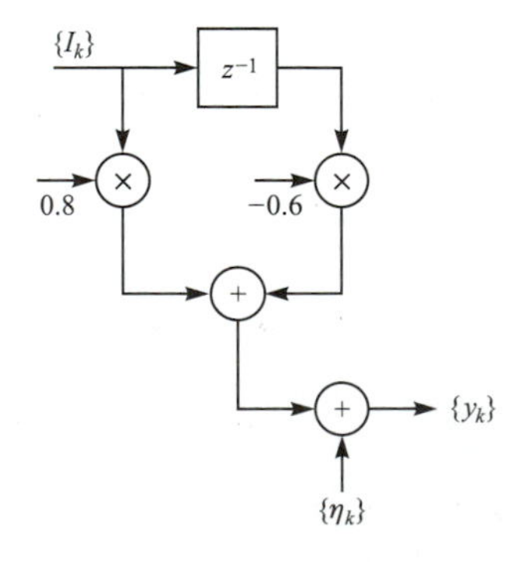

FIGURE P10.24

10.25 A transversal equalizer with K taps has an impulse response

$$e(t) = \sum_{k=0}^{K-1} c_k \delta(t - kT)$$

where T is the delay between adjacent taps, and a transfer function

$$E(z) = \sum_{k=0}^{K-1} c_k z^{-k}$$

The *discrete Fourier transform* (DFT) of the equalizer coefficients $\{c_k\}$ is defined as

$$E_n \equiv E(z)|_{z=e^{j2\pi n/K}} = \sum_{k=0}^{K-1} c_k e^{-j2\pi kn/K}, \qquad n = 0, 1, \ldots, K-1$$

The *inverse DFT* is defined as

$$b_k = \frac{1}{K} \sum_{n=0}^{K-1} E_n e^{j2\pi nk/K}, \qquad k = 0, 1, \ldots, K-1$$

a) Show that $b_k = c_k$, by substituting for E_n in the above expression.

b) From the relations given above, derive an equivalent filter structure having the z transform

$$E(z) = \underbrace{\frac{1 - z^{-K}}{K}}_{E_1(z)} \sum_{n=0}^{K-1} \underbrace{\frac{E_n}{1 - e^{j2\pi n/K} z^{-1}}}_{E_2(z)}$$

c) If $E(z)$ is considered as two separate filters $E_1(z)$ and $E_2(z)$ in cascade, sketch a block diagram for each of the filters, using z^{-1} to denote a unit of delay.

d) In the transversal equalizer, the adjustable parameters are the equalizer coefficients $\{c_k\}$. What are the adjustable parameters of the equivalent equalizer in (b), and how are they related to $\{c_k\}$?

11

Adaptive Equalization

In Chapter 10, we introduced both optimum and suboptimum receivers that compensate for ISI in the transmission of digital information through band-limited, nonideal channels. The optimum receiver employed maximum-likelihood sequence estimation for detecting the information sequence from the samples of the demodulation filter. The suboptimum receivers employed either a linear equalizer or a decision-feedback equalizer.

In the development of the three equalization methods, we implicitly assumed that the channel characteristics, either the impulse response or the frequency response, were known at the receiver. However, in most communication systems that employ equalizers, the channel characteristics are unknown a priori and, in many cases, the channel response is time-variant. In such a case, the equalizers are designed to be adjustable to the channel response and, for time-variant channels, to be adaptive to the time variations in the channel response.

In this chapter, we present algorithms for automatically adjusting the equalizer coefficients to optimize a specified performance index and to adaptively compensate for time variations in the channel characteristics. We also analyze the performance characteristics of the algorithms, including their rate of convergence and their computational complexity.

11.1
ADAPTIVE LINEAR EQUALIZER

In the case of the linear equalizer, recall that we considered two different criteria for determining the values of the equalizer coefficients $\{c_k\}$. One criterion was based on the minimization of the peak distortion at the output of the equalizer, which is defined by Equation 10.2–4. The other criterion was based on the minimization of the mean square error at the output of the equalizer, which is defined by Equation 10.2–25. Below, we describe two algorithms for performing the optimization automatically and adaptively.

11.1.1 The Zero-Forcing Algorithm

In the peak-distortion criterion, the peak distortion $\mathcal{D}(\mathbf{c})$, given by Equation 10.2–22, is minimized by selecting the equalizer coefficients $\{c_k\}$. In general, there is no simple computational algorithm for performing this optimization, except in the special case where the peak distortion at the input to the equalizer, defined as $\mathcal{D}_0$ in Equation 10.2–23, is less than unity. When $\mathcal{D}_0 < 1$, the distortion $\mathcal{D}(\mathbf{c})$ at the output of the equalizer is minimized by forcing the equalizer response $q_n = 0$, for $1 \leqslant |n| \leqslant K$, and $q_0 = 1$. In this case, there is a simple computational algorithm, called the zero-forcing algorithm, that achieves these conditions.

The zero-forcing solution is achieved by forcing the cross correlation between the error sequence $\varepsilon_k = I_k - \hat{I}_k$ and the desired information sequence $\{I_k\}$ to be zero for shifts in the range $0 \leqslant |n| \leqslant K$. The demonstration that this leads to the desired solution is quite simple. We have

$$\begin{aligned} E(\varepsilon_k I_{k-j}^*) &= E[(I_k - \hat{I}_k) I_{k-j}^*] \\ &= E(I_k I_{k-j}^*) - E(\hat{I}_k I_{k-j}^*), \qquad j = -K, \dots, K \end{aligned} \tag{11.1–1}$$

We assume that the information symbols are uncorrelated, i.e., $E(I_k I_j^*) = \delta_{kj}$, and that the information sequence $\{I_k\}$ is uncorrelated with the additive noise sequence $\{\eta_k\}$. For $\hat{I}_k$, we use the expression given in Equation 10.2–41. Then, after taking the expected values in Equation 11.1–1, we obtain

$$E(\varepsilon_k I_{k-j}^*) = \delta_{j0} - q_j, \qquad j = -K, \dots, K \tag{11.1–2}$$

Therefore, the conditions

$$E(\varepsilon_k I_{k-j}^*) = 0, \qquad j = -K, \dots, K \tag{11.1–3}$$

are fulfilled when $q_0 = 1$ and $q_n = 0$, $1 \leqslant |n| \leqslant K$.

When the channel response is unknown, the cross correlations given by Equation 11.1–1 are also unknown. This difficulty can be circumvented by transmitting a known training sequence $\{I_k\}$ to the receiver, which can be used to estimate the cross correlation by substituting time averages for the ensemble averages given in Equation 11.1–1. After the initial training, which will require the transmission of a training sequence of some predetermined length that equals or exceeds the equalizer length, the equalizer coefficients that satisfy Equation 11.1–3 can be determined.

A simple recursive algorithm for adjusting the equalizer coefficients is

$$c_j^{(k+1)} = c_j^{(k)} + \Delta \varepsilon_k I_{k-j}^*, \qquad j = -K, \dots, -1, 0, 1, \dots, K \tag{11.1–4}$$

where $c_j^{(k)}$ is the value of the jth coefficient at time $t = kT$, $\varepsilon_k = I_k - \hat{I}_k$ is the error signal at time $t = kT$, and Δ is a scale factor that controls the rate of adjustment, as will be explained later in this section. This is the *zero-forcing algorithm*. The term $\varepsilon_k I_{k-j}^*$ is an estimate of the cross correlation (ensemble average) $E(\varepsilon_k I_{k-j}^*)$. The averaging operation of the cross correlation is accomplished by means of the recursive first-order difference equation algorithm in Equation 11.1–4, which represents a simple discrete-time integrator.

Following the training period, after which the equalizer coefficients have converged to their optimum values, the decisions at the output of the detector are generally sufficiently reliable so that they may be used to continue the coefficient adaptation process. This is called a *decision-directed mode* of adaptation. In such a case, the cross correlations in Equation 11.1–4 involve the error signal $\tilde{\varepsilon}_k = \tilde{I}_k - \hat{I}_k$ and the detected output sequence $\tilde{I}_{k-j}, j = -K, \ldots, K$. Thus, in the adaptive mode, Equation 11.1–4 becomes

$$c_j^{(k+1)} = c_j^{(k)} + \Delta \tilde{\varepsilon}_k \tilde{I}_{k-j}^* \tag{11.1–5}$$

Figure 11.1–1 illustrates the zero-forcing equalizer in the training mode and the adaptive mode of operation.

The characteristics of the zero-forcing algorithm are similar to those of the least-mean-square (LMS) algorithm, which minimizes the MSE and which is described in detail in the following section.

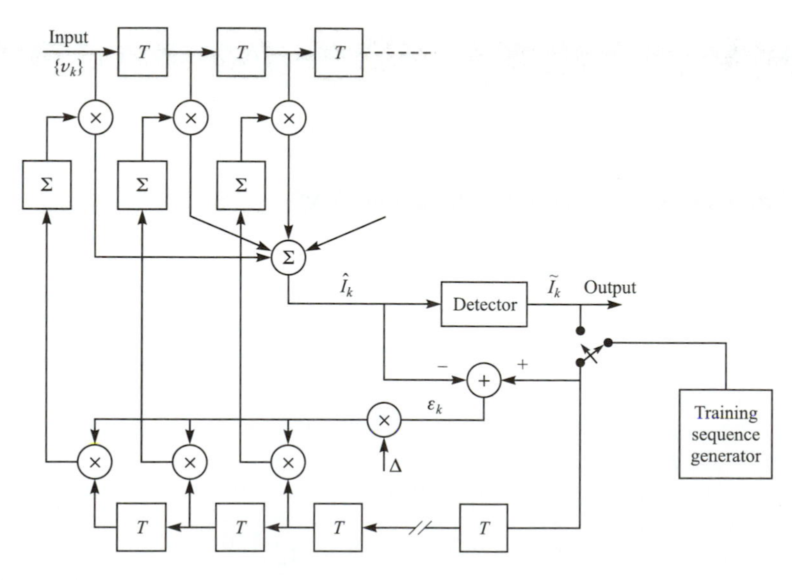

FIGURE 11.1–1
An adaptive zero-forcing equalizer.

11.1.2 The LMS Algorithm

In the minimization of the MSE, treated in Section 10.2.2, we found that the optimum equalizer coefficients are determined from the solution of the set of linear equations, expressed in matrix form as

$$\mathbf{\Gamma C} = \boldsymbol{\xi} \tag{11.1–6}$$

where $\mathbf{\Gamma}$ is the $(2K+1) \times (2K+1)$ covariance matrix of the signal samples $\{v_k\}$, $\mathbf{C}$ is the column vector of $(2K+1)$ equalizer coefficients, and $\boldsymbol{\xi}$ is a $(2K+1)$-dimensional column vector of channel filter coefficients. The solution for the optimum equalizer coefficients vector $\mathbf{C}_{\text{opt}}$ can be determined by inverting the covariance matrix $\mathbf{\Gamma}$, which can be efficiently performed by use of the Levinson–Durbin algorithm described in Appendix A.

Alternatively, an iterative procedure that avoids the direct matrix inversion may be used to compute $\mathbf{C}_{\text{opt}}$. Probably the simplest iterative procedure is the method of steepest descent, in which one begins by arbitrarily choosing the vector $\mathbf{C}$, say as $\mathbf{C}_0$. This initial choice of coefficients corresponds to some point on the quadratic MSE surface in the $(2K+1)$-dimensional space of coefficients. The gradient vector $\mathbf{G}_0$, having the $2K+1$ gradient components $\frac{1}{2}\partial J/\partial c_{0k}$, $k = -K, \ldots, -1, 0, 1, \ldots, K$, is then computed at this point on the MSE surface, and each tap weight is changed in the direction opposite to its corresponding gradient component. The change in the jth tap weight is proportional to the size of the jth gradient component. Thus, succeeding values of the coefficient vector $\mathbf{C}$ are obtained according to the relation

$$\mathbf{C}_{k+1} = \mathbf{C}_k - \Delta \mathbf{G}_k, \qquad k = 0, 1, 2, \ldots \tag{11.1–7}$$

where the gradient vector $\mathbf{G}_k$ is

$$\mathbf{G}_k = \frac{1}{2}\frac{dJ}{d\mathbf{C}_k} = \mathbf{\Gamma C}_k - \boldsymbol{\xi} = -E(\varepsilon_k \mathbf{V}_k^*) \tag{11.1–8}$$

The vector $\mathbf{C}_k$ represents the set of coefficients at the kth iteration, $\varepsilon_k = I_k - \hat{I}_k$ is the error signal at the kth iteration, $\mathbf{V}_k$ is the vector of received signal samples that make up the estimate $\hat{I}_k$, i.e., $\mathbf{V}_k = [v_{k+K} \cdots v_k \cdots v_{k-K}]^t$, and Δ is a positive number chosen small enough to ensure convergence of the iterative procedure. If the minimum MSE is reached for some $k = k_0$, then $\mathbf{G}_k = \mathbf{0}$, so that no further change occurs in the tap weights. In general, $J_{\min}(K)$ cannot be attained for a finite value of k_0 with the steepest-descent method. It can, however, be approached as closely as desired for some finite value of k_0.

The basic difficulty with the method of steepest descent for determining the optimum tap weights is the lack of knowledge of the gradient vector $\mathbf{G}_k$, which depends on both the covariance matrix $\mathbf{\Gamma}$ and the vector $\boldsymbol{\xi}$ of cross correlations. In turn, these quantities depend on the coefficients $\{f_k\}$ of the equivalent discrete-time channel model and on the covariance of the information sequence and the additive noise, all of which may be unknown at the receiver in general. To overcome the difficulty, estimates of the gradient vector may be used. That is, the algorithm for adjusting the tap weight coefficients may be expressed in the form

$$\hat{\mathbf{C}}_{k+1} = \hat{\mathbf{C}}_k - \Delta \hat{\mathbf{G}}_k \tag{11.1–9}$$

where $\hat{\mathbf{G}}_k$ denotes an estimate of the gradient vector $\mathbf{G}_k$ and $\hat{\mathbf{C}}_k$ denotes the estimate of the vector of coefficients.

From Equation 11.1–8 we note that $\mathbf{G}_k$ is the negative of the expected value of the $\varepsilon_k \mathbf{V}_k^*$. Consequently, an estimate of $\mathbf{G}_k$ is

$$\hat{\mathbf{G}}_k = -\varepsilon_k \mathbf{V}_k^* \tag{11.1–10}$$

Since $E(\hat{\mathbf{G}}_k) = \mathbf{G}_k$, the estimate $\hat{\mathbf{G}}_k$ is an unbiased estimate of the true gradient vector $\hat{\mathbf{G}}_k$. Incorporation of Equation 11.1–10 into Equation 11.1–9 yields the algorithm

$$\hat{\mathbf{C}}_{k+1} = \hat{\mathbf{C}}_k + \Delta \varepsilon_k \mathbf{V}_k^* \tag{11.1–11}$$

This is the basic LMS algorithm for recursively adjusting the tap weight coefficients of the equalizer as described by Widrow (1966). It is illustrated in the equalizer shown in Figure 11.1–2.

The basic algorithm given by Equation 11.1–11 and some of its possible variations have been incorporated into many commercial adaptive equalizers that are used in high-speed modems. Three variations of the basic algorithm

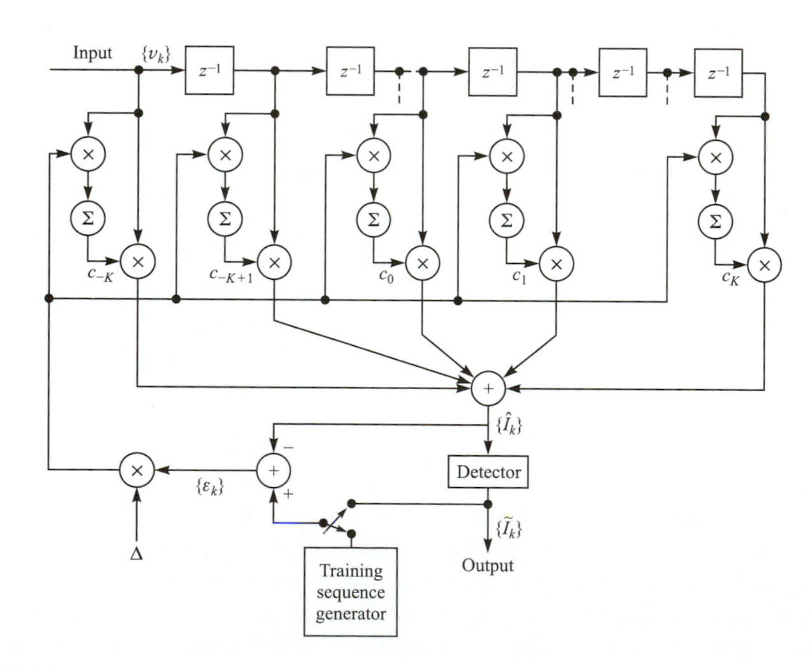

FIGURE 11.1–2
Linear adaptive equalizer based on the MSE criterion.

are obtained by using only sign information contained in the error signal ε_k and/or in the components of $\mathbf{V}_k$. Hence, the three possible variations are

$$c_{(k+1)j} = c_{kj} + \Delta \operatorname{csgn}(\varepsilon_k) v_{k-j}^*, \qquad j = -K, \ldots, -1, 0, 1, \ldots, K \quad (11.1\text{–}12)$$

$$c_{(k+1)j} = c_{kj} + \Delta \varepsilon_k \operatorname{csgn}(v_{k-j}^*), \qquad j = -K, \ldots, -1, 0, 1, \ldots, K \quad (11.1\text{–}13)$$

$$c_{(k+1)j} = c_{kj} + \Delta \operatorname{csgn}(\varepsilon_k) \operatorname{csgn}(v_{k-j}^*), \qquad j = -K, \ldots, -1, 0, 1, \ldots, K \quad (11.1\text{–}14)$$

where $\operatorname{csgn}(x)$ is defined as

$$\operatorname{csgn}(x) = \begin{cases} 1 + j & [\operatorname{Re}(x) > 0, \operatorname{Im}(x) > 0] \\ 1 - j & [\operatorname{Re}(x) > 0, \operatorname{Im}(x) < 0] \\ -1 + j & [\operatorname{Re}(x) < 0, \operatorname{Im}(x) > 0] \\ -1 - j & [\operatorname{Re}(x) < 0, \operatorname{Im}(x) < 0] \end{cases} \quad (11.1\text{–}15)$$

(Note that in Equation 11.1–15, $j \equiv \sqrt{-1}$, as distinct from the index j in Equations 11.1–12 to 11.1–14.) Clearly, the algorithm in Equation 11.1–14 is the most easily implemented, but it gives the slowest rate of convergence to the others.

Several other variations of the LMS algorithm are obtained by averaging or filtering the gradient vectors over several iterations prior to making adjustments of the equalizer coefficients. For example, the average over N gradient vectors is

$$\bar{\hat{\mathbf{G}}}_{mN} = -\frac{1}{N} \sum_{n=0}^{N-1} \varepsilon_{mN+n} \mathbf{V}_{mN+n}^* \quad (11.1\text{–}16)$$

and the corresponding recursive equation for updating the equalizer coefficients once every N iterations is

$$\hat{\mathbf{C}}_{(k+1)N} = \hat{\mathbf{C}}_{kN} - \Delta \bar{\hat{\mathbf{G}}}_{kN} \quad (11.1\text{–}17)$$

In effect, the averaging operation performed in Equation 11.1–16 reduces the noise in the estimate of the gradient vector, as shown by Gardner (1984).

An alternative approach is to filter the noisy gradient vectors by a low-pass filter and use the output of the filter as an estimate of the gradient vector. For example, a simple low-pass filter for the noisy gradients yields as an output

$$\bar{\hat{\mathbf{G}}}_k = w \bar{\hat{\mathbf{G}}}_{k-1} + (1 - w) \hat{\mathbf{G}}_k, \qquad \bar{\hat{\mathbf{G}}}(0) = \hat{\mathbf{G}}(0) \quad (11.1\text{–}18)$$

where the choice of $0 \leqslant w < 1$ determines the bandwidth of the low-pass filter. When w is close to unity, the filter bandwidth is small and the effective averaging is performed over many gradient vectors. On the other hand, when w is small, the low-pass filter has a large bandwidth and, hence, it provides little averaging of the gradient vectors. With the filtered gradient vectors given by Equation 11.1–18 in place of $\mathbf{G}_k$, we obtain the filtered gradient LMS algorithm given by

$$\hat{\mathbf{C}}_{k+1} = \hat{\mathbf{C}}_k - \Delta \bar{\hat{\mathbf{G}}}_k \quad (11.1\text{–}19)$$

In the above discussion, it has been assumed that the receiver has knowledge of the transmitted information sequence in forming the error signal between the desired symbol and its estimate. Such knowledge can be made available during a short training period in which a signal with a known information sequence is transmitted to the receiver for initially adjusting the tap weights. The length of this sequence must be at least as large as the length of the equalizer so that the spectrum of the transmitted signal adequately covers the bandwidth of the channel being equalized.

In practice, the training sequence is often selected to be a periodic pseudorandom sequence, such as a maximum length shift-register sequence whose period N is equal to the length of the equalizer ($N = 2K + 1$). In this case, the gradient is usually averaged over the length of the sequence as indicated in Equation 11.1–16 and the equalizer is adjusted once a period according to Equation 11.1–17. This approach has been called *cyclic equalization*, and has been treated in the papers by Mueller and Spaulding (1975) and Qureshi (1977, 1985). A practical scheme for continuous adjustment of the tap weights may be either a decision-directed mode of operation in which decisions on the information symbols are assumed to be correct and used in place of I_k in forming the error signal ε_k, or one in which a known pseudorandom-probe sequence is inserted in the information-bearing signal either additively or by interleaving in time and the tap weights adjusted by comparing the received probe symbols with the known transmitted probe symbols. In the decision-directed mode of operation, the error signal becomes $\tilde{\varepsilon}_k = \tilde{I}_k - \hat{I}_k$, where $\tilde{I}_k$ is the decision of the receiver based on the estimate $\hat{I}_k$. As long as the receiver is operating at low error rates, an occasional error will have a negligible effect on the convergence of the algorithm.

If the channel response changes, this change is reflected in the coefficients $\{f_k\}$ of the equivalent discrete-time channel model. It is also reflected in the error signal ε_k, since it depends on $\{f_k\}$. Hence, the tap weights will be changed according to Equation 11.1–11 to reflect the change in the channel. A similar change in the tap weights occurs if the statistics of the noise or the information sequence change. Thus, the equalizer is adaptive.

11.1.3 Convergence Properties of the LMS Algorithm

The convergence properties of the LMS algorithm given by Equation 11.1–11 are governed by the step-size parameter Δ. We shall now consider the choice of the parameter Δ to ensure convergence of the steepest-descent algorithm in Equation 11.1–7, which employs the exact value of the gradient.

From Equations 11.1–7 and 11.1–8, we have

$$\begin{aligned} \mathbf{C}_{k+1} &= \mathbf{C}_k - \Delta \mathbf{G}_k \\ &= (\mathbf{I} - \Delta \mathbf{\Gamma})\mathbf{C}_k + \Delta \boldsymbol{\xi} \end{aligned} \tag{11.1–20}$$

where $\mathbf{I}$ is the identity matrix, $\mathbf{\Gamma}$ is the autocorrelation matrix of the received signal, $\mathbf{C}_k$ is the $(2K + 1)$-dimensional vector of equalizer tap gains, and $\boldsymbol{\xi}$ is the vector of cross correlations given by Equation 10.2–45. The recursive relation in

Equation 11.1–20 can be represented as a closed-loop control system as shown in Figure 11.1–3. Unfortunately, the set of $2K+1$ first-order difference equations in Equation 11.1–20 are coupled through the autocorrelation matrix $\mathbf{\Gamma}$. In order to solve these equations and, thus, establish the convergence properties of the recursive algorithm, it is mathematically convenient to decouple the equations by performing a linear transformation. The appropriate transformation is obtained by noting that the matrix $\mathbf{\Gamma}$ is Hermitian and, hence, can be represented as

$$\mathbf{\Gamma} = \mathbf{U}\mathbf{\Lambda}\mathbf{U}'^{*} \tag{11.1–21}$$

where $\mathbf{U}$ is the normalized modal matrix of $\mathbf{\Gamma}$ and $\mathbf{\Lambda}$ is a diagonal matrix with diagonal elements equal to the eigenvalues of $\mathbf{\Gamma}$.

When Equation 11.1–21 is substituted into Equation 11.1–20 and if we define the transformed (orthogonalized) vector $\mathbf{C}_k^o = \mathbf{U}'^{*}\mathbf{C}_k$ and $\boldsymbol{\xi}^o = \mathbf{U}'^{*}\boldsymbol{\xi}$, we obtain

$$\mathbf{C}_{k+1}^o = (\mathbf{I} - \Delta\mathbf{\Lambda})\mathbf{C}_k^o + \Delta\boldsymbol{\xi}^o \tag{11.1–22}$$

This set of first-order difference equations is now decoupled. Their convergence is determined from the homogeneous equation

$$\mathbf{C}_{k+1}^o = (\mathbf{I} - \Delta\mathbf{\Lambda})\mathbf{C}_k^o \tag{11.1–23}$$

We see that the recursive relation will converge provided that all the poles lie inside the unit circle, i.e.,

$$|1 - \Delta\lambda_k| < 1, \qquad k = -K, \ldots, -1, 0, 1, \ldots, K \tag{11.1–24}$$

where $\{\lambda_k\}$ is the set of $2K+1$ (possibly nondistinct) eigenvalues of $\mathbf{\Gamma}$. Since $\mathbf{\Gamma}$ is an autocorrelation matrix, it is positive-definite and, hence, $\lambda_k > 0$ for all k. Consequently convergence of the recursive relation in Equation 11.1–22 is ensured if Δ satisfies the inequality

$$0 < \Delta < \frac{2}{\lambda_{\max}} \tag{11.1–25}$$

where $\lambda_{\max}$ is the largest eigenvalue of $\mathbf{\Gamma}$.

Since the largest eigenvalue of a positive-definite matrix is less than the sum of all the eigenvalues of the matrix and, furthermore, since the sum of the eigenvalues of a matrix is equal to its trace, we have the following simple upper bound on $\lambda_{\max}$:

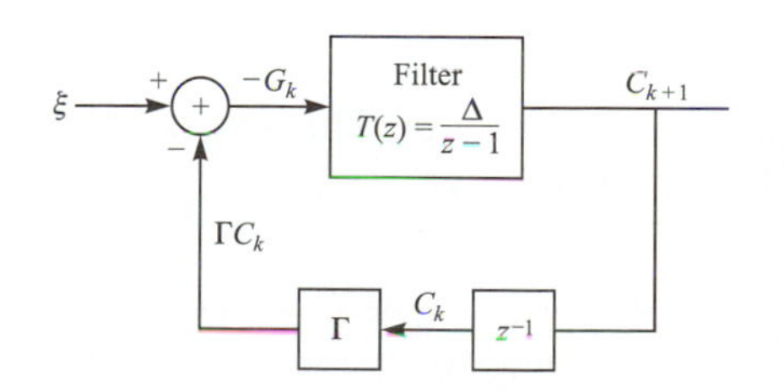

FIGURE 11.1–3
Closed-loop control system representation of the recursive relation in Equation 11.1–20.

$$\lambda_{\max} < \sum_{k=-K}^{K} \lambda_k = \operatorname{tr}\boldsymbol{\Gamma} = (2K+1)\Gamma_{kk} \\ = (2K+1)(x_0 + N_0) \tag{11.1–26}$$

From Equations 11.1–23 and 11.1–24 we observe that rapid convergence occurs when $|1 - \Delta\lambda_k|$ is small, i.e., when the pole positions are far from the unit circle. But we cannot achieve this desirable condition and still satisfy Equation 11.1–25 if there is a large difference between the largest and smallest eigenvalues of $\boldsymbol{\Gamma}$. In other words, even if we select Δ to be near the upper bound given in Equation 11.1–25, the convergence rate of the recursive MSE algorithm is determined by the smallest eigenvalue $\lambda_{\min}$. Consequently, the ratio $\lambda_{\max}/\lambda_{\min}$ ultimately determines the convergence rate. If $\lambda_{\max}/\lambda_{\min}$ is small, Δ can be selected so as to achieve rapid convergence. However, if the ratio $\lambda_{\max}/\lambda_{\min}$ is large, as is the case when the channel frequency response has deep spectral nulls, the convergence rate of the algorithm will be slow.

11.1.4 Excess MSE Due to Noisy Gradient Estimates

The recursive algorithm in Equation 11.1–11 for adjusting the coefficients of the linear equalizer employs unbiased noisy estimates of the gradient vector. The noise in these estimates causes random fluctuations in the coefficients about their optimal values and, thus, leads to an increase in the MSE at the output of the equalizer. That is, the final MSE is $J_{\min} + J_\Delta$, where J_Δ is the variance of the measurement noise. The term J_Δ due to the estimation noise has been termed *excess mean square error* by Widrow (1966).

The total MSE at the output of the equalizer for any set of coefficients $\mathbf{C}$ can be expressed as

$$J = J_{\min} + (\mathbf{C} - \mathbf{C}_{\text{opt}})'^{*}\boldsymbol{\Gamma}(\mathbf{C} - \mathbf{C}_{\text{opt}}) \tag{11.1–27}$$

where $\mathbf{C}_{\text{opt}}$ represents the optimum coefficients, which satisfy Equation 11.1–6. This expression for the MSE can be simplified by performing the linear orthogonal transformation used above to establish convergence. The result of this transformation applied to Equation 11.1–27 is

$$J = J_{\min} + \sum_{k=-K}^{K} \lambda_k E|c_k^o - c_{k\,\text{opt}}^o|^2 \tag{11.1–28}$$

where the $\{c_k^o\}$ are the set of transformed equalizer coefficients. The excess MSE is the expected value of the second term in Equation 11.1–28, i.e.,

$$J_\Delta = \sum_{k=-K}^{K} \lambda_k E|c_k^o - c_{k\,\text{opt}}^o|^2 \tag{11.1–29}$$

It has been shown by Widrow (1970) that the excess MSE is

$$J_\Delta = \Delta^2 J_{\min} \sum_{k=-K}^{K} \frac{\lambda_k^2}{1-(1-\Delta\lambda_k)^2} \tag{11.1–30}$$

The expression in Equation 11.1–30 can be simplified when Δ is selected such that $\Delta\lambda_k \ll 1$ for all k. Then

$$\begin{aligned} J_\Delta &\approx \tfrac{1}{2}\Delta J_{\min} \sum_{k=-K}^{K} \lambda_k \\ &\approx \tfrac{1}{2}\Delta J_{\min} \operatorname{tr} \mathbf{\Gamma} \\ &\approx \tfrac{1}{2}\Delta(2K+1) J_{\min}(x_0+N_0) \end{aligned} \tag{11.1–31}$$

Note that $x_0 + N_0$ represents the received signal plus noise power.

It is desirable to have $J_\Delta < J_{\min}$. That is, Δ should be selected such that

$$\frac{J_\Delta}{J_{\min}} \approx \tfrac{1}{2}\Delta(2K+1)(x_0+N_0) < 1$$

or, equivalently,

$$\Delta < \frac{2}{(2K+1)(x_0+N_0)} \tag{11.1–32}$$

For example, if Δ is selected as

$$\Delta = \frac{0.2}{(2K+1)(x_0+N_0)} \tag{11.1–33}$$

the degradation in the output SNR of the equalizer due to the excess MSE is less than 1 dB.

The analysis given above on the excess mean square error is based on the assumption that the mean value of the equalizer coefficients has converged to the optimum value $\mathbf{C}_{\text{opt}}$. Under this condition, the step size Δ should satisfy the bound in Equation 11.1–32. On the other hand, we have determined that convergence of the mean coefficient vector requires that $\Delta < 2/\lambda_{\max}$. While a choice of Δ near the upper bound $2/\lambda_{\max}$ may lead to initial convergence of the deterministic (known) steepest-descent gradient algorithm, such a large value of Δ will usually result in instability of the LMS stochastic gradient algorithm.

The initial convergence or transient behavior of the LMS algorithm has been investigated by several researchers. Their results clearly indicate that the step size must be reduced in direct proportion to the length of the equalizer as specified by Equation 11.1–32. Hence, the upper bound given by Equation 11.1–32 is also necessary to ensure the initial convergence of the LMS algorithm. The papers by Gitlin and Weinstein (1979) and Ungerboeck (1972) contain analyses of the transient behavior and the convergence properties of the LMS algorithm.

The following example serves to reinforce the important points made above regarding the initial convergence of the LMS algorithm.

EXAMPLE 11.1-1. The LMS algorithm was used to adaptively equalize a communication channel for which the autocorrelation matrix $\mathbf{\Gamma}$ has an eigenvalue spread of $\lambda_{\max}/\lambda_{\min} = 11$. The number of taps selected for the equalizer was $2K + 1 = 11$. The input signal plus noise power $x_0 + N_0$ was normalized to unity. Hence, the upper bound on Δ given by Equation 11.1–32 is 0.18. Figure 11.1–4 illustrates the initial convergence characteristics of the LMS algorithm for $\Delta = 0.045$, 0.09, and 0.115, by averaging the (estimated) MSE in 200 simulations. We observe that by selecting $\Delta = 0.09$ (one-half of the upper bound) we obtain relatively fast initial convergence. If we divide Δ by a factor of 2 to $\Delta = 0.045$, the convergence rate is reduced but the excess mean square error is also reduced, so that the LMS algorithm performs better in steady state (in a time-invariant signal environment). Finally, we note that a choice of $\Delta = 0.115$, which is still far below the upper bound, causes large undesirable fluctuations in the output MSE of the algorithm.

In a digital implementation of the LMS algorithm, the choice of the step-size parameter becomes even more critical. In an attempt to reduce the excess mean square error, it is possible to reduce the step-size parameter to the point where the total mean square error actually increases. This condition occurs when the estimated gradient components of the vector $\varepsilon_k \mathbf{V}_k^*$ after multiplication by the small step-size parameter Δ are smaller than one-half of the least significant bit in the fixed-point representation of the equalizer coefficients. In such a case, adaptation ceases. Consequently, it is important for the step size to be large enough to bring the equalizer coefficients in the vicinity of $\mathbf{C}_{\text{opt}}$. If it is desired to decrease the step size significantly, it is necessary to increase the precision in the equalizer coefficients. Typically, 16 bits of precision may be used for the coefficients, with about 10–12 of the most significant bits used for arithmetic operations in the equalization of the data. The remaining least significant bits are required to provide the necessary precision for the adaptation process. Thus, the scaled estimated gradient components $\Delta\varepsilon\mathbf{V}_k^*$ usually affect only the least-significant bits in any one iteration. In effect, the added precision also allows for the noise to be averaged out, since many incremental changes in the least-significant bits are required before any change occurs in the upper more significant bits used in arithmetic operations for equalizing the data. For an analysis of roundoff errors in a digital implementation of the LMS algorithm, the reader is referred to the papers by Gitlin and Weinstein (1979), Gitlin et al. (1982), and Caraiscos and Liu (1984).

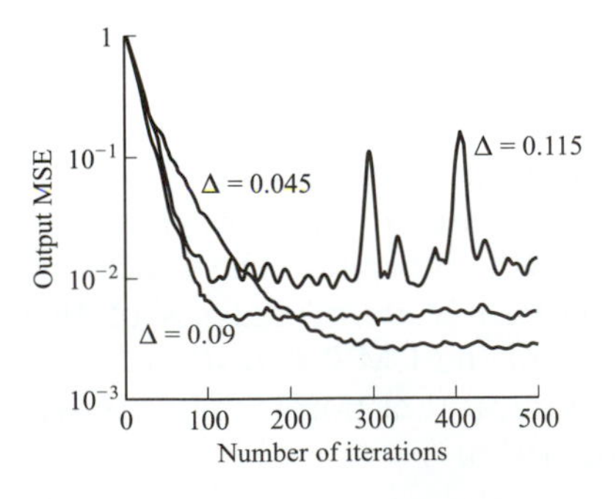

FIGURE 11.1–4
Initial convergence characteristics of the LMS algorithm with different step sizes. (*From* Digital Signal Processing, *by J. G. Proakis and D. G. Manolakis, 1995, Prentice Hall Company. Reprinted with permission of the publisher.*)

As a final point, we should indicate that the LMS algorithm is appropriate for tracking slowly time invariant signal statistics. In such a case, the minimum MSE and the optimum coefficient vector will be time-variant. In other words, $J_{\min}(n)$ is a function of time and the $2(K+1)$-dimensional error surface is moving with the time index n. The LMS algorithm attempts to follow the moving minimum $J_{\min}(n)$ in the $(2K+1)$-dimensional space, but it is always lagging behind due to its use of (estimated) gradient vectors. As a consequence, the LMS algorithm incurs another form of error, called the *lag error*, whose mean square value decreases with an increase in the step size Δ. The total MSE error can now be expressed as

$$J_{\text{total}} = J_{\min}(n) + J_\Delta + J_l \tag{11.1–34}$$

where J_l denotes the mean square error due to the lag.

In any given nonstationary adaptive equalization problem, if we plot the errors J_Δ and J_l as a function of Δ, we expect these errors to behave as illustrated in Figure 11.1–5. We observe that J_Δ increases with an increase in Δ while J_l decreases with an increase in Δ. The total error will exhibit a minimum, which will determine the optimum choice of the step-size parameter.

When the statistical time variations of the signal occur rapidly, the lag error will dominate the performance of the adaptive equalizer. In such a case, $J_l \gg J_{\min} + J_\Delta$, even when the largest possible value of Δ is used. When this condition occurs, the LMS algorithm is inappropriate for the application and one must rely on the more complex recursive least-squares algorithms described in Section 11.4 to obtain faster convergence and tracking.

11.1.5 Accelerating the Initial Convergence Rate in the LMS Algorithm

As we have observed, the initial convergence rate of the LMS algorithm for any given channel characteristic is controlled by the step-size parameter Δ. The initial convergence rate is strongly influenced by the channel spectral characteristics, which are related to the eigenvalues $\{\lambda_n\}$ of the received signal covariance matrix. If the channel amplitude and phase distortions are small, the eigenvalue ratio $\lambda_{\max}/\lambda_{\min}$ is close to unity and, hence, the equalizer converges to its optimum tap

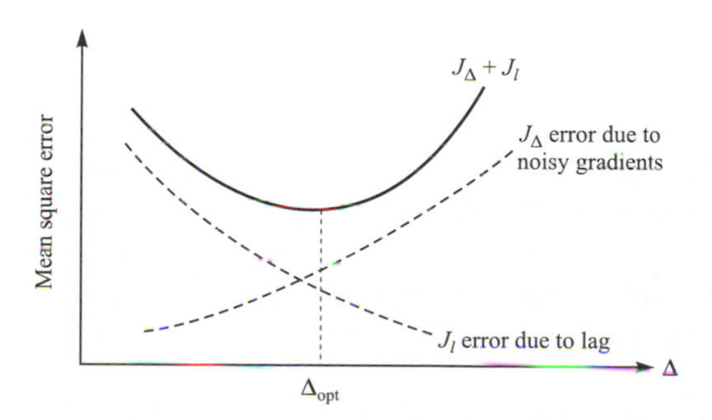

FIGURE 11.1–5
Excess mean square error J_Δ and lag error J_l as a function of the step size. (*From* Digital Signal Processing, *by J. G. Proakis and D. G. Manolakis, 1995, Prentice Hall Company. Reprinted with permission of the publisher.*)

coefficients relatively fast. On the other hand, if the channel exhibits poor spectral characteristics, such as relatively large attenuation in a part of its spectrum, the eigenvalue ratio $\lambda_{\max}/\lambda_{\min} \gg 1$ and, hence, the convergence rate of the LMS algorithm will be slow.

A considerable effort has been spent by researchers on methods to accelerate the initial convergence of the LMS algorithm. A simple remedy is to begin with a large step size, say Δ_0, and reduce the step size as the tap coefficients converge to their optimum values. In other words, we use a sequence of step sizes, $\Delta_0 > \Delta_1 > \Delta_2 > \cdots > \Delta_m \equiv \Delta$, where Δ is the final step size to be used in steady-state operation of the LMS algorithm.

An alternative method for accelerating initial convergence has been proposed and investigated by Chang (1971) and Qureshi (1977). This method is based on introducing additional parameters in the LMS algorithm by replacing the step size with a weighting matrix **W**. In such a case, the LMS algorithm is generalized to the form:

$$\begin{aligned}\hat{\mathbf{C}}_{k+1} &= \hat{\mathbf{C}}_k - \mathbf{W}\hat{\mathbf{G}}_k \\ &= \hat{\mathbf{C}}_k + \mathbf{W}(\mathbf{\Gamma}\hat{\mathbf{C}} - \boldsymbol{\xi}) \\ &= \hat{\mathbf{C}}_k + \mathbf{W}e_k\mathbf{V}_k^*\end{aligned} \tag{11.1–35}$$

where **W** is the weighting matrix. Ideally, $\mathbf{W} = \mathbf{\Gamma}^{-1}$, or if $\mathbf{\Gamma}$ is estimated, then **W** can be set equal to the inverse of the estimate.

When the training sequence for the equalizer is periodic with period N, the covariance matrix $\mathbf{\Gamma}$ is Toeplitz and circulant and its inverse is circulant. In this case, the multiplication by the weighting matrix **W** can be simplified considerably by the implementation of a single finite duration impulse response (FIR) filter with weights equal to the first row of **W**, as indicated by Qureshi (1977). That is, the fast update algorithm that is equivalent to multiplying the gradient vector $\hat{\mathbf{G}}_k$ by **W** is simply implemented as shown in Figure 11.1–6, by inserting the FIR filter with N coefficients $w_0, w_1, \ldots, w_{N-1}$ in the path of the periodic input sequence before it is used for tap coefficient adjustment.

Qureshi (1977) described a method for estimating the weights from the received signal. The basic steps are as follows:

1. Collect one period (N symbols) of received data $v_0, v_1, \ldots, v_{N-1}$ in the equalizer delay line.
2. Compute the N-point discrete Fourier transform (DFT) of $\{v_n\}$ denoted as $\{R_n\}$.
3. Compute the discrete power spectrum $|R_n|^2$. If we neglect the noise, $|R_n|^2$ corresponds to N times the eigenvalues of the circulant covariance matrix of the signal at the input to the equalizer. Then, add N times the estimate of the noise variance σ^2 to $|R_n|^2$.
4. Compute the inverse DFT of the sequence $1/(|R_n|^2 + N\hat{\sigma}^2)$, $n = 0, 1, \ldots, N-1$. This yields the sequence $\{w_n\}$ of filter coefficients for the filter shown in Figure 11.1–6.
5. The algorithm for adjusting the equalizer tap coefficient now becomes

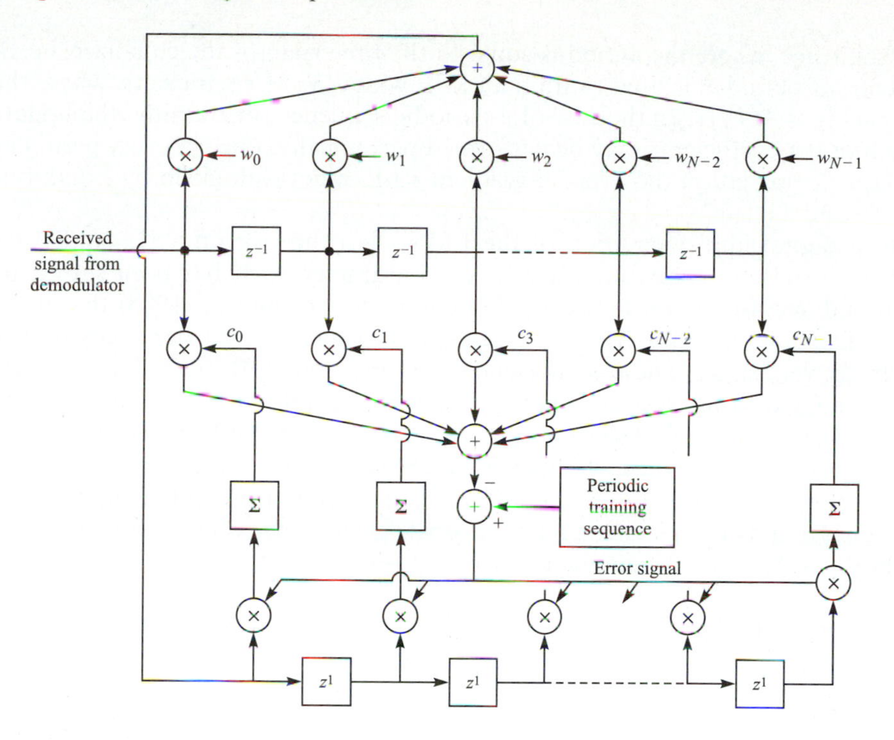

FIGURE 11.1–6
Fast start-up technique for an adaptive equalizer.

$$c_j^{(k+1)} = c_j^{(k)} - e_j \sum_{m=0}^{N-1} w_k v_{k-j-m}^*, \qquad j = 0, 1, \ldots, N-1 \tag{11.1–36}$$

11.1.6 Adaptive Fractionally Spaced Equalizer—The Tap Leakage Algorithm

As described in Section 10.2–4, an FSE is preferable to a symbol rate equalizer (SRE) when the channel characteristics are unknown at the receiver. In such a case, the FSE combines the operations of matched filtering and equalization of intersymbol interference into a single filter. By processing samples at the Nyquist rate, the FSE adapts its coefficients to compensate for any timing phase within a symbol. Thus, its performance is insensitive to the sampling time within a symbol interval, as discussed previously. Consequently, from a performance viewpoint, the FSE is equivalent to a matched filter followed by a symbol rate sampler, which is followed by an SRE.

The LMS algorithm and any of its variants can be used to adjust the coefficients of the FSE adaptively. Suitable training signals for initial adjustment may take the form of an aperiodic pseudorandom sequence or a periodic pseudorandom sequence, where the period is equal to the time span of the equalizer, i.e., a sequence of period P is used to train an FSE with PN/M coefficients, where the tap spacing is MT/N. In the case of a periodic sequence for training, the update of each of the coefficients may be performed periodically, once in every period of the sequence based on the average gradient LMS algorithm given by Equations 11.1–16 and 11.1–17.

In a digital implementation of the LMS algorithm for an FSE, some care must be exercised in selecting the step-size parameter Δ. It has been shown by Gitlin and Weinstein (1981) and further described by Qureshi (1985) that in an FSE, a fraction $(N-M)/N$ of the eigenvalues of the received signal covariance matrix are very small. These small eigenvalues and their corresponding eigenvectors are related to the spectral characteristics of the noise in the frequency band $(1+\beta)/2T \leqslant |f| \leqslant 1/T$. As a consequence, the output MSE becomes insensitive to deviations in the coefficient values corresponding to these eigenvalues. In such cases, errors due to finite precision arithmetic accumulate along the eigenvectors (frequency band) corresponding to the small eigenvalues and eventually cause overflows in the coefficient values, without significantly affecting the overall MSE.

A solution to this problem has been given in the paper by Gitlin et al. (1982). Instead of minimizing the MSE given by Equation 10.2–25, we minimize the performance index

$$J = J_{\mathrm{MSE}} + \mu \sum_{i=-K}^{K} |c_i|^2 \tag{11.1–37}$$

where J_{MSE} is the conventional MSE and μ is a small positive constant. Thus, the ill-conditioning of the received signal covariance matrix is avoided. The minimization of J leads to the following "modified LMS" algorithm.

$$\mathbf{C}_{k+1} = (1 - \Delta\mu)\mathbf{C}_k + \Delta\varepsilon_k \mathbf{V}_k^* \tag{11.1–38}$$

This algorithm is called the *tap-leakage algorithm*.

In adapting the tap coefficients of an FSE, the tap adjustments, as described above, are made periodically either at the symbol rate or slower when a periodic training sequence is transmitted. However, the samples at the input to the FSE occur at a faster rate. For example, if we consider a $T/2$ FSE, there are two samples per information symbol. An interesting question is whether or not it is possible to increase the initial convergence rate of an FSE by adapting its coefficients at the sampling rate. If the tap adjustments are performed at the sampling rate, one must generate additional desired signal values corresponding to sample values that fall between values of the desired symbols. That is, one must design a filter that performs intersymbol interpolation in order to generate the intermediate desired sample sequence. This problem has been considered by Gitlin and Weinstein (1981), Cioffi and Kailath (1984), and Ling (1989). The results given in the paper by Ling provide an answer to the question.

First we note that the initial convergence of the LMS algorithm depends on the number of nontrivial eigenvalues of the autocorrelation matrix of the received signal. This number is equal to the number of independent parameters that are to be optimized. For example, an SRE that has K taps and spans a time interval of KT seconds has K independent parameters to be optimized. In contrast, a $T/2$ complex-valued FSE that spans the same time interval has $2K$ tap coefficients, but its autocorrelation matrix has K nontrivial (and K trivial) eigenvalues and, thus, it has K independent parameters to be optimized. Consequently, the complex-valued $T/2$ FSE that is adapted at the symbol rate has the same convergence rate as the SRE. Now, if the complex-valued FSE employs interpolation to update its coefficients at all time instants $nT/2$, the number of independent parameters to be optimized is $2K$. In this case, there are two autocorrelation matrices, one corresponding to samples at $nT/2$, and the other corresponding to samples at $(nT+1)/2$, and each matrix has K nontrivial eigenvalues. That is, the $T/2$ FSE that employs interpolation adjusts one set of K parameters in one update and the second set of K parameters in the next update. Therefore, the convergence rate of the interpolated FSE will be approximately the same as the convergence rate of the symbol-updated FSE.

In the case of a phase-splitting FSE (PS-FSE), which is implemented at bandpass, with a time span of KT seconds and tap spacing T/N, where $N > 2$, e.g., $N = 3$ or 4, there are KN parameters to be optimized. In this case, Ling (1989) showed that the convergence rate of the PS-FSE was approximately a factor of 2 slower than the convergence rate of the conventional complex-valued FSE, when the PS-FSE is adjusted at the symbol rate. By employing ideal intersymbol interpolation, the convergence rate of the PS-FSE is increased by approximately a factor of 2 compared to symbol rate adjustment of the PS-FSE. Thus, the PS-FSE with intersymbol interpolation achieves the same convergence rate as the conventional complex-valued FSE that is adjusted at the symbol rate.

11.1.7 An Adaptive Channel Estimator for ML Sequence Detection

The ML sequence detection criterion implemented via the Viterbi algorithm as embodied in the metric computation given by Equation 10.1–23 and the probabilistic symbol-by-symbol detection algorithm described in Section 5.1.5 require knowledge of the equivalent discrete-time channel coefficients $\{f_k\}$. To accommodate a channel that is unknown or slowly time varying, one may include a channel estimator connected in parallel with the detection algorithm, as shown in Figure 11.1–7. The channel estimator, which is shown in Figure 11.1–8, is identical in structure to the linear transversal equalizer discussed previously in Section 11.1. In fact, the channel estimator is a replica of the equivalent discrete-time channel filter that models the intersymbol interference. The estimated tap coefficients, denoted by $\{\hat{f}_k\}$, are adjusted recursively to minimize the MSE between the actual received sequence and the output of the estimator. For example, the LMS steepest-descent algorithm in a decision-directed mode of operation is

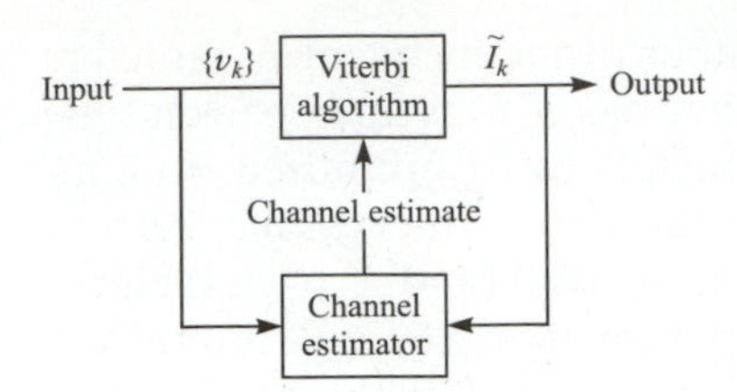

FIGURE 11.1–7
Block diagram of method for estimating the channel characteristics for the Viterbi algorithm.

$$\hat{\mathbf{f}}_{k+1} = \hat{\mathbf{f}}_k + \Delta\varepsilon_k \tilde{\mathbf{I}}_k^* \tag{11.1–39}$$

where $\hat{\mathbf{f}}_k$ is the vector of tap gain coefficients at the kth iteration, Δ is the step size, $\varepsilon_k = v_k - \hat{v}_k$ is the error signal, and $\tilde{\mathbf{I}}_k$ denotes the vector of detected information symbols in the channel estimator at the kth iteration.

We now show that when the MSE between v_k and $\hat{v}_k$ is minimized, the resulting values of the tap gain coefficients of the channel estimator are the values of the discrete-time channel model. For mathematical tractability, we assume that the detected information sequence $\{\tilde{I}_k\}$ is correct, i.e., $\{\tilde{I}_k\}$ is identical to the transmitted sequence $\{I_k\}$. This is a reasonable assumption when the system is operating at a low probability of error. Thus, the MSE between the received signal v_k and the estimate $\hat{v}_k$ is

$$J(\hat{\mathbf{f}}) = E\left(\left|v_k - \sum_{j=0}^{N-1} \hat{f}_j I_{k-j}\right|^2\right) \tag{11.1–40}$$

The tap coefficients $\{\hat{f}_k\}$ that minimize $J(\hat{\mathbf{f}})$ in Equation 11.1–40 satisfy the set of N linear equations

$$\sum_{j=0}^{N-1} \hat{f}_j \phi_{kj} = d_k, \qquad k = 0, 1, \ldots, N-1 \tag{11.1–41}$$

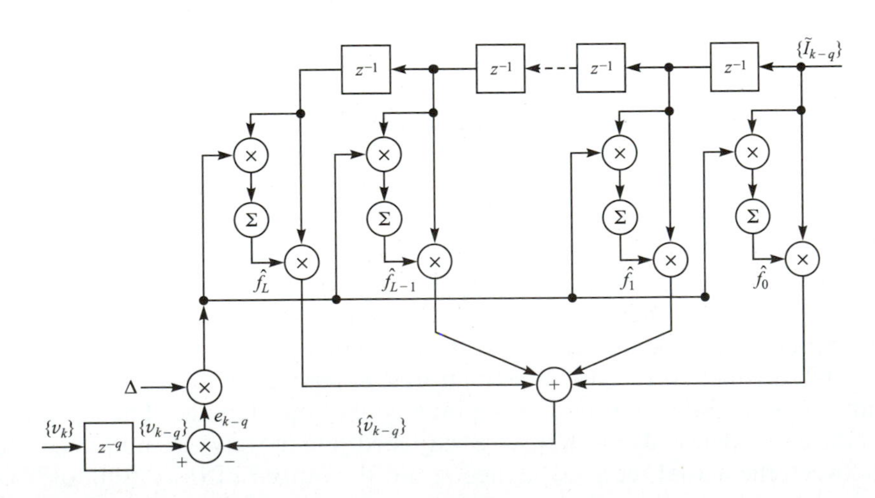

FIGURE 11.1–8
Adaptive transversal filter for estimating the channel dispersion.

where

$$\phi_{kj} = E(I_k I_j^*), \qquad d_k = \sum_{j=0}^{N-1} f_j \phi_j \tag{11.1–42}$$

From Equations 11.1–41 and 11.1–42, we conclude that, as long as the information sequence $\{I_k\}$ is uncorrelated, the optimum coefficients are exactly equal to the respective values of the equivalent discrete-time channel. It is also apparent that when the number of taps N in the channel estimator is greater than or equal to $L+1$, the optimum tap gain coefficients $\{\hat{f}_k\}$ are equal to the respective values of the $\{f_k\}$, even when the information sequence is correlated. Subject to the above conditions, the minimum MSE is simply equal to the noise variance N_0.

In the above discussion, the estimated information sequence at the output of the Viterbi algorithm or the probabilistic symbol-by-symbol algorithm was used in making adjustments of the channel estimator. For start-up operation, one may send a short training sequence to perform the initial adjustment of the tap coefficients, as is usually done in the case of the linear transversal equalizer. In an adaptive mode of operation, the receiver simply uses its own decisions to form an error signal.

11.2 ADAPTIVE DECISION-FEEDBACK EQUALIZER

As in the case of the linear adaptive equalizer, the coefficients of the feedforward filter and the feedback filter in a decision-feedback equalizer (DFE) may be adjusted recursively, instead of inverting a matrix as implied by Equation 10.3–3. Based on the minimization of the MSE at the output of the DFE, the steepest-descent algorithm takes the form

$$\mathbf{C}_{k+1} = \mathbf{C}_k + \Delta E(\varepsilon_k \mathbf{V}_k^*) \tag{11.2–1}$$

where $\mathbf{C}_k$ is the vector of equalizer coefficients in the kth signal interval, $E(\varepsilon_k \mathbf{V}_k^*)$ is the cross correlation of the error signal $\varepsilon_k = I_k - \hat{I}_k$ with $\mathbf{V}_k = [v_{k+K_1} \cdots v_k \; I_{k-1} \cdots I_{k-K_2}]^t$, representing the signal values in the feedforward and feedback filters at time $t = kT$. The MSE is minimized when the cross-correlation vector $E(\varepsilon_k \mathbf{V}_k^*) = 0$ as $k \to \infty$.

Since the exact cross-correlation vector is unknown at any time instant, we use as an estimate the vector $\varepsilon_k \mathbf{V}_k^*$ and average out the noise in the estimate through the recursive equation

$$\hat{\mathbf{C}}_{k+1} = \hat{\mathbf{C}}_k + \Delta \varepsilon_k \mathbf{V}_k^* \tag{11.2–2}$$

This is the LMS algorithm for the DFE.

As in the case of a linear equalizer, we may use a training sequence to adjust the coefficients of the DFE initially. Upon convergence to the (near-) optimum coefficients (minimum MSE), we may switch to a decision-directed mode where

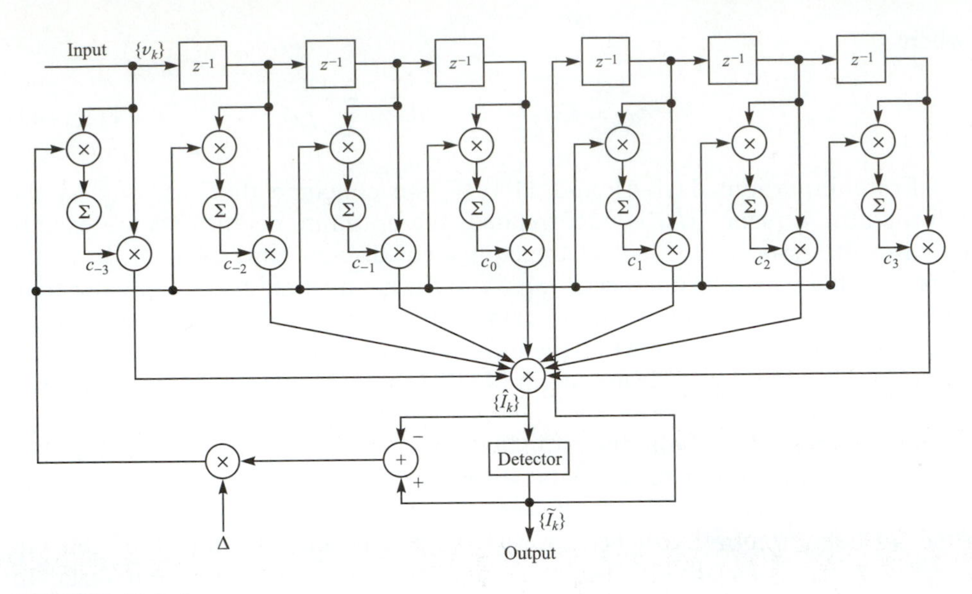

FIGURE 11.2–1
Decision-feedback equalizer.

the decisions at the output of the detector are used in forming the error signal ε_k and fed to the feedback filter. This is the adaptive mode of the DFE, which is illustrated in Figure 11.2–1. In this case, the recursive equation for adjusting the equalizer coefficient is

$$\tilde{\mathbf{C}}_{k+1} = \tilde{\mathbf{C}}_k + \Delta \tilde{\varepsilon}_k \mathbf{V}_k^* \tag{11.2–3}$$

where $\tilde{\varepsilon}_k = \tilde{I}_k - \hat{I}_k$ and $\mathbf{V}_k = [v_{k+K_1} \cdots v_k \tilde{I}_{k-1} \cdots \tilde{I}_{k-K_2}]^t$.

The performance characteristics of the LMS algorithm for the DFE are basically the same as the development given in Sections 11.1.3 and 11.1.4 for the linear adaptive equalizer.

11.3 ADAPTIVE EQUALIZATION OF TRELLIS-CODED SIGNALS

Bandwidth efficient trellis-coded modulation that was described in Section 8.3 is frequently used in digital communications over telephone channels to reduce the required SNR per bit for achieving a specified error rate. Channel distortion of the trellis-coded signal forces us to use adaptive equalization in order to reduce the intersymbol interference. The output of the equalizer is then fed to the Viterbi decoder, which performs soft-decision decoding of the trellis-coded signal.

The question that arises regarding such a receiver is, how do we adapt the equalizer in a data transmission mode? One possibility is to have the equalizer make its own decisions at its output solely for the purpose of generating an error signal for adjusting its tap coefficients, as shown in the block diagram in Figure 11.3–1. The problem with this approach is that such decisions are generally unreliable, since the pre-decoding coded symbol SNR is relatively low. A high error rate would cause a significant degradation in the operation of the equalizer, which would ultimately affect the reliability of the decisions at the output of the decoder. The more desirable alternative is to use the post-decoding decisions from the Viterbi decoder, which are much more reliable, to continuously adapt the equalizer. This approach is certainly preferable and viable when a linear equalizer is used prior to the Viterbi decoder. The decoding delay inherent in the Viterbi decoder can be overcome by introducing an identical delay in the tap weight adjustment of the equalizer coefficients as shown in Figure 11.3–2. The major price that must be paid for the added delay is that the step-size parameter in the LMS algorithm must be reduced, as described by Long et al. (1987, 1989), in order to achieve stability in the algorithm.

In channels with severe ISI, the linear equalizer is no longer adequate for compensating the channel intersymbol interference. Instead, we would like to use a DFE. But the DFE requires reliable decisions in its feedback filter in order to cancel out the intersymbol interference from previously detected symbols. Tentative decisions prior to decoding would be highly unreliable and, hence, inappropriate. Unfortunately, the conventional DFE cannot be cascaded with the Viterbi algorithm in which post-decoding decisions from the decoder are fed back to the DFE.

One alternative is to use the predictive DFE described in Section 10.3.3. In order to accommodate for the decoding delay as it affects the linear predictor, we introduce a periodic interleaver/deinterleaver pair that has the same delay as the Viterbi decoder and, thus, makes it possible to generate the appropriate error signal to the predictor as illustrated in the block diagram of Figure 11.3–3. The way in which a predictive DFE can be combined with Viterbi decoding to equalize trellis-coded signals is described and analyzed by Eyuboglu (1988). This same idea has been carried over to the equalization of fading multipath channels by Zhou et al. (1988, 1990), but the structure of the DFE was modified to use recursive least-squares lattice-type filters, which provide faster adaptation to the time variations encountered in the channel.

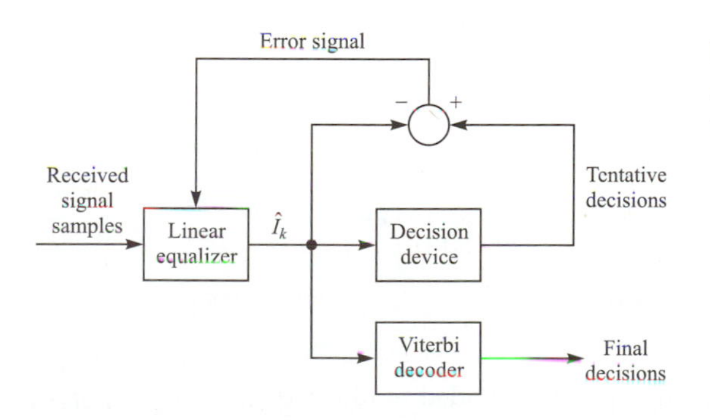

FIGURE 11.3–1
Adjustment of equalizer based on tentative decisions.

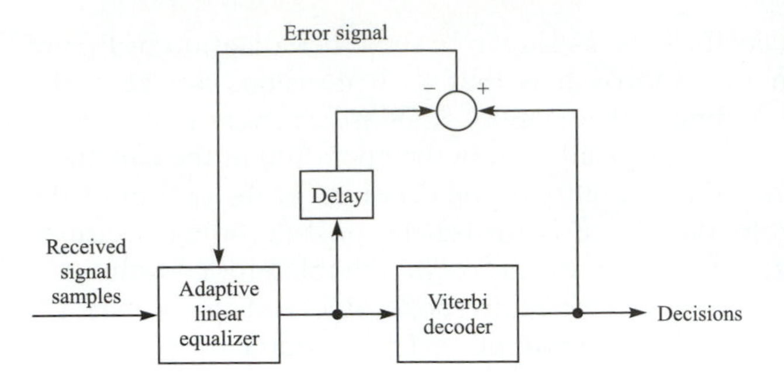

FIGURE 11.3–2
Adjustment of equalizer based on decisions from the Viterbi decoder.

Another approach that is effective in wireline channels, where the channel impulse response is essentially time invariant, is to place the feedback section of the DFE at the transmitter and, thus, eliminate the tail (postcursors) of the channel response prior to transmission. This is the approach previously described in Section 10.3.4, in which the information sequence is precoded using the Tomlinson–Harashima precoding scheme. Generally, this approach is implemented by sending a channel probe signal to measure the channel frequency or impulse response at the receiver and, thus, to inform the transmitter of the channel response in order to synthesize the precoder. An adaptive, fractionally spaced linear equalizer is implemented at the receiver, which serves as the feed forward filter of the DFE and, thus, compensates for any small time variations in the channel response.

Reduced-state Viterbi detection algorithms. From a performance viewpoint, the best method for detecting a TCM signal sequence that is corrupted by ISI is to model the ISI and the trellis code jointly by a single finite state machine and to

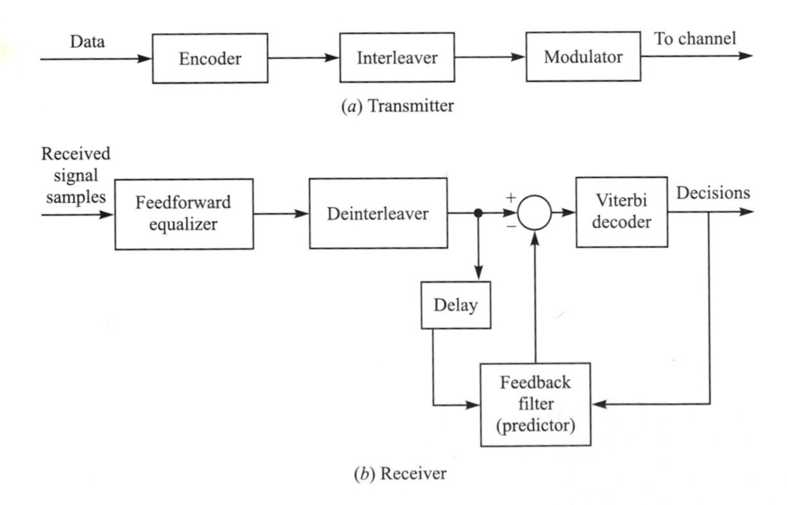

FIGURE 11.3–3
Use of predictive DFE with interleaving and trellis-coded modulation.

use the Viterbi algorithm on the combined trellis, as described in the papers by Chevillat and Eleftheriou (1988, 1989), Eyuboglu et al. (1988, 1989), and Wesolowski (1987b). By using a whitened matched filter (WMF) as described previously for the receiver front end, the model for the combined trellis encoder and ISI channel filter is illustrated in Figure 11.3–4, where the channel filter $F(z)$ is minimum phase. Thus, a TCM encoder that has S states and employs a signal constellation with 2^{m+1} signal points has a combined TCM/ISI trellis that has $S2^{mL}$ states and 2^m transitions (branches) emerging from each state. The states of the combined finite state machine may be denoted as

$$S_n = (I_{n-L}, I_{n-L+1}, \ldots, I_{n-1}, \theta_n) \tag{11.3–1}$$

where $\{I_n\}$ is the information symbol sequence and where θ_n is the encoder state.

The Viterbi decoder operates on the combined ISI and code trellis in the conventional way, by computing the branch metrics

$$\left| v_k - \sum_{i=0}^{L} f_i I_{k-i} \right|^2 \tag{11.3–2}$$

and incrementing the corresponding path metrics.

Clearly, the complexity of the Viterbi detector becomes prohibitively large when the span L of the ISI is large. In such a case, the decoder complexity can be reduced as described in Section 10.4, by truncating the effective channel memory to L_0 terms. With truncation, the combined TCM/ISI trellis has the $S2^{mL_0}$ states

$$S_n^{L_0} = (I_{n-L_0}, I_{n-L_0+1}, \ldots, I_{n-1}, \theta_n) \tag{11.3–3}$$

where $1 \leqslant L_0 \leqslant L$.

Thus, when $L_0 = 1$, the Viterbi algorithm operates directly on the TCM coded trellis and the L ISI terms are estimated and canceled. By selecting $L_0 > 1$, some ISI terms are kept while $L + 1 - L_0$ terms are canceled. To reduce the performance degradation due to tentative decisions in the Viterbi detector, the ISI cancelation is introduced into the branch metric computations using local feedback, as previously described in Section 10.4. Thus, the branch metrics computed in the Viterbi detector take the form

$$\left| v_k - \sum_{i=0}^{L_0-1} f_i I_{k-i} - \sum_{i=L_0}^{L+1} f_i \tilde{I}_{k-i}(S_n^{L_0}) \right|^2 \tag{11.3–4}$$

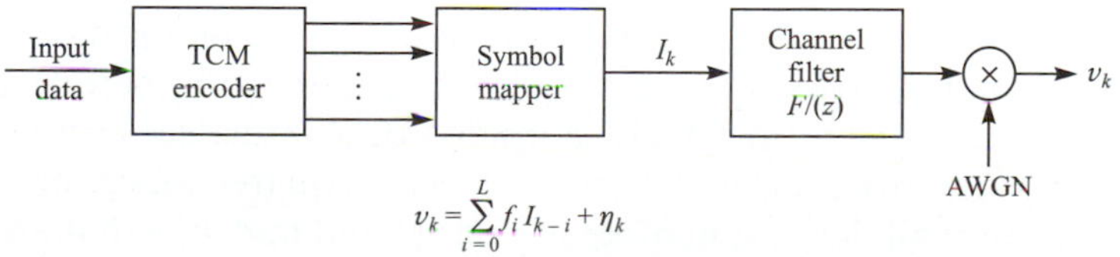

FIGURE 11.3–4
Model of TCM and ISI channel.

where $\tilde{I}_{k-i}(S_n^{L_0})$ denotes the estimated ISI term due to the symbols $\{I_{k-i}, L_0 < i < L\}$ involved in the truncation of the ISI based on local feedback.

In the case of an unknown channel characteristic, both the WMF and the channel estimator of $F(z)$ must be determined adaptively. This may be accomplished by adapting a complex-valued baseband FSE for the WMF and the channel estimator described previously in Section 11.1.7. Thus, a training sequence may be used for initial adjustment and decision-directed estimation may continue following the initial training sequence. The LMS algorithm may be used in both the training and decision-directed modes. Simulation results given by Chevillat and Eleftheriou (1989) demonstrate the superior performance of this adaptive WMF/reduced-state Viterbi detector compared to the combination of a linear equalizer followed by a Viterbi detector.

11.4
RECURSIVE LEAST-SQUARES ALGORITHMS FOR ADAPTIVE EQUALIZATION

The LMS algorithm that we described in Sections 11.1 and 11.2 for adaptively adjusting the tap coefficients of a linear equalizer or a DFE is basically a (stochastic) steepest-descent algorithm in which the true gradient vector is approximated by an estimate obtained directly from the data.

The major advantage of the steepest-descent algorithm lies in its computational simplicity. However, the price paid for the simplicity is slow convergence, especially when the channel characteristics result in an autocorrelation matrix $\mathbf{\Gamma}$ whose eigenvalues have a large spread, i.e., $\lambda_{max}/\lambda_{min} \gg 1$. Viewed in another way, the gradient algorithm has only a single adjustable parameter for controlling the convergence rate, namely, the parameter Δ. Consequently the slow convergence is due to this fundamental limitation. Two simple methods for increasing the convergence rate to some extent were described in Section 11.1.5.

In order to obtain faster convergence, it is necessary to devise more complex algorithms involving additional parameters. In particular, if the matrix $\mathbf{\Gamma}$ is $N \times N$ and has eigenvalues $\lambda_1, \lambda_2, \ldots, \lambda_N$, we may use an algorithm that contains N parameters—one for each of the eigenvalues. The optimum selection of these parameters to achieve rapid convergence is a topic of this section.

In deriving faster converging algorithms, we shall adopt a least-squares approach. Thus, we shall deal directly with the received data in minimizing the quadratic performance index, whereas previously we minimized the expected value of the squared error. Put simply, this means that the performance index is expressed in terms of a time average instead of a statistical average.

It is convenient to express the recursive least-squares algorithms in matrix form. Hence, we shall define a number of vectors and matrices that are needed in this development. In so doing, we shall change the notation slightly. Specifically, the estimate of the information symbol at time t, where t is an integer, from a linear equalizer is now expressed as

$$\hat{I}(t) = \sum_{j=-K}^{K} c_j(t-1)v_{t-j}$$

By changing the index j on $c_j(t-1)$ to run from $j = 0$ to $j = N - 1$ and simultaneously defining

$$y(t) = v_{t+K}$$

the estimate $\hat{I}(t)$ becomes

$$\begin{aligned}\hat{I}(t) &= \sum_{j=0}^{N-1} c_j(t-1)y(t-j) \\ &= \mathbf{C}'_N(t-1)\mathbf{Y}_N(t)\end{aligned} \tag{11.4–1}$$

where $\mathbf{C}_N(t-1)$ and $\mathbf{Y}_N(t)$ are, respectively, the column vectors of the equalizer coefficients $c_j(t-1)$, $j = 0, 1, \ldots, N-1$, and the input signals $y(t-j)$, $j = 0, 1, 2, \ldots, N-1$.

Similarly, in the decision-feedback equalizer, we have tap coefficients $c_j(t)$, $j = 0, 1, \ldots, N-1$, where the first $K_1 + 1$ are the coefficients of the feedforward filter and the remaining $K_2 = N - K_1 - 1$ are the coefficients of the feedback filter. The data in the estimate $\hat{I}(t)$ is $v_{t+K_1}, \ldots, v_{t+1}, \tilde{I}_{t-1}, \ldots, \tilde{I}_{t-K_2}$, where $\tilde{I}_{t-j}$, $1 \leqslant j \leqslant K_2$, denote the decisions on previously detected symbols. In this development, we neglect the effect of decision errors in the algorithms. Hence, we assume that $\tilde{I}_{t-j} = I_{t-j}$, $1 \leqslant j \leqslant K_2$. For notational convenience, we also define

$$y(t-j) = \begin{cases} v_{t+K_1-j} & (0 \leqslant j \leqslant K_1) \\ I_{t+K_1-j} & (K_1 < j \leqslant N-1) \end{cases} \tag{11.4–2}$$

Thus,

$$\begin{aligned}\mathbf{Y}_N(t) &= [y(t) \quad y(t-1) \cdots y(t-N+1)]^t \\ &= [v_{t+K_1} \cdots v_{t+1}\, v_t\, I_{t-1} \cdots I_{t-K_2}]^t\end{aligned} \tag{11.4–3}$$

11.4.1 Recursive Least-Squares (Kalman) Algorithm

The recursive least-squares (RLS) estimation of $\hat{I}(t)$ may be formulated as follows. Suppose we have observed the vectors $\mathbf{Y}_N(n)$, $n = 0, 1, \ldots, t$, and we wish to determine the coefficient vector $\mathbf{C}_N(t)$ of the equalizer (linear or decision-feedback) that minimizes the time-average weighted squared error

$$\mathcal{E}_N^{LS} = \sum_{n=0}^{t} w^{t-n} |e_N(n, t)|^2 \tag{11.4–4}$$

where the error is defined as

$$e_N(n, t) = I(n) - \mathbf{C}'_N(t)\mathbf{Y}_N(n) \tag{11.4–5}$$

and w represents a weighting factor $0 < w < 1$. Thus we introduce exponential weighting into past data, which is appropriate when the channel characteristics are time-variant. Minimization of $\mathcal{E}_N^{LS}$ with respect to the coefficient vector $\mathbf{C}_N(t)$ yields the set of linear equations

$$\mathbf{R}_N(t)\mathbf{C}_N(t) = \mathbf{D}_N(t) \tag{11.4–6}$$

where $\mathbf{R}_N(t)$ is the signal correlation matrix defined as

$$\mathbf{R}_N(t) = \sum_{n=0}^{t} w^{t-n}\mathbf{Y}_N^*(n)\mathbf{Y}_N'(n) \tag{11.4–7}$$

and $\mathbf{D}_N(t)$ is the cross-correlation vector

$$\mathbf{D}_N(t) = \sum_{n=0}^{t} w^{t-n}I(n)\mathbf{Y}_N^*(n) \tag{11.4–8}$$

The solution of Equation 11.4–6 is

$$\mathbf{C}_N(t) = \mathbf{R}_N^{-1}(t)\mathbf{D}_N(t) \tag{11.4–9}$$

The matrix $\mathbf{R}_N(t)$ is akin to the statistical autocorrelation matrix $\mathbf{\Gamma}_N$, while the vector $\mathbf{D}_N(t)$ is akin to the cross-correlation vector ξ_N, defined previously. We emphasize, however, that $\mathbf{R}_N(t)$ is not a Toeplitz matrix. We also should mention that, for small values of t, $\mathbf{R}_N(t)$ may be ill conditioned; hence, it is customary to initially add the matrix $\delta\mathbf{I}_N$ to $\mathbf{R}_N(t)$, where δ is a small positive constant and $\mathbf{I}_N$ is the identity matrix. With exponential weighting into the past, the effect of adding $\delta\mathbf{I}_N$ dissipates with time.

Now suppose we have the solution 11.4–9 for time $t-1$, i.e., $\mathbf{C}_N(t-1)$, and we wish to compute $\mathbf{C}_N(t)$. It is inefficient, and, hence, impractical to solve the set of N linear equations for each new signal component that is received. To avoid this, we proceed as follows. First, $\mathbf{R}_N(t)$ may be computed recursively as

$$\mathbf{R}_N(t) = w\mathbf{R}_N(t-1) + \mathbf{Y}_N^*(t)\mathbf{Y}_N'(t) \tag{11.4–10}$$

We call Equation 11.4–10 the *time-update equation* for $\mathbf{R}_N(t)$.

Since the inverse of $\mathbf{R}_N(t)$ is needed in Equation 11.4–9, we use the matrix-inverse identity

$$\mathbf{R}_N^{-1}(t) = \frac{1}{w}\left[\mathbf{R}_N^{-1}(t-1) - \frac{\mathbf{R}_N^{-1}(t-1)\mathbf{Y}_N^*(t)\mathbf{Y}_N'(t)\mathbf{R}_N^{-1}(t-1)}{w + \mathbf{Y}_N'(t)\mathbf{R}_N^{-1}(t-1)\mathbf{Y}_N^*(t)}\right] \tag{11.4–11}$$

Thus $\mathbf{R}_N^{-1}(t)$ may be computed recursively according to Equation 11.4–11.

For convenience, we define $\mathbf{P}_N(t) = \mathbf{R}_N^{-1}(t)$. It is also convenient to define an N-dimensional vector, called the *Kalman gain vector*, as

$$\mathbf{K}_N(t) = \frac{1}{w + \mu_N(t)}\mathbf{P}_N(t-1)\mathbf{Y}_N^*(t) \tag{11.4–12}$$

where $\mu_N(t)$ is a scalar defined as

$$\mu_N(t) = \mathbf{Y}'_N(t)\mathbf{P}_N(t-1)\mathbf{Y}^*_N(t) \tag{11.4–13}$$

With these definitions, Equations 11.4–11 becomes

$$\mathbf{P}_N(t) = \frac{1}{w}[\mathbf{P}_N(t-1) - \mathbf{K}_N(t)\mathbf{Y}'_N(t)\mathbf{P}_N(t-1)] \tag{11.4–14}$$

Suppose we postmultiply both sides of Equation 11.4–14 by $\mathbf{Y}^*_N(t)$. Then

$$\begin{aligned}\mathbf{P}_N(t)\mathbf{Y}^*_N(t) &= \frac{1}{w}[\mathbf{P}_N(t-1)\mathbf{Y}^*_N(t) - \mathbf{K}_N(t)\mathbf{Y}'_N(t)\mathbf{P}_N(t-1)\mathbf{Y}^*_N(t)] \\ &= \frac{1}{w}\{[w + \mu_N(t)]\mathbf{K}_N(t) - \mathbf{K}_N(t)\mu_N(t)\} \\ &= \mathbf{K}_N(t)\end{aligned} \tag{11.4–15}$$

Therefore, the Kalman gain vector may also be defined as $\mathbf{P}_N(t)\mathbf{Y}_N(t)$.

Now we use the matrix inversion identity to derive an equation for obtaining $\mathbf{C}_N(t)$ from $\mathbf{C}_N(t-1)$. Since

$$\mathbf{C}_N(t) = \mathbf{P}_N(t)\mathbf{D}_N(t)$$

and

$$\mathbf{D}_N(t) = w\mathbf{D}_N(t-1) + I(t)\mathbf{Y}^*_N(t) \tag{11.4–16}$$

we have

$$\begin{aligned}\mathbf{C}_N(t) &= \frac{1}{w}[\mathbf{P}_N(t-1) - \mathbf{K}_N(t)\mathbf{Y}'_N(t)\mathbf{P}_N(t-1)][w\mathbf{D}_N(t-1) + I(t)\mathbf{Y}^*_N(t)] \\ &= \mathbf{P}_N(t-1)\mathbf{D}_N(t-1) + \frac{1}{w}I(t)\mathbf{P}_N(t-1)\mathbf{Y}^*_N(t) \\ &\quad - \mathbf{K}_N(t)\mathbf{Y}'_N(t)\mathbf{P}_N(t-1)\mathbf{D}_N(t-1) \\ &\quad - \frac{1}{w}I(t)\mathbf{K}_N(t)\mathbf{Y}'_N(t)\mathbf{P}_N(t-1)\mathbf{Y}^*_N(t) \\ &= \mathbf{C}_N(t-1) + \mathbf{K}_N(t)[I(t) - \mathbf{Y}'_N(t)\mathbf{C}_N(t-1)]\end{aligned} \tag{11.4–17}$$

Note that $\mathbf{Y}'_N(t)\mathbf{C}_N(t-1)$ is the output of the equalizer at time t, i.e.,

$$\hat{I}(t) = \mathbf{Y}'_N(t)\mathbf{C}_N(t-1) \tag{11.4–18}$$

and

$$e_N(t, t-1) = I(t) - \hat{I}(t) \equiv e_N(t) \tag{11.4–19}$$

is the error between the desired symbol and the estimate. Hence, $\mathbf{C}_N(t)$ is updated recursively according to the relation

$$\mathbf{C}_N(t) = \mathbf{C}_N(t-1) + \mathbf{K}_N(t)e_N(t) \tag{11.4–20}$$

The residual MSE resulting from this optimization is

$$\mathcal{E}_{N\,\min}^{LS} = \sum_{m=0}^{t} w^{t-n}|I(n)|^2 - \mathbf{C}'_N(t)\mathbf{D}^*_N(t) \tag{11.4–21}$$

To summarize, suppose we have $\mathbf{C}_N(t-1)$ and $\mathbf{P}_N(t-1)$. When a new signal component is received, we have $\mathbf{Y}_N(t)$. Then the recursive computation for the time update of $\mathbf{C}_N(t)$ and $\mathbf{P}_N(t)$ proceeds as follows:

- Compute output:

$$\hat{I}(t) = \mathbf{Y}'_N(t)\mathbf{C}_N(t-1)$$

- Compute error:

$$e_N(t) = I(t) - \hat{I}(t)$$

- Compute Kalman gain vector:

$$\mathbf{K}_N(t) = \frac{\mathbf{P}_N(t-1)\mathbf{Y}'_N(t)}{w + \mathbf{Y}'_N(t)\mathbf{P}_N(t-1)\mathbf{Y}^*_N(t)}$$

- Update inverse of the correlation matrix:

$$\mathbf{P}_N(t) = \frac{1}{w}[\mathbf{P}_N(t-1) - \mathbf{K}_N(t)\mathbf{Y}'_N(t)\mathbf{P}_N(t-1)]$$

- Update coefficients:

$$\begin{aligned}\mathbf{C}_N(t) &= \mathbf{C}_N(t-1) + \mathbf{K}_N(t)e_N(t) \\ &= \mathbf{C}_N(t-1) + \mathbf{P}_N(t)\mathbf{Y}^*_N(t)e_N(t)\end{aligned} \tag{11.4–22}$$

The algorithm described by Equation 11.4–22 is called the *RLS direct form* or *Kalman algorithm*. It is appropriate when the equalizer has a transversal (direct-form) structure.

Note that the equalizer coefficients change with time by an amount equal to the error $e_N(t)$ multipled by the Kalman gain vector $\mathbf{K}_N(t)$. Since $\mathbf{K}_N(t)$ is N-dimensional, each tap coefficient in effect is controlled by one of the elements of $\mathbf{K}_N(t)$. Consequently rapid convergence is obtained. In contrast, the steepest-descent algorithm, expressed in our present notation, is

$$\mathbf{C}_N(t) = \mathbf{C}_N(t-1) + \Delta\mathbf{Y}^*_N(t)e_N(t) \tag{11.4–23}$$

and the only variable parameter is the step size Δ.

Figure 11.4–1 illustrates the initial convergence rate of these two algorithms for a channel with fixed parameters $f_0 = 0.26$, $f_1 = 0.93$, $f_2 = 0.26$, and a linear equalizer with 11 taps. The eigenvalue ratio for this channel is $\lambda_{\max}/\lambda_{\min} = 11$. All the equalizer coefficients were initialized to zero. The steepest-descent algorithm was implemented with $\Delta = 0.020$. The superiority of the Kalman algorithm is clearly evident. This is especially important in a time-variant channel. For example, the time variations in the characteristics of an (ionospheric) high-frequency (HF) radio channel are too rapid to be equalized by the gradient

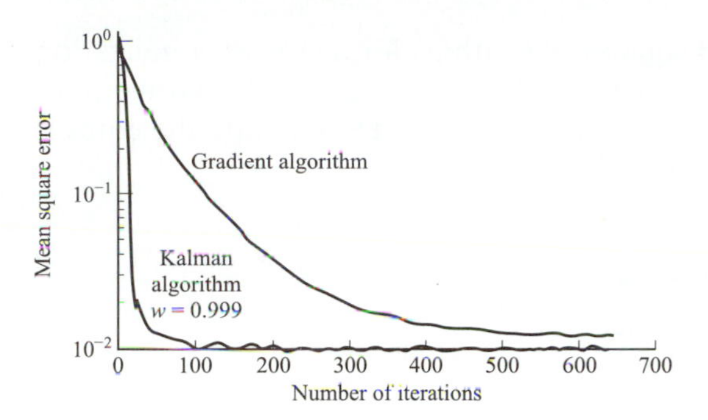

FIGURE 11.4–1
Comparison of convergence rate for the Kalman and gradient algorithms.

algorithm, but the Kalman algorithm adapts sufficiently rapidly to track such variations.

In spite of its superior convergence performance, the Kalman algorithm described above has two disadvantages. One is its complexity. The second is its sensitivity to roundoff noise that accumulates due to the recursive computations. The latter may cause instabilities in the algorithm.

The number of computations or operations (multiplications, divisions, and subtractions) in computing the variables in Equation 11.4–22 is proportional to N^2. Most of these operations are involved in the updating of $\mathbf{P}_N(t)$. This part of the computation is also susceptible to roundoff noise. To remedy that problem, algorithms have been developed that avoid the computation of $\mathbf{P}_N(t)$ according to Equation 11.4–14. The basis of these algorithms lies in the decomposition of $\mathbf{P}_N(t)$ in the form

$$\mathbf{P}_N(t) = \mathbf{S}_N(t)\mathbf{\Lambda}_N(t)\mathbf{S}'_N(t) \tag{11.4–24}$$

where $\mathbf{S}_N(t)$ is a lower-triangular matrix whose diagonal elements are unity, and $\mathbf{\Lambda}_N(t)$ is a diagonal matrix. Such a decomposition is called a *square-root factorization* (see Bierman, 1977). This factorization is described in Appendix D. In a square-root algorithm, $\mathbf{P}_N(t)$ is not updated as in Equation 11.4–14 nor is it computed. Instead, the time updating is performed on $\mathbf{S}_N(t)$ and $\mathbf{\Lambda}_N(t)$.

Square-root algorithms are frequently used in control systems applications in which Kalman filtering is involved. In digital communications, the square-root Kalman algorithm has been implemented in a decision-feedback-equalized PSK modem designed to transmit at high speed over high-frequency radio channels with a nominal 3-kHz bandwidth. This algorithm is described in the paper by Hsu (1982). It has a computational complexity of $1.5N^2 + 6.5N$ (complex-valued multiplications and divisions per output symbol). It is also numerically stable and exhibits good numerical properties. For a detailed discussion of square-root algorithms in sequential estimation, the reader is referred to the book by Bierman (1977).

It is also possible to derive RLS algorithms with computational complexities that grow linearly with the number N of equalizer coefficients. Such algorithms are generally called *fast RLS algorithms* and have been described in the papers by Carayannis et al. (1983), Cioffi and Kailath (1984), and Slock and Kailath (1991).

Another class of recursive least squares algorithms for adaptive equalization are based on the lattice equalizer structure. Below, we derive the lattice filter structure from the transversal filter structure and, thus, demonstrate the equivalence of the two structures.

11.4.2 Linear Prediction and the Lattice Filter

In Chapter 3, we considered the linear prediction of a signal, in the context of speech encoding. In this section, we shall establish the connection between linear prediction and a lattice filter.

The linear prediction problem may be stated as follows: given a set of data $y(t-1), y(t-2), \ldots, y(t-p)$, predict the value of the next data point $y(t)$. The predictor of order p is

$$\hat{y}(t) = \sum_{k=1}^{p} a_{pk} y(t-k) \tag{11.4–25}$$

Minimization of the MSE, defined as

$$\begin{aligned} \mathcal{E}_p &= E[y(t) - \hat{y}(t)]^2 \\ &= E\left[y(t) - \sum_{k=1}^{p} a_{pk} y(t-k)\right]^2 \end{aligned} \tag{11.4–26}$$

with respect to the predictor coefficients $\{a_{pk}\}$ yields the set of linear equations

$$\sum_{k=1}^{p} a_{pk} \phi(k-l) = \phi(l), \qquad l = 1, 2, \ldots, p \tag{11.4–27}$$

where

$$\phi(l) = E[y(t)y(t+l)]$$

These are called the *normal equations* or the *Yule–Walker equations.*

The matrix $\mathbf{\Phi}$ with elements $\phi(k-l)$ is a Toeplitz matrix, and, hence, the Levinson–Durbin algorithm described in Appendix A provides an efficient means for solving the linear equations recursively, starting with a first-order predictor and proceeding recursively to the solution of the coefficients for the predictor of order p. The recursive relations for the Levinson–Durbin algorithm are

$$\begin{aligned} a_{11} &= \frac{\phi(1)}{\phi(0)}, \qquad \mathcal{E}_0 = \phi(0) \\ a_{mm} &= \frac{\phi(m) - \mathbf{A}_m^t \boldsymbol{\phi}_{m-1}^r}{\mathcal{E}_{m-1}} \\ a_{mk} &= a_{m-1k} - a_{mm} a_{m-1m-k} \\ \mathcal{E}_m &= \mathcal{E}_{m-1}(1 - a_{mm}^2) \end{aligned} \tag{11.4–28}$$

for $m = 1, 2, \ldots, p$, where the vectors $\mathbf{A}_{m-1}$ and $\boldsymbol{\phi}_{m-1}^r$ are defined as

$$\begin{aligned} \mathbf{A}_{m-1} &= [a_{m-11} \quad a_{m-12} \cdots a_{m-1m-1}]^t \\ \boldsymbol{\phi}_{m-1}^r &= [\phi(m-1) \quad \phi(m-2) \cdots \phi(1)]^t \end{aligned}$$

The linear prediction filter of order m may be realized as a transversal filter with transfer function

$$A_m(z) = 1 - \sum_{k=1}^{m} a_{mk} z^{-k} \tag{11.4–29}$$

Its input is the data $\{y(t)\}$ and its output is the error $e(t) = y(t) - \hat{y}(t)$. The prediction filter can also be realized in the form of a lattice, as we now demonstrate.

Our starting point is the use of the Levinson–Durbin algorithm for the predictor coefficients a_{mk} in Equation 11.4–29. This substitution yields

$$\begin{aligned} A_m(z) &= 1 - \sum_{k=1}^{m-1} (a_{m-1k} - a_{mm} a_{m-1m-k}) z^{-k} - a_{mm} z^{-m} \\ &= 1 - \sum_{k=1}^{m-1} a_{m-1k} z^{-k} - a_{mm} z^{-m} \left(1 - \sum_{k=1}^{m-1} a_{m-1k} z^{k}\right) \\ &= A_{m-1}(z) - a_{mm} z^{-m} A_{m-1}(z^{-1}) \end{aligned} \tag{11.4–30}$$

Thus we have the transfer function of the mth-order predictor in terms of the transfer function of the $(m-1)$th-order predictor.

Now suppose we define a filter with transfer function $G_m(z)$ as

$$G_m(z) = z^{-m} A_m(z^{-1}) \tag{11.4–31}$$

Then Equation 11.4–30 may be expressed ass

$$A_m(z) = A_{m-1}(z) - a_{mm} z^{-1} G_{m-1}(z) \tag{11.4–32}$$

Note that $G_{m-1}(z)$ represents a transversal filter with tap coefficients $(-a_{m-1m-1}, -a_{m-1m-2}, \ldots, -a_{m-11}, 1)$, while the coefficients of $A_{m-1}(z)$ are exactly the same except that they are given in reverse order.

More insight into the relationship between $A_m(z)$ and $G_m(z)$ can be obtained by computing the output of these two filters to an input sequence $y(t)$. Using z-transform relations, we have

$$A_m(z)Y(z) = A_{m-1}(z)Y(z) - a_{mm}z^{-1}G_{m-1}(z)Y(z) \tag{11.4–33}$$

We define the outputs of the filters as

$$\begin{aligned} F_m(z) &= A_m(z)Y(z) \\ B_m(z) &= G_m(z)Y(z) \end{aligned} \tag{11.4–34}$$

Then Equation 11.4–33 becomes

$$F_m(z) = F_{m-1}(z) - a_{mm}z^{-1}B_{m-1}(z) \tag{11.4–35}$$

In the time domain, the relation in Equation 11.4–35 becomes

$$f_m(t) = f_{m-1}(t) - a_{mm}b_{m-1}(t-1), \qquad m \geqslant 1 \tag{11.4–36}$$

where

$$f_m(t) = y(t) - \sum_{k=1}^{m-1} a_{mk}y(t-k) \tag{11.4–37}$$

$$b_m(t) = y(t-m) - \sum_{k=1}^{m-1} a_{mk}y(t-m+k) \tag{11.4–38}$$

To elaborate, $f_m(t)$ in Equation 11.4–37 represents the error of an mth-order forward predictor, while $b_m(t)$ represents the error of an mth-order backward predictor.

The relation in Equation 11.4–36 is one of two that specifies a lattice filter. The second relation is obtained from $G_m(z)$ as follows:

$$\begin{aligned} G_m(z) &= z^{-m}A_m(z^{-1}) \\ &= z^{-m}[A_{m-1}(z^{-1}) - a_{mm}z^m A_{m-1}(z)] \\ &= z^{-1}G_{m-1}(z) - a_{mm}A_{m-1}(z) \end{aligned} \tag{11.4–39}$$

Now, if we multiply both sides of Equation 11.4–39 by $Y(z)$ and express the result in terms of $F_m(z)$ and $B_m(z)$ using the definitions in Equation 11.4–34, we obtain

$$B_m(z) = z^{-1}B_{m-1}(z) - a_{mm}F_{m-1}(z) \tag{11.4–40}$$

By transforming Equation 11.4–40 into the time domain, we obtain the second relation that corresponds to the lattice filter, namely,

$$b_m(t) = b_{m-1}(t-1) - a_{mm}f_{m-1}(t), \qquad m \geqslant 1 \tag{11.4–41}$$

The initial condition is

$$f_0(t) = b_0(t) = y(t) \tag{11.4–42}$$

The lattice filter described by the recursive relations in Equations 11.4–36 and 11.4–41 is illustrated in Figure 11.4–2. Each stage is characterized by its own multiplication factor $\{a_{ii}\}$, $i = 1, 2, \ldots, m$, which is defined in the Levinson–

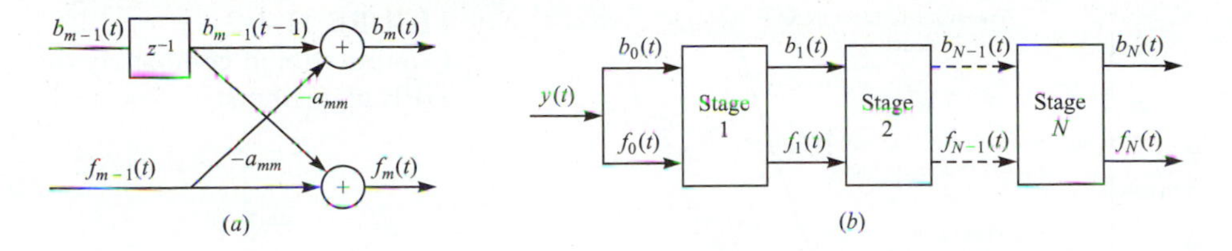

FIGURE 11.4–2
A lattice filter.

Durbin algorithm. The forward and backward errors $f_m(t)$ and $b_m(t)$ are usually called the *residuals*. The mean square value of these residuals is

$$\mathcal{E}_m = E[f_m^2(t)] = E[b_m^2(t)] \tag{11.4–43}$$

$\mathcal{E}_m$ is given recursively, as indicated in the Levinson–Durbin algorithm, by

$$\begin{aligned} \mathcal{E}_m &= \mathcal{E}_{m-1}(1 - a_{mm}^2) \\ &= \mathcal{E}_0 \prod_{i=1}^{m}(1 - a_{ii}^2) \end{aligned} \tag{11.4–44}$$

where $\mathcal{E}_0 = \phi(0)$.

The residuals $\{f_m(t)\}$ and $\{b_m(t)\}$ satisfy a number of interesting properties, as described by Makhoul (1978). Most important of these are the orthogonality properties

$$\begin{aligned} E[b_m(t)b_n(t)] &= \mathcal{E}_m\delta_{mn} \\ E[f_m(t+m)f_n(t+n)] &= \mathcal{E}_m\delta_{mn} \end{aligned} \tag{11.4–45}$$

Furthermore, the cross correlation between $f_m(t)$ and $b_n(t)$ is

$$E[f_m(t)b_n(t)] = \begin{cases} a_{nn}\mathcal{E}_m & (m \geqslant n) \\ 0 & (m < n) \end{cases} \qquad m, n \geqslant 0 \tag{11.4–46}$$

As a consequence of the orthogonality properties of the residuals, the different sections of the lattice exhibit a form of independence that allows us to add or delete one or more of the last stages without affecting the parameters of the remaining stages. Since the residual mean square error $\mathcal{E}_m$ decreases monotonically with the number of sections, $\mathcal{E}_m$ can be used as a performance index in determining where the lattice should be terminated.

From the above discussion, we observe that a linear prediction filter can be implemented either as a linear transversal filter or as a lattice filter. The lattice filter is order-recursive, and, as a consequence, the number of sections it contains can be easily increased or decreased without affecting the parameters of the remaining sections. In contrast, the coefficients of a transversal filter obtained on the basis of the RLS criterion are interdependent. This means that an increase or a decrease in the size of the filter results in a change in all coefficients.

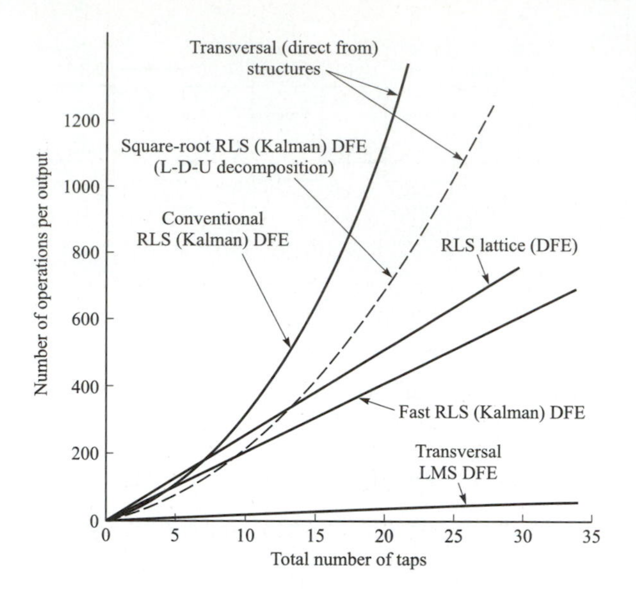

FIGURE 11.4–3
Computational complexity of DFE algorithms.

Consequently, the Kalman algorithm described in Section 11.4.1 is recursive in time but not in order.

Based on least-squares optimization, RLS lattice equalization algorithms have been developed whose computational complexity grows linearly with the number N of filter coefficients (lattice stages). Hence, the lattice equalizer structure is computationally competitive with the direct-form fast RLS equalizer algorithms. For example, Figure 11.4–3 illustrates the computational complexity (number of multiplications and divisions per output symbol) of transversal and lattice, symbol-spaced DFE filter structures. Observe that for equalizer lengths of fewer than 10 taps, the difference in computational complexity among the different structures and algorithms is relatively small. However, as the number of taps increases, the lattice RLS algorithm and the fast (transversal) RLS algorithm are significantly less complex than the conventional and square-root RLS algorithms. Of course, all the RLS algorithms are computationally more complex than the LMS algorithm. RLS lattice algorithms are described in the papers by Morf (1977), Morf and Lee (1978), and Morf et al. (1977a,b,c), Satorius and Alexander (1979), Satorius and Pack (1981), Ling and Proakis (1982, 1984c, 1985, 1986).

RLS lattice algorithms have the distinct feature of being numerically robust to round-off error inherent in digital implementations of the algorithms. A treatment of their numerical properties may be found in the papers by Ling and Proakis (1984a) and Ling et al. (1986a,b).

11.5 SELF-RECOVERING (BLIND) EQUALIZATION

In the conventional zero-forcing or minimum MSE equalizers, we assumed that a known training sequence is transmitted to the receiver for the purpose of initially adjusting the equalizer coefficients. However, there are some applications, such as multipoint communication networks, where it is desirable for the receiver to synchronize to the received signal and to adjust the equalizer without having a known training sequence available. Equalization techniques based on initial adjustment of the coefficients without the benefit of a training sequence are said to be *self-recovering* or *blind*.

Beginning with the paper by Sato (1975), three different classes of adaptive blind equalization algorithms have been developed over the past two decades. One class of algorithms is based on steepest descent for adaptation of the equalizer. A second class of algorithms is based on the use of second- and higher-order (generally, fourth-order) statistics of the received signal to estimate the channel characteristics and to design the equalizer. More recently, a third class of blind equalization algorithms based on the maximum-likelihood criterion have been investigated. In this section, we briefly describe these approaches and give several relevant references to the literature.

11.5.1 Blind Equalization Based on the Maximum-Likelihood Criterion

It is convenient to use the equivalent, discrete-time channel model described in Section 10.1.2. Recall that the output of this channel model with ISI is

$$v_n = \sum_{k=0}^{L} f_k I_{n-k} + \eta_n \tag{11.5–1}$$

where $\{f_k\}$ are the equivalent discrete-time channel coefficients, $\{I_n\}$ represents the information sequence, and $\{\eta_n\}$ is a white Gaussian noise sequence.

For a block of N received data points, the (joint) probability density function of the received data vector $\mathbf{v} = [v_1 \ v_2 \ \cdots \ v_N]^t$ conditioned on knowing the impulse response vector $\mathbf{f} = [f_0 \ f_1 \ \cdots \ f_L]^t$ and the data vector $\mathbf{I} = [I_1 \ I_2 \ \cdots I_N]^t$ is

$$p(\mathbf{v}|\mathbf{f}, \mathbf{I}) = \frac{1}{(2\pi\sigma^2)^N} \exp\left(-\frac{1}{2\sigma^2}\sum_{n=1}^{N}\left|v_n - \sum_{k=0}^{L} f_k I_{n-k}\right|^2\right) \tag{11.5–2}$$

The joint maximum-likelihood estimates of $\mathbf{f}$ and $\mathbf{I}$ are the values of these vectors that maximize the joint probability density function $p(\mathbf{v}|\mathbf{f}, \mathbf{I})$ or, equivalently, the values of $\mathbf{f}$ and $\mathbf{I}$ that minimize the term in the exponent. Hence, the ML solution is simply the minimum over $\mathbf{f}$ and $\mathbf{I}$ of the metric

$$DM(\mathbf{I}, \mathbf{f}) = \sum_{n=1}^{N} \left| v_n - \sum_{k=0}^{L} f_k I_{n-k} \right|^2 \\ = \|\mathbf{v} - \mathbf{A}\mathbf{f}\|^2 \tag{11.5–3}$$

where the matrix $\mathbf{A}$ is called the *data matrix* and is defined as

$$\mathbf{A} = \begin{bmatrix} I_1 & 0 & 0 & \dots & 0 \\ I_2 & I_1 & 0 & \dots & 0 \\ I_3 & I_2 & I_1 & \dots & 0 \\ \vdots & \vdots & \vdots & & \vdots \\ I_N & I_{N-1} & I_{N-2} & \dots & I_{N-L} \end{bmatrix} \tag{11.5–4}$$

We make several observations. First of all, we note that when the data vector $\mathbf{I}$ (or the data matrix $\mathbf{A}$) is known, as is the case when a training sequence is available at the receiver, the ML channel impulse response estimate obtained by minimizing Equation 11.5–3 over $\mathbf{f}$ is

$$\mathbf{f}_{ML}(\mathbf{I}) = (\mathbf{A}'\mathbf{A})^{-1}\mathbf{A}^t\mathbf{v} \tag{11.5–5}$$

On the other hand, when the channel impulse response $\mathbf{f}$ is known, the optimum ML detector for the data sequence $\mathbf{I}$ performs a trellis search (or tree search) by utilizing the Viterbi algorithm for the ISI channel.

When neither $\mathbf{I}$ nor $\mathbf{f}$ are known, the minimization of the performance index $DM(\mathbf{I}, \mathbf{f})$ may be performed jointly over $\mathbf{I}$ and $\mathbf{f}$. Alternatively, $\mathbf{f}$ may be estimated from the probability density function $p(\mathbf{v}|\mathbf{f})$, which may be obtained by averaging $p(\mathbf{v}, \mathbf{f}|\mathbf{I})$ over all possible data sequences. That is,

$$p(\mathbf{v}|\mathbf{f}) = \sum_m p(\mathbf{v}, \mathbf{I}^{(m)}|\mathbf{f}) \\ = \sum_m p(\mathbf{v}|\mathbf{I}^{(m)}, \mathbf{f})P(\mathbf{I}^{(m)}) \tag{11.5–6}$$

where $P(\mathbf{I}^{(m)})$ is the probability of the sequence $\mathbf{I} = \mathbf{I}^{(m)}$, for $m = 1, 2, \dots, M^N$, and M is the size of the signal constellation.

Channel estimation based on average over data sequences. As indicated in the above discussion, when both $\mathbf{I}$ and $\mathbf{f}$ are unknown, one approach is to estimate the impulse response $\mathbf{f}$ after averaging the probability density $p(\mathbf{v}, \mathbf{I}|\mathbf{f})$ over all possible data sequences. Thus, we have

$$p(\mathbf{v}|\mathbf{f}) = \sum_m p(\mathbf{v}|\mathbf{I}^{(m)}, \mathbf{f})P(\mathbf{I}^{(m)}) \\ = \sum_m \left[\frac{1}{(2\pi\sigma^2)^N} \exp\left(-\frac{\|\mathbf{v} - \mathbf{A}^{(m)}\mathbf{f}\|^2}{2\sigma^2} \right) \right] P(\mathbf{I}^{(m)}) \tag{11.5–7}$$

Then, the estimate of $\mathbf{f}$ that maximizes $p(\mathbf{v}|\mathbf{f})$ is the solution of the equation

$$\frac{\partial p(\mathbf{v}|\mathbf{f})}{\partial \mathbf{f}} = \sum_m P(\mathbf{I}^{(m)}) \cdot (\mathbf{A}^{(m)t}\mathbf{A}^{(m)}\mathbf{f} - \mathbf{A}^{(m)t}\mathbf{v}) \exp\left(-\frac{\|\mathbf{v} - \mathbf{A}^{(m)}\mathbf{f}\|^2}{2\sigma^2}\right) = 0 \tag{11.5–8}$$

Hence, the estimate of $\mathbf{f}$ may be expressed as

$$\mathbf{f} = \left[\sum_m P(\mathbf{I}^{(m)})\mathbf{A}^{(m)\prime}\mathbf{A}^{(m)}g(\mathbf{v}, \mathbf{A}^{(m)}, \mathbf{f})\right]^{-1} \times \sum_m P(\mathbf{I}^{(m)})g(\mathbf{v}, \mathbf{A}^{(m)}, \mathbf{f})\mathbf{A}^{(m)\prime}\mathbf{v} \tag{11.5–9}$$

where the function $g(\mathbf{v}, \mathbf{A}^{(m)}, \mathbf{f})$ is defined as

$$g(\mathbf{v}, \mathbf{A}^{(m)}, \mathbf{f}) = \exp\left(-\frac{\|\mathbf{v} - \mathbf{A}^{(m)}\mathbf{f}\|^2}{2\sigma^2}\right) \tag{11.5–10}$$

The resulting solution for the optimum $\mathbf{f}$ is denoted by $\mathbf{f}_{ML}$.

Equation 11.5–9 is a non-linear equation for the estimate of the channel impulse response, given the received signal vector $\mathbf{v}$. It is generally difficult to obtain the optimum solution by solving Equation 11.5–9 directly. On the other hand, it is relatively simple to devise a numerical method that solves for $\mathbf{f}_{ML}$ recursively. Specifically, we may write

$$\mathbf{f}^{(k+1)} = \left[\sum_m P(\mathbf{I}^{(m)})\mathbf{A}^{(m)\prime}\mathbf{A}^{(m)}g(\mathbf{v}, \mathbf{A}^{(m)}, \mathbf{f}^{(k)})\right]^{-1} \times \sum_m P(\mathbf{I}^{(m)})g(\mathbf{v}, \mathbf{A}^{(m)}, \mathbf{f}^{(k)})\mathbf{A}^{(m)\prime}\mathbf{v} \tag{11.5–11}$$

Once $\mathbf{f}_{ML}$ is obtained from the solution of Equation 11.5–9 or 11.5–11, we may simply use the estimate in the minimization of the metric $DM(\mathbf{I}, \mathbf{f}_{ML})$, given by Equation 11.5–3, over all the possible data sequences. Thus, $\mathbf{I}_{ML}$ is the sequence $\mathbf{I}$ that minimizes $DM(\mathbf{I}, \mathbf{f}_{ML})$, i.e.,

$$\min_{\mathbf{I}} DM(\mathbf{I}, \mathbf{f}_{ML}) = \min_{\mathbf{I}} \|\mathbf{v} - \mathbf{A}\mathbf{f}_{ML}\|^2 \tag{11.5–12}$$

We know that the Viterbi algorithm is the computationally efficient algorithm for performing the minimization of $DM(\mathbf{I}, \mathbf{f}_{ML})$ over $\mathbf{I}$.

This algorithm has two major drawbacks. First, the recursion for $\mathbf{f}_{LM}$ given by Equation 11.5–11 is computationally intensive. Second, and, perhaps, more importantly, the estimate $\mathbf{f}_{ML}$ is not as good as the maximum-likelihood estimate $\mathbf{f}_{ML}(\mathbf{I})$ that is obtained when the sequence $\mathbf{I}$ is known. Consequently, the error rate performance of the blind equalizer (the Viterbi algorithm) based on the estimate $\mathbf{f}_{ML}$ is poorer than that based on $\mathbf{f}_{ML}(\mathbf{I})$. Next, we consider joint channel and data estimation.

Joint channel and data estimation. Here, we consider the joint optimization of the performance index $DM(\mathbf{I}, \mathbf{f})$ given by Equation 11.5–3. Since the elements of the impulse response vector $\mathbf{f}$ are continuous and the elements of the data vector $\mathbf{I}$ are discrete, one approach is to determine the maximum-likelihood estimate of $\mathbf{f}$ for each possible data sequence and, then, to select the data sequence that minimizes $DM(\mathbf{I}, \mathbf{f})$ for each corresponding channel estimate. Thus, the channel estimate corresponding to the mth data sequence $\mathbf{I}^{(m)}$ is

$$\mathbf{f}_{ML}(\mathbf{I}^{(m)}) = (\mathbf{A}^{(m)\prime}\mathbf{A}^{(m)})^{-1}\mathbf{A}^{(m)t}\mathbf{v} \tag{11.5–13}$$

For the mth data sequence, the metric $DM(\mathbf{I}, \mathbf{f})$ becomes

$$DM[\mathbf{I}^{(m)}, \mathbf{f}_{ML}(\mathbf{I}^{(m)})] = \|\mathbf{v} - \mathbf{A}^{(m)}\mathbf{f}_{ML}(\mathbf{I}^{(m)})\|^2 \tag{11.5–14}$$

Then, from the set of M^N possible sequences, we select the data sequence that minimizes the cost function in Equation 11.5–14, i.e., we determine

$$\min_{\mathbf{I}^{(m)}} DM[\mathbf{I}^{(m)}, \mathbf{f}_{ML}(\mathbf{I}^{(m)})] \tag{11.5–15}$$

The approach described above is an exhaustive computational search method with a computational complexity that grows exponentially with the length of the data block. We may select $N = L + 1$, and, thus, we shall have one channel estimate for each of the M^L surviving sequences. Thereafter, we may continue to maintain a separate channel estimate for each surviving path of the Viterbi algorithm search through the trellis. This approach to joint channel and data estimation has been called *per-survivor processing* by Raheli et al. (1995).

A similar approach has been proposed by Seshadri (1994). In essence, Seshadri's algorithm is a type of generalized Viterbi algorithm (GVA) that retains $K \geqslant 1$ best estimates of the transmitted data sequence into each state of the trellis and the corresponding channel estimates. In Seshadri's GVA, the search is identical to the conventional Viterbi algorithm (VA) from the beginning up to the Lth stage of the trellis, i.e., up to the point where the received sequence $(v_1, v_2, \ldots, v_L)$ has been processed. Hence, up to the Lth stage, an exhaustive search is performed. Associated with each data sequence $\mathbf{I}^{(m)}$, there is a corresponding channel estimate $\mathbf{f}_{ML}(\mathbf{I}^{(m)})$. From this stage on, the search is modified, to retain $K \geqslant 1$ surviving sequences and associated channel estimates per state instead of only one sequence per state. Thus, the GVA is used for processing the received signal sequence $\{v_n, n \geqslant L + 1\}$. The channel estimate is updated recursively at each stage using the LMS algorithm to further reduce the computational complexity. Simulation results given in the paper by Seshadri (1994) indicate that this GVA blind equalization algorithm performs rather well at moderate signal-to-noise ratios with $K = 4$. Hence, there is a modest increase in the computational complexity of the GVA compared with that for the conventional VA. However, there are additional computations involved with the estimation and updating of the channel estimates $\mathbf{f}(\mathbf{I}^{(m)})$ associated with each of the surviving data estimates.

An alternative joint estimation algorithm that avoids the least-squares computation for channel estimation has been devised by Zervas et al. (1991). In this algorithm, the order for performing the joint minimization of the performance index $DM(\mathbf{I}, \mathbf{f})$ is reversed. That is, a channel impulse response, say $\mathbf{f} = \mathbf{f}^{(1)}$, is selected and then the conventional VA is used to find the optimum sequence for this channel impulse response. Then, we may modify $\mathbf{f}^{(1)}$ in some manner to $\mathbf{f}^{(2)} = \mathbf{f}^{(1)} + \Delta\mathbf{f}^{(1)}$ and repeat the optimization over the data sequences $\{\mathbf{I}^{(m)}\}$.

Based on this general approach, Zervas et al. developed a new ML blind equalization algorithm, which is called a *quantized-channel algorithm*. The algorithm operates over a grid in the channel space, which becomes finer and finer by using the ML criterion to confine the estimated channel in the neighborhood of the original unknown channel. This algorithm leads to an efficient parallel implementation, and its storage requirements are only those of the VA.

11.5.2 Stochastic Gradient Algorithms

Another class of blind equalization algorithms are stochastic-gradient iterative equalization schemes that apply a memoryless non-linearity in the output of a linear FIR equalization filter in order to generate the "desired response" in each iteration.

Let us begin with an initial guess of the coefficients of the optimum equalizer, which we denote by $\{c_n\}$. Then, the convolution of the channel response with the equalizer response may be expressed as

$$\{c_n\} \star \{f_n\} = \{\delta_n\} + \{e_n\} \tag{11.5–16}$$

where $\{\delta_n\}$ is the unit sample sequence and $\{e_n\}$ denotes the error sequence that results from our initial guess of the equalizer coefficients. If we convolve the equalizer impulse response with the received sequence $\{v_n\}$, we obtain

$$\begin{aligned} \{\hat{I}_n\} &= \{v_n\} \star \{c_n\} \\ &= \{I_n\} \star \{f_n\} \star \{c_n\} + \{\eta_n\} \star \{c_n\} \\ &= \{I_n\} \star (\{\delta_n\} + \{e_n\}) + \{\eta_n\} \star \{c_n\} \\ &= \{I_n\} + \{I_n\} \star \{e_n\} + \{\eta_n\} \star \{c_n\} \end{aligned} \tag{11.5–17}$$

In Equation 11.5–17 the term $\{I_n\}$ represents the desired data sequence, the term $\{I_n\} \star \{e_n\}$ represents the residual ISI, and the term $\{\eta_n\} \star \{c_n\}$ represents the additive noise. Our problem is to utilize the deconvolved sequence $\{\hat{I}_n\}$ to find the "best" estimate of a desired response, denoted in general by $\{d_n\}$. In the case of adaptive equalization using a training sequence, $\{d_n\} = \{I_n\}$. In a blind equalization mode, we shall generate a desired response from $\{\hat{I}_n\}$.

The mean square error (MSE) criterion may be employed to determine the "best" estimate of $\{I_n\}$ from the observed equalizer output $\{\hat{I}_n\}$. Since the transmitted sequence $\{I_n\}$ has a non-Gaussian PDF, the MSE estimate is a non-linear transformation of $\{\hat{I}_n\}$. In general, the best estimate $\{d_n\}$ is given by

$$\begin{aligned} d_n &= g(\hat{I}_n) && \text{(memoryless)} \\ d_n &= g(\hat{I}_n, \hat{I}_{n-1}, \ldots, \hat{I}_{n-m}) && (m\text{th-order memory}) \end{aligned} \tag{11.5–18}$$

where $g(\;)$ is a non-linear function. The sequence $\{d_n\}$ is then used to generate an error signal, which is fed back into the adaptive equalization filter, as shown in Figure 11.5–1. Let us consider the non-linear function based on the MSE criterion.

A well-known classical estimation problem is the following. If the equalizer output $\hat{I}_n$ is expressed as

$$\hat{I}_n = I_n + \tilde{\eta}_n \tag{11.5–19}$$

where $\tilde{\eta}_n$ is assumed to be zero-mean Gaussian (the central limit theorem may be invoked here for the residual ISI and the additive noise), $\{I_n\}$ and $\{\tilde{\eta}_n\}$ are statistically independent, and $\{I_n\}$ are statistically independent and identically distributed random variables, then the MSE estimate of $\{I_n\}$ is

$$d_n = E(I_n | \hat{I}_n) \tag{11.5–20}$$

which is a non-linear function of the equalizer output when $\{I_n\}$ is non-Gaussian.

Table 11.5–1 illustrates the general form of existing blind equalization algorithms that are based on LMS adaptation. We observe that the basic difference among these algorithms lies in the choice of the memoryless non-linearity. The most widely used algorithm in practice is the *Godard algorithm*, sometimes also called the *constant-modulus algorithm* (CMA).

It is apparent from Table 11.5–1 that the output sequence $\{d_n\}$ obtained by taking a non-linear function of the equalizer output plays the role of the desired response or a training sequence. It is also apparent that these algorithms are simple to implement, since they are basically LMS-type algorithms. As such, we expect that the convergence characteristics of these algorithms will depend on the autocorrelation matrix of the received data $\{v_n\}$.

With regard to convergence, the adaptive LMS-type algorithms converge in the mean when

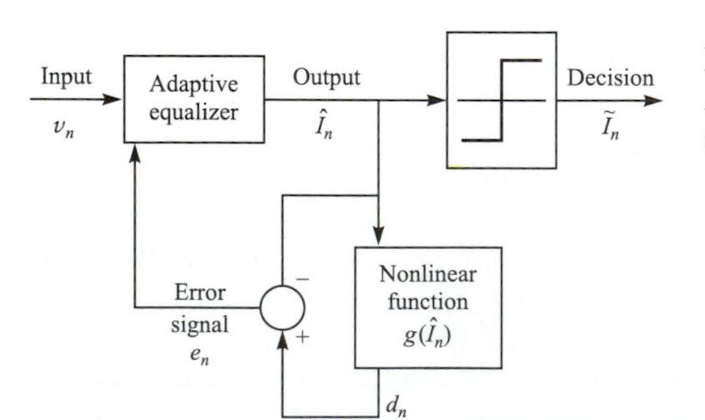

FIGURE 11.5–1
Adaptive blind equalization with stochastic gradient algorithms.

TABLE 11.5–1
Stochastic gradient algorithms for blind equalization

Equalizer tap coefficients	$\{c_n, 0 \leqslant n \leqslant N-1\}$
Received signal sequence	$\{v_n\}$
Equalizer output sequence	$\{\tilde{I}_n\} = \{v_n\} \star \{c_n\}$
Equalizer error sequence	$\{e_n\} = g(\tilde{I}_n) - \tilde{I}_n$
Tap coefficient update equation	$\mathbf{c}_{n+1} = \mathbf{c}_n + \Delta \mathbf{v}_n^* e_n$

Algorithm	**Non-linearity: $g(\tilde{I}_n)$**
Godard	$\dfrac{\hat{I}_n}{\lvert\tilde{I}_n\rvert}(\lvert\tilde{I}_n\rvert + R_2\lvert\hat{I}_n\rvert - \lvert\hat{I}_n\rvert^3)$, $R_2 = \dfrac{E\{\lvert I_n\rvert^4\}}{E\{\lvert I_n\rvert^2\}}$
Sato	ζ csgn $(\hat{I}_n)$, $\zeta = \dfrac{E\{[\mathrm{Re}(I_n)]^2\}}{E\{\lvert\mathrm{Re}(I_n)\rvert\}}$
Benveniste–Goursat	$\hat{I}_n + k_1(\hat{I}_n - I_n) + k_2\lvert\hat{I}_n - \tilde{I}_n\rvert[\zeta \text{ csgn } (\hat{I}_n) - \tilde{I}_n]$, k_1 and k_2 are positive constants
Stop-and-go	$\hat{I}_n + \frac{1}{2}A(\hat{I}_n - \tilde{I}_n) + \frac{1}{2}B(\hat{I}_n - \tilde{I}_n)^*(A, B) = (2, 0)$, $(1, 1)$, $(1, -1)$, or $(0, 0)$, depending on the signs of decision-directed error $\hat{I}_n - \tilde{I}_n$ and the error ζ csgn $(\hat{I}_n) - \tilde{I}_n$

$$E[v_n g^*(\hat{I}_n)] = E[v_n \hat{I}_n^*] \tag{11.5–21}$$

and, in the mean square sense, when (superscript H denotes the conjugate transpose)

$$\begin{aligned} E[\mathbf{c}_n^H v_n g^*(\hat{I}_n)] &= E[\mathbf{c}_n^H v_n \hat{I}_n^*] \\ E[\hat{I}_n g^*(\hat{I}_n)] &= E[\lvert\hat{I}_n\rvert^2] \end{aligned} \tag{11.5–22}$$

Therefore, it is required that the equalizer output $\{\hat{I}_n\}$ satisfy Equation 11.5–22. Note that Equation 11.5–22 states that the autocorrelation of $\{\hat{I}_n\}$ (the right-hand side) equals the cross correlation between $\hat{I}_n$ and a non-linear transformation of $\hat{I}_n$ (left-hand side). Processes that satisfy this property are called *Bussgang* (1952), as named by Bellini (1986). In summary, the algorithms given in Table 11.5–1 converge when the equalizer output sequence $\hat{I}_n$ satisfies the Bussgang property.

The basic limitation of stochastic gradient algorithms is their relatively slow convergence. Some improvement in the convergence rate can be achieved by modifying the adaptive algorithms from LMS-type to RLS-type.

Godard algorithm. As indicated above, the Godard blind equalization algorithm is a steepest-descent algorithm that is widely used in practice when a training sequence is not available. Let us describe this algorithm in more detail, assuming a general QAM signal constellation.

Godard considered the problem of combined equalization and carrier phase recovery and tracking. The carrier phase tracking is performed at baseband, following the equalizer as shown in Figure 11.5–2. Based on this structure, we may express the equalizer output as

$$\hat{I}_k = \sum_{n=-K}^{K} c_n v_{k-n} \tag{11.5–23}$$

and the input to the decision device as $\hat{I}_n \exp(-j\hat{\phi}_k)$, where $\hat{\phi}_k$ is the carrier phase estimate in the kth symbol interval.

If the desired symbol were known, we could form the error signal

$$\varepsilon_k = I_k - \hat{I}_k e^{-j\hat{\phi}_k} \tag{11.5–24}$$

and minimize the MSE with respect to $\hat{\phi}_k$ and $\{c_n\}$, i.e.,

$$\min_{\hat{\phi}_k, \mathbf{C}} E(|I_k - \hat{I}_k e^{-j\hat{\phi}_k}|^2) \tag{11.5–25}$$

This criterion leads us to use the LMS algorithm for recursively estimating $\mathbf{C}$ and ϕ_k. The LMS algorithm based on knowledge of the transmitted sequence is

$$\hat{\mathbf{C}}_{k+1} = \hat{\mathbf{C}}_k + \Delta_c (I_k - \hat{I}_k e^{-j\hat{\phi}_k}) \mathbf{V}_k^* e^{j\hat{\phi}_k} \tag{11.5–26}$$

$$\hat{\phi}_{k+1} = \hat{\phi}_k + \Delta_\phi \mathrm{Im}(I_k \hat{I}_k^* e^{j\hat{\phi}_k}) \tag{11.5–27}$$

where Δ_c and Δ_ϕ are the step-size parameters for the two recursive equations. Note that these recursive equations are coupled together. Unfortunately, these equations will not converge, in general, when the desired symbol sequence $\{I_k\}$ is unknown.

The approach proposed by Godard is to use a criterion that depends on the amount of intersymbol interference at the output of the equalizer but one that is independent of the QAM signal constellation and the carrier phase. For example, a cost function that is independent of carrier phase and has the property that its minimum leads to a small MSE is

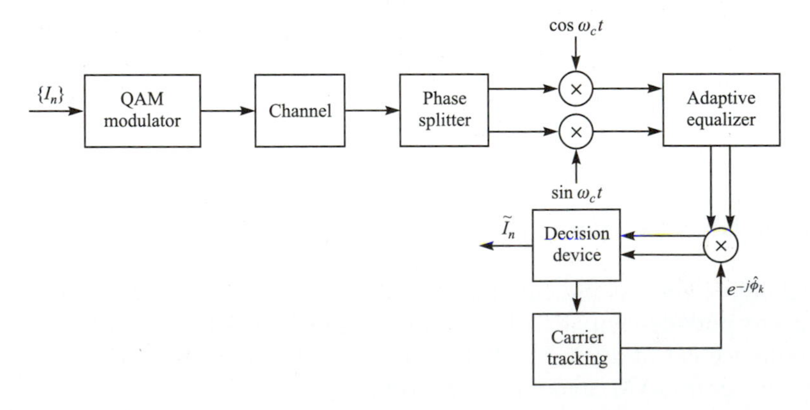

FIGURE 11.5–2
Godard scheme for combined adaptive (blind) equalization and carrier phase tracking.

$$G^{(p)} = E(|\hat{I}_k|^p - |I_k|^p)^2 \tag{11.5–28}$$

where p is a positive and real integer. Minimization of $G^{(p)}$ with respect to the equalizer coefficients results in the equalization of the signal amplitude only. Based on this observation, Godard selected a more general cost function, called the *dispersion of order* p, defined as

$$D^{(p)} = E(|\hat{I}_k|^p - R_p)^2 \tag{11.5–29}$$

where R_p is a positive real constant. As in the case of $G^{(p)}$, we observe that $D^{(p)}$ is independent of the carrier phase.

Minimization of $D^{(p)}$ with respect to the equalizer coefficients can be performed recursively according to the steepest-descent algorithm

$$\mathbf{C}_{k+1} = \mathbf{C}_k - \Delta_p \frac{dD^{(p)}}{d\mathbf{C}_k} \tag{11.5–30}$$

where Δ_p is the step-size parameter. By differentiating $D^{(p)}$ and dropping the expectation operation, we obtain the following LMS-type algorithm for adjusting the equalizer coefficients:

$$\hat{\mathbf{C}}_{k+1} = \hat{\mathbf{C}}_k + \Delta_p \mathbf{V}_k^* \hat{I}_k |\hat{I}_k|^{p-2}(R_p - |\hat{I}_k|^p) \tag{11.5–31}$$

where Δ_p is the step-size parameter and the optimum choice of R_p is

$$R_p = \frac{E(|I_k|^{2p})}{E(|I_k|^p)} \tag{11.5–32}$$

As expected, the recursion in Equation 11.5–31 for $\hat{\mathbf{C}}_k$ does not require knowledge of the carrier phase. Carrier phase tracking may be carried out in a decision-directed mode according to Equation 11.5–27, with $\tilde{I}_k$ substituted in place of I_k.

Of particular importance is the case $p = 2$, which leads to the relatively simple algorithm

$$\begin{aligned} \hat{\mathbf{C}}_{k+1} &= \hat{\mathbf{C}}_k + \Delta_p \mathbf{V}_k^* \hat{I}_k (R_2 - |\hat{I}_k|^2) \\ \hat{\phi}_{k+1} &= \hat{\phi}_k + \Delta_\phi \operatorname{Im}(\tilde{I}_k \hat{I}_k^* e^{j\hat{\phi}_k}) \end{aligned} \tag{11.5–33}$$

where $\tilde{I}_k$ is the output decision based on $\hat{I}_k$, and

$$R_2 = \frac{E(|I_k|^4)}{E(|I_k|^2)} \tag{11.5–34}$$

Convergence of the algorithm given in Equation 11.5–33 was demonstrated in the paper by Godard (1980). Initially, the equalizer coefficients were set to zero except for the center (reference) tap, which was set according to the condition

$$|c_0|^2 > \frac{E|I_k|^4}{2|x_0|^2[E(|I_k|^2)]^2} \tag{11.5–35}$$

which is sufficient, but not necessary, for convergence of the algorithm. Simulation results performed by Godard on simulated telephone channels with typical frequency-response characteristics and transmission rates of 7200–12,000 bits/s indicate that the algorithm in Equation 11.5–31 performs well and leads to convergence in 5000–20,000 iterations, depending on the signal constellation. Initially, the eye pattern was closed prior to equalization. The number of iterations required for convergence is about an order of magnitude greater than the number required to equalize the channels with a known training sequence. No apparent difficulties were encountered in using the decision-directed phase estimation algorithm in Equation 11.5–33 from the beginning of the equalizer adjustment process.

11.5.3 Blind Equalization Algorithms Based on Second- and Higher-Order Signal Statistics

It is well known that second-order statistics (autocorrelation) of the received signal sequence provide information on the magnitude of the channel characteristics, but not on the phase. However, this statement is not correct if the autocorrelation function of the received signal is periodic, as is the case for a digitally modulated signal. In such a case, it is possible to obtain a measurement of the amplitude and the phase of the channel from the received signal. This cyclostationarity property of the received signal forms the basis for a channel estimation algorithm devised by Tong et al. (1994, 1995).

It is also possible to estimate the channel response from the received signal by using higher-order statistical methods. In particular, the impulse response of a linear, discrete-time-invariant system can be obtained explicitly from cumulants of the received signal, provided that the channel input is non-Gaussian. We describe the following simple method, due to Giannakis (1987) and Giannakis and Mendel (1989) for estimation of the channel impulse response from fourth-order cumulants of the received signal sequence. For simplicity, we assume that the received signal sequence is real-valued. The fourth-order cumulant is defined as

$$\begin{aligned} c(v_k, v_{k+m}, v_{k+n}, v_{k+l}) \equiv c_r(m, n, l) &= E(v_k v_{k+m} v_{k+n} v_{k+l}) \\ &\quad - E(v_k\, v_{k+m})E(v_{k+n} v_{k+l}) \\ &\quad - E(v_k\, v_{k+n})E(v_{k+m} v_{k+l}) \\ &\quad - E(v_k\, v_{k+l})E(v_{k+m} v_{k+n}) \end{aligned} \tag{11.5–36}$$

(The fourth-order cumulant of a Gaussian signal process is zero.) Consequently, it follows that

$$c_r(m, n, l) = c(I_k, I_{k+m}, I_{k+n}, I_{k+l}) \sum_{k=0}^{\infty} f_k f_{k+m} f_{k+n} f_{k+l} \tag{11.5–37}$$

For a statistically independent and identically distributed input sequence $\{I_n\}$ to the channel, $c(I_k, I_{k+m}, I_{k+n}, I_{k+l}) = k$, a constant, called the *kurtosis*. Then, if the length of the channel response is $L+1$, we may let $m = n = l = -L$ so that

$$c_r(-L, -L, -L) = kf_L f_0^3 \tag{11.5–38}$$

Similarly, if we let $m = 0$, $n = L$, and $l = p$, we obtain

$$c_r(0, L, p) = kf_L f_0^2 f_p \tag{11.5–39}$$

If we combine Equations 11.5–38 and 11.5–39, we obtain the impulse response within a scale factor as

$$f_p = f_0 \frac{c_r(0, L, p)}{c_r(-L, -L, -L)}, \qquad p = 1, 2, \ldots, L \tag{11.5–40}$$

The cumulants $c_r(m, n, l)$ are estimated from sample averages of the received signal sequence $\{v_n\}$.

Another approach based on higher-order statistics is due to Hatzinakos and Nikias (1991). They have introduced the first polyspectra-based adaptive blind equalization method named the *tricepstrum equalization algorithm* (TEA). This method estimates the channel response characteristics by using the complex cepstrum of the fourth-order cumulants (tricepstrum) of the received signal sequence $\{v_n\}$. TEA depends only on fourth-order cumulants of $\{v_n\}$ and is capable of separately reconstructing the minimum-phase and maximum-phase characteristics of the channel. The channel equalizer coefficients are then computed from the measured channel characteristics. The basic approach used in TEA is to compute the tricepstrum of the received sequence $\{v_n\}$, which is the inverse (three-dimensional) Fourier transform of the logarithm of the trispectrum of $\{v_n\}$. [The *trispectrum* is the three-dimensional discrete Fourier transform of the fourth-order cumulant sequence $c_r(m, n, l)$.] The equalizer coefficients are then computed from the cepstral coefficients.

By separating the channel estimation from the channel equalization, it is possible to use any type of equalizer for the ISI, i.e., either linear, or decision-feedback, or maximum-likelihood sequence detection. The major disadvantage with this class of algorithms is the large amount of data and the inherent computational complexity involved in the estimation of the higher-order moments (cumulants) of the received signal.

In conclusion, we have provided an overview of three classes of blind equalization algorithms that find applications in digital communications. Of the three families of algorithms described, those based on the maximum-likelihood criterion for jointly estimating the channel impulse response and the data sequence are optimal and require relatively few received signal samples for performing channel estimation. However, the computational complexity of the algorithms is large when the ISI spans many symbols. On some channels, such as the mobile radio channel, where the span of the ISI is relatively short, these algorithms are simple to implement. However, on telephone channels, where the ISI spans many symbols but is usually not too severe, the LMS-type (stochastic gradient) algorithms are generally employed.

11.6 BIBLIOGRAPHICAL NOTES AND REFERENCES

Adaptive equalization for digital communications was developed by Lucky (1965, 1966). His algorithm was based on the peak distortion criterion and led to the zero-forcing algorithm. Lucky's work was a major breakthrough, which led to the rapid development of high-speed modems within 5 years of publication of his work. Concurrently, the LMS algorithm was devised by Widrow (1966, 1970), and its use for adaptive equalization for two-dimensional (in-phase and quadrature components) signals was described and analyzed in a tutorial paper by Proakis and Miller (1969).

A tutorial treatment of adaptive equalization algorithms that were developed during the period 1965–1975 is given by Proakis (1975). A more recent tutorial treatment of adaptive equalization is given in the paper by Qureshi (1985). The major breakthrough in adaptive equalization techniques, beginning with the work of Lucky in 1965 coupled with the development of trellis-coded modulation, which was described by Ungerboeck and Csajka (1976), has led to the development of commercially available high-speed modems with a capability of speeds exceeding 30,000 bits/s on telephone channels.

The use of a more rapidly converging algorithm for adaptive equalization was proposed by Godard (1974). Our derivation of the RLS (Kalman) algorithm, described in Section 11.4.1, follows the approach outlined by Picinbono (1978). RLS lattice algorithms for general signal estimation applications were developed by Morf (1977), Morf and Lee (1978), and Morf et al. (1977a,b,c). The applications of these algorithms have been investigated by several researchers, including Makhoul (1978), Satorius and Pack (1981), Satorius and Alexander (1979), and Ling and Proakis (1982, 1984a–c, 1985, 1986). The fast RLS Kalman algorithm for adaptive equalization was first described by Falconer and Ljung (1978). The above references are just a few of the important papers that have been published on RLS algorithms for adaptive equalization and other applications. A comprehensive treatment of RLS algorithms is given in the book by Haykin (1996).

Sato's (1975) original work on blind equalization was focused on PAM (one-dimensional) signal constellations. Subsequently it was generalized to two-dimensional and multidimensional signal constellations in the algorithms devised by Godard (1980), Benveniste and Goursat (1984), Sato et al. (1986), Foschini (1985), Picchi and Prati (1987), and Shalvi and Weinstein (1990). Blind equalization methods based on the use of second- and higher-order moments of the received signal were proposed by Giannakis (1987), Giannakis and Mendel (1989), Hatzinakos and Nikias (1991), and Tong et al. (1994, 1995). The use of the maximum-likelihood criterion for joint channel estimation and data detection has been investigated and treated in papers by Sato (1994), Seshadri (1994),

Ghosh and Weber (1991), Zervas et al. (1991), and Raheli et al. (1995). Finally, the convergence characteristics of stochastic gradient blind equalization algorithms have been investigated by Ding (1990), Ding et al. (1989), and Johnson (1991).

PROBLEMS

11.1 An equivalent discrete-time channel with white Gaussian noise is shown in Figure P11.1

a) Suppose we use a linear equalizer to equalize the channel. Determine the tap coefficients c_{-1}, c_0, c_1 of a three-tap equalizer. To simplify the computation, let the AWGN be zero.

b) The tap coefficients of the linear equalizer in (a) are determined recursively via the algorithm

$$\mathbf{C}_{k+1} = \mathbf{C}_k - \Delta\mathbf{G}_k, \qquad \mathbf{C}_k = [c_{-1k} \quad c_{0k} \quad c_{1k}]^t$$

where $\mathbf{G}_k = \mathbf{\Gamma}\mathbf{C}_k - \boldsymbol{\xi}$ is the gradient vector and Δ is the step size. Determine the range of values of Δ to ensure convergence of the recursive algorithm. To simplify the computation, let the AWGN be zero.

c) Determine the tap weights of a DFE with two feedforward taps and one feedback gap. To simplify the computation, let the AWGN be zero.

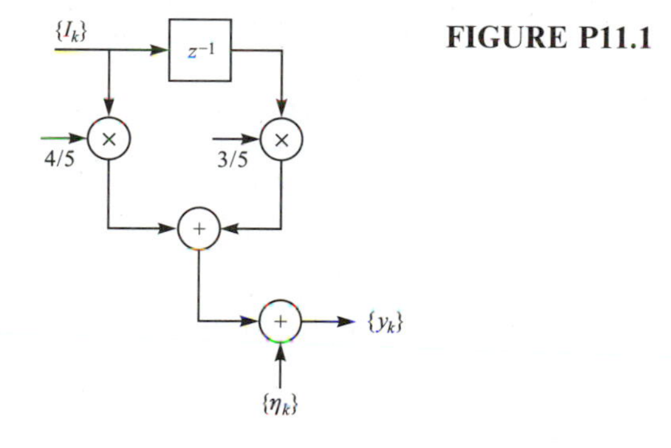

FIGURE P11.1

11.2 Refer to Problem 10.18 and answer the following questions.

a) Determine the maximum value of Δ that can be used to ensure that the equalizer coefficients converge during operation in the adaptive mode.

b) What is the variance of the self-noise generated by the three-tap equalizer when operating in an adaptive mode, as a function of Δ? Suppose it is desired to limit the variance of the self-noise to 10 percent of the minimum MSE for the three-tap equalizer when $N_0 = 0.1$. What value of Δ would you select?

c) If the optimum coefficients of the equalizer are computed recursively by the method of steepest descent, the recursive equation can be expressed in the form

$$\mathbf{C}_{n+1} = (\mathbf{I} - \Delta\mathbf{\Gamma})\mathbf{C}_n + \Delta\boldsymbol{\xi}$$

where $\mathbf{I}$ is the identity matrix. The above represents a set of three coupled first-order difference equations. They can be decoupled by a linear transformation that diagonalizes the matrix $\mathbf{\Gamma}$. That is, $\mathbf{\Gamma} = \mathbf{U}\mathbf{\Lambda}\mathbf{U}^t$ where $\mathbf{\Lambda}$ is the diagonal matrix having the eigenvalues of $\mathbf{\Gamma}$ as its diagonal elements and $\mathbf{U}$ is the (normalized) modal matrix that can be obtained from your answer to Problem 10.18b. Let $\mathbf{C}' = \mathbf{U}^t\mathbf{C}$ and determine the steady-state solution for $\mathbf{C}'$. From this, evaluate $\mathbf{C} = (\mathbf{U}^t)^{-1}\mathbf{C}' = \mathbf{U}\mathbf{C}'$ and, thus, show that your answer agrees with the result obtained in Problem 10.18a.

11.3 When a periodic pseudorandom sequence of length N is used to adjust the coefficients of an N-tap linear equalizer, the computations can be performed efficiently in the frequency domain by use of the discrete Fourier transform (DFT). Suppose that $\{y_n\}$ is a sequence of N received samples (taken at the symbol rate) at the equalizer input. Then the computation of the equalizer coefficients is performed as follows.

a) Compute the DFT of one period of the equalizer input sequence $\{y_n\}$, i.e.,

$$Y_k = \sum_{n=0}^{N-1} y_n e^{-j2\pi nk/N}$$

b) Compute the desired equalizer spectrum

$$C_k = \frac{X_k Y_k^*}{|Y_k|^2}, \qquad k = 0, 1, \ldots, N-1$$

where $\{X_i\}$ is the precomputed DFT of the training sequence.

c) Compute the inverse DFT of $\{C_k\}$ to obtain the equalizer coefficients $\{c_n\}$. Show that this procedure in the absence of noise yields an equalizer whose frequency response is equal to the frequency response of the inverse folded channel spectrum at the N uniformly spaced frequencies $f_k = k/NT$, $k = 0, 1, \ldots, N-1$.

11.4 Show that the gradient vector in the minimization of the MSE may be expressed as

$$\mathbf{G}_k = -E(\varepsilon_k \mathbf{V}_k^*)$$

where the error $\varepsilon_k = I_k - \hat{I}_k$, and the estimate of $\mathbf{G}_k$, i.e.,

$$\hat{\mathbf{G}}_k = -\varepsilon_k \mathbf{V}_k^*$$

satisfies the condition that $E(\hat{\mathbf{G}}_k) = \mathbf{G}_k$.

11.5 The *tap-leakage LMS algorithm* proposed in the paper by Gitlin et al. (1982) may be expressed as

$$\mathbf{C}_N(n+1) = w\mathbf{C}_N(n) + \Delta\varepsilon(n)\mathbf{V}_N^*(n)$$

where $0 < w < 1$, Δ is the step size, and $\mathbf{V}_N(n)$ is the data vector at time n. Determine the condition for the convergence of the mean value of $\mathbf{C}_N(n)$.

11.6 Consider the random process

$$x(n) = gv(n) + w(n), \qquad n = 0, 1, \ldots, M - 1$$

where $v(n)$ is a known sequence, g is a random variable with $E(g) = 0$, and $E(g^2) = G$. The process $w(n)$ is a white noise sequence with

$$\gamma_{ww}(m) = \sigma_w^2 \delta_m$$

Determine the coefficients of the linear estimator for g, that is,

$$\hat{g} = \sum_{n=0}^{M-1} h(n)x(n)$$

that minimize the mean square error.

11.7 A digital transversal filter can be realized in the frequency-sampling form with system function (see Problem 10.25)

$$\begin{aligned} H(z) &= \frac{1 - z^{-M}}{M} \sum_{k=0}^{M-1} \frac{H_k}{1 - e^{j2\pi k/M} z^{-1}} \\ &= H_1(z)H_2(z) \end{aligned}$$

where $H_1(z)$ is the comb filter, $H_2(z)$ is the parallel bank of resonators, and $\{H_k\}$ are the values of the discrete Fourier transform (DFT).

a) Suppose that this structure is implemented as an adaptive filter using the LMS algorithm to adjust the filter (DFT) parameters $\{H_k\}$. Give the time-update equation for these parameters. Sketch the adaptive filter structure.

b) Suppose that this structure is used as an adaptive channel equalizer in which the desired signal is

$$d(n) = \sum_{k=0}^{M-1} A_k \cos \omega_k n, \qquad \omega_k = \frac{2\pi k}{M}$$

With this form for the desired signal, what advantages are there in the LMS adaptive algorithm for the DFT coefficients $\{H_k\}$ over the direct-form structure with coefficients $\{h(n)\}$? [See Proakis (1970).]

11.8 Consider the performance index

$$J = h^2 + 40h + 28$$

Suppose that we search for the minimum of J by using the steepest-descent algorithm

$$h(n + 1) = h(n) - \tfrac{1}{2}\Delta g(n)$$

where $g(n)$ is the gradient.

a) Determine the range of values of Δ that provides an overdamped system for the adjustment process.

b) Plot the expression for J as a function of n for a value of Δ in this range.

11.9 Determine the coefficients a_1 and a_2 for the linear predictor shown in Figure P11.9, given that the autocorrelation $\gamma_{xx}(m)$ of the input signal is

$$\gamma_{xx}(m) = b^{|m|}, \qquad 0 < b < 1$$

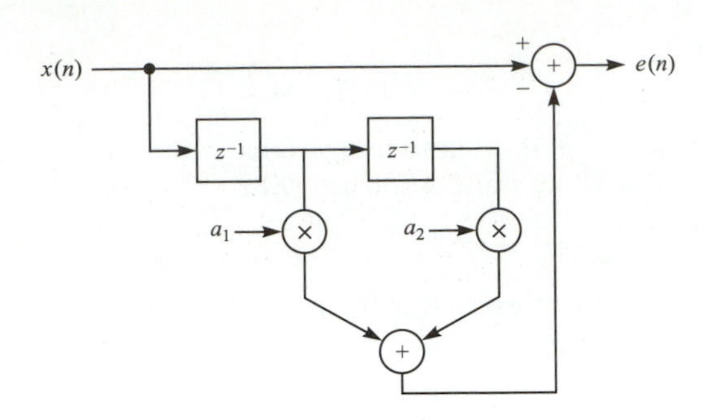

FIGURE P11.9

11.10 Determine the lattice filter and its optimum reflection coefficients corresponding to the linear predictor in Problem 11.9.

11.11 Consider the adaptive FIR filter shown in Figure P11.11. The system $C(z)$ is characterized by the system function

$$C(z) = \frac{1}{1 - 0.9z^{-1}}$$

Determine the optimum coefficients of the adaptive transversal (FIR) filter $B(z) = b_0 + b_1 z^{-1}$ that minimize the mean square error. The additive noise is white with variance $\sigma_w^2 = 0.1$.

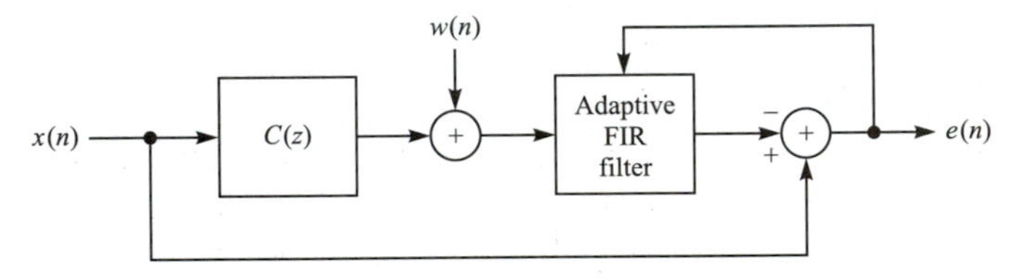

FIGURE P11.11

11.12 An $N \times N$ correlation matrix $\mathbf{\Gamma}$ has eigenvalues $\lambda_1 > \lambda_2 > \cdots > \lambda_N > 0$ and associated eigenvectors $\mathbf{v}_1, \mathbf{v}_2, \ldots, \mathbf{v}_N$. Such a matrix can be represented as

$$\mathbf{\Gamma} = \sum_{i=1}^{N} \lambda_i \mathbf{v}_i \mathbf{v}_i^{*t}$$

a) If $\mathbf{\Gamma} = \mathbf{\Gamma}^{1/2}\mathbf{\Gamma}^{1/2}$, where $\mathbf{\Gamma}^{1/2}$ is the square root of $\mathbf{\Gamma}$, show that $\mathbf{\Gamma}^{1/2}$ can be represented as

$$\mathbf{\Gamma}^{1/2} = \sum_{i=1}^{N} \lambda_i^{1/2} \mathbf{v}_i \mathbf{v}_i^{*t}$$

b) Using this representation, determine a procedure for computing $\mathbf{\Gamma}^{1/2}$.

12

Multichannel and Multicarrier Systems

In some applications, it is desirable to transmit the same information-bearing signal over several channels. This mode of transmission is used primarily in situations where there is a high probability that one or more of the channels will be unreliable from time to time. For example, radio channels such as ionospheric scatter and tropospheric scatter suffer from signal fading due to multipath, which renders the channels unreliable for short periods of time. As another example, multichannel signaling is sometimes employed in military communication systems as a means of overcoming the effects of jamming of the transmitted signal. By transmitting the same information over multiple channels, we are providing signal diversity, which the receiver can exploit to recover the information.

Another form of multichannel communications is multiple carrier transmission, where the frequency band of the channel is subdivided into a number of subchannels and information is transmitted on each of the subchannels. A rationale for subdividing the frequency band of a channel into a number of narrowband channels is given below.

In this chapter, we consider both multichannel signal transmission and multicarrier transmission. We begin with a treatment of multichannel transmission.

12.1 MULTICHANNEL DIGITAL COMMUNICATIONS IN AWGN CHANNELS

In this section, we confine our attention to multichannel signaling over fixed channels that differ only in attenuation and phase shift. The specific model for the multichannel digital signaling system may be described as follows. The signal waveforms, in general, are expressed as

$$s_m^{(n)}(t) = \text{Re}[s_{lm}^{(n)}(t)e^{j2\pi f_c t}], \qquad 0 \leqslant t \leqslant T$$
$$n = 1, 2, \ldots, L, \qquad m = 1, 2, \ldots, M \qquad (12.1\text{–}1)$$

where L is the number of channels and M is the number of waveforms. The waveforms are assumed to have equal energy and to be equally probable a priori. The waveforms $\{s_m^{(n)}(t)\}$ transmitted over the L channels are scaled by the attenuation factors $\{\alpha_n\}$, phase-shifted by $\{\phi_n\}$, and corrupted by additive noise. The equivalent low-pass signals received from the L channels may be expressed as

$$r_l^{(n)}(t) = \alpha_n e^{-j\phi_n} s_{lm}^{(n)}(t) + z_n(t), \quad 0 \leqslant t \leqslant T$$
$$n = 1, 2, \ldots, L, \quad m = 1, 2, \ldots, M \qquad (12.1\text{–}2)$$

where $\{s_{lm}^{(n)}(t)\}$ are the equivalent low-pass transmitted waveforms and $\{z_n(t)\}$ represent the additive noise processes on the L channels. We assume that $\{z_n(t)\}$ are mutually statistically independent and identically distributed Gaussian noise random processes.

We consider two types of processing at the receiver, namely, coherent detection and noncoherent detection. The receiver for coherent detection estimates the channel parameters $\{\alpha_n\}$ and $\{\phi_n\}$ and uses the estimates in computing the decision variables. Suppose we define $g_n = \alpha_n e^{-j\phi_n}$ and let $\hat{g}_n$ be the estimate of g_n. The multichannel receiver correlates each of the L received signals with a replica of the corresponding transmitted signals, multiplies each of the correlator outputs by the corresponding estimates $\{\hat{g}_n^*\}$, and sums the resulting signals. Thus, the decision variables for coherent detection are the correlation metrics

$$CM_m = \sum_{n=1}^{L} \text{Re}\left[\hat{g}_n^* \int_0^T r_l^{(n)}(t) s_{lm}^{(n)*}(t)\, dt\right], \qquad m = 1, 2, \ldots, M \qquad (12.1\text{–}3)$$

In noncoherent detection, no attempt is made to estimate the channel parameters. The demodulator may base its decision either on the sum of the envelopes (envelope detection) or the sum of the squared envelopes (square-law detection) of the matched filter outputs. In general, the performance obtained with envelope detection differs little from the performance obtained with square-law detection in AWGN. However, square-law detection of multichannel signaling in AWGN channels is considerably easier to analyze than envelope detection. Therefore, we confine our attention to square-law detection of the received signals of the L channels, which produces the decision variables

$$CM_m = \sum_{n=1}^{L} \left| \int_0^T r_l^{(n)}(t) s_{lm}^{(n)*}(t)\, dt \right|^2, \qquad m = 1, 2, \ldots, M \qquad (12.1\text{–}4)$$

Let us consider binary signaling first, and assume that $s_{l1}^{(n)}$, $n = 1, 2, \ldots, L$, are the L transmitted waveforms. Then an error is committed if $CM_2 > CM_1$, or, equivalently, if the difference $D = CM_1 - CM_2 < 0$. For noncoherent detection, this difference may be expressed as

$$D = \sum_{n=1}^{L} (|X_n|^2 - |Y_n|^2) \tag{12.1–5}$$

where the variables $\{X_n\}$ and $\{Y_n\}$ are defined as

$$\begin{aligned} X_n &= \int_0^T r_l^{(n)}(t) s_{l1}^{(n)*}(t)\, dt, \qquad n = 1, 2, \ldots, L \\ Y_n &= \int_0^T r_l^{(n)}(t) s_{l2}^{(n)*}(t)\, dt, \qquad n = 1, 2, \ldots, L \end{aligned} \tag{12.1–6}$$

The $\{X_n\}$ are mutually independent and identically distributed Gaussian random variables. The same statement applies to the variables $\{Y_n\}$. However, for any n, X_n and Y_n may be correlated. For coherent detection, the difference $D = CM_1 - CM_2$ may be expressed as

$$D = \tfrac{1}{2} \sum_{n=1}^{L} (X_n Y_n^* + X_n^* Y_n) \tag{12.1–7}$$

where, by definition,

$$\begin{aligned} Y_n &= \hat{g}_n, \qquad n = 1, 2, \ldots, L \\ X_n &= \int_0^T r_l^{(n)}(t)[s_{l1}^{(n)*}(t) - s_{l2}^{(n)*}(t)]\, dt \end{aligned} \tag{12.1–8}$$

If the estimates $\{\hat{g}_n\}$ are obtained from observation of the received signal over one or more signaling intervals, as described in Appendix C, their statistical characteristics are described by the Gaussian distribution. Then the $\{Y_n\}$ are characterized as mutually independent and identically distributed Gaussian random variables. The same statement applies to the variables $\{X_n\}$. As in noncoherent detection, we allow for correlation between X_n and Y_n, but not between X_m and Y_n for $m \neq n$.

12.1.1 Binary Signals

In Appendix B, we derive the probability that the general quadratic form

$$D = \sum_{n=1}^{L} (A|X_n|^2 + B|Y_n|^2 + C X_n Y_n^* + C^* X_n^* Y_n) \tag{12.1–9}$$

in complex-valued Gaussian random variables is less than zero, where A, B, and C are constants. This probability, which is given in Equation B–21 of Appendix B, is the probability of error for binary multichannel signaling in AWGN. A number of special cases are of particular importance.

If the binary signals are antipodal and the estimates of $\{g_n\}$ are perfect, as in coherent PSK, the probability of error takes the simple form

$$P_b = Q(\sqrt{2\gamma_b}) \tag{12.1–10}$$

where

$$\gamma_b = \frac{\mathcal{E}}{N_0}\sum_{n=1}^{L}|g_n|^2 = \frac{\mathcal{E}}{N_0}\sum_{n=1}^{L}\alpha_n^2 \tag{12.1–11}$$

is the SNR per bit. If the channels are all identical, $\alpha_n = \alpha$ for all n and, hence,

$$\gamma_b = \frac{L\mathcal{E}}{N_0}\alpha^2 \tag{12.1–12}$$

We observe that $L\mathcal{E}$ is the total transmitted signal energy for the L signals. The interpretation of this result is that the receiver combines the energy from the L channels in an optimum manner. That is, there is no loss in performance in dividing the total transmitted signal energy among the L channels. The same performance is obtained as in the case in which a single waveform having energy $L\mathcal{E}$ is transmitted on one channel. This behavior holds true only if the estimates $\hat{g}_n = g_n$, for all n. If the estimates are not perfect, a loss in performance occurs, the amount of which depends on the quality of the estimates, as described in Appendix C.

Perfect estimates for $\{g_n\}$ constitute an extreme case. At the other extreme, we have binary DPSK signaling. In DPSK, the estimates $\{\hat{g}_n\}$ are simply the (normalized) signal-plus-noise samples at the outputs of the matched filters in the previous signaling interval. This is the poorest estimate that one might consider using in estimating $\{g_n\}$. For binary DPSK, the probability of error obtained from Equation B–21 is

$$P_b = \frac{1}{2^{2L-1}}e^{-\gamma_b}\sum_{n=0}^{L-1}c_n\gamma_b^n \tag{12.1–13}$$

where, by definition,

$$c_n = \frac{1}{n!}\sum_{k=0}^{L-1-n}\binom{2L-1}{k} \tag{12.1–14}$$

and γ_b is the SNR per bit defined in Equation 12.1–11 and, for identical channels, in Equation 12.1–12. This result can be compared with the single-channel ($L = 1$) error probability. To simplify the comparison, we assume that the L channels have identical attenuation factors. Thus, for the same value of γ_b, the performance of the multichannel system is poorer than that of the single-channel system. That is, splitting the total transmitted energy among L channels results in a loss in performance, the amount of which depends on L.

A loss in performance also occurs in square-law detection of orthogonal signals transmitted over L channels. For binary orthogonal signaling, the expression for the probability of error is identical in form to that for binary DPSK given in Equation 12.1–13, except that γ_b is replaced by $\frac{1}{2}\gamma_b$. That is, binary orthogonal signaling with noncoherent detection is 3 dB poorer than binary DPSK. However, the loss in performance due to noncoherent combination of the signals received on the L channels is identical to that for binary DPSK.

Figure 12.1–1 illustrates the loss resulting from noncoherent (square-law) combining of the L signals as a function of L. The probability of error is not shown, but it can be easily obtained from the curve of the expression

$$P_b = \tfrac{1}{2}e^{-\gamma_b} \tag{12.1–15}$$

which is the error probability of binary DPSK shown in Figure 5.2–12 and then degrading the required SNR per bit, γ_b, by the noncoherent combining loss corresponding to the value of L.

12.1.2 *M*-ary Orthogonal Signals

Now let us consider M-ary orthogonal signaling with square-law detection and combination of the signals on the L channels. The decision variables are given by Equation 12.1–4. Suppose that the signals $s_{l1}^{(n)}(t)$, $n = 1, 2, \dots, L$, are transmitted over the L AWGN channels. Then, the decision variables are expressed as

$$\begin{aligned} CM_1 \equiv U_1 &= \sum_{n=1}^{L} |2\mathcal{E}\alpha_n + N_{n1}|^2 \\ CM_m \equiv U_m &= \sum_{n=1}^{L} |N_{nm}|^2, \qquad m = 2, 3, \dots, M \end{aligned} \tag{12.1–16}$$

where the $\{N_{nm}\}$ are complex-valued zero-mean Gaussian random variables with variance $\sigma^2 = \frac{1}{2}E(|N_{nm}|^2) = 2\mathcal{E}N_0$. Hence U_1 is described statistically as a non-

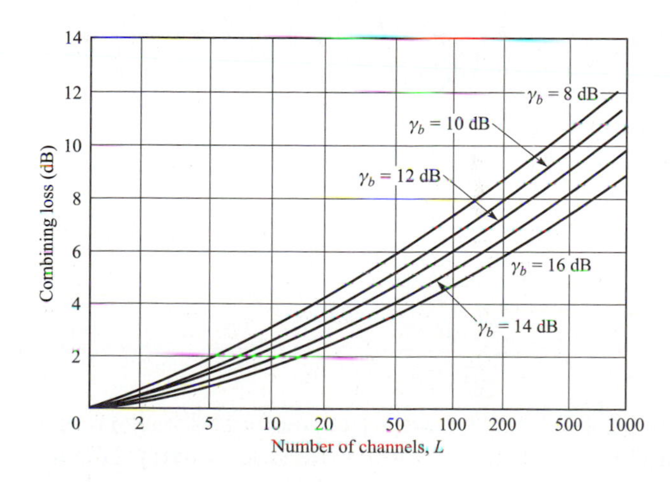

FIGURE 12.1–1 Combining loss in noncoherent detection and combination of binary multichannel signals.

central chi-square random variable with $2L$ degrees of freedom and noncentrality parameter

$$s^2 = \sum_{n=1}^{L} (2\mathcal{E}\alpha_n)^2 = 4\mathcal{E}^2 \sum_{n=1}^{L} \alpha_n^2 \qquad (12.1\text{–}17)$$

Using Equation 2.1–118, we obtain the PDF of U_1 as

$$p(u_1) = \frac{1}{4\mathcal{E}N_0} \left(\frac{u_1}{s^2} \right)^{(L-1)/2} \exp\left(-\frac{s^2 + u_1}{4\mathcal{E}N_0} \right) I_{L-1}\left(\frac{s\sqrt{u_1}}{2\mathcal{E}N_0} \right), \quad u_1 \geqslant 0 \qquad (12.1\text{–}18)$$

On the other hand, the $\{U_m\}$, $m = 2, 3, \ldots, M$, are statistically independent and identically chi-square-distributed random variables, each having $2L$ degrees of freedom. Using Equation 2.1–110, we obtain the PDF for U_m as

$$p(u_m) = \frac{1}{(4\mathcal{E}N_0)^L (L-1)!} u_m^{L-1} e^{-u_m/4\mathcal{E}N_0}, \qquad u_m \geqslant 0$$

$$m = 2, 3, \ldots, M \qquad (12.1\text{–}19)$$

The probability of a symbol error is

$$\begin{aligned} P_M &= 1 - P_c \\ &= 1 - P(U_2 < U_1, U_3 < U_1, \ldots, U_M < U_1) \\ &= 1 - \int_0^\infty [P(U_2 < u_1 | U_1 = u_1)]^{M-1} p(u_1)\, du_1 \end{aligned} \qquad (12.1\text{–}20)$$

But

$$P(U_2 < u_1 | U_1 = u_1) = 1 - \exp\left(-\frac{u_1}{4\mathcal{E}N_0} \right) \sum_{k=0}^{L-1} \frac{1}{k!} \left(\frac{u_1}{4\mathcal{E}N_0} \right)^k \qquad (12.1\text{–}21)$$

Hence,

$$\begin{aligned} P_M &= 1 - \int_0^\infty \left[1 - e^{-u_1/4\mathcal{E}N_0} \sum_{k=0}^{L-1} \frac{1}{k!} \left(\frac{u_1}{4\mathcal{E}N_0} \right)^k \right]^{M-1} p(u_1)\, du_1 \\ &= 1 - \int_0^\infty \left(1 - e^{-v} \sum_{k=0}^{L-1} \frac{v^k}{k!} \right)^{M-1} \left(\frac{v}{\gamma} \right)^{(L=1)/2} e^{-(\gamma+v)} I_{L-1}(2\sqrt{\gamma v})\, dv \end{aligned}$$

$$(12.1\text{–}22)$$

where

$$\gamma = \mathcal{E} \sum_{n=1}^{L} \frac{\alpha_n^2}{N_0}$$

The integral in Equation 12.1–22 can be evaluated numerically. It is also possible to expand the term $(1 - x)^{M-1}$ in Equation 12.1–22 and carry out the integration term by term. This approach yields an expression for P_M in terms of finite sums.

An alternative approach is to use the union bound

$$P_M < (M-1)P_2(L) \tag{12.1–23}$$

where $P_2(L)$ is the probability of error in choosing between U_1 and any one of the $M-1$ decision variables $\{U_m\}$, $m = 2, 3, \ldots, M$. From our previous discussion on the performance of binary orthogonal signaling, we have

$$P_2(L) = \frac{1}{2^{2L-1}} e^{-k\gamma_b/2} \sum_{n=0}^{L-1} c_n \left(\tfrac{1}{2}k\gamma_b\right)^n \tag{12.1–24}$$

where c_n is given by Equation 12.1–14. For relatively small values of M, the union bound in Equation 12.1–23 is sufficiently tight for most practical applications.

12.2 MULTICARRIER COMMUNICATIONS

From our treatment of nonideal linear filter channels in Chapters 10 and 11, we have observed that such channels introduce ISI, which degrades performance compared with the ideal channel. The degree of performance degradation depends on the frequency-response characteristics. Furthermore, the complexity of the receiver increases as the span of the ISI increases.

Given a particular channel characteristic, the communication system designer must decide how to efficiently utilize the available channel bandwidth in order to transmit the information reliably within the transmitter power constraint and receiver complexity constraints. For a nonideal linear filter channel, one option is to employ a single carrier system in which the information sequence is transmitted serially at some specified rate R symbols/s. In such a channel, the time dispersion is generally much greater than the symbol rate, and, hence, ISI results from the nonideal frequency-response characteristics of the channel. As we have observed, an equalizer is necessary to compensate for the channel distortion.

As an example of such an approach, we cite the modems designed to transmit data through voice-band channels in the switched telephone network, which are based on the International Telecommunication Union (ITU) standard V.34. Such modems employ QAM impressed on a single carrier that is selected adaptively along with the symbol rate from a small set of specified values to obtain the maximum throughput at the desired level of performance (error rate). The channel frequency-response characteristics are measured upon initial setup of the telephone circuit, and the symbol rate and carrier frequency are selected based on this measurement.

An alternative approach to the design of a bandwidth-efficient communication system in the presence of channel distortion is to subdivide the available channel bandwidth into a number of subchannels, such that each subchannel is nearly ideal. To elaborate, suppose that $C(f)$ is the frequency response of a nonideal, band-limited channel with a bandwidth W, and that the power spectral density of the additive Gaussian noise is $\Phi_{nn}(f)$. Then, we divide the bandwidth W into $N = W/\Delta f$ subbands of width Δf, where Δf is chosen sufficiently small that $|C(f)|^2/\Phi_{nn}(f)$ is approximately a constant within each subband. Furthermore, we select the transmitted signal power to be distributed in frequency as $P(f)$, subject to the constraint that

$$\int_W P(f)\,df \leqslant P_{\text{av}} \tag{12.2–1}$$

where P_{av} is the available average power of the transmitter. Then, we transmit the data on these N subchannels. Let us evaluate the capacity of the nonideal additive Gaussian noise channel.

12.2.1 Capacity of a Nonideal Linear Filter Channel

Recall that the capacity of an ideal, band-limited, AWGN channel is

$$C = W \log_2\left(1 + \frac{P_{\text{av}}}{WN_0}\right) \tag{12.2–2}$$

where C is the capacity in bits/s, W is the channel bandwidth, and P_{av} is the average transmitted power. In a multicarrier system, with Δf sufficiently small, the subchannel has capacity

$$C_i = \Delta f \log_2\left[1 + \frac{\Delta f P(f_i)|C(f_i)|^2}{\Delta f \Phi_{nn}(f_i)}\right] \tag{12.2–3}$$

Hence, the total capacity of the channel is

$$C = \sum_{i=1}^{N} C_i = \Delta f \sum_{i=1}^{N} \log_2\left[1 + \frac{P(f_i)|C(f_i)|^2}{\Phi_{nn}(f_i)}\right] \tag{12.2–4}$$

In the limit as $\Delta f \to 0$, we obtain the capacity of the overall channel in bits/s as

$$C = \int_W \log_2\left[1 + \frac{P(f)|C(f)|^2}{\Phi_{nn}(f)}\right] df \tag{12.2–5}$$

Under the constraint on $P(f)$ given by Equation 12.2–1, the choice of $P(f)$ that maximizes C may be determined by maximizing the integral

$$\int_W \left\{\log_2\left[1 + \frac{P(f)|C(f)|^2}{\Phi_{nn}(f)}\right] + \lambda P(f)\right\} df \tag{12.2–6}$$

where λ is a Lagrange multiplier, which is chosen to satisfy the constraint. By using the calculus of variations to perform the maximization, we find that the optimum distribution of transmitted signal power is the solution to the equation

$$\frac{1}{P(f) + \Phi_{nn}(f)/|C(f)|^2} + \lambda = 0 \tag{12.2–7}$$

Therefore, $P(f) + \Phi_{nn}(f)/|C(f)|^2$ must be a constant, whose value is adjusted to satisfy the average power constraint in Equation 12.2–1. That is,

$$P(f) = \begin{cases} K - \Phi_{nn}(f)/|C(f)|^2 & (f \in W) \\ 0 & (f \notin W) \end{cases} \tag{12.2–8}$$

This expression for the channel capacity of a nonideal linear filter channel with additive Gaussian noise is due to Shannon (1949). The basic interpretation of this result is that the signal power should be high when the channel SNR $|C(f)|^2/\Phi_{nn}(f)$ is high, and low when the channel SNR is low. This result on the transmitted power distribution is illustrated in Figure 12.2–1. Observe that if $\Phi_{nn}(f)/|C(f)|^2$ is interpreted as the bottom of a bowl of unit depth, and we pour an amount of water equal to P_{av} into the bowl, the water will distribute itself in the bowl so as to achieve capacity. This is called the *water-filling interpretation* of the optimum power distribution as a function of frequency.

It is interesting to note that the channel capacity is smallest when the channel SNR $|C(f)|^2/\Phi_{nn}(f)$ is a constant for all $f \in W$. In this case, $P(f)$ is a constant for all $f \in W$. Equivalently, if the channel frequency response is ideal, i.e., $C(f) = 1$ for $f \in W$, then the worst Gaussian noise power distribution, from the viewpoint of maximizing capacity, is white Gaussian noise.

The above development suggests that multicarrier modulation that divides the available channel bandwidth into subbands of relatively narrow width $\Delta f = W/N$ provides a solution that could yield transmission rates close to capacity. The signal in each subband may be independently coded and modulated at a synchronous symbol rate of $1/\Delta f$, with the optimum power allocation $P(f)$. If Δf is small enough, then $C(f)$ is essentially constant across each subband, so that no equalization is necessary because the ISI is negligible.

Multicarrier modulation has been used in modems for both radio and telephone channels. Multicarrier modulation has also been adopted as a standard for digital audio broadcast applications.

A particularly suitable application of multicarrier modulation is in digital transmission over copper wire subscriber loops. The typical channel attenuation characteristics for such subscriber lines are illustrated in Figure 12.2–2. We observe that the attenuation increases rapidly as a function of frequency. This

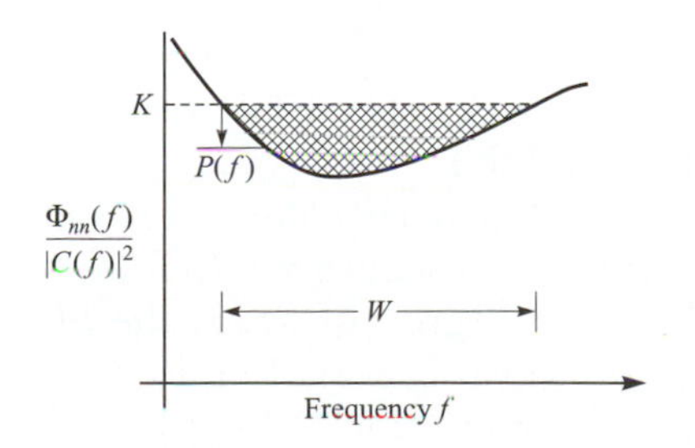

FIGURE 12.2–1
The optimum power distribution based on water-pouring interpretation.

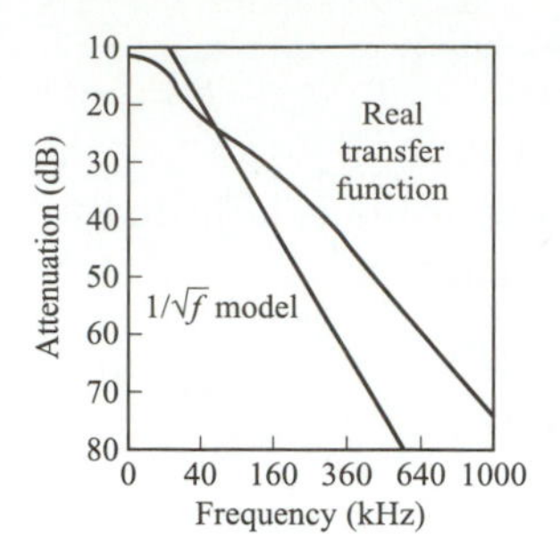

FIGURE 12.2–2
Attenuation characteristic of a 24-gauge 12000 ft. polyethylene-insulated cable loop. [*From Werner (1991)* © *IEEE*.]

characteristic makes it extremely difficult to achieve a high transmission rate with a single modulated carrier and an equalizer at the receiver. The ISI penalty in performance is very large. On the other hand, multicarrier modulation with optimum power distribution provides the potential for a higher transmission rate.

The dominant noise in transmission over subscriber lines is crosstalk interference from signals carried on other telephone lines located in the same cable. The power distribution of this type of noise is also frequency-dependent, which can be taken into consideration in the allocation of the available transmitted power.

A design procedure for a multicarrier QAM system for a nonideal linear filter channel has been given by Kalet (1989). In this procedure, the overall bit rate is maximized, through the design of an optimal power division among the subcarriers and an optimum selection of the number of bits per symbol (sizes of the QAM signal constellations) for each subcarrier, under an average power constraint and under the constraint that the symbol error probabilities for all subcarriers are equal.

Below, we present an implementation of a multicarrier QAM modulator and demodulator that is based on the discrete Fourier transform (DFT) for the generation of the multiple carriers at the transmitter and on the use of the DFT in the demodulation of the received signal.

12.2.2 An FFT-Based Multicarrier System

In this section, we describe a multicarrier communication system that employs the fast Fourier transform (FFT) algorithm to synthesize the signal at the transmitter and to demodulate the received signal at the receiver. The FFT is simply the efficient computational tool for implementing the DFT.

Figure 12.2–3 illustrates a block diagram of a multicarrier communication system. A serial-to-parallel buffer segments the information sequence into frames of N_f bits. The N_f bits in each frame are parsed into $\tilde{N}$ groups, where the ith group is assigned $\tilde{n}_i$ bits, and

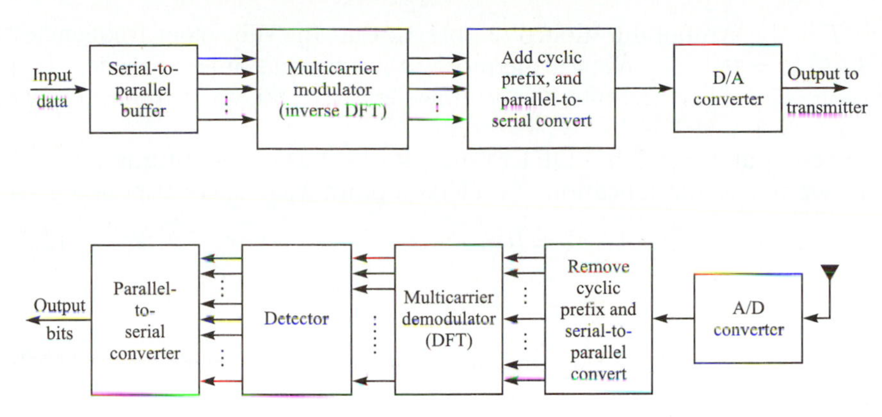

FIGURE 12.2–3
Multicarrier communication system.

$$\sum_{i=1}^{\tilde{N}} \tilde{n}_i = N_f \tag{12.2–9}$$

Each group may be encoded separately, so that the number of output bits from the encoder for the ith group is $n_i \geqslant \tilde{n}_i$.

It is convenient to view the multicarrier modulation as consisting of $\tilde{N}$ independent QAM channels, each operating at the same symbol rate $1/T$, but each channel having a distinct QAM constellation, i.e., the ith channel will employ $M_i = 2^{n_i}$ signal points. We denote the complex-valued signal points corresponding to the information symbols on the subchannels by X_k, $k = 0, 1, \ldots, \tilde{N} - 1$. In order to modulate the $\tilde{N}$ subcarriers by the information symbols $\{X_k\}$, we employ the inverse DFT (IDFT).

However, if we compute the $\tilde{N}$-point IDFT of $\{X_k\}$, we shall obtain a complex-valued time series, which is not equivalent to $\tilde{N}$ QAM-modulated subcarriers. Instead, we create $N = 2\tilde{N}$ information symbols by defining

$$X_{N-k} = X_k^*, \qquad k = 1, \ldots, \tilde{N} - 1 \tag{12.2–10}$$

and $X_0' = \text{Re}(X_0)$, $X_{\tilde{N}} = \text{Im}(X_0)$. Thus, the symbol X_0 is split into two parts, both real. Then, the N-point IDFT yields the real-valued sequence

$$x_n = \frac{1}{\sqrt{N}} \sum_{k=0}^{N-1} X_k e^{j2\pi nk/N}, \qquad n = 0, 1, \ldots, N - 1 \tag{12.2–11}$$

where $1/\sqrt{N}$ is simply a scale factor.

The sequence $\{x_n, 0 \leqslant n \leqslant N - 1\}$ corresponds to the samples of the sum $x(t)$ of $\tilde{N}$ subcarrier signals, which is expressed as

$$x(t) = \frac{1}{\sqrt{N}} \sum_{k=0}^{N-1} X_k e^{j2\pi kt/T}, \qquad 0 \leqslant t \leqslant T \tag{12.2–12}$$

where T is the symbol duration. We observe that the subcarrier frequencies are $f_k = k/T,\ k = 0, 1, \ldots, \tilde{N}$. Furthermore, the discrete-time sequence $\{x_n\}$ in Equation 12.2–11 represents the samples of $x(t)$ taken at times $t = nT/N$ where $n = 0, 1, \ldots, N-1$.

The computation of the IDFT of the data $\{X_k\}$ as given in Equation 12.2–10 may be viewed as multiplication of each data point X_k by a corresponding vector

$$\mathbf{v}_k = [v_{k0} \quad v_{k1} \quad \cdots \quad v_{k(N-1)}] \tag{12.2–13}$$

where

$$v_{kn} = \frac{1}{\sqrt{N}} e^{j(2\pi/N)kn} \tag{12.2–14}$$

as illustrated in Figure 12.2–4. In any case, the computation of the DFT is performed efficiently by the use of the FFT algorithm.

In practice, the signal samples $\{x_n\}$ are passed through a digital-to-analog (D/A) converter whose output, ideally, would be the signal waveform $x(t)$. The output of the channel is the waveform

$$r(t) = x(t) \star c(t) + n(t) \tag{12.2–15}$$

where $c(t)$ is the impulse response of the channel and $\star$ denotes convolution. By selecting the bandwidth Δf of each subchannel to be very small, the symbol duration $T = 1/\Delta f$ is large compared with the channel time dispersion. To be specific, let us assume that the channel dispersion spans $\nu + 1$ signal samples where $\nu \ll N$. One way to avoid the effect of ISI is to insert a time guard band of duration $\nu T/N$ between transmissions of successive blocks.

An alternative method that avoids ISI is to append a cyclic prefix to each block of N signal samples $\{x_0, x_1, \ldots, x_{N-1}\}$. The cyclic prefix for this block of samples consists of the samples $x_{N-\nu}, x_{N-\nu+1}, \ldots, x_{N-1}$. These new samples are appended to the beginning of each block. Note that the addition of the cyclic prefix to the block of data increases the length of the block to $N + \nu$ samples, which may be indexed from $n = -\nu, \ldots, N-1$, where the first ν samples constitute the prefix. Then, if $\{c_n, 0 \leqslant n \leqslant \nu\}$ denotes the sampled channel impulse response, its convolution with $\{x_n, -\nu \leqslant n \leqslant N-1\}$ produces $\{r_n\}$, the received

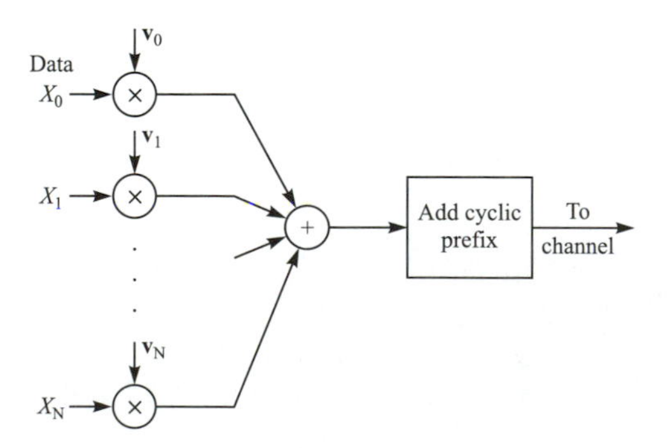

FIGURE 12.2–4
Signal synthesis for multicarrier modulation based on inverse DFT.

sequence. We are interested in the samples of $\{r_n\}$ for $0 \leqslant n \leqslant N-1$, from which we recover the transmitted sequence by using the N-point DFT for demodulation. Thus, the first ν samples of $\{r_n\}$ are discarded.

From a frequency-domain viewpoint, when the channel impulse response is $\{c_n, 0 \leqslant n \leqslant \nu\}$, its frequency response at the subcarrier frequencies $f_k = k/N$ is

$$C_k \equiv C\left(\frac{2\pi k}{N}\right) = \sum_{n=0}^{\nu} c_n e^{-j2\pi nk/N} \tag{12.2–16}$$

Because of the cyclic prefix, successive blocks (frames) of the transmitted information sequence do not interfere and, hence, the demodulated sequence may be expressed as

$$\hat{X}_k = C_k X_k + \eta_k, \qquad k = 0, 1, \ldots, N-1 \tag{12.2–17}$$

where $\{\hat{X}_k\}$ is the output of the N-point DFT demodulator, and η_k is the additive noise corrupting the signal. We note that by selecting $N \gg \nu$, the rate loss due to the cyclic prefix can be rendered negligible.

As shown in Figure 12.2–3, the information is demodulated by computing the DFT of the received signal after it has been passed through an analog-to-digital (A/D) converter. The DFT computation may be viewed as a multiplication of the received signal samples $\{r_n\}$ from the A/D converter by $\mathbf{v}_n^*$, where $\mathbf{v}_n$ is defined in Equation 12.2–13. As in the case of the modulator, the DFT computation at the demodulator is performed efficiently by use of the FFT algorithm.

It is simple matter to estimate and compensate for the channel factors $\{C_k\}$ prior to passing the data to the detector and decoder. A training signal consisting of either a known modulated sequence on each of the subcarriers or unmodulated subcarriers may be used to measure the $\{C_k\}$ at the receiver. If the channel parameters vary slowly with time, it is also possible to track the time variations by using the decisions at the output of the detector or the decoder, in a decision-directed fashion. Thus, the multicarrier system can be rendered adaptive.

By measuring the SNR in each subchannel, one can optimize the transmission rate by allocating the average transmitted power and the number of bits to be carried by each subcarrier. The SNR per subchannel is defined as

$$\mathrm{SNR}_k = \frac{TP_k |C_k|^2}{\sigma_{nk}^2}$$

where T is the symbol duration, P_k is the average power allocated to the kth subchannel, $|C_k|^2$ is the magnitude-square of the frequency response of the kth subchannel, and σ_{nk}^2 is the variance of the noise in the kth subchannel. Based on these SNR measurements, the capacity of each subchannel may be determined as described in Section 12.2.1. Furthermore, system performance may be optimized by selecting the bit and power allocation for each subchannel as described in the papers by Chow et al. (1995) and Fischer and Huber (1996).

Multicarrier QAM of the type described above has been implemented for a variety of applications, including high-speed transmission over telephone lines, such as digital subscriber lines.

Other types of implementation besides the DFT are possible. For example, a digital filter bank that basically performs the DFT may be substituted for the FFT-based implementation when the number of subcarriers is small, e.g., $N \leqslant 32$. For a large number of subcarriers, e.g., $N > 32$, the FFT-based systems are computationally more efficient.

One limitation of the DFT-type modulators and demodulators arises from the relatively large sidelobes in frequency that are inherent in DFT-type filter banks. The first sidelobe is only 13 dB down from the peak at the desired subcarrier. Consequently, the DFT-based implementations are vulnerable to interchannel interference (ICI) unless a full cyclic prefix is used. If ICI is a problem, due to channel anomalies, one may resort to other types of digital filter banks that have much lower sidelobes. In particular, the class of multirate digital filter banks that have the perfect reconstruction property associated with wavelet-based filters appear to be an attractive alternative (see Tzannes et al., 1994; Rizos et al., 1994).

12.2.3 Minimizing Peak-to-Average Ratio in Multicarrier Systems

A major problem with multicarrier modulation is the relatively high peak-to-average power ratio (PAR) that is inherent in the transmitted signal. In general, large signal peaks occur in the transmitted signal when the signals in many of the various subchannels add constructively in phase. Such large signal peaks may result in clipping of the signal voltage in a D/A converter when the multicarrier signal is synthesized digitally, and/or it may saturate the power amplifier and, thus, cause intermodulation distortion in the transmitted signal. When the number N of subcarriers is large, the central limit theorem may be used to model the combined signal on the N subchannels as a zero-mean Gaussian random process. In such a model, the voltage PAR is proportional to $\sqrt{N}$.

To avoid intermodulation distortion, it is common to reduce the power in the transmitted signal and, thus, operate the power amplifier at the transmitter in its linear operating range. This power reduction or "power back-off" results in inefficient operation of the communication system. For example, if the PAR is 10 dB, the power back-off may be as much as 10 dB to avoid intermodulation distortion.

Various methods have been devised to reduce the PAR in multicarrier systems. One of the simplest methods is to insert different phase shifts in each of the subcarriers. These phase shifts can be selected pseudorandomly, or by means of some algorithm, to reduce the PAR. For example, we may have a small set of N stored pseudorandomly selected phase shifts which can be used when the PAR in the modulated subcarriers is large. The information on which set of pseudorandom phase shifts is used in any signal interval can be transmitted to the receiver on one of the N subcarriers. Alternatively, a single set of pseudorandom phase shifts may be employed, where this set is found via computer simulation to reduce the PAR to an acceptable level over the ensemble of possible transmitted data symbols on the N subcarriers.

Another method that can be used to reduce the PAR is to modulate a small subset of the subcarriers with dummy symbols which are selected to reduce the PAR. Since the dummy symbols do not have to be constrained to take amplitude and phase values from a specified signal constellation, the design of the dummy symbols is very flexible. The subcarriers carrying dummy symbols may be distributed across the frequency band. Since modulating subcarriers with dummy symbols results in a lower throughput in data rate, it is desirable to employ only a small percentage of the total subcarriers for this purpose.

As an alternative to allocating subcarriers that are modulated with dummy symbols, one may select a subset of subcarriers that already carry data and expand the signal constellation in such a manner that the data can be correctly detected at the receiver by use of a modulo-q operation, where q is an appropriate integer. For example, if rectangular 16-point QAM is used as the modulation of each subcarrier, a minimally expanded signal constellation for a subset of subcarriers may consist of a 32-point signal constellation that includes the 16 additional points adjacent to the outer points in the original constellation. When the PAR of the original signal constellation exceeds a predetermined amount, the signal point on a selected subcarrier is replaced by a signal point from the minimally expanded set, such that the PAR is reduced. This approach may require several iterations using a different subcarrier each time to reduce the PAR to a desired value. The interested reader may refer to the paper by Tellado and Cioffi (1998), which treats this method.

In a digitally synthesized multicarrier signal, the PAR may be kept within a specified limit by clipping the signal at the D/A converter. The clipping generally causes errors in the transmitted data sequence. In such a case, if the clipping occurs infrequently, the occasional errors may be corrected by introducing a suitable error-correcting code.

Because of its practical importance, the problem of PAR reduction in multicarrier systems has been investigated by many people, and methods other than the ones described above have been considered. The interested reader may refer to the papers by Boyd (1986), Popovic (1991), Jones et al. (1994), Wilkinson and Jones (1995), Wulich (1996), Tarokh and Jafarkhani (2000), and Tellado and Cioffi (1999a).

12.3 BIBLIOGRAPHICAL NOTES AND REFERENCES

Multichannel signal transmission is commonly used on time-varying channels to overcome the effects of signal fading. This topic is treated in some detail in Chapter 14, where we provide a number of references to published work. Of particular relevance to the treatment of multichannel digital communications given in this chapter are the two publications by Price (1962a,b).

There is a large amount of literature on multicarrier digital communication systems. Such systems have been implemented and used for over 35 years. One of the earliest systems, described by Doeltz et al. (1957) and called Kineplex, was used for digital transmission in the HF band. Other early work on multicarrier system design has been reported in the papers by Chang (1966) and Saltzberg (1967). The use of the DFT for modulation and demodulation of multicarrier systems was proposed by Weinstein and Ebert (1971).

Of particular interest in recent years is the use of multicarrier digital transmission for data, facsimile, and video on a variety of channels, including the narrowband (4 kHz) switched telephone network, the 48-kHz group telephone band, digital subscriber lines, cellular radio, and audio broadcast. The interested reader may refer to the many papers in the literature. We cite as examples the papers by Hirosaki (1981), Hirosaki et al. (1986), Chow et al. (1991), and the survey paper by Bingham (1990). The paper by Kalet (1989) gives a design procedure for optimizing the rate in a multicarrier QAM system given constraints on transmitter power and channel characteristics. Finally, we cite the book by Vaidyanathan (1993) and the papers by Tzannes et al. (1994) and Rizos et al. (1994) for a treatment of multirate digital filter banks, and the book by Starr et al. (1999) on the application of multicarrier modulation for digital transmission on digital subscriber lines.

PROBLEMS

12.1 $X_1, X_2, \ldots, X_N$ are a set of N statistically independent and identically distributed real Gaussian random variables with moments $E(X_i) = m$ and var $(X_i) = \sigma^2$.

a) Define

$$U = \sum_{n=1}^{N} X_n$$

Evaluate the SNR of U, which is defined as

$$(\mathrm{SNR})_U = \frac{[E(U)]^2}{2\sigma_U^2}$$

where σ_U^2 is the variance of U.

b) Define

$$V = \sum_{n=1}^{N} X_n^2$$

Evaluate the SNR of V, which is defined as

$$(\mathrm{SNR})_V = \frac{[E(V)]^2}{2\sigma_V^2}$$

where σ_V^2 is the variance of V.

c) Plot $(\mathrm{SNR})_U$ and $(\mathrm{SNR})_V$ versus m^2/σ^2 on the same graph and, thus, compare the SNRs graphically.

d) What does the result in (c) imply regarding coherent detection and combining versus square-law detection and combining of multichannel signals?

12.2 A binary communication system transmits the same information on two diversity channels. The two received signals are

$$r_1 = \pm\sqrt{\mathcal{E}_b} + n_1$$
$$r_2 = \pm\sqrt{\mathcal{E}_b} + n_2$$

where $E(n_1) = E(n_2) = 0$, $E(n_1^2) = \sigma_1^2$ and $E(n_2^2) = \sigma_2^2$, and n_1 and n_2 are uncorrelated Gaussian variables. The detector bases its decision on the linear combination of r_1 and r_2, i.e.,

$$r = r_1 + kr_2$$

a) Determine the value of k that minimizes the probability of error.
b) Plot the probability of error for $\sigma_1^2 = 1$, $\sigma_2^2 = 3$, and either $k = 1$ or k is the optimum value found in (a). Compare the results.

12.3 Assess the cost of the cyclic prefix (used in multicarrier modulation to avoid ISI) in terms of
a) Extra channel bandwidth.
b) Extra signal energy.

12.4 Let $x(n)$ be a finite-duration signal with length N and let $X(k)$ be its N-point DFT. Suppose we pad $x(n)$ with L zeros and compute the $(N + L)$-point DFT, $X'(k)$. What is the relationship between $X(0)$ and $X'(0)$? If we plot $|X(k)|$ and $|X'(k)|$ on the same graph, explain the relationships between the two graphs.

12.5 Show that the sequence $\{x_n\}$ given by Equation 12.2–11 corresponds to the samples of the signal $x(t)$ given by Equation 12.2–12.

12.6 Show that the IDFT of a sequence $\{X_k, 0 \leqslant k \leqslant N-1\}$ can be computed by passing the sequence $\{X_k\}$ through a bank of N linear discrete-time filters with system functions

$$H_n(z) = \frac{1}{1 - e^{j2\pi n/N} z^{-1}}$$

and sampling the filter outputs at $n = N$.

12.7 Plot $P_2(L)$ for $L = 1$ and $L = 2$ as a function of $10 \log \gamma_b$ and determine the loss in SNR due to the combining loss for $\gamma_b = 10$.

13

Spread Spectrum Signals for Digital Communications

Spread spectrum signals used for the transmission of digital information are distinguished by the characteristic that their bandwidth W is much greater than the information rate R in bits/s. That is, the bandwidth expansion factor $B_e = W/R$ for a spread spectrum signal is much greater than unity. The large redundancy inherent in spread spectrum signals is required to overcome the severe levels of interference that are encountered in the transmission of digital information over some radio and satellite channels. Since coded waveforms are also characterized by a bandwidth expansion factor greater than unity and since coding is an efficient method for introducing redundancy, it follows that coding is an important element in the design of spread spectrum signals.

A second important element employed in the design of spread spectrum signals is pseudorandomness, which makes the signals appear similar to random noise and difficult to demodulate by receivers other than the intended ones. This element is intimately related with the application or purpose of such signals.

To be specific, spread spectrum signals are used for

- Combating or suppressing the detrimental effects of interference due to jamming, interference arising from other users of the channel, and self-interference due to multipath propagation.
- Hiding a signal by transmitting it at low power and, thus, making it difficult for an unintended listener to detect in the presence of background noise.
- Achieving message privacy in the presence of other listeners.

In applications other than communications, spread spectrum signals are used to obtain accurate range (time delay) and range rate (velocity) measurements in radar and navigation. For the sake of brevity, we shall limit our discussion to digital communication applications.

In combating intentional interference (jamming), it is important to the communicators that the jammer who is trying to disrupt the communication does not have prior knowledge of the signal characteristics except for the overall channel bandwidth and the type of modulation (PSK, FSK, etc.) being used. If the digital information is just encoded as described in Chapter 8, a sophisticated jammer can easily mimic the signal emitted by the transmitter and, thus, confuse the receiver. To circumvent this possibility, the transmitter introduces an element of unpredictability or randomness (pseudorandomness) in each of the transmitted coded signal waveforms that is known to the intended receiver but not to the jammer. As a consequence, the jammer must synthesize and transmit an interfering signal without knowledge of the pseudorandom pattern.

Interference from the other users arises in multiple-access communication systems in which a number of users share a common channel bandwidth. At any given time, a subset of these users may transmit information simultaneously over the common channel to corresponding receivers. Assuming that all the users employ the same code for the encoding and decoding of their respective information sequences, the transmitted signals in this common spectrum may be distinguished from one another by superimposing a different pseudorandom pattern, also called a *code*, in each transmitted signal. Thus, a particular receiver can recover the transmitted information intended for it by knowing the pseudorandom pattern, i.e., the key, used by the corresponding transmitter. This type of communication technique, which allows multiple users to simultaneously use a common channel for transmission of information, is called *code division multiple access* (CDMA). CDMA will be considered in Sections 13.2 and 13.3.

Resolvable multipath components resulting from time-dispersive propagation through a channel may be viewed as a form of self-interference. This type of interference may also be suppressed by the introduction of a pseudorandom pattern in the transmitted signal, as will be described below.

A message may be hidden in the background noise by spreading its bandwidth with coding and transmitting the resultant signal at a low average power. Because of its low power level, the transmitted signal is said to be "covert." It has a low probability of being intercepted (detected) by a casual listener and, hence, is also called a *low-probability-of-intercept* (LPI) signal.

Finally, message privacy may be obtained by superimposing a pseudorandom pattern on a transmitted message. The message can be demodulated by the intended receivers, who know the pseudorandom pattern or key used at the transmitter, but not by any other receivers who do not have knowledge of the key.

In the following sections, we shall describe a number of different types of spread spectrum signals, their characteristics, and their applications. The emphasis will be on the use of spread spectrum signals for combating, jamming (antijam or AJ signals), CDMA, and LPI. Before discussing the signal design problem, however, we shall briefly describe the types of channel characteristics assumed for the applications cited above.

13.1 MODEL OF SPREAD SPECTRUM DIGITAL COMMUNICATION SYSTEM

The block diagram shown in Figure 13.1–1 illustrates the basic elements of a spread spectrum digital communication system with a binary information sequence at its input at the transmitting end and at its output at the receiving end. The channel encoder and decoder and the modulator and demodulator are basic elements of the system, which were treated in Chapters 5, 7, and 8. In addition to these elements, we have two identical pseudorandom pattern generators, one that interfaces with the modulator at the transmitting end and a second that interfaces with the demodulator at the receiving end. The generators generate a pseudorandom or pseudonoise (PN) binary-valued sequence which is impressed on the transmitted signal at the modulator and removed from the received signal at the demodulator.

Synchronization of the PN sequence generated at the receiver with the PN sequence contained in the incoming received signal is required in order to demodulate the received signal. Initially, prior to the transmission of information, synchronization may be achieved by transmitting a fixed pseudorandom bit pattern that the receiver will recognize in the presence of interference with a high probability. After time synchronization of the generators is established, the transmission of information may commence.

Interference is introduced in the transmission of the information-bearing signal through the channel. The characteristics of the interference depend to a large extent on its origin. It may be categorized as being either broadband or narrowband relative to the bandwidth of the information-bearing signal and as either continuous or pulsed (discontinuous) in time. For example, a jamming signal may consist of one or more sinusoids in the bandwidth used to transmit the information. The frequencies of the sinusoids may remain fixed or they may change with time according to some rule. As a second example, the interference generated in CDMA by other users of the channel may be either broadband or narrowband, depending on the type of spread spectrum signal that is employed to achieve multiple access. If it is broadband, it may be characterized as an

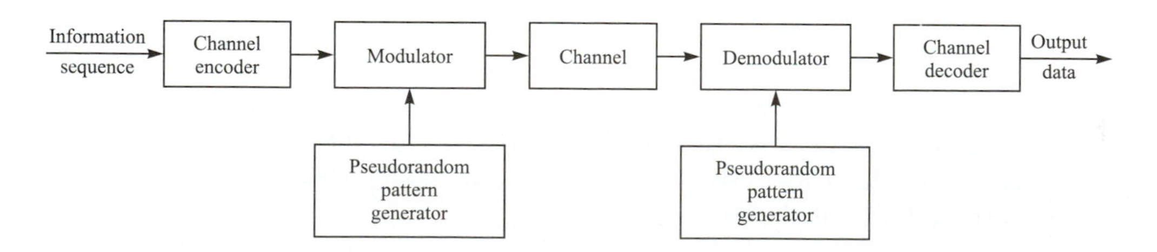

FIGURE 13.1–1
Model of spread spectrum digital communication system.

equivalent additive white Gaussian noise. We shall consider these types of interference and some others in the following sections.

Our treatment of spread spectrum signals will focus on the performance of the digital communication system in the presence of narrowband and broadband interference. Two types of modulation are considered: PSK and FSK. PSK is appropriate in applications where phase coherence between the transmitted signal and the received signal can be maintained over a time interval that is relatively long compared to the reciprocal of the transmitted signal bandwidth. On the other hand, FSK modulation is appropriate in applications where such phase coherence cannot be maintained due to time-variant effects on the communications link. This may be the case in a communications link between two high-speed aircraft or between a high-speed aircraft and a ground terminal.

The PN sequence generated at the modulator is used in conjunction with the PSK modulation to shift the phase of the PSK signal pseudorandomly as described in Section 13.2. The resulting modulated signal is called a *direct sequence* (DS) or a *pseudo-noise* (PN) spread spectrum signal. When used in conjunction with binary or M-ary ($M > 2$) FSK, the pseudorandom sequence selects the frequency of the transmitted signal pseudorandomly. The resulting signal is called a *frequency-hopped* (FH) spread spectrum signal. Although a number of other types of spread spectrum signals will be briefly described, the emphasis of our treatment will be on PN and FH spread spectrum signals.

13.2 DIRECT SEQUENCE SPREAD SPECTRUM SIGNALS

In the model shown in Figure 13.1–1, we assume that the information rate at the input to the encoder is R bits/s and the available channel bandwidth is W Hz. The modulation is assumed to be binary PSK. In order to utilize the entire available channel bandwidth, the phase of the carrier is shifted pseudorandomly according to the pattern from the PN generator at a rate W times/s. The reciprocal of W, denoted by T_c, defines the duration of a pulse, which is called a *chip*; T_c is called the *chip interval*. The pulse is the basic element in a DS spread spectrum signal.

If we define $T_b = 1/R$ to be the duration of a rectangular pulse corresponding to the transmission time of an information bit, the bandwidth expansion factor W/R may be expressed as

$$B_e = \frac{W}{R} = \frac{T_b}{T_c} \tag{13.2–1}$$

In practical systems, the ratio T_b/T_c is an integer,

$$L_c = \frac{T_b}{T_c} \tag{13.2–2}$$

which is the number of chips per information bit. That is, L_c is the number of phase shifts that can occur in the transmitted signal during the bit duration $T_b = 1/R$. Figure 13.2–1a illustrates the relationships between the PN signal and the data signal.

Suppose that the encoder takes k information bits at a time and generates a binary linear (n, k) block code. The time duration available for transmitting the n code elements is kT_b seconds. The number of chips that occur in this time interval is kL_c. Hence, we may select the block length of the code as $n = kL_c$. If the encoder generates a binary convolutional code of rate k/n, the number of chips in the time interval kT_b is also $n = kL_c$. Therefore, the following discussion

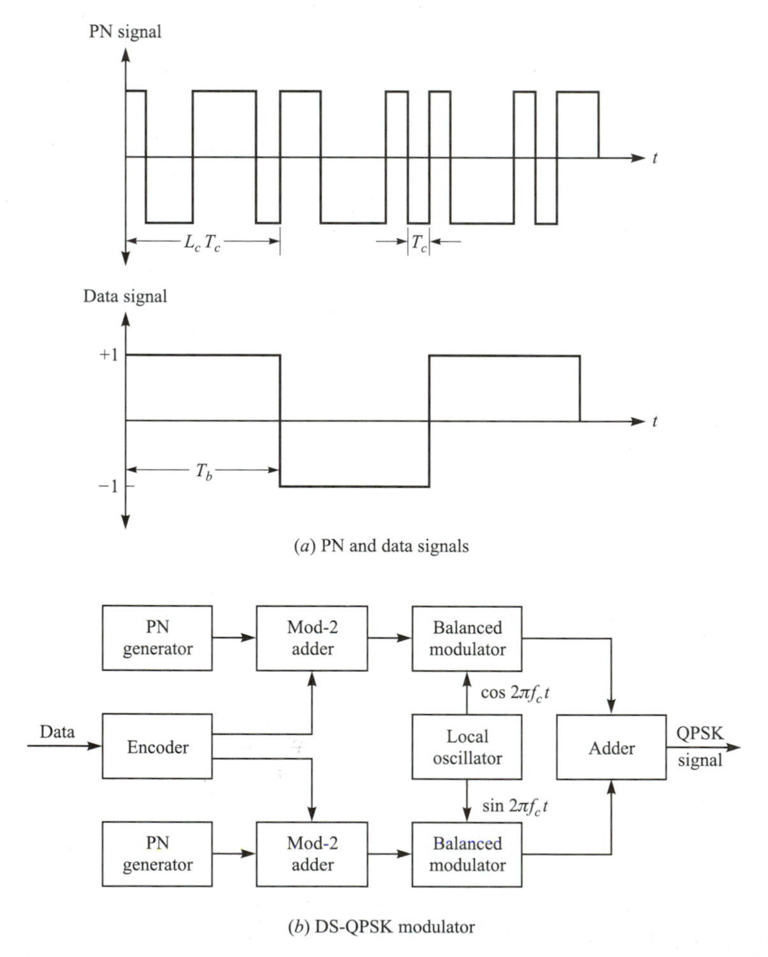

FIGURE 13.2–1
The PN and data signals (*a*) and the QPSK modulator (*b*) for a DS spread spectrum system.

applies to both block codes and convolutional codes. We note that the code rate $R_c = k/n = 1/L_c$.

One method for impressing the PN sequence on the transmitted signal is to alter directly the coded bits by modulo-2 addition with the PN sequence.† Thus, each coded bit is altered by its addition with a bit from the PN sequence. If b_i represents the ith bit of the PN sequence and c_i is the corresponding bit from the encoder, the modulo-2 sum is

$$a_i = b_i \oplus c_i \tag{13.2–3}$$

Hence, $a_i = 1$ if either $b_i = 1$ and $c_i = 0$ or $b_i = 0$ and $c_i = 1$; also $a_i = 0$ if either $b_i = 1$ and $c_i = 1$ or $b_i = 0$ and $c_i = 0$. We may say that $a_i = 0$ when $b_i = c_i$ and $a_i = 1$ when $b_i \neq c_i$. The sequence $\{a_i\}$ is mapped into a binary PSK signal of the form $s(t) = \pm\text{Re}[g(t)e^{j2\pi f_c t}]$ according to the convention

$$g_i(t) = \begin{cases} g(t - iT_c) & (a_i = 0) \\ -g(t - iT_c) & (a_i = 1) \end{cases} \tag{13.2–4}$$

where $g(t)$ represents a pulse of duration T_c seconds and arbitrary shape.

The modulo-2 addition of the coded sequence $\{c_i\}$ and the sequence $\{b_i\}$ from the PN generator may also be represented as a multiplication of two waveforms. To demonstrate this point, suppose that the elements of the coded sequence are mapped into a binary PSK signal according to the relation

$$c_i(t) = (2c_i - 1)g(t - iT_c) \tag{13.2–5}$$

Similarly, we define a waveform $p_i(t)$ as

$$p_i(t) = (2b_i - 1)p(t - iT_c) \tag{13.2–6}$$

where $p(t)$ is a rectangular pulse of duration T_c. Then the equivalent low-pass transmitted signal corresponding to the ith coded bit is

$$\begin{aligned} g_i(t) &= p_i(t)c_i(t) \\ &= (2b_i - 1)(2c_i - 1)g(t - iT_c) \end{aligned} \tag{13.2–7}$$

This signal is identical to the one given by Equation 13.2–4, which is obtained from the sequence $\{a_i\}$. Consequently, modulo-2 addition of the coded bits with the PN sequence followed by a mapping that yields a binary PSK signal is equivalent to multiplying a binary PSK signal generated from the coded bits with a sequence of unit amplitude rectangular pulses, each of duration T_c, and with a polarity which is determined from the PN sequence according to Equation 13.2–6. Although it is easier to implement modulo-2 addition followed by PSK modulation instead of waveform multiplication, it is convenient, for purposes of

---

†When four-phase PSK is desired, one PN sequence is added to the information sequence carried on the in-phase signal component and a second PN sequence is added to the information sequence carried on the quadrature component. In many PN spread spectrum systems, the same binary information sequence is added to the two PN sequences to form the two quadrature components. Thus, a four-phase PSK signal is generated with a binary information stream.

demodulation, to consider the transmitted signal in the multiplicative form given by Equation 13.2–7. A functional block diagram of a four-phase PSK-DS spread spectrum modulator is shown in Figure 13.2–1b.

The received equivalent low-pass signal for the ith code element is

$$\begin{aligned} r_i(t) &= (p_i(t)c_i(t) + z(t), \qquad iT_c \leqslant t \leqslant (i+1)T_c \\ &= (2b_i - 1)(2c_i - 1)g(t - iT_c) + z(t) \end{aligned} \tag{13.2–8}$$

where $z(t)$ represents the interference or jamming signal corrupting the information-bearing signal. The interference is assumed to be a stationary random process with zero mean.

If $z(t)$ is a sample function from a complex-valued Gaussian process, the optimum demodulator may be implemented either as a filter matched to the waveform $g(t)$ or as a correlator, as illustrated by the block diagrams in Figure 13.2–2. In the matched filter realization, the sampled output from the matched filter is multiplied by $2b_i - 1$, which is obtained from the PN generator at the demodulator when the PN generator is properly synchronized. Since

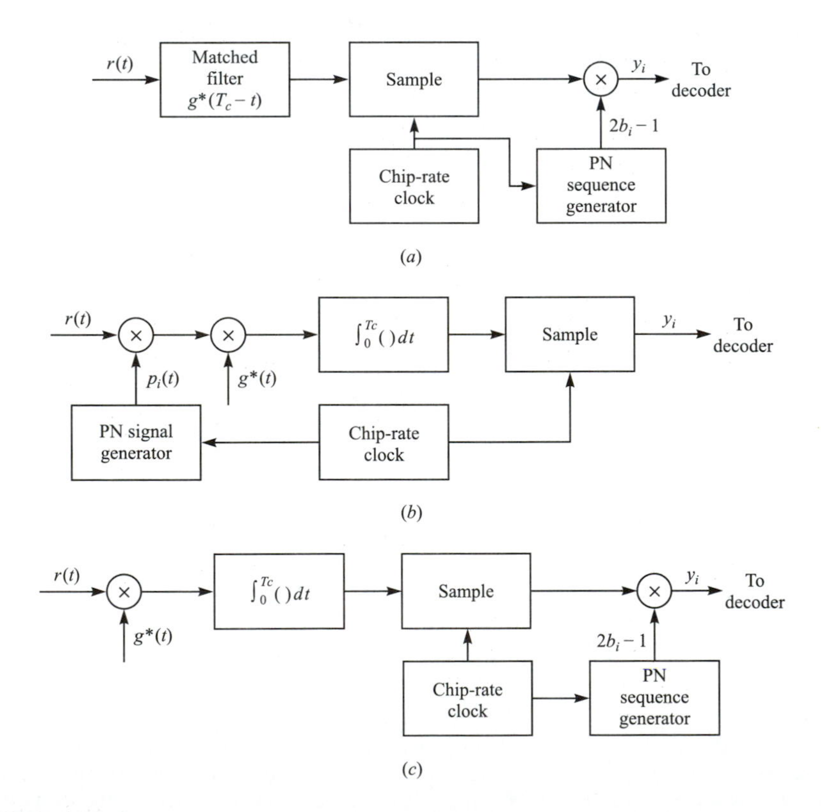

FIGURE 13.2–2
Possible demodulator structures for PN spread spectrum signals.

$(2b_i - 1)^2 = 1$ when $b_i = 0$ and $b_i = 1$, the effect of the PN sequence on the received coded bits is thus removed.

In Figure 13.2–2, we also observe that the cross correlation can be accomplished in either one of two ways. The first, illustrated in Figure 13.2–2b, involves premultiplying $r_i(t)$ with the waveform $p_i(t)$ generated from the output of the PN generator and then cross-correlating with $g^*(t)$ and sampling the output in each chip interval. The second method, illustrated in Figure 13.2–2c, involves cross correlation with $g^*(t)$ first, sampling the output of the correlator and, then, multiplying this output with $2b_i - 1$, which is obtained from the PN generator.

If $z(t)$ is not a Gaussian random process, the demodulation methods illustrated in Figure 13.2–2 are no longer optimum. Nevertheless, we may still use any of these three demodulator structures to demodulate the received signal. When the statistical characteristics of the interference $z(t)$ are unknown a priori, this is certainly one possible approach. An alternative method, which is described later, utilizes an adaptive filter prior to the matched filter or correlator to suppress narrowband interference. The rationale for this second method is also described later.

In Section 13.2.1, we derive the error rate performance of the DS spread spectrum system in the presence of wideband and narrowband interference. The derivations are based on the assumption that the demodulator is any of the three equivalent structures shown in Figure 13.2–2.

13.2.1 Error Rate Performance of the Decoder

Let the unquantized output of the demodulator be denoted by y_j, $1 \leqslant j \leqslant n$. First we consider a linear binary (n, k) block code and, without loss of generality, we assume that the all-zero code word is transmitted.

A decoder that employs soft-decision decoding computes the correlation metrics

$$CM_i = \sum_{j=1}^{n}(2c_{ij} - 1)y_j, \qquad i = 1, 2, \ldots, 2^k \tag{13.2–9}$$

where c_{ij} denotes the jth bit in the ith code word. The correlation metric corresponding to the all-zero code word is

$$\begin{aligned} CM_1 &= 2n\mathcal{E}_c + \sum_{j=1}^{n}(2c_{1j} - 1)(2b_j - 1)v_j \\ &= 2n\mathcal{E}_c - \sum_{j=1}^{n}(2b_j - 1)v_j \end{aligned} \tag{13.2–10}$$

where v_j, $1 \leqslant j \leqslant n$, is the additive noise term corrupting the jth coded bit and $\mathcal{E}_c$ is the chip energy. It is defined as

$$\nu_j = \text{Re}\left\{\int_0^{T_c} g^*(t)z[t + (j-1)T_c]\,dt\right\}, \qquad j = 1, 2, \ldots, n \tag{13.2–11}$$

Similarly, the correlation metric corresponding to code word $\mathbf{C}_m$ having weight w_m is

$$CM_m = 2\mathcal{E}_c n\left(1 - \frac{2w_m}{n}\right) + \sum_{j=1}^{n}(2c_{mj} - 1)(2b_j - 1)\nu_j \tag{13.2–12}$$

Following the procedure used in Section 8.1.4, we shall determine the probability that $CM_m > CM_1$. The difference between CM_1 and CM_m is

$$\begin{aligned} D &= CM_1 - CM_m \\ &= 4\mathcal{E}_c w_m - 2\sum_{j=1}^{n} c_{mj}(2b_j - 1)\nu_j \end{aligned} \tag{13.2–13}$$

Since the code word $\mathbf{C}_m$ has weight w_m, there are w_m nonzero components in the summation of noise terms contained in Equation 13.2–13. We shall assume that the minimum distance of the code is sufficiently large that we can invoke the central limit theorem for the summation of noise components. This assumption is valid for PN spread spectrum signals that have a bandwidth expansion of 10 or more.† Thus, the summation of noise components is modeled as a Gaussian random variable. Since $E(2b_j - 1) = 0$ and $E(\nu_j) = 0$, the mean of the second term in Equation 13.2–13 is also zero.

The variance is

$$\sigma_m^2 = 4\sum_{j=1}^{n}\sum_{i=1}^{n} c_{mi}c_{mj}E[(2b_j - 1)(2b_i - 1)]E(\nu_i\nu_j) \tag{13.2–14}$$

The sequence of binary digits from the PN generator are assumed to be uncorrelated. Hence

$$E[(2b_j - 1)(2b_i - 1)] = \delta_{ij} \tag{13.2–15}$$

and

$$\sigma_m^2 = 4w_m E(\nu^2) \tag{13.2–16}$$

where $E(\nu^2)$ is the second moment of any one element from the set $\{\nu_j\}$. This moment is easily evaluated to yield

$$\begin{aligned} E(\nu^2) &= \int_0^{T_c}\int_0^{T_c} g^*(t)g(\tau)\phi_{zz}(t - \tau)\,dt\,d\tau \\ &= \int_{-\infty}^{\infty} |G(f)|^2\Phi_{zz}(f)\,df \end{aligned} \tag{13.2–17}$$

†Typically, the bandwidth expansion factor in a spread spectrum signal is of the order of 10 to 100 and sometimes higher.

where $\phi_{zz}(\tau) = \frac{1}{2}E[z^*(t)z(t+\tau)]$ is the autocorrelation function and $\Phi_{zz}(f)$ is the power spectral density of the interference $z(t)$.

We observe that when the interference is spectrally flat within the bandwidth† occupied by the transmitted signal, i.e.,

$$\Phi_{zz}(f) = J_0, \qquad |f| \leqslant \tfrac{1}{2}W \tag{13.2–18}$$

the second moment in Equation 13.2–17 is $E(\nu^2) = 2\mathcal{E}_c J_0$, and, hence, the variance of the interference term in Equation 13.2–16 becomes

$$\sigma_m^2 = 8\mathcal{E}_c J_0 w_m \tag{13.2–19}$$

In this case, the probability that $D < 0$ is

$$P_2(m) = Q\left(\sqrt{\frac{2\mathcal{E}_c}{J_0} w_m}\right) \tag{13.2–20}$$

But the energy per coded bit $\mathcal{E}_c$ may be expressed in terms of the energy per information bit $\mathcal{E}_b$ as

$$\mathcal{E}_c = \frac{k}{n}\mathcal{E}_b = R_c\mathcal{E}_b \tag{13.2–21}$$

Witht his substitution, Equation 13.2–20 becomes

$$\begin{aligned} P_2(m) &= Q\left(\sqrt{\frac{2\mathcal{E}_b}{J_0} R_c w_m}\right) \\ &= Q(\sqrt{2\gamma_b R_c w_m}) \end{aligned} \tag{13.2–22}$$

where $\gamma_b = \mathcal{E}_b/J_0$ is the SNR per information bit. Finally, the code word error probability may be upper-bounded by the union bound as

$$P_M \leqslant \sum_{m=2}^{M} Q(\sqrt{2\gamma_b R_c w_m}) \tag{13.2–23}$$

where $M = 2^k$. Note that this expression is identical to the probability of a code word error for soft-decision decoding of a linear binary block code in an AWGN channel.

Although we have considered a binary block code in the derivation given above, the procedure is similar for an (n, k) convolutional code. The result of such a derivation is the following upper bound on the equivalent bit error probability:

$$P_b \leqslant \frac{1}{k} \sum_{d=d_{\text{free}}}^{\infty} \beta_d Q(\sqrt{2\gamma_b R_c d}) \tag{13.2–24}$$

†If the bandwidth of the bandpass channel is W, that of the equivalent low-pass channel is $\frac{1}{2}W$.

The set of coefficients $\{\beta_d\}$ is obtained from an expansion of the derivative of the transfer function $T(D, N)$, as described in Section 8.2.3.

Next, we consider a narrowband interference centered at the carrier (at DC for the equivalent low-pass signal). We may fix the total (average) jamming power to $J_{av} = J_0 W$, where J_0 is the value of the power spectral density of an equivalent wideband interference (jamming signal). The narrowband interference is characterized by the power spectral density

$$\Phi_{zz}(f) = \begin{cases} \dfrac{J_{av}}{W_1} = \dfrac{J_0 W}{W_1} & (|f| \leqslant \frac{1}{2}W_1) \\ 0 & (|f| > \frac{1}{2}W_1) \end{cases} \tag{13.2–25}$$

where $W \gg W_1$.

Substitution of Equation 13.2–25 for $\Phi_{zz}(f)$ into Equation 13.2–17 yields

$$E(v^2) = \frac{J_{av}}{W_1} \int_{-W_1/2}^{W_1/2} |G(f)|^2 \, df \tag{13.2–26}$$

The value of $E(v^2)$ depends on the spectral characteristics of the pulse $g(t)$. In the following example, we consider two special cases.

EXAMPLE 13.2–1. Suppose that $g(t)$ is a rectangular pulse as shown in Figure 13.2–3a and $|G(f)|^2$ is the corresponding energy density spectrum shown in Figure 13.2–3b. For the narrowband interference given by Equation 13.2–25, the variance of the total interference is

$$\begin{aligned} \sigma_m^2 &= 4w_m E(v^2) \\ &= \frac{8\mathcal{E}_c w_m T_c J_{av}}{W_1} \int_{-W_1/2}^{W_1/2} \left(\frac{\sin \pi f T_c}{\pi f T_c} \right)^2 df \\ &= \frac{8\mathcal{E}_c w_m J_{av}}{W_1} \int_{-\beta/2}^{\beta/2} \left(\frac{\sin \pi x}{\pi x} \right)^2 dx \end{aligned} \tag{13.2–27}$$

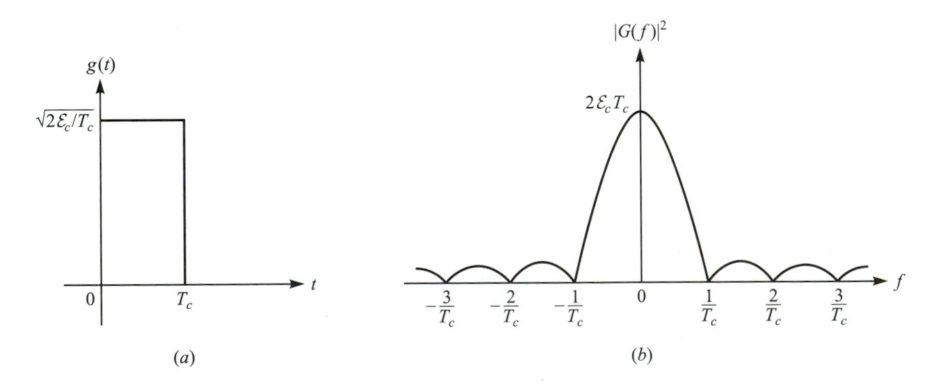

FIGURE 13.2–3
Rectangular pulse and its energy density spectrum.

where $\beta = W_1 T_c$. Figure 13.2–4 illustrates the value of this integral for $0 \leqslant \beta \leqslant 1$. We observe that the value of the integral is upper-bounded by $W_1 T_c$. Hence, $\sigma_m^2 \leqslant 8\mathcal{E}_c w_m T_c J_{\text{av}}$.

In the limit as W_1 becomes zero, the interference becomes an impulse at the carrier. In this case the interference is a pure frequency tone and it is usually called a *continuous wave (CW) jamming signal*. The power spectral density is

$$\Phi_{zz}(f) = J_{\text{av}}\delta(f) \tag{13.2–28}$$

and the corresponding variance for the decision variable $D = CM_1 - CM_m$ is

$$\begin{aligned} \sigma_m^2 &= 4w_m J_{\text{av}}|G(0)|^2 \\ &= 8w_m \mathcal{E}_c T_c J_{\text{av}} \end{aligned} \tag{13.2–29}$$

The probability of a code word error for CW jamming is upper-bounded as

$$P_M \leqslant \sum_{m=2}^{M} Q\left(\sqrt{\frac{2\mathcal{E}_c}{J_{\text{av}} T_c} w_m}\right) \tag{13.2–30}$$

But $\mathcal{E}_c = R_c \mathcal{E}_b$. Furthermore, $T_c \approx 1/W$ and $J_{\text{av}}/W = J_0$. Therefore Equation 13.2–30 may be expressed as

$$P_M \leqslant \sum_{m=2}^{M} Q\left(\sqrt{\frac{2\mathcal{E}_b}{J_0} R_c w_m}\right) \tag{13.2–31}$$

which is the result obtained previously for broadband interference. This result indicates that a CW jammer has the same effect on performance as an equivalent broadband jammer. This equivalence is discussed further below.

EXAMPLE 13.2–2. Let us determine the performance of the DS spread spectrum system in the presence of a CW jammer of average power J_{av} when the transmitted signal pulse $g(t)$ is one-half cycle of a sinusoid as illustrated in Figure 13.2–5, i.e.,

$$g(t) = \sqrt{\frac{4\mathcal{E}_c}{T_c}} \sin \frac{\pi t}{T_c}, \qquad 0 \leqslant t \leqslant T_c \tag{13.2–32}$$

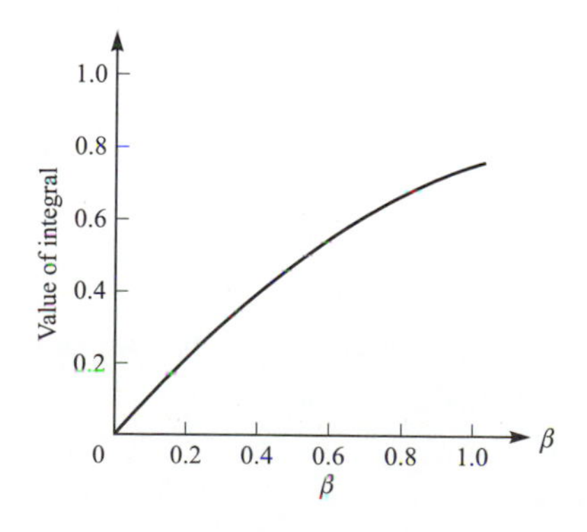

FIGURE 13.2–4
Plot of the value of the integral in Equation 13.2–27.

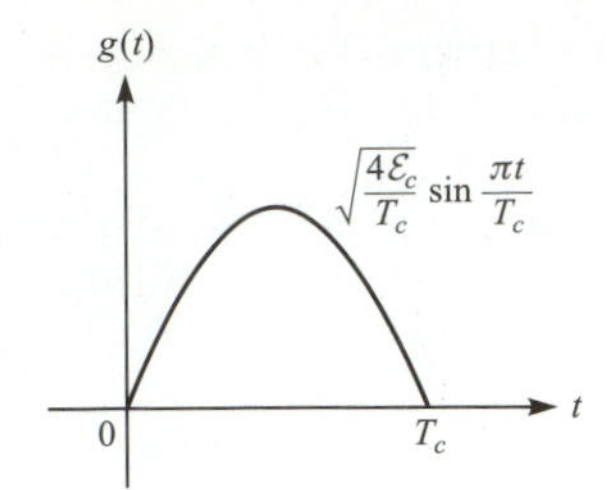

FIGURE 13.2–5
A sinusoidal signal pulse.

The variance of the interference of this pulse is

$$\begin{aligned}\sigma_m^2 &= 4w_m J_{\text{av}}|G(0)|^2 \\ &= \frac{64}{\pi^2}\mathcal{E}_c T_c J_{\text{av}} w_m\end{aligned} \tag{13.2–33}$$

Hence, the upper bound on the code word probability is

$$P_M \leqslant \sum_{m=2}^{M} Q\left(\sqrt{\frac{\pi^2\mathcal{E}_b}{4J_{\text{av}}T_c}R_c w_m}\right) \tag{13.2–34}$$

We observe that the performance obtained with this pulse is 0.9 dB better than that obtained with a rectangular pulse. Recall that this pulse shape when used in offset QPSK results in an MSK signal. MSK modulation is frequently used in DS spread spectrum systems.

The processing gain and the jamming margin. An interesting interpretation of the performance characteristics for the DS spread spectrum signal is obtained by expressing the signal energy per bit $\mathcal{E}_b$ in terms of the average power. That is, $\mathcal{E}_b = P_{\text{av}}T_b$, where P_{av} is the average signal power and T_b is the bit interval. Let us consider the performance obtained in the presence of CW jamming for the rectangular pulse treated in Example 13.2–1. When we substitute for $\mathcal{E}_b$ and J_0 into Equation 13.2–31, we obtain

$$P_M \leqslant \sum_{m=2}^{M} Q\left(\sqrt{\frac{2P_{\text{av}}}{J_{\text{av}}}\frac{T_b}{T_c}R_c w_m}\right) = \sum_{m=2}^{M} Q\left(\sqrt{\frac{2P_{\text{av}}}{J_{\text{av}}}L_c R_c w_m}\right) \tag{13.2–35}$$

where L_c is the number of chips per information bit and $P_{\text{av}}/J_{\text{av}}$ is the signal-to-jamming power ratio.

An identical result is obtained with broadband jamming for which the performance is given by Equation 13.2–23. For the signal energy per bit, we have

$$\mathcal{E}_b = P_{\text{av}}T_b = \frac{P_{\text{av}}}{R} \tag{13.2–36}$$

where R is the information rate in bits/s. The power spectral density for the jamming signal may be expressed as

$$J_0 = \frac{J_{\text{av}}}{W}$$

Using this relation and Equation 13.2–36, the ratio $\mathcal{E}_b/J_0$ may be expressed as

$$\frac{\mathcal{E}_b}{J_0} = \frac{P_{\text{av}}/R}{J_{\text{av}}/W} = \frac{W/R}{J_{\text{av}}/P_{\text{av}}} \tag{13.2–37}$$

The ratio $J_{\text{av}}/P_{\text{av}}$ is the jamming-to-signal power ratio, which is usually greater than unity. The ratio $W/R = T_b/T_c = B_e = L_c$ is just the bandwidth expansion factor, or, equivalently, the number of chips per information bit. This ratio is usually called the *processing gain* of the DS spread spectrum system. It represents the advantage gained over the jammer that is obtained by expanding the bandwidth of the transmitted signal. If we interpret $\mathcal{E}_b/J_0$ as the SNR required to achieve a specified error rate performanace and W/R as the available bandwidth expansion factor, the ratio $J_{\text{av}}/P_{\text{av}}$ is called the *jamming margin* of the DS spread spectrum system. In other words, the jamming margin is the largest value that the ratio $J_{\text{av}}/P_{\text{av}}$ can take and still satisfy the specified error probability.

The performance of a soft-decision decoder for a linear (n, k) binary code, expressed in terms of the processing gain and the jamming margin, is

$$P_M \leqslant \sum_{m=2}^{M} Q\left(\sqrt{\frac{2W/R}{J_{\text{av}}/P_{\text{av}}} R_c w_m}\right) \leqslant (M-1) Q\left(\sqrt{\frac{2W/R}{J_{\text{av}}/P_{\text{av}}} R_c d_{\min}}\right) \tag{13.2–38}$$

In addition to the processing gain W/R and $J_{\text{av}}/P_{\text{av}}$, we observe that the performance depends on a third factor, namely, $R_c w_m$. This factor is the *coding gain*. A lower bound on this factor is $R_c d_{\min}$. Thus the jamming margin achieved by the DS spread spectrum signal depends on the processing gain and the coding gain.

We may express the relationship among these three quantities in dB as

$$(\text{SNR})_{\text{dB}} = \left(\frac{W}{R}\right)_{\text{dB}} + (R_c d_{\min})_{\text{dB}} - \left(\frac{J_{\text{av}}}{P_{\text{av}}}\right)_{\text{dB}} \tag{13.2–39}$$

where the $(\text{SNR})_{\text{dB}}$ is the signal-to-noise ratio required by the receiver to achieve a specified level of performance.

Uncoded DS spread spectrum signals. The performance results given above for DS spread spectrum signals generated by means of an (n, k) code may be specialized to a trivial type of code, namely, a binary repetition code. For this case, $k = 1$ and the weight of the nonzero code word is $w = n$. Thus, $R_c w = 1$ and, hence, the performance of the binary signaling system reduces to

$$\begin{aligned} P_2 &= Q\left(\sqrt{\frac{2\mathcal{E}_b}{J_0}}\right) \\ &= Q\left(\sqrt{\frac{2W/R}{J_{\text{av}}/P_{\text{av}}}}\right) \end{aligned} \tag{13.2–40}$$

Note that the trivial (repetition) code gives no coding gain. It does result in a processing gain of W/R.

EXAMPLE 13.2–3. Suppose that we wish to achieve an error rate performance of 10^{-6} or less with an uncoded DS spread spectrum system. The available bandwidth expansion factor is $W/R = 1000$. Let us determine the jamming margin.

The $\mathcal{E}_b/J_0$ required to achieve a bit error probability of 10^{-6} with uncoded binary PSK is 10.5 dB. The processing gain is $10 \log_{10} 1000 = 30$ dB. Hence the maximum jamming-to-signal power that can be tolerated, i.e., the jamming margin, is

$$10 \log_{10} \frac{J_{\rm av}}{P_{\rm av}} = 30 - 10.5 = 19.5\,\text{dB}$$

Since this is the jamming margin achieved with an uncoded DS spread spectrum system, it may be increased by coding the information sequence.

There is another way to view the modulation and demodulation processes for the uncoded (repetition code) DS spread spectrum system. At the modulator, the signal waveform generated by the repetition code with rectangular pulses, for example, is identical to a unit amplitude rectangular pulse $s(t)$ of duration T_b or its negative, depending on whether the information bit is 1 or 0, respectively. This may be seen from Equation 13.2–7, where the coded chips $\{c_i\}$ within a single information bit are either all 1s or 0s. The PN sequence multiplies either $s(t)$ or $-s(t)$. Thus, when the information bit is a 1, the L_c PN chips generated by the PN generator are transmitted with the same polarity. On the other hand, when the information bit is a 0, the L_c PN chips when multiplied by $-s(t)$ are reversed in polarity.

The demodulator for the repetition code, implemented as a correlator, is illustrated in Figure 13.2–6. We observe that the integration interval in the integrator is the bit interval T_b. Thus, the decoder for the repetition code is eliminated and its function is subsumed in the demodulator.

Now let us qualitatively assess the effect of this demodulation process on the interference $z(t)$. The multiplication of $z(t)$ by the output of the PN generator, which is expressed as

$$w(t) = \sum_i (2b_i - 1)p(t - iT_c)$$

yields

$$v(t) = w(t)z(t)$$

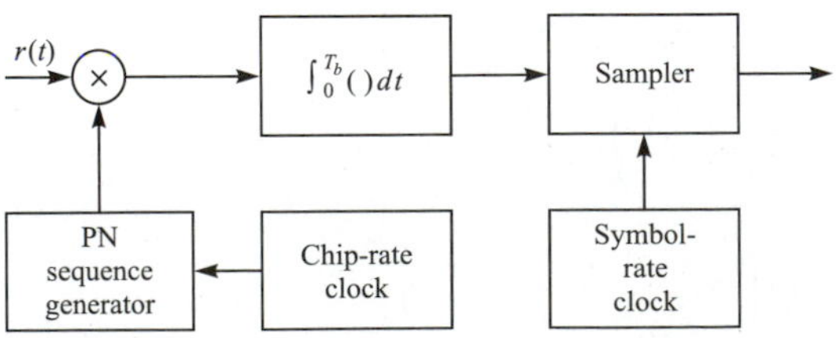

FIGURE 13.2–6
Correlation-type demodulator for a repetition code.

The waveforms $w(t)$ and $z(t)$ are statistically independent random processes each with zero-mean and autocorrelation functions $\phi_{ww}(\tau)$ and $\phi_{zz}(\tau)$, respectively. The product $v(t)$ is also a random process having an autocorrelation function equal to the product of $\phi_{ww}(\tau)$ with $\phi_{zz}(\tau)$. Hence, the power spectral density of the process $v(t)$ is equal to the convolution of the power spectral density of $w(t)$ with the power spectral density of $z(t)$.

The effect of convolving the two spectra is to spread the power in bandwidth. Since the bandwidth of $w(t)$ occupies the available channel bandwidth W, the result of convolution of the two spectra is to spread the power spectral density of $z(t)$ over the frequency band of width W. If $z(t)$ is a narrowband process, i.e., its power spectral density has a width much less than W, the power spectral density of the process $v(t)$ will occupy a bandwidth equal to at least W.

The integrator used in the cross correlation shown in Figure 13.2–6 has a bandwidth approximately equal to $1/T_b$. Since $1/T_b \ll W$, only a fraction of the total interference power appears at the output of the correlator. This fraction is approximately equal to the ratio of bandwidths $1/T_b$ to W. That is,

$$\frac{1/T_b}{W} = \frac{1}{WT_b} = \frac{T_c}{T_b} = \frac{1}{L_c}$$

In other words, the multiplication of the interference with the signal from the PN generator spreads the interference to the signal bandwidth W, and the narrowband integration following the multiplication sees only the fraction $1/L_c$ of the total interference. Thus, the performance of the uncoded DS spread spectrum system is enhanced by the processing gain L_c.

Linear code concatenated with a binary reception code. As illustrated above, a binary repetition code provides a margin against an interference or jamming signal but yields no coding gain. To obtain an improvement in performance, we may use a linear (n_1, k) block or convolutional code, where $n_1 \leqslant n = kL_c$. One possibility is to select $n_1 < n$ and to repeat each code bit n_2 times such that $n = n_1 n_2$. Thus, we can construct a linear (n_1, k) code by concatenating the (n_1, k) code with a binary $(n_2, 1)$ repetition code. This may be viewed as a trivial form of code concatenation where the outer code is the (n_1, k) code and the inner code is the repetition code.

Since the repetition code yields no coding gain, the coding gain achieved by the combined code must reduce to that achieved by the (n_1, k) outer code. It is demonstrated that this is indeed the case. The coding gain of the overall combined code is

$$R_c w_m = \frac{k}{n} w_m, \qquad m = 2, 3, \ldots, 2^k$$

But the weights $\{w_m\}$ for the combined code may be expressed as

$$w_m = n_2 w_m^0$$

where $\{w_m^0\}$ are the weights of the outer code. Therefore, the coding gain of the combined code is

$$R_c w_m = \frac{k}{n_1 n_2} n_2 w_m^0 = \frac{k}{n_1} w_m^0 = R_c^0 w_m^0 \tag{13.2–41}$$

which is just the coding gain obtained from the outer code.

A coding gain is also achieved if the (n_1, k) outer code is decoded using hard decisions. The probability of a bit error obtained with $(n_2, 1)$ repetition code (based on soft-decision decoding) is

$$\begin{aligned} p &= Q\left(\sqrt{\frac{2n_2\mathcal{E}_c}{J_0}}\right) = Q\sqrt{2\frac{\mathcal{E}_b}{J_0}R_c^0}\Bigg) \\ &= Q\left(\sqrt{\frac{2W/R}{J_{\text{av}}/P_{\text{av}}}R_c^0}\right) \end{aligned} \tag{13.2–42}$$

Then the code word error probability for a linear (n_1, k) block code is upper-bounded as

$$P_M \leqslant \sum_{m=t+1}^{n_1} \binom{n_1}{m} p^m (1-p)^{n_1-m} \tag{13.2–43}$$

where $t = \lfloor \frac{1}{2}(d_{\min} - 1) \rfloor$, or as

$$P_M \leqslant \sum_{m=2}^{M} [4p(1-p)]^{w_m^0/2} \tag{13.2–44}$$

where the latter is a Chernoff bound. For an (n_1, k) binary convolutional code, the upper bound on the bit error probability is

$$P_b \leqslant \sum_{d=d_{\text{free}}}^{\infty} \beta_d P_2(d) \tag{13.2–45}$$

where $P_2(d)$ is defined by Equation 8.2–28 for odd d and by Equation 8.2–29 for even d.

Concatenated coding for DS spread spectrum systems. It is apparent from the above discussion that an improvement in performance can be obtained by replacing the repetition code by a more powerful code that will yield a coding gain in addition to the processing gain. Basically, the objective in a DS spread spectrum system is to construct a long, low-rate code having a large minimum distance. This may be best accomplished by using code concatenation. When binary PSK is used in conjunction with DS spread spectrum, the elements of a concatenated code word must be expressed in binary form.

Best performance is obtained when soft-decision decoding is used on both the inner and outer codes. However, an alternative, which usually results in reduced complexity for the decoder, is to employ soft-decision decoding on the inner code and hard-decision decoding on the outer code. The expressions for the error rate performance of these decoding schemes depend, in part, on the type of codes (block or convolutional) selected for the inner and outer codes. For example, the concatenation of two block codes may be viewed as an overall long binary (n, k) block code having a performance given by Equation 13.2–38. The performance of other code combinations may also be readily derived. For the sake of brevity, we shall not consider such code combinations.

13.2.2 Some Applications of DS Spread Spectrum Signals

In this subsection, we shall briefly consider the use of coded DS spread spectrum signals for three specific applications. One is concerned with providing immunity against a jamming signal. In the second, a communication signal is hidden in the background noise by transmitting the signal at a very low power level. The third application is concerned with accommodating a number of simultaneous signal transmissions on the same channel, i.e., CDMA.

Antijamming application. In Section 13.2.1, we derived the error rate performance for a DS spread spectrum signal in the presence of either a narrowband or a wideband jamming signal. As examples to illustrate the performance of a digital communication system in the presence of a jamming signal, we shall select three codes. One is the Golay (24, 12) code, which is characterized by the weight distribution given in Table 8.1–1 and has a minimum distance $d_{\min} = 8$. The second code is an expurgated Golay (24, 11) code obtained by selecting 2048 code words of constant weight 12. Of course, this expurgated code is non-linear. These two codes will be used in conjunction with a repetition code. The third code to be considered is a maximum-length shift-register code.

The error rate performance of the Golay (24, 12) code with soft-decision decoding is

$$P_M \leqslant \left[759Q\left(\sqrt{\frac{8W/R}{J_{\text{av}}/P_{\text{av}}}}\right) + 2576Q\left(\sqrt{\frac{12W/R}{J_{\text{av}}/P_{\text{av}}}}\right) + 759Q\left(\sqrt{\frac{16W/R}{J_{\text{av}}/P_{\text{av}}}}\right) + Q\left(\sqrt{\frac{24W/R}{J_{\text{av}}/P_{\text{av}}}}\right)\right] \tag{13.2–46}$$

where W/R is the processing gain and $J_{\text{av}}/P_{\text{av}}$ is the jamming margin. Since $n = n_1 n_2 = 12W/R$ and $n_1 = 24$, each coded bit is, in effect, repeated $n_2 = W/2R$ times. For example, if $W/R = 100$ (a processing gain of 20 dB), the block length of the repetition code is $n_2 = 50$.

If hard-decision decoding is used, the probability of error for a coded bit is

$$p = Q\left(\sqrt{\frac{W/R}{J_{\text{av}}/P_{\text{av}}}}\right) \tag{13.2–47}$$

and the corresponding probability of a code word error is upper-bounded as

$$P_M \leqslant \sum_{m=4}^{24} \binom{24}{m} p^m (1-p)^{24-m} \tag{13.2–48}$$

As an alternative, we may use the Chernoff bound for hard-decision decoding, which is

$$\begin{aligned} P_M \leqslant\ & 759[4p(1-p)]^4 + 2576[4p(1-p)]^6 \\ & + 759[4p(1-p)]^8 + [4p(1-p)]^{12} \end{aligned} \tag{13.2–49}$$

Figure 13.2–7 illustrates the performance of the Golay (24, 12) code as a function of the jamming margin J_{av}/P_{av}, with the processing gain as a parameter. The Chernoff bound was used to compute the error probability for hard-decision decoding. The error probability for soft-decision decoding is dominated by the term

$$759Q\left(\sqrt{\frac{8W/R}{J_{av}/P_{av}}}\right)$$

and that for hard-decision decoding is dominated by the term $759[4p(1-p)]^4$. Hence, the coding gain for soft-decision decoding† is at most $10 \log 4 = 6$ dB. We note that the two curves corresponding to $W/R = 1000$ (30 dB) are identical in shape to the ones for $W/R = 100$ (20 dB), except that the latter are shifted by 10 dB to the right relative to the former. This shift is simply the difference in processing gain between these two DS spread spectrum signals.

The error rate performance of the expurgated Golay (24, 11) code is upper-bounded as

$$P_M \leqslant 2047Q\left(\sqrt{\frac{11W/R}{J_{av}P_{av}}}\right) \tag{13.2–50}$$

for soft-decision decoding and as‡

$$P_M \leqslant 2047[4p(1-p)]^6 \tag{13.2–51}$$

for hard-decision decoding, where p is given as

$$p = Q\left(\sqrt{\frac{11W/R}{J_{av}/P_{av}}}\right) \tag{13.2–52}$$

†The coding gain is less than 6 dB due to the multiplicative factor of 759, which increases the error probability relative to the performance of the binary uncoded system.

‡We remind the reader that the union bound is not very tight for large signal sets.

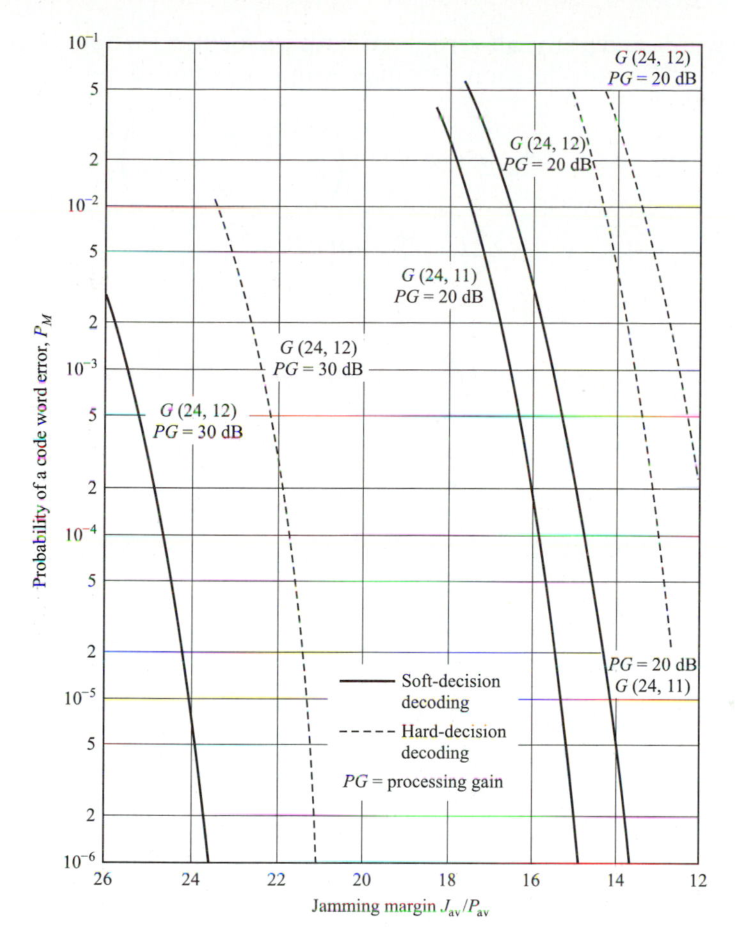

FIGURE 13.2–7
Performance of Golay codes used in DS spread spectrum signal.

The performance characteristics of this code are also plotted in Figure 13.2–7 for $W/R = 100$. We observe that this expurgated Golay (24, 11) code performs about 1 dB better than the Golay (24, 12) code.

Instead of using block code concatenated with a low-rate $(1/n_2)$ repetition code, let us consider using a single low-rate code. A particularly suitable set of low-rate codes is the set of maximum-length shift-register codes described in Section 8.1.3. We recall that for this set of codes,

$$(n, k) = (2^m - 1, m)$$
$$d_{\min} = 2^{m-1} \tag{13.2–53}$$

All code words except the all-zero word have an identical weight of 2^{m-1}. Hence, the error rate for soft-decision decoding is upper-bounded as†

$$\begin{aligned} P_M &\leqslant (M-1)Q\left(\sqrt{\frac{2W/R}{J_{av}/P_{av}} R_c d_{min}}\right) \\ &\leqslant 2^m Q\left(\sqrt{\frac{2W/R}{J_{av}/P_{av}} \frac{m2^{m-1}}{2^m - 1}}\right) \\ &\leqslant 2^m \exp\left(-\frac{W/R}{J_{av}/P_{av}} \frac{m2^{m-1}}{2^m - 1}\right) \end{aligned} \tag{13.2–54}$$

For moderate values of m, $R_c d_{min} \approx \frac{1}{2}m$ and, hence, Equation 13.2–54 may be expressed as

$$P_M \leqslant 2^m Q\left(\sqrt{\frac{W/R}{J_{av}/P_{av}} m}\right) \leqslant 2^m \exp\left(-\frac{mW/R}{2J_{av}/P_{av}}\right) \tag{13.2–55}$$

Hence, the coding gain is at most $10 \log \frac{1}{2}m$.

For example, if we select $m = 10$, then $n = 2^{10} - 1 = 1023$. Since $n = kW/R = mW/R$, it follows that $W/R \approx 102$. Thus, we have a processing gain of about 20 dB and a coding gain of 7 dB. This performance is comparable to that obtained with the expurgated Golay (24, 11) code. Higher coding gains can be achieved with larger values of m.

If hard-decision decoding is used for the maximum-length shift-register codes, the error rate is upper-bounded by the Chernoff bound as

$$P_M \leqslant (M-1)[4p(1-p)]^{d_{min}/2} = (2^m - 1)[4p(1-p)]^{2^{m-2}} \tag{13.2–56}$$

where p is given as

$$p = Q\left(\sqrt{\frac{2W/R}{J_{av}/P_{av} R_c}}\right) = Q\left(\sqrt{\frac{2W/R}{J_{av}/P_{av}} \frac{m}{2^m - 1}}\right) \tag{13.2–57}$$

For $m = 10$, the code word error rate P_M is comparable to that obtained with the expurgated Golay (24, 11) code for hard-decision decoding.

The results given above illustrate the performance that can be obtained with a single level of coding. Greater coding gains can be achieved with concatenated codes.

†The $M = 2^m$ waveforms generated by a maximum-length shift-register code form a simplex set (see Problem 8.20). The exact expression for the error probability, given in Section 5.2.4, may be used for large values of M, where the union bound becomes very loose.

Low-detectability signal transmission. In this application, the signal is purposely transmitted at a very low power level relative to the background channel noise and thermal noise that is generated in the front end of the receiver. If the DS spread spectrum signal occupies a bandwidth W and the spectral density of the additive noise is N_0 W/Hz, the average noise power in the bandwidth W is $N_{\text{av}} = WN_0$.

The average received signal power at the intended receiver is P_{av}. If we wish to hide the presence of the signal from receivers that are in the vicinity of the intended receiver, the signal is transmitted at a low power level such that $P_{\text{av}}/N_{\text{av}} \ll 1$. For example, let us assume that binary PSK is used to transmit the information. The probability of error at the intended receiver may be expressed as

$$\begin{aligned} P_M &< MQ\left(\sqrt{\frac{2\mathcal{E}_b}{N_0} R_c d_{\min}}\right) \\ &< MQ\left(\sqrt{2\left(\frac{W}{R}\right)\left(\frac{P_{\text{av}}}{N_{\text{av}}}\right) R_c d_{\min}}\right) \end{aligned}$$

From this expression, we observe that even though $P_{\text{av}}/N_{\text{av}} \ll 1$, the intended receiver can recover the information-bearing signal with the aid of the processing gain and the coding gain. However, any other receiver that has no prior knowledge of the PN sequence is unable to take advantage of the processing gain and the coding gain. Hence, the presence of the information-bearing signal is difficult to detect. We say that the signal has a *low probability* of being *intercepted* (LPI) and it is called an *LPI signal*.

The probability of error results given in Section 13.2.1 also apply to the demodulation and decoding of LPI signals at the intended receiver.

Code division multiple access. The enhancement in performance obtained from a DS spread spectrum signal through the processing gain and coding gain can be used to enable many DS spread spectrum signals to occupy the same channel bandwidth provided that each signal has its own distinct PN sequence. Thus, it is possible to have several users transmit messages simultaneously over the same channel bandwidth. This type of digital communication in which each user (transmitter–receiver pair) has a distinct PN code for transmitting over a common channel bandwidth is called either *code division multiple access* (CDMA) or *spread spectrum multiple access* (SSMA).

In the demodulation of each PN signal, the signals from the other simultaneous users of the channel appear as an additive interference. The level of interference varies, depending on the number of users at any given time. A major advantage of CDMA is that a large number of users can be accommodated if each transmits messages for a short period of time. In such a multiple access system, it is relatively easy either to add new users or to decrease the number of users without disrupting the system.

Let us determine the number of simultaneous signals that can be supported in a CDMA system.† For simplicity, we assume that all signals have identical average powers. Thus, if there are N_u simultaneous users, the desired signal-to-noise inteference power ratio at a given receiver is

$$\frac{P_{\text{av}}}{J_{\text{av}}} = \frac{P_{\text{av}}}{(N_u - 1)P_{\text{av}}} = \frac{1}{N_u - 1} \tag{13.2–58}$$

Hence, the performance for soft-decision decoding at the given receiver is upper-bounded as

$$P_M \leqslant \sum_{m=2}^{M} Q\left(\sqrt{\frac{2W/R}{N_u - 1} R_c w_m}\right) \leqslant (M-1)Q\left(\sqrt{\frac{2W/R}{N_u - 1} R_c d_{\text{min}}}\right) \tag{13.2–59}$$

In this case, we have assumed that the interference from other users is Gaussian.

As an example, suppose that the desired level of performance (error probability of 10^{-6}) is achieved when

$$\frac{W/R}{N_u - 1} R_c d_{\text{min}} = 20$$

Then the maximum number of users that can be supported in the CDMA system is

$$N_u = \frac{W/R}{20} R_c d_{\text{min}} + 1 \tag{13.2–60}$$

If $W/R = 100$ and $R_c d_{\text{min}} = 4$, as obtained with the Golay (24, 12) code, the maximum number is $N_u = 21$. If $W/R = 1000$ and $R_c d_{\text{min}} = 4$, this number becomes $N_u = 201$.

In determining the maximum number of simultaneous users of the channel, we have implicitly assumed that the PN code sequences are mutually orthogonal and the interference from other users adds on a power basis only. However, orthogonality among a number of PN code sequences is not easily achieved, especially if the number of PN code sequences required is large. In fact, the selection of a good set of PN sequences for a CDMA system is an important problem that has received considerable attention in the technical literature. We shall briefly discuss this problem in Section 13.2.5.

Digital cellular CDMA system based on DS spread spectrum. Direct sequence CDMA has been adopted as one multiple-access method for digital cellular voice communications in North America. This digital cellular communication system was proposed and developed by Qualcomm and has been standardized and designated as IS-95 by the Telecommunications Industry Association (TIA) for use in the 800-MHz and in the 1900-MHz frequency bands.

---

†In this section the interference from other users is treated as a random process. This is the case if there is no cooperation among the users. In Chapter 15 we consider CDMA transmission in which interference from other users is known and is suppressed by the receiver.

The nominal bandwidth used for transmission from a base station to the mobile receivers (forward link) is 1.25 MHz, and a separate channel, also with a bandwidth of 1.25 MHz, is used for signal transmission from mobile receivers to a base station (reverse link). The signals transmitted in both the forward and the reverse links are DS spread spectrum signals having a chip rate of 1.2288×10^6 chips per second (Mchips/s).

Forward link. A block diagram of the modulator for the signals transmitted from a base station to the mobile receivers is shown in Figure 13.2–8. The speech coder is a code-excited linear predictive (CELP) coder which generates data at the variable rates of 9600, 4800, 2400, and 1200 bits/s, where the data rate is a function of the speech activity of the user, in frame intervals of 20 ms. The data from the speech coder is encoded by a rate 1/2, constraint length $K = 9$ convolutional code. For lower speech activity, where the data rates are 4800, 2400, or 1200 bits/s, the output symbols from the convolutional encoder are repeated either twice, four times, or eight times so as to maintain a constant bit rate of 9600 bits/s. At the lower speech activity rates, the transmitter power is reduced by either 3, 6, or 9 dB, so that the transmitted energy per bit remains constant for all speech rates. Thus, a lower speech activity results in a lower transmitter power and, hence, a lower level of interference to other users.

The encoded bits for each frame are passed through a block interleaver, which is needed to overcome the effects of burst errors that may occur in transmission through the channel. The data bits at the output of the block interleaver, which occur at a rate of 19.2 kbits/s, are scrambled by multiplication with the output of a long code (period $N = 2^{42} - 1$) generator running at the chip rate of 1.2288 M chips/s, but whose output is decimated by a factor of 64 to 19.2 kchips/s. The long code is used to uniquely identify a call of a mobile station on the forward and reverse links.

Each user of the channel is assigned a Hadamard (or Walsh) sequence of length 64. There are 64 orthogonal Hadamard sequences assigned to each base station, and, thus, there are 64 channels available. One Hadamard sequence (the all-zero sequence) is used to transmit a pilot signal, which serves as a means for measuring the channel characteristics, including the signal strength and the carrier phase offset. These parameters are used at the receiver in performing phase coherent demodulation. Another Hadamard sequence is used for providing time synchronization. One channel, and possibly more if necessary, is used for paging. That leaves up to 61 channels for allocation to different users.

Each user, using the Hadamard sequence assigned to it, multiplies the data sequence by the assigned Hadamard sequence. Thus, each encoded data bit is multiplied by the Hadamard sequence of length 64. The resulting binary sequence is now spread by multiplication with two PN sequences of length $N = 2^{15}$, so as to create in-phase and quadrature signal components. Thus, the binary data signal is converted to a four-phase signal and both the I and Q components are filtered by baseband spectral shaping filters. Different base stations are identified by different offsets of these PN sequences.The signals for all the 64 channels are transmitted synchronously so that, in the absence of channel

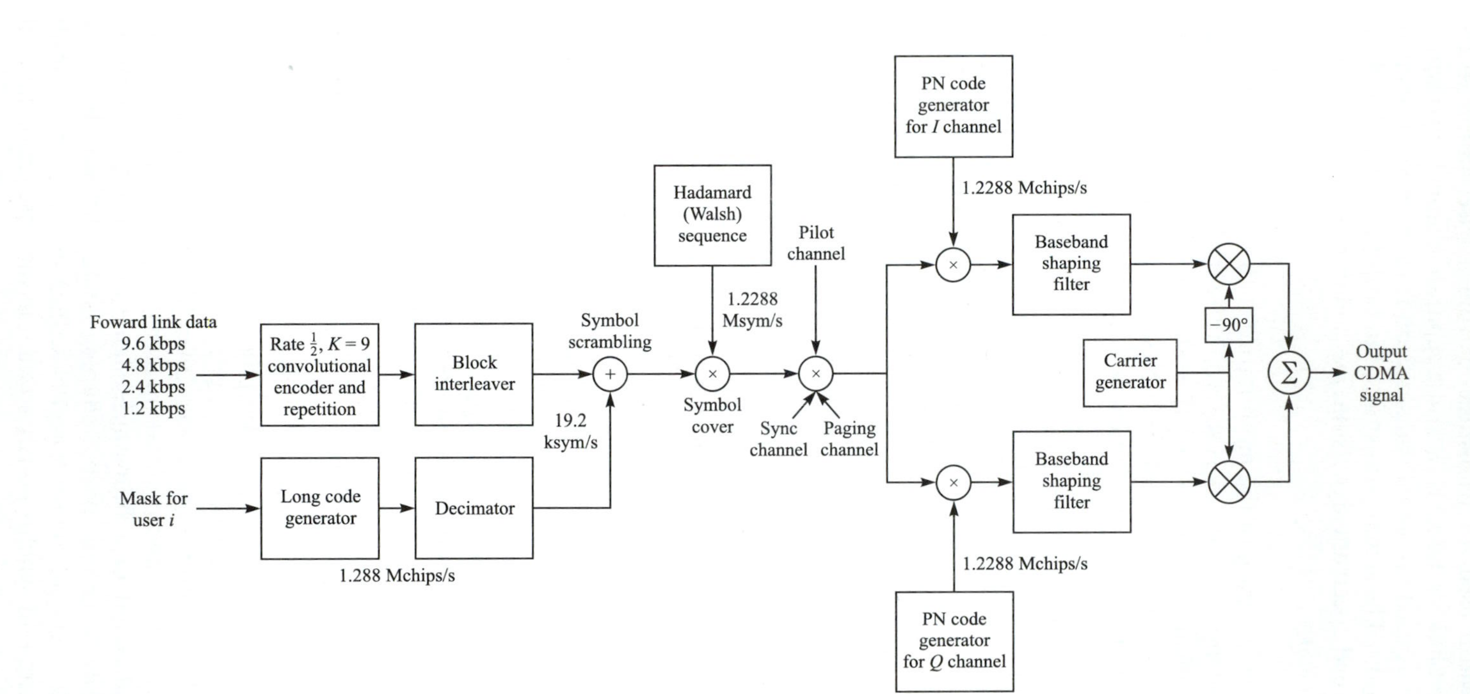

FIGURE 13.2–8
Block diagram of IS-95 forward link.

multipath distortion, the signals of other users received at any mobile receiver do not interfere because of the orthogonality of the Hadamard sequences.

At the receiver, a RAKE demodulator is used to resolve the major multipath signal components, which are then phase-aligned and weighted according to their signal strength using the estimates of phase and signal strength derived from the pilot signal. These components are combined and passed to the Viterbi soft-decision decoder. The RAKE demodulator is described in detail in Chapter 14.

Reverse link. The modulator for the reverse link from a mobile transmitter to a base station is different from that for the forward link. A block diagram of the modulator is shown in Figure 13.2–9. An important consideration in the design of the modulator is that signals transmitted from the various mobile transmitters to the base station are asynchronous and, hence, there is significantly more interference among users. Secondly, the mobile transmitters are usually battery operated and, consequently, these transmissions are power limited. To compensate for these major limitations, a $K = 9$, rate 1/3 convolutional code is used in the reverse link. Although this code has essentially the same coding gain in an AWGN channel as the rate 1/2 code used in the forward link, it has a much higher coding gain in a fading channel, which is the characteristic of digital cellular communication links, as we shall observe in our treatment of communication through fading channels in Chapter 14. As in the case of the forward link, for lower speech activity, the output bits from the convolutional encoder are repeated either two, or four, or eight times. However, the coded bit rate is 28.8 kbits/s.

For each 20-ms frame, the 576 encoded bits are block-interleaved and passed to the modulator. The data is modulated using an $M = 64$ orthogonal signal set using Hadamard sequences of length 64. Thus, a 6-bit block of data is mapped into one of the 64 Hadamard sequences. The result is a bit (or chip) rate of 307.2 kbits/s at the output of the modulator. We note that 64-ary orthogonal modulation at an error probability of 10^{-6} requires approximately 3.5 dB less SNR per bit than binary antipodal signaling.

To reduce interference to other users, the time position of the transmitted code symbol repetitions is randomized so that, at the lower speech activity, consecutive bursts do not occur evenly spaced in time. Following the randomizer, the signal is spread by the output of the long code PN generator, which is running at a rate of 1.2288 Mchips/s. Hence, there are only four PN chips for every bit of the Hadamard sequence from the modulator, so the processing gain in the reverse link is very small. The resulting 1.2288 Mchips/s binary sequence at the output of the multiplier is then further multiplied by two PN sequences of length $N = 2^{15}$, whose rate is also 1.2288 Mchips/s, to create I and Q signals (a QPSK signal) which are filtered by baseband spectral shaping filters and then passed to quadrature mixers. The Q-channel signal is delayed in time by one-half PN chip relative to the I-channel signal prior to the baseband filter. In effect, the signal at the output of the two baseband filters is an offset QPSK signal.

Although the chips are transmitted as an offset QPSK signal, the demodulator employs noncoherent demodulation of the $M = 64$ orthogonal Hadamard

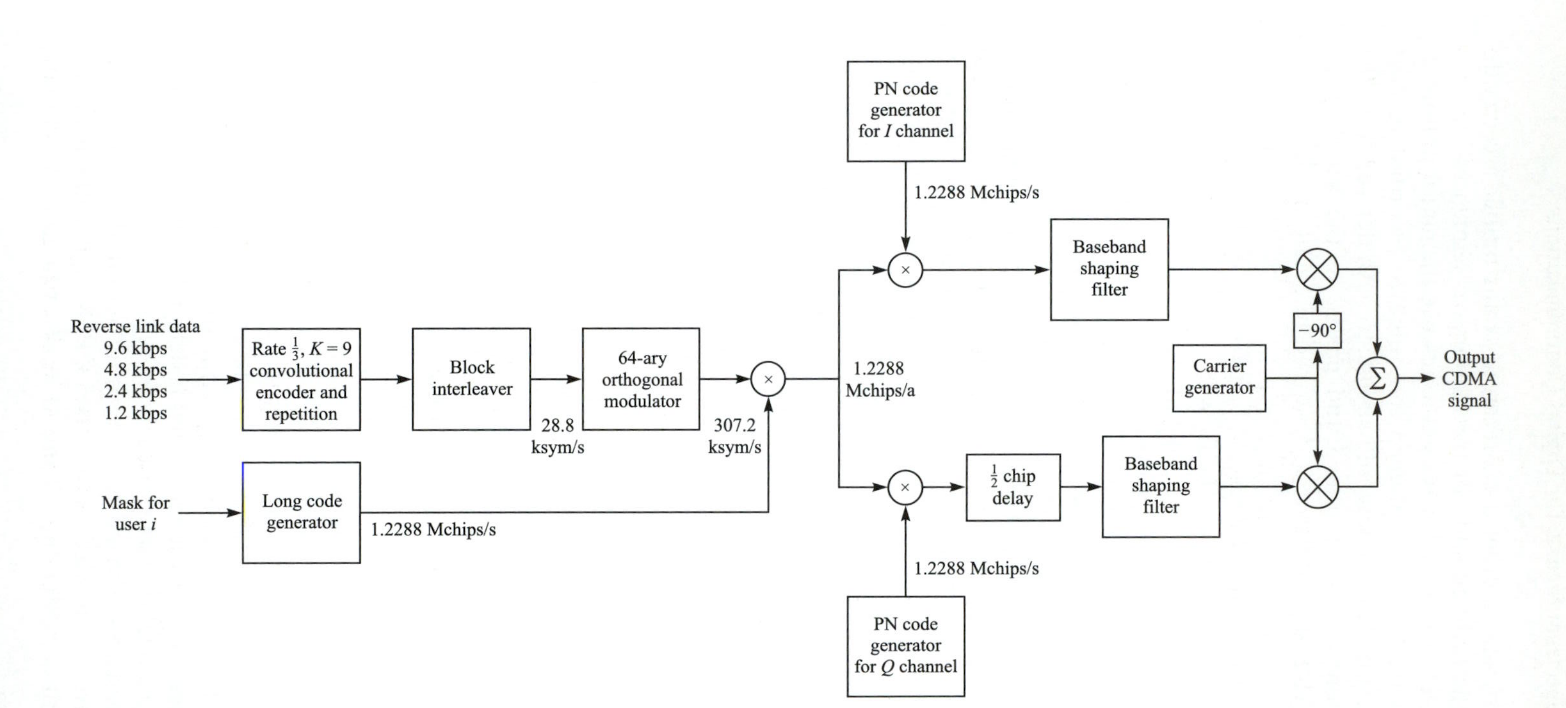

FIGURE 13.2–9
Block diagram of IS-95 reverse link.

waveforms to recover the encoded data bits. A fast Hadamard transform is used to reduce the computational complexity in the demodulation process. The output of the demodulator is then fed to the Viterbi detector, whose output is used to synthesize the speech signal.

13.2.3 Effect of Pulsed Interference on DS Spread Spectrum Systems

Thus far, we have considered the effect of continuous interference or jamming on a DS spread spectrum signal. We have observed that the processing gain and coding gain provide a means for overcoming the detrimental effects of this type of interference. However, there is a jamming threat that has a dramatic effect on the performance of a DS spread spectrum system. That jamming signal consists of pulses of spectrally flat noise that covers the entire signal bandwidth W. This is usually called *pulsed interference* or *partial-time jamming*.

Suppose the jammer has an average power J_{av} in the signal bandwidth W. Hence $J_0 = J_{av}/W$. Instead of transmitting continuously, the jammer transmits pulses at a power J_{av}/α for a percent of the time, i.e., the probability that the jammer is transmitting at a given instant is α. For simplicity, we assume that an interference pulse spans an integral number of signaling intervals and, thus, it affects an integral number of bits. When the jammer is not transmitting, the transmitted bits are assumed to be received error-free, and when the jammer is transmitting, the probability of error for an uncoded DS spread spectrum system is $Q(\sqrt{2\alpha\mathcal{E}_b/J_0})$. Hence, the average probability of a bit error is

$$P_2(\alpha) = \alpha Q(\sqrt{2\alpha\mathcal{E}_b/J_0}) = \alpha Q\left(\sqrt{\frac{2\alpha W/R}{J_{av}/P_{av}}}\right) \tag{13.2–61}$$

The jammer selects the duty cycle α to maximize the error probability. On differentiating Equation 13.2–61 with respect to α, we find that the worst-case pulse jamming occurs when

$$\alpha^* = \begin{cases} \dfrac{0.71}{\mathcal{E}_b/J_0} & (\mathcal{E}_b/J_0 \geqslant 0.71) \\ 1 & (\mathcal{E}_b/J_0 < 0.71) \end{cases} \tag{13.2–62}$$

and the corresponding error probability is

$$P_2 = \begin{cases} \dfrac{0.083}{\mathcal{E}_b/J_0} = \dfrac{0.083 J_{av}/P_{av}}{W/R} & (\mathcal{E}_b/J_0 > 0.71) \\ Q\left(\sqrt{\dfrac{2W/R}{J_{av}/P_{av}}}\right) & (\mathcal{E}_b/J_0 < 0.71) \end{cases} \tag{13.2–63}$$

The error rate performance given by Equation 13.2–61 for $\alpha = 1.0$, 0.1, and 0.01 along with the worst-case performance based on α^* is plotted in Figure 13.2–10. By comparing the error rate for continuous Gaussian noise jamming

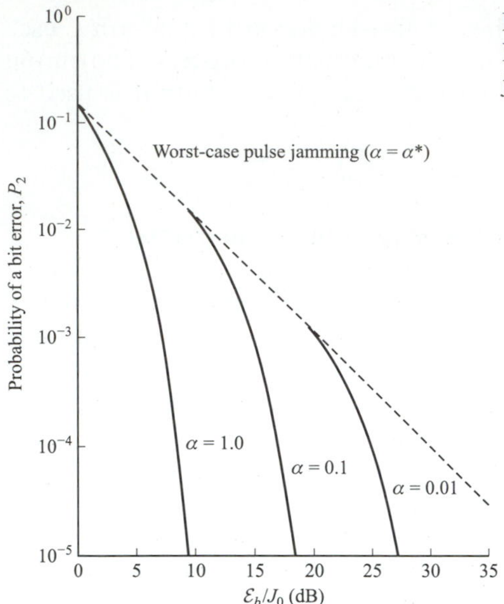

FIGURE 13.2–10
Performance of DS binary PSK with pulse jamming.

with worst-case pulse jamming, we observe a large difference in performance, which is approximately 40 dB at an error rate of 10^{-6}.

We should point out that the above analysis applies when the jammer pulse duration is equal to or greater than the bit duration. In addition, we should indicate that practical considerations may prohibit the jammer from achieving high peak power (small values of α). Nevertheless, the error probability given by Equation 13.2–63 serves as an upper bound on the performance of the uncoded binary PSK in worst-case pulse jamming. Clearly, the performance of the DS spread spectrum system in the presence of such jamming is extremely poor.

If we simply add coding to the DS spread spectrum system, the improvement over the uncoded system is the coding gain. Thus, $\mathcal{E}_b/J_0$ is reduced by the coding gain, which in most cases is limited to less than 10 dB. The reason for the poor performance is that the jamming signal pulse duration may be selected to affect many consecutive coded bits when the jamming signal is turned on. Consequently, the code word error probability is high due to the burst characteristics of the jammer.

In order to improve the performance, we should interleave the coded bits prior to transmission over the channel. The effect of the interleaving, as discussed in Section 8.1.9, is to make the coded bits that are hit by the jammer statistically independent.

The block diagram of the digital communication system that includes interleaving/deinterleaving is shown in Figure 13.2–11. Also shown is the possibility that the receiver knows the jammer state, i.e., that it knows when the jammer is on or off. Knowledge of the jammer state (called *side information*) is sometimes available from channel measurements of noise power levels in adjacent frequency bands. In our treatment, we consider two extreme cases, namely, no knowledge

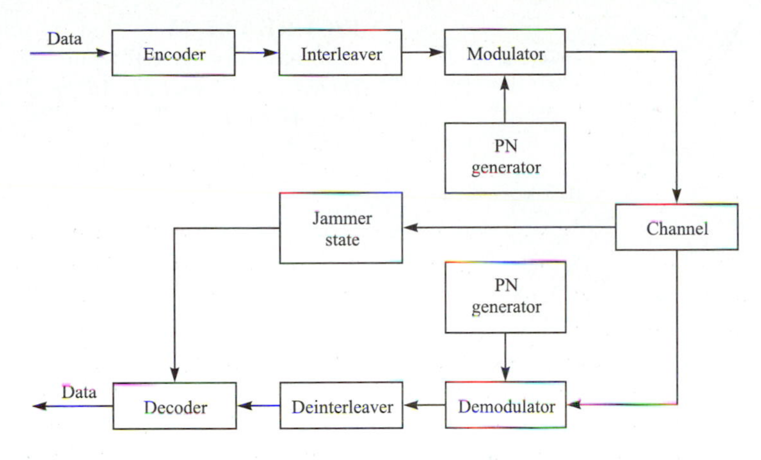

FIGURE 13.2–11
Block diagram of AJ communication system.

of the jammer state or complete knowledge of the jammer state. In any case, the random variable ζ representing the jammer state is characterized by the probabilities

$$P(\zeta = 1) = \alpha, \qquad P(\zeta = 0) = 1 - \alpha$$

When the jammer is on, the channel is modeled as an AWGN with power spectral density $N_0 = J_0/\alpha = J_{\text{av}}/\alpha W$; and when the jammer is off, there is no noise in the channel. Knowledge of the jammer state implies that the decoder knows when $\zeta = 1$ and when $\zeta = 0$, and uses this information in the computation of the correlation metrics. For example, the decoder may weight the demodulator output for each coded bit by the reciprocal of the noise power level in the interval. Alternatively, the decoder may give zero weight (erasure) to a jammed bit.

First, let us consider the effect of jamming without knowledge of the jammer state. The interleaver/deinterleaver pair is assumed to result in statistically independent jammer hits of the coded bits. As an example of the performance achieved with coding, we cite the performance results from the paper of Martin and McAdam (1980). There the performance of binary convolutional codes is evaluated for worst-case pulse jamming. Both hard- and soft-decision Viterbi decoding are considered. Soft decisions are obtained by quantizing the demodulator output to eight levels. For this purpose, a uniform quantizer is used for which the threshold spacing is optimized for the pulse jammer noise level. The quantizer plays the important role of limiting the size of the demodulator output when the pulse jammer is on. The limiting action ensures that any hit on a coded bit does not heavily bias the corresponding path metrics.

The optimum duty cycle for the pulse jammer in the coded system is generally inversely proportional to the SNR, but its value is different from that given by Equation 13.2–62 for the uncoded system. Figure 13.2–12 illustrates graphi-

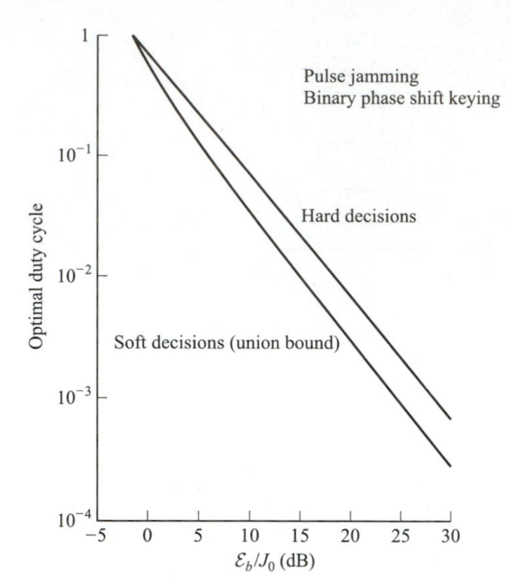

FIGURE 13.2–12
Optimal duty cycle for pulse jammer. [*From Martin and McAdam (1980).* © *1980 IEEE.*]

cally the optimal jammer duty cycle for both hard- and soft-decision decoding of the rate 1/2 convolutional codes. The corresponding error rate results for this worst-case pulse jammer are illustrated in Figures 13.2–13 and 13.2–14 for rate 1/2 codes with constraint lengths $3 \leqslant K \leqslant 9$. For example, note that at $P_2 = 10^{-6}$, the $K = 7$ convolutional code with soft-decision decoding requires $\mathcal{E}_b/J_0 = 7.6$ dB, whereas hard-decision decoding requires $\mathcal{E}_b/J_0 = 11.7$ dB. This 4.1-dB difference in SNR is relatively large. With continuous Gaussian noise, the corresponding SNRs for an error rate of 10^{-6} are 5 dB for soft-decision decoding and 7 dB for hard-decision decoding. Hence, the worst-case pulse jammer has degraded the performance by 2.6 dB for soft-decision decoding and by 4.7 dB for hard-decision decoding. These levels of degradation increase as the constraint length of the convolutional code is decreased. The important point, however, is that the loss in SNR due to jamming has been reduced from 40 dB for the uncoded system to less than 5 dB for the coded system based on a $K = 7$, rate 1/2 convolutional code.

A simpler method for evaluating the performance of a coded anti-jamming (AJ) communication system is to use the cutoff rate parameter R_0 as proposed by Omura and Levitt (1982). For example, with binary-coded modulation, the cutoff rate may be expressed as

$$R_0 = 1 - \log(1 + D_\alpha) \tag{13.2–64}$$

where the factor D_α depends on the channel noise characteristics and the decoder processing. Recall that for binary PSK in an AWGN channel and soft-decision decoding,

$$D_\alpha = e^{-\mathcal{E}_c/N_0} \tag{13.2–65}$$

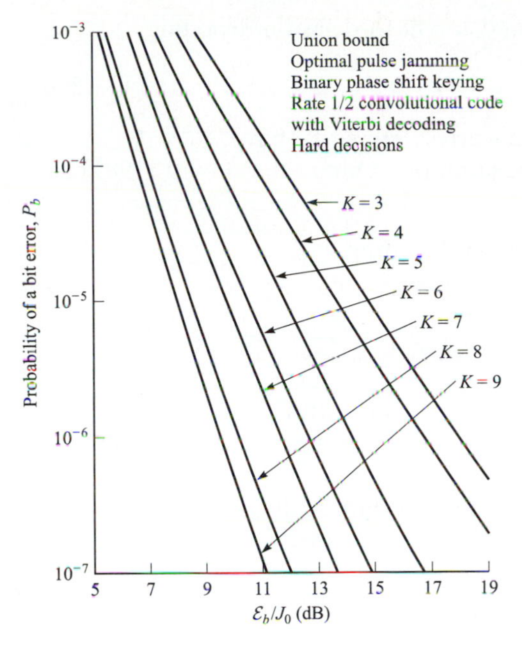

FIGURE 13.2–13
Performance of rate 1/2 convolutional codes with hard-decision Viterbi decoding binary PSK with optimal pulse jamming. [*From Martin and McAdam (1980). © 1980 IEEE.*]

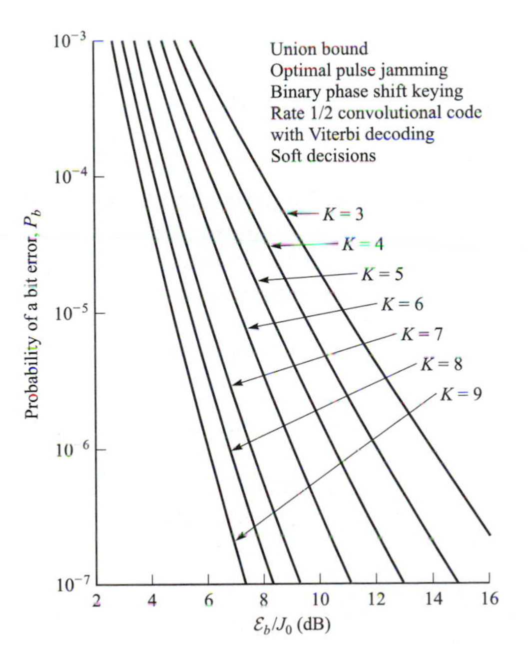

FIGURE 13.2–14
Performance of rate 1/2 convolutional codes with soft-decision Viterbi decoding binary PSK with optimal pulse jamming. [*From Martin and McAdam (1980). © 1980 IEEE.*]

where $\mathcal{E}_c$ is the energy per coded bit; and for hard-decision decoding,

$$D_\alpha = \sqrt{4p(1-p)} \tag{13.2–66}$$

where p is the probability of a coded bit error. Here, we have $N_0 \equiv J_0$.

For a coded binary PSK, with pulse jamming, Omura and Levitt (1982) have shown that

$$D_\alpha = \alpha e^{-\alpha\mathcal{E}_c/N_0} \quad \text{for soft-decision decoding with knowledge of jammer state} \tag{13.2–67}$$

$$D_\alpha = \min_{\lambda \geqslant 0}\{[\alpha \exp(\lambda^2\mathcal{E}_c/N_0/\alpha) + 1 - \alpha] \exp(-2\lambda\mathcal{E}_c)\}$$

$$\text{for soft-decision decoding with no knowledge of jammer state} \tag{13.2–68}$$

$$D_\alpha = \alpha\sqrt{4p(1-p)} \quad \text{for hard-decision decoding with knowledge of the jammer state} \tag{13.2–69}$$

$$D_\alpha = \sqrt{4\alpha p(1-\alpha p)} \quad \text{for hard-decision decoding with no knowledge of the jammer state} \tag{13.2–70}$$

where the probability of error for hard-decision decoding of binary PSK is

$$p = Q\left(\sqrt{\frac{2\alpha\mathcal{E}_c}{N_0}}\right)$$

The graphs for R_0 as a function of $\mathcal{E}_c/N_0$ are illustrated in Figure 13.2–15 for the cases given above. Note that these graphs represent the cutoff rate for the worst-case value of $\alpha = \alpha^*$ that maximizes D_α (minimizes R_0) for each value of $\mathcal{E}_c/N_0$. Furthermore, note that with soft-decision decoding and no knowledge of the jammer state, $R_0 = 0$. This situation results from the fact that the demodulator output is not quantized.

The graphs in Figure 13.2–15 may be used to evaluate the performance of coded systems. To demonstrate the procedure, suppose that we wish to determine the SNR required to achieve an error probability of 10^{-6} with coded binary PSK in worst-case pulse jamming. To be specific, we assume that we have a rate 1/2, $K = 7$ convolutional code. We begin with the performance of the rate 1/2, $K = 7$ convolutional code with soft-decision decoding in an AWGN channel. At $P_2 = 10^{-6}$, the SNR required is found from Figure 8.2–22 to be

$$\frac{\mathcal{E}_b}{N_0} = 5 \text{ dB}$$

Since the code is rate 1/2, we have

$$\frac{\mathcal{E}_c}{N_0} = 2 \text{ dB}$$

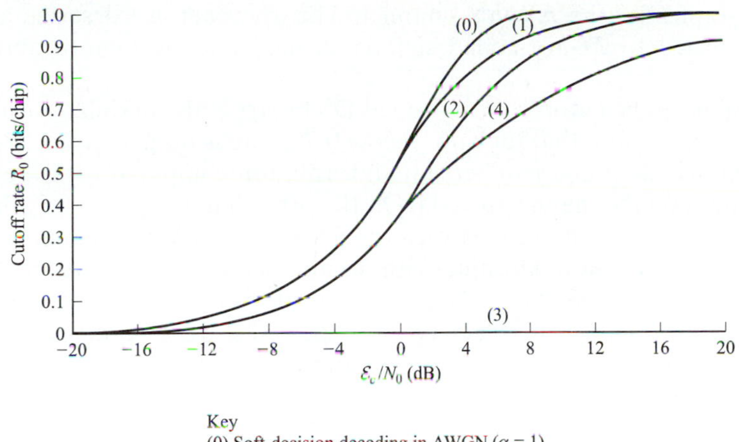

FIGURE 13.2–15
Cutoff rate for coded DS binary PSK modulation. [*From Omura and Levitt (1982).* © *1982 IEEE*].

Now, we go to the graphs in Figure 13.2–15 and find that for the AWGN channel (reference system) with $\mathcal{E}_c/N_0 = 2\,\text{dB}$, the corresponding value of the cutoff rate is

$$R_0 = 0.74 \text{ bits per symbol}$$

If we have another channel with different noise characteristics (a worst-case pulse noise channel) but with the same value of the cutoff rate R_0, then the upper bound on the bit error probability is the same, i.e., 10^{-6} in this case. Consequently, we can use this rate to determine the SNR required for the worst-case pulse jammer channel. From the graphs in Figure 13.2–15, we find that

$$\frac{\mathcal{E}_c}{J_0} = \begin{cases} 10 \text{ dB} & \text{for hard-decision decoding with no knowledge of jammer state} \\ 5 \text{ dB} & \text{for hard-decision decoding with knowledge of jammer state} \\ 3 \text{ dB} & \text{for soft-decision decoding with knowledge of jammer state} \end{cases}$$

Therefore, the corresponding values of $\mathcal{E}_b/J_0$ for the rate 1/2, $K = 7$ convolutional code are 13, 8, and 6 dB, respectively.

This general approach may be used to generate error rate graphs for coded binary signals in a worst-case pulse jamming channel by using corresponding error rate graphs for the AWGN channel. The approach we describe above is easily generalized to M-ary coded signals as indicated by Omura and Levitt (1982).

By comparing the cutoff rate for coded DS binary PSK modulation shown in Figure 13.2–15, we note that for rates below 0.7, there is no penalty in SNR with soft-decision decoding and jammer state information compared with the performance on the AWGN channel ($\alpha = 1$). On the other hand, at $R_0 = 0.7$, there is a 6-dB difference in performance between the SNR in an AWGN channel and that required for hard-decision decoding with no jammer state information. At rates below 0.4, there is no penalty in SNR with hard-decision decoding if the jammer state is unknown. However, there is the expected 2-dB loss in hard-decision decoding compared with soft-decision decoding in the AWGN channel.

13.2.4 Excision of Narrowband Interference in DS Spread Spectrum Systems

We have shown that DS spread spectrum signals reduce the effects of interference due to other users of the channel and intentional jamming. When the interference is narrowband, the cross correlation of the received signal with the replica of the PN code sequence reduces the level of the interference by spreading it across the frequency band occupied by the PN signal. Thus, the interference is rendered equivalent to a lower-level noise with a relatively flat spectrum. Simultaneously the cross correlation operation collapses the desired signal to the bandwidth occupied by the information signal prior to spreading. Consequently, the power in the narrowband interference is reduced by an amount equal to the processing gain.

The interference immunity of a DS spread spectrum communication system corrupted by narrowband interference can be further improved by filtering the signal prior to despreading, where the objective is to reduce the level of the interference at the expense of introducing some distortion on the desired signal. This filtering can be accomplished by exploiting the wideband spectral characteristics of the desired DS signal and the narrowband characteristic of the interference as described below.

To be specific, we consider the demodulator illustrated in Figure 13.2–16. The received signal is passed through a filter matched to the chip pulse $g(t)$. The output of this filter is synchronously sampled every T_c seconds to yield

$$r_j = 2\mathcal{E}_c(2b_j - 1)(2c_{ij} - 1) + \nu_j, \qquad j = 1, 2, \ldots \tag{13.2–71}$$

where $\mathcal{E}_c$ is the energy of the chip pulse, $\{b_j\}$ is the binary-valued PN sequence, and ν_j represents the additive noise and interference term. The additive noise term ν_j will be assumed to consist of two terms, one corresponding to a broadband noise (usually thermal noise) and the other to narrowband interference. Consequently we may express r_j as

$$r_j = s_j + i_j + n_j \tag{13.2–72}$$

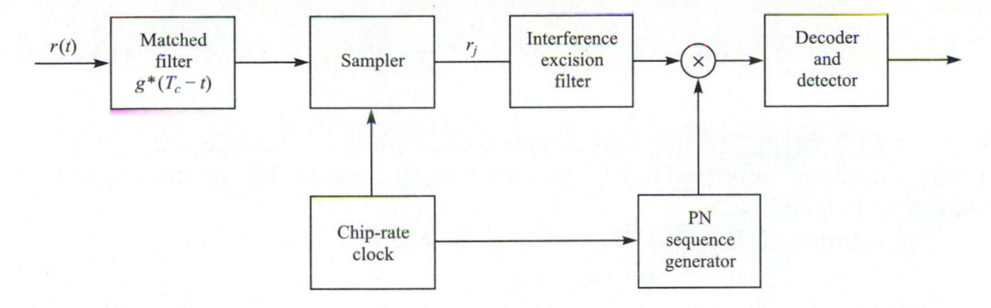

FIGURE 13.2–16
Demodulator for PN spread spectrum signal corrupted by narrowband interference.

where s_j denotes the signal component, i_j the narrowband interference, and n_j the broadband noise.

The received signal sequence $\{r_j\}$ at the output of the sampler is fed to a discrete-time filter that estimates the narrowband interference sequence $\{i_j\}$ and subtracts the estimate $\hat{\imath}_j$ from $\{r_j\}$. This filter may be either linear or non-linear. The resulting signal sequence $\{r_j - \hat{\imath}_j\}$ is then fed to the PN correlator, whose output is passed to the decoder.

Interference estimation and suppression based on linear prediction. The interference component i_j can be estimated from the received signal by passing it through the linear transverse filter. Computationally efficient algorithms based on linear prediction may be used to estimate the interference. Basically, in this method the narrowband interference is modeled as having been generated by passing white noise through an all-pole filter. Hence, the output of this filter is an autoregressive (AR) process. Linear prediction is used to estimate the coefficients of the all-pole model. The estimated coefficients specify an appropriate noise-whitening all-zero (transversal) filter which is used to suppress the narrowband interference.

Let us assume for the moment that the statistics of the sequence $\{i_j\}$ are known and that $\{i_j\}$ is a stationary random sequence. Then, because of the narrowband characteristics of $\{i_j\}$, we can predict i_j from $r_{j-1}, r_{j-2}, \ldots, r_{j-m}$. That is,

$$\hat{\imath}_j = \sum_{l=1}^{m} a_{ml} r_{j-l} \tag{13.2–73}$$

where $\{a_{ml}\}$ are the coefficients of an mth-order linear predictor. It should be emphasized that Equation 13.2–73 predicts the interference but not the signal s_j, because the PN chips are uncorrelated and, hence, s_j is uncorrelated with r_{j-l}, $l = 1, 2, \ldots, m$, where m is less than the length of the PN sequence.

The coefficients in Equation 13.2–73 are determined by minimizing the mean square error between r_j and $\hat{\imath}_j$, with respect to the predictor coefficients. This leads to the set of linear equations, called the Yule–Walker equations,

$$\sum_{l=1}^{m} a_{ml}\phi(k-l) = \phi(k), \qquad k = 1, 2, \ldots, m \tag{13.2–74}$$

where $\phi(k) = E(r_j r_{j+k})$ is the autocorrelation function of the received signal $\{r_j\}$. These equations were previously encountered in Section 3.5, in the context of analog source coding.

The solution of Equation 13.2–74 for the coefficients of the prediction filter requires knowledge of the autocorrelation function $\phi(k)$. In practice, the autocorrelation function of $\{i_j\}$ and, hence, $\{r_j\}$ is usually unknown, and it may also be slowly varying in time (nonstationary interference). In such a case, adaptive algorithms may be used to estimate the narrowband interference. In particular, least-squares-type algorithms, such as the Burg algorithm, are especially effective for estimating the coefficients of the linear prediction filter adaptively, as described in the paper by Ketchum and Proakis (1982).

EXAMPLE 13.2–4. Let us consider a narrowband interference that occupies 20 percent of the spectral band occupied by the PN spread spectrum signal. The average power of the interference is 20 dB above the average power of the signal. The average power of the broadband noise is 20 dB below the average power of the signal. Figure 13.2–17 illustrates the spectral characteristics of a 16-tap and a 29-tap FIR filter when the interference is equally split into four frequency bands. It is apparent that the 29-tap filter has better spectral characteristics. In general, the number of taps in the filter should be about four times the number of interference bands for adequate suppression. It is also apparent that the interference suppression filter acts as a notch filter. In effect, it attempts to whiten the total noise plus interference, so that the power spectral density of these components at its output is approximately flat. While suppressing the interference, the filter also distorts the desired signal by spreading it in time.

Performance improvement with interference suppression. Since the noise plus interference at the output of the suppression filter is spectrally flat, the matched filtering or cross correlation following the suppression filter should be performed with the distorted signal. This may be accomplished by having a filter matched to the interference suppression filter, i.e., a discrete-time filter impulse response $\{-a_{mm}, -a_{mm-1} \ldots - a_{m1}, 1\}$ followed by the PN correlator. In fact, we can combine the interference suppression filter and its matched filter into a single filter having an impulse response

$$\begin{aligned}
h_0 &= -a_{mm} \\
h_k &= -a_{mm-k} + \sum_{l=0}^{k-1} a_{mm-l}a_{mk-l}, \qquad 1 \leqslant k \leqslant m-1 \\
h_m &= 1 + \sum_{l=1}^{m} a_{ml}^2 \\
h_{m+k} &= h_{m-k}, \qquad 0 \leqslant k \leqslant m
\end{aligned} \tag{13.2–75}$$

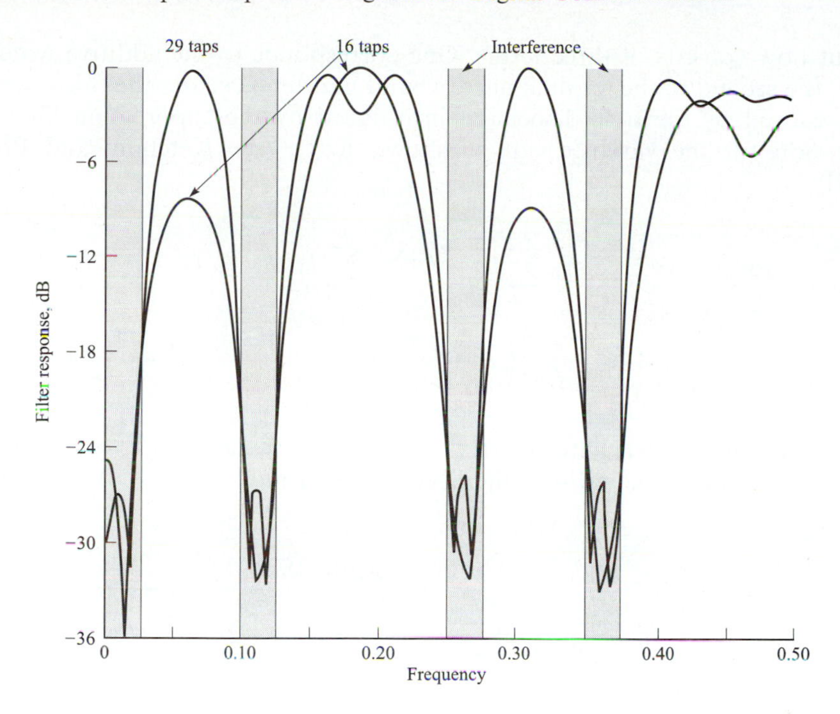

FIGURE 13.2–17
Frequency-response characteristics of 16- and 29-tap filters for four bands of interference.

The combined filter is a linear phase (symmetric) transversal filter with $K = 2m + 1$ taps. The impulse response may be normalized by dividing every term by h_m. Thus the center tap is normalized to unity. In order to demonstrate the effectiveness of the interference suppression filter, we compare the performance of the DS system with and without the suppression filter. The output SNR is a convenient performance index for this purpose. Since the output of the PN correlator is characterized as Gaussian, there is a one-to-one correspondence between the SNR and the probability of error.

Without the suppression filter, the PN correlator output , denoted as U_1, has mean $2\mathcal{E}_c L_c$ and a variance $L_c[2\mathcal{E}_c N_0 + \phi_{ii}(0)]$ where $\phi_{ii}(k)$ is the autocorrelation function of the sequence $\{i_j\}$ and L_c is the number of chips per bit or per symbol. The output SNR is defined as the ratio of the square of the mean to twice the variance. Hence the SNR without the suppression filter is

$$\text{SNR}_{n0} = \frac{\mathcal{E}_c L_c}{N_0 + \phi_{ii}(0)/2\mathcal{E}_c} \tag{13.2–76}$$

With an interference suppression filter having symmetric impulse response as defined in Equation 13.2–75 and normalized such that the center tap is unity, the mean value of the correlator output is also $2\mathcal{E}_c L_c$. However, the variance of the output now consists of three terms. One corresponds to the additive wideband noise, the second to the residual narrowband interference, and the third to a self-noise caused by the time dispersion introduced by the suppression filter. The expression for the variance can be shown to be (see Ketchum and Proakis [1982]):

$$\begin{aligned}\operatorname{var}(U_1) = 2L_c\mathcal{E}_c N_0 \sum_{k=0}^{K} h_k^2 + L_c \sum_{k=0}^{K}\sum_{l=0}^{K} h(l)h(k)\phi_{ii}(k-l) \\ + 4L_c\mathcal{E}_c^2 \sum_{k=0}^{K/2-1}\left(2-\frac{k}{L_c}\right)h_k^2\end{aligned} \tag{13.2–77}$$

Hence the output SNR with the filter is the ratio of the square of the mean to twice the variance. The ratio of the SNR with the filter to the SNR without the filter is

$$\eta_0 = \frac{N_0 + \phi_{ii}(0)/2\mathcal{E}_c}{N_0 \sum_{k=0}^{K} h_k^2 + \frac{1}{2\mathcal{E}_c}\sum_{k=0}^{K}\sum_{l=0}^{K} h(k)h(l)\phi_{ii}(k-l) + 2\mathcal{E}_c \sum_{k=0}^{K/2-1}(2-k/L_c)h_k^2} \tag{13.2–78}$$

This ratio is called the improvement factor resulting from interference suppression. It may be plotted against the normalized SNR per chip without filtering, defined as

$$\frac{\mathrm{SNR}_{n0}}{L_c} = \frac{\mathcal{E}_c}{N_0 + \phi_{ii}(0)/2\mathcal{E}_c} \tag{13.2–79}$$

The resulting graph of η_0 versus SNR_{n0}/L_c is universal in the sense that it applies to any PN spread spectrum system with arbitrary processing gain for a given $\mathcal{E}_c$, N_0, and $\phi_{ii}(0)$.

As an example, the improvement factor in (decibels) is plotted against SNR_{n0}/L_c in Figure 13.2–18 for a single-band equal-amplitude randomly phased sinusoids covering 20 percent of the frequency band occupied by the DS spread spectrum signal. The interference suppression filter consists of a nine-tap suppression filter which corresponds to a fourth-order predictor. These numerical results indicate that the notch filter is very effective in suppressing the interference prior to PN correlation and decoding. As a consequence, the jamming margin of the system is increased.

The use of a linear adaptive FIR filter for suppression of narrowband interference in DS spread spectrum systems has been considered in the literature by many authors. The interested reader is referred to this literature cited in Section 13.6. A practical motivation for excision of narrowband signals from wideband signals is to allow the overlay of narrowband digital cellular systems with wideband CDMA systems.

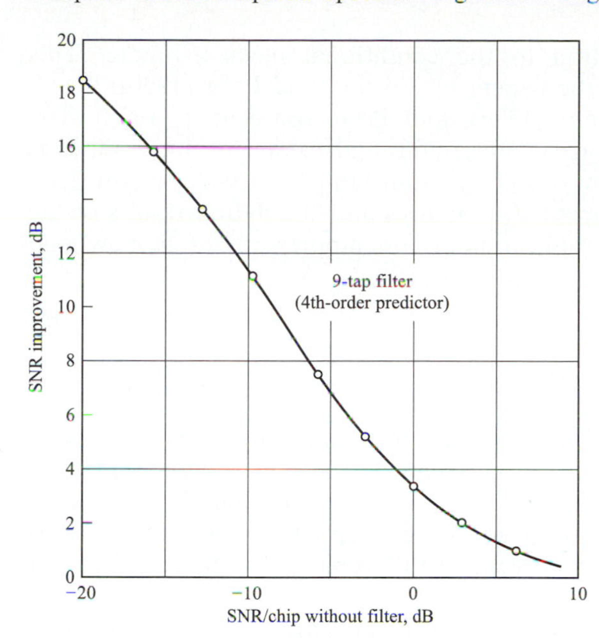

FIGURE 13.2–18
Improvement factor for interference suppression filter in cascade with its matched filter.

Interference estimation and suppression based on non-linear filtering. The linear FIR filter used to predict the narrowband interference, which is modeled as a Gaussian autoregressive (AR) process, is the optimal minimum mean-square-error filter when the signal $\{s_k\}$ and broadband noise $\{n_k\}$ components are Gaussian random processes. However, the DS spread spectrum signal sequence $\{s_k\}$ is non-Gaussian. Consequently, the linear estimation filter is suboptimal, in the sense that it is not the best filter for suppressing the narrowband interference. The optimum estimator for the narrowband interference is non-linear.

By defining the state vector $\mathbf{x}_k$ as

$$\mathbf{x}_k = [i_k \quad i_{k-1} \quad \cdots \quad i_{k-m+1}]^t \tag{13.2–80}$$

where m is the order of the AR model, it is possible to express the state vector and the observation sequence in the state-space form

$$\begin{aligned} \mathbf{x}_k &= \mathbf{\Phi}\mathbf{x}_{k-1} + \mathbf{w}_k \\ \mathbf{r}_k &= \mathbf{H}\mathbf{x}_k + (n_k + s_k) \end{aligned} \tag{13.2–81}$$

where $\mathbf{\Phi}$ is the state transition matrix that depends on the AR model parameters, $\mathbf{w}_k$ is the white Gaussian process driving the AR model, and $\mathbf{H} = [100\ldots0]$. We recall that the minimum mean-square-error estimator for the state at time k given the observations $\mathbf{r}_{k-1} \equiv [r_{k-1}, r_{k-2}, \ldots, r_0]$ is the conditional mean $E(\mathbf{x}_k|\mathbf{r}_{k-1})$. If the signal sequence $\{s_k\}$ and the broadband noise sequence $\{n_k\}$ were Gaussian, the optimum estimator for the state $\mathbf{x}_k$ corresponding to the conditional mean would be the linear predictor obtained from the Kalman filter. Since $\{s_k\}$ is non-Gaussian, the conditional mean estimate is a non-linear function of the abbreviations which, in general, is highly complex. However, it is possible to derive a reduced complexity approximation to the conditional mean estimate. This approach has been described in the papers by Vijayan and Poor (1990), Garth and Poor (1992), Rusch and Poor (1994), and Poor and Rusch (1994). The general configuration of the approximate conditional mean non-linear filter is shown in Figure 13.2–19. The non-linear function tanh(x) provides a soft-decision type feedback signal component. An analysis and simulation results of the performance of this type of non-linear filter for suppression of narrowband interference are given in the papers cited above.

13.2.5 Generation of PN Sequences

The generation of PN sequences for spread spectrum applications is a topic that has received considerable attention in the technical literature. We shall briefly discuss the construction of some PN sequences and present a number of important properties of the autocorrelation and cross-correlation functions of such sequences. For a comprehensive treatment of this subject, the interested reader may refer to the book by Golomb (1967).

By far the most widely known binary PN sequences are the maximum-length shift-register sequences introduced in Section 8.1.3 in the context of coding and suggested again in Section 13.2.2 for use as low-rate codes. A maximum-length shift-register sequence, or m-sequence for short, has length $n = 2^m - 1$ bits and is generated by an m-stage shift register with linear feedback as illustrated in Figure 13.2–20. The sequence is periodic with period n. Each period of the sequence contains 2^{m-1} ones and $2^{m-1} - 1$ zeros.

In DS spread spectrum applications the binary sequence with elements {0, 1} is mapped into a corresponding sequence of positive and negative pulses according to the relation

$$p_i(t) = (2b_i - 1)p(t - iT)$$

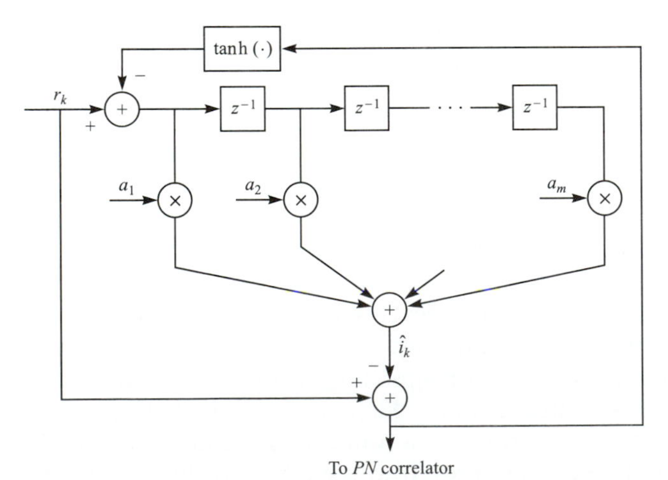

FIGURE 13.2–19
Non-linear excision filter.

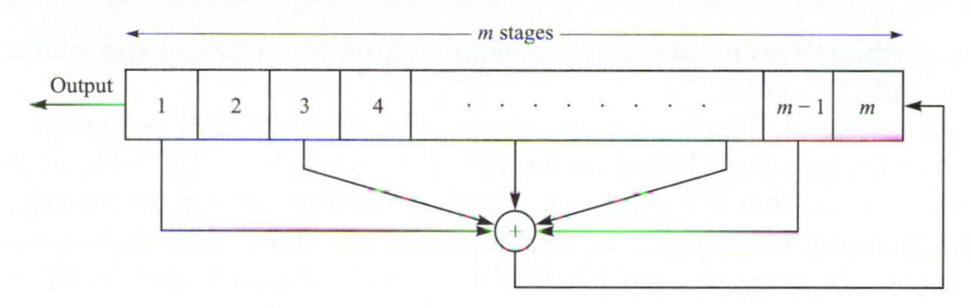

FIGURE 13.2–20
General m-stage shift register with linear feedback.

where $p_i(t)$ is the pulse corresponding to the element b_i in the sequence with elements $\{0, 1\}$. Equivalently, we may say that the binary sequence with elements $\{0, 1\}$ is mapped into a corresponding binary sequence with elements $\{-1, 1\}$. We shall call the equivalent sequence with elements $\{-1, 1\}$ a *bipolar sequence*, since it results in pulses of positive and negative amplitudes.

An important characteristic of a periodic PN sequence is its periodic autocorrelation function, which is usually defined in terms of the bipolar sequence as

$$\phi(j) = \sum_{i=1}^{n} (2b_i - 1)(2b_{i+j} - 1), \qquad 0 \leqslant j \leqslant n - 1 \tag{13.2–82}$$

where n is the period. Clearly, $\phi(j + rn) = \phi(j)$ for any integer value r.

Ideally, a pseudorandom sequence should have an autocorrelation function with the property that $\phi(0) = n$ and $\phi(j) = 0$ for $1 \leqslant j \leqslant n - 1$. In the case of m sequences, the periodic autocorrelation function is

$$\phi(j) = \begin{cases} n & (j = 0) \\ -1 & (1 \leqslant j \leqslant n - 1) \end{cases} \tag{13.2–83}$$

For large values of n, i.e., for long m sequences, the size of the off-peak values of $\phi(j)$ relative to the peak value $\phi(j)/\phi(0) = -1/n$ is small and, from a practical viewpoint, inconsequential. Therefore, m sequences are almost ideal when viewed in terms of their autocorrelation function.

In antijamming applications of PN spread spectrum signals, the period of the sequence must be large in order to prevent the jammer from learning the feedback connections of the PN generator. However, this requirement is impractical in most cases because the jammer can determine the feedback connections by observing only $2m - 1$ chips from the PN sequence. This vulnerability of the PN sequence is due to the linearity property of the generator. To reduce the vulnerability to a jammer, the output sequences from several stages of the shift register or the outputs from several distinct m sequences are combined in a non-linear way to produce a non-linear sequence that is considerably more difficult for the jammer to learn. Further reduction in vulnerability is achieved by frequently changing the feedback connections and/or the number of stages in the shift register according to some prearranged plan formulated between the transmitter and the intended receiver.

In some applications, the cross-correlation properties of PN sequences are as important as the autocorrelation properties. For example, in CDMA, each user is assigned a particular PN sequence. Ideally, the PN sequences among users should be mutually orthogonal so that the level of interference experienced by any one user from transmissions of other users adds on a power basis. However, the PN sequences used in practice exhibit some correlation.

To be specific, we consider the class of m sequences. It is known (Sarwate and Pursley, 1980) that the periodic cross-correlation function between any pair of m sequences of the same period can have relatively large peaks. Table 13.2–1 lists the peak magnitude $\phi_{\max}$ for the periodic cross correlation between pairs of m sequences for $3 \leqslant m \leqslant 12$. The table also shows the number of m sequences of length $n = 2^m - 1$ for $3 \leqslant m \leqslant 12$. As we can see, the number of m sequences of length n increases rapidly with m. We also observe that, for most sequences, the peak magnitude $\phi_{\max}$ of the cross-correlation function is a large percentage of the peak value of the autocorrelation function.

Such high values for the cross correlations are undesirable in CDMA. Although it is possible to select a small subset of m sequences that have relatively smaller cross-correlation peak values, the number of sequences in the set is usually too small for CDMA applications.

PN sequences with better periodic cross-correlation properties than m sequences have been given by Gold (1967, 1968) and Kasami (1966). They are derived from m sequences as described below.

Gold and Kasami proved that certain pairs of m sequences of length n exhibit a three-valued cross-correlation function with values $\{-1, -t(m), t(m) - 2\}$, where

$$t(m) = \begin{cases} 2^{(m+1)/2} + 1 & (\text{odd } m) \\ 2^{(m+2)/2} + 1 & (\text{even } m) \end{cases} \tag{13.2–84}$$

TABLE 13.2–1
Peak cross correlation of *m* sequences and Gold sequences

m	$n = 2^m - 1$	Number of m sequences	Peak cross correlation ϕ_{max}	$\phi_{max}/\phi(0)$	$t(m)$	$t(m)/\phi(0)$
3	7	2	5	0.71	5	0.71
4	15	2	9	0.60	9	0.60
5	31	6	11	0.35	9	0.29
6	63	6	23	0.36	17	0.27
7	127	18	41	0.32	17	0.13
8	255	16	95	0.37	33	0.13
9	511	48	113	0.22	33	0.06
10	1023	60	383	0.37	65	0.06
11	2047	176	287	0.14	65	0.03
12	4095	144	1407	0.34	129	0.03

For example, if $m = 10$, then $t(10) = 2^6 + 1 = 65$ and the three possible values of the periodic cross-correlation function are $\{-1, -65, 63\}$. Hence the maximum cross correlation for the pair of m sequences is 65, while the peak for the family of 60 possible sequences generated by a 10-stage shift register with different feedback connections is $\phi_{\max} = 383$—about a sixfold difference in peak values. Two m sequences of length n with a periodic cross-correlation function that takes on the possible values $\{-1, -t(m), t(m) - 2\}$ are called *preferred sequences*.

From a pair of preferred sequences, say $\mathbf{a} = [a_1 a_2 \cdots a_n]$ and $\mathbf{b} = [b_1 b_2 \cdots b_n]$, we construct a set of sequences of length n by taking the modulo-2 sum of **a** with the n cyclicly shifted versions of **b** or vice versa. Thus, we obtain n new periodic sequences† with period $n = 2^m - 1$. We may also include the original sequences **a** and **b**, and, thus, we have a total of $n + 2$ sequences. The $n + 2$ sequences constructed in this manner are called *Gold sequences*.

EXAMPLE 13.2–5. Let us consider the generation of Gold sequences of length $n = 31 = 2^5 - 1$. As indicated above for $m = 5$, the cross-correlation peak is

$$t(5) = 2^3 + 1 = 9$$

Two preferred sequences, which may be obtained from Peterson and Weldon (1972), are described by the parity polynomials

$$h_1(p) = p^5 + p^3 + 1$$
$$h_2(p) = p^5 + p^4 + p^3 + p + 1$$

The shift registers for generating the two m sequences and the corresponding Gold sequences are shown in Figure 13.2–21. In this case, there are 33 different sequences,

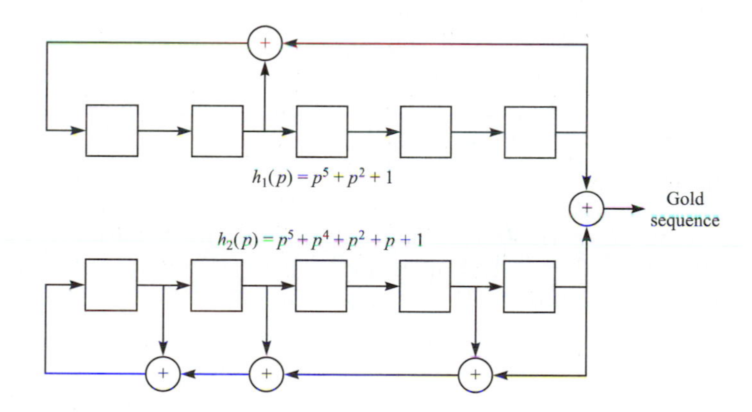

FIGURE 13.2–21
Generation of Gold sequences of length 31.

†An equivalent method for generating the n new sequences is to employ a shift register of length $2m$ with feedback connections specified by the polynomial $h(p) = h_1(p)h_2(p)$, where $h_1(p)$ and $h_2(p)$ are the polynomials that specify the feedback connections of the m-stage shift registers that generate the m sequences **a** and **b**.

corresponding to the 33 relative phases of the two m sequences. Of these, 31 sequences are non-maximal-length sequences.

With the exception of the sequences **a** and **b**, the set of Gold sequences is not comprised of maximum-length shift-register sequences of length n. Hence, their autocorrelation functions are not two-valued. Gold (1968) has shown that the cross-correlation function for any pair of sequences from the set of $n+2$ Gold sequences is three-valued with possible values $\{-1, -t(m), t(m)-2\}$, where $t(m)$ is given by Equation 13.2–84. Similarly, the off-peak autocorrelation function for a Gold sequence takes on values from the set $\{-1, -t(m), t(m)-2\}$. Hence, the off-peak values of the autocorrelation function are upper-bounded by $t(m)$.

The values of the off-peak autocorrelation function and the peak cross-correlation function, i.e., $t(m)$, for Gold sequences is listed in Table 13.2–1. Also listed are the values normalized by $\phi(0)$.

The frequency of occurrence for each of the three possible values of the cross correlation for any pair of Gold sequences may also be of interest to the system designer. In Table 13.2–2, we give the frequency of occurrence of the three values for the case in which m is odd.

It is interesting to compare the peak cross-correlation value of Gold sequences with a known lower bound on the cross correlation between any pair of binary sequences of period n in a set of M sequences. A lower bound derived by Welch (1974) for $\phi_{\max}$ is

$$\phi_{\max} \geqslant n\sqrt{\frac{M-1}{Mn-1}} \tag{13.2–85}$$

which, for large values of n and M, is well approximated as $\sqrt{n}$. For Gold sequences, $M=2^m+1$, $n=2^m-1$ and the lower bound is $\phi_{\max} \approx 2^{m/2}$. This bound is lower by $\sqrt{2}$ for odd m and by 2 for even m relative to $\phi_{\max}=t(m)$ for Gold sequences. Therefore, Gold sequences do not achieve the lower bound.

A procedure similar to that used for generating Gold sequences will generate a smaller set of $M=2^{m/2}$ binary sequences of period $n=2^m-1$, where m is even. In this procedure, we begin with an m sequence **a** and we form a binary sequence **b** by taking every $2^{m/2}+1$ bit of **a**. Thus, the sequence **b** is formed by decimating **a** by $2^{m/2}+1$. It can be verified that the resulting sequence **b** is periodic with period $2^{m/2}-1$. For example, if $m=10$, the period of **a** is $n=1023$ and the peroid of **b** is 31. Hence, if we observe 1023 bits of the sequence **b**,

TABLE 13.2–2
Frequency of occurrence of cross-correlation values for Gold codes of length $n=2^m-1$, m odd

Cross-correlation value	Frequency of occurrence
-1	$2^{n-1}-1$
$-[2^{(m+1)/2}+1]$	$2^{n-2}-2^{(n-3)/2}$
$2^{(m+1)/2}-1$	$2^{n-2}+2^{(n-3)/2}$

we shall see 33 repetitions of the 31-bit sequence. Now, by taking $n = 2^m - 1$ bits of the sequences **a** and **b**, we form a new set of sequences by adding, modulo-2, the bits from **a** and the bits from **b** and all $2^{m/2} - 2$ cyclic shifts of the bits from **b**. By including **a** in the set, we obtain a set of $2^{m/2}$ binary sequences of length $n = 2^m - 1$. These are called *Kasami sequences*. The autocorrelation and cross-correlation functions of these sequences take on values from the set $\{-1, -(2^{m/2} + 1), 2^{m/2} - 1\}$. Hence, the maximum cross-correlation value for any pair of sequences from the set is

$$\phi_{\max} = 2^{m/2} + 1 \qquad (13.2\text{–}86)$$

This value of $\phi_{\max}$ satisfies the Welch lower bound for a set of $2^{m/2}$ sequences of length $n = 2^m - 1$. Hence, the Kasami sequences are optimal.

Besides the well-known Gold and Kasami sequences, there are other binary sequences appropriate for CDMA applications. The interested reader may refer to the work of Scholtz (1979), Olsen (1977), and Sarwate and Pursley (1980).

Finally, we wish to indicate that, although we have discussed the periodic cross-correlation function between pairs of periodic sequences, many practical CDMA systems may use information bit durations that encompass only fractions of a periodic sequence. In such cases, it is the partial-period cross correlation between two sequences that is important. A number of papers deal with this problem, including those by Lindholm (1968), Wainberg and Wolf (1970), Fredricsson (1975), Bekir et al. (1978), and Pursley (1979).

13.3 FREQUENCY-HOPPED SPREAD SPECTRUM SIGNALS

In a *frequency-hopped* (FH) spread spectrum communication system the available channel bandwidth is subdivided into a large number of contiguous frequency slots. In any signaling interval, the transmitted signal occupies one or more of the available frequency slots. The selection of the frequency slot(s) in each signaling interval is made pseudorandomly according to the output from a PN generator. Figure 13.3–1 illustrates a particular FH pattern in the time-frequency plane.

A block diagram of the transmitter and receiver for an FH spread spectrum system is shown in Figure 13.3–2. The modulation is usually either binary or M-ary FSK. For example, if binary FSK is employed, the modulator selects one of two frequencies corresponding to the transmission of either a 1 or a 0. The resulting FSK signal is translated in frequency by an amount that is determined by the output sequence from the PN generator, which, in turn, is used to select a frequency that is synthesized by the frequency synthesizer. This frequency is mixed with the output of the modulator and the resultant frequency-translated signal is transmitted over the channel. For example, m bits from the PN generator may be used to specify $2^m - 1$ possible frequency translations.

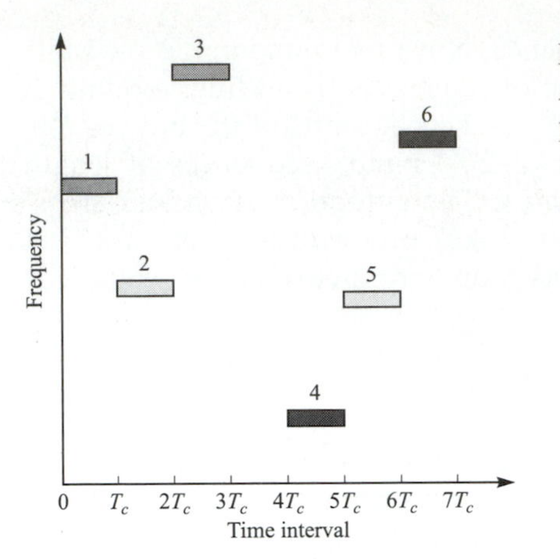

FIGURE 13.3–1
An example of a frequency-hopped (FH) pattern.

At the receiver, we have an identical PN generator, synchronized with the receiver signal, which is used to control the output of the frequency synthesizer. Thus, the pseudorandom frequency translation introduced at the transmitter is removed at the receiver by mixing the synthesizer output with the received signal. The resultant signal is demodulated by means of an FSK demodulator. A signal for maintaining synchronism of the PN generator with the frequency-translated received signal is usually extracted from the received signal.

Although PSK modulation gives better performance than FSK in an AWGN channel, it is difficult to maintain phase coherence in the synthesis of the frequencies used in the hopping pattern and, also, in the propagation of the signal over the channel as the signal is hopped from one frequency to another over a wide bandwidth. Consequently, FSK modulation with noncoherent detection is usually employed with FH spread spectrum signals.

In the FH system depicted in Figure 13.3–2, the carrier frequency is pseudorandomly hopped in every signaling interval. The M information-bearing tones are contiguous and separated in frequency by $1/T_c$, where T_c is the signaling interval. This type of frequency hopping is called *block hopping*.

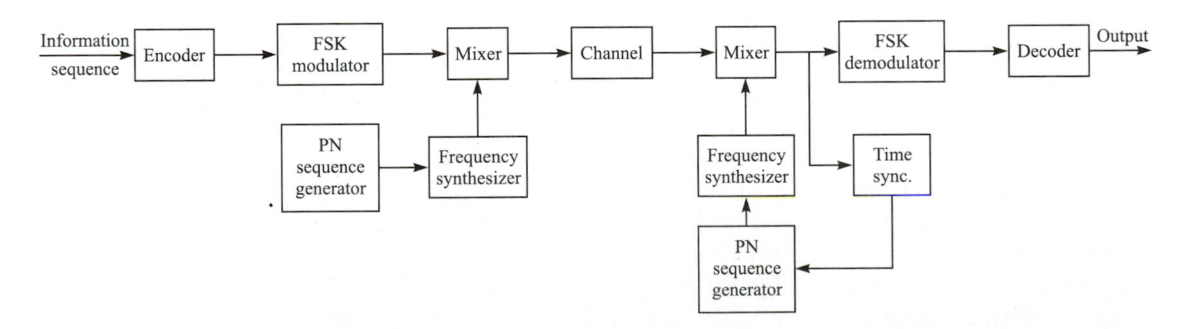

FIGURE 13.3–2
Block diagram of an FH spread spectrum system.

Another type of frequency hopping that is less vulnerable to some jamming strategies is independent tone hopping. In this scheme, the M possible tones from the modulator are assigned widely dispersed frequency slots. One method for accomplishing this is illustrated in Figure 13.3–3. Here, the m bits from the PN generator and the k information bits are used to specify the frequency slots for the transmitted signal.

The FH rate is usually selected to be either equal to the (coded or uncoded) symbol rate or faster than that rate. If there are multiple hops per symbol, we have a fast-hopped signal. On the other hand, if the hopping is performed at the symbol rate, we have a slow-hopped signal.

Fast frequency hopping is employed in AJ applications when it is necessary to prevent a type of jammer, called a *follower jammer,* from having sufficient time to intercept the frequency and retransmit it along with adjacent frequencies so as to create interfering signal components. However, there is a penalty incurred in subdividing a signal into several FH elements because the energy from these separate elements is combined noncoherently. Consequently, the demodulator incurs a penalty in the form of a noncoherent combining loss as described in Section 12.1.

FH spread spectrum signals are used primarily in digital communication systems that require AJ projection and in CDMA, where many users share a common bandwidth. In most cases, an FH signal is preferred over a DS spread spectrum signal because of the stringent synchronization requirements inherent in DS spread spectrum signals. Specifically, in a DS system, timing and synchronization must be established to within a fraction of the chip interval $T_c \approx 1/W$. On the other hand, in an FH system, the chip interval is the time spent in transmitting a signal in a particular frequency slot of bandwidth $B \ll W$. But this interval is approximately $1/B$, which is much larger than $1/W$. Hence the timing requirements in an FH system are not as stringent as in a DS system.

In Sections 13.3.2 and 13.3.3, we shall focus on the AJ and CDMA applications of FH spread spectrum signals. First, we shall determine the error rate performance of an uncoded and a coded FH signal in the presence of broadband AWGN inteference. Then we shall consider a more serious type of interference

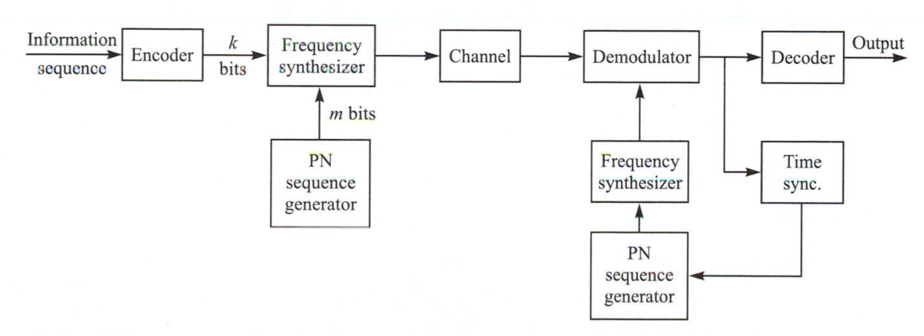

FIGURE 13.3–3
Block diagram of an independent tone FH spread spectrum system.

that arises in AJ and CDMA applications, called *partial-band interference*. The benefits obtained from coding for this type of interference are determined. We conclude the discussion in Section 13.3.3 with an example of an FH CDMA system that was designed for use by mobile users with a satellite serving as the channel.

13.3.1 Performance of FH Spread Spectrum Signals in an AWGN Channel

Let us consider the performance of an FH spread spectrum signal in the presence of broadband interference characterized statistically as AWGN with power spectral density J_0. For binary orthogonal FSK with noncoherent detection and slow frequency hopping (1 hop/bit), the probability of error, derived in Section 5.4.1, is

$$P_2 = \tfrac{1}{2} e^{-\gamma_b/2} \tag{13.3–1}$$

where $\gamma_b = \mathcal{E}_b/J_0$. On the other hand, if the bit interval is subdivided into L subintervals and FH binary FSK is transmitted in each subinterval, we have a fast FH signal. With square-law combining of the output signals from the corresponding matched filters for the L subintervals, the error rate performance of the FH signal, obtained from the results in Section 12.1, is

$$P_2(L) = \frac{1}{2^{2L-1}} e^{-\gamma_b/2} \sum_{i=0}^{L-1} K_i (\tfrac{1}{2}\gamma_b)^i \tag{13.3–2}$$

where the SNR per bit is $\gamma_b = \mathcal{E}_b/J_0 = L\gamma_c$, γ_c is the SNR per chip in the L-chip symbol, and

$$K_i = \frac{1}{i!} \sum_{r=0}^{L-1-i} \binom{2L-1}{r} \tag{13.3–3}$$

We recall that, for a given SNR per bit γ_b, the error rate obtained from Equation 13.3–2 is larger than that obtained from Eequation 13.3–1. The difference in SNR for a given error rate and a given L is called the *noncoherent combining loss*, which was described and illustrated in Section 12.1.

Coding improves the performance of the FH spread spectrum system by an amount, which we call the *coding gain*, that depends on the code parameters. Suppose we use a linear binary (n, k) block code and binary FSK modulation with one hop per coded bit for transmitting the bits. With soft-decision decoding of the square-law-demodulated FSK signal, the probability of a code word error is upper-bounded as

$$P_M \leqslant \sum_{m=2}^{M} P_2(m) \tag{13.3–4}$$

where $P_2(m)$ is the error probability in deciding between the mth code word and the all-zero code word when the latter has been transmitted. The expression for

$P_2(m)$ was derived in Section 8.1.4 and has the same form as Equations 13.3–2 and 13.3–3, with L being replaced by w_m and γ_b by $\gamma_b R_c w_m$, where w_m is the weight of the mth code word and R_c is the code rate. The product $R_c w_m$, which is not less than $R_c d_{\min}$, represents the coding gain. Thus, we have the performance of a block coded FH system with slow frequency hopping in broadband interference.

The probability of error for fast frequency hopping with n_2 hops per coded bit is obtained by reinterpreting the binary event probability $P_2(m)$ in Equation 13.3–4. The n_2 hops per coded bit may be interpreted as a repetition code, which, when combined with a nontrivial (n_1, k) binary linear code having weight distribution $\{w_m\}$, yields an $(n_1 n_2, k)$ binary linear code having weight distribution $\{n_2 w_m\}$. Hence, $P_2(m)$ has the form given in Equation 13.3–2, with L replaced by $n_2 w_m$ and γ_b by $\gamma_b R_c n_2 w_m$, where $R_c = k/n_1 n_2$. Note that $\gamma_b R_c n_2 w_m = \gamma_b w_m k/n_1$, which is just the coding gain obtained from the nontrivial (n_1, k) code. Consequently, the use of the repetition code will result in an increase in the noncoherent combining loss.

With hard-decision decoding and slow frequency hopping, the probability of a coded bit error at the output of the demodulator for noncoherent detection is

$$p = \tfrac{1}{2} e^{-\gamma_b R_c/2} \tag{13.3–5}$$

The code word error probability is easily upper bounded, by use of the Chernoff bound, as

$$P_M \leqslant \sum_{m=2}^{M} [4p(1-p)]^{w_m/2} \tag{13.3–6}$$

However, if fast frequency hopping is employed with n_2 hops per coded bit, and the square-law-detected outputs from the corresponding matched filters for the n_2 hops are added as in soft-decision decoding to form the two decision variables for the coded bits, the bit error probability p is also given by Equation 13.3–2, with L replaced by n_2 and γ_b replaced by $\gamma_b R_c n_2$, where R_c is the rate of the nontrivial (n_1, k) code. Consequently, the performance of the fast FH system in broadband interference is degraded relative to the slow FH system by an amount equal to the noncoherent combining loss of the signals received from the n_2 hops.

We have observed that for both hard-decision and soft-decision decoding, the use of the repetition code in a fast FH system yields no coding gain. The only coding gain obtained comes from the (n_1, k) block code. Hence, the repetition code is inefficient in a fast FH system with noncoherent combining. A more efficient coding method is one in which either a single low-rate binary code or a concatenated code is employed. Additional improvements in performance may be obtained by using nonbinary codes in conjunction with M-ary FSK. Bounds on the error probability for this case may be obtained from the results given in Section 12.1.

Although we have evaluated the performance of linear block codes only in the above discussion, it is relatively easy to derive corresponding performance results for binary convolutional codes. We leave as an exercise for the reader the derivation of the bit error probability for soft-decision Viterbi decoding and hard-decision Viterbi decoding of FH signals corrupted by broadband interference.

Finally, we observe that $\mathcal{E}_b$, the energy per bit, can be expressed as $\mathcal{E}_b = P_{av}/R$, where R is the information rate in bits per second and $J_0 = J_{av}/W$. Therefore, γ_b may be expressed as

$$\gamma_b = \frac{\mathcal{E}_b}{J_0} = \frac{W/R}{J_{av}/P_{av}} \tag{13.3–7}$$

In this expression, we recognize W/R as the processing gain and J_{av}/P_{av} as the jamming margin for the FH spread spectrum signal.

13.3.2 Performance of FH Spread Spectrum Signals in Partial-Band Interference

The partial-band interference considered in this subsection is modeled as a zero-mean Gaussian random process with a flat power spectral density over a fraction α of the total bandwidth W and zero elsewhere. In the region or regions where the power spectral density is nonzero, its value is $\Phi_{zz}(f) = J_0/\alpha$, $0 < \alpha \leqslant 1$. This model of the interference may be applied to a jamming signal or to interference from other users in an FH CDMA system.

Suppose that the partial-band interference comes from a jammer who may select α to optimize the effect on the communication system. In an uncoded pseudorandomly hopped (slow-hopping) FH system with binary FSK modulation and noncoherent detection, the received signal will be jammed with probability α and it will not be jammed with probability $1 - \alpha$. When it is jammed, the probability of error is $\frac{1}{2}\exp(-\mathcal{E}_b\alpha/2J_0)$, and when it is not jammed, the demodulation is error-free. Consequently, the average probability of error is

$$P_2(\alpha) = \tfrac{1}{2}\alpha \exp\left(-\frac{\alpha\mathcal{E}_b}{2J_0}\right) \tag{13.3–8}$$

where $\mathcal{E}_b/J_0$ may also be expressed as $(W/R)/(J_{av}/P_{av})$.

Figure 13.3–4 illustrates the error rate as a function of $\mathcal{E}_b/J_0$ for several values of α. The jammer's optimum strategy is to select the value of α that maximizes the error probability. By differentiating $P_2(\alpha)$ and solving for the extremum with the restriction that $0 \leqslant \alpha \leqslant 1$, we find that

$$\alpha^* = \begin{cases} \dfrac{1}{\mathcal{E}_b/2J_0} = 2\dfrac{J_{av}/P_{av}}{W/R} & (\mathcal{E}_b/J_0 \geqslant 2) \\ 1 & (\mathcal{E}_b/J_0 < 2) \end{cases} \tag{13.3–9}$$

The corresponding error probability for the worst-case partial-band jammer is

$$P_2 = \frac{e^{-1}}{\mathcal{E}_b/J_0} = \left[e\left(\frac{W/R}{J_{av}/P_{av}}\right)\right]^{-1} \tag{13.3–10}$$

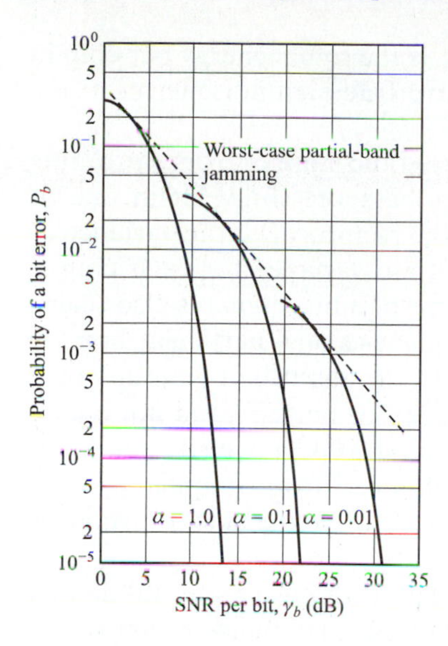

FIGURE 13.3–4
Performance of binary FSK with partial-band interference.

Whereas the error probability decreases exponentially for full-band jamming, we now find that the error probability decreases only inversely with $\mathcal{E}_b/J_0$ for the worst-case partial-band jamming. This result is similar to the error rate performance of binary FSK in a Rayleigh fading channel (see Section 14.3) and to the uncoded DS spread spectrum system corrupted by worst-case pulse jamming (see Section 13.2.3).

As we shall demonstrate below, signal diversity obtained by means of coding provides a significant improvement in performance relative to uncoded signals. This same approach to signal design is also effective for signaling over a fading channel, as we shall demonstrate in Chapter 14.

To illustrate the benefits of diversity in an FH spread spectrum signal with partial-band interference, we assume that the same information symbol is transmitted by binary FSK on L independent frequency hops. This may be accomplished by subdividing the signaling interval into L subintervals, as described previously for fast frequency hopping. After the hopping pattern is removed, the signal is demodulated by passing it through a pair of matched filters whose outputs are square-law-detected and sampled at the end of each subinterval. The square-law-detected signals corresponding to the L frequency hops are weighted and summed to form the two decision variables (metrics), which are denoted as U_1 and U_2.

When the decision variable U_1 contains the signal components, U_1 and U_2 may be expressed as

$$\begin{aligned} U_1 &= \sum_{k=1}^{L} \beta_k |2\mathcal{E}_c + N_{1k}|^2 \\ U_2 &= \sum_{k=1}^{L} \beta_k |N_{2k}|^2 \end{aligned} \tag{13.3–11}$$

where $\{\beta_k\}$ represent the weighting coefficients, $\mathcal{E}_c$ is the signal energy per chip in the L-chip symbol, and $\{N_{jk}\}$ represent the additive Gaussian noise terms at the output of the matched filters.

The coefficients are optimally selected to prevent the jammer from saturating the combiner should the transmitted frequencies be successfully hit in one or more hops. Ideally, β_k is selected to be equal to the reciprocal of the variance of the corresponding noise terms $\{N_k\}$. Thus, the noise variance for each chip is normalized to unity by this weighting and the corresponding signal is also scaled accordingly. This means that when the signal frequencies on a particular hop are jammed, the corresponding weight is very small. In the absence of jamming on a given hop, the weight is relatively large. In practice, for partial-band noise jamming, the weighting may be accomplished by use of an AGC having a gain that is set on the basis of noise power measurements obtained from frequency bands adjacent to the transmitted tones. This is equivalent to having side information (knowledge of jammer state) at the decoder.

Suppose that we have broadband Gaussian noise with power spectral density N_0 and partial-band interference, over αW of the frequency band, which is also Gaussian with power spectral density J_0/α. In the presence of partial-band interference, the second moments of the noise terms N_{1k} and N_{2k} are

$$\sigma_k^2 = \tfrac{1}{2}E(|N_{1k}|^2) = \tfrac{1}{2}E(|N_{2k}|^2) = 2\mathcal{E}_c\left(N_0 + \frac{J_0}{\alpha}\right) \tag{13.3–12}$$

In this case, we select $\beta_k = 1/\sigma_k^2 = [2\mathcal{E}_c(N_0 + J_0/\alpha)]^{-1}$. In the absence of partial-band interference, $\sigma_k^2 = 2\mathcal{E}_c N_0$ and, hence, $\beta_k = (2\mathcal{E}_c N_0)^{-1}$. Note that β_k is a random variable.

An error occurs in the demodulation if $U_2 > U_1$. Although it is possible to determine the exact error probability, we shall resort to the Chernoff bound, which yields a result that is much easier to evaluate and interpret. Specifically, the Chernoff (upper) bound on the error probability is

$$\begin{aligned} P_2 &= P(U_2 - U_1 > 0) \leqslant E\{\exp[\nu(U_2 - U_1)]\} \\ &= E\left\{\exp\left[-\nu \sum_{k=1}^{L} \beta_k(|2\mathcal{E}_c + N_{1k}|^2 - |N_{2k}|^2)\right]\right\} \end{aligned} \tag{13.3–13}$$

where ν is a variable that is optimized to yield the tightest possible bound.

The averaging in Equation 13.3–13 is performed with respect to the statistics of the noise components and the statistics of the weighting coefficients $\{\beta_k\}$, which are random as a consequence of the statistical nature of the interference. Keeping the $\{\beta_k\}$ fixed and averaging over the noise statistics first, we obtain

$$\begin{aligned}P_2(\boldsymbol{\beta}) &= E\left[\exp\left(-\nu\sum_{k=1}^{L}\beta_k|2\mathcal{E}_c + N_{1k}|^2 + \nu\sum_{k=1}^{L}\beta_k|N_{2k}|^2\right)\right] \\ &= \prod_{k=1}^{L} E[\exp(-\nu\beta_k|2\mathcal{E}_c + N_{1k}|^2)]E[\exp(\nu\beta_k|N_{2k}|^2)] \\ &= \prod_{k=1}^{L}\frac{1}{1-4\nu^2}\exp\left(\frac{-4\mathcal{E}_c^2\beta_k\nu}{1+2\nu}\right)\end{aligned} \tag{13.3–14}$$

Since the FSK tones are jammed with probability α, it follows that $\beta_k = [2\mathcal{E}(N_0 + J_0/\alpha)]^{-1}$ with probability α and $(2\mathcal{E}_c N_0)^{-1}$ with probability $1-\alpha$. Hence, the Chernoff bound is

$$\begin{aligned}P_2 &\leqslant \prod_{k=1}^{L}\left\{\frac{\alpha}{1-4\nu^2}\exp\left[\frac{-2\mathcal{E}_c\nu}{(N_0+J_0/\alpha)(1+2\nu)}\right] + \frac{1-\alpha}{1-4\nu^2}\exp\left[\frac{-2\mathcal{E}_c\nu}{N_0(1+2\nu)}\right]\right\} \\ &= \left\{\frac{\alpha}{1-4\nu}\exp\left[\frac{-2\mathcal{E}_c\nu}{(N_0+J_0/\alpha)(1+2\nu)} + \frac{1-\alpha}{1-4\nu^2}\exp\left[\frac{-2\mathcal{E}_c\nu}{N_0(1+2\nu)}\right]\right\}^L\end{aligned} \tag{13.3–15}$$

The next step is to optimize the bound in Equation 13.3–15 with respect to the variable ν. In its present form, however, the bound is messy to manipulate. A significant simplification occurs if we assume that $J_0/\alpha_0, \geqslant N_0$, which renders the second term in Equation 13.3–15 negligible compared with the first. Alternatively, we let $N_0 = 0$, so that the bound on P_2 reduces to

$$P_2 \leqslant \left\{\frac{\alpha}{1-4\nu^2}\exp\left[\frac{-2\alpha\nu\mathcal{E}_c}{J_0(1+2\nu)}\right]\right\}^L \tag{13.3–16}$$

The minimum value of this bound with respect to ν and the maximum with respect to α (worst-case partial-band interference) is easily shown to occur when $\alpha = 3J_0/\mathcal{E}_c \leqslant 1$ and $\nu = \frac{1}{4}$. For these values of the parameters, Equation 13.3–16 reduces to

$$P_2 \leqslant P_2(L) = \left(\frac{4}{e\gamma_c}\right)^L = \left(\frac{1.47}{\gamma_c}\right)^L, \qquad \gamma_c = \frac{\mathcal{E}_c}{J_0} = \frac{\mathcal{E}_b}{LJ_0} \geqslant 3 \tag{13.3–17}$$

where γ_c is the SNR per chip in the L-chip symbol. Equivalently,

$$P_2 \leqslant \left[\frac{1.47(J_{av}/P_{av})}{W/R}\right]^L, \qquad \frac{W/R}{L(J_{av}/P_{av})} \geqslant 3 \tag{13.3–18}$$

The result in Equation 13.3–17 was first derived by Viterbi and Jacobs (1975).

We observe that the probability of error for the worst-case partial-band interference decreases exponentially with an increase in the SNR per chip γ_c. This result is very similar to the performance characteristics of diversity techniques for Rayleigh fading channels (see Section 14.4). We may express the right-hand side of Equation 13.3–17 in the form

$$P_2(L) = \exp[-\gamma_b h(\gamma_c)] \tag{13.3–19}$$

where the function $h(\gamma_c)$ is defined as

$$h(\gamma_c) = -\frac{1}{\gamma_c}\left[\ln\left(\frac{4}{\gamma_c}\right) - 1\right] \tag{13.3–20}$$

A plot of $h(\gamma_c)$ is given in Figure 13.3–5. We observe that the function has a maximum value of $\frac{1}{4}$ at $\gamma_c = 4$. Consequently, there is an optimum SNR per chip of $10 \log \gamma_c = 6$ dB. At the optimum SNR, the error rate is upper-bounded as

$$P_2 \leqslant P_2(L_{\text{opt}}) = e^{-\gamma_b/4} \tag{13.3–21}$$

When we compare the error probability bound in Equation 13.3–21 with the error probability for binary FSK in spectrally flat noise, which is given by Equation 13.3–1, we see that the combined effect of worst-case partial-band interference and the noncoherent combining loss in the square-law combining of the L chips is 3 dB. We emphasize, however, that for a given $\mathcal{E}_b/J_0$, the loss is greater when the order of diversity is not optimally selected.

Coding provides a means for improving the performance of the FH system corrupted by partial-band interference. In particular, if a block orthogonal code is used, with $M = 2^k$ code words and Lth-order diversity per code word, the probability of a code word error is upper-bounded as

$$P_M \leqslant (2^k - 1)P_2(L) = (2^k - 1)\left(\frac{1.47}{\gamma_c}\right)^L = (2^k - 1)\left(\frac{1.47}{k\gamma_b/L}\right)^L \tag{13.3–22}$$

and the equivalent bit error probability is upper-bounded as

$$P_b \leqslant 2^{k-1}\left(\frac{1.47}{k\gamma_b/L}\right)^L \tag{13.3–23}$$

Figure 13.3–6 illustrates the probability of a bit error for $L = 1, 2, 4, 8$ and $k = 1, 3$. With an optimum choice of diversity, the upper bound can be expressed as

$$P_b \leqslant 2^{k-1} \exp(-\tfrac{1}{4}k\gamma_b) = \tfrac{1}{2}\exp[-k(\tfrac{1}{4}\gamma_b - \ln 2)] \tag{13.3–24}$$

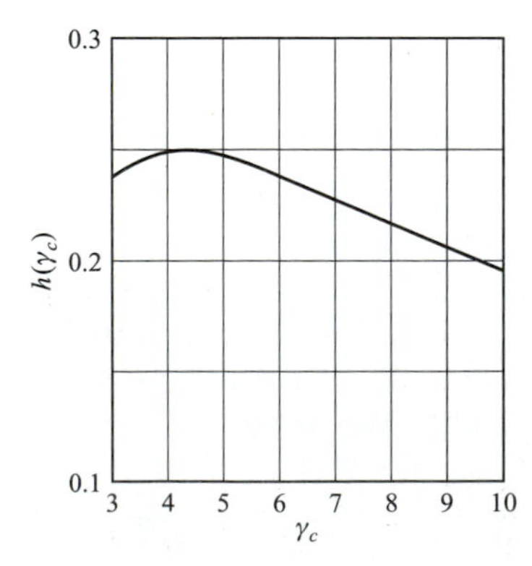

FIGURE 13.3–5
Graph of the function $h(\gamma_c)$.

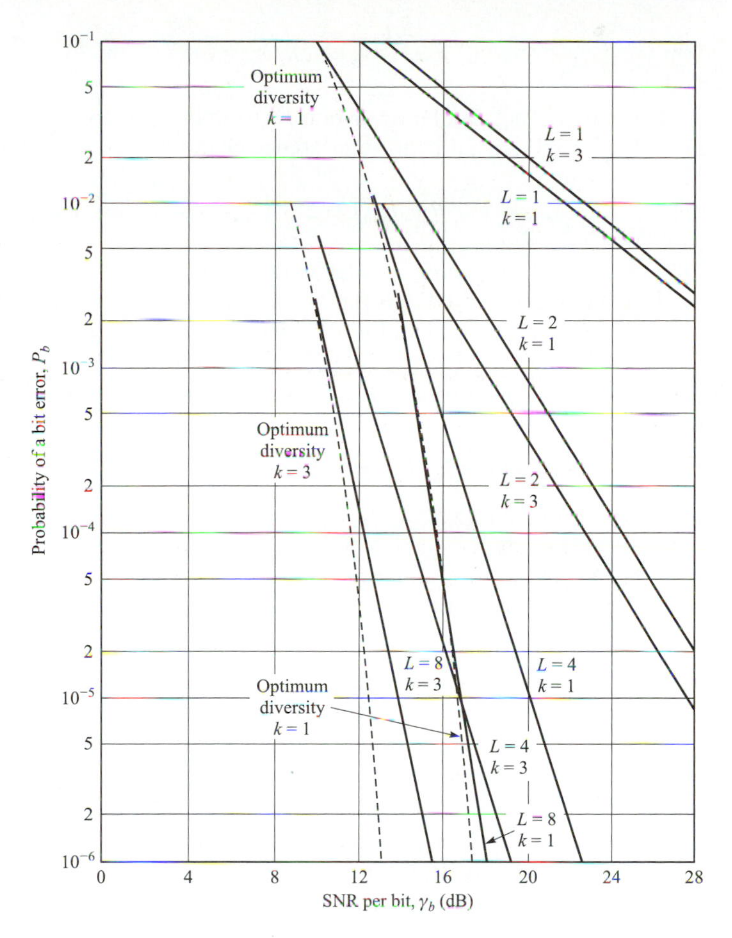

FIGURE 13.3–6
Performance of binary and octal FSK with L-order diversity for a channel with worst-case partial-band interference.

Thus, we have an improvement in performance by an amount equal to $10\log[k(1-2.77/\gamma_b)]$. For example, if $\gamma_b = 10$ and $k = 3$ (octal modulation), then the gain is 3.4 dB, while if $k = 5$, then the gain is 5.6 dB.

Additional gains can be achieved by employing concatenated codes in conjunction with soft-decision decoding. In the example below, we employ a dual-k convolutional code as the outer code and a Hadamard code as the inner code on the channel with partial-band interference.

EXAMPLE 13.3–1. Suppose we use a Hadamard $H(n, k)$ constant weight code with on–off keying (OOK) modulation for each code bit. The minimum distance of the code is $d_{\min} = \frac{1}{2}n$, and, hence, the effective order of diversity obtained with OOK modulation is $\frac{1}{2}d_{\min} = \frac{1}{4}n$. There are $\frac{1}{2}n$ FH tones transmitted per code word. Hence,

$$\gamma_c = \frac{k}{\frac{1}{2}n}\gamma_b = 2R_c\gamma_b \tag{13.3–25}$$

when this code is used alone. The bit error rate performance for soft-decision decoding of these codes for the partial-band interference channel is upper-bounded as

$$P_b \leqslant 2^{k-1}P_2(\tfrac{1}{2}d_{\min}) = 2^{k-1}\left(\frac{1.47}{2R_c\gamma_b}\right)^{n/4} \tag{13.3–26}$$

Now, if a Hadamard (n, k) code is used as the inner code and a rate 1/2 dual-k convolutional code (see Section 8.2.9) is the outer code, the bit error performance in the presence of worst-case partial-band interference is (see Equation 8.2–56)

$$P_b \leqslant \frac{2^{k-1}}{2^k-1}\sum_{m=4}^{\infty}\beta_m P_2(\tfrac{1}{2}md_{\min}) = \frac{2^{k-1}}{2^k-1}\sum_{m=4}^{\infty}\beta_m P_2(\tfrac{1}{2}mn) \tag{13.3–27}$$

where $P_2(L)$ is given by Equation 13.3–17 with

$$\gamma_c = \frac{k}{n}\gamma_b = R_c\gamma_b \tag{13.3–28}$$

Figure 13.3–7 illustrates the performance of the dual-k codes for $k = 5, 4$, and 3 concatenated with the Hadamard $H(20, 5)$, $H(16, 4)$, and $H(12, 3)$ codes, respectively.

In the above discussion, we have focused on soft-decision decoding. On the other hand, the performance achieved with hard-decision decoding is significantly (several decibels) poorer than that obtained with soft-decision decoding. In a concatenated coding scheme, however, a mixture involving soft-decision decoding of the inner code and hard-decision decoding of the outer code represents a reasonable compromise between decoding complexity and performance.

Finally, we wish to indicate that another serious threat in an FH spread spectrum system is partial-band multitone jamming. This type of interference

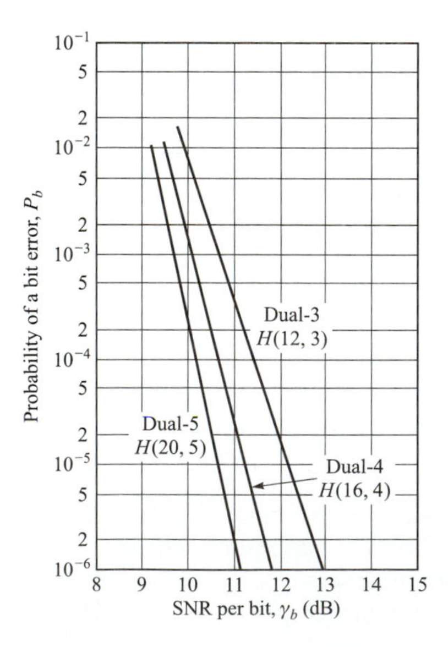

FIGURE 13.3–7
Performance of dual-k codes concatenated with Hadamard codes for a channel with worst-case partial-band interference.

is similar in effect to partial-band spectrally flat noise jamming. Diversity obtained through coding is an effective means for improving the performance of the FH system. An additional improvement is achieved by properly weighting the demodulator outputs so as to suppress the effects of the jammer.

13.3.3 A CDMA System Based on FH Spread Spectrum Signals

In Section 13.2.2, we considered a CDMA system based on the use of DS spread spectrum signals. As previously indicated, it is also possible to have a CDMA system based on FH spread spectrum signals. Each transmitter–receiver pair in such a system is assigned its own pseudorandom FH pattern. Aside from this distinguishing feature, the transmitters and receivers of all the users may be identical in that they may have identical encoders, decoders, modulators, and demodulators.

CDMA systems based on FH spread spectrum signals are particularly attractive for mobile (land, air, sea) users because timing requirements are not as stringent as in a DS spread spectrum signal. In addition, frequency synthesis techniques and associated hardware have been developed that make it possible to frequency-hop over bandwidths that are significantly larger than those currently possible with DS spread spectrum systems. Consequently, larger processing gains are possible with FH. The capacity of CDMA with FH is also relatively high. Viterbi (1978) has shown that with dual-k codes and M-ary FSK modulation, it is possible to accommodate up to $\frac{3}{8}W/R$ simultaneous users who transmit at an information rate R bits/s over a channel with bandwidth W.

One of the earliest CDMA systems based on FH coded spread spectrum signals was built to provide multiple-access tactical satellite communications for small mobile (land, sea, air) terminals each of which transmitted relatively short messages over the channel intermittently. The system was called the *Tactical Transmission System* (TATS), and it is described in a paper by Drouilhet and Bernstein (1969).

An octal Reed–Solomon (7, 2) code is used in the TATS system. Thus, two 3-bit information symbols from the input to the encoder are used to generate a seven-symbol code word. Each 3-bit coded symbol is transmitted by means of octal FSK modulation. The eight possible frequencies are spaced $1/T_c$ Hz apart, where T_c is the time (chip) duration of a single frequency transmission. In addition to the seven symbols in a code word, an eighth symbol is included. That symbol and its corresponding frequency are fixed and transmitted at the beginning of each code word for the purpose of providing timing and frequency synchronization† at the receiver. Consequently, each code word is transmitted in $8T_c$ seconds.

---

†Since mobile users are involved, there is a Doppler frequency offset associated with transmission. This frequency offset must be tracked and compensated for in the demodulation of the signal. The sync symbol is used for this purpose.

TATS was designed to transmit at information rates of 75 and 2400 bits/s. Hence, $T_c = 10$ ms and 312.5 μs, respectively. Each frequency tone corresponding to a code symbol is frequency-hopped. Hence, the hopping rate is 100 hops/s at the 75-bits/s rate and 3200 hops/s at the 2400-bits/s rate.

There are $M = 2^6 = 64$ code words in the Reed–Solomon (7, 2) code and the minimum distance of the code is $d_{\min} = 6$. This means that the code provides an effective order of diversity equal to 6.

At the receiver, the received signal is first dehopped and then demodulated by passing it through a parallel bank of eight matched filters, where each filter is tuned to one of the eight possible frequencies. Each filter output is envelope-detected, quantized to 4 bits (one of 16 levels), and fed to the decoder. The decoder takes the 56 filter outputs corresponding to the reception of each seven-symbol code word and forms 64 decision variables corresponding to the 64 possible code words in the (7, 2) code by linearly combining the appropriate envelope-detected outputs. A decision is made in favor of the code word having the largest decision variable.

By limiting the matched filter outputs to 16 levels, interference (crosstalk) from other users of the channel causes a relatively small loss in performance (0.75 dB with strong interference on one chip and 1.5 dB with strong interference on two chips out of the seven). The AGC used in TATS has a time constant greater than the chip interval T_c, so that no attempt is made to perform optimum weighting of the demodulator outputs as described in Section 13.3.2.

The derivation of the error probability for the TATS signal in AWGN and worst-case partial-band interference is left as an exercise for the reader (Problems 13.23 and 13.24).

13.4 OTHER TYPES OF SPREAD SPECTRUM SIGNALS

DS and FH are the most common forms of spread spectrum signals used in practice. However, other methods may be used to introduce pseudorandomness in a spread spectrum signal. One method, which is analogous to FH, is *time hopping* (TH). In TH, a time interval, which is selected to be much larger than the reciprocal of the information rate, is subdivided into a large number of time slots. The coded information symbols are transmitted in a pseudorandomly selected time slot as a block of one or more code words. PSK modulation may be used to transmit the coded bits.

For example, suppose that a time interval T is subdivided into 1000 time slots of width $T/1000$ each. With an information bit rate of R bits/s, the number of bits to be transmitted in T seconds is RT. Coding increases this number to RT/R_c bits, where R_c is the code rate. Consequently, in a time interval of $T/1000$ s, we must transmit RT/R_c bits. If binary PSK is used as the modulation method, the bit rate is $1000R/R_c$ and the bandwidth required is approximately $W = 1000R/R_c$.

A block diagram of a transmitter and a receiver for a TH spread spectrum system is shown in Figure 13.4–1. Because of the burst characteristics of the transmitted signal, buffer storage must be provided at the transmitter in a TH system, as shown in Figure 13.4–1. A buffer may also be used at the receiver to provide a uniform data stream to the user.

Just as partial-band interference degrades an uncoded FH spread spectrum system, partial-time (pulsed) interference has a similar effect on a TH spread spectrum system. Coding and interleaving are effective means for combating this type of interference, as we have already demonstrated for FH and DS systems. Perhaps the major disadvantage of a TH system is the stringent timing requirements compared not only with FH but, also, with DS.

Other types of spread spectrum signals can be obtained by combining DS, FH, and TH. For example, we may have a hybrid DS/FH, which means that a PN sequence is used in combination with frequency hopping. The signal transmitted on a single hop consists of a DS spread spectrum signal which is demodulated coherently. However, the received signals from different hops are combined noncoherently (envelope or square-law combining). Since coherent detection is performed within a hop, there is an advantage obtained relative to a pure FH system. However, the price paid for the gain in performance is an increase in complexity, greater cost, and more stringent timing requirements.

Another possible hybrid spread spectrum signal is DS/TH. This does not seem to be as practical as DS/FH, primarily because of an increase in system complexity and more stringent timing requirements.

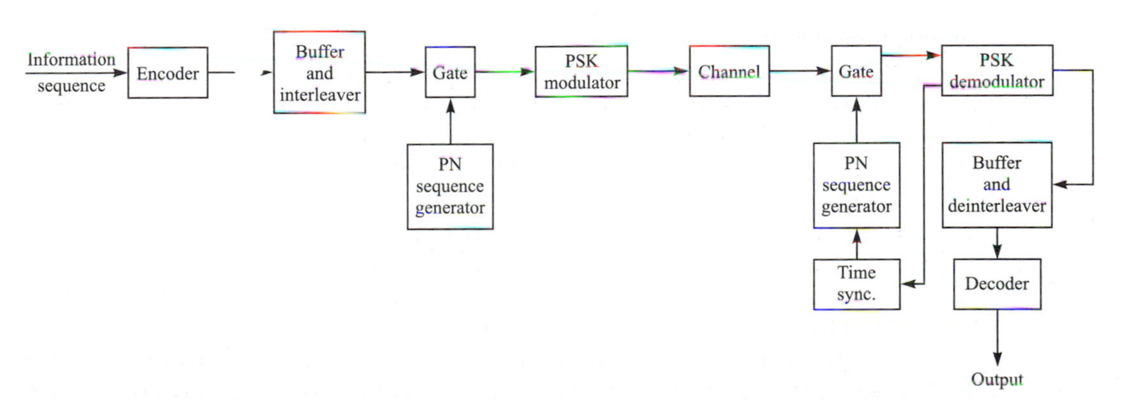

FIGURE 13.4–1
Block diagram of time-hopping (TH) spread spectrum system.

13.5 SYNCHRONIZATION OF SPREAD SPECTRUM SYSTEMS

Time synchronization of the receiver to the received spread spectrum signal may be separated into two phases. There is an initial acquisition phase and a tracking phase after the signal has been initially acquired.

Acquisition. In a direct sequence spread spectrum system, the PN code must be time-synchronized to within a small fraction of the chip interval $T_c \approx 1/W$. The problem of initial synchronization may be viewed as one in which we attempt to synchronize in time the receiver clock to the transmitter clock. Usually, extremely accurate and stable time clocks are used in spread spectrum systems. Consequently, accurate time clocks result in a reduction of the time uncertainty between the receiver and the transmitter. However, there is always an initial timing uncertainty due to range uncertainty between the transmitter and the receiver. This is especially a problem when communication is taking place between two mobile users. In any case, the usual procedure for establishing initial synchronization is for the transmitter to send a known pseudorandom data sequence to the receiver. The receiver is continuously in a search mode looking for this sequence in order to establish initial synchronization.

Let us suppose that the initial timing uncertainty is T_u and the chip duration is T_c. If initial synchronization is to take place in the presence of additive noise and other interference, it is necessary to dwell for $T_d = NT_c$ in order to test synchronism at each time instant. If we search over the time uncertainty interval in (coarse) time steps of $\frac{1}{2}T_c$, then the time required to establish initial synchronization is

$$T_{\text{init sync}} = \frac{T_u}{\frac{1}{2}T_c} NT_c = 2NT_u \tag{13.5–1}$$

Clearly, the synchronization sequence transmitted to the receiver must be at least as long as $2NT_u$ in order for the receiver to have sufficient time to perform the necessary search in a serial fashion.

In principle, matched filtering or cross correlation are optimum methods for establishing initial synchronization. A filter matched to the known data waveform generated from the known pseudorandom sequence continuously looks for exceedence of a predetermined threshold. When this occurs, initial synchronization is established and the demodulator enters the "data receive" mode.

Alternatively, we may use a *sliding correlator* as shown in Figure 13.5–1. The correlator cycles through the time uncertainty, usually in discrete time intervals of $\frac{1}{2}T_c$, and correlates the received signal with the known synchronization sequence. The cross correlation is performed over the time interval NT_c (N chips) and the correlator output is compared with a threshold to determine if the known signal sequence is present. If the threshold is not exceeded, the known reference sequence is advanced in time by $\frac{1}{2}T_c$ seconds and the correlation process is repeated. These operations are performed until a signal is detected or until the

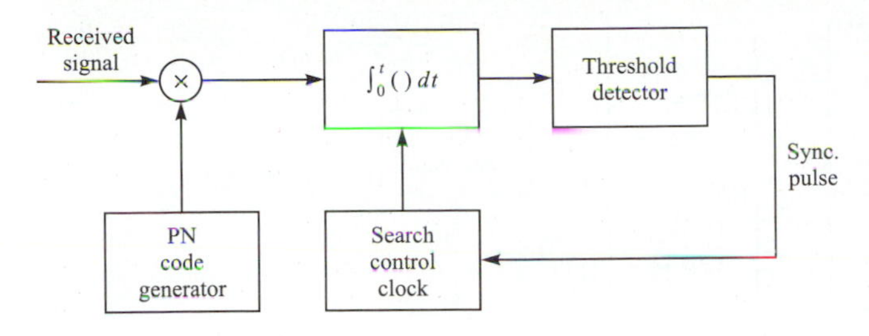

FIGURE 13.5–1
A sliding correlator for DS signal acquisition.

search has been performed over the time uncertainty interval T_u. In the latter case, the search process is then repeated.

A similar process may also be used for FH signals. In this case, the problem is to synchronize the PN code that controls the hopped frequency pattern. To accomplish this initial synchronization, a known FH signal is transmitted to the receiver. The initial acquisition system at the receiver looks for this known FH signal pattern. For example, a bank of matched filters tuned to the transmitted frequencies in the known pattern may be employed. Their outputs must be properly delayed, envelope- or square-law-detected, weighted, if necessary, and added (noncoherent integration) to produce the signal output which is compared with a threshold. A signal present is declared when the threshold is exceeded. The search process is usually performed continuously in time until a threshold is exceeded. A block diagram illustrating this signal acquisition scheme is given in Figure 13.5–2. As an alternative, a single matched-filter–envelope detector pair may be used, preceded by an FH pattern generator and followed by a postdetection integrator and a threshold detector. This configuration, shown in Figure 13.5–3, is based on a serial search and is akin to the sliding correlator for DS spread spectrum signals.

The sliding correlator for the DS signals or its counterpart shown in Figure 13.5–3 for FH signals basically perform a serial search that is generally time-consuming. As an alternative, one may introduce some degree of parallelism by having two or more such correlators operating in parallel and searching over non-overlapping time slots. In such a case, the search time is reduced at the expense of a more complex and costly implementation.

During the search mode, there may be false alarms that occur at the designed false alarm rate of the system. To handle the occasional false alarms, it is necessary to have an additional method or circuit that checks to confirm that the received signal at the output of the correlator remains above the threshold. With such a detection strategy, a large noise pulse that causes a false alarm will cause only a temporary exceedence of the threshold. On the other hand, when a signal is present, the correlator or matched filter output will stay above the threshold for the duration of the transmitted signal. Thus, if confirmation fails, the search is resumed.

Another initial search strategy, called a *sequential search*, has been investigated by Ward (1965) and Ward and Yiu (1977). In this method, the dwell time at each delay in the search process is made variable by employing a correlator

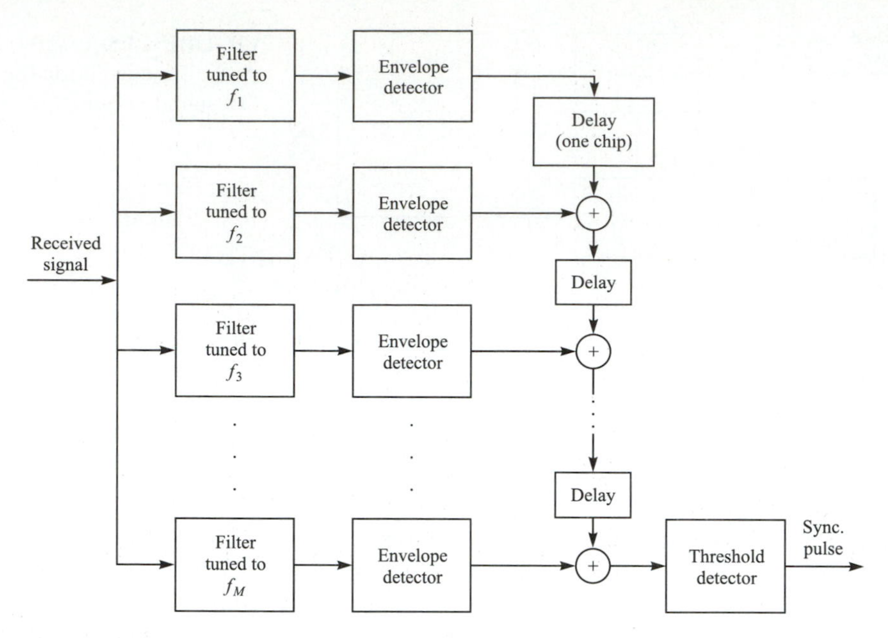

FIGURE 13.5–2
System for acquisition of an FH signal.

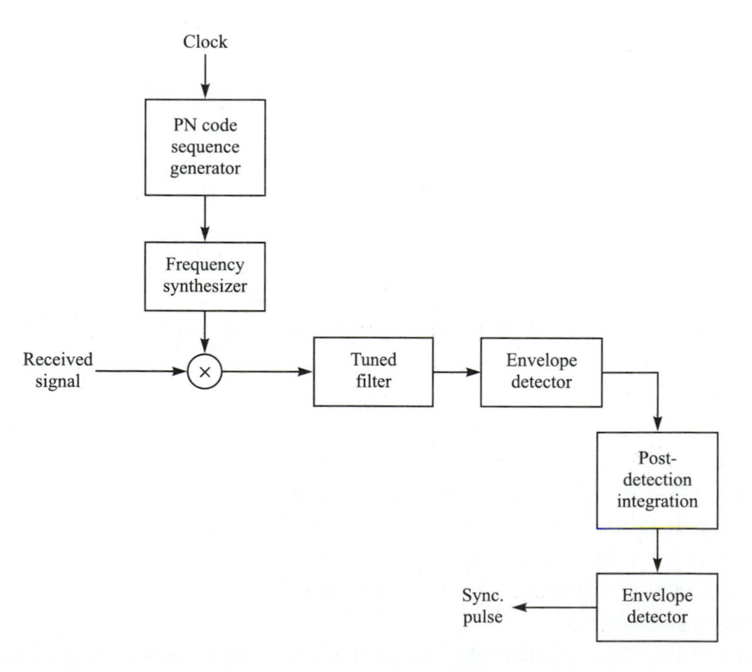

FIGURE 13.5–3
Alternative system for acquisition of an FH signal.

with a variable integration period whose (biased) output is compared with two thresholds. Thus, there are three possible decisions:

1. If the upper threshold is exceeded by the correlator output, initial synchronization is declared established.
2. If the correlator output falls below the lower threshold, the signal is declared absent at that delay and the search process resumes at a different delay.
3. If the correlator output falls between the two thresholds, the integration time is increased by one chip and the resulting output is compared with the two thresholds again.

Hence, steps 1, 2, and 3 are repeated for each chip interval until the correlator output either exceeds the upper threshold or falls below the lower threshold.

The sequential search method falls in the class of sequential estimation methods proposed by Wald (1947), which are known to result in a more efficient search in the sense that the average search time is minimized. Hence, the search time for a sequential search is less than that for the fixed dwell time integrator.

In the above discussion, we have considered only time uncertainty in establishing initial synchronization. However, another aspect of initial synchronization is frequency uncertainty. If the transmitter and/or the receiver are mobile, the relative velocity between them results in a Doppler frequency shift in the received signal relative to the transmitted signal. Since the receiver does not usually know the relative velocity, a priori, the Doppler frequency shift is unknown and must be determined by means of a frequency search method. Such a search is usually accomplished in parallel over a suitably quantized frequency uncertainty interval and serially over the time uncertainty interval. A block diagram of this scheme is shown in Figure 13.5–4. Appropriate Doppler frequency search methods can also be devised for FH signals.

Tracking. Once the signal is acquired, the initial search process is stopped and fine synchronization and tracking begins. The tracking maintains the PN code generator at the receiver in synchronism with the incoming signal. Tracking includes both fine chip synchronization and, for coherent demodulation, carrier phase tracking.

The commonly used tracking loop for a DS spread spectrum signal is the delay-locked loop (DLL) which is shown in Figure 13.5–5. In this tracking loop, the received signal is applied to two multipliers, where it is multiplied by two outputs from the local PN code generator, which are delayed relative to each other by an amount $2\delta \leqslant T_c$. Thus, the product signals are the cross correlations between the received signal and the PN sequence at the two values of delay. These products are band-pass-filtered and envelope- (or square-law-) detected and then subtracted. This difference signal is applied to the loop filter that drives the voltage-controlled clock (VCC). The VCC serves as the clock for the PN code signal generator.

If the synchronism is not exact, the filtered output from one correlator will exceed the other and the VCC will be appropriately advanced or delayed. At the equilibrium point, the two filtered correlator outputs will be equally displaced

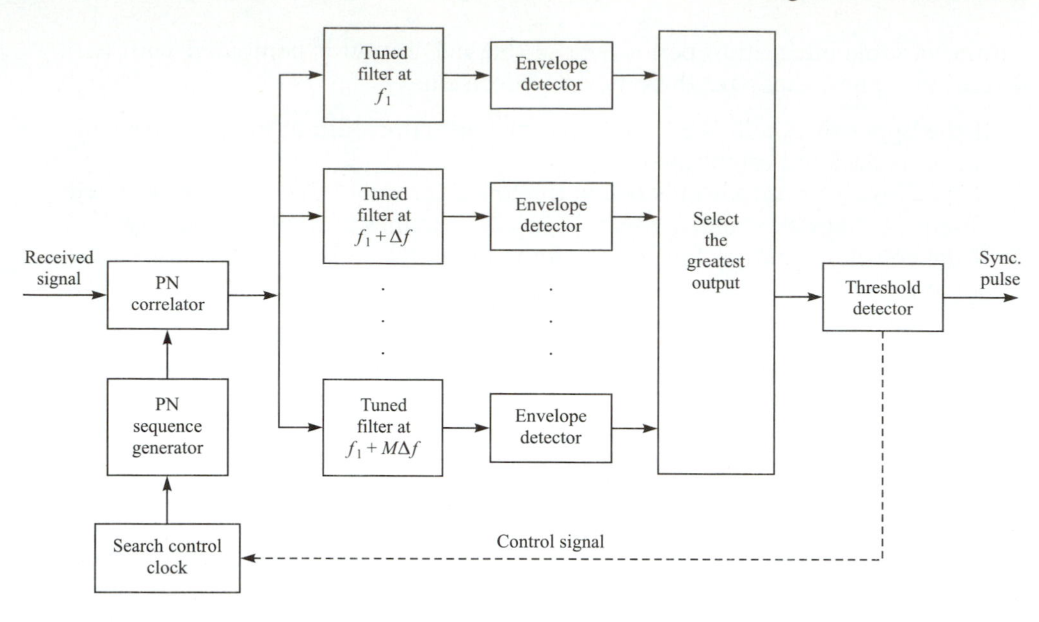

FIGURE 13.5–4
Initial search for Doppler frequency offset in a DS system.

from the peak value, and the PN code generator output will be exactly synchronized to the received signal that is fed to the demodulator. We observe that this implementation of the DLL for tracking a DS signal is equivalent to the early–late gate bit tracking synchronizer previously discussed in Section 6.3.2 and shown in Figure 6.3–5.

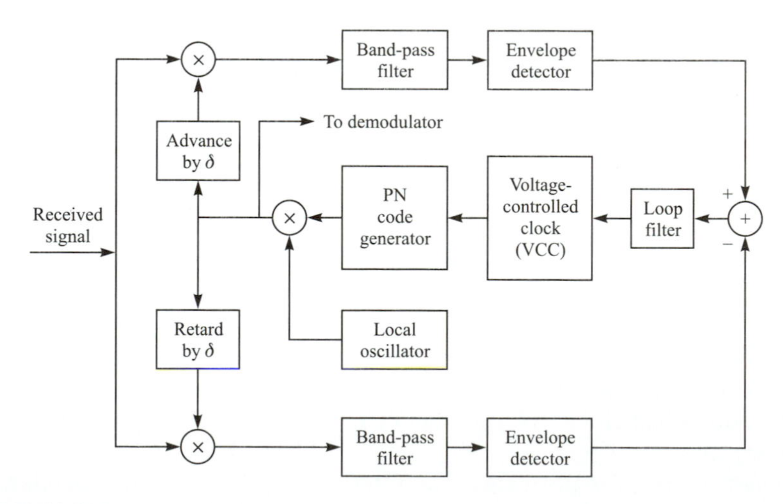

FIGURE 13.5–5
Delay-locked loop (DLL) for PN code tracking.

An alternative method for time tracking a DS signal is to use a *tau-dither loop* (TDL), illustrated by the block diagram in Figure 13.5–6. The TDL employs a single "arm" instead of the two "arms" shown in Figure 13.5–5. By providing a suitable gating waveform, it is possible to make this "single-arm" implementation appear to be equivalent to the "two-arm" realization. In this case, the cross correlation is regularly sampled at two values of delay, by stepping the code clock forward or backward in time by an amount δ. The envelope of the cross correlation that is sampled at $\pm\delta$ has an amplitude modulation whose phase relative to the tau-dither modulator determines the sign of the tracking error.

A major advantage of the TDL is the less costly implementation resulting from elimination of one of the two arms that are employed in the conventional DLL. A second and less apparent advantage is that the TDL does not suffer from performance degradation that is inherent in the DLL when the amplitude gain in the two arms is not properly balanced.

The DLL (and its equivalent, the TDL) generate an error signal by sampling the signal correlation function at $\pm\delta$ off the peak as shown in Figure 13.5–7a. This generates an error signal as shown in Figure 13.5–7b. The analysis of the performance of the DLL is similar to that for the phase-locked loop (PLL) carried out in Section 6.2. If it were not for the envelope detectors in the two arms of the DLL, the loop would resemble a Costas loop. In general, the variance of the time estimation error in the DLL is inversely proportional to the loop SNR, which depends on the input SNR to the loop and the loop bandwidth. Its performance is somewhat degraded as in the squaring PLL by non-linearities inherent in the envelope detectors, but this degradation is relatively small.

A typical tracking technique for FH spread spectrum signals is illustrated in Figure 13.5–8a. This method is also based on the premise that, although initial acquisition has been achieved, there is a small timing error between the received signal and the receiver clock. The band-pass filter is tuned to a single intermediate frequency and its bandwidth is of the order of $1/T_c$, where T_c is the chip interval. Its output is envelope-detected and then multiplied by the clock signal to produce a three-level signal, as shown in Figure 13.5–8b, which drives the loop

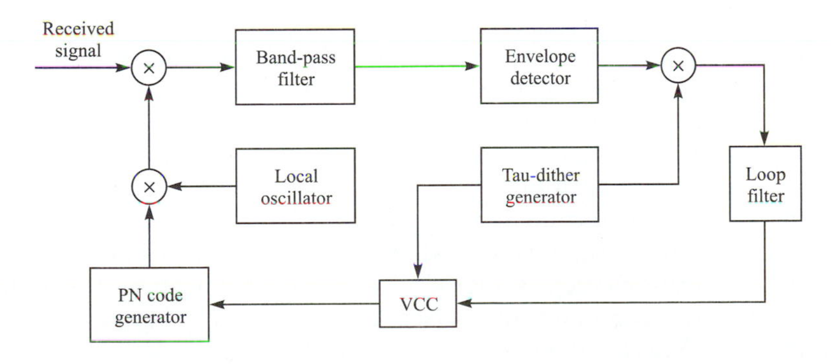

FIGURE 13.5–6
Tau-dither loop (TDL).

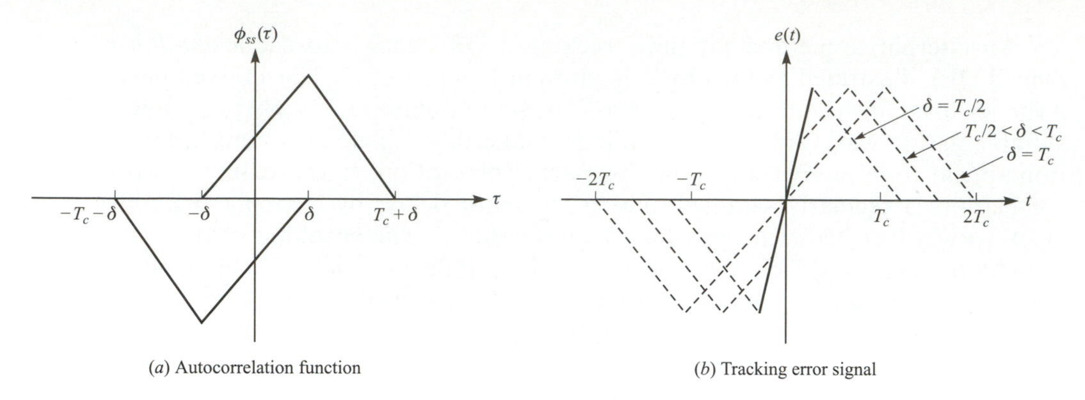

FIGURE 13.5–7
Autocorrelation function and tracking error signal for DLL.

filter. Note that when the chip transitions from the locally generated sinusoidal waveform do not occur at the same time as the transitions in the incoming signal, the output of the loop filter will be either negative or positive, depending on whether the VCC is lagging or advanced relative to the timing of the input signal. This error signal from the loop filter will provide the control signal for adjusting the VCC timing signal so as to drive the frequency synthesized pulsed sinusoid to proper synchronism with the received signal.

13.6
BIBLIOGRAPHICAL NOTES AND REFERENCES

The introductory treatment of spread spectrum signals and their performance that we have given in this chapter is necessarily brief. Detailed and more specialized treatments of signal acquisition techniques, code tracking methods, and hybrid spread spectrum systems, as well as other general topics on spread spectrum signals and systems, can be found in the vast body of technical literature that now exists on the subject.

Historically, the primary application of spread spectrum communications has been in the development of secure (AJ) digital communication systems for military use. In fact, prior to 1970, most of the work on the design and development of spread spectrum communications was classified. Since then, this trend has been reversed. The open literature now contains numerous publications on all aspects of spread spectrum signal analysis and design. Moreover, we have recently seen the application of spread spectrum signaling techniques to commercial communications such as interoffice radio communications (see Pahlavan, 1985), mobile radio communications (see Yue, 1983), and digital cellular communications (see Viterbi, 1995).

A historical perspective on the development of spread spectrum communication systems covering the period 1920–1960 is given in a paper by Scholtz (1982).

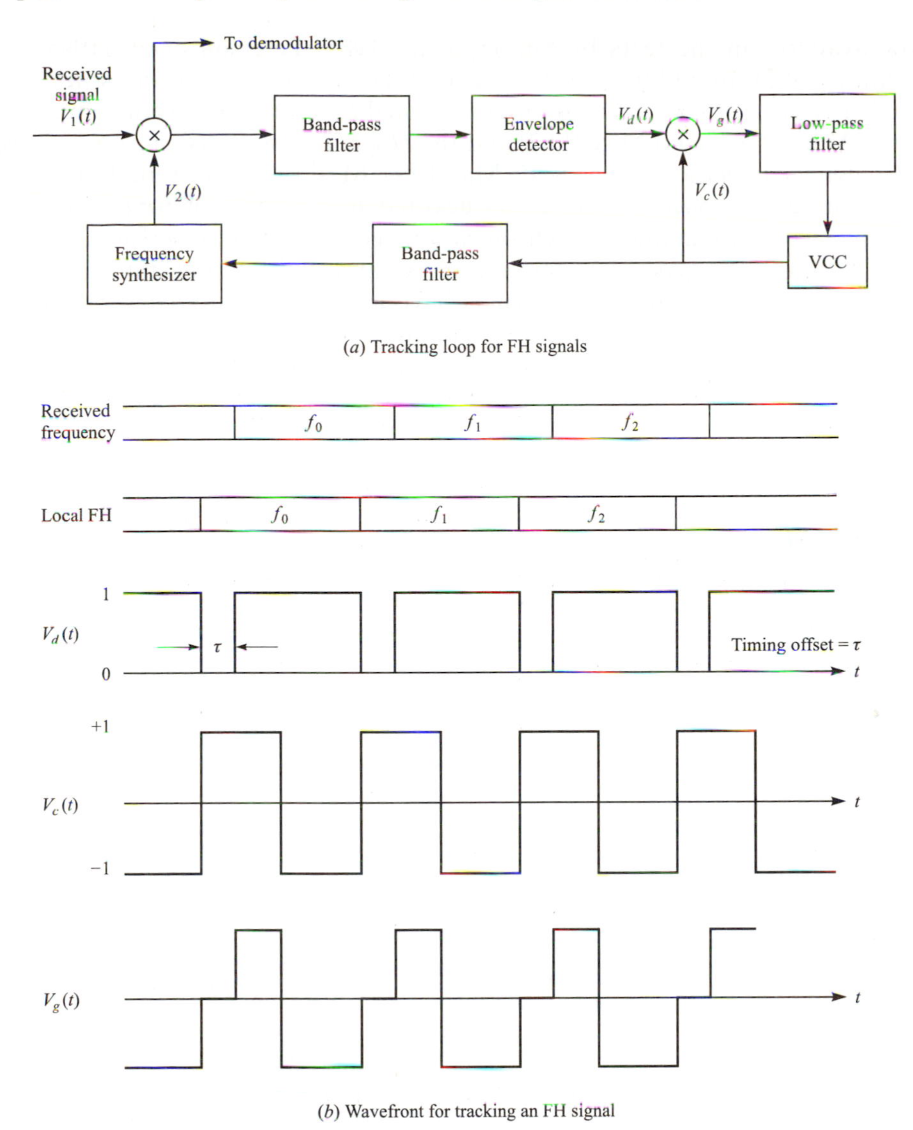

FIGURE 13.5–8
Tracking method for FH signals. [*From Pickholtz et al. (1982). © 1982 IEEE.*]

Tutorial treatments focusing on the basic concepts are found in the papers by Scholtz (1977) and Pickholtz et al. (1982). These papers also contain a large number of references to previous work. In addition, there are two papers by Viterbi (1979, 1985) that provide a basic review of the performance characteristics of DS and FH signaling techniques.

Comprehensive treatments of various aspects of analysis and design of spread spectrum signals and systems, including synchronization techniques are

13.12 Consider the DS spread spectrum signal

$$c(t) = \sum_{n=-\infty}^{\infty} c_n p(t - nT_c)$$

where c_n is a periodic m sequence with a period $N = 127$ and $p(t)$ is a rectangular pulse of duration $T_c = 1\ \mu s$. Determine the power spectral density of the signal $c(t)$.

13.13 Suppose that $\{c_{1i}\}$ and $\{c_{2i}\}$ are two binary (0, 1) periodic sequences with periods N_1 and N_2, respectively. Determine the period of the sequence obtained by forming the modulo-2 sum of $\{c_{1i}\}$ and $\{c_{2i}\}$.

13.14 An $m = 10$ ML shift register is used to generate the pseudorandom sequence in a DS spread spectrum system. The chip duration is $T_c = 1\ \mu s$, and the bit duration is $T_b = NT_c$, where N is the length (period) of the m sequence.

a) Determine the processing gain of the system in dB.

b) Determine the jamming margin if the required $\mathcal{E}_b/J_0 = 10$ and the jammer is a tone jammer with an average power J_{av}.

13.15 An FH binary orthogonal FSK system employs an $m = 15$ stage linear feedback shift register that generates an ML sequence. Each state of the shift register selects one of L non-overlapping frequency bands in the hopping pattern. The bit rate is 100 bits/s and the hop rate is one per bit. The demodulator employs noncoherent detection.

a) Determine the hopping bandwidth for this channel.

b) What is the processing gain?

c) What is the probability of error in the presence of AWGN?

13.16 Consider the FH binary orthogonal FSK system described in Problem 13.15. Suppose that the hop rate is increased to 2 hops/bit. The receiver uses square-law combining to combine the signal over the two hops.

a) Determine the hopping bandwidth for the channel.

b) What is the processing gain?

c) What is the error probability in the presence of AWGN?

13.17 In a fast FH spread spectrum system, the information is transmitted via FSK, with noncoherent detection. Suppose there are $N = 3$ hops/bit, with hard-decision decoding of the signal in each hop.

a) Determine the probability of error for this system in an AWGN channel with power spectral density $\frac{1}{2}N_0$ and an SNR $=$ 13 dB (total SNR over the three hops).

b) Compare the result in (a) with the error probability of an FH spread spectrum system that hops once per bit.

13.18 A slow FH binary FSK system with noncoherent detection operates at $\mathcal{E}_b/J_0 = 10$, with a hopping bandwidth of 2 GHz, and a bit rate of 10 kbits/s.

a) What is the processing gain for the system?

b) If the jammer operates as a partial-band jammer, what is the bandwidth occupancy for worst-case jamming?

c) What is the probability of error for the worst-case partial-band jammer?

13.19 Determine the error probability for an FH spread spectrum signal in which a binary convolutional code is used in combination with binary FSK. The interference on the channel is AWGN. The FSK demodulator outputs are square-law-detected and passed to the decoder, which performs optimum soft-decision Viterbi decoding as described in Section 8.2. Assume that the hopping rate is 1 hop per coded bit.

13.20 Repeat Problem 13.19 for hard-decision Viterbi decoding.

13.21 Repeat Problem 13.19 when fast frequency hopping is performed at a hopping rate of L hops per coded bit.

13.22 Repeat Problemm 13.19 when fast frequency hopping is performed with L hops per coded bit and the decoder is a hard-decision Viterbi decoder. The L chips per coded bit are square-law-detected and combined prior to the hard decision.

13.23 The TATS signal described in Section 13.3.3 is demodulated by a parallel bank of eight matched filters (octal FSK), and each filter output is square-law-detected. The eight outputs obtained in each of seven signal intervals (56 total outputs) are used to form the 64 possible decision variables corresponding to the Reed–Solomon (7, 2) code. Determine an upper (union) bound of the code word error probability for AWGN and soft-decision decoding.

13.24 Repeat Problem 13.23 for the worst-case partial-band interference channel.

13.25 Derive the results in Equations 13.2–62 and 13.2–63 from Equation 13.2–61.

13.26 Show that Equation 13.3–14 follows from Equation 13.3–13.

13.27 Derive Equation 13.3–17 from Equation 13.3–16.

13.28 The parity polynomials for constructing Gold code sequences of length $n = 7$ are

$$h_1(p) = p^3 + p + 1$$
$$h_2(p) = p^3 + p^2 + 1$$

Generate all the Gold codes of length 7 and determine the cross correlations of one sequence with each of the others.

13.29 In Section 13.2.3, we demonstrated techniques for evaluating the error probability of a coded system with interleaving in pulse interference by using the cutoff rate parameter R_0. Use the error probability curves given in Figure P13.29 for rate 1/2 and 1/3 convolutional codes with soft-decision Viterbi decoding to determine the corresponding error rates for a coded system in pulse interference. Perform this computation for $K = 3, 5,$ and 7.

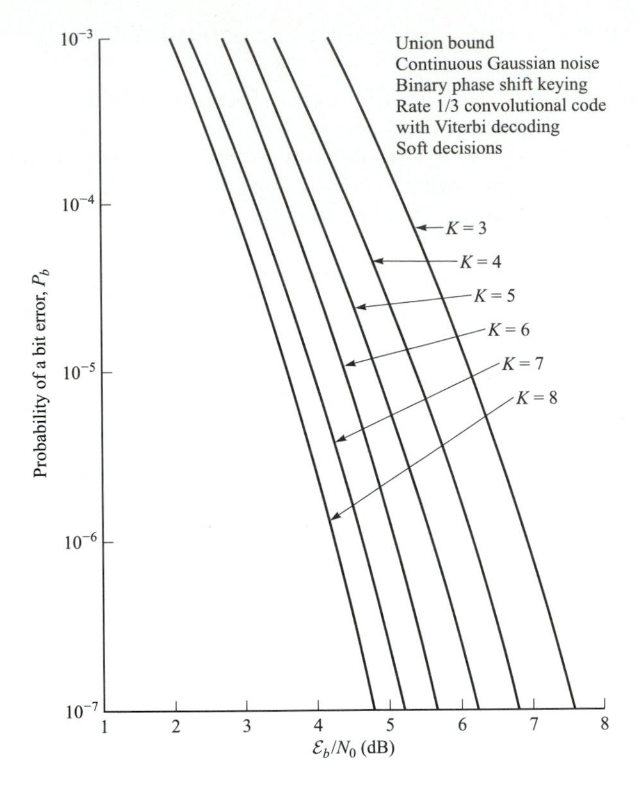

FIGURE P13.29

13.30 In coded and interleaved DS binary PSK modulation with pulse jamming and soft-decision decoding, the cutoff rate is

$$R_0 = 1 - \log_2(1 + \alpha e^{-\alpha\mathcal{E}_c/N_0})$$

where α is the fraction of the time the system is being jammed, $\mathcal{E}_c = \mathcal{E}_b R$, R is the bit rate, and $N_0 \equiv J_0$.

a) Show that the SNR per bit, $\mathcal{E}_b/N_0$, can be expressed as

$$\frac{\mathcal{E}_b}{N_0} = \frac{1}{\alpha R} \ln \frac{\alpha}{2^{1-R_0} - 1}$$

b) Determine the value of α that maximizes the required $\mathcal{E}_b/N_0$ (worst-case pulse jamming) and the resulting maximum value of $\mathcal{E}_b/N_0$.

c) Plot the graph of $10 \log(\mathcal{E}_b/rN_0)$ versus R_0, where $r = R_0/R$, for worst-case pulse jamming and for AWGN ($\alpha = 1$). What conclusions do you reach regarding the effect of worst-case pulse jamming?

13.31 In a coded and interleaved FH q-ary FSK modulation with partial band jamming and coherent demodulation with soft-decision decoding, the cutoff rate is

$$R_0 = \log_2\left[\frac{q}{1 + (q-1)\alpha e^{-\alpha\mathcal{E}_c/2N_0}}\right]$$

where α is the fraction of the band being jammed, $\mathcal{E}_c$ is the chip (or tone) energy, and $N_0 = J_0$.

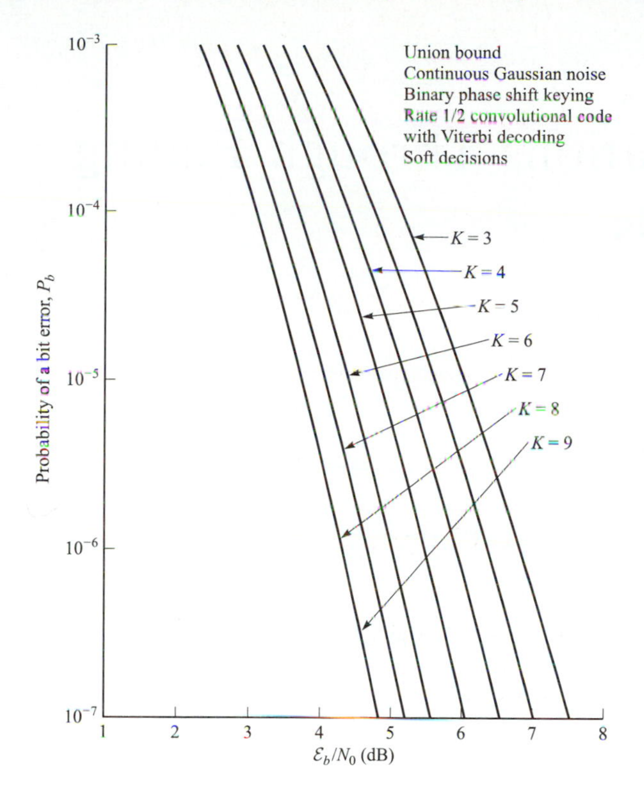

FIGURE P13.29 (*Continued*)

a) Show that the SNR per bit can be expressed as

$$\frac{\mathcal{E}_b}{N_0} = \frac{2}{\alpha R} \ln \frac{(q-1)\alpha}{q2^{-R_0} - 1}$$

b) Determine the value of α that maximizes the required $\mathcal{E}_b/N_0$ (worst-case partial band jamming) and the resulting maximum value of $\mathcal{E}_b/N_0$.

c) Define $r = R_0/R$ in the result for $\mathcal{E}_b/N_0$ from (b), and plot $10\log(\mathcal{E}_b/rN_0)$ versus the normalized cutoff rate $R_0/\log_2 q$ for $q = 2, 4, 8, 16, 32$. Compare these graphs with the results of Problem 13.30c. What conclusions do you reach regarding the effect of worst-case partial band jamming? What is the effect of increasing the alphabet size q? What is the penalty in SNR between the results in Problem 13.30c and q-ary FSK as $q \to \infty$?

14

Digital Communications Through Fading Multipath Channels

The previous chapters have described the design and performance of digital communication systems for transmission on either the classical AWGN channel or a linear filter channel with AWGN. We observed that the distortion inherent in linear filter channels requires special signal design techniques and rather sophisticated adaptive equalization algorithms in order to achieve good performance.

In this chapter, we consider the signal design, receiver structure, and receiver performance for more complex channels, namely, channels having randomly time variant impulse responses. This characterization serves as a model for signal transmission over many radio channels such as shortwave ionospheric radio communication in the 3–30 MHz frequency band (HF), tropsopheric scatter (beyond-the-horizon) radio communications in the 300–3000 MHz frequency band (UHF), and 3000–30,000 MHz frequency band (SHF), and ionospheric forward scatter in the 30–300 MHz frequency band (VHF). The time-variant impulse responses of these channels are a consequence of the constantly changing physical characteristics of the media. For example, the ions in the ionospheric layers that reflect the signals transmitted in the HF band are always in motion. To the user of the channel, the motion of the ions appears to be random. Consequently, if the same signal is transmitted at HF in two widely separated time intervals, the two received signals will be different. The time-varying responses that occur are treated in statistical terms.

We shall begin our treatment of digital signaling over fading multipath channels by first developing a statistical characterization of the channel. Then we shall evaluate the performance of several basic digital signaling techniques for communication over such channels. The performance results will demonstrate the severe penalty in SNR that must be paid as a consequence of the fading characteristics of the received signal. We shall then show that the penalty in SNR can be dramatically reduced by means of efficient modulation/coding and demodulation/decoding techniques.

14.1
CHARACTERIZATION OF FADING MULTIPATH CHANNELS

If we transmit an extremely short pulse, ideally an impulse, over a time-varying multipath channel, the received signal might appear as a train of pulses, as shown in Figure 14.1–1. Hence, one characteristic of a multipath medium is the time spread introduced in the signal that is transmitted through the channel.

A second characteristic is due to the time variations in the structure of the medium. As a result of such time variations, the nature of the multipath varies with time. That is, if we repeat the pulse-sounding experiment over and over, we shall observe changes in the received pulse train, which will include changes in the sizes of the individual pulses, changes in the relative delays among the pulses, and, quite often, changes in the number of pulses observed in the received pulse train as shown in Figure 14.1–1. Moreover, the time variations appear to be unpredictable to the user of the channel. Therefore, it is reasonable to characterize the time-variant multipath channel statistically. Toward this end, let us examine the effects of the channel on a transmitted signal that is represented in general as

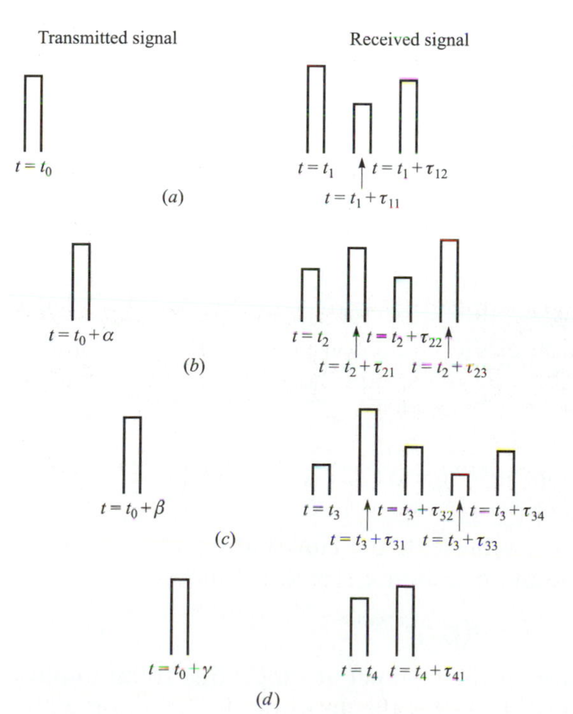

FIGURE 14.1–1
Example of the response of a time-variant multipath channel to a very narrow pulse.

$$s(t) = \text{Re}\left[s_l(t)e^{j2\pi f_c t}\right] \tag{14.1–1}$$

We assume that there are multiple propagation paths. Associated with each path is a propagation delay and an attenuation factor. Both the propagation delays and the attenuation factors are time-variant as a result of changes in the structure of the medium. Thus, the received band-pass signal may be expressed in the form

$$x(t) = \sum_n \alpha_n(t)s[t - \tau_n(t)] \tag{14.1–2}$$

where $\alpha_n(t)$ is the attenuation factor for the signal received on the nth path and $\tau_n(t)$ is the propagation delay for the nth path. Substitution for $s(t)$ from Equation 14.1–1 into Equation 14.1–2 yields the result

$$x(t) = \text{Re}\left(\left\{\sum_n \alpha_n(t)e^{-j2\pi f_c \tau_n(t)} s_l[t - \tau_n(t)]\right\}e^{j2\pi f_c t}\right) \tag{14.1–3}$$

It is apparent from Equation 14.1–3 that the equivalent low-pass received signal is

$$r_l(t) = \sum_n \alpha_n(t)e^{-j2\pi f_c \tau_n(t)} s_l[t - \tau_n(t)] \tag{14.1–4}$$

Since $r_l(t)$ is the response of an equivalent low-pass channel to the equivalent low-pass signal $s_l(t)$, it follows that the equivalent low-pass channel is described by the time-variant impulse response

$$c(\tau; t) = \sum_n \alpha_n(t)e^{-j2\pi f_c \tau_n(t)} \delta[\tau - \tau_n(t)] \tag{14.1–5}$$

For some channels, such as the tropospheric scatter channel, it is more appropriate to view the received signal as consisting of a continuum of multipath components. In such a case, the received signal $x(t)$ is expressed in the integral form

$$x(t) = \int_{-\infty}^{\infty} \alpha(\tau; t)s(t - \tau)d\tau \tag{14.1–6}$$

where $\alpha(\tau; t)$ denotes the attenuation of the signal components at delay τ and at time instant t. Now substitution for $s(t)$ from Equation 14.1–1 into Equation 14.1–6 yields

$$x(t) = \text{Re}\left\{\left[\int_{-\infty}^{\infty} \alpha(\tau; t)e^{-j2\pi f_c \tau} s_l(t - \tau)\, d\tau\right]e^{j2\pi f_c t}\right\} \tag{14.1–7}$$

Since the integral in Equation 14.1–7 represents the convolution of $s_l(t)$ with an equivalent low-pass time-variant impulse response $c(\tau; t)$, it follows that

$$c(\tau; t) = \alpha(\tau; t)e^{-j2\pi f_c \tau} \tag{14.1–8}$$

where $c(\tau; t)$ represents the response of the channel at time t due to an impulse applied at time $t - \tau$. Thus Equation 14.1–8 is the appropriate definition of the equivalent low-pass impulse response when the channel results in continuous multipath and Equation 14.1–5 is appropriate for a channel that contains discrete multipath components.

Now let us consider the transmission of an unmodulated carrier at frequency f_c. Then $s_l(t) = 1$ for all t, and, hence, the received signal for the case of discrete multipath, given by Equation 14.1–4, reduces to

$$\begin{aligned} r_l(t) &= \sum_n \alpha_n(t) e^{-j2\pi f_c \tau_n(t)} \\ &= \sum_n \alpha_n(t) e^{-j\theta_n(t)} \end{aligned} \tag{14.1–9}$$

where $\theta_n(t) = 2\pi f_c \tau_n(t)$. Thus, the received signal consists of the sum of a number of time-variant vectors (phasors) having amplitudes $\alpha_n(t)$ and phases $\theta_n(t)$. Note that large dynamic changes in the medium are required for $\alpha_n(t)$ to change sufficiently to cause a significant change in the received signal. On the other hand, $\theta_n(t)$ will change by 2π rad whenever τ_n changes by $1/f_c$. But $1/f_c$ is a small number and, hence, θ_n can change by 2π rad with relatively small motions of the medium. We also expect the delays $\tau_n(t)$ associated with the different signal paths to change at different rates and in an unpredictable (random) manner. This implies that the received signal $r_l(t)$ in Equation 14.1–9 can be modeled as a random process. When there are a large number of paths, the central limit theorem can be applied. That is, $r_l(t)$ may be modeled as a complex-valued Gaussian random process. This means that the time-variant impulse response $c(\tau; t)$ is a complex-valued Gaussian random process in the t variable.

The multipath propagation model for the channel embodied in the received signal $r_l(t)$, given in Equation 14.1–9, results in signal fading. The fading phenomenon is primarily a result of the time variations in the phases $\{\theta_n(t)\}$. That is, the randomly time variant phases $\{\theta_n(t)\}$ associated with the vectors $\{\alpha_n e^{-j\theta_n}\}$ at times result in the vectors adding destructively. When that occurs, the resultant received signal $r_l(t)$ is very small or practically zero. At other times, the vectors $\{\alpha_n e^{-j\theta_n}\}$ add constructively, so that the received signal is large. Thus, the amplitude variations in the received signal, termed *signal fading*, are due to the time-variant multipath characteristics of the channel.

When the impulse response $c(\tau; t)$ is modeled as a zero-mean complex-valued Gaussian process, the envelope $|c(\tau; t)|$ at any instant t is Rayleigh-distributed. In this case the channel is said to be a *Rayleigh fading channel*. In the event that there are fixed scatterers or signal reflectors in the medium, in addition to randomly moving scatterers, $c(\tau; t)$ can no longer be modeled as having zero-mean. In this case, the envelope $|c(\tau; t)|$ has a Rice distribution and the channel is said to be a *Ricean fading channel*. Another probability distribution function that has been used to model the envelope of fading signals is the Nakagami-m distribution. These fading channel models are considered in Section 14.1.2.

14.1.1 Channel Correlation Functions and Power Spectra

We shall now develop a number of useful correlation functions and power spectral density functions that define the characteristics of a fading multipath channel. Our starting point is the equivalent low-pass impulse response $c(\tau; t)$, which is characterized as a complex-valued random process in the t variable. We assume that $c(\tau; t)$ is wide-sense-stationary. Then we define the autocorrelation function of $c(\tau; t)$ as

$$\phi_c(\tau_1, \tau_2; \Delta t) = \tfrac{1}{2} E[c^*(\tau_1; t)c(\tau_2; t + \Delta t)] \tag{14.1–10}$$

In most radio transmission media, the attentuation and phase shift of the channel associated with path delay τ_1 is uncorrelated with the attenuation and phase shift associated with path delay τ_2. This is usually called *uncorrelated scattering*. We make the assumption that the scattering at two different delays is uncorrelated and incorporate it into Equation 14.1–10 to obtain

$$\tfrac{1}{2} E[c^*(\tau_1; t)c(\tau_2; t + \Delta t)] = \phi_c(\tau_1; \Delta t)\delta(\tau_1 - \tau_2) \tag{14.1–11}$$

If we let $\Delta t = 0$, the resulting autocorrelation function $\phi_c(\tau; 0) \equiv \phi_c(\tau)$ is simply the average power output of the channel as a function of the time delay τ. For this reason, $\phi_c(\tau)$ is called the *multipath intensity profile* or the *delay power spectrum* of the channel. In general, $\phi_c(\tau; \Delta t)$ gives the average power output as a function of the time delay τ and the difference Δt in observation time.

In practice, the function $\phi_c(\tau; \Delta t)$ is measured by transmitting very narrow pulses or, equivalently, a wideband signal and cross-correlating the received signal with a delayed version of itself. Typically, the measured function $\phi_c(\tau)$ may appear as shown in Figure 14.1–2. The range of values of τ over which $\phi_c(\tau)$ is essentially nonzero is called the *multipath spread of the channel* and is denoted by T_m.

A completely analogous characterization of the time-variant multipath channel begins in the frequency domain. By taking the Fourier transform of $c(\tau; t)$, we obtain the time-variant transfer function $C(f; t)$, where f is the frequency variable. Thus,

$$C(f; t) = \int_{-\infty}^{\infty} c(\tau; t)e^{-j2\pi f \tau}\, d\tau \tag{14.1–12}$$

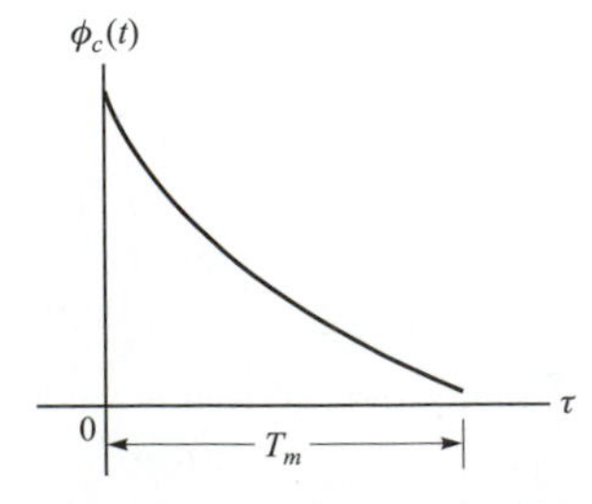

FIGURE 14.1–2
Multipath intensity profile.

If $c(\tau; t)$ is modeled as a complex-valued zero-mean Gaussian random process in the t variable, it follows that $C(f; t)$ also has the same statistics. Under the assumption that the channel is wide-sense-stationary, we define the autocorrelation function

$$\phi_C(f_1, f_2; \Delta t) = \tfrac{1}{2} E[C^*(f_1; t) C(f_2; t + \Delta t)] \tag{14.1–13}$$

Since $C(f; t)$ is the Fourier transform of $c(\tau; t)$, it is not surprising to find that $\phi_C(f_1, f_2; \Delta t)$ is related to $\phi_C(\tau; \Delta t)$ by the Fourier transform. The relationship is easily established by substituting Equation 14.1–12 into Equation 14.1–13. Thus,

$$\begin{aligned}
\phi_C(f_1, f_2; \Delta t) &= \frac{1}{2}\int_{-\infty}^{\infty}\int_{-\infty}^{\infty} E[c^*(\tau_1; t)c(\tau_2; t + \Delta t)] e^{j2\pi(f_1\tau_1 - f_2\tau_2)} d\tau_1 d\tau_2 \\
&= \int_{-\infty}^{\infty}\int_{-\infty}^{\infty} \phi_c(\tau_1; \Delta t)\delta(\tau_1 - \tau_2) e^{j2\pi(f_1\tau_1 - f_2\tau_2)} d\tau_1 d\tau_2 \\
&= \int_{-\infty}^{\infty} \phi_c(\tau_1; \Delta t) e^{j2\pi(f_1 - f_2)\tau_1} d\tau_1 \\
&= \int_{-\infty}^{\infty} \phi_c(\tau_1; \Delta t) e^{-j2\pi\Delta f \tau_1} d\tau_1 \equiv \phi_C(\Delta f; \Delta t)
\end{aligned} \tag{14.1–14}$$

where $\Delta f = f_2 - f_1$. From Equation 14.1–14, we observe that $\phi_C(\Delta f; \Delta t)$ is the Fourier transform of the multipath intensity profile. Furthermore, the assumption of uncorrelated scattering implies that the autocorrelation function of $C(f; t)$ in frequency is a function of only the frequency difference $\Delta f = f_2 - f_1$. Therefore, it is appropriate to call $\phi_C(\Delta f; \Delta t)$ the *spaced-frequency, spaced time correlation function of the channel.* It can be measured in practice by transmitting a pair of sinusoids separated by Δf and cross-correlating the two separately received signals with a relative delay Δt.

Suppose we set $\Delta t = 0$ in Equation 14.1–14. Then, with $\phi_C(\Delta f; 0) \equiv \phi_C(\Delta f)$ and $\phi_c(\tau; 0) \equiv \phi_c(\tau)$, the transform relationship is simply

$$\phi_C(\Delta f) = \int_{-\infty}^{\infty} \phi_c(\tau) e^{-j2\pi\Delta f \tau} d\tau \tag{14.1–15}$$

The relationship is depicted graphically in Figure 14.1–3. Since $\phi_C(\Delta f)$ is an autocorrelation function in the frequency variable, it provides us with a measure of the frequency coherence of the channel. As a result of the Fourier transform relationship between $\phi_C(\Delta f)$ and $\phi_c(\tau)$, the reciprocal of the multipath spread is a measure of the *coherence bandwidth of the channel.* That is,

$$(\Delta f)_c \approx \frac{1}{T_m} \tag{14.1–16}$$

where $(\Delta f)_c$ denotes the coherence bandwidth. Thus, two sinusoids with frequency separation greater than $(\Delta f)_c$ are affected differently by the channel. When an information-bearing signal is transmitted through the channel, if

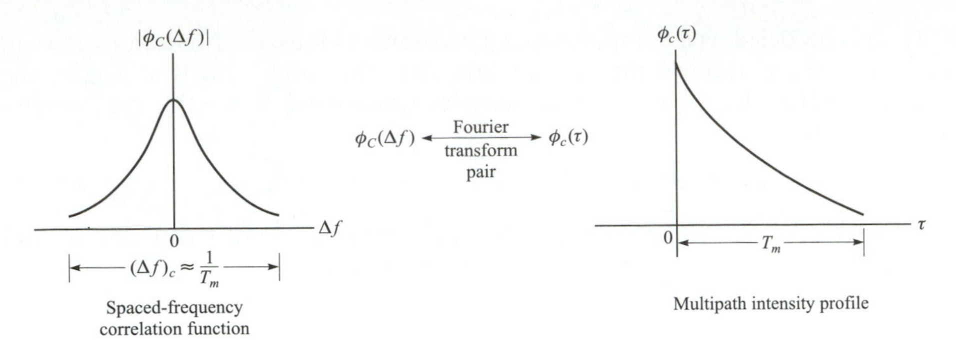

FIGURE 14.1–3
Relationship between $\phi_C(\Delta f)$ and $\phi_c(\tau)$.

$(\Delta f)_c$ is small in comparison to the bandwidth of the transmitted signal, the channel is said to be *frequency-selective*. In this case, the signal is severely distorted by the channel. On the other hand, if $(\Delta f)_c$ is large in comparison with the bandwidth of the transmitted signal, the channel is said to be *frequency-nonselective*.

We now focus our attention on the time variations of the channel as measured by the parameter Δt in $\phi_C(\Delta f;\, \Delta t)$. The time variations in the channel are evidenced as a Doppler broadening and, perhaps, in addition as a Doppler shift of a spectral line. In order to relate the Doppler effects to the time variations of the channel, we define the Fourier transform of $\phi_C(\Delta f;\, \Delta t)$ with respect to the variable Δt to be the function $S_C(\Delta f;\, \lambda)$. That is,

$$S_C(\Delta f;\, \lambda) = \int_{-\infty}^{\infty} \phi_C(\Delta f;\, \Delta t) e^{-j2\pi\lambda\, \Delta t} d\Delta t \tag{14.1–17}$$

With Δf set to zero and $S_C(0;\, \lambda) \equiv S_C(\lambda)$, the relation in Equation 14.1–17 becomes

$$S_C(\lambda) = \int_{-\infty}^{\infty} \phi_C(0;\, \Delta t) e^{-j2\pi\lambda\, \Delta t} d\Delta t \tag{14.1–18}$$

The function $S_C(\lambda)$ is a power spectrum that gives the signal intensity as a function of the Doppler frequency λ. Hence, we call $S_C(\lambda)$ the *Doppler power spectrum of the channel.*

From Equation 14.1–18, we observe that if the channel is time-invariant, $\phi_C(\Delta t) = 1$ and $S_C(\lambda)$ becomes equal to the delta function $\delta(\lambda)$. Therefore, when there are no time variations in the channel, there is no spectral broadening observed in the transmission of a pure frequency tone.

The range of values of λ over which $S_C(\lambda)$ is essentially nonzero is called the *Doppler spread B_d of the channel.* Since $S_C(\lambda)$ is related to $\phi_C(\Delta t)$ by the Fourier transform, the reciprocal of B_d is a measure of the coherence time of the channel. That is,

$$(\Delta t)_c \approx \frac{1}{B_d} \tag{14.1–19}$$

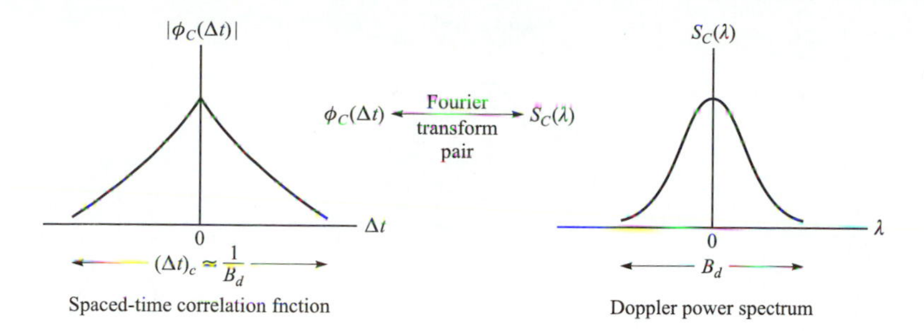

FIGURE 14.1–4
Relationship between $\phi_C(\Delta t)$ and $S_C(\lambda)$.

where $(\Delta t)_c$ denotes the *coherence time*. Clearly, a slowly changing channel has a large coherence time or, equivalently, a small Doppler spread. Figure 14.1–4 illustrates the relationship between $\phi_C(\Delta t)$ and $S_C(\lambda)$.

We have now established a Fourier transform relationship between $\phi_C(\Delta f; \Delta t)$ and $\phi_c(\tau; \Delta t)$ involving the variables $(\tau, \Delta f)$, and a Fourier transform relationship between $\phi_C(\Delta f; \Delta t)$ and $S_C(\Delta f; \lambda)$ involving the variables $(\Delta t, \lambda)$. There are two additional Fourier transform relationships that we can define, which serve to relate $\phi_c(\tau; \Delta t)$ to $S_C(\Delta f; \lambda)$ and, thus, close the loop. The desired relationship is obtained by defining a new function, denoted by $S(\tau; \lambda)$, to be the Fourier transform of $\phi_c(\tau; \Delta t)$ in the Δt variable. That is,

$$S(\tau; \lambda) = \int_{-\infty}^{\infty} \phi_c(\tau; \Delta t) e^{-j2\pi\lambda\,\Delta t} d\Delta t \tag{14.1–20}$$

It follows that $S(\tau; \lambda)$ and $S_C(\Delta f; \lambda)$ are a Fourier transform pair. That is,

$$S(\tau; \lambda) = \int_{-\infty}^{\infty} S_C(\Delta f; \lambda) e^{j2\pi\tau\,\Delta f} d\Delta f \tag{14.1–21}$$

Furthermore, $S(\tau; \lambda)$ and $\phi_C(\Delta f; \Delta t)$ are related by the double Fourier transform

$$S(\tau; \lambda) = \int_{-\infty}^{\infty}\int_{-\infty}^{\infty} \phi_C(\Delta f; \Delta t) e^{-j2\pi\lambda\,\Delta t} e^{j2\pi\tau\,\Delta f}\, d\Delta t\, d\Delta f \tag{14.1–22}$$

This new function $S(\tau; \lambda)$ is called the *scattering function of the channel*. It provides us with a measure of the average power output of the channel as a function of the time delay τ and the Doppler frequency λ.

The relationships among the four functions $\phi_C(\Delta f; \Delta t)$, $\phi_c(\tau; \Delta t)$, $\phi_C(\Delta f; \lambda)$, and $S(\tau; \lambda)$ are summarized in Figure 14.1–5.

EXAMPLE 14.1–1: SCATTERING FUNCTION OF A TROPOSPHERIC SCATTER CHANNEL. The scattering function $S(\tau; \lambda)$ measured on a 150-mi tropospheric scatter link is shown in Figure 14.1–6. The signal used to probe the channel had a time resolution of 0.1 μs. Hence, the time-delay axis is quantized in increments of 0.1 μs. From the graph, we observe that the multipath spread $T_m = 0.7$ μs. On the other

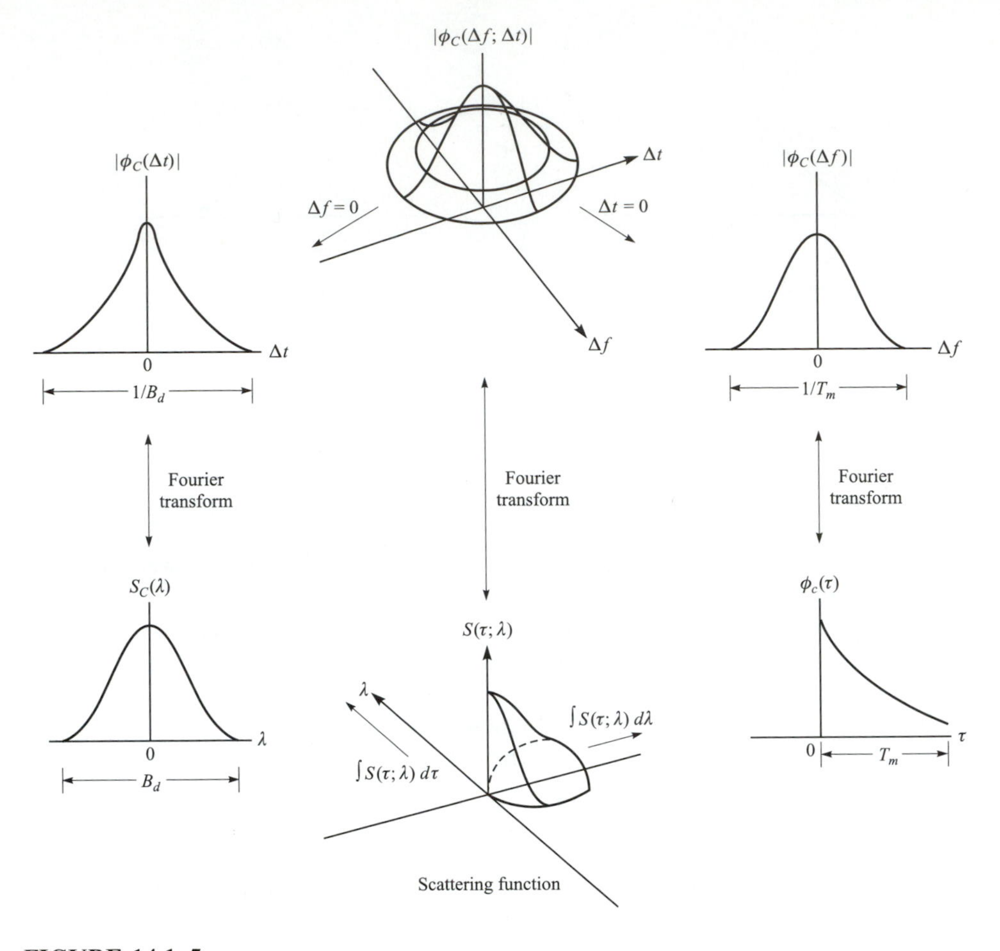

FIGURE 14.1–5
Relationships among the channel correlation functions and power spectra. [*From Green (1962), with permission.*]

hand, the Doppler spread, which may be defined as the 3-dB bandwidth of the power spectrum for each signal path, appears to vary with each signal path. For example, in one path it is less than 1 Hz, while in some other paths it is several hertz. For our purposes, we shall take the largest of these 3-dB bandwidths of the various paths and call that the *Doppler spread.*

EXAMPLE 14.1–2: MULTIPATH INTENSITY PROFILE OF MOBILE RADIO CHANNELS. The multipath intensity profile of a mobile radio channel depends critically on the type of terrain. Numerous measurements have been made under various conditions in many parts of the world. In urban and suburban areas, typical values of multipath spreads range from 1 to 10 μs. In rural mountainous areas, the multipath spreads are much greater, with typical values in the range of 10 to 30 μs. Two models for the multipath intensity profile that are widely used in evaluating system performance for these two types of terrain are illustrated in Figure 14.1–7.

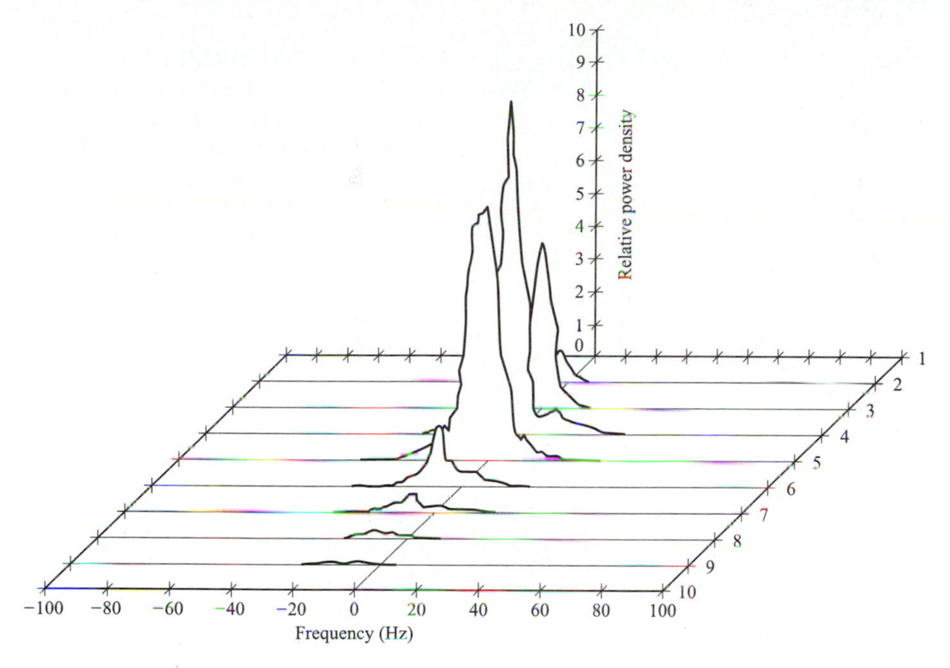

FIGURE 14.1–6
Scattering function of a medium-range tropospheric scatter channel. The taps delay increment is 0.1 μs.

EXAMPLE 14.1–3: DOPPLER POWER SPECTRUM OF MOBILE RADIO CHANNELS. A widely used model for the Doppler power spectrum of a mobile radio channel is the so-called Jakes' model (Jakes, 1974). In this model, the autocorrelation of the time-variant transfer function $C(f;t)$ is given as

$$\begin{aligned}\phi_C(\Delta t) &= \tfrac{1}{2}E[C^*(f;t)C(f;t+\Delta t)]\\ &= J_0(2\pi f_m\,\Delta t)\end{aligned}$$

where $J(\ )$ is the zero-order Bessel function of the first kind and $f_m = vf_0/c$ is the maximum Doppler frequency, where v is the vehicle speed in meters per second (m/s), f_0 is the carrier frequency, and c is the speed of light (3×10^8 m/s). The Fourier transform of this autocorrelation function yields the Doppler power spectrum. That is

$$\begin{aligned}S_c(\lambda) &= \int_{-\infty}^{\infty}\phi_C(\Delta t)e^{-j2\pi\lambda\,\Delta t}\,d\Delta t\\ &= \int_{-\infty}^{\infty}J_0(2\pi f_m\,\Delta t)e^{-j2\pi\lambda\,\Delta t}\,d\Delta t\\ &= \begin{cases}\dfrac{1}{\pi f_m}\dfrac{1}{\sqrt{1-(f/f_m)^2}} & (|f|\leqslant f_m)\\ 0 & (|f|>f_m)\end{cases}\end{aligned}$$

Plots of $\phi_C(\Delta t)$ and $S_c(\lambda)$ are shown in Figure 14.1–8.

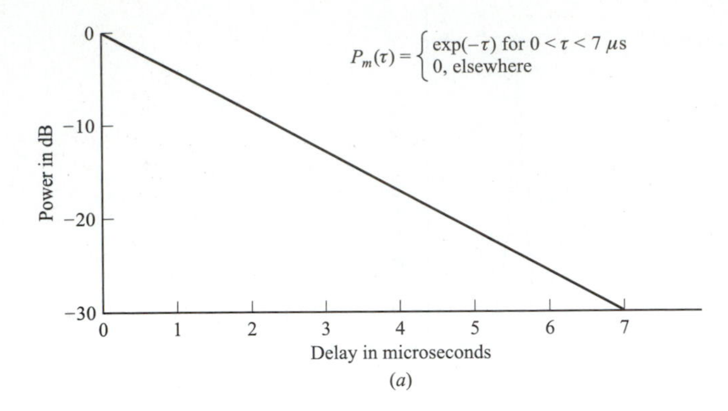

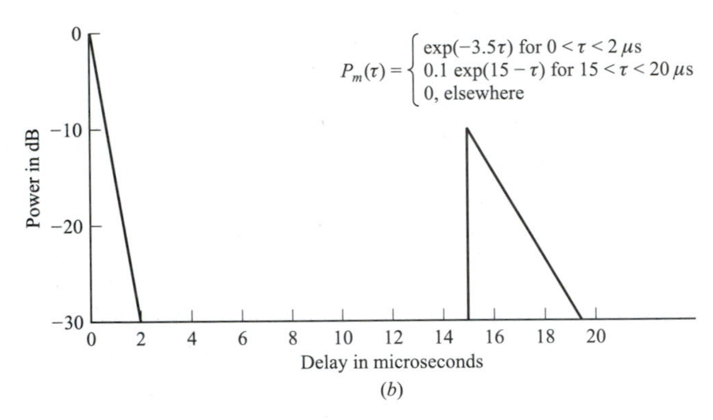

FIGURE 14.1–7 Cost 207 average power delay profiles: (**a**) typical delay profile for suburban and urban areas; (**b**) typical "bad"-case delay profile for hilly terrain. [*From Cost 207 Document 207 TD (86)51 rev 3.*]

14.1.2 Statistical Models for Fading Channels

There are several probability distributions that can be considered in attempting to model the statistical characteristics of the fading channel. When there are a large number of scatterers in the channel that contribute to the signal at the receiver, as is the case in ionospheric or tropospheric signal propagation, application of the central limit theorem leads to a Gaussian process model for the channel impulse response. If the process is zero-mean, then the envelope of the channel response at any time instant has a Rayleigh probability distribution and the phase is uniformly distributed in the interval $(0, 2\pi)$. That is

$$p_R(r) = \frac{2r}{\Omega} e^{-r^2/\Omega}, \qquad r \geqslant 0 \tag{14.1–23}$$

where

$$\Omega = E(R^2) \tag{14.1–24}$$

We observe that the Rayleigh distribution is characterized by the single parameter $E(R^2)$.

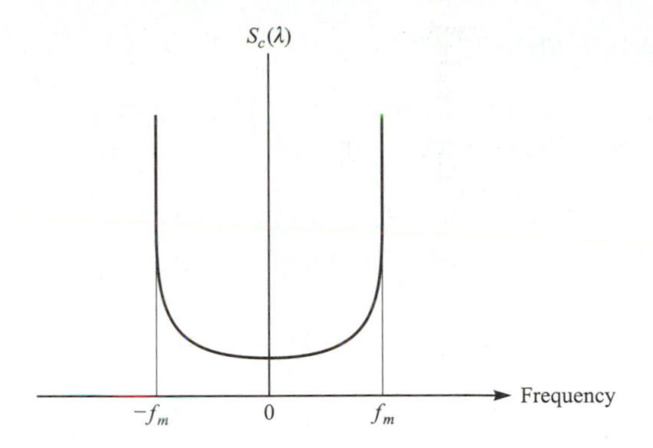

FIGURE 14.1–8
Model of Doppler spectrum for a mobile radio channel.

An alternative statistical model for the envelope of the channel response is the Nakagami-m distribution given by the PDF in Equation 2.1–147. In contrast to the Rayleigh distribution, which has a single parameter that can be used to match the fading channel statistics, the Nakagami-m is a two-parameter distribution, involving the parameter m and the second moment $\Omega = E(R^2)$. As a consequence, this distribution provides more flexibility and accuracy in matching the observed signal statistics. The Nakagami-m distribution can be used to model fading channel conditions that are either more or less severe than the Rayleigh distribution, and it includes the Rayleigh distribution as a special case ($m = 1$). For example, Turin et al. (1972) and Suzuki (1977) have shown that the Nakagami-m distribution is the best fit for data signals received in urban radio multipath channels.

The Rice distribution is also a two-parameter distribution. It may be expressed by the PDF given in Equation 2.1–141, where the parameters are s and σ^2. Recall that s^2 is called the *noncentrality parameter* in the equivalent chi-square distribution. It represents the power in the nonfading signal components, sometimes called *specular components*, of the received signal.

There are many radio channels in which fading is encountered that are basically line-of-sight (LOS) communication links with multipath components arising from secondary reflections, or signal paths, from surrounding terrain. In such channels, the number of multipath components is small, and, hence, the channel may be modeled in a somewhat simpler form. We cite two channel models as examples.

As the first example, let us consider an airplane to ground communication link in which there is the direct path and a single multipath component at a delay t_0 relative to the direct path. The impulse response of such a channel may be modeled as

$$c(\tau; t) = \alpha\delta(\tau) + \beta(t)\delta[\tau - \tau_0(t)] \qquad (14.1\text{–}25)$$

where α is the attenuation factor of the direct path and $\beta(t)$ represents the time-variant multipath signal component resulting from terrain reflections. Often, $\beta(t)$ can be characterized as a zero-mean Gaussian random process. The transfer function for this channel model may be expressed as

$$C(f;t) = \alpha + \beta(t)e^{-j2\pi f \tau_0(t)} \tag{14.1–26}$$

This channel fits the Ricean fading model defined previously. The direct path with attenuation α represents the specular component and $\beta(t)$ represents the Rayleigh fading component.

A similar model has been found to hold for microwave LOS radio channels used for long-distance voice and video transmission by telephone companies throughout the world. For such channels, Rummler (1979) has developed a three-path model based on channel measurements performed on typical LOS links in the 6-GHz frequency band. The differential delay on the two multipath components is relatively small, and, hence, the model developed by Rummler is one that has a channel transfer function

$$C(f) = \alpha[1 - \beta e^{-j2\pi(f-f_0)\tau_0}] \tag{14.1–27}$$

where α is the overall attenuation parameter, β is called a shape parameter which is due to the multipath components, f_0 is the frequency of the fade minimum, and τ_0 is the relative time delay between the direct and the multipath components. This simplified model was used to fit data derived from channel measurements.

Rummler found that the parameters α and β may be characterized as random variables that, for practical purposes, are nearly statistically independent. From the channel measurements, he found that the distribution of β has the form $(1-\beta)^{2.3}$. The distribution of α is well modeled by the lognormal distribution, i.e., $-\log\alpha$ is Gaussian. For $\beta > 0.5$, the mean of $-20\log\alpha$ was found to be 25 dB and the standard deviation was 5 dB. For smaller values of β, the mean decreases to 15 dB. The delay parameter determined from the measurements was $\tau_0 = 6.3$ ns. The magnitude-square response of $C(f)$ is

$$|C(f)|^2 = \alpha^2[1 + \beta^2 - 2\beta\cos 2\pi(f - f_0)\tau_0] \tag{14.1–28}$$

$|C(f)|$ is plotted in Figure 14.1–9 as a function of the frequency $f - f_0$ for $\tau_0 = 6.3$ ns. Note that the effect of the multipath component is to create a deep attenuation at $f = f_0$ and at multiples of $1/\tau_0 \approx 159$ MHz. By comparison, the typical channel bandwidth is 30 MHz. This model was used by Lundgren and Rummler (1979) to determine the error rate performance of digital radio systems.

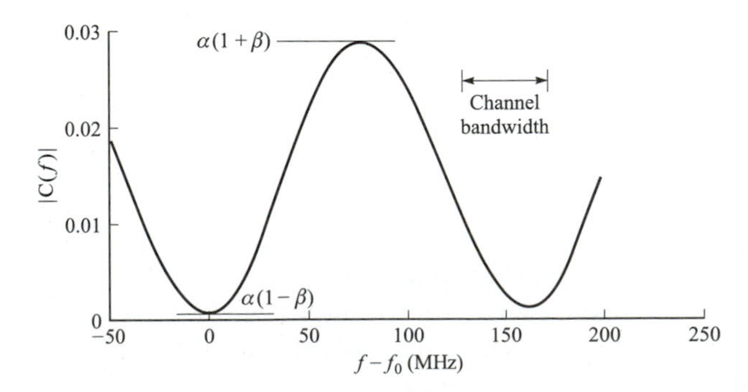

FIGURE 14.1–9
Magnitude frequency response of LOS channel model.

Propagation models for mobile radio channels. In the link budget calculations that were described in Section 5.5.2, we had characterized the path loss of radio waves propagating through free space as being inversely proportional to d^2, where d is the distance between the transmitter and the receiver. However, in a mobile radio channel, propagation is generally neither free space nor line of sight. The mean path loss encountered in mobile radio channels may be characterized as being inversely proportional to d^p, where $2 \leqslant p \leqslant 4$, with d^4 being a worst-case model. Consequently, the path loss is usually much more severe compared to that of free space.

There are a number of factors affecting the path loss in mobile radio communications. Among these factors are base station antenna height, mobile antenna height, operating frequency, atmospheric conditions, and presence or absence of buildings and trees. Various mean path loss models have been developed that incorporate such factors. For example, a model for a large city in an urban area is the Hata model, in which the mean path loss is expressed as

$$\begin{aligned}\text{Loss in dB} = 69.55 + 26.16 \log_{10} f - 13.82 \log_{10} h_t - a(h_r) \\ + (44.9 - 6.55 \log_{10} h_t) \log_{10} d\end{aligned} \tag{14.1–29}$$

where f is the operating frequency in MHz ($150 < f < 1500$), h_t is the transmitter antenna height in meters ($30 < h_t < 200$), h_r is the receiver antenna height in meters ($1 < h_r < 10$), d is the distance between transmitter and receiver in km ($1 < d < 20$), and

$$a(h_r) = 3.2(\log_{10} 11.75 h_r)^2 - 4.97, \qquad f \geqslant 400 \text{ MHz} \tag{14.1–30}$$

Another problem with mobile radio propagation is the effect of shadowing of the signal due to large obstructions, such as large buildings, trees, and hilly terrain between the transmitter and the receiver. Shadowing is usually modeled as a multiplicative and, generally, slowly time varying random process. That is, the received signal may be characterized mathematically as

$$r(t) = A_0 g(t) s(t) \tag{14.1–31}$$

where A_0 represents the mean path loss, $s(t)$ is the transmitted signal, and $g(t)$ is a random process that represents the shadowing effect. At any time instant, the shadowing process is modeled statistically as lognormally distributed. The probability density function for the lognormal distribution is

$$p(g) = \begin{cases} \dfrac{1}{\sqrt{2\pi\sigma^2}\, g} e^{-(\ln g - \mu)^2/2\sigma^2} & (g \geqslant 0) \\ 0 & (g < 0) \end{cases} \tag{14.1–32}$$

If we define a new random variable X as $X = \ln g$, then

$$p(x) = \frac{1}{\sqrt{2\pi\sigma^2}} e^{-(x-\mu)^2/2\sigma^2}, \qquad -\infty < x < \infty \tag{14.1–33}$$

The random variable X represents the path loss measured in dB, μ is the mean path loss in dB, and σ is the standard deviation of the path loss in dB. For typical cellular and microcellular environments, σ is in the range of 5–12 dB.

14.2
THE EFFECT OF SIGNAL CHARACTERISTICS ON THE CHOICE OF A CHANNEL MODEL

Having discussed the statistical characterization of time-variant multipath channels generally in terms of the correlation functions describe in Section 14.1, we now consider the effect of signal characteristics on the selection of a channel model that is appropriate for the specified signal. Thus, let $s_l(t)$ be the equivalent low-pass signal transmitted over the channel and let $S_l(f)$ denote its frequency content. Then the equivalent low-pass received signal, exclusive of additive noise, may be expressed either in terms of the time-domain variables $c(\tau; t)$ and $s_l(t)$ as

$$r_l(t) = \int_{-\infty}^{\infty} c(\tau; t) s_l(t - \tau)\, d\tau \qquad (14.2\text{–}1)$$

or in terms of the frequency functions $C(f; t)$ and $S_l(f)$ as

$$r_l(t) = \int_{-\infty}^{\infty} C(f; t) S_l(f) e^{j2\pi ft}\, df \qquad (14.2\text{–}2)$$

Suppose we are transmitting digital information over the channel by modulating (either in amplitude, or in phase, or both) the basic pulse $s_l(t)$ at a rate $1/T$, where T is the signaling interval. It is apparent from Equation 14.2–2 that the time-variant channel characterized by the transfer function $C(f; t)$ distorts the signal $S_l(f)$. If $S_l(f)$ has a bandwidth W greater than the coherence bandwidth $(\Delta f)_c$ of the channel, $S_l(f)$ is subjected to different gains and phase shifts across the band. In such a case, the channel is said to be *frequency-selective*. Additional distortion is caused by the time variations in $C(f; t)$. This type of distortion is evidenced as a variation in the received signal strength, and has been termed *fading*. It should be emphasized that the frequency selectivity and fading are viewed as two different types of distortion. The former depends on the multipath spread or, equivalently, on the coherence bandwidth of the channel relative to the transmitted signal bandwidth W. The latter depends on the time variations of the channel, which are grossly characterized by the coherence time $(\Delta t)_c$ or, equivalently, by the Doppler spread B_d.

The effect of the channel on the transmitted signal $s_l(t)$ is a function of our choice of signal bandwidth and signal duration. For example, if we select the signaling interval T to satisfy the condition $T \gg T_m$, the channel introduces a negligible amount of intersymbol interference. If the bandwidth of the signal pulse $s_l(t)$ is $W \approx 1/T$, the condition $T \gg T_m$ implies that

$$W \ll \frac{1}{T_m} \approx (\Delta f)_c \qquad (14.2\text{–}3)$$

That is, the signal bandwidth W is much smaller than the coherence bandwidth of the channel. Hence, the channel is *frequency-nonselective*. In other words, all the frequency components in $S_l(f)$ undergo the same attenuation and phase shift in transmission through the channel. But this implies that, within the bandwidth occupied by $S_l(f)$, the time-variant transfer function $C(f; t)$ of the channel is a complex-valued constant in the frequency variable. Since $S_l(f)$ has its frequency content concentrated in the vicinity of $f = 0$, $C(f; t) = C(0; t)$. Consequently, Equation 14.2–2 reduces to

$$\begin{aligned} r_l(t) &= C(0; t) \int_{-\infty}^{\infty} S_l(f) e^{j2\pi ft} \, df \\ &= C(0; t) s_l(t) \end{aligned} \tag{14.2–4}$$

Thus, when the signal bandwidth W is much smaller than the coherence bandwidth $(\Delta f)_c$ of the channel, the received signal is simply the transmitted signal multiplied by a complex-valued random process $C(0; t)$, which represents the time-variant characteristics of the channel. In this case, we say that the multipath components in the received are not resolvable because $W \ll (\Delta f)_c$.

The transfer function $C(0; t)$ for a frequency-nonselective channel may be expressed in the form

$$C(0; t) = \alpha(t) e^{-j\phi(t)} \tag{14.2–5}$$

where $\alpha(t)$ represents the envelope and $\phi(t)$ represents the phase of the equivalent low-pass channel. When $C(0; t)$ is modeled as a zero-mean complex-valued Gaussian random process, the envelope $\alpha(t)$ is Rayleigh-distributed for any fixed value of t and $\phi(t)$ is uniformly distributed over the interval $(-\pi, \pi)$. The rapidity of the fading on the frequency-nonselective channel is determined either from the correlation function $\phi_C(\Delta t)$ or from the Doppler power spectrum $S_C(\lambda)$. Alternatively, either of the channel parameters $(\Delta t)_c$ or B_d can be used to characterize the rapidity of the fading.

For example, suppose it is possible to select the signal bandwidth W to satisfy the condition $W \ll (\Delta f)_c$ and the signaling interval T to satisfy the condition $T \ll (\Delta t)_c$. Since T is smaller than the coherence time of the channel, the channel attenuation and phase shift are essentially fixed for the duration of at least one signaling interval. When this condition holds, we call the channel a *slowly fading channel*. Furthermore, when $W \approx 1/T$, the conditions that the channel be frequency-nonselective and slowly fading imply that the product of T_m and B_d must satisfy the condition $T_m B_d < 1$.

The product $T_m B_d$ is called the *spread factor* of the channel. If $T_m B_d < 1$, the channel is said to be *underspread*; otherwise, it is *overspread*. The multipath spread, the Doppler spread, and the spread factor are listed in Table 14.2–1 for several channels. We observe from this table that several radio channels, including the moon when used as a passive reflector, are underspread. Consequently, it is possible to select the signal $s_l(t)$ such that these channels are frequency-nonselective and slowly fading. The slow-fading condition implies

TABLE 14.2–1
Multipath spread, Doppler spread, and spread factor for several time-variant multipath channels

Type of channel	Multipath duration, s	Doppler spread, Hz	Spread factor
Shortwave ionospheric propagation (HF)	10^{-3}–10^{-2}	10^{-1}–1	10^{-4}–10^{-2}
Ionospheric propagation under distributed auroral conditions (HF)	10^{-3}–10^{-2}	10–100	10^{-2}–1
Ionospheric forward scatter (VHF)	10^{-4}	10	10^{-3}
Tropospheric scatter (SHF)	10^{-6}	10	10^{-5}
Orbital scatter (X band)	10^{-4}	10^{3}	10^{-1}
Moon at max. libration ($f_0 = 0.4\,\text{kmc}$)	10^{-2}	10	10^{-1}

that the channel characteristics vary sufficiently slowly that they can be measured.

In Section 14.3, we shall determine the error rate performance for binary signaling over a frequency-nonselective slowly fading channel. This channel model is, by far, the simplest to analyze. More importantly, it yields insight into the performance characteristics for digital signaling on a fading channel and serves to suggest the type of signal waveforms that are effective in overcoming the fading caused by the channel.

Since the multipath components in the received signal are not resolvable when the signal bandwidth W is less than the coherence bandwidth $(\Delta f)_c$ of the channel, the received signal appears to arrive at the receiver via a single fading path. On the other hand, we may choose $W \gg (\Delta f)_c$, so that the channel becomes frequency-selective. We shall show later that, under this condition, the multipath components in the received signal are resolvable with a resolution in time delay of $1/W$. Thus, we shall illustrate that the frequency-selective channel can be modeled as a tapped delay line (transversal) filter with time-variant tap coefficients. We shall then derive the performance of binary signaling over such a frequency-selective channel model.

14.3 FREQUENCY-NONSELECTIVE, SLOWLY FADING CHANNEL

In this section, we derive the error rate performance of binary PSK and binary FSK when these signals are transmitted over a frequency-nonselective, slowly fading channel. As described in Section 14.2, the frequency-nonselective channel results in multiplicative distortion of the transmitted signal $s_l(t)$. Furthermore, the condition that the channel fades slowly implies that the multiplicative process may be regarded as a constant during at least one signaling interval. Consequently, if the transmitted signal is $s_l(t)$, the received equivalent low-pass signal in one signaling interval is

$$r_l(t) = \alpha e^{-j\phi} s_l(t) + z(t), \qquad 0 \leqslant t \leqslant T \tag{14.3–1}$$

where $z(t)$ represents the complex-valued white Gaussian noise process corrupting the signal.

Let us assume that the channel fading is sufficiently slow that the phase shift ϕ can be estimated from the received signal without error. In that case, we can achieve ideal coherent detection of the received signal. Thus, the received signal can be processed by passing it through a matched filter in the case of binary PSK or through a pair of matched filters in the case of binary FSK. One method that we can use to determine the performance of the binary communication systems is to evaluate the decision variables and from these determine the probability of error. However, we have already done this for a fixed (time-invariant) channel. That is, for a fixed attenuation α, we know the probability of error for binary PSK and binary FSK. From Equation 5.2–5, the expression for the error rate of binary PSK as a function of the received SNR γ_b is

$$P_2(\gamma_b) = Q\left(\sqrt{2\gamma_b}\right) \tag{14.3–2}$$

where $\gamma_b = \alpha^2 \mathcal{E}_b / N_0$. The expression for the error rate of binary FSK, detected coherently, is given by Equation 5.2–10 as

$$P_2(\gamma_b) = Q(\sqrt{\gamma_b}) \tag{14.3–3}$$

We view Equation 14.3–2 and 14.3–3 as conditional error probabilities, where the condition is that α is fixed. To obtain the error probabilities when α is random, we must average $P_2(\gamma_b)$, given in Equations 14.3–2 and 14.3–3, over the probability density function of γ_b. That is, we must evaluate the integral

$$P_2 = \int_0^\infty P_2(\gamma_b) p(\gamma_b)\, d\gamma_b \tag{14.3–4}$$

where $p(\gamma_b)$ is the probability density function of γ_b when α is random.

Rayleight fading. When α is Rayleigh-distributed, α^2 has a chi-square probability distribution with two degrees of freedom. Consequently, γ_b also is chi-square-distributed. It is easily shown that

$$p(\gamma_b) = \frac{1}{\bar{\gamma}_b} e^{-\gamma_b/\bar{\gamma}_b}, \qquad \gamma_b \geqslant 0 \tag{14.3–5}$$

where $\bar{\gamma}_b$ is the average signal-to-noise ratio, defined as

$$\bar{\gamma}_b = \frac{\mathcal{E}_b}{N_0} E(\alpha^2) \tag{14.3–6}$$

The term $E(\alpha^2)$ is simply the average value of α^2.

Now we can substitute Equation 14.3–5 into Equation 14.3–4 and carry out the integration for $P_2(\gamma_b)$ as given by Equations 14.3–2 and 14.3–3. The result of this integration for binary PSK is

$$P_2 = \frac{1}{2}\left(1 - \sqrt{\frac{\bar{\gamma}_b}{1+\bar{\gamma}_b}}\right) \tag{14.3–7}$$

If we repeat the integration with $P_2(\gamma_b)$ given by Equation 14.3–3, we obtain the probability of error for binary FSK, detected coherently, in the form

$$P_2 = \frac{1}{2}\left(1 - \sqrt{\frac{\bar{\gamma}_b}{2+\bar{\gamma}_b}}\right) \tag{14.3–8}$$

In arriving at the error rate results in Equations 14.3–7 and 14.3–8, we have assumed that the estimate of the channel phase shift, obtained in the presence of slow fading, is noiseless. Such an ideal condition may not hold in practice. In such a case, the expressions in Equations 14.3–7 and 14.3–8 should be viewed as representing the best achievable performance in the presence of Rayleigh fading. In Appendix C we consider the problem of estimating the phase in the presence of noise and we evaluate the error rate performance of binary and multiphase PSK.

On channels for which the fading is sufficiently rapid to preclude the estimation of a stable phase reference by averaging the received signal phase over many signaling intervals, DPSK, is an alternative signaling method. Since DPSK requires phase stability over only two consecutive signaling intervals, this modulation technique is quite robust in the presence of signal fading. In deriving the performance of binary DPSK for a fading channel, we begin again with the error probability for a nonfading channel, which is

$$P_2(\gamma_b) = \tfrac{1}{2}e^{-\gamma_b} \tag{14.3–9}$$

This expression is substituted into the integral in Equation 14.3–4 along with $p(\gamma_b)$ obtained from Equation 14.3–5. Evaluation of the resulting integral yields the probability of error for binary DPSK, in the form

$$P_2 = \frac{1}{2(1+\bar{\gamma}_b)} \tag{14.3–10}$$

If we choose not to estimate the channel phase shift at all, but instead employ a noncoherent (envelope or square-law) detector with binary, orthogonal FSK signals, the error probability for a nonfading channel is

$$P_2(\gamma_b) = \tfrac{1}{2}e^{-\gamma_b/2} \tag{14.3–11}$$

When we average $P_2(\gamma_b)$ over the Rayleigh fading channel attenuation, the resulting error probability is

$$P_2 = \frac{1}{2+\bar{\gamma}_b} \tag{14.3–12}$$

The error probabilities in Equations 14.3–7, 14.3–8, 14.3–10, and 14.3–12 are illustrated in Figure 14.3–1. In comparing the performance of the four binary signaling systems, we focus our attention on the probabilities of error for large

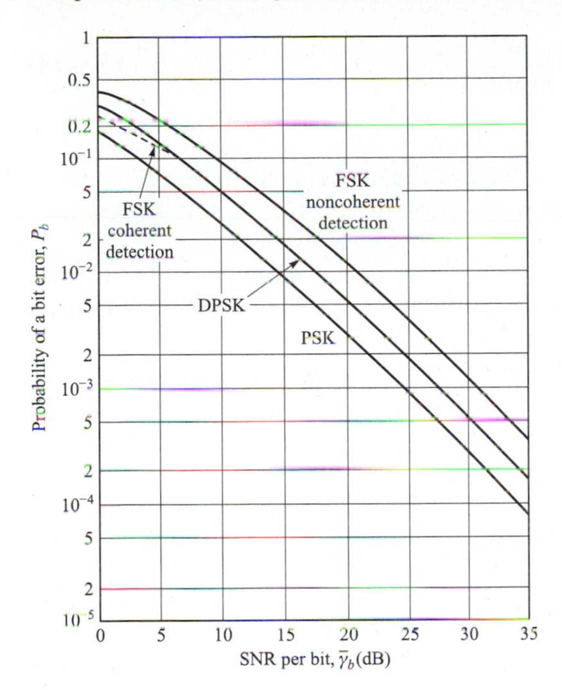

FIGURE 14.3–1
Performance of binary signaling on a Rayleigh fading channel.

SNR, i.e., $\bar{\gamma}_b \gg 1$. Under this condition, the error rates in Equations 14.3–7, 14.3–8, 14.3–10, and 14.3–12 simplify to

$$P_2 \approx \begin{cases} 1/4\bar{\gamma}_b & \text{for coherent PSK} \\ 1/2\bar{\gamma}_b & \text{for coherent, orthogonal FSK} \\ 1/2\bar{\gamma}_b & \text{for DPSK} \\ 1/\bar{\gamma}_b & \text{for noncoherent, orthogonal FSK} \end{cases} \tag{14.3–13}$$

From Equation 14.3–13, we observe that coherent PSK is 3 dB better than DPSK and 6 dB better than noncoherent FSK. More striking, however, is the observtion that the error rates decrease only inversely with SNR. In contrast, the decrease in error rate on a nonfading channel is exponential with SNR. This means that, on a fading channel, the transmitter must transmit a large amount of power in order to obtain a low probability of error. In many cases, a large amount of power is not possible, technically and/or economically. An alternative solution to the problem of obtaining acceptable performance on a fading channel is the use of redundancy, which can be obtained by means of diversity techniques, as discussed in Section 14.4.

Nakagami fading. If α is characterized statistically by the Nakagami-m distribution, the random variable $\gamma = \alpha^2 \mathcal{E}_b/N_0$ has the PDF (see Problem 14.15)

$$p(\gamma) = \frac{m^m}{\Gamma(m)\bar{\gamma}^m} \gamma^{m-1} e^{-m\gamma/\bar{\gamma}} \tag{14.3–14}$$

where $\bar{\gamma} = E(\alpha^2)\mathcal{E}/N_0$.

The average probability of error for any of the modulation methods is simply obtained by averaging the appropriate error probability for a nonfading channel over the fading signal statistics.

As an example of the performance obtained with Nakagami-m fading statistics, Figure 14.3–2 illustrates the probability of error of binary PSK with m as a parameter. We recall that $m = 1$ corresponds to Rayleigh fading. We observe that the performance improves as m is increased above $m = 1$, which is indicative of the fact that the fading is less severe. On the other hand, when $m < 1$, the performance is worse than Rayleigh fading.

Other fading signal statistics. Following the procedure describe above, one can determine the performance of the various modulation methods for other types of fading signal statistics, such as the Rice distribution.

Error probability results for Rice-distributed fading statistics can be found in the paper by Lindsey (1964), while for Nakagami-m fading statistics, the reader may refer to the papers by Esposito (1967), Miyagaki et al. (1978), Charash (1979), Al-Hussaini et al. (1985), and Beaulieu and Abu-Dayya (1991).

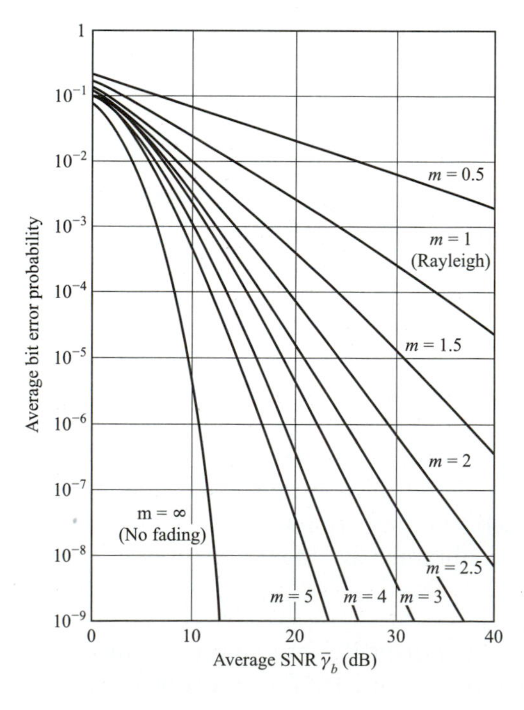

FIGURE 14.3–2
Average error probability for two-phase PSK with Nakagami fading.

14.4 DIVERSITY TECHNIQUES FOR FADING MULTIPATH CHANNELS

Diversity techniques are based on the notion that errors occur in reception when the channel attenuation is large, i.e., when the channel is in a deep fade. If we can supply to the receiver several replicas of the same information signal transmitted over independently fading channels, the probability that all the signal components will fade simultaneously is reduced considerably. That is, if p is the probability that any one signal will fade below some critical value, then p^L is the probability that all L independently fading replicas of the same signal will fade below the critical value. There are several ways in which we can provide the receiver with L independently fading replicas of the same information-bearing signal.

One method is to employ *frequency diversity*. That is, the same information-bearing signal is transmitted on L carriers, where the separation between successive carriers equals or exceeds the coherence bandwidth $(\Delta f)_c$ of the channel.

A second method for achieving L independently fading versions of the same information-bearing signal is to transmit the signal in L different time slots, where the separation between successive time slots equals or exceeds the coherence time $(\Delta t)_c$ of the channel. This method is called *time diversity*.

Note that the fading channel fits the model of a bursty error channel. Furthermore, we may view the transmission of the same information either at different frequencies or in difference time slots (or both) as a simple form of repetition coding. The separation of the diversity transmissions in time by $(\Delta t)_c$ or in frequency by $(\Delta f)_c$ is basically a form of block-interleaving the bits in the repetition code in an attempt to break up the error bursts and, thus, to obtain independent errors. Later in the chapter, we shall demonstrate that, in general, repetition coding is wasteful of bandwidth when compared with nontrivial coding.

Another commonly used method for achieving diversity employs multiple antennas. For example, we may employ a single transmitting antenna and multiple receiving antennas. The latter must be spaced sufficiently far apart that the multipath components in the signal have significantly different propagation delays at the antennas. Usually a separation of a few wavelengths is required between two antennas in order to obtain signals that fade independently.

A more sophisticated method for obtaining diversity is based on the use of a signal having a bandwidth much greater than the coherence bandwidth $(\Delta f)_c$ of the channel. Such a signal with bandwidth W will resolve the multipath components and, thus, provide the receiver with several independently fading signal paths. The time resolution is $1/W$. Consequently, with a multipath spread of T_m seconds, there are $T_m W$ resolvable signal components. Since $T_m \approx 1/(\Delta f)_c$, the number of resolvable signal components may also be expressed as $W/(\Delta f)_c$. Thus, the use of a wideband signal may be viewed as just another method for obtaining frequency diversity of order $L \approx W/(\Delta f)_c$. The optimum receiver for processing the wideband signal will be derived in Section 14.5. It is called a

RAKE correlator or a *RAKE matched filter* and was invented by Price and Green (1958).

There are other diversity techniques that have received some consideration in practice, such as angle-of-arrival diversity and polarization diversity. However, these have not been as widely used as those described above.

14.4.1 Binary Signals

We shall now determine the error rate performance for a binary digital communication system with diversity. We begin by describing the mathematical model for the communication system with diversity. First of all, we assume that there are L diversity channels, carrying the same information-bearing signal. Each channel is assumed to be frequency-nonselective and slowly fading with Rayleigh-distributed envelope statistics. The fading processes among the L diversity channels are assumed to be mutually statistically independent. The signal in each channel is corrupted by an additive zero-mean white Gaussian noise process. The noise processes in the L channels are assumed to be mutually statistically independent, with identical autocorrelation functions. Thus, the equivalent low-pass received signals for the L channels can be expressed in the form

$$r_{lk}(t) = \alpha_k e^{-j\phi_k} s_{km}(t) + z_k(t), \qquad k = 1, 2, \ldots, L, \qquad m = 1, 2 \tag{14.4–1}$$

where $\{\alpha_k e^{-j\phi_k}\}$ represent the attenuation factors and phase shifts for the L channels, $s_{km}(t)$ denotes the mth signal transmitted on the kth channel, and $z_k(t)$ denotes the additive white Gaussian noise on the kth channel. All signals in the set $\{s_{km}(t)\}$ have the same energy.

The optimum demodulator for the signal received from the kth channel consists of two matched filters, one having the impulse response

$$b_{k1}(t) = s_{k1}^*(T - t) \tag{14.4–2}$$

and the other having the impulse response

$$b_{k2}(t) = s_{k2}^*(T - t) \tag{14.4–3}$$

Of course, if binary PSK is the modulation method used to transmit the information, then $s_{k1}(t) = -s_{k2}(t)$. Consequently, only a single matched filter is required for binary PSK. Following the matched filters is a combiner that forms the two decision variables. The combiner that achieves the best performance is one in which each matched filter output is multiplied by the corresponding complex-valued (conjugate) channel gain $\alpha_k e^{j\phi_k}$. The effect of this multiplication is to compensate for the phase shift in the channel and to weight the signal by a factor that is proportional to the signal strength. Thus, a strong signal carries a larger weight than a weak signal. After the complex-valued weighting operation is performed, two sums are formed. One consists of the real parts of the weighted outputs from the matched filters corresponding to a transmitted 0. The second consists of the real part of the outputs from the matched filters corresponding to a transmitted 1. This optimum combiner is called a *maximal ratio combiner* by Brennan (1959). Of course, the realization of this optimum combiner is based on the assumption that the channel attenuations $\{\alpha_k\}$ and the phase shifts $\{\phi_k\}$ are known perfectly. That is, the estimates of the parameters $\{\alpha_k\}$ and $\{\phi_k\}$ contain no noise. (The effect of noisy estimates on the error rate performance of multiphase PSK is considered in Appendix C.)

A block diagram illustrating the model for the binary digital communication system described above is shown in Figure 14.4–1.

Let us first consider the performance of binary PSK with Lth-order diversity. The output of the maximal ratio combiner can be expressed as a single decision variable in the form

$$\begin{aligned} U &= \mathrm{Re}\left(2\mathcal{E}\sum_{k=1}^{L}\alpha_k^2 + \sum_{k=1}^{L}\alpha_k N_k\right) \\ &= 2\mathcal{E}\sum_{k=1}^{L}\alpha_k^2 + \sum_{k=1}^{L}\alpha_k N_{kr} \end{aligned} \tag{14.4–4}$$

where N_{kr} denotes the real part of the complex-valued Gaussian noise variable

$$N_k = e^{j\phi_k}\int_0^T z_k(t)s_k^*(t)\,dt \tag{14.4–5}$$

We follow the approach used in Section 14.3 in deriving the probability of error. That is, the probability of error conditioned on a fixed set of attenuation factors $\{\alpha_k\}$ is obtained first. Then the conditional probability of error is averaged over the probability density function of the $\{\alpha_k\}$.

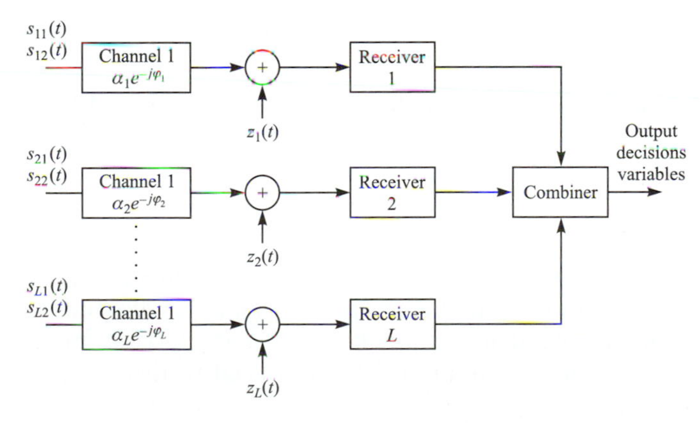

FIGURE 14.4–1
Model of binary digital communication system with diversity.

Rayleigh fading. For a fixed set of $\{\alpha_k\}$ the decision variable U is Gaussian with mean

$$E(U) = 2\mathcal{E} \sum_{k=1}^{L} \alpha_k^2 \tag{14.4–6}$$

and variance

$$\sigma_U^2 = 2\mathcal{E}N_0 \sum_{k=1}^{L} \alpha_k^2 \tag{14.4–7}$$

For these values of the mean and variance, the probability that U is less than zero is simply

$$P_2(\gamma_b) = Q\left(\sqrt{2\gamma_b}\right) \tag{14.4–8}$$

where the SNR per bit, γ_b, is given as

$$\begin{aligned} \gamma_b &= \frac{\mathcal{E}}{N_0} \sum_{k=1}^{L} \alpha_k^2 \\ &= \sum_{k=1}^{L} \gamma_k \end{aligned} \tag{14.4–9}$$

where $\gamma_k = \mathcal{E}\alpha_k^2/N_0$ is the instantaneous SNR on the kth channel. Now we must determine the probability density function $p(\gamma_b)$. This function is most easily determined via the characteristic function of γ_b. First of all, we note that for $L = 1$, $\gamma_b \equiv \gamma_1$ has a chi-square probability density function given in Equation 14.3–5. The characteristic function of γ_1 is easily shown to be

$$\begin{aligned} \psi_{\gamma_1}(jv) &= E(e^{jv\gamma_1}) \\ &= \frac{1}{1 - jv\bar{\gamma}_c} \end{aligned} \tag{14.4–10}$$

where $\bar{\gamma}_c$ is the average SNR per channel, which is assumed to be identical for all channels. That is,

$$\bar{\gamma}_c = \frac{\mathcal{E}}{N_0} E(\alpha_k^2) \tag{14.4–11}$$

independent of k. This assumption applies for the results throughout this section. Since the fading on the L channels is mutually statistically independent, the $\{\gamma_k\}$ are statistically independent, and, hence, the characteristic function for the sum γ_b is simply the result in Equation 14.4–10 raised to the Lth power, i.e.,

$$\psi_{\gamma_b}(jv) = \frac{1}{(1 - jv\bar{\gamma}_c)^L} \tag{14.4–12}$$

But this is the characteristic function of a chi-square-distributed random variable with $2L$ degrees of freedom. It follows from Equation 2.1–10 that the probability density function $p(\gamma_b)$ is

$$p(\gamma_b) = \frac{1}{(L-1)!\bar{\gamma}_c^L} \gamma_b^{L-1} e^{-\gamma_b/\bar{\gamma}_c} \tag{14.4–13}$$

The final step in this derivation is to average the conditional error probability given in Equation 14.4–8 over the fading channel statistics. Thus, we evaluate the integral

$$P_2 = \int_0^\infty P_2(\gamma_b) p(\gamma_b)\, d\gamma_b \tag{14.4–14}$$

There is a closed-form solution for Equation 14.4–14, which can be expressed as

$$P_2 = \left[\tfrac{1}{2}(1-\mu)\right]^L \sum_{k=0}^{L-1} \binom{L-1+k}{k} \left[\tfrac{1}{2}(1+\mu)\right]^k \tag{14.4–15}$$

where, by definition

$$\mu = \sqrt{\frac{\bar{\gamma}_c}{1+\bar{\gamma}_c}} \tag{14.4–16}$$

When the average SNR per channel, $\bar{\gamma}_c$, satisfies the condition $\bar{\gamma}_c \gg 1$, the term $\frac{1}{2}(1+\mu) \approx 1$ and the term $\frac{1}{2}(1-\mu) \approx 1/4\bar{\gamma}_c$. Furthermore,

$$\sum_{k=0}^{L-1} \binom{L-1+k}{k} = \binom{2L-1}{L} \tag{14.4–17}$$

Therefore, when $\bar{\gamma}_c$ is sufficiently large (greater than 10 dB), the probability of error in Equation 14.4–15 can be approximated as

$$P_2 \approx \left(\frac{1}{4\bar{\gamma}_c}\right)^L \binom{2L-1}{L} \tag{14.4–18}$$

We observe from Equation 14.4–18 that the probability of error varies as $1/\bar{\gamma}_c$ raised to the Lth power. Thus, with diversity, the error rate decreases inversely with the Lth power of the SNR.

Having obtained the performance of binary PSK with diversity, we now turn our attention to binary, orthogonal FSK that is detected coherently. In this case, the two decision variables at the output of the maximal ratio combiner may be expressed as

$$\begin{aligned} U_1 &= \operatorname{Re}\left(2\mathcal{E} \sum_{k=1}^{L} \alpha_k^2 + \sum_{k=1}^{L} \alpha_k N_{k1}\right) \\ U_2 &= \operatorname{Re}\left(\sum_{k=1}^{L} \alpha_k N_{k2}\right) \end{aligned} \tag{14.4–19}$$

where we have assumed that signal $s_{k1}(t)$ was transmitted and where $\{N_{k1}\}$ and $\{N_{k2}\}$ are the two sets of noise components at the output of the matched filters. The probability of error is simply the probability that $U_2 > U_1$. This computation is similar to the one performed for PSK, except that we now have twice the noise power. Consequently, when the $\{\alpha_k\}$ are fixed, the conditional probability of error is

$$P_2(\gamma_b) = Q(\sqrt{\gamma_b}) \tag{14.4–20}$$

We use Equation 14.4–13 to average $P_2(\gamma_b)$ over the fading. It is not surprising to find that the result given in Equation 14.4–15 still applies, with $\bar{\gamma}_c$ replaced by $\frac{1}{2}\bar{\gamma}_c$. That is, Equation 14.4–15 is the probability of error for binary, orthogonal FSK with coherent detection, where the parameter μ is defined as

$$\mu = \sqrt{\frac{\bar{\gamma}_c}{2 + \bar{\gamma}_c}} \tag{14.4–21}$$

Furthermore, for large values of $\bar{\gamma}_c$, the performance P_2 can be approximated as

$$P_2 \approx \left(\frac{1}{2\bar{\gamma}_c}\right)^L \binom{2L-1}{L} \tag{14.4–22}$$

In comparing Equation 14.4–22 with Equation 14.4–18, we observe that the 3-dB difference in performance between PSK and orthogonal FSK with coherent detection, which exists in a nonfading, nondispersive channel, is the same also in a fading channel.

In the above discussion of binary PSK and FSK, detected coherently, we assumed that noiseless estimates of the complex-valued channel parameters $\{\alpha_k e^{-j\phi_k}\}$ were used at the receiver. Since the channel is time-variant, the parameters $\{\alpha_k e^{-j\phi_k}\}$ cannot be estimated perfectly. In fact, on some channels, the time variations may be sufficiently fast to preclude the implementation of coherent detection. In such a case, we should consider using either DPSK or FSK with noncoherent detection.

Let us consider DPSK first. In order for DPSK to be a viable digital signaling method, the channel variations must be sufficiently slow so that the channel phase shifts $\{\phi_k\}$ do not change appreciably over two consecutive signaling intervals. In our analysis, we assume that the channel parameters $\{\alpha_k e^{-j\phi_k}\}$ remain constant over two successive signaling intervals. Thus the combiner for binary DPSK will yield as an output the decision variable

$$U = \text{Re}\left[\sum_{k=1}^{L} (2\mathcal{E}\alpha_k e^{-j\phi_k} + N_{k2})(2\mathcal{E}\alpha_k e^{j\phi_k} + N_{k1}^*)\right] \tag{14.4–23}$$

where $\{N_{k1}\}$ and $\{N_{k2}\}$ denote the received noise components at the output of the matched filters in the two consecutive signaling intervals. The probability of error is simply the probability that $U < 0$. Since U is a special case of the general quadratic form in complex-valued Gaussian random variables treated in Appendix B, the probability of error can be obtained directly from the results given in that appendix. Alternatively, we may use the error probability given in 12.1–3, which applies to binary DPSK transmitted over L time-invariant channels, and average it over the Rayleigh fading channel statistics. Thus, we have the conditional error probability

$$P_2(\gamma_b) = (\tfrac{1}{2})^{2L-1} e^{-\gamma_b} \textstyle\sum_{k=0}^{L-1} b_k \gamma_b^k \tag{14.4–24}$$

where γ_b is given by Equation 14.4–9 and

$$b_k = \frac{1}{k!} \sum_{n=0}^{L-1-k} \binom{2L-1}{n} \tag{14.4–25}$$

The average of $P_2(\gamma_b)$ over the fading channel statistics given by $p(\gamma_b)$ in Equation 14.4–13 is easily shown to be

$$P_2 = \frac{1}{2^{2L-1}(L-1)!(1+\bar{\gamma}_c)^L} \sum_{k=0}^{L-1} b_k (L-1+k)! \left(\frac{\bar{\gamma}_c}{1+\bar{\gamma}_c} \right)^k \tag{14.4–26}$$

We indicate that the result in Equation 14.4–26 can be manipulated into the form given in Equation 14.4–15, which applies also to coherent PSK and FSK. For binary DPSK, the parameter μ in Equation 14.4–15 is defined as (see Appendix C)

$$\mu = \frac{\bar{\gamma}_c}{1+\bar{\gamma}_c} \tag{14.4–27}$$

For $\bar{\gamma}_c \gg 1$, the error probability in Equation 14.4–26 can be approximated by the expression

$$P_2 \approx \left(\frac{1}{2\bar{\gamma}_c} \right)^L \binom{2L-1}{L} \tag{14.4–28}$$

Orthogonal FSK with noncoherent detection is the final signaling technique that we consider in this section. It is appropriate for both slow and fast fading. However, the analysis of the performance presented below is based on the assumption that the fading is sufficiently slow so that the channel parameters $\{\alpha_k e^{-j\phi_k}\}$ remain constant for the duration of the signaling interval. The combiner for the multichannel signals is a square-law combiner. Its output consists of the two decision variables

$$\begin{aligned} U_1 &= \sum_{k=1}^{L} |2\mathcal{E}\alpha_k e^{-j\phi_k} + N_{k1}|^2 \\ U_2 &= \sum_{k=1}^{L} |N_{k2}|^2 \end{aligned} \tag{14.4–29}$$

where U_1 is assumed to contain the signal. Consequently the probability of error is the probability that $U_2 > U_1$.

As in DPSK, we have a choice of two approaches in deriving the performance of FSK with square-law combining. In Section 12.1, we indicated that the expression for the error probability for square-law-combined FSK is the same as that for DPSK with γ_b replaced by $\frac{1}{2}\gamma_b$. That is, the FSK system requires 3 dB of additional SNR to achieve the same performance on a time-invariant channel. Consequently, the conditional error probability for DPSK given in Equation 14.4–24 applies to square-law-combined FSK when γ_b is replaced by $\frac{1}{2}\gamma_b$. Furthermore, the result obtained by averaging Equation 14.4–24 over the fading, which is given by Equation 14.4–26, must also apply to FSK with $\bar{\gamma}_c$ replaced by $\frac{1}{2}\bar{\gamma}_c$. But we also stated previously that Equations 14.4–26 and 14.4–15 are equivalent. Therefore, the error probability given in Equation 14.4–15 also applies to square-law-combined FSK with the parameter μ defined as

$$\mu = \frac{\bar{\gamma}_c}{2 + \bar{\gamma}_c} \tag{14.4–30}$$

An alternative derivation used by Pierce (1958) to obtain the probability that the decision variable $U_2 > U_1$ is just as easy as the method described above. It begins with the probability density functions $p(U_1)$ and $p(U_2)$. Since the complex-valued random variables $\{\alpha_k e^{-j\phi_k}\}$, $\{N_{k1}\}$, and $\{N_{k2}\}$ are zero-mean Gaussian-distributed, the decision variables U_1 and U_2 are distributed according to a chi-square probability distribution with $2L$ degrees of freedom. That is,

$$p(U_1) = \frac{1}{(2\sigma_1^2)^L (L-1)!} U_1^{L-1} \exp\left(-\frac{U_1}{2\sigma_1^2}\right) \tag{14.4–31}$$

where

$$\begin{aligned} \sigma_1^2 &= \tfrac{1}{2} E(|2\mathcal{E}\alpha_k e^{-j\phi_k} + N_{k1}|^2) \\ &= 2\mathcal{E}N_0(1 + \bar{\gamma}_c) \end{aligned}$$

Similarly,

$$p(U_2) = \frac{1}{(2\sigma_2^2)(L-1)!} U_2^{L-1} \exp\left(-\frac{U_2}{2\sigma_2^2}\right) \tag{14.4–32}$$

where

$$\sigma_2^2 = 2\mathcal{E}N_0$$

The probability of error is just the probability that $U_2 > U_1$. It is left as an exercise for the reader to show that this probability is given by Equation 14.4–15, where μ is defined by Equation 14.4–30.

When $\bar{\gamma}_c \gg 1$, the performance of square-law-detected FSK can be simplified as we have done for the other binary multichannel systems. In this case, the error rate is well approximated by the expression

$$P_2 \approx \left(\frac{1}{\bar{\gamma}_c}\right)^L \binom{2L-1}{L} \tag{14.4–33}$$

The error rate performance of PSK, DPSK, and square-law-detected orthogonal FSK is illustrated in Figure 14.4–2 for $L = 1$, 2, and 4. The performance is

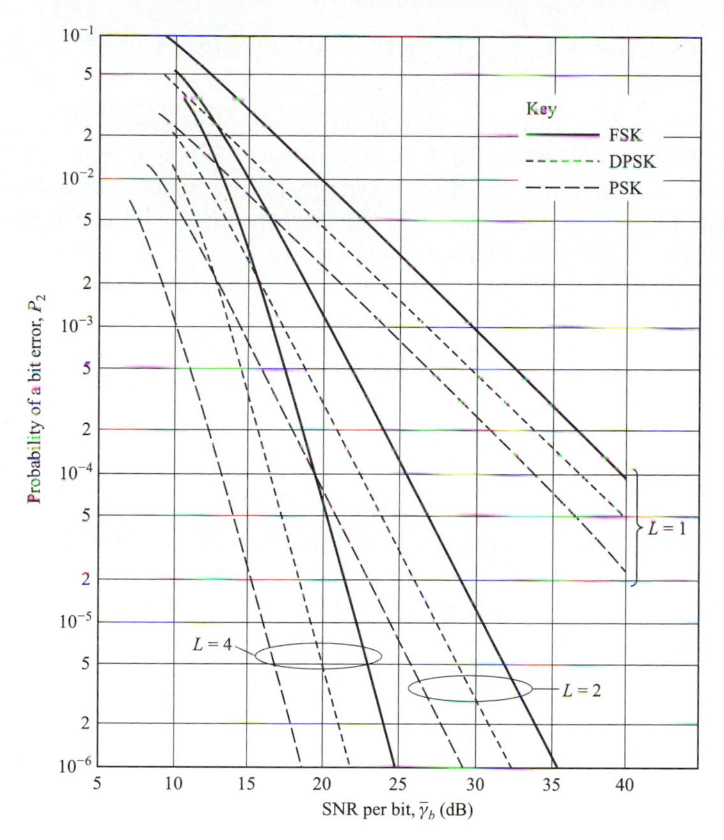

FIGURE 14.4–2
Performance of binary signals with diversity.

plotted as a function of the average SNR per bit, $\bar{\gamma}_b$, which is related to the average SNR per channel, $\bar{\gamma}_c$, by the formula

$$\bar{\gamma}_b = L\bar{\gamma}_c \tag{14.4–34}$$

The results in Figure 14.4–2 clearly illustrate the advantage of diversity as a means for overcoming the severe penalty in SNR caused by fading.

Nakagami fading. It is a simple matter to extend the results of this section to other fading models. We shall briefly consider Nakagami fading. Let us compare the Nakagami PDF for the single-channel SNR parameter $\gamma_b = \alpha^2\mathcal{E}_b/N_0$, previously given by Equation 14.3–14 as

$$p(\gamma_b) = \frac{1}{\Gamma(m)(\bar{\gamma}_b/m)^m}\gamma_b^{m-1}e^{-\gamma_b/(\bar{\gamma}_b/m)} \tag{14.4–35}$$

with the PDF $p(\gamma_b)$ obtained for the L-channel SNR with Rayleigh fading, given by Equation 14.4–13 as

$$p(\gamma_b) = \frac{1}{(L-1)!\bar{\gamma}_c^L} \gamma_b^{L-1} e^{-\gamma_b/\bar{\gamma}_c} \tag{14.4–36}$$

By noting that $\bar{\gamma}_c = \bar{\gamma}_b / L$ in the case of an Lth order diversity system, it is clear that the two PDFs are identical for $L = m =$ integer. When $L = m = 1$, the two PDFs correspond to a single channel Rayleigh fading system. For the case in which the Nakagami parameter $m = 2$, the performance of the single-channel system is identical to the performance obtained in a Rayleigh fading channel with dual ($L = 2$) diversity. More generally, any single-channel system with Nakagami fading in which the parameter m is an integer, is equivalent to an L-channel diversity system for a Rayleigh fading channel. In view of this equivalence, the characteristic function of a Nakagami-m random variable must be of the form

$$\psi_{\gamma_b}(jv) = \frac{1}{(1 - jv\bar{\gamma}_b/m)^m} \tag{14.4–37}$$

which is consistent with the result given in Equation 14.4–12 for the characteristic function of the combined signal in a system with Lth order diversity in a Rayleigh fading channel. Consequently, it follows that a K-channel system transmitting in a Nakagami fading channel with independent fading is equivalent to an $L = Km$ channel diversity in a Rayleigh fading channel.

14.4.2 Multiphase Signals

Multiphase signaling over a Rayleigh fading channel is the topic presented in some detail in Appendix C. Our main purpose in this section is to cite the general result for the probability of a symbol error in M-ary PSK and DPSK systems and the probability of a bit error in four-phase PSK and DPSK.

The general result for the probability of a symbol error in M-ary PSK and DPSK is

$$P_M = \frac{(-1)^{L-1}(1-\mu^2)^L}{\pi(L-1)!} \left(\frac{\partial^{L-1}}{\partial b^{L-1}} \left\{ \frac{1}{b-\mu^2} \left[\frac{\pi}{M}(M-1) - \frac{\mu \sin(\pi/M)}{\sqrt{b-\mu^2\cos^2(\pi/M)}} \cot^{-1} \frac{-\mu\cos(\pi/M)}{\sqrt{b-\mu^2\cos^2(\pi/M)}} \right] \right\} \right)_{b=1} \tag{14.4–38}$$

where

$$\mu = \sqrt{\frac{\bar{\gamma}_c}{1+\bar{\gamma}_c}} \tag{14.4–39}$$

for coherent PSK and

$$\mu = \frac{\bar{\gamma}_c}{1+\bar{\gamma}_c} \tag{14.4–40}$$

for DPSK. Again $\bar{\gamma}_c$ is the average received SNR per channel. The SNR per bit is $\bar{\gamma}_b = L\bar{\gamma}_c/k$, where $k = \log_2 M$.

The bit error rate for four-phase PSK and DPSK is derived on the basis that the pair of information bits is mapped into the four phases according to a Gray code. The expression for the bit error rate derived in Appendix C is

$$P_b = \frac{1}{2}\left[1 - \frac{\mu}{\sqrt{2-\mu^2}}\sum_{k=0}^{L-1}\binom{2k}{k}\left(\frac{1-\mu^2}{4-2\mu^2}\right)^k\right] \tag{14.4–41}$$

where μ is again given by Equations 14.4–39 and 14.4–40 for PSK and DPSK, respectively.

Figure 14.4–3 illustrates the probability of a symbol error of DPSK and coherent PSK for $M = 2$, 4, and 8 with $L = 1$. Note that the difference in performance between DPSK and coherent PSK is approximately 3 dB for all three values of M. In fact, when $\bar{\gamma}_b \gg 1$ and $L = 1$, Equation 14.4–38 is well approximated as

$$P_M \approx \frac{M-1}{(M\log_2 M)[\sin^2(\pi/M)]\bar{\gamma}_b} \tag{14.4–42}$$

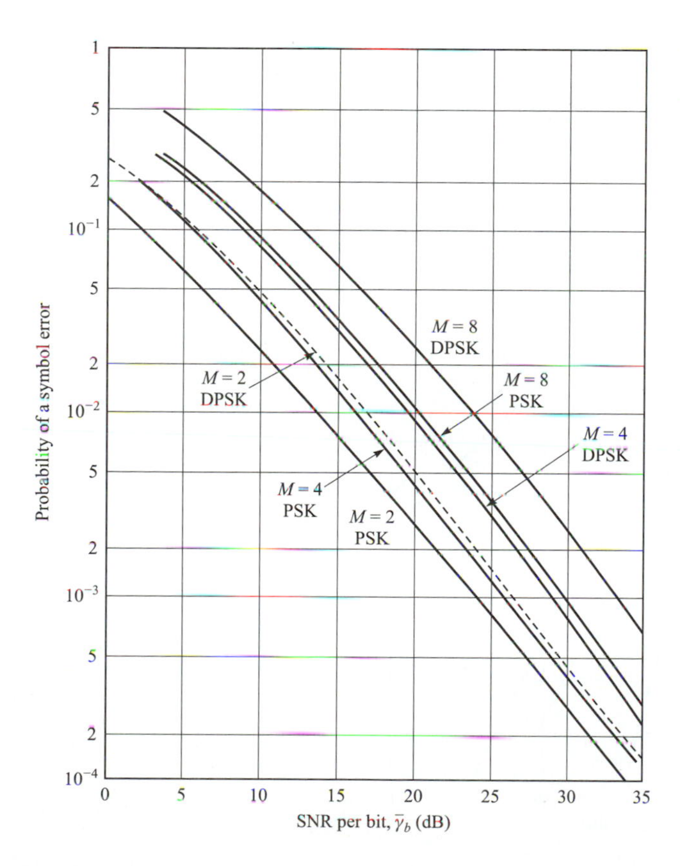

FIGURE 14.4–3
Probability of symbol error for PSK and DPSK for Rayleigh fading.

for DPSK and as

$$P_M \approx \frac{M-1}{(M \log_2 M)[\sin^2(\pi/M)]2\bar{\gamma}_b} \tag{14.4–43}$$

for PSK. Hence, at high SNR, coherent PSK is 3 dB better than DPSK on a Raleigh fading channel. This difference also holds as L is increased.

Bit error probabilities are depicted in Figure 14.4–4 for two-phase, four-phase, and eight-phase DPSK signaling with $L = 1$, 2, and 4. The expression for the bit error probability of eight-phase DPSK with Gray encoding is not given here, but it is available in the paper by Proakis (1968). In this case, we observe that the performances for two- and four-phase DPSK are (approximately) the same, while that for eight-phase DPSK is about 3 dB poorer. Although we have not shown the bit error probability for coherent PSK, it can be demonstrated that two- and four-phase coherent PSK also yield approximately the same performance.

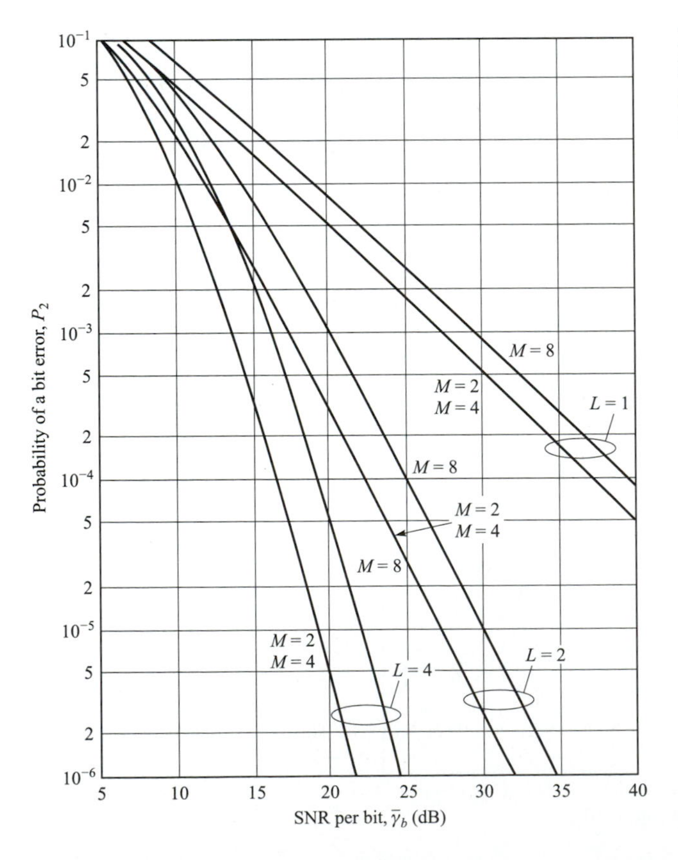

FIGURE 14.4–4
Probability of a bit error for DPSK with diversity for Rayleigh fading.

14.4.3 M-ary Orthogonal Signals

In this subsection, we determine the performance of M-ary orthogonal signals transmitted over a Rayleigh fading channel and we assess the advantages of higher-order signal alphabets relative to a binary alphabet. The orthogonal signals may be viewed as M-ary FSK with a minimum frequency separation of an integer multiple of $1/T$, where T is the signaling interval. The same information-bearing signal is transmitted on L diversity channels. Each diversity channel is assumed to be frequency-nonselective and slowly fading, and the fading processes on the L channels are assumed to be mutually statistically independent. An additive white Gaussian noise process corrupts the signal on each diversity channel. We assume that the additive noise processes are mutually statistically independent.

Although it is relatively easy to formulate the structure and analyze the performance of a maximal ratio combiner for the diversity channels in the M-ary communication system, it is more likely that a practical system would employ noncoherent detection. Consequently, we confine our attention to square-law combining of the diversity signals. The output of the combiner containing the signal is

$$U_1 = \sum_{k=1}^{L} |2\mathcal{E}\alpha_k e^{-j\phi_k} + N_{k1}|^2 \tag{14.4–44}$$

while the outputs of the remaining $M - 1$ combiners are

$$U_m = \sum_{k=1}^{L} |N_{km}|^2, \qquad m = 2, 3, 4, \ldots, M \tag{14.4–45}$$

The probability of error is simply 1 minus the probability that $U_1 > U_m$ for $m = 2, 3, \ldots, M$. Since the signals are orthogonal and the additive noise processes are mutually statistically independent, the random variables $U_1, U_2, \ldots, U_M$ are also mutually statistically independent. The probability density function of U_1 was given in Equation 14.4–31. On the other hand, $U_2, \ldots, U_M$ are identically distributed and described by the marginal probability density function in Equation 14.4–32. With U_1 fixed, the joint probability $P(U_2 < U_1, U_3 < U_1, \ldots, U_m < U_1)$ is equal to $P(U_2 < U_1)$ raised to the $M - 1$ power. Now,

$$\begin{aligned} P(U_2 < U_1) &= \int_0^{U_1} p(U_2)\, dU_2 \\ &= 1 - \exp\left(-\frac{U_1}{2\sigma_2^2}\right) \sum_{k=0}^{L-1} \frac{1}{k!} \left(\frac{U_1}{2\sigma_2^2}\right)^k \end{aligned} \tag{14.4–46}$$

where $\sigma_2^2 = 2\mathcal{E}N_0$. The $M - 1$ power of this probability is then averaged over the probability density function of U_1 to yield the probability of a correct decision. If we subtract this result from unity, we obtain the probability of error in the form given by Hahn (1962)

$$\begin{aligned} P_M &= 1 - \int_0^\infty \frac{1}{(2\sigma_1^2)^L (L-1)!} U_1^{L-1} \exp\left(-\frac{U_1}{2\sigma_1^2}\right) \\ &\quad \times \left[1 - \exp\left(-\frac{U_1}{2\sigma_2^2}\right) \sum_{k=0}^{L-1} \frac{1}{k!} \left(\frac{U_1}{2\sigma_2^2}\right)^k\right]^{M-1} dU_1 \\ &= 1 - \int_0^\infty \frac{1}{(1+\bar{\gamma}_c)^L (L-1)!} U_1^{L-1} \exp\left(-\frac{U_1}{1+\bar{\gamma}_c}\right) \\ &\quad \times \left(1 - e^{-U_1} \sum_{k=0}^{L-1} \frac{U_1^k}{k!}\right)^{M-1} dU_1 \end{aligned} \tag{14.4–47}$$

where $\bar{\gamma}_c$ is the average SNR per diversity channel. The average SNR per bit is $\bar{\gamma}_b = L\bar{\gamma}_c / \log_2 M = L\bar{\gamma}_c/k$.

The integral in Equation 14.4–47 can be expressed in closed form as a double summation. This can be seen if we write

$$\left(\sum_{k=0}^{L-1} \frac{U_1^k}{k!}\right)^m = \sum_{k=0}^{m(L-1)} \beta_{km} U_1^k \tag{14.4–48}$$

where β_{km} is the set of coefficients in the above expansion. Then it follows that Equation 14.4–47 reduces to

$$\begin{aligned} P_M &= \frac{1}{(L-1)!} \sum_{m=1}^{M-1} \frac{(-1)^{m+1} \binom{M-1}{m}}{(1+m+m\bar{\gamma}_c)^L} \\ &\quad \times \sum_{k=0}^{m(L-1)} \beta_{km} (L-1+k)! \left(\frac{1+\bar{\gamma}_c}{1+m+m\bar{\gamma}_c}\right)^k \end{aligned} \tag{14.4–49}$$

When there is no diversity ($L = 1$), the error probability in Equation 14.4–49 reduces to the simple form

$$P_M = \sum_{m=1}^{M-1} \frac{(-1)^{m+1} \binom{M-1}{m}}{1+m+m\bar{\gamma}_c} \tag{14.4–50}$$

The symbol error rate P_M may be converted to an equivalent bit error rate by multiplying P_M with $2^{k-1}/(2^k - 1)$.

Although the expression for P_M given in Equation 14.4–49 is in closed form, it is computationally cumbersome to evaluate for large values of M and L. An alternative is to evaluate P_M by numerical integration using the expression in Equation 14.4–47. The results illustrated in the following graphs were generated from Equation 14.4–47.

First of all, let us observe the error rate performance of M-ary orthogonal signaling with square-law combining as a function of the order of diversity. Figures 14.4–5 and 14.4–6 illustrate the characteristics of P_M for $M = 2$ and 4

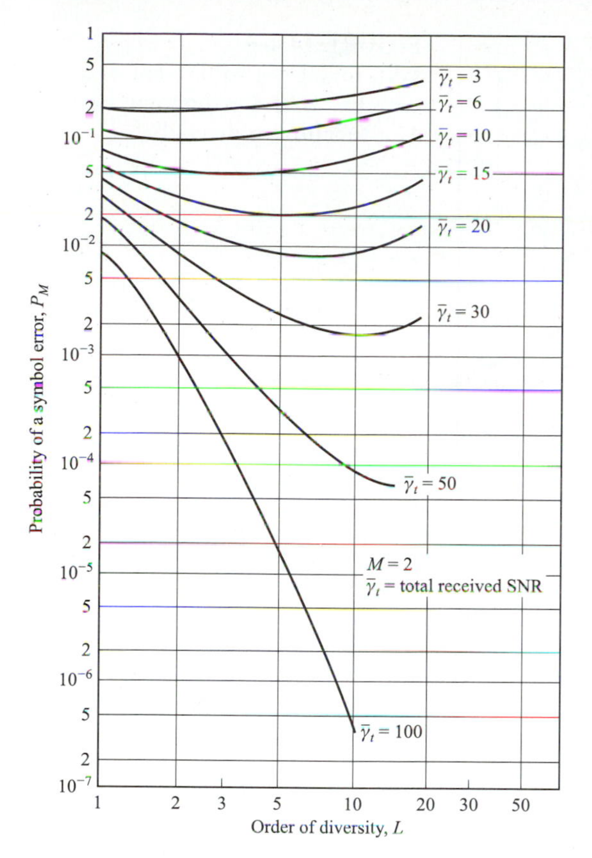

FIGURE 14.4–5
Performance of square-law-detected binary orthogonal signals as a function of diversity.

as a function of L when the total SNR, defined as $\bar{\gamma}_t = L\bar{\gamma}_c$, remains fixed. These results indicate that there is an optimum order of diversity for each $\bar{\gamma}_t$. That is, for any $\bar{\gamma}_t$, there is a value of L for which P_M is a minimum. A careful observation of these graphs reveals that the minimum in P_M is obtained when $\bar{\gamma}_c = \bar{\gamma}_t/L \approx 3$. This result appears to be independent of the alphabet size M.

Second, let us observe the error rate P_M as a function of the average SNR per bit, defined as $\bar{\gamma}_b = L\bar{\gamma}_c/k$. (If we interpret M-ary orthogonal FSK as a form of coding† and the order of diversity as the number of times a symbol is repeated in a repetition code, then $\bar{\gamma}_b = \bar{\gamma}_c/R_c$, where $R_c = k/L$ is the code rate.) The graphs of P_M versus $\bar{\gamma}_b$ for $M = 2, 4, 8, 16, 32$ and $L = 1, 2, 4$ are shown in Figure 14.4–7. These results illustrate the gain in performance as M increases and L increases. First, we note that a significant gain in performance is obtained by increasing L. Second, we note that the gain in performance obtained with an increase in M is relatively small when L is small. However, as L increases, the

---

†In Section 14.6, we show that M-ary orthogonal FSK with diversity may be viewed as a block orthogonal code.

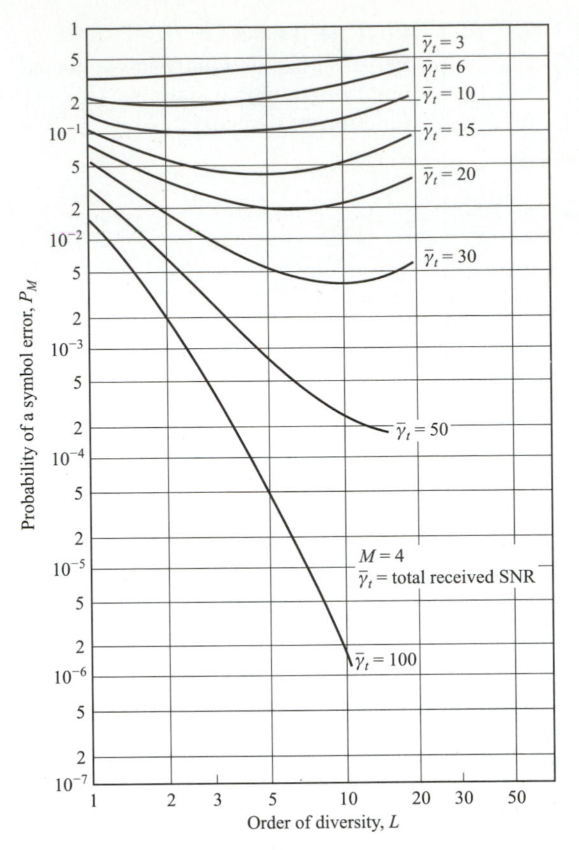

FIGURE 14.4–6
Performance of square-law-detected $M = 4$ orthogonal signals as a function of diversity.

gain achieved by increasing M also increases. Since an increase in either parameter results in an expansion of bandwidth, i.e.,

$$B_e = \frac{LM}{\log_2 M} \tag{14.4–51}$$

the results illustrated in Figure 14.4–7 indicate that an increase in L is more efficient than a corresponding increase in M. As we shall see in Section 14.6, coding is a bandwidth-effective means for obtaining diversity in the signal transmitted over the fading channel.

Chernoff bound. Before concluding this section, we develop a Chernoff upper bound on the error probability of binary orthogonal signaling with Lth-order diversity, which will be useful in our discussion of coding for fading channels, the topic of Section 14.6. Our starting point is the expression for the two decision variables U_1 and U_2 given by Equation 14.4–29, where U_1 consists of the square-law-combined signal-plus-noise terms and U_2 consists of square-law-combined noise terms. The binary probability of error, denoted here by $P_2(L)$, is

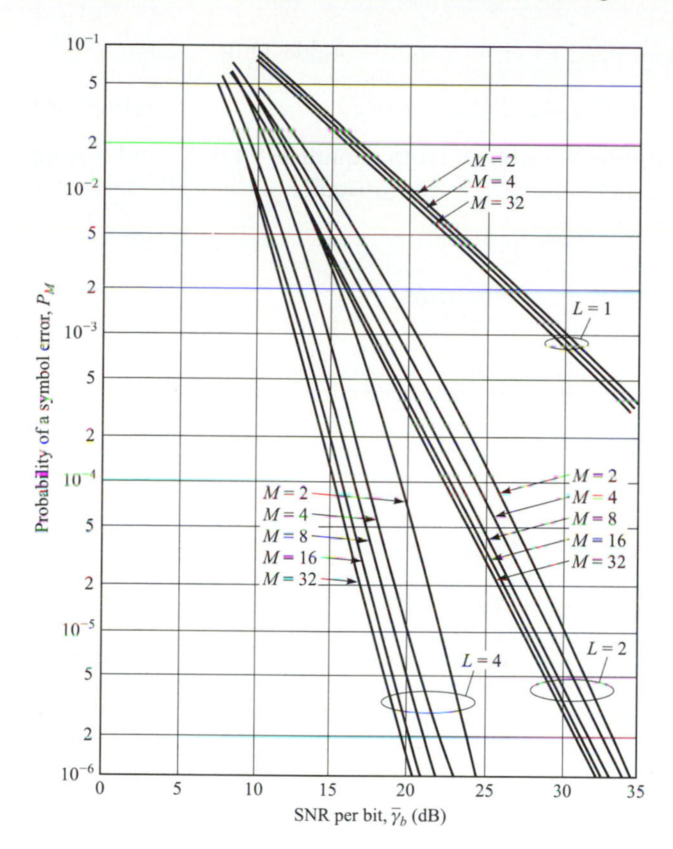

FIGURE 14.4–7 Performance of orthogonal signaling with M and L as parameters.

$$
\begin{aligned}
P_2(L) &= P(U_2 - U_1 > 0) \\
&= P(X > 0) = \int_0^{\infty} p(x)\, dx
\end{aligned}
\tag{14.4–52}
$$

where the random variable X is defined as

$$
X = U_2 - U_1 = \sum_{k=1}^{L} \left(|N_{k2}|^2 - |2\mathcal{E}\alpha_k + N_{k1}|^2\right) \tag{14.4–53}
$$

The phase terms $\{\phi_k\}$ in U_1 have been dropped since they do not affect the performance of the square-law detector.

Let $S(x)$ denote the unit step function. Then the error probability in 14.4–52 can be expressed in the form

$$
P_2(L) = E[S(x)] \tag{14.4–54}
$$

Following the development in Section 2.1.5, the Chernoff bound is obtained by overbounding the unit step function by an exponential function. That is,

$$
S(x) \leqslant e^{\zeta X}, \qquad \zeta \geqslant 0 \tag{14.4–55}
$$

where the parameter ζ is optimized to yield a tight bound. Thus, we have

$$P_2(L) = E[S(x)] \leqslant E(e^{\zeta X}) \tag{14.4–56}$$

Upon substituting for the random variable X from Equation 14.4–53 and noting that the random variables in the summation are mutually statistically independent, we obtain the result

$$P_2(L) \leqslant \prod_{k=1}^{L} E(e^{\zeta|N_{k2}|^2}) E(e^{-\zeta|2\mathcal{E}\alpha_k + N_{k1}|^2}) \tag{14.4–57}$$

But

$$E(e^{\zeta|N_{k2}|^2}) = \frac{1}{1 - 2\zeta\sigma_2^2}, \qquad \zeta < \frac{1}{2\sigma_2^2} \tag{14.4–58}$$

and

$$E(e^{-\zeta|2\mathcal{E}\alpha_k + N_{k1}|^2}) = \frac{1}{1 + 2\zeta\sigma_1^2}, \qquad \zeta > \frac{-1}{2\sigma_1^2} \tag{14.4–59}$$

where $\sigma_2^2 = 2\mathcal{E}N_0$, $\sigma_1^2 = 2\mathcal{E}N_0(1 + \bar{\gamma}_c)$, and $\bar{\gamma}_c$ is the average SNR per diversity channel. Note that σ_1^2 and σ_2^2 are independent of k, i.e., the additive noise terms on the L diversity channels as well as the fading statistics are identically distributed. Consequently, Equation 14.4–57 reduces to

$$P_2(L) \leqslant \left[\frac{1}{(1 - 2\zeta\sigma_2^2)(1 + 2\zeta\sigma_1^2)}\right]^L, \qquad 0 \leqslant \zeta \leqslant \frac{1}{2\sigma_2^2} \tag{14.4–60}$$

By differentiating the right-hand side of Equation 14.4–60 with respect to ζ, we find that the upper bound is minimized when

$$\zeta = \frac{\sigma_1^2 - \sigma_2^2}{4\sigma_1^2\sigma_2^2} \tag{14.4–61}$$

Substitution of Equation 14.4–61 for ζ into Equation 14.4–60 yields the Chernoff upper bound in the form

$$P_2(L) \leqslant \left[\frac{4(1 + \bar{\gamma}_c)}{(2 + \bar{\gamma}_c)^2}\right]^L \tag{14.4–62}$$

It is interesting to note that Equation 14.4–62 may also be expressed as

$$P_2(L) \leqslant [4p(1 - p)]^L \tag{14.4–63}$$

where $p = 1/(2 + \bar{\gamma}_c)$ is the probability of error for binary orthogonal signaling on a fading channel without diversity.

A comparison of the Chernoff bound in Equation 14.4–62 with the exact error probability for binary orthogonal signaling and square-law combining of the L diversity signals, which is given by the expression

$$\begin{aligned} P_2(L) &= \left(\frac{1}{2+\bar{\gamma}_c}\right)^L \sum_{k=0}^{L-1} \binom{L-1+k}{k} \left(\frac{1+\bar{\gamma}_c}{2+\bar{\gamma}_c}\right)^k \\ &= p^L \sum_{k=0}^{L-1} \binom{L-1+k}{k} (1-p)^k \end{aligned} \tag{14.4–64}$$

reveals the tightness of the bound. Figure 14.4–8 illustrates this comparison. We observe that the Chernoff upper bound is approximately 6 dB from the exact error probability for $L = 1$, but, as L increases, it becomes tighter. For example, the difference between the bound and the exact error probability is about 2.5 dB when $L = 4$.

Finally we mention that the error probability for M-ary orthogonal signaling with diversity can be upper-bounded by means of the union bound

$$P_M \leqslant (M-1)P_2(L) \tag{14.4–65}$$

where we may use either the exact expression given in Equation 14.4–64 or the Chernoff bound in Equation 14.4–62 for $P_2(L)$.

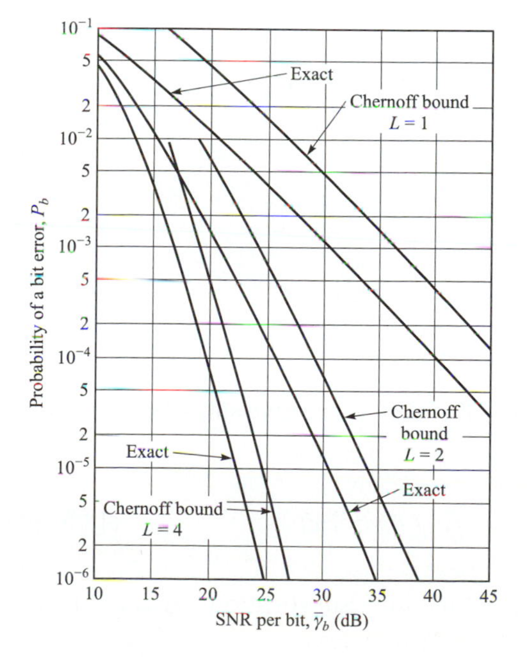

FIGURE 14.4–8
Comparison of Chernoff bound with exact error probability.

14.5 DIGITAL SIGNALING OVER A FREQUENCY-SELECTIVE, SLOWLY FADING CHANNEL

When the spread factor of the channel satisfies the condition $T_m B_d \ll 1$, it is possible to select signals having a bandwidth $W \ll (\Delta f)_c$ and a signal duration $T \ll (\Delta t)_c$. Thus, the channel is frequency-nonselective and slowly fading. In such a channel, diversity techniques can be employed to overcome the severe consequences of fading.

When a bandwidth $W \gg (\Delta f)_c$ is available to the user, the channel can be subdivided into a number of frequency-division multiplexed (FDM) subchannels having a mutual separation in center frequencies of at least $(\Delta f)_c$. Then the same signal can be transmitted on the FDM subchannels, and, thus, frequency diversity is obtained. In this section, we describe an alternative method.

14.5.1 A Tapped-Delay-Line Channel Model

As we shall now demonstrate, a more direct method for achieving basically the same results is to employ a wideband signal covering the bandwidth W. The channel is still assumed to be slowly fading by virtue of the assumption that $T \ll (\Delta t)_c$. Now suppose that W is the bandwidth occupied by the real band-pass signal. Then the band occupancy of the equivalent low-pass signal $s_l(t)$ is $|f| \leqslant \frac{1}{2}W$. Since $s_l(t)$ is band-limited to $|f| \leqslant \frac{1}{2}W$, application of the sampling theorem results in the signal representation

$$s_l(t) = \sum_{n=-\infty}^{\infty} s_l\left(\frac{n}{W}\right) \frac{\sin[\pi W(t - n/W)]}{\pi W(t - n/W)} \tag{14.5–1}$$

The Fourier transform of $s_l(t)$ is

$$S_l(f) = \begin{cases} \dfrac{1}{W} \displaystyle\sum_{n=-\infty}^{\infty} s_l(n/W) e^{-j2\pi f n/W} & (|f| \leqslant \frac{1}{2}W) \\ 0 & (|f| > \frac{1}{2}W) \end{cases} \tag{14.5–2}$$

The noiseless received signal from a frequency-selective channel was previously expressed in the form

$$r_l(t) = \int_{-\infty}^{\infty} C(f; t) S_l(f) e^{j2\pi f t}\, df \tag{14.5–3}$$

where $C(f; t)$ is the time-variant transfer function. Substitution for $S_l(f)$ from Equation 14.5–2 into 14.5–3 yields

$$r_l(t) = \frac{1}{W} \sum_{n=-\infty}^{\infty} s_l(n/W) \int_{-\infty}^{\infty} C(f;t) e^{j2\pi f(t-n/W)}\, df$$
$$= \frac{1}{W} \sum_{n=-\infty}^{\infty} s_l(n/W) c(t - n/W; t) \tag{14.5–4}$$

where $c(\tau; t)$ is the time-variant impulse response. We observe that Equation 14.5–4 has the form of a convolution sum. Hence, it can also be expressed in the alternative form

$$r_l(t) = \frac{1}{W} \sum_{n=-\infty}^{\infty} s_l(t - n/W) c(n/W; t) \tag{14.5–5}$$

It is convenient to define a set of time-variable channel coefficients as

$$c_n(t) = \frac{1}{W} c\left(\frac{n}{W}; t\right) \tag{14.5–6}$$

Then Equation 14.5–5 expressed in terms of these channel coefficients becomes

$$r_l(t) = \sum_{n=-\infty}^{\infty} c_n(t) s_l(t - n/W) \tag{14.5–7}$$

The form for the received signal in Equation 14.5–7 implies that the time-variant frequency-selective channel can be modeled or represented as a tapped delay line with tap spacing $1/W$ and tap weight coefficients $\{c_n(t)\}$. In fact, we deduce from Equation 14.5–7 that the low-pass impulse response for the channel is

$$c(\tau; t) = \sum_{n=-\infty}^{\infty} c_n(t) \delta(\tau - n/W) \tag{14.5–8}$$

and the corresponding time-variant transfer function is

$$C(f; t) = \sum_{n=-\infty}^{\infty} c_n(t) e^{-j2\pi f n/W} \tag{14.5–9}$$

Thus, with an equivalent low-pass-signal having a bandwidth $\frac{1}{2}W$, where $W \gg (\Delta f)_c$, we achieve a resolution of $1/W$ in the multipath delay profile. Since the total multipath spread is T_m, for all practical purposes the tapped delay line model for the channel can be truncated at $L = \lfloor T_m W \rfloor + 1$ taps. Then the noiseless received signal can be expressed in the form

$$r_l(t) = \sum_{n=1}^{L} c_n(t) s_l\left(t - \frac{n}{W}\right) \tag{14.5–10}$$

The truncated tapped delay line model is shown in Figure 14.5–1. In accordance with the statistical characterization of the channel presented in Section 14.1, the time-variant tap weights $\{c_n(t)\}$ are complex-valued stationary random processes. In the special case of Rayleigh fading, the magnitudes $|c_n(t)| \equiv \alpha_n(t)$ are Rayleigh-distributed and the phases $\phi_n(t)$ are uniformly distributed. Since the

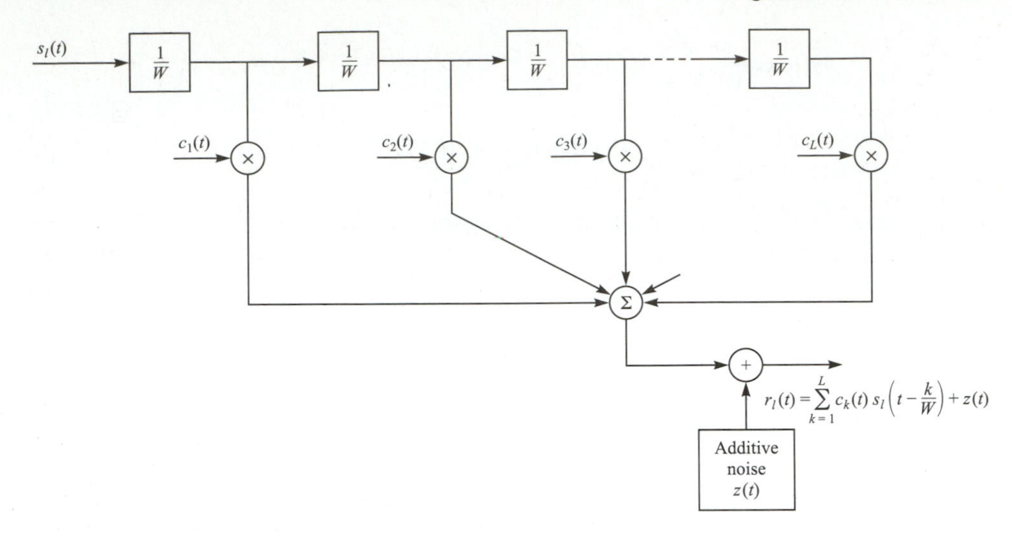

FIGURE 14.5–1
Trapped delay line model of frequency-selective channel.

$\{c_n(t)\}$ represent the tap weights corresponding to the L different delays $\tau = n/W$, $n = 1, 2, \ldots, L$, the uncorrelated scattering assumption made in Section 14.1 implies that the $\{c_n(t)\}$ are mutually uncorrelated. When the $\{c_n(t)\}$ are Gaussian random processes, they are statistically independent.

14.5.2 The RAKE Demodulator

We now consider the problem of digital signaling over a frequency-selective channel that is modeled by a tapped delay line with statistically independent time-variant tap weights $\{c_n(t)\}$. It is apparent at the outset, however, that the tapped delay line model with statistically independent tap weights provides us with L replicas of the same transmitted signal at the receiver. Hence, a receiver that processes the received signal in an optimum manner will achieve the performance of an equivalent Lth-order diversity communication system.

Let us consider binary signaling over the channel. We have two equal-energy signals $s_{l1}(t)$ and $s_{l2}(t)$, which are either antipodal or orthogonal. Their time duration T is selected to satisfy the condition $T \gg T_m$. Thus, we may neglect any intersymbol interference due to multipath. Since the bandwidth of the signal exceeds the coherent bandwidth of the channel, the received signal is expressed as

$$\begin{aligned} r_l(t) &= \sum_{k=1}^{L} c_k(t) s_{li}(t - k/W) + z(t) \\ &= v_i(t) + z(t), \qquad 0 \leqslant t \leqslant T, \qquad i = 1, 2 \end{aligned} \tag{14.5–11}$$

where $z(t)$ is a complex-valued zero-mean white Gaussian noise process. Assume for the moment that the channel tap weights are known. Then the optimum demodulator consists of two filters matched to $v_1(t)$ and $v_2(t)$. The demodulator output is sampled at the symbol rate and the samples are passed to a decision circuit that selects the signal corresponding to the largest output. An equivalent optimum demodulator employs cross correlation instead of matched filtering. In either case, the decision variables for coherent detection of the binary signals can be expressed as

$$\begin{aligned} U_m &= \mathrm{Re}\left[\int_0^T r_l(t) v_m^*(t)\,dt\right] \\ &= \mathrm{Re}\left[\sum_{k=1}^{L}\int_0^T r_l(t) c_k^*(t) s_m^*(t-k/W)\,dt\right], \qquad m=1,2 \end{aligned} \tag{14.5–12}$$

Figure 14.5–2 illustrates the operations involved in the computation of the decision variables. In this realization of the optimum receiver, the two reference signals are delayed and correlated with the received signal $r_l(t)$.

An alternative realization of the optimum demodulator employs a single delay line through which is passed the received signal $r_l(t)$. The signal at each tap is correlated with $c_k^*(t)s_{lm}^*(t)$, where $k = 1, 2, \ldots, L$ and $m = 1, 2$. This receiver structure is shown in Figure 14.5–3. In effect, the tapped delay line demodulator attempts to collect the signal energy from all the received signal paths that fall within the span of the delay line and carry the same information. Its action is somewhat analogous to an ordinary garden rake and, consequently, the name "RAKE demodulator" has been coined for this demodulator structure by Price and Green (1958).

14.5.3 Performance of RAKE Demodulator

We shall now evaluate the performance of the RAKE demodulator under the condition that the fading is sufficiently slow to allow us to estimate $c_k(t)$ perfectly (without noise). Furthermore, within any one signaling interval, $c_k(t)$ is treated as a constant and denoted as c_k. Thus the decision variables in Equation 14.5–12 may be expressed in the form

$$U_m = \mathrm{Re}\left[\sum_{k=1}^{L} c_k^* \int_0^T r(t) s_{lm}^*(t-k/W)\,dt\right], \qquad m=1,2 \tag{14.5–13}$$

Suppose the transmitted signal is $s_{l1}(t)$; then the received signal is

$$r_l(t) = \sum_{n=1}^{L} c_n s_{l1}(t-n/W) + z(t), \qquad 0 \leqslant t \leqslant T \tag{14.5–14}$$

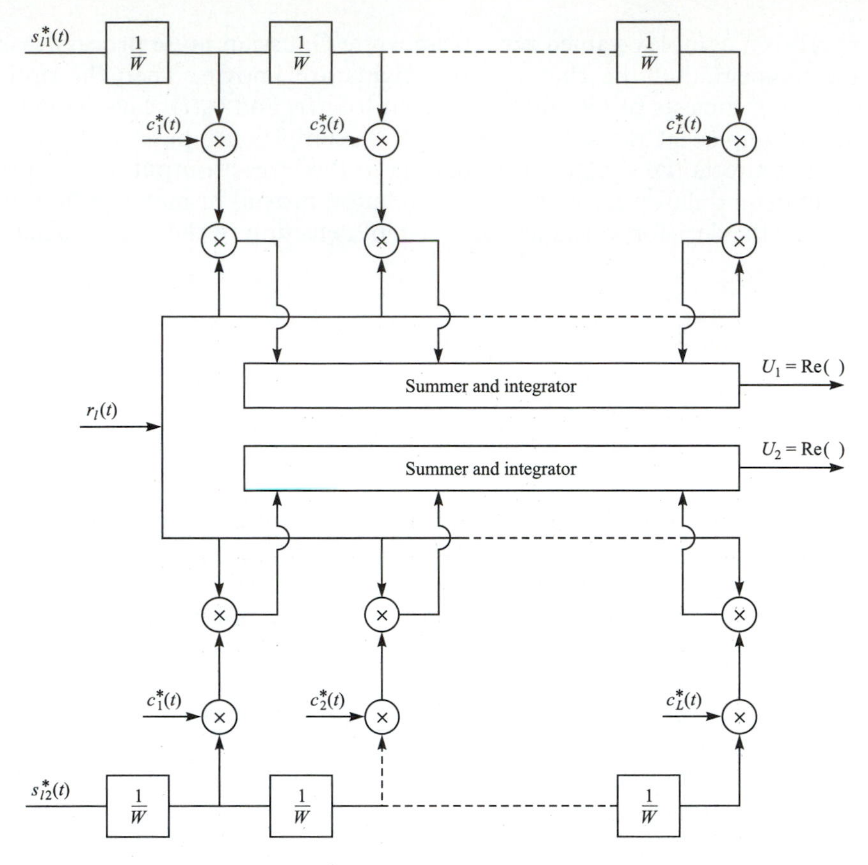

FIGURE 14.5–2
Optimum demodulator for wideband binary signals (delayed reference configuration).

Substitution of Equation 14.5–14 into Equation 14.5–13 yields

$$\begin{aligned} U_m = \operatorname{Re}\left[\sum_{k=1}^{L} c_k^* \sum_{n=1}^{L} c_n \int_0^T s_{l1}(t - n/W) s_{lm}^*(t - k/W)\, dt\right] \\ + \operatorname{Re}\left[\sum_{k=1}^{L} c_k^* \int_0^T z(t) s_{lm}^*(t - k/W)\, dt\right], \qquad m = 1, 2 \end{aligned} \tag{14.5–15}$$

Usually the wideband signals $s_{l1}(t)$ and $s_{l2}(t)$ are generated from pseudorandom sequences, which result in signals that have the property

$$\int_0^T s_{li}(t - n/W) s_{li}^*(t - k/W)\, dt \approx 0, \qquad k \neq n, \qquad i = 1, 2 \tag{14.5–16}$$

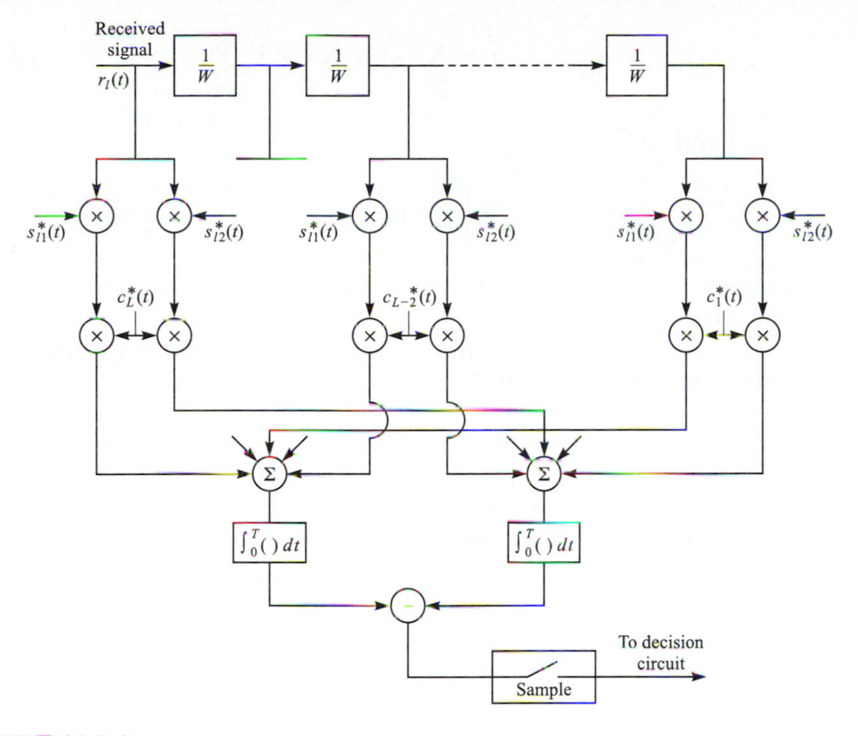

FIGURE 14.5–3
Optimum demodulator for wideband binary signals (delayed received signal configurtion).

If we assume that our binary signals are designed to satisfy this property, then Equation 14.5–15 simplifies to†

$$
\begin{aligned}
U_m = \operatorname{Re}\left[\sum_{k=1}^{L} |c_k|^2 \int_0^T s_{l1}(t - k/W) s_{lm}^*(t - k/W)\, dt\right] \\
+ \operatorname{Re}\left[\sum_{k=1}^{L} c_k^* \int_0^T z(t) s_{lm}^*(t - k/W)\, dt\right], \qquad m = 1, 2
\end{aligned}
\tag{14.5–17}
$$

When the binary signals are antipodal, a single decision variable suffices. In this case, Equation 14.5–17 reduces to

†Although the orthogonality property specified by Equation 14.5–16 can be satisfied by proper selection of the pseudorandom sequences, the cross correlation of $s_{l1}(t - n/W)$ with $s_{li}^*(t - k/W)$ gives rise to a signal-dependent self-noise, which ultimately limits the performance. For simplicity, we do not consider the self-noise term in the following calculations. Consequently, the performance results presented below should be considered as lower bounds (ideal RAKE). An approximation to the performance of the RAKE can be obtained by treating the self-noise as an additional Gaussian noise component with noise power equal to its variance.

$$U_1 = \mathrm{Re}\left(2\mathcal{E}\sum_{k=1}^{L}\alpha_k^2 + \sum_{k=1}^{L}\alpha_k N_k\right) \tag{14.5–18}$$

where $\alpha_k = |c_k|$ and

$$N_k = e^{j\phi_k}\int_0^T z(t)s_l^*(t - k/W)\,dt \tag{14.5–19}$$

But Equation 14.5–18 is identical to the decision variable given in Equation 14.4–4, which corresponds to the output of a maximal ratio combiner in a system with Lth-order diversity. Consequently, the RAKE demodulator with perfect (noiseless) estimates of the channel tap weights is equivalent to a maximal ratio combiner in a system with Lth-order diversity. Thus, when all the tap weights have the same mean-square value, i.e., $E(\alpha_k^2)$ is the same for all k, the error rate performance of the RAKE demodulator is given by Equations 14.4–15 and 14.4–16. On the other hand, when the mean-square values $E(\alpha_k^2)$ are not identical for all k, the derivation of the error rate performance must be repeated since Equation 14.4–15 no longer applies.

We shall derive the probability of error for binary antipodal and orthogonal signals under the condition that the mean-square values of $\{\alpha_k\}$ are distinct. We begin with the conditional error probability

$$P_2(\gamma_b) = Q\left(\sqrt{\gamma_b(1-\rho_r)}\right) \tag{14.5–20}$$

where $\rho_r = -1$ for antipodal signals, $\rho_r = 0$ for orthogonal signals, and

$$\gamma_b = \frac{\mathcal{E}}{N_0}\sum_{k=1}^{L}\alpha_k^2 = \sum_{k=1}^{L}\gamma_k \tag{14.5–21}$$

Each of the $\{\gamma_k\}$ is distributed according to a chi-squared distribution with two degrees of freedom. That is,

$$p(\gamma_k) = \frac{1}{\bar{\gamma}_k}e^{-\gamma_k/\bar{\gamma}_k} \tag{14.5–22}$$

where $\bar{\gamma}_k$ is the average SNR for the kth path, defined as

$$\bar{\gamma}_k = \frac{\mathcal{E}}{N_0}E(\alpha_k^2) \tag{14.5–23}$$

Furthermore, from Equation 14.4–10 we know that the characteristic function of γ_k is

$$\psi_{\gamma_k}(jv) = \frac{1}{1 - jv\bar{\gamma}_k} \tag{14.5–24}$$

Since γ_b is the sum of L statistically independent components $\{\gamma_k\}$, the characteristic function of γ_b is

$$\psi_{\gamma_b}(jv) = \prod_{k=1}^{L} \frac{1}{1 - jv\bar{\gamma}_k} \qquad (14.5\text{–}25)$$

The inverse Fourier transform of the characteristic function in Equation 14.5–25 yields the probability density function of γ_b in the form

$$p(\gamma_b) = \sum_{k=1}^{L} \frac{\pi_k}{\bar{\gamma}_k} e^{-\gamma_b/\bar{\gamma}_k}, \qquad \gamma_b \geqslant 0 \qquad (14.5\text{–}26)$$

where π_k is defined as

$$\pi_k = \prod_{\substack{i=1 \\ i \neq k}}^{L} \frac{\bar{\gamma}_k}{\bar{\gamma}_k - \bar{\gamma}_i} \qquad (14.5\text{–}27)$$

When the conditional error probability in Equation 14.5–20 is averaged over the probability density function given in Equation 14.5–26, the result is

$$P_2 = \tfrac{1}{2} \sum_{k=1}^{L} \pi_k \left[1 - \sqrt{\frac{\bar{\gamma}_k(1 - \rho_r)}{2 + \bar{\gamma}_k(1 - \rho_r)}} \right] \qquad (14.5\text{–}28)$$

This error probability can be approximated as ($\bar{\gamma}_k \gg 1$)

$$P_2 \approx \binom{2L-1}{L} \prod_{k=1}^{L} \frac{1}{2\bar{\gamma}_k(1 - \rho_r)} \qquad (14.5\text{–}29)$$

By comparing Equation 14.5–29 for $\rho_r = -1$ with Equation 14.4–18, we observe that the same type of asymptotic behavior is obtained for the case of unequal SNR per path and the case of equal SNR per path.

In the derivation of the error rate performance of the RAKE demodulator, we assumed that the estimates of the channel tap weights are perfect. In practice, relatively good estimates can be obtained if the channel fading is sufficiently slow, e.g., $(\Delta t)_c/T \geqslant 100$, where T is the signaling interval. Figure 14.5–4 illustrates a method for estimating the tap weights when the binary signaling waveforms are orthogonal. The estimate is the output of the low-pass filter at each tap. At any one instant in time, the incoming signal is either $s_{l1}(t)$ or $s_{l2}(t)$. Hence, the input to the low-pass filter used to estimate $c_k(t)$ contains signal plus noise from one of the correlators and noise only from the other correlator. This method for channel estimation is not appropriate for antipodal signals, because the addition of the two correlator outputs results in signal cancellation. Instead, a single correlator can be employed for antipodal signals. Its output is fed to the input of the low-pass filter after the information-bearing signal is removed. To accomplish this, we must introduce a delay of one signaling interval into the channel estimation procedure, as illustrated in Figure 14.5–5. That is, first the receiver must decide whether the information in the received signal is $+1$ or -1 and, then, it uses the decision to remove the information from the correlator ouput prior to feeding it to the low-pass filter.

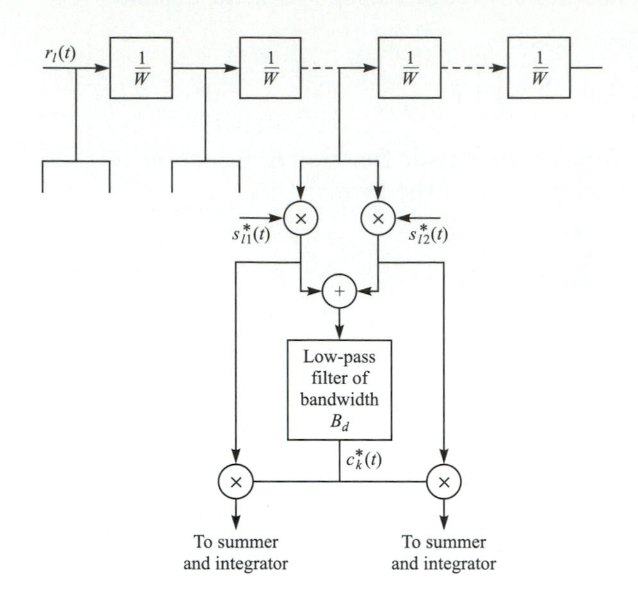

FIGURE 14.5–4
Channel tap weight estimation with binary orthogonal signals.

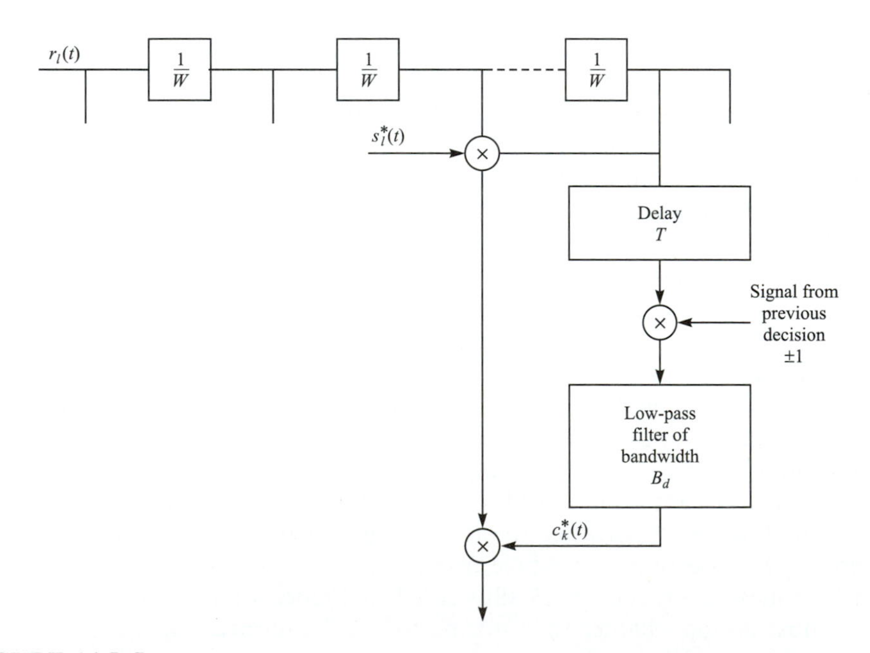

FIGURE 14.5–5
Channel tap weight estimation with binary antipodal signals.

If we choose not to estimate the tap weights of the frequency-selective channel, we may use either DPSK signaling or noncoherently detected orthogonal signaling. The RAKE demodulator structure for DPSK is illustrated in Figure 14.5–6. It is apparent that when the transmitted signal waveform $s_l(t)$ satisfies the orthogonality property given in Equation 14.5–16, the decision variable is identical to that given in Equation 14.4–23 for an Lth-order diversity system. Consequently, the error rate performance of the RAKE demodulator for a binary DPSK is identical to that given in Equation 14.4–15 with $\mu = \bar{\gamma}_c/(1+\bar{\gamma}_c)$, when all the signal paths have the same SNR $\bar{\gamma}_c$. On the other hand, when the SNRs $\{\bar{\gamma}_k\}$ are distinct, the error probability can be obtained by averaging Equation 14.4–24, which is the probability of error conditioned on a time-invariant channel, over the probability density function of γ_b given by Equation 14.5–26. The result of this integration is

$$P_2 = \left(\tfrac{1}{2}\right)^{2L-1} \sum_{m=0}^{L-1} m!\, b_m \sum_{k=1}^{L} \frac{\pi_k}{\bar{\gamma}_k} \left(\frac{\bar{\gamma}_k}{1+\bar{\gamma}_k}\right)^{m+1} \tag{14.5–30}$$

where π_k is defined in Equation 14.5–27 and b_m in Equation 14.4–25.

Finally, we consider binary orthognal signaling over the frequency-selective channel with square-law detection at the receiver. This type of signal is appropriate when either the fading is rapid enough to preclude a good estimate of the channel tap weights or the cost of implementing the channel estimators is high.

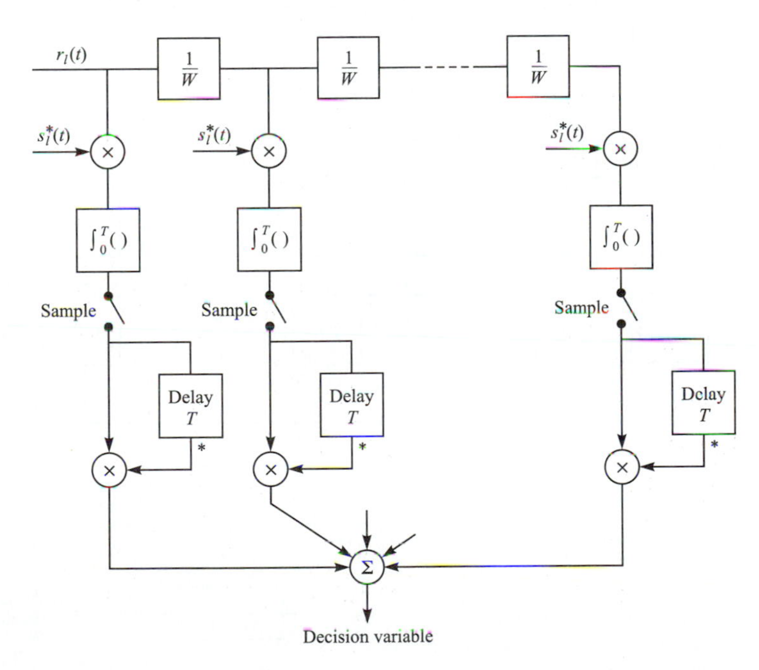

FIGURE 14.5–6
RAKE demodulator for DPSK signals.

The RAKE demodulator with square-law combining of the signal from each tap is illustrated in Figure 14.5–7. In computing its performance, we again assume that the orthogonality property given in Equation 14.5–16 holds. Then the decision variables at the output of the RAKE are

$$\begin{aligned} U_1 &= \sum_{k=1}^{L} |2\mathcal{E}c_k + N_{k1}|^2 \\ U_2 &= \sum_{k=1}^{L} |N_{k2}|^2 \end{aligned} \tag{14.5–31}$$

where we have assumed that $s_{l1}(t)$ was the transmitted signal. Again we observe that the decision variables are identical to the ones given in Equation 14.4–29, which apply to orthogonal signals with Lth-order diversity. Therefore, the performance of the RAKE demodulator for square-law-detected orthogonal signals is given by Equation 14.4–15 with $\mu = \bar{\gamma}_c/(2 + \bar{\gamma}_c)$ when all the signal paths have the same SNR. If the SNRs are distinct, we can average the conditional error probability given by Equation 14.4–24, with γ_b replaced by $\frac{1}{2}\gamma_b$, over the prob-

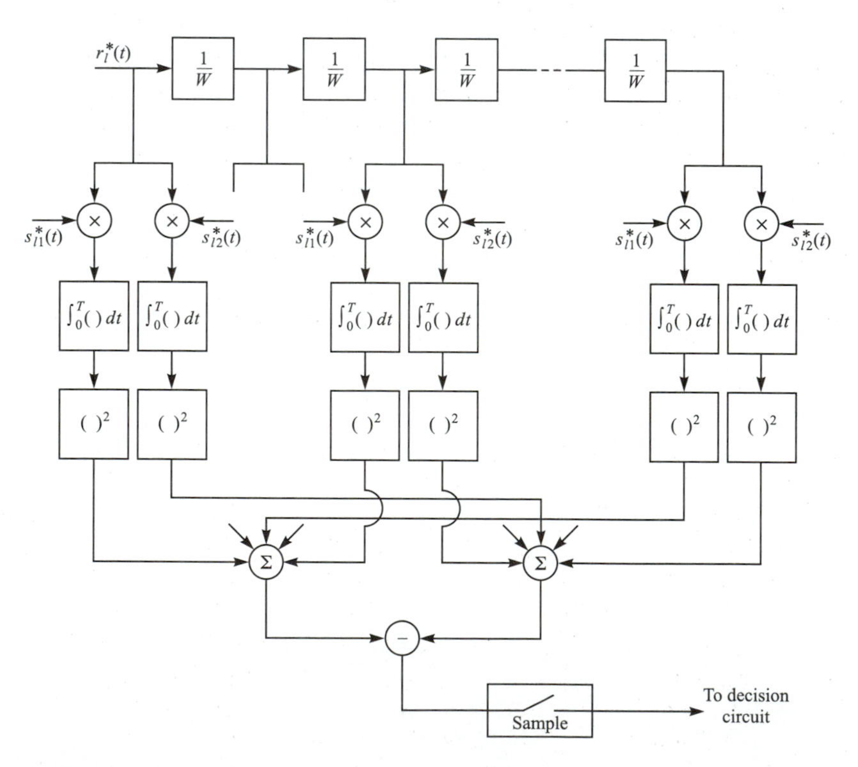

FIGURE 14.5–7
RAKE demodulator for square-law combination of orthogonal signals.

ability density function $p(\gamma_b)$ given in Equation 14.5–26. The result of this averaging is given by Equation 14.5–30, with $\bar{\gamma}_k$ replaced by $\frac{1}{2}\bar{\gamma}_k$.

In the above analysis, the RAKE demodulator shown in Figure 14.5–7 for square-law combining of orthogonal signals is assumed to contain a signal component at each delay. If that is not the case, its performance will be degraded, since some of the tap correlators will contribute only noise. Under such conditions, the low-level, noise-only contributions from the tap correlators should be excluded from the combiner, as shown by Chyi et al. (1988).

The configurations of the RAKE demodulator presented in this section can be easily generalized to multilevel signaling. In fact, if M-ary PSK or DPSK is chosen, the RAKE structures presented in this section remain unchanged. Only the PSK and DPSK detectors that follow the RAKE correlator are different.

14.5.4 Receiver Structures for Channels with Intersymbol Interference

As described above, the wideband signal waveforms that are transmitted through the multipath channels resolve the multipath components with a time resolution of $1/W$, where W is the signal bandwidth. Usually, such wideband signals are generated as direct sequence spread spectrum signals, in which the *PN* spreading sequences are the outputs of linear feedback shift registers, e.g., maximal-length linear feedback shift registers. The modulation impressed on the sequences may be binary PSK, QPSK, DPSK, or binary orthogonal. The desired bit rate determines the bit interval or symbol interval.

The RAKE demodulator that we described above is the optimum demodulator based on the condition that the bit interval $T_b \gg T_m$, i.e., there is negligible ISI. When this condition is not satisfied, the RAKE demodulator output is corrupted by ISI. In such a case, an equalizer is required to suppress the ISI.

To be specific, we assume that binary PSK modulation is used and spread by a PN sequence. The bandwidth of the transmitted signal is sufficiently broad to resolve two or more multipath components. At the receiver, after the signal is demodulated to baseband, it may be processed by the RAKE, which is the matched filter to the channel response, followed by an equalizer to suppress the ISI. The RAKE output is sampled at the bit rate, and these samples are passed to the equalizer. An appropriate equalizer, in this case, would be a maximum-likelihood sequence estimator implemented by use of the Viterbi algorithm or a decision feedback equalizer (DFE). This demodulator structure is shown in Figure 14.5–8.

Other receiver structures are also possible. If the period of the PN sequence is equal to the bit interval, i.e., $LT_c = T_b$, where T_c is the chip interval and L_c is the number of chips per bit, a fixed filter matched to the spreading sequence may be used to process the received signal and followed by an adaptive equalizer, such as a fractionally spaced DFE, as shown in Figure 14.5–9. In this case, the matched filter output is sampled at some multiple of the chip rate, e.g., twice the chip rate, and fed to the fractionally spaced DFE. The feedback filter in the DFE would

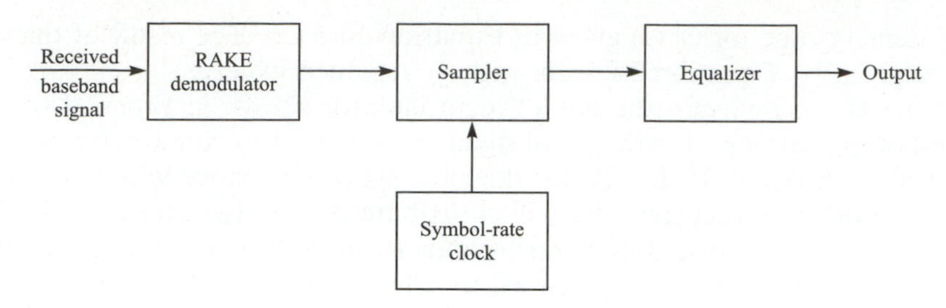

FIGURE 14.5–8
Receiver structure for processing wideband signal corrupted by ISI.

have taps spaced at the bit interval. The adaptive DFE would require a training sequence for adjustment of its coefficients to the channel multipath structure.

An even simpler receiver structure is one in which the spread spectrum matched filter is replaced by a low-pass filter whose bandwidth is matched to the transmitted signal bandwidth. The output of such a filter may be sampled at an integer multiple of the chip rate and the samples are passed to an adaptive fractionally spaced DFE. In this case, the coefficients of the feedback filter in the DFE, with the aid of a training sequence, will adapt to the combination of the spreading sequence and the channel multipath. Abdulrahman et al. (1994) consider the use of a DFE to suppress ISI in a CDMA system in which each user employs a wideband direct sequence spread spectrum signal.

The paper by Taylor et al. (1998) provides a broad survey of equalization techniques and their performance for wireless channels.

14.6 CODED WAVEFORMS FOR FADING CHANNELS

Up to this point, we have demonstrated that diversity techniques are very effective in overcoming the detrimental effects of fading caused by the time-variant dispersive characteristics of the channel. Time- and/or frequency-diversity tech-

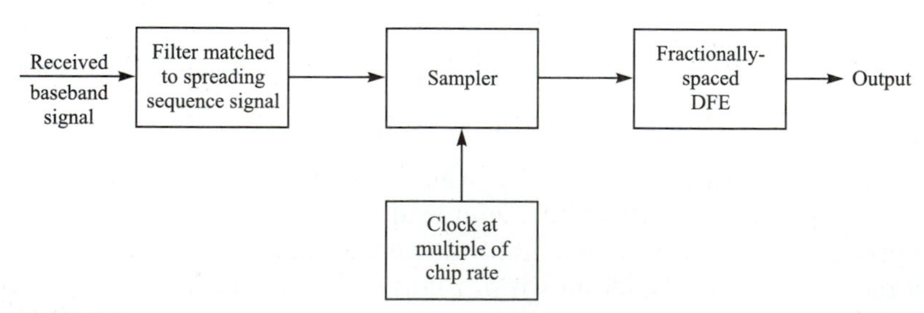

FIGURE 14.5–9
Alternative receiver structure for processing wideband signal corrupted by ISI.

niques may be viewed as a form of repetition (block) coding of the information sequence. From this point of view, the combining techniques described previously represent soft-decision decoding of the repetition code. Since a repetition code is a trivial form of coding, we shall now consider the additional benefits derived from more efficient types of codes. In particular, we demonstrate that coding provides an efficient means of obtaining diversity on a fading channel. The amount of diversity provided by a code is directly related to its minimum distance.

As explained in Section 14.4, time diversity is obtained by transmitting the signal components carrying the same information in multiple time intervals mutually separated by an amount equal to or exceeding the coherence time $(\Delta t)_c$ of the channel. Similarly, frequency diversity is obtained by transmitting the signal components carrying the same information in multiple frequency slots mutually separated by an amount at least equal to the coherence bandwidth $(\Delta f)_c$ of the channel. Thus, the signal components carrying the same information undergo statistically independent fading.

To extend these notions to a coded information sequence, we simply require that the signal waveform corresponding to a particular code bit or code symbol fade independently of the signal waveform corresponding to any other code bit or code symbol. This requirement may result in inefficient utilization of the available time-frequency space, with the existence of large unused portions in this two-dimensional signaling space. To reduce the inefficiency, a number of code words may be interleaved in time or in frequency or both, in such a manner that the waveform corresponding to the bits or symbols of a given code word fade independently. Thus, we assume that the time-frequency signaling space is partitioned into non-overlapping time-frequency cells. A signal waveform corresponding to a code bit or code symbol is transmitted within such a cell.

In addition to the assumption of statistically independent fading of the signal components of a given code word, we also assume that the additive noise components corrupting the received signals are white Gaussian processes that are statistically independent and identically distributed among the cells in the time–frequency space. Also, we assume that there is sufficient separation between adjacent cells so that intercell interference is negligible.

An important issue is the modulation technique that is used to transmit the coded information sequence. If the channel fades slowly enough to allow the establishment of a phase reference, then PSK or DPSK may be employed. If this is not possible, then FSK modulation with noncoherent detection at the receiver is appropriate. In our treatment below, we assume that it is not possible to establish a phase reference or phase references for the signals in the different cells occupied by the transmitted signal. Consequently, we choose FSK modulation with noncoherent detection. In the last part of this section, we briefly treat phase-coherent detection of coded signals.

A model of the digital communication system for which the error rate performance will be evaluated is shown in Figure 14.6–1. The encoder may be binary, nonbinary, or a concatenation of a nonbinary encoder with a binary encoder. Furthermore, the code generated by the encoder may be a block

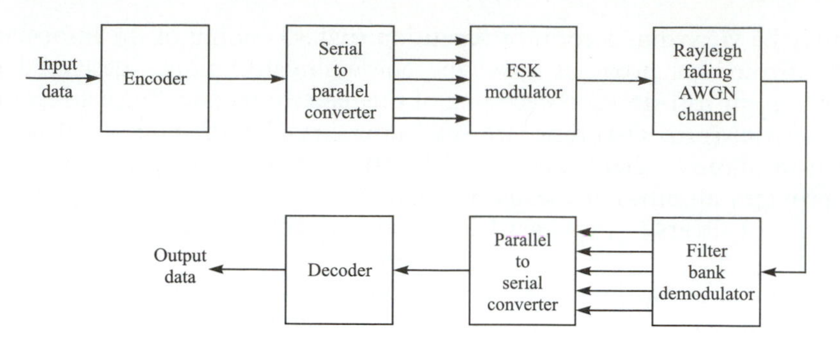

FIGURE 14.6–1
Model of communications system with FSK modulation/demodulation and encoding/decoding.

code, a convolutional code, or, in the case of concatenation, a mixture of a block code and a convolutional code.

In order to explain the modulation, demodulation, and decoding for FSK-type (orthogonal) signals, consider a linear binary block code in which k information bits are encoded into a block of n bits. For simplicity and without loss of generality, let us assume that all n bits of a code word are transmitted simultaneously over the channel on multiple frequency cells. A code word $\mathbf{C}_i$ having bits $\{c_{ij}\}$ is mapped into FSK signal waveforms in the following way. If $c_{ij} = 0$, the tone f_{0j} is transmitted, and if $c_{ij} = 1$, the tone f_{1j} is transmitted. This means that $2n$ tones or cells are available to transmit the n bits of the code word, but only n tones are transmitted in any signaling interval. Since each code word conveys k bits of information, the bandwidth expansion factor for FSK is $B_e = 2n/k$.

The demodulator for the received signal separates the signal into $2n$ spectral components corresponding to the available tone frequencies at the transmitter. Thus, the demodulator can be realized as a bank of $2n$ filters, where each filter is matched to one of the possible transmitted tones. The outputs of the $2n$ filters are detected noncoherently. Since the Rayleigh fading and the additive white Gaussian noises in the $2n$ frequency cells are mutually statistically independent and identically distributed random processes, the optimum maximum-likelihood soft-decision decoding criterion requires that these filter responses be square-law-detected and appropriately combined for each code word to form the $M = 2^k$ decision variables. The code word corresponding to the maximum of the decision variables is selected. If hard-decision decoding is employed, the optimum maximum-likelihood decoder selects the code word having the smallest Hamming distance relative to the received code word.

Although the discussion above assumed the use of a block code, a convolutional encoder can be easily accommodated in the block diagram shown in Figure 14.6–1. For example, if a binary convolutional code is employed, each bit in the output sequence may be transmitted by binary FSK. The maximum-likelihood soft-decision decoding criterion for the convolutional code can be efficiently implemented by means of the Viterbi algorithm, in which the metrics for the surviving sequences at any point in the trellis consist of the square-law-combined outputs for the corresponding paths through the trellis. On the other hand, if hard-decision decoding is employed, the Viterbi algorithm is implemented with Hamming distance as the metric.

14.6.1 Probability of Error for Soft-Decision Decoding of Linear Binary Block Codes

Consider the decoding of a linear binary (n, k) code transmitted over a Rayleigh fading channel, as described above. The optimum soft-decision decoder, based on the maximum-likelihood criterion, forms the $M = 2^k$ decision variables.

$$\begin{aligned} U_i &= \sum_{j=1}^{n} [(1 - c_{ij})|y_{0j}|^2 + c_{ij}|y_{1j}|^2] \\ &= \sum_{j=1}^{n} [|y_{0j}|^2 + c_{ij}(|y_{1j}|^2 - |y_{0j}|^2)],\ i = 1, 2, \ldots, 2^k \end{aligned} \tag{14.6–1}$$

where $|y_{rj}|^2$, $j = 1, 2, \ldots, n$, and $r = 0, 1$ represent the squared envelopes at the outputs of the $2n$ filters that are tuned to the $2n$ possible transmitted tones. A decision is made in favor of the code word corresponding to the largest decision variable of the set $\{U_i\}$.

Our objective in this section is the determination of the error rate performance of the soft-decision decoder. Toward this end, let us assume that the all-zero code word $\mathbf{C}_1$ is transmitted. The average received signal-to-noise ratio per tone (cell) is denoted by $\bar{\gamma}_c$. The total received SNR for the n tones in $n\bar{\gamma}_c$ and, hence, the average SNR per bit is

$$\bar{\gamma}_b = \frac{n}{k}\bar{\gamma}_c = \frac{\bar{\gamma}_c}{R_c} \tag{14.6–2}$$

where R_c is the code rate.

The decision variable U_1 corresponding to the code word $\mathbf{C}_1$ is given by Equation 14.6–1 with $c_{ij} = 0$ for all j. The probability that a decision is made in favor of the mth code word is just

$$\begin{aligned} P_2(m) &= P(U_m > U_1) = P(U_1 - U_m < 0) \\ &= P\left[\sum_{j=1}^{n} (c_{1j} - c_{mj})(|y_{1j}|^2 - |y_{0j}|^2) < 0\right] \\ &= P\left[\sum_{j=1}^{w_m} (|y_{0j}|^2 - |y_{1j}|^2) < 0\right] \end{aligned} \tag{14.6–3}$$

where w_m is the weight of the mth code word. But the probability in Equation 14.6–3 is just the probability of error for square-law combining of binary orthogonal FSK with w_mth-order diversity. That is,

$$P_2(m) = p^{w_m} \sum_{k=0}^{w_m-1} \binom{w_m - 1 + k}{k} (1-p)^k \tag{14.6–4}$$

$$\leqslant p^{w_m} \sum_{k=0}^{w_m-1} \binom{w_m - 1 + k}{k} = \binom{2w_m - 1}{w_m} p^{w_m} \tag{14.6–5}$$

where

$$p = \frac{1}{2+\bar{\gamma}_c} = \frac{1}{2 + R_c \bar{\gamma}_b} \tag{14.6–6}$$

As an alternative, we may use the Chernoff upper bound derived in Section 14.4, which in the present notation is

$$P_2(m) \leqslant [4p(1-p)]^{w_m} \tag{14.6–7}$$

The sum of the binary error events over the $M-1$ nonzero-weight code words gives an upper bound on the probability of error. Thus,

$$P_M \leqslant \sum_{m=2}^{M} P_2(m) \tag{14.6–8}$$

Since the minimum distance of the linear code is equal to the minimum weight, it follows that

$$(2 + R_c \bar{\gamma}_b)^{-w_m} \leqslant (2 + R_c \bar{\gamma}_b)^{-d_{\min}}$$

The use of this relation is conjunction with Equations 14.6–5 and 14.6–8 yields a simple, albeit looser, upper bound that may be expressed in the form

$$P_M < \frac{\sum_{m=2}^{M} \binom{2w_m - 1}{w_m}}{(2 + R_c \bar{\gamma}_b)^{d_{\min}}} \tag{14.6–9}$$

This simple bound indicates that the code provides an effective order of diversity equal to $d_{\min}$. An even simpler bound is the union bound

$$P_M < (M-1)[4p(1-p)]^{d_{\min}} \tag{14.6–10}$$

which is obtained from the Chernoff bound given in Equation 14.6–7.

As an example serving to illustrate the benefits of coding for a Rayleigh fading channel, we have plotted in Figure 14.6–2 the performance obtained with the extended Golay (24,12) code and the performance of binary FSK and quaternary FSK each with dual diversity. Since the extended Golay code requires a total of 48 cells and $k = 12$, the bandwidth expansion factor $B_e = 4$. This is also the bandwidth expansion factor for binary and quaternary FSK with $L = 2$. Thus, the three types of waveforms are compared on the basis of the same

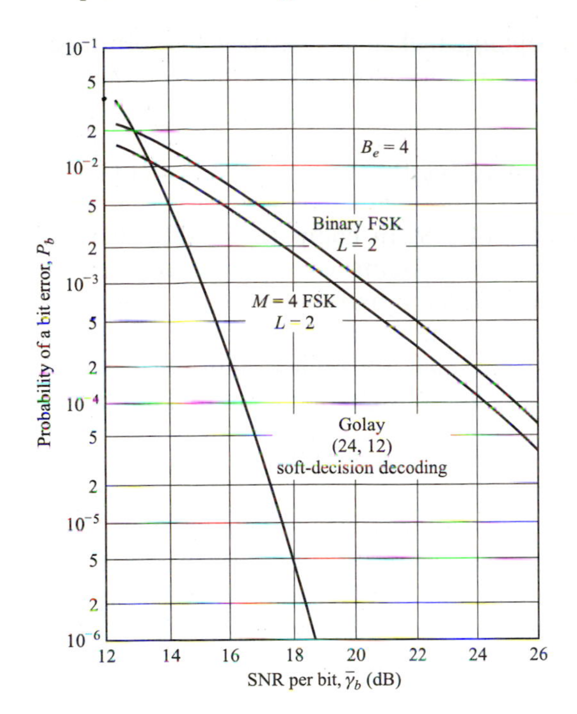

FIGURE 14.6–2
Example of performance obtained with conventional diversity versus coding for $B_e = 4$.

bandwidth expansion factor. Note that at $P_b = 10^{-4}$, the Golay code outperforms quaternary FSK by more than 6 dB, and at $P_b = 10^{-5}$, the difference is approximately 10 dB.

The reason for the superior performanc of the Golay code is its large minimum distance ($d_{\min} = 8$), which translates into an equivalent eighth-order ($L = 8$) diversity. In contrast, the binary and quaternary FSK signals have only second-order diversity. Hence, the code makes more efficient use of the available channel bandwidth. The price that we must pay for the superior performance of the code is the increase in decoding complexity.

14.6.2 Probability of Error for Hard-Decision Decoding of Linear Binary Block Codes

Bounds on the performance obtained with hard-decision decoding of a linear binary (n, k) code have already been given in Section 8.1.5. These bounds are applicable to a general binary-input, binary-output memoryless (binary symmetric) channel, and, hence, they apply without modification to a Rayleigh fading AWGN channel with statistically independent fading of the symbols in the code word. The probability of a bit error needed to evaluate these bounds when binary FSK with noncoherent detection is used as the modulation and demodulation techniques is given by Equation 14.6–6.

soft-decision Viterbi decoding. First of all, Figure 14.6–4 shows the performance of the rate 1/2 convolutional codes for constraint lengths 3, 4, and 5. The bandwidth expansion factor for binary FSK modulation is $B_e = 2n$. Since an increase in the constraint length results in an increase in the complexity of the decoder to go along with the corresponding increase in the minimum free distance, the system designer can weight these two factors in the selection of the code.

Another way to increase the distance without increasing the constraint length of the code is to repeat each output bit m times. This is equivalent to reducing the code rate by a factor of m or expanding the bandwidth by the same factor. The result is a convolutional code that has a minimum free distance of md_{free}, where d_{free} is the minimum free distance of the original code without repetitions. Such a code is almost as good, from the viewpoint of minimum distance, as a maximum free distance, rate $1/mn$ code. The error rate performance with repetitions is upper-bounded by

$$P_b < \frac{1}{k} \sum_{d_{\text{free}}}^{\infty} \beta_d P_2(md) \tag{14.6–14}$$

where $P_2(md)$ is given by Equation 14.6–12. Figure 14.6–5 illustrates the performance of the rate 1/2 codes with repetitions ($m = 1, 2, 3, 4$) for constraint length 5.

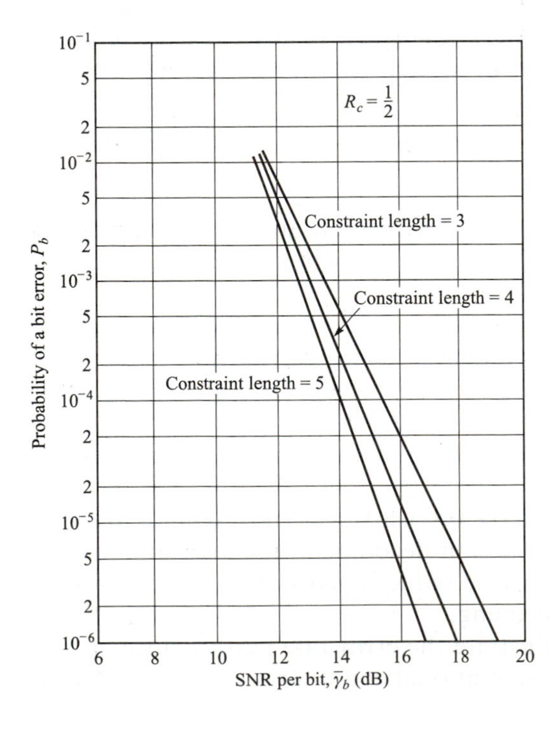

FIGURE 14.6–4
Performance of rate 1/2 binary convolutional codes with soft-decision decoding.

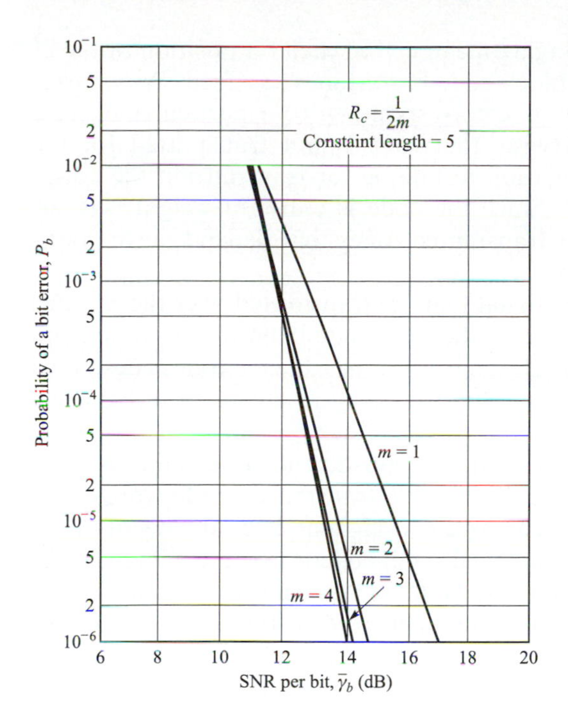

FIGURE 14.6–5
Performance of rate $1/2m$, constraint length 5, binary convolutional codes with soft-decision decoding.

14.6.4 Use of Constant-Weight Codes and Concatenated Codes for a Fading Channel

Our treatment of coding for a Rayleigh channel to this point was based on the use of binary FSK as the modulation technique for transmitting each of the binary digits in a code word. For this modulation technique, all the 2^k code words in the (n, k) code have identical transmitted energy. Furthermore, under the condition that the fading on the n transmitted tones is mutually statistically independent and identically distributed, the average received signal energy for the $M = 2^k$ possible code words is also identical. Consequently, in a soft-decision decoder, the decision is made in favor of the code word having the largest decision variable.

The condition that the received code words have identical average SNR has an important ramification in the implementation of the receiver. If the received code words do not have identical average SNR, the receiver must provide bias compensation for each received code word so as to render it equal energy. In general, the determination of the appropriate bias terms is difficult to implement because it requires the estimation of the average received signal power; hence, the equal-energy condition on the received code words considerably simplifies the receiver processing.

There is an alternative modulation method for generating equal-energy waveforms from code words when the code is constant-weight, i.e., when every code word has the same number of 1s. Note that such a code is non-linear.

Nevertheless, suppose we assign a single tone or cell to each bit position of the 2^k code words. Thus, an (n, k) binary block code has n tones assigned. Waveforms are constructed by transmitting the tone corresponding to a particular bit in a code word if that bit is a 1; otherwise, that tone is not transmitted for the duration of the interval. This modulation technique for transmitting the coded bits is called *on–off keying* (OOK). Since the code is constant-weight, say, w, every coded waveform consists of w transmitted tones that depend on the positions of the 1s in each of the code words.

As in FSK, all tones in the OOK signal that are transmitted over the channel are assumed to fade independently across the frequency band and in time from one code word to another. The received signal envelope for each tone is described statistically by the Rayleigh distribution. Statistically independent additive white Gaussian noise is assumed to be present in each frequency cell.

The receiver employs maximum-likelihood (soft-decision) decoding to map the received waveform into one of the M possible transmitted code words. For this purpose, n matched filters are employed, each matched to one of the n frequency tones. For the assumed statistical independence of the signal fading for the n frequency cells and additive white Gaussian noise, the envelopes of the matched filter outputs are squared and combined to form the M decision variables

$$U_i = \sum_{j=1}^{n} c_{ij} |\, y_j|^2, \qquad i = 1, 2, \ldots, 2^k \tag{14.6–15}$$

where $|\, y_j|^2$ corresponds to the squared envelope of the filter corresponding to the jth frequency, where $j = 1, 2, \ldots, n$.

It may appear that the constant-weight condition severely restricts our choice of codes. This is not the case, however. To illustrate this point, we briefly describe some methods for constructing constant-weight codes. This discussion is by no means exhaustive.

Method 1: Non-linear transformation of a linear code. In general, if in each word of an arbitrary binary code we substitute one binary sequence for every occurrence of a 0 and another sequence for each 1, a constant-weight binary block code will be obtained if the two substitution sequences are of equal weights and lengths. If the length of the sequence is ν and the original code is an (n, k) code, then the resulting constant-weight code will be an $(\nu n, k)$ code. The weight will be n times the weight of the substitution sequence, and the minimum distance will be the minimum distances of the original code times the distances between the two substitution sequences. Thus, the use of complementary sequences when ν is even results in a code with minimum distance $\nu d_{\min}$ and weight $\frac{1}{2}\nu n$.

The simplest form of this method is the case $\nu = 2$, in which every 0 is replaced by the pair 01 and every 1 is replaced by the complementary sequence 10 (or vice versa). As an example, we take as the initial code the (24,12) extended Golay code. The parameters of the original and the resultant constant-weight code are given in Table 14.6–1.

TABLE 14.6–1
Example of constant-weight code formed by Method 1

Code parameters	Original Golay	Constant-weight
n	24	48
k	12	12
M	4096	4096
$d_{\min}$	8	16
w	Variable	24

Note that this substitution process can be viewed as a separate encoding. This secondary encoding clearly does not alter the information content of a code word—it merely changes the form in which it is transmitted. Since the new code word is composed of pairs of bits—one "on" and one "off"—the use of OOK transmission of this code word produces a waveform that is identical to that obtained by binary FSK modulation for the underlying linear code.

Method 2: Expurgation. In this method, we start with an arbitrary binary block code and select from it a subset consisting of all words of a certain weight. Several different constant-weight codes can be obtained from one initial code by varying the choice of the weight w. Since the code words of the resulting expurgated code can be viewed as a subset of all possible permutations of any one code word in the set, the term *binary expurgated permutation modulation* (BEXPERM) has been used by Gaarder (1971) to describe such a code. In fact, the constant-weight binary block codes constructed by the other methods may also be viewed as BEXPERM codes. This method of generating constant-weight codes is in a sense opposite to the first method in that the word length n is held constant and the code size M is changed.The minimum distance for the constant-weight subset will clearly be no less than that of the original code. As an example, we consider the Golay (24, 12) code and form the two different constant-weight codes shown in Table 14.6–2.

Method 3: Hadamard matrices. This method might appear to form a constant-weight binary block code directly, but it actually is a special case of the method of expurgation. In this method, a Hadamard matrix is formed as

TABLE 14.6–2
Examples of constant-weight codes formed by expurgation

Parameters	Original	Constant weight no. 1	Constant weight no. 2
n	24	24	24
k	12	9	11
M	4096	759	2576
$d_{\min}$	8	$\geqslant 8$	$\geqslant 8$
w	Variable	8	12

described in Section 8.1.2, and a constant-weight code is created by selection of rows (code words) from this matrix. Recall that a Hadamard matrix is an $n \times n$ matrix (n even integer) of 1s and 0s with the property that any row differs from any other row in exactly $\frac{1}{2}n$ positions. One row of the matrix is normally chosen as being all 0s.

In each of the other rows, half of the elements are 0s and the other half 1s. A Hadamard code of size $2(n-1)$ code words is obtained by selecting these $n-1$ rows and their complements. By selecting $M = 2^k \leqslant 2(n-1)$ of these code words, we obtain a Hadamard code, which we denote by $H(n, k)$, where each code word conveys k information bits. The resulting code has constant weight $\frac{1}{2}n$ and minimum distance $d_{\min} = \frac{1}{2}n$.

Since n frequency cells are used to transmit k information bits, the bandwidth expansion factor for the Hadamard $H(n, k)$ code is defined as

$$B_e = \frac{n}{k} \quad \text{cells per information bit}$$

which is simply the reciprocal of the code rate. Also, the average SNR per bit, denoted by $\bar{\gamma}_b$, is related to the average SNR per cell, $\bar{\gamma}_c$, by the expression

$$\bar{\gamma}_c = \frac{k}{\frac{1}{2}n}\bar{\gamma}_b = 2\frac{k}{n}\bar{\gamma}_b = 2R_c\bar{\gamma}_b = \frac{2\bar{\gamma}_b}{B_e} \tag{14.6–16}$$

Let us compare the performance of the constant-weight Hadamard codes under a fixed bandwidth constraint with a conventional M-ary orthogonal set of waveforms where each waveform has diversity L. The M orthogonal waveforms with diversity are equivalent to a block orthogonal code having a block length $n = LM$ and $k = \log_2 M$. For example, if $M = 4$ and $L = 2$, the code words of the block orthogonal code are

$$\begin{aligned}
\mathbf{C}_1 &= [1 \quad 1 \quad 0 \quad 0 \quad 0 \quad 0 \quad 0 \quad 0] \\
\mathbf{C}_2 &= [0 \quad 0 \quad 1 \quad 1 \quad 0 \quad 0 \quad 0 \quad 0] \\
\mathbf{C}_3 &= [0 \quad 0 \quad 0 \quad 0 \quad 1 \quad 1 \quad 0 \quad 0] \\
\mathbf{C}_4 &= [0 \quad 0 \quad 0 \quad 0 \quad 0 \quad 0 \quad 1 \quad 1]
\end{aligned}$$

To transmit these code words using OOK modulation requires $n = 8$ cells, and since each code word conveys $k = 2$ bits of information, the bandwidth expansion factor $B_e = 4$. In general, we denote the block orthogonal code as $\mathrm{O}(n, k)$. The bandwidth expansion factor is

$$B_e = \frac{n}{k} = \frac{LM}{k} \tag{14.6–17}$$

Also, the SNR per bit is related to the SNR per cell by the expression

$$\bar{\gamma}_c = \frac{k}{L}\bar{\gamma}_b = M\left(\frac{k}{n}\right)\bar{\gamma}_b = M\frac{\bar{\gamma}_b}{B_e} \tag{14.6–18}$$

Now we turn our attention to the performance characteristics of these codes. First, the exact probability of a code word (symbol) error for M-ary orthogonal signaling over a Rayleigh fading channel with diversity was given in closed form in Section 14.4. As previously indicated, this expression is rather cumbersome to evaluate, especially if either L or M or both are large. Instead, we shall use a union bound that is very convenient. That is, for a set of M orthogonal waveforms, the probability of a symbol error can be upper-bounded as

$$\begin{aligned} P_M &\leqslant (M-1)P_2(L) \\ &= (2^k - 1)P_2(L) < 2^k P_2(L) \end{aligned} \tag{14.6–19}$$

where $P_2(L)$, the probability of error for two orthogonal waveforms, each with diversity L, is given by Equation 14.6–12 with $p = 1/(2+\bar{\gamma}_c)$. The probability of bit error is obtained by multiplying P_M by $2^{k-1}/(2^k - 1)$, as explained previously.

A simple upper (union) bound on the probability of a code word error for the Hadamard $H(n, k)$ code is obtained by noting the probability of error in deciding between the transmitted code word and any other code word is bounded from above by $P_2(\frac{1}{2}d_{\min})$, where $d_{\min}$ is the minimum distance of the code. Therefore, an upper bound on P_M is

$$P_M \leqslant (M-1)P_2(\tfrac{1}{2}d_{\min}) < 2^k P_2(\tfrac{1}{2}d_{\min}) \tag{14.6–20}$$

Thus the "effective order of diversity" of the code for OOK modulation is $\frac{1}{2}d_{\min}$. The bit error probability may be approximated as $\frac{1}{2}P_M$, or slightly overbounded by multiplying P_M by the factor $2^{k-1}/(2^k - 1)$, which is the factor used above for orthogonal codes. The latter was selected for the error probability computations given below.

Figure 14.6–6 illustrates the error rate performance of a selected number of Hadamard codes for several bandwidth expansion factors. The advantage resulting from an increase in the size M of the alphabet (or k, since $k = \log_2 M$) and an increase in the bandwidth expansion factor is apparent from observation of these curves. Note, for example, that the $H(20, 5)$ code when repeated twice results in a code that is denoted by ${}_2H(20, 5)$ and has a bandwidth expansion factor $B_e = 8$. Figure 14.6–7 shows the performance of the Hadamard and block orthogonal codes compared on the basis of equal bandwidth expansion factors. It is observed that the error rate curves for the Hadamard codes are steeper than the corresponding curves for the block orthogonal codes. This characteristic behavior is due simply to the fact that, for the same bandwidth expansion factor, the Hadamard codes provide more diversity than block orthogonal codes. Alternatively, one may say that Hadamard codes provide better bandwidth efficiency than block orthogonal codes. It should be mentioned, however, that at low SNR, a lower-diversity code outperforms a higher-diversity code as a consequence of the fact that, on a Rayleigh fading channel, there is an optimum distribution of the total received SNR among the diversity signals. Therefore, the curves for the block orthogonal codes will cross over the curves for the Hadamard codes at the low-SNR (high-error-rate) region.

Method 4: Concatenation. In this method, we begin with two codes: one binary and the other nonbinary. The binary code is the inner code and is an

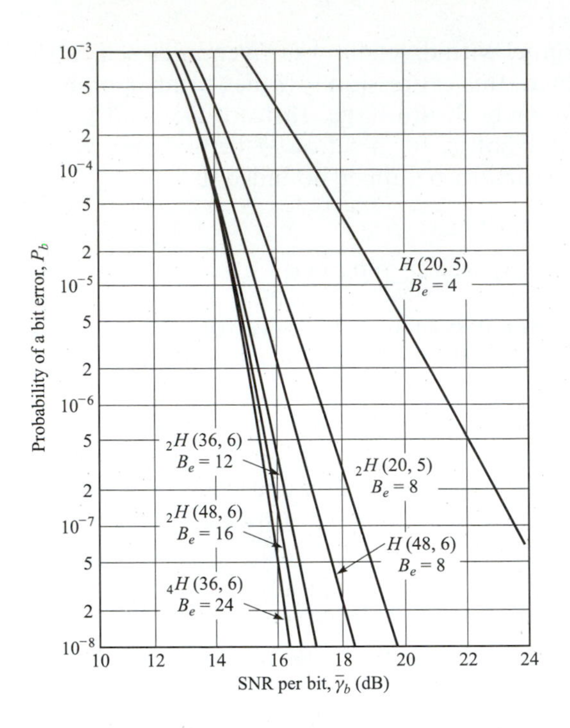

FIGURE 14.6–6
Performance of Hadamard codes.

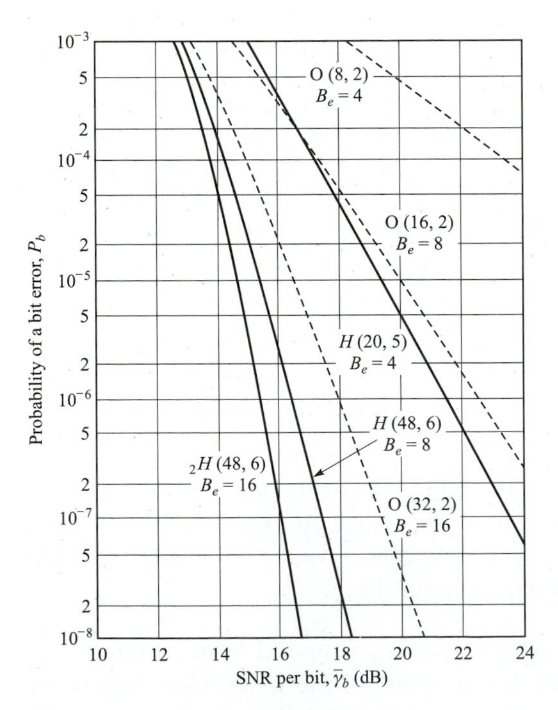

FIGURE 14.6–7
Comparison of performance between Hadamard codes and block orthogonal codes.

(n, k) constant-weight (non-linear) block code. The nonbinary code, which may be linear, is the outer code. To distinguish it from the inner code, we use uppercase letters, e.g., an (N, K) code, where N and K are measured in terms of symbols from a q-ary alphabet. The size q of the alphabet over which the outer code is defined cannot be greater than the number of words in the inner code. The outer code, when defined in terms of the binary inner code words rather than q-ary symbols, is the new code.

An important special case is obtained when $q = 2^k$ and the inner code size is chosen to be 2^k. Then the number of words is $M = 2^{kK}$ and the concatenated structure is an (nN, kK) code. The bandwidth expansion factor of this concatenated code is the product of the bandwidth expansions for the inner and outer codes.

Now we shall demonstrate the performance advantages obtained on a Rayleigh fading channel by means of code concatenation. Specifically, we construct a concatenated code in which the outer code is a dual-k (nonbinary) convolutional code and the inner code is either a Hadamard code or a block orthogonal code. That is, we view the dual-k code with M-ary ($M = 2^k$) orthogonal signals for modulation as a concatenated code. In all cases to be considered, soft-decision demodulation and Viterbi decoding are assumed.

The error rate performance of the dual-k convolutional codes is obtained from the derivation of the transfer function given by Equation 8.2–39. For a rate-1/2, dual-k code with no repetitions, the bit error probability, appropriate for the case in which each k-bit output symbol from the dual-k encoder is mapped into one of $M = 2^k$ orthogonal code words, is upper-bounded as

$$P_b < \frac{2^{k-1}}{2^k - 1} \sum_{m=4}^{\infty} \beta_m P_2(m) \qquad (14.6\text{–}21)$$

where $P_2(m)$ is given by Equation 14.6–12.

For example, a rate-1/2, dual-2 code may employ a 4-ary orthogonal code O(4, 2) as the inner code. The bandwidth expansion factor of the resulting concatenated code is, of course, the product of the bandwidth expansion factors of the inner and outer codes. Thus, in this example, the rate of the outer code is 1/2 and the inner code is 1/2. Hence, $B_e = (4/2)(2) = 4$.

Note that if every symbol of the dual-k is repeated r times, this is equivalent to using an orthogonal code with diversity $L = r$. If we select $r = 2$ in the example given above, the resulting orthogonal code is denoted as O(8, 2) and the bandwidth expansion factor for the rate-1/2, dual-2 code becomes $B_e = 8$. Consequently, the term $P_2(m)$ in Equation 14.6–21 must be replaced by $P_2(mL)$ when the orthogonal code has diversity L. Since a Hadamard code has an "effective diversity" $\frac{1}{2}d_{\min}$, it follows that when a Hadamard code is used as the inner code with a dual-k outer code, the upper bound on the bit error probability of the resulting concatenated code given by Equation 14.6–21 still applies if $P_2(m)$ is replaced by $P_2(\frac{1}{2}md_{\min})$. With these modifications, the upper bound on the bit error probability given by Equation 14.6–21 has been evaluated for rate-1/2, dual-k convolutional codes with either Hadamard codes or block orthogonal codes as inner codes. Thus the resulting concatenated code has a bandwidth exansion factor equal to twice the bandwidth expansion factor of the inner code.

First, we consider the performance gains due to code concatenation. Figure 14.6–8 illustrates the performance of dual-k codes with block orthogonal inner codes compared with the performance of block orthogonal codes for bandwidth expansion factors $B_e = 4$, 8, 16, and 32. The performance gains due to concatenation are very impressive. For example, at an error rate of 10^{-6} and $B_e = 8$, the dual-k code outperforms the orthogonal block code by 7.5 dB. In short, this gain may be attributed to the increased diversity (increase in minimum distance) obtained via code concatenation. Similarly, Figure 14.6–9 illustrates the performance of two dual-k codes with Hadamard inner codes compared with the performance of the Hadamard codes alone for $B_e = 8$ and 12. It is observed that the performance gains due to code concatenation are still significant, but certainly not as impressive as those illustrated in Figure 14.6–8. The reason is that the Hadamard codes alone yield a large diversity, so that the increased diversity arising from concatenation does not result in as large a gain in performance for the range of error rates covered in Figure 14.6–9.

The numerical results given above illustrate the performance advantages in using codes with good distance properties and soft-decision decoding on a Rayleigh fading channel as an alternative to conventional M-ary orthogonal signaling with diversity. In addition, the results illustrate the benefits of code concatenation on such a channel, using a dual-k convolutional code as the outer

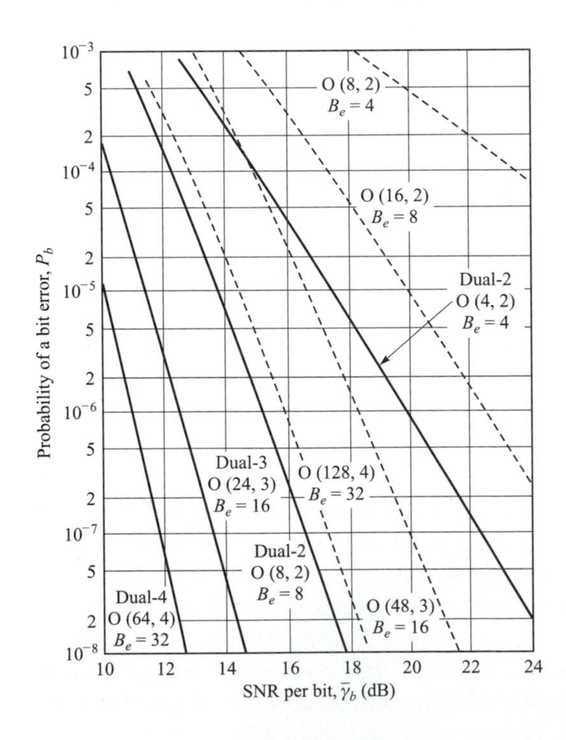

FIGURE 14.6–8
Comparison of performance between block orthogonal codes and dual-k with block orthogonal inner codes.

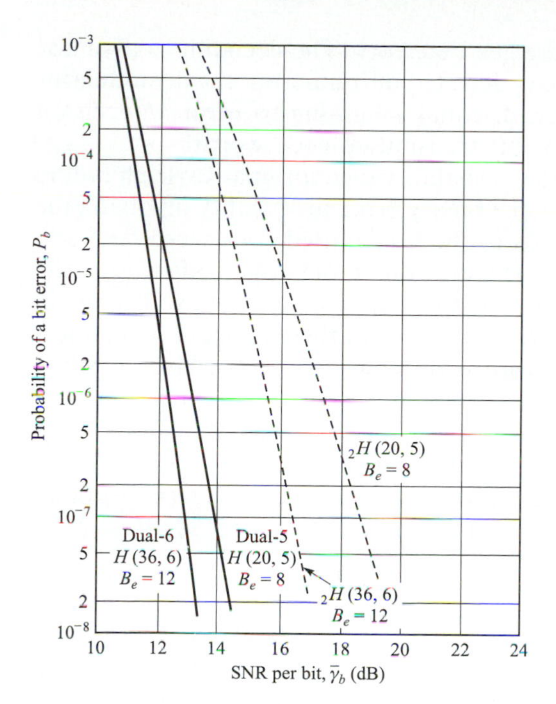

FIGURE 14.6–9
Comparison of performance between Hadamard codes and dual-k codes with Hadamard inner codes.

code and either a Hadamard code or a block orthogonal code as the inner code. Although dual-k codes were used for the outer code, similar results are obtained when a Reed–Solomon code is used for the outer code. There is an even greater choice in the selection of the inner code.

The important parameter in the selection of both the outer and the inner codes is the minimum distance of the resultant concatenated code required to achieve a specified level of performance. Since many codes will meet the performance requirements, the ultimate choice is made on the basis of decoding complexity and bandwidth requirements.

14.6.5 System Design Based on the Cutoff Rate

In the above treatment of coded waveforms, we have demonstrated the effectiveness of various codes for fading channels. In particular, we have observed the benefits of soft-decision decoding and code concatenation as a means for increasing the minimum distance and, hence, the amount of diversity in the coded waveforms. In this subsection, we consider randomly selected code words and derive an upper (union) bound on the error probability that depends on the cutoff rate parameter for the Rayleigh fading channel.

Let us consider the model for the communication system illustrated in Figure 14.6–1. The modulator has a q-ary orthogonal FSK alphabet. Code words of block length n are mapped into waveforms by selecting n tones from the alphabet of q tones. The demodulation is performed by passing the signal through a bank of q matched filters followed by square-law detectors. The decoding is assumed to be soft-decision. Thus, the square-law-detected outputs from the demodulator are appropriately combined (added) with equal weighting to form M decision variables corresponding to the M possible transmitted code words.

To evaluate the union bound on the probability of error in a Rayleigh fading channel with AWGN, we first evaluate the binary error probability involving the decision variable U_1, which corresponds to the transmitted code word, and any of the other $M - 1$ decision variables corresponding to the other code words. Let U_2 be the other decision variable and suppose that U_1 and U_2 have l tones in common. Hence, the contributions to U_1 and U_2 from these l tones are identical and, therefore, cancel out when we form the difference $U_1 - U_2$. Since the two decision variables differ in $n - l$ tones, the probability of error is simply that for a binary orthogonal FSK system with $n - l$ order diversity. The exact form for this probability of error is given by Equation 14.6–4, where $p = 1/(2 + \bar{\gamma}_c)$, and $\bar{\gamma}_c$ is the average SNR per tone. For simplicity, we choose to use the Chernoff bound for this binary event probability, given by Equation 14.6–7, i.e.,

$$P_2(U_1, U_2 \mid l) \leqslant [4p(1-p)]^{n-l} \tag{14.6–22}$$

Now, let us average over the ensemble of binary communication systems. There are q^n possible code words, from which we randomly select two code words. Thus, each code word is selected with equal probability. Then, the probability that two randomly selected code words have l tones in common is

$$P(l) = \binom{n}{l}\left(\frac{1}{q}\right)^l\left(1 - \frac{1}{q}\right)^{n-l} \tag{14.6–23}$$

When we average Equation 14.6–22 over the probability distribution of l given by Equation 14.6–23, we obtain

$$\begin{aligned}\overline{P_2(U_1, U_2)} &= \sum_{l=0}^{n} P_2(U_1, U_2 \mid l)P(l) \\ &\leqslant \sum_{l=0}^{n}\binom{n}{l}\left(\frac{1}{q}\right)^l\left[4\left(1-\frac{1}{q}\right)p(1-p)\right]^{n-l} \\ &\leqslant \left\{\frac{1}{q}[1 + 4(q-1)p(1-p)]\right\}^n\end{aligned} \tag{14.6–24}$$

Finally, the union bound for communication systems that use $M = 2^k$ randomly selected code words is simply

$$\bar{P}_M \leqslant (M-1)\overline{P_2(U_1, U_2)} < M\overline{P_2(U_1, U_2)} \tag{14.6–25}$$

By combining Equation 14.6–24 with Equation 14.6–25, we obtain the upper bound on the symbol error probability as

$$\bar{P}_M < 2^{-n(R_0 - R_c)} \tag{14.6–26}$$

where $R_c = k/n$ is the code rate and R_0 is the cutoff rate defined as

$$R_0 = \log_2 \frac{q}{1 + 4(q-1)p(1-p)} \tag{14.6–27}$$

with

$$p = \frac{1}{2 + \bar{\gamma}_c} \tag{14.6–28}$$

Graphs of R_0 as a function of $\bar{\gamma}_c$ are shown in Figure 14.6–10 for $q = 2$, 4, and 8.

A more interesting form of Equation 14.6–26 is obtained if we express $\bar{P}_M$ in terms of the SNR per bit. In particular, Equation 14.6–26 may be expressed as

$$\bar{P}_M < 2^{-k[\bar{\gamma}_b g(q,\bar{\gamma}_c)-1]} \tag{14.6–29}$$

where, by definition,

$$\begin{aligned} g(q, \bar{\gamma}_c) &= \frac{R_0}{\bar{\gamma}_c} \\ &= \frac{1}{\bar{\gamma}_c} \log_2 \left[\frac{q}{1 + 4(q-1)p(1-p)} \right] \end{aligned} \tag{14.6–30}$$

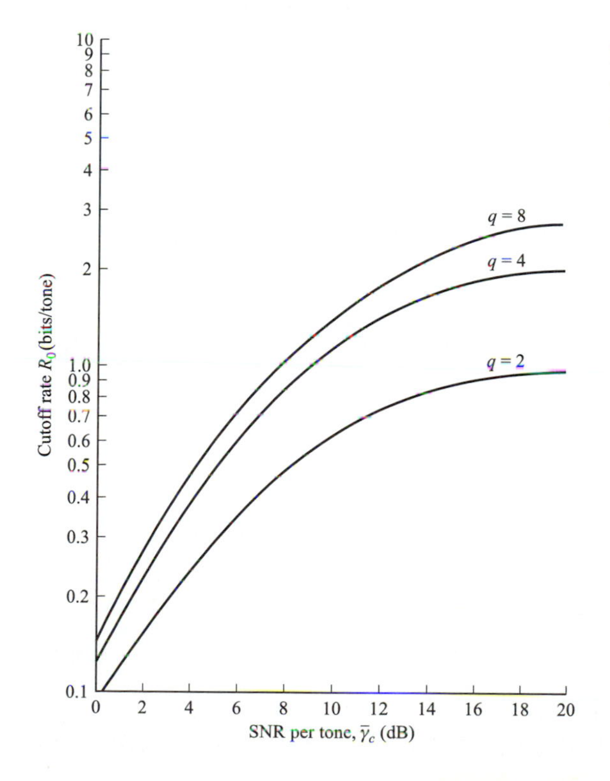

FIGURE 14.6–10
Cutoff rate as a function of $\bar{\gamma}_c$ for Rayleigh fading channel.

Graphs of $g(q, \bar{\gamma}_c)$ as a function of $\bar{\gamma}_c$ are plotted in Figure 14.6–11, with q as a parameter. First, we note that there is an optimum $\bar{\gamma}_c$ for each value of q that minimizes the probability of error. For large q, this value is approximately $\bar{\gamma}_c = 3$ (5 dB), which is consistent with our previous observation for ordinary square-law diversity combining. Furthermore, as $q \to \infty$, the function $g(q, \bar{\gamma}_c)$ approaches a limit, which is

$$\lim_{q\to\infty} g(q, \bar{\gamma}_c) = g_\infty(\bar{\gamma}_c) = \frac{1}{\bar{\gamma}_c} \log_2 \left[\frac{(2+\bar{\gamma}_c)^2}{4(1+\bar{\gamma}_c)} \right] \tag{14.6–31}$$

The value of $g_\infty(\bar{\gamma}_c)$ evaluated at $\bar{\gamma}_c = 3$ is

$$g_\infty(3) = \max_{\bar{\gamma}_c} g_\infty(\bar{\gamma}_c) = 0.215 \tag{14.6–32}$$

Therefore, the error probability in Equation 14.6–29 for this optimum division of total SNR is

$$\bar{P}_M < 2^{-0.215k(\bar{\gamma}_b - 4.65)} \tag{14.6–33}$$

This result indicates that the probability of error can be made arbitrarily small with optimum SNR per code chip, if the average SNR per bit $\bar{\gamma}_b > 4.65$ (6.7 dB).

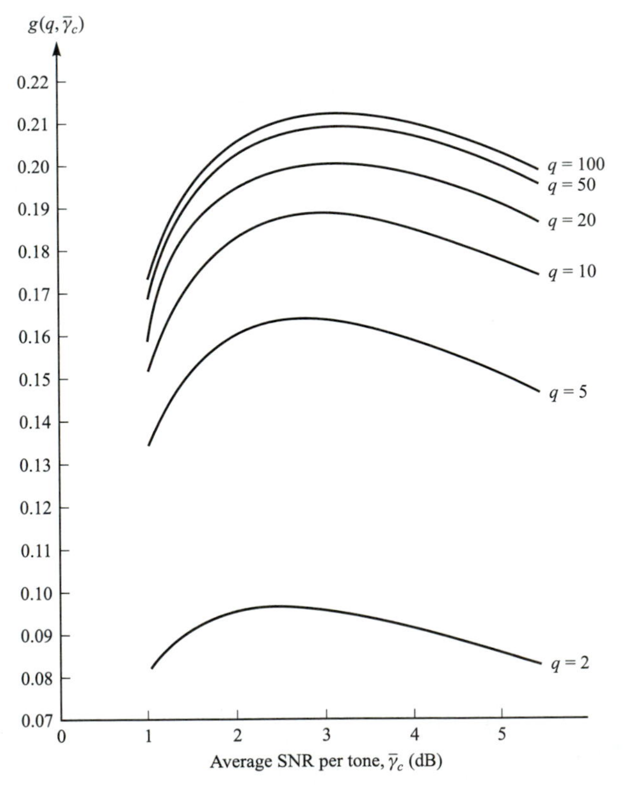

FIGURE 14.6–11
Graph of function $g(q, \bar{\gamma}_c)$.

Even a relatively modest value of $q = 20$ comes close to this minimum value. As seen from Figure 14.6–11, $g(20, 3) = 0.2$, so that $P_M \to 0$, provided $\bar{\gamma}_b > 5$ (7 dB). On the other hand, if $q = 2$, the maximum value of $g(2, \bar{\gamma}_c) \approx 0.096$ and the corresponding minimum SNR per bit is 10.2 dB.

In the case of binary FSK waveforms ($q = 2$), we may easily compare the cutoff rate for the unquantized (soft-decision) demodulator output with the cutoff rate for binary quantization, for which

$$R_Q = 1 - \log\left[1 + \sqrt{4p(1-p)}\right], \qquad q = 2 \tag{14.6–34}$$

as was given in Equation 7.3–6. Figure 14.6–12 illustrates the graphs for R_0 and R_Q. Note that the difference between R_0 and R_Q is approximately 3 dB for rates below 0.3 and the difference increases rapidly at high rates. This loss may be reduced significantly by increasing the number of quantization levels to $Q = 8$ (three bits).

Similar comparisons in the relative performance between unquantized soft-decision decoding and quantized decision decoding can also be made for $q > 2$.

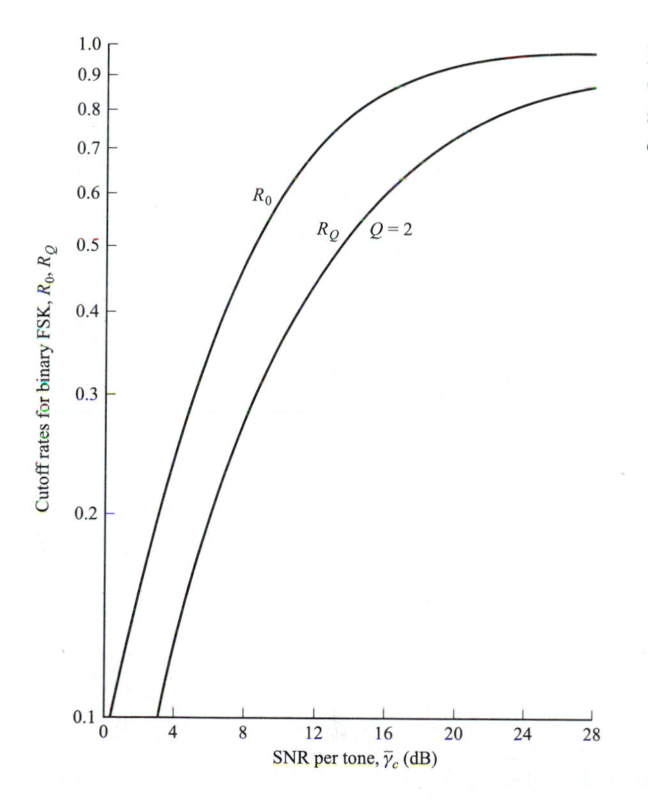

FIGURE 14.6–12
Cutoff rate for (unquantized) soft-decision and hard-decision decoding of coded binary FSK.

14.6.6 Performance of Coded Phase-Coherent Communication Systems—Bit-Interleaved Coded Modulation

Our treatment of coded communication systems for fading channels focused on coded FSK signal waveforms which are detected noncoherently. Such waveforms are generally more robust to channel fading than PSK or QAM signals, since the latter require phase-coherent demodulation. On the other hand, there are channels in which the fading is sufficiently slow to allow for phase-coherent demodulation of PSK and QAM signals. In this section, we briefly consider the performance of coded signals in which the coded bits are transmitted by binary PSK.

The model of the digital communication system with coding is shown in Figure 14.6–13. We assume that the encoder generates an (n, k) linear binary block code.† To ensure that the bits in any code word fade independently the sequence is bit-wise interleaved. We assume that the bits are interleaved with a depth exceeding the coherence time of the channel. Each coded bit is transmitted via binary PSK through a frequency-nonselective fading channel and demodulated coherently at the receiver.

To evaluate the probability of error for the coded system, with soft-decision decoding, we may again use the results on diversity reception with maximal ratio combining that we obtained in Section 4.4.1. In particular, the binary event error probability, assuming that the all-zero code work was transmitted

$$P_2(w_m) = \left(\frac{1-\mu}{2}\right)^{w_m} \sum_{k=0}^{w_m-1} \binom{w_m - 1 + k}{k} \left(\frac{1+\mu}{2}\right)^k \tag{14.6–35}$$

where w_m is the weight of the mth code word and, for a perfect channel gain estimate, i.e., perfect knowledge of the channel fading process,

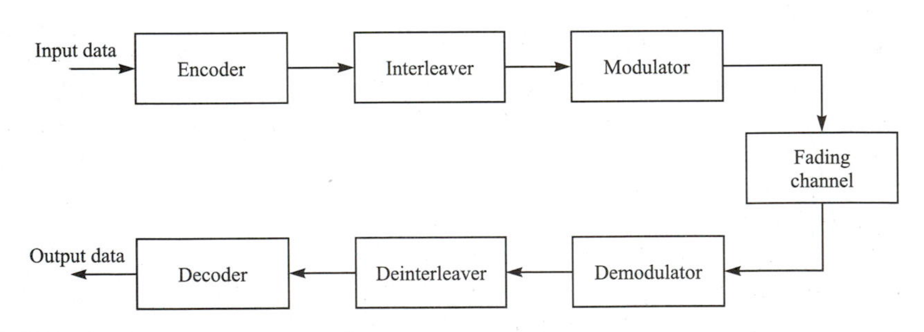

FIGURE 14.6–13
Block diagram of trellis-coded modulation systems.

†A convolutional code may be used in place of the block code.

$$\mu = \sqrt{\frac{\bar{\gamma}_c}{1+\bar{\gamma}_c}} \tag{14.6–36}$$

where $\bar{\gamma}_c$ is the average SNR per coded bit. The union bound may be used to obtain an upper bound on the error probability for a code word error as

$$\begin{aligned} P_M &< \sum_{m=2}^{M} P_2(w_m) < (M-1)P_2(d_{\min}) \\ &< 2^k \binom{2d_{\min}-1}{d_{\min}} \left(\frac{1}{4R_c\bar{\gamma}_b}\right)^{d_{\min}} \end{aligned} \tag{14.6–37}$$

where $M = 2^k$, R_c is the code rate, and $\bar{\gamma}_b$ is the average SNR per bit.

In contrast, for hard-decision decoding the expression for P_M given by Equation 14.6–37 is simply modified by substituting $d_{\min}/2$ for $d_{\min}$.

We observe that the communication system that employs phase-coherent modulation with coding results in a 6-dB improvement in performance compared with a noncoherent system that employs FSK. Furthermore, since PSK requires less bandwidth than FSK, the bandwidth expansion factor for the phase-coherent system is also reduced. Consequently, with bit-wise interleaving, a coded phase-coherent communication system is preferable to one that employs phase-noncoherent demodulation if the channel exhibits sufficient phase stability to support phase-coherent demodulation. The effect of a phase estimation error on the system performance can be easily determined by use of the analysis described in Appendix C.

In the discussion above, we considered the mapping of the bit-interleaved coded sequence into binary PSK modulation. A generalization is to subdivide the bit-interleaved sequence into sub-sequences of m bits and use M-ary ($M = 2^m$) modulation, e.g., M-ary PSK or M-ary QAM for phase-coherent modulation and demodulation. Gray encoding should be used in mapping m-bit sequences into signal points in the M-ary signal constellation. For soft-decision decoding, the error rate performance of the bit-interleaved sequence with M-ary modulation also decreases exponentially with the inverse of the SNR per bit, where the exponent is the minimum (Hamming) distance of the code.

As in the case of coded FSK signals with noncoherent demodulation, code concatenation is highly desirable in a coded phase-coherent system as a practical means for increasing the minimum Hamming distance of the code. The performance analysis of concatenated codes for phase-coherent systems is similar to that already described for systems employing noncoherent demodulation.

14.6.7 Trellis-Coded Modulation

Trellis-coded modulation was described in Section 8.3 as a means for achieving a coding gain on bandwidth-constrained channels, where we wish to transmit at a bit rate-to-bandwidth ratio $R/W > 1$. For such channels, the digital communication system is designed to use bandwidth-efficient multilevel or multiphase modulation (PAM, PSK, DPSK, or QAM), which allows us to achieve an $R/W > 1$. When coding is applied in signal design for a bandwidth-constrained channel, a coding gain is desired without expanding the signal bandwidth. This goal can be achieved, as described in Section 8.3, by increasing the number of signal points in the constellation over the corresponding uncoded system to compensate for the redundancy introduced by the code, and designing the trellis code so that the Euclidean distance in a sequence of transmitted symbols corresponding to paths that merge at any node in the trellis is larger than the Euclidean distance per symbol in an uncoded system.

In contrast, the coding schemes that we have described above in conjunction with FSK or PSK modulation expand the bandwidth of the modulated signal for the purpose of achieving signal diversity. Coupled with FSK modulation, which is not bandwidth-efficient, the coding schemes we have described are inappropriate for use on bandwidth-constrained channels.

In designing trellis-coded signal waveforms for fading channels, we may use the same basic principles that we have learned and applied in the design of conventional coding schemes. In particular, the most important objective in any coded signal design for fading channels is to achieve as large a signal diversity as possible. This implies that successive output symbols from the encoder must be interleaved or sufficiently separated in transmission, either in time or in frequency, so as to achieve independent fading in a sequence of symbols that equals or exceeds the minimum free distance of the trellis code. Therefore, we may represent such a trellis-coded modulation system by the block diagram in Figure 14.6–13, where the encoder is a trellis encoder and the interleaver is viewed broadly as a device that separates the successive coded symbols so as to provide independent fading on each symbol (through frequency or time separation of symbols) in the sequence. The receiver consists of a signal demodulator whose output is deinterleaved and fed to the trellis decoder.

As indicated above, the candidate modulation methods that achieve high bandwidth efficiency are M-ary PSK, DPSK, QAM, and PAM. The choice depends to a large extent on the channel characteristics. If there are rapid amplitude variations in the received signal, QAM and PAM may be particularly vulnerable, because a wideband automatic gain control (AGC) must be used to compensate for the channel variations. In such a case, PSK or DPSK are more suitable, since the information is conveyed by the signal phase and not by the signal amplitude. DPSK provides the additional benefit that carrier phase coherence is required only over two successive symbols. However, there is an SNR degradation in DPSK relative to PSK.

In the design of the trellis code, our objective is to achieve as large a free distance as possible, since this parameter is equivalent to the amount of diversity in the received signal. In conventional Ungerboeck trellis coding, each branch in the trellis corresponds to a single M-ary (PSK, DPSK, QAM) output channel symbol. Let us define the *shortest error event path* as the error event path with the smallest number of nonzero distances between itself and the correct path, and let L be its length. In other words, L is the Hamming distance between the M-ary symbols on the shortest error event path and those in the correct path. Hence, if we assume that the transmitted sequence corresponds to the all-zero path in the trellis, L is the number of branches in the shortest-length path with a nonzero M-ary symbol. In a trellis diagram with parallel paths, the paths are constrained to have a shortest error event length of one branch, so that $L = 1$. This means that such a trellis code provides no diversity in a fading channel and, hence, the probability of error is inversely proportional to the SNR per symbol. Therefore, in conventional trellis coding for a fading channel, it is undesirable to design a code that has parallel paths in its trellis, because such a code yields no diversity. This is the case in a conventional Ungerboeck rate-$m/(m+1)$ trellis code, where we are forced to have parallel paths when the number of states is less than 2^m.

One possible way to increase the minimum free distance and, thus, the order of diversity in the code, is to introduce asymmetry in the signal point constellation. This approach appears to be somewhat effective and has been investigated by Simon and Divsalar (1985), Divsalar and Yuen (1984), and Divsalar et al. (1987).

A more effective way to increase the distance L and, thus, the order of diversity is to employ multiple trellis-coded modulation (MTCM). In MTCM, illustrated in Figure 14.6–14, b inputs bits to the encoder are coded into c output bits, which are then subdivided into k groups, each of m bits, such that $c = km$. Each m-bit group is mapped into an M-ary symbol. Thus, we obtain the M-ary output symbols. The special case $k = 1$ corresponds to the conventional Ungerboeck codes. With k M-ary output symbols, it is possible to design trellis codes with parallel paths having a distance $L = k$. Thus, we can achieve an error probability that decays inversely as $(\mathcal{E}/N_0)^k$.

An important consideration in the design of the decoder for the trellis code is the use of any side information regarding the channel attenuation for each symbol. In the case of FSK modulation with a square-law combiner at the decoder to form the decision metrics, it is not necessary to know the channel attenuation for demodulated symbols. However, with coherent detection, the optimum Euclidean distance metric for each demodulated symbol is of the form $|r_n - \alpha_n s_n|^2$, where α_n is the channel attenuation for the transmitted symbol

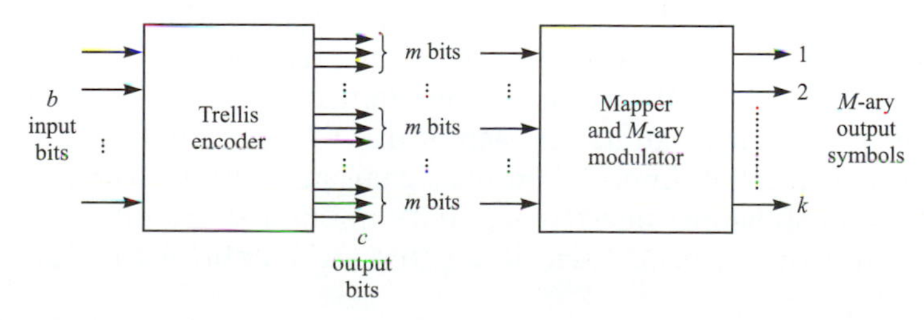

FIGURE 14.6–14
Block diagram of an MTCM transmitter.

s_n and r_n is the demodulation output. Hence, the sum of branch metrics for any given path through the trellis is of the form

$$D(\mathbf{r}, \mathbf{s}^{(i)}) = \sum_n |r_n - \alpha_n s_n^{(i)}|^2$$

where the superscript (i) indicates the ith path through the trellis. Therefore, the estimation of the channel attenuation must be performed in order to realize the optimum trellis decoder. The estimation of the channel attenuation and phase shift, is considered in Appendix C for the case of PSK modulation and demodulation. The effect of the quality of the attenuation and phase estimates on the performance of PSK (uncoded) modulation is also assessed in Appendix C.

A different approach from the conventional Ungerboeck TCM method is considered in the papers by Viterbi et al. (1989), Zehavi (1992), and Caire et al. (1998). Instead of jointly optimizing the encoding and modulation, the two functions are separated. In this approach, a binary convolutional code of a specified constraint length with the largest free Hamming distance is employed. The output of the encoder is bit-wise interleaved and passed to a mapper which maps groups of m bits into PSK or QAM signal constellations. With bit-wise interleaving, the error rate performance of the coded modulation on a Rayleigh fading channel with perfect channel state information decreases inversely as the SNR per bit raised to the power of the minimum free (Hamming) distance of the convolutional code. The paper by Caire et al. (1998) presents an in-depth comparison of the performance of bit-interleaved coded modulation and conventional coded modulation with symbol interleaving for Rayleigh fading channels.

14.7
MULTIPLE-ANTENNA SYSTEMS

Using multiple antennas at the receiver side is a standard method for achieving (spatial) diversity to combat fading without expanding the bandwidth of the transmitted signal. Spatial diversity can also be achieved by using multiple antennas at the transmitter side. For example, it is possible to achieve dual diversity with two transmitting antennas and one receiving antenna (see Problem 14.16).

Multiple transmitting antennas can also be used to create multiple spatial channels and, thus, provide the capability to increase the data rate. In particular, with N transmitting antennas and M receiving antennas,† it is possible to achieve an N-fold increase in the data rate and simultaneously provide Mth-order reception diversity to combat fading for each of the N transmitted signals. As there is no orthogonal structure imposed on the signals at the N transmitting antennas, there will be interchannel interference (ICI) among the spatial channels. In this section, we describe detection algorithms that are suitable for multiple transmit-

†In this case, we assume that $M \geqslant N$.

ting and receiving antennas and evaluate their performance with and without coding.

The general configuration for a communication system that employs multiple transmitting and receiving antennas to increase the data rate is shown in Figure 14.7–1. The data is encoded and interleaved. The interleaver is necessary in the case of encoded data in order to ensure independent fading of the coded bits or symbols. A block of N symbols is converted from serial to parallel, and each symbol is fed to one of the N identical modulators, where each modulator is connected to a (spatially) separate antenna. Thus, the N symbols are transmitted in parallel and received on M spatially separated receiving antennas, where $M \geqslant N$.

We assume that each signal from a transmitting antenna to a receiving antenna undergoes frequency nonselective Rayleigh fading. We also assume that the differences in propagation times of the signals from the N transmitting antennas to the M receiving antennas are small relative to the symbol duration T, so that for practical purposes, the signals from the N transmitting antennas to any receiving antenna are synchronous. Hence, we can represent the equivalent low-pass received signals at the receiving antennas in a signaling interval as

$$r_{lm}(t) = \sum_{n=1}^{N} d_n a_{mn} g(t) + z_m(t), \qquad 0 \leqslant t \leqslant T, \qquad m = 1, 2, \ldots, M \tag{14.7–1}$$

where $g(t)$ is the pulse shape (impulse response) of the modulation filters, a_{mn} is the complex-valued zero-mean Gaussian channel gain between the nth transmit-

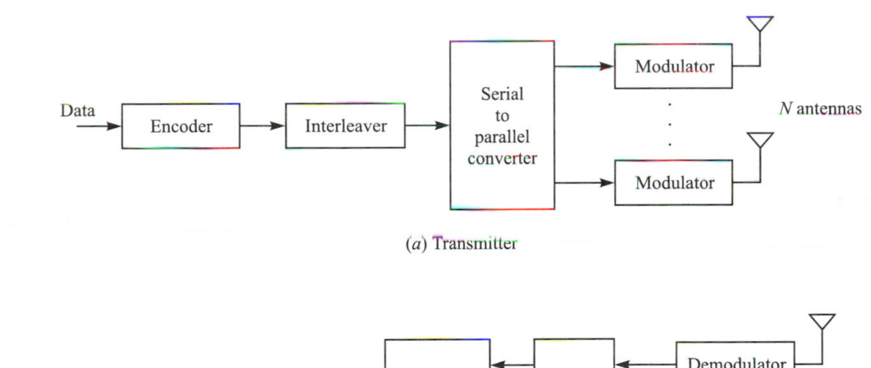

(*a*) Transmitter

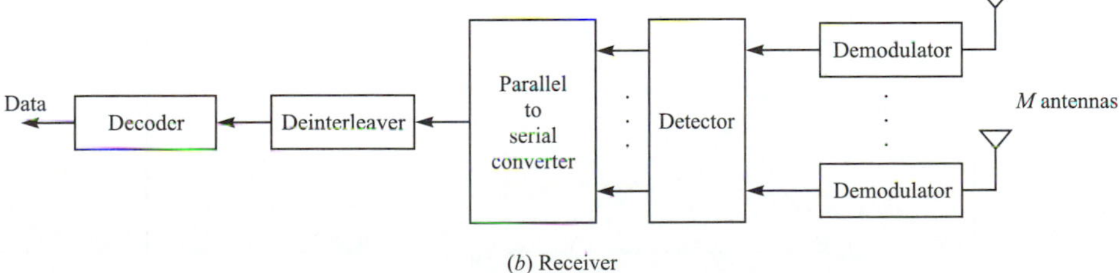

(*b*) Receiver

FIGURE 14.7–1
A communication system with multiple transmitting and receiving antennas.

ting antenna and the mth receiving antenna, d_n is the symbol transmitted on the nth antenna, and $z_m(t)$ is a sample function of an AWGN process. The channel coefficients $\{a_{mn}\}$ are identically distributed, statistically independent from channel to channel and, with interleaving, from coded symbol to coded symbol. The information symbols are drawn from a binary or expanded PSK/QAM signal constellation.

The demodulator for the signal at each of the M receiving antennas consists of a matched filter to the pulse $g(t)$ (or a cross correlator), whose output is sampled at the end of each symbol interval. The output of the demodulator corresponding to the mth antenna in the kth signaling interval can be represented as

$$y_m(k) = \sum_{n=1}^{N} d_n(k) a_{mn}(k) + \eta_m(k), \qquad m = 1, 2, \ldots, M \tag{14.7–2}$$

where the energy of the signal pulse $g(t)$ is normalized to unity and $\eta_m(k)$ is the additive Gaussian noise component. The M soft outputs from the demodulators are passed to the signal detector. For convenience, Equation 14.7–2 may be expressed in matrix form as

$$\mathbf{y} = \mathbf{Ad} \tag{14.7–3}$$

where $\mathbf{y} = [y_1(k), \ldots, y_M(k)]^t$, $\mathbf{d} = [d_1(k), \ldots, d_N(k)]^t$, $\boldsymbol{\eta} = [\eta_1(k), \ldots, \eta_M(k)]^t$, and $\mathbf{A}$ is the $M \times N$ matrix of channel coefficients. Figure 14.7–2 illustrates the model for the multiple transmitter and receiver signals.

Maximum-likelihood detector (MLD). Since the additive noise on the M receiver antennas is (statistically independent) Gaussian, the joint conditional PDF $p(\mathbf{y}|\mathbf{d})$ is Gaussian. Therefore, the MLD selects the symbol vector $\mathbf{d}$ that minimizes the Euclidean distance metric

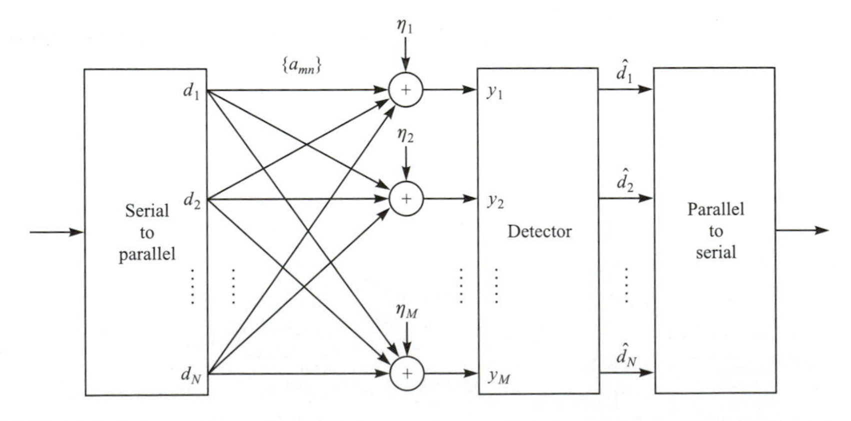

FIGURE 14.7–2
Discrete-time model of the communication system with multiple antennas.

$$\mu(\mathbf{d}) = \sum_{m=1}^{M} \left| y_m(k) - \sum_{n=1}^{N} a_{mn}(k)\, d_n(k) \right|^2 \tag{14.7–4}$$

Minimum mean-square-error (MMSE) detector. The MMSE detector linearly combines the received signal samples $y_1(k), \ldots, y_M(k)$ to form estimates of the transmitted symbols $d_1(k), \ldots, d_N(k)$. The linear combining is represented in matrix form as

$$\hat{\mathbf{d}} = \mathbf{W}\mathbf{y} \tag{14.7–5}$$

where $\mathbf{W}$ is an $N \times M$ weighting matrix, which is selected to minimize the mean square error

$$J(\mathbf{W}) = E[\|\mathbf{e}\|^2] = E[\|\mathbf{d} - \mathbf{W}\mathbf{y}\|^2] \tag{14.7–6}$$

Each value of $\hat{\mathbf{d}}$ is then quantized to the closest transmitted symbol value.

Inverse channel detector (ICD). The ICD also forms an estimate of $\hat{\mathbf{d}}$ by linearly combining the received signal samples. However, in this case we set $M = N$ and the weight matrix $\mathbf{W}$ is selected so that the interchannel interference is completely eliminated, i.e.,

$$\hat{\mathbf{d}} = \mathbf{A}^{-1}\mathbf{y} \tag{14.7–7}$$

Each value of the estimate $\hat{\mathbf{d}}$ is then quantized to the closest transmitted symbol value.

In addition to the maximum-likelihood detector, which is non-linear, there are certainly other non-linear detection algorithms that can be devised whose performance will fall between that of the optimum MLD and the linear ICD and MMSE detectors. However, we shall not consider such detectors here.

The performance of the three detectors in a Rayleigh fading channel with no channel coding, obtained via computer simulation, is illustrated in Figures 14.7–3 and 14.7–4 for $(M, N) = (2, 2)$ and $(3, 2)$, respectively. In both cases, the variances of the channel coefficients are identical and normalized such that

$$\sum_{n,m} E[|a_{mn}|^2] = 1 \tag{14.7–8}$$

The error probability for binary PSK modulation is plotted as a function of the average SNR per bit. With the normalization of the energy in the channel coefficients as given by Equation 14.7–8, the average received signal energy is simply the transmitted signal energy.

The performance results illustrate that the MLD exploits the full diversity of order M available in the received signal and, thus, its performance is comparable to that of a maximal ratio combiner (MRC) of the M received signals, without the presence of interchannel interference. The two linear detectors, the MMSE and the ICD, achieve an error rate which decreases inversely as the SNR raised to the $(M - 1)$ power for $N = 2$, transmit antennas. In general, it can be shown

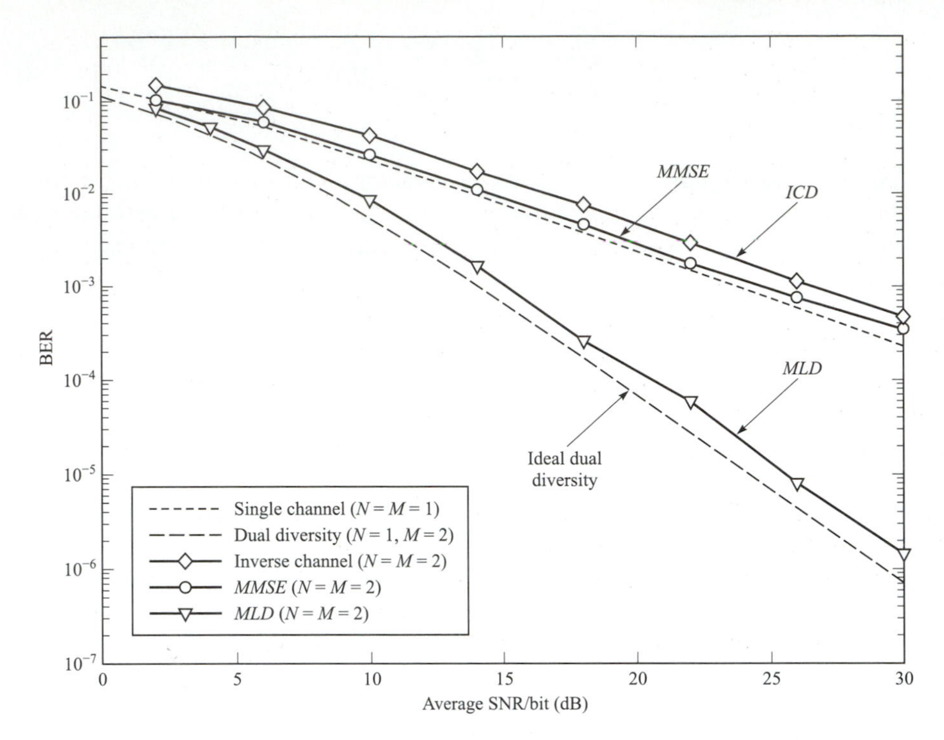

FIGURE 14.7–3
Performance of MLD, MMSE, and inverse channel detectors with $M = 2$ receiving antennas.

(see papers by Nichols et al., 1977 and Winters et al., 1994) that for any $M > N$, the MLD detector achieves a diversity of order M and the linear detectors achieve a diversity of order $M - N + 1$.

The performance of the multiple-antenna system can be further enhanced by the addition of coding with interleaving. Coding provides a means for further increasing the order of diversity. With soft-decision decoding, the order of diversity is increased by a factor of d_{min} (or d_{free}), whereas, if hard-decision decoding is employed, the order of diversity is increased by a factor of $d_{\text{min}}/2$ (or $d_{\text{free}}/2$). We demonstrate the benefits of coding by using a rate $R_c = 1/2$ convolutional code with $d_{\text{free}} = 5$. Figures 14.7–5 and 14.7–6 illustrate the performance of the multiple-antenna system for binary PSK with hard-decision decoding and soft-decision decoding, respectively, for $(M, N) = (2, 2)$ and $(MN) = (3, 2)$. We observe that coding with interleaving further improves the performance of the multiple-antenna system at the expense of a reduction in the data rate by the reciprocal of the code rate. For $M = 3$, $N = 2$, and hard-decision decoding, we observe that the MMSE detector with coding performs almost as well as the MLD with coding. Also, for MLD, we observe the superiority of soft-decision decoding over hard-decision decoding, where, for example, at 10^{-5} the difference in performance is more than 5 dB. This performance advantage is due to the factor-of-

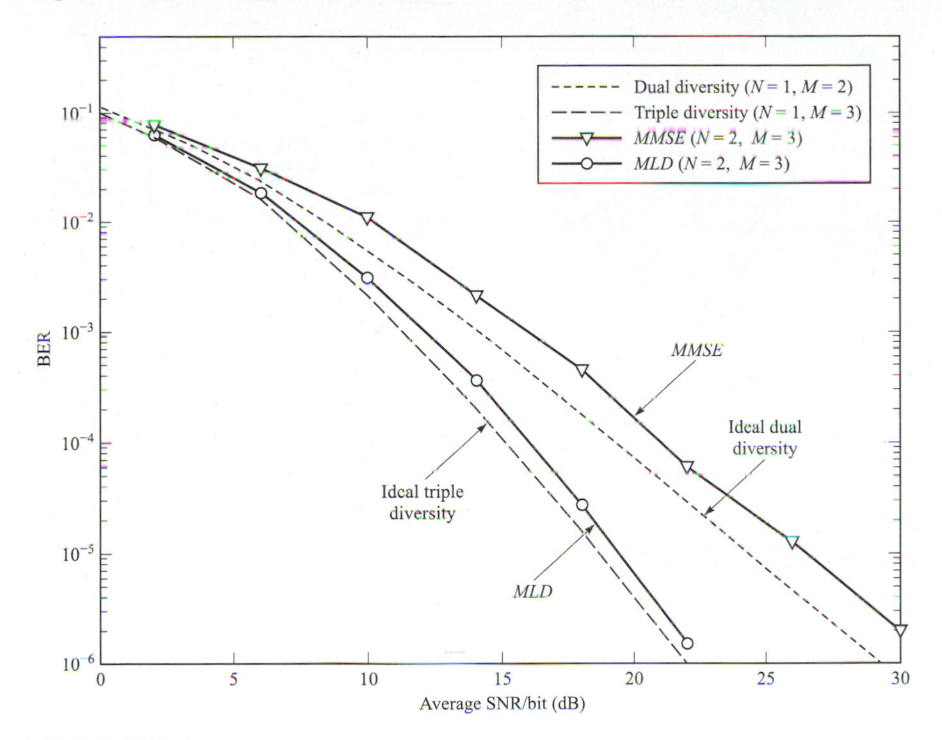

FIGURE 14.7–4
Performance of MLD and MMSE detectors with $M = 3$ receiving antennas.

2 difference in the order of diversity achieved by the two types of decoders. We also observe that there is about a 2-dB degradation at $P_2 = 10^{-5}$ in the performance of the soft-decision decoder compared to the ideal performance of the coded system, which is given in Section 14.6.6. This loss is attributed to the interference resulting from the use of multiple transmitting antennas. For additional performance results on the use of multiple transmitting antennas as a means for increasing the data rate, the reader may refer to the paper by Winters et al. (1994), Foschini and Gans (1998), and Bjerke and Proakis (1999).

Another approach to coding for use with multiple transmitting and receiving antennas is described in the papers by Tarokh et al. (1998, 1999a,b,c) and Alamouti (1998). In these papers, trellis codes are designed that are tailored specifically to the number of transmitting and receiving antennas, with a goal of achieving signal diversity. This class of trellis codes for multiple-antenna systems has been called *space–time codes*.

Figure 14.7–7 illustrates the trellis structure of two space–time trellis codes. One is a 4-PSK space–time code with four states. The two output symbols of the encoder for each input symbol are transmitted simultaneously on two different transmitting antennas, which are spaced sufficiently far apart to yield independent fading. Thus, the 4-PSK space–time trellis code achieves dual diversity and a transmission rate of 2 bits per information symbol. The second code is an 8-

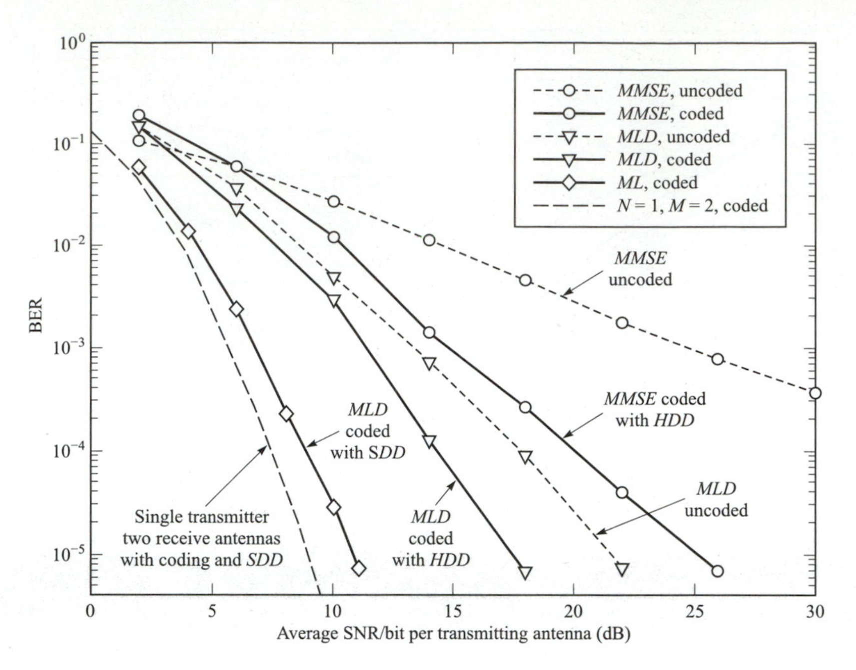

FIGURE 14.7–5
Performance of coded ($R_c = 1/2, d_{\text{free}} = 5$) systems with $N = M = 2$.

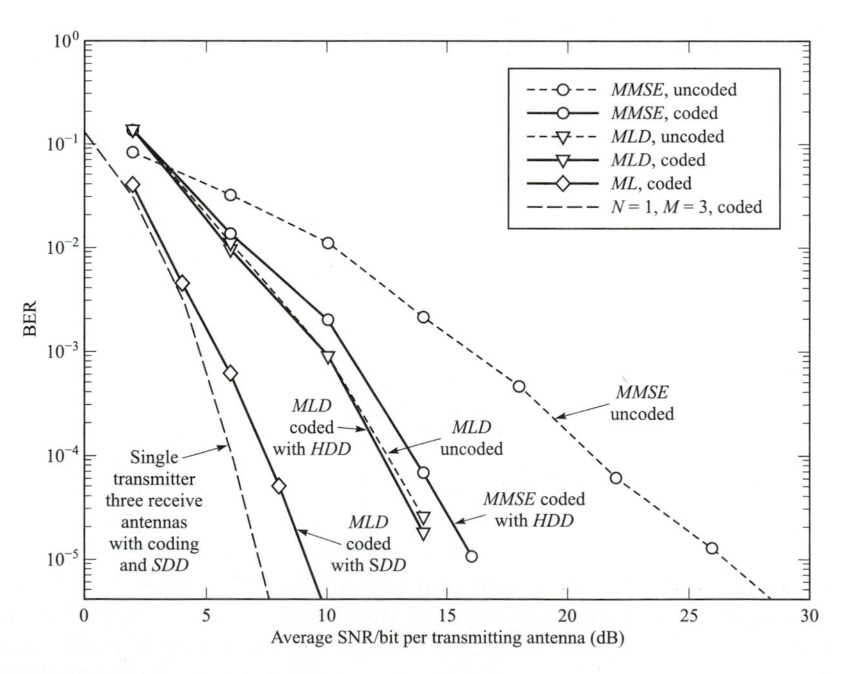

FIGURE 14.7–6
Performance of coded ($R_c = 1/2, d_{\text{free}} = 5$) systems with $N = 2, M = 3$.

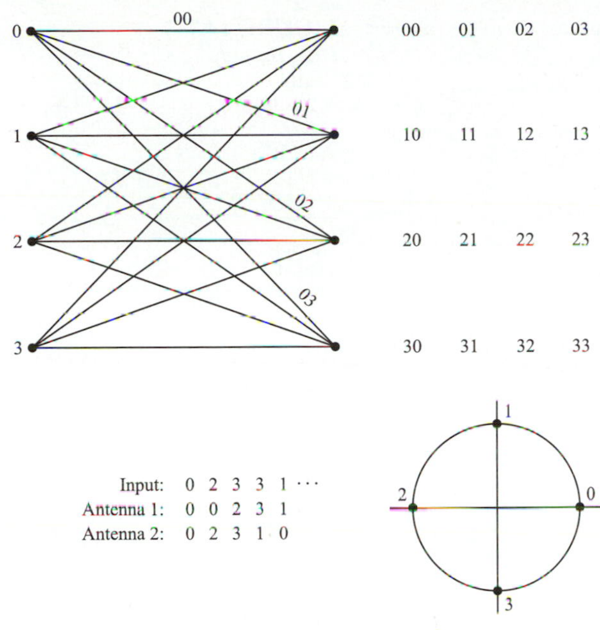

(*a*) 4-PSK, 4-state, space–time code

FIGURE 14.7–7
Two-space–time codes for two transmitting antennas.

PSK space–time trellis code with eight states. As in the case of the 4-PSK code, there are two output symbols for every 3-bit input symbol. The two symbols are transmitted simultaneously, the first on one antenna and the second on the other antenna. Thus, this code achieves a transmission rate of 3 bits per information symbol and yields dual diversity with two transmitting antennas. Additional diversity can be achieved with either code by using multiple antennas at the receiver.

The design of space–time trellis codes is considered in the paper by Tarokh et al. (1998). The design of space–time block codes is treated in the papers by Tarokh et al. (1999b,c).

14.8
BIBLIOGRAPHICAL NOTES AND REFERENCES

In this chapter, we have considered a number of topics concerned with digital communications over a fading multipath channel. We began with a statistical characterization of the channel and then described the ramifications of the channel characteristics on the design of digital signals and on their performance. We observed that the reliability of the communication system is enhanced by the use of diversity transmission and reception. Finally we demonstrated that channel encoding and soft-decision decoding provide a bandwidth-efficient means for obtaining diversity over such channels.

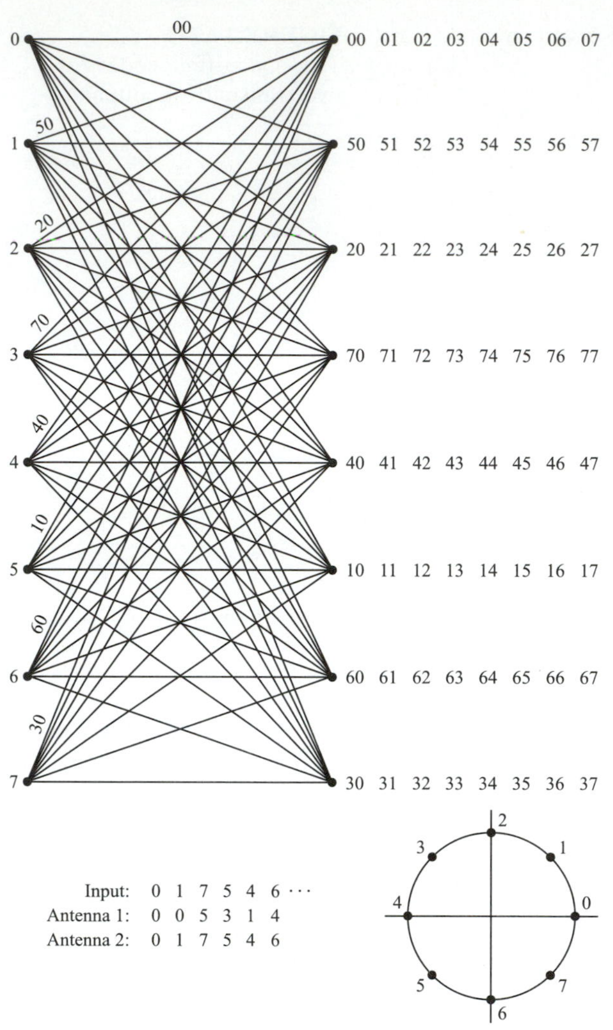

(*b*) 8-PSK, 8-state, space–time code

FIGURE 14.7–7
(*Continued*)

The pionerring work on the characterization of fading multipath channels and on signal and receiver design for reliable digital communciations over such channels was done by Price (1954, 1956). This work was followed by additional significant contributions from Price and Green (1958, 1960), Kailath (1960, 1961), and Green (1962). Diversity transmission and diversity combining techniques under a variety of channel conditions have been considered in the papers by Pierce (1958), Brennan (1959), Turin (1961, 1962), Pierce and Stein (1960), Barrow (1963), Bello and Nelin (1962a,b, 1963), Price (1962,a,b), and Lindsey (1964).

Our treatment of coding for fading channels has relied on contributions from a number of researchers. In particular, the use of dual-k codes with M-ary orthogonal FSK was proposed in publications by Viterbi and Jacobs (1975) and Odenwalder (1976). The importance of coding for digital communications over a fading channel was also emphasized in a paper by Chase (1976). The benefits derived from concatenated coding with soft-decision decoding for a fading channel were demonstrated by Pieper et al. (1978). There, a Reed–Solomon code was used for the outer code and a Hadamard code was selected as the inner code. The performance of dual-k codes with either block orthogonal codes or Hadamard codes as inner codes were investigated by Proakis and Rahman (1979). The error rate performance of maximal free distance binary convolutional codes was evaluated by Rahman (1981). Finally, the derivation of the cutoff rate for Rayleigh fading channels is due to Wozencraft and Jacobs (1965).

Information-theoretic aspects of communication through fading channels are treated in a paper by Biglieri et al. (1998). Also, the design and analysis of turbo codes for Rayleigh fading channels is considered in the paper by Wilson and Hall (1998).

Trellis-coded modulation for fading channels has been investigated by many researchers, whose work was motivated to a large extent by applications to mobile and cellular communications. For example, the paper by Zehavi (1992) considers trellis codes for 8-PSK modulation. The book by Biglieri et al. (1991) gives a tutorial treatment of this topic and contains a large number of references to the technical literature. More recent treatments of coded modulation are given in the papers by Caire et al. (1998) and Salehi and Proakis (1995).

Our treatment of digital communications over fading channels focused primarily on the Rayleigh fading channel model. For the most part, this is due to the wide acceptance of this model for describing the fading effects on many radio channels and to its mathematical tractability. Although other statistical models, such as the Ricean fading model or the Nakagami fading model may be more appropriate for characterizing fading on some real channels, the general approach in the design of reliable communications presented in this chapter carries over. Alouini and Goldsmith (1998), Simon and Alouini (1988, 2000), and Annamalai et al. (1998, 1999) have presented a unified approach to evaluating the error rate performance of digital communication systems for various fading channel models. A more general treatment of wireless communications is given in the book by Rappaport (1996).

PROBLEMS

14.1 The scattering function $S(\tau; \lambda)$ for a fading multipath channel is nonzero for the range of values $0 \leqslant \tau \leqslant 1$ ms and -0.1 Hz $\leqslant \lambda \leqslant 0.1$ Hz. Assume that the scattering function is approximately uniform in the two variables.

a) Give numerical values for the following parameters:

(i) The multipath spread of the channel.

(ii) The Doppler spread of the channel.

(iii) The coherence time of the channel.

(iv) The coherence bandwidth of the channel.
(v) The spread factor of the channel.

b) Explain the meaning of the following, taking into consideration the answers given in (a):
(i) The channel is frequency-nonselective.
(ii) The channel is slowly fading.
(iii) The channel is frequency-selective.

c) Suppose that we have a frequency allocation (bandwidth) of 10 kHz and we wish to transmit at a rate of 100 bits/s over this channel. Design a binary communication system with frequency diversity. In particular, specify
(i) The type of modulation.
(ii) The number of subchannels.
(iii) The frequency separation between adjacent carriers.
(iv) The signaling interval used in your design.
Justify your choice of parameters.

14.2 Consider a binary communication system for transmitting a binary sequence over a fading channel. The modulation is orthogonal FSK with third-order frequency diversity ($L = 3$). The demodulator consists of matched filters followed by square-law detectors. Assume that the FSK carriers fade independently and identically according to a Rayleigh envelope distribution. The additive noises on the diversity signals are zero-mean Gaussian with autocorrelation functions $\frac{1}{2}E[z_k^*(t)z_k(t+\tau)] = N_0\delta(\tau)$. The noise processes are mutually statistically independent.

a) The transmitted signal may be viewed as binary FSK with square-law detection, generated by a repetition code of the form

$$1 \rightarrow \mathbf{C}_1 = [1 \quad 1 \quad 1], \qquad 0 \rightarrow \mathbf{C}_0 = [0 \quad 0 \quad 0]$$

Determine the error rate performance P_{2h} for a hard-decision decoder following the square-law-detected signals.

b) Evaluate P_{2h} for $\bar{\gamma}_c = 100$ and 1000.

c) Evaluate the error rate P_{2s} for $\bar{\gamma}_c = 100$ and 1000 if the decoder employs soft-decision decoding.

d) Consider the generalization of the result in (a). If a repetition code of block length L (L odd) is used, determine the error probability P_{2h} of the hard-decision decoder and compare that with P_{2s}, the error rate of the soft-decision decoder. Assume $\bar{\gamma} \gg 1$.

14.3 Suppose that the binary signal $\pm s_l(t)$ is transmitted over a fading channel and the received signal is

$$r_l(t) = \pm a s_l(t) + z(t), \qquad 0 \leqslant t \leqslant T$$

where $z(t)$ is zero-mean white Gaussian noise with autocorrelation function

$$\phi_{zz}(\tau) = N_0\delta(\tau)$$

The energy in the transmitted signal is $\mathcal{E} = \frac{1}{2}\int_0^T |s_l(t)|^2\, dt$. The channel gain a is specified by the probability density function

$$p(a) = 0.1\delta(a) + 0.9\delta(a-2)$$

a) Determine the average probability of error P_2 for the demodulator that employs a filter matched to $s_l(t)$.
b) What value does P_2 approach as $\mathcal{E}/N_0$ approaches infinity?
c) Suppose that the same signal is transmitted on two statistically *independently fading* channels with gains a_1 and a_2, where

$$p(a_k) = 0.1\delta(a_k) + 0.9\delta(a_k - 2), \qquad k = 1, 2$$

The noises on the two channels are statistically independent and identically distributed. The demodulator employs a matched filter for each channel and simply adds the two filter outputs to form the decision variable. Determine the average P_2.
d) For the case in (c) what value does P_2 approach as $\mathcal{E}/N_0$ approaches infinity?

14.4 A multipath fading channel has a multipath spread of $T_m = 1\,\text{s}$ and a Doppler spread $B_d = 0.01\,\text{Hz}$. The total channel bandwidth at band-pass available for signal transmission is $W = 5\,\text{Hz}$. To reduce the effects of intersymbol interference, the signal designer selects a pulse duration $T = 10\,\text{s}$.
a) Determine the coherence bandwidth and the coherence time.
b) Is the channel frequency selective? Explain.
c) Is the channel fading slowly or rapidly? Explain.
d) Suppose that the channel is used to transmit binary data via (antipodal) coherently detected PSK in a frequency diversity mode. Explain how you would use the available channel bandwidth to obtain frequency diversity and determine how much diversity is available.
e) For the case in (d), what is the *approximate* SNR required per diversity to achieve an error probability of 10^{-6}?
f) Suppose that a wideband signal is used for transmission and a RAKE-type receiver is used for demodulation. How many taps would you use in the RAKE receiver?
g) Explain whether or not the RAKE receiver can be implemented as a coherent receiver with maximal ratio combining.
h) If binary orthogonal signals are used for the wideband signal with square-law postdetection combining in the RAKE receiver, what is the *approximate* SNR required to achieve an error probability of 10^{-6}? (Assume that all taps have the same SNR.)

14.5 In the binary communication system shown in Figure P14.5, $z_1(t)$ and $z_2(t)$ are statistically independent white Gaussian noise processes with zero-mean and identical autocorrelation functions $\phi_{zz}(\tau) = N_0\delta(\tau)$. The sampled values U_1 and U_2 represent the *real parts* of the matched filter outputs. For example, if $s_l(t)$ is transmitted, then we have

$$U_1 = 2\mathcal{E} + N_1$$
$$U_2 = N_1 + N_2$$

where $\mathcal{E}$ is the transmitted signal energy and

$$N_k = \text{Re}\left[\int_0^T s_l^*(t) z_k(t)\, dt\right], \qquad k = 1, 2$$

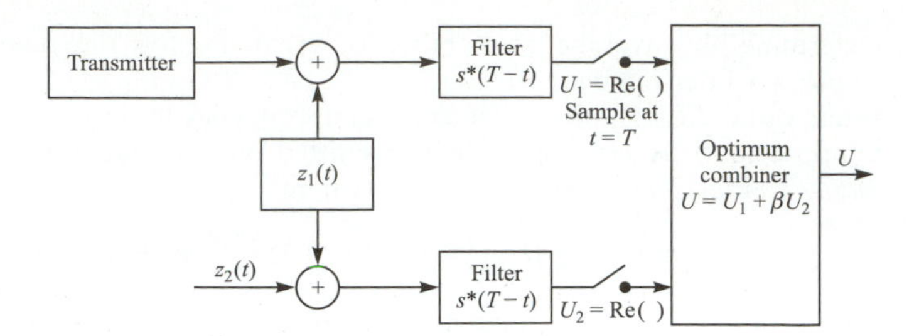

FIGURE P14.5

It is apparent that U_1 and U_2 are correlated Gaussian variables while N_1 and N_2 are independent Gaussian variables. Thus,

$$p(n_1) = \frac{1}{\sqrt{2\pi}\sigma} \exp\left(-\frac{n_1^2}{2\sigma^2}\right)$$

$$p(n_2) = \frac{1}{\sqrt{2\pi}\sigma} \exp\left(-\frac{n_2^2}{2\sigma^2}\right)$$

where the variance of N_k is $\sigma^2 = 2\mathcal{E}N_0$.

a) Show that the joint probability density function for U_1 and U_2 is

$$p(U_1, U_2) = \frac{1}{2\pi\sigma^2} \exp\left\{-\frac{1}{\sigma^2}\left[(U_2 - 2\mathcal{E})^2 - U_2(U_1 - 2\mathcal{E}) + \tfrac{1}{2}U_2^2\right]\right\}$$

if $s(t)$ is transmitted and

$$p(U_1, U_2) = \frac{1}{2\pi\sigma^2} \exp\left\{-\frac{1}{\sigma^2}\left[(U_1 + 2\mathcal{E})^2 - U_2(U_1 + 2\mathcal{E}) + \tfrac{1}{2}U_2^2\right]\right\}$$

if $-s(t)$ is transmitted.

b) Based on the likelihood ratio, show that the optimum combination of U_1 and U_2 results in the decision variable

$$U = U_1 + \beta U_2$$

where β is a constant. What is the optimum value of β?

c) Suppose that $s(t)$ is transmitted. What is the probability density function of U?

d) What is the probability error assuming that $s(t)$ was transmitted? Express your answer as a function for the SNR $\mathcal{E}/N_0$.

e) What is the loss in performance if only $U = U_1$ is the decision variable?

14.6 Consider the model for a binary communication system with diversity as shown in Figure P14.6. The channels have fixed attenuations and phase shifts. The $\{z_k(t)\}$ are complex-valued white Gaussian noise processes with zero-mean and autocorrelation functions

$$\phi_{zz}(t) = \tfrac{1}{2}E[z_k^*(t)z_k(t+\tau)] = N_{0k}\delta(\tau)$$

(Note that the spectral densities $\{N_{0k}\}$ are all different.) Also, the noise processes $\{z_k(t)\}$ are mutually statistically independent. The $\{\beta_k\}$ are complex-valued weighting factors to be determined. The decision variable from the combiner is

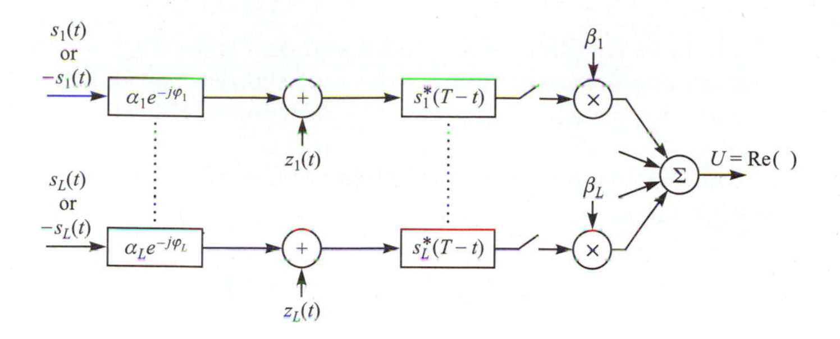

FIGURE P14.6

$$U = \text{Re}\left(\sum_{k=1}^{L} \beta_k U_k\right) \underset{-1}{\overset{1}{\gtrless}} 0$$

a) Determine the PDF $p(U)$ when $+1$ is transmitted.
b) Determine the probability of error P_2 as a function of the weights $\{\beta_k\}$.
c) Determine the values of $\{\beta_k\}$ that minimize P_2.

14.7 Determine the probability of error for binary orthogonal signaling with Lth-order diversity over a Rayleigh fading channel. The PDFs of the two decision variables are given by Equations 14.4–31 and 14.4–32.

14.8 The rate 1/3, $K = 3$, binary convolutional code with transfer function given by Equation 8.2–5 is used for transmitting data over a Rayleigh fading channel via binary PSK.

a) Determine and plot the probability of error for hard-decision decoding. Assume that the transmitted waveforms corresponding to the coded bits fade independently.
b) Determine and plot the probability of error for soft-decision decoding. Assume that the waveforms corresponding to the coded bits fade independently.

14.9 A binary sequence is transmitted via binary antipodal signaling over a Rayleigh fading channel with Lth-order diversity. When $s_l(t)$ is transmitted, the received equivalent low-pass signals are

$$r_k(t) = \alpha_k e^{-j\phi_k} s_l(t) + z_k(t), \qquad k = 1, 2, \ldots, L$$

The fading among the L subchannels is statistically independent. The additive noise terms $\{z_k(t)\}$ are zero-mean, statistically independent, and identically distributed white Gaussian noise processes with autocorrelation function $\phi_{zz}(\tau) = N_0\delta(\tau)$. Each of the L signals is passed through a filter matched to $s_l(t)$ and the output is phase-corrected to yield

$$U_k = \text{Re}\left[e^{j\phi_k} \int_0^T r_k(t) s_l^*(t)\, dt\right], \qquad k = 1, 2, \ldots, L$$

The $\{U_k\}$ are combined by a linear combiner to form the decision variable

$$U = \sum_{k=1}^{L} U_k$$

a) Determine the PDF of U conditional on fixed values for the $\{a_k\}$.
b) Determine the expression for the probability of error when the $\{a_k\}$ are statistically independent and identically distributed Rayleigh random variables.

14.10 The Chernoff bound for the probability of error for binary FSK with diversity L in Rayleigh fading was shown to be

$$P_2(L) < [4p(1-p)]^L = \left[4\frac{1+\bar{\gamma}_c}{(2+\bar{\gamma}_c)^2}\right]^L$$
$$< 2^{-\bar{\gamma}_b g(\bar{\gamma}_c)}$$

where

$$g(\bar{\gamma}_c) = \frac{1}{\bar{\gamma}_c}\log_2\left[\frac{(2+\bar{\gamma}_c)^2}{4(1+\bar{\gamma}_c)}\right]$$

a) Plot $g(\bar{\gamma}_c)$ and determine its approximate maximum value and the value of $\bar{\gamma}_c$ where the maximum occurs.
b) For a given $\bar{\gamma}_b$, determine the optimal order of diversity.
c) Compare $P_2(L)$, under the condition that $g(\bar{\gamma}_c)$ is maximized (optimal diversity), with the error probability for binary FSK and AWGN with no fading, which is

$$P_2 = \tfrac{1}{2}e^{-\gamma_b/2}$$

and determine the penalty in SNR due to fading and noncoherent (square-law) combining.

14.11 A DS spread spectrum system is used to resolve the multipath signal components in a two-path radio signal propagation scenario. If the path length of the secondary path is 300 m longer than that of the direct path, determine the minimum chip rate necessary to resolve the multipath components.

14.12 A baseband digital communication system employs the signals shown in Figure P14.12a for the transmission of two equiprobable messages. It is assumed that the communication problem studied here is a "one-shot" communication problem; that is, the above messages are transmitted just once and no transmission takes place afterward. The channel has no attenuation ($\alpha = 1$), and the noise is AWGN with power spectral density $\frac{1}{2}N_0$.
a) Find an appropriate orthonormal basis for the representation of the signals.
b) In a block diagram, give the precise specifications of the optimum receiver using matched filters. Label the diagram carefully.
c) Find the error probability of the optimum receiver.
d) Show that the optimum receiver can be implemented by using just *one* filter (see the block diagram in Figure P14.12b. What are the characteristics of the matched filter, the sampler and decision device?
e) Now assume that the channel is not ideal but has an impulse response of $c(t) = \delta(t) + \frac{1}{2}\delta(t - \frac{1}{2}T)$. Using the same matched filter as in (d), design an optimum receiver.
f) Assuming that the channel impulse response is $c(t) = \delta(t) + a\delta(t - \frac{1}{2}T)$, where a is a random variable uniformly distributed on [0, 1], and using the same matched filter as in (d), design the optimum receiver.

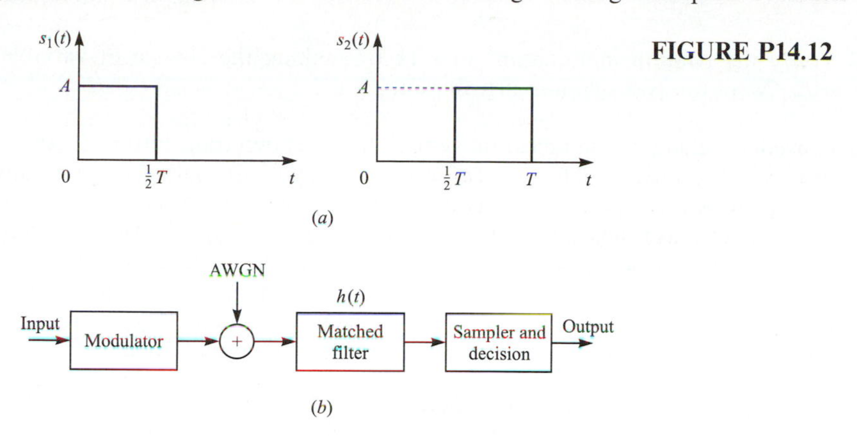

FIGURE P14.12

14.13 A communication system employs dual antenna diversity and binary orthogonal FSK modulation. The received signals at the two antennas are

$$r(t) = \alpha_1 s(t) + n_1(t)$$
$$r_2(t) = \alpha_2 s(t) + n_2(t)$$

where α_1 and α_2 are statistically iid Rayleigh random variables, and $n_1(t)$ and $n_2(t)$ are statistically independent, zero-mean and white Gaussian random processes with power-spectral density $\frac{1}{2}N_0$. The two signals are demodulated, squared, and then combined (summed) prior to detection.

a) Sketch the functional block diagram of the entire receiver, including the demodulator, the combiner, and the detector.

b) Plot the probability of error for the detector and compare the result with the case of no diversity.

14.14 The two equivalent low-pass signals shown in Figure P14.14 are used to transmit a binary sequence. The equivalent low-pass impulse response of the channel is $h(t) = 4\delta(t) - 2\delta(t - T)$. To avoid pulse overlap between successive transmissions, the transmission rate in bits/s is selected to be $R = 1/2T$. The transmitted signals are equally probable and are corrupted by additive zero-mean white Gaussian noise having an equivalent low-pass representation $z(t)$ with an autocorrelation function

$$\phi_{zz}(\tau) = \tfrac{1}{2}E[z^*(t)z(t+\tau)] = N_0\delta(\tau)$$

a) Sketch the two possible equivalent low-pass noise-free *received* waveforms.

b) Specify the optimum receiver and sketch the equivalent low-pass impulse responses of all filters used in the optimum receiver. Assume *coherent detection* of the signals.

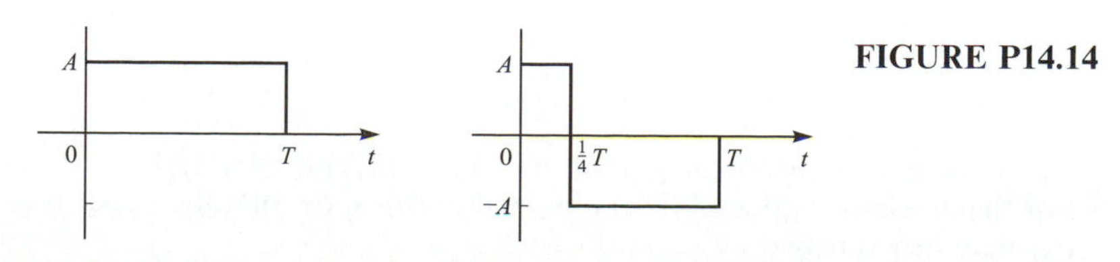

FIGURE P14.14

14.15 Verify the relation in Equation 14.3–14 by making the change of variable $\gamma = \alpha^2 \mathcal{E}_b / N_0$ in the Nakagami-m distribution.

14.16 Consider a digital communication system that uses two transmitting antennas and one receiving antenna. The two transmitting antennas are sufficiently separated so as to provide dual spatial diversity in the transmission of the signal. The transmission scheme is as follows: If s_1 and s_2 represent a pair of symbols from either a one-dimensional or a two-dimensional signal constellation, which are to be transmitted by the two antennas, the signal from the first antenna over two signal intervals is (s_1, s_2^*) and from the second antenna the transmitted signal is $(s_2, -s_1^*)$. The signal received by the single receiving antenna over the two signal intervals is

$$r_1 = h_1 s_1 + h_2 s_2 + n_1$$
$$r_2 = h_1 s_2^* - h_2 s_1^* + n_2$$

where (h_1, h_2) represent the complex-valued channel path gains, which may be assumed to be zero-mean, complex Gaussian with unit variance and statistically independent. The channel path gains (h_1, h_2) are assumed to be constant over the two signal intervals and known to the receiver. The terms (n_1, n_2) represent additive white Gaussian noise terms that have zero-mean and variance σ^2 and uncorrelated.

a) Show how to recover the transmitted symbols (s_1, s_2) from (r_1, r_2) and achieve dual diversity reception.
b) If the energy in the pair (s_1, s_2) is $(\mathcal{E}_s, \mathcal{E}_s)$ and the modulation is binary PSK, determine the probability of error.
c) Repeat (b) if the modulation is QPSK.

14.17 As shown in Section 14.6.6, an upper bound on the code word error probability in a Rayleigh fading channel for coherent detection of the binary-PSK-coded bits is given by Equation 14.6–37.

a) Beginning with the expression in Equation 14.6–35, show that an upper bound on the code word error probability for binary PSK, DPSK, and FSK (with noncoherent detection) and soft-decision decoding may be expressed as

$$P_M < 2^k \binom{2d_{\min} - 1}{d_{\min}} \left(\frac{1}{\beta R_c \bar{\gamma}_b} \right)^{d_{\min}}$$

where $\beta = 4$ for PSK, $\beta = 2$ for DPSK, and $\beta = 1$ for FSK.

b) Express the bound on P_M as

$$P_M < 2^k \binom{2d_{\min} - 1}{d_{\min}} e^{-d_{\min} R_c \bar{\gamma}_b f(\bar{\gamma}_c)}$$

where

$$f(\bar{\gamma}_c) = \frac{1}{\bar{\gamma}_c} \ln \beta \bar{\gamma}_c$$

and show that the value of $\bar{\gamma}_c$ that maximizes $f(\bar{\gamma}_c)$ is $\bar{\gamma}_c = e/\beta$.

c) For the maximum value of $f(\bar{\gamma}_c)$ obtained in (b), show that the upper bound on P_M may be expressed as

$$P_M < \binom{2d_{\min} - 1}{d_{\min}} e^{-k(\beta d_{\min} \bar{\gamma}_b / ne - \ln 2)}$$

and solve for the minimum value of $\bar{\gamma}_b$ that ensures that the bound goes to zero as k is increased to infinity. Evaluate the minimum $\bar{\gamma}_b$ for PSK, DPSK, and FSK, when $d_{\min}/n = 1/2$.

15

Multiuser Communications

Our treatment of communication systems up to this point has been focused on a single communication link involving a transmitter and a receiver. In this chapter, the focus shifts to multiple users and multiple communication links. We explore the various ways in which the multiple users access a common channel to transmit information. The multiple access methods that are described in this chapter form the basis for current and future wireline and wireless communication networks, such as satellite networks, cellular and mobile communication networks, and underwater acoustic networks.

15.1 INTRODUCTION TO MULTIPLE ACCESS TECHNIQUES

It is instructive to distinguish among several types of multiuser communication systems. One type is a multiple access system in which a large number of users share a common communication channel to transmit information to a receiver. Such a system is depicted in Figure 15.1–1. The common channel may be the uplink in a satellite communication system, or a cable to which are connected a set of terminals that access a central computer, or some frequency band in the radio spectrum that is used by multiple users to communicate with a radio receiver. For example, in a mobile cellular communication system, the users are the mobile transmitters in any particular cell of the system and the receiver resides in the base station of the particular cell.

A second type of multiuser communication system is a broadcast network in which a single transmitter sends information to multiple receivers as depicted in Figure 15.1–2. Examples of broadcast systems include the common radio and TV broadcast systems, as well as the downlinks in a satellite system.

The multiple access and broadcast networks are probably the most common multiuser communication systems. A third type of multiuser system is a store-

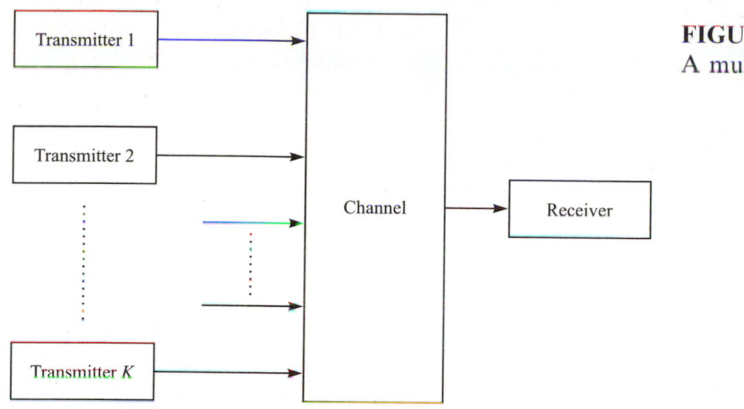

FIGURE 15.1–1
A multiple access system

and-forward network, as depicted in Figure 15.1–3. Yet a fourth type is the two-way communication system shown in Figure 15.1–4.

In this chapter, we focus on multiple access methods for multiuser communications. In general, there are several different ways in which multiple users can send information through the communication channel to the receiver. One simple method is to subdivide the available channel bandwidth into a number, say K, of frequency non-overlapping subchannels, as shown in Figure 15.1–5, and to assign a subchannel to each user upon request by the users. This method is generally called *frequency-division multiple access* (FDMA) and is commonly used in wireline channels to accommodate multiple users for voice and data transmission.

Another method for creating multiple subchannels for multiple access is to subdivide the duration T_f, called the *frame duration*, into, say, K non-overlapping subintervals, each of duration T_f/K. Then each user who wishes to transmit information is assigned to a particular time slot within each frame. This multiple access method is called *time-division multiple access* (TDMA) and it is frequently used in data and digital voice transmission.

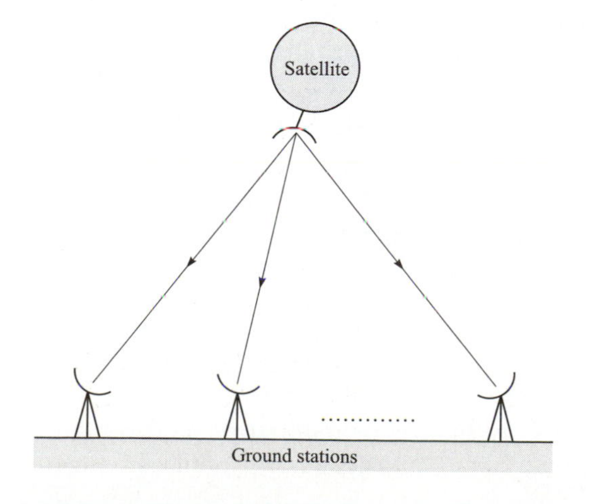

FIGURE 15.1–2
A broadcast network.

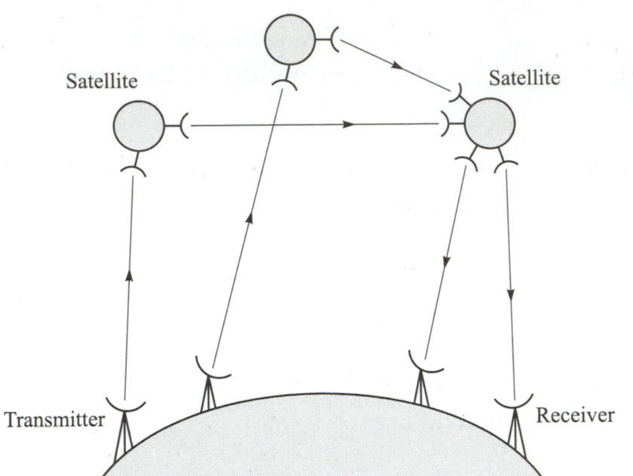

FIGURE 15.1–3
A store-and-forward communication network with satellite relays.

We observe that in FDMA and TDMA, the channel is basically partitioned into independent single-user subchannels. In this sense, the communication system design methods that we have described for single-user communication are directly applicable and no new problems are encountered in a multiple access environment, except for the additional task of assigning users to available channels.

The interesting problems arise when the data from the users accessing the network is bursty in nature. In other words, the information transmissions from a single user are separated by periods of no transmission, where these periods of silence may be greater than the periods of transmission. Such is the case generally with users at various terminals in a computer communication network. To some extent, this is also the case in mobile cellular communication systems carrying digitized voice, since speech signals typically contain long pauses.

In such an environment where the transmission from the various users is bursty and low-duty-cycle, FDMA and TDMA tend to be inefficient because a certain percentage of the available frequency slots or time slots assigned to users do not carry information. Ultimately, an inefficiently designed multiple access system limits the number of simultaneous users of the channel.

An alternative to FDMA and TDMA is to allow more than one user to share a channel or subchannel by use of direct-sequence spread spectrum signals. In this method, each user is assigned a unique code sequence or *signature sequence*

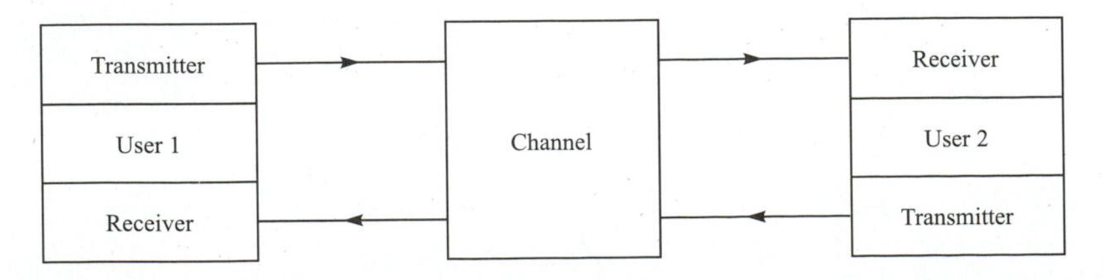

FIGURE 15.1–4
A two-way communication channel.

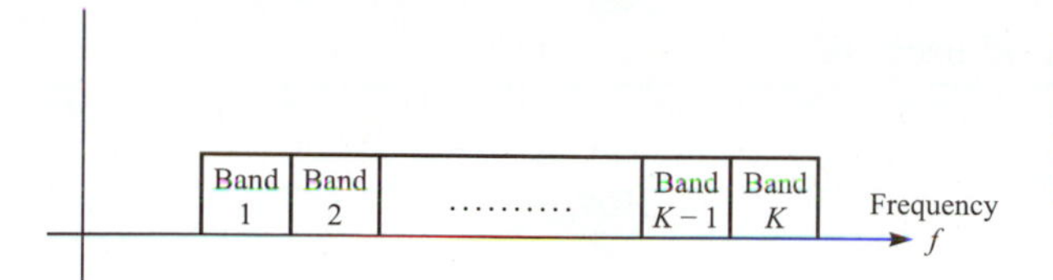

FIGURE 15.1–5
Subdivision of the channel into non-overlapping frequency bands.

that allows the user to spread the information signal across the assigned frequency band. Thus signals from the various users are separated at the receiver by cross correlation of the received signal with each of the possible user signature sequences. By designing these code sequences to have relatively small cross-correlations, the crosstalk inherent in the demodulation of the signals received from multiple transmitters is minimized. This multiple access method is called *code division multiple access* (CDMA).

In CDMA, the users access the channel in a random manner. Hence, the signal transmissions among the multiple users completely overlap both in time and in frequency. The demodulation and separation of these signals at the receiver is facilitated by the fact that each signal is spread in frequency by the pseudorandom code sequence. CDMA is sometimes called *spread spectrum multiple access* (SSMA).

An alternative to CDMA is nonspread random access. In such a case, when two users attempt to use the common channel simultaneously, their transmissions collide and interefere with each other. When that happens, the information is lost and must be retransmitted. To handle collisions, one must establish protocols for retransmission of messages that have collided. Protocols for scheduling the retransmission of collided messages are described below.

15.2 CAPACITY OF MULTIPLE ACCESS METHODS

It is interesting to compare FDMA, TDMA, and CDMA in terms of the information rate that each multiple access method achieves in an ideal AWGN channel of bandwidth W. Let us compare the capacity of K users, where each user has an average power $P_i = P$, for all $1 \leqslant i \leqslant K$. Recall that in an ideal band-limited AWGN channel of bandwidth W, the capacity of a single user is

$$C = W \log_2\left(1 + \frac{P}{WN_0}\right) \qquad (15.2\text{–}1)$$

where $\frac{1}{2}N_0$ is the power spectral density of the additive noise.

In FDMA, each user is allocated a bandwidth W/K. Hence, the capacity of each user is

$$C_k = \frac{W}{K} \log_2\left[1 + \frac{P}{(W/K)N_0}\right] \qquad (15.2\text{–}2)$$

and the total capacity for the K users is

$$KC_K = W\log_2\left(1 + \frac{KP}{WN_0}\right) \tag{15.2–3}$$

Therefore, the total capacity is equivalent to that of a single user with average power $P_{\text{av}} = KP$.

It is interesting to note that for a fixed bandwidth W, the total capacity goes to infinity as the number of users increases linearly with K. On the other hand, as K increases, each user is allocated a smaller bandwidth (W/K) and, consequently, the capacity per user decreases. Figure 15.2–1 illustrates the capacity C_K per user normalized by the channel bandwidth W, as a function of $\mathcal{E}_b/N_0$, with K as a parameter. This expression is given as

$$\frac{C_K}{W} = \frac{1}{K}\log_2\left[1 + K\frac{C_K}{W}\left(\frac{\mathcal{E}_b}{N_0}\right)\right] \tag{15.2–4}$$

A more compact form of Equation 15.2–4 is obtained by defining the normalized total capacity $C_n = KC_K/W$, which is the total bit rate for all K users per unit of bandwidth. Thus, Equation 15.2–4 may be expressed as

$$C_n = \log_2\left(1 + C_n\frac{\mathcal{E}_b}{N_0}\right) \tag{15.2–5}$$

or, equivalently,

$$\frac{\mathcal{E}_b}{N_0} = \frac{2^{C_n} - 1}{C_n} \tag{15.2–6}$$

The graph of C_n versus $\mathcal{E}_b/N_0$ is shown in Figure 15.2–2. We observe that C_n increases as $\mathcal{E}_b/N_0$ increases above the minimum value of ln 2.

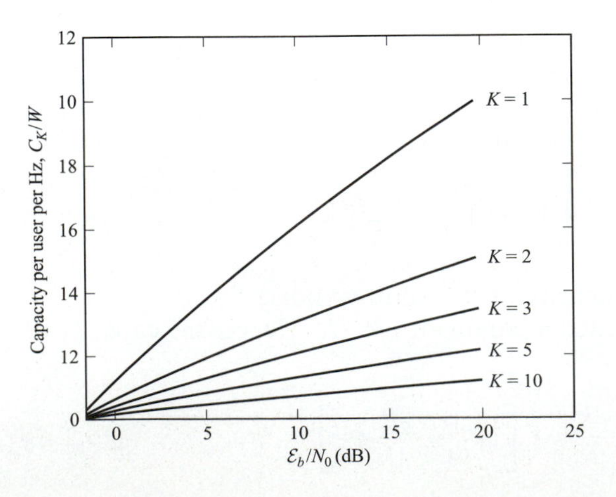

FIGURE 15.2–1
Normalized capacity as a function of $\mathcal{E}_b/N_0$ for FDMA.

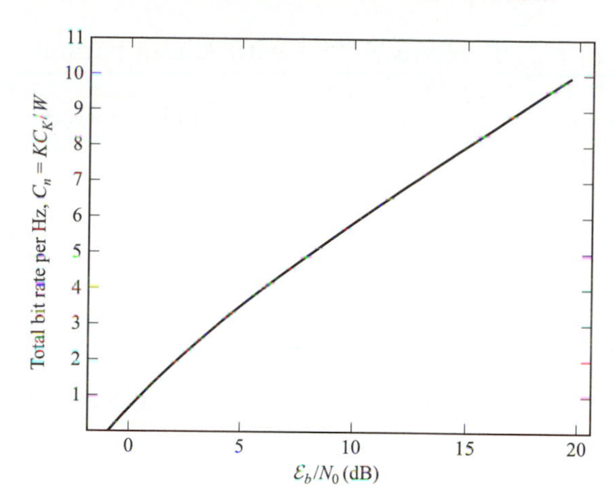

FIGURE 15.2–2
Total capacity per hertz as a function of $\mathcal{E}_b/N_0$ for FDMA.

In a TDMA system, each user transmits for $1/K$ of the time through the channel of bandwidth W, with average power KP. Therefore, the capacity per user is

$$C_K = \left(\frac{1}{K}\right) W \log_2\left(1 + \frac{KP}{WN_0}\right) \tag{15.2–7}$$

which is identical to the capacity of an FDMA system. However, from a practical standpoint, we should emphasize that, in TDMA, it may not be possible for the transmitters to sustain a transmitter power of KP when K is very large. Hence, there is a practical limit beyond which the transmitter power cannot be increased as K is increased.

In a CDMA system, each user transmits a pseudorandom signal of a bandwidth W and average power P. The capacity of the system depends on the level of cooperation among the K users. At one extreme is noncooperative CDMA, in which the receiver for each user signal does not know the spreading waveforms of the other users, or chooses to ignore them in the demodulation process. This is called *single-user detection*. Hence, the other users' signals appear as interference at the receiver of each user. In this case, the multiuser receiver consists of a bank of K single-user matched filters. If we assume that each user's pseudorandom signal waveform is Gaussian, then each user signal is corrupted by Gaussian interference of power $(K-1)P$ and additive Gaussian noise of power WN_0. Therefore, the capacity per user for single-user detection is

$$C_K = W \log_2\left[1 + \frac{P}{WN_0 + (K-1)P}\right] \tag{15.2–8}$$

or, equivalently,

$$\frac{C_K}{W} = \log_2\left[1 + \frac{C_K}{W}\frac{\mathcal{E}_b/N_0}{1 + (K-1)(C_K/W)(\mathcal{E}_b/N_0)}\right] \tag{15.2–9}$$

Figure 15.2–3 illustrates the graph of C_K/W versus $\mathcal{E}_b/N_0$, with K as a parameter.

For a large number of users, we may use the approximation $\ln(1+x) \leqslant x$. Hence,

$$\frac{C_K}{W} \leqslant \frac{C_K}{W} \frac{\mathcal{E}_b/N_0}{1 + K(C_K/W)(\mathcal{E}_b/N_0)} \log_2 e \tag{15.2–10}$$

or, equivalently, the normalized total capacity $C_n = KC_K/W$ is

$$\begin{aligned} C_n &\leqslant \log_2 e - \frac{1}{\mathcal{E}_b/N_0} \\ &\leqslant \frac{1}{\ln 2} - \frac{1}{\mathcal{E}_b/N_0} < \frac{1}{\ln 2} \end{aligned} \tag{15.2–11}$$

In this case, we observe that the total capacity does not increase with K as in TDMA and FDMA.

On the other hand, suppose that the K users cooperate by transmitting synchronously in time, and the multiuser receiver knows the spreading waveforms of all users and jointly demodulates and detects all the users' signals. This is called *multiuser detection*. Thus, each user is assigned a rate R_i, $1 \leqslant i \leqslant K$, and a code book containing a set of 2^{nR_i} code words of power P. In each signal interval, each user selects an arbitrary code word, say $\mathbf{X}_i$, from its own code book and all users transmit their code words simultaneously. Thus, the decoder at the receiver observes

$$\mathbf{Y} = \sum_{i=1}^{K} \mathbf{X}_i + \mathbf{Z} \tag{15.2–12}$$

where $\mathbf{Z}$ is an additive noise vector. The optimum decoder looks for the K code words, one from each code book, that have a vector sum closest to the received vector $\mathbf{Y}$ in Euclidean distance.

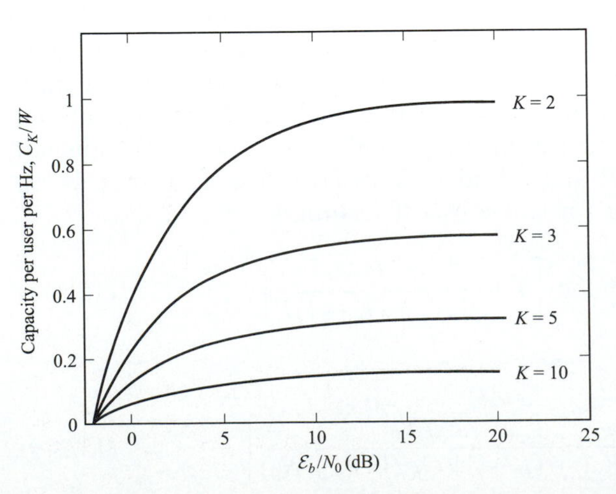

FIGURE 15.2–3 Normalized capacity as a function of $\mathcal{E}_b/N_0$ for noncooperative CDMA.

The achievable K-dimensional rate region for the K users in an AWGN channel, assuming equal power for each user, is given by the following equations:

$$R_i < W \log_2\left(1 + \frac{P}{WN_0}\right), \qquad 1 \leqslant i \leqslant K \tag{15.2–13}$$

$$R_i + R_j < W \log_2\left(1 + \frac{2P}{WN_0}\right), \qquad 1 \leqslant i, j \leqslant K \tag{15.2–14}$$

$$\vdots$$

$$R_{\text{SUM}}^{\text{MU}} = \sum_{i=1}^{K} R_i < W \log_2\left(1 + \frac{KP}{WN_0}\right) \tag{15.2–15}$$

where $R_{\text{SUM}}^{\text{MU}}$ is the total rate achieved by the K users by employing multiuser detection. In the special case when all the rates are identical, the inequality 15.2–15 is dominant over the other $K-1$ inequalities. It follows that if the rates $\{R_i, 1 \leqslant i \leqslant K\}$ for the K cooperative synchronous users are selected to fall in the capacity region specified by the inequalities given above, then the probabilities of error for the K users tend to zero as the code block length n tends to infinity.

From the above discussion, we conclude that the sum of the rates of the K users $R_{\text{SUM}}^{\text{MU}}$ goes to infinity with K. Therefore, with cooperative synchronous users, the capacity of CDMA has a form similar to that of FDMA and TDMA. Note that if all the rates in the CDMA system are selected to be identical to R, then Equation 15.2–15 reduces to

$$R < \frac{W}{K} \log_2\left(1 + \frac{KP}{WN_0}\right) \tag{15.2–16}$$

which is identical to the rate constraint in FDMA and TDMA. In this case, CDMA does not yield a higher rate than TDMA and FDMA. However, if the rates of the K users are selected to be unequal such that the inequalities 15.2–13 to 15.2–15 are satisfied, then it is possible to find the points in the achievable rate region such that the sum of the rates for the K users in CDMA exceeds the capacity of FDMA and TDMA.

EXAMPLE 15.2–1. Consider the case of two users in a CDMA system that employs coded signals as described above. The rates of the two users must satisfy the inequalities

$$R_1 < W \log_2\left(1 + \frac{P}{WN_0}\right) \tag{15.2–17}$$

$$R_2 < W \log_2\left(1 + \frac{P}{WN_0}\right) \tag{15.2–18}$$

$$R_1 + R_2 < W \log_2\left(1 + \frac{2P}{WN_0}\right) \tag{15.2–19}$$

where P is the average transmitted power of each user and W is the signal bandwidth.

The capacity region for the two-user CDMA system with coded signal waveforms has the form illustrated in Figure 15.2–4, where

$$C_i = W \log_2\left(1 + \frac{P_i}{WN_0}\right), \qquad i = 1, 2$$

are the capacities corresponding to the two users with $P_1 = P_2 = P$. We note that if user 1 is transmitting at capacity C_1, user 2 can transmit up to a maximum rate

$$\begin{aligned} R_{2m} &= W \log_2\left(1 + \frac{2P}{WN_0}\right) - C_1 \\ &= W \log_2\left(1 + \frac{P}{P + WN_0}\right) \end{aligned} \tag{15.2–20}$$

which is illustrated in Figure 15.2–4 as point A. This result has an interesting interpretation. We note that rate R_{2m} corresponds to the case in which the signal from user 1 is considered as an equivalent additive noise in the detection of the signal of user 2. On the other hand, user 1 can transmit at capacity C_1, since the receiver knows the transmitted signal from user 2 and, hence, it can eliminate its effect in detecting the signal of user 1.

Because of symmetry, a similar situation exists if user 2 is transmitting at capacity C_2. Then user 1 can transmit up to a maximum rate $R_{1m} = R_{2m}$, which is illustrated in Figure 15.2–4 as point B. In this case, we have a similar interpretation as above, with an interchange in the roles of user 1 and user 2.

The points A and B are connected by a straight line, which is defined by Equation 15.2–19. It is easily seen that this straight line is the boundary of the achievable rate region, since any point on the line corresponds to the maximum rate $W \log_2 (1 + 2P/WN_0)$, which can be obtained by simply time sharing the channel between the two users.

In the next section, we consider the problem of signal detection for a multiuser CDMA system and assess the performance and the computational complexity of several receiver structures.

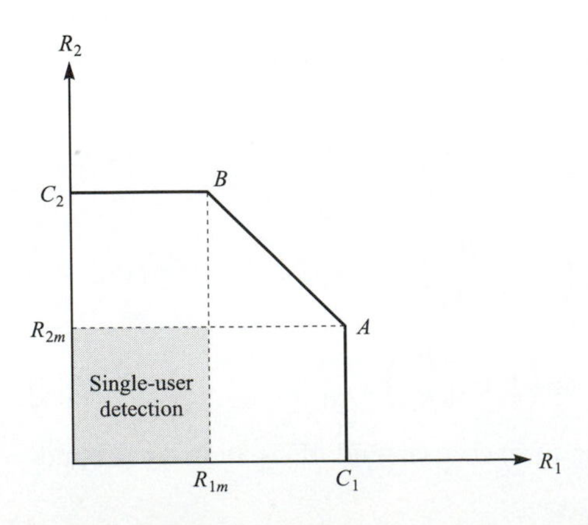

FIGURE 15.2–4
Capacity region of two-user CDMA multiple access Gaussian channel.

15.3 CODE-DIVISION MULTIPLE ACCESS

As we have observed, TDMA and FDMA are multiple access methods in which the channel is partitioned into independent, single-user subchannels, i.e., non-overlapping time slots or frequency bands, respectively. In CDMA, each user is assigned a distinct signature sequence (or waveform), which the user employs to modulate and spread the information-bearing signal. The signature sequences also allow the receiver to demodulate the message transmitted by multiple users of the channel, who transmit simultaneously and, generally, asynchronously.

In this section, we treat the demodulation and detection of multiuser CDMA signals. We shall see that the optimum maximum-likelihood detector has a computational complexity that grows exponentially with the number of users. Such a high complexity serves as a motivation to devise suboptimum detectors having lower computational complexities. Finally, we consider the performance characteristics of the various detectors.

15.3.1 CDMA Signal and Channel Models

Let us consider a CDMA channel that is shared by K simultaneous users. Each user is assigned a signature waveform $g_k(t)$ of duration T, where T is the symbol interval. A signature waveform may be expressed as

$$g_k(t) = \sum_{n=0}^{L-1} a_k(n) p(t - nT_c), \qquad 0 \leqslant t \leqslant T \tag{15.3–1}$$

where $\{a_k(n), 0 \leqslant n \leqslant L-1\}$ is a pseudonoise (PN) code sequence consisting of L chips that take values $\{\pm 1\}$, $p(t)$ is a pulse of duration T_c, and T_c is the chip interval. Thus, we have L chips per symbol and $T = LT_c$. Without loss of generality, we assume that all K signature waveforms have unit energy, i.e.,

$$\int_0^T g_k^2(t)\, dt = 1 \tag{15.3–2}$$

The cross correlations between pairs of signature waveforms play an important role in the metrics for the signal detector and on its performance. We define the following cross correlations, where $0 \leqslant \tau \leqslant T$ and $i < j$,

$$\rho_{ij}(\tau) = \int_{\tau_c}^{T} g_i(t) g_j(t - \tau)\, dt \tag{15.3–3}$$

$$\rho_{ji}(\tau) = \int_0^{\tau} g_i(t) g_j(t + T + \tau)\, dt \tag{15.3–4}$$

The cross correlations in Equations 15.3–3 and 15.3–4 apply to asynchronous transmissions among the K users. For synchronous transmission, we need only $\rho_{ij}(0)$.

For simplicity, we assume that binary antipodal signals are used to transmit the information from each user. Hence, let the information sequence of the kth user be denoted by $\{b_k(m)\}$, where the value of each information bit may be ± 1. It is convenient to consider the transmission of a block of bits of some arbitrary length, say N. Then, the data block from the kth user is

$$\mathbf{b}_k = [b_k(1) \cdots b_k(N)]^t \tag{15.3–5}$$

and the corresponding equivalent low-pass, transmitted waveform may be expressed as

$$s_k(t) = \sqrt{\mathcal{E}_k} \sum_{i=1}^{N} b_k(i) g_k(t - iT) \tag{15.3–6}$$

where $\mathcal{E}_k$ is the signal energy per bit. The composite transmitted signal for the K users may be expressed as

$$\begin{aligned} s(t) &= \sum_{k=1}^{K} s_k(t - \tau_k) \\ &= \sum_{k=1}^{K} \sqrt{\mathcal{E}_k} \sum_{i=1}^{N} b_k(i) g_k(t - iT - \tau_k) \end{aligned} \tag{15.3–7}$$

where $\{\tau_k\}$ are the transmission delays, which satisfy the condition $0 \leqslant \tau_k < T$ for $1 \leqslant k \leqslant K$. Without loss of generality, we assume that $0 \leqslant \tau_1 \leqslant \tau_2 \leqslant \cdots \leqslant \tau_K < T$. This is the model for the multiuser transmitted signal in an asynchronous mode. In the special case of synchronous transmission, $\tau_k = 0$ for $1 \leqslant k \leqslant K$.

The transmitted signal is assumed to be corrupted by AWGN. Hence, the received signal may be expressed as

$$r(t) = s(t) + n(t) \tag{15.3–8}$$

where $s(t)$ is given by Equation 15.3–7 and $n(t)$ is the noise, with power spectral density $\frac{1}{2}N_0$.

15.3.2 The Optimum Receiver

The optimum receiver is defined as the receiver that selects the most probable sequence of bits $\{b_k(n), 1 \leqslant n \leqslant N, 1 \leqslant k \leqslant K\}$ given the received signal $r(t)$ observed over the time interval $0 \leqslant t \leqslant NT + 2T$. First, let us consider the case of synchronous transmission; later, we shall consider asynchronous transmission.

Synchronous transmission. In synchronous transmission, each (user) interferer produces exactly one symbol which interferes with the desired symbol. In additive white Gaussian noise, it is sufficient to consider the signal received in one signal interval, say $0 \leqslant t \leqslant T$, and determine the optimum receiver. Hence, $r(t)$ may be expressed as

$$r(t) = \sum_{k=1}^{K} \sqrt{\mathcal{E}_k} b_k(1) g_k(t) + n(t), \qquad 0 \leqslant t \leqslant T \tag{15.3–9}$$

The optimum maximum-likelihood receiver computes the log-likelihood function

$$\Lambda(\mathbf{b}) = \int_0^T \left[r(t) - \sum_{k=1}^{K} \sqrt{\mathcal{E}_k} b_k(1) g_k(t) \right]^2 dt \tag{15.3–10}$$

and selects the information sequence $\{b_k(1),\ 1 \leqslant k \leqslant K\}$ that minimizes $\Lambda(\mathbf{b})$. If we expand the integral in Equation 15.3–10, we obtain

$$\begin{aligned}\Lambda(\mathbf{b}) = &\int_0^T r^2(t)\,dt - 2\sum_{k=1}^{K} \sqrt{\mathcal{E}_k} b_k(1) \int_0^T r(t) g_k(t)\,dt \\ &+ \sum_{j=1}^{K}\sum_{k=1}^{K} \sqrt{\mathcal{E}_j \mathcal{E}_k} b_k(1) b_j(1) \int_0^T g_k(t) g_j(t)\,dt\end{aligned} \tag{15.3–11}$$

We observe that the integral involving $r^2(t)$ is common to all possible sequences $\{b_k(1)\}$ and is of no relevance in determining which sequence was transmitted. Hence, it may be neglected. The term

$$r_k = \int_0^T r(t) g_k(t)\,dt, \qquad 1 \leqslant k \leqslant K \tag{15.3–12}$$

represents the cross correlation of the received signal with each of the K signature sequences. Instead of cross correlators, we may employ matched filters. Finally, the integral involving $g_k(t)$ and $g_j(t)$ is simply

$$\rho_{jk}(0) = \int_0^T g_j(t) g_k(t)\,dt \tag{15.3–13}$$

Therefore, Equation 15.3–11 may be expressed in the form of correlation metrics

$$C(\mathbf{r}_K, \mathbf{b}_K) = 2\sum_{k=1}^{K} \sqrt{\mathcal{E}_k} b_k(1) r_k - \sum_{j=1}^{K}\sum_{k=1}^{K} \sqrt{\mathcal{E}_j \mathcal{E}_k} b_k(1) b_j(1) \rho_{jk}(0) \tag{15.3–14}$$

These correlation metrics may also be expressed in vector inner product form as

$$C(\mathbf{r}_K, \mathbf{b}_K) = 2\mathbf{b}_K' \mathbf{r}_K - \mathbf{b}_K' \mathbf{R}_s \mathbf{b}_K \tag{15.3–15}$$

where

$$\mathbf{r}_K = [r_1 \quad r_2 \quad \cdots \quad r_K]', \qquad \mathbf{b}_K = [\sqrt{\mathcal{E}_1} b_1(1) \ldots \sqrt{\mathcal{E}_K} b_K(1)]'$$

and $\mathbf{R}_s$ is the correlation matrix, with elements $\rho_{jk}(0)$. It is observed that the optimum detector must have knowledge of the received signal energies in order to compute the correlation metrics. Figure 15.3–1 depicts the optimum multiuser receiver.

There are 2^K possible choices of the bits in the information sequence of the K users. The optimum detector computes the correlation metrics for each sequence and selects the sequence that yields the largest correlation metric. We observe that the optimum detector has a complexity that grows exponentially with the number of users, K.

In summary, the optimum receiver for symbol-synchronous transmission consists of a bank of K correlators or matched filters followed by a detector that computes the 2^K correlation metrics given by Equation 15.3–15 corresponding to the 2^K possible transmitted information sequences. Then, the detector selects the sequence corresponding to the largest correlation metric.

Asynchronous transmission. In this case, there are exactly two consecutive symbols from each interferer that overlap a desired symbol. We assume that the receiver knows the received signal energies $\{\mathcal{E}_k\}$ for the K users and the transmission delays $\{\tau_k\}$. Clearly, these parameters must be measured at the receiver or provided to the receiver as side information by the users via some control channel.

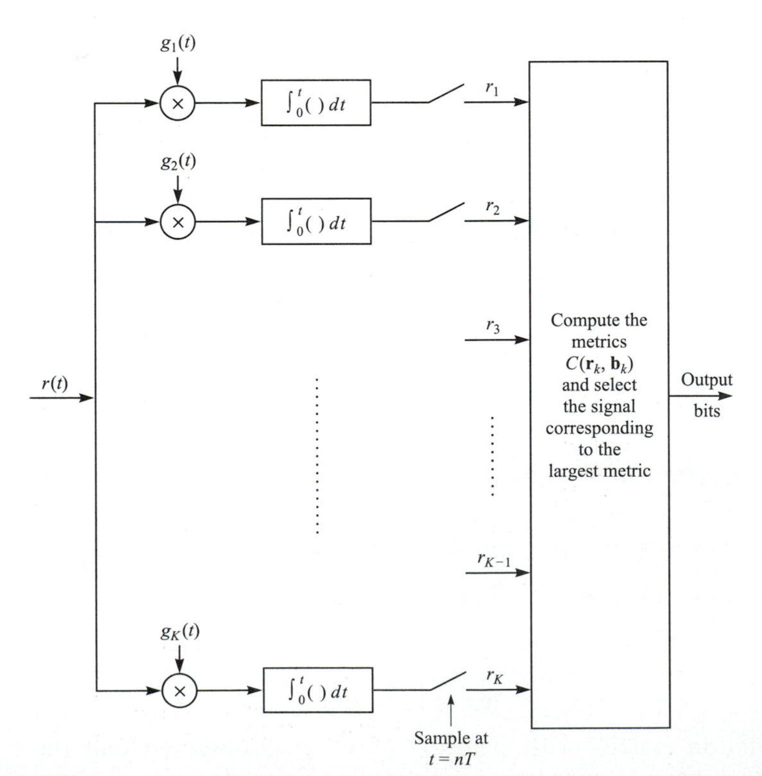

FIGURE 15.3–1
Optimum multiuser receiver for synchronous transmission.

The optimum maximum-likelihood receiver computes the log-likelihood function

$$\begin{aligned}\Lambda(\mathbf{b}) &= \int_0^{NT+2T}\left[r(t) - \sum_{k=1}^{K}\sqrt{\mathcal{E}_k}\sum_{i=1}^{N} b_k(i)g_k(t-iT-\tau_k)\right]^2 dt \\ &= \int_0^{NT+2T} r^2(t)\,dt - 2\sum_{k=1}^{K}\sqrt{\mathcal{E}_k}\sum_{i=1}^{N} b_k(i)\int_0^{NT+2T} r(t)g_k(t-iT-\tau_k)\,dt \\ &\quad + \sum_{k=1}^{K}\sum_{l=1}^{K}\sqrt{\mathcal{E}_k\mathcal{E}_l}\sum_{i=1}^{N}\sum_{j=1}^{N} b_k(i)b_l(j)\int_0^{NT+2T} g_k(t-iT-\tau_k)g_l(t-jT-\tau_l)\,dt\end{aligned} \tag{15.3–16}$$

where **b** represents the data sequences from the K users. The integral involving $r^2(t)$ may be ignored, since it is common to all possible information sequences. The integral

$$r_k(i) \equiv \int_{iT+\tau_k}^{(i+1)T+\tau_k} r(t)g_k(t-iT-\tau_k)\,dt, \qquad 1 \leqslant i \leqslant N \tag{15.3–17}$$

represents the outputs of the correlator or matched filter for the kth user in each of the signal intervals. Finally, the integral

$$\begin{aligned}&\int_0^{NT+2T} g_k(t-iT-\tau_k)g_l(t-jT-\tau_l)\,dt \\ &\qquad = \int_{-iT-\tau_k}^{NT+2T-iT-\tau_k} g_k(t)g_l(t+iT-jT+\tau_k-\tau_1)\,dt\end{aligned} \tag{15.3–18}$$

may be easily decomposed into terms involving the cross correlation $\rho_{kl}(\tau) = \rho_{kl}(\tau_l - \tau_k)$ for $k \leqslant 1$ and $\rho_{ik}(\tau)$ for $k > 1$. Therefore, we observe that the log-likelihood function may be expressed in terms of a correlation metric that involves the outputs $\{r_k(i), 1 \leqslant k \leqslant K, \leqslant i \leqslant N\}$ of K correlators or matched filters—one for each of the K signature sequences. Using vector notation, it can be shown that the NK correlator or matched filter outputs $\{r_k(i)\}$ can be expressed in the form

$$\mathbf{r} = \mathbf{R}_N\mathbf{b} + \mathbf{n} \tag{15.3–19}$$

where, by definition

$$\begin{aligned}\mathbf{r} &= [\mathbf{r}^t(1) \quad \mathbf{r}^t(2) \quad \cdots \quad \mathbf{r}^t(N)]' \\ \mathbf{r}(i) &= [r_1(i) \quad r_2(i) \quad \cdots \quad r_K(i)]'\end{aligned} \tag{15.3–20}$$

$$\begin{aligned}\mathbf{b} &= [\mathbf{b}^t(1) \quad \mathbf{b}^t(2) \quad \cdots \quad \mathbf{b}^t(N)]' \\ \mathbf{b}(i) &= [\sqrt{\mathcal{E}_1}b_1(i) \quad \sqrt{\mathcal{E}_2}b_2(i) \quad \cdots \quad \sqrt{\mathcal{E}_K}b_K(i)]'\end{aligned} \tag{15.3–21}$$

$$\begin{aligned}\mathbf{n} &= [\mathbf{n}^t(1) \quad \mathbf{n}^t(2) \quad \cdots \mathbf{n}^t(N)]' \\ \mathbf{n}(i) &= [n_1(i) \quad n_2(i) \quad \cdots \quad n_K(i)]'\end{aligned} \tag{15.3–22}$$

$$\mathbf{R}_N = \begin{bmatrix} \mathbf{R}_a(0) & \mathbf{R}_a'(1) & \mathbf{0} & \cdots & \cdots & \mathbf{0} \\ \mathbf{R}_a(1) & \mathbf{R}_a(0) & \mathbf{R}_a'(1) & \mathbf{0} & \cdots & \mathbf{0} \\ \vdots & \vdots & \vdots & \vdots & \vdots & \vdots \\ \mathbf{0} & \mathbf{0} & \mathbf{0} & \mathbf{R}_a(1) & \mathbf{R}_a(0) & \mathbf{R}_a'(1) \\ \mathbf{0} & \mathbf{0} & \mathbf{0} & \mathbf{0} & \mathbf{R}_a(1) & \mathbf{R}_a(0) \end{bmatrix} \tag{15.3–23}$$

and $\mathbf{R}_a(m)$ is a $K \times K$ matrix with elements

$$R_{kl}(m) = \int_{-\infty}^{\infty} g_k(t - \tau_k) g_l(t + mT - \tau_l)\, dt \tag{15.3–24}$$

The Gaussian noise vectors $\mathbf{n}(i)$ have zero-mean and autocorrelation matrix

$$E[\mathbf{n}(k)\mathbf{n}'(j)] = \tfrac{1}{2} N_0 \mathbf{R}_a(k - j) \tag{15.3–25}$$

Note that the vector $\mathbf{r}$ given by Equation 15.3–19 constitutes a set of sufficient statistics for estimating the transmitted bits $b_k(i)$.

If we adopt a block processing approach, the optimum *ML* detector must compute 2^{NK} correlation metrics and select the K sequences of length N that correspond to the largest correlation metric. Clearly, such an approach is much too complex computationally to be implemented in practice, especially when K and N are large. An alternative approach is *ML* sequence estimation employing the Viterbi algorithm. In order to construct a sequential-type detector, we make use of the fact that each transmitted symbol overlaps at most with $2K - 2$ symbols. Thus, a significant reduction in computational complexity is obtained with respect to the block size parameter N, but the exponential dependence on K cannot be reduced.

It is apparent that the optimum *ML* receiver employing the Viterbi algorithm involves such a high computational complexity that its use in practice is limited to communication systems where the number of users is extremely small, e.g., $K < 10$. For larger values of K, one may consider a sequential-type detector that is akin to either the sequential decoding or the stack algorithms described in Chapter 8. Below, we consider a number of suboptimum detectors whose complexity grows linearly with K.

15.3.3 Suboptimum Detectors

In the above discussion, we observed that the optimum detector for the K CDMA users has a computational complexity, measured in the number of arithmetic operations (additions and multiplications/divisions) per modulated symbol, that grows exponentially with K. In this subsection we describe suboptimum detectors with computational complexities that grow linearly with the number of users, K. We begin with the simplest suboptimum detector, which we call the conventional (single-user) detector.

Conventional single-user detector. In conventional single-user detection, the receiver for each user consists of a demodulator that correlates (or match-filters) the received signal with the signature sequence of the user and passes the correlator output to the detector, which makes a decision based on the single correlator output. Thus, the conventional detector neglects the presence of the other users of the channel or, equivalently, assumes that the aggregate noise plus interference is white and Gaussian.

Let us consider synchronous transmission. Then, the output of the correlator for the kth user for the signal in the interval $0 \leqslant t \leqslant T$ is

$$r_k = \int_0^T r(t) g_k(t)\, dt \tag{15.3–26}$$

$$= \sqrt{\mathcal{E}_k} b_k(1) + \sum_{\substack{j=1 \\ j \neq k}}^{K} \sqrt{\mathcal{E}_j} b_j(1) \rho_{jk}(0) + n_k(1) \tag{15.3–27}$$

where the noise component $n_k(1)$ is given as

$$n_k(1) = \int_0^T n(t) g_k(t)\, dt \tag{15.3–28}$$

Since $n(t)$ is white Gaussian noise with power spectral density $\frac{1}{2}N_0$, the variance of $n_k(1)$ is

$$E[n_k^2(1)] = \tfrac{1}{2}N_0 \int_0^T g_k^2(t)\, dt = \tfrac{1}{2}N_0 \tag{15.3–29}$$

Clearly, if the signature sequences are orthogonal, the interference from the other users given by the middle term in Equation 15.3–27 vanishes and the conventional single-user detector is optimum. On the other hand, if one or more of the other signature sequences are not orthogonal to the user signature sequence, the interference from the other users can become excessive if the power levels of the signals (or the received signal energies) of one or more of the other users is sufficiently larger than the power level of the kth user. This situation is generally called the *near–far problem* in multiuser communications, and necessitates some type of power control for conventional detection.

In asynchronous transmission, the conventional detector is more vulnerable to interference from other users. This is because it is not possible to design signature sequences for any pair of users that are orthogonal for all time offsets. Consequently, interference from other users is unavoidable in asynchronous transmission with the conventional single-user detection. In such a case, the near–far problem resulting from unequal power in the signals transmitted by the various users is particularly serious. The practical solution generally requires a power adjustment method that is controlled by the receiver via a separate communication channel that all users are continuously monitoring. Another option is to employ one of the multiuser detectors described below.

Decorrelating detector. We observe that the conventional detector has a complexity that grows linearly with the number of users, but its vulnerability to the near–far problem requires some type of power control. We shall now devise another type of detector that also has a linear computational complexity but does not exhibit the vulnerability to other-user interference.

Let us first consider the case of symbol-synchronous transmission. In this case, the received signal vector $\mathbf{r}_K$ that represents the output of the K matched filters is

$$\mathbf{r}_K = \mathbf{R}_s \mathbf{b}_K + \mathbf{n}_K \tag{15.3–30}$$

where $\mathbf{b}_K = [\sqrt{\mathcal{E}_1} b_1(1) \quad \sqrt{\mathcal{E}_2} b_2(1) \quad \cdots \quad \sqrt{\mathcal{E}_K} b_K(1)]'$ and the noise vector with elements $\mathbf{n}_K = [n_1(1) \; n_2(1) \cdots n_K(1)]'$ has a covariance

$$E(\mathbf{n}_K \mathbf{n}'_K) = \frac{N_0}{2} \mathbf{R}_s \tag{15.3–31}$$

Since the noise is Gaussian, $\mathbf{r}_K$ is described by a K-dimensional Gaussian PDF with mean $\mathbf{R}_s \mathbf{b}_K$ and covariance $\mathbf{R}_s$. That is,

$$p(\mathbf{r}_K | \mathbf{b}_K) = \frac{1}{\sqrt{(N_0 \pi)^K \det \mathbf{R}_s}} \exp\left[-\frac{1}{N_0}(\mathbf{r}_K - \mathbf{R}_s \mathbf{b}_K)' \mathbf{R}_s^{-1} (\mathbf{r}_K - \mathbf{R}_s \mathbf{b}_K)\right] \tag{15.3–32}$$

The best linear estimate of $\mathbf{b}_K$ is the value of $\mathbf{b}_K$ that minimizes the likelihood function

$$\Lambda(\mathbf{b}_K) = (\mathbf{r}_K - \mathbf{R}_s \mathbf{b}_K)' \mathbf{R}_s^{-1} (\mathbf{r}_K - \mathbf{R}_s \mathbf{b}_K) \tag{15.3–33}$$

The result of this minimization yields

$$\mathbf{b}_K^0 = \mathbf{R}_s^{-1} \mathbf{r}_K \tag{15.3–34}$$

Then, the detected symbols are obtained by taking the sign of each element of $\mathbf{b}_K^0$, i.e.,

$$\hat{\mathbf{b}}_K = \operatorname{sgn}(\mathbf{b}_K^0) \tag{15.3–35}$$

Figure 15.3–2 illustrates the receiver structure. Since the estimate $\mathbf{b}_K^0$ is obtained by performing a linear transformation on the vector of correlator outputs, the computational complexity is linear in K.

The reader should observe that the best (maximum-likelihood) linear estimate of $\mathbf{b}_K$ given by Equation 15.3–34 is different from the optimum non-linear ML sequence detector that finds the best discrete-valued $\{\pm 1\}$ sequence that maximizes the likelihood function. It is also interesting to note that the estimate $\mathbf{b}_K^0$ is the best linear estimate that maximizes the correlation metric given by Equation 15.3–15.

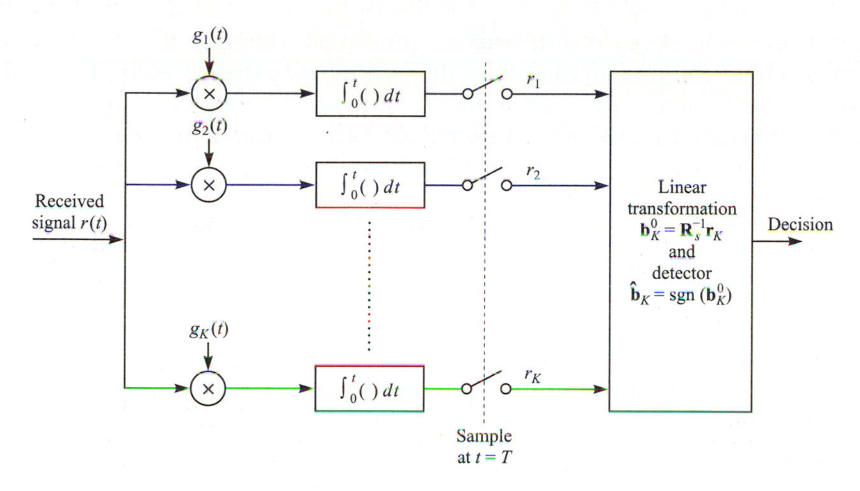

FIGURE 15.3–2
Receiver structure for decorrelation receiver.

An interesting interpretation of the detector that computes $\mathbf{b}_K^0$ as in Equation 15.3–34 and makes decisions according to Equation 15.3–35 is obtained by considering the case of $K = 2$ users. In this case,

$$\mathbf{R}_s = \begin{bmatrix} 1 & \rho \\ \rho & 1 \end{bmatrix} \tag{15.3–36}$$

$$\mathbf{R}_s^{-1} = \frac{1}{1-\rho^2}\begin{bmatrix} 1 & -\rho \\ -\rho & 1 \end{bmatrix} \tag{15.3–37}$$

where

$$\rho = \int_0^T g_1(t) g_2(t)\,dt \tag{15.3–38}$$

Then, if we correlate the received signal

$$r(t) = \sqrt{\mathcal{E}_1} b_1 g_1(t) + \sqrt{\mathcal{E}_2} b_2 g_2(t) + n(t) \tag{15.3–39}$$

with $g_1(t)$ and $g_2(t)$, we obtain

$$\mathbf{r}_2 = \begin{bmatrix} \sqrt{\mathcal{E}_1} b_1 + \rho\sqrt{\mathcal{E}_2} b_2 + n_1 \\ \rho\sqrt{\mathcal{E}_1} b_1 + \sqrt{\mathcal{E}_2} b_2 + n_2 \end{bmatrix} \tag{15.3–40}$$

where n_1 and n_2 are the noise components at the output of the correlators. Therefore,

$$\begin{aligned} \mathbf{b}_2^0 &= \mathbf{R}_s^{-1}\mathbf{r}_2 \\ &= \begin{bmatrix} \sqrt{\mathcal{E}_1} b_1 + (n_1 - \rho n_2)/(1-\rho^2) \\ \sqrt{\mathcal{E}_2} b_2 + (n_2 - \rho n_1)/(1-\rho^2) \end{bmatrix} \end{aligned} \tag{15.3–41}$$

This is a very interesting result, because the transformation $\mathbf{R}_s^{-1}$ has eliminated the interference components between the two users. Consequently, the near–far problem is eliminated and there is no need for power control.

It is interesting to note that a result similar to Equation 15.3–41 is obtained if we correlate $r(t)$ given by Equation 15.3–39 with two modified signature waveforms

$$g_1'(t) = g_1(t) - \rho g_2(t) \tag{15.3–42}$$

$$g_2'(t) = g_2(t) - \rho g_1(t) \tag{15.3–43}$$

This means that, by correlating the received signal with the modified signature waveforms, we have tuned out or *decorrelated* the multiuser interference. Hence, the detector based on Equation 15.3–34 is called a *decorrelating detector*.

In asynchronous transmission, the received signal at the output of the correlators is given by Equation 15.3–19. Hence, the log-likelihood function is given as

$$\Lambda(\mathbf{b}) = (\mathbf{r} - \mathbf{R}_N\mathbf{b})'\mathbf{R}_N^{-1}(\mathbf{r} - \mathbf{R}_N\mathbf{b}) \tag{15.3–44}$$

where $\mathbf{R}_N$ is defined by Equation 15.3–23 and $\mathbf{b}$ is given by Equation 15.3–21. It is relatively easy to show that the vector $\mathbf{b}$ that minimizes $\Lambda(\mathbf{b})$ is

$$\mathbf{b}^0 = \mathbf{R}_N^{-1}\mathbf{r} \tag{15.3–45}$$

This is the ML estimate of $\mathbf{b}$ and it is again obtained by performing a linear transformation of the outputs from the bank of correlators of matched filters.

Since $\mathbf{r} = \mathbf{R}_N\mathbf{b} + \mathbf{n}$, it follows from Equation 15.3–45 that

$$\mathbf{b}^0 = \mathbf{b} + \mathbf{R}_N^{-1}\mathbf{n} \tag{15.3–46}$$

Therefore, $\mathbf{b}^0$ is an unbiased estimate of $\mathbf{b}$. This means that the multiuser interference has been eliminated, as in the case of symbol-synchronous transmission. Hence, this detector for asynchronous transmission is also called a *decorrelating detector*.

A computationally efficient method for obtaining the solution given by Equation 15.3–45 is the square-root factorization method described in Appendix D. Of course, there are many other methods that may be used to invert the matrix $\mathbf{R}_N$. Iterative methods to decorrelate the signals have also been explored.

Minimum mean-square-error detector. In the above discussion, we showed that the linear ML estimate of $\mathbf{b}$ is obtained by minimizing the quadratic log-likelihood function in Equation 15.3–44. Thus, we obtained the result given by Equation 15.3–45, which is an estimate derived by performing a linear transformation on the outputs of the bank of correlators or matched filters.

Another, somewhat different, solution is obtained if we seek the linear transformation $\mathbf{b}^0 = \mathbf{A}\mathbf{r}$, where the matrix $\mathbf{A}$ is to be determined so as to minimize the mean square error (MSE)

$$\begin{aligned} J(\mathbf{b}) &= E[(\mathbf{b} - \mathbf{b}^0)'(\mathbf{b} - \mathbf{b}^0)] \\ &= E[(\mathbf{b} - \mathbf{Ar})'(\mathbf{b} - \mathbf{Ar}] \end{aligned} \tag{15.3–47}$$

where the expectation is with respect to the data vector $\mathbf{b}$ and the additive noise $\mathbf{n}$. The optimum matrix $\mathbf{A}$ may be found by forcing the error $(\mathbf{b} - \mathbf{Ar})$ to be orthogonal to the data vector $\mathbf{r}$. Thus,

$$\begin{aligned} E[(\mathbf{b} - \mathbf{Ar})\mathbf{r}'] &= \mathbf{0} \\ E(\mathbf{br}') - \mathbf{A}E(\mathbf{rr}') &= \mathbf{0} \end{aligned} \tag{15.3–48}$$

Let us consider the case of synchronous transmission. We have

$$E(\mathbf{b}_K\mathbf{r}'_K) = E(\mathbf{b}_K\mathbf{b}'_K)\mathbf{R}'_s = \mathbf{DR}'_s \tag{15.3–49}$$

and

$$\begin{aligned} E(\mathbf{r}_K\mathbf{r}'_K) &= E[(\mathbf{R}_s\mathbf{b}_K + \mathbf{n}_K)(\mathbf{R}_s\mathbf{b}_K + \mathbf{n}_K)'] \\ &= \mathbf{R}_s\mathbf{DR}'_s + \frac{N_0}{2}\mathbf{R}'_s \end{aligned} \tag{15.3–50}$$

where $\mathbf{D}$ is a diagonal matrix with diagonal elements $\{\mathcal{E}_k, 1 \leqslant k \leqslant K\}$. By substituting Equation 15.3–49 and 15.3–50 into Equation 15.3–48 and solving for $\mathbf{A}$, we obtain

$$\mathbf{A}^0 = \left(\mathbf{R}_s + \frac{N_0}{2}\mathbf{D}^{-1}\right)^{-1} \tag{15.3–51}$$

Then,

$$\mathbf{b}_K^0 = \mathbf{A}^0\mathbf{r}_K \tag{15.3–52}$$

and

$$\hat{\mathbf{b}}_K = \operatorname{sgn}(\mathbf{b}_K^0) \tag{15.3–53}$$

Similarly, for asynchronous transmission, it can be shown that the optimum choice of $\mathbf{A}$ that minimizes $J(\mathbf{b})$ is

$$\mathbf{A}^0 = (\mathbf{R}_N + \tfrac{1}{2}N_0\mathbf{I})^{-1} \tag{15.3–54}$$

and, hence,

$$\mathbf{b}^0 = (R_N + \tfrac{1}{2}N_0\mathbf{I})^{-1}\mathbf{r} \tag{15.3–55}$$

The output of the detector is then $\hat{\mathbf{b}} = \operatorname{sgn}(\mathbf{b}^0)$.

The estimate given by Equation 15.3–52 or 15.3–55 is called the *minimum MSE* (MMSE) estimate of $\mathbf{b}$. Note that when $\frac{1}{2}N_0$ is small compared with the diagonal elements of $\mathbf{R}_N$, the MMSE solution approaches the ML solution given by Equation 15.3–45. On the other hand, when the noise level is large compared with the signal level in the diagonal elements of $\mathbf{R}_N$, $\mathbf{A}^0$ approaches the identity matrix (scaled by $\frac{1}{2}N_0$). In this low-SNR case, the detector basically ignores the interference from other users, because the additive noise is the dominant term. It should also be noted that the MMSE criterion produces a biased estimate of $\mathbf{b}$. Hence, there is some residual multiuser interference.

To perform the computations that lead to the values of $\mathbf{b}$, we solve the set of linear equations

$$(\mathbf{R}_N + \tfrac{1}{2}N_0\mathbf{I})\mathbf{b} = \mathbf{r} \tag{15.3–56}$$

This solution may be computed efficiently using a square-root factorization of the matrix $\mathbf{R}_N + \frac{1}{2}N_0\mathbf{I}$ as indicated above. Thus, to detect NK bits requires $3NK^2$ multiplications. Therefore, the computational complexity is $3K$ multiplications per bit, which is independent of the block length N and is linear in K.

We observe that both the decorrelating detector and the MMSE detector exhibit the desirable property of being near-far resistant. In fact, in the case of the decorrelating detector, the interference from other users is completely eliminated.

We also observe that both the decorrelating detector and the MMSE detector described above involve performing linear transformations on a block of data obtained from K correlators or matched filters. The linear transformations are akin to the linear equalization of intersymbol interference treated in Chapter 10. In fact, the decorrelating detector is akin to the zero-forcing linear equalizer, and the MMSE detector is akin to the linear MMSE equalizer. Consequently, these multiuser detectors for asynchronous transmission can be implemented by employing a tapped-delay-line filter with adjustable coefficients for each user and selecting the filter coefficients to either eliminate the interuser interference or to minimize the MSE for each user signal. Thus, the received information bits are estimated sequentially with finite delay, instead of as a block.

A decision-feedback-type filter can be used instead of a linear filter to implement the multiuser detector that processes the data sequentially. In particular, Xie et al. (1990b) demonstrated that the transmitted bits may be recovered sequentially from the received signal by employing a form of a decision-feedback equalizer with finite delay. Hence, there is a similarity between the detection of signals corrupted by ISI in a single-user communication system and the detection of signals in a multiuser system with asynchronous transmission.

Other types of multiuser detectors. Because of the widespread interest in the development of commercial CDMA communication systems, the design of multiuser detection algorithms continues to be a very active area of research. Our treatment in this chapter focused on the optimum MLSE algorithm, suboptimum linear (MMSE and decorrelating detection) algorithms, and non-linear successive interference cancellation algorithms based on hard decisions.

In addition to these relatively simple algorithms, a number of more complex algorithms have been described in the literature that are appropriate for time-dispersive channels which result in ISI. In addition, one may assume that knowledge of the signature waveforms of the other users is not available to a user receiver. Hence, a user receiver is confronted with both ISI and multiple access interference (MAI). In such a scenario, it is possible to design adaptive interference suppression algorithms that are akin to equalization algorithms previously described in Chapters 10 and 11.

Adapative algorithms for suppressing ISI and MAI in multiuser CDMA systems are described in the papers by Abdulrahman et al. (1994), Honig (1998), Miller (1995, 1996), Rapajic and Vucetic (1994), and Mitra and Poor (1995). In some cases, the adaptive algorithms are designed to converge without the use of any training symbols. Such algorithms are called *blind multiuser detection algorithms*. Examples of such blind algorithms are described in the papers by Honig et al. (1995), Madhow (1998), Wang and Poor (1998a,b), Bensley and Aazhang (1996).

The use of multiple transmitting and/or receiving antennas in CDMA systems provides each user with the opportunity to employ spatial filtering in addition to temporal filtering to reduce ISI and MAI. Blind multiuser detection algorithms for multiple antenna systems have been described by Wang and Poor (1999).

In general, the signals transmitted by the various users in a CDMA communication system are coded, either using a single level of coding or a concatenated code. Instead of separating the signal processing of the demodulator from the decoder, a better strategy is to use soft-information metrics from the decoder to enhance the suppression of the MAI and ISI at the demodulator. Thus, one can devise turbo-type iterative demodulation-decoding algorithms for suppressing MAI and ISI. Such algorithms for coded CDMA systems have been described in the papers by Reed et al. (1998), Moher (1998), Alexander et al. (1999), and Wang and Poor (1999).

15.3.4 Successive Interference Cancellation

Another multiuser detection technique is called successive interference cancellation (SIC). This technique is based on removing the interfering signal waveforms from the received signal, one at a time as they are detected. One approach is to demodulate the users in the order of decreasing received powers. Thus, the user having the strongest received signal is demodulated first. After a signal has been demodulated and detected, the detected information is used to subtract the signal of the particular user from the received signal.

When making a decision about the transmitted information of the kth user, we assume that the decisions of users $k+1, \ldots, K$ are correct and neglect the presence of users $1, \ldots, k-1$. Therefore, the decision for the information bit of the kth user, for synchronous transmission, is

$$\hat{b}_k = \operatorname{sgn}\left[r_k - \sum_{j=k+1}^{K} \sqrt{\mathcal{E}_j}\rho_{jk}(0)\hat{b}_j \right] \tag{15.3–57}$$

where r_k is the output of the correlator or matched filter corresponding to the kth user's signature sequence.

The approach based on demodulating the user signals in the order of decreasing received powers does not take into account the cross correlations among users. An alternative approach is to demodulate the user signals according to the powers at the outputs of the cross correlators or matched filters, i.e., according to the correlation metrics

$$E\left\{\left[\int_0^T g_k(t)r(t)\,dt\right]^2\right\} = \mathcal{E}_k + \sum_{j\neq k} \mathcal{E}_j \rho_{jk}^2(0) + \frac{N_0}{2} \tag{15.3–58}$$

which applies to the case of synchronous transmission.

We make the following observations regarding the SIC of multiuser interference. First of all, SIC requires that we estimate the received signal powers of the users in order to cancel the interference. Estimation errors result in residual multiuser interference, which causes a degradation in performance. Secondly, the interference from users whose signals are weaker than the user signal being detected is treated as additive interference. Thirdly, the computational complexity in the demodulation of a user information bit is linear in the number of users. Finally, the delay in demodulating the weakest user increases linearly with the number of users.

SIC is easily generalized to asynchronous signal transmission. In this case, both the user signal strengths and the time delays must be estimated.

Finally, we note that the SIC multiuser detector given in Equation 15.3–57 is also a suboptimum detector, since the signals of weaker users are treated as additive interference. The jointly optimum interference canceller for synchronous transmission may be defined as the detector which computes the decisions $\hat{b}_k$ as

$$\hat{b}_k = \text{sgn}\left[\mathrm{r_k} - \sum_{\mathrm{j\neq k}} \sqrt{\mathcal{E}_\mathrm{j}}\rho_\mathrm{jk}(0)\hat{\mathrm{b}}_\mathrm{j}\right] \tag{15.3–59}$$

Multistage interference cancellation (MIC). Multiuser detection based on MIC is a technique that employs multiple iterations in detecting the user bits and cancelling the interference. The method is easily described by means of an example.

EXAMPLE 15.3–1: TWO USERS AND SYNCHRONOUS TRANSMISSION. For the first stage of the detector, we may use the SIC detector or any of the suboptimum detectors. For example, suppose we use the decorrelating detector in the first stage.

First stage (decorrelating detector):

$$\hat{b}_1 = \text{sgn}(r_1 - \rho r_2)$$
$$\hat{b}_2 = \text{sgn}(r_2 - \rho r_1)$$

Second stage:

$$\hat{\hat{b}}_1 = \text{sgn}\left(r_1 - \sqrt{\mathcal{E}_2}\hat{b}_2\rho\right)$$
$$\hat{\hat{b}}_2 = \text{sgn}\left(r_2 - \sqrt{\mathcal{E}_1}\hat{b}_1\rho\right)$$

Third stage:

$$\hat{\hat{\hat{b}}}_1 = \text{gn}\left(r_1 - \sqrt{\mathcal{E}_2}\hat{\hat{b}}_2\rho\right)$$
$$\hat{\hat{\hat{b}}}_2 = \text{sgn}\left(r_2 - \sqrt{\mathcal{E}_1}\hat{\hat{b}}_1\rho\right)$$

The computations may be terminated when there is no change in the decisions over two successive iterations.

Successive interference cancellation and multistage interference cancellation are two types of multiple access interference cancellation techniques that have received considerable attention by many researchers. For reference, we include the papers by Varanasi and Aazhang (1990), Patel and Holtzman (1994), Buehrer et al. (1996, 1999), and Divsalar et al. (1998).

We should indicate that the MIC is a suboptimum detector and does not converge to the jointly optimum multiuser detector defined above.

15.3.5 Performance Characteristics of Detectors

The bit error probability is generally the desirable performance measure in multiuser communications. In evaluating the effect of multiuser interference on the performance of the detector for a single user, we may use as a benchmark the probability of a bit error for a single-user receiver in the absence of other users of the channel, which is

$$P_k(\gamma_k) = Q(\sqrt{2\gamma_k}) \tag{15.3–60}$$

where $\gamma_k = \mathcal{E}_k/N_0$, $\mathcal{E}_k$ is the signal energy per bit, and $\frac{1}{2}N_0$ is the power spectral density of the AWGN.

In the case of the optimum detector for either synchronous or asynchronous transmission, the probability of error is extremely difficult and tedious to evaluate. In this case, we may use Equation 15.3–60 as a lower bound and the performance of a suboptimum detector as an upper bound.

Let us consider, first, the suboptimum, conventional single-user detector. For synchronous transmission, the output of the correlator for the kth user is given by Equation 15.3–27. Therefore, the probability of error for the kth user, conditional on a sequence $\mathbf{b}_i$ of bits from other users, is

$$P_k(\mathbf{b}_i) = Q\left(\sqrt{2\left[\sqrt{\mathcal{E}_k} + \sum_{\substack{j=1 \\ j \neq k}}^{K} \sqrt{\mathcal{E}_j} b_j(1) \rho_{jk}(0)\right]^2 \Big/ N_0}\right) \tag{15.3–61}$$

Then, the average probability of error is simply

$$P_k = \left(\tfrac{1}{2}\right)^{K-1} \sum_{i=1}^{2^{K-1}} P_k(\mathbf{b}_i) \tag{15.3–62}$$

The probability in Equation 15.3–62 will be dominated by the term that has the smallest argument in the Q function. The smallest argument will result in an SNR of

$$(\text{SNR})_{\min} = \frac{1}{N_0}\left[\sqrt{\mathcal{E}_k} - \sum_{\substack{j=1 \\ j \neq k}}^{K} \sqrt{\mathcal{E}_j} |\rho_{jk}(0)|\right]^2 \tag{15.3–63}$$

Therefore,

$$\left(\tfrac{1}{2}\right)^{K-1} Q\left(\sqrt{2(\text{SNR})_{\min}}\right) < P_k < Q\left(\sqrt{2(\text{SNR})_{\min}}\right) \tag{15.3–64}$$

A similar development can be used to obtain bounds on the performance for asynchronous transmission.

In the case of a decorrelating detector, the other-user interference is completely eliminated. Hence, the probability of error may be expressed as

$$P_k = Q\left(\sqrt{\mathcal{E}_k / \sigma_k^2}\right) \tag{15.3–65}$$

where σ_k^2 is the variance of the noise in the kth element of the estimate $\mathbf{b}^0$.

EXAMPLE 15.3–2. Consider the case of synchronous, two-user transmission, where $\mathbf{b}_2^0$ is given by Equation 15.3–41. Let us determine the probability of error.

The signal component for the first term in Equation 15.3–41 is $\sqrt{\mathcal{E}_1}$. The noise component is

$$n = \frac{n_1 - \rho n_2}{1 - \rho^2}$$

where ρ is the correlation between the two signature signals. The variance of this noise is

$$\begin{aligned} \sigma_1^2 &= \frac{E[(n_1 - \rho n_2)]^2}{(1 - \rho^2)^2} \\ &= \frac{1}{1 - \rho^2} \frac{N_0}{2} \end{aligned} \tag{15.3–66}$$

and

$$P_1 = Q\left(\sqrt{\frac{2\mathcal{E}_1}{N_0}(1 - \rho^2)}\right) \tag{15.3–67}$$

A similar result is obtained for the performance of the second user. Therefore, the noise variance has increased by the factor $(1-\rho^2)^{-1}$. This noise enhancement is the price paid for the elimination of the multiuser interference by the decorrelating detector.

The error rate performance of the MMSE detector is similar to that for the decorrelating detector when the noise level is low. For example, from Equation 15.3–55, we observe that when N_0 is small relative to the diagonal elements of the signal correlation matrix $\mathbf{R}_N$,

$$\mathbf{b}^0 \approx \mathbf{R}_N^{-1}\mathbf{r} \tag{15.3–68}$$

which is the solution for the decorrelating detector. For low multiuser interference, the MMSE detector results in a smaller noise enhancement compared with the decorrelating detector, but has some residual bias resulting from the other users. Thus, the MMSE detector attempts to strike a balance between the residual interference and the noise enhancement.

An alternative to the error probability as a figure of merit that has been used to characterize the performance of a multiuser communication system is the ratio of SNRs with and without the presence of interference. In particular, Equation 15.3–60 gives the error probability of the kth user in the absence of other-user interference. In this case, the SNR is $\gamma_k = \mathcal{E}_k/N_0$. In the presence of multiuser interference, the user that transmits a signal with energy $\mathcal{E}_k$ will have an error probability P_k that exceeds $P_k(\gamma_k)$. The *effective SNR* γ_{ke} is defined as the SNR required to achieve the error probability

$$P_k = P_k(\gamma_{ke}) = Q(\sqrt{2\gamma_{ke}}) \tag{15.3–69}$$

The *efficiency* is defined as the ratio γ_{ke}/γ_k and represents the performance loss due to the multiuser interference. The desirable figure of merit is the *asymptotic efficiency*, defined as

$$\eta_k = \lim_{N_0 \to 0} \frac{\gamma_{ke}}{\gamma_k} \tag{15.3–70}$$

This figure of merit is often simpler to compute than the probability of error.

EXAMPLE 15.3–3. Consider the case of two symbol-synchronous users with signal energies $\mathcal{E}_1$ and $\mathcal{E}_2$. Let us determine the asymptotic efficiency of the conventional detector.

In this case, the probability of error is easily obtained from Equation 15.3–61 and Equation 15.3–62 as

$$P_1 = \tfrac{1}{2}Q\left(\sqrt{2(\sqrt{\mathcal{E}_1}+\rho\sqrt{\mathcal{E}_2})^2/N_0}\right) + \tfrac{1}{2}Q\left(\sqrt{2(\sqrt{\mathcal{E}_1}-\rho\sqrt{\mathcal{E}_2})^2/N_0}\right)$$

However, the asymptotic efficiency is much easier to compute. It follows from the definition of Equation 15.3–70 and from Equation 15.3–61 that

$$\eta_1 = \left[\max\left(0, 1-\sqrt{\frac{\mathcal{E}_2}{\mathcal{E}_1}}|\rho|\right)\right]^2$$

A similar expression is obtained for η_2.

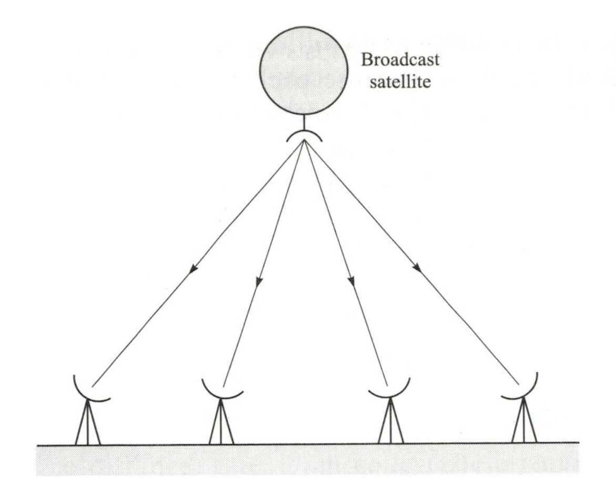

FIGURE 15.4–2
Broadcast system.

users can monitor the satellite transmissions and, thus, establish whether or not their packets have been transmitted successfully.

There are basically two types of ALOHA systems: *synchronized or slotted* and *unsynchronized or unslotted.* In an unslotted ALOHA system, a user may begin transmitting a packet at any arbitrary time. In a slotted ALOHA, the packets are transmitted in time slots that have specified beginning and ending times.

We assume that the start time of packets that are transmitted is a Poisson point process having an average rate of λ packets/s. Let T_p denote the time duration of a packet. Then, the normalized channel traffic G, also called the *offered channel traffic*, is defined as

$$G = \lambda T_p \tag{15.4–1}$$

There are many channel access protocols that can be used to handle collisions. Let us consider the one due to Abramson (1973). In Abramson's protocol, packets that have collided are retransmitted with some delay τ, where τ is randomly selected according to the PDF

$$p(\tau) = \alpha e^{-\alpha\tau} \tag{15.4–2}$$

where α is a design parameter. The random delay τ is added to the time of the initial transmission and the packet is retransmitted at the new time. If a collision occurs again, a new value of τ is randomly selected and the packet is retransmitted with a new delay from the time of the second transmission. This process is continued until the packet is transmitted successfully. The design parameter α determines the average delay between retransmissions. The smaller the value of α, the longer the delay between retransmissions.

Now, let λ', where $\lambda' < \lambda$, be the rate at which packets are transmitted successfully. Then, the normalized channel throughput is

$$S = \lambda' T_p \tag{15.4–3}$$

We can relate the channel throughput S to the offered channel traffic G by making use of the assumed start time distribution. The probability that a packet will not overlap a given packet is simply the probability that no packet begins T_p seconds before or T_p seconds after the start time of the transmitted packet. Since the start time of all packets is Poisson-distributed, the probability that a packet will not overlap is $\exp(-2\lambda T_p) = \exp(-2G)$. Therefore,

$$S = Ge^{-2G} \tag{15.4–4}$$

This relationship is plotted in Figure 15.4–3. We observe that the maximum throughput is $S_{\max} = 1/2e = 0.184$ packets per slot, which occurs at $G = \frac{1}{2}$. When $G > \frac{1}{2}$, the throughput S decreases. The above development illustrates that an unsynchronized or unslotted random access method has a relatively small throughput and is inefficient.

Throughput for slotted ALOHA. To determine the throughput in a slotted ALOHA system, let G_i be the probability that the ith user will transmit a packet in some slot. If all the K users operate independently and there is no statistical dependence between the transmission of the user's packet in the current slot and the transmission of the user's packet in previous time slots, the total (normalized) offered channel traffic is

$$G = \sum_{i=1}^{K} G_i \tag{15.4–5}$$

Note that, in this case, G may be greater than unity.

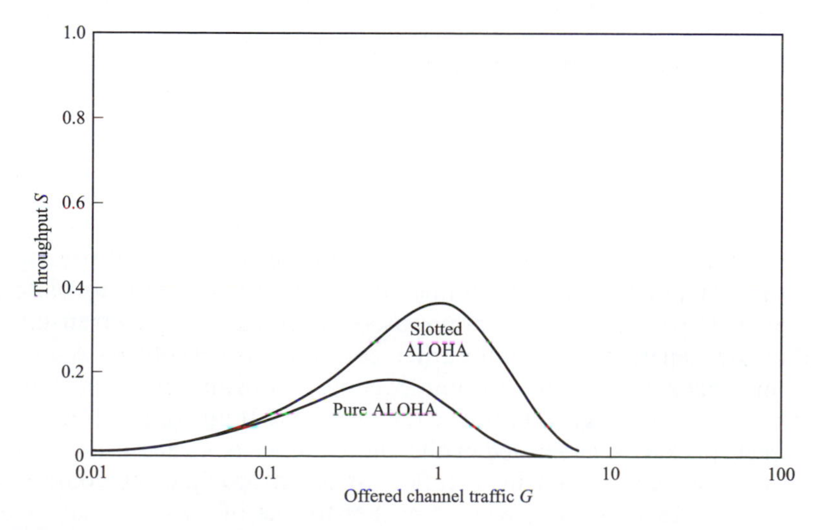

FIGURE 15.4–3
Throughput in ALOHA systems.

Now, let $S_i \leqslant G_i$ be the probability that a packet transmitted in a time slot is received without a collision. Then, the normalized channel throughput is

$$S = \sum_{i=1}^{K} S_i \tag{15.4–6}$$

The probability that a packet from the ith user will not have a collision with another packet is

$$Q_i = \prod_{\substack{j=1 \\ j \neq i}}^{K} (1 - G_j) \tag{15.4–7}$$

Therefore,

$$S_i = G_i Q_i \tag{15.4–8}$$

A simple expression for the channel throughput is obtained by considering K identical users. Then,

$$S_i = \frac{S}{K}, \qquad G_i = \frac{G}{K}$$

and

$$S = G\left(1 - \frac{G}{K}\right)^{K-1} \tag{15.4–9}$$

Then, if we let $K \to \infty$, we obtain the throughput

$$S = Ge^{-G} \tag{15.4–10}$$

This result is also plotted in Figure 15.4–3. We observe that S reaches a maximum throughput of $S_{\max} = 1/e = 0.368$ packets per slot at $G = 1$, which is twice the throughput of the unslotted ALOHA system.

The performance of the slotted ALOHA system given above is based on Abramson's protocol for handling collisions. A higher throughput is possible by devising a better protocol.

A basic weakness in Abramson's protocol is that it does not take into account the information on the amount of traffic on the channel that is available from observation of the collisions that occur. An improvement in throughput of the slotted ALOHA system can be obtained by using a tree-type protocol devised by Capetanakis (1979). In this algorithm, users are not allowed to transmit new packets that are generated until all earlier collisions are resolved. A user can transmit a new packet in a time slot immediately following its generation, provided that all previous packets that have collided have been transmitted successfully. If a new packet is generated while the channel is clearing the previous collisions, the packet is stored in a buffer. When a new packet collides with another, each user assigns its respective packet to one of two sets, say A or B, with equal probability (by flipping a coin). Then, if a packet is put in set A, the user transmits it in the next time slot. If it collides again, the user will again randomly assign the packet to one of two sets and the process of transmission is repeated. This process continues until all packets contained in set A are transmitted successfully. Then, all packets in set B are transmitted following the same procedure. All the users monitor the state of the channel, and, hence, they know when all the collisions have been serviced.

When the channel becomes available for transmission of new packets, the earliest generated packets are transmitted first. To establish a queue, the time scale is subdivided into subintervals of sufficiently short duration such that, on average, approximately one packet is generated by a user in a subinterval. Thus, each packet has a "time tag" that is associated with the subinterval in which it was generated. Then, a new packet belonging to the first subinterval is transmitted in the first available time slot. If there is no collision, then a packet from the second subinterval is transmitted, and so on. This procedure continues as new packets are generated and as long as any backlog of packets for transmission exists. Capetanakis has demonstrated that this channel access protocol achieves a maximum throughput of 0.43 packets per slot.

In addition to throughput, another important performance measure in a random access system is the average transmission delay in transmitting a packet. In an ALOHA system, the average number of transmissions per packet is G/S. To this number we may add the average waiting time between transmissions and, thus, obtain an average delay for a successful transmission. We recall from the above discussion that in the Abramson protocol, the parameter α determines the average delay between retransmissions. If we select α small, we obtain the desirable effect of smoothing out the channel load at times of peak loading, but the result is a long retransmission delay. This is the trade-off in the selection of α in Equation 5.4–2. On the other hand, the Capetanakis protocol has been shown to have a smaller average delay in the transmission of packets. Hence, it outperforms Abramson's protocol in both average delay and throughput.

Another important issue in the design of random access protocols is the stability of the protocol. In our treatment of ALOHA-type channel access protocols, we implicitly assumed that for a given offered load, an equilibrium point is reached where the average number of packets entering the channel is equal to the average number of packets transmitted successfully. In fact, it can be demonstrated that any channel access protocol, such as the Abramson protocol, that does not take into account the number of previous unsuccessful transmissions in establishing a retransmission policy is inherently unstable. On the other hand, the Capetanakis algorithm differs from the Abramson protocol in this respect and has been proved to be stable. A thorough discussion of the stability issues of random access protocols is found in the paper by Massey (1988).

15.4.2 Carrier Sense Systems and Protocols

As we have observed, ALOHA-type (slotted and unslotted) random access protocols yield relatively low throughput. Furthermore, a slotted ALOHA system requires that users transmit at synchronized time slots. In channels where transmission delays are relatively small, it is possible to design random access protocols that yield higher throughput. An example of such a protocol is *carrier sensing* with collision detection, which is used as a standard Ethernet protocol in local area networks. This protocol is generally known as *carrier sense multiple access with collision detection* (CSMA/CD).

The CSMA/CD protocol is simple. All users listen for transmissions on the channel. A user who wishes to transmit a packet seizes the channel when it senses that the channel is idle. Collisions may occur when two or more users sense an idle channel and begin transmission. When the users that are transmitting simultaneously sense a collision, they transmit a special signal, called a *jam signal*, that serves to notify all users of the collision and abort their transmissions. Both the carrier sensing feature and the abortion of transmission when a collision occurs result in minimizing the channel downtime and, hence, yield a higher throughput.

To elaborate on the efficiency of CSMA/CD, let us consider a local area network having a bus architecture, as shown in Figure 15.4–4. Consider two users U_1 and U_2 at the maximum separation, i.e., at the two ends of the bus, and let τ_d be the propagation delay for a signal to travel the length of the bus. Then, the (maximum) time required to sense an idle channel is τ_d. Suppose that U_1 transmits a packet of duration T_p. User U_2 may seize the channel τ_d seconds later by using carrier sensing and begins to transmit. However, user U_1 would not know of this transmission until τ_d seconds after U_2 begins transmission. Hence, we may define the time interval $2\tau_d$ as the (maximum) time interval to detect a collision. If we assume that the time required to transmit the jam signal is negligible, the CSMA/CD protocol yields a high throughput when $2\tau_d \ll T_p$.

There are several possible protocols that may be used to reschedule transmissions when a collision occurs. One protocol is called *nonpersistent CSMA*, a second is called *1-persistent CSMA*, and a generalization of the latter is called *p*-persistent CSMA.

Nonpersistent CSMA. In this protocol, a user that has a packet to transmit senses the channel and operates according to the following rule.

(a) If the channel is idle, the user transmits a packet.
(b) If the channel is sensed busy, the user schedules the packet transmission at a later time according to some delay distribution. At the end of the delay interval, the user again senses the channel and repeats steps (a) and (b).

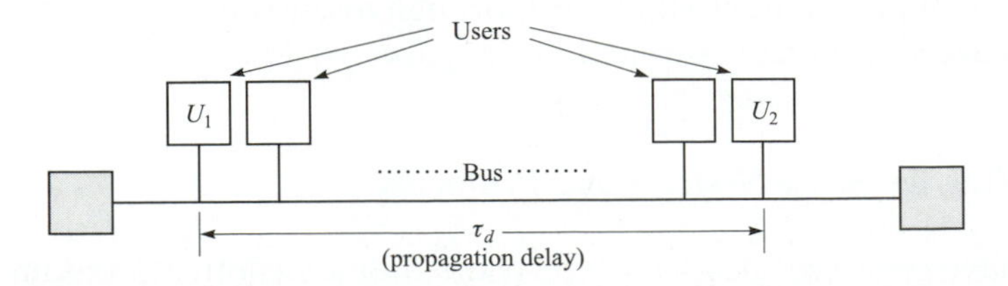

FIGURE 15.4–4
Local area network with bus architecture.

1-Persistent CSMA. This protocol is designed to achieve high throughput by not allowing the channel to go idle if some user has a packet to transmit. Hence, the user senses the channel and operates according to the following rule.

(a) If the channel is sensed idle, the user transmits the packet with probability 1.
(b) If the channel is sensed busy, the user waits until the channel becomes idle and transmits a packet with probability one. Note that in this protocol, a collision will always occur when more than one user has a packet to transmit.

p-Persistent CSMA. To reduce the rate of collisions in 1-persistent CSMA and increase the throughput, we should randomize the starting time for transmission of packets. In particular, upon sensing that the channel is idle, a user with a packet to transmit sends it with probability p and delays it by τ with probability $1-p$. The probability p is chosen in a way that reduces the probability of collisions while the idle periods between consecutive (non-overlapping) transmissions is kept small. This is accomplished by subdividing the time axis into minislots of duration τ and selecting the packet transmission at the beginning of a minislot. In summary, in the p-persistent protocol, a user with a packet to transmit proceeds as follows.

(a) If the channel is sensed idle, the packet is transmitted with probability p, and with probability $1-p$ the transmission is delayed by τ seconds.
(b) If at $t=\tau$, the channel is still sensed to be idle, step (a) is repeated. If a collision occurs, the users schedule retransmission of the packets according to some preselected transmission delay distribution.
(c) If at $t=\tau$, the channel is sensed busy, the user waits until it becomes idle, and the operates as in steps (a) and (b) above.

Slotted versions of the above protocol can also be constructed.

The throughput analysis for the nonpersistent and the p-persistent CSMA/CD protocols has been performed by Kleinroch and Tobagi (1975), based on the following assumptions:

1. The average retransmission delay is large compared with the packet duration T_p.
2. The interarrival times of the point process defined by the start times of all the packets plus retransmissions are independent and exponentially distributed.

For the nonpersistent CSMA, the throughput is

$$S = \frac{Ge^{-aG}}{G(1+2a)+e^{-aG}} \tag{15.4–11}$$

where the parameter $a=\tau_d/T_p$. Note that as $a\to 0$, $S\to G/(1+G)$. Figure 15.4–5 illustrates the throughput versus the offered traffic G, with a as a parameter. We observe that $S\to 1$ as $G\to\infty$ for $a=0$. For $a>0$, the value of $S_{\max}$ decreases.

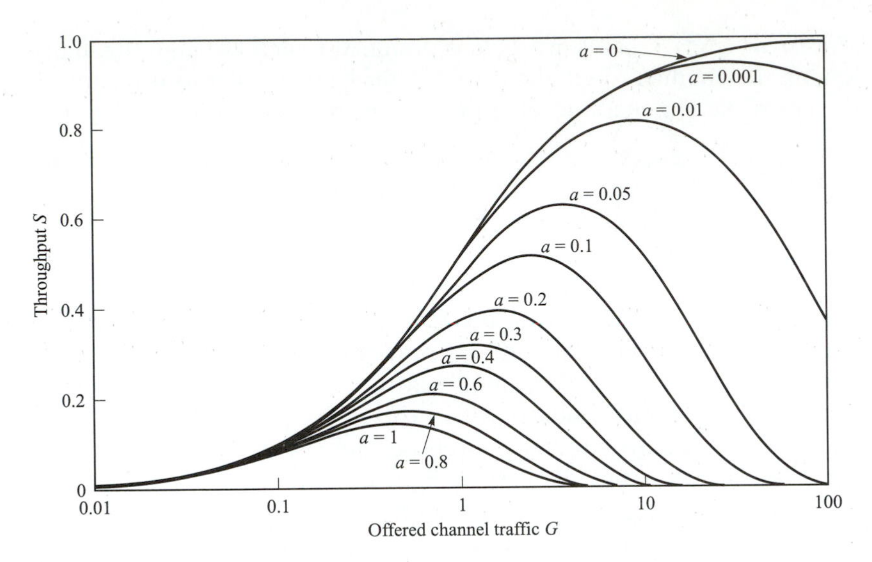

FIGURE 15.4–5
Throughput in nonpersistent CSMA. [*From Kleinrock and Tobagi* (*1975*), © *IEEE*.]

For the 1-persistent protocol, the throughput obtained by Kleinrock and Tobagi (1975) is

$$S = \frac{G[1 + G + aG(1 + G + \frac{1}{2}aG)]e^{-G(1+2a)}}{G(1 + 2a) - (1 - e^{-aG}) + (1 + aG)e^{-G(1+a)}} \tag{15.4–12}$$

In this case,

$$\lim_{a \to 0} S = \frac{G(1 + G)e^{-G}}{G + e^{-G}} \tag{15.4–13}$$

which has a smaller peak value than the nonpersistent protocol.

By adopting the p-persistent protocol, it is possible to increase the throughput relative to the 1-persistent scheme. For example, Figure 15.4–6 illustrates the throughput versus the offered traffic with $a = \tau_d/T_p$ fixed and with p as a parameter. We observe that as p increases toward unity, the maximum throughput decreases.

The transmission delay was also evaluated by Kleinrock and Tobagi (1975). Figure 15.4–7 illustrates the graphs of the delay (normalized by T_p) versus the throughput S for the slotted nonpersistent and p-persistent CSMA protocols. Also shown for comparison is the delay versus throughput characteristic of the ALOHA slotted and unslotted protocols. In this simulation, only the newly generated packets are derived independently from a Poisson distribution. Collisions and uniformly distributed random retransmissions are handled without further assumptions. These simulation results illustrate the superior performance of the p-persistent and the nonpersistent protocols relative to the ALOHA protocols. Note that the graph label "optimum p-persistent" is obtained by

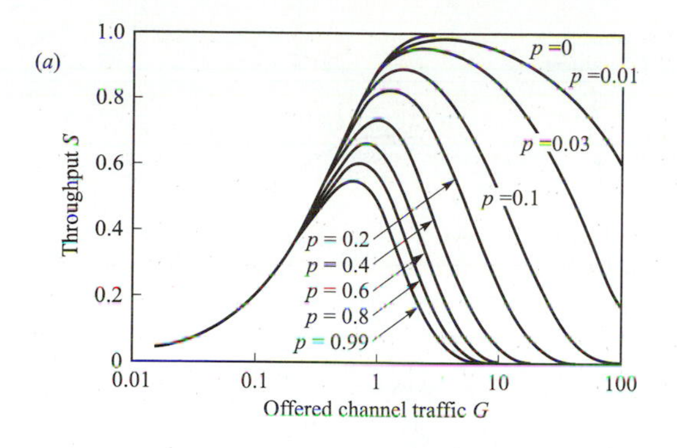

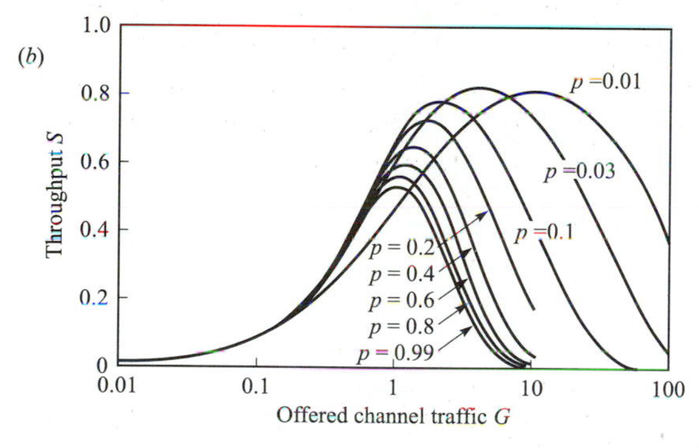

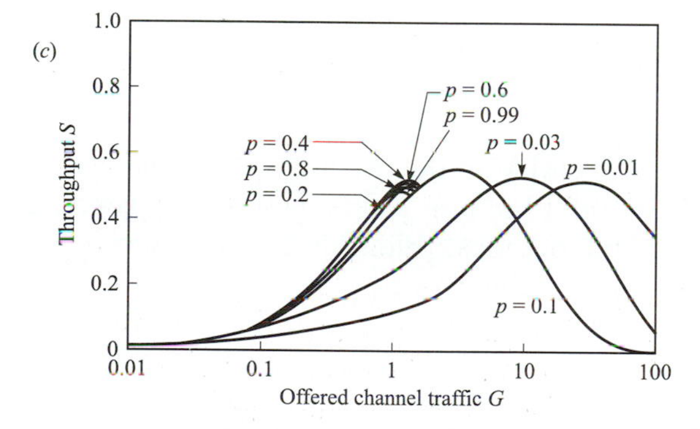

FIGURE 15.4–6
Channel throughput in p-persistent CSMA: (**a**) $a = 0$; (**b**) $a = 0.01$; (**c**) $a = 0.1$. [*From Kleinrock and Tobagi (1975), © IEEE.*]

finding the optimum value of p for each value of the throughput. We observe that for small values of the throughput, the 1-persistent ($p = 1$) protocol is optimal.

15.5 BIBLIOGRAPHICAL NOTES AND REFERENCES

FDMA was the dominant multiple access scheme that has been used for decades in telephone communication systems for analog voice transmission. With the advent of digital speech transmission using PCM, DPCM, and other speech

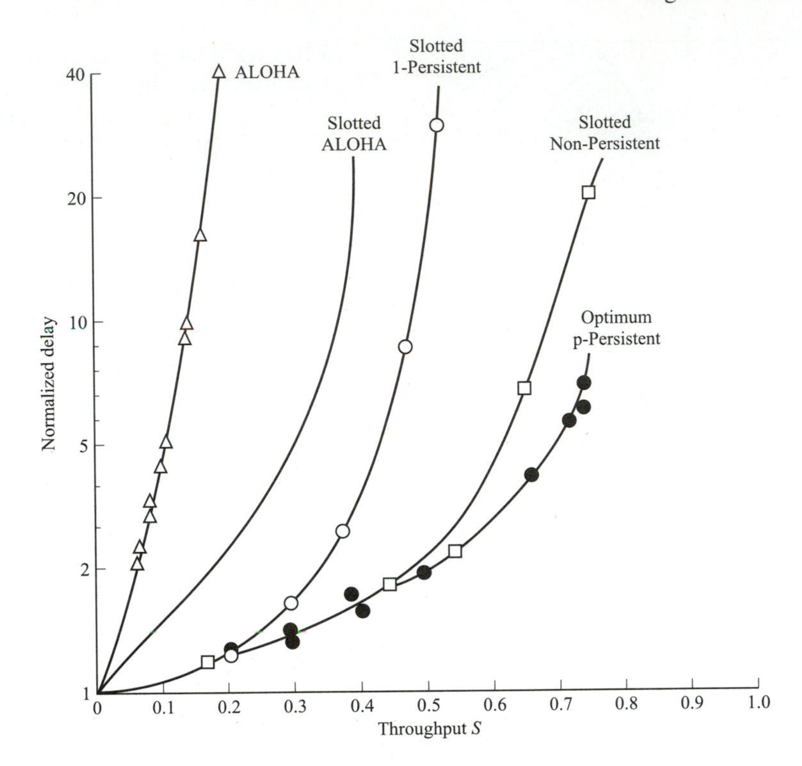

FIGURE 15.4–7
Throughput versus delay from simulation ($a = 0.01$). [*From Kleinrock and Tobagi (1975), © IEEE.*]

coding methods, TDMA has replaced FDMA as the dominant multiple access scheme in telecommunications. CDMA and random access methods, in general, have been developed over the past three decades, primarily for use in wireless signal transmission and in local area wireline networks.

Multiuser information theory deals with basic information-theoretic limits in source coding for multiple sources, and channel coding and modulation for multiple access channels. A large amount of literature exists on these topics. In the context of our treatment of multiple access methods, the reader will find the papers by Cover (1972), El Gamal and Cover (1980), Bergmans and Cover (1974), Hui (1984), Cover (1998), and the book by Cover and Thomas (1991) particularly relevant. The capacity of a cellular CDMA system has been considered in the paper by Gilhousen et al. (1991).

Signal demodulation and detection for multiuser communications has received considerable attention in recent years. The reader is referred to the papers by Verdu (1986a,b,c, 1989), Lupas and Verdu (1990), Xie et al. (1990a,b), Poor and Verdu (1988), Zhang and Brady (1993), Madhow and Honig (1994), Zvonar and Brady (1995), Viterbi (1990), Varanasi (1999), and the books by Verdu (1998), Viterbi (1995), and Garg et al. (1997). Earlier work on signal design and demodulation for multiuser communications is found in the papers by Van Etten (1975, 1976), Horwood and Gagliardi (1975), and Kaye and George (1970).

The ALOHA system, which was one of the earliest random access systems, is treated in the papers by Abramson (1970, 1977) and Roberts (1975). These papers contain the throughput analysis for unslotted and slotted systems. More recently, Abramson (1994), considers an ALOHA system that employs spread spectrum signals and provides a link to CDMA systems. Stability issues regarding the ALOHA protocols may be found in the papers by Carleial and Hellman (1975), Ghez et al. (1988), and Massey (1988). Stable protocols based on tree algorithms for random access channels were first given by Capetanakis (1979). The carrier sense multiple access protocols that we described are due to Kleinrock and Tobagi (1975). Finally, we mention the IEEE Press book edited by Abramson (1993), which contains a collection of papers dealing with multiple access communications.

PROBLEMS

15.1 In the formulation of the CDMA signal and channel models described in Section 15.3.1, we assumed that the received signals are real. For $K > 1$, this assumption implies phase synchronism at all transmitters, which is not very realistic in a practical system. To accommodate the case where the carrier phases are not synchronous, we may simply alter the signature waveforms for the K users, given by Equation 15.3–1, to be complex-valued, of the form

$$g_k(t) = e^{j\theta_k} \sum_{n=0}^{L-1} a_k(n) p(t - nT_c), \qquad 1 \leqslant k \leqslant K$$

where θ_k represents the constant phase offset of the kth transmitter as seen by the common receiver.

a) Given this complex-valued form for the signature waveforms, determine the form of the optimum ML receiver that computes the correlation metrics analogous to Equation 15.3–15.

b) Repeat the derivation for the optimum ML detector for asynchronous transmission that is analogous to Equation 15.3–19.

15.2 Consider a TDMA system where each user is limited to a transmitted power P, independent of the number of users. Determine the capacity per user, C_K, and the total capacity KC_K. Plot C_K and KC_K as functions of $\mathcal{E}_b/N_0$ and comment on the results as $K \to \infty$.

15.3 Consider an FDMA system with $K = 2$ users, in an AWGN channel, where user 1 is assigned a bandwidth $W_1 = \alpha W$ and user 2 is assigned a bandwidth $W_2 = (1 - \alpha)W$, where $0 \leqslant \alpha \leqslant 1$. Let P_1 and P_2 be the average powers of the two users.

a) Determine the capacities C_1 and C_2 of the two users and their sum $C = C_1 + C_2$ as a function of α. On a two-dimensional graph of the rates R_2 versus R_1, plot the graph of the points (C_2, C_1) as α varies in the range $0 \leqslant \alpha \leqslant 1$.

b) Recall that the rates of the two users must satisfy the conditions

$$R_1 < W_1 \log_2\left(1 + \frac{P_1}{W_1 N_0}\right)$$
$$R_2 < W_2 \log_2\left(1 + \frac{P_2}{W_2 N_0}\right)$$
$$R_1 + R_2 < W \log_2\left(1 + \frac{P_1 + P_2}{W N_0}\right)$$

Determine the total capacity C when $P_1/\alpha = P_2/(1-\alpha) = P_1 + P_2$, and, thus, show that the maximum rate is achieved when $\alpha/(1-\alpha) = P_1/P_2 = W_1/W_2$.

15.4 Consider a TDMA system with $K = 2$ users in an AWGN channel. Suppose that the two transmitters are peak-power-limited to P_1 and P_2, and let user 1 transmit for 100α percent of the available time and user-2 transmit $100(1-\alpha)$ percent of the time. The available bandwidth is W.

a) Determine the capacities C_1, C_2, and $C = C_1 + C_2$ as functions of α.

b) Plot the graph of the points (C_2, C_1) as α varies in the range $0 \leqslant \alpha \leqslant 1$.

15.5 Consider a TDMA system with $K = 2$ users in an AWGN channel. Suppose that the two transmitters are average-power-limited, with powers P_1 and P_2. User 1 transmits $100a$ per cent of the time and user 2 transmits $100(1-\alpha)$ percent of the time. The channel bandwidth is W.

a) Determine the capacities C_1, C_2, and $C = C_1 + C_2$ as functions of α.

b) Plot the graph of the points (C_2, C_1) as α varies in the range $0 \leqslant \alpha \leqslant 1$.

c) What is the similarity between this solution and the FDMA system in Problem 15.3?

15.6 Consider a two-user, synchronous CDMA transmission system, where the received signal is

$$r(t) = \sqrt{\mathcal{E}_1} b_1 g_1(t) + \sqrt{\mathcal{E}_2} b_2 g_2(t) + n(t), \qquad 0 \leqslant t \leqslant T$$

and $(b_1, b_2) = (\pm 1, \pm 1)$. The noise process $n(t)$ is zero-mean Gaussian and white, with spectral density $N_0/2$. The demodulator for $r(t)$ is shown in Figure P15.6.

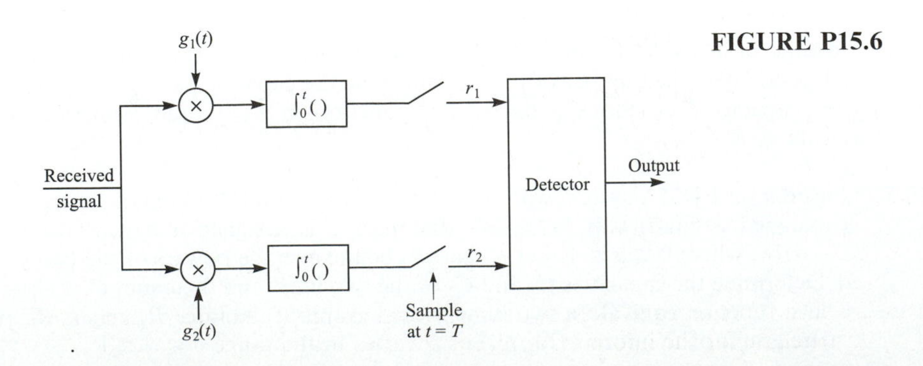

FIGURE P15.6

a) Show that the correlator outputs r_1 and r_2 at $t = T$ may be expressed as

$$r_1 = \sqrt{\mathcal{E}_1} b_1 + \sqrt{\mathcal{E}_2} b_2 \rho + n_1$$
$$r_2 = \sqrt{\mathcal{E}_1} b_1 \rho + \sqrt{\mathcal{E}_2} b_2 + n_2$$

b) Determine the variances of n_1 and n_2 and the covariance of n_1 and n_2.
c) Determine the joint PDF $p(r_1, r_2 | b_1, b_2)$.

15.7 Consider the two-user, synchronous CDMA transmission system described in Problem 15.6. The conventional single-user detector for the information bits b_1 and b_2 gives the outputs

$$b_1 = \operatorname{sgn}(r_1)$$
$$b_2 = \operatorname{sgn}(r_2)$$

Assuming that $P(b_1 = 1) = P(b_2 = 1) = \frac{1}{2}$, and b_1 and b_2 are statistically independent, determine the probability of error for this detector.

15.8 Consider the two-user, synchronous CDMA transmission system described in Problem 15.6. $P(b_1 = 1) = P(b_2 = 1) = \frac{1}{2}$ and $P(b_1, b_2) = P(b_1)P(b_2)$. The *jointly optimum detector* makes decisions based on the maximum a posteriori probability (MAP) criterion. That is, the detector computes

$$\max_{b_1, b_2} P[b_1, b_2 | r(t), 0 \leqslant t \leqslant T]$$

a) For the equally likely information bits (b_1, b_2) show that the MAP criterion is equivalent to the maximum-likelihood (ML) criterion

$$\max_{b_1, b_2} p[r(t), 0 \leqslant t \leqslant T | b_1, b_2]$$

b) Show that the ML criterion in (a) leads to the jointly optimum detector that makes decisions on b_1 and b_2 according to the following rule:

$$\max_{b_1, b_2} \left(\sqrt{\mathcal{E}_1} b_1 r_1 + \sqrt{\mathcal{E}_2} b_2 r_2 - \sqrt{\mathcal{E}_1 \mathcal{E}_2} \rho b_1 b_2 \right)$$

15.9 Consider the two-user, synchronous CDMA transmission system described in Problem 15.6. $P(b_1 = 1) = P(b_2 = 1) = \frac{1}{2}$ and $P(b_1, b_2) = P(b_1)P(b_2)$. The individually optimum detector makes decisions based on the MAP criterion. That is, the detector computes the a posteriori probabilities.

$$P[b_1 | r(t), 0 \leqslant t \leqslant T] = P[b_1, b_2 = 1 | r(t), 0 \leqslant t \leqslant T] + P[b_1, b_2 = -1 | r(t), 0 \leqslant t \leqslant T]$$

and

$$P[b_2 | r(t), 0 \leqslant t \leqslant T] = P[b_1 = 1, b_2 | r(t), 0 \leqslant t \leqslant T] + P[b_1 = -1, b_2 | r(t), 0 \leqslant t \leqslant T]$$

a) Show that an equivalent test statistic for this individually optimum MAP detector for the information bit b_1 is

$$\max_{b_1}\left\{\frac{\sqrt{\mathcal{E}_1}r_1}{N_0}b_1+\ln\cosh\left(\frac{\sqrt{\mathcal{E}_2}r_2-\sqrt{\mathcal{E}_1\mathcal{E}_2}\rho b_1}{N_0}\right)\right\}$$

b) By substituting $b_1=1$ and $b_1=-1$ into the expression in (a), show that the test statistic in (a) is equivalent to selecting b_1 according to the relation

$$\hat{b}_1=\text{sgn}\left[r_1-\frac{N_0}{2\sqrt{\mathcal{E}_1}}\ln\frac{\cosh(\sqrt{\mathcal{E}_2}r_2+\sqrt{\mathcal{E}_1\mathcal{E}_2}\rho)/N_0}{\cosh(\sqrt{\mathcal{E}_2}r_2-\sqrt{\mathcal{E}_1\mathcal{E}_2}\rho)/N_0}\right]$$

15.10 Show that the asymptotic efficiency of the conventional single-user detector in a CDMA system with K users transmitting synchronously is

$$\eta_k=\left[\max\left\{0,1-\sum_{j\neq k}\sqrt{\frac{\mathcal{E}_j}{\mathcal{E}_k}}|\rho_{jk}(0)|\right\}\right]^2$$

15.11 Consider the jointly optimum detector defined in Problem 15.8 for the two-user, synchronous CDMA system. Show that the (symbol) error probability for this detector may be upper bounded as

$$Pe<Q\left(\sqrt{\frac{2\mathcal{E}_1}{N_0}}\right)+\frac{1}{2}Q\left(\sqrt{\frac{\mathcal{E}_1+\mathcal{E}_2-2\sqrt{\mathcal{E}_1\mathcal{E}_2}|\rho|}{N_0/2}}\right)$$

15.12 Consider the jointly optimum detector defined in Problem 15.8 for the two-user, synchronous CDMA system.

a) Show that the asymptotic efficiency for this detector for user 1

$$\eta_1=\min\left\{1,1+\frac{\mathcal{E}_2}{\mathcal{E}_1}-2\sqrt{\frac{\mathcal{E}_2}{\mathcal{E}_1}}|\rho|\right\}$$

b) Plot and compare the asymptotic efficiencies of the jointly optimum detector and the conventional single-user detector for $\rho=0.1$ and $\rho=0.2$.

15.13 Consider the two-user synchronous CDMA system in Problem 5.6. Determine the probability of error for each user that employs a decorrelating detector when $\mathcal{E}_1\neq\mathcal{E}_2$.

15.14 Consider a two-user synchronous CDMA system where the received signal is given in Problem 5.6. Each user employs the minimum MSE detector specified by Equations 15.3–51 to 15.3–53.

a) Determine the linear transformation matrix $\mathbf{A}^0$ for the two users.

b) Show that the MMSE detector approaches the decorrelating detector as $N_0\to 0$.

c) Show that the MMSE detector approaches the conventional single-user detector as $N_0\to\infty$.

15.15 Consider the asynchronous communication system shown in Figure P15.15. The two receivers are not colocated, and the white noise processes $n^{(1)}(t)$ and $n^{(2)}(t)$ may be considered to be independent. The noise processes are identically distributed, with power spectral density σ^2 and zero-mean. Since the receivers are not colo-

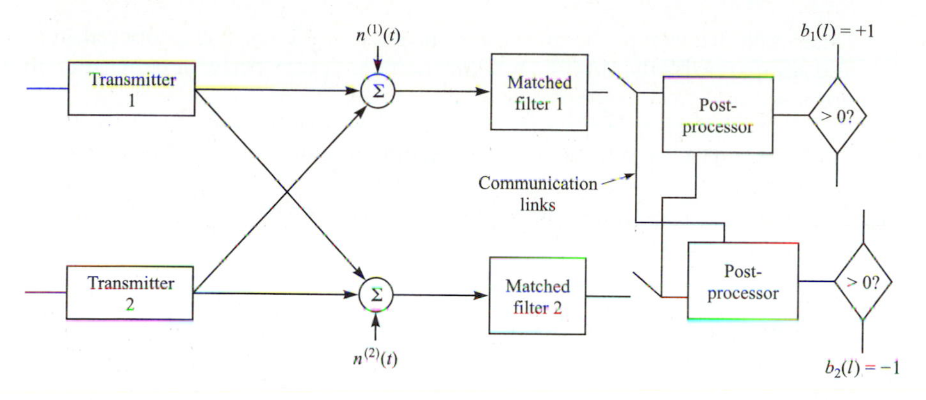

FIGURE P15.15

cated, the relative delays between the users are not the same—denote the relative delay of user k at receiver i by $\tau_k^{(i)}$. All other signal parameters coincide for the receivers, and the received signal at receiver i is

$$r^{(i)}(t) = \sum_{k=1}^{2} \sum_{l=-\infty}^{\infty} b_k(l) s_k(t - lT - \tau_k^{(i)}) + n^{(i)}(t)$$

where s_k has support on $[0, T]$. You may assume that the receiver i has full knowledge of the waveforms, energies, and relative delays $\tau_1^{(i)}$ and $\tau_2^{(i)}$. Although receiver i is eventually interested only in the data from transmitter i, note that there is a free communication link between the sampler of one receiver, and the postprocessing circuitry of the other. Following each postprocessor, the decision is attained by threshold detection. In this problem, you will consider options for postprocessing and for the communication link in order to improve performance.

a) What is the bit error probability for users 1 and 2 of a receiver pair that does not utilize the communication link and does not perform postprocessing? Use the following notation:

$$y_k(l) = \int s_k(t - lT - \tau_k^{(k)}) r^{(k)}(t)\, dt$$

$$\rho_{12}^{(i)} = \int s_1(t - \tau_1^{(i)}) s_2(t - \tau_2^{(i)})\, dt$$

$$\rho_{21}^{(i)} = \int s_1(t - \tau_1^{(i)}) s_2(t + T - \tau_2^{(i)})\, dt$$

$$w_k = \int s_k^2(t - \tau_k^{(1)})\, dt = \int s_k^2(t - \tau_k^{(2)})\, dt$$

b) Consider a postprocessor for receiver 1 that accepts $y_2(l-1)$ and $y_2(l)$ from the communication link and implements the following postprocessing on $y_1(l)$

$$z_l(l) = y_1(l) - \rho_{21}^{(1)} \mathrm{sgn}[y_2(l-1)] - \rho_{12}^{(1)} \mathrm{sgn}[y_2(l)].$$

Determine an exact expression for the bit error rate for user 1.

c) Determine the asymptotic multiuser efficiency of the receiver proposed in (b), and compare with that in (a). Does this receiver always perform better than that proposed in (a)?

15.16 In a pure ALOHA system, the channel bit rate is 2400 bits/s. Suppose that each terminal transmits a 100-bit message every minute on the average.
a) Determine the maximum number of terminals that can use the channel.
b) Repeat (a) if slotted ALOHA is used.

15.17 An alternative derivation for the throughput in a pure ALOHA system may be obtained from the relation $G = S + A$, where A is the average (normalized) rate of retransmissions. Show that $A = G(1 - e^{-2G})$ and then solve for S.

15.18 For a Poisson process, the probability of k arrivals in a time interval T is

$$P(k) = \frac{e^{-\lambda T}(\lambda T)^k}{k!}, \qquad k = 0, 1, 2, \ldots$$

a) Determine the average number of arrivals in the interval T.
b) Determine the variance σ^2 in the number of arrivals in the interval T.
c) What is the probability of at least one arrival in the interval T?
d) What is the probability of exactly one arrival in the interval T?

15.19 Refer to Problem 15.18. The average arrival rate is $\lambda = 10$ packets/s. Determine
a) The average time between arrivals.
b) The probability that another packet will arrive within 1 s; within 100 ms.

15.20 Consider a pure ALOHA system that is operating with a throughput $S = 0.1$ and packets are generated with a Poisson arrival rate λ. Determine
a) The value of G.
b) The average number of attempted transmissions to send a packet.

15.21 Consider a CSMA/CD system in which the transmission rate on the bus is 10 Mbits/s. The bus is 2 km and the propagation delay is $5\,\mu$s/km. Packets are 1000 bits long. Determine
a) The end-to-end delay τ_d.
b) The packet duration T_p.
c) The ratio τ_d/T_p.
d) The maximum utilization of the bus and the maximum bit rate.

APPENDIX A

The Levinson–Durbin Algorithm

The Levinson–Durbin algorithm is an order recursive method for determining the solution to the set of linear equations

$$\mathbf{\Phi}_p \mathbf{a}_p = \boldsymbol{\phi}_p \tag{A–1}$$

where $\mathbf{\Phi}_p$ is a $p \times p$ Toeplitz matrix, $\mathbf{a}_p$ is the vector of predictor coefficients expressed as

$$\mathbf{a}'_p = [a_{p1} \;\; a_{p2} \;\; \cdots \;\; a_{pp}]$$

and $\boldsymbol{\phi}_p$ is a p-dimensional vector with elements

$$\boldsymbol{\phi}'_p = [\phi(1) \;\; \phi(2) \;\; \cdots \;\; \phi(p)]$$

For a first-order ($p = 1$) predictor, we have the solution

$$\begin{aligned} \phi(0)a_{11} &= \phi(1) \\ a_{11} &= \phi(1)/\phi(0) \end{aligned} \tag{A–2}$$

The residual mean square error (MSE) for the first-order predictor is

$$\begin{aligned} \mathcal{E}_1 &= \phi(0) - a_{11}\phi(1) \\ &= \phi(0) - a_{11}^2\phi(0) \\ &= \phi(0)(1 - a_{11}^2) \end{aligned} \tag{A–3}$$

In general, we may express the solution for the coefficients of an mth-order predictor in terms of the coefficients of the $(m-1)$th-order predictor. Thus, we express $\mathbf{a}_m$ as the sum of two vectors, namely,

$$\mathbf{a}_m = \begin{bmatrix} a_{m1} \\ a_{m2} \\ \vdots \\ a_{mm} \end{bmatrix} = \begin{bmatrix} \mathbf{a}_{m-1} \\ \hline 0 \end{bmatrix} + \begin{bmatrix} \mathbf{d}_{m-1} \\ \hline k_m \end{bmatrix} \tag{A–4}$$

where the vector $\mathbf{d}_{m-1}$ and the scalar k_m are to be determined. Also, $\mathbf{\Phi}_m$ may be expressed as

$$\mathbf{\Phi}_m = \left[\begin{array}{c|c} \mathbf{\Phi}_{m-1} & \boldsymbol{\phi}^r_{m-1} \\ \hline \boldsymbol{\phi}^{r'}_{m-1} & \phi(0) \end{array}\right] \tag{A–5}$$

where $\boldsymbol{\phi}^r_{m-1}$ is just the vector $\boldsymbol{\phi}_{m-1}$ in reverse order.

Now

$$\left[\begin{array}{c|c} \mathbf{\Phi}_{m-1} & \boldsymbol{\phi}^r_{m-1} \\ \hline \boldsymbol{\phi}^{r'}_{m-1} & \phi(0) \end{array}\right] \left(\begin{bmatrix} \mathbf{a}_{m-1} \\ \hline 0 \end{bmatrix} + \begin{bmatrix} \mathbf{d}_{m-1} \\ \hline k_m \end{bmatrix} \right) = \begin{bmatrix} \boldsymbol{\phi}_{m-1} \\ \hline \phi(m) \end{bmatrix} \tag{A–6}$$

From Equation A–6, we obtained two equations. The first is the matrix equation

$$\mathbf{\Phi}_{m-1}\mathbf{a}_{m-1} + \mathbf{\Phi}_{m-1}\mathbf{d}_{m-1} + k_m\boldsymbol{\phi}^r_{m-1} = \boldsymbol{\phi}_{m-1} \tag{A–7}$$

But $\mathbf{\Phi}_{m-1}\mathbf{a}_{m-1} = \boldsymbol{\phi}_{m-1}$. Hence, Equation A–7 simplifies to

$$\mathbf{\Phi}_{m-1}\mathbf{d}_{m-1} + k_m\boldsymbol{\phi}^r_{m-1} = \mathbf{0} \tag{A–8}$$

This equation has the solution

$$\mathbf{d}_{m-1} = -k_m\mathbf{\Phi}^{-1}_{m-1}\boldsymbol{\phi}^r_{m-1} \tag{A–9}$$

But $\boldsymbol{\phi}^r_{m-1}$ is just $\boldsymbol{\phi}_{m-1}$ in reverse order. Hence, the solution in Equation A–9 is simply $\mathbf{a}_{m-1}$ in reverse order multiplied by $-k_m$. That is,

$$\mathbf{d}_{m-1} = -k_m \begin{bmatrix} a_{m-1m-1} \\ a_{m-1m-2} \\ \vdots \\ a_{m-1\ 1} \end{bmatrix} \tag{A–10}$$

The second equation obtained from Equation A–6 is the scalar equation

$$\boldsymbol{\phi}^{r'}_{m-1}\mathbf{a}_{m-1} + \boldsymbol{\phi}^{r'}_{m-1}\mathbf{d}_{m-1} + \phi(0)k_m = \phi(m) \tag{A–11}$$

We eliminate $\mathbf{d}_{m-1}$ from Equation A–11 by use of Equation A–10. The resulting equation gives us k_m. That is,

$$
\begin{aligned}
k_m &= \frac{\phi(m) - \boldsymbol{\phi}_{m-1}^{r'} \mathbf{a}_{m-1}}{\phi(0) - \boldsymbol{\phi}_{m-1}^{r'} \boldsymbol{\Phi}_{m-1}^{-1} \boldsymbol{\phi}_{m-1}^{r}} \\
&= \frac{\phi(m) - \boldsymbol{\phi}_{m-1}^{r'} \mathbf{a}_{m-1}}{\phi(0) - \mathbf{a}_{m-1}' \boldsymbol{\phi}_{m-1}} \\
&= \frac{\phi(m) - \boldsymbol{\phi}_{m-1}^{r'} \mathbf{a}_{m-1}}{\mathcal{E}_{m-1}}
\end{aligned}
\tag{A–12}
$$

where $\mathcal{E}_{m-1}$ is the residual MSE given as

$$
\mathcal{E}_{m-1} = \phi(0) - \mathbf{a}_{m-1}' \boldsymbol{\phi}_{m-1} \tag{A–13}
$$

By substituting Equation A–10 for $\mathbf{d}_{m-1}$ in Equation A–4, we obtain the order-recursive relation

$$
a_{mk} = a_{m-1k} - k_m a_{m-1m-k}, \qquad k = 1, 2, \ldots, m-1, \qquad m = 1, 2, \ldots, p \tag{A–14}
$$

and

$$
a_{mm} = k_m
$$

The minimum MSE may also be computed recursively. We have

$$
\mathcal{E}_m = \phi(0) - \sum_{k=1}^{m} a_{mk} \phi(k) \tag{A–15}
$$

Using Equation A–14 in Equation A–15, we obtain

$$
\mathcal{E}_m = \phi(0) - \sum_{k=1}^{m-1} a_{m-1k} \phi(k) - a_{mm} \left[\phi(m) - \sum_{k=1}^{m-1} a_{m-1m-k} \phi(k) \right] \tag{A–16}
$$

But the term in square brackets in Equation A–16 is just the numerator of k_m in Equation A–12. Hence,

$$
\begin{aligned}
\mathcal{E}_m &= \mathcal{E}_{m-1} - a_{mm}^2 \mathcal{E}_{m-1} \\
&= \mathcal{E}_{m-1}(1 - a_{mm}^2)
\end{aligned}
\tag{A–17}
$$

APPENDIX B

Error Probability for Multichannel Binary Signals

In multichannel communication systems that employ binary signaling for transmitting information over the AWGN channel, the decision variable at the detector can be expressed as a special case of the general quadratic form

$$D = \sum_{k=1}^{L} (A|X_k|^2 + B|Y_k|^2 + CX_kY_k^* + C^*X_k^*Y_k) \tag{B–1}$$

in complex-valued Gaussian random variables. A, B, and C are constants; X_k and Y_k are a pair of correlated complex-valued Gaussian random variables. For the channels considered, the L pairs $\{X_k, Y_k\}$ are mutually statistically independent and identically distributed.

The probability of error is the probability that $D < 0$. This probability is evaluated below.

The computation begins with the characteristic function, denoted by $\psi_D(jv)$, of the general quadratic form. The probability that $D < 0$, denoted here as the probability of error P_b, is

$$P_b = P(D < 0) = \int_{-\infty}^{0} p(D)\, dD \tag{B–2}$$

where $p(D)$, the probability density function of D, is related to $\psi_D(jv)$ by the Fourier transform, i.e.,

$$p(D) = \frac{1}{2\pi} \int_{-\infty}^{\infty} \psi_D(jv) e^{-jvD}\, dv$$

Hence,

$$P_b = \int_{-\infty}^{0} dD \frac{1}{2\pi} \int_{-\infty}^{\infty} \psi_D(jv) e^{-jvD}\, dv \tag{B–3}$$

Let us interchange the order of integration and carry out first the integration with respect to D. The result is

$$P_b = -\frac{1}{2\pi j}\int_{-\infty+j\varepsilon}^{\infty+j\varepsilon} \frac{\psi_D(jv)}{v}\, dv \tag{B–4}$$

where a small positive number ε has been inserted in order to move the path of integration away from the singularity at $v = 0$ and which must be positive in order to allow for the interchange in the order of integration.

Since D is the sum of statistically independent random variables, the characteristic function of D factors into a product of L characteristic functions, with each function corresponding to the individual random variables d_k, where

$$d_k = A|X_k|^2 + B|Y_k|^2 + CX_k Y_k^* + C^* X_k^* Y_k$$

The characteristic function of d_k is

$$\psi_{d_k}(jv) = \frac{v_1 v_2}{(v + jv_1)(v - jv_2)} \exp\left[\frac{v_1 v_2(-v^2\alpha_{1k} + jv\alpha_{2k})}{(v + jv_1)(v - jv_2)}\right] \tag{B–5}$$

where the parameters v_1, v_2, α_{1k}, and α_{2k} depend on the means $\bar{X}_k$ and $\bar{Y}_k$ and the second (central) moments μ_{xx}, μ_{yy}, and μ_{xy} of the complex-valued Gaussian variables X_k and Y_k through the following definitions ($|C|^2 - AB > 0$) :

$$\begin{aligned}
v_1 &= \sqrt{w^2 + \frac{1}{4(\mu_{xx}\mu_{yy} - |\mu_{xy}|^2)(|C|^2 - AB)}} - w \\
v_2 &= \sqrt{w^2 + \frac{1}{4(\mu_{xx}\mu_{yy} - |\mu_{xy}|^2)(|C|^2 - AB)}} + w \\
w &= \frac{A\mu_{xx} + B\mu_{yy} + C\mu_{xy}^* + C^*\mu_{xy}}{4(\mu_{xx}\mu_{yy} - |\mu_{xy}|^2)(|C|^2 - AB)} \\
\alpha_{1k} &= 2(|C|^2 - AB)(|\bar{X}_k|^2\mu_{yy} + |\bar{Y}_k|^2\mu_{xx} - \bar{X}_k^*\bar{Y}_k\mu_{xy} - \bar{X}_k\bar{Y}_k^*\mu_{xy}^*) \\
\alpha_{2k} &= A|\bar{X}_k|^2 + B|\bar{Y}_k|^2 + C\bar{X}_k^*\bar{Y}_k + C^*\bar{X}_k\bar{Y}_k^* \\
\mu_{xy} &= \tfrac{1}{2}E[(X_k - \bar{X}_k)(Y_k - \bar{Y}_k)^*]
\end{aligned} \tag{B–6}$$

Now, as a result of the independence of the random variables d_k, the characteristic function of D is

$$\begin{aligned}
\psi_D(jv) &= \prod_{k=1}^{L} \psi_{d_k}(jv) \\
\psi_D(jv) &= \frac{(v_1 v_2)^L}{(v + jv_1)^L (v - jv_2)^L} \exp\left[\frac{v_1 v_2(jv\alpha_2 - v^2\alpha_1)}{(v + jv_1)(v - jv_1)}\right]
\end{aligned} \tag{B–7}$$

where

$$\alpha_1 = \sum_{k=1}^{L} \alpha_{1k}, \qquad \alpha_2 = \sum_{k=1}^{L} \alpha_{2k} \tag{B–8}$$

The result B–7 is substituted for $\psi_D(jv)$ in Equation B–4, and we obtain

$$P_b = -\frac{(v_1 v_2)^L}{2\pi j} \int_{-\infty+j\varepsilon}^{\infty+j\varepsilon} \frac{dv}{v(v+jv_1)^L (v-jv_2)^L} \exp\left[\frac{v_1 v_2 (jv\alpha_2 - v^2 \alpha_1)}{(v+jv_1)(v-jv_2)}\right] \quad \text{(B–9)}$$

This integral is evaluated as follows.

The first step is to express the exponential function in the form

$$\exp\left(-A_1 + \frac{jA_2}{v+jv_1} - \frac{jA_3}{v-jv_2}\right)$$

where one can easily verify that the constants A_1, A_2, and A_3 are given as

$$\begin{aligned} A_1 &= \alpha_1 v_1 v_2 \\ A_2 &= \frac{v_1^2 v_2}{v_1 + v_2}(\alpha_1 v_1 + \alpha_2) \\ A_3 &= \frac{v_1 v_2^2}{v_1 + v_2}(\alpha_1 v_2 - \alpha_2) \end{aligned} \quad \text{(B–10)}$$

Second, a conformal transformation is made from the v plane onto the p plane via the change in variable

$$p = -\frac{v_1}{v_2}\frac{v - jv_2}{v + jv_1} \quad \text{(B–11)}$$

In the p plane, the integral given by Equation B–9 becomes

$$P_b = \frac{\exp[v_1 v_2(-2\alpha_1 v_1 v_2 + \alpha_2 v_1 - \alpha_2 v_2)/(v_1+v_2)^2]}{(1+v_2/v_1)^{2L-1}} \frac{1}{2\pi j}\int_\Gamma f(p)\,dp \quad \text{(B–12)}$$

where

$$f(p) = \frac{[1+(v_2/v_1)p]^{2L-1}}{p^L(1-p)} \exp\left[\frac{A_2(v_2/v_1)}{v_1+v_2}p + \frac{A_3(v_1/v_2)}{v_1+v_2}\frac{1}{p}\right] \quad \text{(B–13)}$$

and Γ is a circular contour of radius less than unity that encloses the origin.

The third step is to evaluate the integral

$$\begin{aligned} \frac{1}{2\pi j}\int_\Gamma f(p)\,dp &= \frac{1}{2\pi j}\int_\Gamma \frac{[1+(v_2/v_1)p]^{2L-1}}{p^L(1-p)} \\ &\quad \times \exp\left[\frac{A_2(v_2/v_1)}{v_1+v_2}p + \frac{A_3(v_1/v_2)}{v_1+v_2}\frac{1}{p}\right]dp \end{aligned} \quad \text{(B–14)}$$

In order to facilitate subsequent manipulations, the constants $a \geqslant 0$ and $b \geqslant 0$ are introduced and defined as follows:

$$\tfrac{1}{2}a^2 = \frac{A_3(v_1/v_2)}{v_1+v_2}, \qquad \tfrac{1}{2}b^2 = \frac{A_2(v_2/v_1)}{v_1+v_2} \quad \text{(B–15)}$$

Let us also expand the function $[1+(v_2/v_1)p]^{2L-1}$ as a binomial series. As a result, we obtain

$$\frac{1}{2\pi j}\int_{\Gamma} f(p)\,dp = \sum_{k=0}^{2L-1}\binom{2L-1}{k}\left(\frac{v_2}{v_1}\right)^k \times \frac{1}{2\pi j}\int_{\Gamma}\frac{p^k}{p^L(1-p)}\exp\left(\frac{\frac{1}{2}a^2}{p}+\frac{1}{2}b^2 p\right)dp \tag{B–16}$$

The contour integral given in Equation B–16 is one representation of the Bessel function. It can be solved by making use of the relations

$$I_n(ab) = \begin{cases} \dfrac{1}{2\pi j}\left(\dfrac{a}{b}\right)^n \displaystyle\int_{\Gamma}\frac{1}{p^{n+1}}\exp\left(\frac{\frac{1}{2}a^2}{p}+\frac{1}{2}b^2 p\right)dp \\ \dfrac{1}{2\pi j}\left(\dfrac{b}{a}\right)^n \displaystyle\int_{\Gamma} p^{n-1}\exp\left(\frac{\frac{1}{2}a^2}{p}+\frac{1}{2}b^2 p\right)dp \end{cases}$$

where $I_n(x)$ is the nth-order modified Bessel function of the first kind and the series representation of Marcum's Q function in terms of Bessel functions, i.e.,

$$Q_1(a,b) = \exp[-\tfrac{1}{2}(a^2+b^2)] + \sum_{n=0}^{\infty}\left(\frac{a}{b}\right)^n I_n(ab)$$

First, consider the case $0 \leqslant k \leqslant L-2$ in Equation B–16. In this case, the resulting contour integral can be written in the form†

$$\frac{1}{2\pi j}\int_{\Gamma}\frac{1}{p^{L-K}(1-p)}\exp\left(\frac{\frac{1}{2}a^2}{p}+\frac{1}{2}b^2 p\right)dp = Q_1(a,b)\exp[\tfrac{1}{2}(a^2+b^2)] + \sum_{n=1}^{L-1-k}\left(\frac{b}{a}\right)^n I_n(ab) \tag{B–17}$$

Next, consider the term $k = L-1$. The resulting contour integral can be expressed in terms of the Q function as follows:

$$\frac{1}{2\pi j}\int_{\Gamma}\frac{1}{p(1-p)}\exp\left(\frac{\frac{1}{2}a^2}{p}+\frac{1}{2}b^2 p\right)dp = Q_1(a,b)\exp[\tfrac{1}{2}(a^2+b^2)] \tag{B–18}$$

†This contour integral is related to the generalized Marcum Q function, defined as

$$Q_m(a,b) = \int_b^{\infty} x(x/a)^{m-1}\exp[-\tfrac{1}{2}(x^2+a^2)]I_{m-1}(ax)\,dx, \qquad m \geqslant 1$$

in the following manner:

$$Q_m(a,b)\exp[\tfrac{1}{2}(a^2+b^2)] = \frac{1}{2\pi j}\int_{\Gamma}\frac{1}{p^m(1-p)}\exp\left(\frac{\frac{1}{2}a^2}{p}+\frac{1}{2}b^2 p\right)dp$$

Finally, consider the case $L \leqslant k \leqslant 2L-1$. We have

$$\begin{aligned}
\frac{1}{2\pi j}\int_\Gamma \frac{p^{k-L}}{1-p}\exp\left(\frac{\frac{1}{2}a^2}{p}+\tfrac{1}{2}b^2 p\right)dp \\
&= \sum_{n=0}^{\infty}\frac{1}{2\pi j}\int_\Gamma p^{k-L+n}\exp\left(\frac{\frac{1}{2}a^2}{p}+\tfrac{1}{2}b^2 p\right)dp \qquad \text{(B–19)}\\
&= \sum_{n=k+1-L}^{\infty}\left(\frac{a}{b}\right)^n I_n(ab) = Q_1(a,b)\exp[\tfrac{1}{2}(a^2+b^2)] - \sum_{n=0}^{k-L}\left(\frac{a}{b}\right)^n I_n(ab)
\end{aligned}$$

Collecting the terms that are indicated on the right-hand side of Equation B–16 and using the results given in Equations B–17 to B–19, the following expression for the contour integral is obtained after some algebra:

$$\begin{aligned}
\frac{1}{2\pi j}\int_\Gamma f(p)\,dp &= \left(1+\frac{v_2}{v_1}\right)^{2L-1}\{\exp[\tfrac{1}{2}(a^2+b^2)]Q_1(a,b) - I_0(ab)\}\\
&\quad + I_0(ab)\sum_{k=0}^{L-1}\binom{2L-1}{k}\left(\frac{v_2}{v_1}\right)^k\\
&\quad + \sum_{n=1}^{L-1} I_n(ab)\sum_{k=0}^{L-1-n}\binom{2L-1}{k}\left[\left(\frac{b}{a}\right)^n\left(\frac{v_2}{v_1}\right)^k - \left(\frac{a}{b}\right)^n\left(\frac{v_2}{v_1}\right)^{2L-1-k}\right] \qquad \text{(B–20)}
\end{aligned}$$

Equation B–20 in conjunction with Equation B–12 gives the result for the probability of error. A further simplification results when one uses the following identity, which can easily be proved:

$$\exp\left[\frac{v_1 v_2}{(v_1+v_2)^2}(-2\alpha_1 v_1 v_2 + \alpha_2 v_1 - \alpha_2 v_2)\right] = \exp[-\tfrac{1}{2}(a^2+b^2)]$$

Therefore, it follows that

$$\begin{aligned}
P_b &= Q_1(a,b) - I_0(ab)\exp[-\tfrac{1}{2}(a^2+b^2)]\\
&\quad + \frac{I_0(ab)\exp[-\frac{1}{2}(a^2+b^2)]}{(1+v_2/v_1)^{2L-1}}\sum_{k=0}^{L-1}\binom{2L-1}{k}\left(\frac{v_2}{v_1}\right)^k + \frac{\exp[-\frac{1}{2}(a^2+b^2)]}{(1+v_2/v_1)^{2L-1}}\\
&\quad \times \sum_{n=1}^{L-1} I_n(ab)\sum_{k=0}^{L-1-n}\binom{2L-1}{k}\\
&\quad \times \left[\left(\frac{b}{a}\right)^n\left(\frac{v_2}{v_1}\right)^k - \left(\frac{a}{b}\right)^n\left(\frac{v_2}{v_1}\right)^{2L-1-k}\right], \qquad L>1\\
P_b &= Q_1(a,b) - \frac{v_2/v_1}{1+v_2/v_1}I_0(ab)\exp[-\tfrac{1}{2}(a^2+b^2)], \qquad L=1 \qquad \text{(B–21)}
\end{aligned}$$

This is the desired expression for the probability of error. It is now a simple matter to relate the parameters a and b to the moments of the pairs $\{X_k, Y_k\}$. Substituting for A_2 and A_3 from Equation B–10 into Equation B–15, we obtain

$$a = \left[\frac{2v_1^2 v_2(\alpha_1 v_2 - \alpha_2)}{(v_1 + v_2)^2}\right]^{1/2}$$
$$b = \left[\frac{2v_1 v_2^2(\alpha_1 v_1 + \alpha_2)}{(v_1 + v_2)^2}\right]^{1/2} \tag{B–22}$$

Since v_1, v_2, α_1, and α_2 have been given in Equations B–6 and B–8 directly in terms of the moments of the pairs X_k and Y_k, our task is completed.

APPENDIX C

Error Probabilities for Adaptive Reception of M-Phase Signals

In this appendix, we derive probabilities of error for two- and four-phase signaling over an L-diversity-branch time-invariant Gaussian noise channel and for M-phase signaling over an L-diversity-branch Rayleigh fading additive Gaussian noise channel. Both channels corrupt the signaling waveforms transmitted through them by introducing additive white Gaussian noise and an unknown or random multiplicative gain and phase shift in the transmitted signal. The receiver processing consists of cross-correlating the signal plus noise received over each diversity branch by a noisy reference signal, which is derived either from the previously received information-bearing signals or from the transmission and reception of a pilot signal, and adding the outputs from all L-diversity branches to form the decision variable.

C.1

MATHEMATICAL MODEL FOR AN M-PHASE SIGNALING COMMUNICATION SYSTEM

In the general case of M-phase signaling, the signaling waveforms at the transmitter are†

$$s_n(t) = \mathrm{Re}[s_{ln}(t)e^{j2\pi f_c t}]$$

where

$$s_{ln}(t) = g(t)\exp\left[j\frac{2\pi}{M}(n-1)\right], \qquad n = 1, 2, \ldots, M, \qquad 0 \leqslant t \leqslant T \quad \text{(C–1)}$$

and T is the time duration of the signaling interval.

†The complex representation of real signals is used throughout. Complex conjugation is denoted by an asterisk.

Consider the case in which one of these M waveforms is transmitted, for the duration of the signaling interval, over L channels. Assume that each of the channels corrupts the signaling waveform transmitted through it by introducing a multiplicative gain and phase shift, represented by the complex-valued number g_k, and an additive noise $z_k(t)$. Thus, when the transmitted waveform is $s_{ln}(t)$, the waveform received over the kth channel is

$$r_{lk}(t) = g_k s_{ln}(t) + z_k(t), \qquad 0 \leqslant t \leqslant T, \qquad k = 1, 2, \ldots, L \qquad \text{(C–2)}$$

The noises $\{z_k(t)\}$ are assumed to be sample functions of a stationary white Gaussian random process with zero-mean and autocorrelation function $\phi_z(\tau) = N_0\delta(\tau)$, where N_0 is the value of the spectral density. These sample functions are assumed to be mutually statistically independent.

At the demodulator, $r_{lk}(t)$ is passed through a filter whose impulse response is matched to the waveform $g(t)$. The output of this filter, sampled at time $t = T$, is denoted as

$$X_k = 2\mathcal{E} g_k \exp\left[j\frac{2\pi}{M}(n-1) \right] + N_k \qquad \text{(C–3)}$$

where $\mathcal{E}$ is the transmitted signal energy per channel and N_k is the noise sample from the kth filter. In order for the demodulator to decide which of the M phases was transmitted in the signaling interval $0 \leqslant t \leqslant T$, it attempts to undo the phase shift introduced by each channel. In practice, this is accomplished by multiplying the matched filter output X_k by the complex conjugate of an estimate $\hat{g}_k$ of the channel gain and phase shift. The result is a weighted and phase-shifted sampled output from the kth-channel filter, which is then added to the weighted and phase-shifted sampled outputs from the other $L - 1$ channel filters.

The estimate $\hat{g}_k$ of the gain and phase shift of the kth channel is assumed to be derived either from the transmission of a pilot signal or by undoing the modulation on the information-bearing signals received in previous signaling intervals. As an example of the former, suppose that a pilot signal, denoted by $s_{pk}(t)$, $0 \leqslant t \leqslant T$, is transmitted over the kth channel for the purpose of measuring the channel gain and phase shift. The received waveform is

$$g_k s_{pk}(t) + z_{pk}(t), \qquad 0 \leqslant t \leqslant T$$

where $z_{pk}(t)$ is a sample function of a stationary white Gaussian random process with zero-mean and autocorrelation function $\phi_p(\tau) = N_0\delta(\tau)$. This signal plus noise is passed through a filter matched to $s_{pk}(t)$. The filter output is sampled at time $t = T$ to yield the random variable $X_{pk} = 2\mathcal{E}_p g_k + N_{pk}$, where $\mathcal{E}_p$ is the energy in the pilot signal, which is assumed to be identical for all channels, and N_{pk} is the additive noise sample. An estimate of g_k is obtained by properly normalizing X_{pk}, i.e., $\hat{g}_k = g_k + N_{pk}/2\mathcal{E}_p$.

On the other hand, an estimate of g_k can be obtained from the information-bearing signal as follows. If one knew the information component contained in the matched filter output, then an estimate of g_k could be obtained by properly normalizing this output. For example, the information component in the filter output given in Equation C–3 is $2\mathcal{E}g_k \exp[j(2\pi/M)(n-1)]$, and, hence, the estimate is

$$\hat{g}_k = \frac{X_k}{2\mathcal{E}} \exp\left[-j\frac{2\pi}{M}(n-1)\right] = g_k + \frac{N'_k}{2\mathcal{E}}$$

where $N'_k = N_k \exp[-j(2\pi/M)(n-1)]$ and the PDF of N'_k is identical to the PDF of N_k. An estimate that is obtained from the information-bearing signal in this manner is called a *clairvoyant estimate*. Although a physically realizable receiver does not possess such clairvoyance, it can approximate this estimate by employing a time delay of one signaling interval and by feeding back the estimate of the transmitted phase in the previous signaling interval.

Whether the estimate of g_k is obtained from a pilot signal or from the information-bearing signal, the estimate can be improved by extending the time interval over which it is formed to include several prior signaling intervals in a way that has been described by Price (1962,a,b). As a result of extending the measurement interval, the signal-to-noise ratio in the estimate of g_k is increased. In the general case where the estimation interval is the infinite past, the normalized *pilot signal estimate* is

$$\hat{g}_k = g_k + \sum_{i=1}^{\infty} c_i N_{pki} \Big/ 2\mathcal{E}_p \sum_{i=1}^{\infty} c_i \tag{C–4}$$

where c_i is the weighting coefficient on the subestimate of g_k derived from the ith prior signal interval and N_{pki} is the sample of additive Gaussian noise at the output of the filter matched to $s_{pk}(t)$ in the ith prior signaling interval. Similarly, the clairvoyant estimate that is obtained from the information-bearing signal by undoing the modulation over the infinite past is

$$\hat{g}_k = g_k + \sum_{i=1}^{\infty} c_i N_{ki} \Big/ 2\mathcal{E} \sum_{i=1}^{\infty} c_i \tag{C–5}$$

As indicated, the demodulator forms the product between $\hat{g}_k^*$ and X_k and adds this to the products of the other $L-1$ channels. The random variable that results is

$$\begin{aligned} z &= \sum_{k=1}^{L} X_k \hat{g}_k^* = \sum_{k=1}^{L} X_k Y_k^* \\ &= z_r + jz_i \end{aligned} \tag{C–6}$$

where, by definition, $Y_k = \hat{g}_k$, $z_r = \text{Re}(z)$, and $z_i = \text{Im}(z)$. The phase of z is the decision variable. This is simply

$$\theta = \tan^{-1}\left(\frac{z_i}{z_r}\right) = \tan^{-1}\left[\text{Im}\left(\sum_{k=1}^{L} X_k Y_k^*\right) \Big/ \text{Re}\left(\sum_{k=1}^{L} X_k Y_k^*\right)\right] \tag{C–7}$$

C.2
CHARACTERISTIC FUNCTION AND PROBABILITY DENSITY FUNCTION OF THE PHASE θ

The following derivation is based on the assumption that the transmitted signal phase is zero, i.e., $n = 1$. If desired, the PDF of θ conditional on any other transmitted signal phase can be obtained by translating $p(\theta)$ by the angle $2\pi(n-1)/M$. We also assume that the complex-valued numbers $\{g_k\}$, which characterize the L channels, are mutually statistically independent and identically distributed zero-mean Gaussian random variables. This characterization is appropriate for slowly Rayleigh fading channels. As a consequence, the random variables (X_k, Y_k) are correlated, complex-valued, zero-mean, Gaussian, and statistically independent, but identically distributed with any other pair (X_i, Y_i).

The method that has been used in evaluating the probability density $p(\theta)$ in the general case of diversity reception is as follows. First, the characteristic function of the joint probability distribution function of z_r and z_i, where z_r and z_i are two components that make up the decision variable θ, is obtained. Second, the double Fourier transform of the characteristic function is performed and yields the density $p(z_r, z_i)$. Then the transformation

$$r = \sqrt{z_r^2 + z_i^2}, \qquad \theta = \tan^{-1}\left(\frac{z_i}{z_r}\right) \tag{C-8}$$

yields the joint PDF of the envelope r and the phase θ. Finally, integration of this joint PDF over the random variable r yields the PDF of θ.

The joint characteristic function of the random variables z_r and z_i can be expressed in the form

$$\psi(jv_1, jv_2) = \left[\frac{\dfrac{4}{m_{xx}m_{yy}(1-|\mu|^2)}}{\left(v_1 - j\dfrac{2|\mu|\cos\varepsilon}{\sqrt{m_{xx}m_{yy}}(1-|\mu|^2)}\right)^2 + \left(v_2 - j\dfrac{2|\mu|\sin\varepsilon}{\sqrt{m_{xx}m_{yy}}(1-|\mu|^2)}\right)^2 + \dfrac{4}{m_{xx}m_{yy}(1-|\mu|^2)^2}}\right] \tag{C-9}$$

where, by definition,

$$\begin{aligned} m_{xx} &= E(|X_k|^2), && \text{identical for all } k \\ m_{yy} &= E(|Y_k|^2), && \text{identical for all } k \\ m_{xy} &= E(X_k Y_k^*), && \text{identical for all } k \\ \mu &= \frac{m_{xy}}{\sqrt{m_{xx}m_{yy}}} = |\mu| e^{-j\varepsilon} \end{aligned} \tag{C-10}$$

The result of Fourier-transforming the function $\psi(jv_1, jv_2)$ with respect to the variables v_1 and v_2 is

$$p(z_r, z_i) = \frac{(1-|\mu|^2)^L}{(L-1)!\pi 2^L}\left(\sqrt{z_r^2+z_i^2}\right)^{L-1} \times \exp[|\mu|(z_r\cos\varepsilon + z_i\sin\varepsilon)]K_{L-1}\left(\sqrt{z_r^2+z_i^2}\right) \tag{C–11}$$

where $K_n(x)$ is the modified Hankel function of order n. Then the transformation of random variables, as indicated in Equation C–8 yields the joint PDF of the envelope r and the phase θ in the form

$$p(r, \theta) = \frac{(1-|\mu|^2)^L}{(L-1)!\pi 2^L} r^L \exp[|\mu| r\cos(\theta-\varepsilon)]K_{L-1}(r) \tag{C–12}$$

Now, integration over the variable r yields the marginal PDF of the phase θ. We have evaluated the integral to obtain $p(\theta)$ in the form

$$p(\theta) = \frac{(-1)^{L-1}(1-|\mu|^2)^L}{2\pi(L-1)!}\left\{\frac{\partial^{L-1}}{\partial b^{L-1}}\left[\frac{1}{b-|\mu|^2\cos^2(\theta-\varepsilon)} + \frac{|\mu|\cos(\theta-\varepsilon)}{[b-|\mu|^2\cos^2(\theta-\varepsilon)]^{3/2}}\cos^{-1}\left(-\frac{|\mu|\cos(\theta-\varepsilon)}{b^{1/2}}\right)\right]\right\}\Bigg|_{b=1} \tag{C–13}$$

In this equation, the notation

$$\frac{\partial^L}{\partial b^L} f(b, \mu)\Bigg|_{b=1}$$

denotes the Lth partial derivative of the function $f(b, \mu)$ evaluated at $b = 1$.

C.3

ERROR PROBABILITIES FOR SLOWLY RAYLEIGH FADING CHANNELS

In this section, the probability of a character error and the probability of a binary digit error are derived for M-phase signaling. The probabilities are evaluated via the probability density function and the probability distribution function of θ.

The probability distribution function of the phase. In order to evaluate the probability of error, we need to evaluate the definite integral

$$P(\theta_1 \leqslant \theta \leqslant \theta_2) = \int_{\theta_1}^{\theta_2} p(\theta)\, d\theta$$

where θ_1 and θ_2 are limits of integration and $p(\theta)$ is given by Equation C–13. All subsequent calculations are made for a real cross-correlation coefficient μ. A

Evaluation of the cross-correlation coefficient. The expressions for the probabilities of error given above depend on a single parameter, namely, the cross-correlation coefficient μ. The clairvoyant estimate is given by Equation C–5, and the matched filter output, when signal waveform $s_{l1}(t)$ is transmitted, is $X_k = 2\mathcal{E}g_k + N_k$. Hence, the cross-correlation coefficient is

$$\mu = \frac{\sqrt{\nu}}{\sqrt{(\bar{\gamma}_c^{-1} + 1)(\bar{\gamma}_c^{-1} + \nu)}} \tag{C–22}$$

where, by definition,

$$\nu = \left|\sum_{i=1}^{\infty} c_i\right|^2 \Big/ \sum_{i=1}^{\infty} |c_i|^2$$
$$\bar{\gamma}_c = \frac{\mathcal{E}}{N_0} E(|g_k|^2), \qquad k = 1, 2, \ldots, L \tag{C–23}$$

The parameter ν represents the effective number of signaling intervals over which the estimate is formed, and $\bar{\gamma}_c$ is the average SNR per channel.

In the case of differential phase signaling, the weighting coefficients are $c_1 = 1$, $c_i = 0$ for $i \neq 1$. Hence, $\nu = 1$ and $\mu = \bar{\gamma}_c/(1 + \bar{\gamma})_c)$.

When $\nu = \infty$, the estimate is perfect and

$$\lim_{\nu \to \infty} \mu = \sqrt{\frac{\bar{\gamma}_c}{\bar{\gamma}_c + 1}}$$

Finally, in the case of a pilot signal estimate, given by Equation C–4 the cross-correlation coefficient is

$$\mu = \left[\left(1 + \frac{r+1}{r\bar{\gamma}_t}\right)\left(1 + \frac{r+1}{\nu\bar{\gamma}_t}\right)\right]^{-1/2} \tag{C–24}$$

where, by definition.

$$\bar{\gamma}_t = \frac{\mathcal{E}_t}{N_0} E(|g_k|^2)$$
$$\mathcal{E}_t = \mathcal{E} + \mathcal{E}_p$$
$$r = \mathcal{E}/\mathcal{E}_p$$

The values of μ given above are summarized in Table C–1.

C.4
ERROR PROBABILITIES FOR TIME-INVARIANT AND RICEAN FADING CHANNELS

In Section C.2, the complex-valued channel gains $\{g_k\}$ were characterized as zero-mean Gaussian random variables, which is appropriate for Rayleigh fading

TABLE C–1
Rayleigh fading channel

Type of estimate	Cross-correlation coefficient μ
Clairvoyant estimate	$\dfrac{\sqrt{\nu}}{\sqrt{(\bar{\gamma}_c^{-1}+1)(\hat{\gamma}_c^{-1}+\nu)}}$
Pilot signal estimate	$\dfrac{\sqrt{r\nu}}{(r+1)\sqrt{\left(\dfrac{1}{\bar{\gamma}_t}+\dfrac{r}{r+1}\right)\left(\dfrac{1}{\bar{\gamma}_t}+\dfrac{\nu}{r+1}\right)}}$
Differential phase signaling	$\dfrac{\bar{\gamma}_c}{\bar{\gamma}_c+1}$
Perfect estimate	$\sqrt{\dfrac{\bar{\gamma}_c}{\bar{\gamma}_c+1}}$

channels. In this section, the channel gains $\{g_k\}$ are assumed to be nonzero-mean Gaussian random variables. Estimates of the channel gains are formed by the demodulator and are used as described in Section C.1. Moreover, the decision variable θ is defined again by Equation C–7. However, in this case, the Gaussian random variables X_k and Y_k, which denote the matched filter output and the estimate, respectively, for the kth channel, have nonzero-means, which are denoted by $\bar{X}_k$ and $\bar{Y}_k$. Furthermore, the second moments are

$$
\begin{aligned}
m_{xx} &= E(|X_k - \bar{X}_k|^2), && \text{identical for all channels} \\
m_{yy} &= E(|Y_k - \bar{Y}_k|^2), && \text{identical for all channels} \\
m_{xy} &= E[(X_k - \bar{X}_k)(Y_k^* - \bar{Y}_k^*)], && \text{identical for all channels}
\end{aligned}
$$

and the normalized covariance is defined as

$$\mu = \frac{m_{xy}}{\sqrt{m_{xx}m_{yy}}}$$

Error probabilities are given below only for two- and four-phase signaling with this channel model. We are interested in the special case in which the fluctuating component of each of the channel gains $\{g_k\}$ is zero, so that the channels are time-invariant. If, in addition to this time invariance, the noises between the estimate and the matched filter output are uncorrelated, then $\mu = 0$.

In the general case, the probability of error for two-phase signaling over L statistically independent channels characterized in the manner described above can be obtained from the results in Appendix B. In its most general form, the expression for the binary error rate is

$$P_2 = Q_1(a,b) - I_0(ab)\exp[-\tfrac{1}{2}(a^2-b^2)]$$
$$+\frac{I_0(ab)\exp[-\frac{1}{2}(a^2+b^2)]}{[2/(1-\mu)]^{2L-1}}\sum_{k=0}^{L-1}\binom{2L-1}{k}\left(\frac{1+\mu}{1-\mu}\right)^k$$
$$+\frac{\exp[-\frac{1}{2}(a^2+b^2)]}{[2/(1-\mu)]^{2L-1}}$$
$$\times\sum_{k=1}^{L-1}I_n(ab)\sum_{k=0}^{L-1-n}\binom{2L-1}{k}\left[\left(\frac{b}{a}\right)^n\left(\frac{1+\mu}{1-\mu}\right)^k-\left(\frac{a}{b}\right)^n\left(\frac{1+\mu}{1-\mu}\right)^{2L-1-k}\right]\quad (L \geqslant 2)$$
$$P_2 = Q_1(a,b) - \tfrac{1}{2}(1+\mu)I_0(ab)\exp[-\tfrac{1}{2}(a^2+b^2)]\quad (L=1) \qquad \text{(C–25)}$$

where, by definition,

$$a = \left(\frac{1}{2}\sum_{k=1}^{L}\left|\frac{\bar{X}_k}{\sqrt{m_{xx}}}-\frac{\bar{Y}_k}{\sqrt{m_{yy}}}\right|^2\right)^{1/2}$$
$$b = \left(\frac{1}{2}\sum_{k=1}^{L}\left|\frac{\bar{X}_k}{\sqrt{m_{xx}}}+\frac{\bar{Y}_k}{\sqrt{m_{yy}}}\right|^2\right)^{1/2} \qquad \text{(C–26)}$$
$$Q_1(a,b) = \int_b^\infty x\exp[-\tfrac{1}{2}(a^2+x^2)]I_0(ax)\,dx$$

$I_n(x)$ is the modified Bessel function of the first kind and of order n.

Let us evaluate the constants a and b when the channel is time-invariant, $\mu = 0$, and the channel gain and phase estimates are those given in Section C.1. Recall that when signal $s_1(t)$ is transmitted, the matched filter output is $X_k = 2\mathcal{E}g_k + N_k$. The clairvoyant estimate is given by Equation C–5. Hence, for this estimate, the moments are $\bar{X}_k = 2\mathcal{E}_{gk}$, $\bar{Y}_k = g_k$, $m_{xx} = 4\mathcal{E}N_0$, and $m_{yy} = N_0/\mathcal{E}\nu$, where $\mathcal{E}$ is the signal energy, N_0 is the value of the noise spectral density, and ν is defined in Equation C–23. Substitution of these moments into Equation C–26 results in the following expressions for a and b:

$$a = \sqrt{\tfrac{1}{2}\gamma_b}|\sqrt{\nu}-1|$$
$$b = \sqrt{\tfrac{1}{2}\gamma_b}|\sqrt{\nu}+1| \qquad \text{(C–27)}$$
$$\gamma_b = \frac{\mathcal{E}}{N_0}\sum_{k=1}^{L}|g_k|^2$$

This is a result originally derived by Price (1962).

The probability of error for differential phase signaling can be obtained by setting $\nu = 1$ in Equation C–27.

Next, consider a pilot signal estimate. In this case, the estimate is given by Equation C–4 and the matched filter output is again $X_k = 2\mathcal{E}g_k + N_k$. When the moments are calculated and these are substituted into Equation C–26, the following expressions for a and b are obtained:

$$
\begin{aligned}
a &= \sqrt{\frac{\gamma_t}{2}}\left|\sqrt{\frac{\nu}{r+1}} - \sqrt{\frac{r}{r+1}}\right| \\
b &= \sqrt{\frac{\gamma_t}{2}}\left(\sqrt{\frac{\nu}{r+1}} + \sqrt{\frac{r}{r+1}}\right)
\end{aligned}
\qquad \text{(C–28)}
$$

where

$$
\begin{aligned}
\gamma_t &= \frac{\mathcal{E}_t}{N_0}\sum_{k=1}^{L}|g_k|^2 \\
\mathcal{E}_t &= \mathcal{E} + \mathcal{E}_p \\
r &= \mathcal{E}/\mathcal{E}_p
\end{aligned}
$$

Finally, we consider the probability of a binary digit error for four-phase signaling over a time-invariant channel for which the condition $\mu = 0$ obtains. One approach that can be used to derive this error probability is to determine the PDF of θ and then to integrate this over the appropriate range of values of θ. Unfortunately, this approach proves to be intractable mathematically. Instead, a simpler, albeit roundabout, method may be used that involves the Laplace transform. In short, the integral in Equation 14.4–14 of the text that relates the error probability $P_2(\gamma_b)$ in an AWGN channel to the error probability P_2 in a Rayleigh fading channel is a Laplace transform. Since the bit error probabilities P_2 and P_{4b} for a Rayleigh fading channel, given by Equations C–18 and C–21, respectively, have the same form but differ only in the correlation coefficient, it follows that the bit error probabilities for the time-invariant channel also have the same form. That is, Equation C–25 with $\mu = 0$ is also the expression for the bit error probability of a four-phase signaling system with the parameters a and b modified to reflect the difference in the correlation coefficient. The detailed derivation may be found in the paper by Proakis (1968). The expressions for a and b are given in Table C–2.

TABLE C–2
Time-invariant channel

Type of estimate	*a*	*b*
	Two-phase signaling	
Clairvoyant estimate	$\sqrt{\frac{1}{2}\gamma_b}\lvert\sqrt{\nu}-1\rvert$	$\sqrt{\frac{1}{2}\gamma_b}(\sqrt{\nu}+1)$
Differential phase signaling	0	$\sqrt{2\gamma_b}$
Pilot signal estimate	$\sqrt{\frac{\gamma_t}{2}}\left\lvert\sqrt{\frac{\nu}{r+1}}-\sqrt{\frac{r}{r+1}}\right\rvert$	$\sqrt{\frac{\gamma_t}{2}}\left(\sqrt{\frac{\nu}{r+1}}+\sqrt{\frac{r}{r+1}}\right)$
	Four-phase signaling	
Clairvoyant estimate	$\sqrt{\frac{1}{2}\gamma_b}\left\lvert\sqrt{\nu+1+\sqrt{\nu^2+1}}-\sqrt{\nu+1-\sqrt{\nu^2+1}}\right\rvert$	$\sqrt{\frac{1}{2}\gamma_b}\left(\sqrt{\nu+1+\sqrt{\nu^2+1}}+\sqrt{\nu+1-\sqrt{\nu^2+1}}\right)$
Differential phase signaling	$\sqrt{\frac{1}{2}\gamma_b}\left(\sqrt{2+\sqrt{2}}-\sqrt{2-\sqrt{2}}\right)$	$\sqrt{\frac{1}{2}\gamma_b}\left(\sqrt{2+\sqrt{2}}+\sqrt{2-\sqrt{2}}\right)$
Pilot signal estimate	$\sqrt{\frac{\gamma_t}{4(r+1)}}\left\lvert\sqrt{\nu+r+\sqrt{\nu^2+r^2}}-\sqrt{\nu+r-\sqrt{\nu^2+r^2}}\right\rvert$	$\sqrt{\frac{\gamma_t}{4(r+1)}}\left(\sqrt{\nu+r+\sqrt{\nu^2+r^2}}+\sqrt{\nu+r-\sqrt{\nu^2+r^2}}\right)$

APPENDIX D

Square-Root Factorization

Consider the solution of the set of linear equations

$$\mathbf{R}_N \mathbf{C}_N = \mathbf{U}_N \tag{D–1}$$

where $\mathbf{R}_N$ is an $N \times N$ positive-definite symmetric matrix, $\mathbf{C}_N$ is an N-dimensional vector of coefficients to be determined, and $\mathbf{U}_N$ is an arbitrary N-dimensional vector. The equations in D–1 can be solved efficiently by expressing $\mathbf{R}_N$ in the factored form

$$\mathbf{R}_N = \mathbf{S}_N \mathbf{D}_N \mathbf{S}_N^t \tag{D–2}$$

where $\mathbf{S}_N$ is a lower triangular matrix with elements $\{s_{ik}\}$ and $\mathbf{D}_N$ is a diagonal matrix with diagonal elements $\{d_k\}$. The diagonal elements of $\mathbf{S}_N$ are set to unity, i.e., $s_{ii} = 1$. Then we have

$$r_{ij} = d \sum_{k=1}^{j} s_{ik} d_k s_{jk}, \qquad 1 \leqslant j \leqslant i-1, \qquad i \geqslant 2 \tag{D–3}$$

$$r_{11} = d_1$$

where $\{r_{ij}\}$ are the elements of $\mathbf{R}_N$. Consequently, the elements $\{s_{ik}\}$ and $\{d_k\}$ are determined from Equation D–3 according to the equations

$$d_1 = r_{11}$$

$$s_{ij} d_j = r_{ij} - \sum_{k=1}^{j-1} s_{ik} d_k s_{jk}, \qquad 1 \leqslant j \leqslant i-1, \qquad 2 \leqslant i \leqslant N \tag{D–4}$$

$$d_i = r_{ii} - \sum_{k=1}^{i-1} s_{ik}^2 d_k, \qquad 2 \leqslant i \leqslant N$$

Thus, Equation D–4 define $\mathbf{S}_N$ and $\mathbf{D}_N$ in terms of the elements of $\mathbf{R}_N$.

The solution to Equation D–1 is performed in two steps. With Equation D–2 substituted into Equation D–1 we have

$$\mathbf{S}_N \mathbf{D}_N \mathbf{S}_N^t \mathbf{C}_N = \mathbf{U}_N$$

Let

$$\mathbf{Y}_N = \mathbf{D}_N \mathbf{S}_N^t \mathbf{C}_N \tag{D–5}$$

Then

$$\mathbf{S}_N \mathbf{Y}_N = \mathbf{U}_N \tag{D–6}$$

First we solve Equation D–6 for $\mathbf{Y}_N$. Because of the triangular form of $\mathbf{S}_N$, we have

$$\begin{aligned} y_1 &= u_1 \\ y_i &= u_i - \sum_{j=1}^{i-1} s_{ij} y_j, \qquad 2 \leqslant i \leqslant N \end{aligned} \tag{D–7}$$

Having obtained $\mathbf{Y}_N$, the second step is to compute $\mathbf{C}_N$. That is,

$$\begin{aligned} \mathbf{D}_N \mathbf{S}_N^t \mathbf{C}_N &= \mathbf{Y}_N \\ \mathbf{S}_N^t \mathbf{C}_N &= \mathbf{D}_N^{-1} \mathbf{Y}_N \end{aligned}$$

Beginning with

$$c_N = y_N / d_N \tag{D–8}$$

the remaining coefficients of $\mathbf{C}_N$ are obtained recursively as follows:

$$c_i = \frac{y_i}{d_i} - \sum_{j=i+1}^{N} s_{ji} c_j, \qquad 1 \leqslant i \leqslant N-1 \tag{D–9}$$

The number of multiplications and divisions required to perform the factorization of $\mathbf{R}_N$ is proportional to N^3. The number of multiplications and divisions required to compute $\mathbf{C}_N$, once $\mathbf{S}_N$ is determined, is proportional to N^2. In contrast, when $\mathbf{R}_N$ is Toeplitz the Levinson–Durbin algorithm should be used to determine the solution of Equation D–1, since the number of multiplications and divisions is proportional to N^2. On the other hand, in a recursive least-squares formulation, $\mathbf{S}_N$ and $\mathbf{D}_N$ are not computed as in Equation D–3, but they are updated recursively. The update is accomplished with N^2 operations (multiplications and divisions). Then the solution for the vector $\mathbf{C}_N$ follows the steps of Equations D–5 to D–9. Consequently, the computational burden of the recursive least-squares formulation is proportional to N^2.

References and Bibliography

Abdulrahman, A., Falconer, D. D., and Sheikh, A. U. (1994). "Decision Feedback Equalization for CDMA in Indoor Wireless Communications," *IEEE J. Select. Areas Commun.*, vol. 12, pp. 698–706, May.

Abend, K. and Fritchman, B. D. (1970). "Statistical Detection for Communication Channels with Intersymbol Interference," *Proc. IEEE*, pp. 779–785, May.

Abramson, N. (1963). *Information Theory and Coding*, McGraw-Hill, New York.

Abramson, N. (1970). "The ALOHA System—Another Alternative for Computer Communications," *1970 Fall Joint Comput. Conf., AFIDS Conf. Proc.*, vol. 37, pp. 281–285, AFIPS Press, Montvale, N.J.

Abramson, N. (1977). "The Throughput of Packet Broadcasting Channels," *IEEE Trans. Commun*, vol. COM-25, pp. 117–128, January.

Abramson, N. (1993). *Multiple Access Communications*, IEEE Press, New York.

Abramson, N. (1994). "Multiple Access in Wireless Digital Networks," *Proc. IEEE*, vol. 82, pp. 1360–1369, September.

Adler, R. L., Coppersmith, D., and Hassner, M. (1983). "Algorithms for Sliding Block Codes," *IEEE Trans. Inform. Theory*, vol. IT-29, pp. 5–22, January.

Alamouti, A. (1998). "A Simple Transmitter Diversity Scheme for Wireless Communications," *IEEE J. Selected Areas Commun.*, vol. JSAC-16, pp. 1451–1458, October.

Alexander, P. D., Reed, M. C., Asenstorfer, J. A., and Schlegel, C. B. (1999). "Iterative Multiuser Interference Reduction: Turbo CDMA," *IEEE Trans. Commun.*, vol. 47, pp. 1008–1014, July.

Al-Hussaini, E. and Al-Bassiouni, A.A.M. (1985). "Performance of MRC Diversity Systems for the Detection of Signals with Nakagami Fading," *IEEE Trans. Commun*, vol. COM-33, pp. 1315–1319, December.

Alouini, M. and Goldsmith, A. (1998). "A Unified Approach for Calculating Error Rates of Linearly Modulated Signals over Generalized Fading Channels," *Proc. IEEE ICC'98*, pp. 459–464, Atlanta, GA.

Altekar, S. A. and Beaulieu, N. C. (1993). "Upper Bounds on the Error Probability of Decision Feedback Equalization," *IEEE Trans. Inform. Theory*, vol. IT-39, pp. 145–156, January.

Anderberg, M. R. (1973). *Cluster Analysis for Applications*, Academic, New York.

Anderson, J. B., Aulin, T., and Sundberg, C. W. (1986). *Digital Phase Modulation*, Plenum, New York.

Anderson, R. R. and Salz, J. (1965). "Spectra of Digital FM," *Bell Syst. Tech. J.*, vol. 44, pp. 1165–1189, July–August.

Annamalai, A., Tellambura, C., and Bhargara, V. K. (1999). "A Unified Approach to Performance Evaluation of Diversity Systems on Fading Channels," in *Wireless Multimedia Network Technologies*, chap. 17, R. Ganesh ed., Kluwer Academic Publishers, Boston, MA.

Annamalai, A., Tellambura, C. and Bhargara, V. K. (1998). "A Unified Analysis of MPSK and MDPSK with Diversity Reception in Different Fading Environments," *IEEE Electr. Lett.*, vol. 34, pp. 1564–1565, August.

Ash, R. B. (1965). *Information Theory*, Interscience, New York.

Aulin, T. (1980). "Viterbi Detection of Continuous Phase Modulated Signals," *Nat Telecommun. Conf. Record*, pp. 14.2.1–14.2.7, Houstin, TX, November.

Aulin, T., Rydbeck, N., and Sundberg, C. W. (1981). "'Continuous Phase Modulation—Part II: Partial Response Signaling," *IEEE Trans. Commun.*, vol. COM-29, pp. 210–225, March.

Aulin, T., Sundberg, C. W., and Svensson, A. (1981). "Viterbi Detectors with Reduced Complexity for Partial Response Continuous Phase Modulation," *Conf. Record NTC'81*, pp. A7.61–A7.6.7, New Orleans, LA.

Aulin, T. and Sundberg, C. W. (1981). "Continuous Phase Modulation—Part I: Full Response Signaling," *IEEE Trans. Commun*, vol. COM-29, pp. 196–209, March.

Aulin, T. and Sundberg, C. W. (1982a). "On the Minimum Euclidean Distance for a Class of Signal Space Codes," *IEEE Trans. Inform. Theory*, vol. IT-28, pp. 43–55, January.

Aulin, T. and Sundberg, C. W. (1982b). "Minimum Euclidean Distance and Power Spectrum for a Class of Smoothed Phase Modulation Codes with Constant Envelope," *IEEE Trans. Commun.*, vol. COM-30, pp. 1721–1729, July.

Aulin, T. and Sundberg, C. W. (1984). "CPM—An Efficient Constant Amplitude Modulation Scheme," *Int. J. Satellite Commun*, vol. 2, pp. 161–186.

Austin, M. E. (1967). "Decision-Feedback Equalization for Digital Communication Over Dispersive Channels," MIT Lincoln Laboratory, Lexington, MA. Tech. Report No. 437, August.

Bahl, L. R., Cocke, J., Jelinek, F., and Raviv, J. (1974). "Optimal Decoding of Linear Codes for Minimizing Symbol Error Rate" *IEEE Trans. Inform. Theory*, vol. IT-20, pp. 284–287, March.

Barrow, B. (1963). "Diversity Combining of Fading Signals with Unequal Mean Strengths," *IEEE Trans. Commun. Syst.*, vol. CS-11, pp. 73–78, March.

Bauch, G. and Franz, V. (1998). "Iterative Equalization and Decoding for the GSM System," *Proc. VTC '98*, pp. 2262–2266, April.

Bauch, G., Khorram, H., and Hagenauer, J. (1997). "Iterative Equalization and Decoding in Mobile Communications Systems," *Proc. European Personal Mobile Commun. Conf. (EPMCC'77)*, pp. 307–312, September.

Beare, C. T. (1978). "The Choice of the Desired Impulse Response in Combined Linear-Viterbi Algorithm Equalizers," *IEEE Trans. Commun.*, vol. 26, pp. 1301–1307, August.

Beaulieu, N. C. (1990). "An Infinite Series for the Computation of the Complementary Probability Distribution Function of a Sum of Independent Random Variables and Its Application to the Sum of Rayleigh Random Variables," *IEEE Trans. Commun.*, vol. COM-38, pp. 1463–1474, September.

Beaulieu, N. C. (1994). "Bounds on Recovery Times of Decision Feedback Equalizers," *IEEE Trans. Commun*, vol. 42, pp. 2786–2794, October.

Beaulieu, N. C. and Abu-Dayya, A. A. (1991). "Analysis of Equal Gain Diversity on Nakagami Fading Channels," *IEEE Trans. Commun*, vol. COM-39, pp. 225–234, February.

Bégin, G. and Haccoun, D. (1989). "High-Rate Punctured Convolutional Codes: Structure, Properties and Construction Technique," *IEEE Trans. Commun.*, vol. 37, pp. 1381–1385, December.

Bégin, G., Haccoun, D., and Paguin, C. (1990). "Further Results on High-Rate Punctured Convolutional Codes for Viterbi and Sequential Decoding," *IEEE Trans. Commun.*, vol. 38, pp. 1922–1928, November.

Bekir, N. E., Scholtz, R. A., and Welch, L. R. (1978). "Partial-Period Correlation Properties of PN Sequences," *1978 Nat. Telecommun. Conf. Record*, pp. 35.1.1–25.1.4, Birmingham, Alabama, November.

Belfiore, C. A. and Park, J. H., Jr. (1979). "Decision-Feedback Equalization," *Proc. IEEE*, vol. 67, pp. 1143–1156, August.

Bellini, J. (1986). "Bussgang Techniques for Blind Equalization," *Proc. GLOBECOM'86*, pp. 46.1.1–46.1.7, Houston, TX, December.

Bello, P. A. and Nelin, B. D. (1962a). "Predetection Diversity Combining with Selectivity Fading Channels," *IRE Trans. Commun Syst.*, vol. CS-10, pp. 32–42, March.

Bello, P. A. and Nelin, B. D. (1962b). "The Influence of Fading Spectrum on the Binary Error Probabilities of Incoherent and Differentially Coherent Matched Filter Receivers," *IRE Trans. Commun. Syst.*, vol. CS-10, pp. 160–168, June.

Bello, P. A. and Nelin, B. D. (1963). "The Effect of Frequency Selective Fading on the Binary Error Probabilities of Incoherent and Differentially Coherent Matched Filter Receivers," *IEEE Trans. Commun. Syst.*, vol. CS-11, pp. 170–186, June.

Benedetto, S., Ajmone Marsan, M., Albertengo, G., and Giachin, E. (1988). "Combined Coding and Modulation: Theory and Applications," *IEEE Trans. Inform. Theory*, vol. 34, pp. 223–236, March.

Benedetto, S., Divsalar, D., Montorsi, G., and Pollara, F. (1998). "Serial Concatenation of Interleaved Codes: Performance Analysis, Design and Iterative Decoding," *IEEE Trans. Inform. Theory*, vol. 44, pp. 909–926, May.

Benedetto, S., Mondin, M., and Montorsi, G. (1994). "Performance Evaluation of Trellis-Coded Modulation Schemes," *Proc. IEEE*, vol. 82, pp. 833–855, June.

Benedetto, S. and Montorsi, G. (1996). "Unveiling Turbo Codes: Some Results on Parallel Concatenated Coding Schemes," *IEEE Trans. Inform. Theory*, vol. 42, pp. 409–428, March.

Bennett, W. R. and Davey, J. R. (1965). *Data Transmission*, McGraw-Hill, New York.

Bennett, W. R. and Rice, S. O. (1963). "Spectral Density and Autocorrelation Functions Associated with Binary Frequency-Shift Keying," *Bell Syst. Tech. J.*, vol. 42, pp. 2355–2385, September.

Bensley, S. E. and Aazhang, B. (1996). "Subspace-Based Channel Estimation for Code-Division Multiple Access Communication Systems," *IEEE Trans. Commun.*, vol. 44, pp. 1009–1020, August.

Benveniste, A. and Goursat, M. (1984). "Blind Equalizers," *IEEE Trans. Commun.*, vol. COM-32, pp. 871–883, August.

Berger, T. (1971). *Rate Distortion Theory*, Prentice-Hall, Englewood Cliffs, NJ.

Berger, T. and Gibson, J. D. (1998). "Lossy Source Coding," *IEEE Trans. Inform. Theory*, vol. 44, pp. 2693–2723, October.

Berger, T. and Tufts, D. W. (1967). "Optimum Pulse Amplitude Modulation, Part I: Transmitter-Receiver Design and Bounds from Information Theory," *IEEE Trans. Inform. Theory*, vol. IT-13, pp. 196–208.

Bergmans, J. W. M. (1995). "Efficiency of Data-Aided Timing Recovery Techniques," *IEEE Trans. Inform. Theory*, vol. 41, pp. 1397–1408, September.

Bergmans, J. W. M., Rajput, S. A., and Van DeLaar, F. A. M. (1987). "On the Use of Decision Feedback for Simplifying the Viterbi Detector," *Philips J. Research*, vol. 42, no. 4, pp. 399–428.

Bergmans, P. P. and Cover, T. M. (1974). "Cooperative Broadcasting," *IEEE Trans. Inform. Theory*, vol. IT-20, pp. 317–324, May.

Berlekamp, E. R. (1968). *Algebraic Coding Theory*, McGraw-Hill, New York.

Berlekamp, E. R. (1973). "Goppa Codes," *IEEE Trans. Inform. Theory*, vol. IT-19, pp. 590–592.

Berlekamp, E. R. (1974). *Key Papers in the Development of Coding Theory*, IEEE Press, New York.

Berrou, C. and Glavieux, A. (1996). "Near Optimum Error-Correcting Coding and Decoding: Turbo Codes," *IEEE Trans. Commun.*, vol. 44, pp. 1261–1271.

Berrou, C., Glavieux, A., and Thitimajshima, P. (1993). "Near Shannon Limit Error-Correcting Coding and Decoding: Turbo Codes," *Proc. IEEE Int. Conf. Commun.*, pp. 1064–1070, May, Geneva, Switzerland.

Bierman, G. J. (1977). *Factorization Methods for Discrete Sequential Estimation*, Academic, New York.

Biglieri, E., Divsalar, D., McLane, P. J., and Simon, M. K. (1991). *Introduction to Trellis-Coded Modulation with Applications*, Macmillan, New York.

Biglieri, E., Proakis, J. G., and Shamai, S. (1998). "Fading Channels: Information-Theoretic and Communications Aspects," *IEEE Trans. Inform. Theory*, vol. 44, pp. 2619–2692, October.

Bingham, J. A. C. (1990). "Multicarrier Modulation for Data Transmission: An Idea Whose Time Has Come," *IEEE Commun. Mag.*, vol. 28, pp. 5–14, May.

Bjerke, B. A. and Proakis, J. G. (1999). "Multiple Antenna Diversity Techniques for Transmission over Fading Channels," *Proc. WCNC'99*, September, New Orleans, LA.

Blahut, R. E. (1983). *Theory and Practice of Error Control Codes*, Addison-Wesley, Reading, MA.

Blahut, R. E. (1987). *Principles and Practice of Information Theory*, Addison-Wesley, Reading, MA.

Blahut, R. E. (1990). *Digital Transmission of Information*, Addison-Wesley, Reading, MA.

Bose, R. C. and Ray-Chaudhuri, D. K. (1960a). "On a Class of Error Correcting Binary Group Codes," *Inform, Control*, vol. 3, pp. 68–79, March.

Bose, R. C. and Ray-Chaudhuri, D. K. (1960b). "Further Results in Error Correcting Binary Group Codes,' "*Inform. Control*, vol. 3, pp. 279–290, September.

Boyd, S. (1986). "Multitone Signals with Low Crest Factor," *IEEE Trans. Circuits and Systems*, vol. CAS-33, pp. 1018–1022.

Brennan, D. G. (1959). "Linear Diversity Combining Techniques," *Proc. IRE.*, vol. 47, pp. 1075–1102.

Bucher, E. A. (1980). "Coding Options for Efficient Communications on Non-Stationary Channels," *Rec. IEEE Int. Conf. Commun.*, pp. 4.1.1–4.1.7.

Buehrer, R. M., Nicoloso, S. P., and Gollamudi, S. (1999). "Linear versus Nonlinear Interference Cancellation," *J. Commun. and Networks*, vol. 1, pp. 118–133, June.

Buehrer, R. M. and Woerner, B. D. (1966). "Analysis of Multistage Interference Cancellation for CDMA Using an Improved Gaussian Approximation," *IEEE Trans. Commun.*, vol. 44, pp. 1308–1316, October.

Burton, H. O. (1969). "A Class of Asymptotically Optimal Burst Correcting Block Codes," *Proc. ICCC*, Boulder, CO, June.

Bussgang, J. J. (1952). "Crosscorrelation Functions of Amplitude-Distorted Gaussian Signals," MIT RLE Tech. Report 216.

Buzo, A., Gray, A. H., Jr., Gray, R. M., and Markel, J. D. (1980). "Speech Coding Based Upon Vector Quantization," *IEEE Trans. Acoust., Speech, Signal Processing*, vol. ASSP-28, pp. 562–574, October.

Cahn, C. R. (1960). "Combined Digital Phase and Amplitude Modulation Communication Systems," *IRE Trans. Common. Syst.*, vol. CS-8, pp. 150–155, September.

Cain, J. B., Clark, G. C., and Geist, J. M. (1979). "Punctured Convolutional Codes of Rate $(n-1)/n$ and Simplified Maximum Likelihood Decoding," *IEEE Trans. Inform. Theory*, vol. IT-25, pp. 97–100, January.

Caire, G., Taricco, G., and Biglieri, E. (1998). "Bit-Interleaved Coded Modulation," *IEEE Trans. Inform. Theory*, vol. 44, pp. 927–946, May.

Calderbank, A. R. (1998). "The Art of Signalling: Fifty Years of Coding Theory," *IEEE Trans. Inform. Theory*, vol. 44, pp. 2561–2595, October.

Calderbank, A. R. and Sloane, N. J. A. (1987). "New Trellis Codes Based on Lattices and Cosets," *IEEE Trans. Inform. Theory*, vol. IT-33, pp. 177–195. March.

Campanella, S. J. and Robinson, G. S. (1971). "A Comparison of Orthogonal Transformations for Digital Speech Processing," *IEEE Trans. Commun.*, vol. COM-19, pp. 1045–1049, December.

Campopiano, C. N. and Glazer, B. G. (1962). "A Coherent Digital Amplitude and Phase Modulation Scheme," *IRE Trans. Commun. Syst.*, vol. CS-10, pp. 90–95, June.

Capetanakis, J. I. (1979). "Tree Algorithms for Packet Broadcast Channels," *IEEE Trans. Inform. Theory*, vol. IT-25, pp. 505–515, September.

Caraiscos, C. and Liu, B. (1984). "A Roundoff Error Analysis of the LMS Adaptive Algorithm," *IEEE Trans. Acoust., Speech, Signal Processing*, vol. ASSP-32, pp. 34–41, January.

Carayannis, G., Manolakis, D. G., and Kalouptsidis, N. (1983). "A Fast Sequential Algorithm for Least-Squares Filtering and Prediction," *IEEE Trans. Acoust., Speech, Signal Processing*, vol. ASSP-31, pp. 1394–1402, December.

Carayannis, G., Manolakis, D. G., and Kalouptsidis, N. (1986). " A Unified View of Parametric Processing Algorithms for Prewindowed Signals," *Signal Processing*, vol. 10, pp. 335–368, June.

Carleial, A. B. and Hellman, M. E. (1975). "Bistable Behavior of ALOHA-Type Systems," *IEEE Trans. Commun.*, vol. COM-23, pp. 401–410, April 1975.

Carlson, A. B. (1975). *Communication Systems*, McGraw-Hill, New York.

Chang, D. Y., Gersho, A., Ramamurthi, B., and Shohan, Y. (1984). "Fast Search Algorithms for Vector Quantization and Pattern Matching," *Proc. IEEE Int. Conf. Acoust., Speech, Signal Processing*, paper 9.11, San Diego, CA, March.

Chang, R. W. (1966). "Synthesis of Band-Limited Orthogonal Signals for Multichannel Data Transmission," *Bell Syst. Tech. J.*, vol. 45, pp. 1775–1796, December.

Chang, R. W. (1971). "A New Equalizer Structure for Fast Start-up Digital Communication," *Bell Syst. Tech. J.*, vol. 50, pp. 1969–2001.

Charash, U. (1979). "Reception Through Nakagami Fading Multipath Channels with Random Delays," *IEEE Trans. Commun.*, vol. COM-27, pp. 657–670, April.

Chase, D. (1972). "A Class of Algorithms for Decoding Block Codes with Channel Measurement Information," *IEEE Trans. Inform. Theory*, vol. IT-18, pp. 170–182, January.

Chase, D. (1976). "Digital Signal Design Concepts for a Time-Varying Ricean Channel," *IEEE Trans. Commun.*, vol. COM-24, pp. 164–172, February.

Chevillat, P. R. and Eleftheriou, E. (1989). "Decoding of Trellis-Encoded Signals in the Presence of Intersymbol Interference and Noise," *IEEE Trans. Commun.*, vol. 37, pp. 669–676, July.

Chevillat, P. R. and Eleftheriou, E. (1988). "Decoding of Trellis-Coded Signals in the Presence of Intersymbol Interference and Noise," *Conf. Rec. ICC'88*, pp. 23.1.1–23.1.6, June, Philadelphia, PA.

Chien, R. T. (1964). "Cyclic Decoding Procedures for BCH Codes," *IEEE Trans. Inform. Theory*, vol. IT-10, pp. 357–363, October.

Chow, J. S., Tu, J. C., and Cioffi, J. M. (1991). "A Discrete Multitone Transceiver System for HDSL Applications," *IEEE J. Selected Areas Commun.*, vol. SAC-9, pp. 895–908, August.

Chow, J. S., Cioffi, J. M., and Bingham, J. A. C. (1995). "A Practical Discrete Multitone Transceiver Loading Algorithm for Data Transmission over Spectrally Shaped Channels," *IEEE Trans. Commun.*, vol. 43, pp. 773–775, February/March/April.

Chyi, G. T., Proakis, J. G., and Keller, C. M. (1988). "Diversity Selection/Combining Schemes with Excess Noise-Only Diversity Reception Over a Rayleigh-Fading Multipath Channel." *Proc. Conf. Inform. Sci. Syst.*, Princeton University Princeton, NJ, March.

Cioffi, J. M. and Kailath, T. (1984a). "Fast Recursive-Least Squares Transversal Filters for Adaptive Filtering," *IEEE Trans. Acoust., Speech, Signal Processing*, vol. ASSP-32, pp. 304–337, April.

Cioffi, J. M. and Kailath, T. (1984b). "An Efficient Exact-Least-Squares Fractionally Spaced Equalizer Using Intersymbol Interpolation," *IEEE J. Selected Areas Commun.*, vol. 2, pp. 743–756, September.

Clark, A. P., Abdullah, S. N., Jayasinghe, S. J., and Sun, K. H. (1985). "Pseudobinary and Pseudoquaternary Detection Processes for Linearly Distorted Multilevel QAM Signals," *IEEE Trans. Commun.*, vol. COM-33, pp. 639–645, July.

Clark, A. P. and Clayden, M. (1984). "Pseudobinary Viterbi Detector," *Proc. IEE*, vol. 131, part F, pp. 280–218, April.

Cook, C. E., Ellersick, F. W., Milstien, L. B., and Schilling, D. L. (1983). *Spread Spectrum Communications*, IEEE Press, New York.

Costas, J. P. (1956). "Synchronous Communications," *Proc. IRE*, vol. 44, pp. 1713–1718, December.

Costello, D. J., Jr., Hagenauer, J., Imai, H., and Wicker, S. B. (1998). "Applications of Error-Control Coding," *IEEE Trans. Inform. Theory*, vol. 44, pp. 2531–2560, October.

Cover, T. M. (1972). "Broadcast Channels," *IEEE Trans. Inform. Theory*, vol. IT-18, pp. 2–14, January.

Cover, T. M. (1998). "Comments on Broadcast Channels," *IEEE Trans. Inform. Theory*, vol. 44, pp. 2524–2530, October.

Cover, T. and Thomas, J. (1991). *Elements of Inform. Theory*, Wiley, New York.

Cramér, H. (1946). *Mathematical Methods of Statistics*, Princeton University Press, Princeton, NJ.

Daneshgaran, F. and Mondin, M. (1999). "Design of Interleavers for turbo codes: Iterative Interleaver Growth Algorithms of Polynomial Complexity," *IEEE Trans. Inform. Theory*, vol. 45, pp. 1845–1859, September.

Daut, D. G., Modestino, J. W., and Wismer, L. D. (1982). "'New Short Constraint Length Convolutional Code Construction for Selected Rational Rates," *IEEE Trans. Inform. Theory*, vol. IT-28, pp. 793–799, September.

Davenport, W. B., Jr. (1970). *Probability and Random Processes*, McGraw-Hill, New York.

Davenport, W. B., Jr. and Root, W. L. (1958). *Random Signals and Noise*, McGraw-Hill, New York.

Davisson, L. D. (1973). "Universal Noiseless Coding," *IEEE Trans. Inform. Theory*, vol. IT-19, pp. 783–795.

Davisson, L. D., McEliece, R. J. Pursley, M. B., and Wallace, M. S. (1981). "Efficient Universal Noiseless Source codes," *IEEE Trans. Inform. Theory*, vol. IT-27, pp. 269–279.

deBuda, R. (1972). "Coherent Demodulation of Frequency Shift Keying with Low Deviation Ratio," *IEEE Trans. Commun.*, vol. COM-20, pp. 429–435, June.

Deller, J. P., Proakis, J. G., and Hansen, H. L. (2000). *Discrete-Time Processing of Speech Signals*, IEEE Press, New York.

Ding, Z. (1990). *Application Aspects of Blind Adaptive Equalizers in QAM Data Communications*, Ph.D. Thesis, Department of Electrical Engineering, Cornell University.

Ding, Z., Kennedy, R. A., Anderson, B. D. O., and Johnson, C. R. (1989). "Existence and Avoidance of Ill-Convergence of Godard Blind Equalizers in Data Communication Systems," *Proc. 23rd Conf. on Inform. Sci. Systems.*, Baltimore, MD.

Divsalar, D., Simon, M. K., and Raphelli, D. (1998). "Improved Parallel Interference Cancellation," *IEEE Trans. Commun.*, vol. 46, pp. 258–268, February.

Divsalar, D., Simon, M. K., and Yuen, J. H. (1987). "Trellis Coding with Asymmetric Modulation," *IEEE Trans. Commun.*, vol. COM-35, pp. 130–141, February.

Divsalar, D. and Yuen, J. H. (1984). "Asymmetric MPSK for Trellis Codes," *Proc. GLOBECOM'84*, pp. 20.6.1–20.6.8, Atlanta, GA, November.

Dixon, R. C. (1976). *Spread Spectrum Techniques*, IEEE Press, New York.

Dobrushin, R.L. and Lupanova, O.B. (1963). *Papers in Information Theory and Cybernetics* (in Russian), Edited by Dobrushin and Lupanova, Izd. Inostr. Lit., Moscow.

Doelz, M. L., Heald, E. T., and Martin, D. L. (1957). "Binary Data Transmission Techniques for Linear Systems," *Proc. IRE*, vol. 45, pp. 656–661, May.

Douillard, C., Jézéquel, M., Berrou, C., Picart, A., Didier, P., and Glavieux, A. (1995). "Iterative Correction of Intersymbol Interference: Turbo-equalization," *ETT European Trans. Telecommun.* vol. 6, pp. 507–511, September/October.

Drouilhet, P. R., Jr. and Bernstein, S. L. (1969). "TATS—A Bandspread Modulation-Demodulation System for Multiple Access Tactical Satellite Communication," *1969 IEEE Electronics and Aerospace Systems (EASCON) Conv. Record*, Washington, DC, pp. 126–132, October 27–29.

Duel-Hallen, A. and Heegard, C. (1989). "Delayed Decision-Feedback Sequence Estimation," *IEEE Trans. Commun.*, vol. 37, pp. 428–436, May.

Duffy, F. P. and Tratcher, T. W. (1971). "Analog Transmission Performance on the Switched Telecommunications Network," *Bell Syst. Tech. J.*, vol. 50, pp. 1311–1347, April.

Duman, T. M. and Salehi, M. (1997). "New Performance Bounds for Turbo codes," *Proc. GLOBECOM'97*, pp. 634–638, November, Phoenix, AZ.

Durbin, J. (1959). "Efficient Estimation of Parameters in Moving-Average Models," *Biometrika*, vol. 46, parts 1 and 2, pp. 306–316.

Duttweiler, D. L. Mazo, J. E., and Messerschmitt, D. G. (1974). "Error Propagation in Decision-Feedback Equalizers," *IEEE Trans. Inform. Theory*, vol. IT-20, pp. 490–497, July.

Eleftheriou, E. and Falconer, D. D. (1987). "Adapative Equalization Techniques for HF Channels," *IEEE J. Selected Areas Commun.*, vol. SAC-5, pp. 238–247, February.

El Gamal, A. and Cover, T. M. (1980). "Multiple User Information Theory," *Proc. IEEE*, vol. 68, pp. 1466–1483, December.

Elias, P. (1954). "Error-Free Coding," *IRE Trans. Inform. Theory*, vol. IT-4, pp. 29–37, September.

Elias, P. (1955). "Coding for Noisy Channels," *IRE Convention Record*, vol. 3, part 4, pp. 37–46.

Esposito, R. (1967). "Error Probabilities for the Nakagami Channel," *IEEE Trans. Inform. Theory*, vol. IT-13, pp. 145–148, January.

Eyuboglu, M. V. (1988). "Detection of Coded Modulation Signals on Linear, Severely Distorted Channels Using Decision-Feedback Noise Prediction with Interleaving," *IEEE Trans. Commun.*, vol. COM-36, pp. 401–409, April.

Eyuboglu, M. V. and Qureshi, S. U. H. (1989). "Reduced-State Sequence Estimation for Coded Modulation on Intersymbol Interference Channels," *IEEE J. Selected Areas Commun.*, vol. 7, pp. 989–955, August.

Eyuboglu, M. V., Qureshi, S. U., and Chen, M. P. (1988). "Reduced-State Sequence Estimation for Trellis-Coded Modulation on Intersymbol Interference Channels," *Proc. GLOBEROM '88*, pp., November, Hollywood, FL.

Eyuboglu, M. V. and Qureshi, S. U. (1988). "Reduced-State Sequence Estimation with Set Partitioning and Decision Feedback," *IEEE Trans. Commun.* vol. 36, pp. 13–20, January.

Falconer, D. D. (1976). "Jointly Adaptive Equalization and Carrier Recovery in Two-Dimensional Digital Communication Systems," *Bell Syst. Tech. J.*, vol. 55, pp. 317–334, March.

Falconer, D. D. and Ljung, L. (1978). "Application of Fast Kalman Estimation to Adaptive Equalization," *IEEE Trans. Commun.*, vol. COM-26, pp. 1439–1446, October.

Falconer, D. D. and Magee, F. R. (1973). "Adaptive Channel Memory Truncation for Maximum Likelihood Sequence Estimation," *Bell Syst. Tech. J.*, vol. 52, pp. 1541–1562, November.

Falconer, D. D. and Salz, J. (1977). "Optimal Reception of Digital Data Over the Gaussian Channel with Unknown Delay and Phase Jitter," *IEEE Trans. Inform. Theory*, vol. IT-23, pp. 117–126, January.

Fano, R. M. (1961). *Transmission of Information*, MIT Press, Cambridge, MA.

Fano, R. M. (1963). " A Heuristic Discussion of Probabilistic Decoding," *IEEE Trans. Inform. Theory*, vol. IT-9, pp. 64–74, April.

Feinstein, A. (1958). *Foundations of Information Theory*, McGraw-Hill, New York.

Fire, P. (1959). "A Class of Multiple-Error-Correcting Binary Codes for Non-Independent Errors," Sylvania Report No. RSL-E-32, Sylvania Electronic Defense Laboratory, Mountain view, CA, March.

Fischer, R. F. H. and Huber, J. B. (1996). "A New Loading Algorithm for Discrete Multitone Transmission," *Proc. IEEE GLOBECOM'96*, pp. 724–728, November, London.

Flanagan, J. L., et al. (1979). "Speech Coding," *IEEE Trans. Commun.*, vol. COM-27, pp. 710–736, April.

Forney, G. D., Jr. (1965). "On Decoding BCH Codes," *IEEE Trans. Inform. Theory*, vol. IT-11, pp. 549–557, October.

Forney, G. D., Jr. (1966a). *Concatenated Codes*, MIT Press, Cambridge, MA.

Forney, G. D., Jr. (1966b). "Generalized Minimum Distance Decoding," *IEEE Trans. Inform. Theory*, vol. IT-12, pp. 125–131, April.

Forney, G. D., Jr. (1968). "Exponential Error Bounds for Erasure, List, and Decision-Feedback Schemes," *IEEE Trans. Inform. Theory*, vol. IT-14, pp. 206–220, March.

Forney, G. D., Jr. (1970a). "Coding and Its Application in Space Communications," *IEEE Spectrum*, vol. 7, pp. 47–58, June.

Forney, G. D., Jr. (1970b). "Convolutional Codes I: Algebraic Structure," *IEEE Trans. Inform. Theory*, vol. IT-16, pp. 720–738, November.

Forney, G. D., Jr (1971). "Burst Correcting Codes for the Classic Bursty Channel," *IEEE Trans. Common. Tech.*, vol. COM-19, pp. 772–781. October.

Forney, G. D., Jr. (1972). "Maximum-Likelihood Sequence Estimation of Digital Sequences in the Presence of Intersymbol Interference." *IEEE Trans. Inform. Theory*, vol. IT-18, pp. 363–378, May.

Forney, G. D., Jr. (1974). "Convolutional Codes III: Sequential Decoding," *Inform. Control*, vol. 25, pp. 267–297, July.

Forney, G. D., Jr. (1988). "Coset Codes I: Introduction and Geometrical Classification," *IEEE Trans. Inform. Theory*, vol. IT-34, pp. 671–680, September.

Forney, G. D., Jr., Gallager, R. G., Lang, G. R., Longstaff, F. M., and Qureshi, S. U. (1984). "Efficient Modulation for Band-Limited Channels," *IEEE J. Selected Areas Commun.*, vol. SAC-2, pp. 632–647, September.

Forney, G. D., Jr. and Ungerboeck, G. (1998). "Modulation and Coding for Linear Gaussian Channels," *IEEE Trans. Inform. Theory*, vol. 44, pp. 2384–2415, October.

Foschini, G. J. (1977). "A Reduced State Variant of Maximum Likelihood Sequence Detection Attaining Optimum Performance for High Signal-to-Noise Ratios," *IEEE Trans. Inform. Theory*, vol. 23, pp. 605–609.

Foschini, G. J. (1984). "Contrasting Performance of Faster-Binary Signaling with QAM," *Bell Syst. Tech. J.*, vol. 63, pp. 1419–1445, October.

Foschini, G. J. (1985). "Equalizing Without Altering or Detecting Data, *Bell Syst. Tech. J.*, vol. 64, pp. 1885–1911, October.

Foschini, G. J. and Gans, M. J. (1998). "On Limits of Wireless Communications in a Fading Environment When Using Multiple Antennas," *Wireless Personal Commun.* pp. 311–335, June.

Foschini, G. J., Gitlin, R. D., and Weinstein, S. B. (1974). " Optimization of Two-Dimensional Signal Constellations in the Presence of Gaussian Noise," *IEEE Trans. Commun.*, vol. COM-22, pp. 28–38, January.

Franaszek, P. A. (1968). "Sequence-State Coding for Digital Transmission," *Bell Syst. Tech. J.*, vol. 27, p. 143.

Franaszek, P. A. (1969). "On Synchronous Variable Length Coding for Discrete Noiseless Channels," *Inform. Control.*, vol. 15, pp. 155–164.

Franaszek, P. A. (1970). "Sequence-State Methods for Run-Length-Limited Coding," *IBM J. Res. Dev.*, pp. 376–383, July.

Franks, L. E. (1969). *Signal Theory*, Prentice-Hall, Englewood Cliff, NJ.

Franks, L. E. (1983). "'Carrier and Bit Synchronization in Data Communication—A Tutorial Review," *IEEE Trans. Commun.*, vol. COM-28, pp. 1107–1121, August.

Franks, L. E. (1981). "Synchronization Subsystems: Analysis and Design," in *Digital Communications, Satellite/Earth Station Engineering*, K. Feher (ed.), Prentice-Hall, Englewood Cliffs, NJ.

Franks, L. E. (1980). "Carrier and Bit Synchronization in Data Communication—A Tutorial Review," *IEEE Trans. Commun.*, vol. COM-28, pp. 1107–1120, August.

Fredricsson, S. (1974). "Optimum Transmitting Filter in Digital PAM Systems with a Viterbi Detector," *IEEE Trans. Inform. Theory*, vol. 20, pp. 479–489.

Fredricsson, S. (1975). "Pseudo-Randomness Properties of Binary Shift Register Sequences," *IEEE Trans. Inform. Theory*, vol. IT-21, pp. 115–120, January.

Freiman, C. E. and Wyner, A. D. (1964). "Optimum Block Codes for Noiseless Input Restricted Channels," *Inform. Control*, vol. 7, pp. 398–415.

Frenger, P., Orten, P., Ottosson, T., and Svensson, A. (1998). "Multirate convolutional Codes:, Tech. Report No. 21, Communication Systems Group, Department of Signals and Systems, Chalmers University of Technology, Goteborg, Sweden, April.

Gaarder, N. T. (1971). "Signal Design for Fast-Fading Gaussian Channels," *IEEE Trans. Inform. Theory*, vol. IT-17, pp. 247–256, May.

Gabor, A. (1967). "'Adaptive Coding for Self Clocking Recording," *IEEE Trans. Electronic Comp.* vol. EC-16, p. 866.

Gallager, R. G. (1965). "Simple Derivation of the Coding Theorem and Some Applications," *IEEE Trans. Inform. Theory*, vol. IT-11, pp. 3–18, January.

Gallager, R. G. (1968). *Information Theory and Reliable Communication*, Wiley, New York.

Gaarder, F. M. (1979). *Phaselock Techniques*, Wiley, New York.

Gardner, W. A. (1984). "Learning Characteristics of Stochastic-Gradient Descent Algorithms: A General Study, Analysis, and Critique," *Signal Processing*, vol. 6, pp. 113–133, April.

Garg, V. K., Smolik, K., and Wilkes, J. E. (1997). *Applications of CDMA in Wireless/Personal Communications*, Prentice-Hall, Upper Saddle River, NJ.

Garth, L. M. and Poor, H. V. (1992). "Narrowband Interference Suppression in Impulsive Channels," *IEEE Trans. Aerospace and Electronic Sys.*, vol. 28, pp. 81–89, January.

George, D. A., Bowen, R. R., and Storey, J. R. (1971). "An Adaptive Decision-Feedback Equalizer," *IEEE Trans. Commun. Tech.*, vol. COM-19, pp. 281–293, June.

Gersho, A. (1982). "On the Structure of Vector Quantizers," *IEEE Trans. Inform. Theory*, vol. IT-28, pp. 157–166, March.

Gersho, A. and Gray, R. M. (1992). *Vector Quantization and Signal Compression*, Kluwer Academic Publishers, Boston.

Gersho, A. and Lawrence, V. B. (1984). "Multidimensional Signal Constellations for Voiceband Data Transmission," *IEEE J. Selected Areas Commun*, vol. SAC-2, pp. 687–702, September.

Gerst, I. and Diamond, J. (1961). "The Elimination of Intersymbol Interference by Input Pulse Shaping," *Proc. IRE*, vol. 53, July.

Ghez, S., Verdu, S., and Schwartz, S. C. (1988). "Stability Properties of Slotted Aloha with Multipacket Reception Capability," *IEEE Trans. Autom. Control*, vol. 33, pp. 640–649, July.

Ghosh, M. and Weber, C. L. (1991). "Maximum Likelihood Blind Equalization," *Proc. 1991 SPIE Conf.*, San Diego, CA, July.

Giannakis, G. B. (1987). "Cumulants: A Powerful Tool in Signal Processing," *Proc. IEEE*, vol. 75, pp. 1333–1334, September.

Giannakis, G. B. and Mendel, J. M. (1989). "Identification of Nonminimum Phase Systems Using Higher-Order Statistics," *IEEE Trans. Acoust., Speech and Signal Processing*, vol. 37, pp. 360–377, March.

Gilbert, E. N. (1952). "A Comparison of Signaling Alphabets," *Bell Syst. Tech. J.*, vol. 31, pp. 504–522, May.

Gilhousen, K. S., Jacobs, I. M., Podovani, R., Viterbi, A. J., Weaver, L. A., and Wheatley, G. E. III (1991). "On the Capacity of a Cellular CDMA System," *IEEE Trans. Vehicular Tech*, vol. 40, pp. 303–312, May.

Gitlin, R. D., Meadors, H. C., and Weinstein, S. B. (1982). "The Tap Leakage Algorithm: An Algorithm for the Stable Operation of a Digitally Implemented Fractionally Spaced, Adaptive Equalizer," *Bell Syst. Tech. J.*, vol. 61, pp. 1817–1839, October.

Gitlin, R. D. and Weinstein, S. B. (1979). "On the Required Tap-Weight Precision for Digitally Implemented Mean-Squared Equalizers," *Bell Syst. Tech. J.*, vol. 58, pp. 301–321, February.

Gitlin, R. D. and Weinstein, S. B. (1981). "Fractionally-Spaced Equalization: An Improved Digital Transversal Equalizer," *Bell Syst. Tech. J.*, vol. 60, pp. 275–296, February.

Glave, F. E. (1972). "An Upper Bound on the Probability of Error due to Intersymbol Interference for Correlated Digital Signals," *IEEE Trans. Inform. Theory*, vol. IT-18, pp. 356–362, May.

Goblick, T. J., Jr. and Holsinger, J. L. (1967). "Analog Source Digitization: A Comparison of Theory and Practice," *IEEE Trans. Inform. Theory*, vol. IT-13, pp. 323–326, April.

Godard, D. N. (1974). "Channel Equalization Using a Kalman Filter for Fast Data Transmission," *IBM J. Res. Dev.*, vol. 18, pp. 267–273, May.

Godard, D. N. (1980). "Self-Recovering Equalization and Carrier Tracking in Two-Dimensional Data Communications Systems," *IEEE Trans. Commun.*, vol. COM-28, pp. 1867–2875, November.

Golay, M. J. E. (1949). "Note on Digital Coding," *Proc. IRE*, vol. 37, p. 657, June.

Gold, R. (1967). "Optimal Binary Sequences for Spread Spectrum Multiplexing," *IEEE Trans. Inform. Theory*, vol. IT-13, pp. 619–621, October.

Gold, R. (1968). "Maximal Recursive Sequences with 3-Valued Recursive Cross Correlation Functions," *IEEE Trans. Inform. Theory*, vol. IT-14, pp. 154–156, January.

Golomb, S. W. (1967). *Shift Register Sequences*, Holden-Day, San Francisco, CA.

Goppa, V. D. (1970). "New Class of Linear Correcting Codes," *Probl. Peredach. Inform.*, vol. 6, pp. 24–30.

Goppa, V. D. (1971). "Rational Presentation of Codes and (L, g)-codes," *Probl. Peredach. Inform.*, vol 7, pp. 41–49.

Gray, R. M. (1975). "Sliding Block Source Coding," *IEEE Trans. Inform. Theory*, vol. IT-21, pp. 357–368, July.

Gray, R. M. (1990). *Source Coding Theory*, Kluwer Academic Publishers, Boston.

Gray, R. M. and Neuhoff, D. L. (1998). "Quantization," *IEEE Trans. Inform. Theory*, vol. 44, pp. 2325–2383, October.

Greefkes, J. A. (1970). "A Digitally Companded Delta Modulation Modem for Speech Transmission," *Proc. IEEE Int. Conf. on Commun*, pp. 7.33–7.48, June.

Green, P. E., Jr. (1962). "Radar Astronomy Measurement Techniques," MIT Lincoln Laboratory, Lexington, MA, Tech. Report No. 282, December.

Gronemeyer, S. A. and McBride, A. L. (1976). "MSK and Offset QPSK Modulation," *IEEE Trans. Commun*, vol. COM-24, pp. 809–820, August.

Gupta, S. C. (1975). "Phase-Locked Loops," *Proc. IEEE*, vol. 63, pp. 291–306, February.

Haccoun, D. and Bégin, G. (1989). "High-Rate Punctured Convolutional Codes for Viterbi and Sequential Decoding," *IEEE Trans. Commun.*, vol. 37, pp. 1113–1125, November.

Hagenauer, J. (1988). "Rate Compatible Punctured Convolutional Codes and Their Applications," *IEEE Trans. Commun.*, vol. 36, pp. 389–400, April.

Hagenauer, J. and Hoeher, P. (1989). "A Viterbi Algorithm with Soft-Decision Outputs and its Applications," *Proc. IEEE GLOBECOM Conf.*, pp. 1680–1686, November, Dallas, TX.

Hagenauer, J., Offer, E., Méasson, C., and Mörz, M. (1999). "Decoding and Equalization with Analog Non-Linear Networks," *European Trans. Telecommun.*, vol. 10, pp. 659–680, November/December.

Hagenauer, J., Offer, E., and Papke, L. (1996). "Iterative Decoding of Binary Block and Convolutional Codes," *IEEE Trans. Inform. Theory*, vol. IT-42, pp. 429–445, March.

Hagenauer, J., Seshadri, N., and Sundberg, C.-E. (1990). "The Performance of Rate-Compatible Punctured Convolutional Codes for Digital Mobile Radio," *IEEE Trans. Commun.*, vol. 38, pp. 966–980, July.

Hahn, P. M. (1962). "Theoretical Diversity Improvement in Multiple Frequency Shift Keying," *IRE Trans. Commun. Syst.*, vol. CS-10, pp. 177–184, June.

Hamming, R. W. (1950). "'Error Detecting and Error Correcting Codes," *Bell Syst. Tech. J.,* vol. 29, pp. 147–160, April.

Hamming, R. W. (1986). *Coding and Information Theory*, Prentice-Hall, Englewood Cliffs, NJ.

Hancock, J. C. and Lucky, R. W. (1960). "Performance of Combined Amplitude and Phase-Modulated Communication Systems," *IRE Trans. Commun. syst.*, vol. CS-8, pp. 232–237, December.

Harashima, H. and Miyakawa, H. (1972). "Matched-Transmission Technique for Channels with Intersymbol Interference," *IEEE Trans. Commun.*, vol. COM-20, pp. 774–780.

Hartley, R. V. (1928). "Transmission of Information," *Bell Syst. Tech. J.*, vol. 7, p. 535.

Hatzinakos, D. and Nikias, C. L. (1991). "Blind Equalization Using a Tricepstrum-Based Algorithm," *IEEE Trans. Commun*, vol. COM-39, pp. 669–682, May.

Haykin, S. (1996). *Adaptive Filter Theory*, 3rd ed., Prentice-Hall: Upper Saddle River, NJ.

Hecht, M. and Guida, A. (1969). "Delay Modulation," *Proc. IEEE*, vol. 57, pp. 1314–1316, July.

Heegard, C. and Wicker, S. B. (1990). *Turbo Coding*, Kluwer Academic Publishers, Boston, MA.

Heller, J. A. (1968). "Short Constraint Length Convolutional Codes," Jet Propulsion Laboratory, California Institute of Technology, Pasadena, CA *Space Program Summary* 37–54, vol. 3, pp. 171–174, December.

Heller, J. A. (1975). "Feedback Decoding of Convolutional Codes," in *Advances in Communication Systems*, vol. 4, A. J. Viterbi (ed.), Academic, New York.

Heller, J. A. and Jacobs, I. M. (1971). "Viterbi Decoding for Satellite and Space Communication," *IEEE Trans. Commun. Tech.*, vol. COM-19, pp. 835–848, October.

Helstrom, C. W. (1955). "The Resolution of Signals in White Gaussian Noise," *Proc. IRE*, vol. 43, pp. 1111–11187, September.

Helstrom, C. W. (1968). *Statistical Theory of Signal Detection*, Pergamon, London.

Helstrom, C. W., (1991). *Probability and Stochastic Processes for Engineers*, Macmillan, New York.

Hildebrand, F. B. (1961). *Methods of Applied Mathematics, Prentice-Hall, Englewood Cliffs, NJ.*

Hirosaki, B. (1981). "An Orthogonality Multiplexed QAM System Using the Discrete Fourier Transform," *IEEE Trans. Commun.*, vol. COM-29, pp. 982–989, July.

Hirosaki, B., Hasegawa, S., and Sabato, A. (1986). "Advanced Group-Band Modem Using Orthogonally Multiplexed QAM Techniques," *IEEE Trans. Commun.*, vol. COM-34, pp. 587–592, June.

Ho, E. Y. and Yeh, Y. S. (1970). "A New Approach for Evaluating the Error Probability in the Presence of Intersymbol Interference and Additive Gaussian Noise," *Bell Syst. Tech. J.*, vol. 49, pp. 2249–2265, November.

Hocquenghem, A. (1959). "'Codes Correcteurs d'Erreurs," *Chiffres*, vol. 2, pp. 147–156.

Hole, K. J. (1988). "New Short Constraint Length Rate $(n = 1)/n$ Punctured Convolutional Codes for Soft-Decision Viterbi Decoding," *IEEE Trans. Inform. Theory*, vol. 34, pp. 1079–1081, September.

Holmes, J. K. (1982). *Coherent Spread Spectrum Systems*, Wiley-Interscience, New York.

Honig, M. L. (1998). "Adaptive Linear Interference Suppression for Packet DS-CDMA," *European Trans. Telecommun. (ETT)*, vol. 9, pp. 173–181, March–April.

Honig, M. L., Madhow, U., and Verdu, S. (1995). "Blind Adaptive Multiuser Detection," *IEEE Trans. Inform. Theory*, vol. 41, pp. 944–960, July.

Horwood, D. and Gagliardi, R. (1975). "Signal Design for Digital Multiple Access Communications," *IEEE Trans. Commun.*, vol. COM-23, pp. 378–383, March.

Hsu, F. M. (1982). "Square-Root Kalman Filtering for High-Speed Data Received over Fading Dispersive HF Channels," *IEEE Trans. Inform. Theory*, vol. IT-28, pp. 753–763, September.

Huffman, D. A. (1952). "A Method for the Construction of Minimum Redundancy Codes," *Proc. IRE*, vol. 40, pp. 1098–1101, September.

Hui, J. Y. N. (1984). "Throughput Analysis for Code Division Multiple Accessing of the Spread Spectrum Channel," *IEEE J. Selected Areas Commun.*, vol. SAC-2, pp. 482–486, July.

Im, G. H. and Un, C. K. (1987). "A Reduced Structure of the Passband Fractionally-Spaced Equalizer," *Proc. IEEE*, vol. 75, pp. 847–849, June.

Immink, K. A. S. (1990). "Runlength-Limited Sequences," *Proc. IEEE*, vol. 78, pp. 1745–1759, November.

Immink, K. A. S., Siegel, P. H., and Wolf, J. K. (1998). "Codes for Digital Recorder," *IEEE Trans. Inform. Theory*, vol. 44, pp. 2260–2299, October.

Itakura, F. (1975). "Minimum Prediction Residual Principle Applied to Speech Recognition," *IEEE Trans. Acoust., Speech, Signal Processing*, vol. ASSP-23, pp. 67–72, February.

Itakura, F. and Saito, S. (1968). "Analysis Synthesis Telephony Based on the Maximum-Likelihood Methods," *Proc. 6th Int. Congr. Acoust.*, Tokyo, Japan, pp. C17–C20.

Jacobs, I. M. (1974). "Practical Applications of Coding," *IEEE Trans. Inform. Theory*, vol. IT-20, pp. 305–310, May.

Jacoby, G. V. (1977). "A New Look-ahead Code for Increased Data Density," *IEEE Trans. Magnetics*, vol. MAG-13, pp. 1202–1204.

Jakes, W. C. (1974). *Microwave Mobile Communications*, Wiley, New York.

Jayant, N. S. (1970). "Adaptive Delta Modulation with a One-Bit Memory," *Bell Syst. Tech. J.*, pp. 321–342, March.

Jayant, N. S. (1974). "Digital Coding of Speech Waveforms: PCM, DPCM, and DM Quantizers," *Proc. IEEE*, vol. 62, pp. 611–632, May.

Jayant, N. S. (1976). *Waveform Quantization and Coding*, IEEE Press, New York.

Jayant, N. S. (1990). "High Quality Coding of Telephone Speech and Wideband Audio," *IEEE Commun. Mag.* vol. 28, pp. 10–20, January.

Jayant, N. S. and Noll, P. (1984). *Digital Coding of Waveforms*, Prentice-Hall, Englewood Cliffs, NJ.

Jelinek, F. (1968). *Probabilistic Information Theory*, Mc-Graw-Hill, New York.

Jelinek, F. (1969). "Fast Sequential Decoding Algorithm Using a Stack," *IBM J. Res. Dev., vol. 13, pp. 675–685, November.*

Johannesson, R. and Zigangirov, K. S. (1999). *Fundamentals of Convolutional Coding*, IEEE Press, New York.

Johnson, C. R. (1991). "Admissibility in Blind Adaptive Channel Equalization," *IEEE Control Syst. Mag.*, pp. 3–15, January.

Jones, A. E., Wilkinson, T. A., and Barton, S. K. (1994). "Block Coding Scheme for Reduction of Peak-to-Mean Envelope Power Ratio of Multicarrier Transmission Schemes," *Electr. Lett.*, vol. 30, pp. 2098–2099, December.

Jones, S. K., Cavin, R. K., and Reed, W. M. (1982). "Analysis of Error-Gradient Adaptive Linear Equalizers for a Class of Stationary-Dependent Process," *IEEE Trans. Inform. Theory*, vol. IT-28, pp. 318–329, March.

Jordan, K. L., Jr. (1966). '"The Performance of Sequential Decoding in Conjunction with Efficient Modulation," *IEEE Trans. Commun. Syst.*, vol. CS-14, pp. 283–287, June.

Justesen, J. (1972). '"A Class of Constructive Asymptotically Good Algebraic Codes," *IEEE Trans. Inform. Theory*, vol. IT-18, pp. 652–656, September.

Kailath, T. (1960). "Correlation Detection of Signals Perturbed by a Random Channel," *IRE Trans. Inform. Theory*, vol. IT-6, pp. 361–366, June.

Kailath, T. (1961). "Channel Characterization: Time-Variant Dispersive Channels, in *Lectures on Communication System Theory*, chap. 6, E. Baghdady (ed.), McGraw-Hill, New York.

Kalet, I. (1989). "The Multitone Channel," *IEEE Trans. Commun.*, vol. COM-37, pp. 119–124, February.

Karabed, R. and Siegel, P. H. (1991). "Matched-Spectral Null Codes for Partial-Response Channels," *IEEE Trans. Inform. Theory*, vol. IT-37, pp. 818–855, May.

Kasami, T. (1966). "Weight Distribution Formula for Some Class of Cyclic Codes," Coordinated Science Laboratory, University of Illinois, Urbana, IL, Tech. Report No. R-285, April.

Kawas Kalet, G. (1989). "Simple Coherent Receivers for Partial Response Continuous Phase Modulation," *IEEE J. Selected Areas Commun.*, vol. 7, pp. 1427–1436, December.

Kaye, A. R. and George, D. A. (1970). "Transmission of Multiplexed PAM Signals over Multiple Channel and Diversity Systems," *IEEE Trans. Commun.*, vol. COM-18, pp. 520–525, October.

Kelly, E. J., Reed, I. S., and Root, W. L. (1960). "The Detection of Radar Echoes in Noise, Pt. I." *J. SIAM*, vol. 8, pp. 309–341, September.

Ketchum, J. and Proakis, J. G. (1982). "Adaptive Algorithms for Estimating and Suppressing Narrowband Interference in PN Spread Spectrum Systems," *IEEE Trans. Commun.*, vol. COM-30, pp. 913–924, May.

Kleinrock, L. and Tobagi, F. A. (1975). "Packet Switching in Radio Channels: Part I—Carrier Sense Multiple-Access Modes and Their Throughput-Delay Characteristics," *IEEE Trans. Commun.*, vol. COM-23, pp. 1400–1416, December.

Klovsky, D. and Nikolaev, B. (1978) *Sequential Transmission of Digital Information in the Presence of Intersymbol Interference*, Mir Publishers, Moscow.

Kobayashi, H. (1971). "Simultaneous Adaptive Estimation and Decision Algorithm for Carrier Modulated Data Transmission Systems," *IEEE Trans. Commun. Tech.*, vol. COM-19, pp. 268–280, June.

Kolmogorov, A.N. (1939). "Sur l'interpolation et extrapolation des suites stationaires," *Comptes Rendus de l'Académie des Sciences*, vol. 208, p. 2043.

Kotelnikov, V. A. (1947). "The Theory of Optimum Noise Immunity," Ph.D. Dissertation, Molotov Energy Institute, Moscow. [Translated by R. A. Silverman, McGraw-Hill, New York.]

Kretzmer, E. R. (1966). "Generalization of a Technique for Binary Data Communication," *IEEE Trans. Commun. Tech.*, vol. COM-14, pp. 67–68, February.

Larsen, K. J. (1973). "Short Convolutional Codes with Maximal Free Distance for Rates 1/2, 1/3, and 1/4," *IEEE Trans. Inform. Theory*, vol. IT-19, pp. 371–372, May.

Laurent, P. A. (1986). "Exact and Approximate Construction of Digital Phase Modulations by Superposition of Amplitude Modulated Pulses," *IEEE Trans. Commun.*, vol. COM-34, pp. 150–160, February.

Lee, P. J. (1988). "Construction of Rate $(n-1)/n$ Punctured Convolutional Codes with Minimum Required *SNR* Criterion," *IEEE Trans. Commun.*, vol. 36, pp. 1171–1174, October.

Lee, W. U. and Hill, F. S. (1977). "A Maximum-Likelihood Sequence Estimator with Decision-Feedback Equalizer," *IEEE Trans. Commun.*, vol. 25, pp. 971–979, September.

LeGoff, S., Glavieux, A., and Berrou, C. (1994). "Turbo-codes and High Spectral Efficiency Modulation," *Proc. Int. Conf. Commun.* (*ICC '94*), pp. 645–649, May, New Orleans, LA.

Lender, A. (1963). "The Duobinary Technique for High Speed Data Transmission," *AIEE Trans. Commun. Electronics*, vol. 82, pp. 214–218.

Leon-Garcia, A. (1994). *Probability and Random Processes for Electrical Engineering*, Addison-Wesley, Reading, MA.

Levinson, N. (1947). "The Wiener RMS (Root Mean Square) Error Criterion in Filter Design and Prediction," *J. Math. and Phys.*, vol. 25, pp. 261–278.

Lin, S. and Costello, D. J., Jr. (1983). *Error Control Coding: Fundamentals and Applications*, Prentice-Hall, Englewood Cliffs, NJ.

Linde, Y., Buzo, A., and Gray, R. M. (1980). "An Algorithm for Vector Quantizer Design." *IEEE Trans. Commun.*, vol. COM-28, pp. 84–95, January.

Lindell, G. (1985). "On Coded Continuous Phase Modulation," Ph.D. Dissertation, Telecommunication Theory, University of Lund, Lund, Sweden, May.

Lindholm, J. H. (1968). "An Analysis of the Pseudo-Randomness Properties of Subsequences of Long *m*-Sequencies," *IEEE Trans. Inform. Theory*, vol. IT-14, pp. 569-576, July.

Lindsey, W. C. (1964). "Error Probabilities for Ricean Fading Multichannel Reception of Binary and *N*-Ary Signals," *IEEE Trans. Inform. Theory*, vol. IT-10, pp. 339–350, October.

Lindsey, W. C. (1972). *Synchronization Systems in Communications*, Prentice-Hall, Englewood Cliffs, NJ.

Lindsey, W. C. and Chie, C. M. (1981). "A Survey of Digital Phase-Locked Loops," *Proc. IEEE*, vol. 69, pp. 410–432.

Lindsey, W. C. and Simon, M. K. (1973). *Telecommunication Systems Engineering*, Prentice-Hall, Englewood Cliffs, NJ.

Ling, F. (1988). "Convergence Characteristics of LMS and LS Adaptive Algorithms for Signals with Rank-Deficient Correlation Matrices," *Proc. Int. Conf. Acoust., Speech, Signal Processing*, New York, 25.D.4.7, April.

Ling, F. (1989). "On Training Fractionally-Spaced Equalizers Using Intersymbol Interpolation," *IEEE Trans. Commun.*, vol. 37, pp. 1096–1099, October.

Ling, F., Manolakis, D. G., and Proakis, J. G. (1986a). "Finite, Word-Length Effects in Recursive Least Squares Algorithms with Application to Adaptive Equalization," *Annales des Telecommunications*, vol. 41, pp. 1–9, May/June.

Ling, F., Manolakis, D. G., and Proakis, J. G. (1986b). "Numerically Robust Least-Squares Lattice-Ladder Algorithms with Direct Updating of the Reflection Coefficients," *IEEE Trans. Acoust., Speech, Signal Processing*, vol. ASSP-34, pp. 837–845, August.

Ling, F. and Proakis, J. G. (1982). Generalized Least Squares Lattice and Its Applications to DFE," *Proc. 1982, IEEE Int. Conf. on Acoust. Speech, Signal Processing*, Paris, France, May.

Ling, F. and Proakis, J. G. (1984a), "Numerical Accuracy and Stability: Two Problems of Adaptive Estimation Algorithms Caused by Round-Off Error," *Proc. Int. Conf. Acoust., Speech, Signal Processing*, pp. 30.3.1–30.3.4, San Diego, CA. March.

Ling, F. and Proakis, J. G. (1984b). "Nonstationary Learning Characteristics of Least Squares Adaptive Estimation Algorithms," *Proc. Int. Conf. Acoust, Speech, Signal Processing*, pp. 3.7.1–3.7.4, San Diego, CA, March.

Ling, F. and Proakis, J. G. (1984c). "A Generalized Multichannel Least-Squares Lattice Algorithm with Sequential Processing Stages," *IEEE Trans. Acoust., Speech, Signal Processing*, vol. ASSP-32, pp. 381–389, April.

Ling, F. and Proakis, J. G. (1985). "Adaptive Lattice Decision-Feedback Equalizers—Their Performance and Application to Time-Variant Multipath Channels," *IEEE Trans. Commun*, vol. COM-33, pp. 348–356, April.

Ling, F. and Proakis, J. G. (1986). "A Recursive Modified Gram–Schmidt Algorithm," *IEEE Trans. Acoust., Speech, Signal Processing*, vol. ASSP-34, pp. 829–836, August.

Ling, F. and Qureshi, S. U. H. (1986). "Lattice Predictive Decision-Feedback Equalizer for Digital Communication Over Fading Multipath Channels," *Proc. GLOBECOM '86*, Houston, TX, December.

Ling, F. and Qureshi, S. U. H. (1990). "Convergence and Steady State Behavior of a Phase-Splitting Fractionally Spaced Equalizer," *IEEE Trans. Commun.* vol. 38, pp. 418–425, April.

Ljung, S. and Ljung, L. (1985). "Error Propagation Properties of Recursive Least-Squares Adaptation Algorithms," *Automatica*, vol. 21, pp. 159–167.

Lloyd, S. P. (1982). "Least Squares Quantization in PCM," *IEEE Trans. Inform. Theory*, vol. IT-28, pp. 129–137, March.

Loeve, M. (1955). *Probability Theory*, Van Nostrand, Princeton, NJ.

Long, G., Ling, F., and Proakis, J. G. (1987). "Adaptive Transversal Filters with Delayed Coefficient Adaptation," *Proc. Int. Conf. Acoust., Speech, Signal Processing*, Dallas, TX, March.

Long, G., Ling, F., and Proakis, J. G. (1988a). "Fractionally-Spaced Equalizers Based on Singular-Value Decomposition," *Proc. Int. Conf. Acoust., Speech, Signal Processing*, New York, 25.D.4.10, April.

Long, G., Ling, F., and Proakis, J. G. (1988b). "Applications of Fractionally-Spaced Decision-Feedback Equalizers to HF Fading Channels," *Proc. MILCOM*, San Diego, CA, October.

Long, G., Ling, F., and Proakis, J. G. (1989). "The LMS Algorithm with Delayed Coefficient Adaptation," *IEEE Trans. Acoust., Speech, Signal Processing*, vol. ASSP-37, October.

Lu, J., Letaief, K. B., Chuang, J. C., and Liou, M. L. (1999). "M-PSK and M-QAM BER Computation Using Signal-Space Concepts," *IEEE Trans. Commun.*, vol. 47, pp. 181–184, February.

Lucky, R. W. (1965). "Automatic Equalization for Digital Communications, *Bell Syst. Tech. J.,* vol. 44, pp. 547–588, April.

Lucky, R. W. (1966). "Techniques for Adaptive Equalization of Digital Communication," *Bell Syst. Tech. J.*, vol. 45, pp. 255–286.

Lucky, R. W. and Hancock, J. C. (1962). "On the Optimum Performance of N-ary Systems Having Two Degrees of Freedom," *IRE Trans. Commun. Syst.*, vol. CS-10, pp. 185–192, June.

Lucky, R. W., Salz, J., and Weldon, E. J., Jr. (1968). *Principles of Data Communication*, McGraw-Hill, New York.

Lugannani, R. (1969). "Intersymbol Interference and Probability of Error in Digital Systems," *IEEE Trans. Inform. Theory*, vol, IT-15, pp. 682–688, November.

Lundgren, C. W. and Rummler, W. D. (1979). "Digital Radio Outage Due to Selective Fading—Observation vs. Prediction from Laboratory Simulation," *Bell Syst. Tech. J.*, vol. 58, pp. 1074–1100, May/June.

Lupas, R. and Verdu, S. (1989). "Linear Multiuser Detectors for Synchronous Code-Division Multiple-Access Channels," *IEEE Trans. Inform. Theory*, vol. IT-35, pp. 123–136, January.

Lupas, R. and Verdu, S. (1990). "Near-Far Resistance of Multiuser Detectors in Asynchronous Channels," *IEEE Trans. Commun.*, vol. COM-38, pp. 496–508, April.

MacKenchnie, L. R. (1973). "Maximum Likelihood Receivers for Channels Having Memory," Ph.D. Dissertation, Department of Electrical Engineering, University of Notre Dame, Notre Dame, IN, January.

MacWilliams, F. J. and Sloane, J. J. (1977). *The Theory of Error Correcting Codes*, North Holland, New York.

Madhow, U. (1998). "Blind Adaptive Interference Suppression for Direct Sequence CDMA," *Proc. IEEE*, vol. 86, pp. 2049–2069, October.

Madhow, U. and Honig, M. L. (1994). "MMSE Interference Suppression for Direct-Sequence Spread-Spectrum CDMA," *IEEE Trans. Commun.*, vol. 42, pp. 3178–3188, December.

Magee, F. R. and Proakis, J. G. (1973). "Adaptive Maximum-Likelihood Sequence Estimation for Digital Signaling in the Presence of Intersymbol Interference," *IEEE Trans. Inform. Theory*, vol. IT-19, pp. 120–124, January.

Makhoul, J. (1978). "A Class of All-Zero Lattice Digital Filters: Properties and Applications," *IEEE Trans. Acoust., Speech, Signal Processing*, vol. ASSP-26, pp. 304–314, August.

Makhoul, J., Roucos, S., and Gish, H. (1985). "Vector Quantization in Speech Coding," *Proc. IEEE*, vol. 73, pp. 1551–1587, November.

Martin, D. R. and McAdam, P. L. (1980). "Convolutional Code Performance with Optimal Jamming," *Conf. Rec. Int. Conf. Commun.*, pp. 4.3.1–4.3.7, May.

Massey, J. L. (1963). *Threshold Decoding*, MIT Press Cambridge, MA.

Massey, J. L. (1965). "Step-by-Step Decoding of the BCH Codes," *IEEE Trans. Inform. Theory*, vol. IT-11, pp. 580–585, October.

Massey, J. L. (1988). "Some New Approaches to Random Access Communications," *Performance '87*, pp. 551–569. [Reprinted 1993 in *Multiple Access Communications*, N. Abramson (ed.), IEEE Press, New York.]

Massey, J. L. and Sain, M. (1968). "Inverses of Linear Sequential Circuits," *IEEE Trans. Comput.*, vol. C-17, pp. 330–337, April.

Matis, K. R. and Modestino, J. W. (1982). "Reduced-State Soft-Decision Trellis Decoding of Linear Block Codes," *IEEE Trans. Inform. Theory*, vol. IT-28, pp. 61–68, January.

Max, J. (1960). "Quantizing for Minimum Distortion," *IRE Trans. Inform. Theory*, vol. IT-6, pp. 7–12, March.

Mazo, J. E. (1975). "Faster-Than-Nyquist Signaling," *Bell Syst. Tech. J.*, vol. 54, pp. 1451–1462, October.

Mazo, J. E. (1979). "On the Independence Theory of Equalizer Convergence," *Bell Syst. Tech. J.,* vol. 58, pp. 963–993, May.

McMahon, M. A. (1984). *The Making of a Profession—A Century of Electrical Engineering in America*, IEEE Press, New York.

Mengali, U. (1977). "Joint Phase and Timing Acquisition in Data Transmission," *IEEE Trans. Commun.*, vol. COM-25, pp. 1174–1185, October.

Mengali, U. and D'Andrea, A. N. (1997). *Synchronization Techniques for Digital Receivers*, Plenum Press, New York.

Mengali, U. and Morelli, M. (1995). "Decomposition of M-ary CPM Signals into PAM Waveforms," *IEEE Trans. Inform. Theory*, vol. 41, pp. 1265–1275, September.

Meyers, M. H. and Franks, L. E. (1980). "Joint Carrier Phase and Symbol Timing for PAM Systems," *IEEE Trans. Commun.*, vol. COM-28, pp. 1121–1129, August.

Meyr, H. and Ascheid, G. (1990). *Synchronization in Digital Communications*, Wiley Interscience, New York.

Meyr, H., Moenclaey, M., and Fechtel, S. A. (1998). *Digital Commun. Receivers*, Wiley, New York.

Miller, K. S. (1964). *Multidimensional Gaussian Distributions*, Wiley, New York.

Miller, S. L. (1996). "Training Analysis of Adaptive Interference Suppression for Direct-Sequence CDMA Systems," *IEEE Trans. Commun.*, vol. 44, pp. 488–495, April.

Miller, S. L. (1995). "An Adaptive Direct-Sequence Code-Division Multiple Access Receiver for Multiuser Interference Rejection," *IEEE Trans. Commun.*, vol. 43, pp. 1746–1755, Feb./March/April.

Millman, S. (ed.) (1984). *A History of Engineering and Science in the Bell System—Communication Sciences (1925–1980)*, AT&T Bell Laboratories.

Milstein, L. B. (1988). "Interference Rejection in Spread Spectrum Communications," *Proc. IEEE*, vol. 76, pp. 657–671, June.

Mitra, U. and Poor, H. V. (1995). "Adaptive Receiver Algorithm for Near-Far Resistant CDMA," *IEEE Trans. Commun.*, vol. 43, pp. 1713–1724, April.

Miyagaki, Y., Morinaga, N., and Namekawa, T. (1978). "Error Probability Characteristics for CPSK Signal Through m-Distributed Fading Channel," *IEEE Trans. Commun.*, vol. COM-26, pp. 88–100, January.

Moher, M. (1998). "An Iterative Multiuser Decoder for Near-Capacity Communications," *IEEE Trans. Commun.*, vol. 46, pp. 870–880, July.

Moon, J. and Carley, L. R. (1988). "Partial Response Signaling in a Magnetic Recording Channel," vol. MAG-24, pp. 2973–2975, November.

Monsen, P. (1971). "Feedback Equalization for Fading Dispersive Channels," *IEEE Trans. Inform. Theory*, vol. IT-17, pp. 56–64, January.

Morf, M. (1977). "Ladder Forms in Estimation and System Identification," *Proc. 11th Annual Asilomar Conf. on Circuits, Systems and Computers*, Monterey, CA, Nov. 7–9.

Morf, M., Dickinson, B., Kailath, T., and Vieira, A. (1977a). "Efficient Solution of Covariance Equations for Linear Prediction," *IEEE Trans. Acoust., Speech, Signal Processing*, vol. ASSP-25, pp. 429–433, October.

Morf, M. and Lee, D. (1978). "Recursive Least Squares Ladder Forms for Fast Parameter Tracking," *Proc. 1978 IEEE Conf. on Decision and Control*, San Diego, CA, pp. 1362–1367, January 12.

Morf, M., Lee, D., Nickolls, J., and Vieira, A. (1977b). "A Classification of Algorithms for ARMA Models and Ladder Realizations," *Proc. 1977 IEEE Int. Conf on Acoustics, Speech, Signal Processing*, Hartford, CT, pp. 13–19, May.

Morf, M., Vieira, A., and Lee, D. (1977c). "Ladder Forms for Identification and Speech Processing," *Proc. 1977 IEEE conf. on Decision and Control*, New Orleans, LA, pp. 1074–1078, December.

Mueller, K. H. and Muller, M. S. (1976). "Timing Recovery in Digital Synchronous Data Receivers," *IEEE Trans. Commun*, vol. COM-24, pp. 516–531, May.

Mueller, K. H. and Spaulding, D. A. (1975). "Cyclic Equalization—A New Rapidly Converging Equalization Technique for Synchronous Data Communications," *Bell Sys. Tech. J.*, vol. 54, pp. 369–406, February.

Mueller, K. H. and Werner, J. J. (1982). "A Hardware Efficient Passband Equalizer Structure for Data Transmission," *IEEE Trans. Commun.*, vol. COM-30, pp. 438–541, March.

Muller, D. E. (1954). "Application of Boolean Algebra to Switching Circuit Design and to Error Detection," *IRE Trans. Comput., vol. EC-3, pp. 6–12, September.*

Mulligan, M. G. (1988). "Multi-Amplitude Continuous Phase Modulation with Convolutional Coding," Ph.D. Dissertation, Department of Electrical and Computer Engineering, Northeastern University, June.

Nakagami, M. (1960). "The m-Distribution—A General Formula of Intensity Distribution of Rapid Fading," in *Statistical Methods of Radio Wave Propagation*, W. C. Hoffman (ed.), pp. 3–36, Pergamon Press, New York.

Natali, F. D. and Walbesser, W. J. (1969). "Phase-Locked Loop Detection of Binary PSK Signals Utilizing Decision Feedback," *IEEE Trans. Aerospace Electronic Syst.*, vol. AES-5, pp. 83–90, January.

Neyman, J. and Pearson, E. S. (1933). "On the Problem of the Most Efficient Tests of Statistical Hypotheses," *Phil. Trans. Roy. Soc. London, Series A*, vol. 231, pp. 289–337.

Nichols, H., Giordano, A. and Proakis, J. G. (1977). "MLD and MSE Algorithms for Adaptive Detection of Digital Signals in the Presence of Interchannel Interference," *IEEE Trans. Inform. Theory*, vol. IT-23, pp. 563–575, September.

North, D. O. (1943). "An Analysis of the Factors Which Determine Signal/Noise Discrimination in Pulse-Carrier Systems," RCA Tech. Report No. 6 PTR-6C.

Nyquist, H. (1924). "Certain Factors Affecting Telegraph Speed," *Bell Syst. Tech. J.*, vol. 3, p. 324.

Nyquist, H. (1928). "Certain Topics in Telegraph Transmission Theory," *AIEE Trans.*, vol. 47, pp. 617–644.

Odenwalder, J. P. (1970). "Optimal Decoding of Convolutional Codes," Ph.D. Dissertation, Department of Systems Sciences, School of Engineering and Applied Sciences, University of California, Los Angeles.

Odenwalder, J. P. (1976). "Dual-k Convolutional Codes for Noncoherently Demodulated Channels," *Proc. Int. Telemetering Conf.*, vol. 12, pp. 165–174, September.

Olsen, J. D. (1977). "Nonlinear Binary Sequences with Asymptotically Optimum Periodic Cross Correlation," Ph.D. Dissertation, University of Southern California, December.

Omura, J. (1971). "Optimal Receiver Design for Convolutional Codes and Channels with Memory Via Control Theoretical Concepts," *Inform. Sci.*, vol, 3, pp. 243–266.

Omura, J. K. and Levitt, B. K. (1982). "Code Error Probability Evaluation for Antijam Communication Systems," *IEEE Trans. Commun.*, vol. COM-30, pp. 896–903, May.

Osborne, W. P. and Luntz, M. B. (1974). "Coherent and Noncoherent Detection of CPSK," *IEEE Trans. Commun.*, vol. COM-22, pp. 1023–1036, August.

Paaske, E. (1974). "Short Binary Convolutional Codes with Maximal Free Distance for Rates 2/3 and 3/4," *IEEE Trans. Inform. Theory*, vol. IT-20, pp. 683–689, September.

Paez, M. D. and Glisson, T. H. (1972). "Minimum Mean Squared Error Quantization in Speech PCM and DPCM Systems," *IEEE Trans. Commun.*, vol. COM-20, pp. 225–230, April.

Pahlavan, K. (1985). "Wireless Communications for Office Information Networks," *IEEE Commun. Mag.*, vol. 23, pp. 18–27, June.

Palenius, T. (1991). "On Reduced Complexity Noncoherent Detectors for Continuous Phase Modulation," Ph.D. Dissertation, Telecommunication Theory, University of Lund, Lund, Sweden.

Palenius, T. and Svensson, A. (1993). "Reduced Complexity Detectors for Continuous Phase Modulation Based on Signal Space Approach," *European Trans. Telecommun.*, vol. 4, pp. 51–63, May/June.

Papoulis, A. (1984). *Probability, Random Variables, and Stochastic Processes*, McGraw-Hill, New York.

Patel, P. and Holtzman, J. (1994). "Analysis of a Simple Successive Interference Cancellation Scheme in a DS/CDMA System," *IEEE J. Select. Areas Commun.*, vol. 12, pp. 796–807, 1994.

Paul, D. B. (1983). "An 800 bps Adaptive Vector Quantization Vocoder Using a Perceptual Distance Measure," *Proc. IEEE Int. Conf. Acoust., Speech, Signal Processing*, Boston, MA, pp. 73–76, April.

Pearson, K. (1965). *Tables of the Incomplete Γ-Function*, Cambridge University Press, London.

Peebles, P. Z. (1987). *Probability, Random Variables, and Random Signal Principles*, McGraw-Hill, New York.

Peterson, R. L., Ziemer, R. E., and Borth, D. E. (1995). *Introduction to Spread Spectrum Communications*, Prentice-Hall, Upper Saddle, River, NJ.

Peterson, W. W. (1960). "Encoding and Error-Correction Procedures for Bose–Chaudhuri Codes," *IRE Trans. Inform. Theory*, vol. IT-6, pp. 459–470, September.

Peterson, W. W. and Weldon, E. J., Jr. (1972). *Error-Correcting Codes*, 2nd ed., MIT Press, Cambridge, MA.

Picchi, G. and Prati, G. (1987). "Blind Equalization and Carrier Recovery Using a Stop-and-Go Decision Directed Algorithm," *IEEE Trans. Commun.*, vol. COM-35, pp. 877–887, September.

Picinbono, B. (1978). "Adaptive Signal Processing for Detection and Communication," in *Communication Systems and Random Process Theory*, J. K. Skwirzynski (ed.), Sijthoff & Nordhoff, Alphen aan den Rijn, The Netherlands.

Pickholtz, R. L., Schilling, D. L., and Milstein, L. B. (1982). "Theory of Spread Spectrum Communications—A Tutorial," *IEEE Trans. Commun.*, vol. COM-30, pp. 855–884, May.

Pieper, J. F., Proakis, J. G., Reed, R. R., and Wolf, J. K. (1978). "Design of Efficient Coding and Modulation for a Rayleigh Fading Channel," *IEEE Trans. Inform. Theory*, vol. IT-24, pp. 457–468, July.

Pierce, J. N. (1958). "Theoretical Diversity Improvement in Frequency-Shift Keying," *Proc. IRE*, vol. 46, pp. 903–910, May.

Pierce, J. N. and Stein, S. (1960). " Multiple Diversity with Non-Independent Fading," *Proc. IRE*, vol. 48, pp. 89–104, January.

Plotkin, M. (1960). "Binary Codes with Specified Minimum Distance," *IRE Trans. Inform. Theory*, vol. IT-6, pp. 445–450, September.

Poor, H. V. and Rusch, L. A. (1994). "Narrowband Interference Suppression in Spread Spectrum CDMA," *IEEE Personal Commun.*, vol. 1, pp. 14–27, Third Quarter.

Poor, H. V. and Verdu, S. (1988). "Single-User Detectors for Multiuser Channels," *IEEE Trans. Commun.*, vol. 36, pp. 50–60, January.

Popovic, B. M. (1991). "Synthesis of Power Efficient Multitone Signals with Flat Amplitude Spectrum," *IEEE Trans. Commun.*, vol. 39, pp. 1031–1033, July.

Price, R. (1954). "The Detection of Signals Perturbed by Scatter and Noise," *IRE Trans. Inform. Theory*, vol. PGIT-4, pp. 163–170, September.

Price, R. (1956). "Optimum Detection of Random Signals in Noise, with Application to Scatter-Multipath Communication," *IRE Trans. Inform. Theory*, vol. IT-2, pp. 125–135, December.

Price, R. (1962a). "Error Probabilities for Adaptive Multichannel Reception of Binary Signals," MIT Lincoln Laboratory, Lexington, MA, Techn. Report No. 258, July.

Price, R. (1962b). "Error Probabilities for Adaptive Multichannel Reception of Binary Signals," *IRE Trans. Inform. Theory*, vol. IT-8, pp. 305–316, September.

Price, R. (1972). "Nonlinearly Feedback-Equalized PAM vs. Capacity," *Proc. 1972 IEEE Int. Conf. on Commun.* Philadelphia, PA, pp. 22.12–22.17, June.

Price, R. and Green, P. E., Jr. (1958). "A Communication Technique for Multipath Channels," *Proc. IRE*, vol. 46, pp. 555–570, March.

Price, R. and Green, P. E., Jr. (1960). "Signal Processing in Radar Astronomy—Communication via Fluctuating Multipath Media," MIT Lincoln Laboratory, Lexington, MA, Tech. Report No. 234, October.

Proakis, J. G. (1968). "Probabilities of Error for Adaptive Reception of *M*-Phase Signals," *IEEE Trans. Commun. Tech.*, vol. COM-16, pp. 71–81, February.

Proakis, J. G. (1970). "Adaptive Digital Filters for Equalization of Telephone Channels," *IEEE Trans. Audio and Electroacoustics*, vol. AU-18, pp. 195–200, June.

Proakis, J. G. (1975). "Advances in Equalization for Intersymbol Interference," in *Advances in Communication Systems*, vol. 4, A. J. Viterbi (ed.), Academic, New York.

Proakis, J. G. (1998). "Equalization Techniques for High-Density Magnetic Recording," *IEEE Signal Processing Mag.*, vol. 15, pp. 73–82, July.

Proakis, J. G., Drouilhet, P. R., Jr., and Price, R. (1964). "Performance of Coherent Detection Systems Using Decision-Directed Channel Measurement," *IEEE Trans. Commun. Syst.*, vol. CS-12, pp. 54–63, March.

Proakis, J. G. and Ling, F. (1984). "'Recursive Least Squares Algorithms for Adaptive Equalization of Time-Variant Multipath Channels," *Proc. Int. Conf. Commun.* Amsterdam, The Netherlands, May.

Proakis, J. G. and Manolakis, D. G. (1995). *Introduction to Digital Processing*, Prentice-Hall, Upper Saddle River, NJ.

Proakis, J. G. and Miller, J. H. (1969). "Adaptive Receiver for Digital Signaling through Channels with Intersymbol Interference," *IEEE Trans. Inform. Theory*, vol. IT-15, pp. 484–497, July.

Proakis, J. G. and Rahman, I. (1979). "Performance of Concatenated Dual-k Codes on a Rayleigh Fading Channel with a Bandwidth Constraint," *IEEE Trans. Commun.*, vol. COM-27, pp. 801–806, May.

Pursley, M. B. (1979). "On the Mean-Square Partial Correlation of Periodic Sequences," *Proc. 1979 Conf. Inform. Science and Systems*, Johns Hopkins University, Baltimore, MD., pp. 377–379, March.

Qureshi, S. U. H. (1976). "Timing Recovery for Equalized Partial Response Systems," *IEEE Trans. Commun.*, vol. COM-24, pp. 1326–1331, December.

Qureshi, S. U. H. (1977). "Fast Start-up Equalization with Periodic Training Sequences," *IEEE Trans. Inform. Theory*, vol. IT-23, pp. 553–563, September.

Qureshi, S. U. H. (1985). "Adaptive Equalization," *Proc. IEEE*, vol. 53, pp. 1349–1387, September.

Qureshi, S. U. H. and Forney, G. D., Jr. (1977). "Performance and Properties of a $T/2$ Equalizer," *Natl. Telecom. Conf. Record*, pp. 11.1.1–11.1.14, Los Angeles, CA. December.

Rabiner, L. R. and Schafer, R. W. (1978). *Digital Processing of Speech Signals*, Prentice-Hall, Englewood Cliffs, NJ.

Raheli, R., Polydoros, A., and Tzou, C. K. (1995). "Per-Survivor Processing: A General Approach to MLSE in Uncertain Environment," *IEEE Trans. Commun.*, vol. 43, pp. 354–364, Feb./March/April.

Rahman, I. (1981). "Bandwidth Constrained Signal Design for Digital Communication over Rayleigh Fading Channels and Partial Band Interference Channels," Ph.D. Dissertation, Department of Electrical Engineering, Northeastern University, Boston, MA.

Ramsey, J. L. (1970). "Realization of Optimum Interleavers," *IEEE Trans. Inform. Theory*, vol. IT-16, pp. 338–345.

Rapajic, P. B. and Vucetic, B. S. (1994). "Adaptive Receiver Structures for Asynchronous CDMA Systems," *IEEE J. Select. Areas Commun*, vol. 12, pp. 685–697, May.

Raphaeli, D. and Zarai, Y. (1998). "Combined Turbo Equalization and Turbo Decoding," *IEEE Commun. Letters*, vol. 2, pp. 107–109, April.

Rappaport, T. S. (1996). *Wireless Commun.*, Prentice-Hall, Upper Saddle River, NJ.

Reed, I. S. (1954). "A Class of Multiple-Error Correcting Codes and the Decoding Scheme," *IRE Trans. Inform.*, vol. IT-4, pp. 38–49, September.

Reed, I. S. and Solomon, G. (1960). "Polynomial Codes Over Certain Finite Fields," *SIAM J.*, vol. 8, pp. 300–304, June.

Reed, M. C., Schlegel, C. B., Alexander, P. D., and Asenstorfer, J. A. (1998). "Iterative Multiuser Detection for CDMA with FEC: Near Single User Performance," *IEEE Trans. Commun.*, vol. 46, pp. 1693–1699, December.

Rimoldi, B. E. (1989). "Design of Coded CPFSK Modulation Systems for Bandwidth and Energy Efficiency," *IEEE Trans. Commun.*, vol. 37, pp. 897–905, September.

Rimoldi, B. E. (1988). "A Decomposition Approach to CPM," *IEEE Trans. Inform. Theory*, vol. 34, pp. 260–270, March.

Rizos, A. D., Proakis, J. G., and Nguyen, T. Q. (1994). "Comparison of DFT and Cosine Modulated Filter Banks in Multicarrier Modulation," *Proc. Globecom'94*, pp. 687–691, San Francisco, CA, November.

Roberts, L. G. (1975). "Aloha Packet System with and without Slots and Capture," *Comp. Commun. Rev.*, vol. 5, pp. 28–42, April.

Robertson, P. and Wörz, T. (1998). "Bandwidth-Efficient Turbo Trellis-Coded Modulation Using Punctured Component Codes," *IEEE J. Selected Areas, Commun.*, vol. 16, pp. 206–218, February.

Roucos, S., Schwartz, R., and Makhoul, J. (1982). "Segment Quantization for Very-Low-Rate Speech Coding," *Proc. Int. Conf. Acoust., Speech, Signal Processing*, Paris, France, pp. 1565–1569, May.

Rowe, H. E. and Prabhu, V. K. (1975). "Power Spectrum of a Digital Frequency Modulation Signal," *Bell Syst. Tech. J.*, vol. 54, pp. 1095–1125, July/August.

Rummler, W. D. (1979). "A New Selective Fading Model: Application to Propagation Data," *Bell Syst. Tech. J.*, vol. 58, pp. 1037–1071, May/June.

Rusch, L. A. and Poor, H. V. (1994). "Narrowband Interference Suppression in CDMA Spread Spectrum Communications," *IEEE Trans. Commun.*, vol. 42, pp. 1969–1979, April.

Ryder, J. D. and Fink, D. G. (1984). *Engineers and Electronics*, IEEE Press, New York.

Salehi, M. and Proakis, J. G. (1995). "Coded Modulation Techniques for Cellular Mobile Systems," in *Worldwide Wireless Communications*, F. S. Barnes (ed.), pp. 215–238, International Engineering Consortion, Chicago, IL.

Saltzberg, B. R. (1967). "Performance of an Efficient Parallel Data Transmission System," *IEEE Trans. Commun.*, vol. COM-15, pp. 805–811, December.

Saltzberg, B. R. (1968). "Intersymbol Interference Error Bounds with Application to Ideal Bandlimited Signaling," *IEEE Trans. Inform. Theory*, vol. IT-14, pp. 563–568, July.

Salz, J. (1973). "Optimum Mean-Square Decision Feedback Equalization," *Bell Syst. Tech. J.*, vol. 52, pp. 1341–1373, October.

Salz, J., Sheehan, J. R., and Paris, D. J. (1971). "Data Transmission by Combined AM and PM," *Bell Syst. Tech. J.*, vol. 50, pp. 2399–2419, September.

Sarwate, D. V. and Pursley, M. B. (1980). "Crosscorrelation Properties of Pseudorandom and Related Sequences," *Proc. IEEE*, vol. 68, pp. 593–619, May.

Sato, Y. (1975). "A Method of Self-Recovering Equalization for Multilevel Amplitude-Modulation Systems," *IEEE Trans. Commun*, vol. COM-23, pp. 679–682, June.

Sato, Y. et al. (1986). "Blind Suppression of Time Dependency and Its Extension to Multi-Dimensional Equalization," *Proc. ICC'86*, pp. 46.4.1–46.4.5.

Sato, Y. (1994). "Blind Equalization and Blind Sequence Estimation," *IEICE Trans. Commun.*, vol. E77-b, pp. 545–556, May.

Satorius, E. H. and Alexander, S. T. (1979). "Channel Equalization Using Adaptive Lattice Algorithms," *IEEE Trans. Commun.*, vol. COM-27, pp. 899–905, June.

Satorius, E. H. and Pack, J. D. (1981). "Application of Least Squares Lattice Algorithms to Adaptive Equalization," *IEEE Trans. Commun.*, vol. COM-29, pp. 136–142, February.

Savage, J. E. (1966). "Sequential Decoding—The Computation Problem," *Bell Syst. Tech. J.*, vol. 45, pp. 149–176, January.

Schlegel, C. (1997). *Trellis Coding*, IEEE Press, New York.

Scholtz, R. A. (1977). "The Spread Spectrum Concept," *IEEE Trans. Commun.*, vol. COM-25, pp. 748–755, August.

Taylor, D. P., Vitetta, G. M., Hart, B. D., and Mammala, A. (1998). "Wireless Channel Equalization," *European Trans. Telecommun. (ETT)*, vol. 9, pp. 117–143, March/April.

Tellado, J. and Cioffi, J. M. (1998). "Efficient Algorithms for Reducing PAR in Multicarrier Systems," *Proc. 1998 IEEE Int. Symp. Inform. Theory*, p. 191, August 16–21, Cambridge, MA. Also in *Proc. 1998 GLOBECOM*, Nov. 8–12, Sydney, Australia.

Titsworth, R. C. and Welch, L. R. (1961). "Power Spectra of Signals Modulated by Random and Pseudorandom Sequences," *JPL Tech. Rep. 32–140*, October 10.

Thomas, C. M., Weidner, M. Y., and Durrani, S. H. (1974). "Digital Amplitude-Phase-Keying with M-ary Alphabets," *IEEE Trans. Commun.*, vol. COM-22, pp. 168–180, February.

Tomlinson, M. (1971). "A New Automatic Equalizer Employing Modulo Arithmetic," *Electr. Lett.*, vol. 7, pp. 138–139.

Tong, L., Xu, G., Hassibi, B., and Kailath, T. (1995). "Blind Channel Identification Based on Second-Order Statistics: A Frequency-Domain Approach," *IEEE Trans. Inform. Theory*, vol. IT-41, pp. 329–334, January.

Tong, L., Xu, G., and Kailath, T. (1994). "Blind Identification and Equalization Based on Second-Order Statistics," *IEEE Trans. Inform. Theory*, vol. IT-40, pp. 340–349, March.

Tufts, D. W. (1965). "Nyquist's Problem—The Joint Optimization of Transmitter and Receiver in Pulse Amplitude Modulation," *Proc. IEEE*, vol. 53, pp. 248–259, March.

Turin, G. L. (1961). "On Optimal Diversity Reception," *IRE Trans. Inform. Theory*, vol. IT-7, pp. 154–166, July.

Turin, G. L. (1962). "On Optimal Diversity Reception II," *IRE Trans. Commun. Syst.*, vol. CS-12, pp. 22–31, March.

Turin, G. L. et al. (1972). "Simulation of Urban Vehicle Monitoring Systems," *IEEE Trans. Vehicular Tech.*, pp. 9–16, February.

Tyner, D. J. and Proakis, J. G. (1993). "Partial Response Equalizer Performance in Digital Magnetic Recording Channels," *IEEE Trans. Magnetics*, vol. 29, pp. 4194–4208, November.

Tzannes, M. A., Tzannes, M. C., Proakis, J. G., and Heller, P. N. (1994). "DMT Systems, DWMT Systems and Digital Filter Banks," *Proc. Int. Conf. Commun.*, pp. 31–315, New Orleans, LA, May 1–5.

Ungerboeck, G. (1972). "Theory on the Speed of Convergence in Adaptive Equalizers for Digital Communication," *IBM J. Res. Dev.*, vol. 16, pp. 546–555, November.

Ungerboeck, G. (1974). "Adaptive Maximum-Likelihood Receiver for Carrier-Modulated Data-Transmission Systems," *IEEE Trans. Commun.*, vol. COM-22, pp. 624–636, May.

Ungerboeck, G. (1976). "Fractional Tap-Spacing Equalizer and Consequences for Clock Recovery in Data Modems," *IEEE Trans. Commun.*, vol. COM-24, pp. 856–864, August.

Ungerboeck, G. (1982). "Channel Coding with Multilevel/Phase Signals," *IEEE Trans. Inform. Theory*, vol. IT-28, pp. 55–67, January.

Ungerboeck, G. (1987). "Trellis-Coded Modulation with Redundant Signal Sets, Parts I and II," *IEEE Commun. Mag.*, vol. 25, pp. 5–21, February.

Ungerboeck, G. and Csajka, I. (1976). "On Improving Data-Link Performance by Increasing the Channel Alphabet and Introducing Sequence Coding," *1976 Int. Conf. Inform. Theory, Ronneby, Sweden*, June.

Vaidyanathan, P. P. (1993). *Multirate Systems and Filter Banks*, Prentice-Hall, Englewood Cliffs, NJ.

Van Etten, W. (1975). "An Optimum Linear Receiver for Multiple Channel Digital Transmission Systems," *IEEE Trans. Commun.*, vol. COM-23, pp. 828–834, August.

Van Etten, W. (1976). "Maximum Likelihood Receiver for Multiple Channel Transmission Systems," *IEEE Trans. Commun.*, vol. COM-24, pp. 276–283, February.

Van Trees, H. L. (1968). *Detection, Estimation, and Modulation Theory, Part I,* Wiley, New York.

Varanasi, M. K. (1999). "Decision Feedback Multiuser Detection: A Systematic Approach," *IEEE Trans. Inform. Theory*, vol. 45, pp. 219–240, January.

Varanasi, M. K. and Aazhang, B. (1990). "Multistage Detection in Asynchronous Code-Division Multiple Access Communications," *IEEE Trans. Commun.*, vol. 38, pp. 509–519, April.

Varsharmov, R. R. (1957). "Estimate of the Number of Signals in Error Correcting Codes," *Doklady Akad. Nauk, S.S.S.R.*, vol. 117, pp. 739–741.

Verdu, S. (1986a). "Minimum Probability of Error for Asynchronous Gaussian Multiple-Access Channels," *IEEE Trans. Inform. Theory*, vol. IT-32, pp. 85–96, January.

Verdu, S. (1986b). "Multiple-Access Channels with Point-Process Observation: Optimum Demodulation," *IEEE Trans. Inform. Theory*, vol. IT-32, pp. 642–651, September.

Verdu, S. (1986c). "Optimum Multiuser Asymptotic Efficiency," *IEEE Trans. Commun.*, vol. COM-34, pp. 890–897, September.

Verdu, S. (1989). "'Recent Progress in Multiuser Detection," *Advances in Communications and Signal Processing*, Springer-Verlag, Berlin. [Reprinted in *Multiple Access Communications,* N. Abramson (ed.), IEEE Press, New York.]

Verdu, S. (1998). *Multiuser Detection*, Cambridge University Press, New York.

Verdu, S. (1998). "Fifty Years of Information Theory," *IEEE Trans. Inform. Theory*, vol. 44, pp. 2057–2078, October.

Vermeulen, F. L. and Hellman, M. E. (1974). "Reduced-State Viterbi Decoders for Channels with Intersymbol Interference," *Conf. Rec. ICC '74*, pp. 37B.1–37B.4, June, Minneapolis, MN.

Vijayan, R. and Poor, H. V. (1990). "Nonlinear Techniques for Interference Suppression in Spread Spectrum Systems," *IEEE Trans. Commun*, vol. 38, pp. 1060–1065, July.

Viterbi, A. J. (1966). *Principles of Coherent Communication*, McGraw-Hill, New York.

Viterbi, A. J. (1967). "Error Bounds for Convolutional Codes and an Asymptotically Optimum Decoding Algorithm," *IEEE Trans. Inform. Theory*, vol. IT-13, pp. 260–269, April.

Viterbi, A. J. (1969). "Error Bounds for White Gaussian and Other Very Noisy Memoryless Channels with Generalized Decision Regions," *IEEE Trans. Inform. Theory*, vol., IT-15, pp. 279–287, March.

Viterbi, A. J. (1971). "Convolutional Codes and Their Performance in Communication Systems," *IEEE Trans. Commun. Tech.*, vol. COM-19, pp. 751–772, October.

Viterbi, A. J. (1978). "A Processing Satellite Transponder for Multiple Access by Low-Rate Mobile Users," *Proc. Fourth Int. Conf. on Digital Satellite Communications*, Montreal, Canada, pp. 166–174, October.

Viterbi, A. J. (1979). "Spread Spectrum Communication—Myths and Realities," *IEEE Commun. Mag.*, vol. 17, pp. 11–18, May.

Viterbi, A. J. (1985). "When Not to Spread Spectrum—A Sequel," *IEEE Commun. Mag.*, vol. 23, pp. 12–17, April.

Viterbi, A. J. (1995). *CDMA: Principles of Spread Spectrum Communications*, Addison-Wesley, Reading, MA.

Viterbi, A. J. (1990). "Very Low Rate Convolutional Codes for Maximum Theoretical Performance of Spread-Spectrum Multiple-Access Channels," *IEEE J. Selected Areas Commun.*, vol. 8, pp. 641–649, May.

Viterbi, A. J. and Jacobs, I. M. (1975). "Advances in Coding and Modulation for Noncoherent Channels Affected by Fading, Partial Band, and Multiple-Access Interference," in *Advances in Communication Systems*, vol. 4, A. J. Viterbi (ed.), Academic, New York.

Viterbi, A. J. and Omura, J. K. (1979). *Principles of Digital Communication and Coding*, McGraw-Hill, New York.

Viterbi, A. J., Wolf, J. K., Zehavi, E. and Padovani, R. (1989). "A Pragmatic Approach to Trellis-Coded Modulation," *IEEE Commun. Mag.*, vol. 27, pp. 11–19, July.

Wainberg, S. and Wolf, J. K. (1970). "Subsequences of Pseudo-Random Sequences," *IEEE Trans. Commun. Tech.*, vol. COM-18, pp. 606–612, October.

Wainberg, S., and Wolf, J. K. (1973). "Algebraic Decoding of Block Codes Over a q-ary Input, Q-ary Output Channel, $Q > q$," *Inform. Control*, vol. 22, pp. 232–247, April.

Wald, A. (1947). *Sequential Analysis*, Wiley, New York.

Wang, X. and Poor, H. V. (1998a). "Blind Equalization and Multiuser Detection for CDMA Communications in Dispersive Channels," *IEEE Trans. Commun.*, vol. 46, pp. 91–103, January.

Wang, X. and Poor, H. V. (1998b). "Blind Multiuser Detection: A Subspace Approach," *IEEE Trans. Inform. Theory*, vol. 44, pp. 91–103, January.

Wang, X. and Poor, H. V. (1999). "Iterative (Turbo) Soft Interference Cancellation and Decoding for Coded CDMA," *IEEE Trans. Commun.*, vol. 47, pp. 1046–1061, July.

Wang, X. and Wicker, S. B. (1996). "A Soft-Output Decoding Algorithm for Concatenated Systems," *IEEE Trans. Inform. Theory*, vol. 42, pp. 543–553, March.

Ward, R. B. (1965). "Acquisition of Pseudonoise Signals by Sequential Estimation," *IEEE Trans. Commun. Tech.*, vol. COM-13, pp. 474–483, December.

Ward, R. B. and Yiu, K. P. (1977). "Acquisition of Pseudonoise Signals by Recursion-Aided Sequential Estimation," *IEEE Trans. Commun.*, vol. COM-25, pp. 784–794, August.

Weber, W. J., III, Stanton, P. H., and Sumida, J. T. (1978). "A Bandwidth Compressive Modulation System Using Multi-Amplitude Minimum-Shift Keying (MAMSK)," *IEEE Trans. Commun.*, vol. COM-26, pp. 543–551, May.

Wei, L. F. (1984a). "Rotationally Invariant Convolutional Channel Coding with Expanded Signal Space, Part I: 180°," *IEEE J. Selected Areas Commun.*, vol. SAC-2, pp. 659–671, September.

Wei, L. F. (1984b). "Rotationally Invariant Convolutional Channel Coding with Expanded Signal Space, Part II: Nonlinear Codes," *IEEE J. Selected Areas Commun.*, vol. SAC-2, pp. 672–686, September.

Wei, L. F. (1987). "Trellis-Coded Modulation with Multi-Dimensional Constellations," *IEEE Trans. Inform. Theory*, vol. IT-33, pp. 483–501, July.

Weinstein, S. B. and Ebert, P. M. (1971). "Data Transmission by Frequency-Division Multiplexing Using the Discrete Fourier Transform," *IEEE Trans. Commun.*, vol. COM-19, pp. 628–634, October.

Welch, L. R. (1974). "Lower Bounds on the Maximum Cross Correlation of Signals," *IEEE Trans. Inform. Theory*, vol. IT-20, pp. 397–399, May.

Weldon, E. J., Jr. (1971). "'Decoding Binary Block Codes on Q-ary Output Channels," *IEEE Trans. Inform. Theory*, vol. IT-17, pp. 713–718, November.

Werner, J. J. (1991). "The HDSL Environment," *IEEE Journal on Selected Areas in Communications*, vol. 9, pp. 785–800, August.

Wesolowski, K. (1987a). "An Efficient DFE and ML Suboptimum Receiver for Data Transmission over Dispersive Channels Using Two-Dimensional Signal Constellations," *IEEE Trans. Commun.*, vol. COM-35, pp. 336–339, March.

Wesolowski, K. (1987b). "Efficient Digital Receiver Structure for Trellis-Coded Signals Transmitted Through Channels with Intersymbol Interference," *Electronics Lett.*, pp. 1265–1267, November.

Wicker, S. B. (1995). *Error Control Systems for Digital Communication and Storage*, Prentice-Hall, Upper Saddle River, NJ.

Wicker, S. B. and Bhargava, V. K. (1994). *Reed Solomon Codes and their Applications*, IEEE Press, New York.

Widrow, B. (1966). "Adaptive Filters, I: Fundamentals," Stanford Electronics Laboratory, Stanford University, Stanford, CA, Tech Report No. 6764-6, December.

Widrow, B. (1970). "Adaptive Filters," in *Aspects of Network and System Theory*, R. E. Kalman and N. DeClaris (eds.), Holt, Rinehart and Winston, New York.

Wiener, N. (1949). *The Extrapolation, Interpolation, and Smoothing of Stationary Time Series with Engineering Applications*, Wiley, New York. (Reprint of original work published as an MIT Radiation Laboratory Report in 1942.)

Wilkinson, T. A. and Jones, A. E. (1995). "Minimization of the Peak-to-Mean Envelope Power Ratio of Multicarrier Transmission Schemes by Block Coding," *Proc. IEEE Vehicular Tech. Conf.*, pp. 825–829, July.

Wilson, S. G. and Hall, E. K. (1998). "Design and Analysis of Turbo Codes on Rayleigh Fading Channels," *IEEE J. Selected Areas Commun.*, vol. 16, pp. 160–174, February.

Winters, J. H., Salz, J., and Gitlin, R. D. (1994). "The Impact of antenna Diversity on the Capacity of Wireless Communication Systems," *IEEE Trans. Commun.*, vol. COM-42, pp. 1740–1751, Feb./March/April.

Wintz, P. A. (1972). "Transform Picture Coding," *Proc. IEEE*, vol. 60, pp. 880–920, July.

Wolf, J. K. (1978). "Efficient Maximum Likelihood Decoding of Linear Block Codes Using a Trellis," *IEEE Trans. Inform. Theory*, vol. IT-24, pp. 76–81, January.

Wozencraft, J. M. (1957). "Sequential Decoding for Reliable Communication," *IRE Nat. Conv. Rec.*, vol. 5, pt. 2, pp. 11–25.

Wozencraft, J. M. and Jacobs, I. M. (1965). *Principles of Communication Engineering*, Wiley, New York.

Wozencraft, J. M. and Kennedy, R. S. (1966). "Modulation and Demodulation for Probabilistic Decoding," *IEEE Trans. Inform. Theory*, vol. IT-12, pp. 291–297, July.

Wozencraft, J. M. and Reiffen, B. (1961). *Sequential Decoding*, MIT Press, Cambridge, MA.

Wulich, D. (1996). "Reduction of Peak-to-Mean Ratio of Multicarrier Modulation Using Cyclic Coding," *Electr. Lett.*, vol. 32, pp. 432–433, February.

Wyner, A. D. (1965). "Capacity of the Band-Limited Gaussian Channel," *Bell. Syst. Tech. J.*, vol. 45, pp. 359–371, March.

Xie, Z., Rushforth, C. K., and Short, R. T. (1990a). "Multiuser Signal Detection Using Sequential Decoding," *IEEE Trans. Commun.*, vol. COM-38, pp. 578–583, May.

Xie, Z., Short, R. T., and Rushforth, C. K. (1990b). "A Family of Suboptimum Detectors for Coherent Multiuser Communications," *IEEE J. Selected Areas Commun.*, vol. SAC-8, pp. 683–690, May.

Yao, K. (1972). "On Minimum Average Probability of Error Expression for Binary Pulse-Communication System with Intersymbol Interference," *IEEE Trans. Inform. Theory*, vol. IT-18, pp. 528–531, July.

Yao, K. and Tobin, R. M. (1976). "Moment Space Upper and Lower Error Bounds for Digital Systems with Intersymbol Interference," *IEEE Trans. Inform. Theory*, vol. IT-22, pp. 65–74, January.

Yasuda, Y., Kashiki, K., and Hirata, Y. (1984). "High-Rate Punctured Convolutional Codes for Soft-Decision Viterbi Decoding," *IEEE Trans. Commun.*, vol. COM-32, pp. 315–319, March.

Yue, O. (1983). "Spread Spectrum Mobile Radio 1977–1982," *IEEE Trans. Vehicular Tech.*, vol. VT-32, pp. 98–105, February.

Zehavi, E. (1992). "8-PSK Trellis Codes for a Rayleigh Channel," *IEEE Trans. Commun.*, vol. 40, pp. 873–884, May.

Zelinski, P. and Noll, P. (1977). "Adaptive Transform Coding of Speech Signals," *IEEE Trans. Acoustics, Speech, Signal Processing*, vol. ASSP-25, pp. 299–309, August.

Zervas, E., Proakis, J. G., and Eyuboglu, V. (1991). "A Quantized Channel Approach to Blind Equalization," *Proc. ICC'91*, Chicago, IL, June.

Zhang, X. and Brady, D. (1993). "Soft-Decision Multistage Detection of Asynchronous AWGN Channels," *Proc. 31st Allerton Conf. on Commun., Contr., Comp.* Allerton, IL, October.

Zhou, K. and Proakis, J. G. (1988). "Coded Reduced-Bandwidth QAM with Decision-Feedback Equalization," *Conf. Rec. IEEE Int. Conf. Commun.*, Philadelphia, PA, pp. 12.6.1–12.6.5, June.

Zhou, K., Proakis, J. G., and Ling, F. (1990). "Decision-Feedback Equalization of Time-Dispersive Channels with Coded Modulation," *IEEE Trans. Commun.*, vol. COM-38, pp. 18–24 January.

Zigangirov, K. S. (1966). "Some Sequential Decoding Procedures," *Probl. Peredach. Inform.*, vol. 2, pp. 13–25.

Ziv, J. (1985). "Universal Quantization," *IEEE Trans. Inform. Theory*, vol. 31, pp. 344–347.

Ziv, J. and Lempel, A. (1977). "A Universal Algorithm for Sequential Data Compression," *IEEE Trans. Inform. Theory*, vol. IT-23, pp. 337–343.

Ziv, J. and Lempel, A. (1978). "Compression of Individual Sequences via Variable-Rate Coding," *IEEE Trans. Inform. Theory*, vol. IT-24, pp. 530–536.

Zvonar, Z. and Brady, D. (1995). "Differentially Coherent Multiuser Detection in Asynchronous CDMA Flat Rayleigh Fading Channels," *IEEE Trans. Commun.*, vol. COM-43, pp. 1252–1255, February/March/April.

INDEX